OPTICAL NETWORKS

Optical Networks

Debasish Datta

Former Professor, Electronics and Electrical Communication Engineering, Indian Institute of Technology, Kharagpur

OXFORD
UNIVERSITY PRESS

Great Clarendon Street, Oxford, OX2 6DP,
United Kingdom

Oxford University Press is a department of the University of Oxford.
It furthers the University's objective of excellence in research, scholarship,
and education by publishing worldwide. Oxford is a registered trade mark of
Oxford University Press in the UK and in certain other countries

First published 2021
First published in paperback 2023

Published in the United States of America by Oxford University Press
198 Madison Avenue, New York, NY 10016, United States of America

British Library Cataloguing in Publication Data
Data available

Library of Congress Cataloging in Publication Data
Data available

ISBN 978–0–19–883422–9 (Hbk.)
ISBN 978–0–19–289048–1 (Pbk.)

Printed and bound in the UK by
TJ Books Limited

Dedicated to my parents.

Preface

The last six decades have witnessed outstanding developments in the area of optical communication systems and networks. With the emergence of modern optical transmission and networking technologies, accompanied by the ceaseless developments in the area of wireless communication systems, the field of telecommunication networking has undergone unprecedented changes in respect of network bandwidth, geographical coverage, and variety of services. Following the invention of the laser as an optical source in the 1960s and then optical fiber as the optical transmission medium, optical networking made its humble beginning during the mid-1970s with experimentation on local/metropolitan-area networks (LANs/MANs). These studies led to a variety of optical LANs/MANs: passive-star-based optical LANs, fiber-distributed digital interface (FDDI) on ring, and distributed-queue dual bus (DQDB), later followed by Gigabit Ethernet and its successors with higher speed.

During the late 1970s and early 1980s, while using digital telephony over the public-switched telephone network (PSTN), long-haul intercity microwave links gradually turned out to be inadequate in respect of available transmission bandwidth. Coincidentally, the technology for optical-fiber transmission systems arrived around the same time and became a potential candidate to replace the erstwhile long-haul microwave communication links in PSTN. This development promised quantum leaps in the transmission speed/bandwidth as well as in the span of telecommunication networks, and soon ushered in (during the late 1980s) the first set of optical networking standards for PSTN, such as synchronous optical network (SONET) and synchronous digital hierarchy (SDH), to be deployed in intercity telecommunication links using single-wavelength optical fiber transmission systems. However, optical fibers, offering enormously wide transmission spectrum, nearing 40 THz spanning over three transmission windows, were far from being fully utilized with only single-wavelength transmission, and this limitation necessitated worldwide investigations on the feasibility of multiwavelength transmission through optical fibers using the wavelength-division multiplexing (WDM) technology.

The synergy of microelectronics, communication, and computing, along with the variety of transmission media including copper wires and cables, radio, and optical fibers, led to a seamless growth of telecommunication services across society and penetration therein. In particular, the arrival of the Internet with its increasing demand from one and all, wireless connectivity at the user-end, vast footprints of satellites, and enormous bandwidth of optical fibers serving the long-haul, metro, and also partly the access segment, significantly changed the ways one can communicate in our society and live one's life today. Among all these enabling technologies, networking solutions based on the optical fiber transmission today form an indispensable part of the modern telecommunication network to meet the growing bandwidth demands in its various hierarchical segments.

In view of all this, the field of optical networking deserves an in-depth deliberation in the form of a comprehensive book, that will capture the past, present, and the ensuing developments, with a balanced emphasis on both networking and transmission aspects. Such a book would be useful at graduate level and would also provide the base material helping the engineers and researchers to quickly grasp the ongoing developments and contribute through further investigations. The present book is expected to serve this purpose with its fifteen chapters and three appendices, as discussed in the following.

The book is organized in four parts, with Part I giving an overview of optical networking and its enabling technologies. Part II deals with optical networks employing single-wavelength transmission, while Part III presents all forms of WDM-based optical networks. Finally, Part IV describes some selected topics on optical networks. The contents of all the chapters are discussed below in further detail.

Part I consists of two chapters. Chapter 1 begins with a background of today's telecommunication network and the roles of optical fibers therein. Next, the chronology of developments in telecommunication networks is presented from the days of PSTN offering the plain old telephone service (POTS), followed by the divestiture of Bell Laboratories and subsequent developments of the demonopolized regime of telecommunication networks. Thereafter, the chapter presents the salient features of two generations of optical networks for various network segments, including single-wavelength and WDM-based LANs/MANs, access networks, metro and long-haul networks, datacenters, along with the forthcoming/yet-to-mature forms of optical networks: elastic optical networks and optical packet/burst-switched networks. Finally, the chapter concludes with a brief discussion on the possible network architectures with the evolving optical-networking technologies. Chapter 2 presents a comprehensive description of various enabling technologies for optical networks, including optical fibers, a large number of optical, optoelectronic and electro-optic devices, as well as the basic network elements using these devices, such as, optical add/drop multiplexer, wavelength-selective switch, optical crossconnect etc., as the building blocks used in optical networks.

Part II presents four chapters on optical networks employing single-wavelength transmission in various networking segments. Chapter 3, the first chapter in Part II, presents optical LANs/MANs, covering passive-star-based LANs, FDDI, DQDB, different versions of high-speed Ethernet (1/10/40/100 Gbps), and storage-area networks (SANs). Chapter 4 deals with optical access networks, starting with the chronology of developments in the area of access networks, followed by a comprehensive description of the optical-access network architectures using passive optical networks (PONs). The salient features of the physical topology and the various TDMA/TDM-based transmission schemes used for the up/downstream traffic in PONs are described in details. The chapter concludes with brief descriptions on the various PON standards, such as, EPON, GPON, and 10Gbps PONs. Chapter 5, the final chapter of Part II, presents the SONET/SDH networks, covering fundamental concepts of network synchronization, followed by the framing and multiplexing techniques and operational features in SONET/SDH networks. Subsequently, the chapter presents the techniques for transmission of packet-switched data through circuit-switched SONET/SDH networks using specialized techniques, such as general framing procedure (GFP), virtual concatenation (VCAT), and link capacity adjustment scheme (LCAS). Further, the optical transport networking (OTN) standard is presented for the integration of high-speed SONET/SDH and data networks, whose frames/packets are encapsulated into the OTN frames to realize a universal transport platform. Finally, the chapter presents the packet-switched version of optical ring networks, known as resilient packet ring (RPR).

Part III has four chapters dealing with different forms of WDM-based optical networks. To start with, Chapter 6 presents WDM LANs/MANs, beginning with the ways one can utilize the potential of WDM in optical LANs/MANs, followed by the descriptions of the early noteworthy experimentations on WDM LANs/MANs. Thereafter, the chapter presents in detail the two principal architectures of WDM LANs/MANs using single-hop and multihop logical topologies, realized through broadcast-and-select transmission over optical passive-star couplers. Chapter 7 presents the WDM-based access networks as an extension of TDM-based PONs with WDM transmission. Several configurations of WDM-based PONs are presented, including WDM PONs using only WDM transmission in both directions and TDM-WDM (TWDM) PONs combining WDM with TDM for enhanced bandwidth utilization. Chapter 8 deals with the WDM metro networks, covering the basic architectures using point-to-point WDM as well as wavelength-routing techniques. The design methodologies for WDM metro networks transporting circuit-switched SONET/SDH traffic over point-to-point WDM and wavelength-routed rings are presented in detail, using linear programming (LP) as well as heuristic schemes. The chapter finally presents the testbeds developed for the packet-switched WDM metro rings to improve bandwidth utilization in optical fibers, as compared to the circuit-switched WDM metro rings, followed by a performance analysis of circuit/packet-switched WDM metro rings to estimate and compare the bandwidth utilization in the respective cases. Chapter 9 deals with the mesh-configured WDM long-haul networks, presenting the underlying design issues, node configurations, and the basic features of the offline design and online operations using wavelength-routed transmission. Thereafter, the chapter presents various offline design methodologies, based on an appropriate mix of LP-based and heuristic schemes. The need of wavelength conversion in such networks plus various possible node configurations using wavelength conversion are explained, and a performance analysis is presented to examine the impact of wavelength conversion in wavelength-routed long-haul networks. Finally, some of the useful online routing and wavelength assignment techniques are described for the operational wavelength-routed long-haul networks.

Part IV deals with some selected topics in six chapters. First, Chapter 10 deals with the various transmission impairments and the power-consumption issues in the optical networks. Chapter 11 deals with the survivable WDM networks, presenting various protection and restoration techniques in the events of network failures. Chapter 12 looks at network management and control schemes, including generalized multi-protocol label switching (GMPLS), automatically switched optical network (ASON) and software-defined optical network (SDN/SDON). Chapter 13 deals with datacenters, where a comprehensive overview of electrical, optical, and hybrid (electrical-cum-optical) intra-datacenter networks is presented followed by some heuristic design methodologies for long-haul wavelength-routed optical networks hosting datacenters at suitable locations, where various objects are made available with varying popularity levels. Chapter 14 presents the emerging developments in the area of elastic optical networks using a flexible grid for better utilization of the available optical fiber spectrum. Finally, Chapter 15 presents the basic concepts and operational features of optical packet- and burst-switched networks.

Three appendices are included at the end of the book to help the readers who need to learn the fundamentals on the three useful topics that are referred to in the main text in various chapters. Appendix A deals with the basics of the LP formulation and the Simplex algorithm to solve the LP-based optimization problems. Appendix B deals with the various noise processes encountered in the optical communication systems. Appendix C presents the fundamentals of queuing theory and the applications of these concepts in evaluating the performance of open networks of queues.

Further, a solution manual for the chapter-end exercise problems in the book has been developed. The interested instructors may request the publisher to get access to the solution manual.

Acknowledgments

It is my great pleasure to acknowledge the help received from several persons for writing this book. To begin with, I wish to acknowledge the help from Aneek Adhya and Goutam Das of IIT Kharagpur, with whom I had a series of useful discussions on various aspects of this book. I greatly appreciate the help from Pranabendu Gangopadhyay of IIT Kharagpur for several in-depth discussions on Chapter 2 dealing with the technologies used in optical networks. Next, I express my sincere thanks to the anonymous reviewers, whose insightful comments have helped me in deciding the contents of the book. I thank Somnath Maity, former Chief General Manager in Bharat Sanchar Nigam Limited, India, who helped me in perceiving a realistic picture of optical-networking practice in India. Further, Partha Goswami from IIT Kharagpur, with his experience in campus networking, helped me with useful inputs on today's campus LANs.

I also take this opportunity to acknowledge the help received from all of my former/ current students, who have worked with me over the years in this area. In particular, I want to express my sincere thanks to Sushovan Das, Dibbendu Roy, Sadananda Behera, and Upama Vyas for their valuable help towards solving some of the exercise problems for the book. Special thanks are due to Sushovan Das for the series of useful discussions I had with him on the chapters dealing with WDM LANs, WDM metro networks, and datacenters.

I must also thank Katherine Ward and Francesca McMahon of Oxford University Press, and their former colleague Harriet Konoshi, without whose constant encouragement and support this book wouldn't have seen the light of the day. I also express my sincere thanks to Elakkia Bharathi and her team members at SPi Global, who have helped me a lot during the preparation of the final manuscript.

Finally, I wish to thank my wife Rumjhum and our daughter Arunima, whose emotional support was a pillar of strength for me during this entire project. Indeed, this acknowledgment would remain incomplete without special thanks to Rumjhum, who tirelessly went through the entire manuscript for preliminary proofreading and tried her best to keep me in good health and spirits all through this long and arduous journey.

Debasish Datta
Kolkata, India
ddatta@ece.iitkgp.ac.in

Contents

Part I
Introduction

Optical Networks: An Overview

1

1.1 Background

Today's telecommunication network serves as an indispensable platform in our society for the exchange of information in its various possible forms, viz. voice, video, and data, enabling us to live our life in this modern era. Networking practices have evolved relentlessly over the last two centuries with a wide range of technologies, calling for the accommodation of heterogeneous networking equipment and transmission media. The synergy of microelectronics with telecommunication and computing has brought in novel networking solutions offering a variety of telecommunication services for users. In particular, the arrival of the Internet with its overwhelming popularity, use of wireless connectivity for the users in the last mile of telecommunication networks, the vast footprints of satellites, and the enormous bandwidth of optical fibers, have significantly changed the ways one can communicate in our society and live one's life today.

Transmission media for telecommunication networks can be broadly divided into two categories: guided or wired, and radio or wireless. The various media for wired transmission include copper wire pairs, copper coaxial cables, and optical fibers. Copper wire pairs have been traditionally used in subscriber or local loops (also referred to as last mile) of the public switched-telephone networks (PSTNs), and later also in local/metropolitan-area computer networks (LANs/MANs), while coaxial cables have been used in cable TV networks, PSTNs, and also in LANs/MANs. The long-haul links in the earlier generation of PSTNs were realized using the line-of-sight radio links at microwave frequencies, generally using multiple hops due to the earth's curvature. Wireless transmission has also been employed in various other ways and scales, such as for satellite communication offering large coverage over a nation or multiple nations, cellular communication for mobile telephone networks (between the mobile users and base stations), wireless LANs (e.g., Wi-Fi) in a campus or at home, etc.

In the area of wired transmission, ground-breaking developments took place during the 1960s and 1970s with the arrival of lasers (Maiman 1960) as optical sources, and optical fibers (Kao and Hockham 1966; Werts 1966; Hecht 2005) as a viable optical transmission medium. The optical fibers were realized as thin cylindrical waveguides with two silica-based concentric layers, viz. core and cladding, having slightly different refractive indices (Fig. 1.1), which were able to function as a guided transmission medium with extremely low loss ($\simeq$ 0.2 to 0.5 dB/km) and enormous bandwidth (almost 40,000 GHz). On the other hand, semiconductor lasers and light-emitting diodes (LEDs) could be fabricated as optical sources, that could emit appropriate wavelengths falling in the low-loss windows of the optical fibers and could be modulated at high speed. The wavelengths transmitted by the optical sources, having traversed through the optical fibers, could be photodetected at the

Optical Networks. Debasish Datta, Oxford University Press (2021). © Debasish Datta.
DOI: 10.1093/oso/9780198834229.003.0001

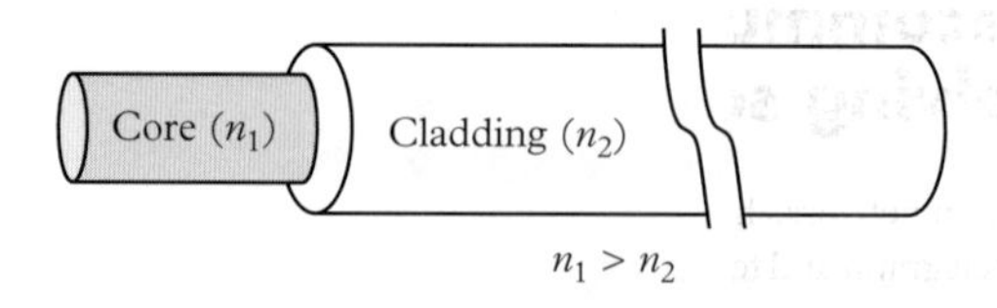

Figure 1.1 *Basic optical fiber structure with two concentric cylindrical layers: core and cladding. The core diameter can vary typically from a few micrometers to a few tens of micrometers for different types of optical fibers. The cladding diameter is mostly standardized at 125 μm to be compatible with standard connectors.*

destination end by using photodiodes with the appropriate wavelength window, enabling end-to-end optical communication at high transmission rate (or speed) (~ 1 Gbps) over tens of kilometers with acceptable transmission quality. With time, the transmission rate and distance of the optical links steadily increased, along with the development of high-power lasers with narrow spectral width and optical fibers with enhanced transmission characteristics, thereby supporting the ever-increasing network traffic in various network segments.

Initially, optical-fiber transmission was explored for enhancing the transmission rate and size of the existing copper-based LANs, and subsequently employed in the long-haul inter-city links in PSTNs. The emergence of pulse code modulation (PCM) to transmit digitized voice paved the way to receive signals with much better quality, but pushed up the necessary bandwidth from 4 kHz for the telephone-quality analog voice signal to a much larger bandwidth, commensurate with the 64-kbps PCM bit stream. Consequently, the provision of high-speed transmission of time-division multiplexed PCM signals using line-of-sight radio transmission systems in PSTNs became infeasible due to the limited bandwidth of the microwave links operating at carrier frequencies in the GHz range. It was at this juncture, with the long-haul bandwidth demands on the rise towards Gbps and thus gradually approaching the carrier frequency itself, the optical fiber-transmission systems with carrier frequencies in THz range arrived, and proved to be a great savior for telecommunication industry.

With enormous bandwidth and extremely low loss, optical fiber communication systems provided such dramatic savings in equipment and operational costs, that all-out deployment of optical fibers in the high-traffic routes became economically the most viable. Thus, optical fibers captured the market of long-haul links for national and transnational networks (terrestrial and transoceanic links) as well as for metro networks. Optical fibers were also found useful as an effective replacement for the bandwidth-constrained copper cables in LANs/MANs for large academic campuses and business organizations. More recently, optical fibers have entered the access segment in the form of broadband access networks, providing a wide range of services for end-users. Furthermore, optical fibers are also going to play significant roles in providing high-speed backhaul/fronthaul support for mobile communication networks. However, the huge bandwidth of optical fibers remained under-utilized with single-wavelength optical transmission, until the introduction of wavelength-division multiplexing (WDM) technology, which we will discuss later in this chapter. A holistic view of the traditional PSTN and its evolution toward today's telecommunication networks using optical-fiber transmission are presented in the following.

1.2 Telecommunication networks: evolving scenario

Telecommunication networks have grown ceaselessly over the past two centuries, ever since the invention of telegraph and telephony. The growth has been multi-dimensional in terms of network capacity, coverage, and services, so much so that describing its overall architecture in a comprehensive manner is quite a task. Nevertheless, in the following, we try to capture the chronology of the major developments in the area of telecommunication networks, starting from the early days of PSTN.

Historically, PSTN was first deployed to offer a basic telephone service, named as the *plain-old telephone service* (POTS) in the USA about two years after the great invention of the telephone by Alexander Graham Bell in 1876. POTS was made available to the public from a centrally located switching office, known as the central office (CO), in New Haven, Connecticut, and the access area covered by a CO was called its local or subscriber loop. The COs were initially operated manually and later upgraded to automated switching systems, which were replicated in other areas and were interconnected at switching offices at a higher level through long-distance analog communication links (trunks) using coaxial cables, and later upgraded with line-of-sight microwave links. Following this hierarchical topology, five levels of switching offices emerged in the USA: *regional centers* as the switching offices at level 5 (the highest level in hierarchy), *sectional centers* as the switching offices at level 4, *primary centers* as the switching offices at level 3, *toll centers* as the switching offices at level 2, and COs as the switching offices at level 1, the lowest level in hierarchy. Further, at the lowest level, a few COs were interconnected together through another local switching unit, called a *tandem switch*, and the local loops covered by the COs connected to a tandem switch were together called *local access and transport areas* (LATAs), where between any two subscribers the telephone calls were toll-free. The LATAs with the constituent COs were interconnected through the inter-city trunks and switching offices at higher levels, covering the entire nation geographically by using the *inter-exchange carrier* (IXC) network, as shown in Fig. 1.2. This simple PSTN architecture continued to operate as long as the telephone industry in the USA remained as a monopoly of the American Telegraph and Telephone (AT&T) company.

In terms of today's networking terminologies, the above type of PSTN was essentially *circuit-switched*, wherein the commodity (voice) to be switched was analog while the switching functionalities were fundamentally digital. The automated switching offices (replacing the manual COs) were realized by using step-by-step electromechanical devices (invented by Almon B. Strowger), which were subsequently improved by using the cross bar switching technology. Later on, motivated by the *digitization of voice* by PCM, these networks were transformed into integrated digital networks (IDNs), wherein the switching functions (being digital in nature) and the digitized voice were integrated into the same digital platform using computerized electronic switching systems (ESS) at the switching offices. However, at this stage, the *last mile* of the network between the telephone subscribers and the line cards of COs remained analog because of various practical constraints. Thus, the analog voice signals from the subscribers were sampled and digitized using PCM at COs, and thereafter time-multiplexed and switched through the various levels of switching offices digitally. Attempts were also made to convert the last mile of IDN into digital format, leading to the integrated service digital network (ISDN), which was introduced at a later stage but had to take a back seat eventually due to its high cost and the arrival of other solutions which were

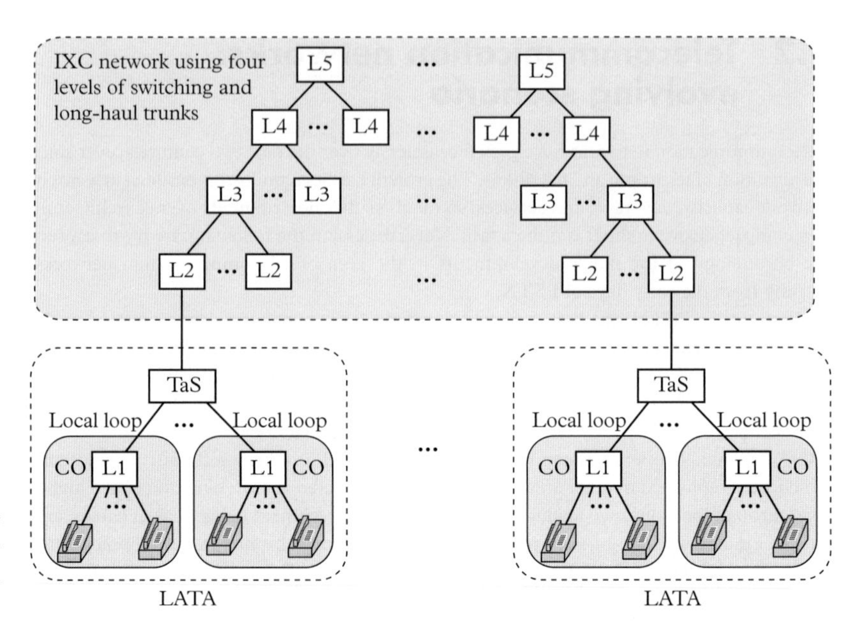

Figure 1.2 *PSTN architecture in the USA before deregulation of AT&T. Li's represent the switching offices at level i, with L1's as the COs, and TaS represents the tandem switch in each exchange area, i.e., LATA.*

more cost-effective. Similar PSTN architectures were adopted in various other nations with varying levels of penetration, depending on the needs of the society therein.

However, even before the arrival of IDN in PSTN, gradually a different scenario started emerging for the telephone networks in the USA from the 1930s (Economides 2013). Till then, the telecommunication networking had not become competitive with enough market force, and hence was a regulated business community, with the sole monopoly for AT&T supported by Western Electric Company as the manufacturing counterpart of AT&T, and Bell Labs as AT&T's research wing. Subsequently, the patents of the telecommunication equipment started expiring in the late 1930s while telephony also started gaining popularity, thereby motivating private enterprises to manufacture telecommunication products for setting up their own independent telecommunication networks, particularly at local levels. However, AT&T continued to enjoy its monopoly in the long-haul segment till 1970s. Meanwhile, the private telecommunication companies also started setting up their own long-haul networks, leading to the coexistence of parallel telecommunication networks across the country. At this juncture, AT&T confronted them by denying interconnections between the subscribers from AT&T and non-AT&T networks, which forced many subscribers and organizations to subscribe to two telecommunication companies to get around the problem of isolation between disconnected telecommunication networks. This eventually led to legal issues with antitrust lawsuits, mainly the one between the USA and Western Electric in 1949, and the other between the USA and AT&T in 1974.

Eventually, after a long-drawn legal process, AT&T had to go through the process of *divestiture*, initiated in 1984, which was completed in 1996, with the creation of some local companies, called Baby Bells, and each Baby Bell company took charge of one specific region of the country, while the long-haul networking stayed with AT&T. The impact of the divestiture gradually took effect in various other countries, leading to the emergence of a scenario of coexistent network operators without an overall monopoly of one single company. These developments led to an architectural change in the old PSTN (shown

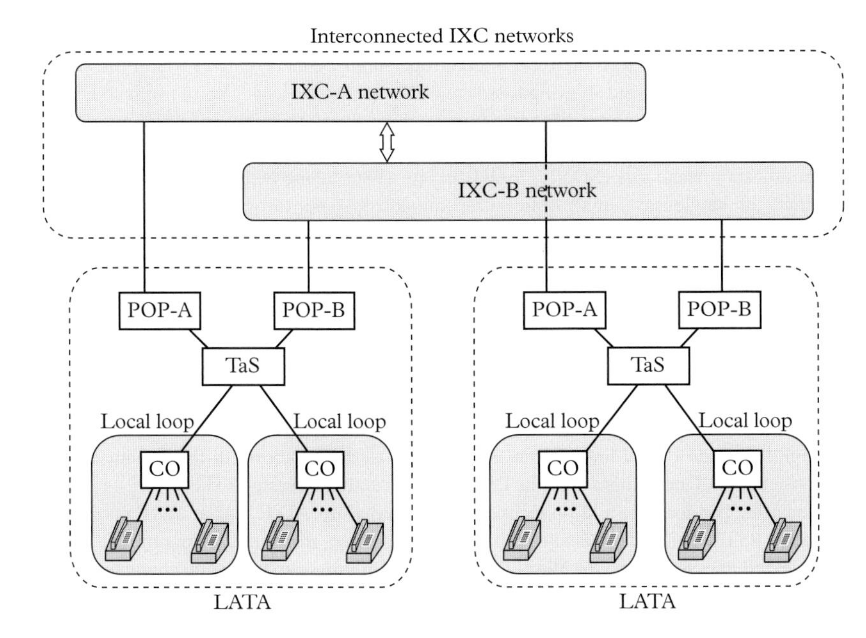

Figure 1.3 *Post-divestiture PSTN architecture with two IXCs.*

earlier in Fig. 1.2), wherein multiple IXCs were allowed to coexist and each LATA had to ensure equal access to different IXCs through an additional class of switching hubs, called point-of-presence (PoP), which were placed within each LATA for each IXC, between the higher-level switch connected to the particular LATA and its tandem switch (Fig. 1.3). This architectural transformation started a new chapter in the history of telecommunication industry, leading to a plethora of innovations and a reduction in tariff for common people. We discuss briefly this transformation process of the telecommunication networks in the following.

With the monopoly of one single organization gone, several telecommunication companies started developing networking technologies for offering different types of services from one given access segment, triggering a series of innovations and providing competition. For example, the PSTN operators in the access segment offered data connections from residential as well as business offices with landline telephones using dial-up connections, ISDN, and digital subscriber line schemes, such as, asymmetric digital subscriber line (ADSL) and its variants. Similarly, the cable TV operators offered cable-telephony and cable-modem services for homes, while autonomous LANs also brought in voice-over-IP services along with various messaging and online services. Thus, the practice of providing *one service from one platform* of the access networks moved toward the paradigm of offering multiple services from any given access-networking platform.

Alongside the above developments, when the optical sources and fibers became available in 1970s, initially some universities and research groups started exploring the potential of optical fiber communication for developing high-speed LANs with larger coverage area. Then in the 1980s, as mentioned earlier, the digitized-voice (i.e., PCM) transmission along with increased traffic volume called for the use of optical fibers for bandwidth enhancement in the long-haul links of the PSTN. In particular, the new standards of optical communication, known as the *synchronous optical network* (SONET) and *synchronous digital hierarchy* (SDH) emerged, which were similar to and compatible with each other, but originated respectively from the USA and Europe, almost concurrently. Around the same time, the data traffic started showing trends to catch up with the legacy voice traffic, but was found manageable with the newly introduced optical fiber communication systems (with single-wavelength links). Thus, with the divestiture of AT&T in 1984, arrival of optical communication standards (SONET/SDH) in late 1980s, along with the popularity of the Internet, the single-wavelength optical transmission systems turned out to be the most appropriate solution for the metro and long-haul telecommunication networks.

Driven by the needs of data communications between the distant organizational offices, the long-haul PSTN links were utilized through leased-line connections, which *circuit-switched* the packets between the *packet-switching* LANs located at different geographical areas. Through this process, distant LANs got interconnected through the PSTN links, leading to what were called wide-area computer networks, or simply *wide-area networks* (WANs). In other words, the data communication between distant autonomous LANs was set up using the omnipresent public network (i.e., PSTN) by *leasing* the WAN connectivity through long-haul PSTN links. Thus, using PSTN as the backbone, different networking options evolved. One of them used the PSTN infrastructure offering POTS as well as a data service using the available add-on technologies, e.g., dial-up, ADSL etc. at users' premises, while the PSTN backbone was also utilized to interconnect the distant autonomous LANs using leased lines leading to the WAN connectivity. With these architectural characteristics, effectively the voice and data networks continued to share the network resources and coexist over the same legacy PSTN.

However, the needs for telecommunication services never ceased to grow thereafter, being more so for the data traffic, eventually calling for better utilization of the bandwidth available in optical fibers. In particular, though the optical fibers offered about 40,000 GHz of bandwidth, single-wavelength transmission could only utilize a small fraction of the same ($\leq$ 100 Gbps), and the best utilization of this enormous resource was only possible through concurrent transmission on multiple non-overlapping wavelengths using WDM technology. Consequently, in a few years' time around the mid-1990s, use of WDM technologies in the optical-fiber transmission systems became imperative, particularly for long-haul and metro networks, and extensive research activities were carried out in this direction. Gradually, the network operators started introducing WDM technology in the national/transnational long-haul and metro segments. However, in the access segment, the copper and wireless continued to be the popular physical media, although the optical fibers started entering this segment as well, wherever the available bandwidth and network reach needed further enhancement. Along with further growth in data traffic, parallel developments also took place from the telecommunication companies to set up long-haul optical networks for data services. The constraints of provisioning WAN connections through legacy PSTN started becoming relaxed by such ventures, and the telecommunication networks assumed a hybrid and complex architecture, wherein the fiber-enabled PSTN continued to serve as the long-haul and metro networks.

In Fig. 1.4 we present a contemporary telecom networking hierarchy with its three segments: long-haul, metro, and access, wherein excepting the access segment, optical transmission using WDM has now been a widely accepted technology. The long-haul networks, though initially realized as interconnected SONET/SDH rings, gradually started adopting mesh topology, while for metro segment, the networking topology has been ring, mostly with SONET/SDH-based transmission. Both the segments have been augmented over time, using WDM transmission and high per-wavelength transmission rates, along with WDM-compliant network protocols. At metro level, the WDM ring networks have been mostly divided into two tiers, with the higher one, called the metro-core, connecting with

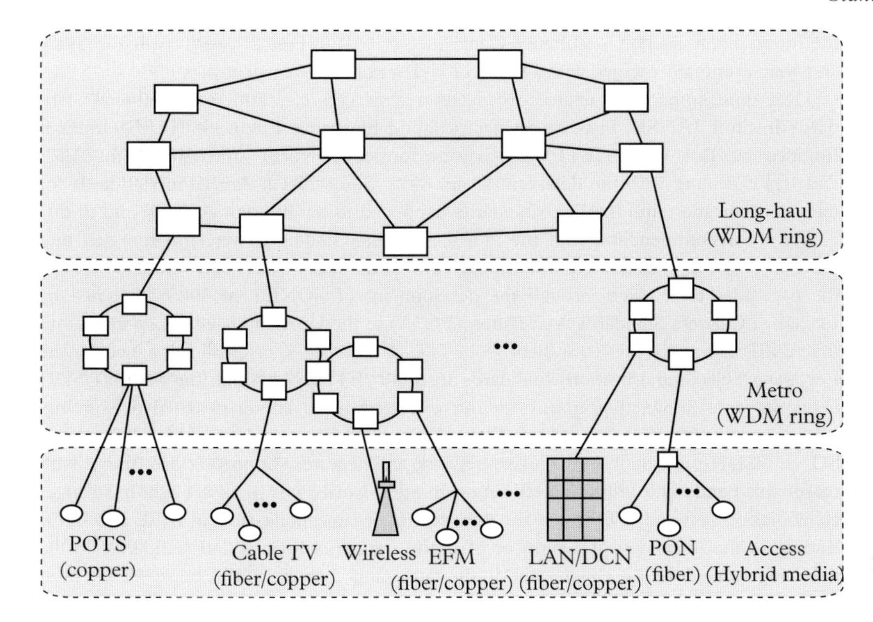

Figure 1.4 *Today's telecom network architecture: a generic form.*

the long-haul network. The metro-core ring is used to interconnect several smaller fiber-optic rings at the lower tier, called metro-edge rings, connecting the access segment for reaching the end-users in the last mile. However, the access segment continues to be a mix of various media: POTS and ADSL on copper wire, cable TV on coaxial cable and fiber, campus LAN and datacenter networks (DCNs) on copper wire and fiber, mobile access on wireless, wireless LAN, passive optical network (PON) on fiber, and so on.

1.3 Standards for telecommunication networks

The standards for telecommunication networks have been an essential part for all the developments, as the networking standards play one of the most significant roles in enabling the manufacturers to develop compatible networking products, so that the various network operators could avail competitive pricing for the necessary products and interoperate between themselves. As a result, various standard-making organizations came up and worked over several decades to present the telecommunication industry with the essential standards, which keep evolving seamlessly to support the ongoing developments in the field of telecommunications.

Historically, through the initiative of Napoleon III, 20 member states (currently 189) from around the world met in Paris in 1865 to constitute the first-ever telecommunication standard committee, called the International Telegraph Union (ITU), wherefrom its successor, the International Telecommunication Union (also abbreviated to ITU), evolved and now plays the central role in forming most of the telecommunication standards followed around the world. The ITU was subsequently incorporated as a Specialized Agency in the United Nations in 1947 following the Second World War. Thereafter, the International Telegraph and Telephone Consultative Committee (CCITT in the French language) and the International Radio Consultative Committee (CCIR in French) were created in 1956, and were eventually merged with ITU as ITU-T and ITU-R, respectively.

Over time, some other related organizations emerged, e.g., American National Standards Institute (ANSI), Institute of Electrical and Electronic Engineers (IEEE), Internet Engineering Task Force (IETF), which were formed in North America with the IEEE and IETF dealing with the data-centric networks. Some North American standards for telecommunication and the ITU standards happened to be different at times, and in due course the recommendations of the North American standards were incorporated into the CCITT/ITU-T standards. Interestingly, one of the noteworthy joint ventures between the two sides took place through the development of SONET in 1988 (designed by Exchange Carriers Standards Association (ECSA) in the USA and later incorporated into the ANSI) along with its equivalent from ITU-T, i.e., SDH, originally designed by the European Telecommunications Standards Institute (ETSI). Although the SONET/SDH networks were designed primarily for the voice-optimized circuit-switched networking, subsequent growth of packet-switched data traffic called for a transition of the voice-centric SONET/SDH towards more inclusive network architectures supporting data traffic with reasonable bandwidth-utilization efficiency. In other words, this trend brought in stronger collaboration between ITU-T and the data-centric organizations, such as IEEE and IETF, leading to the formation of a number of useful industrial forums and standards. In the following sections in this chapter and in the remaining chapters of this book, we will keep referring to some of the relevant telecommunication standards, as necessary.

1.4 Single-wavelength optical networks

The potential of optical fibers was initially explored with single-wavelength transmission in the 1970s for enhancing the speed and size of the erstwhile LANs. From the mid/late 1980s, optical fibers started being considered also for the high-speed long-haul digital transmission links in PSTN with single-wavelength transmission between the COs. Some of the early attempts to develop optical LANs include the ring-configured optical LAN, called Hallo, from the University of California at Irvine (Farber 1975), the passive-star based optical LAN, called Fibernet, from Xerox Palo Alto Research Center (Rawson and Metcalfe 1978), and an augmented version of the pre-existing copper-based Cambridge Ring using optical fibers from the University of Cambridge (Hunkin and Litchfield 1983). Further developments in the fiber-optic LANs took place in late 1980s, which we discuss later in this section.

Gradually the bandwidth crunch in microwave links led to the use of optical fibers in long-haul links in PSTNs. However, optical transmission systems had to deviate from the erstwhile time-division multiplexing (TDM) technique adopted for digital telephony – plesiochronous digital hierarchy (PDH) – and develop a more robust TDM technique, to alleviate the problem of clock jitter in high-speed bit streams transmitted over optical fibers. This new generation of TDM standard was named as SONET in North America (Ballart and Chiang 1989; ANSI SONET X3T12 1995; Kartalopoulos 2004), while a similar multiplexing standard (and compatible with SONET) was introduced, called as SDH, for the European and some other countries (ITU-T G.803 2000; Kartalopoulos 2004), which were mainly realized in ring topologies (sometimes linear) using a class of network element/equipment called add-drop multiplexers (ADMs). The SONET/SDH standards were primarily designed for voice traffic along with some options to integrate data traffic

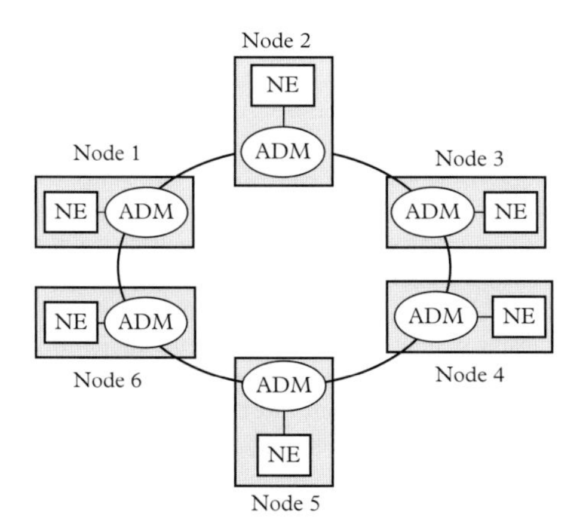

Figure 1.5 *A typical SONET/SDH ring using ADMs. NEs represent the network equipment of various clients, viz., PDH, IP, Ethernet etc.*

in the form of Ethernet, Internet protocol (IP), asynchronous transfer mode (ATM) data flows. A typical SONET/SDH-based ring network is shown in Fig. 1.5. The SONET/SDH-based optical fiber networks brought in a major change in the way the metro and long-haul networks were set up, opening up great possibilities with the potential to support higher bandwidth demands.

In due course, with the proliferation of LANs and the needs for their interconnection when located in distant places, the SDH/SONET networks with the options to carry data traffic alongside voice was harnessed to offer leased (circuit-switched) connections to interconnect LANs as a carrier of data traffic thereby forming WANs over PSTN. Thus, the circuit-switched optical networks using SDH/SONET standard were used to provide support for the long-distance packet-switched data communications between LANs, by using circuit-switching of *bursty* packet-switched data streams, but with limited bandwidth utilization in optical fibers. Later, in order to enhance the efficiency for the bursty data traffic in metro networks, a packet-switched version of optical ring networks was developed, named as the resilient packet ring (RPR), wherein intelligent protocols were designed to carry both data and real-time traffic streams for enhanced bandwidth utilization in optical fibers (Yuan *et al.* 2004).

The early attempts at upgrading the copper-based LANs using optical fibers continued as well, but in doing so some specific challenges had to be faced. In the bus topology, when used in the optical LANs, tapping loss at the network nodes turned out to be significant and progressive in nature along the bus length. Thus, the traditional media-access control (MAC) protocol for bus topology using *carrier-sense multiple access scheme with collision detection* (CSMA/CD), appeared infeasible as an optical packet coming from a distant node to a given node would have a weaker power level as compared to the packet emerging from the given local node, thereby making collision detection rather challenging. However, the star topology of optical fiber was found compatible and logically could be similar (passive broadcast) to a copper bus without progressive loss. However, the CSMA/CD-based protocols with high-speed operation and longer propagation delay across the network (due to the larger network size enabled by the low losses of optical fibers) would result in smaller values for the ratio of the packet duration to the round-trip propagation delay in a LAN, thereby causing frequent collisions leading to poor LAN throughput. As an alternative solution to this problem on the bus topology, one could go for dual or folded-bus topologies with collision-avoidance protocols, eventually leading to the distributed-queue dual-bus (DQDB) network, standardized as a MAN from IEEE, operating at a data rate of 100 Mbps (Tran-Gia and Stock 1989).

However, optical fiber in the ring topology, which was earlier tried in the augmented Cambridge Ring, was more compatible, as the ring can offer collision-free operation and regenerates signal by optical-electronic-optical (OEO) conversion at every node. However, the traditional IEEE token-ring protocol in optical-fiber rings also faced the problem of low throughput, as the optical ring size turned out to be much bigger than the optical packet durations with high transmission rates (as compared to the copper rings). Thus *one-packet-at-a-time* protocol (i.e., no packet allowed to enter the ring until the last packet gets back home) was found to be inefficient, as each short packet (shrunk at high bit rate) would traverse the large network span without allowing any other packet to share the remaining fiber space. This aspect was addressed in an improved version of the ring-based fiber-optic LAN/MAN – fiber-distributed digital interface (FDDI) – by using a modified capacity allocation scheme at 100 Mbps (Ross 1986).

Subsequently, in a race to compete with 100-Mbps FDDI, 10 Mbps Ethernet was upgraded to 100 Mbps fast Ethernet with great success, eventually winning over the DQDB and FDDI standards. Thereafter, the Ethernet community went on for further speed enhancements with 1, 10, 40 and 100-Gbps versions, by using *switched full-duplex* optical fiber transmissions (Frazier and Johnson 1999; Cunningham 2001; D'Ambrosia 2009; Roese and Braun 2010). However, most importantly, the lower ends (user side) were kept compatible with the legacy interfaces with lower transmission rates (10/100 Mbps), while at higher levels of network hierarchy, large bandwidth and low loss of optical fibers were harnessed with switched transmission, instead of continuing with the CSMA/CD-based *shared-transmission* technology.

Following the overwhelming growth of data networks, accompanied by the popularity of high-speed Ethernet versions, a noteworthy development took place for providing a convergent optical networking platform for ≥ 1 Gbps Ethernet versions and IP traffic along with the legacy SONET/SDH traffic streams at similar data rates, leading to the optical transport network (OTN) standard (ITU-T G.709 2016). The other optical networks developed over smaller network spans include the storage area networks (SANs), such as HIPPI, ESCON, and Fiber channel. Further developments in this direction include the datacenters employing high-speed optical interconnections between the servers and switches therein, which are emerging as a major storehouse of data and computing power with the capability to offer cloud-based services (Xia *et al.* 2017).

In the access segment, TDM-based passive optical networks (TDM-PONs, or simply PONs) were developed offering broadband services to the end-users in homes, large building complexes, offices, academic campuses, and business organizations (Kramer *et al.* 2012). As shown in Fig. 1.6, PONs (using two separate wavelengths for upstream and downstream transmissions) employ passive optical splitter/combiner at a remote node (RN), and the RN is connected to the access hub, called optical line terminal (OLT), through a feeder segment, while on the user side (called collection and distribution segment) the remaining ports of the RN are connected to the optical networking units (ONUs), placed at or close to the customers' premises. Thus, a PON attempts to bring the fiber as close to the end-users as possible, hence being called the fiber-to-the-home (FTTH), fiber-to-the-building (FTTB), or fiber-to-the-curb (FTTC) network, determined by the locations of the ONUs, which are also collectively referred to as FTTx networks, with x representing the possible locations of the ONUs (i.e., home, building, or curb located at a central place in a group/community of users).

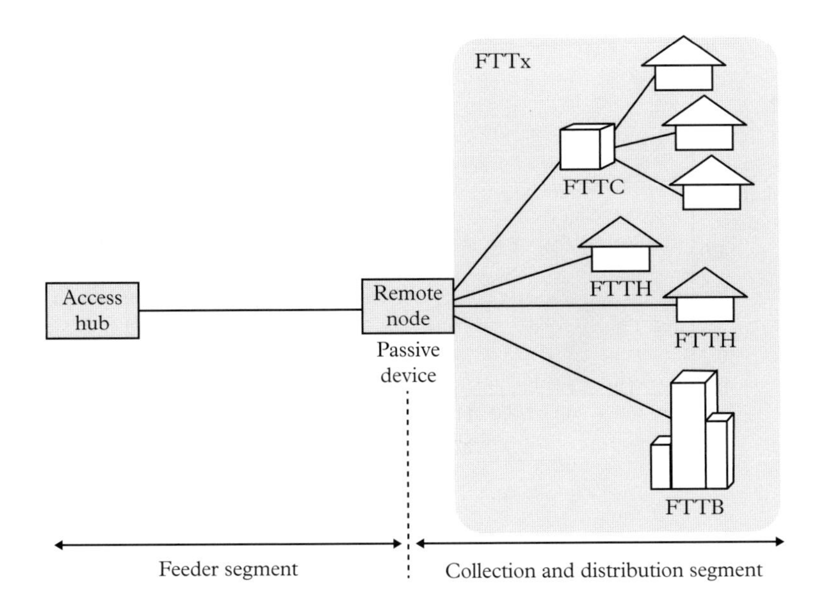

Figure 1.6 *Illustration of a PON, showing the possible FTTx connections to the end-users.*

1.5 Wavelength-division multiplexing

As indicated earlier, in order to utilize the full capacity of optical fibers, it gradually became necessary to employ concurrent transmissions at multiple wavelengths, enabled by WDM devices. Today, WDM technology plays a significant role in utilizing the bandwidth of optical fibers in telecommunication networks. As shown in Fig. 1.7, the typical high-quality optical fibers can operate over a wide transmission spectrum, which is generally divided into three windows – first, second and third window – governed by the loss mechanisms in optical fibers. The third window is the widest, with a wavelength span of 200 nm located around 1550 nm (giving a bandwidth of 25,000 GHz), while the other two narrower windows are located around 1300 nm and 900 nm, with the three windows together offering an enormous transmission bandwidth of $\simeq$ 40,000 GHz. These three windows, more particularly the third and second windows (due to smaller losses), can be utilized effectively through concurrent (WDM) transmission from multiple optical sources with non-overlapping wavelengths or channels. Before going into further details on the WDM networking aspects, we address below a frequently-asked question as to why we use the term WDM, and why not frequency-division multiplexing (FDM) for concurrent multi-wavelength transmission.

To answer the above question, we consider a device schematic of an optical demultiplexer, as shown in Fig. 1.8, whose task is to receive a WDM signal with M wavelengths at the input port and forward each input wavelength to one specific output port, based on the principle of constructive/destructive interference between the lightwaves of each wavelength. In other words, each input lightwave (wavelength) gets spread over (typically due to a diffraction phenomenon, as in a double-slit experiment or in some grating device) the cross-section of the device in multiple directions. Thus, the multiple (diffracted) lightwave components from each input wavelength are made to travel through different path lengths (dashed lines) to each output port, and at the *desired output port* for a given wavelength w_1 (say), the *optical* path difference $\Delta L_1^{opt}(ij)$ between any two paths, say paths i and j with a geometrical

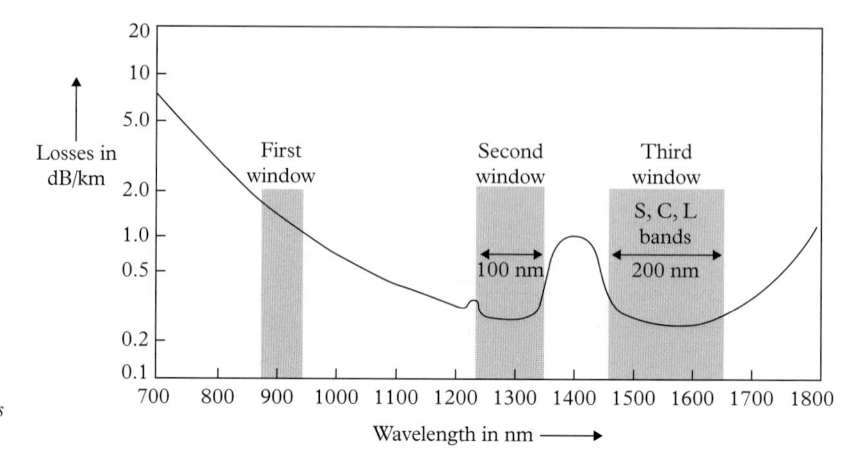

Figure 1.7 *A representative plot of the loss in optical fibers versus wavelength.*

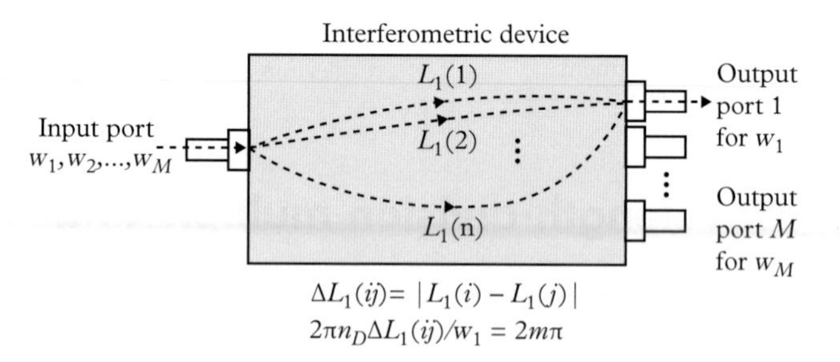

Figure 1.8 *Schematic diagram of an optical demultiplexer.*

path difference $\Delta L_1(ij)$ (Fig. 1.8), has to be an even multiple of π for a constructive interference, i.e.,

$$\Delta L_1^{opt}(ij) = \frac{2\pi \Delta L_1(ij)}{w_1/n_D} = 2m\pi, \tag{1.1}$$

with m as an integer and n_D representing the effective refractive index of the device medium. The above expression implies that $\Delta L_1(ij)$ must be proportional to the wavelength w_1, i.e.,

$$\Delta L_1(ij) = \frac{mw_1}{n_D} \tag{1.2}$$

and hence the choice of the physical dimensions and placements of output ports of the demultiplexer would be *wavelength-centric*, based on the wavelengths under consideration.

Thus, it is presumably the wavelength-centric design approach, that has oriented the device physicists in visualizing such devices as wavelength demultiplexers (or multiplexers in the reverse direction), and the underlying task of optical signal-processing as wavelength-division multiplexing (i.e., WDM), rather than FDM. Note that, in electronic circuits, such spectral demultiplexers are realized by frequency-selective bandpass filters, in contrast to the usage of wavelength-selective interferometric means in optical devices.

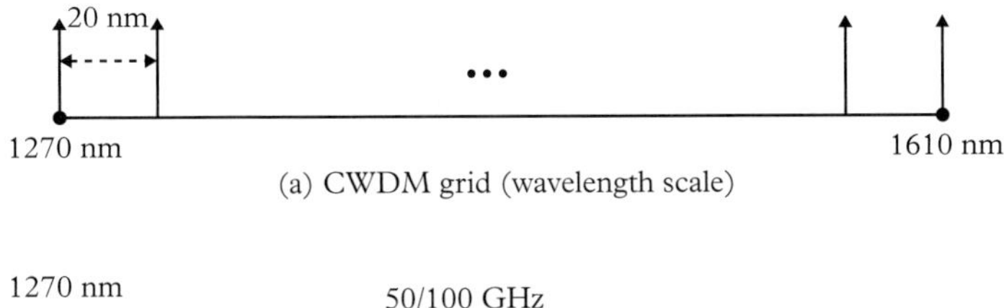

(a) CWDM grid (wavelength scale)

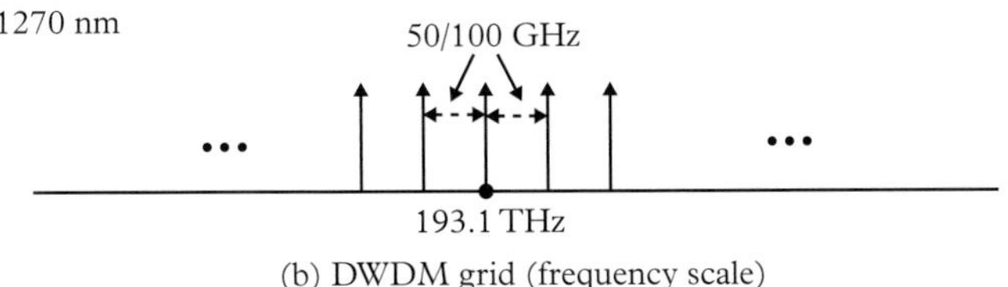

(b) DWDM grid (frequency scale)

Figure 1.9 *WDM channel grids recommended by ITU-T: (a) CWDM wavelength grid, (b) DWDM frequency grid.*

Governed by ITU recommendations, the WDM channels (wavelengths or frequencies) can be chosen following one of the two possible WDM grids: dense WDM (DWDM) and coarse WDM (CWDM). The earlier ITU-T recommendation for DWDM systems (ITU-T G.694.1 2002) was to use WDM channels in the third window, centered around 193.1 THz (1552.52 nm) with a 100-GHz channel spacing on each side, which was later modified with the same central frequency but using 50-GHz channel spacing. With the recent developments on the elastic optical networks (EON) with a flexible frequency grid (for better bandwidth utilization of optical fibers), the DWDM standard has been further revised with narrower channel spacing of 12.5 GHz with the same central frequency, i.e., 193.1 THz. However, the CWDM systems were recommended (ITU-T G.694.2 2003) with 12 wavelengths over 1270–1610 nm (i.e., including the second and third windows of optical fiber) with 20-nm channel spacing. The ITU-T standards for the CWDM and DWDM grids are illustrated in Fig. 1.9, where the CWDM systems can accommodate 18 channels over the range of 1270–1610 nm and the DWDM systems can choose channel frequencies from C and L bands in the third window[1] with the total number of channels determined by the evolving technology of optical amplifiers to accommodate a wider range of channels for amplification. With a typical optical-amplifier bandwidth of 35 nm in C band, a DWDM system using a 50-GHz grid will be able to transmit about 80 channels concurrently. Thus, with each channel carrying a data stream at a speed in the range of 40–100 Gbps, the DWDM system would be able to offer a total transmission capacity of 3.2–8 Tbps over an optical fiber.

The CWDM systems use the WDM channels placed with equal or uniform wavelength spacing across the spectrum, while the DWDM systems choose the channels with equal frequency spacing. In the CWDM systems, the WDM channels being placed with equal wavelength spacing (= 20 nm), the inter-channel frequency spacings would become unequal, governed by the relation: $|\Delta f| = |\Delta \mathrm{w}|c/\mathrm{w}^2$, with f and w representing the frequency and wavelength, respectively, and c as the velocity of light in free space. However, in DWDM systems, the channel-packing needs to be tight (but with necessary margins for various uncontrollable features, e.g., frequency misalignment between the lasers and optical filters, laser frequency drift etc.), thus calling for a channel spacing that would be close and related directly to the modulation bandwidth of the WDM channels. Modulation bandwidths being

[1] Note that the third spectral window in optical fibers is divided into three bands, called S band (shorter wavelength region), C band (central wavelength region), and L band (longer wavelength region).

decided by the channel transmission rates, the DWDM channel placement is therefore made with equal frequency spacing, thus leading to a uniform frequency grid.

Besides WDM, some other multiplexing techniques have been explored for optical networking, one of them being optical time-division multiplexing (OTDM) (Barry *et al.* 1996), where ultra-narrow optical pulses are generated, modulated, and time-division multiplexed optically using optical delay lines flanked by a splitter-combiner combination. Though OTDM transmission can exploit the fiber bandwidth through narrow pulse transmission, the device technologies for OTDM multiplexing/demultiplexing are complex and yet to mature, and thus presently cannot offer an acceptable commercial alternative to the WDM systems. The other multiplexing technique uses optical code-division multiplexing (OCDM), which is also yet to mature enough for commercial use in optical networks (O'Mahony *et al.* 2006).

1.6 WDM optical networks

Initially WDM transmission was tried in optical LANs/MANs to enhance the speed and coverage areas, and more effective use of WDM was made thereafter in the metro and long-haul networks using the SONET/SDH technology. Extensive studies have been carried out on the WDM-based LANs/MANs in various research laboratories and universities (Mukherjee 1992a; Mukherjee 1992b; Brackett 1990). Generally, the WDM LANs/MANs (or simply WDM LANs) were formed around optical passive-star couplers, with each node employing broadcast-and-select transmission over one or more suitable wavelengths. The transmitters and receivers in the network nodes used tunable or fixed receivers, leading to different types of WDM LANs, known as single-hop and multihop networks. However, from the commercial viewpoint WDM LANs could not make much impact in the long run, while the technologies that emerged from this research were later found useful for the WDM-based long-haul, metro, and access network segments.

For the metro[2] and long-haul networks, while responding to the increasing traffic, the existing optical networks went initially through the deployment of WDM technology using point-to-point WDM links. However, with the point-to-point WDM links, the growth in IP traffic started overloading the IP routers (or other client network elements) in handling the aggregate network traffic for all WDM channels with OEO conversion for each wavelength and hop-by-hop forwarding of traffic at the intermediate nodes. The burden of routers in the electronic domain could be reduced by using some novel optical technologies, enabling the intermediate nodes in a WDM network to carry out the routing functionality in the optical domain itself (Chlamtac *et al.* 1992). Such networks, generally known as wavelength-routed optical networks (WRONs), can be realized in various topological forms, including mesh, ring, interconnected rings, etc.

Figure 1.10 illustrates a typical WRON, realized over a mesh topology with wavelength-selective optical bypass operation at intermediate nodes, by using optical crossconnects (OXCs), where the end-to-end all-optical connections are called *lightpaths*. Such end-to-end lightpaths between the node pairs (in particular, between the respective client NEs, e.g., IP routers, Ethernet switches, SONET/SDH ADMs etc.) are realized through multiple fiber

[2] Note that there is a subtle difference between MANs and metro networks. MANs by definition cover metropolitan areas, and such networks generally remain under the control of an autonomous organization. However, the networking infrastructures that are set up over metropolitan areas with free access to individual users are different from the autonomous MANs (also considered as larger versions of LANs) and are popularly referred to as the public-domain metropolitan or simply *metro* networks.

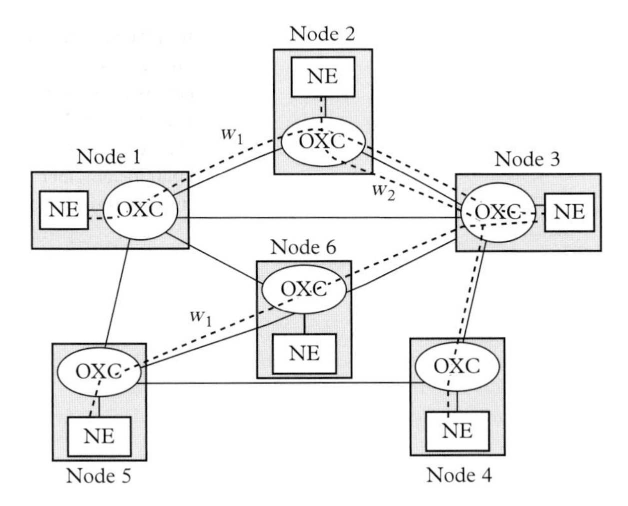

Figure 1.10 *An example 6-node WRON with mesh topology using OXCs. Three lightpaths (shown by dashed lines) are set up: one between the nodes 1 and 3 on wavelength* w_1*, through node 2 as an intermediate node, another between nodes 2 and 4 on wavelength* w_2*, through node 3, and the third between the nodes 5 and 3 on wavelength* w_1*, through node 6.*

links by using reconfigurable OXCs, which make use of optical multiplexers, demultiplexers, and switches. However, OXCs can also be realized using OEO conversion for various reasons (e.g., to refresh signal quality, relax wavelength-continuity constraint, en route traffic grooming on the pass through lightpaths etc.), but with increased hardware complexity. In Fig. 1.10, three lightpaths (shown by the dashed lines) are set up: a lightpath between the nodes 1 and 3 through the OXC of node 2 on wavelength w_1, another lightpath between nodes 2 and 4 through the OXC of node 3 on wavelength w_2, and the third lightpath between the nodes 5 and 3 through the OXC of node 6 on wavelength w_1. Note that, for the lightpath between the nodes 2 and 4 one has to choose wavelength w_2 to avoid wavelength-overlap with the lightpath between the nodes 1 and 3, while for the lightpath between the nodes 5 and 3 one can choose again wavelength w_1 without any wavelength-overlap.

As in a mesh topology, the wavelength-routing functionality can also be realized in the ring-based metro networks (Fig. 1.11), where the OXCs are replaced with fixed/reconfigurable optical ADMs (OADMs/ ROADMs). Note that the OADMs, while performing the wavelength-routing function for passthrough traffic, do not have to perform crossconnect operation, as the nodes in a ring need to operate with a physical degree of two only. Thus, having arrived at one of the two ports of an OADM, the passthrough traffic has the only option to get forwarded to the other port (unlike, having the choices of multiple outgoing ports in OXCs with nodal degrees exceeding two), thereby making the OADM hardware simpler than that of OXCs.

One important device in the long-haul WDM networks is the optical amplifier, mainly an erbium-doped fiber amplifier (EDFA), which is used in appropriate locations in a network to compensate for the losses incurred in the network nodes and fiber segments (Desurvire 1994). In particular, EDFAs are typically placed immediately before and after a WDM node and as in-line amplifiers after every $\simeq$ 80 km in a long fiber link. These optical amplifiers, having a bandwidth of about 35 nm (and somewhat more for the improved versions) in the third transmission window of optical fiber, can greatly simplify the network design by amplifying all the WDM signals in one go. For example, with in-line EDFAs operating over its 35-nm wide amplification band, a DWDM system with the ITU-T 100/50-GHz frequency grid, can transmit about 40/80 WDM channels through a long-haul link spanning a few hundred kilometers. However, an upper limit on such EDFA-amplified long-haul links

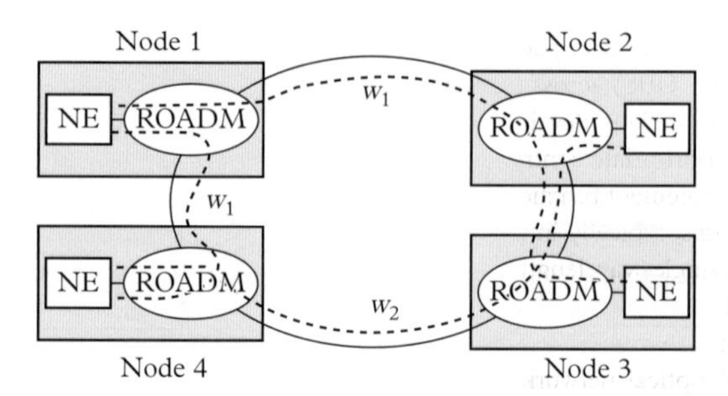

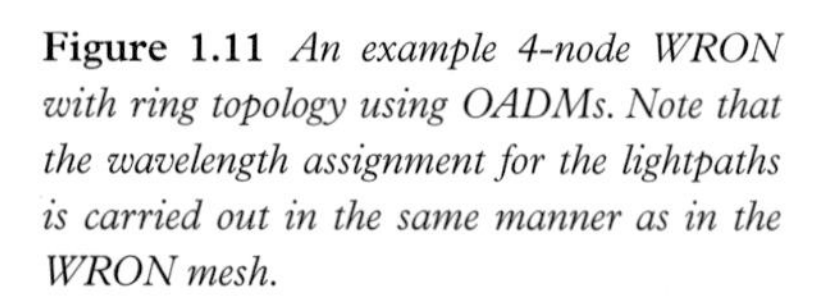
Figure 1.11 *An example 4-node WRON with ring topology using OADMs. Note that the wavelength assignment for the lightpaths is carried out in the same manner as in the WRON mesh.*

is imposed by the accumulation of the noise components generated in each EDFA, therefore needing OEO regeneration at some intermediate nodes, as necessary.

Currently, an important research topic in the area of WRONs deals with the flexibility of the frequency-grid for WDM transmission (Gerstel *et al.* 2012). In particular, based on the finest WDM grid of ITU-T standard, i.e., 12.5-GHz channel spacing, a notion of *flexible or elastic grid* is being explored in EONs, where one can aggregate a number of 12.5-GHz frequency slots, to realize a flexible $n \times 12.5$ GHz channel bandwidth as per the incoming connection requests. However, such flexibility in the bandwidth selection in WRONs can only be achieved at the cost of more complex bandwidth-variable WDM technologies.

As in the case of RPR developed for the single-wavelength packet-switched ring network, extensive studies have also been carried out on the packet-switched WDM ring networks for metro segment (Herzog *et al.* 2004). The testbeds on WDM packet-switched metro rings include MAWSON, HORNET, Ringo, Ringostar, etc. These studies reported various possible ways to realize packet-switched WDM rings, by using novel MAC protocols and node architectures. However, these networks, notwithstanding their novel data-friendly architectures, are yet to mature as commercial networks.

The capacity and the span of optical access networks can also be enhanced significantly by using WDM over PON topology, leading to WDM PONs (Kramer *et al.* 2012). WDM PONs can use a WDM device at the RN, such as arrayed-waveguide grating (AWG), to employ user-specific channels for the ONUs at the customer premises. However, following the AWG at RN, WDM-PONs can also use splitters/combiners for each wavelength to accommodate more users per wavelength by using TDM/TDMA. Furthermore, one can also make such WDM-PONs more flexible by replacing the AWG at the RN by a splitter/combiner to realize more flexible and bandwidth-efficient WDM PONs, where optical amplifiers may have to be used in the feeder segment to restore the power budget due to excessive power-splitting introduced to support large number of ONUs.

Further, as mentioned earlier, datacenters are evolving fast to serve as the centralized storehouses of data and computing resources, typically located in a building complex, employing huge number of interconnected servers, in the range of thousands to hundreds of thousands and even more in the foreseeable future (Xia *et al.* 2017). In most of the datacenters, the interconnections between the servers need to be augmented with optical-networking support, reducing significantly the network latency and power consumption.

The other noteworthy research efforts in the area of optical WDM networks include the use of optical packet and burst-switching techniques, leading to optical packet-switched (OPS) (Guillemot *et al.* 1998) and optical burst-switched (OBS) networks (Yoo 2006), which are yet to mature as a viable technology for commercial application. The OPS and OBS schemes aim to improve the bandwidth utilization in optical WDM networks handling large volume of bursty data traffic (as compared to the circuit-switched WRONs), by switching/routing of optical packets (for OPS) or aggregated optical packets, i.e., optical packet-bursts (for OBS) across the optical network.

Alongside the various developments as described above, the task of operating optical networks with appropriate control and management functionalities had to be explored. The control and management functionalities of networks have been embedded traditionally in all forms of networks, typically in a distributed manner. For optical networks, SONET/SDH standards with single-wavelength transmission have incorporated these functionalities as an entity, referred to as operations administration and maintenance (OAM), which are carried out in a distributed manner through the respective overhead bytes in SONET/SDH frames. For the WDM optical networks, individual WDM channels, while using SONET/SDH standard, also come under the purview of OAM. However, the WDM optical networks need to exercise separate control and management functionalities, by using different class of standards, e.g., generalized multiprotocol label switching (GMPLS), automatically switched optical network (ASON), which are also operated in a distributed manner. More recently, another standard for this purpose has evolved, referred to as software-defined optical network (SDON) (as an extension of software-defined network (SDN)), which is operated in a centralized manner. As we shall see in the following, choice of the standard for network control and management will also determine the overall architecture of an optical network.

1.7 Architectural options in optical networks

From the preceding discussions, it is evident that optical networks can be operated with several possible options for the overall network architectures. As the preliminary single-wavelength LANs were all conceived as packet-switched networks, they could directly fit into the traditional five-layer TCP/IP protocol suite with the fiber-specific MAC protocols for the link layer, as in FDDI, DQDB, or different versions of Gigabit Ethernet. However, SONET/SDH networks, though primarily designed for voice-centric circuit-switching service, also have had the options to circuit-switch the packet-switched traffic using TDM slots over the metro and long-haul networks, implying that SONET/SDH could support both kinds of traffic (circuit and packet-switched) over the fiber. This led to an inclusive architecture permitting both circuit and packet-switched traffic streams from the client layers, such as PDH, Ethernet, IP and ATM traffic streams, to pass through the SONET/SDH networks, as shown in Fig. 1.12. In reality, the ability of SONET/SDH networks evolved over time to carry some other real- and non-real-time traffic including video as well as data streams from other packet-switched client networks (e.g., video, SAN etc.), leading to a more inclusive version of the layered architecture.

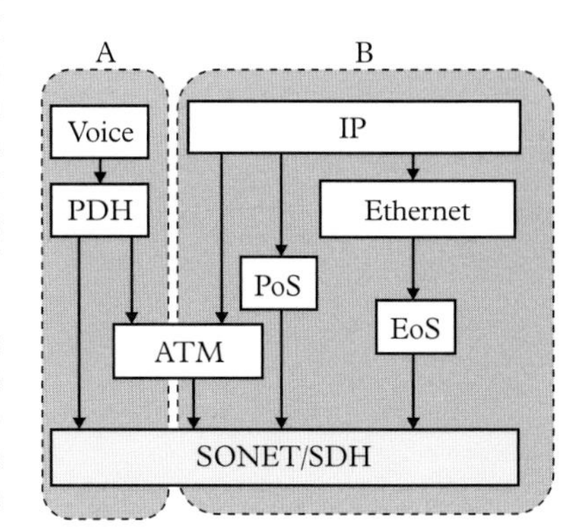

Figure 1.12 *SONET/SDH-based layered architecture for single-wavelength optical fiber transmission. Note that, the traffic streams belonging to categories A and B are circuit-multiplexed within the SONET/SDH frames using TDM.*

Note from Fig. 1.12 that all the client traffic flows can be visualized into two categories: category A and category B. The traffic flows belonging to category A represent the voice traffic that is circuit-switched typically (*i*) through PDH and SONET/SDH or (*ii*) through PDH, ATM (virtual circuit-switching), and SONET/SDH. However, the traffic flows in category B are originally packet-switched traffic which are first passed through a suitable adaptation interface, e.g., packet-over-SONET/SDH (PoS) or Ethernet-over-SONET/SDH (EoS), and thereafter circuit-switched through the SONET/SDH layer. Finally, viewing from the SONET/SDH layer, we observe that the traffic categories A and B are eventually *circuit-multiplexed* together through the respective TDM slots in a SONET/SDH bit stream.

In WDM-based optical networks, with manifold increase in the network capacity due to concurrent transmission of a large number of wavelengths (and associated complexities), several architectural options have emerged to support the variety of traffic classes and client layers. In particular, the use of WDM in optical networks led to a distinct additional layer (in comparison to the single-wavelength networks), called the *optical layer*, which deals with the appropriate wavelength selection, multiplexing/demultiplexing of wavelengths at network nodes, and optical amplification of the WDM signals on their way to the destination nodes.

Figure 1.13 illustrates some of the candidate architectural options for WDM networking, where the optical layer (also called the *WDM layer*) appears at the bottom with three sublayers, through which all possible traffic streams from the respective clients must pass. In the optical layer, looking downward, all the client signals get assigned with appropriate wavelengths in the *optical channel selection* sublayer, optically multiplexed in the *optical channel multiplexer* sublayer, and amplified by the optical amplifiers operating across the network in the *optical amplifier* sublayer. Equivalent operations are carried out in the reverse direction (upward) while receiving a WDM channel. As shown in Fig. 1.13, some candidate architectural options (though not exhaustive) are as follows.

- PDH-over-SONET/SDH-over-WDM: This stack of layers implies that the PDH frames are packed into the SONET/SDH frames, which are subsequently transmitted through the optical layer using an appropriate wavelength (WDM channel).
- IP-over-SONET/SDH-over-WDM using the PoS interface: In this case, the IP packets are packed into the SONET/SDH frames using PoS as the intermediate layer, and then transmitted through the optical layer using an appropriate wavelength.
- IP-over-ATM-over-SONET/SDH-over-WDM: When the IP traffic needs to go through the ATM layer as a convergent platform (by asynchronously multiplexing all kinds of services), the IP packets are split into ATM cells, which are subsequently transmitted through the SONET/SDH-over-WDM protocol suite as above.
- PDH-over-ATM-over-SONET/SDH-over-WDM: This stack of layers is equivalent to the above case (i.e., IP-over-ATM-over-SONET/ SDH-over-WDM), wherein the PDH frames (instead of IP packets) are split into ATM cells. The ATM cells are thereafter transmitted through the SONET/SDH-over-WDM protocol suite.
- Ethernet-over-SONET/SDH-over-WDM using the EoS interface: In this case, the Ethernet frames are packed into the SONET/SDH frames using EoS as the intermediate layer and transmitted through the optical layer using an appropriate wavelength.
- x-over-OTN-over-WDM: As mentioned earlier, OTN is a standard that serves as a convergence layer, through which some of the higher-layer packets/frames can be encapsulated into a universal format and passed on to the optical layer for onward transmission. Three important clients (x) for OTN are SONET/SDH, Ethernet, and IP. Due to the huge popularity of Ethernet, high-speed Ethernet clients adopt this option to make use of the optical layer for onward WDM transmission. Thus, Ethernet-over-OTN-over-WDM has been a useful protocol suite for connecting the distant high-speed Ethernet LANs through long-haul WDM networks, and is popularly referred to as *carrier-grade Ethernet*.
- IP-over-GMPLS-over-WDM: In this case, IP packets bypass the SONET/SDH layer, and are instead passed through the GMPLS layer and transmitted over an appropriate wavelength, where the GMPLS protocol may have to resort to using ASON protocol for inter-domain networking, as necessary. Further, as mentioned earlier, research efforts are also being made to migrate from the distributed control-plane regime of GMPLS/ASON to SDON, with SDON executing the network control and management operations in a centralized manner.

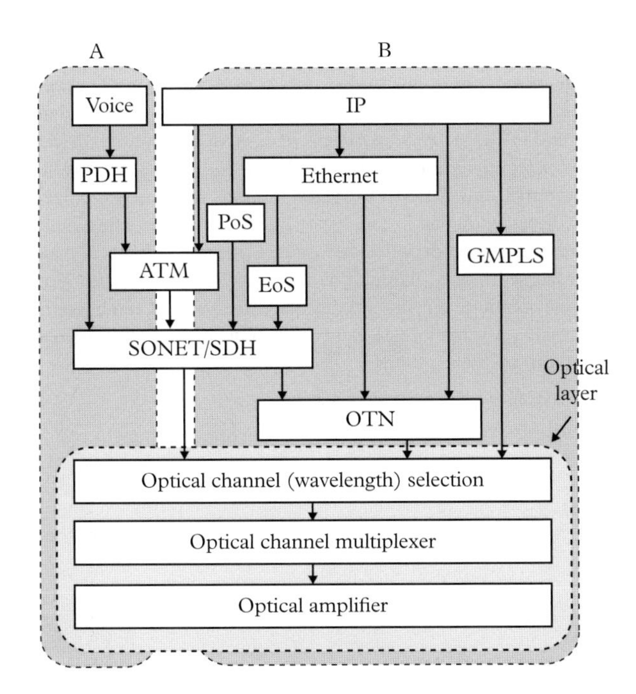

Figure 1.13 *Architectural options for optical WDM networks, with the traffic categories A and B having the same implications as in Fig.1.12.*

As shown earlier, in Fig. 1.12, in this case also (i.e., in Fig. 1.13) the incoming traffic flows can be notionally divided into categories A and B, which are eventually wavelength-multiplexed through the optical layer. Further, note that the packet or burst-switched WDM networks, if found viable for commercial use in future, would have to adopt an appropriate protocol suite, such as, packet/burst-over-GMPLS-over-WDM, thereby bypassing the circuit-switched SONET/SDH layer (as in IP-over-GMPLS-over-WDM architectural option) for enhanced bandwidth utilization. Similarly, appropriate modifications would be needed in the optical layer to accommodate a flexible optical-frequency grid as and when the EON technologies become commercially viable.

1.8 Summary

In this chapter, we presented an overview of optical networks, beginning with the background of today's telecommunication network and the roles of optical fibers therein. Next, we discussed the chronology of developments in telecommunication networks starting from the days of PSTN offering POTS as the basic service, followed by the divestiture of Bell Laboratories and subsequent developments of the demonopolized regime of telecommunication networks. Thereafter, we described the salient features of the two generations of optical networks for various network segments.

The first-generation optical networks, employing single-wavelength transmission, were described with an introduction to the various networking segments: optical LANs/MANs, SONET/SDH networks for metro and long-haul networks, OTN standard covering packet and circuit-switched traffic over a convergent platform, packet-switched ring using RPR, PONs for access segment, SANs and datacenters. Next, we presented the second-generation optical networks employing concurrent multiwavelength transmission over optical fibers using WDM technologies. In doing so, first we introduced the basic concept of WDM along with the basic features of the ITU-recommended frequency grid for dense and coarse WDM (CWDM and DWDM) systems. Then we discussed the variety of WDM networks for different network segments: access, metro and long-haul, including WDM LAN, WDM ring and mesh networks, and WDM PONs. As an improvisation of the fixed-grid WDM networks, we also discussed the basic concept of EON, which uses a flexible frequency grid for better bandwidth utilization in optical networks. Basic ideas of optical packet and burst-switching were discussed briefly, though the respective technologies are yet to mature as commercially viable solutions. Finally, we discussed the various candidate network layers that are stacked to set up viable optical network architectures, both for single-wavelength and WDM-based transmission systems.

Technologies for Optical Networking

2

The technologies used in optical networks have evolved seamlessly over the past six decades. Optical fibers with extremely low loss and enormous bandwidth are used as the transmission medium in optical networks, while semiconductor lasers and LEDs serve as optical sources, and photodetectors, such as *pin* and avalanche photodiodes, are used to receive the optical signal at the destination nodes. The transmitted optical signal has to pass through a variety of network elements, which in turn need a wide range of passive and active devices, carrying out the necessary networking functionalities. For WDM optical networks, many of these tasks need to be accomplished in the optical domain itself in a wavelength-selective manner, calling for various types of WDM-based networking elements. In this chapter, we present a comprehensive description of the optical and optoelectronic devices that are used in setting up modern optical networks.

2.1 Optical networking: physical-layer perspective

With extremely low loss and enormous bandwidth, optical fiber stands today as the most attractive wireline transmission medium, so much so it has already conquered the long-haul and metro segments of telecommunication networks, and has also penetrated considerably into the access segment. A wide range of passive and active photonic components operate together to make the optical transmission systems (physical layer) in telecommunication networks operate through optical fiber links.

During the initial phase of development, optical transmission systems employed a single wavelength, with continuous effort toward decreasing the fiber loss and increasing the transmission rate from a few tens of Mbps to the rates in the Gbps range, which helped in developing high-speed inter-city telecommunication links as well as high-speed LANs/MANs with larger network spans. However, the switching and routing of traffic over single-wavelength optical networks remained confined to the class of network elements operating in the electrical domain (e.g., electronic switches, routers, ADMs, digital crossconnects). For the long-haul intercity links, the feasibility of enhanced transmission span was also established by exploring coherent transmission of lightwaves, which could improve the receiver sensitivity by 15 dB or more, with an enhancement of un-repeatered link lengths by $\geq$ 70 km. Furthermore, with the potential of $\simeq 40,000$ GHz transmission bandwidth of optical fbers, single-wavelength transmission could not utilize the full capacity, thereby ushering in the WDM transmission systems.

Initially, WDM networks remained confined to point-to-point communication links, which gradually evolved with time to realize the switching and routing of lightwaves at intermediate nodes in the optical domain itself. Further, the amplification of lightwaves for all co-propagating wavelengths in an optical fiber using a novel and simple device – doped

Optical Networks. Debasish Datta, Oxford University Press (2021). © Debasish Datta.
DOI: 10.1093/oso/9780198834229.003.0002

fiber-optical amplifiers – was demonstrated, which significantly simplified the regeneration hardware using OEO conversion for each co-propagating wavelength in a fiber. At this stage, the optical networks graduated from being merely *electronically switched/routed* networks to the genre of optical networks exercising the capability of switching, routing, and regeneration in optical domain itself, along with its per-wavelength transmission speed increasing steadily from a few Gbps to 100 Gbps and more.

For single-wavelength non-coherent optical transmission, both LEDs and lasers are used as intensity-modulated (IM) optical sources while the receiver employs a *pin* or avalanche photodiode for direct detection (DD), and thus the non-coherent transmissions are functionally based on IM-DD transceivers. The choice of optical sources (LED or laser) for the IM-DD systems is governed by the span and speed of the transmission link. Thus, in single-wavelength systems, for covering longer distances at high speed, lasers are preferred and typically OEO regeneration is employed at intermediate nodes. However, with transmission rates exceeding a certain limit, typically beyond 10 Gbps, one needs to adopt coherent or nearly coherent transmission schemes, by employing traditional modulation techniques, as used in radio communication systems.

Typical single-wavelength IM-DD binary optical transmission links without and with OEO regeneration are illustrated in Fig. 2.1. In a practical network-setting, the linear optical fiber link shown in the figure is used to form a ring topology with all the nodes using OEO conversion in metro segments, while the long-haul networks make use of interconnected rings or mesh topologies. Further, tree topologies are used with optical splitters/combiners in the access segments. In the ring and mesh topologies, electronic switching/routing is employed (with OEO conversion) at intermediate nodes for realizing the end-to-end connections. Single-wavelength coherent transmission systems are far more complex than IM-DD systems, and a discussion on the coherent systems is presented toward the end of this chapter.

For WDM networking, a variety of WDM devices needs to be used, including optical multiplexers (OMUXs) and demultiplexers (ODMUXs), along with optical filters and switches, optical add-drop multiplexers (OADMs), optical crossconnects (OXCs), optical amplifiers (OAs) etc., as shown using a few representative examples in Fig. 2.2. While in single-wavelength networks, the networking functionalities are carried out in the electrical domain at the intermediate nodes, in WDM networks a variety of controlled optical devices are used for routing/switching the wavelengths across the network. In these devices, the lightwave traverses in the optical domain, while the control of its path through the switching devices (OXCs, OADMs etc.) takes place through appropriate transducers

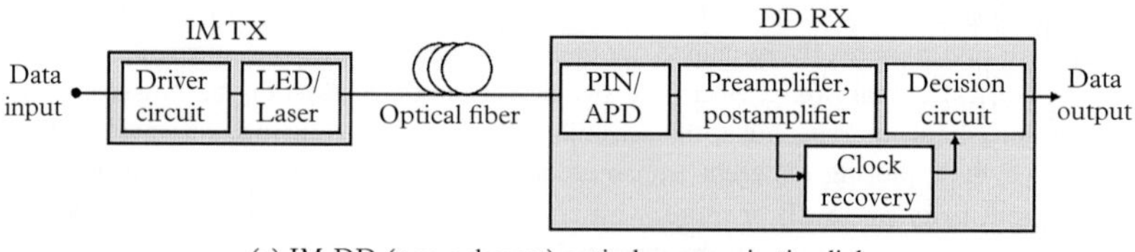

(a) IM-DD (non-coherent) optical communication link.

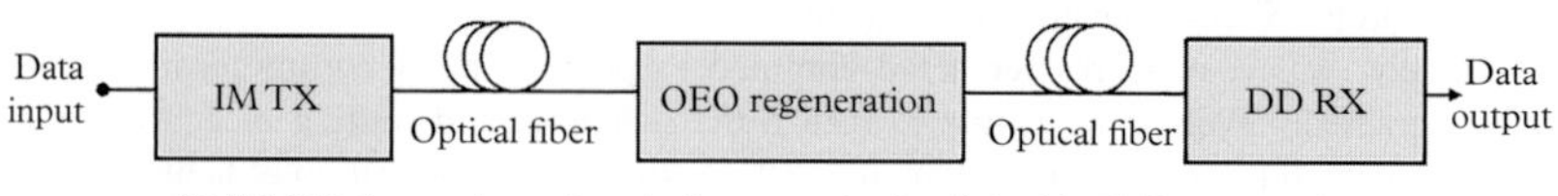

(b) IM-DD (non-coherent) optical communication link with OEO regeneration.

Figure 2.1 *Single-wavelength optical transmission using intensity modulation (IM) and direct detection (DD).*

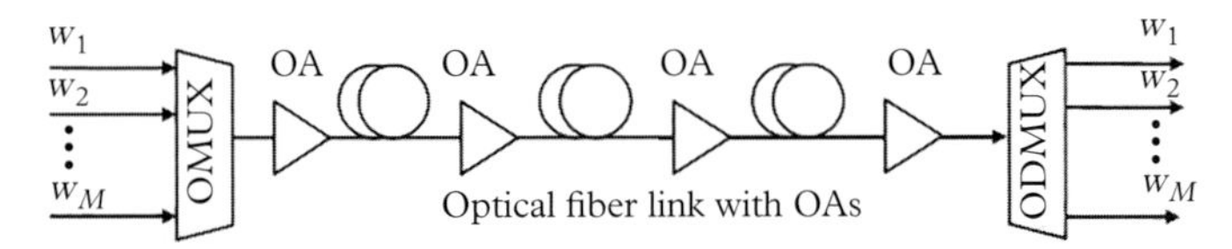

(a) Point-to-point WDM transmission with optical amplifiers (OAs).

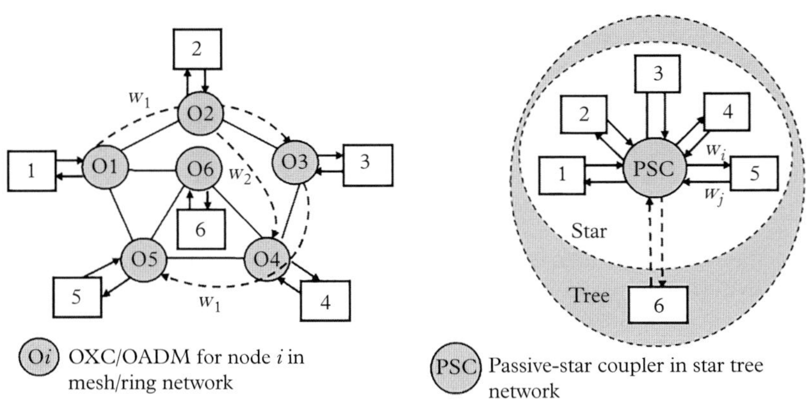

(b) WDM networking: mesh (becomes a ring with node 6 removed) and PSC-based star tree (with node 6 added to the star as a central hub with a feeder fiber).

Figure 2.2 *WDM optical transmission over various networking topologies. Note that when the mesh is changed to ring by removing node 6, the OXCs are replaced by OADMs with only two fiber ports. Further, the PSC in star/tree topologies operates as a passive combiner and a splitter as well, while viewing from any node.*

(e.g., electro-optic, thermo-optic, accousto-optic, electromechanical etc.). As shown in Fig. 2.2(a), a linear point-to-point WDM link would make use of an OMUX at the transmitting end, and an ODMUX at the receiving end, while in the middle it might need optical regeneration using OAs with their separation not exceeding typically 80 km.

As in single-wavelength networks, for practical WDM networks one needs to extend the linear segments to adopt a variety of topological formations – mesh, ring, star, tree – where multiple wavelengths are used in various possible ways. For example, in mesh topology shown in the figure, the connection from node 1 to node 3 is realized using transmission over wavelength w_1, which is bypassed at node 2 using an OXC or an OADM (OADMs being typically used for the ring topologies).[1] The two other connections are set up on wavelengths w_2 and w_1, such that they do not *clash* with each other, but are reused wherever feasible. The other approaches to employ WDM would be to use a star or a tree topology, where for star topology each node transmits some wavelengths which are broadcast through a passive-star coupler (PSC), and the destined nodes select the appropriate wavelengths using optical filters.

Following the above introduction to the basic physical-layer needs of optical networks, we next proceed to describe the salient features of the important components/devices that enable the optical networks to operate at various segments, i.e., long-haul backbone, metro, and access.

[1] Note that multi-degree OADMs (i.e., with a degree > 2) can also be used in WDM mesh topologies, typically formed using interconnected WDM rings.

2.2 Optical fibers

Optical fibers represent a class of cylindrical waveguides for the incident light on the end face to propagate along the longitudinal axis of the cylindrical structure. In particular, as shown in Fig. 2.3, an optical fiber employs an inner dielectric cylinder, called a *core* with a refractive index n_1, surrounded by a cylindrical *cladding* with a refractive index n_2, such that $n_1 > n_2$. With $n_1 > n_2$, as the light propagates through the optical fiber along the longitudinal direction, the light rays undergo total internal reflection (TIR) at the core–cladding interface and thus remain confined in the core region during the course of its propagation. The materials used for the core and cladding are made from silica glass, with appropriate dopings to realize the desired refractive indices n_1 and n_2.

The phenomenon of TIR is illustrated in Fig. 2.4, using cross-sectional and longitudinal views of an optical fiber. In the cross-sectional view, the refractive index profile (RIP) of the fiber is shown as a *step* function between the core centre and the core–cladding interface on both sides of the centre, wherein n_1 at the top of the RIP represents the refractive index in the core, while the base of the RIP is the cladding refractive index n_2. The fibers having such RIP are called *step-index* fibers, with $\Delta = (n_1 - n_2)/n_1$ representing the core–cladding refractive-index difference (or simply index difference), normalized by the core refractive index n_1. In Fig. 2.4, two light rays are shown to be incident on the end surface of core (air-to-core interface) making angles θ and θ' with the longitudinal axis, which get refracted into the core region, following Snell's law of refraction. For the ray incident at an angle θ with the longitudinal axis, the refracted ray is incident on the core–cladding interface making an angle of incidence ψ, and the value of ψ exceeds the critical angle for the core–cladding interface, leading to TIR and the light thereafter keeps propagating along the fiber with zig-zag path in a series of TIRs. However, the light ray with an incidence angle θ' reaches the core–cladding interface at an angle of incidence lower than the critical angle and hence escapes from the core to the cladding region without any TIR. The second category of rays do not contribute to the propagation of light. As we shall see later, there exists a maximum limit of θ, called the acceptance angle θ_A, beyond which no incident light ray is accepted by the fiber for onward propagation. In other words, one can construct, at the air-to-core interface, a cone around the longitudinal axis with a half angle θ_A, beyond which no incident ray will be able to propagate through the fiber.

Based on the basic material and structural parameters (e.g., the values of n_1, n_2 and the core radius a), optical fibers exhibit certain characteristic features which are important for optical transmission systems, such as the amount of power that can be coupled into an optical fiber and the maximum transmission rate that an optical fiber link can support. Using these features, optical fibers can be categorized broadly into two types: step-index and graded-index fibers. In step-index fibers (as shown in Fig. 2.4), the refractive index across the cross-section abruptly changes at the core–cladding interface, while in graded-index fibers the refractive index of the core gradually decreases from its maximum value at the center of the core to the base value at the core–cladding interface. In both cases, the choices of core radius and the refractive indices in core and cladding regions can be varied

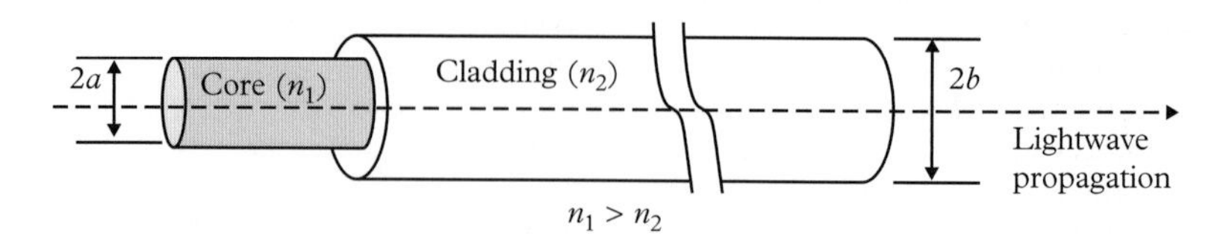

Figure 2.3 *Optical fiber. A protective coating is used around the cladding region without any significant role for the light propagation in the fiber, and hence not shown here.*

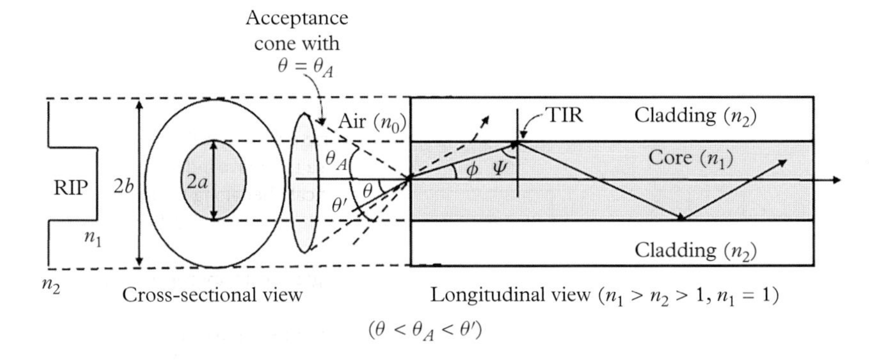

Figure 2.4 *Illustration of TIR in optical fibers. Note that, for the light ray with $\theta(< \theta_A)$ as the angle of incidence on the core–cladding interface, TIR takes place as ψ exceeds the critical angle. On the other hand, the other light ray (dashed line) with $\theta'(>\theta_A)$ as the angle of incidence does not satisfy the condition for TIR and thus escapes from the core to the cladding region.*

to obtain different propagation characteristics of optical fibers, leading to *single-mode* and *multi-mode* optical fibers. In particular, with larger values of a/w (with w as the operating wavelength) and the refractive-index difference Δ, one can have a class of optical fibers, where the launched optical power gets divided into a large number of modes, leading to the multi-mode fibers. Each mode implies a certain pattern of electromagnetic fields in the optical fiber, with most of these modes having some equivalence to a group or congruence of light rays. On the other hand, with smaller values of a/w and Δ, one can design the single-mode optical fibers, where only the fundamental mode can propagate.

Figure 2.5 illustrates different types of optical fibers, using step-index and graded-index RIPs with single and multi-mode propagation. As shown in the figure, typical values for cladding diameter ($2b$) in today's optical fibers is 125 μm, while the core diameter ($2a$) for single-mode optical fibers would be in the range of 8–10 μm (Fig. 2.5(a)), and multi-mode fibers can have core diameters as large as 50 μm or more (Fig. 2.5(b)). Regarding the RIP, multi-mode fibers would in general have an index difference Δ in the range of 1–3%, while for single-mode fibers Δ can be as small as 0.2%. Note that, in the case of multi-mode fibers, the axial and non-axial lines represent the shortest and longest rays, respectively. The difference in lengths between the longest and shortest rays (corresponding to the slowest and fastest modes) leads to intermodal dispersion, causing pulse spreading, which can be reduced in graded-index optical fibers, as shown in Fig. 2.5(c). We discuss later on the various dispersion mechanisms in optical fibers in detail. However, for single-mode fibers, the axial line (a dashed line) implies only the direction of propagation and not exactly a ray, as in this case the ray theory loses most of its significance due to small core diameter. We shall get back to the choice of the structural dimensions, the values of Δ, and RIP variations for different types of fibers, when dealing with the propagation phenomena in optical fibers.

As the light propagates along the fiber length, it suffers power loss for various reasons, forming some specific spectral windows, where the signal *loss* or *attenuation* remains below acceptable limits. Further, due to the structural and material properties of optical fiber and also due to the non-ideal spectral properties of optical sources, an optical pulse propagating along the fiber length spreads in the time domain due to some phenomena, collectively known as *dispersion*. Due to the loss and dispersion, the received optical power at the destination gets reduced (due to loss) along with stretched pulses (due to dispersion) leading to inter-symbol interference (overlapping in time) between the adjacent symbols in a propagating bit stream. While the loss mechanisms in optical fibers are determined by the absorption and scattering properties of the materials used in optical fibers, the dispersion mechanisms are governed by the propagation characteristics of optical fibers, which are in

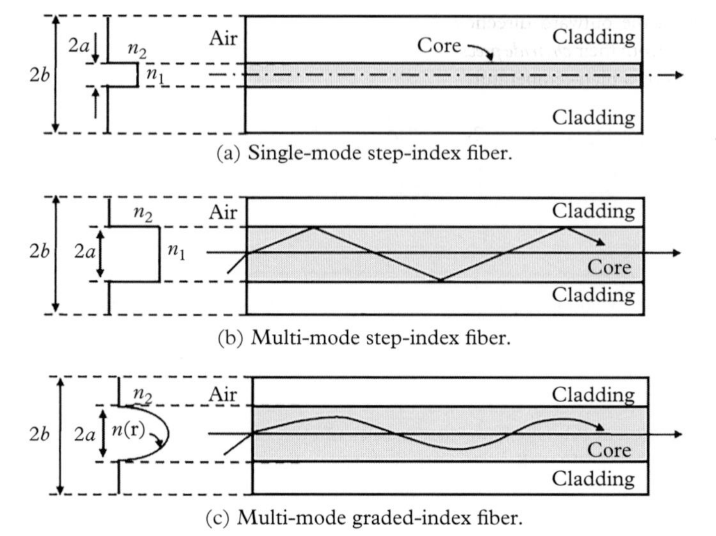

Figure 2.5 *Different types of optical fibers with their RIPs and relative core/cladding dimensions. Generally the cladding diameter (2b) for all the types is kept the same so that the same type of connectors can be used for all of them, while core diameter varies significantly between single-mode and multi-mode fibers.*

turn determined by the structural properties (e.g., core diameter, RIP, structural irregularity) and the dispersive (wavelength-sensitive) properties of the materials used in optical fibers.

In view of the above, we first discuss the materials and techniques used to manufacture optical fibers. Then, with the knowledge of the materials used, we describe in detail the loss mechanisms and light propagation models in optical fibers using ray and wave theories. Subsequently, the two models are used to explain the various dispersion mechanisms in optical fibers. Further, we describe nonlinear effects in optical fibers, which can affect the propagation characteristics in optical fibers carrying large optical power. Finally, we wrap up the section on optical fibers with a discussion of the possible remedies to address the problems created by dispersion and nonlinearity in optical fibers.

2.2.1 Fiber materials and manufacturing process

Before describing the loss mechanisms in optical fibers, we discuss briefly the basic manufacturing process and the materials used. In general, fiber manufacturing is carried out in a two-step process. In the first step, a thick tube, called a *preform*, with a hollow cylindrical form (with an outer diameter much larger than that of the actual fiber) is fabricated with concentric layers of higher (inner side) and lower (outer side) refractive indices by using a vapor deposition process. In the second step, the preform is pulled out at high temperature to form thin fibers with appropriate RIP.

The methods to manufacture preforms have evolved over time. One of the most used methods to make preforms is based on chemical vapor deposition (CVD), which was later improved and named modified CVD (MCVD). Using an MCVD-based processing, the preform is fabricated by depositing the vapors of pure silica and a dopant on a removable thin *bait* rod, typically an alumina rod with a diameter $\simeq$ 30 mm. Initially, on the bait rod a mixture-gas of pure silica and the dopants suitable for increasing the refractive index (e.g., GeO_2, P_2O_5) are deposited to realize a refractive index, that is slightly higher than that of pure silica (= 1.458 at 850 nm). As the width of the deposited layer grows, the concentration of the dopant is reduced gradually, thereby decreasing the refractive index radially in the outward direction. The vapors of the silica and the dopant materials are prepared from their *chloride precursors*, i.e., $SiCl_4$ (for silica) and $GeCl_4$ or $POCl_3$ (for GeO_2 or P_2O_5) by a high-temperature oxidization process, such as,

$$SiCl_4 + O_2 \Rightarrow SiO_2 + 2Cl_2 \text{ for silica as core,}$$
$$GeCl_4 + O_2 \Rightarrow GeO_2 + 2Cl_2 \text{ for } GeO_2 \text{ as dopant}$$
$$2POCl_3 + \tfrac{3}{2}O_2 \Rightarrow P_2O_5 + 3Cl_2 \text{ for } P_2O_5 \text{ as dopant.}$$

One can also realize the refractive indices below that of silica, by using the vaporized version of B_2O_3 as the dopant by oxidizing its chloride precursor BCl_3, while initially starting the deposition process with only silica vapor.

Having completed the vaporization process, the bait rod is pulled out to have the primary version of the preform in a hollow cylindrical form, which is then sintered at 1500°C to remove the bubbles, if any. In the final step, known as the *fiber-drawing process*, the preform, while being hung vertically, is heated at around 1800–1900°C using an appropriate heating furnace and pulled down simultaneously to collapse it into the thinned-down thread of glass fiber. Note that the axial hole created earlier by the removal of the bait rod practically disappears at this stage. The glass fibers are then covered with protective coating and, in practice, several such fibers are bundled together with adequate packing material and a strengthening metallic wire to finally obtain a multi-fiber (also known as multi-core) optical fiber cable. Single-core fiber cables are also made for the in-house interconnections between optical devices as well as for carrying out laboratory experiments.

2.2.2 Loss mechanisms

The loss mechanisms in optical fibers can be categorized into three major types: intrinsic losses, extrinsic losses, and bending losses. As described above, optical fibers are made from pure silica glass along with suitable dopants for core/cladding regions to realize the desired value of Δ. The silica glass exhibits intrinsic losses caused by some of its fundamental characteristics. However, the silica glass used in manufacturing optical fibers carries some residual impurities, in spite of the utmost care being taken while producing the preform. The losses in optical fibers that are caused by impurities are called extrinsic losses. Furthermore, losses can also take place due to structural irregularities in optical fibers developed during the manufacturing process, such as micro-bending. There is another component of loss, though negligible as compared to the other components, occurring from the atomic defects and bubbles (even after running the bubble-removal step), developed during the fiber manufacturing process.

All the above losses in optical fibers take place in certain parts of the ultraviolet-to-infrared spectral region, leaving out some specific windows with low losses, known as *spectral windows* of optical fibers. Optical sources make use of these windows to set up high-speed connections in optical transmission systems. We describe below the three main categories of the loss components in optical fibers in further details.

Intrinsic losses

Intrinsic losses in optical fibers can be categorized into three types: intrinsic absorption in the ultraviolet and infrared regions of the optical spectrum, and the scattering (Rayleigh scattering) of light in the backward direction during propagation.

Pure silica glass has a large absorption peak in the ultraviolet region at 140 nm, caused by the interaction between the photons in the propagating lightwave and the tightly bound electrons in the oxygen ions of silica (SiO_2). This binding creates an energy bandgap of 8.9 eV, and the bound electrons get excited from the valence band by ultraviolet light at a wavelength $w = 140$ nm. The tail of this absorption spectrum extends into the near-infrared region above 700 nm, leading to an absorption loss that decays fast with increasing wavelength. On the other side of the spectrum, there are a few absorption peaks at infrared wavelengths due to the vibration of silica molecules caused by photons at the wavelengths $w = 8000$, 4400, 3800 and 3200 nm, with 8000 nm being the fundamental tone, and the rest representing the combination and overtone wavelengths. This absorption phenomenon creates a loss tail extending far downward in the wavelength scale, eventually fading out to a very low value beyond 1650 nm.

Next, we consider another intrinsic loss, caused by the scattering of light in optical fibers. In particular, the silica glass is a non-crystalline amorphous material having inhomogeneous density, which scatters the propagating light in the backward direction, falling into the category of Rayleigh scattering. Rayleigh scattering is also exacerbated due to the dopants added to the silica, e.g., GeO_2, P_2O_5, fluorine, B_2O_3, for increasing (or decreasing) the refractive index. The loss due to Rayleigh scattering is inversely proportional to the fourth power of wavelength (i.e., $\propto w^{-4}$), thereby reducing dramatically at the long-wavelength region (above 1300 nm). The losses due to the ultraviolet absorption and the Rayleigh scattering overlap in the near-infrared region, though the ultraviolet absorption remains comparatively much smaller with respect to the Rayleigh-scattering loss.

Thus, the three intrinsic losses create a *natural valley* of low loss in optical-fiber loss spectrum, approximately between 850 and 1650 nm. However, as we shall see below, within this wide range of wavelength, some extrinsic losses can *pop up* to divide the available spectral valley into multiple smaller spectral windows.

Extrinsic losses

Extrinsic losses are incurred in optical fibers due to the impurities that are absorbed through fiber manufacturing process as well as the added dopants. The impurities include transition metal ions, such as Fe^{2+} and Cr^{3+}, along with hydroxyl (OH^-) ions from water absorption. The transition metal ions can increase the total loss by about 1 dB/km with an impurity concentration of 1 part per billion (ppb). On the other hand, hydroxyl ions have the peak absorption located at 2700 nm, while its overtones encroach into the valley (850–1650 nm) of the spectrum formed by the intrinsic losses, and are located at 1390, 1250, and 950 nm. With an impurity concentration of 1 part per million (ppm), hydroxyl ions cause a loss as large as 30 dB/km at 1390 nm. Hence, for low-loss fibers, it is essential to keep the hydroxyl concentration below a few 10's of ppb. The dopants also contribute to the total loss incurred in optical fibers. For example, germanium-doping in the core can increase the loss by about 0.05 dB/km, which is not a negligible contribution for the present-day optical fibers trying to achieve a total loss $\simeq 0.2$ dB/km.

Bending losses

During the manufacturing process, optical fibers develop some microscopic variations in the cross-sectional size at some locations, which are known as micro-bending sites. Sites with micro-bending change the boundary conditions locally, creating higher-order modes which cannot propagate much farther, leading to a loss of power (going into the cladding region), as micro-bending loss. Also, while laying the fiber cable, the fiber might get bent into a small radius of curvature; such bent spots would also cause power loss for the same reason as with micro-bending. From the ray-theoretic viewpoint, these bending spots would make the light rays incident on the core–cladding interface at an angle less than the critical

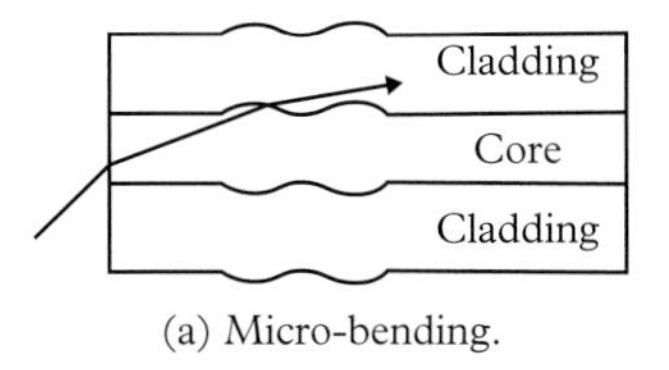

(a) Micro-bending.

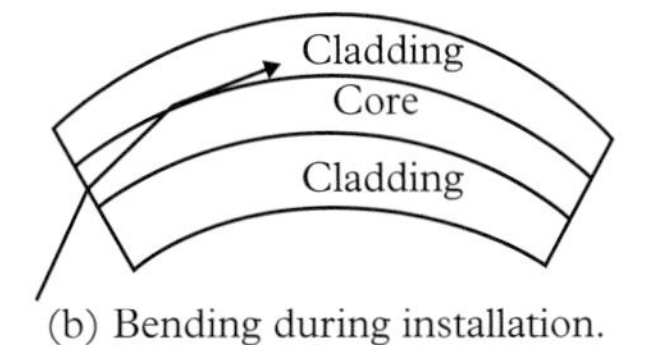

(b) Bending during installation.

Figure 2.6 *Two different bending scenarios in optical fibers: (a) micro-bending taking place during the manufacturing process, (b) bending introduced during the installation process.*

angle, thereby getting lost into the cladding. The two different scenarios of bending losses in optical fibers are illustrated in Fig. 2.6.

Loss-versus-wavelength plot and spectral windows

The various loss components in optical fibers are shown as functions of wavelength in Fig. 2.7, to visualize the relative contributions of the loss components on the overall loss-wavelength profile. As shown in the figure, Rayleigh scattering, and to a minor extent the tail of the ultraviolet absorption peak, decide the lower end of the spectral valley of optical fibers starting at wavelengths > 850 nm, while on the larger wavelength side, the infrared absorption starts increasing from 1600 nm onward. In between, the overtones of hydroxyl ions come in at 1390, 1250, and 950 nm, thereby splitting the 850–1650 nm valley into practically three segments, called spectral windows. In effect, there appears a wavelength span around 900 nm, which has the largest loss and is known as the first window, while the wavelength span around 1300 nm with a lower loss is known as the second window. The next span – indeed the most important one, with the smallest loss and widest bandwidth – is called the third window, with the minimum loss in the range of 1550 ± 100 nm, paving the way for high-speed DWDM transmission over optical fibers. Note that, in the early-generation optical fibers, the hydroxyl ions used to be present in larger proportions along with some other transition-metal impurities, making the overall loss much higher than found in present-day commercially available optical fibers.

A typical variation of the overall fiber loss with wavelength in today's single-mode optical fibers is shown in Fig. 2.8. As evident from the plot, with improved technologies, impurities as well as hydroxyl ions have been eliminated so much so that, the first and second windows appear as a continuum, while between the second and third windows, the impact of OH peak is practically not visible for fibers with low-OH content (solid line). The first window loss around 900 nm has the largest magnitude ($\simeq$ 2–2.5 dB) and is not used much in today's optical networks. The second window around 1300 nm has a lower loss ($\simeq$0.4–0.5 dB) and is used in optical networks selectively, more when we need fewer wavelengths over relatively shorter links. The third window offers a loss as low as 0.2–0.25 dB in the range of 1550 ± 100 nm, enabling high-speed DWDM transmission over optical fibers. The 200 nm span of the third window around 1550 nm leads to a bandwidth $|\Delta f| = |\Delta w|c/w^2 \simeq 200 \times 10^{-9} \times 3 \times 10^8/(1550 \times 10^{-9})^2 \simeq 25,000$ GHz.

The third transmission window of optical fibers with the minimum loss offers the most significant spectral span in WDM transmission systems. Further, in this window, there is a central portion with bandwidth of 35 nm around $\simeq$ 1550 nm, which can be optically amplified using fiber-based optical amplifiers. Thus, the third window is further segmented into three bands, S, C, and L, with the C band representing the central band as above, and S and L representing the left-side (shorter wavelengths with respect to the C band) and the right-side (longer wavelengths with respect to C band) bands on the two sides of the C band (see Fig. 2.9). The wavelengths placed in the C band (1530–1565 nm) will have the privilege of being periodically amplified by the in-line fiber-based optical amplifiers along the long-haul links.

Figure 2.7 *Various loss components in optical fibers. RS: Rayleigh scattering, UVA: ultraviolet absorption, IRA: infrared absorption, OH1, OH2, and OH3: absorption spectra for the first, second, and third overtones of hydroxyl ions, respectively. In present-day optical fibers, the hydroxyl content has been almost eliminated to the extent that the humps due to the OH peaks have become practically invisible in the loss-versus-wavelength profile for low-OH fibers, as shown in Fig. 2.8.*

Spectral valley from intrinsic losses (RS, UVA, IRA)
First window
Second window
Third window
RS
UVA
OH3
OH2
OH1
IRA
Losses in dB/km
0 1 2 3 4 5 6
700 800 900 1000 1100 1200 1300 1400 1500 1600 1700 1800
Wavelength in nm

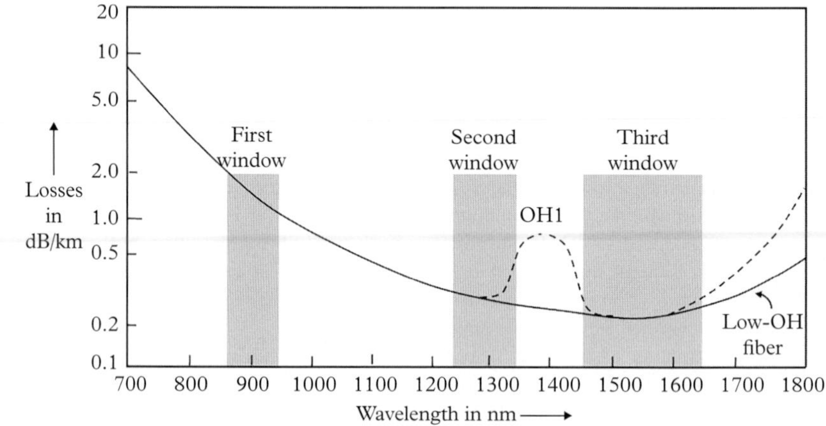

Figure 2.8 *Variation of overall loss (in log(dB/km) scale) with wavelength in today's single-mode optical fibers. With advanced fiber-manufacturing technology, the impact of OH ions and other impurities (OH1 peak shown by the dotted hump) has now been practically removed, offering thereby a single* bowl-like *low-OH loss-versus-wavelength profile (solid line) for optical fibers.*

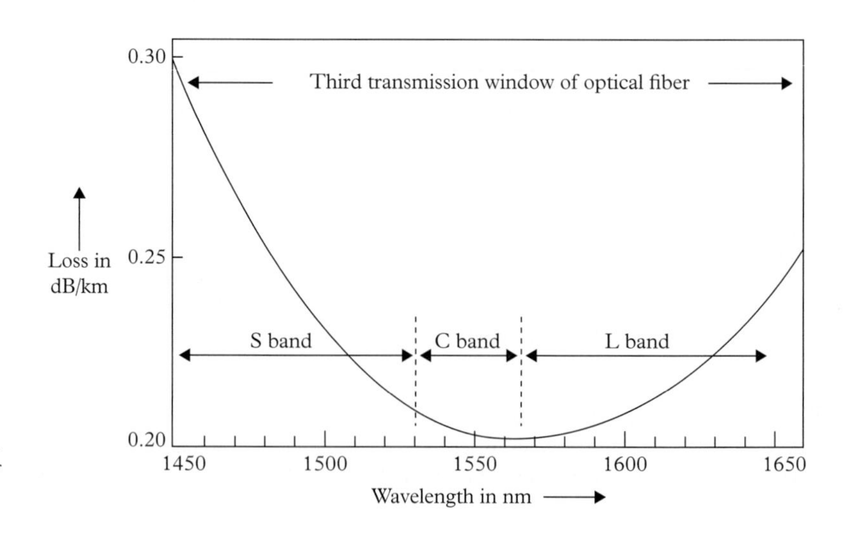

Figure 2.9 *S, C (1530–1565 nm), and L bands in the third transmission window of optical fibers.*

2.2.3 Propagation characteristics: two models

The propagation of light in optical fibers can be explained by using geometrical optics based on the ray theory or by the electromagnetic wave theory (or simply wave theory). However, the ray theory loses its significance when the core diameter $2a$ becomes comparable to the propagating wavelength w. Nevertheless, some of the important features of optical fibers can be readily explained using the ray theory, which can also be analyzed using the wave theory, though the reverse may not always be true, particularly for single-mode optical fibers. In the following we consider both the approaches (ray theory and wave theory) to develop a comprehensive understanding of the propagation of light in optical fibers.

Ray theory model

The ray theory models the propagating light as rays, with a ray representing a normal to the planar wavefront of the propagating light in the direction of propagation. Our earlier discussion with reference to Fig. 2.4 has been based on the ray theory applied for the step-index optical fibers, where the incident ray on the fiber-end surface moves into the core region and keeps propagating, guided by the basic laws of geometric optics. In particular, we consider here only the meridional rays, i.e., the rays that lie on the longitudinal planes passing through the fiber axis. Though there can be some rays that follow helical paths around the fiber axis, we don't consider them in the ray theory model as these rays do not matter for the features of our immediate interest. By using the ray theory, we next examine some important characteristics, which are fundamental for understanding the functioning of optical fibers.

As illustrated in Fig. 2.4, we consider the light ray that is incident on the air-to-core interface making an incidence angle θ, with the air having a refractive index $n_0 = 1$. Note that, at the core–cladding interface this light ray is incident with an angle ψ, which exceeds the critical angle and hence undergoes TIR and remains confined to the core region, thus contributing to the light propagation in the optical fiber. Using the law of refraction and simple geometry, one can relate the angles ϕ, ψ inside the core region and the incidence angle θ at the air-to-core interface as

$$n_1 \sin\phi = \sin\theta, \psi = \frac{\pi}{2} - \phi. \tag{2.1}$$

Note that, the ray incident on the core–cladding interface will experience TIR only if ψ exceeds the critical angle ψ_{cr}, given by $n_1 \sin\psi_{cr} = n_2 \sin(\frac{\pi}{2})$, i.e.,

$$\psi_{cr} = \arcsin\frac{n_2}{n_1}. \tag{2.2}$$

Thus, the maximum allowable value of ϕ that will enable TIR-based propagation along the fiber length would be given by

$$\phi_{max} = \frac{\pi}{2} - \psi_{cr} = \frac{\pi}{2} - \arcsin\frac{n_2}{n_1}. \tag{2.3}$$

Using the above value of ϕ_{max} along with the expression of $\sin\theta$ from Eq. 2.1, one can express the maximum allowable value for θ as

$$\sin\theta_{max} = n_1 \sin\phi_{max} = \sqrt{n_1^2 - n_2^2}. \tag{2.4}$$

From the above equation, we obtain $\theta_{max} = \arcsin\sqrt{n_1^2 - n_2^2}$, which is known as the acceptance angle θ_A of optical fiber (as shown in Fig. 2.4) and the corresponding cone with the half-angle θ_A around the fiber axis at the air-to-core interface would be the acceptance cone for the incident light. No light ray incident on the core-to-cladding interface with an angle of incidence $\theta > \theta_A$ (i.e., outside the acceptance cone) will be able to propagate along the optical fiber. The acceptance angle leads to an important parameter of optical fiber, known as the numerical aperture (NA), given by

$$\text{NA} = \sin\theta_A = \sin\theta_{max} = \sqrt{n_1^2 - n_2^2} \simeq n_1\sqrt{2\Delta}, \tag{2.5}$$

which represents the fraction of the total emitted power that will be captured by a given optical fiber into its core from a *Lambertian optical source*, having a normalized radiation pattern, given by $I(\theta) = \cos\theta$. Note that the approximation for the NA formula in the above equation (i.e., $\text{NA} \simeq n_1\sqrt{2\Delta}$) holds good with $\Delta \ll 1$. Practical optical sources will have a sharper radiation pattern as compared to a Lambertian source, implying that the percentage of power captured by an optical fiber from a given emitted power from the laser or LED will be more (much more for lasers with extremely sharp radiation patterns) than one would get by multiplying the radiated power from the source by NA. Notwithstanding this difference, NA stands as a universal reference parameter, when comparing different types of optical fibers in terms of the percentage of emitted optical power from a source, captured within the fiber core for subsequent propagation.

In Fig. 2.4, the acceptance angle θ_A forms a cone at the point of incidence of light at the air-to-core interface, within which one could imagine infinite number of rays that would be accepted by the core to propagate. However, in reality, there will be only a finite number of rays that would be able to propagate successfully along the fiber, with specific discrete values of $\theta \in [0, \theta_A]$. With the ray theory, based on the planar-wave propagation, this phenomenon is attributed to the fact that the plane waves resulting from a given incident ray from its subsequent TIRs can undergo *constructive interference* only for some specific discrete values of $\theta < \theta_A$. The rays that satisfy this criterion would be successful in carrying optical power through the fiber length. As we shall see in the wave theory model of the fiber, these rays would have an equivalence with some of the modes in the cylindrical waveguide representing an optical fiber. Thus, the optical fibers that allow multiple modes (rays) are called as multi-mode fibers, while one can design an optical fiber allowing only one mode to propagate, leading to single-mode fiber. Note that the core diameter in the single-mode optical fibers needs to be much smaller than that in the multi-mode fibers (typically, 8–10 μm for single-mode fibers and 50 μm or more for multi-mode fibers); we shall get back to this aspect while describing the wave-theory model.

As mentioned earlier, in the graded-index multi-mode fibers, intermodal dispersion can be reduced by using a graded-index RIP. For graded-index fibers, the refractive index $n(r)$ in the RIP can be expressed as a function of fiber radius r, given by

$$\begin{aligned} n(r) &= n_1\left[1 - 2\Delta\left(\frac{r}{a}\right)^{\alpha}\right]^{1/2} \quad \text{for} \quad 0 \leq r \leq a \\ &= n_1\sqrt{1-2\Delta} \simeq n_1(1-\Delta) = n_2 \quad \text{for} \quad r \geq a \end{aligned} \tag{2.6}$$

The parameter α determines the shape of the RIP, and the most commonly used value for α is 2. Note that α goes to ∞ to transform the graded-index profile into a step-index profile.

As shown earlier in part (c) of Fig. 2.5, the path of any off-axis meridional ray in a graded-index fiber goes through continuous bending as it moves ahead, due to the decreasing refractive index in the radial direction from the axis. In effect, the decreasing refractive index across the radius creates an infinite number of concentric layers with different refractive indices, and at each interface between two such layers (with outer layer having lower refractive index than the inner one) there occurs a refraction, making the rays bend continuously. This *continuous* refraction process during propagation reduces the length of the path traversed by all the rays, except the axial ray. Thus, unlike the step-index fibers shown in Fig. 2.5(b), the differential optical path length between the axial ray and the longest meridional ray in the graded-index fibers is reduced *physically* due to the continuous bending of the latter (with the TIR occurring at a radius $r < a$) and also *optically* as the longest ray passes through the core regions where the off-axis refractive indices are more than n_2 (and hence closer to n_1, as compared to the step-index fibers). This two-fold effect helps bring down considerably the intermodal dispersion in the graded-index fibers as compared to the step-index fibers.

Wave theory model

As mentioned earlier, the ray theory doesn't hold good when the core diameter $2a$ becomes comparable to the operating wavelength w. With smaller core diameter the propagating light doesn't find enough space around to keep its wavefront *planar* due to the tight boundary conditions imposed by the small cross-sectional area of the core-cladding interface. In such cases, one needs to use a more exhaustive analytical model based on (electromagnetic) wave theory, which is indeed applicable to the cases of fibers with large core diameter as well.

The wave theory model for the step-index optical fibers is described using the (r, θ, z) cylindrical coordinate system, as shown in Fig. 2.10, with r as the radial coordinate at a given point in the core/cladding region, θ as the angle between the given point and a reference plane passing through the fiber axis (x-z plane), and z representing the longitudinal coordinate of the point along the fiber axis. With this framework, one can use Maxwell's equations to find the expressions for all the electric and magnetic fields in the optical fiber, i.e., $E_r^x, E_\theta^x, E_r^x, H_r^x, H_\theta^x$, and H_r^x (all being functions of r, θ, and z), with E and H representing the electric and magnetic fields, and the superfix x representing the core or cladding region. Using this framework, next we carry out an analytical exercise to develop some insight into the single-mode and multi-mode propagations in optical fibers (Keiser 2008; Senior 1996; Cherin 1983).

Using Maxwell's equations, it is possible to express the electric and magnetic field components, E_z and H_z, in the core and cladding regions. Without elaborating the further

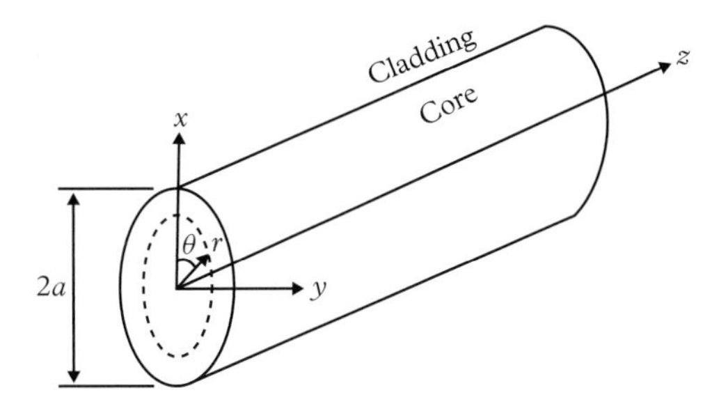

Figure 2.10 *Cylindrical coordinate system with (r, θ, z) coordinates for optical fibers with the underlying Cartesian (x, y, z) framework. The angle θ is measured from the x-z plane as the reference.*

details of the preliminary analytical steps, we next express the defining wave equations for E_z and H_z for a cylindrical waveguide as (Cherin 1983)

$$\frac{\partial^2 E_z}{\partial r^2} + \frac{1}{r}\frac{\partial E_z}{\partial r} + \frac{1}{r^2}\frac{\partial^2 E_z}{\partial \theta^2} + q^2 E_z = 0, \tag{2.7}$$

$$\frac{\partial^2 H_z}{\partial r^2} + \frac{1}{r}\frac{\partial H_z}{\partial r} + \frac{1}{r^2}\frac{\partial^2 H_z}{\partial \theta^2} + q^2 H_z = 0, \tag{2.8}$$

where $q^2 = \omega^2 - \beta^2$, with ω representing the angular frequency given by $\omega = 2\pi f = 2\pi c/w$. Next, in order to find solutions to these equations, we first express E_z (which is also applicable to H_z) in terms of the cylindrical coordinates (r, θ, z) as

$$E_z = AR(r)\Theta(\theta)\exp\{j(\omega t - \beta z)\}, \tag{2.9}$$

where the three separate functions are chosen in product form (a well-known technique, known as *separation of variables*), viz., $R(r)$ for r-dependence, $\Theta(\theta)$ for θ-dependence, and exponential function of time t and z representing the longitudinal propagation mechanism, with β as the propagation constant and A as the amplitude of E_z.

Note that the field in a cylindrical structure at a point r, θ, z should be the same as the field at the point at $r, \theta + 2\pi, z$, implying that $\Theta(\theta)$ should be a periodic function of θ in an optical fiber. Hence, one can express $\Theta(\theta)$ as

$$\Theta(\theta) = \exp(j\nu\theta), \tag{2.10}$$

where ν represents an integer to ensure the periodicity. Substituting Eq. 2.10 in Eq. 2.9, we obtain E_z as

$$E_z = AR(r)\exp(j\nu\theta)\exp\{j(\omega t - \beta z)\}. \tag{2.11}$$

Using Eq. 2.11 in Eq. 2.7, and removing the common factors (i.e., θ and z-dependent factors) we obtain the wave equation for E_z in terms of the radial distance r as

$$\frac{\partial^2 R(r)}{\partial r^2} + \frac{1}{r}\frac{\partial R(r)}{\partial r} + \left(q^2 - \frac{\nu^2}{r^2}R(r)\right) = 0, \tag{2.12}$$

which represents the well-known Bessel equation for the radial field distribution function $R(r)$. The solutions for $R(r)$ can be chosen from the family of Bessel functions. Following similar steps, an equivalent equation can also be arrived at for H_z.

The choices of Bessel functions for the core and cladding regions are different due to the conditions necessary for propagation. In the core region, Bessel functions of the first kind and ν-th order suit the requirements, as these functions are finite at any value of r, while exhibiting an oscillatory trend with varying r. The oscillatory nature of these functions with r makes them unsuitable in the cladding region, as the fields outside the core region must decay monotonically for propagation to materialize. For the cladding region, the choice of the modified Bessel functions of the third kind (also called Hankel functions) serves the purpose with its monotonically decaying profile with increasing r, while these functions can not be chosen for the core region as they approach infinity at $r = 0$. Thus, the core and cladding regions must choose two different solutions of Bessel equations: (i) Bessel functions of the first kind $J_\nu(\rho r)$ for the core region ($r \leq a$) and (*ii*) modified Bessel functions of the third kind $K_\nu(\gamma r)$ for the cladding region ($r \geq a$), where ρ and γ are obtained from the wave equations (Equation 2.7 or 2.8) as

$$\begin{aligned} \rho^2 &= \omega^2\mu\epsilon_1 - \beta^2 = k_1^2 - \beta^2, \\ \gamma^2 &= \beta^2 - \omega^2\mu\epsilon_2 = \beta^2 - k_2^2, \end{aligned} \tag{2.13}$$

with ϵ_1 and ϵ_2 as the dielectric constants of the core and cladding regions and μ as the permeability.

Using the above observations on Eq. 2.11, we obtain the expressions for all the z-components in the core and cladding regions as

$$\begin{aligned} E_z^{core}(r \leq a) &= AJ_\nu(\rho r)\exp(j\nu\theta)\exp\{j(\omega t - \beta z)\} \\ H_z^{core}(r \leq a) &= BJ_\nu(\rho r)\exp(j\nu\theta)\exp\{j(\omega t - \beta z)\} \\ E_z^{clad}(r \geq a) &= CK_\nu(\gamma r)\exp(j\nu\theta)\exp\{j(\omega t - \beta z)\} \\ H_z^{clad}(r \geq a) &= DK_\nu(\gamma r)\exp(j\nu\theta)\exp\{j(\omega t - \beta z)\}, \end{aligned} \tag{2.14}$$

where A, B, C, and D are the amplitudes of the respective fields, which are related to each other through the constraints of the boundary conditions at the core–cladding interface in a manner such that the total power carried by these fields is equal to the power captured by the optical fiber. Note that using Maxwell's equations one can also find out the other set of fields in the core and cladding regions.

Optical fibers have a fundamental difference with the metallic waveguides, as the tangential components of electromagnetic fields on the metallic walls must vanish to zero. However, an optical waveguide having the *dielectric-to-dielectric* interface (between the core and cladding regions), does not force the tangential electromagnetic fields to zero at the core–cladding interface, and hence makes them *continuous* across the interface. This feature manifests itself as a *soft* boundary between the core and cladding, which in turn allows the electromagnetic fields to reach the cladding region. For a successful propagation, the vestige of these fields intruding from the core into the cladding region must decay sharply beyond the core–cladding interface. In the following, our analysis seeks for appropriate solutions for the electromagnetic fields in optical fibers, complying with this criterion.

In a cylindrical optical waveguide, the tangential field components, i.e., z- and ϕ-components of electric and magnetic fields, must be continuous at $r = a$, leading to the following four homogeneous equations

$$\begin{aligned} E_z^{core}(r = a) - E_z^{clad}(r = a) &= 0 \\ E_\phi^{core}(r = a) - E_\phi^{clad}(r = a) &= 0 \\ H_z^{core}(r = a) - H_z^{clad}(r = a) &= 0 \\ H_\phi^{core}(r = a) - H_\phi^{clad}(r = a) &= 0, \end{aligned} \tag{2.15}$$

where $E_z^x(r = a), E_\phi^x(r = a)$ are the z- and ϕ-components, respectively, of the electric fields in the x-region at $r = a$, with x representing the core ($r \leq a$) or cladding ($r \geq a$) region, and similarly $H_z^x(r = a), H_\phi^x(r = a)$ are the respective magnetic fields. Substituting the field expressions from Eq. 2.14 in Eq. 2.15, one can obtain a set of four homogeneous equations (right-hand sides being zero) with the four unknown variables (i.e., the field amplitudes A, B, C, and D), given by

$$\begin{aligned} y_{11}A + y_{12}B + y_{13}C + y_{14}D &= 0 \\ y_{21}A + y_{22}B + y_{23}C + y_{24}D &= 0 \\ y_{31}A + y_{32}B + y_{33}C + y_{34}D &= 0 \\ y_{41}A + y_{42}B + y_{43}C + y_{44}D &= 0, \end{aligned} \tag{2.16}$$

where the coefficients y_{ij} are determined by the field expressions given in Eq. 2.14 (for example, $y_{12} = y_{14} = 0$ for the first line of the above equation set). The above set of equations involving the variables A, B, C, and D can be expressed in a matrix form, given by

$$\mathbf{Y}\{ABCD\}^T = 0, \tag{2.17}$$

with $\mathbf{Y} = \{y_{ij}\}$ representing a 4×4 matrix obtained from Eq. 2.16, and $\{ABCD\}^T$ represents a column vector composed of the field amplitudes.

The implication of the above expressions defined by Equations 2.16 and 2.17 for A, B, C, and D is that they can have non-trivial (i.e., non-zero) solutions, if and only if, the determinant of the matrix $\mathbf{Y}$ goes to zero. In other words, with the homogeneous equations (i.e., right-hand side of Eq. 2.17 being a null vector), all the column vectors in $\mathbf{Y}$ are linearly dependent, representing four *parallel* hyperplanes in four-dimensional space. In such cases, non-trivial solutions would exist only when they fall upon each other being *coplanar* while passing through the origin, thereby making the determinant zero. This will lead to only specific linear relations between them without any absolute values, and the absolute values will finally be governed by their relative values decided by the above condition and the total power launched into the fiber. Equating the determinant $|\mathbf{Y}|$ to zero will eventually lead to the characteristic equation for the optical fiber, producing an infinite number of solutions theoretically, corresponding to all the possible modes in the optical fiber, from which some specific modes will be able to propagate, as governed by the constraints imposed by the fiber parameters.

Using the expressions for the tangential field components, and equating the determinant $|\mathbf{Y}|$ to zero, one can obtain the characteristic equation for step-index optical fibers as (Keiser 2008; Cherin 1983)

$$(\mathbf{J}_\nu + \mathbf{K}_\nu)(k_1^2\mathbf{J}_\nu + k_2^2\mathbf{K}_\nu) = \left(\frac{\beta\nu}{a}\right)\left(\frac{1}{\rho^2} + \frac{1}{\gamma^2}\right), \tag{2.18}$$

where $\mathbf{J}_\nu$ and $\mathbf{K}_\nu$ are given by

$$\mathbf{J}_\nu = \frac{J'_\nu(\rho a)}{\rho J_\nu(\rho a)}, \ \mathbf{K}_\nu = \frac{K'_\nu(\gamma a)}{\gamma K_\nu(\gamma a)}. \tag{2.19}$$

The characteristic equation (Eq. 2.18) determines the permitted values for the propagation constant β, which can assume only *discrete* values, corresponding to the possible modes in a step-index optical fiber. Furthermore, these discrete values of β would be limited within $k_1 = n_1 k$ and $k_2 = n_2 k$, which are the free-space propagation constants for the core and cladding materials respectively, i.e.,

$$k_1 > \beta > k_2. \tag{2.20}$$

The subsequent procedure for solving Eq. 2.18 is quite long in the present context, and hence we discuss here the end results on the appropriate values of β for wave propagation in step-index optical fibers. In metallic waveguides, electromagnetic waves propagate either with transverse electric fields (TE modes) or with transverse magnetic fields (TM modes). However, optical waveguides with soft boundary also allow hybrid modes, called HE and EH modes, with various possible components of the fields. One such hybrid mode, viz., HE_{11} mode (we discuss later the implications of this mode), turns out to be the fundamental mode having a zero cutoff wavelength. This implies that the HE_{11} mode would be able to propagate in an optical fiber with any value of wavelength. For all other modes, there would exist a cutoff wavelength, above which those modes won't be able to propagate. Thus, if the operating wavelength can be higher than the cutoff wavelength of the modes with the *lowest non-zero* cutoff wavelength, the fiber would function as a single-mode fiber operating with only the HE_{11} mode. As we shall see soon, such fibers would result in minimum possible dispersion, thereby ensuring maximum possible transmission speed in an optical network. Next, we discuss the implications of cutoff conditions in the step-index optical fibers.

Cutoff conditions

The cutoff conditions in optical waveguides are reached when the electromagnetic fields completely escape from the core to cladding region. This condition can be analytically identified from the nature of variation of the fields in the cladding region, governed by the modified Bessel function for $r > a$ as (Keiser 2008, Cherin 1983)

$$K_\nu(\gamma r) \propto \frac{\exp(-\nu r)}{\sqrt{\nu r}}. \tag{2.21}$$

Hence, with large values of γ, $K_\nu(\gamma r)$ decays fast with r, and the fields remain tightly confined to the core region. On the other hand, with decreasing γ, the fields start moving further into the cladding region, and finally, at $\gamma = 0$, the fields get detached from the core and propagation of light ceases to exist. The wavelength (frequency) at which γ goes to zero is called the cutoff wavelength (frequency), given by

$$\gamma = 0 = \sqrt{\beta_c^2 - k_{2c}^2} = \sqrt{\beta_c^2 - \omega_c^2 \mu \epsilon_2}, \tag{2.22}$$

where k_{2c} is the value of k_2 (see Eq. 2.13) at the cutoff angular frequency $\omega_c = 2\pi f_c = 2\pi c/w_c$. From the above, we obtain the propagation constant β at cutoff as

$$\beta_c^2 = k_{2c}^2 = \omega_c^2 \mu \epsilon_2 = 4\pi^2 f_c^2 \mu \epsilon_2. \tag{2.23}$$

In the core region, using Eq. 2.13 we therefore get the value of ρ at cutoff frequency as

$$\rho_c^2 = k_{1c}^2 - \beta_c^2 = 4\pi^2 f_c^2 \mu \epsilon_1 - 4\pi^2 f_c^2 \mu \epsilon_2 = 4\pi^2 f_c^2 \mu (\epsilon_1 - \epsilon_2). \tag{2.24}$$

Finally, using Eq. 2.24, we obtain the expressions for the cutoff frequency and wavelength (f_c, w_c) as

$$f_c = \frac{c}{w_c} = \frac{\rho_c}{2\pi \sqrt{\mu(\epsilon_1 - \epsilon_2)}}. \tag{2.25}$$

Thus, the cutoff frequency $f_c = 0$ (i.e., the cutoff wavelength $w_c \to \infty$), when ρ_c goes to zero which, as we shall see soon, is possible for the hybrid mode HE_{11}, allowing thereby any frequency or wavelength to propagate through the optical fiber. In order to determine the cutoff frequencies and wavelengths for specific modes, one needs to find out their characteristic equations, from which the values of ρ_c (cutoff conditions) are obtained and used in Eq. 2.25. With these observations, we next determine the solutions (i.e., β_c values) for specific modes: TE, TM, and hybrid modes.

TE and TM modes

For the TE and TM modes with $\nu = 0$, the characteristic equation can be simplified to (Cherin 1983)

$$J_0(\rho_c a) = 0. \tag{2.26}$$

The solutions or roots of the above equation, i.e., ρ_c, lead to the TE and TM modes with $\nu = 0$, known as TE_{0m} and TM_{0m} modes, where the first number in the suffix (i.e., $\nu = 0$) implies uniformity of $\Theta(\theta)$ around the fiber axis, and the second integer m represents the order of the roots of the characteristic equation (Eq. 2.26). Thus, TE_{01} mode represents a TE mode with $\nu = 0$ and the smallest root of Eq. 2.26, and so on. These values of ρ_c are substituted in Eq. 2.25 to obtain the values of cutoff frequencies and wavelengths for TE and TM modes with $\nu = 0$ in a given step-index optical fiber. Characteristic equations and the corresponding cutoff criteria for the higher-order TE and TM modes can also be found out following similar steps, albeit with a more complex process for simplifying the characteristic equations.

Hybrid modes

For hybrid $HE_{\nu m}$ modes, the characteristic equation can be obtained as

$$J_\nu(\rho_c a) = 0, \tag{2.27}$$

with $\rho_c a$ given by

$$\rho_c a = x_{\nu m} \quad \text{for} \quad \nu = 1, 2, 3, \cdots \tag{2.28}$$

Note that, for the $EH/HE_{\nu m}$ modes with $\nu > 1$, the process to find cutoff criteria is not considered here, as the corresponding cutoff wavelengths are higher than those of HE_{11} and TE/TM_{01} modes. Therefore, in order to get the overall picture of the relevant modes, we next compute the cutoff wavelengths for the HE_{11} mode with the largest cutoff wavelength and for the modes (TE/TM_{01}) with the next-lower cutoff wavelength.

Cutoff wavelengths for HE_{11} and TE_{01}/TM_{01} modes:

The electromagnetic fields of the HE_{11} mode are governed by the function $J_1(x)$. From the standard plot of $J_1(x)$ vs. x, the first or the smallest root of $J_1(x)$ takes place at $x = 0$. Hence, the cutoff condition for the HE_{11} mode can be expressed as

$$\rho_c a = 0. \tag{2.29}$$

With $\rho_c^2 = 4\pi^2 f_c^2 \mu(\epsilon_1 - \epsilon_2)$ from Eq. 2.24, we observe that the cutoff frequency f_c for HE_{11} mode is simply zero, meaning that even infinitely large wavelengths can propagate with HE_{11} mode through step-index optical fiber. In other words, HE_{11} does not exhibit any cutoff phenomenon with zero cutoff frequency leading to infinite cutoff wavelength.

Through inspection of the plots of all $J_\nu(x)$'s vs. x (Cherin 1983), the next (smaller) cutoff wavelength occurs for the TE_{01} and TM_{01} modes, whose fields are functions of $J_1(x)$, with the smallest root of $J_1(x)$ occurring at $x = 2.405$. Hence, these modes (nearest to the HE_{11} mode) must be cut off to realize the single-mode operation of fiber. Therefore, by assuming $\rho_c a = 2.405$, and using Eq. 2.25, the cutoff frequency f_c for the TE_{01} and TM_{01} modes is obtained as

$$f_c = \frac{2.405}{2\pi a\sqrt{\mu(\epsilon_1 - \epsilon_2)}} \quad \text{for} \quad \mathrm{TE}_{01}, \mathrm{TM}_{01} \quad \text{modes.} \tag{2.30}$$

The above result enables us to design the parameters for optical fibers with single-mode propagation. We consider this aspect in the following.

Single-mode propagation

As discussed above, in order to allow only single-mode propagation, the operating frequency must be smaller than the above cutoff frequency (or the operating wavelength must be larger than the cutoff wavelength), i.e.,

$$f < \frac{2.405}{2\pi a\sqrt{\mu(\epsilon_1 - \epsilon_2)}} \quad \text{or} \quad w > \frac{2\pi ca}{2.405}\sqrt{\mu(\epsilon_1 - \epsilon_2)}. \tag{2.31}$$

Noting that $\sqrt{\mu(\epsilon_1 - \epsilon_2)} = \sqrt{\mu\epsilon_0(n_1^2 - n_2^2)} = \mathrm{NA}/c$ with ϵ_0, representing the dielectric constant of free space, the above condition can be simplified as

$$f < \frac{2.405c}{\mathrm{NA}} \quad \text{or} \quad w > \frac{2\pi a}{2.405}\mathrm{NA}. \tag{2.32}$$

At this stage, we define a well-known parameter for optical fibers, called the V parameter, given by

$$V = \left(\frac{2\pi a}{w}\right)\mathrm{NA}, \tag{2.33}$$

and express the above condition (Eq. 2.32) for single-mode operation in terms of V parameter as

$$V < 2.405, \tag{2.34}$$

which implies that both the core diameter and NA should be small enough for a given operating wavelength, so that the V parameter remains below 2.405. As mentioned before, typically the single-mode fibers use core diameters in the range of 8–10μm with the refractive index difference Δ lying in the range 0.2–1%. Considering the operating wavelength w as 1.55 μm, along with $2a = 8$ μm, $\Delta = 0.2\%$ and $n_1 = 1.45$, and assuming that for small values of Δ ($\ll 1$), $\mathrm{NA} = \sqrt{n_1^2 - n_2^2} \simeq n_1\sqrt{2\Delta}$, we obtain V parameter as

$$V = \frac{2\pi \times (4 \times 10^{-6})}{1.55 \times 10^{-6}} \times (1.45\sqrt{2 \times 0.002}) = 1.487, \tag{2.35}$$

ensuring single-mode propagation in the given optical fiber.

Wave theory – overall perspective

The analysis of optical fibers using the wave theory helps in visualizing the phenomenon of mode-based propagation of lightwaves (i.e., electromagnetic waves in the infrared optical spectrum) with the soft boundary at the core–cladding interface. Hence, we review in the following the major aspects of the above analysis for a comprehensive understanding of the propagation phenomena in optical fibers, which will help us later in addressing the dispersion mechanisms and assessing the transmission capacity in terms of the maximum possible transmission rates over optical fibers.

To start with, a discussion on the rays vis-a-vis the modes would be in order. As mentioned earlier, in the context of the ray theory, all the rays considered for studying fiber characteristics were considered to be meridional rays passing through the fiber axis. In reality, these rays correspond to the TE and TM modes. On the other hand, the hybrid modes lead to helical or skew rays spiraling around the fiber axis, and the simple approach using meridional rays can not be used to explain them.

Next, in Fig. 2.11 we present the plot of the number of modes N_m as a function of V parameter, along with a table with the exact values of V for the modes shown in the plot. As evident from the figure, with the increase in the V parameter, the higher-order modes get excited. As expected, the plot assumes a *staircase-like* form, as the onset of propagation of a given mode can only take place at a discrete value of V parameter, governed by the specific root of the Bessel function of the respective order. As mentioned earlier, the plot shows two types of hybrid modes – HE and EH modes – where the first letter (E or H) indicates the type of characteristic equation followed by the respective hybrid mode. As evident from the plot, in the range of $0 < V < 2.405$, only one mode, i.e., the HE_{11} mode, can propagate. As V goes above 2.405, the TE_{01}, TM_{01} modes come in, and soon after at $V = 2.42$ (not shown separately in the plot due to small difference between 2.405 and 2.42) the HE_{21} mode joins with the last three modes. Thus, beyond $V = 2.405$ with increase in V various modes join, leading to multi-mode operation.

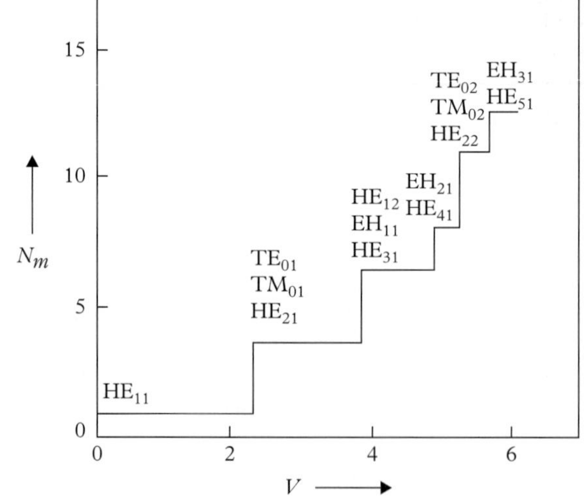

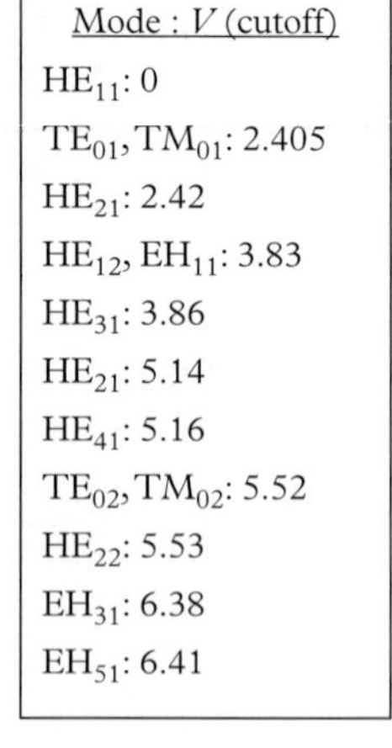

Mode : V (cutoff)
HE_{11}: 0
TE_{01}, TM_{01}: 2.405
HE_{21}: 2.42
HE_{12}, EH_{11}: 3.83
HE_{31}: 3.86
HE_{21}: 5.14
HE_{41}: 5.16
TE_{02}, TM_{02}: 5.52
HE_{22}: 5.53
EH_{31}: 6.38
EH_{51}: 6.41

Figure 2.11 *Variation of the number of modes N_m versus V parameter. The plot gives a staircase-like growth of N_m with V, while the associated table shows the exact cutoff values of V parameters for the modes shown in the plot. Note that, in the plot the modes with close values of V are clubbed together with one single step of increment for N_m (after Cherin 1983).*

Note that though the single-mode operation offers high transmission capacity over long distance (due to the absence of intermodal dispersion), one must use a laser as the optical source, as the power coupling from an LED (having broader radiation pattern) into the single-mode fiber with small NA (and hence small acceptance angle) becomes inadequate. Furthermore, the spectral spread of an LED leads to unacceptable chromatic dispersion in single-mode fibers (discussed later in more detail), preventing high-speed transmission. However, multi-mode fibers are preferred for shorter links operating at moderate transmission rates, while permitting cheaper LEDs as sources to transmit power through the fibers with larger values of NA. In order to get this benefit, multi-mode fibers are designed with a large number of modes to offer adequate acceptance angles for LEDs. In such cases, the number of modes N_m in a multi-mode optical fiber can be estimated in terms of the V parameter as (Keiser 2008)

$$N_m \simeq \frac{V^2}{2}. \tag{2.36}$$

Thus, a typical multi-mode optical fiber with $2a = 50$ μm, $n_1 = 1.45$ and $\Delta = 2\%$, while operating with a wavelength $w = 1.55$ μm, will have a $V = \frac{2\pi a}{w}\text{NA} = 29.388$, allowing as many as $N_m \simeq 29.388^2/2 = 432$ modes. Indeed, with this large value of N_m, intermodal dispersion needs to be estimated to check whether the fiber would support the needed transmission rate over the given link length.

Finally, note that, unlike the metallic wave-guides, multi-mode propagation in optical fibers presents a unique propagation scenario, wherein the hybrid modes (absent in metallic waveguides) combine together, giving rise to various distinct clusters or sets of modes, called as *linearly polarized modes* (LP modes), typically in the context of the weakly guiding fibers (i.e., with $\Delta \ll 1$). Moreover, all the hybrid modes in these LP modes, with the perfect circularity (if possible) of the fiber core, can exist in pairs with orthogonally polarized fields depending on the polarization(s) of light being fed from the optical source, and yet can co-propagate with the same propagation characteristics. As we shall see later, this condition is not fulfilled in practical optical fibers due to structural irregularities (e.g., non-circular core, fiber bending etc.), thereby enforcing different propagation constants for the two co-propagating orthogonally polarized versions of the same modes; this phenomenon in optical fibers is known as birefringence, which we discuss in further detail while discussing the various types of dispersion in optical fibers. Further, the LP modes are helpful in visualizing the field formations in optical fibers, and developing deeper insight. On this aspect, extensive studies have been made and reported in the literature during the early days of research on optical fibers (Cherin 1983).

2.2.4 Dispersion mechanisms

The dispersion mechanisms in optical fibers can be classified in three broad categories: *intermodal* or *modal dispersion*, *intramodal* or *chromatic dispersion*, and *polarization-mode dispersion*. Intermodal dispersion takes place as different modes in the optical fibers propagate with different group velocities, while the spectral spread of optical sources causes intramodal dispersion within each mode, and polarization-mode dispersion takes place due to the birefringence property resulting from the structural irregularities of the optical fibers. In the following, we discuss the three types of dispersions in further detail.

Intermodal dispersion

Intermodal dispersion can be readily estimated by using the ray-theory model, wherein the meridional rays can be mapped into the equivalent propagating modes. With discrete number of rays propagating in a multi-mode optical fiber, when an optical pulse is incident on the end surface of the optical fiber within an angle θ_A, it would get split into a number of rays, each making different angle (θ) with the longitudinal axis. As a result, each ray (i.e., plane wave) traveling with a velocity c/n_1 will take a different time to reach the destination end of the optical fiber, spreading the received pulse over a time duration, which would be larger than the original pulse width. The pulse spread will be equal to the difference between the propagation times of the rays along the longest and shortest (axial) paths (T_{max} and T_{min} for an optical-fiber segment of length L, respectively). The spread of the optical pulse per unit length of the fiber due to this phenomenon is called *intermodal* or *modal dispersion*, denoted as D_{mod}. Noting that the velocity of light in the core is c/n_1, and using the illustration shown in Fig. 2.12 and the value of ϕ_{max} from Eq. 2.3, we express D_{mod} as

$$D_{mod} = \frac{T_{max} - T_{min}}{L} = \frac{1}{L}\left(\frac{Ln_1/c}{\cos\phi_{max}} - Ln_1/c\right) \simeq \frac{n_1\Delta}{c}. \tag{2.37}$$

In an optical fiber link transmitting at a bit rate of $r_b = 1/T_b$ (with T_b as the bit interval) over a link length L, the pulse spread due to the intermodal dispersion, i.e., LD_{mod}, should not exceed $\frac{T_b}{2} = \frac{1}{2r_b}$, implying that $r_b \leq \frac{1}{2LD_{mod}}$. Note that, in reality, multi-mode fibers would exhibit somewhat less intermodal dispersion due to a phenomenon called *mode mixing*, where the propagating modes of the lightwaves keep *moving* between the various permissible modes (slower/faster) due to the fiber imperfections (Keiser 2008).

Chromatic dispersion

The fundamental cause of chromatic dispersion is the finite spectral width of optical sources (i.e., the chromaticity of the optical-source spectrum), as the different frequencies/wavelengths of the optical spectrum travel with different velocities in an optical fiber even within a given mode, causing the transmitted optical pulses to spread in time.

In free space, a monochromatic lightwave travels with the velocity c which is independent of the source wavelength. However, when a lightwave travels within an unbounded medium, the velocity (more precisely, the phase velocity v_p) of light becomes $v_p = c/n$ (n being the refractive index of the medium) and hence wavelength-dependent, when n turns out to be a function of wavelength w. Furthermore, when the lightwave becomes chromatic due to the modulation from the information-bearing data stream and finite spectral spread of the optical source, the resulting envelope of the lightwave can no longer move with the phase velocity. In this scenario, one can conceive the bandpass spectrum of the chromatic lightwave as a combination of a large number of infinitesimal wavelength bands or groups, where each

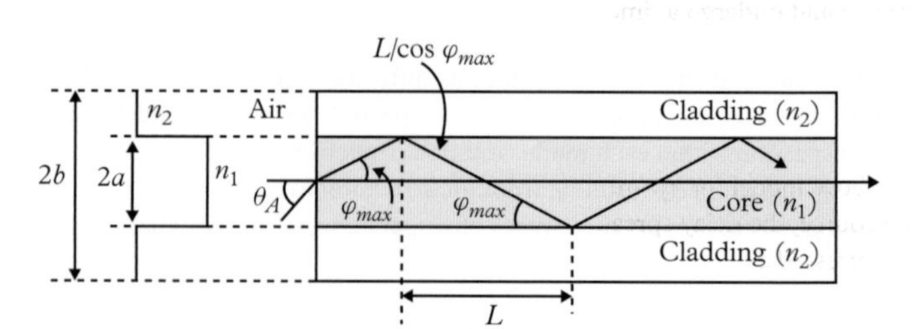

Figure 2.12 *Illustration of intermodal dispersion in a step-index multi-mode optical fiber, using the shortest (axial) and the longest meridional rays. Note that, the longest ray is incident on the core end-surface with an angle equaling to the acceptance angle θ_A of the optical fiber.*

group can travel with a specific group velocity v_g. The significance of v_g plays the central role in the chromatic dispersion mechanism.

We illustrate the concept of v_g using an example of two propagating electromagnetic waves $E_1(z,t) = A\cos(\omega_1 t - \beta_1 z)$ and $E_2(z,t) = A\cos(\omega_2 t - \beta_2 z)$ with two closely spaced frequencies $\omega_1 = \omega_0 - \delta\omega$ and $\omega_2 = \omega_0 + \delta\omega$ and corresponding propagation constants $\beta_1 = \beta_0 - \delta\beta$ and $\beta_2 = \beta_0 + \delta\beta$, respectively. These two closely spaced frequencies ($2\delta\omega \ll \omega_0$) form a wavelength group with the central frequency located at ω_0. The resultant electric field of these two waves becomes $E(z,t) = E_1(z,t) + E_2(z,t) = A\cos(\delta\omega t - \delta\beta z)\cos(\omega t - \beta z) = A\cos\psi_g \cos\psi_p$, with $\psi_g = \delta\omega t - \delta\beta z$ and $\psi_p = \omega t - \beta z$. Note that, while the cos ψ_p factor represents a wave propagating with a frequency ω_0, the cos ψ_g factor is in contrast another wave with much lower frequency behaving like a *modulating envelope* of the high-frequency wave as a carrier. With ψ_g and ψ_p representing the phase of the two waves, their respective velocities can be determined as follows.

For the high-frequency carrier wave, consider a location $z + \Delta z$ at time $t + \Delta t$, where and when one gets back the same value of the phase ψ_p, i.e., $\omega t - \beta z = \omega(t + \Delta t) - \beta(z + \Delta z)$ = a constant (Ψ, say). This observation leads to the fact that the phase Ψ has traveled a distance Δz in time Δt, implying that the carrier at ω_0 has traveled with a phase velocity $v_p = \Delta z/\Delta t = \omega/\beta$. Using a similar approach, one can readily show (with $\delta\omega \to 0$) that the wavelength group has traveled with a group velocity v_g, given by

$$v_g = \frac{1}{\partial\beta/\partial\omega}. \tag{2.38}$$

Note that, when β becomes a function of ω (which is so in optical fibers), the group velocity would be different for different groups of frequencies/wavelengths, leading to chromatic dispersion and consequent pulse (envelope) spreading. Moreover, the chromatic dispersion has an intrinsic dichotomy, as explained in the following.

First, we consider an *unbounded* medium with the core material having a refractive index n_1 as a function of wavelength, making $\beta = 2\pi n_1/w$ a function of wavelength too. Hence, for an un-bounded medium also, a lightwave with finite spectral width will undergo chromatic dispersion due to the material property (wavelength dependence), and this type of chromatic dispersion is called *material dispersion*. Next, we observe that, in optical fibers (which is a medium bounded by the core-cladding interface), even if the refractive indices of the core and cladding regions become independent of w, the parameter a/w and hence the propagation constant of the fiber, which is a function of a/w, becomes dependent on w, leading to another type of chromatic dispersion, referred to as *waveguide dispersion*. Consequently, different parts of the lightwave spectrum would arrive at the fiber end at different instants of time, causing a delay spread for the lightwave envelope due to the combined effect of the material and waveguide dispersion mechanisms.

The expression for v_g (Eq. 2.38) implies that, a spectral group in the wavelength interval $[w, w + \delta w]$ would undergo a time delay τ_g after traversing a fiber length L, given by

$$\tau_g = \frac{L}{v_g} = L\frac{\partial\beta}{\partial\omega}. \tag{2.39}$$

Using the above model for τ_g, and assuming a linear variation of τ_g over the spectral width Δw of the source, the delay spread $\Delta\tau$ for an optical pulse after traversing the fiber length L can be expressed as

$$\Delta\tau = \frac{\partial\tau_g}{\partial w}\Delta w. \tag{2.40}$$

The chromatic dispersion D_{ch} of a given mode (or a fiber, if it is designed as a single-mode fiber) is defined as the value of the above delay spread for unit spectral width (typically measured in nm) of the source transmitting over a kilometer of fiber. Hence, by using $\Delta\tau$ from Eq. 2.40, we express D_{ch} as

$$D_{ch} = \frac{\Delta\tau}{\Delta w L} = \frac{1}{L}\frac{\partial \tau_g}{\partial w} = \frac{1}{L}\frac{\partial}{\partial w}\left(L\frac{\partial \beta}{\partial \omega}\right) = \frac{\partial}{\partial w}\left(\frac{\partial \beta}{\partial \omega}\right). \tag{2.41}$$

Noting that $\delta\omega = \left(-\frac{2\pi c}{w^2}\right)\delta w$ we obtain the expression for D_{ch} in terms of the derivatives of w as

$$D_{ch} = -\frac{1}{2\pi c}\frac{\partial}{\partial w}\left(w^2\frac{\partial \beta}{\partial w}\right) = -\frac{1}{2\pi c}\left(2w\frac{\partial \beta}{\partial w} + w^2\frac{\partial^2 \beta}{\partial w^2}\right). \tag{2.42}$$

D_{ch} can also be expressed, using Eq. 2.41 and the relation between δw and $\delta\omega$, in a different form,

$$D_{ch} = \frac{\partial}{\partial w}\left(\frac{\partial \beta}{\partial \omega}\right) = -\frac{2\pi c}{w^2}\frac{\partial^2 \beta}{\partial \omega^2} = -\frac{2\pi c}{w^2}\beta_2, \tag{2.43}$$

where $\beta_2 = \frac{\partial^2 \beta}{\partial \omega^2}$ is known as the group velocity dispersion (GVD) in optical fibers. Note that, in the present model, we have used a linear (first-order) approximation in Eq. 2.40 for evaluating D_{ch}, resulting in an expression dependent only on β_2. More accurate model can be arrived at by considering yet higher-order expansion leading to the dependence of D_{ch} on β_n, with $n > 2$.

Generically, D_{ch} includes the contributions for material as well as waveguide dispersions in an intertwined (and hence inseparable) manner as β is a complex function of a, n_1, and w. However, with the narrow spectral spread of the lightwaves (compared to the operating wavelength/frequency) one can approximate D_{ch}, in practice, by a linear combination of the two types of chromatic dispersions as

$$D_{ch} \simeq D_{mat} + D_{wg}. \tag{2.44}$$

In the above expression, D_{mat} represents the material dispersion in absence of any waveguiding structure, implying a core with $a \to \infty$. However, D_{wg} represents the waveguide dispersion for an optical fiber with its materials having refractive indices that are independent of wavelength.

For evaluating D_{mat}, one can simply assume that $\beta = 2\pi n_1/w$, which ensures that the lightwave propagates through an infinitely large medium of silica with a refractive index n_1, which is a function of w. Using this simplified model of β in Eq. 2.42, we express D_{mat} as

$$D_{mat} = -\frac{1}{2\pi c}\frac{\partial}{\partial w}\left[w^2\frac{\partial(2\pi n_1/w)}{\partial w}\right] = -\frac{w}{c}\frac{\partial^2 n_1}{\partial w^2}. \tag{2.45}$$

The above result has a major significance for optical communication systems, as the second-order derivative of refractive index in silica vanishes at a specific wavelength. In particular, $\frac{\partial^2 n_1}{\partial w^2} = 0$ at $w = 1270$ nm for pure silica glass, implying thereby that the material dispersion D_{mat} goes to zero at this wavelength for a silica fiber. However, the total chromatic dispersion becomes zero at a different wavelength owing to the additive waveguide dispersion component D_{wg} (see Eq. 2.44).

Waveguide dispersion in optical fibers can be estimated from Eq. 2.43 by assuming that the refractive indices in the expression of β remain constant against wavelength variation, which implies that β is assumed to vary with w only due to the structural constraints. The evaluation of $\frac{\partial}{\partial w}(w^2 \frac{\partial \beta}{\partial w})$ for a given mode is a fairly complex process, where one has to use the solution for β from the characteristic equation for the mode under consideration and thereafter determine its second-order derivative with the constant refractive-index assumption. As this process is outside the scope of this book, we refer the interested readers to (Keiser 2008; Cherin 1983). In the following, we discuss the salient features of chromatic dispersion, resulting from the interplay of its two components.

Figure 2.13 shows some representative plots of chromatic dispersion and its two components (i.e., D_{ch}, D_{mat}, D_{wg}) as the functions of wavelength w in nanometer, wherein the respective dispersion components are *measured* in picoseconds per unit spectral spread (one nanometer) of optical spectrum over one kilometer of fiber (i.e, in ps/nm/km). It is evident from the plots that the material dispersion generally shows stronger variation with wavelength as compared to the waveguide dispersion. As mentioned earlier, for a pure silica core, the material dispersion goes to zero at 1270 nm, and the waveguide dispersion, when added to the material dispersion, the zero-dispersion wavelength for the overall chromatic dispersion moves from 1270 nm to ~1300 nm with a = 5 μm (Fig. 2.13), as governed by the core diameter. The material-dispersion plot can be shifted toward higher wavelength using appropriate doping as shown in the figure (GeO_2 in the present example). This effort toward bringing down the dispersion at the wavelengths with minimum loss (i.e., around 1550 nm) can also be supplemented by decreasing the core diameter, as the decrease in waveguide dispersion with smaller a (i.e., increase in magnitude) pulls down the total chromatic dispersion at 1550 nm. However, in this process with reduced core diameter, power launching from lasers into optical fibers becomes critical.

To have a feel of the chromatic dispersion in a single-mode optical fiber link, consider a 100-km link operating at 10 Gbps with a wavelength w = 1550 nm. Assuming that the total chromatic dispersion $D_{ch} \simeq$ 10 ps/nm/km, and that the spectral spread of the source

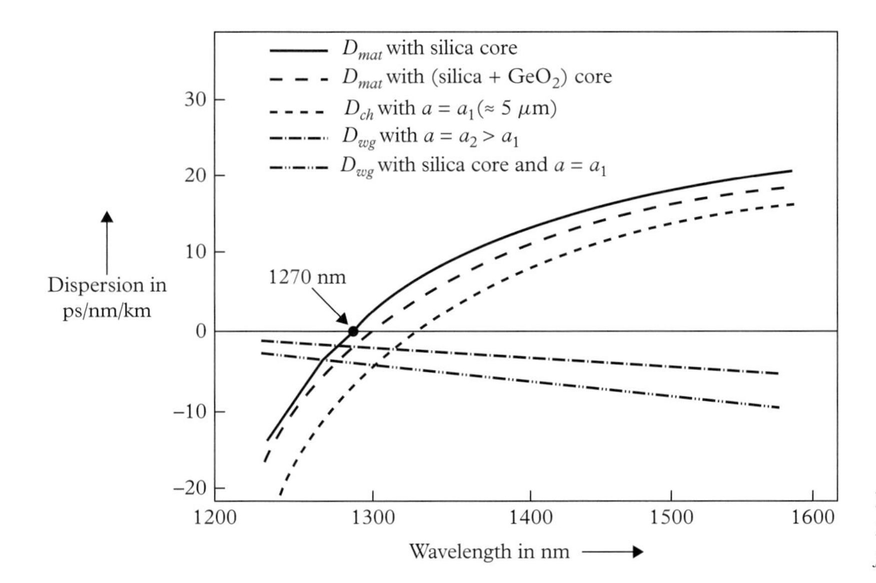

Figure 2.13 *Representative plots of D_{ch}, D_{mat}, and D_{wg} for step-index single-mode fibers.*

$\delta w = 0.1$ nm, we get the pulse spreading due to chromatic dispersion as $\Delta\tau_g = 10 \times 0.1 \times 100 = 100$ ps, which is just equal to the bit duration of a 10 Gbps transmission link, while in practice it should not exceed 50 ps for the same link.[2] This calls for further reduction of the chromatic dispersion in single-mode fibers by appropriate means, so that the zero-dispersion wavelength ($\simeq 1300$ nm in the present example with silica core and $a = 5$ μm) moves to the third window with minimum fiber loss. As mentioned earlier, the zero-crossing point for the material dispersion can be moved upward toward 1550 nm by using appropriate dopants (e.g., GeO_2) and narrower core sizes, along with specially designed RIPs. We shall discuss later in this section the various design considerations to control the chromatic dispersion in single-mode optical fibers.

Polarization-mode dispersion

As discussed earlier, single-mode optical fibers having perfectly circular cores can ideally support and carry through two orthogonally polarized fundamental modes simultaneously with identical propagation characteristics. However, each of these two modes can undergo transformation at any location along the fiber, due to the inherent birefringence of optical fibers resulting from structural irregularities. Any such deviation of the core cross-section from circularity at a fiber location will transform the two co-propagating orthogonally polarized modes into another pair of orthogonally polarized modes with different orientations, thereby changing the states of polarization (SOPs) of the propagating lightwave, as shown in Fig. 2.14. As evident from the figure, the SOPs of the orthogonal modes might get reoriented by an angle $\Delta\theta$ after traversing a given segment of a single-mode optical fiber. In practice, the propagating lightwave keeps encountering this kind of structural irregularity along an optical fiber in an unpredictable manner, caused by minute deformations of cross-sectional contour of the core and variation in the RIP, developed during the fiber-manufacturing process, as well as, from external factors such as fiber bending, twisting, anisotropic stresses on the fiber, at times being time-varying in nature due to environmental challenges.

While moving through a stretch of optical fiber exhibiting some structural irregularity, a given pair of orthogonally polarized modes experience different boundary conditions due to the non-circularity of the core, and thus move with different group velocities. At different randomly appearing stretches exhibiting such irregularities, the power in an optical pulse during its journey through an optical fiber gets repeatedly split and reconstructed from one pair of orthogonally polarized modes to another. Thus, an input optical pulse keeps getting spread over time, as the two orthogonally polarized modes in each mode pair propagate with different group velocities due to the birefringence property of the fiber in the various imperfect stretches of the fiber, leading eventually to another dispersion mechanism in optical fibers, known as *polarization-mode dispersion* (PMD).

The delay spread $\Delta\tau_{pmd}$ caused by PMD, as shown in Fig. 2.14, is a random variable, and hence estimated statistically for a given fiber with its average value $\bar{D}_{pmd/\sqrt{km}}$ expressed

[2] The spectrum of the emitted light from an optical source will be practically determined by the combined effect of the spectral spread due to the modulation of light by the data stream as well as the laser linewidth due to the inherent phase noise (see Eq. B.57 in Appendix B). Thus, with a high transmission rate $\geq$ 10 Gbps) the overall spectral spread of the emitted light will be dominated by the modulation bandwidth, rather than by the laser linewidth which has come down to much lower values (MHz range) with present-day laser technologies. However, notwithstanding this aspect, the laser linewidth will continue to remain important as this will determine the receiver bit-error rate (BER) performance, when applying multilevel modulation schemes involving phase modulation, i.e., QPSK, QAM, and OFDM. In particular, when using any of these phase-modulation schemes, the phase noise in the lasers should not cause any significant phase variation within a symbol (combination of a few consecutive bits, such as 2 bits for QPSK) interval, thereby demanding that the laser linewidth due to phase noise remain much smaller compared to the inverse of the symbol intervals used in the multi-level modulation schemes.

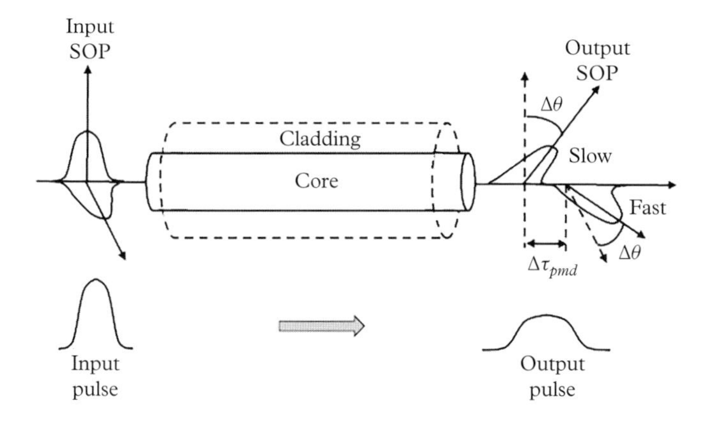

Figure 2.14 *Illustration of birefringence in a single-mode optical fiber, where the two orthogonally polarized modes undergo reorientation by an angle $\Delta\theta$ with a delay spread $\Delta\tau_{pmd}$ due to PMD. Note that the effect of PMD is shown to have manifested in the outgoing optical pulse from the fiber, which has been spread over time as compared to the input pulse.*

in ps/$\sqrt{\text{km}}$ as its unit. Note that the unit is chosen to be proportional to the square-root of the distance, as the effect of PMD also gets averaged out to some extent, as an optical pulse during its long journey through fiber keeps moving back and forth between the slow and fast modes, thereby preventing a linear growth of PMD. In practice, the impact of PMD is small in magnitude as compared to the other dispersion mechanisms, typically in the range of 0.5–1 ps/$\sqrt{\text{km}}$, and hence not much *felt* in shorter and low-speed optical fiber links. For example, over a 50 km fiber link with $\bar{D}_{pmd/\sqrt{km}} = 0.5$ ps/$\sqrt{\text{km}}$, the pulse spreading would turn out to be $\Delta\tau_{pmd} = 0.5 \times \sqrt{50} = 3.54$ ps, which would be manageable with a 10 Gbps transmission corresponding to bit durations of 100 ps. However, for high-speed ($\geq$ 10 Gbps) long-distance ($\geq$ 100 km) links, the impact of PMD would no longer remain negligible as compared to the chromatic dispersion. In order to address this problem, multi-level modulation schemes need to be employed (e.g., QPSK, QAM) to stretch the symbol intervals to an integral multiple of bit intervals, thereby absorbing the impact of various dispersion mechanisms within the increased symbol durations.

2.2.5 Fiber nonlinearities

The performance of high-speed WDM transmission systems may be significantly affected by fiber nonlinearities, when the power carried by the lightwaves in optical fibers exceeds a certain limit. There are different types of nonlinear effects in optical fibers, depending on the ways the incident lightwaves interact with the propagating medium. These nonlinear effects might also lead to some useful technologies that can be utilized in optical networks in certain situations.

Fiber nonlinearities can be categorized broadly in two types: inelastic scattering phenomena and Kerr nonlinearity. Inelastic scattering could be mainly of two types: stimulated Raman scattering (SRS) and stimulated Brillouin scattering (SBS). In both cases, the incident lightwave, called a pump, generates a second lightwave with lower frequency, referred to as a Stokes wave, giving away part of its energy to the molecules of the medium. Kerr nonlinearity is the fallout of the dependence of the refractive index of silica on the incident optical power, which manifests itself in a number of nonlinear phenomena, but without transferring any energy to the medium (and hence viewed as an elastic process).

In general, the nonlinear effects in optical fibers recede after a certain distance due to the power loss in the fiber, and typically the manifestation of the nonlinear effects take place within an effective length L_{eff}, given by

$$L_{eff} = \frac{1 - \exp(-\alpha L)}{\alpha}, \tag{2.46}$$

where L represents the actual length of the fiber, and L_{eff} is the length, that the signal would traverse through if it had a constant power over that length (i.e., $z \in L_{eff}$) and zero beyond (i.e., $z > L_{eff}$). Thus, if $\alpha L \ll 1$, then $L_{eff} \simeq L$. However, if $\alpha L \gg 1$, then $L \simeq 1/\alpha$ (i.e., $\simeq$ 22 km with a fiber loss of $\alpha = 0.2$ dB/km ($\simeq 0.046$ neper/km)) in the third spectral window of the optical fiber.

Concerning the impact of nonlinearity, another important parameter of optical fibers is the effective cross-sectional area A_{eff}, which is essentially governed by the cross-sectional field distribution $F(r, \theta)$ in the fiber, given by (Agrawal 2001)

$$A_{eff} = \frac{[\int_r \int_\theta |F(r,\theta)|^2 r dr d\theta]^2}{\int_r \int_\theta |F(r,\theta)|^4 r dr d\theta}. \tag{2.47}$$

The above expression implies how tightly the optical power is bounded in the core. Thus, the effective intensity of the light in a fiber core is given by $I = P/A_{eff}$, with P as the total power flowing across the fiber cross-section. The value of A_{eff} will depend on the actual core diameter and the RIP of the fiber.

In the following, we describe the various types of nonlinearities and their manifestations in optical transmission systems.

Inelastic scattering

As mentioned earlier, the inelastic scattering phenomena in optical fibers – SRS and SBS – extract power from the incident lightwaves, that can take place if the total incident power in an optical fiber exceeds certain specific limits, and these power limits are different for SRS and SBS. The basic difference between the two scattering phenomena is that, in SRS the pump photons interact with the optical phonons, while in SBS the acoustic phonons participate with the pump photons. We describe these two inelastic scattering phenomena in the following.

Stimulated Raman scattering:

In SRS the incident lightwave interacts with a silica molecule in the medium (optical fiber in the present context), to create a Stokes wave at a lower frequency, and the difference in energy between the pump and the Stokes wave is absorbed by the silica molecule, thereby exciting a vibrational state in it. Viewed quantum-mechanically, this amounts to scattering of the incoming photon by a molecule in the medium to generate another photon with lower energy, while moving up itself (the scattering molecule) to a higher-energy vibrational state, representing an optical phonon. For a given pump (incident lightwave) frequency ω_p (angular frequency), the vibrational energy of the molecule determines the frequency ω_p of the Stokes wave as $\omega_p - \omega_s = \Omega_R$, with Ω_R as the Raman shift. Note that, due to the isotropic nature of silica molecules, SRS occurs in both backward and forward directions.

Although the complete analytical model of SRS is quite complex, the basic physics behind SRS can be described by using a simple differential equation:

$$\frac{\partial I_s}{\partial z} = g_R I_p I_s - \alpha_s I_s, \tag{2.48}$$

with I_s as the Stokes intensity, I_p as the pump intensity, g_R as the Raman gain coefficient, and α_s representing the fiber loss for the Stokes frequency. In Eq. 2.48, the first term on

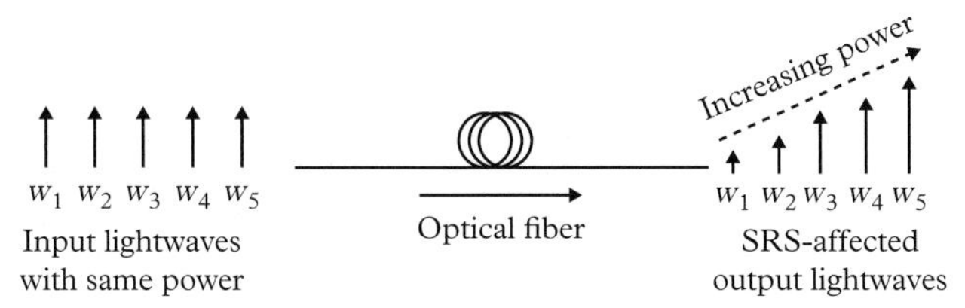

Figure 2.15 *Illustration of SRS in WDM transmission system.*

the right-hand side represents the growth of Stokes intensity with z, and the second term represents the loss of the Stokes wave with z, so the left side, as a sum of the two counteracting terms, gives the net growth rate of the Stokes wave along the fiber axis. In order to represent the pump depletion due to SRS and fiber loss, one can form another differential equation, which we don't consider here for brevity. The above equation gives an insight and leads to approaches for assessing how the incident optical power, SRS gain, and fiber loss can together affect the performance of optical transmission systems.

In single-wavelength transmission systems, the impact of SRS remains practically negligible due to the high power requirement $\simeq$ 1 W (evaluated using Eq. 2.48) for the incident lightwave in optical fibers (Agrawal 2001). In WDM transmission systems, when the lightwaves with multiple wavelengths co-propagate through the fiber, and the aggregate power exceeds a certain limit, due to SRS the power gets transferred from the higher-energy (lower-wavelength) lightwaves to the lower-energy (higher-wavelength) lightwaves, as illustrated in Fig. 2.15. This leads to SRS-induced inter-channel crosstalk between the WDM channels.

Experimentally, it is observed that g_R in Eq. 2.48 exhibits a *gain spectrum* spanning from the pump frequency ω_p down to the frequencies in the range $[\omega_p, \omega_p - \Delta\omega]$, with $\Delta\omega$ representing the bandwidth of the gain spectrum (Chraplyvy 1984). As shown in the figure, the gain coefficient g_R in the SRS gain spectrum for silica monotonically increases with frequency upto a maximum value $g_R(max)$ ($\simeq 6 \times 10^{-14}$ m/W at the pump wavelength 1550 nm). Thereafter, g_R falls quite abruptly and passes through much-subdued lower peaks at 18 THz, 24 THz, and 32 THz, eventually decaying to negligible values beyond 40 THz. The variation of g_R with $\delta\omega$ is generally approximated as a triangular function (see Fig. 2.16) for analytical purpose. This implies that in a WDM system the lightwave frequencies covered by this bandwidth of the triangular gain spectrum may be affected by SRS. We discuss the implication of this aspect in the following.

As indicated earlier (Fig. 2.15), in a WDM transmission system, the channel with the highest frequency (i.e., lowest wavelength) becomes most vulnerable to SRS as it transfers energy to all other channels, that fall within $\Delta\omega$ of the SRS gain spectrum. The total power carried by the fiber will determine the extent of the SRS that would take place, while the use of a larger number of channels would imply that the lightwave with the highest-frequency channel will be driven by the SRS process to the maximum extent in transferring its power to all the remaining lightwaves. This leads to the fact that the product of the total WDM power (as this will decide the strength of the SRS) and the number of channels (deciding how many channels would be *sucking* power from the largest-frequency channel through SRS) would have to be constrained within an upper-bound, so that the power loss from the highest-frequency channel remains confined to a specified limit, say 0.5 dB. Using the triangular approximation for the SRS gain spectrum (Fig. 2.16), i.e.,

$$g_R = g_R(max)\delta\omega/\Delta\omega \text{ for } 0 \leq \delta\omega \leq \Delta\omega, \tag{2.49}$$

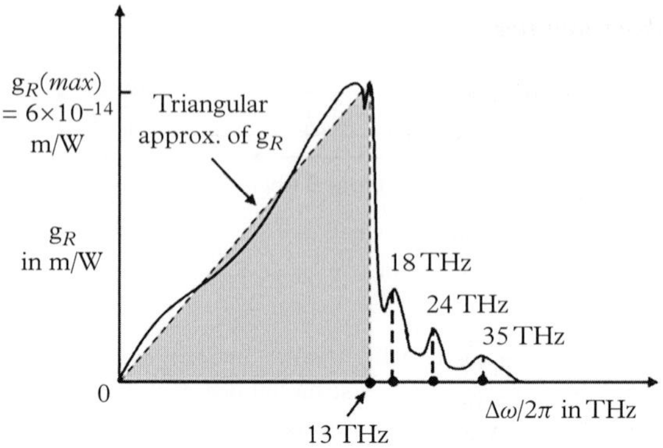

Figure 2.16 *SRS gain spectrum of silica with 1550 nm as the pump wavelength. Triangular approximation of the gain spectrum:* $g_R \simeq g_R(max)\delta\omega/\Delta\omega$ *(Chraplyvy 1984, ©IEEE).*

and the 0.5 dB limit for the SRS-caused power loss, one can reach a handy design formula for controlling the impact of SRS in a WDM system (with zero dispersion in the fibers), given by (Ramaswami *et al.* 2010; Chraplyvy 1984)

$$P_T \Delta w L_{eff} < 40,000 \text{ mW.nm.km}, \tag{2.50}$$

where P_T is the total power carried by the M WDM channels, each with a power of P (i.e., $P_T = MP$) and $\Delta w = (M-1)\delta w$ is the wavelength span covered by the entire WDM transmission window with δw ($\simeq$ 0.8nm $\equiv$ 100 GHz) representing the channel spacing in terms of wavelength (with the assumption that Δw falls within the spectral bandwidth of g_R). Further, since the impact of SRS would effectively last up to the effective length of the fiber due to the fiber loss, the left side of the above expression comprises L_{eff}, instead of the full length L of the fiber (presuming that $L > L_{eff}$).

With in-line optical amplifiers placed along an optical fiber link, in each fiber span (usually with a length exceeding L_{eff}) between any two successive amplifiers the effect of SRS reappears, making a periodic contribution along the link. While designing long-haul WDM links, one therefore needs to reduce the separation between the in-line optical amplifiers, as much as possible without exceeding the affordable limit on the number of optical amplifiers in a link, and sum up all these contributions to take into account the cumulative effect of SRS from the various optically amplified fiber spans. However, the polarization mismatch (already taken into account in Eq. 2.50) between the lightwaves and the dispersion in fibers ameliorate the impact of SRS, and the latter (dispersion) in turn relaxes the right-hand side of Eq. 2.49 by a factor of two, i.e., from 40,000 mW.nm.km. to 80,000 mW.nm.km. Using this constraint, one can readily estimate the power threshold in a WDM transmission system caused by SRS. For example, in a WDM transmission system with 64 channels placed 100 GHz apart and transmitted over a 20 km fiber segment, one needs to keep P below 1.24 mW to achieve the 0.5 dB limit on the power penalty due to SRS.

Finally, on the flip side, SRS can also be used for amplifying a lightwave by injecting a strong pump wave into an optical fiber segment at an appropriate frequency within the SRS gain spectrum. Such optical amplifiers are known as Raman amplifiers, which we will discuss later in this chapter.

Stimulated Brillouin scattering

In SBS, with strong lightwaves, the glass molecules in the optical fiber get compressed, leading to *electrostriction*, which in turn generates an acoustic wave in the fiber. The acoustic wave modulates the refractive index of the fiber material, thereby inducing a refractive-index grating, which causes backscattering of the lightwave through Bragg diffraction. The backscattered light, known as a Stokes wave, undergoes a downshift in frequency due to the Doppler effect, brought in by the movement of the grating at the acoustic velocity. Note that, unlike SRS, SBS generates only backward scattering.

Viewing the SBS process quantum-mechanically, an injected pump photon into the optical fiber gets engaged into SBS to create a scattered photon (Stokes photon) and an acoustic phonon, and the energy as well as the momentum of all the three participating *particles* should be conserved in the scattering event. Denoting ω_p and ω_s as the angular frequencies of the pump and Stokes photons, one can express the energy conservation criterion as $\omega_p = \omega_s - \Omega_B$, with Ω_B as the Brillouin shift representing the frequency of the SBS-induced acoustic phonon. Similarly, for the wave vectors of the pump and Stokes waves, the criterion for the conservation of momentum must be satisfied. Unlike SRS, the bandwidth of SBS spectrum is much smaller, i.e., $\simeq 20$ MHz at 1550 nm as compared to 13 THz in SRS, but with a much higher gain coefficient $g_B \simeq 5 \times 10^{-11}$ m/W, which remains practically independent of the lightwave frequency. The scattering phenomenon in SBS can be explained with the differential equation (Eq. 2.48) as used in SRS, with g_R replaced by g_B.

The impact of SBS on an optical transmission system depends mainly on g_B, along with the effective length L_{eff} and the effective cross-sectional area A_{eff} of the fiber. Further, the polarization states of the pump and Stokes waves determine the strength of SBS, and so also the relative value of the spectral width $\delta\omega = 2\pi\delta f$ of the optical source with respect to the SBS bandwidth $\Delta\omega_B = 2\pi\Delta f_B \simeq 2\pi \times 20$ MHz. In other words, better polarization match between the pump and Stokes waves intensifies the impact of SBS, bringing down the threshold power. However, the larger spectral width of the source spreads the process spectrally, thereby bringing up the threshold power of SBS. Considering all these aspects, one can find out a comprehensive relation to express the threshold power for SBS (Ramaswami *et al.* 2010; Smith 1972) as

$$P_{th} = \frac{21\eta_p A_{eff}}{g_B L_{eff}}\left(1 + \frac{\delta\omega}{\Delta\omega_B}\right), \tag{2.51}$$

where η_p represents the polarization-matching factor (= 1 for full match, 2 otherwise). For an optical source with a spectral width of 100 MHz and with the worst-case scenario in respect of polarisation matching (i.e., $\eta_p = 1$), one gets $P_{th} = 7.8$ mW, which would go down by 3 dB to 3.9 mW with $\eta_p = 2$, in a practical scenario.

In SBS, the Stokes wave, propagating only in the backward direction, causes pump depletion and the backscattered light might reach the transmitter end and needs to be avoided using optical circulators. Further, SBS hardly exhibits any interaction between different wavelengths, unless their separation is < 20 MHz, which is indeed not possible in practical WDM systems. However, with low threshold power, WDM systems may become sensitive to SBS, and special care needs to be taken with lower transmission rates due to the small spectral spread from low-speed modulation. In such cases, on-off keying creating laser chirps or deliberate spectral-spreading with additional phase modulation in the transmitter (called spectral dithering) can be helpful in reducing the impact of SRS. The additional chromatic dispersion arising from the deliberate spectral spreading may not be of much concern for lower transmission rates, and in any case, one can resort to dispersion management techniques to combat the problem.

Kerr nonlinearity

When a lightwave in an optical fiber carries a power above a certain limit, the dielectric constant of silica in the fiber core exhibits a nonlinear dependence on the propagating electric field, leading to the Kerr effect. In particular, high power in the propagating lightwave produces anharmonic motion of the electrons bound in the silica molecules, which in turn makes the induced electric polarization vary nonlinearly with the electric field. This underlying phenomenon changes the refractive index of the medium in proportion to the light intensity (i.e., optical power per unit area of cross-section). In order to examine this nonlinear effect, we first discuss the propagation model in a linear medium, and thereafter extend this for the nonlinear case (Agrawal 2001).

For a transmission medium (silica in the present context) operating in the linear regime, one can express the induced electric flux density D_L as a function of the propagating electric field E as

$$D_L = \epsilon_0 E + P_L, \tag{2.52}$$

where P_L (or more precisely $P_L(z,t)$, when it is necessary to represent its evolving nature with time and space) is the linear dielectric polarization resulting from the convolution of E (again as $E(z,t)$ wherever necessary, with the same reasoning as for $P_L(z,t)$) in the given medium with the first-order susceptibility $\chi^{(1)}(t)$, given by

$$P_L(z,t) = \epsilon_0 \chi^{(1)}(t) * E(z,t). \tag{2.53}$$

Note that, the convolution operation in the above expression indicates that the temporal spread of $\chi^{(1)}(t)$ (intrinsically representing the impulse response of the medium) makes $P_L(z,t)$ dependent on the present and past values of $E(z,t)$. Taking the Fourier transform on both sides of Eq. 2.53, we therefore obtain

$$\tilde{P}_L(z,\omega) = \epsilon_0 \tilde{\chi}^{(1)}(\omega)\tilde{E}(z,\omega), \tag{2.54}$$

with $\tilde{P}_L(z,\omega)$, $\tilde{\chi}^{(1)}(\omega)$ and $\tilde{E}(z,\omega)$ as the Fourier transforms of $P_L(z,t)$, $\chi^{(1)}(t)$ and $E(z,t)$, respectively. Making use of the above expression of $\tilde{P}_L(z,t)$ in Eq. 2.52, one can express $\tilde{D}_L(z,\omega)$ as

$$\begin{aligned} \tilde{D}_L(z,\omega) &= \epsilon_0 \tilde{E}(z,\omega) + \epsilon_0 \tilde{\chi}^{(1)}(\omega)\tilde{E}(z,\omega) \\ &= \epsilon_0 [1 + \tilde{\chi}^{(1)}(\omega)]\tilde{E}(z,\omega) \\ &= \epsilon_0 \epsilon_r(\omega)\tilde{E}(z,\omega) \\ &= \epsilon_0 n^2(\omega)\tilde{E}(z,\omega), \end{aligned} \tag{2.55}$$

with $n(\omega)$ and $\epsilon_r(\omega)$ as the refractive index and relative dielectric constant (permittivity) of the medium, respectively, which are related through the first-order susceptibility as

$$n^2(\omega) = \epsilon_r(\omega) = 1 + \tilde{\chi}^{(1)}(\omega). \tag{2.56}$$

Making use of the above results, one can therefore write the linear propagation equation as

$$\nabla^2 E + \beta^2 E = 0, \quad \beta = \frac{\omega n(\omega)}{c}, \tag{2.57}$$

which we need to modify to bring in the impact of the Kerr effect in the medium. However, when the medium exhibits the Kerr effect, the corresponding wave equation becomes very complex. In the following, we present a practical approach to modifying the linear wave equation itself to take into account the impact of the Kerr effect on the wave propagation.

In general, there can be various orders of nonlinearities in a medium affecting its dielectric constant, determined by second- and higher-order susceptibilities of the medium. However, silica molecules being symmetric in nature, the second-order susceptibility vanishes to zero, while the third-order susceptibility is the only significant higher-order nonlinearity that produces a tangible impact in silica. Further, unlike the first-order susceptibility, the third-order susceptibility $\chi^{(3)}$ doesn't exhibit any tangible temporal spread in its response (i.e., resembles closely a delta function in time), thereby enabling us to express the nonlinear polarization $P_{NL}(z,t)$ as $P_{NL}(z,t) = \epsilon_0 \chi^{(3)} E^3(z,t)$ (i.e., without using convolution). Representing $E(z,t)$ as a monochromatic lightwave given by the electric field $E\cos(\omega t - \beta z)$, one can therefore express $P_{NL}(z,t)$ as

$$\begin{aligned}
P_{NL}(z,t) &= \epsilon_0 \chi^{(3)} E^3(z,t) \\
&= \epsilon_0 \chi^{(3)} E^3 \cos^3(\omega t - \beta z) \\
&= \epsilon_0 \chi^{(3)} E^3 \left[\frac{3}{4}\cos(\omega t - \beta z) + \frac{1}{4}\underbrace{\cos(3\omega t - 3\beta z)}_{neglected}\right] \\
&= \frac{3}{4}\epsilon_0 \chi^{(3)} E^3 \cos(\omega t - \beta z) \\
&= \epsilon_0 \left(\frac{3}{4}\chi^{(3)} E^2\right) \underbrace{E\cos(\omega t - \beta z)}_{E(z,t)} \\
&= \epsilon_0 \epsilon_{NL} E(z,t).
\end{aligned} \tag{2.58}$$

In the above expression, $\epsilon_{NL} = \frac{3}{4}\chi^{(3)} E^2$ represents the nonlinear equivalent of the relative dielectric constant $\epsilon_r(\omega)$ of the linear medium, wherein the term with the frequency 3ω (in the third line of Eq. 2.58) is neglected due to the lack of phase matching in silica.

By using the above result, the wave equation in the presence of Kerr effect can be simplified with some practical assumptions. In particular, we consider that the lightwave carries a *slowly time-varying envelope* (i.e., modulation), such that the intensity of light remains fairly *constant* over a given *observation interval* (typically a bit or a symbol duration), which allows us to assume that ϵ_{NL} remains constant in the present model. With this assumption, one can therefore obtain the modified relative dielectric constant $\hat{\epsilon}(\omega)$ of the medium due to nonlinearity as the sum of $\epsilon_r(\omega)$ and ϵ_{NL} as

$$\hat{\epsilon}(\omega) = \epsilon_r(\omega) + \epsilon_{NL} = n^2(\omega) + \epsilon_{NL} = 1 + \tilde{\chi}^{(1)}(\omega) + \epsilon_{NL} = \hat{n}^2(\omega), \tag{2.59}$$

where $\hat{n}(\omega)$ represents the modified refractive index of the medium operating under the nonlinear regime, leading to the modified propagation constant $\hat{\beta}$ of the nonlinear medium, given by $\hat{\beta} = \hat{n}(\omega)\omega/c$. Therefore, $\hat{\beta}$ can now be used in place of β in the linear wave equation (Eq. 2.57) to obtain the quasi-linear version of the nonlinear wave equation as

$$\nabla^2 E + \hat{\beta}^2 E = 0, \quad \hat{\beta} = \frac{\hat{n}(\omega)\omega}{c} = \frac{\omega\sqrt{\epsilon_r(\omega) + \epsilon_{NL}}}{c} = \frac{\omega\sqrt{\hat{\epsilon}_r(\omega)}}{c}, \tag{2.60}$$

where ϵ_{NL}, being a constant, behaves as a *small perturbation* to the propagation constant of the linear wave equation. This substitution ($\hat{\beta}$ in place of β) becomes extremely useful in representing a nonlinear phenomenon by a *quasi-linear* differential equation, which becomes possible because the electric field E in ϵ_{NL} is assumed to remain constant during an observation interval.

Next, with the observation that $\chi^{(3)} \ll \chi^{(1)}$ in silica, one can simplify the expression for $\hat{n}(\omega)$ from Eq. 2.59 as

$$\hat{n}(\omega) = \sqrt{\hat{\epsilon}_r(\omega)} = \sqrt{n^2(\omega) + \frac{3}{4}\chi^{(3)}E^2} \simeq n(\omega)\left(1 + \frac{3}{8}\frac{\chi^{(3)}}{n^2(\omega)}E^2\right). \tag{2.61}$$

Note that the intensity I of the propagating lightwave is the power per unit area developed by the electric field E across the material impedance $\eta = \sqrt{\mu_0/\epsilon} = \frac{1}{cn(\omega)\epsilon_0}$, given by $I = \frac{E^2}{2}\eta = \frac{\epsilon_0 n(\omega)cE^2}{2}$. Using this result in Eq. 2.61, we obtain the expression for the refractive index $\hat{n}(\omega)$ for the nonlinear medium as

$$\hat{n}(\omega) = n(\omega) + \left(\frac{3\chi^{(3)}}{4\epsilon_0 n^2(\omega)c}\right) \times \left(\frac{\epsilon_0 n(\omega)cE^2}{2}\right) = n(\omega) + \bar{n}(\omega)I, \tag{2.62}$$

with $\bar{n}(\omega)$ as the *nonlinear refractive-index coefficient* (in $\mu m^2/W$) of the medium governed by its third-order susceptibility. Using Equations 2.62 and 2.60, we express $\hat{\beta}$ as

$$\hat{\beta} = \hat{n}(\omega)\omega/c = \frac{\big(n(\omega) + \bar{n}(\omega)I\big)\omega}{c} = \beta + \frac{\bar{n}(\omega)\omega}{c}I = \beta + \xi I, \tag{2.63}$$

where $\xi = \bar{n}(\omega)\omega/c$. Note that, due to the dependence of propagation constant $\hat{\beta}$ on the light intensity I brought in by the Kerr effect, the lightwaves undergo phase variation during the course of propagation through optical fibers.

Making use of the above feature of the Kerr effect, we next explain its various manifestations in optical fibers: self-phase modulation (SPM) for single-wavelength as well as WDM systems, and cross-phase modulation (XPM) and four-wave mixing (FWM) for WDM systems.

Self-phase modulation

Equation 2.63 implies that, in presence of the Kerr effect, the propagation constant $\hat{\beta}$ of a lightwave varies instantaneously with light intensity I. The instantaneous variation of the propagation constant enforced by the light intensity brings in a *spatial-cum-temporal* variation of phase shift incurred by the propagating lightwave pulses along the optical fiber length. Thus, when observed at a given instant of time t, a spatial variation of the propagation constant takes place within the length spanned by a propagating optical pulse, and this *snapshot* of the spatial phase variation keeps moving and varying in shape *en block* in space and time. However, when viewed at a given location z along the fiber axis, the different parts of an optical pulse (i.e., a THz sinusoid with a pulse-shaped baseband envelope) pass through the cross-section at the given location z with their light intensity varying with time. Through the space-time dependence of the propagation constant, the pulse-modulated lightwave undergoes a temporal phase variation, thereby causing a *phase modulation* of the optical carrier during the pulse interval itself.

The above self-induced phase modulation due to the Kerr effect is known as *self-phase modulation* (SPM), which in effect causes a variation of the instantaneous frequency (i.e., the rate of change of phase with time) within the pulse, resulting in frequency *chirping*. In the presence of SPM-induced chirping, the shift in the instantaneous angular frequency (i.e., frequency chirping) can be determined from Eq. 2.63, by estimating the derivative of the instantaneous phase shift of the lightwave in the presence of Kerr nonlinearity. The phase shift incurred by the lightwave at time t after traversing a distance $z > L_{eff}$ through the fiber can be expressed as

$$\phi(z,t) = z\beta + L_{eff}\xi I(t). \tag{2.64}$$

Thus, the angular frequency chirping $\delta\omega(t)$ at the distance z can be expressed as

$$\delta\omega(t) = \frac{\partial\phi(z,t)}{\partial t} = L_{eff}\xi\frac{\partial I(t)}{\partial t}, \tag{2.65}$$

where the time-derivative of linear propagation constant β, i.e., $\partial\beta/\partial t$, is assumed to be zero. Note that, when the long-haul links employ a series of in-line optical amplifiers for compensating the fiber losses, the above formulation of $\phi(z,t)$ involving L_{eff} needs to be modified to obtain the realistic picture, as the optical power will get restored several times on the way to its original level, thereby bringing in the nonlinear effects over and over again.

To examine the impact of SPM-induced chirping on the information-bearing envelopes of the propagating lightwaves, we consider a Gaussian function $p_g(t)$ as a reference shape of lightwave pulses, which is launched at the fiber input at $z = 0$. The Gaussian pulse $p_g(t)$, as shown in Fig. 2.17, is expressed as

$$p_g(t) = A_0 \exp\left(\frac{-t^2}{2T_p^2}\right)\cos\omega t, \tag{2.66}$$

where A_0 represents the amplitude of the Gaussian pulse, being carried as an envelope by the lightwave of a frequency ω, with the $\frac{1}{e}$-intensity half-width as T_p. Note that, the instantaneous intensity $I(t)$ of the lightwave is proportional to the square of the pulse function $p_g(t)$. The shape of the optical pulse and the constituent frequency variation (SPM-induced chirping) keep evolving under the joint influence of GVD and the Kerr effect in optical fiber. However, in the following, our immediate interest is to examine the impact of SPM-induced chirping only.

In order to account for the chirping effect due to SPM on the above Gaussian pulse, we use a simplistic approach by replacing the $\cos\omega t$ term by $\cos(\omega t + \int_{-\infty}^{t}\delta\omega(t)dt)$ in Eq. 2.66, and obtain the expression of the chirped optical pulse as

$$p_g^{ch}(t) = A_0 \exp\left(\frac{-t^2}{2T_p^2}\right)\cos\left\{\omega t + \int_{-\infty}^{t}\delta\omega(t)dt\right\}. \tag{2.67}$$

This expression can now be used to visualize and estimate the effect of chirping on the Gaussian pulse, as illustrated in Fig. 2.17. As shown in the figure, without GVD, the pulse develops chirping *only*, without any pulse broadening, which is illustrated further by the representative variations of the instantaneous phase $\phi(t)$, and the corresponding

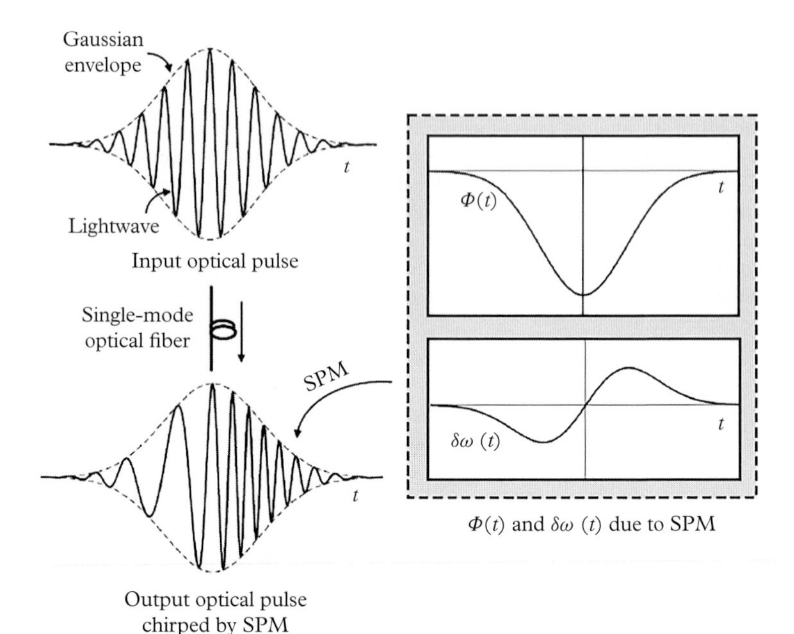

Figure 2.17 *Chirping in a Gaussian pulse due to SPM only.*

instantaneous frequency increment $\delta\omega(t)$ along with the input and output pulses. Note that, the phase lag of the lightwave first increases in magnitude (though with a negative sign due to the lag) as the pulse intensity goes up and hence the incremental frequency (proportional to phase-derivative) of the propagating pulse, initially goes down as the slope of $p_g(t)$ increases. Thereafter, the slope of the phase plot decreases when the intensity slope starts reducing before reaching the peak. Hence, the incremental frequency reaches its most negative value at this point, and then its magnitude decreases toward zero as the phase plot moves toward its peak (corresponding to the peak of the pulse shape) with zero slope. After crossing the peak of the pulse, what follows is just the mirror image of the previous episode. Consequently, the optical frequency inside the envelope gets chirped initially with lower and later with higher frequencies during the pulse interval.

Further, a Gaussian lightwave pulse affected by GVD *only* due to the pulse modulation (i.e., with a monochromatic source and without any SPM) is shown in Fig. 2.18, wherein only the pulse spreading takes place without any chirping. In the linear propagation regime, pulse spreading for Gaussian pulses without SPM and with $\delta w = 0$ (i.e., monochromatic source), can be expressed as (Agrawal 2001)

$$T_p(z) = T_p\sqrt{1 + \left(\frac{\beta_2 z}{T_p^2}\right)^2} = \sqrt{T_p^2 + (\beta_2 z)^2}. \tag{2.68}$$

This expression is obtained by solving the linear wave equation in the presence of pulse modulation, which implies that the pulse duration T_p and the pulse spreading $\beta_2 z$ due to chromatic dispersion add on the root-mean-square basis. This expression has to be further augmented for the nonlinear wave propagation to take care of the SPM, leading to a nonlinear Schrödinger equation (Agrawal 2001). Further extension of this analysis is necessary to take into account the effect of the phase noise of optical sources (which manifests itself as the non-zero spectral width of the sources) along with the spectral

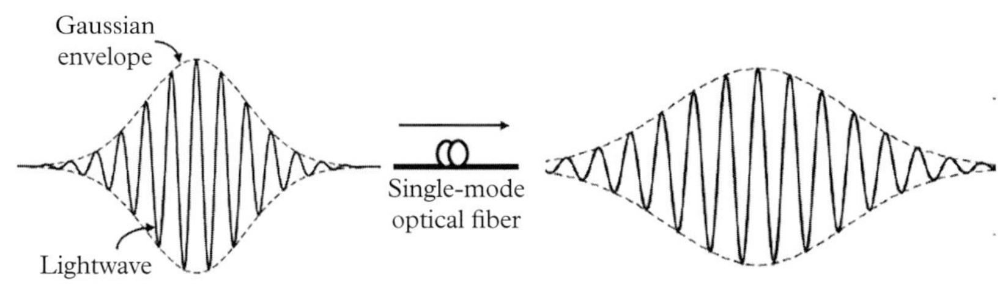

Figure 2.18 *Pulse spreading in a Gaussian pulse due to GVD only.*

spreading due to modulation and SPM-induced chirping. Note that, in line with the observation made from Eq. 2.68, a simple approach for the system designers has been in vogue for assessing the effects of various pulse-spreading mechanisms on the system design in the linear propagation regime. In this approach, the various pulse-spreading components (e.g., pulse spreading due to dispersion mechanisms, transmitter and receiver rise times, etc.) are added up on a root-mean-square basis to obtain the overall system rise time, which must remain well below the bit interval in an optical bit stream. We shall discuss this issue again in Chapter 10 when addressing the system rise-time budget in optical communication links.

In the combined presence of SPM-induced chirping from strong optical pulses and the spectral spread of optical source, the fiber would exhibit complex features, including frequency chirping due to SPM along with GVD-caused changes in the pulse shapes due to the source chromaticity as well as the instantaneous frequency variation from chirping. In particular, both effects (SPM-induced chirping and spectral width of the transmitted light) may or may not act in the same direction for the GVD, leading to the possibilities of pulse spreading, pulse compression, or even maintaining a fixed pulse duration during the course of the propagation. The third scenario (maintaining the same pulse shape throughout the fiber) can be realized with critical choices of launched pulse shape, power, and fiber parameters, and the propagation mechanism to achieve this special feature is known as soliton propagation.

Cross-phase modulation

Cross-phase modulation (XPM) takes place between the co-propagating lightwaves of different wavelengths in WDM systems. In a WDM system, if the aggregate optical power of all wavelengths in an optical fiber exceeds a specific limit, then along with the SPM occurring on the individual channels, the optical pulses in each channel get phase-modulated due to the power carried by the other channels in the optical fiber, thereby spreading the effect of the Kerr nonlinearity from one channel to the other, hence the name XPM. In other words, the intensity variation in some channel, while generating SPM for the same channel, concurrently creates XPM for other channels.

As observed in the case of SPM, in a given channel (channel j, say), the phase modulation caused by SPM and the XPM caused by the other channels on a given channel at a distance $z \geq L_{eff}$, can be expressed as (Agrawal 2001)

$$\Delta\phi_j = L_{eff}\xi\left(\frac{\partial I_j}{\partial t} + 2\sum_{i(\neq j)\in M}\frac{\partial I_i}{\partial t}\right), \tag{2.69}$$

where I_j's are the light intensities carried by the respective channels.

The impact of XPM can be reduced significantly in WDM systems by increasing the channel separation. However, in a typical WDM system, the lightwaves on different wavelengths with their respective pulsed envelopes won't be able to propagate in phase due to the finite chromatic dispersion in the fiber. In other words, these lightwaves would walk away from each other, thereby not getting adequate scope to take effect. However, the fibers with a very small and flat dispersion profile in the third window may give rise to XPM, and thus it is desirable to have some minimum amount of dispersion in the optical fibers in this window to avoid the deleterious effects of XPM (and also FWM, as described in the following).

Four-wave mixing

Due to Kerr nonlinearity, when multiple wavelengths co-propagate along the fiber in a WDM link, the induced polarization might exhibit cubic nonlinearity with the electric fields, implying thereby that any three propagating lightwaves would interact to generate a fourth lightwave. This process of fourth lightwave generation is known as FWM, and the frequency of the fourth lightwave might also fall upon (i.e., overlap with) the existing lightwave frequencies, leading to interference. We explain the basic features of FWM in the following.

Consider a WDM link carrying M wavelengths/frequencies with the electric fields of the propagating lightwaves, given by

$$\begin{aligned} E_1(z,t) &= A_1\cos(\omega_1 t-\beta_1 z)=A_1\cos\psi_1 \\ E_2(z,t) &= A_2\cos(\omega_2 t-\beta_2 z)=A_2\cos\psi_2 \\ &\cdots \\ E_i(z,t) &= A_i\cos(\omega_i t-\beta_i z)=A_i\cos\psi_i \\ &\cdots \\ E_M(z,t) &= A_M\cos(\omega_M t-\beta_M z)=A_M\cos\psi_M \end{aligned} \tag{2.70}$$

where $\psi_i = \omega_i t - \beta_i z$, and A_i, ω_i and β_i represent the amplitudes, frequencies, and propagation constants of the respective lightwaves. While these lightwaves co-propagate through the fiber, their aggregate power might exceed a certain limit, which would in turn cause the Kerr nonlinearity to take effect. The induced nonlinear polarization $P_{NL}^M(z,t)$ from the M propagating lightwaves can be represented as

$$\begin{aligned} P_{NL}(z,t) &= \epsilon_0\chi^{(3)}\left[\sum_{1=1}^{M}E_i(z,t)\right]^3 \\ &= \epsilon_0\chi^{(3)}\sum_{i=1}^{M}A_i\cos\psi_i\sum_{j=1}^{M}A_j\cos\psi_j\sum_{j=1}^{M}A_k\cos\psi_k \\ &= \epsilon_0\chi^{(3)}\left[\sum_{i=1}^{M}A_i^3\cos^3\psi_i+\sum_{i=1}^{M}A_i^2\cos^2\psi_i\sum_{j\neq i}^{M}A_j\cos\psi_j+\right. \\ &\left.\sum_{i\neq j,k}^{M}\sum_{j\neq i,k}^{M}\sum_{k\neq i,j}^{M}A_i\cos\psi_iA_j\cos\psi_jA_k\cos\psi_k\right] \\ &= X(i=j=k)+Y(i=j\neq k)+Z(i\neq j\neq k), \end{aligned} \tag{2.71}$$

where X, Y, and Z represent three groups of terms with varying relations between the indices i, j, and k, as indicated in their parentheses. In particular, X represents the group of terms with $i = j = k$, given by

$$
\begin{aligned}
X &= \epsilon_0 \chi^{(3)} \sum_{i=1}^{M} A_i^3 \cos^3 \psi_i \\
&= \frac{1}{4} \epsilon_0 \chi^{(3)} \sum_{i=1}^{M} A_i^3 (3 \cos \psi_i + \cos 3\psi_i) \\
&= \underbrace{X_1}_{SPM} + X_2,
\end{aligned}
\tag{2.72}
$$

where $X_2 = \frac{1}{4}\epsilon_0 \chi^{(3)} \sum_{i=1}^{M} A_i^3 \cos 3\psi_i = \frac{1}{4}\epsilon_0 \chi^{(3)} \sum_{i=1}^{M} A_i^3 \cos 3(\omega_i t - \beta_i z)$ in the last line is composed of the terms with the frequencies $3\omega_i$, which are neglected because of their non-phase-matched nature. However, all terms in X_1 propagate through the optical fiber, but are affected by SPM, because no other lightwaves (i.e, A_j or A_k) contribute to its amplitude (i.e., $\propto A_i^3$).

Y and Z in Eq. 2.71 represent the groups of terms, having at least one frequency different from the two others, and hence lead to the generation of various other frequencies in the form of $\omega_i \pm \omega_j \pm \omega_j$. The various terms in Y are expressed as

$$
\begin{aligned}
Y &= 3\epsilon_0 \chi^{(3)} \sum_{i=1}^{M} A_i^2 \cos^2 \psi_i \sum_{j \neq i}^{M} A_j \cos \psi_j \\
&= \frac{3}{2} \epsilon_0 \chi^{(3)} \sum_{i=1}^{M} A_i^2 (1 + \cos 2\psi_i) \sum_{j \neq i}^{M} A_j \cos \psi_j) \\
&= \underbrace{\frac{3}{2} \epsilon_0 \chi^{(3)} \sum_{i=1}^{M} \sum_{j \neq i}^{M} A_i^2 A_j \cos \psi_i +}_{XPM} \\
&\quad \underbrace{\frac{3}{4} \epsilon_0 \chi^{(3)} \sum_{i=1}^{M} \sum_{j \neq i}^{M} A_i^2 A_j \cos(2\psi_i - \psi_j) +}_{FWM} \\
&\quad \frac{3}{4} \epsilon_0 \chi^{(3)} \sum_{i=1}^{M} \sum_{j \neq i}^{M} A_i^2 A_j \cos(2\psi_i + \psi_j) \\
&= Y_1 + Y_2 + Y_3.
\end{aligned}
\tag{2.73}
$$

In the above expression for Y, the terms in Y_3 with the factor of $\cos(2\psi_i + \psi_j)$ would be non-phase-matched and hence won't be able to propagate in the optical fiber. However, the terms in Y_1 and Y_2, generated from the Kerr effect, will be able to participate in the propagation. Note that each of these terms is generated from the nonlinear interaction of only two independent lightwaves with third-order nonlinearity, as reflected in their respective amplitudes $\frac{3}{2}\epsilon_0 \chi^{(3)} A_i^2 A_j$ or $\frac{3}{4}\epsilon_0 \chi^{(3)} A_i^2 A_j$ (i.e., involving A_i and A_j). However, each term in Y_1 carries the impact of nonlinearity from ω_j's on ω_i's only on the amplitude and not on the frequency, indicating thereby the occurrence of the XPM phenomenon. However, the Y_2-terms clearly represent the FWM interactions as their frequencies appear as the linear combination of ω_j and ω_i, since $2\psi_i - \psi_j = (2\omega_i - \omega_j)t - (2\beta_i - \beta_j)z$. As mentioned above, these FWM components appear with two participating lightwaves having the same frequencies, leading to what is known as degenerate FWM interaction. We will also come across the other class of FWM terms from Z with all the participating frequencies being different from each other. From Eq. 2.71, we therefore express the various terms in Z as

$$
\begin{aligned}
Z &= 6\epsilon_0\chi^{(3)} \sum_{i=1}^{M}\sum_{j\neq i}^{M}\sum_{k\neq j}^{M} A_iA_jA_k \cos\psi_i \cos\psi_j \cos\psi_k \qquad (2.74)\\
&= \frac{3}{2}\epsilon_0\chi^{(3)} \sum_{i=1}^{M}\sum_{j\neq i}^{M}\sum_{k\neq j}^{M} A_iA_jA_k \cos(\psi_i + \psi_j + \psi_k)\\
&\quad + \underbrace{\frac{3}{2}\epsilon_0\chi^{(3)} \sum_{i=1}^{M}\sum_{j\neq i}^{M}\sum_{k\neq j}^{M} A_iA_jA_k \cos(\psi_i + \psi_j - \psi_k)}_{FWM}\\
&\quad + \underbrace{\frac{3}{2}\epsilon_0\chi^{(3)} \sum_{i=1}^{M}\sum_{j\neq i}^{M}\sum_{k\neq j}^{M} A_iA_jA_k \cos(\psi_i - \psi_j + \psi_k)}_{FWM}\\
&\quad + \underbrace{\frac{3}{2}\epsilon_0\chi^{(3)} \sum_{i=1}^{M}\sum_{j\neq i}^{M}\sum_{k\neq j}^{M} A_iA_jA_k \cos(\psi_i - \psi_j - \psi_k)}_{FWM}\\
&= Z_1 + Z_2 + Z_3 + Z_4.
\end{aligned}
$$

As before, the terms of Z_1 cannot propagate, while all the terms of Z_2, Z_3, and Z_4 are generated from the FWM interaction between the respective lightwaves with three different frequencies. Next, we collect together all the FWM terms from X, Y, and Z, and express the comprehensive FWM contributions to the nonlinear polarization as

$$
P_{NL}^F(z,t) = Y_2 + Z_2 + Z_3 + Z_4. \qquad (2.75)
$$

For illustration, we consider three frequencies $\omega_1 < \omega_2 < \omega_3$, all with the same amplitude A. Further, we consider two possible scenarios: (i) frequencies with equal spacing, i.e., $\omega_3 - \omega_2 = \omega_2 - \omega_1 = \Delta\omega$, and (ii) frequencies with unequal spacing, i.e., $\omega_3 - \omega_2 \neq \omega_2 - \omega_1$. In both cases, nine FWM frequencies are generated as follows:

- *Group of six frequencies from Y_2 with $i = j \neq k$:* $\omega_{112} = 2\omega_1 - \omega_2$, $\omega_{113} = 2\omega_1 - \omega_3$, $\omega_{221} = 2\omega_2 - \omega_1$, $\omega_{223} = 2\omega_2 - \omega_3$, $\omega_{331} = 2\omega_3 - \omega_1$, $\omega_{332} = 2\omega_3 - \omega_2$;

- *Group of three frequencies from Z_2, Z_3* and *Z_4 with $i \neq j \neq k$:* $\omega_{123} = \omega_1 + \omega_2 - \omega_3$, $\omega_{231} = \omega_2 + \omega_3 - \omega_1$, $\omega_{312} = \omega_3 + \omega_1 - \omega_2$.

Note that, among the nine FWM frequencies, the last three frequencies, i.e., $\omega_{123}, \omega_{231}$, and ω_{312}, are generated from the three different lightwaves each with a unique frequency. However, each of the first six FWM frequencies, i.e., $\omega_{112}, \omega_{113}, \omega_{221}, \omega_{223}, \omega_{331}$, and ω_{332}, is generated from only two of the three participating lightwaves, leading to degenerate FWM components. For the last three frequencies ($\omega_{123}, \omega_{231}$, and ω_{312}), all the three lightwaves having different frequencies exhibit weaker interaction between themselves. This makes the first group of six frequencies (i.e., the degenerate FWM terms) stronger than the second group, although the degenerate components are accompanied by a weighting of $\frac{3}{4}$, which is half of the weighting ($\frac{3}{2}$) for the non-degenerate components. The impact of degeneracy is taken into consideration by using an additional factor, called the phase-matching efficiency (defined later). Using the above results, the strength of the dielectric polarizations for the various FWM components can be expressed as

$$\begin{aligned} \mathrm{P}_{NL}^{F}(ijk) &= \frac{\epsilon_0 \chi^{(3)}}{4} d_{ijk} A_i A_j A_k \cos(\psi_i + \psi_j - \psi_k) \\ &= \frac{\epsilon_0 \chi^{(3)}}{4} d_{ijk} A_i A_j A_k \cos\{(\omega_i + \omega_j - \omega_k)t - (\beta_i + \beta_j - \beta_k)z\}, \end{aligned} \tag{2.76}$$

where d_{ijk} is the degeneracy factor, given by

$$\begin{aligned} d_{ijk} &= 3, \text{ for } i = j \\ &= 6, \text{ for } i \neq j. \end{aligned} \tag{2.77}$$

Correspondingly the optical power in each FWM component can be obtained as

$$P_{ijk} = \left(\frac{\omega_{ijk} \bar{n} d_{ijk}}{3cA_{eff}}\right)^2 P_i P_j P_k L^2. \tag{2.78}$$

In a practical scenario, the phase-matching between the participating waves won't be perfect, thereby bringing down the FWM power. Furthermore, due to power loss, the overall effect won't persist beyond the effective length L_{eff} of the fiber. With these observations, the above expression for P_{ijk} is modified as (Chraplyvy 1984)

$$P_{ijk} = \eta_{ijk} \left(\frac{\omega_{ijk} \bar{n} d_{ijk}}{3cA_{eff}}\right)^2 P_i P_j P_k L_{eff}^2, \tag{2.79}$$

where η_{ijk} represents the phase-matching efficiency (as mentioned earlier), which is a strong function of the phase difference $\Delta\beta_{ijk}$ between the participating FWM lightwaves, given by $\Delta\beta_{ijk} = \beta_i + \beta_j - \beta_k - \beta_{ijk}$, with β_x representing the propagation constant of wavelength w_x.

Next, we consider the implications of equal/unequal spacing of the original or *parent* lightwave frequencies. If the spacings are equal, then the frequencies of several FWM components will coincide with the parent lightwaves, thereby creating crosstalk from the interferences. For example, with $\omega_1 = \omega_0$, $\omega_2 = \omega_0 + \Delta\omega$ and $\omega_3 = \omega_0 + 2\Delta\omega$, one would get $\omega_{312} = \omega_3 + \omega_1 - \omega_2 = (\omega_0 + 2\Delta\omega) + (\omega_0) - (\omega_0 + \Delta\omega) = \omega_0 + \Delta\omega = \omega_2$. However, while the frequency spacing is unique for each adjacent pair of parent frequencies, none of the FWM components would coincide with the parent frequencies.

The above features of FWM are illustrated in Fig. 2.19 for the three participating lightwaves with equal power levels. As evident from the figure, if one intends to avoid completely the interference from FWM components, the frequency spacing between the adjacent channels must change from pair to pair to become unique, while their minimum value must be large enough to carry the sidebands of the modulated carrier. This approach would force the channel spacings to increase steadily, which can severely limit the available number of channels in the fiber transmission window, thereby significantly reducing the overall network capacity. In this regard, a pragmatic approach is needed to control rather than avoid completely the FWM interferences in WDM networks. We address this design issue in Chapter 10, while discussing the impact of transmission impairments in optical networks.

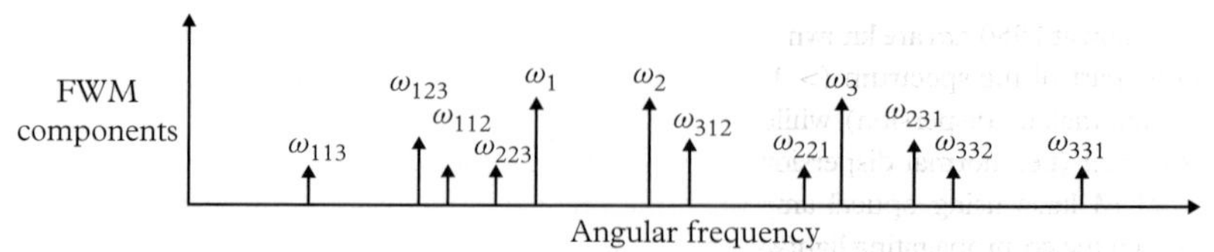

(a) Non-uniform frequency spacing (no FWM component falls on the original lightwave frequencies).

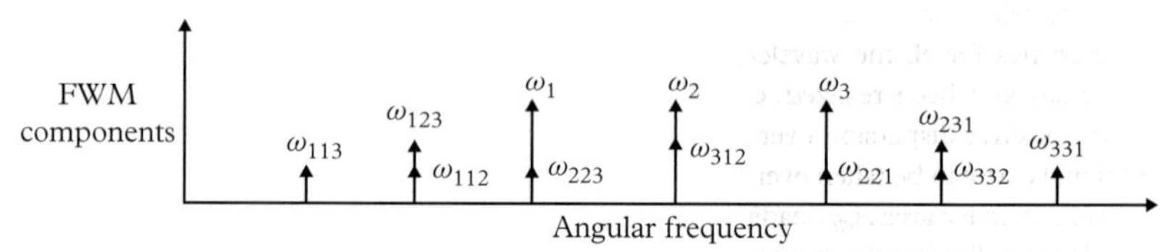

(b) Uniform frequency spacing (three out of nine FWM components fall on the three original lightwave frequencies; some other FWM components also have same frequencies, e.g., $\omega_{123} = \omega_{112}$ and $\omega_{231} = \omega_{332}$).

Figure 2.19 *FWM spectrum with non-uniform and uniform frequency spacings.*

2.2.6 Controlling dispersion and nonlinear effects

The dispersion mechanisms and nonlinear effects in optical fibers may degrade the communication system performance significantly, and hence one needs to design the optical fibers appropriately to keep these transmission impairments under control. Further, care needs to be taken to compensate for the residual/accumulated effects of dispersion in a link using suitable dispersion-compensating techniques. In this section, we discuss briefly the special quality fibers that can alleviate the impact of dispersion and nonlinear effects in optical fibers as well as some special optical components for dispersion compensation.

Before we describe the fibers needed to control dispersion and/or nonlinearities, a brief discussion on some related issues would be worthwhile. As discussed earlier, the GVD in optical fibers is given by $D_{ch} = -\frac{2\pi c}{w^2}\beta_2$. An optical fiber is considered to be working in the *normal dispersion regime* when $\beta_2 > 0$, else the fiber is said to be operating in the *anomalous dispersion regime*. Since $D_{ch} \propto -\beta_2$, the wavelengths in Fig. 2.13 lying above the zero-crossing point of chromatic dispersion operate in the anomalous dispersion regime. Further, the impact of all kinds of nonlinearities in optical fibers can be ameliorated by increasing the value of A_{eff}, while maintaining the single-mode propagating condition in the fibers. This feature can be realized by exploring different shapes of RIPs, where the core field is allowed to spread over a larger cross-sectional area.

Several approaches have been explored to reduce the chromatic dispersion in optical fibers. In general, optical fibers are designed so that the GVD remains confined to an acceptable limit for the operating wavelengths. As discussed earlier (in the section on chromatic dispersion), one can move the point of zero chromatic dispersion of a fiber from $\sim$ 1300 nm to the higher values entering the third window near 1550 nm, by reducing the core diameter and also by trying out different core dopants and RIPs.

The fibers that are designed with zero chromatic dispersion at 1300 nm are called 1300-nm-optimized fibers, which are realized relatively more easily and used for transmission in the second transmission window. For transmission over the third window, one needs to preferably engineer both core diameter and RIP or else only reducing the core diameter might make the acceptance angle too small for the optical sources. Fibers realized for zero dispersion at 1550 nm are known as dispersion-shifted fibers (DSFs), where the wavelengths in one part of the spectrum (> 1550 nm) traverse the fiber with positive dispersion (i.e., with anomalous dispersion) while the wavelengths in the other part go through negative dispersion (i.e., normal dispersion). In order to control the impact of fiber nonlinearities in WDM links using optical amplifiers (offering gains in C band), the phase-matching between the co-propagating lightwaves has to be kept low, thereby demanding some minimal dispersion over the C band wavelengths, and such fibers are called non-zero dispersion-shifted fibers (NZ-DSF) with a small positive dispersion ($\simeq$ 5 ps/nm/km at 1550 nm) within the said wavelength range. However, NZ-DSF doesn't ensure uniform dispersion characteristics for all the wavelengths in C band, and to address this issue another type of fiber has also been realized, called dispersion-flattened fibers, with a small and nearly uniform positive dispersion over a wide range of wavelengths from 1300 nm to 1600 nm. Additional care can be taken over the non-linearities, by further modifying the optical fibers with large effective area A_{eff}, leading to what is known as large-effective-area fibers (LEAFs) from Corning. Similar fibers were also made by Lucent Technologies, named as TrueWave XL fibers.

Figure 2.20 shows the RIPs of some of these special quality fibers, viz., 1300-nm-optimized, DSF, dispersion-flattened, and large-effective-area DSF, while in Fig. 2.21 we present the typical plots of dispersion versus wavelengths for four types of fibers: 1300-nm-optimized, DSF, dispersion flattened fiber and LEAF DSF.

In various optical networks around the world, optical fibers have already been in place and need therefore dispersion management with higher transmission speeds with add-on schemes to compensate for the pulse spreading that takes place in long-haul optical fiber links. In such cases, the interplay between the normal and anomalous dispersion mechanisms can be exploited for dispersion compensation by using dispersion compensating fibers (DCFs). For example, the pulse spreading caused in the existing terrestrial fibers due to anomalous dispersion in the given wavelength window can be compensated by additional fiber segments, specially designed to exhibit normal dispersion in the same window. In the long-haul terrestrial optical fiber links, typically such DCF rolls can precede an EDFA

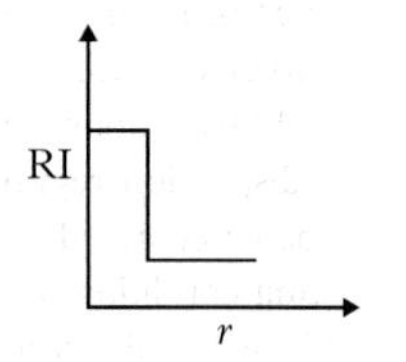

(a) 1300-nm-optimized fiber with standard step index profile.

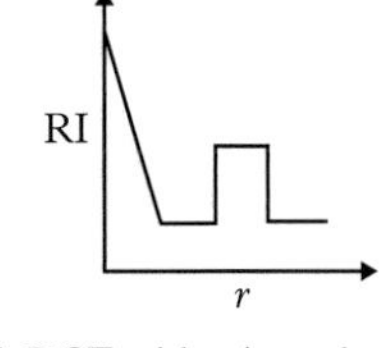

(b) DSF with triangular and annular ring.

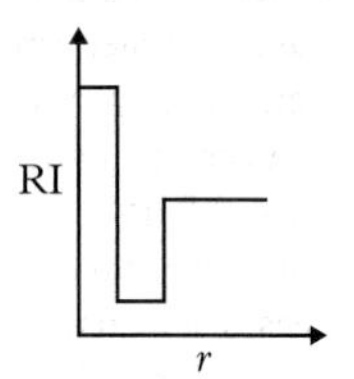

(c) Dispersion-flattened fiber with double-clad profile.

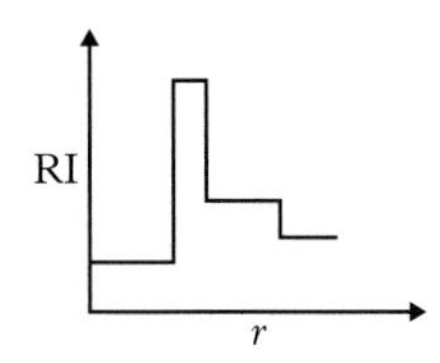

(d) Large-effective-area DSF with depressed-core profile.

Figure 2.20 *Illustration of the RIPs of the special quality single-mode optical fibers: (a) 1300 nm optimized fiber, (b) DSF, (c) dispersion flattened fiber, and (d) large effective area DSF. RI: refractive index, r = radial distance in fiber core from the fiber axis.*

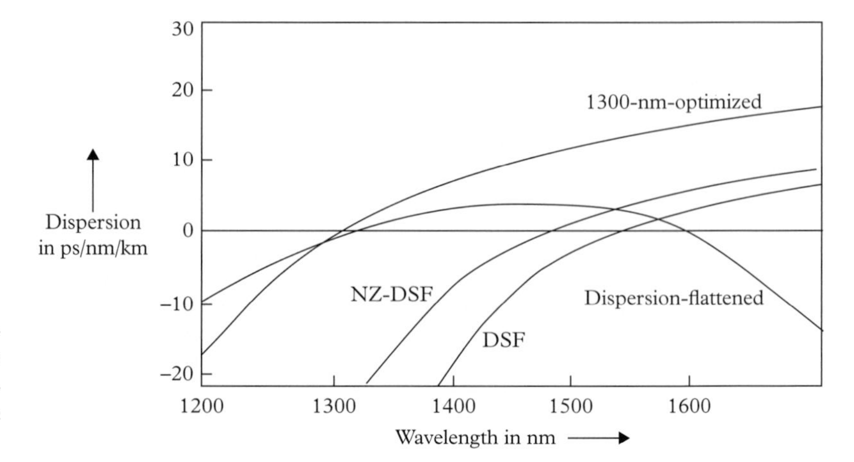

Figure 2.21 *Plots of dispersion vs. wavelength for different types of fibers: 1300 nm optimized fiber, DSF, NZ-DSF and dispersion flattened fibers with non-zero dispersion between 1300 and 1600 nm.*

at regular intervals (typically 80 km), thereby compensating the accumulated dispersion in each amplified fiber segment. Note that for the undersea optical fiber communication systems the optical fibers are generally used with negative dispersion, thereby necessitating DCFs with positive dispersion. Further, dispersion compensation using DCFs brings in some non-negligible signal loss, and an alternative to this solution is to use specially designed FBGs, which in general offer more dispersion compensation over narrower spectral range (and vice versa), thus with an operational constraint of dispersion-bandwidth product.

Interplay between the fiber nonlinearities and chromatic dispersion in the optical fibers gives rise to some more challenges. Recall the SPM-induced chirping in optical pulses due to the Kerr effect, which causes additional spectral spread over and above that due to the spectral width of the source and its modulation. However, the chirping due to SPM turns out to be positive, and its deleterious effect can be ameliorated by negative dispersion. Normally the NZ-DSF and dispersion flattened fibers operate mostly with positive dispersion in the third window (particularly in C band), and hence the designer needs to control the optical power so that SPM-induced chirping does not become a problem. Or else, it is also possible to tame the effect of SPM-chirping as well as chromatic dispersion by using optical fibers operating with negative dispersion in the respective wavelength range in WDM links.

For high-speed optical fiber links, the impact of polarization-mode dispersion might become non-negligible, which can be controlled by using polarization-aware schemes at the receiving end. In a typical polarization-aware scheme, one can split the received optical signal into two orthogonal polarizations and slow down the faster polarization using an optical delay line and combine both polarizations thereafter to compensate for the polarization-mode dispersion. However, given that the SOPs of the propagating lightwaves vary randomly with time, the above scheme needs to assess dynamically the differential delay between the two polarizations to configure the optical delay line from time to time, making the scheme much complex. However, polarization mode dispersion doesn't represent much of a challenge for moderate bit rates, while the high-speed links (typically, 40/100 Gbps and beyond) opt for multi-level modulation schemes thereby extending the symbol duration so that the pulse spreading due to the polarization-mode dispersion gets nearly absorbed therein.

2.3 Optical couplers

Optical couplers can be realized in various forms and represent a useful passive component for optical networks. The simplest form of optical couplers is a 2 × 2 coupler, which can be realized either by fusing two fibers together, or by using an optical integrated circuit, as shown in Fig. 2.22.

In the configuration shown in Fig. 2.22(a), the two fibers are first held together and twisted. Thereafter, the twisted fiber-pair is fused while being pulled from both sides. Consequently, the fused and elongated fibers take a biconically tapered shape with a narrow *waist* in the central part. The power injected at any one of the two input ports gets coupled to both output ports with a ratio, that will depend on the length and diameter of the waist region. One useful version of such couplers (or splitters) is 3 dB coupler, where the power incident at any input port gets divided equally into the two output ports. Note that, in a 3 dB coupler, the actual power P_o (in dBm) at any output port will be lesser than half of the input power P_i (in dBm) due to some insertion loss IL (in dB) within the passive device, i.e., $P_o = P_i - (3 + IL)$. Similar functionality can also be realized using an optical integrated circuit based on coupled planar waveguides, as shown in Fig. 2.22(b).

Furthermore, it is important to note that in 2 × 2 couplers the lightwaves coupled to the output ports undergo unequal phase shifts. In particular, for a 3 dB coupler, the lightwave coupled from the incident port (say, input port 1) to the output port 2 (i.e., the output port that is directly connected through fiber/waveguide to the input port 2) will undergo an additional $\pi/2$ phase shift, as compared to the lightwave coming from the input port 1 to the output port 1 (i.e., the output port that is directly connected through fiber/waveguide to the input port 1). This phase shift takes place during the coupling process and is in conformity with energy conservation between the input and output ports.

While both types of 2×2 couplers can be used as a building block for developing different types of couplers/splitters/combiners with larger numbers of ports, the fused couplers, being drawn out of fibers with circular cross-section, are polarization insensitive and hence can be more readily used for distributing and combining optical signals in a network-setting. However, the 2 × 2 couplers using planar wave-guides are especially useful for developing Mach–Zehnder interferometers (MZIs) and the entire class of devices that make use of MZIs, such as modulators, filters, etc. In such devices, planar waveguides help in maintaining polarization, which is important for realizing interference-based signal processing, albeit with polarization losses while being coupled from optical fibers to planar waveguides.

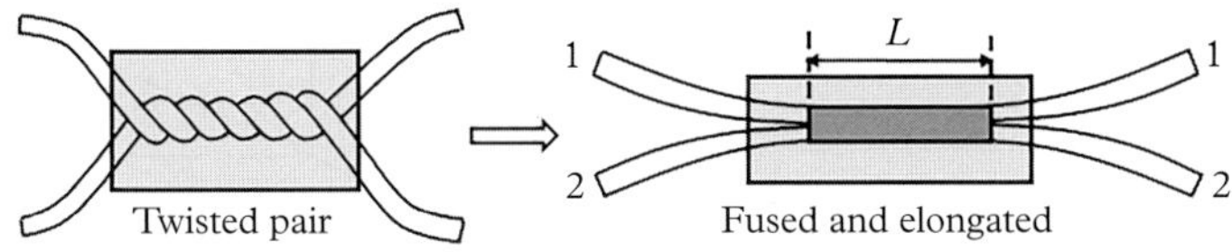

(a) 2×2 optical coupler using fused biconical tapering of twisted pair of fibers.

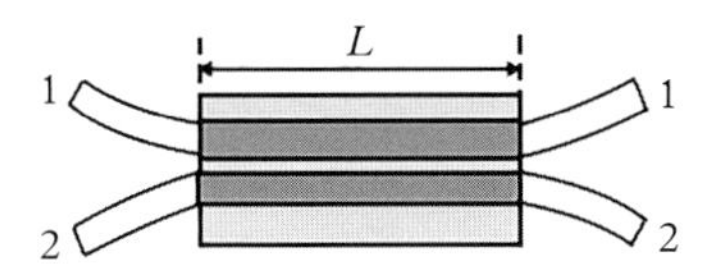

(b) 2×2 optical coupler using integrated optics, based on planar waveguides.

Figure 2.22 2 × 2 *optical couplers using fused-silica fibers and integrated optics.*

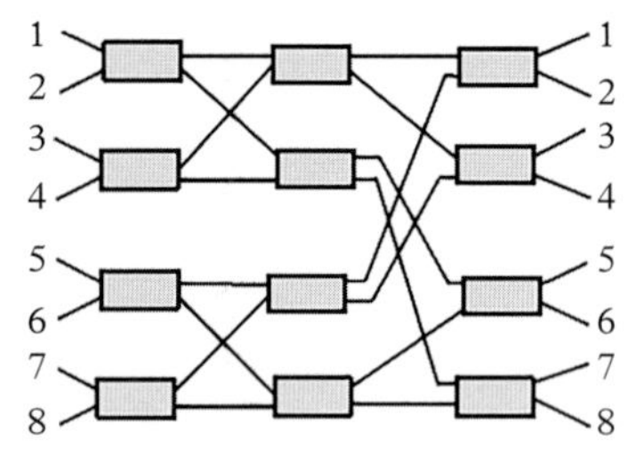

Figure 2.23 *8×8 optical star coupler using 2×2 couplers as building block. Note that, the number of stages required is $k = \log_2 N = 3$, and the total number of 2×2 couplers is $\frac{N}{2} \times \log_2 N = 12$.*

Usually, a 2×2 optical coupler is used as a basic building block for constructing various devices, e.g., $N \times N$ star coupler, $1 \times N$ splitter and $N \times 1$ power combiner, and various other combinations that are required to form a network topology or for the nodes used in a network. For example, in PONs a $1 \times N$ splitter is used for both distribution and collection of traffic to/from the user-end. Further, the same device becomes useful in building $N \times N$ passive star couplers, which are used in WDM LANs and various other broadcast-based network devices/topologies. Figure 2.23 shows the schematic of an $N \times N$ star coupler using k columns of 2×2 coupler, supporting $N(= 8)$ ports given by $N = 2^k$. In the star couplers using 2×2 couplers, the output power P_o (in dBm) at each output port would be given by $P_o = P_i - (10 \log N + \log_2 N \times IL)$, as the $N \times N$ coupler would employ $k = \log_2 N$ columns of 2×2 couplers, with each of them causing an insertion loss denoted by IL (in dB) and P_i representing the input power in dBm.

2.4 Isolators and circulators

In some optical and optoelectronic devices, isolators are needed at their output ports to prevent undesired signals that might get reflected back into the device. In particular, lasers and optical amplifiers are highly sensitive to the reflected signals at their output ports, and isolators as two-port devices are used to prevent such reflected signals. Circulators also support only one-way traversal of lightwaves, but employing three or four ports. Thus, isolators and circulators are non-reciprocal devices, where the *Faraday effect* plays an important role.

The Faraday effect takes place in optically active media such as quartz crystals, calcite etc., where the SOP of a linearly polarized light keeps rotating during propagation, as illustrated in Fig. 2.24. In the example shown in Fig. 2.24, the input SOP ($\theta^{in} = 0$) of the propagating lightwave evolves through proportional rotation with the distance traversed, thereby showing up with a SOP of $\theta' > \theta^{in}$ at an intermediate point of the optically active cylindrical waveguide. Finally, the lightwave emerges with a SOP of $\theta^{out} > \theta' > \theta^{in}$. This feature of optically active media is used to realize a Faraday cell or rotator, which in turn helps in making isolators and circulators.

Figure 2.25 shows the block schematic of an optical isolator, using a Faraday rotator (realized using a crystal that can rotate the input SOP) flanked by two polarizers at the

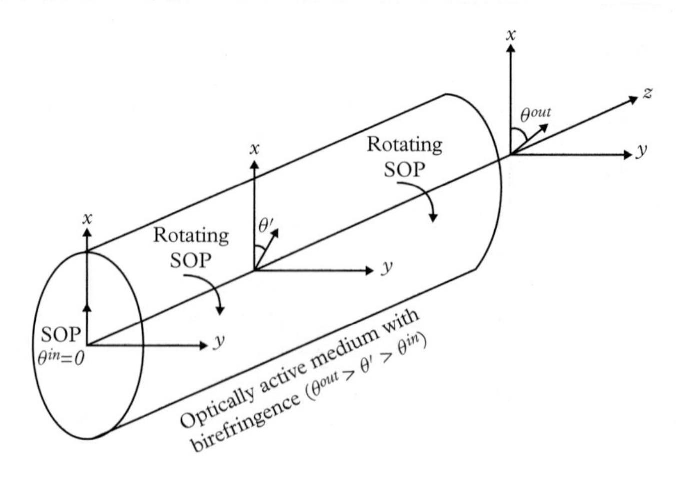

Figure 2.24 *Illustration of Faraday rotation in an optically active propagating medium.*

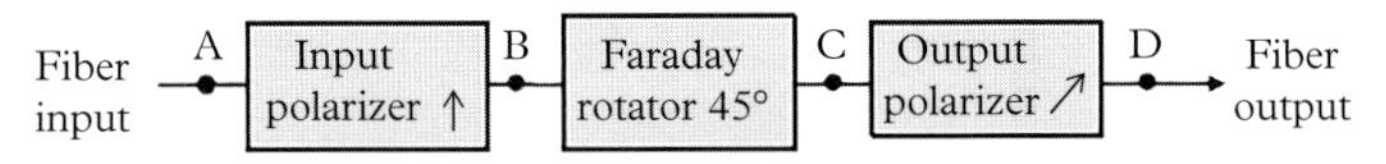

(a) Isolator schematic with 45° Faraday rotator.

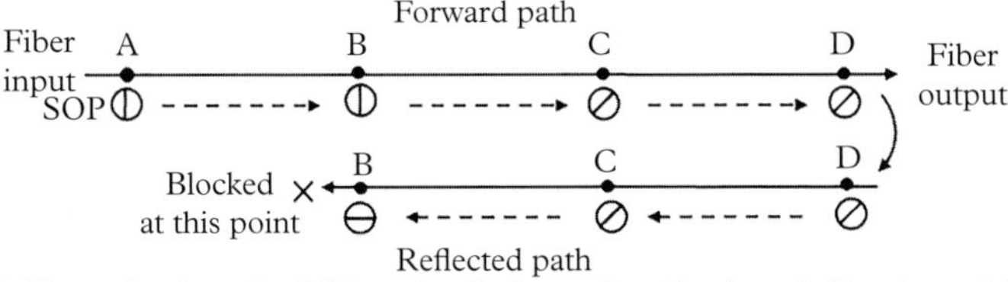

(b) Illustrating how the SOP evolves in forward and backward directions of the isolator in part (a); the locatioins A, B, C, and D are the same as those in part (a).

Figure 2.25 *Illustration of an optical isolator using Faraday rotator and polarizers.*

input and output sides. The input lightwave enters through an optical fiber with a vertical polarization into a polarizer that supports the same SOP. The polarizer therefore passes on the lightwave with vertical SOP to the Faraday rotator, which rotates the polarization by 45°, and passes on the lightwave to the output polarizer which permits the rotated SOP. Hence, the lightwave passes through the output polarizer and reaches the output port. If a part of this lightwave gets reflected back from the connector connecting to a fiber or some other device, it passes through the output polarizer again in the reverse direction toward the Faraday rotator. Thereafter, the SOP of the reflected lightwave gets rotated again by 45° to attain a horizontal polarization, which eventually gets *blocked* by the input polarizer (supporting only vertical polarization), thereby realizing the *one-way functionality* of an isolator.

The schematic diagram of a circulator using three ports is shown in Fig. 2.26, which also uses similar devices as in isolators. As shown in the figure, the circulator allows clockwise propagation of light and thus the lightwave incident at port 1 exits from port 2 and does not get to reach port 3. On the other hand, the lightwave entering the device from port 2 also moves in the clockwise direction and exits from port 3, without any chance of reaching port 1. A similar property also applies to port 3, wherefrom an input lightwave can only reach port 1. Such devices become useful in various network nodes for some critical functionalities: add-drop multiplexers, bi-directional optical amplifiers, etc.

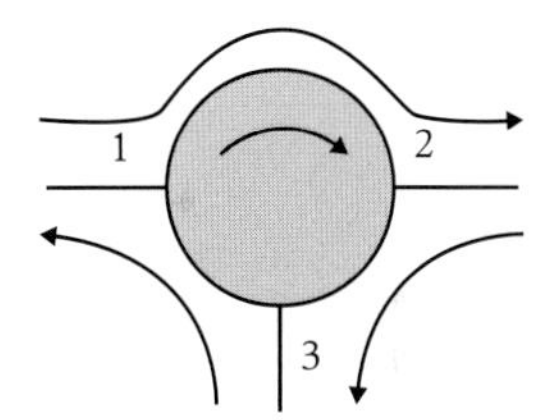

Figure 2.26 *Schematic diagram of a three-port optical circulator.*

2.5 Grating

A grating is a useful signal-processing device in optics, with its principle of operation being based on the interference between multiple lightwaves. The interfering lightwaves are generated from the *same light source*, and passed through different paths using an appropriate grating device, thereby undergoing different phase shifts. Thereafter, these lightwaves are combined together on a given plane of observation. If the relative phase difference between the different paths is an integral multiple of 2π, the lightwaves at a given location on the plane of observation interfere constructively with high brightness, or else they might undergo partially or fully destructive interference with lesser or zero intensity, respectively.

A grating can be conceived as an extension of the well-known *double-slit* experiment in optics, where the light from a monochromatic optical source passes through two closely spaced slits and gets diffracted in all directions. When the diffracted lightwaves are received on the plane of observation (a screen, say) a fringe pattern is formed on the screen with

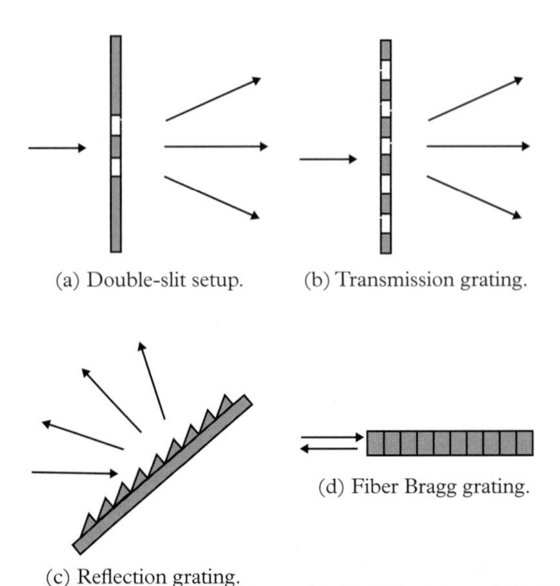

Figure 2.27 *Double-slit setup and some typical grating configurations.*

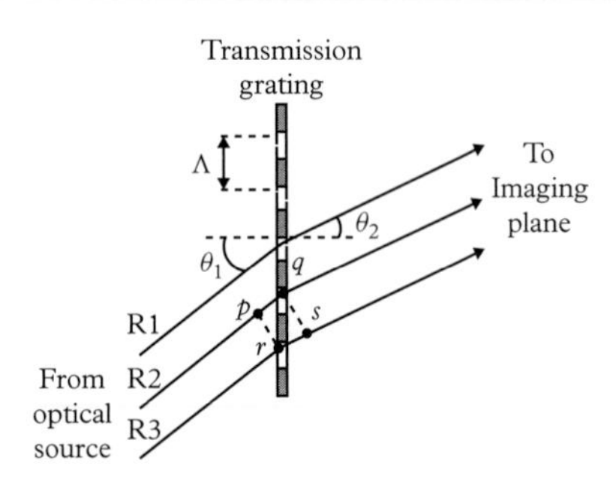

Figure 2.28 *Illustration of the transmission grating with a pitch* Λ, *using the three rays, R1, R2, and R3. The length of the ray R2 between the points p and q is* $pq = \Lambda \sin\theta_1$, *and the length of the ray R3 between the points r and s is* $rs = \Lambda \sin\theta_2$. *Note that, on the source side the ray R2 lags behind the ray R3 by pq in reaching the grating plane, while on the imaging side the ray R2 leads over the ray R3 by rs in reaching the imaging plane. This observation implies that the net phase difference between the two rays, R2 and R3, is proportional to the difference of the two ray-segments pq and rs, i.e.,* $(pq - rs)$, *leading to the formation of Eq. 2.80.*

bright and dark lines due to constructive and destructive interference. However, the fringe may not be formed at all, if the differential delay between the different paths exceeds the *coherence time* of the light source, beyond which the phase of the lightwave changes following a *random-walk* process. In a grating, several such *slits* or equivalent diffracting elements are placed in a periodic manner on a plane to obtain an interference pattern. The features of such interference patterns observed using grating-based devices are wavelength-sensitive and extremely useful for the network elements used in WDM networks.

Figure 2.27 illustrates the basic double-slit experimental setup and its subsequent extensions to form different types of grating configurations. In the double-slit configuration, as shown in (a), one can realize the slits by cutting slots in a metallic sheet, which functions as a transmission-based diffraction setup. For the transmission grating of (b), one can extend the double-slit configuration with more slits placed periodically on the metallic sheet, or by using a dielectric sheet with periodically varying refractive index. In the reflection grating shown in (c), one can implement a diffracting surface over a dielectric sheet by realizing wedges, again in a periodic manner, while in the fiber Bragg grating of (d), a small segment of optical fiber is used, where the refractive index is varied periodically along the axis. Before describing the specific grating configurations used for WDM networks, we explain in the following the salient features of a transmission grating configuration.

Consider a transmission grating, as shown in Fig. 2.28 (the same as that shown in Fig. 2.27(b), and reconsidered here for analysis), operating with a pitch (spatial period) of Λ, whereupon a WDM light is incident making an angle θ_1 with the normal to the plane of the grating and gets diffracted in all possible directions toward the imaging plane. We focus on the two rays R2 and R3 (or the equivalent planar lightwaves) diffracted in the direction making an angle θ_2 with the normal to the plane of the grating. On the imaging plane, these two lightwaves interfere with each other with a phase difference $\Delta\phi_i$ for a given wavelength w_i, given by

$$\Delta\phi_i = \frac{2\pi}{w_i}\big(\Lambda \sin\theta_1 - \Lambda \sin\theta_2\big), \tag{2.80}$$

where the first term in the parentheses on the right side represents the optical path difference on the source side of the grating, and the second term is the optical path difference on the imaging side of the grating.

To obtain a constructive interference for the wavelength w_i in the example shown in Fig. 2.28, the total phase difference must be an integral multiple of 2π, given by

$$\Delta\phi_i = \frac{2\pi}{w_i}\big(\Lambda \sin\theta_1 - \Lambda \sin\theta_2\big) = 2k\pi, \tag{2.81}$$

with k as an integer, leading to the grating equation, expressed as

$$\Lambda\big(\sin\theta_1 - \sin\theta_2\big) = kw_i. \tag{2.82}$$

Thus, a given wavelength w_i will undergo constructive interference and hence appear with a bright spot along the direction of angle θ_2 on the imaging plane, if the above condition (Eq. 2.82) is satisfied with a specific value of k. Since k can have different integer values, the bright spot of constructive interference will appear for various possible discrete values of θ_2 ($\theta_2(k, w_i), k = 1, 2, 3, \cdots$ for each w_i), leading to the various orders of interference. Thus, all the constituent wavelengths (i.e., $w_1, w_2, \cdots, w_i, \cdots, w_M$) in the input WDM signal will be diffracted in unique directions $\theta_2(k, w_i), i = 1, 2, 3, \cdots, M$ for a specific k, and the whole set will keep repeating periodically with different values of k. Note that the maximum power will be emerging for $k = 0$. However, with $k = 0$ for a given θ_1, all the wavelengths will be emerging from the grating along the same angle, i.e.,

$$\theta_2(0, w_1) = \theta_2(0, w_2) = \cdots = \theta_2(0, w_i) = \cdots = \theta_2(0, w_M) = \theta_2, \tag{2.83}$$

implying that a practical grating should be so designed that the image plane captures a mode with $k \neq 0$, so that different wavelengths can exit the grating along different angles, enabling wavelength demultiplexing.

The basic functionality of a grating, as described above, manifests itself in various ways for different types of grating configurations. We discuss below two grating configurations, Bragg grating and arrayed-waveguide grating, which are extremely useful for optical WDM networks.

2.5.1 Bragg grating

Bragg grating is realized with a simple configuration using a suitable propagation medium for light, such as a planar waveguide or an optical fiber. In any of these two propagation media, one can realize periodic variation of the refractive index along the direction of propagation with a specific pitch Λ, to reflect and transmit wavelengths selectively (Fig. 2.29).

As shown in part (a) of Fig. 2.29, the incident WDM lightwave moving in the forward direction gets diffracted backward from the crests of the periodically varying refractive index in a planar waveguide. These diffracted lightwaves in the backward direction may or may not interfere constructively for the wavelength of the lightwave under consideration. If the pitch Λ of the periodic variation of refractive index *written* in the medium is such that $(2\pi \times 2\Lambda)/(w_i/n_{eff}) = 2\pi$, i.e.,

$$w_i = 2n_{eff}\Lambda, \tag{2.84}$$

with n_{eff} as the effective refractive index of the medium, then the various backward-diffracted components of the wavelength w_i from all the crests of the grating will interfere

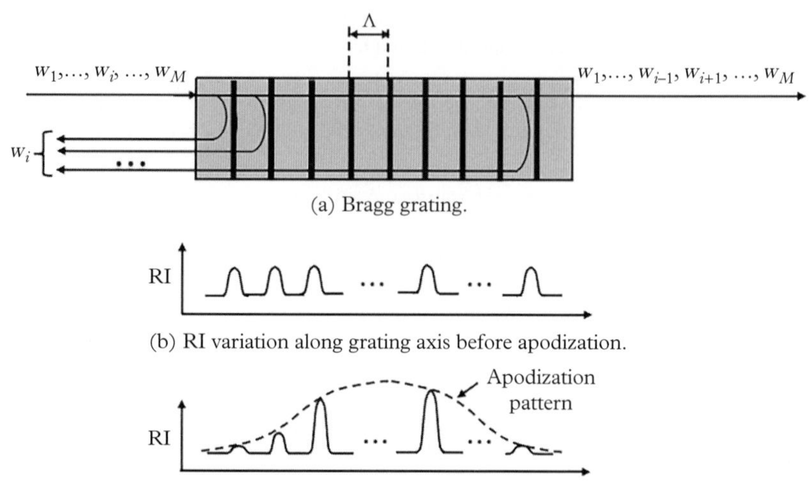

Figure 2.29 *Illustration of Bragg grating on a planar waveguide and the apodization profile of the refractive index along the grating axis.*

constructively, implying that the wavelength w_i would be sent back to the input end of the grating, while the remaining wavelengths will move ahead in the forward direction. However, the Bragg grating also diffracts backward other wavelengths partially, which needs to be minimized by modifying the grating pattern by a process called *apodization*. In particular, instead of having the same refractive index variation in each period across the Bragg grating length, the maximum value of the refractive index at the crests are slowly tapered down towards both ends, while the pitch remains the same throughout (Fig. 2.29 (b) and (c)). This in effect reduces the amplitude of the reflected waves at wavelengths other than the desired one.

Fiber Bragg grating

A Bragg grating can be realized in optical fibers as well, known as a fiber Bragg grating (FBG), by making the fiber material photosensitive by doping GeO_2 into silica, which is next exposed to ultraviolet radiation. By periodically varying the intensity of ultraviolet light exposure (using the fringes formed by the interference of two ultraviolet lightwaves) across the fiber length, one can *write* the grating profile therein along with the necessary tapered-periodic profile of refractive index for apodization.

As we shall see later in subsequent chapters, the FBG offers itself as a versatile passive device for optical networking elements: optical filtering, demultiplexing, add-drop operation, fiber dispersion management, etc.

2.5.2 Arrayed waveguide grating

Arrayed waveguide grating (AWG) is formed again by using the principle of interference between multiple paths of light from the same optical source, but not using the diffraction phenomenon. An AWG, though it can generically operate as an $N \times K$ passive device, usually it is employed in $1 \times N$, $N \times 1$, or $N \times N$ configuration.

To begin with, we consider an $N \times N$ AWG as shown in Fig. 2.30, where the incident lightwave at an input port is split through a dielectric slab as a coupler, and the

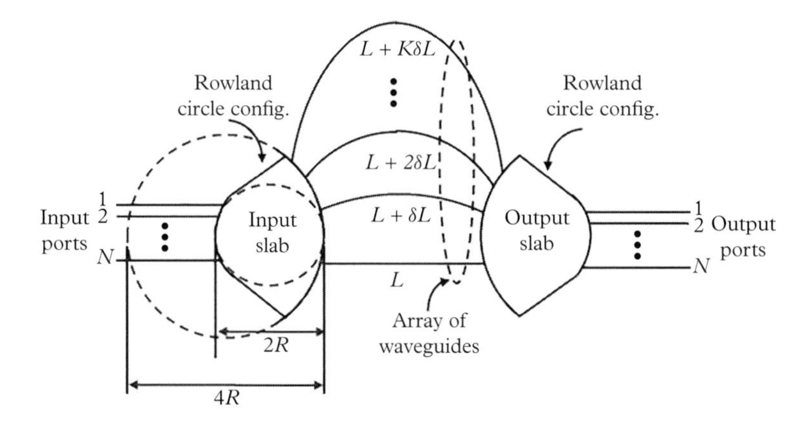

Figure 2.30 *Illustration of an $N \times N$ AWG.*

split-components are passed concurrently through an array of waveguides with incrementally increasing lengths, i.e., $L, L+\Delta L, \cdots, L+k\Delta L, \cdots, L+K\,\Delta L$, with k as an integer $\in [0, K]$. The lightwaves emerging from these waveguides are combined in a similar dielectric slab coupler placed in a laterally reversed position with respect to the one at the input side, and passed on to all the output ports.

The two dielectric slabs on the two sides, designed using a specific geometry, known as Rowland's circle, have some specific features to facilitate the AWG functionality. In particular, a Rowland circle at the input end is constructed as a convex-shaped slab created by using two circles of radii R and $2R$, where the smaller circle is *nested* within the larger circle in the right half of the latter. Thus, for the input slab, the Rowland circle uses the left-side arc of the smaller circle with radius R and the right-side arc of the larger circle with radius $2R$, which are connected by two lines at the top and bottom of the slab. The slab on the output side is simply the mirror-image of the slab on the input side. In the input slab, the central input port is aligned with the centers of both circles (with radii R and $2R$) and hence equidistant from the left ends of all the arrayed waveguides, enabling almost uniform power distribution to the waveguides. However, for the other input ports, the split components of the input lightwave might reach the inputs of the arrayed waveguides with some asymmetry in power distribution. At the output side of the AWG, all the lightwaves after passing through the arrayed waveguides are incident on the left-side arc of the larger circle (radius $2R$). For these lightwaves, the right-side arc of the smaller circle (radius R) forms the focal plane, thereby facilitating the incoming lightwaves to be focused at the output ports of the AWG.

The split optical lightwaves in the input dielectric slab traverse through different paths (waveguides) and reach the output ports after passing through the output slab, and the path difference between any pair of these paths reaching the same output port determines the nature of interference between the respective lightwaves. For a given wavelength at an input port, all the split components from the input end are designed to meet with constructive interference, only at one specific output port attached to the right-side arc of the smaller circle in the output slab. Thus, from a given input port, each wavelength undergoes constructive interference at a unique output port, which amounts to demultiplexing of a WDM input signal into its constituent wavelengths, each appearing at a specific output port (Fig. 2.31(a)). In the reverse direction, the same feature makes an AWG a candidate for an $N \times 1$ optical multiplexer or combiner. Further, in $N \times N$ mode, these features are ensured for each input/output port, implying that a given wavelength at an input port reaches a unique output port without any overlap with other signals at the same wavelength from some other

$w_1, w_2, \ldots, w_N$ AWG as $1 \times N$ or $N \times 1$ device w_1 w_2 w_N

→ $1 \times N$ (left to right) for demultiplexing

← $N \times 1$ (right to left) for multiplexing/combining

(a) AWG offering $1 \times N$ or $N \times 1$ functionality.

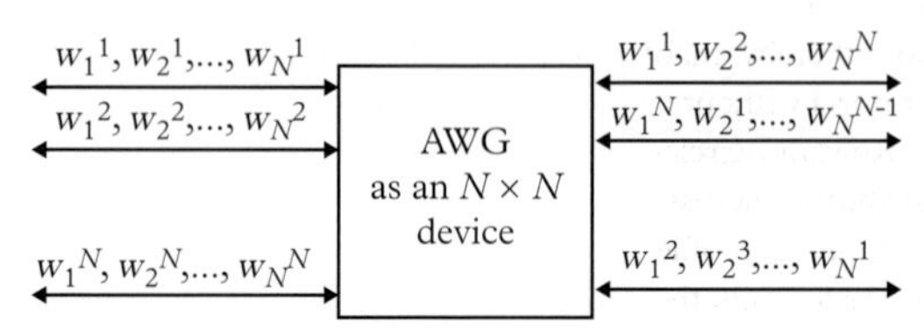

(b) AWG offering $N \times N$ static optical crossconnect functionality.

Figure 2.31 *Illustration of various functionalities in an $N \times N$ AWG.*

input ports (Fig. 2.31(b)). This leads to a static WDM crossconnect functionality across the input and output ports in an AWG.

For designing an AWG with the above features, one needs to estimate the path differences between the split lightwaves from one input port going through input/output slabs and different waveguides, eventually meeting with each other at a given output port. In order to trace and estimate the lengths of these paths, consider a pair of input and output ports, say the input or source port s and the output or destination port d. Also consider the two lightpaths between the port-pair (s, d) passing through the two adjacent waveguides, say the kth and $(k+1)$th waveguides in the waveguide array. The *optical path lengths* traversed by these two lightpaths at the wavelength w_i can be expressed as

$$\begin{aligned} \rho_{s,d}^{k}(i) &= \frac{2\pi}{w_i}\left[n_s l_{s,k} + n_w(L + k\delta L) + n_s l_{d,k}\right] \\ \rho_{s,d}^{k+1}(i) &= \frac{2\pi}{w_i}\left[n_s l_{s,k+1} + n_w\{L + (k+1)\delta L\} + n_s l_{d,k+1}\right], \end{aligned} \tag{2.85}$$

where n_s and n_w represent the effective refractive indices of the input/output slab couplers and the arrayed waveguides, respectively. Further, $l_{s,k}$ and $l_{d,k}$ represent the path lengths traversed by the lightwaves (through the kth waveguide) in the input and output slabs, respectively. Using the above two optical path lengths, we obtain the optical path difference $\Delta\rho$ between the two lightpaths as

$$\begin{aligned} \Delta\rho &= \rho_{s,d}^{k+1}(i) - \rho_{s,d}^{k}(i) \\ &= \frac{2\pi}{w_i}\left[n_s(l_{s,k+1} - l_{s,k}) + n_w\delta L + n_s(l_{d,k+1} - l_{d,k})\right], \end{aligned} \tag{2.86}$$

which has to be equated to an integral multiple of 2π, i.e., $2n\pi$ with n as an integer for constructive interference, leading to the AWG design equation, given by

$$n_s(l_{s,k+1} - l_{s,k}) + n_w\delta L + n_s(l_{d,k+1} - l_{d,k}) = nw_i. \tag{2.87}$$

With the input and output slabs constructed in the form of Rowland circles and appropriate choices for n_s, n_w, δL, and locations of the input and output waveguides, it is possible to satisfy the above condition for constructive interference, which would enable the AWG to achieve the desired connectivities between its input and output ports on appropriate wavelengths, as shown in Fig. 2.31. Note that, due to the integer n on the right-hand side of the above expression (Eq. 2.87), the connectivity for a given wavelength w_i will again repeat at w_i + FSR, with FSR representing the free spectral range for the AWG. Hence, the wavelengths in a WDM input signal should be chosen accordingly, so that the wavelength span covered by the wavelengths from w_1 to w_M fall within the FSR of the AWG under consideration, thereby enabling the wavelengths of each input port to emerge from the distinct output ports without any spectral overlap with the others.

In spite of the Rowland-circle-based configuration of the input and output slabs, the optical power distribution across the output ports can not be perfectly uniform, and it typically follows a tapered profile from the central port, that can be approximated by a Gaussian function. As a result, the size of AWGs in terms of port count gets limited as the power levels at the output ports toward the two edges are reduced, leading to poor signal-to-noise ratio (SNR) and higher BER at these output ports. Moreover, the width and spacing of the input and output waveguides have to be designed judiciously to minimize possible crosstalk from the neighbouring ports. Interested readers are referred to (Smit and Dam 1996) for further details on the design issues of AWG.

2.6 Fabry–Perot interferometer

The Fabry–Perot interferometer (FPI), also called an *etalon*, is a classical device in optics, built with a suitable optical propagating medium, bounded by two partially reflecting walls, called *facets*, at each end. The operation of an FPI is solely based on the interferences between the reflected lightwaves from the facets placed at each end of the propagating medium.

Figure 2.32 shows the basic schematic of an FPI, where two partially reflecting/transmitting facets are placed at the ends of a segment of propagating medium of length L and an effective refractive index n_{eff}. The input lightwave with a wavelength w_i (say) enters the interferometer *cavity* (formed by the two facets) through the left facet and passes through the cavity toward the right facet. Due to the partially reflecting nature of the facet, a part of it exits the cavity on the right side and the rest is bounced back into the cavity to get reflected back again by the left facet. This component of lightwave co-propagates with the lightwave entering from the left side and reach the right facet. This process goes on, with the intensity of the lightwave reducing after every reflection. However, if the wavelength of the lightwave and the cavity length are so chosen that the exiting lightwaves interfere constructively, i.e.,

$$\frac{2\pi n_{eff} \times 2L}{w_i} = 2k\pi, \tag{2.88}$$

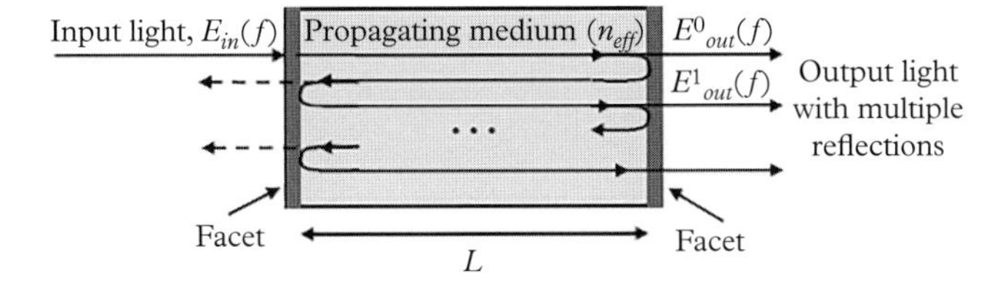

Figure 2.32 *FPI schematic with a cavity of length L and a propagating medium with an effective refractive index n_{eff}.*

with k as an integer, i.e., if $w_i = 2n_{eff}L/k$, then the Fabry–Perot output delivers a large power for the same wavelength, while the other wavelengths fail to produce as much power due to partial or fully destructive interference. The wavelength that gets selected by the cavity is called the resonant wavelength, as the cavity goes into resonance for that wavelength. Again, due to the inherent periodic nature of the process, an FPI also exhibits the feature of FSR (as in AWG), and thus it can handle the wavelengths only in the range of its FSR, while being used to select or *filter* a specific wavelength for a WDM input signal.

Owing to the partial reflections from the facets, the periodic transmission function of an FPI can significantly vary in respect of the sharpness in selecting/rejecting the wavelengths when used to filter a WDM input signal. Assuming symmetric power reflectivity R and absorption coefficient A for the two facets, the returned electric field into the cavity is reduced by the ratio $\sqrt{1-A-R}$, and the delay incurred in each one-way trip is $\tau_d = n_{eff}L/c$. Moreover, the first traversal through the cavity incurs a phase shift of $2\pi f\tau_d$ for the lightwave, while each of the subsequent round trips causes a phase shift of $4\pi f\tau_d$. Using these observations, one can construct an infinite-series model for the exiting electric field E_{out} associated with the light output, expressed in terms of the incident electric field E_i as (Green 1993)

$$\begin{aligned} E_{out}(f) &= \sum_{i=0}^{\infty} E_{out}^{i}(f) \\ &= \sum_{i=0}^{\infty} E_{in}(f)(1-A-R)R^2 \exp[-j2\pi f(\tau_d + 2i\tau_d)], \end{aligned} \tag{2.89}$$

which leads to the transfer function $H_{FP}(f)$ of FPI, given by $H_{FP}(f) = \frac{E_{out}(f)}{E_{in}(f)}$. The magnitude-square of the transfer function, i.e., the power transfer function, is obtained from $H_{FP}(f)$ as

$$T_{FP}(f) = |H_{FP}(f)|^2 = \frac{(1-\frac{A}{1-R})^2}{1+\left(\frac{2\sqrt{R}}{1-R}\sin 2\pi f\tau_d\right)^2}, \tag{2.90}$$

which is known as the Airy function. To have an insight into the frequency response of FPIs, we present some representative plots of $T_{FP}(f)$ versus normalized frequency f/FSR in Fig. 2.33 for two extreme values of R, with zero absorption loss in the facets (i.e., $A = 0$). The plots exhibit the expected periodicity with the normalized frequency, and the selectivity of the transfer function around each peak becomes higher (sharper) as R is increased from a very low (= 5%) to a very high value (= 90%). The selectivity of $T_{FP}(f)$ is expressed in terms of its *full-width half-maximum* (FWHM) bandwidth around a peak, and it is desirable that, for a given FSR, the transfer function peaks are sharp enough, i.e., the FSR is large enough with respect to FWHM to accommodate a reasonably large number of wavelengths in a WDM signal. To have a measure of this feature, the ratio of FSR to FWHM turns out to be a useful parameter of FPIs, and is known as the *finesse*, given by

$$F_{FP} = \frac{\text{FSR}}{\text{FWHM}} = \frac{\pi\sqrt{R}}{1-R}, \tag{2.91}$$

implying that large reflectivity leads to high finesse values. As we shall see later, FPI or etalon plays a significant role in designing various optical components that are used in optical WDM networks.

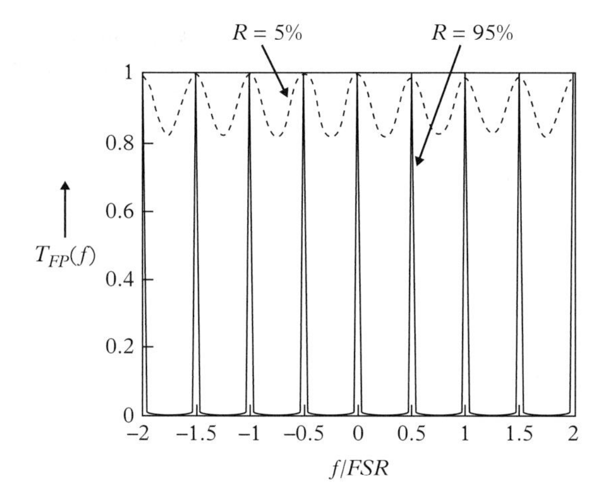

Figure 2.33 *Plots of power transfer function versus normalized frequency for an FPI.*

2.7 Mach–Zehnder interferometer

Another interferometric device, known as the Mach–Zehnder interferometer (MZI), plays an important role in designing a variety of optical components. As shown in Fig. 2.34(a), MZI is a 2 × 2 device, usually realized with planar waveguides for optical networking components. The input lightwave from one input port, say input port 1, is first split using a 3 dB coupler into two paths and the two split components are reunited using another 3 dB coupler after passing through the two MZI arms having different path lengths. In the example shown in the figure, the upper arm is longer than the lower arm by a length ΔL, such that the phase shift in the upper arm due to the longer physical length is higher than that of the lower arm by $\Delta\phi$, given by

$$\Delta\phi = \frac{2\pi n_{eff}}{w}\Delta L. \tag{2.92}$$

Further, note that, each 3 dB coupler also introduces a $\pi/2$ phase shift to the output port, corresponding to the coupled waveguide along with the power distribution between the two ports.

In view of the above features, an MZI can be conceived as a cascade of three functional blocks (see Fig. 2.34(b)): two 3 dB couplers on the two sides and the pair of MZI waveguides in the middle. Hence, the overall *scattering matrix* S^{MZI} (representing the ratios of the electric fields received at all the ports for the given input fields) of the MZI can be determined by multiplying the scattering matrices of the three blocks S^c_{in}, $S^{\Delta\phi}$, and S^c_{out} for the input coupler, intermediate waveguide pair for differential phase-shifting, and the output coupler, respectively. We therefore express S^{MZI} as

$$S^{MZI} = S^c_{in} S^{\Delta\phi} S^c_{out}, \tag{2.93}$$

with S^c_{in} and S^c_{out} expressed (see the discussion on the 2 × 2 optical coupler, shown in Fig. 2.22) as

$$S^c_{in} = S^c_{out} = \begin{bmatrix} \frac{1}{\sqrt{2}} & \frac{\exp(j\pi/2)}{\sqrt{2}} \\ \frac{\exp(j\pi/2)}{\sqrt{2}} & \frac{1}{\sqrt{2}} \end{bmatrix}, \tag{2.94}$$

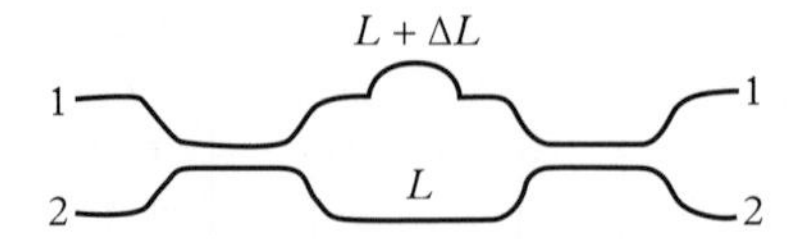

(a) MZI device using planar waveguides.

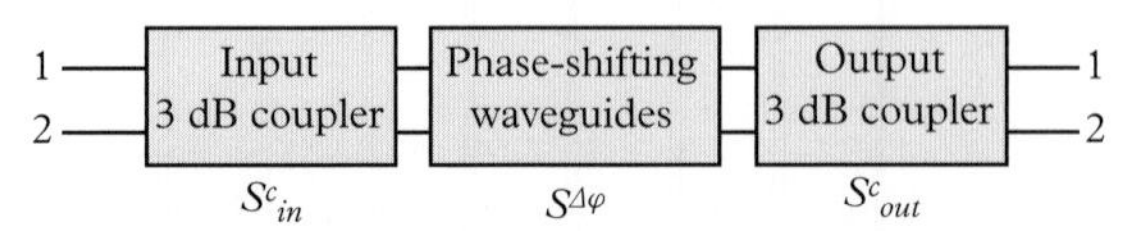

(b) MZI model using three scattering matrices.

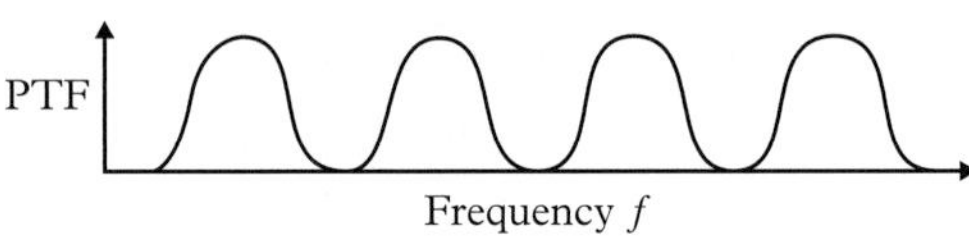

(c) MZI power transfer function (PTF).

Figure 2.34 *MZI device schematic, analytical model, and power transfer function (i.e.,* T_{11}^{MZI} *or* T_{12}^{MZI}*).*

and $S^{\Delta\phi}$ given by

$$S^{\Delta\phi} = \begin{bmatrix} \exp(j\Delta\phi) & 0 \\ 0 & 1 \end{bmatrix}. \tag{2.95}$$

Using Equations 2.94 and 2.95 in Eq. 2.93, and simplifying the resulting expression, one can obtain the power transfer functions T_{11}^{MZI} and T_{12}^{MZI} from input port 1 to output ports 1 and 2, respectively (generally one of the input ports in MZI is not used) as

$$\begin{aligned} T_{11}^{MZI} &= |S_{11}^{MZI}|^2 = \sin^2(\pi f \tau_{\Delta L}) \\ T_{12}^{MZI} &= |S_{12}^{MZI}|^2 = \cos^2(\pi f \tau_{\Delta L}), \end{aligned} \tag{2.96}$$

where $\tau_{\Delta L} = n_{eff}\Delta L/c$. The above expressions imply that both the PTFs are periodic in frequency with a period of $\delta f = 1/\tau_{\Delta L}$, as shown in Fig. 2.34(c). Thus, an MZI behaves as a periodic filtering device with an FSR = δf. However, for a large FSR, the sharpness reduces significantly, and one can use cascaded stages of MZI with different periods to get around this problem.

2.8 Optical sources

Primary requirements for the optical sources to be used in the transmitters of optical fiber links are the appropriate emission wavelengths matching with the transmission windows of optical fibers and the flexibility to be modulated by digital or analog signals with the required transmission speed/bandwidth. While gas lasers have high power, they are bulky and cannot operate over the wide range of optical frequencies used for optical fiber communication, and this is more of a problem for WDM systems. Furthermore, it is not possible to employ direct modulation on these lasers, nor can one integrate them with the transmitter electronics. However, semiconductor-based optical sources, viz., LEDs and lasers, support direct modulation, are available in small sizes, and are easily integrable with the electronic circuitry in the transmitter. These optical sources, lasers in particular, need to have adequate power with a sharp radiation pattern and narrow power spectrum, when needed to transmit through the single-mode optical fibers over long-haul links. Further, the sources should have high temperature stability and long life, along with tunability over a given range of wavelengths, when used in the WDM networks. We summarize these features in the following.

- emitted optical power: large enough to support the necessary link length
- radiation pattern: commensurate with the fiber acceptance angle
- peak emission wavelength/frequency: matching the desired transmission window of the fiber
- spectral width: small enough to keep the dispersion under control
- switching speed (digital transmission) or modulation bandwidth (analog transmission): large enough to support the modulating baseband signal
- tunability: fast-tunable within a given spectral window, when used as a tunable source in a WDM system
- temperature stability: important for lasers; needs special care
- life: long enough to enable the link to survive for several years

In general, LEDs have smaller power, broader spectrum and radiation pattern, and are cheaper than lasers. Due to large spectral width and broad radiation pattern, LEDs are not suitable for long-haul high-speed communication links and also cannot be used for single-mode fibers. Further, LEDs with large spectral width cannot be used for coherent optical communication and are not usable in tunable transmitters. Notwithstanding these limitations, LEDs serve well for optical fiber links with shorter distances operating with moderate speed/bandwidth, without much concern for temperature instability.

However, lasers are available with high power, narrow spectral width, and sharp radiation pattern but with high sensitivity to temperature, and practical transmitters using lasers need to employ appropriate driver circuitry for controlling the temperature as well as the power output. With their high power, narrow spectral width and sharp beams, lasers are best suited for single-mode fibers and can support long-haul communication links at high transmission rates. Furthermore, lasers can be used for coherent optical communication, are tunable, and are most useful for WDM transmission.

2.8.1 Source materials

Both LEDs and lasers are basically *pn* junction diodes, operated in the forward-biased mode, where the forward-bias current works as the electrical pump, turning itself into optical power through the electro-optic conversion process. However, not every diode emits light, and the diodes for optical sources need to use specific materials to emit light at specific wavelengths. The first and foremost criterion for any semiconductor used as an optical source is that they should have a *direct bandgap* between the valence and conduction bands. The electrons from the valence band move up to the conduction band with forward bias, and emit light while falling back to the valence band. However, the emission becomes much unlikely if the bottom-most point of the conduction band and the topmost point of the valence band are not aligned vertically in respect of momentum K, as shown in Fig. 2.35 illustrating

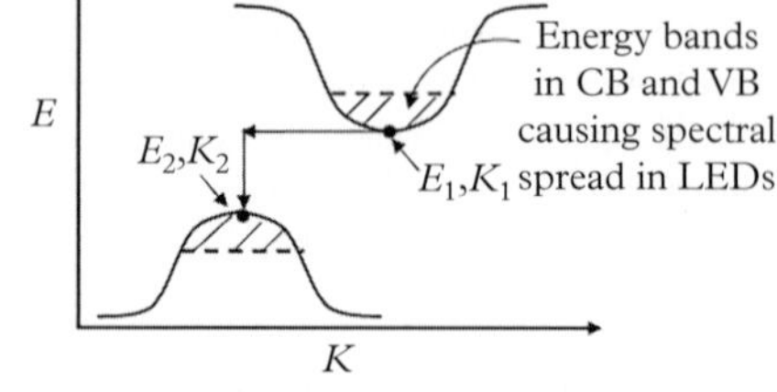

(a) Indirect-bandgap semiconductor.

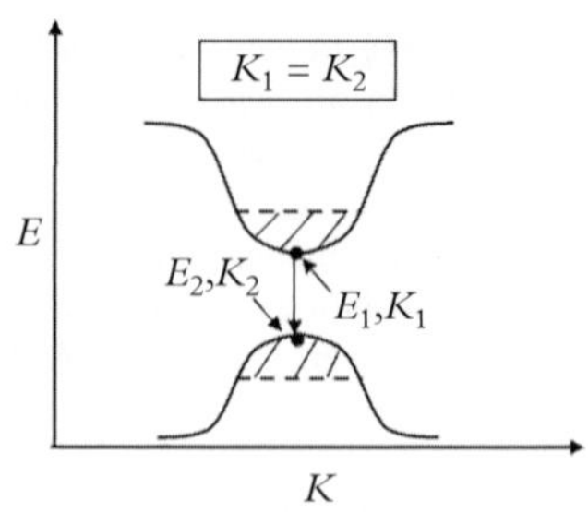

(b) Direct-bandgap semiconductor.

Figure 2.35 *E-K diagrams for indirect bandgap and direct bandgap semiconductors. CB: conduction band, VB: valence band*

the EK-diagram of indirect and direct-bandgap semiconductors. In other words, in any such quantum-mechanical process, both energy E and momentum K must be conserved. Conservation of energy results from the photon emission with an energy $hf = E_1 - E_2$, and conservation of momentum takes place with a phonon generation with a momentum $\Delta K = K_1 - K_2$. However, the probability of *simultaneous* occurrence of both the photon and the phonon is rather low in indirect bandgap semiconductors, making them unable to produce any perceptible light emission. However, for the direct-bandgap semiconductors ΔK being zero, generation of a phonon does not come into the picture, thereby making the photon emission a successful process with high probability.

Silicon and germanium being indirect-bandgap semiconductors cannot emit light, and thus for optical sources one needs to use compound semiconductors having direct bandgaps, along with the right choice of the bandgap E_g for the desired wavelength w, given by $w = \frac{1.24}{E_g}$ μm $= \frac{1.24}{E_g} \times 10^3$ nm, with E_g in eV. The following materials are examples of the direct-bandgap compound semiconductors, used for optical sources in different wavelength ranges:

- Binary alloy GaAs: $E_g = 1.424$ eV, $w = 870$ nm (first window)
- Quaternary alloy $In_{1-x}Ga_xAs_yP_{1-y}$, with x and y representing the gallium and arsenic mole fractions in the respective pairs of group III/V semiconductors, having $x \in [0, 0.47]$. The bandgap E_g with $y \simeq 2.2x$ is expressed as a function of y, given by

$$E_g = 1.35 - 0.72y + 0.12y^2. \tag{2.97}$$

This quaternary alloy of group III-V compounds can be used for all windows by varying x and y. Some typical examples are as follows:

◇ $x = 0, y = 0$: $E_g = 1.35$ eV, $w = 919$ nm (first window),

◇ $x = 0.26, y = 0.572$: $E_g = 0.9774$ eV, $w = 1269$ nm (second window),

◇ $x = 0.45, y = 0.990$: $E_g = 0.7548$ eV, $w = 1643$ nm (third window).

In general, the above quaternary compounds are developed from the basic binary compound InP through subsequent deposition steps, and for such compounds to work together the lattice parameter of the constituent III-V compounds need to match. In the following, we describe the basic principles and salient features of LEDs and lasers, that are relevant for developing optical transmitters.

2.8.2 LEDs

LEDs are made from *pn* junctions based on the spontaneous emission of photons. As indicated earlier, the forward current through an LED moves the electrons up from the valence band to the conduction band, and these electrons subsequently fall down *spontaneously* to the valence band, releasing photons with the frequency f, given by

$$f = \frac{c}{w} = \frac{cE_g}{1.24} \times 10^6 \quad \text{Hz.} \tag{2.98}$$

Since the topmost point of the valence band and the bottom-most point of the conduction band have a continuum of allowed energy levels below and above themselves respectively, the energy difference $E_1 - E_2$ of the *spontaneously emitted* photons get spread over these possible levels (see the shaded regions of the valence and conduction bands in Fig. 2.35), resulting in a continuum of possible photon frequencies and hence a large spectral width of the emitted light, typically ranging from a few tens of nanometer to as large as $\sim$ 150–200 nm. Note that, in lasers, the choice of wavelength is constrained by an internal feedback mechanism, thereby producing a much narrower spectrum (discussed in the following section). The LED output power typically ranges over a few hundreds of microwatts, supporting a transmission speed limited to a few hundreds of Mbps. Moreover, the radiation pattern being broad, LEDs are generally used with the multi-mode fibers. These features make LEDs suitable for optical fiber links covering short to moderate distances (typically for LANs), from both the received power and dispersion considerations.

In order to enhance the optical power output, LEDs employ a specific configuration, called double-heterojunction (DH), which was originally conceived for lasers (Fig. 2.36). In a DH configuration, a pair of p and n layers forms a basic *pn* junction; however, between these two layers an additional layer (an n layer in Fig. 2.36) is inserted. The additional layer (shown as n' layer in the figure) uses a slightly different refractive index (higher) and E_g (lower) as compared to the adjoining p and n layers for *confining* the photons and the charge carriers within this layer itself. With this arrangement, the additional layer, called the *active layer*, generates large number of photons with high quantum efficiency, leading to enhanced optical output power as compared to the ordinary *pn* junctions. These three layers forming the two heterojunctions on two sides of the active layer are called DH layers (DHLs), as shown in Fig. 2.36. Note that the above requirements call for specific combinations of compound semiconductors suitable for DH design. The forward bias current is fed through the metalization layers (MLs) on both sides through the DHLs. The striped area on the top ML and the geometry of the insulation layer (IL) (using SiO_2) define the width of the active area (AA) in the active layer for photon generation, which has a longitudinal stretch from one end to the other forming a rectangular parallelepiped segment in the active layer.

The generated photons from the spontaneous emission in the active layer are taken out, either through one of the two edges of the active-layer segment along the axis (which is normal to the direction of forward current) or from one of the two broad surfaces of the AA (which is normal to the axis of the active layer), thereby leading to two types of LEDs: edge-emitting LEDs and surface-emitting (also called Burrus emitter) LEDs. Figure 2.36 shows a typical configuration for an edge-emitting DH LED. The light output from the LED is fed to a small fiber chord (fixed permanently), which is then connected to the end surface of the actual transmission fiber using a fiber connector. LEDs working with spontaneous emission are very stable against temperature variation and are modulated by varying the forward current, with typical forward currents in the range of $\simeq$ 50–100 mA. The carrier lifetimes

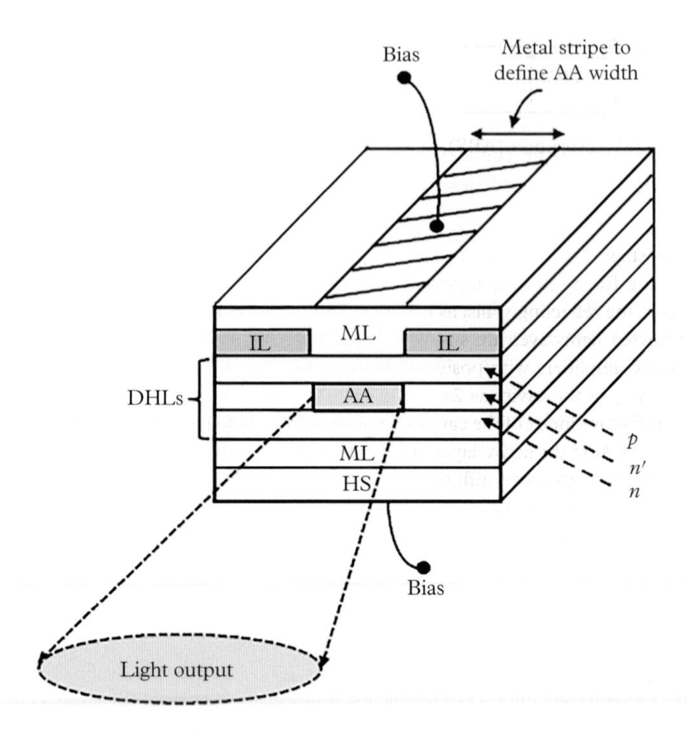

Figure 2.36 *Edge-emitting double-heterojunction structure of LED. ML: metalization layer, IL: insulation layer, AA: active area, HS: heat sink, DHLs: double-heterojunction layers.*

and the junction capacitance, along with the forward current, determine the switching speed of LEDs.

2.8.3 Semiconductor lasers

While using the same materials and *pn* junction as a DH structure, lasers differ from LEDs significantly in respect of several performance features, which are realized by using *stimulated emission* of photons (instead of the spontaneous emission used in LEDs). In particular, in a laser, the DH structure shown earlier for LEDs (Fig. 2.36) is engineered by preparing the two end faces of the active layer as partial reflectors, so that a fraction of the lightwaves generated from the release of photons keep coming back into the active layer from the partially reflecting ends over and over again, thereby leading to an optical feedback. The reflecting walls are typically realized by using the *cleaved* ends of the semiconductor substrate on the crystal planes normal to the longitudinal axis. Thus the active layer of the laser bounded by the cleavage planes gets transformed into a rectangular resonant cavity, with an active cross-sectional area in the vicinity of a few μm^2.

As shown in Fig. 2.37, the basic emission process starts with an *absorption* process, through which initially some electrons are *pumped up* to move from an energy level (E_1, say) in the valence band to another (E_2, say) in the conduction band. Absorbed electrons keep returning through *spontaneous emission* of photons to the valence band. As the forward current is increased further (well beyond what is typically used for LEDs), a large number (population) of electrons move up to the conduction band in the cavity space, so much so that it overcompensates the underlying loss process in the medium, thereby creating a *population inversion* from the valence band to the conduction band.

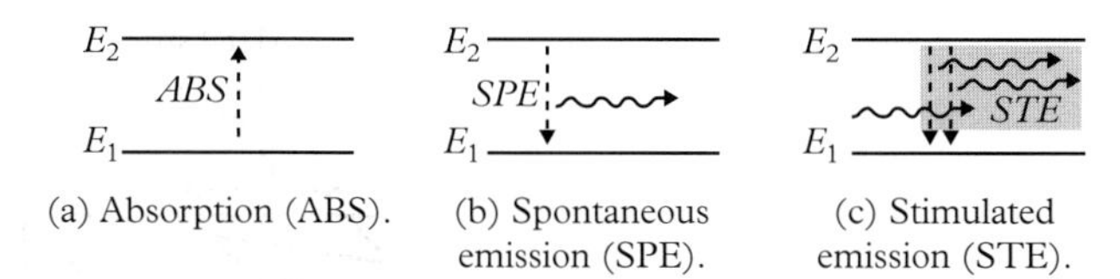

Figure 2.37 *Illustration of semiconductor laser operation.*

At this stage, large numbers of photons are generated by spontaneous emission in the cavity, and a fraction of the lightwave associated with these photons come back into the cavity from the reflecting walls, as observed earlier in a Fabry–Perot cavity. However, only those reflected lightwaves are sustained in the cavity (similar to the sustained oscillation in electronic amplifiers with positive feedback), whose wavelengths w_i's are related to the axial length L_c of the cavity as $2\pi \times (2L_c n_c/w_i) = 2\pi i$ (with i as an integer and n_c as the effective refractive index of the cavity). In other words, these lightwaves undergo *constructive interference*, making the active layer of the laser behave as a resonant cavity.

The reflected lightwaves influence the phase and frequency of the fields associated with the excited electrons *waiting* in the conduction band for subsequent photon emission. The sustained lightwaves, corresponding to the fed-back photons with the wavelengths w_i's, stimulate the excited electrons in the conduction band, and in turn the excited electrons release photons in sync with the fed-back photons and fall down to their ground states in the valence band, resulting in *stimulated emission* of photons. The fields of the lightwaves associated with these photons (stimulating and stimulated) add up coherently with the same frequency and phase, and thus keep growing to build up large optical power with narrow spectral spread. However, some photons keep emitting spontaneously, contributing to the phase noise and spectral spread of the resultant lightwave produced in the laser. In the interference process taking place in the resonant cavity, only the wavelengths selected by the abovementioned constraint, i.e.,

$$w_i = 2L_c n_c/i, \tag{2.99}$$

can exist, which are discrete in nature but large in number, resulting in as many longitudinal modes. However, finally the number of modes gets pruned by the gain spectrum of the cavity, leading to multi-mode (and single-mode) lasers with much fewer modes.

Laser modes and spectrum

The longitudinal modes in a laser are decided by the interference constraint in Eq. 2.99 and the overall gain spectrum of the medium in the active layer. The medium in the active layer has its own pump-induced gain, longitudinal propagation loss and confinement factors, which work together to produce an overall medium gain spectrum $g_m(w)$ as a function of the wavelength w. The combined effect of interference constraint and the gain spectrum is illustrated in Fig. 2.38, where we find that the modes are separated by a fixed spacing δw, centered around the peak-emission wavelength w_0 of the gain spectrum.

The wavelength spacing between the two adjacent modes in a laser can be estimated by considering two successive modes, say ith and $(i-1)$th modes. The wavelengths for these two modes are expressed as

$$\begin{aligned} w_i &= 2L_c n_c/i \\ w_{i-1} &= 2L_c n_c/(i-1). \end{aligned} \tag{2.100}$$

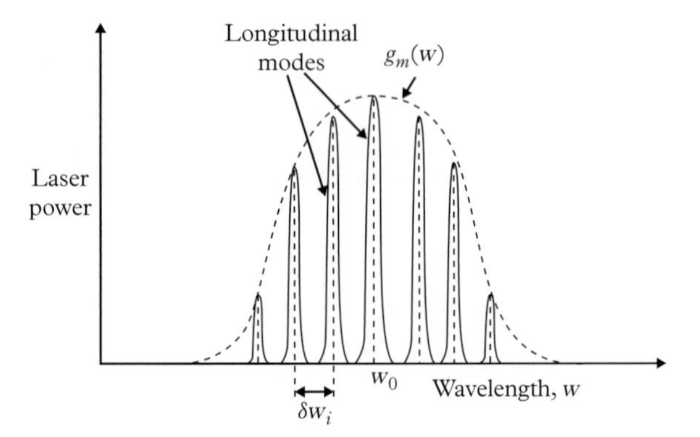

Figure 2.38 *Laser spectrum with multiple longitudinal modes.*

Using the above equations, one can express the mode numbers i and $(i-1)$ in terms of the respective modal frequencies $f_i = c/w_i$ and $f_{i-1} = c/w_{i-1}$ as

$$i = \frac{2L_c n_c}{w_i} = \frac{2L_c n_c f_i}{c} \tag{2.101}$$

$$i-1 = \frac{2L_c n_c}{w_{i-1}} = \frac{2L_c n_c f_{i-1}}{c},$$

leading to $\delta f_i = f_i - f_{i-1}$, given by

$$\delta f_i = \frac{c}{2L_c n_c} = \delta f, \tag{2.102}$$

implying that the modal frequencies are equally spaced. The wavelength spacing δw_i can be obtained using the relation $\delta f = (c/w^2)\delta w$ (ignoring the negative sign in the expression in the present context) as

$$\delta w_i = \frac{w_i^2}{2L_c n_c} \simeq \frac{w_0^2}{2L_c n_c} = \delta w, \tag{2.103}$$

where the approximate expression using w_0 is justifiable as the percentage variation of the w_i's over the laser spectrum is rather small. For DWDM transmission, one needs to reduce the number of modes to realize single-mode lasers by making the laser cavity more selective in respect of the resonant frequencies. We discuss in the following some useful techniques to realize single-mode lasers.

Single-mode lasers

Single-mode lasers can be realized by using frequency-selective feedback mechanisms in many ways, either within the laser cavity or by attaching an external cavity with frequency-selective reflecting walls (Suematsu and Arai 2000; Suematsu 2014). We discuss some of these single-mode lasers in the following.

One important class of single-mode lasers employs distributed feedback of the generated lightwaves, instead of having feedback only from the end faces of the cavity. These lasers are generically known as distributed feedback (DFB) lasers. In DFB lasers, the distributed feedback mechanism is employed in the cavity by using corrugation on one of its broader walls along the longitudinal dimension (Fig. 2.39), operating as a *Bragg grating*. Thus, the

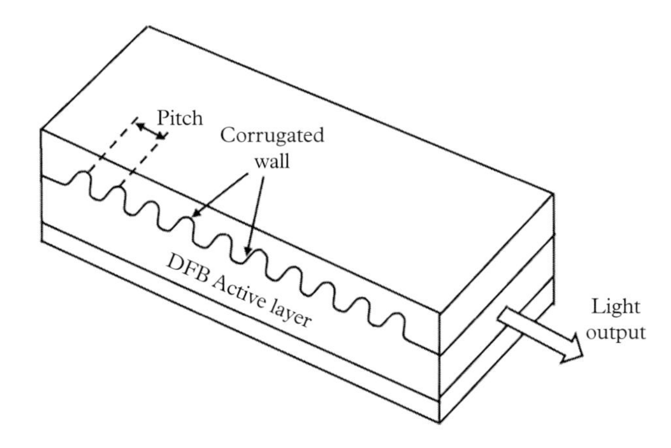

Figure 2.39 *Corrugated active layer of a DFB laser.*

cavity keeps reflecting back the lightwaves from each *crest* of its periodically corrugated internal wall across the length of the cavity, and with this arrangement only the lightwaves with specific wavelengths can experience constructive interference. For example, if the pitch of the corrugation is made half of the desired wavelength w_i (say) divided by the effective refractive index n_c of the medium, then the round-trip path of the reflected waves at a given point turns out to be integral multiples of the wavelength w_i leading to constructive interference, while the other wavelengths are mostly suppressed due to phase mismatch between their reflected versions. This configuration, along with having one of the two ends kept mostly reflective, leads to the realization of a single-mode DFB laser. Note that, with the Bragg grating configuration used on the walls of the cavity, one can realize different emission wavelengths by choosing different pitch lengths of the corrugation.

Another useful realization of the single-mode lasers with distributed feedback, known as distributed Bragg reflector (DBR) lasers, uses a Bragg grating in a distributed manner as shown in Fig. 2.40. In particular, the DBR configuration employs the corrugation on the wall near the end surfaces of the cavity on both sides and not at the center, thus distributing the Bragg grating region in two segments, while the central part of the cavity uses a planar wall. The central part controls the gain mechanism by sending the forward bias through this region only (called a gain segment), while the two corrugated Bragg grating segments on the two sides of the cavity make the wavelength selection (called a wavelength selection segment) in a passive manner, i.e., without any direct forward bias current through this region. This configuration keeps the two mechanisms (gain and frequency selection) fairly independent, which in turn offers a flexible design approach for the laser.

Two other methods to realize single-mode lasers are shown in Fig. 2.41. One of them (Fig. 2.41(a)), known as an *external-cavity laser*, employs an external cavity attached to the original cavity of a semiconductor laser, so that the two cascaded cavities can make a sharper selection of a frequency. Another type of single-mode lasers (Fig. 2.41(b)) uses one of the surfaces of the cavity for emission (instead of one of the edges), with a cavity that is formed with vertical deposition of layers on a substrate material, and hence called a *vertical-cavity surface-emitting laser* (VCSEL). The speciality of the VCSEL configuration is that the vertical deposition process enables realizing a small cavity length L_c sandwiched between two highly reflective wavelength-selective layers (e.g., Bragg gratings) at the top and bottom, which in turn places the modal frequencies far apart, since $\delta f_i \propto \frac{1}{L_c}$ (see Eq. 2.102). Thus the VCSEL configuration helps the cavity gain spectrum by selecting

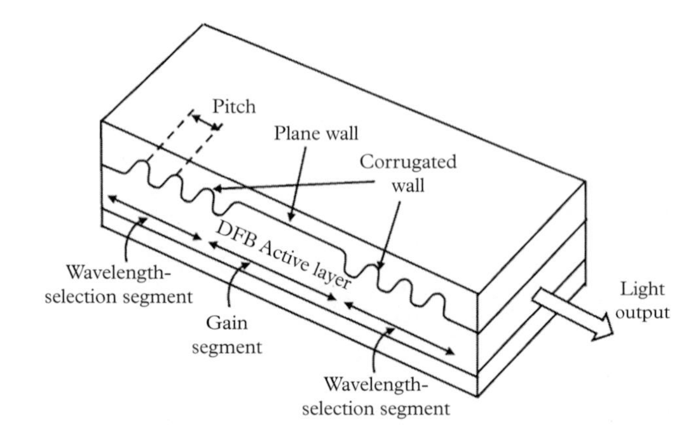

Figure 2.40 *Selectively corrugated active layer of a DBR laser.*

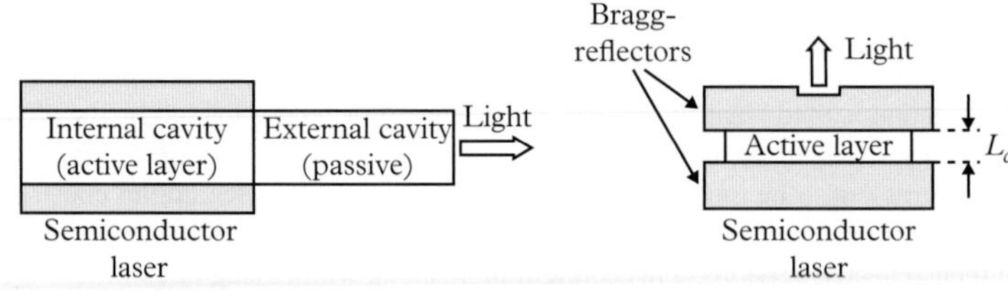

Figure 2.41 *Single-mode lasers using external-cavity and vertical-cavity surface-emitting configurations.*

only one mode, leading to single-mode operation. Given these alternatives, one needs to choose the laser type based on various underlying parameters: cost, design complexity, compactness, temperature sensitivity, compatibility to high-speed modulation, etc.

Furthermore, in semiconductor lasers, some useful improvements can be made by confining quantum-mechanically the active layer, which in effect transforms the smooth E–K diagram of semiconductors into a staircase-shaped formation, but enveloped by the original parabolic shapes (Green 1993; Arakawa and Yariv 1986). This feature can be realized by decreasing at least one of the two dimensions of its cross-section, leading to multiple quantum wells (MQW) where the energy states get confined to multiple discrete levels for the carriers. With this feature, MQW-based semiconductor lasers offer lower spectral width, lower threshold current, better linearity (P–I_F variation), higher yield in fabrication process and high temperature stability, thus superseding the earlier Fabry–Perot configurations.

Tunable lasers

Tunable lasers represent an extremely useful form of optical source for WDM networks. Although a WDM network can operate over a large optical transmission window, such as the third window (200 nm), or a part of it, such as the C band (35 nm), no node in a network in practice would simultaneously use all the available wavelengths (e.g., $\simeq$ 32 or 64 for C band, presuming a channel spacing of 100 or 50 GHz, respectively). Thus, a laser should desirably have the option to transmit over one among a few given wavelengths, that can be chosen from the given wavelength range in operation across the network. Thus a node may need to have a few lasers and each one of them should be able to *tune* across the entire or some part of the network spectrum; such lasers are called *tunable lasers.*

The tunable-laser technology has evolved a lot over the years and several choices are available for implementation. In the InGaAsP/InP fixed-wavelength lasers, one can realize any operating wavelength within the entire third window, while to make them tunable over the entire range, special design techniques are necessary. One of them is based on the external cavity lasers, wherein the *optical length* of the external cavity can be varied by using a suitable mechanism based on mechanical, acoustic, or electro-optic control. In mechanical control, the external cavity is formed by extending the active-layer-based cavity with a rotating reflection grating placed beside the semiconductor laser. The rotating cavity selects the wavelength that will be reflected back and will find itself in constructive interference in the active layer, thereby realizing the tuning of emission wavelength. The mechanical control can obviously be effected over a wide range of wavelengths ($\sim$ 500 nm), but it will be sluggish in its operation with a tuning time ranging over a few milliseconds to 100 ms, governed by the *distance* between the present and the target wavelengths. In acousto-optic external cavities, the refractive index is changed dynamically by using sound waves, thereby giving faster tuning time ($\sim$ 10μs) along with a large tuning range ($\sim$ 700 nm). However, electro-optic tuning realizes much faster tuning ($\leq$ 10 ns) by injecting current in the external cavity, although with a much shorter tuning range ($\leq$ 10 nm).

The other class of tunable lasers, preferred over the external cavity lasers for the compactness of design, is the bias-current-controlled DBR configuration of Fig. 2.40, which can control the wavelength-selection segments at the two ends of the active layer by using independent forward bias current. This part of the forward bias current is varied only when the frequency needs to be changed, while the forward bias current for the central region varies with the modulation at a much faster rate commensurate with the data transmission speed. Such DBR lasers are called multi-segment wavelength-tunable DBR (MS-WT DBR) lasers, offering fast tuning ($\leq$ 10 ns), but with a rather small tuning range in the order of 5 nm.

In WDM networks, one needs to have a fast as well as wide tuning range for more flexible design and operation, preferably over the S, C, and L bands in the third window, spanning over $\pm$ 100 nm around 1550 nm. Several research groups have worked toward the development of such lasers, mostly extending the basic idea of MS-WT DBR lasers. We discuss below some of the useful developments in this direction: the sampled grating DBR (SG-DBR) and superstructure grating DBR (SSG-DBR) lasers.

As described earlier, DBR lasers employ two separate Bragg grating regions at the two ends of the active layer. In particular, in SG-DBR lasers (Fig. 2.42), the Bragg grating pattern is interrupted spatially by *multiplying* the periodic grating pattern $g(x)$ by a *spatial sampling function* $s(x)$, such that at some intermediate parts the corrugated layer is replaced by planar wall as in $g_s(x)$. Such an interrupted grating profile is known as *sampled grating*, leading to the SG-DBR configuration. The pitches of two segments of the sampled grating at the two ends, i.e., Λ_1 and Λ_2, are kept slightly different (i.e., $g_s(x)$ assumes two different patterns, $g_{s1}(x)$ and $g_{s2}(x)$, at the two ends), and between these two sampled grating regions, one gain segment and one phase segment are placed, one after another. Each of these segments in the SG-DBR configuration is driven by an independent forward bias current: I_{sg1} and I_{sg2} for the two sampled grating segments, I_g for the gain segment and I_p for the phase segment. By controlling the currents over the two sampled grating regions with dissimilar pitches and the phase segment, the laser realizes a convolution of the reflected lightwaves, thereby offering a continuous tuning of wavelength with single-mode operation over a reasonably wide range of $\simeq$ 57 nm, which can cover the C band as well as some parts of the S and L bands of the third window (Jayaraman *et al.* 1993). In SSG-DBR lasers, further improvisation is made on

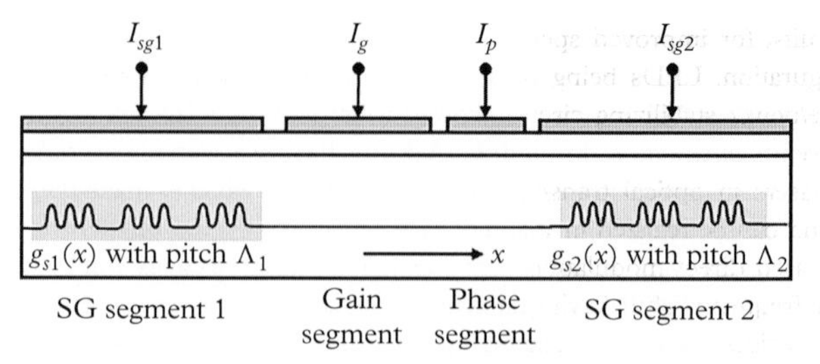

(a) Longitudinal side-view of an SG-DBR laser ($\Lambda_1 \neq \Lambda_2$).

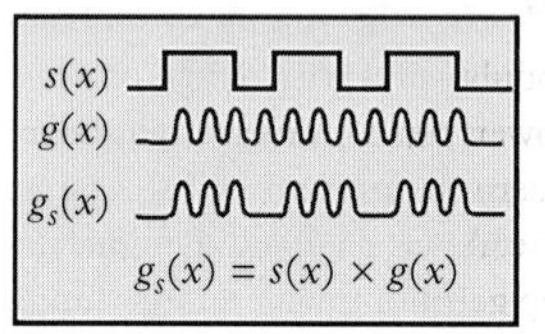

(b) Spatial sampling mechanism in the SG segment.

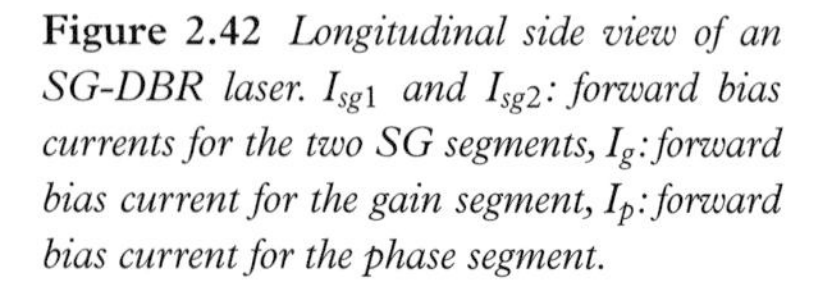

Figure 2.42 *Longitudinal side view of an SG-DBR laser. I_{sg1} and I_{sg2}: forward bias currents for the two SG segments, I_g: forward bias current for the gain segment, I_p: forward bias current for the phase segment.*

SG-DBR lasers, by using a superstructure for the sampled grating zones. In SSG structure, grating regions are chirped spatially, thereby increasing the tuning range significantly to $\simeq$100 nm, while retaining the tuning speed within 10 ns.

At times, introducing a tunability feature in lasers increases their hardware complexity and cost, and in such situations an easier alternative is preferred where one can use an array of lasers, with each of them being fixed-tuned at one particular wavelength, so the node can select and switch on a few of them at a time to transmit its ingress traffic over the network. Laser arrays can be realized by using VCSELs along with a MEMS-based optical switching mechanism to choose the desired laser from the given array of lasers.

2.8.4 Optical transmitters

Transmitters in optical communication systems can use LEDs or lasers, depending on the needs of transmitting power, spectral width, radiation pattern, modulation schemes, bit rate, link length, etc. In general, for short-haul and low-speed non-coherent communication links using multi-mode fibers, LED becomes the appropriate choice using IM (i.e., direct modulation) of light. However, for long-haul single-mode fiber links, lasers offer the right choice with their high power, narrow spectral width, and sharp radiation pattern. In particular, WDM transmission systems with the need of well-defined transmission bandwidth within the WDM frequency grid, demand a narrow spectrum of optical sources for both non-coherent (i.e., IM) as well as coherent modulation schemes, e.g., quadrature phase-shift keying (QPSK), quadrature amplitude modulation (QAM), or orthogonal frequency-division multiplexing (OFDM). For such high-speed non-coherent and coherent transmission schemes, lasers qualify as the indisputable choice for optical sources.

IM-based non-coherent transmission of light from LEDs can be realized with typical transistor-based switches operating in on-off mode with additional care to reduce the switching time in the driving transistors and LEDs, where the LED current is switched between low and high levels to obtain an optical output signal with binary modulation. One can also use an emitter-coupled current-switching scheme, as in emitter-coupled logic

(ECL) circuits, for improved speed, as the transistors are not driven into saturation in ECL configuration. LEDs being based on spontaneous emission need not employ any temperature/power stabilizing circuits, thereby making the LED transmitters simple and cost-effective.

Use of lasers in optical transmitters requires more complex hardware as compared to LEDs, and can be realized in various ways, which can be broadly categorized into two types: IM-based direct modulation by varying the forward-bias current (which can also be used for frequency-shift keying (FSK)), and external modulation by using additional modulating devices, the latter scheme being more useful for coherent transmission systems. In the following, we discuss both modulation techniques for lasers.

Direct modulation of lasers

In order to carry out direct modulation (i.e., IM) of lasers, it is important to know the nature of variation of laser output power versus forward bias current, as shown in Fig. 2.43. The output optical power P of a laser initially increases slowly with the forward current I_F, as in an LED with spontaneous noise. As the current increases, the laser goes through population inversion, and the photons keep getting generated more and more. Soon after, many of these photons are reflected back from the cavity end-faces and interfere constructively, leading to a sharp increase in the output power with increasing forward current. The forward current at which this happens (due to the stimulated emission), exhibiting a sharp increase in the slope of P-I_F plot, is called the threshold current I_{th} of the laser, and at this point the laser is said to have been *lased*. As the forward bias current keeps increasing, along with the increase in the emitted power, the laser spectrum gets narrower with fewer longitudinal modes, as shown in

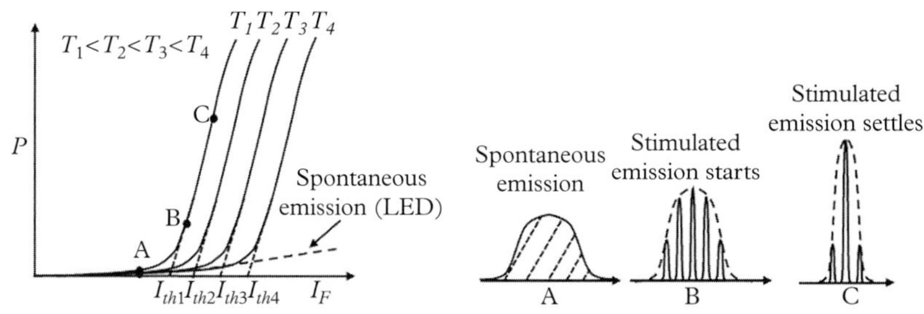

(a) Laser output power versus forward bias current for different temperatures.

(b) Evolution of laser spectrum with increasing forward current at different points for $T = T_1$.

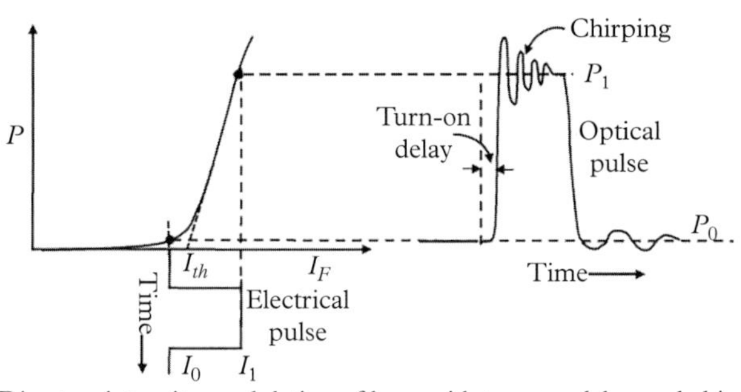

(c) Direct or intensity modulation of laser with turn-on delay and chirping.

Figure 2.43 *Representative plots of optical power output versus forward bias current in lasers for different temperatures. Typically, the maximum laser power output P_1 can be as large as 10 mW, with a forward-bias current over 100 mA, while the threshold currents would be mostly in the range of 30–50 mA.*

Fig. 2.43. The laser threshold current can increase significantly with the laser temperature T, following an exponential relation, given by

$$I_{th}(T) \propto \exp\left(\frac{T}{T_0}\right), \tag{2.104}$$

where $I_{th}(T)$ is the threshold current at temperature T and T_0 is an equivalent temperature of the laser, representing the *relative temperature insensitivity* of the laser, decided by its physical parameters (both T and T_0 are the temperatures in °K). It is desirable to have lasers with T_0 as large as possible, and it typically varies from $\sim$ 120°K in the first spectral window to $\sim$ 60°K in the third spectral window of optical fibers. Figure 2.43 shows three such plots in part (a) with three temperatures, T_1, T_2, and T_3, with the three threshold currents I_{th1}, I_{th2}, and I_{th3}, respectively. Part (b) of the figure shows how the laser spectrum from the spontaneous emission stage (at point A in the P–I_F plot for $T = T_1$) evolves to the ones at the points B and C with stimulated emission at the increased laser currents.

Figure 2.43(c) shows how the laser current can be directly modulated by varying the forward bias current to obtain optical pulses, but with a *turn-on delay* along with transient intensity fluctuations in the optical pulses. The transient intensity fluctuation leads to frequency chirping in the optical pulses, which in turn causes pulse broadening due to the chromatic dispersion in the optical fibers. For reducing the turn-on delay, lasers need to be pre-biased (for binary *zero* transmission) at a forward bias current above zero, while not being too close to the threshold current for stable operation.

The illustration in Fig. 2.43(a) indicates that, the P–I_F plot shifts toward the right side with increasing temperature, thereby reducing the optical power output for a given forward bias current. To restore the power output, one needs to increase the forward bias current which will in turn cause more heating of the laser resulting further increase in laser temperature, thereby leading to *thermal runaway* of the laser causing permanent damage. This necessitates to exercise the power as well as temperature controls of lasers, so that the laser offers a reasonably constant power output over time, without any thermal runaway.

Note that, the direct modulation of lasers using IM is fundamentally different from the amplitude modulation of the lightwave, as in IM the forward bias current controls the number of emitted photons, i.e., the light intensity (or power), which is the *square-and-average* of the instantaneous amplitude of the respective lightwave. Furthermore, in an IM-based transmitter, each time a laser transmits a *binary one* following the transmission of a binary zero, it needs to restart its lasing process (as the laser is turned off or *nearly* turned off using a small prebias current during the preceding zero transmission), thereby missing the phase continuity of lightwave from the last binary one (separated by one or more zeros). This makes the IM of lasers indeed a non-coherent modulation scheme. In other words, since the laser has to lase again afresh, the phase trail of the previous lightwave doesn't necessarily match with the phase epoch of the freshly generated (and hence uncorrelated) lightwave. On the other hand, while restarting, the laser also takes a while to lase, causing a *turn-on delay* (as shown earlier in Fig. 2.43(b)). Following the turn-on delay, due to the momentary refractive index variation in the cavity, the laser frequency varies before settling down at the desired frequency, leading to the laser chirping, as discussed earlier (see Fig. 2.43(c)).

Through IM, one can also realize frequency modulation of lasers, as any change in injection (forward bias) current changes the cavity refractive index, thereby bringing in frequency modulation. However, the variation in the bias current also brings a small variation in laser output power, which may not be desirable when the transmitter needs to transmit a constant-envelope FSK lightwave with large frequency deviation. Also note that in optical communication systems in general, analog IM hasn't been a very popular choice.

However, the subcarrier modulation (SCM) systems have been useful in some situations, where the binary data stream first modulates a microwave/millimeter-wave subcarrier, and several such non-overlapping (in frequency domain) analog (modulated) microwave carriers are multiplexed and used to modulate the light intensity directly.

External modulation of lasers

In view of the basic limitations of IM, especially for high-speed transmission ($\geq$ 10 Gbps), it is preferable to replace IM by external modulation, which can be used to modulate amplitude, phase, or both (as in QAM). External modulators can produce coherent modulation, as the laser operates in CW mode and hence never gets switched off during data transmission.

Figure 2.44 shows the schematic diagrams for phase shift keying (PSK) and amplitude shift keying (ASK) modulation schemes using photonic waveguides with electro-optic control. In part (a), Ti-diffused $LiNbO_2$ waveguides are used to introduce the necessary phase shift (0 or π) for PSK using two electrodes. In part (b), an MZI is used for ASK modulation. The input light, having been fed the input port 1, is first split into two parts by a 3 dB coupler in the MZI. Before the two parts reach the second 3 dB coupler, an *additional phase shift* is introduced electro-optically through the electrodes in the upper arm (with respect to the lower arm) when the data is a binary zero, or else the additional phase shift in the upper arm is set at zero. Governed by the differential phase shifts in the two waveguide arms of the MZI, the lightwave at the output port 1 becomes zero for binary-zero data, owing to destructive interference, while for the binary one data the output assumes the full power from the constructive interference.

By utilizing the basic devices, such as, the phase shifters and MZIs, one can mix and match ASK and PSK modulation to realize bandwidth-efficient multi-level modulation schemes such as QPSK and QAM. The external modulators used in such cases can also be integrated with the semiconductor laser, making the transmitter compact and stable.

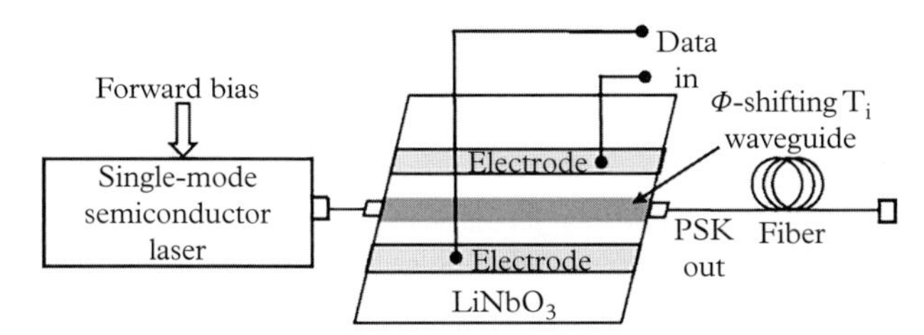

(a) External PSK modulation using Ti-$LiNbO_3$ waveguides.

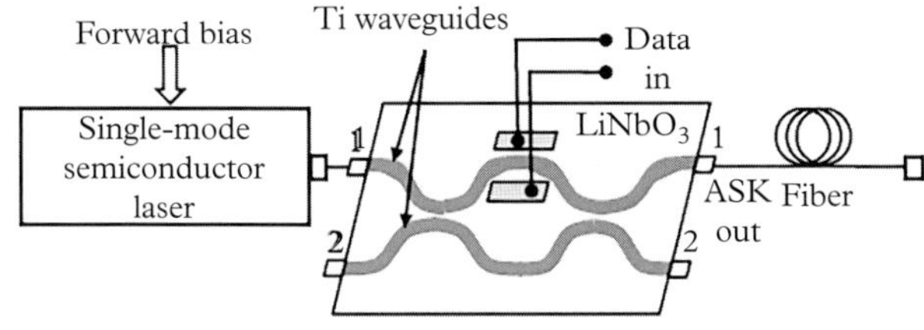

(b) External ASK modulation using MZI with Ti-$LiNbO_3$ waveguides.

Figure 2.44 *External modulation of lasers.*

2.9 Photodetectors

Optical receivers need to employ suitable optoelectronic devices to translate the received optical signal into an equivalent electrical signal through photodetection. Photodetectors need to operate in the desired wavelength range with adequate speed and efficiency to convert the received light into an equivalent electrical signal. There are different types of photodetectors: *pin* photodiodes, avalanche photodiodes (APDs), which are semiconductor-based devices, and photomultipliers using photocathodes and electron multipliers based on vacuum-tube technology. Photomultipliers with large size and high-voltage operation are not suitable for optical fiber communication systems. On the other hand, *pin* photodiodes are the most handy form of photodetectors, while APDs offer additional internal gain, as compared to *pin* diodes, for the photodetected signal, albeit with much larger reverse bias voltage and additional noise contribution from the avalanche multiplication process. In the following we describe the operational principles and salient features of *pin* photodiodes and APDs.

2.9.1 *pin* photodiodes

pin photodiodes, or simply *pin* diodes, employ basically *pn* junctions, where a nearly intrinsic layer (i-layer) is introduced between the p- and n-layers. As shown in Fig. 2.45, the *pin* diode is kept reverse biased and the incoming optical signal is received at its front-end surface. The photons in the received optical signal impinge on the end surface (of the p-layer in Fig. 2.45) and move through the device creating hole-electron pairs. The movement of hole-electron pairs across the diode develops a photocurrent flowing through the resistance R used in the biasing circuit. The photocurrent is thus converted into a voltage across R, which is amplified thereafter, so that the amplified electrical signal has adequate amplitude for further signal processing in the receiver.

The incident photons on the *pin* diode entering from the front-end surface (Fig. 2.45) get partially absorbed in the p-type material, before reaching the *pi* junction. The i-layer uses *lightly doped* n-type semiconductor, and thus the depletion layer developed across the *pi* junction due to reverse bias becomes short in length in the p-region and long in the i-region, with a strong electric field operating across the junction. The photons having arrived at the junction are absorbed and generate hole-electron pairs. The holes and electrons, so produced, are swept away in opposite directions swiftly by the prevailing strong electric field across the depletion layer, before they get the chance to recombine. Flows of these charge carriers constitute the photocurrent, and in order to raise the efficiency of this process, one needs to use longer i-layer along with shorter p-layer. Longer i-layer enhances the photodetection efficiency, while shorter p-layer reduces the loss of incident photons due to

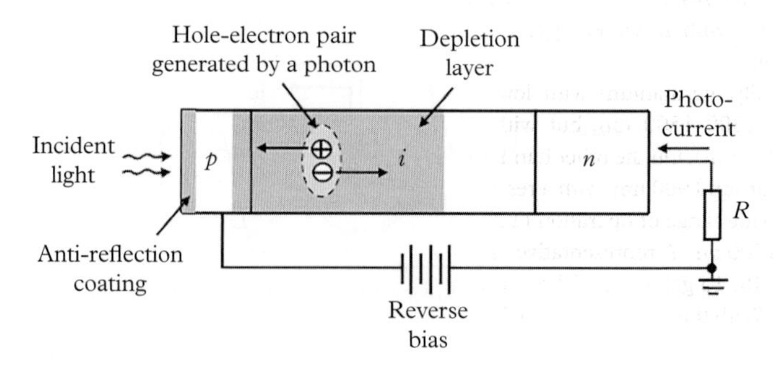

Figure 2.45 *Schematic diagram of a reverse biased pin diode generating photocurrent from the incident light, with R as the biasing resistance.*

the absorption of light in the p-layer. However, too long an i-layer introduces a large transit time for the carriers, thereby degrading the device speed, implying that an optimum choice for the length of the i-layer needs to be made to achieve high photodetection efficiency along with the needed speed of operation.

Materials used in *pin* diodes are chosen depending on the operating wavelength range. The parameter that decides the overall efficiency of photodetection is the *responsivity* R_w, given by

$$R_w = \frac{I}{P}, \tag{2.105}$$

where I represents the photocurrent generated from an incident optical power P. Assuming that a rectangular optical pulse with an instantaneous optical power P and a duration of T has been incident onto the end-face of a *pin* diode, resulting in a pulse of photocurrent of amplitude I over the same time duration T, one can express I and P as

$$P = \frac{\Lambda hf}{T} \quad \text{and} \quad I = \frac{\eta \Lambda q}{T}, \tag{2.106}$$

where h represents Planck's constant, Λ is the number of photons received during T, q is the electronic charge, and η represents the quantum efficiency of the photodetector. Using Eq. 2.106 in Eq. 2.105, we obtain the expression for R_w as

$$R_w = \frac{\eta q}{hf}. \tag{2.107}$$

As evident from the above expression, the responsivity R_w of a *pin* diode depends on the quantum efficiency η and the operational wavelength $w = c/f$, and η is governed by the semiconductor material used in the device and its structural features. Overall, R_w exhibits a bandpass-like behavior over a specific wavelength range, governed by two opposing phenomena. In particular, η depends significantly on the absorption (loss) in the front layer of the device and the reflection from the front surface. Absorption loss in the front layer (p-layer in Fig. 2.45) during the passage of light from the front-end surface to the depletion layer increases at lower wavelengths, leading to a lower cutoff wavelength w_{min}. The fraction of light actually reaching the depletion layer can be enhanced by thinning down the front layer along with an antireflection coating on the front-end surface. Typically, η varies in the range of 40–95% for various materials. On the other hand, from the relation between the operating wavelength w and the bandgap E_g of the semiconductor material used, one can obtain $w_{max} = 1/f_{min} = hc/E_g$, implying that the incident photons must have at least an energy $= hf_{min}$, leading to a maximum permissible (i.e., upper-cutoff) wavelength w_{max}. Generally, the responsivity falls sharply above w_{max} as the photons with higher wavelengths (and hence with lesser energy) are not able to effect any transition for photocurrent generation.

Typically, germanium with low E_g, offers a passband spanning over a wavelength range of 1300–1500 nm, but with a low responsivity ($\simeq$ 0.45 A/W in the interval of 1350–1400 nm). On the other hand, silicon with higher value of E_g performs well in the first window, around 900 nm, with a responsivity of $\simeq$ 0.7 A/W. For longer wavelengths, InGaAs offers a wide range of operation (1300–1600 nm) with a responsivity as high as 0.9–1 A/W around 1500 nm. A representative plot of R_w versus w for *pin* diodes using InGaAs is shown in Fig. 2.46. To get a feel of the *pin* parameters, consider a *pin* diode with a responsivity of 0.9 A/W. With this *pin* diode, an incident optical signal with a power of 1 μW (i.e., −30 dBm)

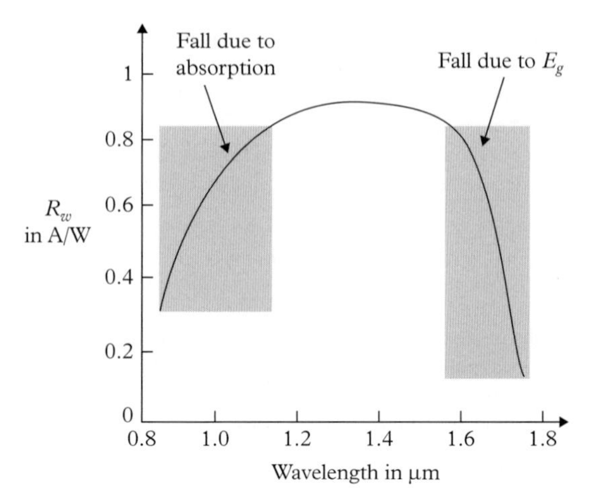

Figure 2.46 *Plot of responsivity versus wavelength for an InGaAs pin diode.*

for binary one transmission and zero watt for binary zero transmission, will lead to a photocurrent swing with an amplitude between 0 and 0.9 μA, which needs to be amplified by using amplifier chain in the optical receiver, needing typically ∼ 100 mV voltage swing to be detectable by a comparator in the presence of noise. Note that, in contrast with the thermal noise experienced in electrical receivers, optical receivers get contaminated additionally by signal-dependent shot noise, the origin of which remains in the source and *not in the receiver*. We address this aspect while describing the noise processes encountered in optical receivers in Appendix B, and make use of the noise models to estimate BERs in Chapter 10, dealing with the transmission impairments in optical communication systems.

2.9.2 Avalanche photodiodes

APDs also employ a reverse-biased *pn* junction, but with a somewhat different structure (using an additional layer) as compared to *pin* diodes, and let the reverse-biased junction enter the avalanche-breakdown region by using a large reverse voltage. Consequently, the primary photocurrent gets amplified internally due to the avalanche-gain mechanism, though the gain turns out to be random in nature, thereby adding a certain amount of additional noise, and one needs to optimize the bias voltage such that the overall SNR also gets maximized and enhanced compared to the *pin* diodes.

Figure 2.47 shows the basic structure of an APD, where the reach-through APD configuration is used with four layers $p^+i(\pi)pn^+$, developed from a heavily doped p^+ substrate. The intrinsic (i) layer, in reality, gets lightly doped by p-type impurity and referred to as π-layer. When the APD is biased with a low reverse-bias voltage, the entire bias voltage practically appears across the pn^+ junction. With increasing bias voltage, when the peak field across the pn^+ junction approaches the minimum electric field E_{min} required for avalanche breakdown, the depletion layer across the pn^+ junction extends toward the π-region (hence the name reach-through APD). Under this condition, the light enters through the p^+-layer and gets absorbed in the π-layer as in a *pin* diode, creating the primary charge carriers, i.e., the hole-electron pairs. Soon after, the electrons start to pass through the pn^+ junction, highly reverse-biased to operate in avalanche breakdown mode. As a result, the electrons, while passing through this junction under high electric field (as shown in Fig. 2.47 in the plot of electric field versus the distance across the APD), get multiplied in

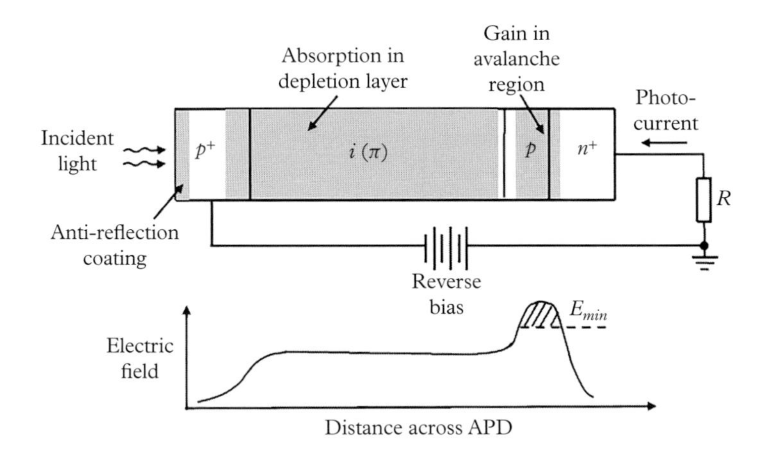

Figure 2.47 *Schematic diagram of a reverse-biased APD. E_{min}: minimum electric field needed for avalanche breakdown.*

number due to the avalanche breakdown mechanism, producing a considerable gain over the primary photocurrent. However, the carrier multiplication factor or gain g due to avalanche breakdown is not a deterministic process, thereby having a mean gain $E[g] = G$ along with a mean-squared value $E[g^2]$, the latter accounting for the additional noise in the receiver, referred to as *excess noise*.

APD mean gain G is defined as the ratio of the APD output current I_{apd} to the primary photocurrent I_p, as

$$G = \frac{I_{apd}}{I_p}, \tag{2.108}$$

with I_p measured as the APD current with low reverse bias (effectively being the primary photocurrent), when the avalanche breakdown process remains dormant, i.e., with $G = 1$. The responsivity of APDs gets enhanced by the mean gain G with respect to the *pin* diodes, and hence can be expressed as

$$R_{apd} = G\frac{\eta q}{hf}. \tag{2.109}$$

The mean-squared value of the APD gain plays an important role in determining the receiver performance, and is expressed as

$$E[G^2] = G^2F(G) = G^2\Big[kG + \Big(2 - \frac{1}{G}\Big)(1-k)\Big], \tag{2.110}$$

where $F(G) = kG + \big(2 - \frac{1}{G}\big)(1-k)$ represents a device parameter, known as excess noise factor in APD, and k represents the ionization ratio in APD. An approximate empirical expression for $F(G)$ is often used in receiver analysis, given by $F(G) = G^x$, with the parameter x varying in the range of [0,1], governed by the APD material.

2.9.3 Optical receivers

The basic configuration of a binary optical receiver for the IM-DD single-wavelength transmission, as shown in Fig. 2.48, consists of a suitable photodetector followed by an

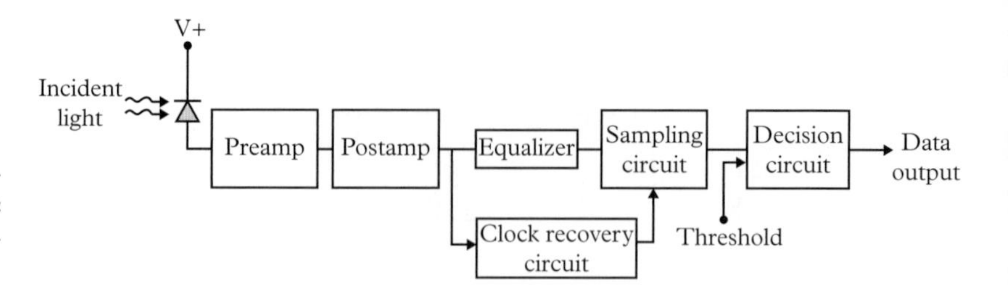

Figure 2.48 *Schematic diagram of an optical receiver. Preamp: preamplifier (generally a transimpedance amplifier), Postamp: postamplifier (chain of voltage amplifiers).*

amplifier-chain along with an equalizer to retrieve the binary signal with adequate amplitude and appropriate shape for decision-making. The amplified and equalized signal is thereafter sampled at each bit interval by using a clock recovered from the amplified signal by using a parallel clock-recovery module. The sampled signal is next passed through the decision circuit with an appropriate decision threshold to retrieve the binary data stream.

As mentioned earlier, in a typical case, the received optical power for a binary 1 in a *pin* diode receiver in the range of -30 dBm (i.e., 1 μW) and zero Watt for a binary 0 would produce a photocurrent swing of about 0.7–0.9 μA (corresponding to a *pin* responsivity in the range of 0.7–0.9 A/W), which needs to be raised to a voltage level in the order of 100 mV. This requirement implies that the amplifier chain following the photodetector should offer a voltage-to-current gain in the form of a transimpedance = 100 mV/0.8 μA = 125,000 Ω (assuming that $R_w = 0.8$ A/W). In order to attain the bandwidth requirement for high-speed transmissions, it is customary to distribute this large transimpedance gain between a low-noise transimpedance amplifier as a pre-amplifier, followed by a chain of voltage amplifiers as a post-amplifier block.

The transimpedance amplifier, usually a voltage-shunt feedback amplifier with a feedback resistance R_F, converts the current swing $\Delta I = I_1 - I_0$ (I_1 and I_0 being the photocurrents corresponding to the received optical power levels for binary one and zero transmissions, respectively) into a voltage swing $\Delta V = \Delta I \times R_F$, which is in general much smaller than what is required for the decision circuit. For example, with an $R_F = 50\ \Omega$, one will get $\Delta V = 0.8 \times 10^{-6} \times 50 = 40\ \mu$V, which is far below the distinguishable voltage swing (typically $\sim$ 100 mV). The remaining task of raising the small voltage swing at the preamplifier output to a distinguishable level (at the input of the decision circuit) is accomplished by the postamplifier. Along with this arrangement, one also needs to provide automatic gain control in the postamplifier to attain a satisfactory dynamic range in the receiver. In some situations, the received power would vary significantly from time to time, typically with a variability between the incoming groups or bursts of packets, and in such cases the receiver needs to adjust the decision threshold dynamically with additional circuits in a packet-by-packet manner, leading to what is known as a burst-mode receiver.

For a WDM transmission system, the receiver would employ an optical filter to extract the optical signal in the desired channel, and thereafter employ similar configuration as discussed above for the single-wavelength reception. However, for coherent receivers for the demodulation of lightwaves transmitted with QPSK/FSK/QAM/OFDM modulation would require much more complex hardware and are briefly discussed later in this chapter.

2.10 Optical filters

Optical filters play significant roles in WDM transmission systems, both for receiving specific wavelengths and for wavelength multiplexing/demultiplexing tasks in the network nodes. We describe below some of the useful options to realize the optical filters.

2.10.1 Fabry–Perot filters

As seen in our earlier discussions, FPIs offer one useful option for optical filtering in transmission mode, but with the constraint of FSR. In other words, if the FSR is smaller than the aggregate bandwidth of a WDM signal (i.e., including all WDM channels), the same FPI filter would select more than one frequency at a time. In order to get around this problem, one can cascade multiple Fabry–Perot filters with different values of FSR so that, over the given span of an input WDM signal, the cascaded assembly picks up only one frequency. Note that the Fabry–Perot section that has the lowest FSR will offer the sharpest selectivity (smallest bandwidth), while the ones with larger FSRs will prune the other frequencies also passed by the first stage, thereby leading to a filter assembly with narrow bandwidth while avoiding the problem of small FSR. Figure 2.49 illustrates this design approach with multiple Fabry–Perot filtering stages with different values of FSR.

2.10.2 Filters based on fiber Bragg grating

As discussed earlier, an FBG, unlike the FPI-based filters, works in reflection mode, i.e., it reflects the desired wavelength instead of transmitting it in the forward direction. As we shall see later, optical filters based on FBG find useful applications in tunable optical filters and add-drop operation for the wavelengths in a node operating in a WDM network.

2.10.3 MZI-based filters

MZIs can be used as optical filters, but with coarser resolution compared to MDTFs (discussed below). Following the same principle as in cascaded FPIs, one can also construct such bandpass filters, which would be useful to separate the wavelengths placed with wider spacings. A typical construction of an MZI-based filter is shown in Fig. 2.50. As shown in the

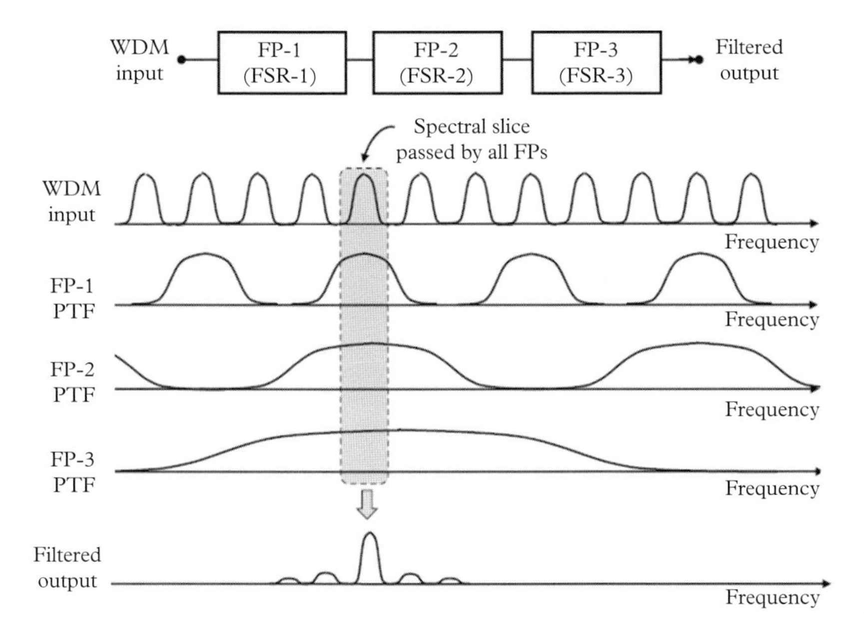

Figure 2.49 *Fabry–Perot filter with three cascaded stages. FP: Fabry–Perot, PTF: power transfer function.*

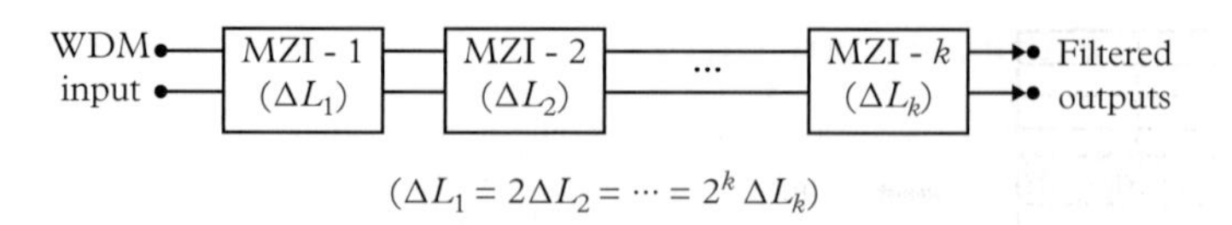

Figure 2.50 *k-stage MZI-based filter.*

figure, different cascaded stages of MZIs need to use different values of ΔL for the different phase shifts to realize large FSR with narrow and sharp passband, much the same way as in FPI-based filters shown earlier in Figure 2.49, though all the power transfer functions in this case would follow the *squared-sinusoidal/cosinusoidal* functions instead of the sharper Airy functions in the case of FPI-based filters.

2.10.4 Multilayer dielectric thin-film filters

A useful design approach, which evolved from the concept of the cascaded FPI-based filters, uses multiple layers of dielectric thin-film slabs (Yariv and Yeh 2007; Knittl 1976), and hence is called a multilayer dielectric thin-film (MDTF) filter. Multiple layers are developed by using appropriate deposition techniques from a substrate, and each of these layers uses a specific refractive index and width. The desired wavelength passes through the entire stack of layers, while the rest of the wavelengths are reflected backward. Theoretically, it is possible to realize a desired transfer function with a flat-top passband, large enough to accommodate the spectral spread and drifts/offsets of lasers along with sharp transitions (skirts) between the passband and the adjoining stopbands by using an MDTF configuration.

However, in practice the constraint on the availability of arbitrary refractive indices usually leads to a binary choice of the materials with high (H) and low (L) refractive index, typically by using TiO_2 (refractive index $n_H \simeq 2.3$) and SiO_2 (refractive index $n_L \simeq 1.46$), respectively, and the width of these H and L layers is chosen to be a quarter of the desired wavelength, i.e., $w/4$. A stack of two such H or L layers with the total thickness of $w/2$ behaves like a resonating FPI, called a cavity, which in turn offers a periodic power transfer function. In order to get around the periodicity of the transfer function, a practical MDTF configuration uses a sequence of H and L layers, which usually starts with the glass (G) layer followed by a specific sequence of H and L layers, leading to the formation of a few cavities (i.e., HH or LL layers) and intermediate quarter-wave layers (i.e., alternating H and L layers), finally being terminated again with a G layer. Note that, one single-cavity configuration fails to give the necessary specifications: flat passband and sharp skirt between the passband and stopbands. Thus, a typical sequence of the H and L layers for an MDTF filter uses a multi-cavity configuration, starting with the alternating layers of H and L, followed by a cavity, i.e., HH or LL layers, again followed by the alternating layers of H and L and a cavity, and so on, with the entire sequence sandwiched by G layers at both ends. Thus, the sequence turns out typically as $G(HL)^i HLL(HL)^j HLL \cdots HLL(HL)^m HG$, wherein $i, j, \cdots, m$ are integers ranging typically between 5 and 15, and each set of LL (or HH) implies a cavity. Single-cavity as well as multiple-cavity configurations of MDTF are illustrated in Fig. 2.51.

Cascading of multiple cavities along with intermediate alternating H and L layers, makes the passband almost flat while realizing a sharp skirt of the power transfer function beyond the passband. This technique leads to a highly stable and compact design of optical filters with the desired power transfer function. Typically, for a WDM channel in the third window with 50/100 GHz channel width, the passband ripple should remain confined within 0.5 dB, with the stopband rejection being in the range of ≥ 50 dB. Due to the underlying flexibility

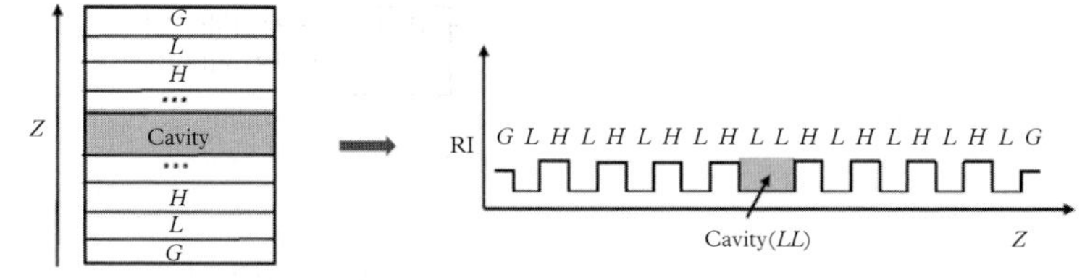

(a) Typical single-cavity MDTF.

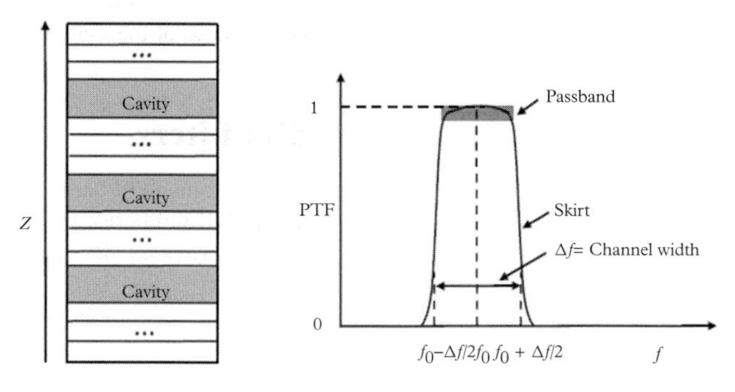

(b) Three-cavity MDTF.

(c) Respresentative power transfer function of a multi-cavity MDTF (tyically, passband variation ≤ 0.5 dB, stopband rejection ≥ 50 dB).

Figure 2.51 *MDTF configuration and frequency response.*

of the design in respect of the number and the sequence of layers and cavities, one can thus synthesize an excellent design complying with the needs of DWDM channels in optical networks.

2.10.5 Tunable filters

In WDM networks at all hierarchical levels, tunability of optical filters turns out to be an important feature to make the network agile in a dynamic traffic scenario. The tunability of optical filters is assessed by the tuning range and speed of the device used, which would vary with the underlying technologies. The basic frequency selection mechanisms includes mechanical, thermo-optic, acousto-optic, and electro-optic controls of various frequency-selective devices, such as stretched Bragg gratings, FPI cavities with movable facets, liquid-crystal-based and acousto-optic FPIs, etc. In the following, we describe some of these devices in brief.

Stretched Bragg gratings

One possible option to realize the tunable filters is to utilize the FBGs by *stretching* the grating pitch of the fiber mechanically using thermo-mechanical, piezo-electric, or stepper-motor-based methods. The thermo-mechanical and stepper-motor-based methods are relatively slow (a few ms of tuning time), but offer good tuning range (a few 100s of nm), while the piezo-electric method is fast (tuning time $\leq$ 10 ns) and precise, although with a limited tuning range of about 15 nm.

MZI-chains with electric control of refractive index

MZI chains offer a viable option to realize tunable filters, where the refractive index of the semiconductor material used as the substrate is controlled electrically. Injection of current into the semiconductor through electrodes can cause fast changes in its refractive index, thereby offering fast tunability, but again over a limited tuning range. Typically, the electrically controlled MZI chains can cover a tuning range of about 15 nm with extremely short tuning time in the range of a few ns.

MDTFs with thermo-optic control of refractive index

MDTFs can also help in realizing tunable optical filters, by thermally controlling the refractive index of the dielectric films in the multi-layer stack. However, thermo-optic control is a slow process, leading to a sluggish tuning option. Typically, such filters can offer a good tuning range up to 50–60 nm, but with large tuning time of about 10 ms.

MEMS-based electromechanical control of FPIs

Another useful method is to vary the cavity length of FPIs by making the reflecting facets movable. Micro-electromechanical system (MEMS) (to be discussed later in detail) are utilized to provide the mobility-feature to the FPI facets. In other words, MEMS-based mirrors are placed in the positions of the facets, which are actuated electrically, giving a wide tuning range ($\approx$ 500 nm), but with a sluggish tuning time in the range of a few ms.

Acousto-optic tunable filters

Acousto-optic controllability of frequency selective optical devices leads to yet another kind of tunable optical filter, capable of realizing fast tunability (a few μs) over a large tuning range ($\simeq$ 250 nm). Through acoustic-wave excitation (using transducers) on planar waveguides, the medium refractive index is changed periodically through localized condensation and rarefication of the propagating medium. This leads to a creation of refractive index variation as in a Bragg grating, whose pitch can be changed by controlling the frequency of the acoustic signal. However, the acousto-optic tunable filters can not have a sharp transfer function, and thus are not recommended for DWDM systems.

Liquid crystal based FPIs

Liquid crystal based FPIs are another type of tunable filter, which can be tuned over a range of about 50 nm with a tuning time in the order of a few μs. In this type of tunable filters, the cavity is kept filled with liquid crystal, whose refractive index can be modulated by an electrical input, leading to the electro-optic control of the cavity length and hence the resonant wavelength. We discuss more on the properties of liquid crystals in the following section on optical switches.

2.11 Optical switches

Optical switching plays a major role in optical networks to realize the desired switching functionality in network nodes. An optical switch can have a variety of requirements, depending on its usage in the network. As discussed in Chapter 1, there are broadly two different switching regimes in today's networks – circuit and packet switching – and their design challenges are different in high-speed optical networks.

For example, in a circuit-switched optical network, a connection can stay for hours to months, and hence switching speed doesn't pose any technological challenge. However the switches at network nodes should ensure a *nonblocking configuration* so that high-speed connection requests do not face internal blocking within a switch. However, an optical switch to be used in packet-switched networks must perform fast switching, as the packet sizes shrink considerably at high transmission speeds. Considering the widespread deployment of Ethernet across the globe, let's take the case of the smallest Ethernet packet size, which happens to be 60 bytes, i.e. 480 bits. With a transmission speed of 10 Gbps, the size of the packet goes down to $480 \times \frac{1}{10}$ ns = 48 ns. Hence, an optical packet-switching device should be typically capable of switching packets with a switching time not exceeding ~ 5 ns (i.e., one order lower than 48 ns).

In practice, an optical switch in a network node is developed using a large number of small switching units as building blocks, and the speed of these switching units eventually determines the switching/reconfiguration time of the nodes. A switching unit with a switching time in the order of a few ms is going to be well accepted for optical circuit-switched networks, but would perform miserably for the optical packet-switched networks. Moreover, with the switching units having super-fast switching speed, one needs to check whether the other desirable performance specifications of a switch – acceptable switch crosstalk, insertion loss, internal blocking probability, cost, etc. – are being met. In the following, we describe some of the important switching technologies used for the basic optical switching units, followed by sections on a number of candidate switch architectures using these switching units as the building blocks.

2.11.1 Switching technologies

There are a number of useful technologies for realizing the basic switching units as building blocks, including electro-optic, liquid-crystal, and thermo-optic devices. We describe in the following the operational features of these devices.

Electro-optic switch

The fastest building block in an $N \times N$ optical switch can be realized as a 2×2 electro-optic switch using Ti-diffused planar waveguides on a $LiNbO_3$ substrate. Switches, with the electrical control input, can switch between the *cross* and *bar* states with a fast switching time in the range of just a few tens of picoseconds. A typical 2×2 electro-optically controlled Ti-$LiNbO_3$ switch is illustrated in Fig. 2.52(a), wherein the control voltage can change the effective refractive index of the $LiNbO_3$ substrate, thereby changing the boundary conditions for the propagating modes in the waveguides.

In Fig. 2.52, for a control voltage (typically, in the range of 0–15 volts for the two switching states) that ensures the bar state, input lightwaves fed to the input ports 1 and 2 propagate along the respective waveguides and reach the output ports 1 and 2, respectively. However, for a control voltage ensuring cross state, the refractive index changes in the substrate in a manner such that the lightwave propagating in one waveguide leaks out and migrates to the other waveguide, thereby moving the incident lightwave at the input port 1 to the output port 2. Concurrently by the symmetry of the geometry, the lightwave fed to the input 2 also reaches the output port 1. However, the migration of the lightwaves from one waveguide to the other depends on the length of the waveguide, which is wavelength-specific. Further, if the length is increased, the lightwaves would get back to the previous waveguides, leading to a periodic migration process. Thus the switch length must be cut to proper size, governed by the wavelength that it needs to handle.

Despite being the fastest switching device (switching time $\simeq 10$ ps), electro-optic switches suffer from large crosstalk as shown in Fig. 2.52(b) by the dashed lines. Typically the crosstalk remains around 30 dB down from the desired signal, which can accumulate

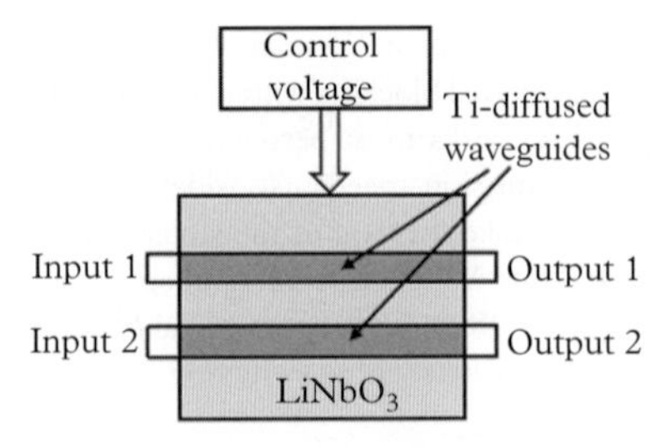

(a) 2×2 Ti-LiNbO$_3$ switch schematic.

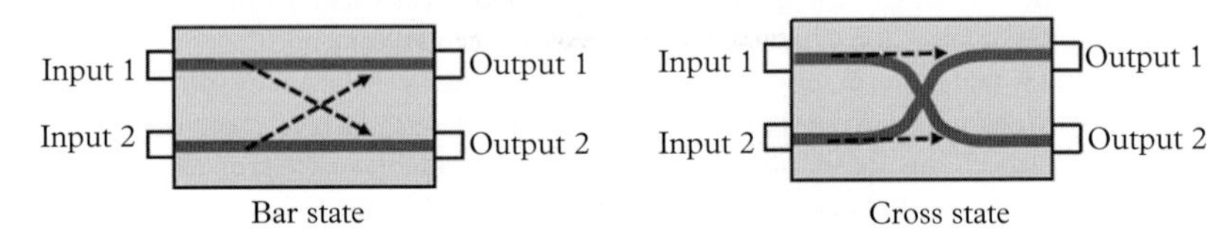

(b) Switching states of the 2×2 Ti-LiNbO$_3$ switch, with dashed arrows representing the crosstalk paths.

Figure 2.52 *Basic 2 × 2 electro-optic switch.*

significantly when a lightpath passes through multiple switching nodes. Further, along with the insertion loss on the higher side (typically, 6–8 dB), these switches, due to the planar-waveguide-based fabrication, suffer from polarization dependent loss (PDL) ($\simeq$ 1 dB) as the input lightwaves can have random SOPs, in contrast with the planar waveguides supporting only the electric field polarizations normal to the surface of the substrate.

Liquid crystal switch

Another means of realizing a basic switching unit is to employ a liquid-crystal cell as the switching device. Liquid crystals represent a specific state of matter, typically observed in certain organic compounds, e.g., hexyl-cyanobiphenyl compound and a few others. In particular, the liquid crystal state of such compounds lies between the crystalline solid state and the state of an amorphous liquid. In the liquid crystal state the constituent molecules, typically having shapes of oval rods or flat discs, remain aligned along a specific direction, which can be changed by applying voltage across the material (Fig. 2.53(a)). As the orientation changes with the applied voltage, the material permits/transforms (i.e., re-orients) the input lightwave field to exit with a *specific polarization* aligned with the prevalent molecular orientation. This specific property of the liquid crystal compound is harnessed in realizing the liquid-crystal-based optical switching units. In other words, in a liquid crystal cell, by using a control voltage, one can change the orientation of the suspended liquid molecules in its solvent, thereby allowing one specific polarization to pass through the liquid crystal compound.

At a given input port of a liquid crystal switch (Fig. 2.53(b)), the incoming lightwave at input port from a fiber (with its inherent birefringence) is split into horizontal (H) and vertical (V) polarizations by the polarization beam splitter and fed into the two parallel liquid crystal cells (in the two arms of a 1 × 2 switching unit), which are controlled by the respective control voltages. The polarizations of the two polarized lights (V and H) passing through the two liquid crystal arms may or may not be changed to the other polarizations (H and V), i.e., by 90°, depending on the control voltage. Thereafter, the outputs from the liquid crystal

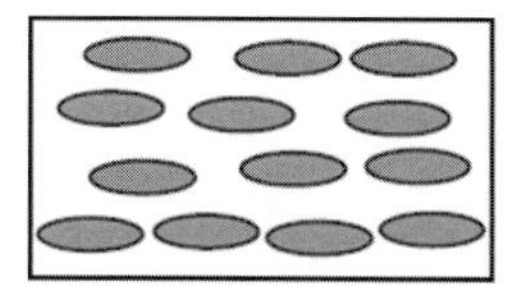

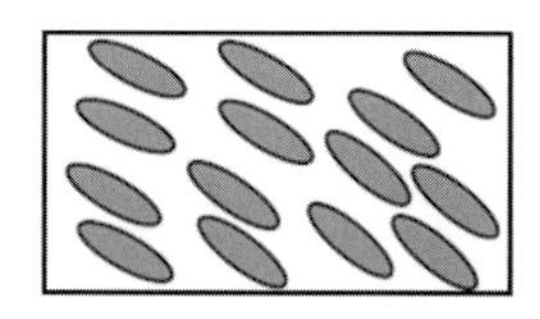

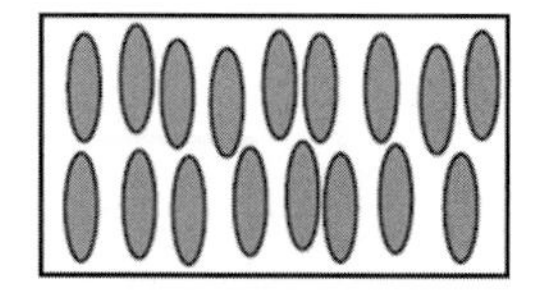

(a) Different possible orientations of liquid-crystal molecules (oval-shaped rods).

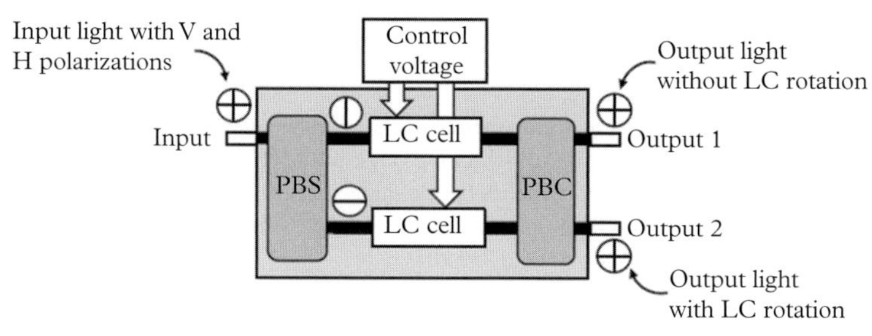

(b) Liquid-crystal-based 1 × 2 optical switching unit. Output signal appears at output 1 or 2 depending on the rotations undergone in the LC cells.

Figure 2.53 *Basic 1 × 2 liquid crystal switch. LC: liquid crystal, PBS: polarization beam splitter, PBC: polarization beam combiner.*

arms are combined together in a polarization beam combiner. The combined output appears at one of the two output ports, depending on how their polarizations were rotated by the liquid crystal cells in the two parallel arms.

The liquid crystal switches are inherently polarization insensitive, as it splits any given input polarization from the fiber into V and H polarizations and treats both polarizations as per the switching need, thereby resulting in a very small PDL ($\simeq$ 0.1 dB). Along with this feature, the reconfiguration time of these switches has of late gone much below the previously reported milliseconds mark to around 30–50 ns (Geis *et al.* 2010), while crosstalk performance is moderate ($\sim$ 30–35 dB). However, these switches cannot be used for the nodes with large port counts owing to the hardware complexity for large-scale integration.

Thermo-optic switch

Another option to realize basic optical switching units is to use thermo-optic control on the switching device, laid down following the same configuration as in Fig. 2.52(a), but with a substrate material whose refractive index can be changed by varying the temperature. In particular, by using thermal control, the refractive index of the substrate is changed, so that the lightwave gets switched to the other arm for realizing the cross-state of the switch. However, this switching device exhibits sluggish switching in the range of a few milliseconds with a moderate crosstalk performance ($\sim$ 35–40 dB), and is not scalable enough to suit the nodes with large port count.

MEMS-based switch

For large optical switches, micro-electro mechanical system (MEMS) offers a very effective technology with mechanically actuated silicon mirrors (Kim *et al.* 2003). A typical silicon mirror used in MEMS-based switches is shown in Fig. 2.54. A MEMS-based switching unit is basically a tiny rotatable mirror, dealing with only one lightwave at a time without having any crossbar functionality. As shown in the figure, the tiny reflecting element can

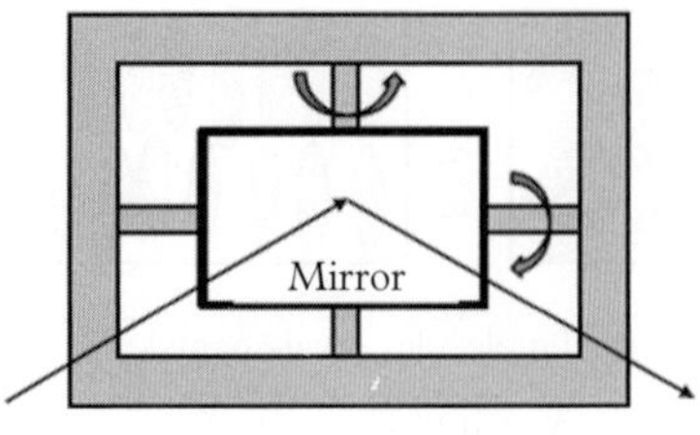

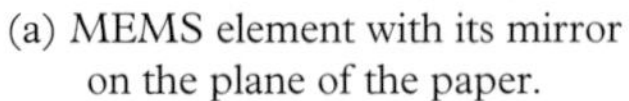
(a) MEMS element with its mirror on the plane of the paper.

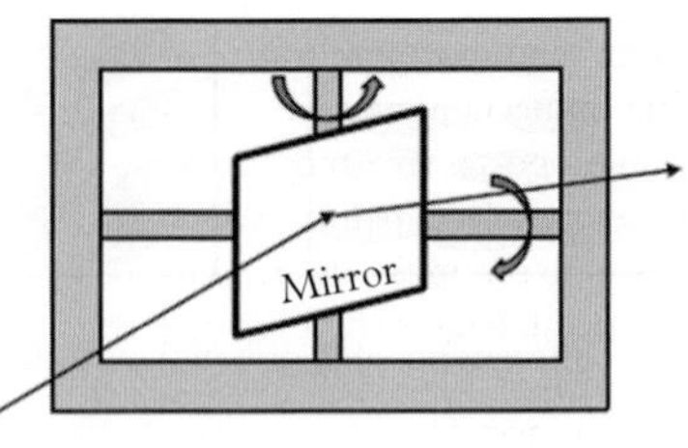

(b) MEMS element with its mirror turned right around the vertical axis.

Figure 2.54 *Basic MEMS switch with two possible orientations (parts (a) and (b))* switching *the incident light ray along two different directions.*

be rotated around both horizontal and vertical axes, thereby offering a two-dimensional steering power in deflecting (and hence switching) an input lightwave to the desired output direction. However, MEMS-based switches have a sluggish reconfiguration time ($\sim$ 10 ms), as compared to the other switching options, while offering the best crosstalk performance ($\sim$ 55–60 dB) with very low PDL ($\sim$ 0.3–0.5 dB).

2.11.2 Switch architectures

Optical switches for network nodes can be built with various possible architectures by using one of the above switching units as the building block. We describe in the following some of these switch architectures that can be used in the nodes of optical networks.

Switch architectures are broadly categorized into two types: blocking and nonblocking architectures. A switch is said to be of blocking type, when an input signal at a given input port can not reach an output port, even if that output port is not being used for any connection. Such incidents of blocking take place when the input ports do not have an adequate alternative path to reach the output ports within the switch. Note that a connection request across a network may also get blocked owing to the unavailability of a route through the given network. However, this instance of blocking doesn't consider the possibility of blocking within a switch hosted by a network node. To preclude the network blocking scenario while discussing the performance of a switch, the blocking encountered within a switch due to inadequate *internal* connectivity is also referred to as *internal blocking*; however, for the sake of simplicity we will continue to refer to the internal blocking simply as blocking in the present context. In reality, the probability of blocking within a switch would be much lower than the across-the-network blocking. Furthermore, in high-speed optical backbones with a huge investment, no one would accept a connection blocking for a 10 Gbps connection request due to a switch having inadequate internal connectivity, thereby implying that our objective in the present context would be to seek switch configurations that would be of non-blocking nature. In view of the above, we consider henceforth the candidate switching architectures which are nonblocking in nature.

2.11.3 Nonblocking switch architectures

Nonblocking architectures of switches can be categorized into various types: wide-sense nonblocking, strictly nonblocking and rearrangeably nonblocking architectures. In the wide-sense nonblocking architecture, a connection between a given input and an unused output port can be made without re-routing the existing connections, which in turn needs a specific routing mechanism so that no future connection request for a free output port gets blocked. In the strictly nonblocking architectures, if an output is free, it would be reachable from any given input port requiring a new connection to that output port, regardless of how the existing connections were set up. However, for the rearrangeably nonblocking architectures, if an output port is free and if an input port looking for a connection to that output port can not find an available path, it would be possible to rearrange the existing connections to find out a path to set up the desired connection.

In general, large switch architectures can employ single-stage or multi-stage switching scheme. One of the classical switch architectures used earlier in the telephone networks is the *crossbar* switch employing single-stage switching, which offers a strictly nonblocking architecture. Further, the Clos and Spanke architectures also offer strictly nonblocking switching functionality, but with multi-stage switching schemes. As we shall see soon, multi-stage switching has some merits over the single-stage switching, particularly in respect of the hardware complexity. However, the Beneš architecture, again a multi-stage architecture, falls in the rearrangeably nonblocking category. A specific combination of the Spanke and Beneš architectures leads to wide-sense nonblocking connectivity, to as Spanke-Beneš architecture. In the following, we discuss some of the strictly nonblocking switch architectures.

Clos architecture

As mentioned above, Clos architecture uses multi-stage switching configuration. Multi-stage configurations offer better scalability as compared to their single-stage counterpart, i.e., crossbar switches. In the crossbar architecture the number N_{cp} of crosspoints (i.e., the basic 2×2 switching devices) increases as $N_{cp} = N^2$ (Fig. 2.55), while for its multi-stage counterparts N_{cp} scales up at a slower rate, e.g., $N_{cp} \simeq 4N\sqrt{2N}$ for Clos architecture, which we explain in the following. First, we consider some example multistage architectures, and show under what condition such architectures can be strictly nonblocking, and then describe the Clos architecture.

Consider a 9×9 switch with three stages, as shown in Fig. 2.56, and assume that a new connection is to be set up between the input port 9 and the output port 5, given that there are already three ongoing connections between the pairs of input-output ports 5–6, 7–2, and 8–9. It is evident that, due to the existing three connections, the new connection cannot be set up because the switching blocks 1, 2, and 3 in the central stage are the candidate *en route* switching blocks, which cannot be used without disrupting one of the existing three connections, even though port 5 at the third stage is free. Thus, the three-stage switch fails

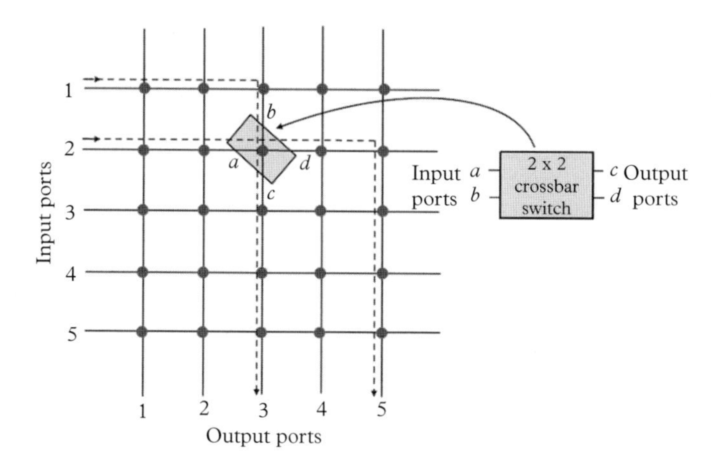

Figure 2.55 *A 5×5 crossbar switch with $5^2 = 25$ crosspoints. Every crosspoint represents a 2×2 crossbar switch as shown on the right side.*

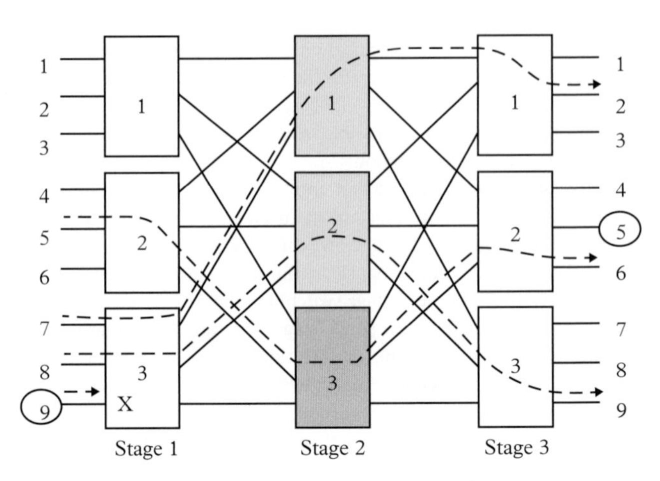

Figure 2.56 *An example multistage 9 × 9 switch exhibiting connection blocking. Note that the connection request from input port 9 to output port 5 cannot be set up (i.e., it is blocked), as the* en route *switching blocks for this connection (i.e., blocks 1, 2, and 3 in the central stage) cannot be used due to the existing connections (dashed lines).*

to become nonblocking, mainly due to the paucity of switching blocks in the second stage. The logical step forward is to dilate the second stage vertically, i.e., to have more switching blocks than those in the first or the third stage.

In view of the above, the problem of blocking is addressed by increasing the number of switching blocks in the second stage as shown in Fig. 2.57(a), where N input ports are divided into r groups, with each group having $n = N/r$ input ports. Each group of n ports is fed into an $n \times k$ switching block (which can be realized using the switching technologies described earlier), implying thereby that the first stage would consist of r switching blocks, each with n input ports and k output ports. The second stage has k switching blocks, each switch block having a symmetric $r \times r$ configuration, with $k > r$, implying a dilation in the central stage (second stage). The final or the third stage is configured as a mirror image of the first stage, thus having again r number of $k \times n$ switching blocks.

With the above switch configuration, first we consider a specific multi-stage switch architecture with the number of switching blocks in the central stage as $k = 2n - 1$, wherein the number of switching blocks in the first/third stage is assumed to be $r = N/n < k$. In this switch, consider one of the $n \times (2n - 1)$ switching blocks at the first stage, say the switching block i, as shown in Fig. 2.57(b), where one of the n inputs needs to set up a connection to one of the n output ports of the switching block j in the third stage. This $(2n - 1) \times n$ switching block in the third stage (i.e., the switching block j) can support n outgoing connections, out of which one should be free due to the above connection request between the switching blocks i and j. Thus, in the worst case, from the switching block i in the first stage the remaining $n - 1$ input ports might want to reach the third stage for as many connections, each through a separate switching block in the second (i.e., central) stage to ensure that these connections are not blocked by any other connections. At the same time, the block j in the third stage, through its remaining $n - 1$ output ports, might have to support $n - 1$ connection requests from another set of $n - 1$ switching blocks in the central stage, again to avoid blocking due to the other possible connections. This amounts to having at least $k = 2(n - 1) + 1 = 2n - 1$ switching blocks in the central stage to set up a strictly nonblocking architecture. Hence, the multi-stage switch architecture in part (b) of Fig. 2.57 having exactly $k = 2n - 1$, becomes strictly nonblocking.

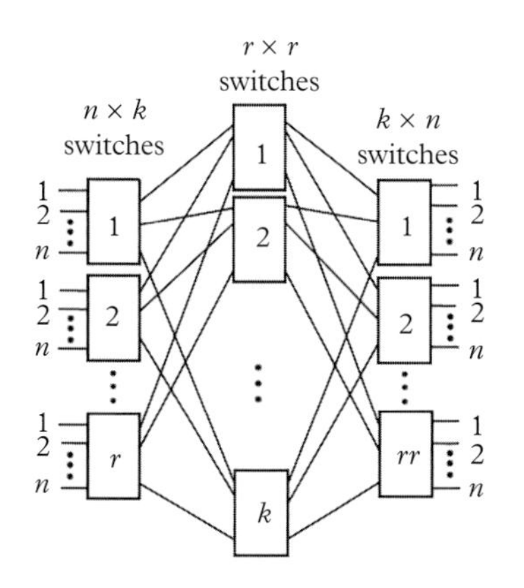

(a) Three-stage $N \times N$ switch architecture with dilated central stage; $N = nr$, $k > r$.

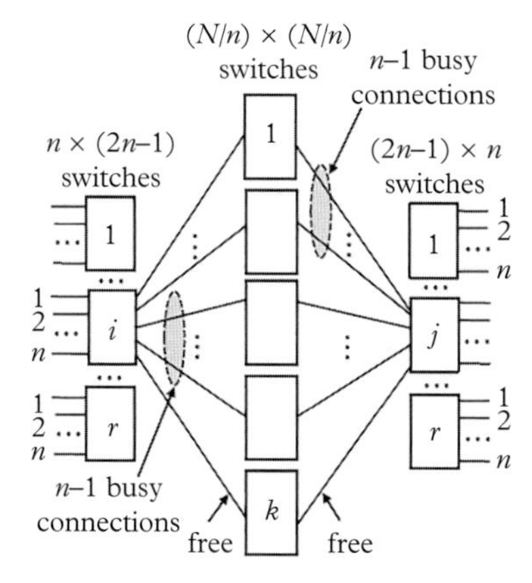

(b) Three-stage $N \times N$ nonblocking Clos architecture with dilated central stage; $N = nr$, $k = 2n–1$.

Figure 2.57 *$N \times N$ multistage switch architectures with the dilated central stage ($N = nr$): (a) generic configuration (b) illustration of the nonblocking condition for the Clos switch.*

This switch architecture was proposed by Clos in a pioneering work on strictly non-blocking switch architecture, which has thereafter been known as Clos architecture (Clos 1953).

Next we examine the hardware requirements of the Clos switch. Note that the number of crosspoints in each of the N/n switching blocks in the first, as well as in the third stage, is $n \times k = n \times (2n - 1)$. Similarly, the number of crosspoints in each of the $k = 2n - 1$ switching blocks in the second stage is $\frac{N}{n} \times \frac{N}{n}$. Hence, the number of crosspoints (i.e., the number of basic switching units) in the Clos architecture is the sum of the crosspoints in all the switching blocks in the three stages, given by

$$\begin{aligned} N_{cp} &= 2 \times \left(\frac{N}{n}\right) \times \left\{n \times (2n-1)\right\} + \\ &\quad (2n-1) \times \left(\frac{N}{n} \times \frac{N}{n}\right) \\ &= 2N(2n-1) + (2n-1)\left(\frac{N}{n}\right)^2 \\ &= \frac{4N^2 - 2Nr}{r} + 2Nr - r^2, \text{ with } r = N/n. \end{aligned} \tag{2.111}$$

Note that $r = N/n$ is a design variable and needs to be optimized to minimize the number of crosspoints. By differentiating N_{cp} with respect to r and equating $\partial N_{cp}/\partial r$ to zero, one can obtain the optimum value of r as $r_{opt} = \sqrt{2N}$, leading to the minimum number of crosspoints needed in a Clos architecture, given by

$$N_{cp} = 4N(\sqrt{2N} - 1) \simeq 4N\sqrt{2N}, \tag{2.112}$$

which, as mentioned earlier, indeed scales up at a slower rate as compared to N^2 in the crossbar switches.

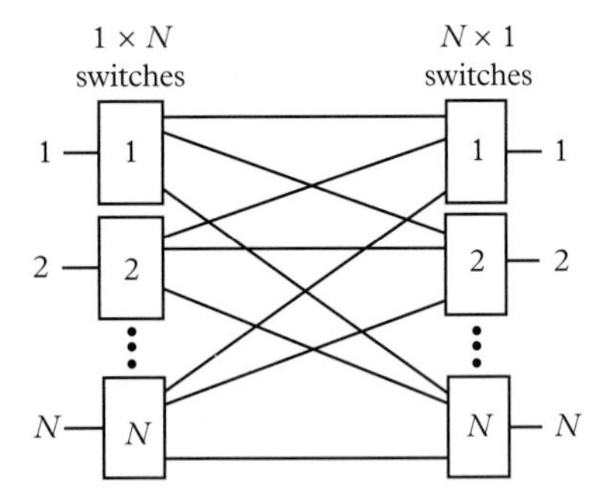

Figure 2.58 *N × N Spanke's switch architecture.*

Spanke's architecture

Spanke's architecture also offers a strictly nonblocking configuration, which is based on $1 \times N$ and $N \times 1$ switching blocks, as shown in Fig. 2.58. In particular, this architecture needs N number of $1 \times N$ switching blocks on the input side and N number of $N \times 1$ switching blocks on the output side, thereby needing $2N$ basic switching units. All of the switching blocks in this architecture can be realized by using 2×2 or 1×2 switching units as well as from the MEMS-based switches. As we shall see, a MEMS-based realization of Spanke's architecture presents a useful implementation scheme for optical switches with large port counts, which are much needed in WDM networks.

Switches Using MEMS in Spanke's architecture

The input and output stages of the Spanke's architecture can be realized using two two-dimensional arrays of MEMS-based switching elements, as shown in Fig. 2.59. The incoming lightwave on a specific wavelength is made incident on a specific rotatable MEMS-mirror in the input array. The MEMS-mirror in the input array is rotated appropriately using a control signal, so the lightwave incident on it gets forwarded to the desired MEMS-mirror in the output array, realizing the $1 \times N$ switching operation in the Spanke's switch. The MEMS-mirror in the output array is in turn rotated appropriately to reach the desired output port, realizing thereby the reverse functionality, i.e., the $N \times 1$ switching operation.

The MEMS-based switches have large reconfiguration time in the range of a few milliseconds (≤ 10 ms), and hence can be used in optical networks if the switching requirements are not critical in respect of reconfiguration time, particularly when dealing with circuit-switched connections across the network. Furthermore, the two-dimensional arrays with large numbers of MEMS mirrors are readily implementable with today's MEMS technology with high signal-to-crosstalk ratio (55–60 dB), thereby facilitating the design of large optical switches with port counts in the range of 250–1000. Furthermore, the MEMS-based switches are practically insensitive to the polarization of light, with the PDL remaining in the range of 0.3–0.5 dB, which is much lower than switches built on the planar integrated optics technology.

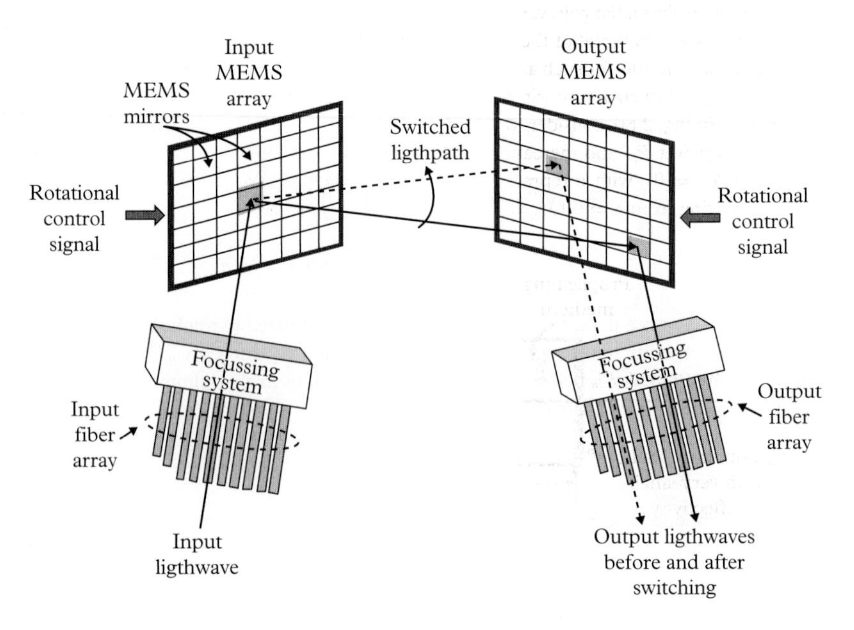

Figure 2.59 *Two-array MEMS-based switch architecture.*

2.12 Optical amplifiers

Optical amplifiers can be used for different purposes in optical networks, e.g., optical amplification of WDM signals along long optical fiber links, loss compensation in network nodes, optical gating/switching of optical signals, wavelength conversion, etc. Optical amplifiers can be of different types: semiconductor optical amplifiers (SOAs), doped-fiber amplifiers using rare earth elements as the doping material, or Raman amplifiers. While SOAs have flexibility in respect of the bandwidth of amplification, they are temperature sensitive and can get into nonlinear operation; this is a limitation in some cases, but useful in others. Doped fiber amplifiers, more frequently used with the doping of erbium ions in silica fibers and hence known as erbium-doped fiber amplifiers (EDFAs), are generally used as optical amplifiers in WDM optical links for loss compensation in the C band. While EDFAs offer simple implementation, the available bandwidth in this type of amplifier is limited by the atomic structure of the doping elements. However, EDFAs have also been realized in the C and L bands together by using some special techniques. Raman amplifiers utilize Raman effect and can amplify over larger bandwidth than EDFAs, but needing multiple pump sources (lasers) with large power. In the following we describe the operation of these optical amplifiers.

2.12.1 SOA

In electronics, amplifiers are typically the prelude towards implementing the oscillators, and the same is true with SOAs when compared with lasers. In particular, the stimulation (read *amplification*) in the active layer using pump currents along with the optical feedback from the reflecting facets leads to the lightwave generation from the lasers. In SOA, by using the end facets with very small reflections, the frequency-selective feedback doesn't get to take place, and hence a lightwave once injected into the active layer of the device, passes through the same and exits with an optical amplification. Note that, with higher reflectivity from the facets, the gain response of SOA would assume a ripple-like form (as in FPIs) with an envelope that follows the gain profile of the material. Thus, with antireflection coatings on the facets, one can realize a flat gain variation (i.e., without ripple) with the operating frequency owing to *single-pass* traversal of the lightwave through the active layer. In the following, we describe a generic model for such amplifiers, referred to as traveling-wave amplifier, by using a simple Fabry–Perot structure with *negligible* reflectivity.

As shown in Fig. 2.60, consider the Fabry–Perot structure of length L, where a lightwave enters with a power P_{in} and thereafter exits with a power P_{out} with a gain $G = P_{out}/P_{in} = P(z = L)/P(z = 0)$, with z representing the longitudinal coordinate of the Fabry–Perot structure varying from $z = 0$ to $z = L$. Assume that $P(z)$ represents the power incident at

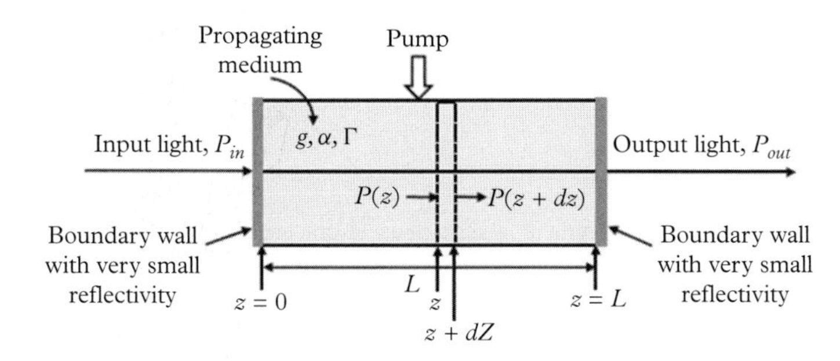

Figure 2.60 *A generic model of a traveling-wave (single-pass) optical amplifier.*

the input of an infinitesimal slice of the medium between z and $z + dz$, which comes out of the slice with an enhanced power denoted as $P(z) + dP(z)$. The net internal gain in the slice of the medium is given by $(g\Gamma - \alpha)dz$, with g as the intrinsic medium gain per unit length, α as the loss per unit length, and Γ as the confinement factor for the light in the medium. Assuming that the power growth $dP(z)$ in the given medium slice equals the *product* of the power $P(z)$ incident on the medium slice and the net internal gain therein $(= (g\Gamma - \alpha)dz)$, one can express $dP(z)$ as

$$dP(z) = P(z) \times (g\Gamma - \alpha)dz, \tag{2.113}$$

which is in effect a first-order linear differential equation, i.e., $\frac{dP(z)}{dz} = P(z)(g\Gamma - \alpha)$, with its solution given by

$$P(z) = P(0)\exp[(g\Gamma - \alpha)z]. \tag{2.114}$$

From this equation and representing the overall amplifier gain as $G = P(z = L)/P(z = 0)$, one can express G (and its value in dB) in terms of the basic device parameters as

$$G = \exp[(g\Gamma - \alpha)L] \text{ and } G_{dB} = 10\log G. \tag{2.115}$$

The above expressions for the amplifier (power) gain are generic in nature and we can use them for different kinds of traveling-wave amplifiers, with the respective values of the device parameters, i.e., g, Γ, α and L. In the case of SOAs, g would by and large depend on the forward (pump) current and the semiconductor medium, α on the loss characteristics of the semiconductor medium, Γ will be governed by the confinement factor obtainable from the heterostructure configuration, and L would represent the length of the active layer (i.e., medium) between the facets. However, for other types of traveling-wave amplifiers (e.g., EDFAs), for accurate analysis, one may have to modify the model to consider the effects that are not prevalent in SOAs, such as the pump deletion in doped fibers during the course of propagation from the input to the output end. Nevertheless, this model gives an insight into the gain phenomenon in all kinds of traveling-wave optical amplifiers.

As in lasers, the SOAs also exhibit high temperature sensitivity and need appropriate temperature control. Further, when amplifying the WDM signals, the forward bias current serves as the *shared pump* for all the WDM channels. Hence, when the signal in one WDM channel enters the SOA and gets amplified, the gain in other channel(s) gets reduced (as the same pump is shared as the common resource among all the channels) thereby exhibiting a crosstalk effect. This has a stronger effect in SOAs than in EDFAs, as the carrier life time ($\simeq$ 1 ms) in semiconductors is much shorter than that of the doped fiber amplifiers ($\simeq$ 11 ms in EDFAs). Furthermore, the active layer in SOAs being built in the form of planar waveguide, light input from the fiber with circular cross-section (and hence birefringent) undergoes PDL reducing thereby the SOA gain. In practice, SOAs find more useful applications inside the WDM nodes, which we shall discuss later in this chapter.

2.12.2 EDFA

By doping erbium ions in a segment of silica fiber, several energy bands are created within the propagating medium, where the higher energy levels populated by the charge carriers are pushed upward from the ground state by using a pump laser. The discrete energy levels of the erbium ions are spread due to a phenomenon, called Stark splitting, which facilitates the amplification of input optical signals over a band of wavelengths, determined by the

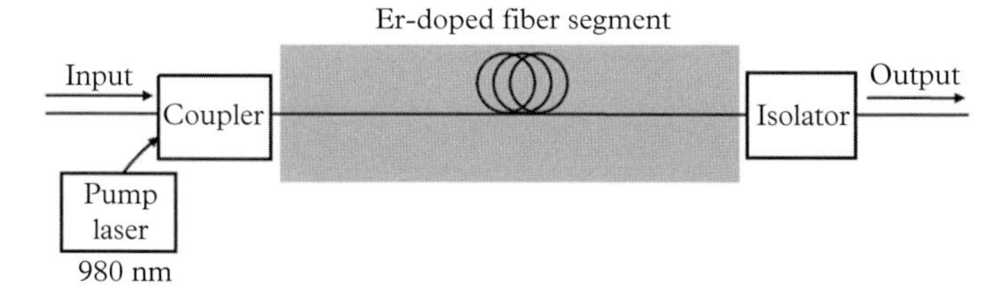

Figure 2.61 *Basic EDFA schematic.*

above energy spread. When an input lightwave passes through the doped fiber segment, the charge carriers already pumped up to the higher energy levels get stimulated by the traveling lightwave and return to the ground level, releasing energy in the form of photons at the same frequency, thereby offering an amplification of the incident lightwave.

A schematic diagram of an EDFA is shown in Fig. 2.61, where a segment of erbium-doped fiber (typically, a few tens of meters in length) is connected to the output fiber through an isolator to prevent the reflected light from gaining entry into the doped fiber. The pump power comes in from a laser diode, which is connected through a coupler to the input fiber. The input fiber is thereafter connected to the erbium-doped fiber segment, functioning as a gain medium for the input lightwave. Note that, in EDFAs the problem of PDL doesn't arise (unlike SOAs) as this amplifier is realized using the doped fiber with circular cross-section, which is also connected to the circular fibers on the input/output ends, thereby making the setup oblivious to the problem of polarization mismatch. In regard to Eq. 2.115, the values of g, Γ, and α for EDFAs are determined by the power injected by the pump laser, RIP of the fiber, and the loss in the erbium-doped fiber segment, respectively.

Though EDFA offers an extremely useful form of optical amplifier for loss compensation in long-haul optical links, it has a limited operational bandwidth ($\simeq$ 35 nm). To get an insight into this aspect, we need to look at the energy-level diagram of EDFA. Note that, erbium is a rare-earth element with large-sized atoms, that are difficult to dope in silica. With the help of alumina as a co-dopant, ionized erbium atoms (Er^{3+}) are generally doped to get adequate concentration. The ionized erbium atoms have several energy levels, as shown in Fig. 2.62, and the transition dynamics of the charge carriers between these energy levels through the absorption and stimulation processes sets up the process of optical amplification. In particular, the pump power is absorbed and the carriers move to a higher state. Thereafter, these excited carriers fall back to the ground state with the emission of light (photons), which can be stimulated by the incoming lightwave, leading to optical amplification.

As shown in Fig. 2.62, there are several energy levels (and the energy bands around them) above the ground level E_0 up to the highest energy level E_M (corresponding to an equivalent wavelength of 450 nm). However, for an operational EDFA, the energy levels E_0, E_1, E_2, and E_3 play important roles. By using an appropriate laser as the pump, charge carriers are pushed upward to the energy bands around E_2 or E_3, wherefrom the carriers fall below to the energy levels around E_1. The basic difference between the level E_2 or E_3 and the level E_1 is that the carrier life times τ in the energy levels around E_2 and E_3 are relatively small ($\tau \simeq 1$ ms), while the energy levels around E_1 exhibit a much longer life time ($\tau \simeq 11$ ms) and hence are called *metastable states*. Moreover, the difference between the level E_1 and the ground level E_0, i.e., $E_1 - E_0$ (in eV) $= 1.24/w$ (w in μm), corresponds to the wavelength $w \simeq 1550$ nm, which is close to the center of the C band in optical fibers. As a result, the charge carriers pumped up to the higher levels (around E_2 or E_3, in this case) stay there for a short duration of about 1 ms, and subsequently fall back to the levels

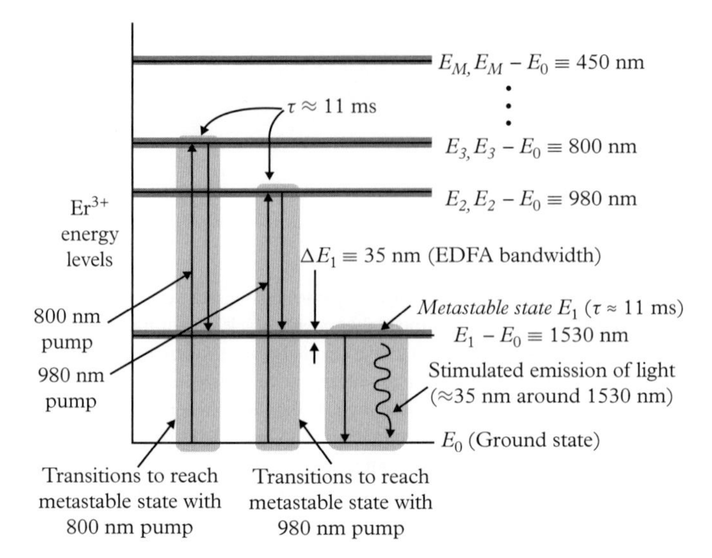

Figure 2.62 *Energy-level diagram of erbium ions in EDFA.* $E_i - E_0$ *(in eV) =* $1.24/w_i$*, with* w_i *in* μm.

around E_1, where they stay for a much higher lifetime of 11 ms, thereby offering themselves with a high likelihood of possible stimulated emission from the passing input lightwave.

Since the difference $E_1 - E_0$ corresponds to the wavelength $w = 1550$ nm, the emitted photons get stimulated by the incoming lightwave having the wavelengths around 1500 nm, thereby providing an optical amplification in the same band. Moreover, the spread of the energy level E_1, i.e., ΔE_1 ($\simeq 35$ nm), is conducive to the amplification of the wavelengths around 1550 nm with an amplifier bandwidth of $\simeq 35$ nm, i.e., over the C band in the third window. Typically, the pump laser is chosen for the energy level E_2, operating at 980 nm, though lasers operating at other wavelengths, such as, 800 nm (corresponding to E_3) and even 1480 nm (corresponding to the upper edge of the energy band around the energy level E_1) can also be used. The typical spectral variation of EDFA gain is shown in Fig. 2.63 for different pump powers. As evident from the plots, the lower end of the spectrum has a *hump* at $\simeq 1530$ nm, followed by flat region in the C band and a sharp fall in the S band, where the hump at the lower end of the gain spectrum is due to the higher energy-level density at the E_1 level of EDFAs at around 1530 nm. However, note that the gain spectrum above 1565 nm (i.e., in the L band) falls significantly (by about 4–5 dB), but remains mostly flat thereafter (not shown) within the L band. As we shall see later, this particular feature of the gain spectrum is important to extend the EDFA bandwidth beyond the C band.

EDFAs are generally used as optical amplifiers in three possible locations in optical networks. One such location immediately follows a node, compensating for the losses in the passive components in the node as a *post-amplifier*, before the lightwaves are launched into the outgoing fiber link. Another usage of EDFAs is to periodically compensate for the fiber losses in long-haul links, typically placed at intervals of 80 km, and hence known as *in-line amplifiers*. The last potential usage of EDFAs is found as a *preamplifier*, placed just in front of a receiving node, thereby restoring the received optical signal power to an adequate level before entering the node.

Some of the challenges while using EDFAs include amplifier gain saturation with increasing input power, nonuniform passband limited within 35 nm around 1550 nm (C band) and amplified spontaneous emission (ASE) noise. We discuss these aspects in the following with possible remedies.

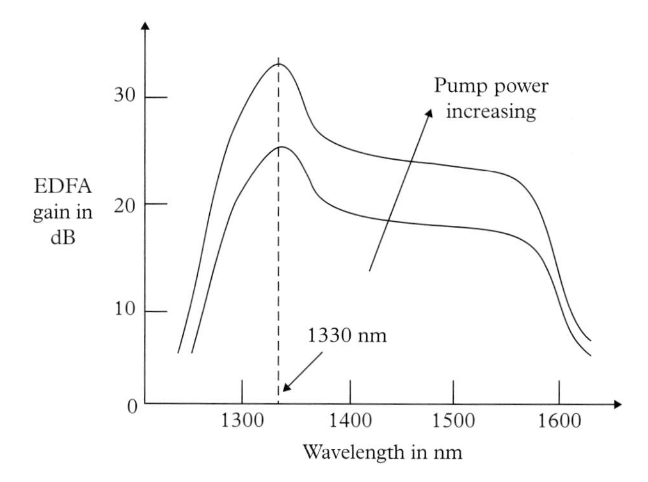

Figure 2.63 *Representative plots of EDFA gain versus wavelength.*

In an EDFA, the pump power serves as a fixed resource to provide amplification to the input power. With increased input power, the amplified output power from an EDFA tends to get saturated due to the limited available resource, thereby causing a reduced power gain. In order to have some insight into this phenomenon in EDFA, we examine below the gain saturation process in EDFAs, by using a simple analytical approach based on the model discussed earlier in the section of SOA.

First, we go back to Eq. 2.113 representing the basic gain process in the traveling-wave optical amplifiers, where the gain g was assumed to be a constant. As mentioned earlier, in the case of EDFAs this model needs some modification to include the effect of *pump depletion* that takes place along the direction of propagation in the doped fiber while contributing to the ongoing amplification process. However, to get an insight through a simpler model, we ignore this aspect in the present analysis, which would give a reasonable estimate of the gain, albeit on the higher side. With this observation, in order to capture the impact of gain saturation phenomenon in EDFAs, the overall gain $(g\Gamma - \alpha)$ in Eq. 2.113 is now *replaced* by a gain function $g(z)$ at the location z along the doped-fiber axis, given by

$$g(z) = \frac{g_0}{1 + P_z/Q}, \tag{2.116}$$

where g_0 represents the overall EDFA gain with negligibly small input power (and hence the unsaturated gain), and Q represents a parameter of the amplifier with the dimension of power. Note that the assumed model implies that higher values for Q would make the amplifier gain less sensitive to the input power variation. Using this model for the gain function, we modify Eq. 2.113 as

$$dP(z) = P(z)\frac{g_0}{1 + P_z/Q}dz, \tag{2.117}$$

thereby leading to the integral equation given by

$$\int_0^L g_0 dz = \int_{P_{in}}^{P_{out}} \frac{1 + P_z/Q}{P_z} dP(z). \tag{2.118}$$

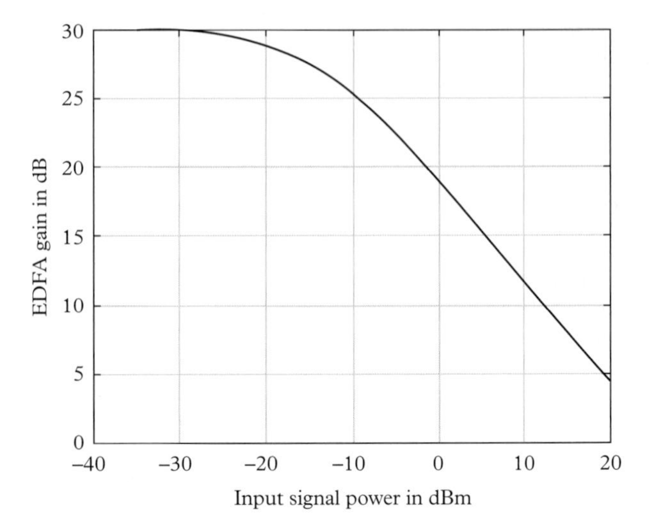

Figure 2.64 *Representative plot of EDFA gain versus input optical power, exhibiting gain saturation. G_0 = 30 dB, Q = 15 dBm.*

Solving the above equation, and using $g_0L = G_0$ as the overall amplifier gain with small input power, one can obtain the EDFA gain G_{sat} in the presence of gain saturation, as

$$G_{sat} = \frac{P_{out}}{P_{in}} = 1 + \frac{Q}{P_{in}} \ln \frac{G_0}{G_{sat}}, \tag{2.119}$$

which is a transcendental equation and needs to be solved by using an appropriate numerical method. Figure 2.64 presents a typical plot of EDFA gain versus input power, showing how the EDFA gain gradually goes down with the increase in input optical power.

A practical way to estimate the gain variation is to increase the input power, which would reduce the gain from G_0 to $G_{sat} = G_0/2$ (i.e., a 3 dB fall in gain), corresponding to an output power denoted as P_{out}^{sat} (say). Substituting this value of G_{sat} in the above equation, and replacing P_{in} by $P_{out}^{sat}/G_{sat} = 2P_{out}^{sat}/G_0$, we obtain from Eq. 2.119

$$G_0/2 = 1 + \frac{G_0 Q}{2P_{out}^{sat}} \ln 2. \tag{2.120}$$

Assuming that $G_0 \gg 1$, the above expression leads to the estimate of the saturating output power P_{out}^{sat} of the EDFA as

$$P_{out}^{sat} \simeq Q \ln 2, \tag{2.121}$$

implying that the parameter Q determines the saturating power level which would be typically in the range of 10–20 dBm in a practical EDFA.

In WDM transmission systems, when some lightpaths are added in or removed from a fiber link, EDFAs exhibit two kinds of gain fluctuations: steady-state and transient fluctuations. The steady-state gain fluctuations take place due to the variations in gain saturation, as implied in Eq. 2.119, and need to be controlled (typically, within 1–1.5 dB) by using some in-built feedback mechanisms in the EDFA design, such as through pump-power control or by using a co-propagating control lightpath on each fiber link with variable transmit power to dynamically balance the total power carried in the fiber (Shehadeh *et al.* 1996). Furthermore, any introduction or removal of connections can create *transient swings* of EDFA gains, and in one of the possible methods the gain transients can be controlled by adding some *dummy bits* at the both ends of a connection duration with *power tapering* (Chaitanya and Datta 2009). Such power tapers would in turn prevent the abrupt power changes at the EDFA input, thereby reducing the transient gain fluctuations. This method is, however, applicable for circuit-switched connections with long durations, and one needs to employ more proactive methods in the cases of packet or burst-switched networks.

As shown earlier in Fig. 2.63, the EDFA-gain spectrum exhibits a peak at the lower wavelength end ($\simeq$ 1532 nm) toward the S band. This peak can be flattened by including an optical filter in the doped fiber itself, which would selectively attenuate the specific band of wavelengths in the range of the spectral hump. Typically, an MDTF can be integrated with the doped fiber itself for this purpose, with long FBG being another alternative. One practical version of EDFAs makes use of two cascaded stages to optimize the performance in respect of gain and noise generation, where the first stage makes use of a forward pump at 980 nm and the second stage uses a backward pump at 1480 nm. The first stage has a lower gain and hence generates lower noise power, while the second stage offers larger gain, which helps in reducing the overall noise figure of the amplifier. A gain-flattening filter (GFF) is inserted in this configuration between the two stages (see Fig. 2.65) to obtain an acceptable passband with gain uniformity.

As mentioned earlier, on the longer wavelength end (i.e., in L band), EDFA gain falls significantly but remains mostly flat. This feature of the EDFA can be utilized to extend its bandwidth to cover the L band with appropriate engineering of the basic configuration, by using one of the two possible options. In one of these options, one can use two parallel traveling wave arms in the amplifier for C and L bands, where one arm uses the basic EDFA fiber for the C band, while the other arm uses a modified version of the EDFA for the L band. In particular, the arm used for the L band needs to use longer doped fiber segment to enhance the smaller intrinsic gain of EDFA along with a stronger pump power (see Fig. 2.66(a)). The input WDM signal is split into C and L bands using optical demultiplexer, and the two wavebands are passed through the two respective amplifying arms, following which the amplified signals of the two wavebands are combined in a multiplexer/combiner to obtain the amplified version of the input WDM signal. However, in this approach, it becomes hard to get seamless amplification of the input optical signal over the entire wavelength range from 1530 to 1630 nm (i.e., covering C and L bands) as the filtering devices used for separating the two wavebands in the demultiplexer create *spectral holes* between the two bands.

In the second option, one can make use of different fiber materials, typically tellurite fibers (i.e., TeO_2 used instead of SiO_2, as the hosting fiber material for erbium-doping), which leads to a seamless amplification of WDM signals over a wide range from 1530 nm to 1610 nm. A schematic diagram of EDFA using three stages of erbium-doped Tellurite

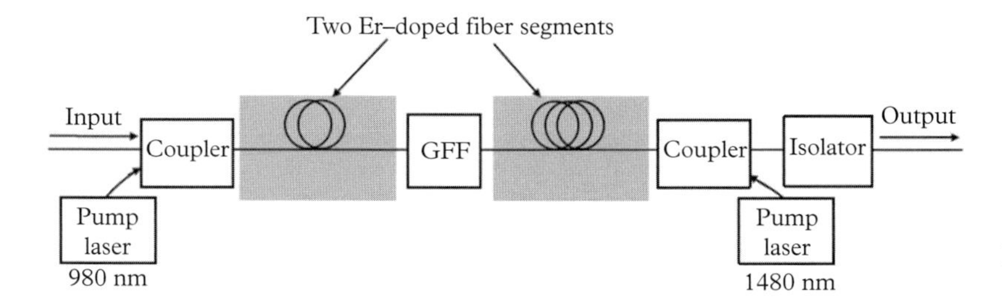

Figure 2.65 *Two-stage EDFA schematic. GFF: gain-flattening filter.*

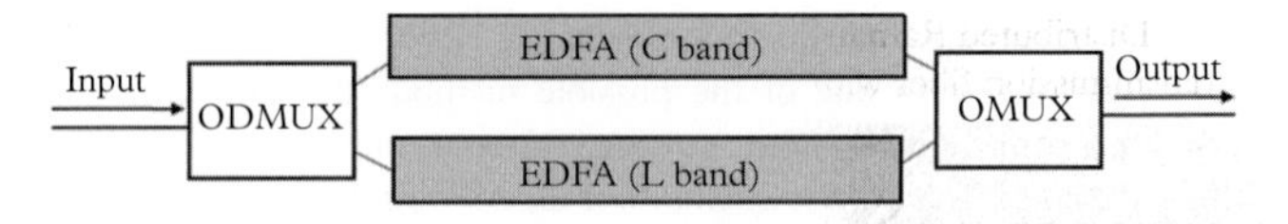

(a) Split-band configuration for wideband EDFA.

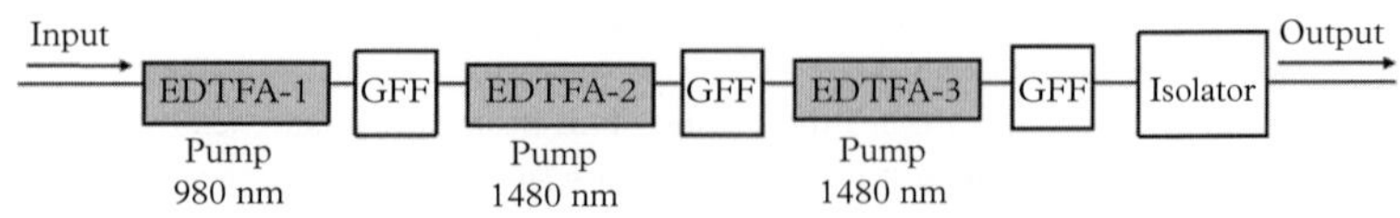

(b) Wideband EDFA using three stages of erbium-doped telurite fiber amplifiers (EDTFAs).

Figure 2.66 *Wideband EDFA configurations for amplification in C and L bands.*

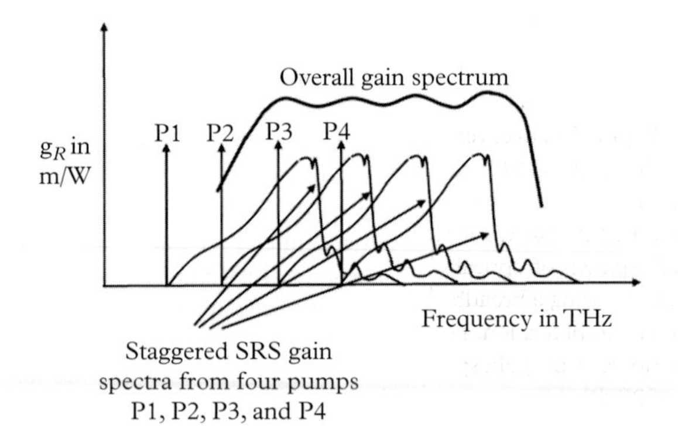

Figure 2.67 *Illustration of the gain-flattening scheme in Raman amplifiers using multiple pump lasers.*

fiber amplifiers (EDTFAs) is shown in Fig. 2.66(b), where three cascaded amplifying stages along with 980 nm pump (for the first stage) and 1480 nm pumps (for the second and third stages) are used to realize amplification ($\simeq$ 24 dB) over C and L bands seamlessly (Ono *et al.* 2002; Yamada *et al.* 1998). Following each stage, one GFF is inserted, using suitable optical filter (FBG, MDTF) to flatten the overall gain spectrum.

2.12.3 Raman amplifier

Operation of Raman amplifiers is based on the SRS phenomenon, as described earlier in this chapter. In contrast with EDFA, Raman amplifiers use normal optical fibers, where the silica absorbs the pump photons and releases photons that are Stokes-shifted to longer wavelengths (i.e., with lower frequencies and energies), with the remaining energy transformed into phonons. Raman gain can be realized either in a localized or distributed manner. In the localized or lumped realization, a spool of fiber (typically a few tens of meters) is used along with appropriate pump lasers. In distributed realization, the amplification of the propagating lightwave can take place over a few tens of kilometers, with the pump lasers feeding power into the propagating fiber, generally in the backward direction.

As shown earlier in Fig. 2.16, the gain spectrum of SRS is approximately triangular in nature, which needs to be flattened over the desired wavelength range. In order to have a flat (or nearly flat) gain over the desired wavelength band (usually C and L bands), one needs to employ multiple pump lasers at different wavelengths, leading to many staggered gain spectra, which overlap and combine together to offer an acceptable gain variation

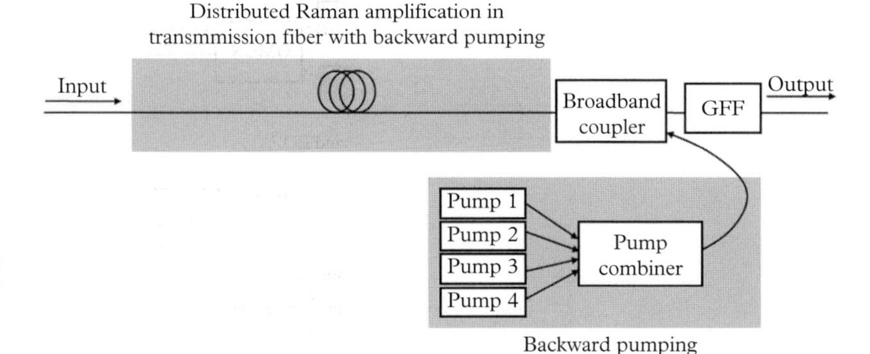

Figure 2.68 *Schematic block diagram of a Raman amplifier using multiple pumps.*

(see Fig. 2.67). In practice, to cover C and L bands with a distributed amplification scheme, pump lasers in the range of 1420–1500 nm are used as shown in Fig. 2.68, where a pump combiner is used to combine the output powers of four lasers in the stated wavelength range (typically, at 1425, 1445, 1465, and 1485 nm). The combined pump power is passed through a small segment of optical fiber, which in turn connects to the transmission fiber from the output end using a broadband optical coupler, leading to backward pumping with distributed Raman amplification. Following the broadband optical coupler, a GFF is used to clean up the ripple in the gain spectrum of the amplifier, thereby offering an overall gain in the vicinity of 20 dB.

2.13 Wavelength multiplexers/ demultiplexers

The devices for wavelength multiplexing/demultiplexing can be realized in a number of ways, with varying performance features. One of the early realizations of these devices was made using reflection grating supported by bulk optics (Ishio *et al.* 1984), as shown in Fig. 2.69(a). This configuration, known as Littrow configuration, employs a focusing arrangement (lenses) through which a multiplexed signal is received by a reflection grating. The grating diffracts the light in different directions with constructive interference along a specific angle for each wavelength, thereby offering an *angularly dispersive* mechanism to demultiplex the incoming wavelengths. The diffracted lightwaves pass through the same focusing lenses and are received by an array of output fibers, each fiber receiving one unique wavelength, and the same device can also operate as multiplexer in the reverse direction. Angularly dispersive wavelength demultiplexers can also be realized by using transmission gratings (known as Czerry-Turner configuration), though with bulkier optical hardware.

Wavelength demultiplexers in Littrow configuration can also be realized by using a graded index (GRIN) lens in the form of a cylindrical rod, as shown in Fig. 2.70, or by using planar waveguide structure with integrated optics. As shown in Fig. 2.69(b), in such configurations it is not possible to realize *flat-top* passbands, nor can one obtain *sharp* transitions between the passband and the stopbands on both sides for the adjacent channels. However, for each WDM channel, particularly in DWDM systems, these features are important for wavelength demultiplexers to capture the entire (maximum possible) bandwidth of the lightwaves transmitted from the lasers having spectral spread due to

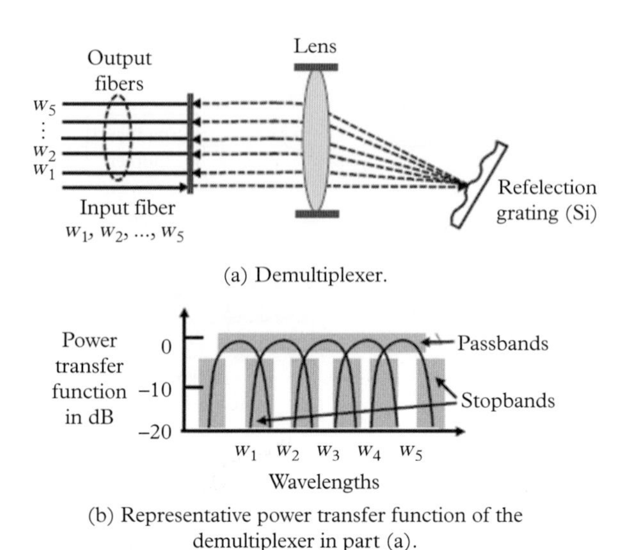

Figure 2.69 *Optical multiplexer/demultiplexer (shown in demultiplexing mode) using reflection grating in Littrow configuration. Note that the passbands are not flat enough and the stopbands lack sharpness, making this class of demultiplexers unsuitable for DWDM systems.*

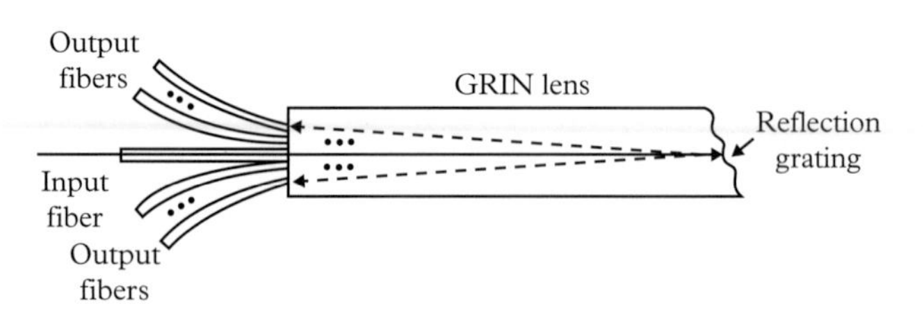

Figure 2.70 *Optical multiplexer/demultiplexer using a reflection grating with a GRIN rod lens in Littrow configuration.*

the modulation and phase noise, as well as unavoidable frequency offsets and drifts, while minimizing crosstalk contributions from the adjacent WDM channels.

Wavelength demultiplexers using optical filters based on MDTF can be realized to support the above features, which become crucial in DWDM systems. As discussed earlier, with a large number of thin-film layers and binary choice of refractive indices for the various layers, MDTFs can be used to synthesize the desired power transfer functions (with flat passbands and sharp transitions between the passband and stopbands). Hence, using a combination of such filters, one can implement wavelength demultiplexers for closely placed channels. Figure 2.71 illustrates a five-wavelength MDTF-based optical demultiplexer/multiplexer, where the MDTF filters for the respective wavelengths are placed on the two sides of a glass slab. In the demultiplexer mode of operation, the multiplexed wavelengths enter the device at the input port (port 1), which uses a GRIN lens for focusing the incoming light into the glass slab. Thereafter, the WDM input traverses the glass slab and falls on the first MDTF filter attached with a GRIN lens to pass one wavelength (w_1, say) to exit from port 2 and reflect back the rest of the wavelengths to the next MDTF/GRIN-lens combination at port 3. The wavelengths reflected from port 2 are extracted similarly, one by one, through the respective MDTF/GRIN lens combinations at the subsequent output ports. As mentioned before, the demultiplexer can also be used as a multiplexer in the reverse mode of operation.

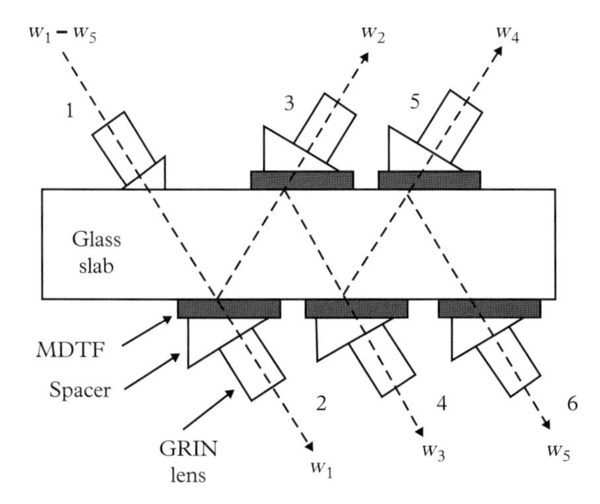

Figure 2.71 *Optical multiplexer/demultiplexer using MDTF filters.*

Furthermore, AWGs can be used in 1:N and N:1 configurations to support the operations of demultiplexer and multiplexer operations, respectively. Mach-Zehnder chains are also suitable as demultiplexing/multiplexing devices for small number of wavelengths.

2.14 Wavelength converters

While setting up an all-optical connection in a WDM network through multiple fiber links, one single wavelength may not be free on all the fiber links along the desired route between the source and destination nodes. As shown in Fig. 2.72, the connection request from node 1 to node 4 via nodes 2 and 3 finds the wavelength w_i free on link 1-2 and link 3-4, while on link 2-3 the wavelength w_j is free instead of w_i. In such cases, one can terminate the connection at node 2 and retransmit on the wavelength w_j, but with the necessary burden of OEO conversion (at nodes 2 and 3) along with the loss of the end-to-end all-optical transparency of the connection. Use of wavelength converters in such cases would convert w_i into w_j on the *fly* at node 2 and similarly convert w_j into w_i at node 3, thereby obviating the need for OEO conversions. Wavelength converters can be realized through various mechanisms, as discussed next (Yoo 1996; Durhuus *et al.* 1996).

One simple method to realize wavelength converters is to use an SOA with cross-gain modulation (CGM) as shown in Fig. 2.73, where the SOA is fed with the incoming lightwave pulses (carrying a binary data stream) on wavelength w_i along with another unmodulated lightwave on wavelength w_j with low power, usually referred to as *probe* wave. When the data pulses on w_i go high for binary ones, the SOA gets saturated and the gain falls to a low level, thereby passing on little power to the output port. However, when the data pulses go low, the lightwave on w_j come out at the SOA output with high power. Overall, this leads to the CGM functionality, where the data pulses of high-power lightwave on w_i control the gain for the lightwave on w_j, and thereby reproduce the data stream of w_i on w_j at the SOA output (albeit with an inversion). The output of the SOA is thereafter passed through an optical filter tuned to w_j, to remove the residual lightwave on w_i, if any. However, the modulated lightwave on w_j might suffer from incomplete extinction for binary zeros as the SOA may

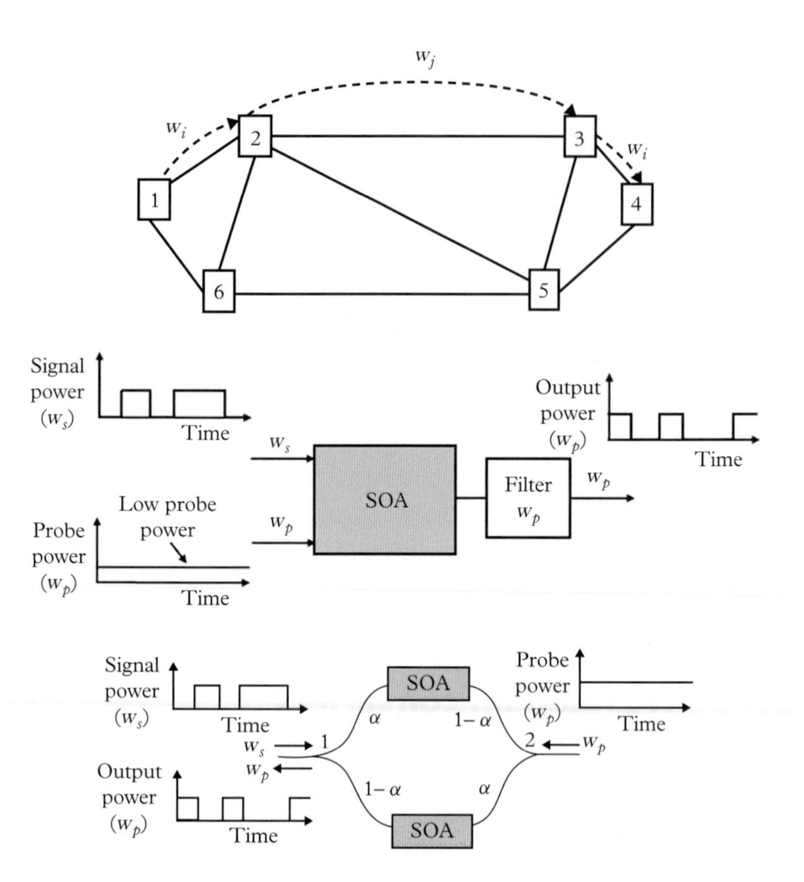

Figure 2.72 *Illustration of wavelength blocking for all-optical connections.*

Figure 2.73 *Wavelength conversion using CGM through an SOA.*

Figure 2.74 *Wavelength conversion using SOAs in MZI-based configuration.*

not get saturated enough during the binary one pulses of the input signal on w_i. Further, the carrier density modulation in SOAs might cause phase distortion in the probe output.

Another method for implementing wavelength converters is to use two SOAs in an MZI of two equally-long arms, as shown in Fig. 2.74, where the incoming lightwave on w_i carrying the data stream and the probe wave on w_p, coming from opposite directions are used to interact with each other through the MZI configuration hosting two SOAs. In particular, the active layers of the SOAs undergo *carrier-density modulation* owing to the modulated lightwave on w_i, which manifests itself as *cross-phase modulation* on the probe wave on w_p. As shown in Fig. 2.74, the incoming signal (i.e., modulated lightwave with binary ones and zeros) enters port 1, while the unmodulated probe lightwave is fed from port 2, and the power splitting ratios of MZI on both sides are designed to be asymmetric. Looking from port 1, the upper arm receives a fraction α (< 1) of the incoming power, while the lower arm receives the fraction $(1 - \alpha)$ of the same. On the other end, the splitting ratio gets reversed at port 2, i.e., at port 2 the splitting ratio is made $(1-\alpha) : \alpha$ for the upper/lower arms. As we shall see in the following, the asymmetry in the splitting ratio plays a major role in realizing the cross-phase modulation between the input and probe waves. Note that the SOAs in this configuration operate in a bidirectional manner.

In the absence of an incoming signal (i.e., during the reception of binary zero), the probe lightwave on w_p arrives at port 1 from port 2 through the two arms with amplification. When

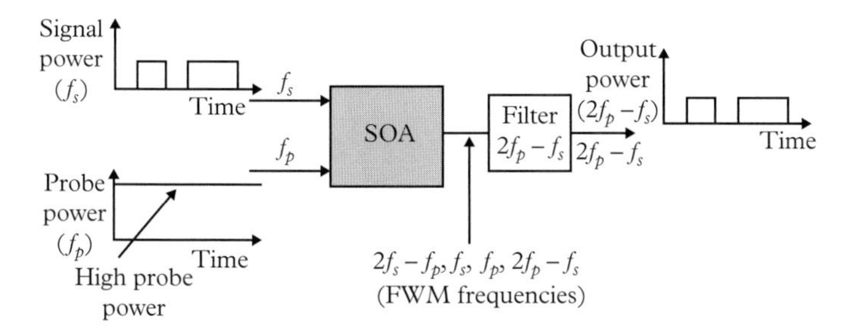

Figure 2.75 *Wavelength conversion using FWM in an SOA.*

the binary one pulse of the modulated lightwave on w_i enters port 1, it affects the carrier densities in the two SOAs in different magnitudes because the two SOAs in the two arms receive different amounts of optical power on w_i (α and $(1 - \alpha)$ times the input power). If the parameters of the setup are properly chosen, the differential phase shift between the two arms can be made an odd multiple of π, leading to a destructive interference of the probe waves on w_p coming from port 2 to port 1 via the two MZI arms, thereby producing a binary zero at port 1 for the probe wave. Thus, at port 1, by using a circulator, one would be able to receive the probe wave (coming from port 2) having zero power for a binary one in the received input lightwave, and a binary one otherwise, thereby offering a complementary modulation (i.e., inverted data stream) on w_j. However, one can also adjust the MZI parameters to receive the same data on w_p at port 1 as brought in by the input lightwave on w_i (i.e., without inversion). The major advantages of this configuration are that the setup does not need large power and the extinction of lightwave during zero transmission can be much better than what is realizable in the CGM-based implementation.

One can also realize wavelength converters using FWM in an SOA, as shown in Fig. 2.75, where a nonlinear interaction, i.e., FWM, takes place between the input and probe lightwaves. Probe power is kept high, and from the FWM several frequencies are generated, from which the one with the frequency $2f_p - f_s$ is filtered out carrying the data stream, as in the input signal on frequency f_s. Note that, in this case, we describe the phenomenon using frequencies instead of wavelengths as the FWM frequencies are linearly related to the input frequencies.

2.15 Wavelength-selective switches

Wavelength-selective switches (WSSs) offer a useful functionality for the nodes used in WDM networks. In its basic form, WSS offers $1 \times N$ demultiplexing of its WDM input, followed by a switching operation so that each demultiplexed signal can be switched to any desired output port of the device. The basic schematic diagram of a WSS is shown in Fig. 2.76, where the demultiplexing part operates in the way we discussed earlier for Littrow-configured demultiplexers using a reflection grating. Thereafter, the demultiplexed wavelengths are forwarded to a MEMS switching array instead of directing them to the output ports. The MEMS elements reflect each demultiplexed wavelength back to the reflection grating along a specific angle, and the grating in turn forwards each wavelength from the MEMS to a specific output port by reusing the Littrow lens system. Since the MEMS mirrors can be rotated, the demultiplexed wavelength incident on a MEMS mirror can be directed to any one of the output ports of the WSS, which is not possible in traditional demultiplexers as discussed earlier. This feature of WSS makes it useful in realizing various

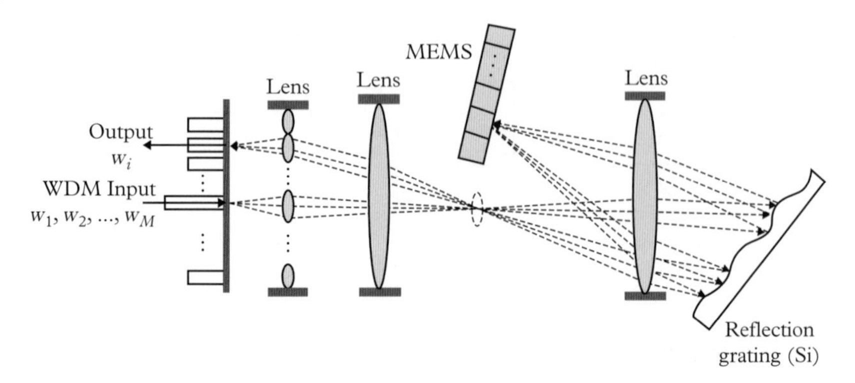

Figure 2.76 *Basic WSS schematic.*

types of WDM nodes, such as reconfigurable optical add-drop multiplexers (ROADMs) and optical crossconnects (OXCs).

Some compact realizations of WSSs are possible, such as by using an AWG along with a few switches and wavelength couplers used in loopback paths. In particular, an $N \times N$ AWG is used as a $1 \times k$ ($k < N$) demultiplexer, while using its spare input/output ports for feeding back lightwaves from some of the output ports to some of the input ports through the switches and wavelength couplers (Yoshida *et al.* 2014), such that the fed-back signals are subsequently routed to the desired output ports of AWG with another additional passage through the AWG. Through this feedback mechanism, an incoming wavelength can be switched to different possible output ports at different times (i.e., flexible switching, unlike the simple demultiplexers), as governed by the network need. However, this version of WSS, being realized using the planar-waveguide technology, will suffer from the polarization-dependent losses, preventing thereby its use for large WDM networks. A more versatile realization of WSSs, the bandwidth-variable WSS configuration, is presented in Chapter 14 on elastic optical networks.

2.16 Optical add-drop multiplexers

As mentioned earlier in this chapter (see Fig. 2.2), optical add-drop multiplexers (OADMs) are generally used in network nodes using WDM over a ring topology. Using OADMs in a WDM ring, one can drop some wavelengths and add the same wavelengths for onward transmission in a node, while the remaining wavelengths are passed on as transit or passthrough wavelengths. OADMs can be of two types: static OADM or simply OADM, and reconfigurable OADM (ROADM).

Figure 2.77 shows the schematic diagrams of the two types of OADM. In Fig. 2.77(a) an OADM (i.e., static OADM) uses optical demultiplexers (ODMUXs) and multiplexers (OMUXs), where the wavelengths w_3 and w_4 can be dropped and added, while the wavelengths w_1 and w_2 are passed through the OADM over the ring to which the OADM is connected. Figure 2.77(b) shows a ROADM, where any wavelength can be chosen as drop/add or as passthrough wavelength by using 2×2 optical switches, depending on the network needs.

OADMs can also be realized using FBG along with a circulator and a directional coupler for the drop and add operations, respectively. Further, ROADMs can employ MEMS switches, as shown in Fig. 2.78, where the incoming WDM signal in a node is first

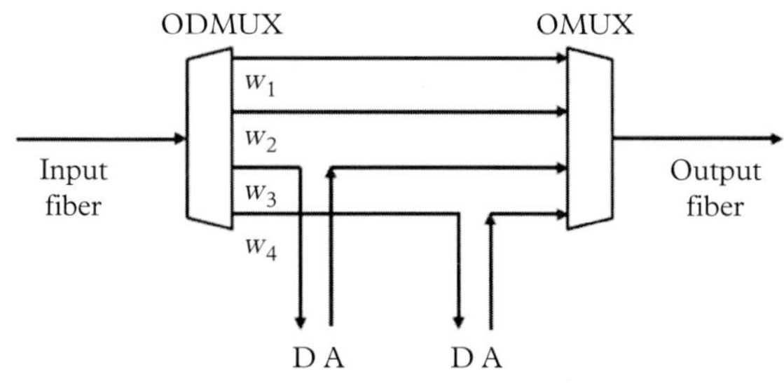

(a) Static OADM (or simply OADM) schematic.

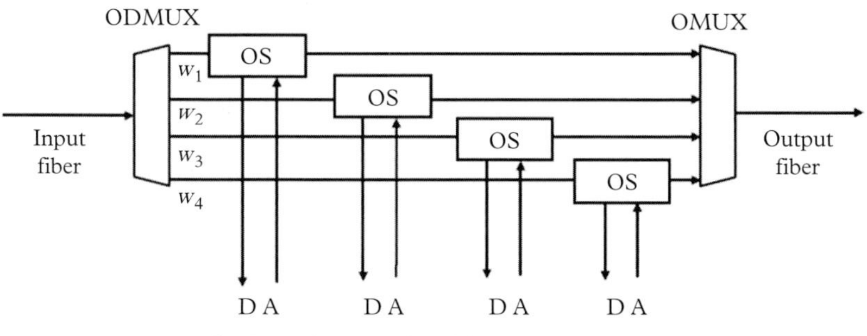

(b) Reconfigurable OADM (ROADM) schematic.

Figure 2.77 *Schematic diagrams of OADMs and ROADMs using optical multiplexers/demultiplexers along with 2 × 2 switches for reconfigurable functionality (in ROADMs). OS: optical switch, A: add port, D: drop port.*

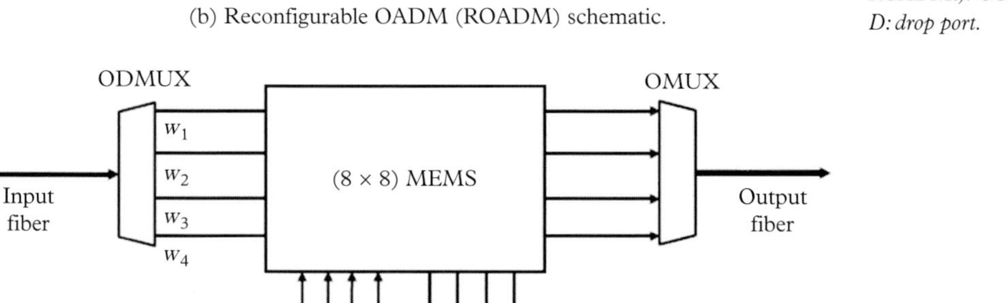

Figure 2.78 *Schematic diagram of a ROADM using MEMS.*

demultiplexed, and thereafter all of them are switched by an $M \times M$ MEMS switch for add-drop and passthrough functionalities, with M representing the number of wavelengths. The MEMS outputs are subsequently multiplexed into the output fiber for onward transmission. One can also use the other switching technologies for realizing ROADMs, as discussed earlier.

So far we have discussed OADMs/ROADMs for ring topologies, where the nodes have a degree of two only. However, as the network expands, the nodes on a ring may have degrees more than two in order to interconnect with other rings, leading to multi-degree ROADMs (MD-ROADMs).

MD-ROADMs use optical splitters in place of demultiplexers at the input end and WSSs at the output end, with at least four ports, typically called as *north, south, east,* and *west* ports. Use of optical splitters (OSPs) and WSSs makes the MD-ROADMs more versatile, offering

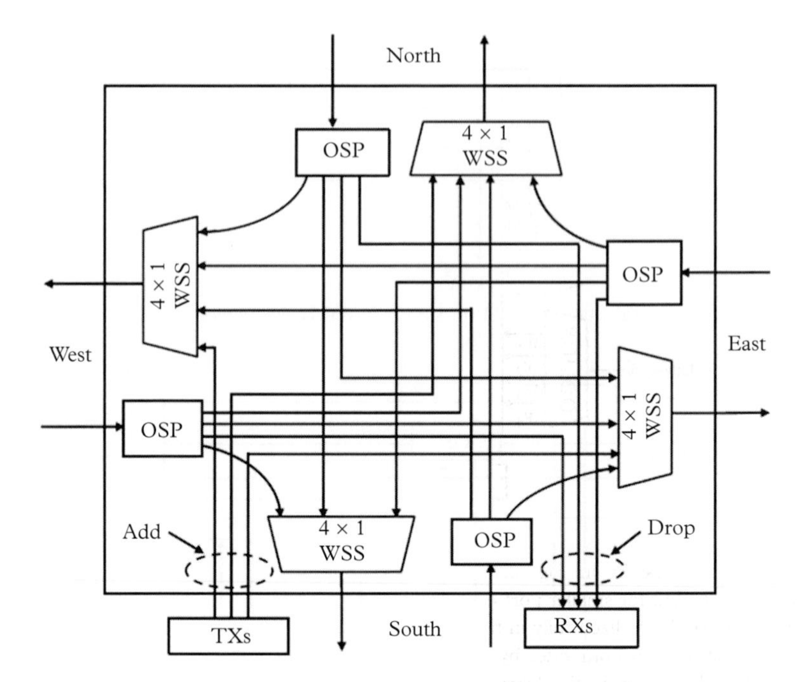

Figure 2.79 *Schematic diagram of a colorless-directionless MD-ROADM using OSPs and WSSs. Such MD-ROADMs or similar ROADMs (i.e., with D = 2) can also be made contentionless with minor modification, as discussed in the text.*

colorless and *directionless* features. Figure 2.79 illustrates one such colorless-directionless MD-ROADM with four ports. The OSP at each input port *broadcasts* all WDM signals to all other ports (making the ROADM directionless) while the WSS at each outgoing port can select any wavelength dynamically (making the ROADM colorless). In other words, by using OSP-WSS combination, from a WDM input signal at a given port, flexible allocation can be made for any incoming wavelength (color) at a given port to exit from any other port. However, there remains a possibility of contention between two signals of the same wavelength arriving from two different input ports at the same outgoing port (during the reconfiguration process), which can however be resolved by providing multiple add/drop options at the local port. As we shall see in Chapter 14, one can also use bandwidth-variable WSSs to add-drop variable-bandwidth signals for the WDM networks using flexible frequency grid.

2.17 Optical crossconnects

ROADMs are designed in general for the ring topology with a nodal degree of two, thereby needing to forward the transit as well as locally added signals to one single and hence a unique output port. However, in mesh topology the nodes need to have larger nodal degree, and with this feature the node needs to make a *choice* of forwarding the incoming signals from one port to one of the remaining ports (more than one), thereby bringing in the need of *crossconnect* functionality between each input port to one of the remaining multiple output ports. The optical devices that carry out this operation are known as an optical crossconnect (OXC)

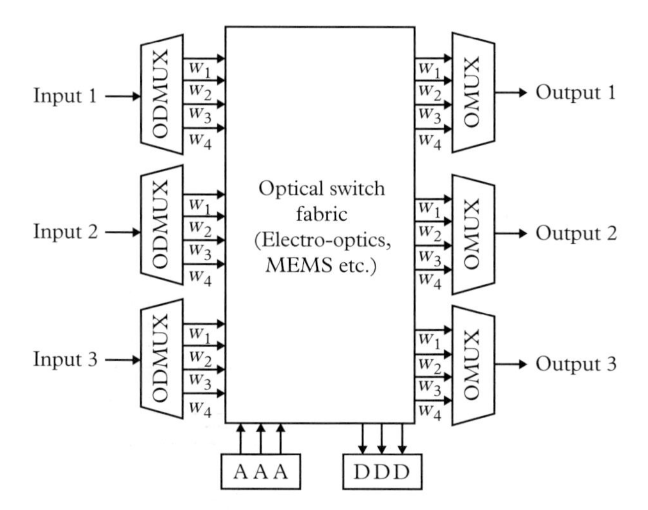

Figure 2.80 *Schematic diagram of an all-optical OXC.*

(as illustrated earlier in Fig. 2.2), although MD-ROADMs can also realize similar functionality, but for smaller port counts (discussed later in further detail). Crossconnect operation can be realized fully in the optical domain by using a suitable optical switching module flanked on both sides by optical demultiplexers and multiplexers, as shown in Fig. 2.80, where the switching fabric can employ one of the candidate switching technologies, e.g., electro-optic, MEMS, and others.

The OXCs using optical switching without any OEO conversion are called all-optical OXCs offering an end-to-end lightpath for a connection request. As mentioned earlier, one can also realize OXC functionality using the splitter-WSS combinations (as in MD-ROADMs), though for large number of ports the splitters would reduce the transmit power level for the passthrough lightpaths, thereby reducing their optical reach. However, in order to get around such problems, whether due to the use of MD-ROADMs or all-optical (and hence lossy) OXCs for the long-haul connections at high speed, one can employ optical regeneration using optical amplifiers leading to the first level of regeneration (1R) scheme, wherein the accumulated ASE noise from the optical amplifiers might degrade the quality of reception in respect of the receiver BER. On the other hand, the OXCs with OEO regeneration and electrical switching provide the opportunity to regenerate, reshape, and retime (3R) the passthrough signal with better physical-layer performance, but at the cost of considerable increase in the node hardware. Thus, the connections employing a 1R type of regeneration offer *all-optical transparency* across the network, though at the cost of higher BER in some cases, while the 3R-based OEO regeneration offers best receiver performance at the cost of increased hardware. There is also an intermediate level of regeneration, called 2R regeneration, which skips the retiming stage (i.e., bit synchronization using clock recovery), but doesn't extract the full benefit of costly OEO-based regeneration. In another type of OXC, all input wavelengths, after being optically demultiplexed, are OEO-converted into 1300 nm (and hence undergo 2R/3R regeneration) and forwarded locally within the OXC to an optical switch operating on one single wavelength (typically, 1300 nm). Having carried out the switching at 1300 nm, the switch output signals are again converted into the electrical domain and retransmitted on the desired wavelengths through the multiplexer stage. We discuss these OXC configurations in further detail in Chapter 9 for long-haul backbone WDM networks.

2.18 Optical fiber communication systems

Optical fiber communication systems deal at large with the transmission and reception of modulated lightwaves over different possible formations of physical-layer connectivity, such as simple point-to-point fiber links, wavelength-routed connections over multiple fiber links, and one-to-many broadcast transmissions, which are realized over a wide range of network topologies, such as simple linear segments, rings, meshes, stars, and trees. Indeed, the transmission of data streams using optical sources over optical fibers through intermediate nodes (if any) and reception of the transmitted signals at the optical receivers with proper signal quality is not a trivial task, as it involves the entire gamut of complex technologies, as discussed in the foregoing. In this section, we briefly describe the various possible means that can be adopted to make such communication systems work for point-to-point links (i.e., signal transmission between the transmitting and receiving nodes through optical fibers without any intermediate nodes or optical amplification), with a note that in a practical scenario, especially for WDM networks, the received signal will go through numerous types of transmission impairments (at intermediate nodes and optical amplifiers) which we deal with in Chapter 10. Furthermore, note that, in the following, we consider only *digital* optical fiber communication systems.

Digital optical fiber communication systems can be broadly categorized into two types: *non-coherent* and *coherent* systems. In non-coherent systems, an LED or a laser undergoes IM, where their forward bias current gets modulated proportionately by the digital baseband signal, typically a binary data stream. This in turn modulates the output light intensity (i.e., instantaneous optical power output and hence the number of photons per bit) of the source. The IM lightwave from the optical source travels through the optical fiber network and is finally photodetected in an optical receiver at the destination node through a *pin* diode or APD. The photodiode (*pin* or APD) produces a photocurrent in proportion to the received light intensity through the photodetection process implying DD-based reception of photons, as each photon generates ideally one hole-electron pair thereby leading to the photocurrent. The entire process of transmission and reception of the IM-DD system nowhere pays any attention to the coherence of the lightwave – neither during modulation, nor during detection – and hence such IM-DD systems are called *non-coherent optical communication systems*. For a long time, IM-DD systems have been the mainstay, satisfying mostly the expected physical-layer performance criteria.

However, non-coherent systems can not attain the best performance that an optical communication system can offer, particularly in respect of the received power required for a given BER specification. The minimum power required to attain a given BER is known as *receiver sensitivity* which can be improved (i.e., reduced) significantly by using a *coherent optical communication system*. In coherent systems, as discussed in Section 2.8.4 on laser modulation schemes, one can modulate phase, frequency, or amplitude, or a combination of them by using appropriate photonic devices, e.g., MZI-based external modulators and others. At the destination end, the receiver needs to employ an appropriate scheme to extract the phase, frequency, or amplitude, or their combinations, in the received lightwave.

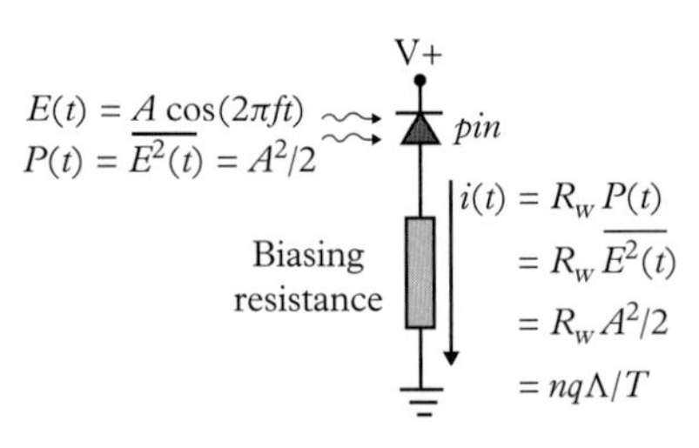

Figure 2.81 *Illustration of the* square-and-average *model for a photodetector.*

Intrinsically, a photodetector, be it *pin* diode or APD, is a *photon counter*, as the photocurrents are proportional to the number of photons arriving in a given bit interval. Since the photon count is proportional to the power or intensity of the lightwave, one can derive a deeper insight for the photodetection process from the fact that the power of a lightwave is the *square-and-average* of the electromagnetic wave (lightwave) associated with the photon stream that is received by the photodetector (see Fig. 2.81). In other words, if the received lightwave is represented by a phase-modulated electric field $E(t) = A\cos\{2\pi ft + \phi(t)\}$ in $t \in [0, T]$ with $\phi(t)$ as the phase term representing the modulating signal, then the associated power $P(t)$ and the average number of photons Λ received in an interval T can be related as

$$P(t) = \overline{E^2(t)} = \frac{A^2}{2} = \frac{hf\Lambda}{T}, \tag{2.122}$$

where the $\cos(4\pi ft)$ term is removed in the averaging operation. The resulting photocurrent $i(t)$ is therefore expressed as

$$i(t) = R_w P(t) = R_w \frac{A^2}{2} = \frac{\eta q}{hf} \times \frac{hf\Lambda}{T} = \frac{\eta q \Lambda}{T}. \tag{2.123}$$

However, if the received signal is added to a continuous-wave lightwave obtained from a local laser, functioning as a local oscillator (LO) in the receiver, given by $E_{LO}(t) = B\cos(2\pi f_{LO}t)$, then the photocurrent $i(t)$ will be given by

$$\begin{aligned} i(t) &= R_w\overline{[E(t) + E_{LO}(t)]^2} \\ &= R_w\frac{A^2}{2} + R_w\frac{B^2}{2} + R_w AB\cos\{2\pi(f - f_{LO})t + \phi(t)\}, \end{aligned} \tag{2.124}$$

which, after bandpass filtering, leads to an intermediate frequency (IF) current $i_{IF}(t)$ at $f_{IF} = f \sim f_{LO}$, given by

$$i_{IF}(t) = R_w AB\cos\{2\pi f_{IF}t + \phi(t)\}. \tag{2.125}$$

Equation 2.125 reveals an important aspect of the photodetection process, that its square-and-average operation over the combined electromagnetic wave $E(t) + E_{LO}(t)$ produces an IF waveform carrying the modulating signal $\phi(t)$ in its phase, which is an *observable* signal in the electrical domain (e.g., viewable on an oscilloscope screen) with f_{IF} being designable to fall within the bandwidth of realizable electronic devices. In other words, the heterodyne operation orchestrated by the photodetector along with the local laser and the IF filter, enables the receiver to process and estimate the phase, frequency, and amplitude of the received lightwave, which were otherwise unfathomable (or *invisible* electronically) in the non-coherent IM-DD systems.

Once the received optical signal gets translated down to the electronic IF frequency, one can employ any of the traditional demodulation techniques, enabling the receiver to extract its phase, frequency, or amplitude (or some combination of them) representing the information-bearing bit stream. In order to carry out the demodulation of the IF signal, one needs the transmitting and the local lasers to be highly coherent, so that the IF frequency does not drift over time due to the phase noise in the two lasers, which is why such systems are called coherent systems. In the presence of large phase noise in noncoherent sources (e.g., LEDs) the IF filter must have proportionately large bandwidth to capture the IF signal with jittery IF frequency, which will in turn increase the receiver noise. Further, note that the coherent receiver would also work with $f_{IF} = 0$, leading to a homodyne receiver; however, heterodyne receivers with non-zero IF frequency lead to the more practical form of coherent receivers as in traditional radio receivers. Typically, coherent heterodyne systems can enhance the receiver sensitivity by ≥ 10 dB, implying an increase in the unrepeatered link lengths by over 50 km (with a 0.2 dB/km optical fiber).

Notwithstanding the potential of coherent optical communication system, realization of its transceivers, more specifically the receivers, involves complex optical hardware. For

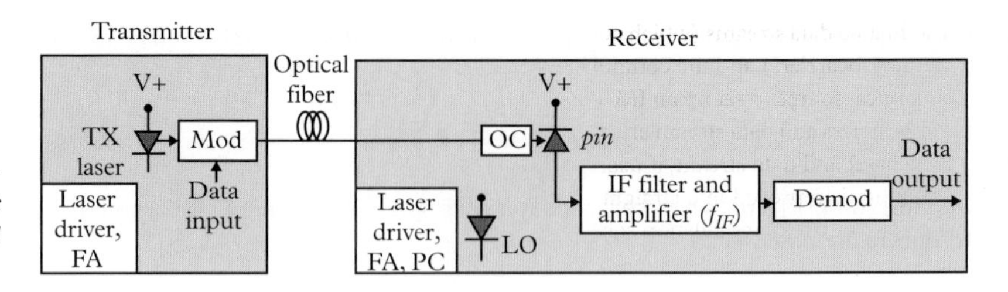

Figure 2.82 *Schematic diagram of a coherent optical communication system. Mod: modulator, Demod: demodulator, OC: optical coupler.*

example, even with coherent lasers, one needs to arrange for frequency alignment (FA) between the incoming lightwave and the local laser output in the receiver as well as in the transmitter, along with a polarization control (PC) module in the receiver to ensure that the SOPs of the incoming lightwaves from the optical fiber and the local laser remain aligned. Figure 2.82 shows a block schematic of a coherent optical communication system, where the FA and PC modules are shown along with the laser drivers in the insets of the transmitter and receiver. However, with the choice of the differential version of digital phase-modulation schemes – DPSK, DQPSK etc. – the design becomes simpler as the receiver derives the phase reference of the carrier from the delayed version of the incoming lightwave, thereby obviating the needs for LO, FA, and PC, albeit with some tolerable degradation in the receiver sensitivity.

In the late 1980s the research activities on the coherent WDM optical communication systems assumed considerable importance all over the world. However, after the arrival of EDFA as optical amplifiers for amplifying all WDM channels in one go, the problem of receiver sensitivity of IM-DD systems was significantly overcome, thereby pushing back the erstwhile research activities in the area of coherent optical communication systems. However, in recent years, the interest in coherent communication systems has been rejuvenated for the evolving needs of higher speed and bandwidth utilization in optical WDM networks. With the 50-GHz WDM grid, the transmission of 40 Gbps, 100 Gbps, and even higher speeds brings the need for bandwidth-efficient *multilevel modulation* of optical carriers, which require mostly phase-cum-amplitude modulation schemes (e.g., QAM, OFDM). For realizing such transmission systems, one needs to revisit the techniques used in coherent optical communication systems and utilize them appropriately as per the system requirements (e.g., speed, distance, BER, etc.). In Chapter 14, when describing the elastic optical networks (EONs), we will revisit some of the relevant issues on this matter, especially while discussing the realization of optical OFDM techniques with a much finer optical grid (typically 12.5 GHz).

The nature of the optical transceivers, be it for IM-DD or coherent communication system, would remain similar as for single-wavelength and WDM transmissions, albeit with appropriate optical filters placed in front of the optical receivers to select the desired wavelength in the WDM-enabled network nodes. With reference to WDM networks, another important aspect would be dispersion and nonlinearity management, as discussed earlier in Section 2.2.6. This issue will become crucial in particular for the long-haul WDM networks, along with the accumulated noise and crosstalk components from the *en route* optical amplifiers and intermediate nodes. We shall address these issues in Chapter 10, dealing with transmission impairments in optical WDM communication systems.

Another useful mode of optical communication, used in optical networks, employs subcarrier modulation and multiplexing. In this scheme, high electronic frequencies in the microwave/millimeter-wave range are modulated by data streams, and thereafter the data-modulated subcarriers are frequency-division multiplexed (FDM), at times also with some unmodulated data streams (which occupy the baseband spectrum and hence do not interfere with the subcarriers) and the composite FDM signal is used to intensity-modulate the light of an optical source to set up an IM-DD system. On the receiving end, after photodetection, the subcarriers and data stream are demultiplexed with a bank of filters (with a lowpass filter for the baseband data stream, if used, and bandpass filters for subcarriers), from which the data stream is retrieved in a straightforward manner, and the subcarriers are demodulated using appropriate demodulation schemes.

2.19 Summary

In this chapter, we presented an overview of optical fibers and different types of optical and optoelectronic components that are used in today's optical networks. Beginning with a brief physical-layer perspective of optical networks, first the optical fibers were considered, describing the fundamental principles of propagation of light, along with a full discussion of the loss, dispersion mechanisms, and nonlinear effects in optical fibers. Propagation mechanism in fibers was examined using ray theory as well as electromagnetic wave propagation, leading to the concepts of numerical aperture, acceptance angle, V parameter, and the conditions for single-mode propagation in optical fibers. Various dispersion mechanisms in optical fibers – intermodal, chromatic and polarization-mode dispersions – were explained and analyzed with appropriate analytical models. Basic concepts of various fiber nonlinearities – SBS, SRS, Kerr effect, and Kerr-effect-induced SPM, XPM, and FWM – were described, and their impact on lightwave propagation in optical fibers were examined.

Next, the operating principles and salient features of the semiconductor-based optical sources (LEDs and lasers) were described along with the essential properties of the semiconductors used, structural geometry, modulation characteristics and transmitter configurations for the direct and external modulation schemes. The operating principles and basic features of the photodiodes, *pin* and APD, were described, along with the typical receiver configuration. Various basic optical devices, such as isolators, circulators, couplers, splitters, interferometers, gratings, filters, multiplexers/demultiplexers, switches, and optical amplifiers were described and explained with suitable illustrations. Using these devices, the configurations and operational features of the important WDM-based network elements, such as (R)OADM, OXC and WSS, were presented and explained using suitable illustrations. Finally, the chapter concluded with a brief description of the point-to-point non-coherent (i.e., IM-DD) and coherent optical communication systems, that serve as the essential physical-layer links in optical networks.

EXERCISES

(2.1) A step-index multi-mode optical fiber has a refractive-index difference $\Delta = 1\%$ and a core refractive index of 1.5. If the core radius is 25 μm, find out the approximate number of propagating modes in the fiber, while operating with a wavelength of 1300 nm.

(2.2) A step-index multi-mode optical fiber has a cladding with the refractive index of 1.45. If it has a limiting intermodal dispersion of 35 ns/km, find its acceptance angle. Also calculate the maximum possible data transmission rate that the fiber would support over a distance of 5 km.

(2.3) Consider that a step-index multi-mode optical fiber receives optical power from a Lambertian source with the emitted intensity pattern given by $I(\theta) = I_0 \cos\theta$, where θ is the angle subtended by an incident light ray from the source with the fiber axis. The total power emitted by the source is 1 mW, while the power coupled into the fiber is found to be -4 dBm. Derive the relation between the launched power and the numerical aperture of the optical fiber. If the refractive index of the core is 1.48, determine the refractive index of the cladding.

(2.4) Consider a 20 km single-mode optical fiber with a loss of 0.5 dB/km at 1330 nm and 0.2 dB/km at 1550 nm. Presuming that the optical fiber is fed with an optical power that is large enough to force the fiber towards exhibiting nonlinear effects, determine the effective lengths of the fiber in the two operating conditions. Comment on the results.

(2.5) Consider an optical communication link operating at 1550 nm over a 60 km optical fiber having a loss of 0.2 dB/km. Determine the threshold power for the onset of SBS in the fiber. Given: SBS gain coefficient $g_B = 5 \times 10^{-11}$ m/W, effective area of cross-section of the fiber A_{eff} = 50 μm^2, SBS bandwidth = 20 MHz, laser spectral width = 200 MHz.

(2.6) Consider an optical communication link operating at 1550 nm over a 60 km optical fiber having a loss of 0.2 dB/km. The effective area of cross-section of the fiber A_{eff} = 50 μm^2, where an optical power of 0 dBm is launched. Determine the nonlinear phase shift introduced by SPM in the fiber. Given: $\tilde{n}(\omega) = 2.6 \times 10^{-20}$ m^2/W.

(2.7) Determine the Bragg wavelength of an integrated-optic grating with a period of 527 nm and an effective refractive index of 1.47. Sketch a block schematic using a Bragg grating and other relevant components for an optical add-drop multiplexer for three wavelengths at a given node in a WDM ring network.

(2.8) A laser cavity (i.e., an active layer in DH configuration) has a length L with a medium loss α dB per unit length, which is pumped with a gain g per unit length using an appropriate forward bias current. The walls on two ends of the cavity are designed with reflectivities r_1 and r_2, and the cavity has a confinement factor of Γ. Determine the condition to be satisfied by the pumped cavity to function as a laser. Give a typical sketch of g as a function of wavelength, and explain its impact on the laser spectrum.

(2.9) What is the fundamental difference between the spectral spreads in lasers due to the phase noise and modulation? Determine the spectral spread of a 1550 nm laser transmitting at 10 Gbps with the unmodulated linewidths of (i) 200 MHz and (ii) 0.08 nm.

(2.10) The threshold current density of a stripe-geometry AlGaAs laser is 3000 A/cm^2 at 15°C. Estimate the required threshold current at 50°C, when the laser characteristic temperature T_0 = 170°K and the contact stripe of the laser has a size (area) of 20 μm × 100 μm.

(2.11) Consider that a binary optical signal is incident from a fiber onto a photodetector. Presuming that the incident light has a duality in its nature (particle and wave), give examples for the manifestation of both the forms of light on the performance of a digital optical receiver.

(2.12) An APD operates at a wavelength of 900 nm with 95% quantum efficiency. Consider that an incident light with a power of -30 dBm has produced a photocurrent of 15 μA at the APD output. Determine the mean avalanche gain of the APD.

(2.13) If the received optical power in a *pin*-based optical receiver is -20 dBm and the preamplifier needs a voltage swing of 1 mV at its output, calculate the value of the feedback resistance needed for the preamplifier. Given: the optical transmitter uses a laser with perfect extinction, *pin* diode responsivity = 0.8 A/W.

(2.14) Draw a block schematic for a strictly nonblocking 8×8 optical switch, employing 2×2 electro-optic switching elements as the building block in Spanke's architecture. Estimate the total insertion loss of a similar $N \times N$ (with $N = 2^k$) optical switch in terms of its number of stages and the losses incurred in all the passive devices during the traversal of a lightpath from an input port to an output port.

(2.15) Using the results on gain saturation in EDFA (Eq. 2.119), estimate the decrease in EDFA gain when its input power increases from −20 dBm to −10 dBm. Given: $Q = 10$ dBm. Discuss how the gain saturation in EDFAs can affect the performance of an optical link in a WDM network and suggest some possible remedy.

Part II

Single-Wavelength Optical Networks

Optical Local/Metropolitan and Storage-Area Networks

3

The first generation of local/metropolitan area networks (LANs/MANs) used various types of copper-based media, spread out typically across a building or a campus under one autonomous administration. With the arrival of optical-fiber transmission, considerable developments took place to enhance the speed and size of these copper-based LANs and MANs, by using optical fibers over various possible topologies, e.g., bus, ring, and star. In this chapter we begin with the earlier experiments carried out on optical LANs/MANs in various research groups, and thereafter present some of the standardized optical LANs/MANs, such as distributed queue dual bus (DQDB), fiber-distributed digital interface (FDDI), and different versions of high-speed Ethernet (1/10/40/100 Gbps). Finally, we describe some of the storage-area networks (SANs) augmented by optical fiber transmission, where locally networked clusters of servers and storage devices are accessed by workstations through LANs, such as HIPPI, ESCON, and fiber channel.

3.1 Optical fibers in local/metropolitan-area networks

Development of local area computer networks or simply local area networks (LANs) has been a long and evolving process with a wide range of architectures using copper, wireless, and fiber-based transmission media. Use of a LAN in an organization for sharing various resources, be it in a business office or an academic campus, is now a basic necessity to run the system efficiently. Furthermore, the need to share information and computing resources between distant LANs also became a necessity, which was realized by interconnecting such LANs through leased lines from PSTN. These developments led to the emergence of wide area networks (WANs) operating at national and international levels. LANs of larger size, called metropolitan area networks (MANs) were also developed, spanning across metropolitan areas. However, these MANs have functioned practically as large-area LANs confined within autonomous organizations. On the other hand, the networking infrastructures set up in metropolitan areas with free access to individuals are different from the autonomous MANs and are viewed as public-domain metropolitan or simply *metro* networks.

With the arrival of optical communication technology, various attempts were made to speed up and extend the geographical spans of the erstwhile copper-based LANs and MANs. These efforts led to the development of varieties of optical LANs/MANs with higher speed and larger size, as compared to their copper-based counterparts. Furthermore, alongside the developments of optical versions of LANs and MANs, the standards for optical communication, SONET and SDH, emerged for high-speed digital communication links, and started serving the metro as well as the long-haul networks (see Chapter 5).

Optical Networks. Debasish Datta, Oxford University Press (2021). © Debasish Datta.
DOI: 10.1093/oso/9780198834229.003.0003

For optical LANs/MANs, several architectures were examined with various physical topologies, such as ring, bus, and star. Some of the early optical LANs explored in various research groups include Halo at the University of California, Irvine (Farber 1975), the Cambridge Ring at the University of Cambridge (Hunkin and Litchfield 1983), and Ethernet-based passive and active-star networks at the Xerox Palo Alto Research Center (Rawson and Metcalfe 1978; Rawson 1985). Following these initial experiments, several standards for optical LANs/MANs were developed, such as distributed-queue dual bus (DQDB) (IEEE 802.6) (Tran-Gia and Stock 1989; Zuckerman and Potter 1989), fiber-distributed digital interface (FDDI) (ANSI X3T9.5) (Ross 1986; Jain 1990), and gigabit Ethernet (GbE) series: GbE, 10GbE, 40GbE, and 100GbE (IEEE 802.3z, 802.3ae, and 802.3ba) (Frazier and Johnson 1999; D'Ambrosia 2009).

During the course of above developments, some of the earlier versions of optical LANs/MANs were unable to stay in the competitive market; however, the various concepts used and the challenges faced while developing these networks proved useful for the subsequent LAN/MAN design and implementations. In this chapter, we therefore first describe some of the earlier developments of the optical LANs/MANs – FDDI and DQDB – and then consider the more successful optically-augmented LANs/MANs (i.e., GbE series) with their lowest-level connections at the user-end typically using copper cables.[1] We also describe the operation of some storage-area networks (SANs), which employ locally networked clusters of servers and storage devices, which are accessed by the computers through a LAN in the same organization or even from the distant LANs through metro/long-haul networks.

3.2 Choice of physical topologies and MAC protocols

Before describing the optical LANs/MANs, we examine whether the physical topologies and MAC protocols used in the copper-based LANs would also be suitable for optical LANs/MANs. First, we consider the bus topology that was used in the earlier versions of Ethernet-based LANs. In the coaxial cables used in Ethernet, the electrical signal is transmitted and received bidirectionally at the network nodes with high input-impedance voltage taps. With these taps draining little current into the tapping nodes, signal can propagate both ways along the bus without much loss over a reasonable distance (e.g., 500 meter in the 10Base5 Ethernet[2]), which allows the media access control (MAC) protocol to operate with the carrier-sense multiple-access scheme using collision detection (CSMA/CD). In particular, any signal coming from a distant node is comparable in strength with the signal transmitted at the local node, and hence the local node can easily detect the presence of two signals at its receiver with the almost-doubled signal level (due to overlap), leading to collision detection. However, the CSMA/CD protocol imposes a restriction on the maximum size of the network, as the round-trip propagation delay in the bus should not exceed the minimum packet size for the detection of collisions by the local transmitting

[1] Note that, hereafter, we shall be using interchangeably the terms, LAN and MAN, with the understanding that an optical LAN is expected to cover a large area (as compared to their electrical counterparts), at times being spread over an area large enough to consider the LAN as a MAN. For example, FDDI and DQDB can be large enough in size (i.e., maximum possible size) for being termed as MANs, though they might operate as LANs when set up over a small campus or building-complex. On the other hand, native Ethernet started as a copper-based LAN, but later got transformed into the GbE series covering much larger areas using optical fibers as well as copper cables (i.e., hybrid media), which would also qualify to be termed as MANs.

[2] We discuss later in Section 3.5 on the original and subsequent versions of Ethernet in details.

nodes, so that it can receive the distant signal (already transmitted) before the end of transmission of its current packet and make the collision *detectable*. Thus the CSMA/CD scheme enforces the discipline of *one-packet-at-a-time* transmission across the whole network span.

In an optical fiber bus, the above criterion for successful functioning of the CSMA/CD protocol became difficult to satisfy. First, the packets *shrink* in time due to high-speed transmission and the bus length is *increased* for larger area coverage by utilizing the low-loss potential of optical fibers. Moreover, the optical fiber taps are unidirectional in nature as shown in Fig. 3.1, and thus each bidirectional tapping-unit needs to apply two directional couplers for the transmitting and receiving functionalities (Fig. 3.2). In order to have bidirectional transmission and reception, an optical fiber bus must use two such taps in a cascade along with circulators for the transmitting ports, causing more loss at each node. Thus, at each node the passive taps cause non-negligible power losses for the propagating optical signal, as it keeps getting partially coupled into the intermediate tapping nodes (unlike in the low-loss high-impedance voltage taps in a copper bus) along with insertion losses in the tapping devices. This leads to a progressive power loss of the signal while traversing the intermediate nodes along the fiber bus. Thus, a given node along an optical bus may receive signals of widely varying power levels from the nodes located at different places on both sides. This aspect makes the CSMA/CD protocol extremely difficult to operate in fiber, as the collision detection mechanism at each node uses a fixed threshold level to detect whether multiple transmissions are going on concurrently at the same spot. Consequently, optical buses cannot fit well with the legacy CSMA/CD-based Ethernet, both for large round-trip delay (compared to the shrunk durations of packets at high speed) and progressive tapping losses over long fiber spans.

Optical LANs/MANs using ring topology can avoid collision altogether as in the legacy token-passing rings, and don't have the problem of progressive power loss, as each node in a ring being active functionally uses OEO conversion. However, optical rings face a different kind of challenge, as one needs to exploit the full benefit of optical transmission using enhanced network size and speed. As shown in Fig. 3.3, with much higher transmission speed (as compared to the copper-ring speeds of a few Mbps), the packet durations in an optical ring get shrunk by several times, occupying a small part of the physical length of the optical ring at any given instant of time. As in a long high-speed optical bus, this problem gets exacerbated by the fact that the physical size of the ring (circumference) also has to increase desirably for utilizing the low-loss potential of optical fibers. Thus, the transmitting node (with the conventional token-ring protocol) has to wait for the return of the transmitted shrunk packet around the entire optical ring before releasing the free token for the neighbor. During this period, the entire ring space (and hence bandwidth) remains unutilized by any other nodes in the network. Consequently, these two problems (shorter packet duration and longer propagation delay) along with one-packet-at-a-time MAC protocol across the entire ring called for some appropriate changes in the token-passing MAC protocol.

For the third option of using star topology, one could employ a PSC (see Chapter 2), placed at a central location with respect to the network nodes. In this topology, with all

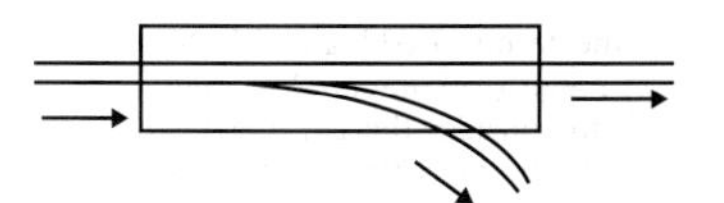

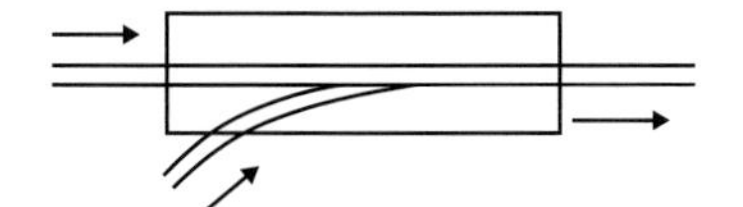

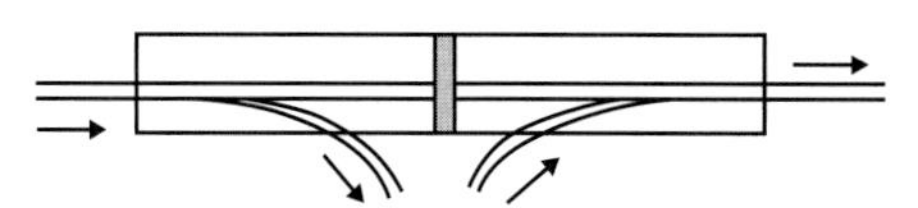

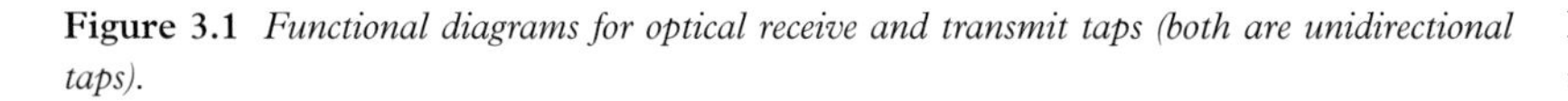

Figure 3.1 *Functional diagrams for optical receive and transmit taps (both are unidirectional taps).*

Figure 3.2 *Functional diagram for unidirectional transmit-receive optical tap.*

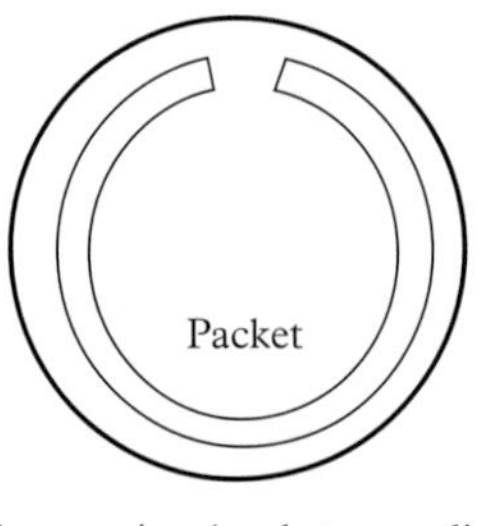

Copper ring (packet spreading over the entire ring)

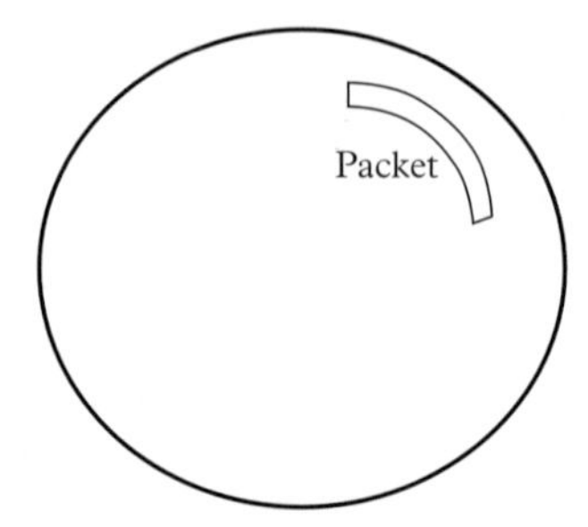

Optical ring (shrunk packet occupying a small space in the bigger ring)

Figure 3.3 *Copper versus optical rings.*

the nodes being at comparable distances from the centrally located PSC, the strength of the received signals from the various nodes is of the same order at any given node, therefore making collision detection a feasible option. However, even then it gets hard to realize a full-fledged CSMA/CD MAC protocol over the PSCs as the large optical network diameter makes the round-trip delay much larger than the shrunk packet sizes at high speeds ($\geq$ 1 Gbps).

In view of the above, replacing the copper cables by optical fibers in the legacy LANs didn't appear to be a straightforward task, and hence several architectures were explored for extracting the benefits of optical fiber transmission in LANs/MANs (in terms of speed and size of network) by judiciously changing the old MAC protocols. We describe in the following some of the standardized optical LANs/MANs: DQDB, FDDI, and the GbE series.

3.3 Distributed-queue dual bus (DQDB)

As discussed earlier, an optical fiber as a single bus, using CSMA/CD as MAC protocol, does not offer a suitable topology for LAN/MAN implementation. The difficulty of realizing optical bus topology led to the use of its two variants (dual-bus and folded-bus) in optical LANs/MANs with the MAC protocols avoiding collisions. Investigations on these topologies and related MAC protocols started in early 1980s with copper-based cables even before the optical LANs/MANs were conceived with bus topology – Expressnet with folded-bus topology, Fasnet with dual-bus topology etc. – all of them attempting to realize high-speed communication over longer distances than typical LANs with collision-free MAC protocols (Tobagi and Fine 1983). In these networks, the MAC protocols employed some distributed conflict-free round-robin schemes, which unlike CSMA/CD scheme, could efficiently carry multiple packets at a time without collision. The above topologies and the MAC protocols were found suitable for the optical buses as suitable replacements of the coaxial copper cables (and more so for the dual-bus Fasnet architecture), leading to the development of the DQDB MAN (Tran-Gia and Stock 1989; Zuckerman and Potter 1989; IEEE DQDB 802.6 1990).

3.3.1 Physical topology and basic features

DQDB network was introduced as an optical MAN standard from IEEE (IEEE 802.6) at a transmission speed of 44.736 Mbps, to serve over long fiber spans ($\leq$ 160 km). The physical topology and some basic features of DQDB are described in the following.

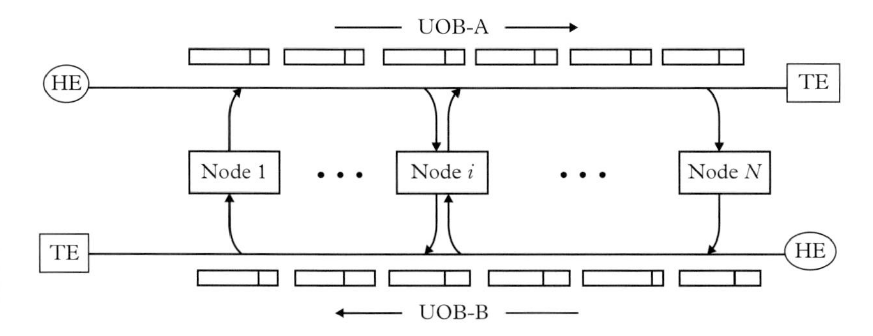

Figure 3.4 *DQDB schematic.*

As shown in Fig. 3.4, DQDB employs two unidirectional optical buses (UOBs) with the data streams flowing through them in the form of fixed-size slots (53 bytes, and hence ATM compatible) in opposite directions, i.e., by using UOB-A and UOB-B for the left-to-right and the right-to-left traffic flows, respectively. On each bus, transmission takes place from the head-end (HE) side toward the terminating end (TE) and the node nearest to HE (i.e., node 1 for UOB-A and node N for UOB-B) transmits periodically fixed-size slots to be used by itself and the downstream nodes. Thus, all the nodes in DQDB operate synchronously with the passing-by slots, and each node employs OEO regeneration on both buses and maintains a *distributed queue* (DQ), which stores its own transmission requests as well as the transmission requests sent from the downstream nodes through UOA-B. DQDB supports both asynchronous and synchronous communications (e.g., digitized voice, video) between the nodes by using the queue-arbitrated (QA) and pre-arbitrated (PA) access schemes, respectively. Some slots are kept with fixed allocations for the PA-based services, while for the QA-based data transfers, the access of each node to the passing-by free slots is determined by using the DQ status. Since both the buses remain operational concurrently, the overall capacity of the network is twice the capacity of a single bus. In the following, we describe the MAC protocol used in DQDB.

3.3.2 MAC Protocol

In order to explain the QA-based MAC protocol of DQDB, we consider one of the two UOBs, say UOB-A. Note that, node 1 being nearest to the HE in UOB-A, generates fixed-size slots periodically on UOB-A. Each slot uses a one-byte access control (AC) field with busy (B) and request (R) bits and a few more, followed by a 52-byte data field. Each node (say, node i) stores in its DQ the transmission requests from all the nodes on its right-hand side (i.e., all downstream nodes) received through UOB-B along with its own transmission requests for UOB-A. Thus, if node i wants to transmit a data packet (called a *segment* in the DQDB terminology) to a downstream node j with $i < j \leq N$ on UOB-A, then it needs to notify all of its upstream nodes (i.e., node 1 through node $(i-1)$) on UOB-B.

The use of DQ in each node with local and upstream transmission requests (with first-come-first-serve discipline) prevents each node from using a passing-by free slot, if its own request waits behind the requests of the upstream nodes in the DQ, thereby bringing in some fairness in the access mechanism. However, a certain amount of unfairness would still remain due to the finite propagation delay along the bus. In other words, when a local request in a node and a request from an upstream node are generated at the same instant of time, the local request will always get preference as compared to the request from the upstream node due to the finite propagation delay between the two nodes. In the following, we describe a typical DQDB node configuration that employs this scheme for

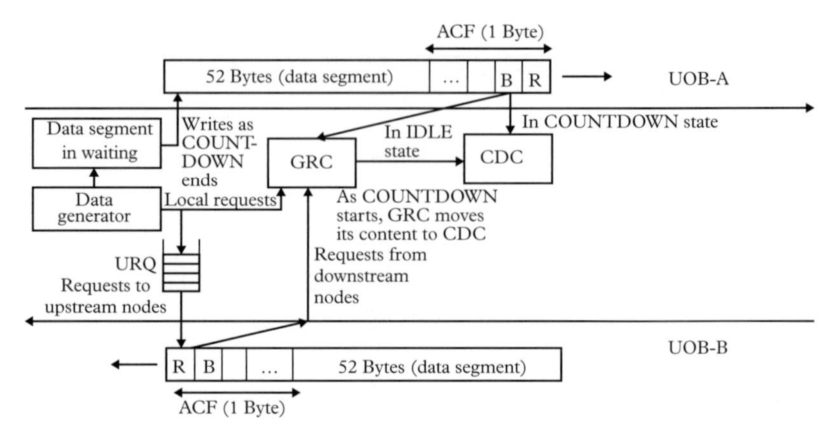

Figure 3.5 *Typical DQDB node configuration for data transmission on UOB-A. GRC: global request counter, CDC: countdown counter, URQ: upstream request queue, ACF: AC field.*

data transmission on UOB-A. The same configuration is also used (duplicated) for data transmission on UOB-B, with the transmission requests being sent through UOB-A.

Figure 3.5 illustrates the node configuration in DQDB, showing all the functional blocks that are essential for data transmission on UOB-A. As shown in the figure, every node uses two counters: (*i*) global requests counter (GRC) keeping a count of all the requests coming from downstream nodes, (*ii*) countdown counter (CDC) used to know when the given node should transmit its data segment over UOB-A. These counters, along with an upstream request queue (URQ), are initially reset to zero. As indicated earlier, in order to use a free data slot on UOB-A, the node under consideration, say node i, has to send a request to its upstream nodes (i.e., to the nodes 1 through $(i-1)$) by setting the request bit R in the AC field of a passing-by slot on UOB-B in the reverse direction. For this task, the upstream request may have to wait in URQ for a while to get an available slot passing by with R = 0 in the AC field.

Every node can have two possible states: IDLE and COUNTDOWN. In the IDLE state, a node has no data to transmit, and continuously monitors both UOBs, and thus CDC has no function to perform. However, the node in IDLE state increments its GRC count by one, when it reads a set R bit (i.e., R = 1) in a passing by slot on UOB-B, and decrements the GRC count by one for each free passing-by slot (i.e., B = 0) on UOB-A. Thus, the count value of GRC at node i represents the number of downstream nodes waiting to transmit data segments and determines how many free slots (with B = 0) should be allowed to pass by before accessing a free slot for its own data transmission.

Alternatively, node i enters a COUNTDOWN state from IDLE state upon arrival of a request from its host computer (shown as data generator in the figure) to transmit a data segment to a downstream node, say node j ($j > i$), on UOB-A. Thereafter, it moves the content of GRC into CDC and thus clears the content of GRC. At the same time, it sends a request on UOB-B to its upstream nodes (i.e., to node 1 through node $(i-1)$) by setting the request bit R in the ACF part of a passing-by slot on UOB-B in the reverse direction. For this task, it may have to wait in the URQ to get a slot passing by with R = 0. An upstream node, say node k ($k < i$), whenever it reads this request bit, it increments the value of its GRC count by one. In this process, node i registers its claim in the GRCs of all the upstream nodes. Further, node i keeps incrementing GRC for each new request coming from UOB-B and keeps decrementing CDC for every free slot passing by on UOB-A. When the CDC count goes to zero, node i accesses the next free slot on UOB-A to transmit its data segment.

Thereafter, node i goes back to IDLE state. At this stage, the node is allowed to register its next request corresponding to its data segment (if any).

The GRCs and CDCs (working through the ACFs) in all the nodes form together a DQ system to bring a reasonable fairness to the network. However, as mentioned earlier, notwithstanding the DQ formation in GRCs and CDCs, DQDB has a certain amount of unfairness, particularly for large network sizes. The distributed protocol of DQDB can also be extended for multiple priority levels and can support four priority levels, where for each priority a separate request queue is formed and a separate request bit is used in the AC field (not shown in Fig. 3.5). Furthermore, to ensure effective bandwidth sharing, the DQDB standard employs a bandwidth-balancing mechanism to occasionally skip the use of free QA-slots (i.e., the slots moving with $B = 0$). For PA-based isochronous services, the slot generator at node 1 (for UOB-A) takes responsibility for sending an adequate number of slots not available (NA) for QA-based services. Whenever such an NA slot is generated, it sets up a virtual channel identifier (VCI) in the slot header which takes care of the circuit-switched connections for node pairs needing isochronous communication.

In spite of significant effort toward improvement of DQDB network, it could not stay competitive and gave way to the other choices for the optical LANs/MANs: FDDI and GbE. However, some of the ideas used in designing DQDB network, such as the use of dual-bus topology and the fairness-aware MAC protocol, are expected to be useful in designing other networks.

3.4 Fiber-distributed digital interface (FDDI)

Some of the early optical LANs used ring topology, such as Halo and the Cambridge ring. Halo was the result of straightforward use of optical fibers and optoelectronic devices as replacement of the pre-existing copper ring at the University of California, Irvine, and called the Irvine Ring. The Cambridge Ring was developed at the University of Cambridge, where a special monitor station was included to frame packets during initialization of the LAN and monitor the ring performance. The line transmission rate used was moderate (10 Mbps) and thus the ring did not have the problem of small packets moving around a long ring. However, this problem seemed inevitable with the optical ring topology employing enhanced transmission rates and larger sizes. To get around this problem, FDDI was introduced as a standard from ANSI for the ring-based optical LANs/MANs (Ross 1986; Jain 1990; ANSI FDDI X3T9.5 1989), where the MAC protocol was changed for better utilization of the overall capacity of optical ring.

3.4.1 Physical topology and basic features

FDDI uses two counter-propagating rings at a data rate of 100 Mbps, and like DQDB it can support asynchronous as well as synchronous services. A total of 1000 physical connections (stations) and a fiber length (ring circumference) of 200 km can be supported by FDDI, thus qualifying it as a MAN. The two counter-propagating rings offer a fail-safe operation, and its nodes (stations in the FDDI terminology) are categorized in two classes (class A and class B stations) depending on the requirement of operational reliability against the link failures. Class A stations are connected physically to both the counter-propagating rings. Class B stations, with lesser priority for fail-safe operation, connect to one of the two rings and their connections are made through wiring concentrators (Fig. 3.6). For example, for a link-cut X on the ring between the stations 1 and 2, the failure can be isolated by two loopback paths at these two nodes, and then the two rings together using loopback operations can form a

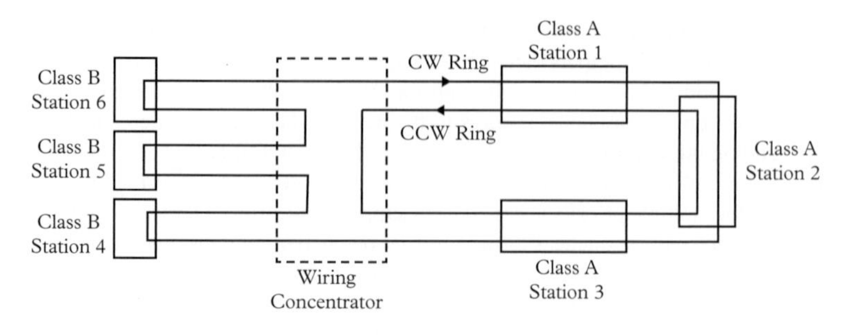

Figure 3.6 *Example FDDI network using two counter-propagating rings (CW: clockwise, CCW: counter-clockwise) and two classes of nodes (class A and class B).*

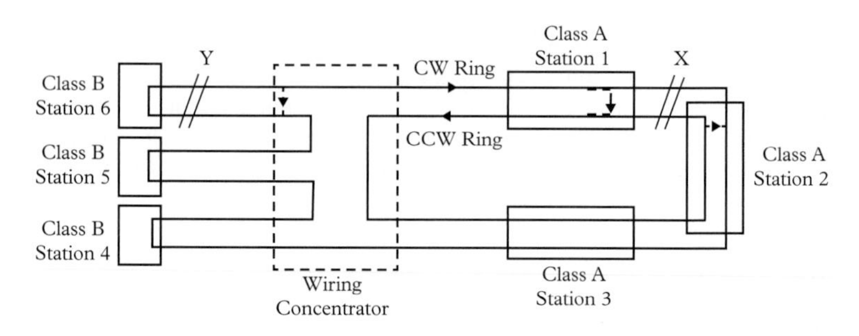

Figure 3.7 *Illustration of fail-safe operation in FDDI network with two link cuts, X and Y.*

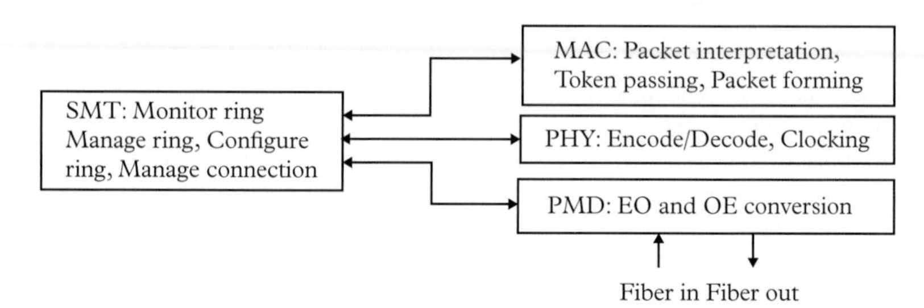

Figure 3.8 *FDDI Protocol stack.*

ring passing through all the nodes. However, this will not be possible if another cable fault Y takes place between stations 1 and 6, when station 6 has to be bypassed for the ring to remain operational. (Fig. 3.7).

The overall operation of FDDI is governed by a protocol stack having four layers: station management (SMT), media access control (MAC), physical (PHY), and physical-medium dependent (PMD) (Fig. 3.8). The SMT layer looks after the following functions: monitoring and management of ring, configuration of ring, and connection management. The functions of PHY and PMD layers are shown in Fig. 3.9. The PMD uses an LED as the optical source (1330 nm) with the frame sizes $\leq$ 4500 bytes. The data transmission employs 4B5B line encoding for ease of clock recovery and DC stability, which increases the final transmission rate to 125 Mbps. The LED output is transmitted through the optical switch, which is used for loopback operation, when needed for failsafe operation. At the receiving mode, the incoming optical signal received from the ring is photodetected and digitized using the recovered clock. Next the recovered digital signal is passed through elastic buffer and line decoder to obtain the data output.

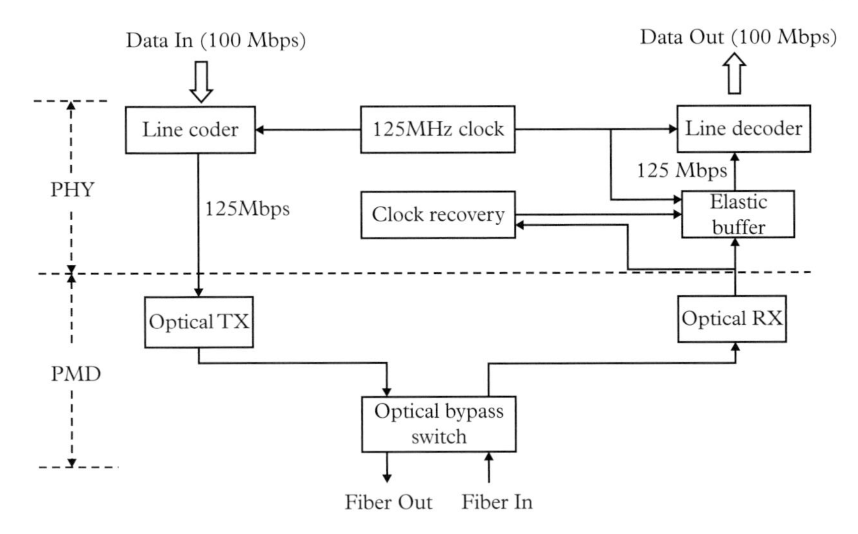

Figure 3.9 *PHY and PMD layers of an FDDI station.*

3.4.2 MAC protocol

In the IEEE 802.5 token-ring protocol, once a token is captured by a station, it is released for the next station only after its own transmitted packet (if transmitted) returns to itself after traversing through the entire ring. With this arrangement in copper rings, a transmitted data packet would mostly spread spatially over the entire ring (at the end of transmission of its last bit) due to lower transmission rate and smaller size of the ring, leading to full utilization of the ring space/bandwidth. However, as shown earlier in Fig. 3.3, this mechanism prevents an optical ring (operating at higher speed with larger size) from utilizing the full capacity of the optical ring due to the shrunk packets and larger ring size. This problem is addressed in FDDI by allowing every station to release the captured token immediately after its own data packet (frame in the FDDI terminology) transmission, instead of holding the token until its own transmitted data frame returns to itself. Effectively, this arrangement transforms the MAC protocol from the one-frame-at-a-time model to the one that allows multiple frames to coexist and traverse the ring at any moment of time.

The MAC-layer operation in FDDI is carried out with data frames and tokens using specific formats as shown in Fig. 3.10. The various fields of the formats shown in the figure are as follows. PA represents the preamble of both token and data frames, consisting of idle-line-state symbols of 64 bits (16 symbols of four bits) occurring at maximum frequency with alternating 1's and 0's, which helps in synchronizing the token and the frame with the clock of each station. A sequence delimiter (SD) is used following PA with a sequence of two 4-bit sequences ($2\times4 = 8$ bits) to recognize the boundary between PA and the rest of the FDDI token or frame. The frame control (FC) field in the token or data frame consists of 8 bits indicating whether it is a synchronous or asynchronous frame, the lengths of addresses (16 or 48 bits), and other necessary control information. The field represented by DA carries destination address, which may be of 16 or 48 bits. Similarly, the field SA carries source address with again 16 or 48 bits. The frame check sequence (FCS) field has a 32-bit cyclic redundancy check (CRC). The ending delimiter (ED) consists of two symbols, i.e., eight bits, while the frame status (FS) field has at least three symbols with the following functions: checking whether (*i*) the addressed station has recognized the address, (*ii*) the frame has been copied, and (*iii*) any station has detected an error in the frame.

FDDI token format:

PA	SD	FC	ED

FDDI data frame format:

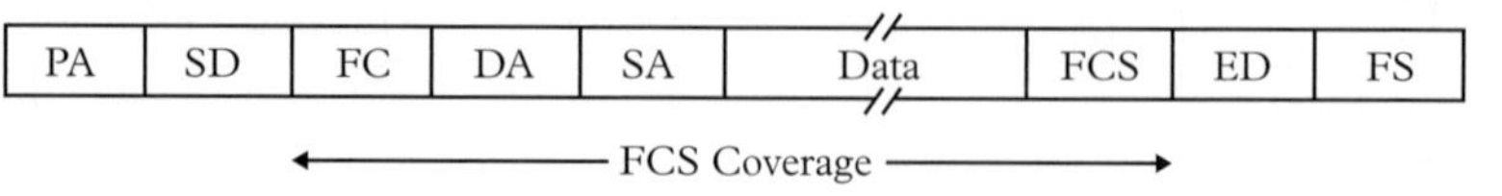

Figure 3.10 *FDDI token and data frame formats.*

In an operational FDDI network, a station waiting to transmit a data frame captures a free token as soon as it receives. Thereafter it starts transmitting its data frame, and the token is released immediately following the completion of the data frame transmission. When a node does not have any data frame to transmit, it regenerates and relays the received token toward the next downstream node. Each node has to remove its own transmitted data frame as and when it comes back after traversing the entire ring, which is referred to as a *source-stripping* mechanism. FDDI has a flexible and distributed capacity (number of bytes allowed to transmit in a given frame) allocation scheme for its nodes and makes use of a number of timers for this purpose. In the following we briefly describe the capacity allocation scheme of FDDI.

During the initialization process, all stations negotiate a target token rotation time ($TTRT$), which represents an estimate of the time duration within which a token should come back to a given node after its last release. $TTRT$ can be expressed as

$$TTRT \geq D + F_{max} + T_{token} + SA_T, \tag{3.1}$$

with D as the total packet propagation time in the ring, F_{max} as the maximum time allotted for asynchronous data frames (sum of frame lengths from all the nodes during one full rotation), T_{token} as the transmission time of a token, and SA_T as the total time allotted for synchronous traffic from all the stations (say N stations). D can be expressed as

$$D = \sum_{i=1}^{N-1} D_{i,i+1}, \tag{3.2}$$

with $D_{i,i+1}$ as the delay incurred by a frame in moving from node i to node $i+1$ including the propagation delay and the processing time in one node. Thus, D can also be represented as

$$D = \frac{L}{(c/n)} + NT_{proc}, \tag{3.3}$$

where L is the length of ring, c is the velocity of light, n is the refractive index of fiber core, N is the number of stations in the ring, and T_{proc} represents the processing time at a node. SA_T can be expressed as the sum of the allotted synchronous transmission times (quotas) of each node (SA_i), given by

$$SA_T = \sum_{i=1}^{N} SA_i. \tag{3.4}$$

Each node has three timers (realized using counters): the token rotation timer to keep record of token rotation time (TRT) around the ring, and the token holding timer for storing token holding time (THT) allowed for the given node and late counter (LC). When the node begins its operation following initialization, TRT is set to $TTRT$ and the node can experience two situations: early return of the token or late return of the token as compared to $TTRT$.

We consider these two cases in the following. Early return takes place when $TRT < TTRT$, i.e., the token arrives at the transmitting station before the scheduled time with a variable defined as earliness $E = TTRT - TRT$. The node sets THT to $E = TTRT - TRT$ and sets LC to 0. Then the node transmits synchronous data first and then for the duration of $E = TTRT - TRT$ it is allowed asynchronous data transmission. Late return is encountered when $TRT \geq TTRT$, i.e., the token arrives at the transmitting station just when $TRT = TTRT$ or after the target time $TTRT$ is exceeded with the earliness E having a zero or a negative value, respectively. Consequently, THT is set to 0 implying that the node cannot send any asynchronous data and LC is set to 1. Thus, with $LC = 1$, no asynchronous data transmission takes place, and only synchronous data transmission is allowed, governed by the pre-decided quota (SA_i). Actual values of TRT will indeed vary from node to node and from one visit cycle of token to another (one visit cycle representing one round of the token in the ring), as governed by the traffic and $TTRT$. Thus, the synchronous transmission is always allowed as per the pre-decided quota for all stations, while the asynchronous transmission is allowed at a given station only when the token returns early. Thus, the TRT at a station directly influences the TRT at the next downstream station.

Using the above capacity allocation scheme, FDDI keeps allocating frame sizes for asynchronous service in a distributed manner to the various stations. Note that, driven by this scheme, the greediness of some station(s) in one cycle (to transmit longer frames) would curtail the asynchronous quota for all stations in the very next cycle, thereby setting up a dynamic control in a distributed manner. With this observation, the overall capacity utilization efficiency η_c can be expressed for N stations as (Jain 1990)

$$\eta_c = \frac{N(TTRT - D)}{(N \times TTRT + D)}. \tag{3.5}$$

Similarly, the maximum access delay can be expressed as

$$t_d = (N - 1)TTRT + 2D. \tag{3.6}$$

As evident from Eq. 3.5, with a $TTRT$ chosen close to the ring latency D, the efficiency η_c will become very low. On the other hand, η_c will increase with the increase in $TTRT$ beyond D, but with an increase in the maximum access delay t_d (Eq. 3.6), and this would call for a tradeoff while setting up the value of $TTRT$.

3.5 Gigabit Ethernet series: 1 Gbps to 100 Gbps

There are several factors that have made Ethernet today the favorite option for corporate as well as residential LANs (and MANs for large-area corporate networks): cost, scalability, reliability, backward compatibility, and widely available management tools (Sömmer *et al.* 2010). Ethernet was developed as a LAN technology at 3 Mbps (precisely, 2.94 Mbps) on coaxial cable, which continued thereafter to evolve over four decades to become the most-used technology not only for LANs/MANs, but also in facilitating data transport through the

SONET/SDH/OTN-based (see Chapter 5) metro, and long-haul carrier networks. As the transmission speed in Ethernet reached and moved beyond 1 Gbps (e.g., 10 Gbps, 40 Gbps, 100 Gbps) (Frazier and Johnson 1999; D'Ambrosia 2009; Roese and Braun 2010)[3] the role of optical fiber transmission became more and more necessary in the relevant parts of the network. In the following, first we discuss the evolutionary developments in Ethernet technologies, and then describe the salient features of Ethernet of different generations, particularly for speeds $\geq$ 1 Gbps.

3.5.1 As it evolved

Ethernet started its journey at Xerox Palo Alto Research Center in 1972, when Robert Metcalfe and his group developed a LAN technology at a speed of 2.94 Mbps to interconnect computers and the peripheral devices through a coaxial cable using Aloha-based MAC protocol (Spurgeon 2000). The initial version was improved later using the now-well-known CSMA/CD MAC protocol (Metcalfe and Boggs 1975), and a patent was filed in 1977 (Metcalfe *et al.* 1977). Later, in 1980, DEC, Intel, and Xerox developed a modified version, named DEC-Intel-Xerox (DIX) Ethernet II, operating at 10 Mbps. In 1985, Ethernet was eventually standardized by IEEE as 802.3 CSMA/CD LAN, and thereby started its long journey with the series of higher-speed versions alongside the continuous growth of the traffic volume with time.

Commercial Ethernet started with the 10Base5 technology, where thick co-axial copper cables were chosen as the transmission medium to carry 10 Mbps data. 10Base5 Ethernet offered a passive cable span of 500 m with an upper limit of five such segments (i.e., 2.5 km) using repeaters. The upper limit of 2.5 km related to the round-trip delay constraint in CSMA/CD protocol (discussed earlier in Section 3.2), which had to be $\leq$ the minimum Ethernet frame duration at 10 Mbps (i.e., 51.2 μs with 512 bits). Subsequently, a cheaper version, i.e., 10Base2 Ethernet, was developed with thin coaxial cables, with a passive cable span of $\simeq$ 200 m and an upper limit of 1 km with repeaters. Subsequently, 10BaseT Ethernet was introduced with twisted wire pairs forming a star topology around a hub having a radius of 100 m, where the hub provided a structured and scalable wiring with increased reliability. In 10BaseT, the collision detection would physically take place within the hub itself. When a given port of a hub received any Ethernet frame, it would broadcast the same to the other ports, leading to collision when such broadcasting took place from multiple ports of the hub.

The hub-centric 10BaseT Ethernet LANs could also be interconnected using multiple hubs at higher level, all sharing a single collision domain (Fig. 3.11). In this configuration, even if one node wanted to communicate with another node in the same hub itself, its frames would be broadcast to all other hubs, thereby reaching the nodes on other hubs causing undesired collisions. This issue constrained the number of levels of hubs (typically two or three) (as in Fig. 3.11) to control the number of collisions in multiple-hub 10BaseT Ethernet.

In order to address this problem, subsequently the concept of bridged-Ethernet was introduced, where an Ethernet having multiple hubs could be broken into a few smaller hub-based Ethernet groups. In this configuration, the highest-level hubs of all the small Ethernet groups, i.e., the apex hubs of each group (e.g., Hub-3, Hub-4, and Hub-5 in Fig. 3.11) were interconnected with each other using a device called a bridge (by replacing Hub-6), which could break one large collision domain down into multiple collision domains of smaller

[3] To support further growth in traffic, the technologies and architectural issues of GbEs transmitting at 400 Gbps and 1 Tbps are also currently being explored by different research groups (Winzer 2010; Wei *et al.* 2015).

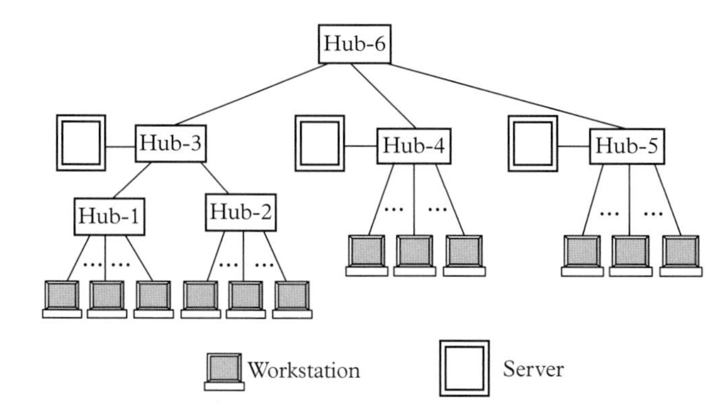

Figure 3.11 *Ethernet formation with hubs. FE: fast Ethernet.*

size. For communications between two nodes belonging to two different groups under two different hubs, the bridge could employ switching selectively (instead of broadcasting to all groups) of the Ethernet frames from one port (connected to one group, hosting the source node) to the specific port which was connected to the apex hub of another group housing the destined node only. Thus, the prevention of broadcasting of an Ethernet frame from a given node of one group to the other not-destined groups could avoid unnecessary collisions, and the collision domains were reduced in size. The bridges were later extended to accommodate lager number of ports, leading to Ethernet switches that made a part of the LAN working as a switched Ethernet, avoiding the collision process. However, at lower level the collision domains could function using the hub/bus-based Ethernet segments.

In the meantime, in order to stay in competition with the 100-Mbps fiber-optic LANs/MANs – DQDB and FDDI – the Ethernet community moved to upgrade the native 10-Mbps Ethernet to offer a 100-Mbps LAN service by using improvised transmission along with hub/switch-based networking. However, in a hub-based CSMA/CD LAN, in order to increase the transmission rate from 10 to 100 Mbps, one had to bring down the upper limit on the network size by a factor of 10 due to the round-trip delay-constraint of CSMA/CD protocol, thereby allowing at the most 2000/10 = 200 m of network diameter. Coincidentally, the hub-based 10-Mbps Ethernet architecture had already been designed by that time with a radius of 100 m (implying a node-to-node distance via hub, i.e., a network diameter of 200 m) for more flexible and reliable physical topology, thereby enabling successful adoption of 100-Mbps Ethernet within the same framework, albeit with some modifications in the communication links to increase the transmission rate from 10 to 100 Mbps. Further, for superior performance, hubs were gradually replaced by switches to avoid CSMA/CD transmission protocol altogether, though the newer version of Ethernet, called fast Ethernet, was kept fully compatible with the native 10-Mbps Ethernet. In particular, the pre-existing computing/networking devices (i.e., computers and hubs) of native Ethernet were made compatible with fast Ethernet bridges/switches using 10/100 Mbps interface cards to set up the communication speed at 10 or 100 Mbps using appropriate message exchanges through a process called *auto-negotiation*. Having gone up to 100 Mbps in this manner, the 100-Mbps fast Ethernet introduced by the Ethernet community competed with and eventually outperformed both FDDI and DQDB, as it provided a smooth backward compatibility to the pre-existing slower Ethernet-based network installations.

Table 3.1 presents the basic features of the various versions of Ethernet technology with the speeds of 10 and 100 Mbps, where three types of 100-Mbps Ethernet are shown: 100BaseT4 using Category-3 (Cat3) unshielded twisted wire pairs (UTPs) with maximum

Table 3.1 *Different versions of Ethernet from 10 Mbps to 100 Mbps.*

Ethernet Name	Cable Type	Maximum length	Terminologies
10Base5	Thick coaxial	Without repeater: 500 m With repeaters (5): 500 m × 5 = 2.5 km	10: 10 Mbps, Base: Baseband 5: 500 m
10Base2	Thin coaxial	Without repeater: 200 m With repeaters (5): 200 m × 5 = 1 km	2: 200 m
10BaseT	Twisted pair	100 m	T: two Cat3 UTP, 10 Mbps
100BaseT4	Twisted pair	100 m	4: four Cat3 UTP, 100 Mbps
100BaseTX	Twisted pair	100 m	X: four Cat5 UTP, 100 Mbps
100BaseF	Optical fiber	2 km	F: fiber, 100 Mbps (switched Ethernet)

spread (radius) of 100 meters, 100BaseTX using Category 5 (Cat5) UTPs again with a radius of 100 meters, and 100BaseF using optical fibers over 2 km, with 100BaseF used only for the switched-Ethernet at 100 Mbps.

By late 1990s, with the need for interconnecting 10/100 Mbps Ethernet segments with higher-speed links, the IEEE 802.3 Ethernet committee started working on the standard of GbE (IEEE 802.3z) operating at 1 Gbps. The GbE networks, introduced in 1998, used GbE switches and optical fibers (for longer distances) as well as high-quality twisted pairs and coaxial cables, thereby offering a high-speed backbone to interconnect lower-speed Ethernet clusters within a large campus/organization. Gbps links were used in selected segments of Ethernet LANs with high traffic, but with the compatibility of GbE with the older versions of Ethernet by using some modifications in the MAC protocol, such as carrier extension and frame bursting (discussed later).

After the development of GbE, within a few years the need for yet-higher speeds was felt, following which the IEEE 802.3 team engaged themselves in exploring higher transmission speeds beyond 1 Gbps, leading to 10 Gbps Ethernet (10GbE) in 2003. Unlike GbE with options for both shared and switched Ethernet operations, the 10GbE LAN (IEEE 802.3ae) with significantly shrunk packet durations and the consequent reduction in the collision domain size, started operating only with the full-duplex links by using 10GbE switches, thereby leaving the regime of CSMA/CD-based shared Ethernet completely.

Until the development of 10GbE LANs, a scaling factor of 10:1 in the network speed was followed for the technology upgrades. However, in due course, in order to assess the growing needs of further speed-enhancement, the IEEE 802.3 working group formed the higher-speed study group (HSSG) in 2006. Interestingly, it was found that the need for higher speed in Ethernet was different for the network aggregation points and the server ports. In particular, HSSG observed that the bandwidth requirement for network aggregation was getting doubled every 18 months, while that for the server ports was doubling up at a slower rate, i.e., in every 24 months. From these observations, HSSG recommended two new speeds simultaneously for the next-level Gigabit Ethernet: 40 Gbps Ethernet (40GbE) for servers and 100 Gbps Ethernet (100GbE) for network aggregation, enabling data transport over long-haul network. This recommendation led to the formation of the IEEE P802.3ba Ethernet task force. In 2010, IEEE standards were established for 40GbE and 100GbE to serve the respective demands, which were in conformity with SONET/SDH and OTN for data transport through metro and long-haul networks, and indeed both of them adopted the switched Ethernet mode using full-duplex transmission.

In the following, we discuss the architectures and basic operational features of the x-GbE series, with x ranging from 1 Gbps to 100 Gbps. The first in the x-GbE series, i.e., GbE, had to take care of backward compatibility in respect of shared vs. switched models of Ethernet using half- and full-duplex links, respectively. However, from 10 Gbps onward, all the high-speed versions of x-GbE were designed with the switched-Ethernet architecture using full-duplex links, thereby leaving altogether the CSMA/CD MAC protocol.

3.5.2 GbE

GbE (i.e., 1 Gbps Ethernet) offered a cost-effective solution to address the demands for larger speed beyond 100 Mbps, while ensuring backward compatibility with 10/100 Mbps versions of Ethernet (Frazier and Johnson 1999). Moreover, while enhancing the speed by ten times with respect to 100 Mbps fast Ethernet, the GbE design caused a cost escalation by $\sim$ 3 times only, thus offering a highly cost-effective upgrade. We describe below the various features of the GbE architecture including its MAC protocol and physical-layer characteristics.

3.5.2.1 GbE architecture

The protocol stack of the GbE architecture is shown in Fig. 3.12, comprising the two major bottom-most layers of the five-layer protocol suite: MAC and physical (PHY) layers, while maintaining the original Ethernet frame format for backward-compatibility by using some intermediate layer/interface/sublayers. In the PHY layer, the full-duplex point-to-point links (without CSMA/CD) became the most predominant communication mode between GbE devices over long distances ($\leq$ 5 km). However, the CSMA/CD-based shared-access connection over half-duplex point-to-point links was also adapted for connecting with 10/100 Mbps devices of Ethernet to make the GbE switches backward compatible with the pre-existing segments in an Ethernet-based LAN.

As shown in Fig. 3.12, the MAC layer is followed below by a layer, called the reconciliation layer, through which the lower layers can interact with the MAC layer in a manner such that

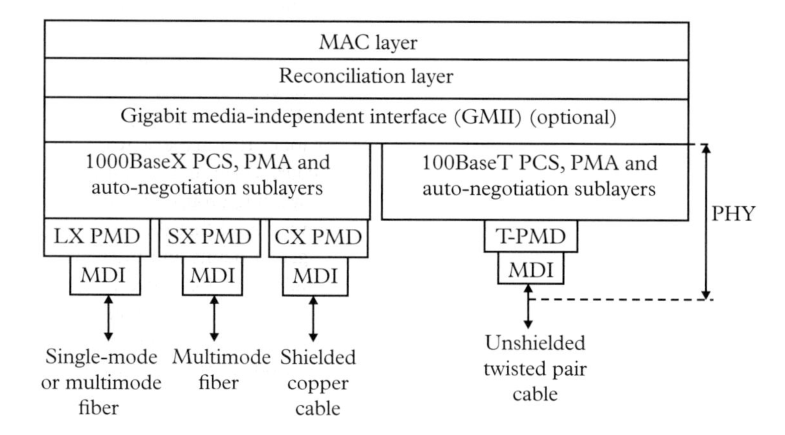

Figure 3.12 *GbE protocol layers.*

the MAC layer gets the *feel* of the native Ethernet, i.e., 802.3. Further, GbE employs a media-independent sublayer, called generalized media-independent interface (GMII) between the reconciliation and PHY layers, though passing through GMII sublayer is not mandatory, particularly when the connection is to be set up through optical fibers or shielded coaxial cables. However, when a GbE connection is set up through a UTP cable, the connection must engage with the GMII sublayer.

Below the GMII sublayer, there are two ways to reach the physical media ports (1000BaseX and 1000BaseT), and in both cases there are three PHY sublayers (shown together in one box): physical coding sublayer (PCS), physical medium attachment (PMA) sublayer and auto-negotiation sublayer. On the left side, 1000BaseX takes effect for the optical fiber and STP copper cable, while on the right side 1000baseT goes through the UTP cable. The PCS sublayer provides the encoding logic, performs multiplexing and synchronization in downward (transmit) direction, and carries out alignment of codes and demultiplexing and decodes the upward (incoming) data stream. The PMA sublayer prepares the outgoing signal from GMII in bit-serial form with mBnB (8B10B) line coding (for ease of clock recovery and to ensure a stable DC component in the signal), and the *auto-negotiation* sublayer needs to negotiate for the speed and half/full-duplex connections, wherever needed. The two sublayers perform equivalent operations in the reverse direction for the incoming signal.

The physical-medium dependent (PMD) sublayer serially transmits the line-coded signal and passes on to the media-dependent interface (MDI) sublayer representing the connectors appropriate for the specific media choices (or receives with the equivalent operations carried out in the reverse direction). The PMD sublayer can operate with four possible media: LX-PMD for long-range single-mode or multi-mode fiber, SX-PMD for short range multi-mode fiber, CX-PMD for copper-based STP cable, and T-PMD for copper-based UTP cable. The connectors used in the MDI sublayer are used in accordance with the four media choices.

An example realization of an Ethernet LAN using GbE backbone along with lower-speed Ethernet clusters is shown in Fig. 3.13, where a few 1 Gbps links have been used as backbones to interconnect the GbE switches, and some of these switches are in turn connected to the 10/100 Mbps Ethernet clusters. The figure shows how seven GbE switches (one as campus distributor (CD), two as building distributors (BDs), and four as floor distributors (FDs)) operate as a backbone to interconnect the central servers and different 10/100 Mbps Ethernet clusters (with their respective workstations and group servers) in different buildings or clusters of the organization. In the following, we discuss the MAC protocols used for half- and full-duplex communications in GbE networks. As evident from the current networking trends, with the development of 10/40/100 GbE standards, even the lowest-level connections in Ethernet-based LANs/MANs have now by and large adopted the switched GbE-connections.

3.5.2.1.1 Half-duplex operation The 1 Gbps links used in a GbE can be realized with full-duplex copper cables (UTP, STP) or optical fibers, depending on the distance, although the optical fiber links are more scalable for longer distances. Even though the GbE backbone is based on switched Ethernet and mostly uses full-duplex connections, in order to ensure backward compatibility, the GbE links have had to retain the option of CSMA/CD-based MAC protocol, particularly when the devices across a given link make use of half-duplex network ports.[4]

[4] As mentioned earlier, today's Ethernet-based LANs have turned out to be increasingly switch-based using full-duplex connections, and thus half-duplex connections are no longer found much in use. Nevertheless, we discuss half-duplex operation here to get an insight into how the MAC protocols have been engineered in the past to ensure backward compatibility, which in turn might be relevant again for future developments in a different context.

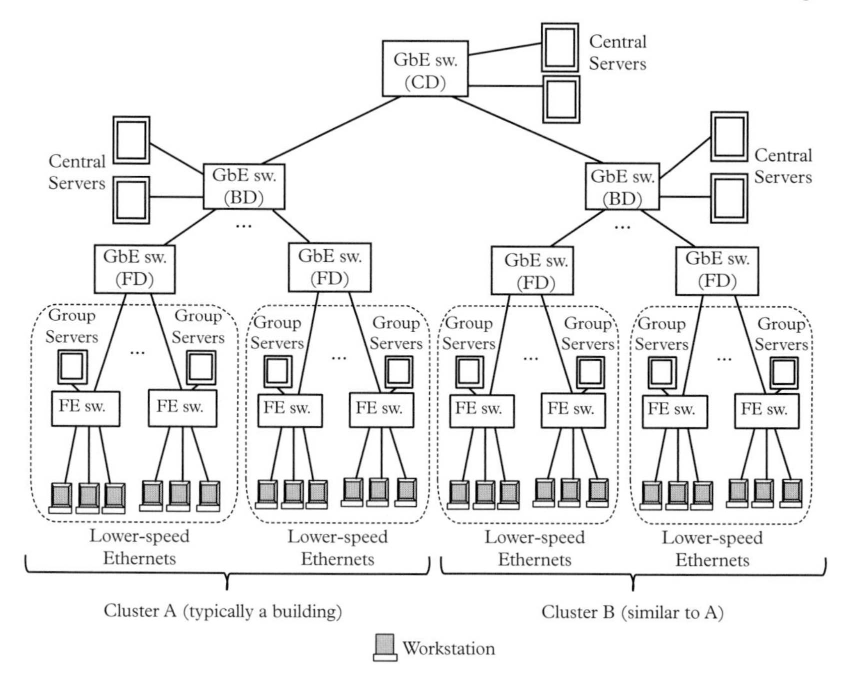

Figure 3.13 *Example of Ethernet-based campus LAN using GbE backbone and lower-speed Ethernet clusters.*

Note that, with 1 Gbps transmission, Ethernet frames will shrink by a factor of 10 with respect to 100 Mbps speed and by a factor of 100 with respect to 10 Mbps speed. Hence, with half-duplex connection, the link lengths must be reduced proportionately by the factors of 100 and 10, respectively, thereby making the network size too small. Thus, with the half-duplex connection operating with CSMA/CD, even a moderate link length (> 25 m) will violate the constraint on the upper-bound of round-trip delay ($\leq$ minimum frame duration) in the link, and both ends will be transmitting shrunk frames without knowing that many of the short-transmitted frames would eventually encounter collisions at the other end.

In order to get around the above problem, the shrunk Ethernet frames at 1 Gbps are extended or padded with additional dummy bits so that the frames get an acceptable duration for collision detection when operating on a half-duplex link. When the padded frames at 1 Gbps arrive at the other end, the dummy extension bits are deleted and thus the frames get back their original form. In particular, 512 bits (minimum) are replaced by a *slot* of 512 bytes (minimum) or 4096 bits at the 1 Gbps transmitter end, with the information-bearing bits padded with an adequate number of dummy bits. This process of padding with dummy bits is known as carrier extension, as shown in Fig. 3.14.

However, the carrier extension is not efficient enough at times to address this problem, particularly when the link length is higher in a large LAN and there are many packets to be transmitted with small sizes. This problem is addressed by a scheme, called *frame bursting*, used along with carrier extension. When a node has a number of frames to transmit over the half-duplex 1 Gbps link, the first frame is treated with carrier extension as before. Thereafter, the remaining frames are transmitted almost back-to-back, albeit with some small inter-frame gaps, until the overall burst size reaches 1500 bytes (controlled by a burst timer). Note that, with 1500 bytes ($1500 \times 8 = 12,000$ bits) as the burst size, the burst duration

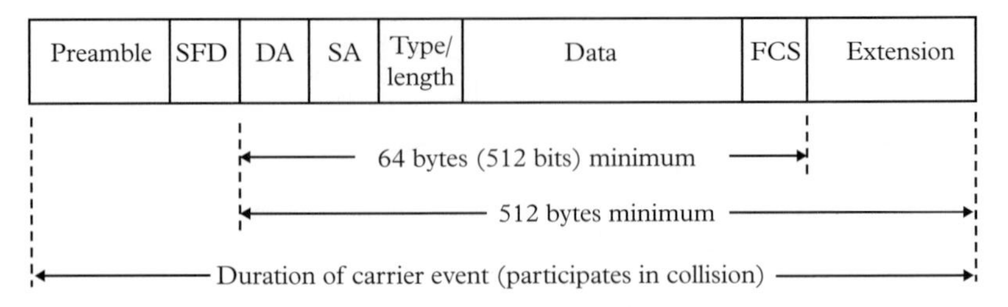

Figure 3.14 *Carrier extension scheme in GbE (SFD: start frame delimiter, DA: destination address, SA: source address, FCS: frame check sequence).*

at 1 Gbps becomes $T_{burst} = 12,000 \times 10^{-9}$ sec = 12 μs. This implies that the half-duplex transmission over optical fiber will be able to support a round-trip delay of 12 μs, implying a maximum permissible optical link length $L_{max} = (12 \times 10^{-6}) \times (2 \times 10^5)/2 = 1.2$ km (with the velocity of light as 2×10^5 km/sec in optical fiber). However, the full-duplex 1-Gbps links will no longer go by the CSMA/CD protocol, thereby obviating the needs of the carrier extension and frame-bursting schemes.

Full-duplex operation

A full-duplex GbE connection allows simultaneous two-way communications between a pair of GbE devices using copper cable or fiber as the medium, which increases the link capacity twofold, i.e., to 2 Gbps. Furthermore, since the transmission is no longer based on CSMA/CD, the constraint on round-trip delay disappears, thereby allowing much longer link lengths wherever needed for the Ethernet-based LANs across large campuses.

Owing to the finite buffers in the GbE switches, the GbE communication under full-duplex mode needs to adopt an appropriate flow-control mechanism. In particular, the transmitting module of each full-duplex switching node sends inter-frame gaps as guard times. Over and above these gaps, the receiving end uses *pause frames* to convey to the transmitting end to throttle its transmission of frames. This becomes necessary as the switches have limited buffers to store the incoming frames from all the ports with full capacity, and thus may get overwhelmed with the incoming traffic and lose the incoming frames owing to buffer overflow. The pause frames are sent to the transmitting end with a timing value (in multiples of 512 bits), indicating that the transmitting end should stall its transmission of frames during the time interval notified through the pause frame. If the receiving end finds itself out of congestion before the completion of the notified pause interval, it sends another pause frame with a zero value for the pause interval, so that the transmitting-end can resume its transmission. Further, note that at higher speeds (10 Gbps onward) full-duplex transmission is employed without any need to have backward compatibility.

3.5.2.2 GbE physical layer

As discussed earlier, the physical layer of GbE can use different types of media: optical fibers, twisted pairs, and coaxial copper cables. As shown in Table 3.2, the standard for GbE transmission was introduced as 1000Base-X, with two categories: 1000Base-LX and 1000Base-SX for long and short transmission distances, respectively. 1000BaseLX GbE operates with lasers transmitting at 1300 nm over (*i*) multi-mode fibers for distances up to 550 m, and (*ii*) single-mode fibers for distances up to 5 km. These links are intended

Table 3.2 *GbE physical-layer specifications.*

Media type	Fiber, copper wire	Optical source	Distance covered	Remarks
Optical fiber: 1000Base-LX	Multi-mode fiber, single-mode fiber	Laser 1300 nm	550 m, 5 km	L: Long range optical link for campus backbone
Optical fiber: 1000Base-SX	Multi-mode fiber	Laser 850 nm	220 m, 500 m	Short range optical link for in-building backbone
Copper wire: 1000Base-T	UTP cable (Cat5)	—	100 m	For horizontal wiring
Copper wire: 1000Base-CX	STP cable	—	25 m	For patch cords in server room

to serve as backbones for a large campus network. 1000BaseSX GbE operates with the lasers emitting at 850 nm over optical multi-mode fibers for shorter in-building backbones of 220–500 m. There are also two types of GbE links using twisted pairs: 1000Base-T and 1000Base-CX. The 1000Base-T uses unshielded twisted-pair (UTP) cables of Cat5 copper wires over 100 m for horizontal wiring purposes, while 1000Base-CX is used within a machine room with shielded twisted-pair (STP) cables over a span of 25 m.

With reference to Fig. 3.13, the links between the GbE switches for CD and BD will generally be long and use 1000Base-LX. However, the links between GbE switches for BD and FD will be shorter and will typically use 1000Base-SX. The connections between GbE switches for FD and the lower-speed switches (floor wiring) can employ optical or copper links with 10/100/1000 Mbps options, but these links might have to operate in half-duplex mode, whenever necessary. Though not in vogue these days, the lower-speed switches would operate at 10/100Mbps to set up connections with the computing devices (workstations, servers etc.) as shown earlier in Fig. 3.13, while the GbE switches, if connected to lower-speed devices ($<$ 1 Gbps), will need to bring down their transmission speed to 10 or 100 Mbps through an auto-negotiation process.

3.5.3 10GbE

From 10GbE onward, owing to the shrunk pulses and small-size collision domains, all higher versions of Ethernet have to employ full-duplex transmission using the switched Ethernet architecture, and the major challenges come from the design of the transmission system. Note that the transmission rate of 10GbE matches closely with SONET/SDH OC-192 rate (9.58464 Gbps), implying that 10GbE is well-suited for transmission through metro and long-haul backbones with some reasonable rate adjustment by using the optical transport network (OTN) as the convergent platform (see Chapter 5). We describe below the various features of 10GbE architecture including its MAC protocol and physical-layer characteristics (Cunningham 2001).

3.5.3.1 10GbE Architecture

The architecture of 10GbE is similar to the protocol stack, shown earlier in Fig. 3.12 for GbE, but without any need to negotiate for half/full-duplex operations. Further, in order to keep itself backward compatible, the frame format remains the same as used in GbE and the lower versions of Ethernet.

One important issue arises in 10GbE from the fact that the transmission rate of 10GbE (i.e., 10 Gbps) is slightly higher than the nearest SONET/SDH rate, i.e., OC-192 rate

(= 9.58464 Gbps), leading to buffer overflow when being transmitted over WANs based on SONET/SDH. However, it is preferred by the Ethernet community to accommodate exactly 10 GbEs for 10:1 scaling of transmission speed. With these differing preferences, an intermediate solution is reached by invoking a rate-adaptation (i.e., flow control) mechanism, similar to the flow-throttling scheme used in GbE. In particular, 10GbE uses a *word-by-word hold scheme*, wherein after receiving a request from the physical layer, the MAC layer pauses by sending a 32-bit word of data for a pre-specified time duration, thereby reducing the effective flow rate down to a value commensurate with OC-192.

3.5.3.2 10GbE physical layer

As already mentioned, the physical layer of 10GbE employs only full-duplex connections, mostly on single-mode/multi-mode optical fibers, and uses copper cables for very short distances. Accordingly, the standards of 10GbE physical layer are divided into four categories: (*i*) 10GBase-LX for long (in the context of LANs/MANs) distances (5–15 km) using multi-mode or single-mode fiber, (*ii*) 10GBase-SX for short distances (100–300 m) using multi-mode fibers, (*iii*) 10GBase-EX for extended-range communication ($\geq$ 40 km) using single-mode fibers, and (*iv*) 10GBase-CX for the distances within 20 m using shielded twisted pair of copper wires. Table 3.3 summarizes these features of 10GbE physical layer along with the laser wavelengths used for the LX, SX, and EX ranges.

3.5.4 40GbE and 100GbE

As mentioned earlier, driven by the different growth of demands for server ports (for computing applications) and the aggregation points (for networking applications), 40GbE and 100GbE were recommended, respectively, by IEEE at the same time. Thus the architectures for these two versions were developed concurrently (D'Ambrosia 2009; Roese and Braun 2010). We discuss briefly their MAC protocols and physical-layer characteristics in the following.

3.5.4.1 40/100GbE architecture

Figure 3.15 presents the architectures of 40GbE and 100GbE. Overall architectural features follow those of 10GbE, but with the additional options of two sublayers in PHY layer – forward error correction (FEC) and auto-negotiation sublayers within the PHY layer – which are used selectively, based on the physical media types. The MAC layer after reconciliation communicates through two different GMII layers, XLGMII and CGMII, for 40 and 100 Gbps transmission, respectively. Transmission media for 40GbE and 100GbE are described in the following.

Table 3.3 *10GbE physical-layer specifications.*

Media type	Fiber, copper wire	Optical source	Distance covered
Optical fiber: 10GBase-LX	Multi-mode fiber, single-mode fiber	Laser 1300 nm	5–15 km
Optical fiber: 10GBase-SX	Multi-mode fiber	Laser 850 nm, 1300 nm	100–300 m
Copper wire: 10GBase-EX	Single-mode fiber	Laser 1500 nm	$\geq$ 40 km
Copper wire: 10GBase-CX	STP cable	—	$\leq$ 20 m

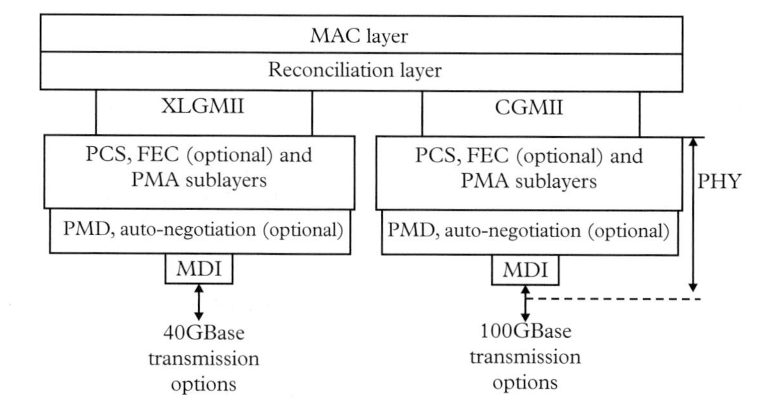

Figure 3.15 *40/100GbE protocol layers.*

Table 3.4 *Physical-layer media for 40/100GbE.*

PMD ports	Reach	40GbE	100GbE	Remarks
40GBase-KR4	At least 1 m on backplane	√	NA	4 × 10 Gbps channels on backplane
40/100GBase-CR4/10	At least 10 m on copper cable	√	√	4/10 × 10 Gbps coaxial cables
40/100GBase-SR4/10	At least 100 m on multimode optical fiber	√	NA	4/10 × 10 Gbps with CWDM transmission on multi-mode fiber
40GBase-LR4	At least 10 km on single-mode optical fiber	√	NA	4 × 10 Gbps with CWDM transmission on single-mode fiber
100GBase-LR4	At least 10 km on single-mode optical fiber	NA	√	4 × 25 Gbps with CWDM transmission single-mode fibers
100GBase-ER4	At least 40 km on single-mode optical fiber	NA	√	4 × 25 Gbps with CWDM transmission on single-mode fibers

3.5.4.2 40/100GbE physical layer

The physical layer for 40/100GbE can be of various types, as employed by the following PMD sublayers:

- 40GBase-KR4: Four 10 Gbps channels for backplane transmission at an aggregate rate of 40 Gbps over at least 1 m,
- 40/100GBase-CR4/10: Four/ten pairs of coaxial cables for 40/100 GbE, respectively, each cable carrying data at 10 Gbps over at least 10 m,
- 40/100GBase-SR4/10: Four/ten optical fibers at 850 nm for 40/100 GbE, respectively, each fiber carrying data at 10 Gbps over at least 100 m,
- 40GBase-LR4: Four 10 Gbps CWDM channels in 1310 nm window with aggregate rate of 40 Gbps, using single-mode fiber over at least 10 km,

- 100GBase-LR4: Four 25 Gbps DWDM channels in 1310 nm window with aggregate rate of 100 Gbps, using single-mode fiber over at least 10 km,
- 100GBase-ER4: Four 25 Gbps DWDM channels in 1310 nm window, using single-mode fiber over at least 40 km.

Thus, for 40 and 100 Gbps transmission, multiple channels are chosen, either with concurrent transmission over multiple electrical cables (or backplane channels) or with C/DWDM transmission. Overall, five target distances were considered in the design: 1 m, 100 m, 1 km, 10 km, and 40 km. Table 3.3 summarizes these specifications.

3.6 Storage-area networks

The enormous volume of data that is generated today by the various business sectors is making the data storage capability and fast access to that data a priority to operate a business efficiently. The storage area networks (SANs) belong to a special class of locally networked devices, which are set up to store and transfer huge volumes of data using a large number of disk arrays, tape libraries, and high-end servers, supported by high-speed interconnections within an organization. A SAN brings together several storage islands using a high-speed network, which can be accessed by the applications running on any networked servers. Unlike the pre-existing direct-attached storages (DASs), SANs remove storage from the individual servers and consolidate storage elements in a location over a networked platform, from where it can be accessed by any application. This in turn helps to improve storage utilization and avoid providing additional storage devices, thereby saving both money and space.

SANs can also be set up in multiple buildings at different locations for storing and transferring large amounts of data, helping disaster management as well. In the early phase of such computing premises, various computing and storage devices were connected using point-to-point links, leading to impractical mesh connectivity which was based on proprietary designs. The concept of SAN was introduced by the standards committees with a switched-connectivity to all the servers and other peripheral devices with simple and scalable connectivity along with the options to connect the SAN to the workstations in a LAN by interfacing SAN servers to the LAN. Figure 3.16 gives a basic schematic of a SAN, which employs a central switching unit, through which all the servers and the peripheral devices are connected in a circuit-switched manner, and the SAN service can be accessed by the workstations through the LAN set up in the organization. Note that a given SAN can be extended to other SANs as well as to the remotely located clients through WAN connectivity.

Furthermore, over the years the needs for large and centralized storage capacity along with high-performance computing power have increased massively as compared to SANs, leading to large computing and storage facilities, called datacenters, hosting thousands to hundreds of thousand of servers in a single site. In this section, we describe the salient features of some of the well-known SAN architectures supported by optical networking, and later in Chapter 13 we present the ongoing developments in the area of datacenters employing electrical as well as optical switching technologies.

There have been some useful SAN technologies, such as, fiber channel (FC) (Clark 1999) and enterprise serial connection (ESCON) (Calta *et al.* 1992) using optical transmission, along with the relatively older version, called high performance parallel interface (HIPPI) using predominantly copper cables with some fiber-based augmentation (Tolmie

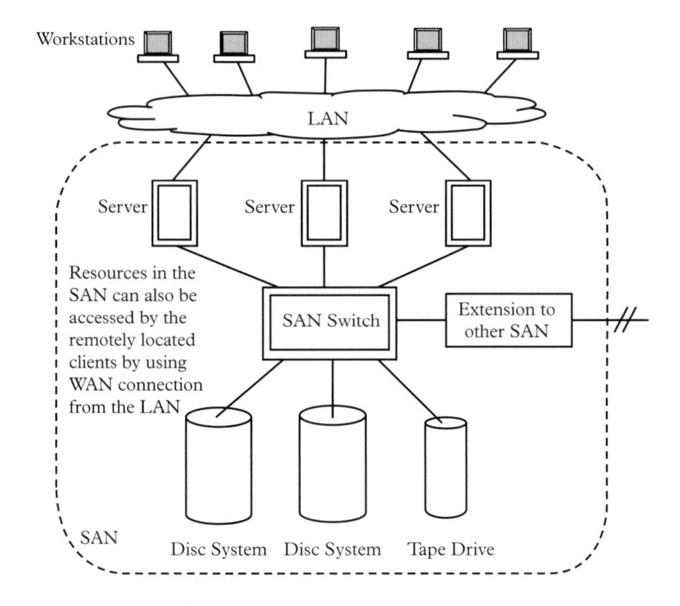

Figure 3.16 *Basic SAN configuration with a centralized switching unit.*

and Renwick 1993). Commercial HIPPI first came out in 1980s and used a crossbar-based SAN switch, using parallel communication with all the storage devices and servers using multiple twisted-pair copper cables. Originally it was standardized with a transmission rate of 100 Mbyte/s (i.e., 800 Mbps) over 50 parallel twisted-pair copper cables, which was limited to 25 meter span. Later on, the older version of HIPPI was improved using 100 twisted-pair cables for transmission at 1.2 Gbyte/s over the same distance, which was further augmented using serial optical fiber transmission for long-distance connections (about 10 km) with other SANs, along with the pre-existing local 100 Mbyte/s and 1.2 Gbyte/s interfaces.

ESCON was introduced by IBM, also in 1980s, with a serial fiber-optic transmission system, initially offering a data rate of 10 Mbyte/s (80 Mbps). It was proposed to replace the pre-existing slower, shorter, and cumbersome copper-based parallel transmission system of interconnecting mainframes and attached devices, known as the bus-and-tag system. ESCON offered simpler connections over a much larger distance (a few kilometers) as compared to copper-wire based SANs. Later, the data rate was increased further to 17 Mbyte/s (136 Mbps) for distances ranging from 3 km (using LED and multimode fiber) to 20 km (using laser and single-mode fiber). The actual transmission rate increased to 200 Mbps due to the overhead bytes and the 8B/10B line encoding used for clock recovery. ESCON used dynamically modifiable switches called ESCON Directors. The links to connect an ESCON Director with the associated servers and storage devices in the SAN used optical fibers with the specifications as mentioned above. The fiber-optic link lengths in an ESCON can be extended up to 60 km to connect to a remote SAN by using a chain of ESCON Directors through OEO regeneration.

ESCON was later outperformed by a faster technology, called fiber channel (FC), which was also designed over optical fibers but with much higher data rates: 1 Gbps, 2 Gbps, and 10 Gbps. As compared to ESCON, FC offered a more effective SAN realization, with higher speed as well as more variations in its connectivity, such as point-to-point (PP), ring with arbitrated loop (AL), and mesh using switched fabric (SF). A typical FC-based SAN

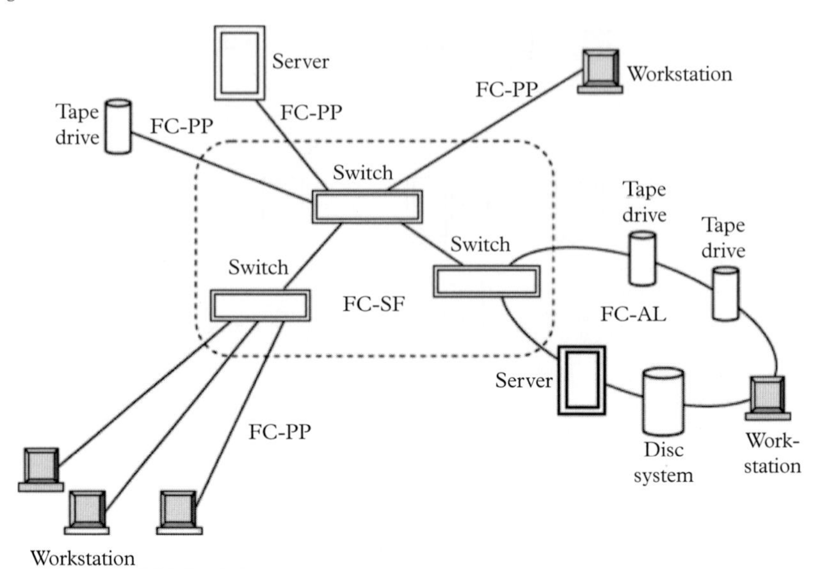

Figure 3.17 *Representative FC-based SAN configuration.*

realization is shown in Fig. 3.17, capturing all the three possible configurations in one. The FC-SF formation is realized with three switches, which brings together the FC-AL loop and FC-PP connections to the workstations, storage devices, and servers.

3.7 Summary

In this chapter, we presented several types of optical LANs/MANs with different physical topologies and MAC protocols, and discussed briefly about SANs. First, we examined the fundamental design issues, while replacing the copper-based transmission links by optical fibers in the traditional LAN/MAN topologies using CSMA/CD or token-based MAC protocols. Next, we discussed the possible remedies and presented the topological features and MAC protocols for the two earlier versions of optical LANs/MANs: DQDB and FDDI. DQDB was based on dual-bus topology (IEEE standard 802.16) while using the notion of a distributed queue in each node to get around the fairness problem in the network. By contrast, FDDI (ANSI X3T9.5) was realized on two counter-propagating fiber rings, with a flexible capacity-allocation scheme.

Next, we considered today's favorite LAN/MAN technology, augmented seamlessly with time by making discreet use of optical fiber transmission wherever needed, such as the series of Gigabit Ethernet standards, including 1-Gbps Ethernet or GbE, 10GbE, 40GbE, and 100GbE. First, we described the evolutionary process that the Ethernet technology has gone through from the days of its native 3 Mbps version, eventually leading to the present-day Gigabit Ethernet series. The basic architectures and physical-layer features were described for all the four versions of the GbE series. Finally, we presented a brief description on SANs: HIPPI, ESCON, and FC, and the readers are referred to Chapter 13 on datacenters, presenting the more recent and the fast-evolving networking trends in this area with vast storage and computing resources.

EXERCISES

(3.1) Discuss with suitable illustrations why the fiber-optic bus or ring topology cannot use traditional *one-packet-at-a-time* MAC protocols, such as CSMA/CD or token-ring protocol, as used in IEEE standards for copper-based LANs. Also indicate the difficulty in using CSMA/CD protocol in passive-star-based optical LANs.

(3.2) Obtain an expression for the progressive power loss in an optical fiber bus connecting N nodes to set up a LAN, with each node tapping power from and transmitting (coupling) power into the fiber bus using a *transmit-receive coupler* (Fig.3.2). Using this formulation, evaluate the dynamic range needed in the receivers used for the network nodes connected to the fiber bus with $N = 5$ and 10. If the receiver dynamic range cannot be more than 10 dB, comment on the feasibility of the LANs in the two cases. Given: connector loss = 1 dB, power tapping for the receiver in the transmit-receive coupler = 10%, insertion loss in the waveguide of each transmit-receive coupler = 0.3 dB, loss at the transmitter-coupling point in each transmit-receive coupler = 0.2 dB, fiber loss = 0.2 dB/km, distance between two adjacent nodes = 1 km.

(3.3) Sketch a block schematic for a possible realization of an 8×8 transmissive passive-star coupler using 2×2 3-dB power splitters as the building block. Assume that, each 2×2 building block has an insertion loss of 1.5 dB (including connectors) apart from its 3-dB power-splitting loss and the LAN diameter is 5 km. Estimate the end-to-end signal loss in a passive-star-based optical LAN using this device. If each node transmits a power of 0 dBm and its receiver sensitivity is -23 dBm, check whether the network would offer a feasible power budget. Given: fiber loss = 0.2 dB/km, connector loss = 1 dB.

(3.4) Consider that a passive-star optical LAN operates at 1 Gbps with a fixed packet size of 1500 bytes. Calculate the maximum possible diameter of the network that would support CSMA/CD as an acceptable MAC protocol. Repeat the calculation for 10 Gbps transmission and comment on the result. (Given: refractive index of fiber core = 1.5).

(3.5) Consider a passive-star optical LAN operating with pure ALOHA as the MAC protocol with a large number of nodes. Assuming that the packet arrival process in the network follows a Poisson distribution with a traffic Λ, determine the expression for the probability that a successful transmission takes place after k attempts. Using this result, determine the average number of attempts to be made for a successful packet transmission.

(3.6) Consider two identical passive-star optical LANs as in Exercise 3.5 (LAN-A and LAN-B, say), each of them working independently using pure ALOHA with a traffic Λ. Next, connect optically one free port of the passive-star coupler used in LAN-A to one free port of the passive-star coupler of LAN-B using an optical fiber. Determine the expression for the throughput of the combined network and compare the same with the throughput of one single LAN (i.e., LAN-A or LAN-B).

(3.7) Assume that, in Exercise 3.6, the two free ports of the two LANs are now connected through an active device (as in a *bridge*), which can convert the optical packets coming from both LANs into electrical domain and store them in a buffer, with the capability to read each packet header from LAN A and LAN B and switch the received packet from one LAN to another, if necessary, and ignore the packets that are destined to some node in the same LAN. What will be the throughput of the combined network?

(3.8) Consider a 10 km long DQDB network with 31 equally distant nodes, where the 1st and 31st nodes have negligible distances from their respective nearest head-ends. Estimate the amount of unfairness (in terms of the time to access the bus) between the 15th and 27th nodes for packet transmission along the unidirectional bus operating from the 1st node to the 31st node. What will be the impact on unfairness between these two nodes (i.e., between 15th and 27th nodes) if we accommodate 30 more nodes in the same network (maintaining equal separation between nodes and the same bus lengths)? Given: refractive index of fiber core = 1.5.

(3.9) In a DQDB network, the terminating end of the upper bus has become defective and hence taken out. Discuss the impact of removal of the termination on each bus and on the overall network operation.

(3.10) Consider an optical LAN based on the ring topology with 100 nodes. Each node in the LAN employs OEO conversion and requires a processing time of 1 μs for each packet that passes by. If the ring operates at a transmission speed of 100 Mbps, and each packet consists of 2000 bytes, determine a suitable value of the ring circumference in km, for which the standard token-based MAC protocol (i.e., IEEE 802.5) will be able to operate efficiently. Given: refractive index of the fiber core = 1.5.

(3.11) Estimate the capacity utilization and maximum delay incurred in an FDDI ring with the following values of ring length L (circumference), number of nodes N, and TTRT setting:

a) $L = 4$ km, $N = 20$, TTRT = 5 ms and 50 ms,

b) $L = 50$ km, $N = 100$, TTRT = 5 ms and 50 ms.

Given: processing time at each node $T_{proc} = 1$ μs.

(3.12) Discuss why the hubs were introduced in 10 Mbps Ethernet in place of coaxial cables as the transmission lines. Indicate the limitations faced subsequently by the hub-based Ethernet architecture and explain how this difficulty was overcome by using switched-Ethernet architecture with bridges/switches.

(3.13) Consider a 100BaseT (i.e., hub-based) fast Ethernet, wherein one has set up by mistake the twisted-pair links between the computers of end-users and the hub with the link lengths ≥ 150 m. Discuss the problems (with appropriate calculations) that the LAN will face in running the CSMA/CD MAC protocol with this network setting.

(3.14) The IEEE standard committee had to introduce 40/100GbE architectures together, though the earlier Ethernet upgrades (e.g., 100 Mbps, 1 Gbps, and 10 Gbps versions) took place one at a time with the speed-enhancement factor of 10. Discuss the underlying issues that had driven the committee to make such a choice.

Optical Access Networks

4

Access networks have evolved relentlessly over time in diverse directions. For example, PSTN initially provided only landline voice service in the last mile and later offered data-access using dial-up, ISDN, and ADSL technologies. Similarly, the cable TV networks and mobile communication systems introduced data and voice/video services, while LANs and Wi-Fi networks also evolved offering voice and video services. Following these developments, the demands for bandwidth and reach in access segment increased and led to the optical-access networking solutions using passive optical networks (PONs). With two wavelengths for the upstream/downstream transmissions over tree topology, PONs use TDM for the downstream transmission to the users and TDMA in the upstream with dynamic bandwidth allocation (DBA) scheme. This chapter focuses on PONs, presenting the building blocks and the candidate MAC protocols using appropriate DBA schemes. Finally, the chapter presents the available PON standards, e.g., EPON, GPON, 10G EPON, and XG-PON.

4.1 Access network: as it evolved

Traditionally, an access network for POTS, popularly called a subscriber loop, used to be formed around a telephone exchange located at a CO, to provide PSTN access to the surrounding homes and business organizations. Today, with a broader sense of *access*, the task of an access segment lies in collecting/distributing voice, video, and data from/to the various types of network users. During the last few decades, the access segment has gone through significant changes with the ever-increasing demand for bandwidth from the users across all sections of society by using various possible transmission media, e.g., copper, wireless, and fiber. The arrival of the Internet stimulated these developments all the more for providing Internet services to homes and offices over the pre-existing voice/video-centric (e.g., telephone and cable TV) access segments. Initially, these developments remained fragmented and disjointed with little interaction between them. In other words, PSTN offered POTS, cable TV network distributed TV programs, and wireless communication systems offered telephone services to individual mobile users, while LANs and their wireless equivalents falling in the category of organizational access segment provided only data access. With time, following the deregulation in the telecommunication industry, this fragmented scenario started disappearing and various technologies came in to support multiple services over single platform, thereby providing voice, data, and video from the same access networking set-up. In other words, this transformation process enabled each networking platform (e.g., PSTN, cable TV, mobile communication systems, wired/wireless LANs) to offer a broad range of services, including voice, data, and video. We briefly describe this evolution process in the following.

The extension of the voice-only access (i.e., POTS) of PSTN for the data service took effect by using *modems*, a combination of modulator and demodulator for data transmission and reception, respectively. In a modem, the data in binary form would modulate suitable

Optical Networks. Debasish Datta, Oxford University Press (2021). © Debasish Datta.
DOI: 10.1093/oso/9780198834229.003.0004

carrier waveforms within the telephone-quality voice spectrum (4 kHz) of POTS using frequency modulation (i.e., frequency-shift keying or FSK) which would be transmitted through the telephone network as *look-alike* analog telephone signals, and finally be received at the destination and demodulated by a modem back to the original binary signal. For the first time, this effort was made way back in 1940 by George Stibitz in USA, who connected a teletype machine from the city of New Hampshire to a computer in New York City through telephone lines. Such modems also found use in SAGE air defense system across USA and Canada.

The idea of transmitting data over the telephone networks was also explored in early 1960s by a group of deaf persons in USA, namely, James C. Marsters, Robert H. Weitbrecht and Andrew Saks, who on seeing the marvel of telephony, wanted to communicate with their deaf friends located in distant places (Lang 2000). However, being challenged by their handicap, they couldn't communicate through the voice-only telephone network, which was being used by others in the community. Driven by sheer passion, they eventually managed to share messages over the PSTN itself using indigenous add-on instruments over the telephones along with teletype machines. In this effort, as in the above modems, messages with English alphabet were coded into binary data using a teletype machine, which was in turn used to modulate the frequency of a carrier, and received at the destination end through PSTN, though this solution was not readily acceptable to AT&T and some legal issues came up against trying non-standard equipment over the AT&T-controlled PSTN. Eventually, the first commercial modem was marketed by AT&T in 1962, named as Bell 103, enabling full-duplex data communication over PSTN at the rates up to 300 bps. During late 1960s, Paul Taylor combined Western Union teletype machines with modems and built non-profitable devices and supplied them to the homes of many in the deaf community in St. Louis, Missouri, which later evolved into the USA's first local telephone relay systems for the deaf community.

Gradually, the modems became a popular device for exchanging data over PSTN and found extensive use in the community, more so after the arrival of Internet. Interestingly, much later in 2009, AT&T received the James C. Marsters (one of those three deaf persons) Promotion Award from TDI (Telecommunications for the Deaf, Inc.) for its efforts to enhance the access of the telecom service by handicapped citizens using modems, although they resisted initially the same effort from Marsters and his friends in 1960s!

Note that, by using the above modems one could *dial up* a number and set up a data connection between two telephone subscribers, but at the cost of blocking their own voice connections during the period of data transmission. This type of dial-up connection over PSTN was followed up by more efficient technologies, such as asymmetric digital subscriber line (ADSL), integrated services digital network (ISDN) etc., where voice and data connections could co-exist over the same transmission medium. ISDN was an effort to offer data and voice connections for residences and offices, all in digital format over PSTN, while ADSL and other members of digital subscriber line (DSL) family kept the voice in analog form and incorporated data on the same copper cable as PSTN. Deregulation in the telecom industry fueled these developments further, with cable TV also providing cable modem and telephony services.

The above developments eventually led to the new generation of access segment offering a broad range of services, which was in due course termed *broadband access networks*. However, with the growing popularity and penetration of the Internet and ever increasing applications from wide-ranging telephone/computing devices, such as smartphones, desktops, laptops, etc., broadband access networks needed to be scaled up with more bandwidth. At this stage, employing optical fibers in the access segment of telecommunication network indeed became worthwhile. Initially, cable TV network providers came forward in this

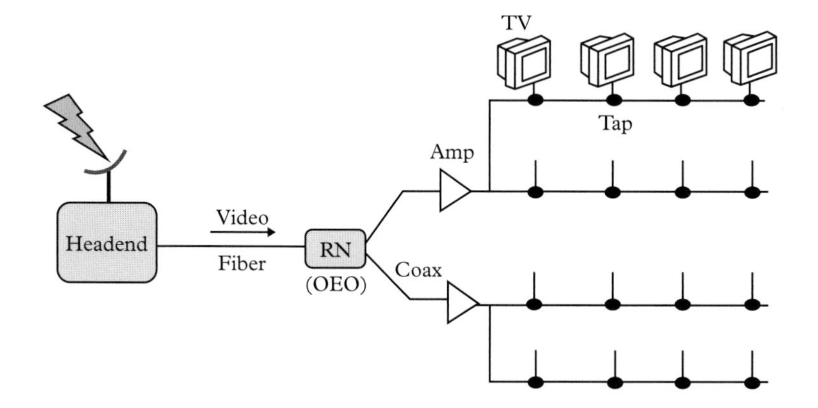

Figure 4.1 *HFC network using tree topology for video distribution.*

direction and upgraded the feeder segment of their networks with optical fibers, leading to hybrid fiber-coax (HFC) networks offering larger network span and bandwidth. An example HFC-based cable TV network is shown in Fig. 4.1, which was however a unidirectional video distribution network. Thereafter, optical fiber was introduced for bidirectional access networks, leading to broadband optical access network architectures.

4.2 Optical access network architectures

While the current metro, regional, and national networks predominantly use optical fibers, the access segments can have various possible choices for the transmission media: copper, wireless, and optical fiber. The optical access solutions in the form of point-to-multipoint configuration do not have the feature of mobility as in wireless, but offer huge bandwidth and can support much larger coverage area, e.g., 20 km and beyond. There are several possible topologies for optical access networks. Figure 4.2 shows some typical topologies, where the hub employs an optical line terminal (OLT) to establish a centralized connectivity with all the optical network units (ONUs) at the customer premises.

The topologies in Fig. 4.2 include two basic configurations: point-to-point configuration (resembling a star around the central hub) as shown in part (a), and tree configuration as shown in parts (b) and (c). The point-to-point configuration offers high per-user speed and supports large coverage area as compared to the other options, but with the complexity of a large number of transceivers at the OLT. The tree topology of (b) uses a feeder fiber up to a remote node (RN), where the signal is converted into the electrical domain for both upstream (ONUs to OLT) and downstream (OLT to ONUs) transmissions. The RN, typically located at a curb, distributes/collects information for upstream/downstream traffic.

The third option shown in part (c) uses a passive tree topology, popularly known as passive optical network (PON), which replaces the active curb switch of the previous configuration by a passive power splitter/combiner as the RN, thereby offering a more reliable access topology. Note that, PONs employ single-wavelength transmission (i.e., no WDM) in each direction, i.e., for upstream and downstream; however, it uses two different wavelengths, w_u and w_d (say), for upstream and downstream transmissions, respectively, as in the feeder fiber both traffic propagate concurrently in opposite directions. Note that, the downstream transmission in a PON employs one-to-many TDM transmission, while the ONUs on the upstream use time-division multiple access (TDMA) with appropriate synchronization, and thus functionally this configuration is also referred to as TDM PON.

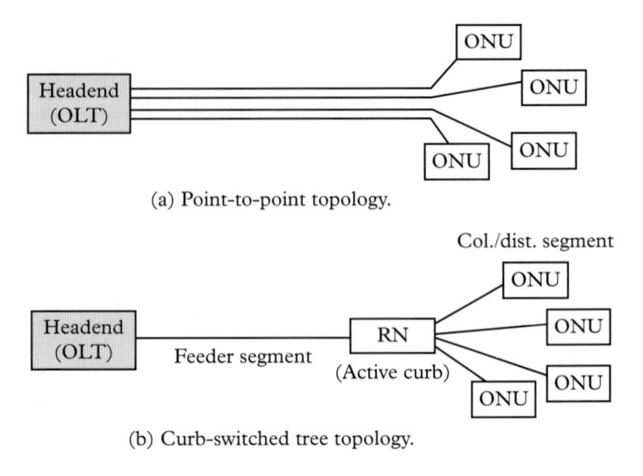

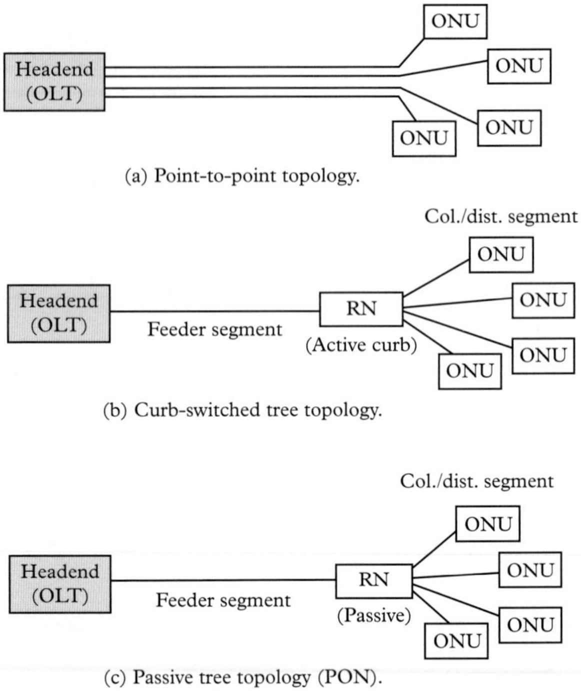

Headend
(OLT)
ONU
ONU
ONU
ONU
Col./dist. segment
Headend
(OLT)
Feeder segment
RN
(Active curb)
ONU
ONU
ONU
ONU
Col./dist. segment
Headend
(OLT)
Feeder segment
RN
(Passive)
ONU
ONU
ONU
ONU

(a) Point-to-point topology.

(b) Curb-switched tree topology.

(c) Passive tree topology (PON).

Figure 4.2 *Some example topologies for the optical access networks.*

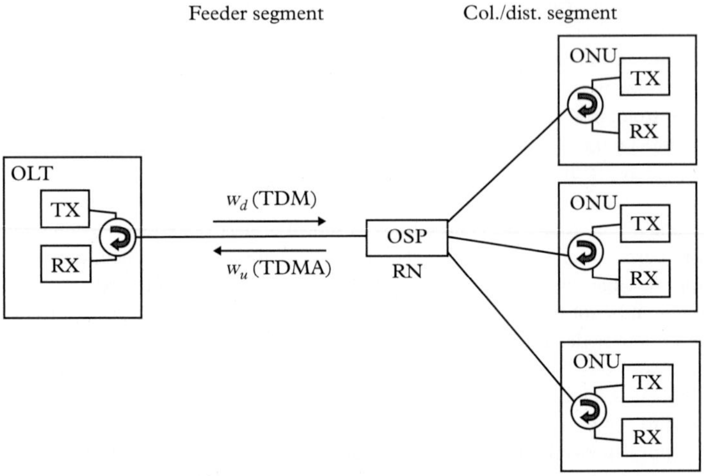

Feeder segment
Col./dist. segment
OLT
TX
RX
w_d (TDM)
w_u (TDMA)
OSP
RN
ONU
TX
RX
ONU
TX
RX
ONU
TX
RX

Figure 4.3 *PON configuration in detail, with optical splitter (OSP) functioning as the RN, and the circulators at the ONUs and the OLT for separating the up/downstream signals. Note that the downstream operates over wavelength w_d with one-to-many TDM transmission, while the upstream operates over wavelength w_u with many-to-one TDMA transmission.*

In the ONUs and the OLT, optical circulators are used to separate the two lightwaves for the transmitters and receivers. These details of the PON configuration are illustrated in Fig. 4.3.

Owing to the widespread usage of Ethernet (see Chapter 3) in access as well as other network segments (metro and long-haul), the Ethernet community adopted PON within its broad framework of access segment, known as, Ethernet in the first mile (EFM)[1] (IEEE EPON 802.ah 2004). Further, besides using optical fibers for Ethernet, EFM can also employ full-duplex copper cables over limited distances. Upon reaching a customer's premises, which might be a central place in a home, an apartment complex, or an office building, onward connections to the end-users are typically provided using full-duplex CAT6 wiring, which is finally terminated on home/office routers with the Ethernet LAN and Wi-Fi options. In the following sections, we describe the various operational features of PONs and some of the PON standards.

4.3 Passive optical networks (PONs)

The PON topology shown in Fig. 4.2(c) uses only one RN, while in general PONs can employ multiple RNs for larger coverage, although with additional complexity at the OLT end, as shown in Fig. 4.4.

In a typical PON, each ONU sends data over a specific non-overlapping time slot to the OLT through RN followed by the *feeder* fiber, while a receiving ONU selects the desired signal from the OLT sent through RN by *listening* to the appropriate time slot. PONs can have varying realizations, such as fiber-to-the-home (FTTH), fiber-to-the-building (FTTB), or fiber-to-the-curb (FTTC), collectively called FTTx networks (Fig. 4.5). The ONUs are located at the users' premises for FTTH, while for FTTB and FTTC, centralized locations are used in buildings or community clusters to provide ONUs, which in turn connect various customer groups through copper and/or wireless networks. Due to these versatile architectural options, PONs are becoming equally suitable for all kinds of residential as well as organizational access networks.

Note too that, as compared to other topological formations of network, tree-based PONs have some conspicuous sensitivity to fiber-cut. In particular, the survivability of a PON greatly depends on the uninterrupted existence of the feeder segment, which can be protected with dual-feeder configurations, as shown in Fig. 4.6. Survivability of PONs can be further enhanced by using a backup OLT and protection fibers in the distribution segment.

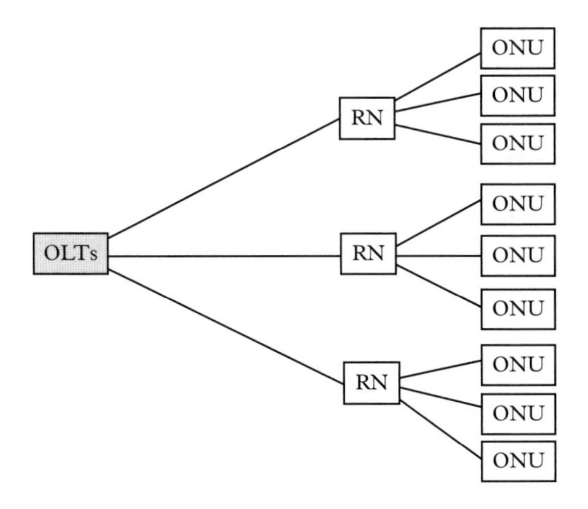

Figure 4.4 *PON with multiple RNs.*

[1] Note that, in EFM, the access segment is viewed as the first mile from the subscriber end, though in traditional sense it represents the last mile of network.

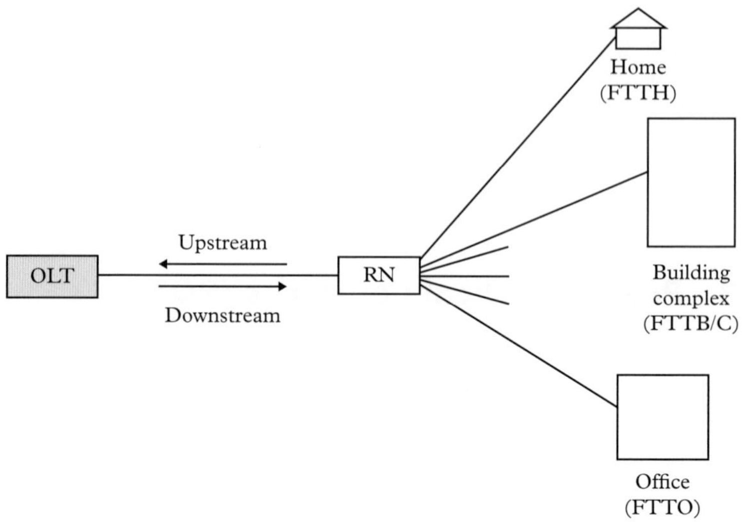

Figure 4.5 *PON supporting FTTx architectures (FTTH/B/C/O).*

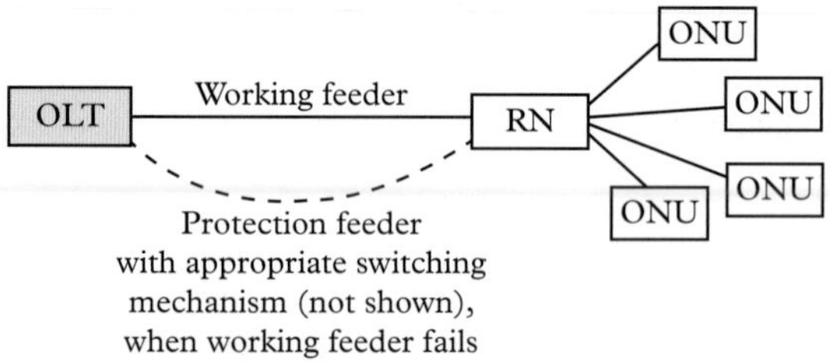

Figure 4.6 *PON with dual-feeder configuration for protection. Various other protection schemes for PONs are considered in further detail in Chapter 11 on network survivability.*

Another survivable formation for PONs is the multi-fiber ring topology, where the function of the feeder segment is realized over a ring, thereby offering network survivability against fiber-cut due to the closed-loop feature of the ring connectivity. We address the survivability issues for PONs and other segments of optical networks in Chapter 11.

4.3.1 Design challenges in PONs

PONs, with tree configuration using a single linear feeder segment, need to handle the upstream and downstream traffic streams in different ways. In particular, for one-to-many (OLT to ONUs) downstream transmission, it employs a broadcast transmission with the OLT located at the hub, distributing traffic to all the ONUs using TDM. However, the upstream packets from the ONUs to OLT need to flow with an appropriate MAC protocol for realizing many-to-one transmission without interfering with each other in the feeder segment. In other words, the upstream packets from all ONUs need to be time-division multiplexed in a distributed manner (i.e., by using TDMA) without any contention. Since PONs usually do not have ONU-to-ONU direct communications, the MAC protocol to realize the TDMA transmission from ONUs has to be coordinated from the OLT only through an appropriate polling-based scheme. Moreover, the MAC protocol needs to ensure fairness between the ONUs in respect of bandwidth allocation for upstream transmission from each ONU. With the bandwidth demands from the ONUs being time-varying in

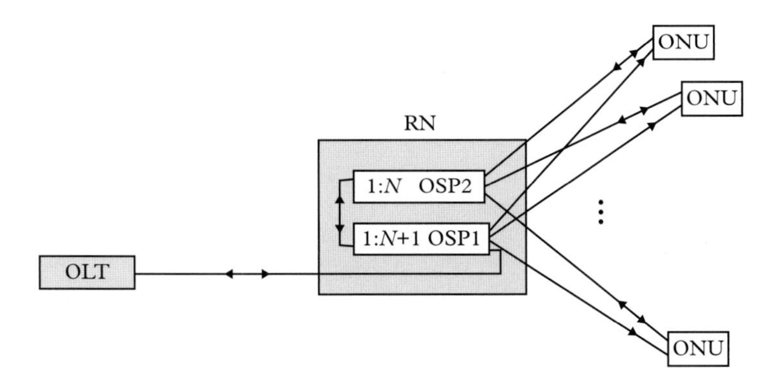

Figure 4.7 *PON with localized inter-ONU communication for higher bandwidth utilization.*

nature, the allocation process has to judiciously employ a dynamic bandwidth allocation scheme (DBA) for upstream transmission.

As mentioned above, the coordination needed for scheduling upstream communications from ONUs without any overlap/contention is carried out from the OLT through a polling process which causes a significant latency due to the round-trip time (RTT) between the OLT and the ONUs. However, this problem can be addressed by introducing some changes in the RN and the distribution segment (Foh *et al.* 2004) (Fig. 4.7). As shown in the figure, the RN uses a pair of optical splitters/combiners (OSP1 and OSP2), and each ONU is connected to both OSPs through two separate fibers. The two OSPs are also connected at one end. The downstream signal from the OLT first goes through OSP1 with $(1 : N + 1)$ configuration and is subsequently fed back to OSP2 with $1 : N$ configuration to reach all the ONUs. On the other hand, the upstream signals from the ONUs arrive first at the OSP2 with $1 : N$ configuration, which are next fed back to OSP1. From OSP1 the upstream signals go to the OLT as well as to all the ONUs including the sender ONU. This provides a localized broadcast between the ONUs, and the collisions that might occur between the upstream signals are taken care of by using appropriate MAC protocols, e.g., CSMA/CD, Aloha, or their variants. Though this configuration has some benefit in respect of avoiding the RTT-induced polling delay (between the ONUs via OLT), it needs additional hardware and increases the number of distribution fibers by a factor of two, and thus hasn't been popular so far.

Physical-layer design in PONs also needs special attention. The ONUs in a PON are in general located at different distances from the RN, leading to unequal received power levels at the OLT from different ONUs. This makes the design of the OLT receiver critical, as it needs to use different threshold levels for bit detection from different packet bursts coming in from different ONUs in quick succession (albeit with some guard time between the successive bursts of packets) with higher/lower threshold levels for nearer/farther ONUs. Thus, the OLT has to dynamically change its receiver threshold between the bursty data packets. Hence, the receiver in OLT, called the burst-mode receiver, needs to derive the threshold information from a set of preamble bits prefixed to each packet, which are also needed for clock recovery from the received signal. The receivers in different ONUs would also have to operate with different received power levels, though they don't need any burst-to-burst dynamic adjustment of the threshold over time.

Further, PONs with splitter/combiner (i.e., OSP) at the RN have the inherent problems of scalability due to the power-splitting loss at the RN. Moreover, with a large number of ONUs, the bandwidth of one single wavelength needs to be shared among all ONUs using TDM for the downstream transmission and TDMA for the upstream transmission. The capacity of the PONs can be greatly enhanced by using WDM transmission over PON topology, leading to the second-generation PONs. Later, we shall be dealing with WDM PONs in Chapter 7.

PONs need to interface with different types of clients, such as Ethernet, ATM, and TDM, and are categorized accordingly. In particular, PONs using Ethernet as a layer-2 client are named as Ethernet PONs (EPONs), and the PONs supporting ATM are called ATM PONs (APONs), while another version that evolved from APON is GPON, representing Gigabit-capable PON. The standards established by ITU-T for PONs include G.983 for APON (called broadband PON or BPON) and G.984 for GPON, where the GPON encapsulation method (GEM) is used to support Ethernet along with ATM, TDM, and IP/MPLS. On the other hand, IEEE has established the IEEE standard 802.3ah for EPON, or more precisely 1Gbps-EPON. However, as discussed in Chapter 3, all over the world, the success of Ethernet has overwhelmed its competitors, with about 90% of data traffic being generated from Ethernet clients, and thus Ethernet-compatible PONs (i.e., both EPONs and GPONs) have attained high popularity as compared to BPONs designed only for ATM. The inherent inefficiency of small cells in ATM, originally designed for network convergence, has been the principal reason behind much higher popularity of EPON/GPON over BPON. Overall, the bandwidth utilization in all these PONs depends greatly on the DBA scheme chosen for the upstream transmission. In the following, we describe the basic operational features of the DBA schemes, which have been adopted in different commercial versions of PONs.

4.3.2 Dynamic bandwidth allocation in PONs

The DBA schemes in a PON play a significant role in realizing bandwidth utilization of the optical fiber during upstream transmission. Designing an efficient DBA in a PON depends on several underlying issues, such as the size of the network, including the geographical span and the number of ONUs, the traffic pattern from the ONUs, and the choice of multiaccess scheme. Typically, a PON covers distances (between the OLT and the farthest ONU) up to 20 km, which leads to the one-way propagation delay of 20 km $\times$ 5 μs/km = 100 μs. Thus, the PONs will have a round-trip time (RTT) of about 200 μs, which will have a significant impact on the performance of a given DBA scheme.

As mentioned earlier, the downstream communication is one-to-many (OLT to ONUs) in nature and hence can manage with simple broadcast protocol from the OLT. However, the upstream communication is many-to-one (ONUs to OLT) having a *common path* through the feeder fiber, and hence is likely to suffer collisions in the time domain unless their transmissions are scheduled appropriately. The DBA schemes must address this problem, along with the guarantee of fairness in the access demands (number of bytes) granted to the constituent ONUs, and these issues are typically managed within the ambit of layer-2 operations in a network.

The basic task of a DBA scheme in a PON is to carry out TDMA-based distributed time-division multiplexing of the upstream packets from the ONUs. In order to realize an appropriate operational protocol, the DBA algorithm should ensure that the upstream packets from the ONUs are time-multiplexed without any temporal overlap. The traffic from each ONU, which on the user-end connects to either a single user or a multi-user LAN, such as Ethernet or Wi-Fi (see Fig. 4.8), is expected to be bursty and time-varying in nature. Given this scenario, the challenge of OLT lies in centrally carrying out remote scheduling for the upstream packets from ONUs, so that the fiber capacity is allocated to ONUs with high bandwidth utilization and minimum possible access delay. The OLT needs to allocate bandwidth to each ONU cyclically in a round-robin manner, thereby needing a *cycle time*, which can be made adaptive with the variation of traffic flows, or static with time. This leads to two categories of DBA: DBA using adaptive cycle time and DBA using a

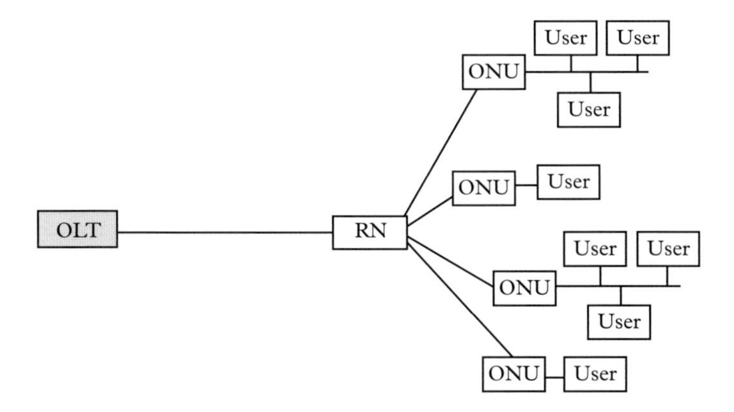

Figure 4.8 *PON with ONUs supporting single as well as multiple users.*

pre-computed fixed cycle time; the former category is adopted in EPONs, while GPONs use the latter category. For both EPONs and GPONs, the DBAs generally employ interleaved (multi-thread) polling, which is an improvization of the standard (single-thread) polling scheme. In the following, we describe these polling schemes for realizing DBAs in PONs.

4.3.3 Standard polling with adaptive cycle time

In a DBA scheme using standard (single-thread) polling, the OLT keeps polling each ONU to know the ONU's bandwidth demand (i.e., the number of bytes), and sending the granted number of bytes as a message, called GATE, to the ONU. As mentioned earlier, typically an ONU is connected on the user-end to one or a number of devices through a LAN or Wi-Fi. Hence, the ONU being driven by the active users/devices, at a time scheduled by the OLT, asks for a number of bytes (synonymous with bandwidth demand) for upstream transmission determined by its transmit-buffer occupancy. Thereafter, having received the grant from the GATE message sent by the OLT, each ONU transmits the upstream packet(s) of the granted size to the OLT along with its transmit-buffer occupancy (i.e., the next demand for the upstream transmission) as a piggy-backed message, called REPORT. This polling process for one ONU is carried out for all other ONUs sequentially over a cycle time, which must allow each ONU to transmit its upstream data once.

The above cycle time keeps repeating in a round-robin manner and can vary from cycle to cycle due to the time-varying bandwidth demands from the ONUs. Thus, at a time one single *thread of communication* between the OLT and one ONU keeps taking place, i.e., either on upstream or on downstream, leading to a single-thread polling scheme, which is known as the basic or standard polling scheme. As we shall see later, in *multiple-thread* or *interleaved* polling, the OLT can transmit to one ONU in downstream direction, while simultaneously receiving (thus using interleaved communications in two directions) the already granted bytes from another ONU. For both cases (i.e., single-thread and multiple-thread schemes) the upstream and downstream transmissions keep using two different wavelengths, w_u and w_d, respectively.

A single-thread DBA scheme for PONs is illustrated in Fig. 4.9, where the horizontal axis represents time and the vertical axis represents the distances of ONUs from the OLT. The time axis reflects the cumulative effect of propagation time, packet-transmission time, and packet-processing time on the overall delay performance of the network. In order to explain the polling mechanism, we assume that the ONUs have already registered in the network

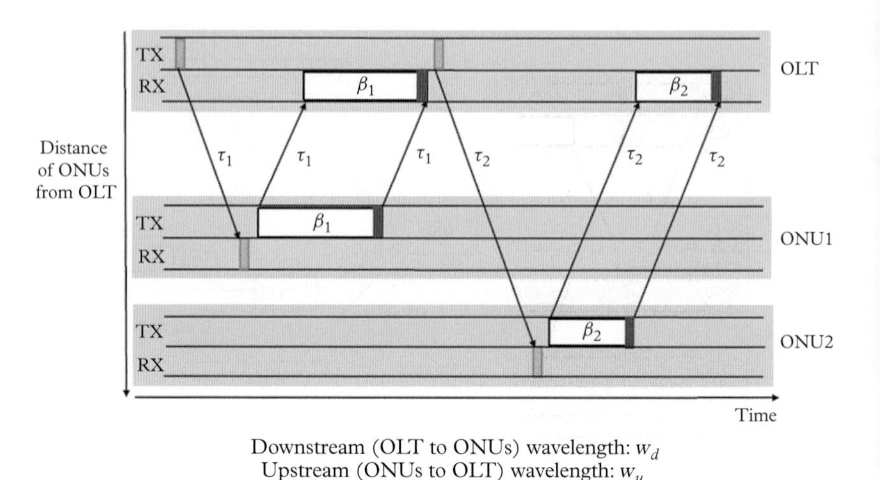

Figure 4.9 *Single-thread DBA scheme for PON.*

and gone through a proper initialization process (discussed later). As shown in the figure, on the extreme left end of the time axis, the OLT sends a GATE message (the corresponding REPORT message with bandwidth request being assumed to have been received during the previous cycle time) carrying the grant for ONU1 (say, β_1 bytes), which reaches ONU1 after a propagation delay τ_1. Thereafter, with a small processing delay, ONU1 transmits β_1 bytes, and appends at the end a REPORT message carrying its transmit-buffer occupancy. The OLT, upon receiving the data bytes and the REPORT message of ONU1 after the propagation time τ_1, responds to the REPORT message of ONU2 (received in the earlier cycle) and transmits the grant through the GATE message for ONU2 granting β_2 bytes. ONU2 receives the GATE message after the propagation time τ_2, and transmits the granted data bytes and the REPORT message in the same way as followed by ONU1, and thereafter starts the next DBA cycle (for the present 2-ONU example).

4.3.4 Interleaved polling with adaptive cycle time

As indicated earlier, in order to reduce the latency of long cycle time in the single-thread standard polling scheme, the OLT in a PON needs to employ interleaving in the polling process, thereby accommodating multiple threads of communications. In other words, when the OLT receives data bytes from an ONU, say ONU*j*, it can also send the GATE message with the granted bytes to another ONU, say ONU*k*, with a timing such that, soon after receiving the grant, ONU*k* can start its granted packet transmission without any contention with others or without creating a large time gap (except for the guard time) between the two successive packet transmissions in the upstream. This DBA scheme is known as *interleaved polling with adaptive cycle time* (IPACT), as illustrated in Fig. 4.10 (Kramer *et al.* 2002). As shown in the figure, the OLT transmits its GATE message to ONU2, while it keeps receiving the granted upstream bytes from ONU1 piggy-backed with the REPORT message (by using two different wavelengths in two directions). In this manner, the DBA carries out interleaved transmissions with the cycle times, which keep changing from cycle to cycle owing to the varying bandwidth demands from the ONUs – hence the name IPACT.

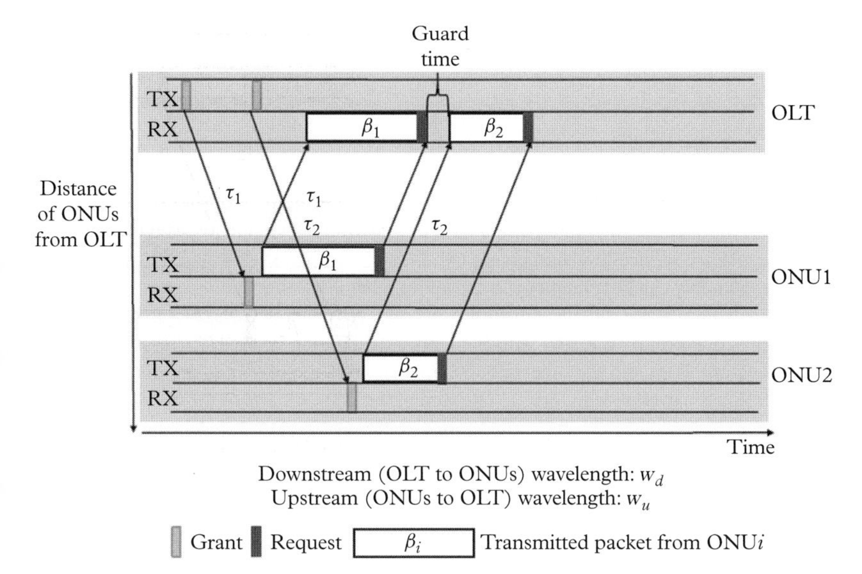

Figure 4.10 *Interleaved DBA scheme for a PON.*

Grant sizing and scheduling in IPACT-based DBA schemes

The cycle times in various DBA schemes might increase significantly, if most of the ONUs keep asking for large grant sizes in their REPORT messages. This greedy trend can be prevented by setting up a control mechanism in the OLT over the grant sizes allocated to the ONUs, leading to what is referred to as *grant sizing*. Furthermore, the OLT/ONUs can also control the access sequence of the ONUs/ONU-users by changing the order of the upstream messages within a cycle time, which is known as *grant scheduling*. We discuss these two schemes in the following.

Grant sizing

Grant sizing can be effected in several possible ways for a PON, by using different types of control schemes as follows (McGarry *et al.* 2008):

- free or gated grant-sizing (i.e., without any control),
- fixed grant-sizing,
- limited grant-sizing,
- limited grant-sizing with excess distribution,
- predictive grant-sizing.

In the free or gated grant-sizing scheme, as the name indicates, there is no control over the demands, and the OLT grants whatever is demanded by each ONU from its existing buffer occupancy, say β_i bytes without waiting for the demands of other ONUs, leading to an online grant-sizing scheme. In the fixed grant-sizing scheme, again an online scheme, the OLT imposes an upper limit for each ONU on the grant size, say B^{max} bytes, so that no ONU can transmit a packet beyond the maximum size. As in any authoritarian scheme, this type of allocation policy leaves some of the ONUs starved, while others do not use the maximum quota. In view of this, one can employ other flexible schemes, such as limited or predictive grant-sizing schemes as follows.

In the limited grant-sizing scheme, the grant for an ONU, say ONUi, is chosen as the minimum of β_i and B_i^{max}, given by

$$B_i = min\{\beta_i, B_i^{max}\}, \tag{4.1}$$

where B_i^{max} represents the upper limit of grant size for ONUi. This scheme does not allow any ONU to hog the fiber bandwidth, as with $\beta_i > B_i^{max}$, the grant size gets limited to B_i^{max}. Moreover, this scheme can fix the upper limit of grant independently for each ONU, based on the respective service-level agreement (SLA).

More flexibility can be brought in for the above scheme by using limited grants (LG) with *excess distribution*. In this scheme, during an ongoing cycle, the OLT classifies the ONUs into two groups, one group ($G1$, say) with the ONUs demanding $\beta_i < B_i^{max}$ (representing the underloaded ONUs) and the other group ($G2$, say) with the ONUs asking for $\beta_i \geq B_i^{max}$ (representing the overloaded ONUs). The sum of the unused parts of the maximum grants from the underloaded ONUs, i.e., $\sum_{i \in G1}(B_i^{max} - \beta_i)$ ($= E_T$, say), is estimated at the end of the cycle, and the same is distributed as excess load to the overloaded ONUs in the next cycle using an appropriate policy. Hence, the OLT needs to wait till the end of each cycle to receive all ONU REPORTs, leading to a class of offline schemes with longer cycle time.

The excess capacity available from the underloaded group of ONUs can be distributed to the overloaded ONUs using a few candidate schemes, such as

- LG with demand-driven excess distribution (LG-DED),
- LG with equitable excess distribution (LG-EED),
- LG with weighted excess distribution (LG-WED).

In the LG-DED scheme, an ONU from the overloaded group gets an excess grant ε_i^{DED} in proportion to its original request β_i, given by

$$\varepsilon_i^{DED} = \frac{\beta_i}{\sum_{i \in G2} \beta_i} E_T = \frac{\beta_i}{\sum_{i \in G2} \beta_i} \sum_{i \in G1} (B_i^{max} - \beta_i). \tag{4.2}$$

In the LG-EED scheme, the entire excess capacity E_T is equally distributed among all the ONUs. Hence, the excess grant ε_i^{LG-EED} is same for all ONUs, given by

$$\varepsilon_i^{EED} = \frac{1}{N_{G2}} E_T = \frac{1}{N_{G2}} \sum_{i \in G1} (B_i^{max} - \beta_i), \tag{4.3}$$

where N_{G2} represents the number of ONUs in the overloaded group $G2$. Grant sizing can also opt for weighted distribution of excess capacity, with prior weights assigned to each ONU, leading to the LG-WED scheme. Thus, with each ONU assigned a specific weight, say α_i, for the ith ONU, the excess grant ε_i^{WED} for the LG-WED scheme can be expressed as

$$\varepsilon_i^{WED} = \frac{\alpha_i}{\sum_{i \in G2} \alpha_i} E_T = \frac{\alpha_i}{\sum_{i \in G2} \alpha_i} \sum_{i \in G1} (B_i^{max} - \beta_i). \tag{4.4}$$

Note that in all of the above ED schemes, the estimated grant for an ONU ($= (B_i^{max} + \varepsilon_i^X)$, with X representing all the three schemes, LG-DED, LG-EED, and LG-WED) might exceed the actual request β_i in some cases, and in such cases the OLT has to reduce the estimated grant ($B_i^{max} + \varepsilon_i^X$) to the actually requested grant β_i to avoid the wastage of bandwidth.

Notwithstanding the merits of the limited grant-sizing schemes, one basic issue is left out, concerning the data accumulation at each ONU buffer, after sending the buffer occupancy value β_i in the REPORT message to the OLT, thereby calling for appropriate predictive schemes. By using a predictive scheme, the ONUs can estimate *ahead of time*, the actual value for the buffer occupancy (say, $\beta_i' \geq \beta_i$) that would be *seen* in reality by each ONU when it actually starts transmitting its next packet in the forthcoming cycle. One of the simplest and most elegant ways of carrying out this task is to use a linear growth model using the past experience in the previous cycle. In this model, one assumes a linear growth of the buffer occupancy over the time elapsed between the instant of sending a REPORT and the start time of the next grant transmission from the ONU. Interested readers are referred to (McGarry *et al.* 2008; Luo and Ansari 2005) for further detail on such schemes.

Grant scheduling

Prioritizing the ONUs within a given cycle can also be exercised by the OLT, leading to the inter-ONU grant scheduling schemes. Furthermore, even within an ONU dealing with multiple users, it is possible to schedule the users in an ONU with preset priorities, leading to the intra-ONU grant scheduling schemes. The two candidate options for inter-ONU scheduling are:

- longest-request first (LRF),
- earliest-request first (ERF).

In the LRF scheme, the longest request (i.e., the maximum grant size) is served first by the OLT and so on. On the other hand, ERF chooses the earliest request first, which lets the ONU send the earliest REPORT message to the OLT, to transmit first and so on. For the LRF-based scheduling, it is necessary to receive all the REPORT messages in a cycle, and hence such a scheme operates at the end of a cycle (i.e., functions as an offline scheme), thereby increasing the length of the cycle time. Note that, in reality, there might be some underloaded ONUs in a given cycle, for which the OLT need not estimate the grant sizes. In such cases, the OLT can send GATE messages to the underloaded ONUs without waiting for the ongoing cycle to end, and this improvement can help in controlling reasonably the cycle time in the LRF-based scheduling scheme (also applicable to LG-ED grant-sizing schemes).

Intra-ONU grant scheduling deals with multiple queues in an ONU dealing with as many users, and is usually executed at the ONU end. There are again a few candidate options for intra-ONU grant scheduling, as follows:

- strict priority (SP),
- weighted fair queuing (WFQ),
- start-time fair queuing (STFQ).

In SP scheduling, each queue is assigned a fixed priority, which might turn out to be unfair at times when a higher-priority queue happens to preempt an operational queue with lower priority. WFQ scheduling is a realistic version of a benchmark yet unrealizable scheme, known as the general processor sharing (GPS) scheme (Parekh and Gallager 1993; Keshav 1997). In order to get an insight into WFQ scheduling, we discuss first the underlying principle of the GPS-based scheduling scheme.

In a typical GPS-based scheme, there would be Q (say) queues being served by one server, just as we visualize in an ONU while dealing with multiple users. While handling such a multiple-queue single-server processing system, GPS splits each packet arriving at a queue into *infinitesimal pieces*. The server visits Q queues cyclically in a round-robin manner, serving an infinitesimal piece of the waiting packet in each queue (serving one in each visit), and keeps skipping the empty queues. In other words, the queues which have lower arrival rates (and hence found empty frequently) get the desired service readily, while the queues with higher packet arrival rates get their packets served with effectively higher service rates because the server gets to serve the non-empty queues in every cycle as compared to the empty queues, thereby achieving a fairness in the queuing system for both underloaded as well as overloaded queues. Such a scheme is said to offer *max-min fair share* processing, as both *max* and *min* demands get a *fair share* of the available server speed (i.e., service rate).

However, the fairness that comes along with the GPS scheme is not implementable in practice because the packets cannot be split into infinitesimal pieces in a practical network setting. In view of this, GPS is generally transformed into some realistic versions, one of them being the packet-by-packet GPS, also called as WFQ (as defined earlier) scheduling, which we describe in the following.

In the WFQ version of GPS, some simplifications are incorporated in the processing scheme. In the first step, for the arrived packets in the queues, WFQ computes the time instants $t_F(i)$ when its scheduler would have finished the service (i.e., complete transmissions) for those packets *if the GPS scheme was followed*. Thereafter, the server serves (transmits) the packets in the increasing order of service finish-time $t_F(i)$. Thus, WFQ mimics GPS internally, and thereafter arranges the transmission order as above by looking at the values of $t_F(i)$, which are actually replaced by rounded-up integers, called *finish numbers*, as the actual values of the time instants bear no significance for forming the ordered sequence.

Note that the computation of the finish numbers is carried out in reality by splitting the packets into their constituent bits, rather than infinitesimal (unimplementable) pieces. In particular, the scheduler finds out the *round numbers* for a packet (i.e., the number of rounds the queue is to be visited in the round-robin process to finish the packet transmission), and the finish numbers are computed using the round numbers along with the packet durations in terms of the number of bits in a packet. However, the WFQ scheme can be further simplified by using the start-time fair queuing (STFQ) system or its modified and more simplified version, modified STFQ (M-STFQ) scheme. Interested readers are referred to (McGarry *et al.* 2008) for further detail of the STFQ and M-STFQ schemes.

Overall, one needs to check the complexity of the scheme for intra-ONU scheduling and its delay performance with different classes of traffic. SP is the simplest scheme while being unfair to lower-priority requests, and WFQ and its variants can address the class-based performance issue, albeit with increased complexity in the ONU design. However, one can also carry out the intra-ONU scheduling at the OLT to avoid design complexity of ONUs, but the OLT-based schemes would be a challenging option while scaling up the network to long reach (with large RTT) and large number of ONUs.

4.3.5 Interleaved polling with fixed cycle time

The interleaved polling schemes discussed in the foregoing have been useful in EPONs, where the cycle time varies with time while adapting to the dynamics of upstream traffic. However, as mentioned earlier, another alternative to such DBA schemes is to set up a fixed cycle time, as used in GPON, and the bandwidth provisioning for each ONU is carried out with a philosophy of proportional bandwidth provisioning along with the consideration of

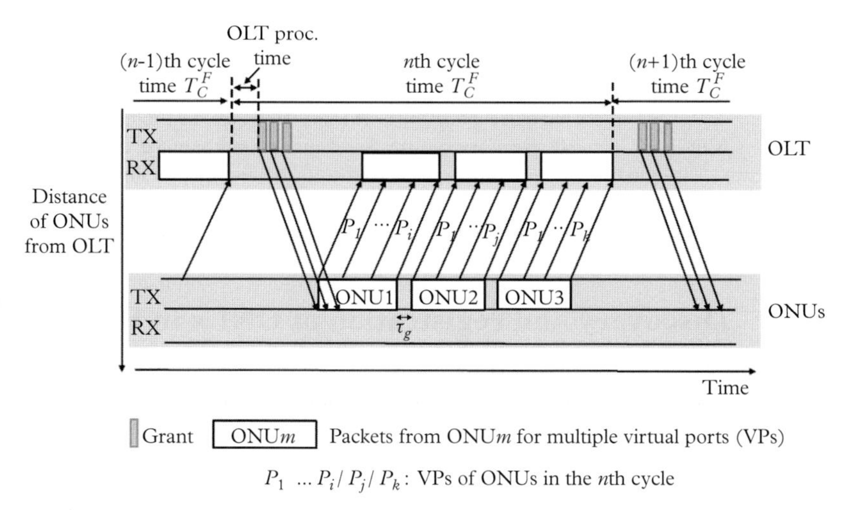

Figure 4.11 *DBA scheme with fixed cycle time for three ONUs, which are assumed to be at the same distance from the OLT (for simplicity), while being placed at different locations. Also, note that, although the grants from the OLT are sent at regular intervals δ_g, only the last grant sent in a cycle time is used by the ONUs. Hence, the intermediate grants are not shown to avoid confusion.*

priorities, if any. With this arrangement, an ONU can move on to the next cycle to get the remaining part of its requested bandwidth, if in the present cycle the requested bandwidth can not be granted fully. We examine below a typical DBA scheme of this category, following the approach used in (Lee *et al.* 2013).

Figure 4.11 illustrates a typical DBA scheme using interleaved polling with fixed cycle time. In general, during a given fixed cycle time T_C^F, the OLT keeps receiving requests in the REPORT messages from the ONUs and estimates at the end of the cycle, as to how much bandwidth each ONU can be granted so that the total grant does not exceed the cycle time (T_C^F) in the next cycle. Further, the DBA ensures that every ONU gets a share of the available bandwidth (i.e., the cycle time), commensurate with its requested bandwidth and the class of request (if any) as well. As mentioned earlier, each ONU may have multiple users, and Fig. 4.11 illustrates this feature using multiple virtual ports (VPs) for the ONUs, with each VP in an ONU sending its own packets P_i's.

In general, the grants from the OLT are transmitted on the downstream at a regular interval δ_g ($\ll T_C^F$), and the ONUs, having finished the transmission of the last grant, look forward for the latest set of grants from the OLT. In other words, the cycle time T_C^F being much longer than the time interval δ_g between the OLT updates on the grants for the ONUs, a given ONU picks up the latest update only, carrying the start time and the duration granted to the same ONU and transmits its data bytes accordingly. (See Fig. 4.11, where only the last grants are shown in a cycle time to avoid clumsiness in the diagram). With the cycle time being fixed (at T_C^F), the bandwidth requested by an ONU needs to be controlled/rationed using a mechanism to allocate the bandwidth that is proportional to the respective demand and honors as well the priority/class, if any. We discuss below this issue in further detail.

Consider that there are N ONUs, and r represents the upstream transmission rate from the ONUs, which makes the byte duration $T_B = 8/r$. Further, assume that ONUi has a demand for $\beta_{i,n}$ bytes, during the nth cycle for the forthcoming cycle, i.e., cycle $(n + 1)$. With these inputs, ONUi is granted a bandwidth (number of bytes) for the $(n+1)$th cycle, given by

$$B_i^{n+1} = \frac{\beta_i^n}{\sum_{i\in N}\beta_i^n}\left(\frac{T_C^F - N\tau_g}{T_B}\right) = \frac{r\beta_i^n}{8\sum_{i\in N}\beta_i^n}\left(T_C^F - N\tau_g\right), \tag{4.5}$$

where τ_g represents the guard time (not considered explicitly in the adaptive case, though it can be accommodated there as well) between the two consecutive upstream packet transmissions. Note that, this policy leads to a scheme similar in spirit to the limited grant sizing with ED, used in the adaptive cycle-time schemes. However, in the earlier case, the ONUs with lower demands were granted the full demands and only the total excess amount was distributed to the rest. In the present scheme, the requests of all ONUs are resized proportionately in each cycle time. Further, this scheme can also be extended for priority/class-based allocation (Lee *et al.* 2013) by granting weighted bandwidth to the ONUs during each cycle time.

4.3.6 Discovery and registration of ONUs in PONs

Discovering and registering a new ONU, or an offline ONU that was switched off for a while, is an important task in a PON, enabling an ONU to start participating in an operational PON. The discovery and registration process is carried out from the OLT following an automatic procedure as follows.

In general, the OLT periodically sends out some discovery messages on downstream to which the new/offline ONUs must respond with a request for registration. There might be contentions in the ensuing process, which is resolved using an appropriate contention resolution scheme. After an ONU request (sent for registration) is successfully received by the OLT, an ONU identification number is issued by the OLT to the respective ONU. Thereafter, all communications between the OLT and ONU in both directions use this ONU ID. Different commercial PONs have their own specific schemes befitting their steady-state operational MAC protocols, to carry out the registration of ONUs. In the following, while presenting the overall architectures of EPON and GPON, we will discuss the relevant features of the registration process used therein.

4.4 EPON, GPON and 10 Gbps PONs

EPON and GPON have emerged almost concurrently from two different communities – IEEE and ITU-T, respectively – and competed with each other over the years. Both are designed over tree topology and comply with 1 Gbps Ethernet (GbE), though GPON has been more generic in respect of accepting non-Ethernet clients as well (and hence relatively more complex than EPON). GPON also supports video distribution from the OLT through coarse WDM on downstream by using a separate wavelength. These standards have also moved later to 10 Gbps PON, called as XG-PON, with “X” implying 10 in Roman. Using the WDM technology, various research groups around the globe are now exploring the idea of extending the 1G/10G technologies toward WDM and TDM-WDM PONs with much enhanced capacity and reach, collectively referred to as next-generation PON (NG-PON), which we consider later in Chapter 7. In the following, we describe briefly the salient features of EPON, GPON, and 10 Gbps PONs.

4.4.1 EPON

EPONs are designed for supporting only Ethernet clients, developed by the IEEE Ethernet Forum (IEEE EPON 802.ah 2004). The basic architecture of EPON is based on tree topology with 1 Gbps transmission speed for upstream as well as downstream traffic (Fig. 4.12). The wavelength for the downstream transmission is 1490 nm, while the upstream transmission uses 1550 nm. Though the initial split ratio for EPON was specified

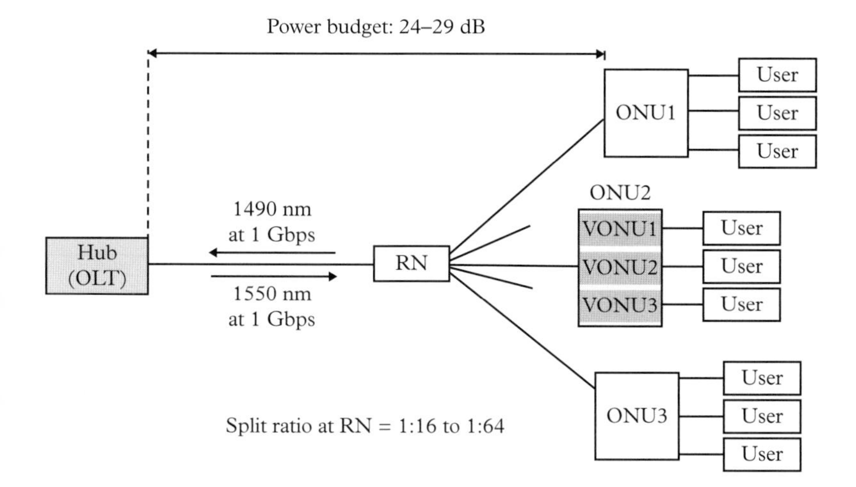

Figure 4.12 *Basic features of EPON.*

to be 1:16, gradually it was raised to 1:32 and 1:64 and in some cases, with the power budget increased from its initial value of 24 dB to 29 dB.

As expected, EPON continued with the spirit of *backward compatibility* with the pre-existing Ethernet frame format and named its protocol for the one-to-multipoint connectivity as multipoint-control protocol (MPCP). However, its preamble field was modified for the burst-mode receptions at the OLT. The ONUs in EPON send their upstream bursts of packets with inter-packet gaps (IPGs), and each burst is preceded by an IDLE code (as a preamble), a bit sequence that is friendly for clock recovery and threshold estimation. Each packet in EPON, be it for upstream or downstream transmission, carries a logical link ID (LLID) of the respective ONU, which represents the logical address of the virtual entity in the ONU (VONU) as the user identity. Note that each ONU in a PON is likely to serve multiple users, thereby leading to multiple LLIDs/VONUs for a given ONU.

The operation of the MPCP MAC protocol is illustrated in Fig. 4.13 with reference to the protocol stack of EPON, both at the OLT as well as at an ONU. The MPCP MAC is shown to operate as a sublayer in the data-link layer (layer-2) of the protocol stack for the OLT and an ONU. On the top of the MPCP MAC sublayer, the operation, administration and management (OAM) sublayer is placed to connect the MAC client (Ethernet), which in turn appears below the stack of the higher layers of LAN (layer-3 upward). Below the MPCP MAC sublayer, Ethernet MAC layer takes effect and interfaces with the physical layer (layer-1) of the network. The sublayers of the physical layer are the same as those described in Chapter 3 for Gigabit Ethernet.

Note that the OLT can instantiate multiple MAC *entities*, while an ONU opens up one such MAC entity at a time. This asymmetry is a topological need for any MPCP protocol, as the OLT needs to communicate concurrently with multiple ONUs during each cycle time with several point-to-point sessions running in parallel, while any ONU can only have one point-to-point communication with the OLT. Further, from the OLT a broadcast-specific MAC entity is also instantiated, as the OLT needs to send some information to all the ONUs from time to time, which are referred to as single-copy broadcast (SCB) messages.

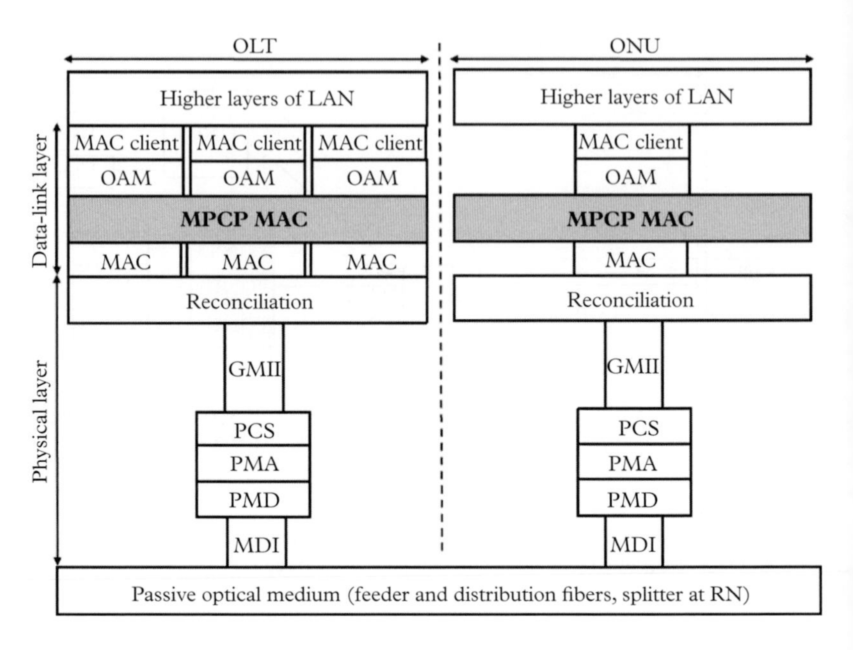

Figure 4.13 *EPON protocol stack using MPCP MAC as the sublayer in data-link layer (layer-2) at the OLT and ONUs. Note that, in this example, the OLT instantiates three MAC entities, while the ONU has one MAC instantiation only. GMII: Gigabit media-independent interface; PCS: physical coding sublayer; PMA: physical media attachment; PMD: physical media dependent sublayer; MDI: media dependent interface.*

The main functionalities of the MPCP MAC are carried out by three signaling protocols:

- discovery processing,
- REPORT processing,
- GATE processing.

Discovery and registration of the new/offline ONUs are carried out in EPON as follows. The OLT keeps periodically carrying out the discovery process in the form of GATE messages on downstream, by asking the unregistered ONUs to respond with an appropriate message for registration in the network. Upon receiving the discovery message, each unregistered ONU sends a request for registration using a message, called REGISTER REQ. Having received this message and approving the registration, the OLT registers the ONU in the network and confirms the same to the ONU using a REGISTER message. Finally, the ONU responds to the OLT with an acknowledgment message, called REGISTER ACK. This completes the registration process in EPON. Note that, an ONU in an EPON can register as a single ONU, or it can register itself as multiple VONUs to the OLT. Based on how an ONU has registered, the OLT treats each VONU in one single (physical) ONU separately as an independent ONU having a unique LLID, with due consideration to their SLAs and the necessary management operations.

REPORT processing includes the generation (at ONUs) and collection (at the OLT) of REPORT messages, using which the ONUs send their bandwidth requirements to the OLT. GATE processing includes the generation (at the OLT) and collection (at ONUs) of GATE messages, using which multiple packets from the ONUs are multiplexed (TDMA) without contention. However, the grant allocations for the ONUs follow the internal DBA scheme (i.e., IPACT) which does not fall under the jurisdiction of the standard, and is implemented in practice by the network providers.

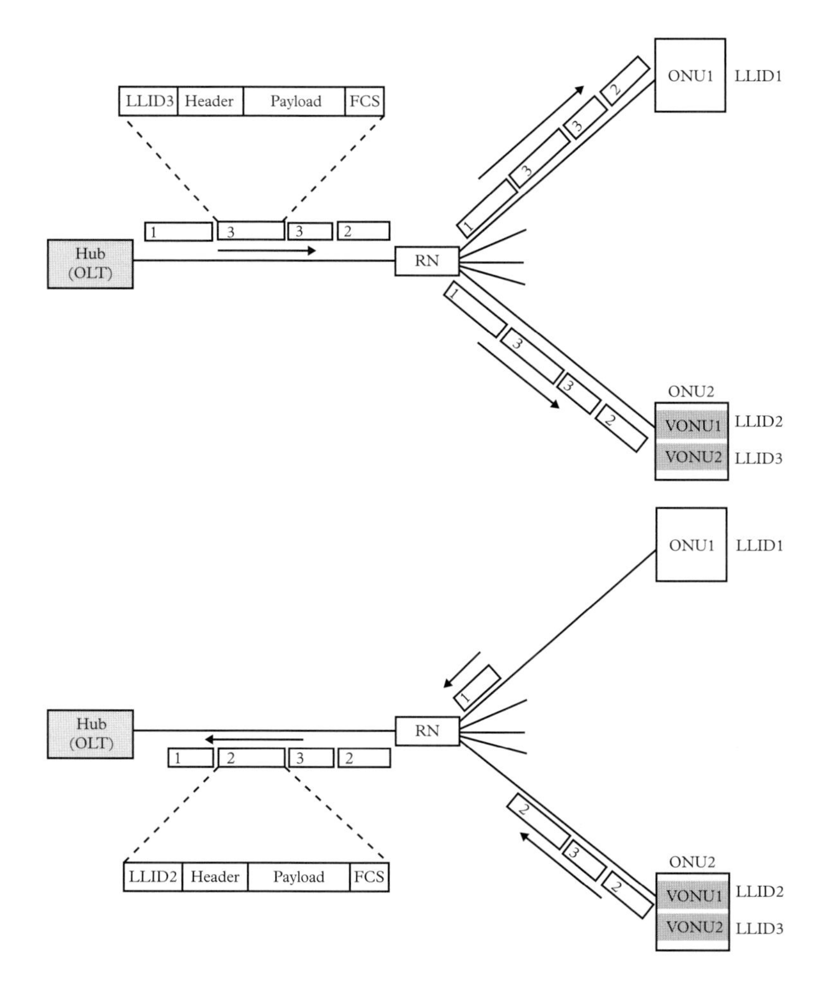

Figure 4.14 *EPON downstream transmission with frame structure. Note that, the downstream transmission is from one to many using TDM, and the gaps between the frames represent the IPGs.*

Figure 4.15 *EPON upstream transmission with frame structure. Note that, the upstream transmission is from many to one using TDMA, which is accomplished by tight synchronization and the appropriate DBA scheme, orchestrated from the OLT.*

Clock-based synchronization is another important task to be carried out in a PON, as it uses TDM in downstream and TDMA in upstream, both needing tight synchronization between the OLT and ONUs in the time domain. In particular, the OLT and ONUs employ 32-bit counters, which get incremented every 16 ns, and these counter readings represent the local time stamps, embedded in the time-stamp field of the MPCP data units. Through the clock-synchronization process, ONUs get aligned in time with the OLT clock. With this arrangement, the OLT can measure the values of RTT of all the ONUs using the return messages from them and can schedule collision-free upstream transmission from ONUs by timing the respective GATE messages in accordance with the corresponding RTT values. Further, EPON as a one-to-multipoint network needs to address the security of data for individual ONUs, which necessitates the use of data encryption. EPON also employs forward error correction (FEC) on its transmitted frames (though an option in the standard), based on Reed–Solomon channel coding using RS (255,239) code.

Figures 4.14 and 4.15 present the EON frame structures for upstream and downstream transmissions. The various fields of the upstream and downstream frames take care of the tasks to be carried out, as discussed above. These figures also illustrate the concept of

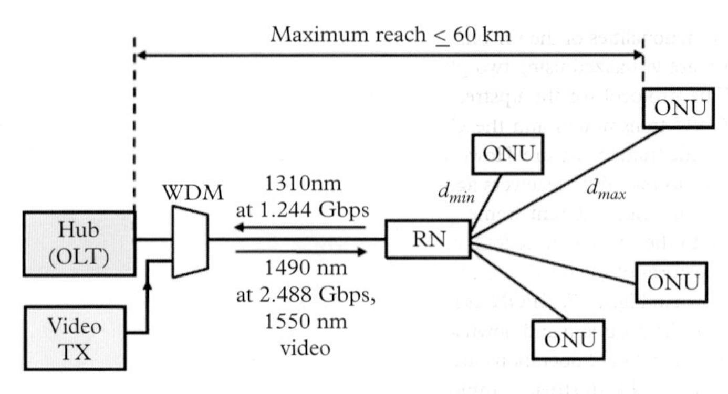

Figure 4.16 *Basic features of GPON.*

LLID-based ONUs/VONUs by showing both categories of ONUs: one ONU (ONU1) using single LLID, and another ONU (ONU2) using multiple LLIDs, with each LLID in ONU2 representing one VONU. Note that, the LLID field would be preceded by the preamble bits (not shown) for clock recovery and threshold estimation.

Finally, we discuss briefly the role of the OAM sublayer. This sublayer is used for link-layer management in EPON, so that the OLT can manage all the necessary operations with the ONUs from its remote location. The OAM functionality comes into play for an ONU, soon after it gets discovered by the OLT, which in turn keeps retrieving the various Ethernet variables of the remote ports, ensuring a smooth network operation.

4.4.2 GPON

GPON was developed by ITU-T as the standard G.984, where the nominal bit rates are 2.48832 Gbps for the downstream transmission and 1.24416 Gbps and 2.48832 Gbps for the upstream transmission. The basic features of GPON are illustrated in Fig. 4.16, where the bidirectional data communication takes place between the OLT and the ONUs at 1490 nm and 1310 nm for the downstream and upstream traffic streams, respectively. Furthermore, a provision for one-way video distribution is also offered on a third wavelength at 1550 nm, providing a similar service as in the HFC-based cable-TV networks. GPON can physically cover upto 60 km of reach, with 20 km of maximum differential reach between the ONUs, which needs to be supported by adequate receiver dynamic range in the OLT and ONUs. The maximum split ratio supported by GPON is 1:128, corresponding to a splitting loss of 21 dB.

The GPON standard, i.e., the ITU-T G.984 series, came out with four constituent parts, G.984.1, G.984.2, G.982.3, and G984.4, representing the basic architecture, physical medium dependent (PMD) layer, transmission convergence (TC) layer, and GPON management, respectively (ITU-T GPON general 2003; ITU-T GPON PMD 2003; ITU-T GPON convergence 2005; ITU-T GPON management 2005; Effenberger *et al.* 2007). Note that, unlike EPON, GPON needs to have a convergence framework to accept different clients, such as Ethernet, IP, MPLS, ATM, and TDM, which in turn necessitates the TC layer in the protocol stack of GPON. In the following, we discuss the operation of GPON for the upstream and downstream transmissions using the protocol stack shown in Fig. 4.17.

The functionalities of the various layers and sublayers of the protocol stack, as shown in Fig. 4.17, are visualized using two planes: user plane and control and management (C/M) plane. The protocol for the upstream transmission from ONUs to the OLT takes effect in the ONU transmitters and the OLT receiver in the downward direction in Fig. 4.17, using specific frame structure. However, the downstream protocols take effect from the OLT transmitter to the ONU receivers again through the same protocol layers in the downward direction, but with different frame structure and processing steps. In the following, with reference to the various protocol layers and sublayers shown in Fig. 4.17, we describe the relevant features of the upstream and downstream transmissions in a GPON.

As shown in Fig. 4.17, GPON uses two layers within the ambit of data-link layer (layer-2) to carry out its upstream and downstream operations: GPON-TC (GTC) layer and PMD layer. However, GTC operations are asymmetric in nature for upstream and downstream transmissions, with upstream employing distributed multiplexing of ONU frames using TDMA, and the downstream using localized TDM from the OLT end. The GTC layer in the user plane is split into two sublayers: TC adaptation (TCA) sublayer and framing sublayer, which keep operating with the support from the C/M plane.

First, we consider the upstream transmission from an ONU to the OLT. With reference to a given ONU, while looking downward in the stack of these two sublayers, the TCA sublayer employs an adaptation process on the ingress packets from the various clients, called the GPON encapsulation method (GEM), and thereafter forwards the encapsulated packets to the underlying framing sublayer. Thus, the packetized traffic from different client networks (e.g., Ethernet, IP, MPLS, TDM) get encapsulated into a common format, leading to a convergent solution for GPON. Adaptation of ATM, while being itself a convergence platform, is included in the GPON architecture for being BPON-compatible (shown by the dashed box in the TCA sublayer), but it has practically fallen out of the race with its competitors owing to its inherent bandwidth-inefficiency. The lower sublayer of GTC, i.e., the framing sublayer, interfaces with the PMD layer and forms the complete GTC frame structure for upstream transmission.

The framing sublayer deals with the frame synchronization, FEC, MAC operation, and physical-layer operation, administration and management (PLOAM). PLOAM covers several TC layer management functions: ONU activation, ONU management and

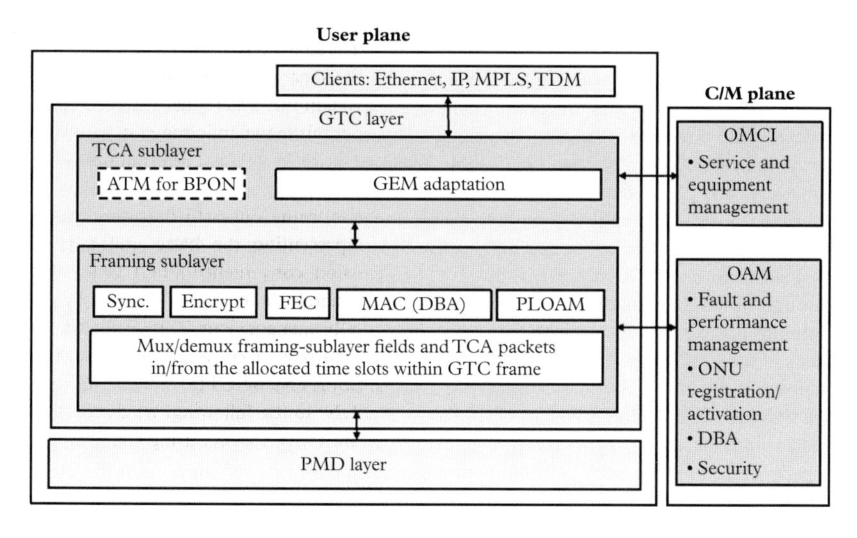

Figure 4.17 *GPON protocol stack.*

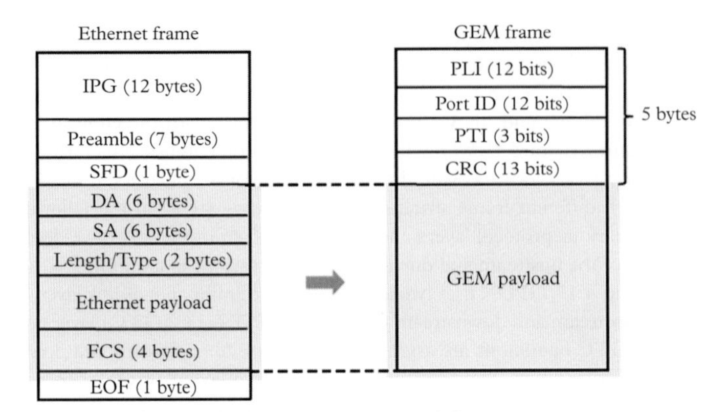

Figure 4.18 *Ethernet-over-GEM in GPON.*

control channel (OMCC) establishment, encryption, etc. The framing sublayer adds the corresponding headers when operating for the upstream traffic.

At an ONU the roles of GEM for upstream transmission in the TCA sublayer, take effect for Ethernet, TDM, IP and MPLS through Ethernet-over-GEM, TDM-over-GEM, IP-over-GEM, and MPLS-over-GEM encapsulations, respectively, with specific adaptation schemes. Figure 4.18 presents the Ethernet-over-GEM adaptation scheme, where some of the overhead fields in each Ethernet frame are removed (as these become redundant with GPON's own overhead fields), and the remaining part is inserted into the GEM frame in place of the GEM payload. In particular, the inter-packet gap (IPG), preamble, start frame delimiter (SFD), and end-of-frame (EoF) fields are deleted from the received Ethernet frame, while the others, including fields for the source and destination addresses (SA and DA), type/length field, and the Ethernet payload, are moved into the GEM-payload field.

Adaptation of the other clients – TDM, IP and MPLS – in the GTC layer are encapsulated into the GEM frames, as shown in Fig. 4.19, where the entire set of data bytes in an incoming packet/frame (i.e., without deleting any part) are moved into the payload field of the GEM frame. In both cases (i.e., Figures 4.18 and 4.19), on the top of the GEM payload, four fields are added by the GTA sublayer – payload length indicator (PLI), identity of a port (Port-ID) in an ONU (an ONU can have multiple ports as in EPON), payload type indicator (PTI), and cyclic redundancy code (CRC) for FEC – and these four fields altogether take five bytes with the breakdown as shown in the two figures. Furthermore, in all TC adaptation schemes, the entire frame is encapsulated with the preambles for clock recovery and threshold estimation, by using PLOu and PCBd fields, as discussed later.

As indicated earlier, the two GTC sublayers in the user plane, i.e., TCA and framing sublayers, are controlled by the C/M plane. Logically there exists a user data plane (U-plane, not shown explicitly) working alongside the C/M plane, and both work together to run the GPON. The user plane activities are carried out using the GEM-encapsulated payload from the TCA sublayer, with the C/M plane interacting with the GEM process through its ONU management and control interface (OMCI). The headers used in the framing sublayer mostly deal with the operation, administration, and management (OAM) block of the C/M plane. Thus, the roles of the C/M plane are divided into two parts: OMCI interworking with TCA sublayer for GEM encapsulation and the OAM interacting with the framing sublayer. In particular, OMCI looks after the management of various services such as Internet access, TDM voice, voice-over-IP, IPTV, etc. It also looks after the physical aspects of equipment, such as equipment configuration and power supply. On the other

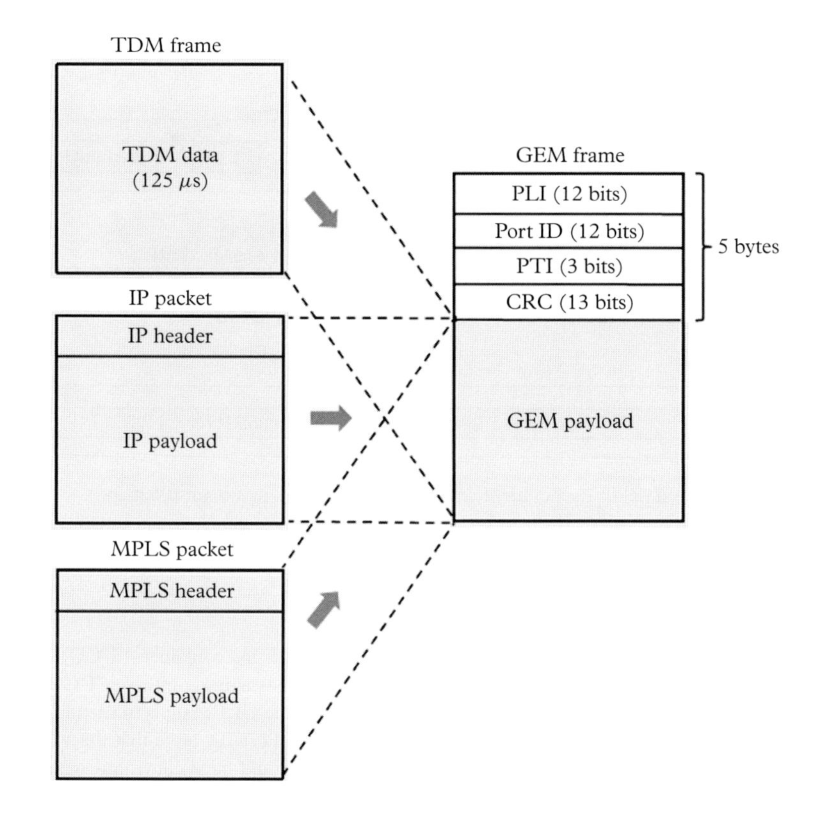

Figure 4.19 *IP/MPLS/TDM-over-GEM in GPON.*

hand, OAM deals with the GTC and PMD configurations, performance management, ONU activation, DBA operation, and security issues.

The GTC upstream output is passed downward through the PMD layer by adding the physical layer overhead for upstream (PLOu). Finally, the GTC upstream frames are transmitted in bursts with a fixed 125 μs duration, including the GEM frames along with the framing-sublayer and PMD header fields. Using an appropriate DBA scheme (with fixed cycle time) with the support from the C/M plane, upstream frames are time-multiplexed (using TDMA) without contention to reach the OLT end. Contention-free TDMA on the upstream is achieved by ranging and delay equalization processes, which are carried out during the activation of the ONUs from the OLT end. The OLT for the upstream packets plays the reverse role with the same protocol stack, but governed by the prior downstream grants already sent by itself through the GATE messages.

Next, we discuss the roles of PLOAM in the context of ONU activation. The ONUs are assigned identification numbers (ONU-IDs) by the OLT through the PLOAM messaging scheme during ONU activation process. Moreover, an ONU can instantiate multiple traffic-bearing entities, identified as allocation IDs (Alloc-IDs), like the LLIDs in EPON. These entities, i.e., Alloc-IDs, are in turn allocated upstream transmission grants. Note that the ONU-ID of an ONU is the default Alloc-ID for the same ONU, and over and above this ID, an ONU can have more Alloc-IDs through PLOAM messaging. The traffic-bearing entities

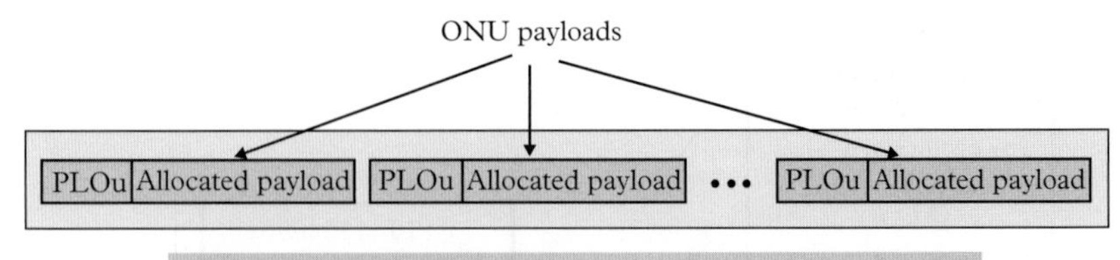

PLOu fields: Preamble, delimiter, BIP, ONU-ID, Ind
Header fields within each allocated payload: PLOAMu, DBRu

(a) GPON upstream frame structure (125 μs)

PCBd | Downstream time-multiplexed GTC payloads for ONUs

PCBd fields: Psync, Ident, PLOAMd, BIP, Plend, upstream BWmap

(b) GPON downstream frame structure (125 μs)

Figure 4.20 *Upstream/downstream frame structures in GPON.*

(corresponding to the Alloc-IDs) are mapped with the transmission containers (T-CONTs), through ONU management and control channel (OMCC) messaging. Any given T-CONT in an ONU can represent a set of logical connections, and the OLT assigns to each logical connection in a T-CONT a unique identity, called a GEM Port-ID. Note that the VPs in our earlier discussion on the DBAs with fixed cycle time would be synonymous with the T-CONTs in GPON.

The downstream transmission process uses localized TDM from the OLT. Thus the OLT generates GEM frames for the ONUs, and thereafter multiplexes and broadcasts these packets through its PMD to all ONUs with fixed 125 μs time frames with appropriate overhead field, called as physical control block downstream (PCBd). Each ONU, based on its own identity (ONU-ID), selectively processes only the packets that are addressed to itself. In the downstream operation, GTC has a unique role (absent in upstream operation) for synchronizing the entire GPON to a single 8 kHz clock, which is transmitted on downstream from the OLT.

The above protocols for the upstream and downstream transmissions are realized using appropriate frame structures, as shown in Fig. 4.20. Each upstream frame has a duration of 125 μs, and thus can accommodate 19,440 bytes for the transmission rate of 1.24416 Gbps, and 38,880 bytes for the transmission rate of 2.48832 Gbps. Each upstream frame can carry multiple transmission bursts from one or more ONUs. As shown in Fig. 4.20(a), each upstream burst in a frame consists of a PLOu section as the header consisting of several fields: preamble, delimiter following the preamble, bit-interleaved parity (BIP), ONU-ID, and indication (Ind) bytes providing the real-time ONU status reports. Each allocated payload section has a GTC overhead with PLOAMu and DBRu fields followed by the respective GTC paylod carrying the GEM payload and framing bytes. The downstream frames are also 125 μs long with the time-multiplexed GTC payload for all ONUs each preceded by the PCBd section playing the role of header for each GTC payload (Fig. 4.20(b)). The PCBd consists of several fields: Psync for synchronization, Ident for large framing structure, PLOAMd for carrying the downstream PLOAM message, BIP, Plend for the payload length, and upstream BWmap for the bandwidth allocation to each T-CONT. Both of the upstream and downstream frames are processed following the respective protocols, as described in the foregoing, and transmitted from the ONUs and OLT, respectively, to run the GPON based on fixed cycle times.

4.4.3 10 Gbps PONs

In an effort to enhance the capacity of GPON and EPON, ITU-T and IEEE came up with their next-level standards, viz., 10G-PON or XG-PON (ITU-T 10GPON 2012) and 10G EPON (IEEE 802.3av) respectively, which follow the similar architectural features to their predecessors, but operating at higher transmission rates.

IEEE's 10G EPON is an extension to the existing EPON to support enhanced speeds, with symmetric as well as asymmetric data rates. However, ITU-T has offered a completely new set of specifications for higher speeds in GPONs: XG-PON1 (with asymmetric rates for upstream and downstream transmissions) and XG-PON2 (with symmetric transmission rates). The asymmetric 10G-PON, i.e., XG-PON1, operates at 10 Gbps on downstream and at 2.5 Gbps on upstream. The symmetric 10G-PON, i.e., XG-PON2, transmits 10 Gbps also for upstream, but it needs to use expensive burst-mode lasers in ONUs to realize the upstream data rate. In Chapter 7, we extend the present discussions on PONs (or more precisely on TDM PONs) to describe the next versions of PONs employing WDM as well as mix-and-match of WDM and TDM toward achieving optical-access solutions offering much enhanced performance features.

4.5 Summary

In this chapter, we dealt with the optical access networks, employing PON configuration over tree topology. Having discussed how the access networks have evolved over time, we first presented different network configurations for realizing optical access networks. In PONs, while the downstream traffic requires one-to-many TDM transmission from the OLT to ONUs through the RN(s), the upstream traffic with many-to-one transmissions from the ONUs to OLT via RN(s) needs appropriate MAC protocol, arbitrated from the OLT end. The OLT operates with a specific polling scheme, granting the requests for bandwidth from the ONUs, upon receiving which the ONUs get to transmit some or all of their packets accumulated in the buffers to the OLT. The entire process, referred to as the DBA scheme, goes on cyclically over time in a round-robin manner, creating a latency in transmission determined by the propagation time through the network, ingress traffic at ONUs, and the efficiency of the DBA scheme in use. In general, DBA schemes in PONs can use fixed or adaptive cycle time.

Several DBA schemes with interleaved polling using adaptive as well as fixed cycle time were described. The basic mechanism for each scheme was described in detail, with reference to the common PON standards, e.g., EPON and GPON, with EPON using IPACT for its polling scheme over adaptive cycle time, and GPON using fixed cycle time with weighted-granting of requests in each cycle commensurate with the requests sent from the ONUs. Finally, the PON standards (EPON, GPON, and 10G PONs) were presented, and the salient features of the respective network architectures were explained.

EXERCISES

(4.1) Discuss the various candidate topologies for realizing optical access networks and indicate why PON has evolved as the favored option. Also discuss the basic differences between the EPON and GPON architectures.

(4.2) Consider a PON operating at 1 Gbps (in both directions) with a maximum reach of 20 km, using single-stage RN. Assume that the transmitters in the PON transmit an optical power of 3 dBm and the receiver sensitivity is -27 dBm. The optical splitter in the RN is realized using multiple stages of 2×2 splitters as the building block, with each having an insertion loss of 0.3 dB. Calculate the received power levels for three different values of the number of ONUs in the PON: (a) 32, (b) 64, and (c) 128. Indicate which value(s) for N will offer a feasible power budget for the PON. Given: fiber loss = 0.2 dB/km, connector loss = 1 dB.

(4.3) Consider the PON with the largest number of ONUs from Exercise 4.2 that can offer a feasible power budget. For this PON, assume that the average grant size requested from each ONU is 5000 bytes, and that all the ONUs are equidistant (i.e., 20 km) from the OLT. Calculate the average cycle time of the PON. Given: guard time = 1 μs, velocity of light in optical fiber = 2×10^8 m/s.

(4.4) A PON operates at 1 Gbps both ways with three ONUs. The REPORT and GATE messages use 60 bytes. The processing time for reading these messages is 3 ns. Initially, the OLT sends GATE messages using broadcast transmission to all ONUs at time t ($t = 0$, say) with zero grant sizes. In the ensuing cycle, the OLT polls only after REPORT from the last ONU arrives. Assume that the ONUs 1, 2, and 3 are located at 10 km, 15 km, and 20 km from the OLT and have the queue sizes of 8000, 10,000, and 20,000 bytes, respectively, before sending their REPORT messages. Having sent the REPORT messages, the ONUs wait for their GATE messages from the OLT, and start data transmission immediately after receiving the respective GATE messages. The OLT employs DBA using interleaved-polling scheme with a guard time of 1 μs, starting with the nearest ONU. Find out the scheduling times for the GATE messages at the OLT to the three ONUs, and the cycle time for the following grant-sizing schemes:

a) fixed scheme, granting 10,000 bytes to each ONU,

b) gated scheme.

Given: velocity of light in optical fiber = 2×10^8 m/s.

(4.5) Consider again the PON of the above exercise (Exercise 4.4) and assume that the DBA employs:

a) limited scheme with the maximum grant size of 9000 bytes,

b) limited scheme with excess distribution (using maximum grant size of 9000 bytes).

Find out the grant sizes, scheduling times for the GATE messages at the OLT, and the cycle time for the two schemes.

(4.6) Consider a DBA framework where the ONUs can be polled only after receiving the REPORT from the last ONU in a cycle. It is understood that a walk-time of at least an RTT is wasted between the grant cycles. Apart from polling the ONU with the least RTT first, what could be a possible grant-sizing strategy to reduce the delay experienced by the worst-affected packets due to long walk time?

(4.7) Consider a PON employing IPACT-based DBA with the gated grant-sizing scheme for N ONUs. The following parameters are defined to model the gated grant-sizing scheme.

τ_j: RTT/2 for ONU_j,

λ: average arrival rate in bits per second at an ONU from the user-end,

T: bit interval,

b_R: size (bits) of REPORT message,

b_G: size (bits) of GATE message,

τ_g: guard time,

t_i^j: time of arrival of the ith transmission from ONU_j at the OLT.

q_i^j: queue length (bits) of ONU$_j$ after completion of the ith data transmission,

g_i^j: grant size (bits) allocated by the OLT to ONU$_j$ for the ONU's ith transmission (including the REPORT message).

Using the above definitions, determine the expressions for:

a) grant size g_i^j in terms of q_{i-1}^j, which is the queue length after completion of the $(i-1)$th data transmission,

b) queue length q_i^j in terms of λ, t_i^j, t_{i-1}^j and g_i^j, g_{i-1}^j for $i > 1$

c) initial transmission time t_1^j in terms of t_1^{j-1},

d) transmission time t_i^j in terms of t_{i-1}^j and g_{i-1}^j if the PON has only one ONU.

(4.8) Consider a single-ONU PON, where the average delay D_{av} incurred by a packet from the instant of arrival at the ONU from the user-end until its delivery to the OLT is estimated by an $M/G/1$ queuing system, with the assumption that the RTT is relatively small compared to D_{av}. If the packet arrival at the ONU from the user-end follows Poisson distribution with a mean arrival rate of λ and the arriving packets have an average length L_{av} and a variance σ_L^2, determine the expression for D_{av}. Ignore guard time.

SONET/SDH, OTN, and RPR

5

With the emergence of technologies for high-speed transmission through optical fibers, the pre-existing TDM-based transmission systems using plesiochronous digital hierarchy (PDH) appeared unsuitable for achieving network synchronization. Various efforts to address this problem from the international standard organizations led to the development of synchronous optical network (SONET) and synchronous digital hierarchy (SDH) as the two equivalent standards for deploying circuit-switched optical networks. Several bandwidth-efficient techniques were also developed to carry packet-switched data over SONET/SDH networks, offering some useful data-over-SONET/SDH architectures. Subsequently, with the increasing transmission rates for SONET/SDH and Ethernet-based LANs, a convergent networking platform, called optical transport network (OTN), was developed. With the ever-increasing volume of bursty data traffic, a standard for packet-switching ring networks, called resilient packet ring (RPR), was also developed as an alternative to SONET/SDH ring in the metro segment for better bandwidth utilization in optical fibers. In this chapter, we first present the SONET/SDH networks and the techniques for supporting the data traffic therein, followed by a description of the basic concepts and salient features of the OTN and RPR networks.

5.1 Arrival of SONET/SDH as a standard

Optical fiber communication systems emerged in the mid-1980s as a timely solution for the growing demands of long-haul high-speed digital communication links in PSTNs. This in turn necessitated the development of an international standard for digital optical communication systems. The standard assumed paramount importance for the manufacturing of various subsystems: optical transmitters, receivers, fibers, and other related components and equipment. Notwithstanding the early efforts for developing some proprietary optical communication systems, international standards were essential to enable the manufacturers to develop compatible networking products, so that the various network operators could obtain competitive prices for the necessary products. Unfortunately, the existing standards for copper and radio-based communication systems were not scalable to cope with the high transmission rates of optical-fiber communication systems, and thus the various international organizations – Bellcore, ANSI, CCITT/ITU-T – started working towards developing universal optical communication standards. This led to a new generation of circuit-switched TDM standards in late 1980s. One of them, called synchronous optical network (SONET), was developed by ANSI and mainly used in North America (Ballart and Chiang 1989; ANSI SONET X3T12 1995; Kartalopoulos 2004). For the rest of the world, CCITT/ITU-T developed a similar multiplexing standard, called synchronous digital hierarchy (SDH) (Kartalopoulos 2004; ITU-T G.803 2000). The SONET/SDH-based optical fiber networks brought a great change in the way the metro and long-haul networks were deployed, thereby opening up great possibilities with the potential to provide much higher transmission rates through optical fibers.

Optical Networks. Debasish Datta, Oxford University Press (2021). © Debasish Datta.
DOI: 10.1093/oso/9780198834229.003.0005

Subsequently, various techniques were explored to carry packet-switched data traffic (typically with transmission rates in the range of 10–1000 Mbps) through the circuit-switched SONET/SDH networks, leading to developments of several data-over-SONET/SDH architectures: general framing procedure (GFP), virtual concatenation (VCAT), link-capacity allocation scheme (LCAS), etc. With the deployment of networks using even higher transmission rates ($\geq$ 10 Gbps), both for the SONET/SDH and Ethernet-based networks, a convergent networking standard became essential, leading to the standard known as optical transport network (OTN). Further, with the ever-increasing volume of data traffic, a standard for packet-switching ring networks, called resilient packet ring (RPR), was also developed for the metro segment. In this chapter, we first present the SONET/SDH networks along with the techniques for supporting packet-switched data traffic. Thereafter, we describe the basic concepts and salient features of the OTN and the RPR-based networks.

5.2 Synchronization issues in telecommunication networks

In telecommunications networks, the need for synchronization can be broadly divided into two categories, *link synchronization* and *network synchronization*. Link synchronization is needed in point-to-point digital communication links between two nodes, and deals with the synchronization of the receiving node with the incoming signal from the transmitting node in respect of the carrier, bit, and frame. For the carrier synchronization, the receiving node must ensure that its local oscillator is synchronized with the carrier of the incoming signal in respect of its phase and frequency, if the incoming signal is modulated (and hence a bandpass signal) and needs coherent demodulation. However, for the communication links transmitting directly a baseband digital signal or employing non-coherent demodulation of the received bandpass signal, the receiver won't need to carry out any carrier synchronization. However, bit and frame synchronization are the essential steps for any point-to-point digital communication link, with the bit synchronization subsystem carrying out *clock recovery* from the received baseband signal to sample the bits at the right instants of time for decision-making, and the frame synchronization subsystem identifying the frame boundaries from the received bit stream.

However, the network synchronization, which must function across the entire network, is an entirely different task, where a given node needs to synchronize its *local clock* with the clocks used by the baseband data streams arriving from the various other nodes, so that the node under consideration can read, add/drop, and switch the received as well as local bit streams, without any *slip* between the various incoming signals and the local clock. The main problem in carrying out the network synchronization is that the signals coming from the various other nodes would be generated using different clock generators at the respective transmitting nodes, and the local clock may not be able to synchronize with all these clocks simultaneously, unless they are derived from one single clock or get a chance to mutually control themselves for realizing synchronization. Thus, the fundamental problem lies in the fact that, when using independent clock generators, even if their mean rates or frequencies are fixed at the same value, their instantaneous rates would vary with time independently, leading to uncontrolled relative timing jitters between themselves. We discuss the various aspects of the network-synchronization problem in the following.

Generically, there are three basic strategies for network synchronization: using independent clock generators at various nodes across the network with full *plesiochrony* (a Greek word, meaning *anarchy*), hierarchical master-slave (HMS) synchronization, and mutual synchronization (Bregni 1998; Bellamy 2003). In a plesiochrony-based system, as the name suggests, no node can synchronize its local clock with all the incoming signals from other nodes at a time (which is obviously not a feasible option). Hence, each node in a plesiochronous network attempts to *patch up* or *manipulate* the difference in bit timings of incoming and local signals during the multiplexing process, if any, by using some additional bits or pulses. The HMS scheme relies on one or more *master* clocks placed judiciously at some nodes, such that the remaining nodes can derive the clock as *slaves* from the master. The mutual synchronization scheme is based on the collaboration between the clocks at different nodes without any notion of master-slave relationship, offering in principle a democratic and reliable scheme, though its implementation across a network is indeed a challenging task (Bregni 1998).

Figure 5.1 shows an example network to explain the network synchronization issues, where five nodes, or more precisely switching nodes (SNs), are interconnected to form a network at the next higher level with respect to the COs (COs being the lowest-level switch in PSTN), and each SN is connected to several COs to exchange the basic tributaries, such as T1s or E1s (for brevity, only SN1 is shown to have been connected with COs in Fig. 5.1).[1] With multiple bit streams or tributaries arriving at SN1, the respective line cards at the input ports of SN1 can recover clocks from and synchronize with each one of them separately using well-established clock-recovery schemes, and store the frames at the respective buffers.

When the above network runs with the plesiochronous synchronization scheme, the clocks of the arriving tributaries are independent of each other (notwithstanding the fact that they all have almost the same mean transmission rate), and hence, have uncorrelated timing jitters. Thus, performing the real-time switching operations on them at SN1 becomes significantly complex, thereby needing an appropriately designed *manipulative* scheme, so that the data streams from different input ports can be dropped, added, and switched without any temporal *slip* between them. Note that, similar problem of plesiochrony would also be observed in the SNs while receiving signals from other SNs (i.e., between the nodes operating at higher levels of switching hierarchy) but with higher transmission rates and hence needing more complex hardware. With the other two schemes, this problem is eased due to the master-slave or mutual interactions between the nodes, albeit with additional arrangements.

Figure 5.2 illustrates the three network synchronization schemes with the eight-node example network shown in part (a). Part (b) of the figure with no logical links between

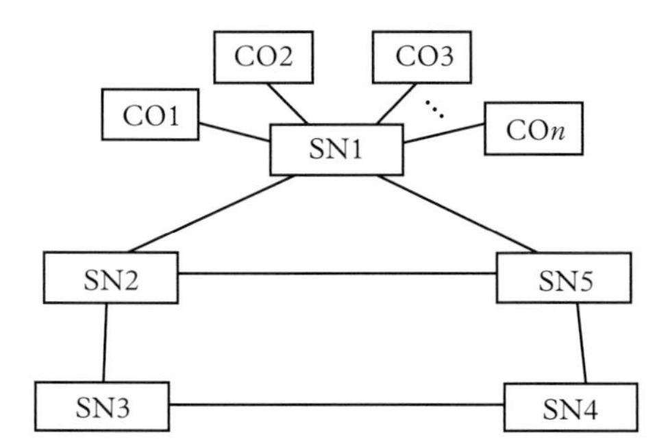

Figure 5.1 *Networking with the switching and edge nodes, where the edge nodes function as COs. The COs for only one switching node (SN) are shown for simplicity.*

[1] Note that, as discussed in Chapter 1, there will be a few more levels of SNs in the PSTN hierarchy. However, without any loss of generality, we don't consider them in the figure to keep the discussion lucid.

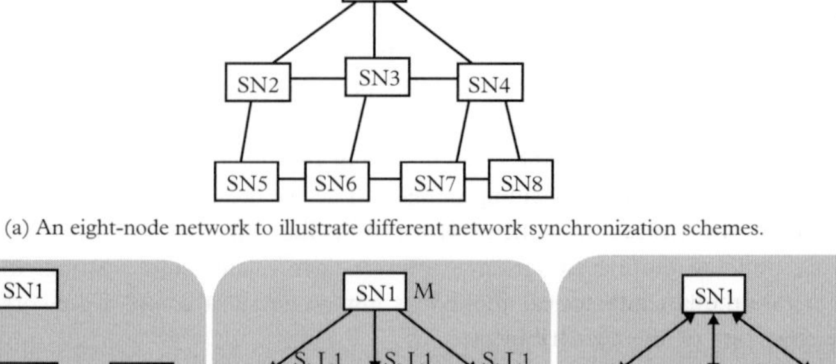

(a) An eight-node network to illustrate different network synchronization schemes.

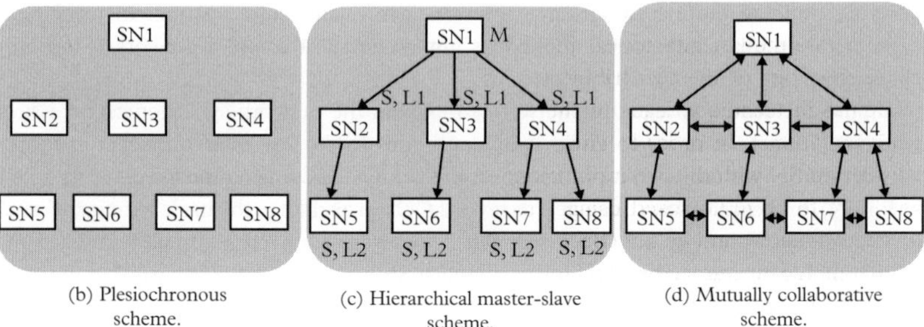

(b) Plesiochronous scheme.

(c) Hierarchical master-slave scheme.

(d) Mutually collaborative scheme.

Figure 5.2 *Illustration of three network synchronization schemes: plesiochronous, HMS, and mutually collaborative. M: master, S.Li: slave at level i. Note that in the plesioschronous scheme, there is no logical connectivity between the nodes in regard to the network synchronization, while for the HMS scheme, the clock information flows downward (shown by the downward arrows) from the master level to levels 1 and 2. For the mutually collaborative scheme, all nodes can interact with each other, as shown by the bidirectional arrows.*

the SNs, implies that there is no coordination between the clocks at the various nodes representing the plesiochronous scheme. Part (c), representing the HMS scheme, indicates that the reference clock flows down from a master (could be more than one, placed in different regions) with the downward arrows indicating the direction of flow of clock information. In this case, the slaves in the third row (level 2) receive indirect synchronization from the master through the slaves at level 1, hence with more jitter as compared to the slaves in the second row (level 1), which receive the clock information from the master directly. Generally, the slaves obtain the clock by line-timing (i.e., by aligning in time domain using an appropriate timing-recovery scheme) with the higher-level slave or a master. The logical tree-connectivity can be realized on other physical topology as well, such as in rings or meshes. In practical networks having varying topologies, there can be more than one source of clock available to a node, and it needs to choose one of them, depending on the in-line jitter introduced by the transmission path traversed. In Fig. 5.2(d) for mutually collaborative synchronization scheme, the connections between the nodes are shown with bidirectional arrows, implying that all the nodes collaborate with each other to mutually fine-tune their clock timings. The plesiochronous and the HMS schemes are the ones used in practice for PDH and SONET/SDH-based networks, respectively, and hence we discuss below the basic principles of these two schemes in further detail.

5.3 Network synchronization in PDH-based networks

The plesiochronous network synchronization scheme, as adopted in the PDH standard, is designed around the fact that an SN would accumulate slightly different numbers of bits over a given time frame in the buffers at its different input ports, while receiving the tributary bit streams from the other nodes (using the same mean clock rate but with independent timing jitters). As a result, the faster tributary will bring in a higher number of bits than the slower ones in their respective SN buffer over a given frame duration (generally, 125 μs). When these received bits from different buffers are to be multiplexed together for a higher-level tributary, the buffers of each lower-level input tributary may not have the same number of bits for multiplexing.

In order to address the problem of plesiochronous networks, the output is synchronized to one of them (the faster one), while adjustment is made for others. Thus, for a given two-tributary example, the faster tributary will have a specific number of bits in the buffer, while the slower one will let one of its allocated bits go with a *stuffed* pulse or bit (equaling zero or one, as governed by the rule set up across the network), but leading to an ambiguity as to whether this stuffed bit is the real bit or not. In order to resolve the ambiguity at the receiving end, one more additional pulse or bit per tributary is added in the multiplexed frame before the last bits appear for the two tributaries, indicating which one of the last information bits of the two tributaries is a valid entry or represents a stuffed bit. This manipulation technique leads to the PDH multiplexing scheme, known as *pulse-stuffing* or *bit-justification*. In order to illustrate the overall PDH-based multiplexing scheme, we consider an SN with the two input tributaries, A and B, with a given frame time (125 μs), which are to be multiplexed using pulse-stuffing into one bit stream in the same frame duration, as shown in Fig. 5.3 (Bellamy 2003).

In each 125-μs frame, the PDH multiplexer shown in Fig. 5.3, will have to accommodate at its output $(2n+2)$ bits for the two incoming tributaries, assuming that the default number of bits per frame for each tributary is n. Thus, at the cost of a minor increase in the transmission rate with the additional two bits (i.e., by the ratio of $\frac{2n+2}{2n} = 1 + 1/n$), the multiplexed frame would help in getting around the problem of plesiochrony between the independent clocks. However, the clock generators used in the source nodes should have a specified upper bound for their jitter variances (so that one additional bit can take care of the possible slips for all practical purposes), ensured by using accurately-tuned highly stable *oscillators*. In Fig. 5.3, SB represents the stuffed bit/pulse for the slower input tributary B. Furthermore, as mentioned above, two more bits, i.e., VA and VB, are added to clarify the status of the last two bits in the multiplexed frame, with VA and VB being zero when both tributaries carry valid bits, or else VA or VB would be made one. For the present example, the clock of tributary A is faster, and hence the last bit of tributary A in the multiplexed frame is a valid bit (= A4, indicated with VA = 0). For the tributary B, VB = 1 and SB is therefore a stuffed bit which will be ignored at the receiving end.

The PDH framing scheme differs in the ANSI and ITU-T standards in respect of transmission rates. For the ANSI standard followed in North America and Japan, the basic multiplexed bit stream, called *digital signal* 1 (DS1), also referred to as T1, is formed by byte-interleaving 24 voice channels, each using 8-bit PCM signal (DS0), along with one bit for frame synchronization, leading to a frame size of $24 \times 8 + 1 = 193$ bits. This frame is transmitted over the inter-sampling interval of 125 μs, leading to a transmission rate of

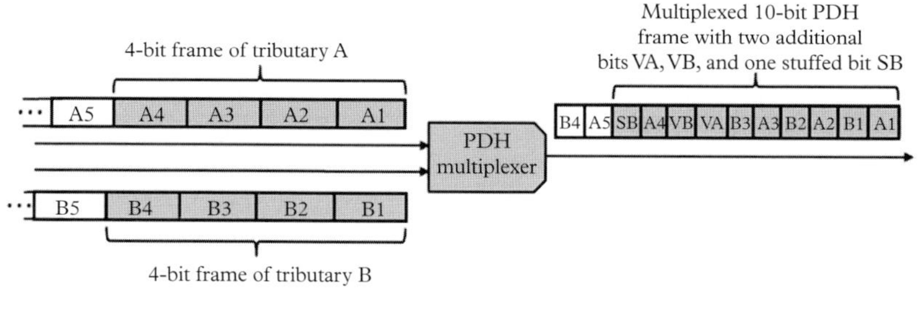

Figure 5.3 *Pulse/bit stuffing schematic in PDH frames for two input tributaries. All the frames shown before and after multiplexing have the same duration, generally 125 μs. Note that tributary B being slower needs a stuffed bit (SB) in the PDH frame, which is identified by VB = 1, while VA remains 0 for the faster tributary A.*

$193/(125 \times 10^{-6}) = 1.544$ Mbps. Note that DS1 is generated at the CO level and hence does not need any pulse-stuffing. The higher-order PDH tributaries, i.e., DS2 or T2 onward, employ pulse-stuffing and hence the transmission rate does not scale up proportionately with the number of voice channels. Similarly for the European standard, the basic signal, called E1, is generated through byte interleaving of 30 8-bit PCM signals with a transmission rate of 2.048 Mbps. As in AT&T's PDH levels, the higher-order PDH tributaries of ITU-T standard, i.e., E2 onward, do not also scale up proportionately with the number of channels owing to pulse-stuffing. Tables 5.1 and 5.2 present the transmission rates for the different PDH levels for both AT&T and ITU-T standards.

The two PDH standards (i.e., AT&T and ITU-T), while taking care of the plesiochronous clocks, disturb the *positional significance* of the bits in the output frame of PDH multiplexer, unlike the way it happens at the basic levels of multiplexing, i.e., in T1/E1 frames, where the pulse-stuffing is not needed. When these PDH frames are further multiplexed at the SNs, the pulse-stuffing process is carried out to form the PDH frames for the next higher transmission rate. Consequently, as the transmission rates go up at the higher levels of hierarchy in the PSTN, identifying the specific bits for a specific 64 kbps signal (bit stream) from the high-level PDH signals by their positions in the multiplexed PDH frames becomes difficult. The best way to access a specific 64 kbps bit stream (for add-drop or switching operations) would be to demultiplex a high-level PDH frame way down to the basic T1/E1 frames (which are not affected by the pulse stuffing) by decoding the stuffed pulses at every higher level and pick up the PCM words from the right set of the contiguous 8 bit positions from the T1/E1 frames only. This amounts to climbing down a *multiplexing*

Table 5.1 *AT&T PDH signal designations and transmission rates.*

AT&T digital signal number	Number of voice channels	Transmission rate in Mbps
DS-1	24	1.544
DS1C	48	3.152
DS-2	96	6.312
DS-3	672	44.736
DS-4	4032	274.16

Table 5.2 *ITU-T PDH signal numbers and transmission rates.*

ITU-T digital signal number	Number of voice channels	Transmission rate in Mbps
E1	30	2.408
E2	120	8.448
E3	480	34.368
E4	1920	139.264
E5	7680	365.148

mountain (the mountain height representing the complexity of the multiplexer/demultiplexer hardware) for getting access to the desired 64 kbps connection. This complexity has nevertheless been manageable in copper and radio communication systems, where bit rates could not exceed a few 100s of Mbps. However, with optical communication links operating at much higher transmission rates exceeding 1 Gbps, such hardware-based manipulation (i.e., pulse stuffing/de-stuffing at every stage of multiplexing/demultiplexing) turned out to be infeasible, and the telecommunication community across the world had to find an alternative, leading to the SONET/SDH-based networking architectures.

5.4 Network synchronization in SONET/SDH-based networks

Note that, historically, the concept of HMS synchronization was used even before the arrival of PDH and SONET/SDH in US networks, as designed by AT&T and the US Independent Telephone Association. However, following the post-divestiture breakup of the nation-wide network, the HMS architecture of network synchronization was changed over to the plesiochronous (PDH) architecture, although the PDH frames used for the lower-rate tributaries had to be eventually packed into the HMS-based SONET/SDH frames for high transmission rates in optical fibers. We discuss this aspect later in further detail.

The SONET/SDH networks, as mentioned earlier, realize network synchronization using a *byte-interleaved* multiplexing technique, based on the HMS scheme. This is achieved by using a set of highly stable *reference* clock generators, which are judiciously placed across the network as masters, though with varying accuracy of clock frequency (which in turn manifests itself as timing jitter). Typically, the SONET/SDH networks employ multiple reference clocks in ring topologies with different levels of accuracy in respect of the clock frequency (Crossett and Krisher 1992). The masters with highest accuracy are realized using the primary reference sources (PRSs) with Stratum 1 standard in SONET or primary reference clocks (PRCs) with G.811 standard in SDH. The next-level reference clocks for SONET (referred to as Stratum 2, Stratum 3, and minimum accuracy) and SDH (referred to as G.812T, G.812L, and G.81S) differ in respect of their specifications and thus exact equivalence cannot be drawn.

The Stratum 1 and G.811 references (of SONET and SDH, respectively) are derived from the cesium atomic clock, and hence offer the unique set of specifications, such that the same accuracy of $\pm 1 \times 10^{-11}$, i.e., $\pm 10^{-5}$ ppm can prevail at this level for SONET and SDH standards worldwide. The other SONET clock references have the following specifications, viz., Stratum 2 accuracy: $\pm 1.6 \times 10^{-8}$ ($\pm 1.6 \times 10^{-2}$ ppm); Stratum 3 accuracy: $\pm 4.6 \times 10^{-6}$ (± 4.6 ppm); Stratum 4 or minimum clock accuracy: $\pm 32 \times 10^{-6}$ (± 32 ppm). In the situations when the reference clock frequencies drift (within the bounds specified by the respective frequency accuracies), the SONET/SDH standards use a novel idea of *pointers* to allow the incoming data streams to *float* and adjust their starting positions within the payload field (called *envelope* in SONET/SDH), thereby adjusting the effective number of bytes carried by the SONET/SDH frames. This feature leads to an in-built dynamic rate adaptation mechanism in SONET/SDH standard, which we discuss later in further detail.

Guided by the master-slave hierarchy, any SONET network element (NE) or node can operate with three possible mechanisms of network synchronization with the available PRSs: it can (*i*) obtain an external timing using a T1 or DS1 (24-channel 1.544 Mbps bit stream) tributary from a Stratum 3 (or better) clock of a CO (typically, the COs are synchronized with T1 timing references of a Stratum 1 source), (*ii*) obtain *line-timing* from an optical interface, i.e., from the incoming high-speed optical signals (which would be

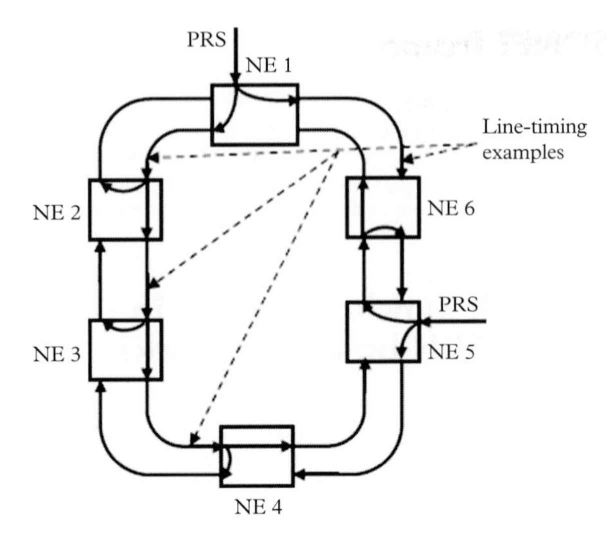

Figure 5.4 *Clock flows in a SONET ring using the master-slave network synchronization scheme with two PRS clocks from BITS (after Crossett and Krisher 1992).*

carrying embedded DS1 tributaries derived from PRSs), or (*iii*) operate in free-running mode with support from an internal clock generator satisfying the SONET's minimum clock accuracy. These basic synchronization options can further be combined into various network configurations.

An example scenario for a SONET ring using the master-slave network synchronization scheme is shown in Fig. 5.4. A SONET-based ring topology with six nodes (NEs) is considered, where each NE has its own internal Stratum 3 clock generator (not shown explicitly), and NE 1 and NE 5 receive Stratum 1 PRS clocks from DS1 tributaries arriving from the external COs. Using these two Stratum 1 PRS clocks, NE 1 and NE 2 transmit higher-speed SONET frames onward, and the other NEs line-tune themselves with the Stratum 1 references (with an acknowledgment sent as a feedback to the NE, wherefrom it gets the clock), albeit with some deterioration in quality as experienced in the NEs 2, 3, 4, and 6. However, the quality of derived clock remains within the permissible limits, and never goes beyond the Stratum 3 specifications.

Moreover, for the survivability against ring failures (e.g., fiber cuts), each NE has direct/line-timed access to two PRSs. Note that, though the physical topology of the network under consideration is a single ring, the two Stratum 1 clocks are distributed through line-timing using two *non-overlapping logical trees*. However, when there is a failure (for example, a cut in the inner ring between NEs 2 and 3), NE 3 will lose access to PRS from NE 1, and consequently the synchronization network will have to be reconfigured along with internal free-running clock in NE 3 (with the minimum accuracy of SONET). However, in doing so, care needs to be taken that no *timing loops* are formed in any combination of primary and secondary timing resources. The network-synchronization management for such issues is carried out through appropriate messages and this aspect remains independent of the connection restoration scheme in the network.

Note that, though SONET and SDH are operational in different parts of the world and are compatible to each other, we will continue our discussions mostly with reference to SONET, while some issues common to SONET and SDH are discussed, wherever needed.

5.5 SONET frame structure

The basic SONET frame, called the synchronous transport signal (STS) level-1, i.e., STS-1, is transmitted at a bit rate of 51.84 Mbps. Every other higher transmission rate of SONET (i.e., for STS-N) is an integer multiple of STS-1 rate, i.e., $N \times 51.84$ Mbps, with N as an integer. This is in sharp contrast to the PDH transmission rates, which are not integral multiples of the lowest level bit rate owing to pulse-stuffing at each level. Every STS-N frame is transmitted with the duration of 125 μs, in conformity with the inter-sample period of 64 kbps PCM and brings in backward compatibility with the PDH tributaries, which are also transmitted using 125 μs frames.

5.5.1 Basic SONET frame: STS-1

An STS-1 frame has 9 rows and 90 columns of bytes (i.e., $9 \times 90 = 810$ bytes = 6480 bits), where the rows are transmitted from top (row 1) to bottom (row 9) and the bytes in every row are transmitted from left to right (Fig. 5.5). Transmission of 6480 bits over 125 μs leads to the bit rate of $6480/(125 \times 10^{-6})$ bps = 51.84 Mbps. Each STS-1 frame comprises some overhead bytes and a payload field following a specific structural form. In the following, we describe the procedure used to construct an STS-1 frame.

The SONET frames, while traversing from a source to a destination node, typically traverse through three types of fiber segments, which are operated and managed by three different groups of overhead bytes. The end-to-end span of a connection in a SONET-based network between the source and destination nodes is called a *path*, while any segment along a path between two nodes where add-drop multiplexing of tributaries takes place is called a *section* and finally any segment (shortest ones) responsible for only regenerating weak signals is called a *line*. The first 3 columns of 9 rows (i.e., $3 \times 9 = 27$ bytes) carry the overhead bytes for section and line, and are called the section overhead (SOH) and line overhead (LOH)

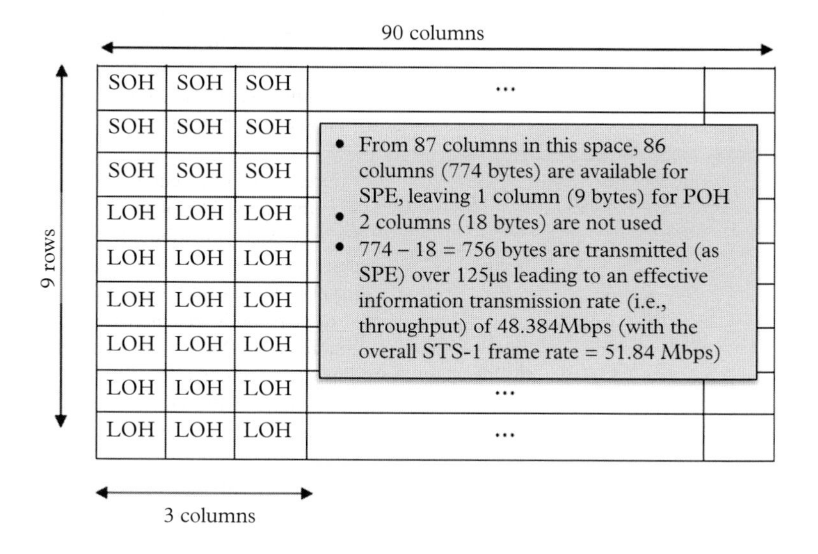

Figure 5.5 *STS-1 frame with $9 \times 90 = 810$ bytes, transmitted in 125 μs at 51.84 Mbps; POH would appear with the respective SPE and is not shown in this figure. Note that the rows in a frame are transmitted from top to bottom, and the bytes and bits in a row are transmitted from left to right.*

bytes, respectively. The SOH bytes are used only at the nodes used for signal regeneration during the traversal of the STS-1 frame through the network, while the LOH bytes are used at all the nodes except the regenerator nodes.

Besides the three columns for the overhead bytes, the remaining 87 columns of 9 rows (i.e., $87 \times 9 = 783$ bytes) carry the payload, called the synchronous payload envelope (SPE), along with one column (9 bytes) of path overhead (POH) having the information necessary for the end-to-end connectivity. The POH bytes are used only at the end nodes of a path. The STS-1 bit stream is finally *scrambled* for ensuring enough data transitions (for efficient clock recovery) and transmitted with electro-optic (EO) conversion using an optical transmitter. The transmitted optical signal has a transmission rate of 51.84 Mbps, called the *optical carrier* at level 1 (OC-1). Optical signals for higher rates (STS-N) are termed OC-N. The functionalities of the three types of overhead bytes, i.e., SOH, LOH, and POH, are discussed later in further detail.

In a SONET node transmitting STS-1 frames the incoming frames of PDH tributaries are written into the SPE without *unpacking* the respective PDH tributary frames. Thus, the lower-rate tributaries are packed, just as they are, into the subrate envelopes within the SPE, which are called virtual tributaries (VTs). An SPE is divided into 7 equal-size VT groups (VTGs) and a given VT group contains an integral multiple of the same type of VT frames (e.g., all DS1, all E1, all DS1C, or all DS2 frames, but with different numbers). For example, a VTG can consist of

- 4 VT1.5 frames, from a DS1 or T1 input at 1.544 Mbps,
- or 3 VT2 frames, from an E1 input at 2.408 Mbps,
- or 2 VT3 frames, from a DS1C input at 3.152 Mbps,
- or 1 VT6 frame, from a DS3 input at 6.312 Mbps,

Table 5.3 *VTs and VTGs in SONET STS-1 frame. Each VTG transmits at the same rate, and the increased bit rate for a VTx is calculated as* $r_x = (r \times c)/T_f$, *with* $T_f = 125$ μs *representing the inter-sampling period of a DS0 (64 kbps) PCM signal. Note that, the outgoing clock rate of each VT in a SONET node is enhanced with respect to the incoming PDH rates, i.e.,* $r_x > r_p$, *and thus all* r_v*'s become identical (i.e., 6.912 Mbps).*

ANSI/ ITU-T	VTx in SPE: r rows, c columns	VTs in a VTG (n_v)	No. of VTs in SPE with 7 VTGs (n_g)	PDH bit rate (r_p)	Raised VT bit rate in STS-1 ($r_x = 8rc/T_F$)	VTG bit rate ($r_v = n_v r_x$)	STS-1 bit rate ($r_s = 7r_v$)
DS1, ANSI	VT1.5: $r = 9, c = 3$	4 VTs	$4 \times 7 =$ 28 VTs	1.544 Mbps	1.728 Mbps	6.912 Mbps	48.384 Mbps
E1, ITU-T	VT2: $r = 9, c = 4$	3 VTs	$3 \times 7 =$ 21 VTs	2.048 Mbps	2.304 Mbps	6.912 Mbps	48.384 Mbps
DS1C, ANSI	VT3: $r = 9, c = 6$	2 VTs	$2 \times 7 =$ 14 VTs	3.152 Mbps	3.456 Mbps	6.912 Mbps	48.384 Mbps
DS2, ANSI	VT6: $r = 9, c = 12$	1 VT	$1 \times 7 =$ 7 VTs	6.312 Mbps	6.912 Mbps	6.912 Mbps	48.384 Mbps

but not any mix of different types of VTs, i.e., a VTx and a VTy with $x \neq y$ can not be mixed in a VTG. The effective payload transmission rate in STS-1 is governed by the number of bytes that can be accommodated in the SPE, i.e., in 86 columns (excluding 4 columns for SOH, LOH, and POH bytes) of 9 rows amounting to $86 \times 9 = 774$ bytes. However, 2 columns (i.e., 18 bytes) remain unused, and thus STS-1 actually can use $774 - 18 = 756$ bytes or $756 \times 8 = 6048$ bits to transmit over 125 μs, leading to an *effective transmission rate* of $6048/(125 \times 10^{-6})$ bps = 48.384 Mbps.

In order to *pack* different incoming VTGs in the same SONET frame, the effective output transmission rates for the various tributaries get adjusted, such that the aggregate rate of all the VTGs carried by a SPE equals 48.384 MBps. Note that all the incoming VT rates are determined independently by the PDH-based multiplexing schemes, which does not allow any integral relationship with the basic DS0 transmission rate (64 kbps) owing to the pulse-stuffing mechanism. Thus, when the STS-1 transmission rate (51.84 Mbps) is assigned to the entire frame, the constituent VTs (VT1.5, E1, DS1C, and DS2) get transmitted at *slightly faster rates* as compared to their original PDH rates.

How the outgoing VT rates are changed (increased) is illustrated in Table 5.3, using the numbers of rows (r) and columns (c) that are allotted for a given VT during the total STS-1 frame duration of 125 μs. As shown in the table, the raised VT rates (r_x's) are 1.728 Mbps, 2.340 Mbps, 3.456 Mbps, and 6.912 Mbps, corresponding to their original PDH rates (r_p's): 1.544 Mbps (VT1.5), 2.048 Mbps (VT2), 3.152 Mbps (VT3), and 6.312 Mbps (VT6), respectively. Note that, the fourth raised rate r_x (for DS2) is integrally related to the first three raised rates (for DS1, E1, and DS1C), and the final VTG transmission rate for all VTGs is $r_v = n_v r_x = 6.912$ Mbps, with the effective STS-1 transmission rate r_s being 48.384 Mbps for all of them.

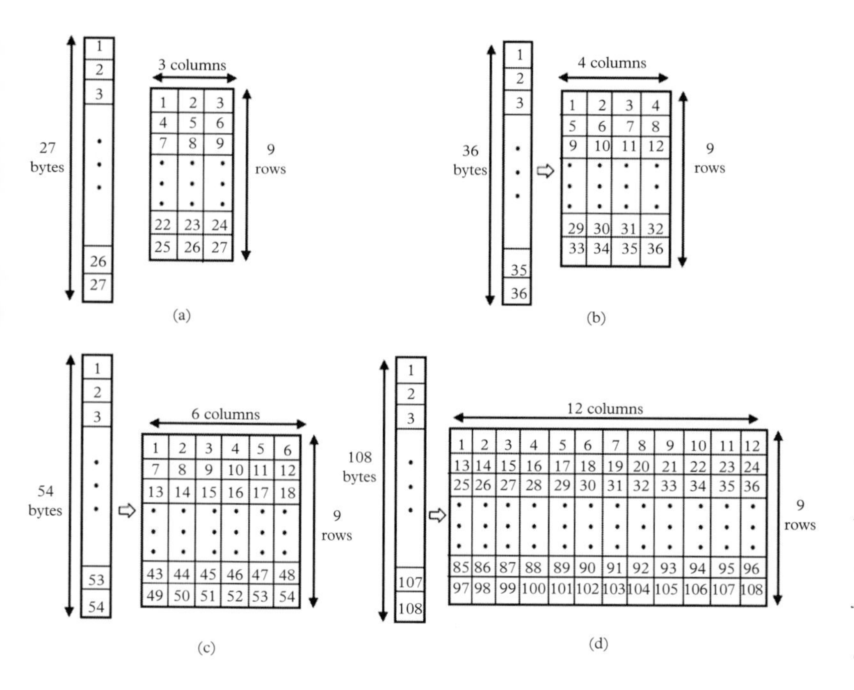

Figure 5.6 *Mapping of VT bytes into STS-1-compatible format. (a) VT1.5 (27 bytes) is formed using 9 rows and 3 columns of bytes in an STS-1 frame with the effective transmission rate of 1.728 Mbps, 4 VT1.5's constitute a VT group of VT1.5, (b) VT2 (with 36 bytes) is formed using 9 rows and 4 columns of bytes in an STS-1 frame with the effective transmission rate of 2.304 Mbps, 3 VT2's constitute a VT group of VT2, (c) VT3 (with 54 bytes) is formed using 9 rows and 6 columns of bytes in an STS-1 frame with the effective transmission rate of 3.456 Mbps, 4 VT3's constitute a VT group of VT3, (d) VT6 (with 108 bytes) is formed using 9 rows and 12 columns of bytes in an STS-1 frame with the effective transmission rate of 6.912Mbps, 1 VT6 constitutes a VT group of VT6.*

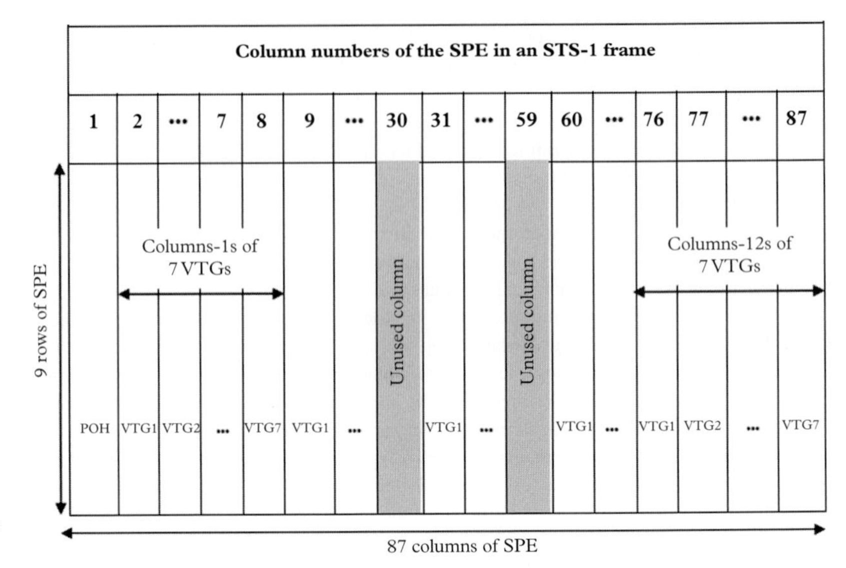

Figure 5.7 *Placement of 7 VT groups (VTGs) in the SPE of an STS-1 frame.*

Figure 5.6 illustrates the formations of different VTs in an STS-1 frame, as governed by the row and column allocations in Table I. As shown in the figure, each VTG will have 12 columns with four options: (*i*) four VT1.5 frames each with 3 columns, (*ii*) three VT2 frames each with 4 columns, (*iii*) two VT3 frames each with 6 columns or (*iv*) one VT6 frame with 12 columns. Seven such VTGs (i.e., $7 \times 12 = 84$ columns) are mapped into the SPE payload of STS-1 frame, along with one column for POH and two unused columns (30th and 59th), as shown in Fig. 5.7.

5.5.2 Multiplexing hierarchy for SONET frames

As discussed in the foregoing, the basic SONET multiplexing starts with the formation of the STS-1 frame from the incoming tributaries: DS1, E1, DS1c, and DS2. Higher-level SONET frames can be formed from the basic frames, as well as directly from some higher-rate PDH tributaries and other data streams (for data-networking over SONET). For example, next to DS2, the higher-rate PDH tributary in the ANSI standard is DS3 with a bit rate of 44.736 Mbps, which can be directly adopted as the SPE payload of the STS-1 frame. The multiplexing hierarchy for STS-1 and the higher-order SONET frames (STS-N) is illustrated in Fig. 5.8.

For the DS1, E1, DS1C, and DS2 signals, first the respective VTx frames (with $x = 1.5$, 2, 3, and 6, respectively) are formed along with their POH bytes, which are subsequently packed into the respective VTGs, with four VT1.5s, three VT2s, two VT3s or only one VT6 forming a VTG. Each VTG is thereafter placed in an SPE followed by the insertion of LOH and SOH bytes, leading to the formation of STS-1 frames. The STS-1 frames are next scrambled for enhanced data transitions (for the ease of clock recovery in the receiving nodes) and transmitted through EO conversion as an OC-1 optical signal.

However, the higher-rate PDH signals, such as E4 (139.264 Mbps) of ITU-T or ATM signal (149.760 Mbps) are mapped into a multiple number of STS-1 frames and then concatenated together to form an STS-3c signal, finally leading to the transmission of an

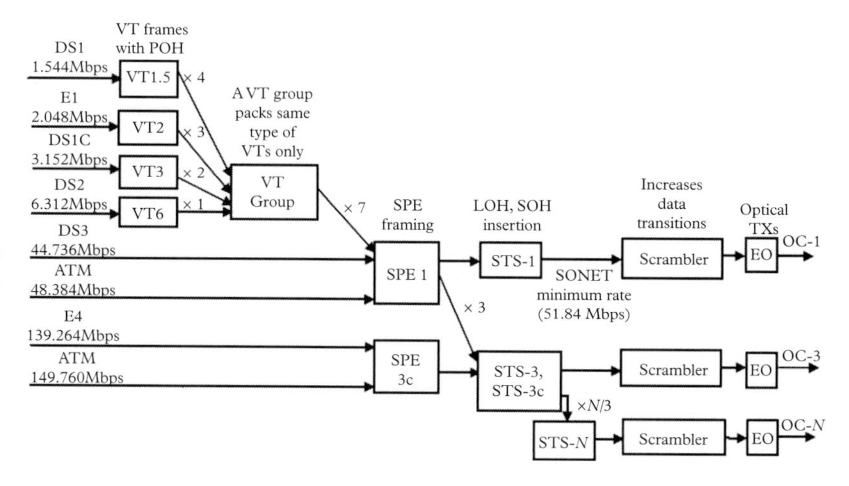

Figure 5.8 *SONET multiplexing hierarchy and OC-N (for OC-1, 3, N) signal generation scheme.*

OC-3 signal. $N/3$ of the STS-3/STS-3c frames can also be byte-multiplexed to obtain an STS-N frame which is thereafter transmitted as an OC-N signal.

Table 5.4 presents the various STS-N/OC-N designations and the respective bit rates along with their equivalents in SDH, where the SDH frames are referred to as synchronous transport modules (STM). There is no equivalent of STS-1 in SDH, and SONET and SDH rates converge from 155.52 Mbps upward, which is the rate used by STM-1 and STS-3 corresponding to OC-3 optical signal. An STS-N signal is scrambled to ensure sufficient data transitions, which helps the clock recovery process at the receiving node. The scrambled STS-N frames thereafter go through EO conversion and are transmitted as an OC-N signal over optical fiber.

Table 5.4 *SONET/SDH and optical signal designations and their bit rates.*

SONET signal designation (electrical)	SDH signal designation (electrical)	Bit rate in Mbps	Optical signal designation
STS-1	No SDH	51.84	OC-1
STS-3	STM-1	155.52	OC-3
STS-12	STM-4	622.08	OC-12
STS-24	STM-8	1244.16	OC-24
STS-48	STM-16	2488.32	OC-48
STS-96	STM-32	4976.64	OC-96
STS-192	STM-64	9953.28	OC-192
STS-768	STM-256	39813.12	OC-768
STS-1920	STM-640	99532.80	OC-1920

5.5.3 SONET layers and overhead bytes

The OH bytes in STS frames are used for operations, administration, and management (OAM) of the entire network. In order to discuss the roles of the OH bytes, i.e., POH, LOH, and SOH bytes, one needs to understand the implications of the SONET layers: section, line, and path, as shown in a segment of SONET-based network in Fig. 5.9. The SONET nodes located at the two ends of the connection shown between nodes i and $i+6$, play the role of the source and destination nodes and the others function as the intermediate nodes. The remaining part of the network, which can use a ring or a linear topology, is not shown explicitly. The SONET nodes, generally referred to as add-drop multiplexers (ADMs), need to carry out a variety of tasks, including signal regeneration, local add-drop operation, forwarding of the transit signal, and various other OAM functionalities.

As mentioned earlier, a typical SONET connection between a given source-destination pair, as shown in Fig. 5.9, is composed of three types of segments corresponding to the three types of SONET layers. The longest one between the source and destination nodes, i.e., the entire connection span, is called a path which is divided into two types of segments, called line and section, and each of these segments has its own group of overhead bytes – POH, LOH, and SOH bytes, respectively.

In Fig. 5.9, the ADMs at the source and destination nodes, i.e., at the nodes i and $i+6$, perform the roles of a path-terminating equipment (PTE) for the path under consideration, and inserts/extracts the payload into/from the SONET frame. If this connection is set up in a linear topology (i.e., not a ring) between nodes i and $i+6$, then these two nodes can also employ terminal multiplexers (TMs), as they don't need to handle any transit traffic (ADMs and TMs are discussed later in further detail). The POH, LOH, and SOH bytes are inserted/extracted along with the payload, and processed in the two PTEs at nodes i and $i+6$. The POH bytes are used to exchange the path-related information between the two nodes and are not processed at the intermediate ADMs, while the LOH and SOH bytes are used to perform the link and section-related functions. The path under consideration goes through five (in this example) ADMs at the intermediate nodes, and these ADMs operate functionally either as line terminating equipment (LTE) or section terminating equipment (STE) for the concerned path. The ADMs operating as LTEs deal with the LOH and SOH bytes for the links between them and perform the tasks of add-drop-forward as well as regeneration along with other associated OAM tasks. Each STE, again an ADM but operating only as a regenerator for the concerned path, performs the OEO-based signal regeneration, and deals with the SOH bytes only for the same path.

As shown in Fig. 5.10, every layer (section, line, and path) in a SONET-based network has specific bytes to make the OAM architecture functional. In Tables 5.5, 5.6, and 5.7, we present the functions of each byte for all three sets of OH bytes, viz., SOH, LOH, POH, with reference to the framing arrangement shown earlier in Fig. 5.5. Note that, every byte in this frame is transmitted over a time interval of 125 μs, leading to an effective byte transmission

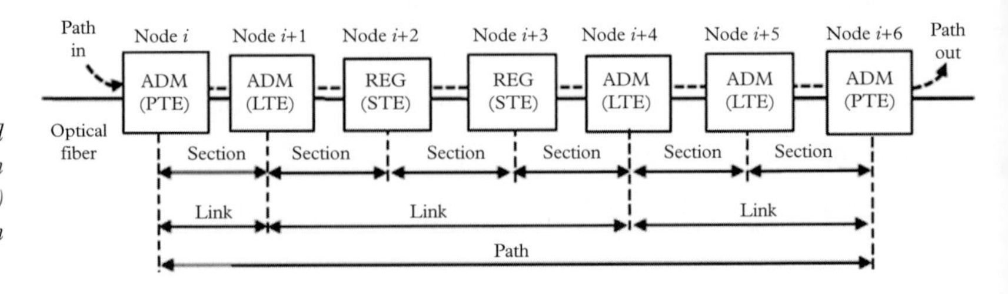

Figure 5.9 *Illustration of path, section, and link in a SONET-based network connection with its various participating nodes (ADMs) playing the roles of PTE, LTE, and STE, with a path being ≥ a line ≥ a section.*

SOH	Framing A1	Framing A2	STS-1 ID C1
	BIP-8 B1	Orderwire E1	User F1
	Data Com D1	Data Com D2	Data Com D3
LOH	Pointer H1	Pointer H2	Pointer Action H3
	BIP-8 B2	APS K1	APS K2
	Data Com D4	Data Com D5	Data Com D6
	Data Com D7	Data Com D8	Data Com D9
	Data Com D10	Data Com D11	Data Com D12
	Growth Z1	Growth Z2	Orderwire E2

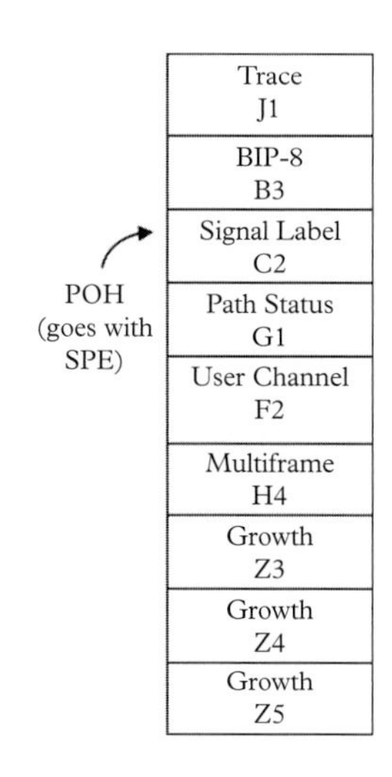

Figure 5.10 *SONET OH bytes (SOH and LOH bytes called together as transport overhead (TOH) bytes).*

rate of 8 bits/125 μs = 64 kbps. In fact, some of the OH bytes do carry 64 kbps PCM channels as well for orderwire communications.

The example connection, shown in Fig. 5.9, is considered again in Fig. 5.11 to illustrate the process of OH-bytes processing along the path. In particular, this figure shows how the three different types of OH bytes are inserted, processed, and reinserted in the respective nodes for a given SONET path. The POH bytes are inserted at the source PTE and processed only at the destination PTE. The LOH bytes are inserted at the source PTE, but are subsequently processed and reinserted at the intermediate nodes playing the role of

Table 5.5 *Functions of SOH bytes.*

Bytes and rows	Functions
A1, A2 (first row)	Framing bytes to mark the beginning of an STS-1 frame; A1 uses the sequence (11110110) and A2 uses the sequence (00101000)
C1 (first row)	STS-1 ID byte to identify the sequence number of the STS-1 frame in a given bigger, i.e., STS-N frame, which multiplexes several STS-1 frames
B1 (second row)	Bit-interleaved parity byte (BIP-8) using even parity with cyclic redundancy code 8 (CRC-8) over previous STS-N frame (after scrambling) for error monitoring at section level
E1 (second row)	64 kbps PCM voice channel (local) for orderwire at section level
F1 (second row)	For users
D1–D3 (third row)	For alarm, maintenance, and administration at section level

Table 5.6 *Functions of LOH bytes.*

Bytes and rows	Functions
H1–H3 (fourth row)	Pointer bytes used for starting address (H1, H2) of SPE in an STS-1 frame and byte stuffing using H3 (discussed later)
B2 (fifth row)	BIP-8 parity byte using CRC-8 (introduced after scrambling, as in B1) for error monitoring at line level
K1, K2 (fifth row)	Two bytes for automatic protection switching (APS) at line level
D1–D12 (sixth, seventh and eighth rows)	For alarm, maintenance, and administration at line level
Z1, Z2 (ninth row)	Reserved for future
E2 (ninth row)	64 kbps PCM voice channel for orderwire at line level

Table 5.7 *Functions of POH Bytes.*

Bytes and rows	Functions
J1 (first row)	One byte (i.e., a 64 kbps channel) in each STS-1 frame, termed as trace byte, is used to send repetitively a 64 byte fixed-length string, so that the terminal receiving node at the other end of the path can check the integrity of the path in a continuous manner
B3 (second row)	BIP-8 parity byte providing end-to-end error monitoring at path level
C2 (third row)	STS path signal label to identify the type of payload being carried
G1 (fourth row)	Status byte to carry maintenance signals
F2 (fifth row)	64 kbps channel for path user
Z3, Z4, and Z5 (seventh, eighth, and ninth rows)	Future use

LTE. The SOH bytes are also inserted at the source PTE, but processed and reinserted at *all* intermediate nodes along the path including the ones playing the roles of LTE as well as STE. Finally, all the overhead bytes (POH, LOH, and SOH bytes) are processed at the destination PTE for termination of the path therein. Typically, a SONET ring is designed to restore the failed connections as well (e.g., failures due to fiber cut, node failure, etc.) within 50 ms, by utilizing the appropriate overhead bytes. We discuss later the in-built survivability mechanisms of SONET rings in Chapter 12.

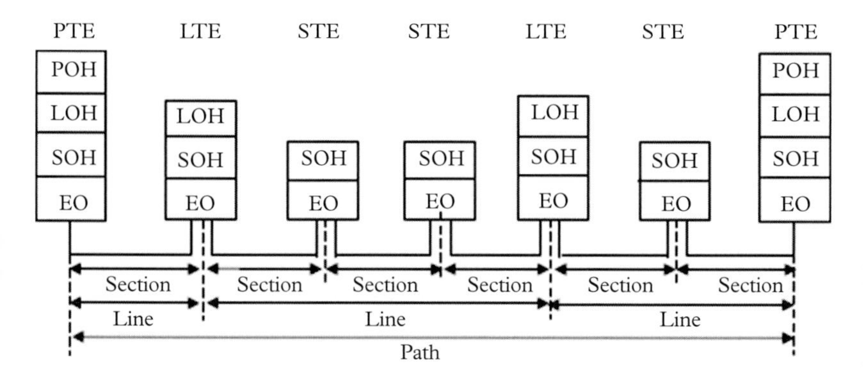

Figure 5.11 *Processing of OH bytes of specific layers in different nodes (in respective roles: as PTE/LTE/STE) in a path over a SONET network, as shown earlier in Fig.5.9.*

5.5.4 Payload positioning and pointer justification

As mentioned earlier, while inserting a tributary (corresponding to a VT) at a SONET node, the POH bytes for its path are attached with the STS-1 payload (SPE) itself. In particular, the nine POH bytes are inserted in the first column of the SPE, and the first byte of the POH column, i.e., J1 byte, can be located at any position in the SPE, as defined by the values of the two pointer bytes, H1 and H2 (in the field of LOH). Since there are $87 \times 9 = 783$ bytes in an SPE excluding the first three OH columns, one single pointer byte (maximum value being $2^8 = 256$) does not give adequate capacity for addressing the 783 possible byte locations. Thus, two pointer bytes, H1 and H2, are used for the addressing purpose, but with an over-provisioning as $2^{16} \gg 783$.

When the SPE gets ready for insertion into an STS-1 frame at the source node (an ADM or a TM), it might get the first byte of the fourth column (earliest available byte in the STS-1 frame, corresponding to a pointer value of zero). However, at an intermediate node (ADM), it might arrive at an instant which may not coincide in phase therein with the beginning of an outgoing SONET frame. In order to prevent any delay in such situation, it becomes necessary to insert the SPE somewhere in the middle of the outgoing SONET frame, without waiting for the beginning of the next SONET frame. Thus, the address defined by H1 and H2 may fall anywhere in the middle of the outgoing SONET frame (see Fig. 5.12). This in turn will cause a spill-over of the corresponding SPE from frame n to frame $n + 1$ (i.e., beyond the last or the 9th row of the nth frame). Moreover, within a given frame, due to the mid-frame starting location, the SPE would also have some columns spilling beyond the 90th column of the STS-1 frame. This part is folded back on the left side of the first column of the SPE, and is inserted from the row, just below the first row of the SPE. This process of assigning the starting point by H1 and H2 and the consequent spill-over and fold-back phenomena, beyond the 9th row and the 90th column of the STS-1 frame, respectively, are illustrated in Fig. 5.12.

Along with the allocation scheme for the SPE-starting point by H1 and H2 bytes, the network may have to adjust this assigned position, depending on the frequency alignment between the clocks being used by the node and the incoming tributary. In order to address this issue, SONET employs a timing adjustment with a granularity of *one single byte*. This implies that the starting point of the given SPE at the node may be shifted by one byte to the left or to the right on the same row, depending on whether the incoming signal has a faster or slower clock with respect to the node clock, respectively. The node makes an assessment from an internal buffer, where the incoming data stream is stored using the recovered clock

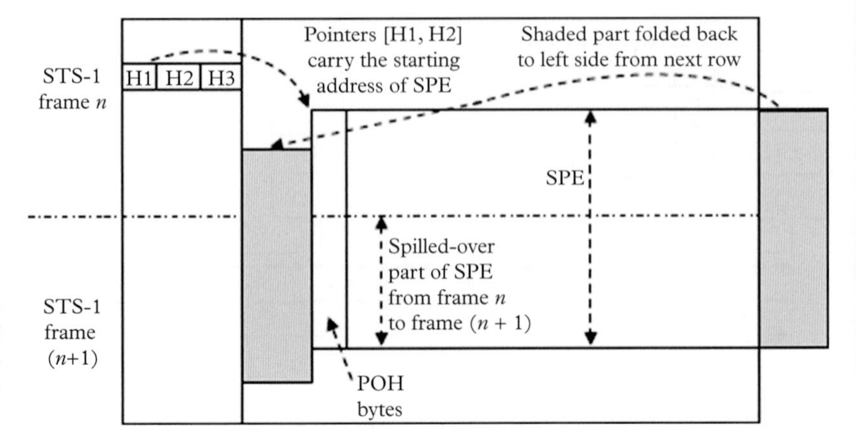

Figure 5.12 *Starting location for an SPE in an STS-1 frame using pointer bytes (H1 and H2) and consequent manipulation using spill-over and fold-back operations. H3 is used for byte stuffing.*

of the same signal, and the buffer is emptied out at a rate governed by the node clock. If the buffer queue is stationary, it implies that the clock rates are aligned. However, if the buffer queue shows a growing trend of the queue length, it indicates that the incoming signal has a clock which is faster than the node clock and vice versa.

Having estimated the buffer-growth rate (negative, zero, or positive) as above, the value of H3 byte is determined by the node. For an incoming STS-1 with a slower clock, a byte is stuffed following H3 byte, delaying the starting instant of the SPE by one byte with respect to the location corresponding to the instant of arrival of the tributary at the node. This effectively realizes a slower payload-transmission rate and this adjustment is called *positive byte-stuffing* (see Fig. 5.13). Along with this byte-stuffing, the pointer location is also incremented by increasing the value of [H1, H2] by one with respect to the earlier setting. However, for an incoming STS-1 with a faster clock (detected by an increasing trend in the queue length in the buffer), H3 is used to carry one byte of data allowing an extra byte before the position assigned by H1 and H2 bytes (i.e., effectively the starting point is advanced in time) to accommodate the higher data rate of the incoming tributary, and this procedure is called *negative byte-stuffing* (see Fig. 5.14). In this case, the pointer value [H1, H2] is decremented by one. Once an appropriate setting is made for H1, H2, and H3, the next three following frames (i.e., four consecutive frames including the frame where the pointer adjustment is made) are allowed for continuous transmission. We explain the implication of this arrangement in the following.

As explained in the foregoing, both positive and negative byte-stuffing schemes enable the STS-1 frame to absorb some frequency misalignment (i.e., relative clock-frequency jitter) between the clock of the incoming signal and the node clock. It is therefore worthwhile to estimate to what extent this arrangement can absorb the frequency mismatch between the two clocks (Bellamy 2003). When both the clocks are in perfect synchronization, the effective transmission rate of the full STS-1 payload (with 9 rows $\times$ 87 columns = 783 bytes) over the four STS-1 frames is given by

$$r_{PL} = (783 \times 4 \times 8)/(4 \times 125 \times 10^{-6}) \text{ bps} = 50.112 \text{ Mbps}. \tag{5.1}$$

When the incoming signal is found to be slower, the SONET node applies positive byte-stuffing and thus delays the payload transmission by one byte. With each frame having the duration of 125 μs, the effective payload transmission rate over the four frames is therefore

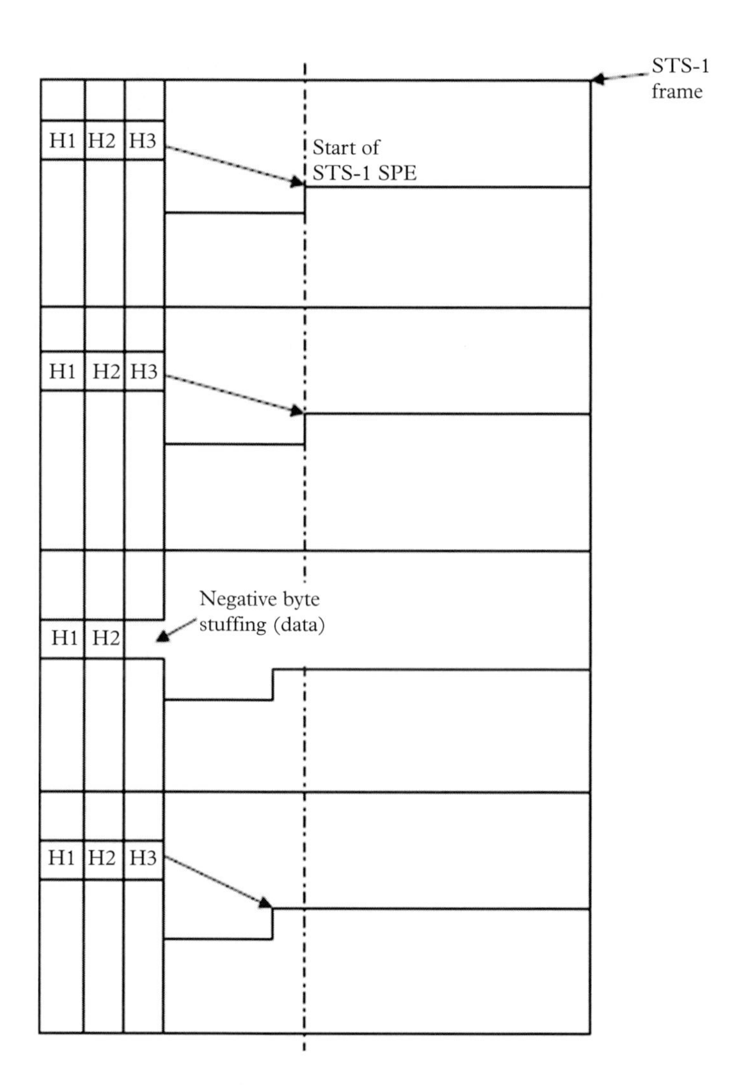

Figure 5.13 *Pointer justification with negative byte stuffing in SONET (after Bellamy 2003).*

reduced to $r_{PL}(\text{min}) = [(783 \times 4 - 1) \times 8]/(4 \times 125 \times 10^{-6})$ bps = 50.096 Mbps. Thus, the positive byte-stuffing would be able to absorb a maximum negative (as the incoming clock is slower) frequency misalignment Δr_{PL}^{-ve}, given by

$$\Delta r_{PL}^{-ve} = r_{PL}(\text{min}) - r_{PL} = -0.016 \text{ Mbps}. \quad (5.2)$$

Normalizing Δr_{PL}^{-ve} with respect to r_{PL}, we obtain the normalized value for the negative frequency misalignment, $\Delta r_{PL}^{-ve}(\text{norm})$, as

$$\Delta r_{PL}^{-ve}(\text{norm}) = \frac{-0.016}{50.112} = -0.0003193 \simeq -320 \text{ ppm}. \quad (5.3)$$

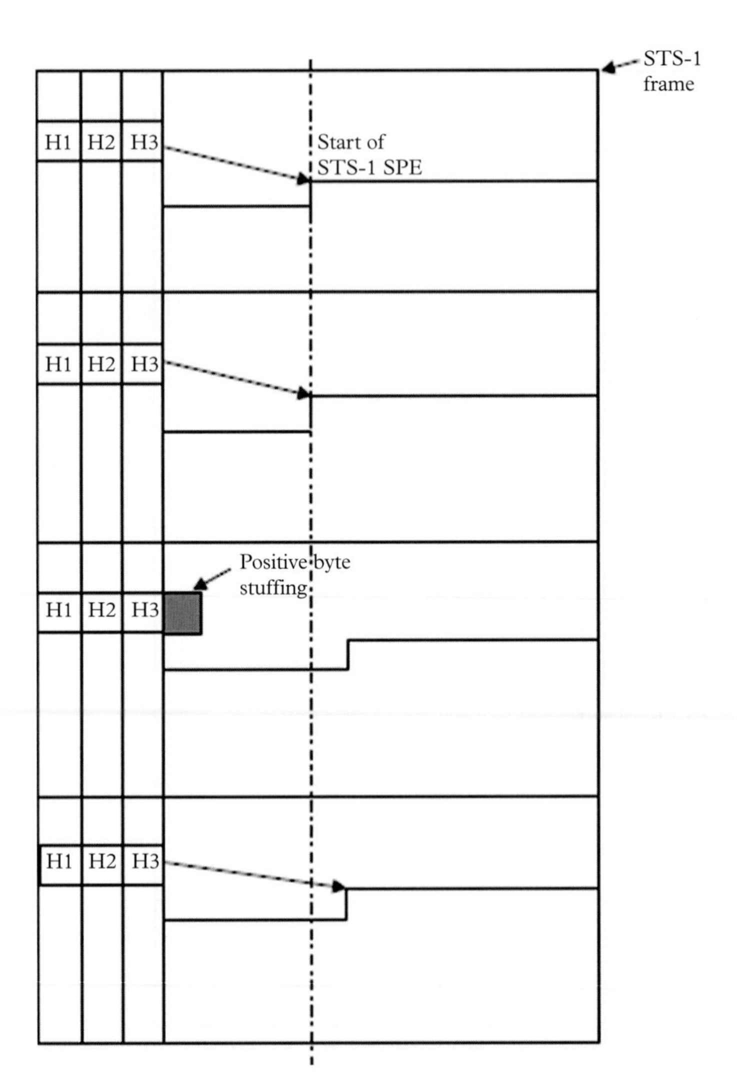

Figure 5.14 *Pointer justification with positive byte stuffing in SONET (after Bellamy 2003).*

When the incoming STS-1 signal is faster, one additional data byte (negative byte stuffing) is accommodated in the first SONET frame. This leads to the transmission of $(783 \times 4 + 1)$ payload bytes over four frames at the rate $r_{PL}(\text{max}) = [(783 \times 4 + 1) \times 8]/(4 \times 125 \times 10^{-6})\text{bps} = 50.128$ Mbps, offering a capacity to absorb a maximum positive frequency misalignment Δr_{PL}^{+ve}, given by

$$\Delta r_{PL}^{+ve} = r_{PL}(\text{max}) - r_{PL} = +0.016 \text{ Mbps}. \tag{5.4}$$

As before, by normalizing Δr_{PL}^{+ve} with respect to r_{PL}, we obtain the normalized value for the positive frequency misalignment, $\Delta r_{PL}^{+ve}(\text{norm})$, as

$$\Delta r_{PL}^{+ve}(\text{norm}) = \frac{+0.016}{50.112} = +0.0003193 \simeq +320 \text{ ppm}. \tag{5.5}$$

It therefore implies that, by using pointer justification, the network can absorb a total frequency misalignment of Δr_{PL}, given by

$$\Delta r_{PL} = |\Delta r_{PL}^{+ve}(\text{norm})| + |\Delta r_{PL}^{-ve}(\text{norm})| = (2 \times 320) \quad \text{ppm} = 640 \quad \text{ppm}, \tag{5.6}$$

including positive and negative byte-stuffing. As discussed earlier, even the lowest-level clocks in SONET nodes have much tighter frequency-jitter specification of $\pm$ 32 ppm, and hence the SONET frames with the pointer-justification scheme rarely falls out of synchronization. We present below an example calculation using a simple statistical model for the frequency (or rate) jitter of the clocks, which helps in appreciating the resilience of SONET standard to the clock-jitter process.

First, we develop a relation between the clock-rate jitter specification in ppm and the corresponding value of the rms rate jitter σ_r. Usually the clock-rate jitter is a random variable (δr, say), that can be represented by a zero-mean Gaussian distribution $p(\delta r)$ with a standard deviation σ_r. However, for practical purposes, generally the peak-to-peak clock-rate jitter Δr_{PP} is used as the clock-rate jitter specification. In particular, Δr_{PP} represents the range of the clock rate (around the mean) within which it remains confined with a high probability $P_{\Delta r}$, ($\geq$ 95%, say). Using $P_{\Delta r} = 0.95$ ($\equiv$ 95%) along with the Gaussian distribution of the jitter, one can express the relation between σ_r and Δr_{PP} as (see Fig.5.15)

$$P_{\Delta r} = 0.95 = \int_{-\Delta r_{PP}/2}^{+\Delta r_{PP}/2} p(\delta r) d(\delta r), \tag{5.7}$$

with $p(\delta f)$ given by,

$$p(\Delta r) = \frac{1}{\sqrt{2\pi\sigma_r^2}} \exp\left(-\frac{\delta r^2}{2\sigma_r^2}\right). \tag{5.8}$$

From Equations (5.7) and (5.8), we get a handy relation between σ_r and Δr_{PP}, given by

$$\Delta r_{PP} = 3.92\sigma_r. \tag{5.9}$$

Thus a 64 ppm peak-to-peak rate jitter of a clock (i.e., with Δr_{PP} = 64 ppm or $\pm$ 32 ppm) will correspond to a standard deviation or an rms rate jitter σ_r = 64/3.92 ppm = 16.33 ppm. With this value of σ_r, we next assess the robustness of the SONET standard in the presence of the clock-rate jitter.

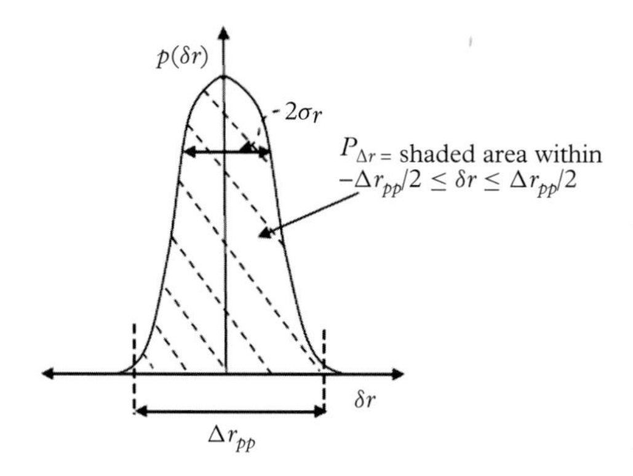

Figure 5.15 *Estimation of $P_{\Delta r}$ from the statistics of δr.*

With the Gaussian model of the clock-rate jitter, as given in Eq. 5.8, we express the probability of loss of synchronization in a SONET-based optical network as the area P_{loss}^{sync} ($=1 - P_{\Delta r}$) of the Gaussian density function of δr (Fig. 5.15), which remains beyond $\delta r = \pm\Delta r_{PL}/2$ (i.e., in the range $-\Delta r_{PL}/2 > r > +\Delta r_{PL}/2$), implying that

$$P_{loss}^{sync} = 1 - \int_{-\Delta r_{PL}/2}^{+\Delta r_{PL}/2} p(\delta r) d(\delta r) \tag{5.10}$$

$$= 1 - \int_{-\Delta r_{PL}/2}^{+\Delta r_{PL}/2} \frac{1}{\sqrt{2\pi\sigma_r^2}} \exp\left(-\frac{\delta r^2}{2\sigma_r^2}\right) d(\delta r).$$

From the above expression with σ_r = 16.33 ppm (corresponding to the $\pm$ 32-ppm peak-to-peak clock-rate jitter of the minimum accuracy SONET clock) and $\Delta r_{PL}/2 = 320$ ppm, we obtain $P_{loss}^{sync} = 1.58 \times 10^{-85}$, which indeed proves that the SONET standard offers an extremely robust network synchronization framework.

5.5.5 Higher-order SONET frames

The SONET frames at higher rates (STS-N, say) are generated by byte-multiplexing N frame-aligned STS-1 frames, frame alignment being necessary for byte-multiplexing. With frame alignment, the SOH and LOH bytes of all the constituent STS-1s are grouped together in the first $3N$ columns in an STS-N frame (with 9 rows and $90N$ columns) for offering a convenient access (right at the beginning) to the receiving node. However, the payloads of all the constituent STS-1s carry with them their independent POH bytes. Figure 5.16 shows an example of an STS-3 frame, formed using the above approach.

As indicated earlier, another type of higher-rate SONET framing is used for high-speed broadband connection, called STS-Nc frames ("c" implying concatenation), wherein the phase and frequency of N constituent STS-1s are locked together, and the entire payload is

A1	A1	A1	A2	A2	A2	C1	C1	C1	
B1	B1	B1	E1	E1	E1	F1	F1	F1	
D1	D1	D1	D2	D2	D2	D3	D3	D3	
H1	H1	H1	H2	H2	H2	H3	H3	H3	Three SPEs (three independent payloads along with the respective POH columns)
B2	B2	B2	K1	K1	K1	K2	K2	K2	
D4	D4	D4	D5	D5	D5	D6	D6	D6	
D7	D7	D7	D8	D8	D8	D9	D9	D9	
D10	D10	D10	D11	D11	D11	D12	D12	D12	
Z1	Z1	Z1	Z2	Z2	Z2	E2	E2	E2	

Figure 5.16 *Example STS-3 frame.*

A1	A1	A1	A2	A2	A2	C1	C1	C1	
B1	NU	NU	E1	NU	NU	F1	NU	NU	
D1	NU	NU	D2	NU	NU	D3	NU	NU	
H1	H1	H1	H2	H2	H2	H3	H3	H3	One SPE (payload along with its POH)
B2	B2	B2	K1	NU	NU	K2	NU	NU	
D4	NU	NU	D5	NU	NU	D6	NU	NU	
D7	NU	NU	D8	NU	NU	D9	NU	NU	
D10	NU	NU	D11	NU	NU	D12	NU	NU	
S1	Z1	Z1	Z2	Z2	M1	E2	NU	NU	

Figure 5.17 *Example STS-3c frame (NU: not used).*

transported end-to-end over one single path with one column of POH. This is accomplished by the use of a concatenation indicator (CI), with the concatenation process implying that it places the constituent STS-1 payload zones one after another with the same path and keeps them locked in that form until the destination node. The CI is used in the locations of pointers for 2nd and subsequent STS-1 frames in the STS-Nc frame, as shown in Fig. 5.17 for an STS-3c frame. By using this method of concatenation, multiple STS-1s can be used for the integrated bigger payload of STS-Nc. In SONET, although there are various permissible realizations of STS-Nc with different values of N, STS-3c and STS-12c are used more frequently.

5.6 Network elements in SONET

In order to deploy the networking infrastructure of SONET, it requires a variety of NEs to be interconnected using optical fibers: TMs, ADMs, and DCSs (Alwayn 2004). The TMs essentially work as PTEs for linear segments, while the ADMs can function as PTEs, LTEs, and STEs. These SONET NEs are used in various ways along with the DCSs supporting various possible topological formations in SONET-based networks. In the following, we describe some example configurations for TM, ADM, and DCS.

Figure 5.18 presents the block schematic of a typical TM, where it functions as an aggregation/distribution unit for various types of PDH tributaries and data streams, i.e., DSn, ATM bit streams along with SONET frames (STS-n), wherever necessary. The various bit rates that are used in this equipment are as follows.

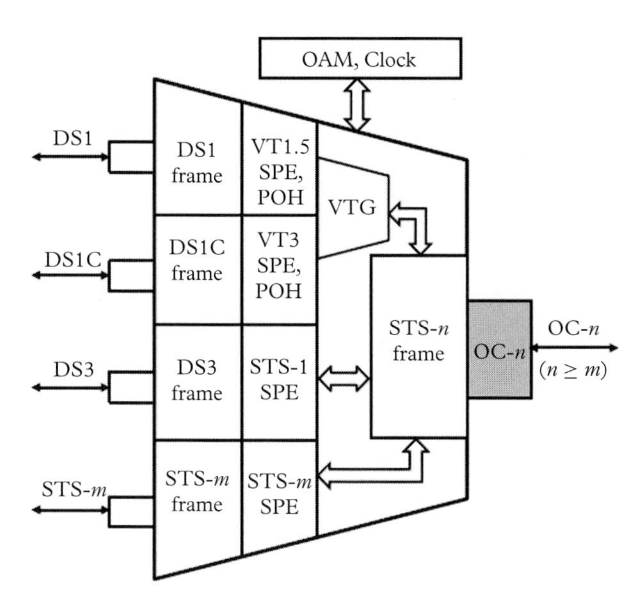

Figure 5.18 *Typical SONET terminal multiplexer (TM).*

Input PDH bit rates:

- DS1 – 1.544 Mbps (24 DS0s)
- DS2 – 6.312 Mbps (96 DS0s)
- DS3 – 44.736 Mbps (28 DS1s)

Output SONET bit rates:

- STS1, OC1 – 51.84 Mbps (28 DS1s or 1 DS3)
- STS3, OC3 – 155.52 Mbps (84 DS1s or 3 DS3s)
- STS12, OC12 – 622.08 Mbps (336 DS1s or 12 DS3s)
- STS48, OC48 – 2488.32 Mbps (1344 DS1s or 48 DS3s)
- STS192, OC192 – 9953.28 Mbps (5376 DS1s or 192 DS3s)

While performing the aggregation in a TM, PDH tributaries are stored and mapped onto the appropriate STS-n-compatible VTs or STSn frames. Thus a TM performs the necessary functions using the OH bytes of all the three layers, SOH, LOH, and POH bytes, assisted by the appropriate OAM and clock interfaces. However, POH bytes are introduced also in ADMs if a path originates/terminates therein. As shown in Fig. 5.18 and governed by the multiplexing hierarchy (Fig. 5.8), the input DS1 and DS1C streams are multiplexed and mapped into VT1.5 and VT3 respectively, which are subsequently grouped to form the respective VTGs for the formation of STS-1 frames. However, each DS3 stream is directly mapped onto one STS-1 frame, and thereafter all STS-1 frames are mapped into higher-order STS-n frames. The lower-order STS-m inputs (with $m < n$) are also mapped onto higher-order STS-n payloads, and the resultant STS-n frames that are formed from all the input streams are finally scrambled and transmitted using the OC-n transmitter of the transceiver port. The reverse functionality works while receiving the OC-n optical signal from the fiber end with OE conversion carried out by optical receiver modules (the other half of the OC-n transceivers). Note that all these functionalities are also included in the ADMs, and are carried out by an ADM while operating as a PTE for a given path.

An example configuration of ADM is shown in Fig. 5.19, which uses two fiber ports, each one using an OC-n transceiver at the desired SONET transmission rate. At the receiver of a given port at one end, it receives OC-n signal and performs add-drop and passthrough operations. In doing so, the ADM has to process the overhead bytes using OAM/clock interface and regroup the VTs in accordance with the add-drop and

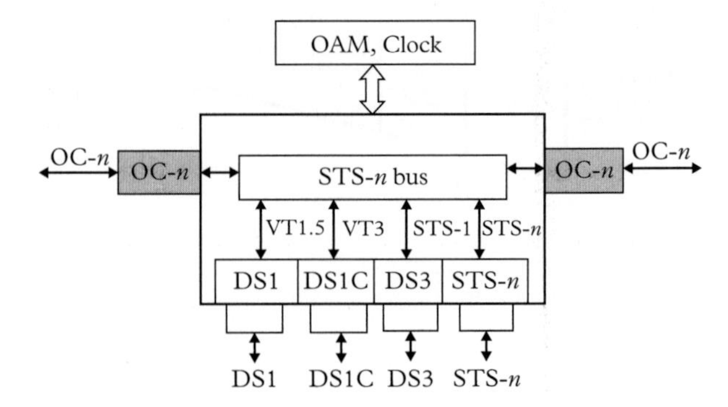

Figure 5.19 *Typical SONET ADM.*

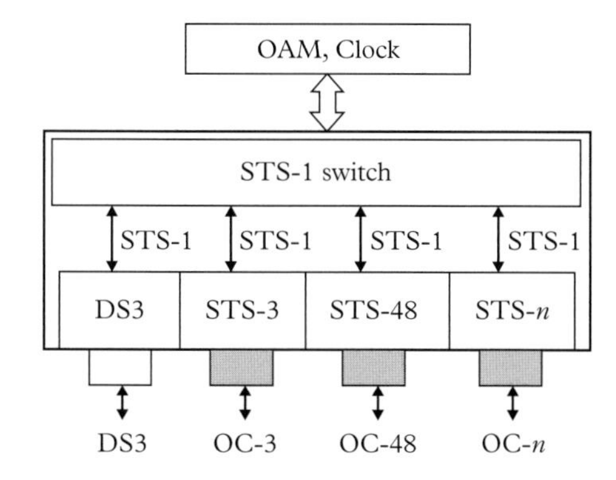

Figure 5.20 *Typical broadband DCS.*

passthrough requirements. Thereafter, the ADM transmits the multiplexed STSn signal at OC-n rate from the fiber port on the other end. The ADMs for lower-speed rings (e.g., OC-3, OC-12) need to add/drop more PDH tributaries, while the ADMs for higher-speed rings (e.g., OC-48, OC-192) will be carrying out add-drop operations with lower-speed SONET/SDH frames (e.g., OC-3, OC-12).

DCSs can be of two different categories: broadband DCS and wideband DCS. Figure 5.20 shows a typical broadband DCS configuration, dealing with the relevant crossconnect functionalities between the SONET rings. Broadband DCSs carry out crossconnect operations for the DS3 and OC-3 through OC-n signals by using an STS-1 switching matrix. On the other hand, the wideband DCSs can perform crossconnect operation with finer granularity, dealing with even the lowest-order PDH tributary to the various levels of OC-n signals as well (not shown).

5.7 Network configurations using SONET

SONET-based networks can be configured using some basic topologies: point-to-point (PP), linear, ring and hub, and a mix-and-match of these basic topologies. A ring is a two-connected mesh, where each node can connect to the remaining network through two independent links, thereby increasing the reliability significantly with respect to the linear configuration. SONET rings are extensively used in both metro and long-haul segments, with interconnected rings for the latter to cover larger areas. Sometimes rings may need linear extension to reach some customer premises where ring connectivity cannot be offered. Proprietary PP links may also be used in some enterprise-network settings. With the evolving trends of today's telecommunication networks, optical networks are also migrating from ring to mesh topology along with WDM transmission technology, where the underlying transmission systems have by and large followed SONET/SDH standards so far, though the other competing standards have evolved with improved performance, particularly for accommodating the growing volume of data traffic.

Figure 5.21 presents the four basic topologies of SONET/SDH infrastructure. The PP links with two nodes don't need any ADM in ADM-mode as all connections need to be sourced and terminated at the two nodes without any passthrough functionality and hence these nodes can simply use TMs or ADMs in TM mode. However, a linear topology with multiple nodes can also use TMs at the two ends, while the intermediate nodes must employ ADMs. In the ring topology, all the nodes employ ADMs, while the hub topology typically

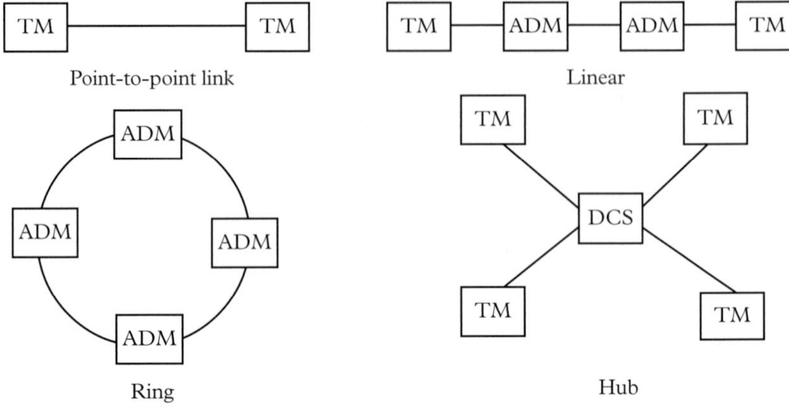

Figure 5.21 *Four basic SONET topologies (regenerators may be needed on a link, depending on its length).*

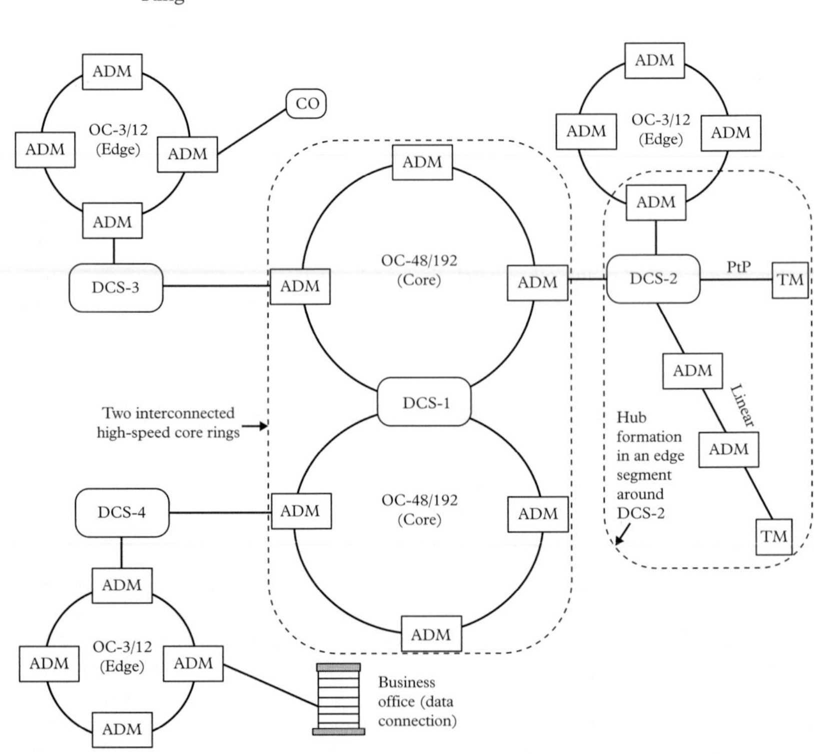

Figure 5.22 *Representative network formation using SONET infrastructure.*

employs a DCS at the center, interconnecting multiple TMs or ADMs, the latter being generally connected to neighbouring rings (not shown). As mentioned before, ADMs can operate as PTE, LTE, and STE, while TMs can only operate as PTEs.

Next we consider some typical network formations based on the SONET framework, as shown in Fig. 5.22. In general, SONET provides connectivity in metro and long-haul segments with interconnected rings. The metro segments are these days split into two levels, metro-core ring and metro-edge ring. For the metro-edge ring, it generally employs lower-rate OCs (e.g., OC-3, OC-12) interfacing with the access networks, while the higher-rate OCs (e.g., OC-48, OC-192, and so on) are used in metro-core rings, flanked by metro-edge and long-haul networks.

Figure 5.22 illustrates a representative SONET-based network spanning across different levels of metro networks. The metro-edge segments are shown using SONET rings with OC-3/12 transmission over ring, along with hub, linear, and point-to-point links, while the metro-core rings employ ring topology using OC-48/192 transmission. Further, the DCS, denoted as DCS-1, interconnects two OC-48/192 core rings and would be grooming and switching DS-3 signals as a wideband DCS between these two higher-speed rings. At lower rates, three more DCSs are employed in this network. DCS-2 forms a hub topology interconnecting a linear segment, a PP segment, one edge ring, and one core ring, and DCS-3 interconnects an upper core ring and one edge ring where the edge ring shows a connection to a CO in a PSTN. Finally, DCS-4 interconnects the lower core ring and an edge ring, where the edge ring extends a connection to a business office dealing with data connection.

SONET and SDH standards offer resilient networks designed with in-built survivability measures against network failures using appropriate protection schemes, e.g., unidirectional path-switched ring (UPSR) and bidirectional line-switched ring (BLSR). These schemes are described in Chapter 11.

5.8 Data transport over SONET

By the late 1990s, data traffic superseded voice traffic in telecommunication networks, which in turn generated an increased need of long-distance connectivity between the geographically separated LANs through optical networks. Initial solutions were obtained over the SONET-based networks by using PDH tributaries as leased lines for the transport of various packet-based data streams (e.g., IP, ATM, Ethernet), as shown in Fig. 5.23. As such, ATM was already included in the original multiplexing hierarchy of SONET (see Fig. 5.8), and thus provided a convergent solution to transport voice and data through its own adaptation layer, called ATM adaptation layer (AAL). However, the IP-over-ATM-over-SONET architecture could not provide an efficient data transport service for heavy traffic (through the AAL5 layer of ATM) owing to the processing latency of small ATM cells (53 bytes) with 10% overhead bytes, popularly referred to as *cell tax*. This necessitated the introduction of more efficient data-over-SONET (DoS) protocols, which were not originally included in the SONET/SDH standards. Besides some proprietary schemes, some early effort was made to develop standards to transport data using techniques, such as IP packets-over-SONET/SDH (PoS) and frame-relay-over-SONET/SDH, with HDLC encapsulation for layer-2 services.

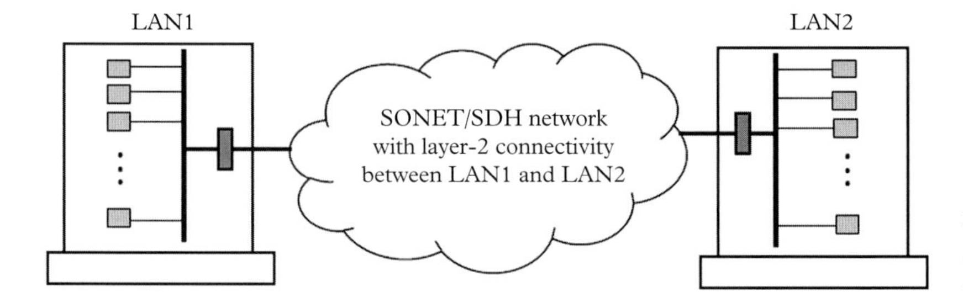

Figure 5.23 *Interconnecting two LANs through SONET/SDH with long-distance layer-2 connectivity.*

However, due to the overwhelming popularity of Ethernet with its wide range of available speeds, strong needs emerged for connecting Ethernet LANs directly with layer-2 connectivity (reducing the processing latency in IP routers and ATM switches) over metro/long-haul optical links. This led to the development of some DoS protocols for direct transport of Ethernet-over-SONET (EoS), offering better performance in respect of delay and overall cost. Eventually, the EoS evolved as a key concept for connecting Ethernet LANs using SONET/SDH networks, leading to one of the carrier Ethernet technologies, known as Metro Ethernet for layer-2 connections across metro networks. In the following, we describe the EoS architecture in further detail.

In general, two basic problems were faced for transmitting packet-based data traffic (e.g., IP, frame relay, Ethernet) over circuit-switched TDM channels of SONET. First, the packet-based data streams are bursty, and each packet has its own stand-alone overhead bytes along with inter-packet gaps, which are not relevant in SONET networks. Secondly, TDM channels in SONET networks carry data at some fixed data rates (OC-N), which don't match with the data rates of packet-based networks; for example, many of the data rates of Ethernet (e.g., 10 Mbps, 100 Mbps, 1 Gbps) don't match with the SONET data rates – 51.84 Mbps and its integral multiples.

Let us look in more detail at the first problem, in the context of Ethernet. Note that Ethernet generates bursty variable-length data packets (called frames in the Ethernet terminology) with overhead bits: preamble bits for packet-by-packet clock recovery, starting delimiter bits, address bits, etc., along with some minimum inter-frame spaces. However, the preamble and starting delimiter bits and also the inter-frame spaces don't remain relevant when the Ethernet payloads are to be packed into the payloads of the STS frames. In order to bring these two different transmission paradigms (asynchronous bursty packets and synchronous TDM bit stream) together, a frame adaptation technique was introduced, known as generic framing procedure (GFP) (Scholten *et al.* 2002; Cavendish *et al.* 2002; Dutta *et al.* 2008; ITU-T GFP G.7041 2016).

The GFP scheme can be categorized into two types: GFP-frame (GFP-F) and GFP-transparent (GFP-T). GFP-F employs deterministic overheads for Ethernet frames or IP packets using store-and-forward operation, while GFP-T transparently maps 8B/10B line-coded blocks for Fiber Channel (FC), Enterprise System Connect (ESCON), Fiber Connection (FICON), etc. with low latency (i.e., minimal packetization and buffering delays). The various possible ways of mapping the data streams from the client layers for voice, IP, SANs, and video traffic onto the SONET/SDH payload through GFP/ATM are illustrated in a layered flow diagram in Fig. 5.24. As shown in the diagram, the basic traffic streams carry voice, data from voice lines (PDH), IP/MPLS traffic through Ethernet/RPR, SAN traffic using ESCON, FC, FICON, and video traffic using digital visual interface (DVI), which are subsequently passed through appropriate intermediate layers: ATM, frame relay (FR), PoS, high-level data link control (HDLC) and GFP (aided by inverse multiplexing of incoming traffic and link-capacity allocation schemes, discussed later), and eventually mapped onto SONET/SDH payloads. In the present discussion, considering the popularity of Ethernet, we consider only the EoS transmission in further detail, focusing on the Ethernet-over-GFP-over-SONET/SDH protocol stack.

For mapping Ethernet frames into GFP-F frames (Fig. 5.25), the preamble and starting delimiter bits, as well as the inter-frame gaps, are removed (all these bits are restored at destination end while mapping GFP frames to Ethernet frames), and the remaining bytes carrying the source and destination addresses, length/type of the frame, and payload data (along with padding if any) are retained in the GFP-F frame. Further, some additional bytes are added: payload identifier (PLI) bytes, core header error control (cHEC) bytes, type

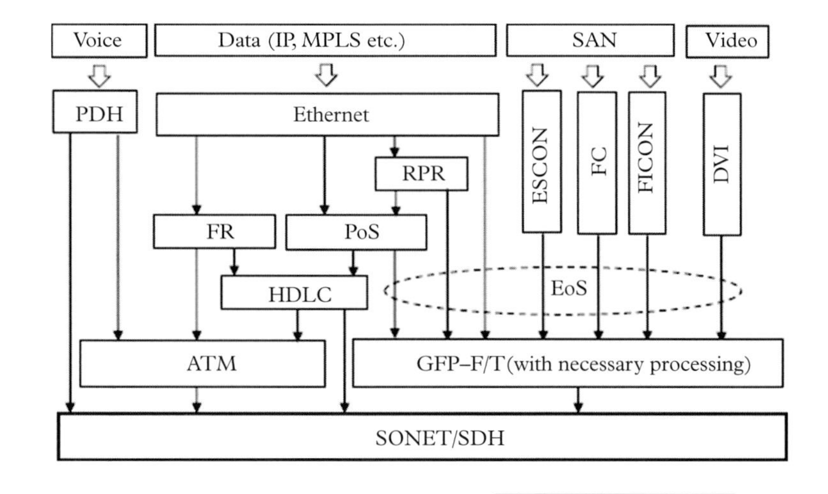

Figure 5.24 *Adaptation of voice, data, storage, and video traffic over SONET/SDH networks.*

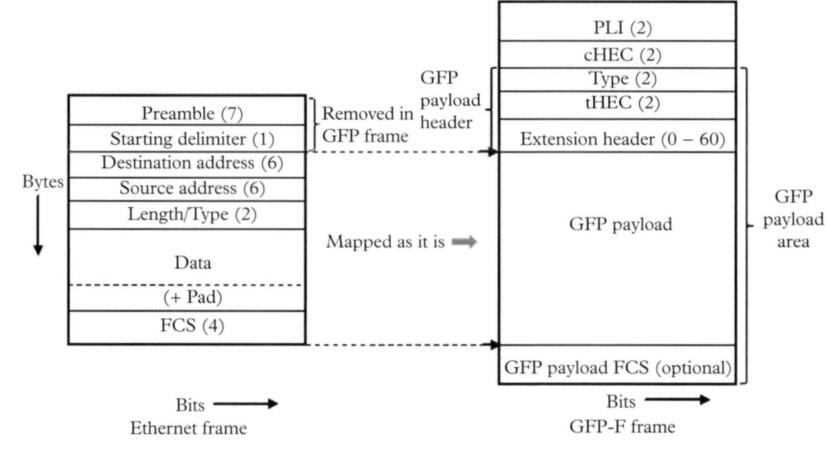

Figure 5.25 *Mapping of Ethernet frame to GFP-F frame. Arrows following the bits and bytes indicate the sequence of transmission, i.e., bytes are transmitted from top to bottom and bits are transmitted from left to right; the numbers in parentheses in the fields indicate the numbers of bytes in the respective fields.*

bytes, type header error control (tHEC) bytes, and GFP extension header bytes. The various roles of GFP headers, as shown in Fig. 5.25, are discussed in the following.

PLI has two bytes indicating the size of the GFP payload area. The PLI values in the range of [0–3] are kept for internal usage of GFP, and the corresponding frames are called GFP control frames, while all other PLI values and the respective frames are designated for GFP client frames. cHEC has two bytes and represents a cyclic redundancy check (CRC) sequence (CRC-16) to protect the core header from error. Below cHEC, Type is a two-byte field of the payload header, indicating the content and format of the payload information. tHEC is another HEC sequence which protects the preceding type field. The presence of the extension header and the optional payload frame-check sequence (FCS) (for the protection of payload error) is specified in the type field. Further note that, along with Ethernet frames, GFP-F also serves for mapping of other client signals: PPP/IP packets or any other HDLC-framed protocol data units (PDUs).

While mapping the frames of a given Ethernet data stream into a SONET payload, one can make use of the appropriate combination of lower-order containers (e.g.,

VT1.5's/VT2's), or higher-order tributaries (e.g., STS-1's/STS-3c's) for efficient capacity (maximum available bandwidth) utilization. Note that just transmitting a 10 Mbps Ethernet data stream using one dedicated STS-1 payload over OC-1 carrier at 51.84Mbps will lead to significant bandwidth wastage. In particular, given an STS-1 frame having the effective data rate of 48.384Mbps, an incoming 10 Mbps Ethernet stream will be able to utilize the available capacity with a capacity-utilization efficiency η_{CU}, given by

$$\eta_{CU} = \frac{10\,\text{Mbps}}{48.384\,\text{Mbps}} \times 100 \approx 21\%\,\text{only}, \tag{5.11}$$

thereby causing a large bandwidth wastage of 79% (Fig. 5.26).

To get around the above problem, a 10 Mbps Ethernet stream can be split into a group of seven VT1.5 frames (denoted as VT1.5-7v) and transported on independent routes over SONET, leaving out the remaining part for other purpose. This will lead to a much-enhanced capacity-utilization efficiency, given by

$$\eta_{CU} = \frac{10\,\text{Mbps}}{(1.728 - 0.128) \times 7\,\text{Mbps}} \times 100\% \approx 89\%, \tag{5.12}$$

where we have considered the effective transmission rate of a VT1.5 as 1.728 Mbps (see Table 5.5) and subtracted the effective speed of the two delimiter bytes to be used in each VT1.5 frame, amounting to $(2 \times 0.064) = 0.128$ Mbps (as each byte in an STS-1 frame consumes 64 kbps = 0.064 Mbps).

Similarly, a 100 Mbps fast Ethernet stream can be transported using VT1.5-64v or STS-1-2v, instead of using STS-3c link for higher capacity utilization. Also, with two incoming GbE streams, it would be worthwhile to split it and fit them into two STS-1-21v groups, instead of sending them through one STS-48 frame. The use of two sets of STS-1-21v will also leave six STS-1's for other applications, such as voice traffic over PDH. Even sending the two GbE streams through one STS-48 will also be less efficient as compared to the above proposition.

In view of the above, network operators could split the incoming data stream (i.e., perform inverse multiplexing), with large payload, into a group of smaller multiple tributaries of SONET signals, and thereafter transport the members of this group through independent routes over the SONET-based network, thereby ensuring enhanced capacity utilization. This technique for efficient capacity utilization is called virtual concatenation (VCAT) of incoming data streams onto appropriate VTx's/VCy's and STS-N's/STM-K's for SONET/SDH

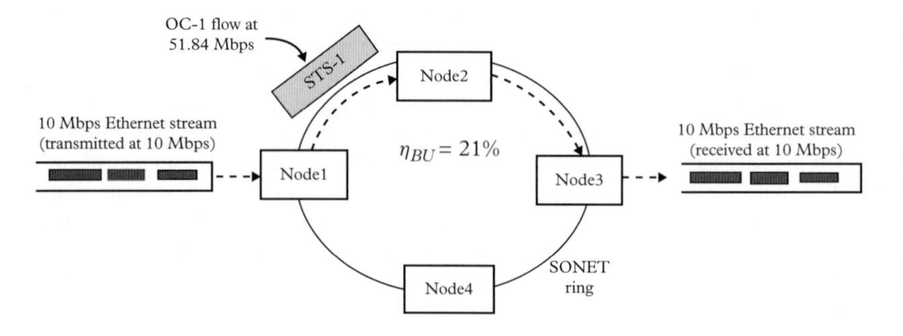

Figure 5.26 *Inefficient (21%) capacity utilization (or 79% bandwidth wastage) while transmitting 10-Mbps Ethernet through OC-1 SONET ring.*

networks (Cavendish *et al.* 2002; Dutta *et al.* 2008).[2] The concatenation process remains virtual in the sense that the different members in a VCAT group may travel through different paths through the SONET/SDH infrastructure (usually a ring or interconnected multiple rings and linear networks), though they remain logically concatenated and are reassembled sequentially in the destination node. However, different members traversing different routes may experience different delays. This issue needs to be addressed while inverse-multiplexing the incoming traffic into smaller containers, so that the network can reach a trade-off between the capacity utilization and the differential delay constraint.

A basic VCAT mapping scheme is shown in Fig. 5.27 for some example cases: 10 Mbps Ethernet, 100 Mbps fast Ethernet, and 1 Gbps Ethernet (i.e., GbE). The results of VCAT in SONET for these Ethernets are summarized in Table 5.8. As discussed earlier, Table 5.8 shows that for 10 Mbps Ethernet bandwidth efficiency increases from 21% (with STS-1) to 89% with a VCAT operation using VT1.5-7v. Similar trends are also observed for Fast Ethernet and GbE LANs.

While VCAT provides the flexibility of creating multipath SONET conduits of different sizes, the source node needs to have flexibility to adjust the capacity of a VCAT group dynamically in accordance with the network traffic and overall status including failure scenarios. This is realized using a technique called *link-capacity adjustment scheme* (LCAS),

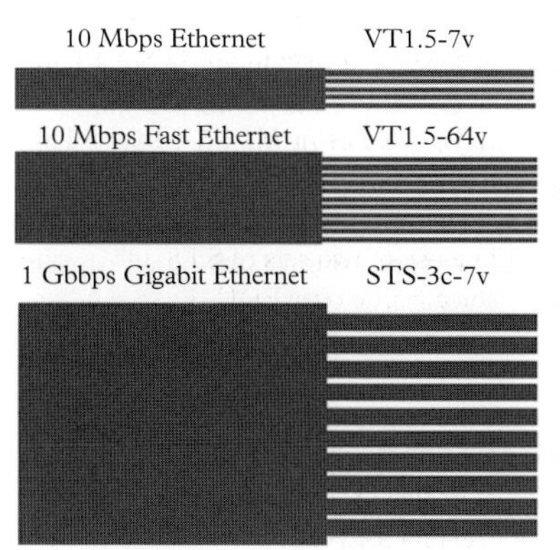

- **10 Mbps Ethernet**: Mapped into SONET SPE with a VCAT group (VCG) of 7 VT1.5 blocks.
- **100 Mbps Fast Ethernet**: Mapped into SONET SPE with VCG of 64 VT1.5 blocks.
- **1 Gbps Gigabit Ethernet (GbE)**: Mapped into SONET SPE with 7 STS-3c frames.

Pro: High bandwidth utilization
Con: Each member in a VCG is routed independently – so they can reach the destination node through different routes causing differential delay.

Figure 5.27 *VCAT group (VCG) formations from various Ethernet streams for improving capacity utilization from SONET.*

Table 5.8 *Use of VCAT for improving capacity-utilization efficiency* η_{CU}.

Ethernet speed	Direct SONET-payload mapping and η_{CU}	SONET-VCAT-based mapping and η_{CU}
10 Mbps Ethernet	STS-1; $\eta_{CU} \approx 21\%$	VT1.5-7v; $\eta_{CU} \approx 89\%$
100 Mbps Ethernet (fast Ethernet)	STS-3c; $\eta_{CU} \approx 67\%$	VT1.5-64v; $\eta_{CU} \approx 98\%$
1 Gbps Ethernet (GbE)	STS-48; $\eta_{CU} \approx 43\%$	STS-3c-7v, STS-1-21v; $\eta_{CU} \approx 98\%$

[2] Note that in SDH the PDH tributaries are transformed into virtual containers (VCs) which are equivalent to VTs in SONET.

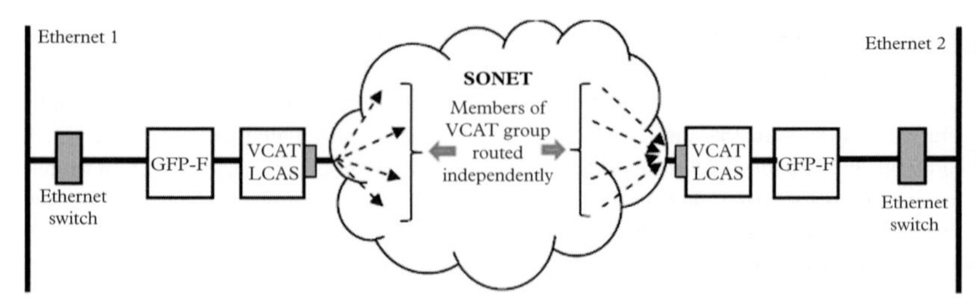

Figure 5.28 *GFP-F/VCAT/LCAS over SONET/SDH network.*

which can automatically increase/decrease VCAT link capacity without affecting the network traffic (hitless adjustment), through the exchange of messages using a network-wide distributed control operation (a two-way handshake protocol) (Cavendish *et al.* 2002; Dutta *et al.* 2008). For example, consider an EoS scenario, where a GbE stream is found to be partially filled. This is quite as expected and often the effective data rate would be around 300 Mbps (typically), though with the maximum possible rate of 1 Gbps. Thus, to reduce the wastage of capacity, LCAS can nominally allocate STS-3c-7v (capacity = 338.688 Mbps) with adequate buffering, and carry out flow control to address the instantaneous transmission rates upper-bounded by the maximum capacity of 1 Gbps. This way, the data rate in EoS can be adapted to the needed data rates at different instants of time within reasonably short durations in the range of a few seconds or minutes. Moreover, VCAT and LCAS do not call for any changes inside the SONET domain of the network. Figure 5.28 presents a representative block diagram for EoS-based transport of Ethernet traffic using GFP-F/VCAT/LCAS scheme. As shown in this figure, Ethernet traffic generated from an Ethernet LAN is passed through the GFP framing process and thereafter split using the VCAT scheme and transmitted over SONET/SDH network using LCAS-based capacity adjustment process.

5.9 Optical transport network

Even with the developments of the data-transport mechanisms using GFP, VCAT, and LCAS over SONET as the carrier, the phenomenal growth of data traffic, particularly from Ethernet-based LANs, overwhelmed these DoS architectures and motivated the telecommunication community to explore a unified carrier platform, leading to the standard called the optical transport network (OTN) (ITU-T G.709 2016). The OTN standard was introduced as a convergent optical networking platform, that would encapsulate and carry inclusively the traffic streams from different potential clients, e.g., Ethernet, IP, Fiber Channel, and SONET, at the higher transmission rates. Note that, in VCAT the Ethernet rates were much below the SONET rates (e.g., the lowest-rate 10 Mbps Ethernet was much below the lowest-rate 51.84 Mbps SONET), thereby calling for virtual concatenation of the lower-rate Ethernet flows, and the associated improvisation schemes. However, with the arrival of GbEs, the lowest rate of GbE (i.e., 1 Gbps) was not that small, as compared to its nearest OC-48 SONET rate at 2.488 Gbps, and more importantly the next higher rates were almost the same for the Ethernet and SONET streams, such as, 10GbE vis-a-vis OC-192 SONET at 9.953 Gbps, or 40GbE vis-a-vis OC-768 SONET at 39.81 Gbps. This aspect motivated the telecommunication community to look for a smarter universal optical transport standard for high-speed connections, more so when the WDM-based optical networks became a reality, with high-speed long-haul all-optical connections, which eventually called for the OTN standard.

In OTN, the respective frames from the clients remain unpacked (as in SONET) during the encapsulation process with their independent synchronization schemes, e.g., preamble-based burst synchronization in Ethernet and pointer-based perpetual synchronization for SONET. At the lowest level of transmission rate, OTN takes up two of 1 Gbps Ethernet (1GbE) streams or one OC-48 SONET stream to form its basic frame, designated as optical channel transport unit 1 (OTU1), where appropriate encapsulations are added with 1GbE or OC-48 frames, leading to the OTU1 rate of 2.67 Gbps. Multiplexing of higher-rate Ethernet and SONET streams as clients led to the next three OTN rates, leading to OTU2, OTU3, and OTU4, generically represented as OTUk, with $k = 1, 2, 3, 4$. Considering the candidate GbEs and SONET as the potential clients, OTUk's for various levels (upto $k = 4$) are assigned with the following bit rates, including the respective overhead bytes and the associated bytes for FEC:

- OTU1: bit rate = 2.67 Gbps, supporting two 1GbEs or one OC-48 (2.488 Gbps),
- OTU2: bit rate = 10.71 Gbps, supporting one 10GbE or one OC-192 (9.953 Gbps),
- OTU3: bit rate = 43.02 Gbps, supporting one 40GbE or one OC-768 (39.813 Gbps),
- OTU4: bit rate = 111.81 Gbps, supporting one 100GbE or one OC-1920 (99.533 Gbps).

In addition to the above four levels, there are two more over-clocked versions of OTU2 and OTU3, viz., OTU2e (*e* for over-clocked signal) and OTU3e, transmitting at 11.22 Gbps and 44.33 Gbps, respectively. Interested readers are referred to the review article on OTN in (Roese and Braun 2010) for further details of this aspect.

The layered architecture of OTN is illustrated in Fig. 5.29, where each layer adds its overhead bytes. For each k, the overhead channel data unit (ODU) layer appears next (below) to the client layer for adding/removing the ODU overhead bytes which are used to adapt the client signals into the OTN frame structure. However, the OTU layer functions in a manner similar to the line and section layers of SONET using the LOH and SOH bytes.

Following the above layered architecture, the OTN frame structure is designed as illustrated in Fig. 5.30. As shown in the figure, an OTN frame comprises bytes arranged in four rows and 4080 columns, and this structure remains the same for the higher multiplexing

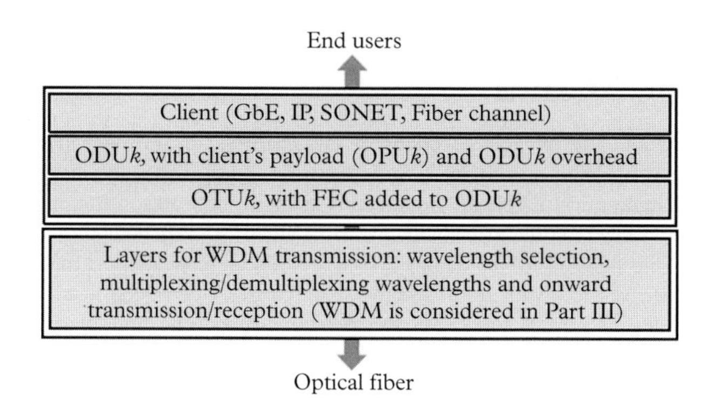

Figure 5.29 *Layered architecture of OTN.*

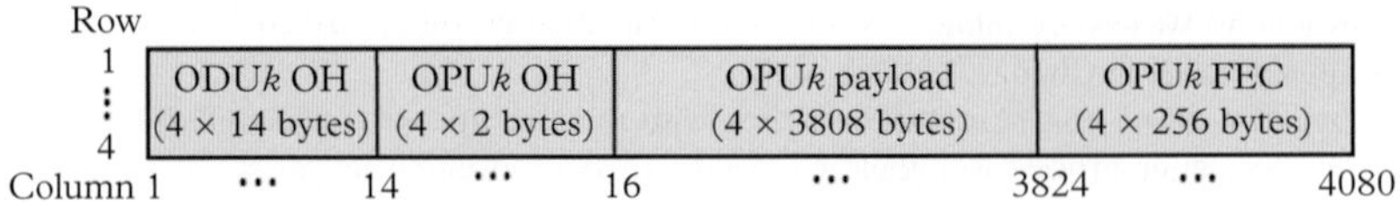

Row	ODUk OH	OPUk OH	OPUk payload	OPUk FEC
1 ⋮ 4	(4 × 14 bytes)	(4 × 2 bytes)	(4 × 3808 bytes)	(4 × 256 bytes)

Column 1 ··· 14 ··· 16 ··· 3824 ··· 4080

Figure 5.30 *OTN frame structure. OH: overhead.*

levels, designated as OTUk with k representing the level of multiplexing hierarchy. In the first 14 columns from the left, i.e., $14 \times 4 = 56$ bytes are ODUk overhead (OH) bytes, added as encapsulation, taking care of the client connectivity, connection monitoring, and protection. Thereafter, two columns, i.e., $2 \times 4 = 8$ bytes are added as overhead channel payload unit (OPUk OH) bytes, dealing with the mapping of the client's payload into a frame structured into the OTN frame. Next, follows the payload field with 3808 columns with $3808 \times 4 = 15,232$ bytes, called the optical channel payload unit (OPUk payload). Finally comes the field for FEC with 256 columns (OPUk FEC), amounting to $256 \times 4 = 1024$ bytes, serving as the overhead for (235-245) Reed–Solomon code to reduce the BER for the entire OTUk frame.

For WDM systems (considered in Part III), higher-level OTUk's (as in SONET) need to comply with a bit rate within the ITU-T-recommended channel widths (100/50 GHz). Thus, the modulation bandwidth needed for OTU3 and OTU4 with bit rates of 43.02 and 111.81 Gbps turned out to be critical for 50/100 GHz channel width, more so due to the problems of finite linewidth and frequency drift/misalignment in lasers. Hence, as in SONET (for OC-768), OTN had to employ multi-level modulation schemes for the higher bit rates to shrink the modulation bandwidth well below 100/50 GHz. For example, with differential QPSK (DQPSK) modulation, a 100 Gbps data stream could modulate a laser with a symbol rate of 50 Gigabauds/sec and this rate would get further reduced to 25 Gigabauds/sec with polarization-shift keying, thereby requiring around 25 GHz of modulation bandwidth. However, this rate will be pushed up to around 30 GHz (yet well within 50 or 100 GHz channel width) owing to the use of FEC codes for meeting the BER specification. In summary, OTN offers a *packet/circuit-switching-friendly* universal backbone transmission mode over optical fibers, while ensuring backward compatibility with its major high-speed clients, e.g., xGbE's and OCy's, with $x = 1, 10, 40, 100$ and $y = 48, 192, 768, 1920$ signifying different bit rates of GbE and SONET bit streams in Gbps, respectively, as well as offering reliable high-speed backhaul networking for 3G, 4G, and forthcoming 5G mobile communication systems.

5.10 RPR: packet-based metro network over ring

In Chapter 3 and the foregoing sections of this chapter, we have considered some of the optical LANs/MANs and SONET/SDH networks. As discussed in Chapter 3, optical MANs and large optical LANs are not suitable as metro networks in the public domain. Metro networks need to offer availability over large areas for the general public, along with fast recovery mechanisms against failure (resilience), and it must support both packet and circuit-switched services with the respective quality-of-service (QoS) needed. In regard to the suitability of various topologies for metro networks, one finds that large optical star topology leads to wasteful fiber deployment. Bus topology offers limited network connectivity and does not guarantee fail-safe operation. However, ring topology needs minimal cabling (two-connected mesh) and covers well a metro area along with the potential for failure-proof operation (resilience) through its closed-loop connectivity. With these issues in mind, we next examine in this context the suitability of the LAN/MAN architectures considered earlier in Chapter 3.

To start with, we consider GbE networks, which today offer the most popular business networking solution. GbE as a packet-centric network does not require static bandwidth allocation between node pairs and can operate over a large area and support high speed. However, the Ethernet-based networks employ spanning-tree protocol for routing purposes (between the GbE switches), where the problem of loop formation is avoided by disabling some physical link(s). This leads to the under-utilization of the physical topology leading to poor transmission efficiency, as the blocked links hinder data transport through shortest paths. Moreover, GbE operating over a spanning tree, does not have any natural potential (as in a ring) for fast recovery against failure. In fact, with a failed link, GbE needs to recompute its spanning tree for reconfiguration, which can take a long time approaching hundreds of milliseconds. Furthermore, GbE is designed primarily as a packet-centric network providing at the most some simple traffic prioritization rules, but without ensuring much the QoS of circuit-switched services (e.g., guaranteed bandwidth, delay, and delay jitter).

On the other hand, the optical networks using SONET/SDH-based ring topology support circuit-switched TDM transmission over a metro area along with the desired metro availability for the public, and resilience as well. However, these voice-optimized transport networks lack the bandwidth efficiency needed for carrying bursty data traffic, as compared to the packet-centric LANs/MANs, such as DQDB, FDDI, or GbE. In reality, SONET/SDH networks set up point-to-point circuit-switched connections (TDM time slots) between the node pairs even for the streams of bursty packet-switched data, which once setup cannot be used by any other nodes even if the given circuit-switched connection goes under-utilized (i.e., the free time slots in the point-to-point TDM connection cannot be utilized by others).

In FDDI networks, though realized over ring topology, transmitted data packets are not removed from the ring until they return to the source nodes. This feature, known as source-stripping, keeps the ring-space engaged by the transmitted packets even after they are received by the destined nodes, thereby preventing the subsequent nodes from accessing the ring. Moreover, FDDI employs two counter-rotating rings, but utilizes one of them in the absence of any failure in the network. DQDB, employing bus topology, lacks fail-safe operation and has the problem of the size-dependent unfairness in MAC protocols despite the use of distributed queues.

In order to address the above limitations of the existing MANs and SONET/SDH networks, some ring-based packet-switched metro networks were proposed, e.g., meta-ring and resilient packet ring (RPR) networks. As mentioned earlier, in today's networking scenario, metro networks are split into two parts: metro-core and metro-edge, with the former connecting to a long-haul backbone and latter bridging between metro-core and access. The SONET/SDH-based circuit-switched metro-core networks, having nearly steady flows of traffic can perform with acceptable bandwidth utilization in optical fibers, while the metro-edge ring, being closer to the end-users of the access segment, needs to handle bursty traffic flows more than in the metro-core segment and hence calls for packet-centric transmission, wherever necessary. The RPR emerged as a bandwidth-efficient metro-edge solution for both packet-switched and delay-sensitive circuit-switched services, along with the necessary resilience and availability of the network (Davik *et al.* 2004; Yuan *et al.* 2004; IEEE RPR 802.17 2004). In particular, RPR combined the better aspects of both sides: resilience and availability of SONET/SDH ring, and packet-based bandwidth-efficient operation of older versions of optical LANs/MANs over the resilient fiber ring. Moreover, RPR traffic can also be carried over SONET/SDH as well as GbE networks wherever necessary. In the following, we discuss the basic features of RPR, its node architecture, MAC protocol and other relevant issues in further details.

5.10.1 Basic features of RPR

RPR employs two counter-propagating packet-switched fiber rings (Fig. 5.31) unlike the circuit-switched SONET/SDH rings, supporting up to 255 nodes over a maximum circumference of 200 km. Both rings are utilized simultaneously without keeping one as a spare for network protection. The circuit-switched traffic streams (e.g., voice, video) are also packetized and transmitted along with the data packets in a packet-switched manner. The total bandwidth available over the two fiber rings, called outer and inner ringlets (designated as Ringlet 0 and Ringlet 1 in Fig. 5.31, respectively), are utilized with high bandwidth efficiency by employing packet-based transmission for both packet-switched as well as circuit-switched services. The bandwidth efficiency is achieved mainly because end-to-end circuit-switched connections between node pairs are avoided altogether. However, notwithstanding the packet-based transmission over the ringlets, the QoS requirements for the circuit-switched services (e.g., voice, video) are taken care of by prioritized scheduling at RPR nodes.

The inner and outer ringlets in RPR carry both information-bearing packets and control messages, where the data traffic and the corresponding control messages flow in opposite directions over the two different ringlets. For example, the control messages for the data traffic on outer ringlet are carried by the inner ringlet in the opposite direction and vice versa. Instead of using the spanning-tree algorithm of Ethernet for routing, RPR employs shortest-path routing between its node pairs, and thus chooses one of the two ringlets depending on which ringlet offers the shortest path. Along with shortest-path routing, RPR employs *destination-stripping* of packets (unlike FDDI), thereby allowing spatial reuse of fiber space/time by the nodes subsequent to the destination node with significant enhancement in capacity utilization. As shown in Fig. 5.31, packet A enters Ringlet 0 at node 8 (as the source or ingress node) and exits from node 2 (as destination or egress node), leading to destination-stripping, which frees the rest of Ringlet 0 for concurrent use (also referred to as spatial reuse). This enables packet B also to enter Ringlet 0 concurrently at node 4 and exit Ringlet 0 at node 6.

The above feature of spatial reuse and concurrent access over disjoint segments of both ringlets significantly increases the throughput of the dual-ring RPR network. In order to examine this aspect, we consider that an RPR has N nodes, and each node is *always ready* to send packets to all $(N-1)$ destinations with equal probability and employs shortest path

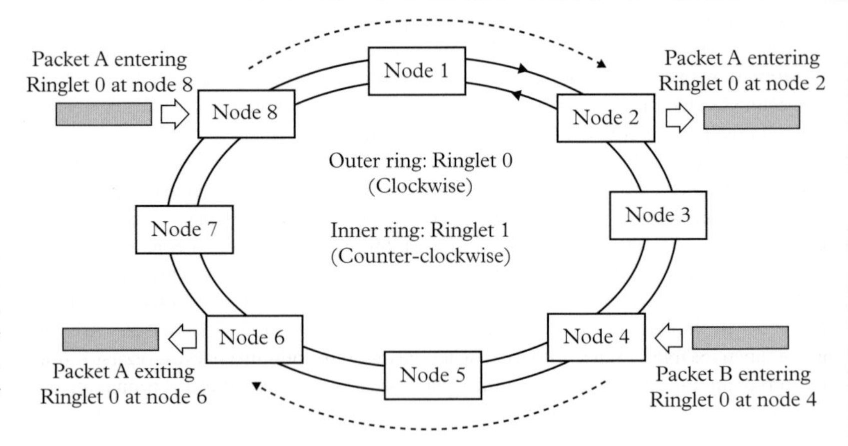

Figure 5.31 *A representative physical topology for an RPR network with two counter-propagating ringlets. The example scenario illustrates how the destination stripping of transmitted packets helps in reusing the fiber ring (enabling concurrent transmissions of packets A and B over Ringlet 0).*

routing (i.e., no connection will span over more than $\frac{N}{2}$ nodes). With this assumption, it turns out that the maximum connection span is $\frac{N}{2}$ (in terms of the number of nodes covered by a connection), and all the connections being equiprobable, the average connection span is $\frac{1}{2} \times \frac{N}{2} = \frac{N}{4}$. This leads to the fact that the spatial reuse enables four nodes to transmit concurrently over each ringlet in one direction. Thus the RPR as a whole can achieve a high capacity, which is, on the average, (with uniform traffic assumption) eight times a single link capacity and four times more than the capacity of bidirectional rings without spatial reuse, such as, FDDI.

Furthermore, RPR ensures resilience with a restoration time not exceeding 50 ms (the same as in SONET/SDH rings). In particular, when a single node or a bidirectional link fails, RPR sets up an alternate path within the stipulated time (i.e., within 50 ms). This is realized by harnessing the dual-ring RPR topology, where the traffic is moved between the surviving nodes, by steering away the route to the ring, working in the opposite direction.

5.10.2 RPR node architecture and MAC protocol

RPR ensures QoS for circuit-switched services over its packet-switched architecture by employing class-based MAC protocol. In particular, RPR operates with multiple classes of traffic: Class A, Class B, and Class C, which are treated in accordance with their QoS requirements. Class A meets the QoS requirements of the circuit-switched traffic, thereby ensuring a guaranteed data rate with specified upper bounds for the end-to-end delay (latency) and delay jitter. Class B offers predictable upper-bounds for latency and delay jitter, while Class C traffic gets the best-effort service.

Class A has two subclasses: subclass A0 and subclass A1, both of which offer guaranteed circumference-independent delay jitter. However, for subclass A0, bandwidth is kept absolutely reserved, so that even if this bandwidth goes unused, it can not be reclaimed by other classes. For subclass A1 the bandwidth is also guaranteed, but if the allocated bandwidth goes unused, it can be reclaimed by some of the lower-priority classes. Class B ensures guaranteed bandwidth but with circumference-dependent delay jitter, and also has two subclasses: subclass B-CIR (class B with committed information rate) and subclass B-EIR (class B with excess information rate). Subclass B-CIR ensures a committed bandwidth as in subclass A1. However, the subclass B-CIR, being a subclass of class B, operates with circumference-dependent jitter in contrast to subclass A1, as the latter guarantees (along with subclass A0) a circumference-independent delay jitter. However, subclass B-CIR has the feature of supporting reclaimable bandwidth as in subclass A1, i.e., the lower-priority classes (subclass B-EIR and class C) can reclaim the unused bandwidth of subclasses A1 and B-CIR in an opportunistic manner. Thus, subclass B-EIR and class C form a group, called the fairness eligible (FE) group. The overall class-based bandwidth provisioning scheme of RPR is illustrated in Fig.5.32.

A typical node architecture of RPR is shown in Fig. 5.33, which is used for one of the two ringlets, and thus each RPR node houses two such units for the two ringlets. Each node, based on the class-based priorities, deals with two different types of traffic flows: the transit traffic flow needing passthrough functionality, and the local ingress (add) and egress (drop) traffic flows needing add-drop functionality. As shown in Fig. 5.33, the RPR node employs a class-based scheduler, which schedules the transit as well as the ingress packets, based on a fairness algorithm. The incoming traffic to the node through the receive link is examined in the functional block, called a checker, which strips off the destined packets (egress traffic) to drop them into the node, and forwards the transit traffic to the transit queuing module (also known as the insertion buffer module). The transit queuing module has either one queue, called primary transit queue (PTQ), or two queues, PTQ and secondary transit

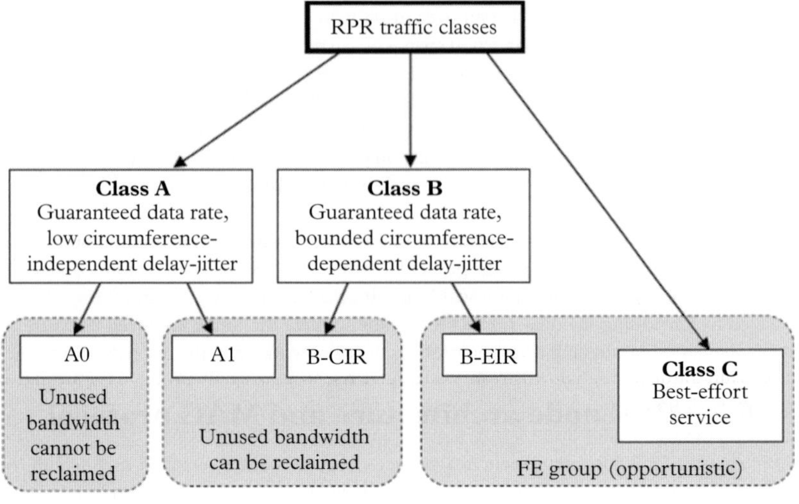

Figure 5.32 *Different classes of traffic in RPR.*

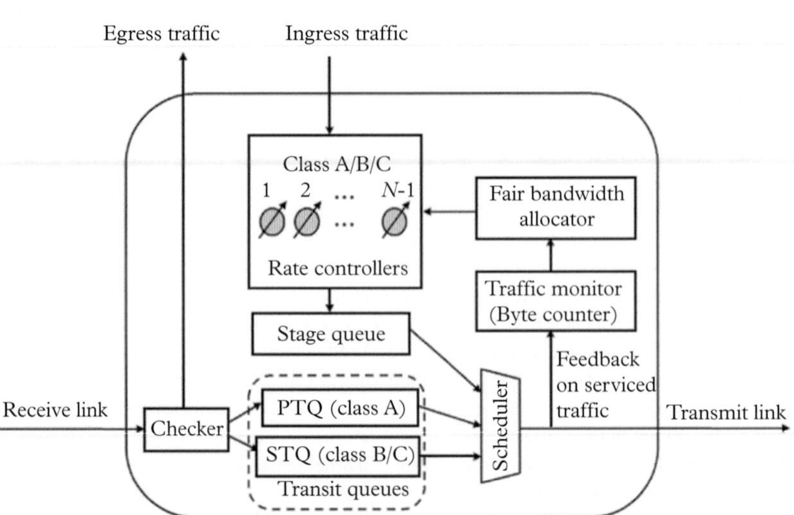

Figure 5.33 *Dual-queue node architecture for an N-node RPR; fairness-aware bandwidth allocation on feedback path applies to FE group only (Yuan et al. 2004, ©IEEE).*

queue (STQ). In a single-queue node, the scheduler exercises strict priority for the transit traffic over the local ingress traffic. In the dual-queue nodes (having PTQ and STQ), PTQ serves for class A transit traffic and STQ serves for class B/C transit traffic. The scheduler in this case first serves the class A transit traffic from PTQ. Once PTQ is served (or if it is empty) and STQ is partially full, then the scheduler serves the local (ingress) traffic belonging to class A. However if STQ is found to be full, the scheduler first serves STQ so that there is no loss of transit traffic. Thus, a class A transit packet will experience a delay, equal to the propagation delay in the ring plus some occasional delay while waiting in PTQs for the already originated (ingress) packets at intermediate nodes.

Regardless of the class, RPR never discards any packets for resolving any congestion. In particular, the packets of classes A and B reach their destination with the respective committed/upper-bounded delay and jitter values, while class C packets, even though being offered the lowest priority, never get lost but may experience high delays at times. As PTQ should not delay class A packets, it usually has a buffer having the capacity to hold one or two class A packets. However, with lower-priority classes, STQ must have much larger size to prevent loss of packets.

As shown in Fig. 5.33, the ingress traffic of all classes at a given node is first passed through one of the $(N-1)$ traffic controllers designated for each of the possible $(N-1)$ destination nodes in an N-node RPR network. These controllers along with the scheduler operate with a *fairness algorithm* based on the feedback about the serviced (i.e., at scheduler output) outgoing packets through the traffic monitor (employing a byte counter). These controllers throttle (control) the data rates of the ingress traffic classes in conformity with the QoS of the transit as well as ingress-traffic demands by using the fair bandwidth (or rate) allocator. Note that, the transit queues (PTQ and STQ), while ensuring lossless flow of upstream traffic through downstream nodes, give rise to a *starvation* problem for the ingress traffic in all nodes, more so indeed for the traffic belonging to FE group. To address this issue, the fairness algorithm plays a central role in a distributed manner (in all the nodes) by ensuring QoS-aware bandwidth provisioning for the FE traffic throughout the entire network. We describe the fairness algorithm of RPR in the following.

5.10.3 RPR fairness algorithm

The main objective of the fairness algorithm in RPR is to fairly distribute the unallocated as well as the reclaimable rate or bandwidth (allocated to but not used by subclasses A1 and B-CIR) among the various nodes for transmitting their ingress packets belonging to subclass B-EIR and class C (i.e., FE group). Before describing the fairness algorithm, we discuss how the bandwidth is first reserved for all the nodes for the highest-priority traffic, i.e., subclass A0, on a given ringlet. This task is carried out through a broadcast mechanism over the other ringlet (operating in opposite direction) as follows.

Each node broadcasts its need of rate reservation for subclass A0 to all other nodes on the other ringlet by using control messages. In this process, each node also receives the same broadcast messages from all other nodes in the ring. Having completed this process, each node finalizes the rate that it should reserve for its subclass A0 traffic. Thereafter, the bandwidth remaining available over the ringlet is used by the other classes of traffic. In particular, the subclasses A1 and B-CIR are next considered for bandwidth allocation, and subsequently the bandwidth to be allocated for the FE group of all the nodes is determined by the fairness algorithm. In doing so, it becomes important to assess the dynamics of traffic congestion around the RPR ringlet under consideration. The process is identical for both ringlets, where the control messages and data packets always flow in two counter-propagating ringlets. As mentioned earlier, if we consider Ringlet 0 for data transport, then Ringlet 1 carries control messages, and vice versa.

When the outgoing transmit link of a node is exhausted, i.e., the bandwidth demand over the link exceeds the link capacity C, the corresponding link and node are considered to be *congested*, and at this point the fairness algorithm of RPR comes into force. In such situations, RPR must fairly distribute the available bandwidth between the congested nodes intending to transmit their respective ingress packets belonging to the FE group. For single-queue nodes (i.e., in absence of STQ), the ingress packets (even class A packets) may have to wait for a long time as the transit packets must not be lost. However, in double-queue nodes, the class A ingress packets need to wait only when STQ (which has longer buffer size as compared to PTQ) gets filled up. This also implies that the class B/C transit packets would have longer waiting time at intermediate nodes. In effect, any congestion observed at a node is encountered owing to its own ingress traffic demand as well as the transit traffic from its upstream nodes on the same ringlet. The nodes and links facing this situation are collectively termed the *congestion domain*.

In order to have faster communication with the congested nodes when a congestion domain develops, the fairness messages (using control packets) proposing the local estimates of *fair rates* for the nodes under consideration are transmitted on upstream over the other ringlet operating in the opposite direction. Using these local fair-rate estimates from all the respective upstream nodes in the congestion domain, each congested node re-estimates its fair rate following a *fairness algorithm* and throttles accordingly its own ingress FE traffic using the rate controllers. After a convergence period, all the nodes set their rates to the respective fair rates and thereby the congestion gets controlled. When the state of congestion ebbs away, the nodes that were affected by the congestion keep periodically enhancing their sending rates to acquire their full bandwidth share.

The fairness algorithm employed at each congested node can choose one of the two modes of operation: aggressive mode (AM) or conservative mode (CM). AM is used in dual-queue nodes with PTQ and STQ in the transit queuing block, while CM is employed in single-queue nodes (i.e., without STQ). For executing the fairness algorithm (for AM or CM), some measurements are carried out in the congested nodes by using byte-counting in its traffic monitor on the serviced traffic (at the scheduler output) over a time duration, called the aging interval T_{age}. In particular, the traffic monitor measures two variables, called forward rate $R_{fwd}(n)$ and add rate $R_{add}(n)$, representing the numbers of bytes counted at the scheduler output in node n during T_{age} for the transit and ingress traffic belonging to FE category, respectively.

Congestion detection and rate estimation in AM

Consider a congested dual-queue node (node n, say), employing AM for fair rate estimation for its ingress traffic belonging to FE category. Such a node can stop the ingress FE traffic only when STQ gets overloaded. STQ is considered to be overloaded when its queue length $L_{STQ}(n)$ reaches 1/8th or 12.5% of the full buffer size. This overload limit of $L_{STQ}(n)$ for AM is referred to as lower threshold, denoted by $L_{th}^{AM}(low)$. The node is also assumed to be aware of the aggregate reserved rate R_{res} of the network-wide class A traffic (through prior broadcast control messages). The rest of the permissible traffic rate of RPR (per ringlet) is designated as the unreserved traffic rate R_{unres}, given by

$$R_{unres} = C - R_{res}. \tag{5.13}$$

Using these parameters and variables along with the measured values of its forward and add rates $R_{fwd}(n)$, $R_{add}(n)$, respectively, node n has to first identify whether it is congested or not by checking the following conditions:

$$L_{STQ}(n) \geq L_{th}^{AM}(low), \text{ or } R_{fwd}(n) + R_{add}(n) \geq R_{unres}. \tag{5.14}$$

When the node experiences either of the above two conditions (i.e., left side exceeds or equals to right side), it considers itself to be congested and chooses its local fair rate $R_{loc}^{fair}(n)$ as its own serviced add rate, given by

$$R_{loc}^{fair}(n) = R_{add}(n), \tag{5.15}$$

and transmits this rate to all the upstream congested nodes (by multicasting) over the ringlet operating in the reverse direction (i.e., to the upstream nodes). Note that if node n transmits a local fair rate which is lower than the local fair rate of an upstream congested node (say node $(n-1)$), then this upstream congested node (node $(n-1)$) will throttle its own add rate to this lower value. In fact, all the congested nodes in a congestion domain will throttle to the lowest received bid of rate to get out of the congestion phase. Having come out of the congestion in this manner, they will again start increasing their add rates, which might lead to oscillatory swings of local add rates around the true fair rate. This will eventually bring in a dynamically balanced state in the RPR in all congestion domains (note: more than one such domain may coexist over each ringlet) with all the local FE add rates moving closely around the true fair rate.

Congestion detection and rate estimation in CM

As mentioned earlier, the single-queue nodes (i.e., having PTQ only) employ CM-based fair rate allocation. Any such node uses an *access timer* to measure the time interval between two successive FE transmissions of the node. Such nodes employ stricter priority for transit traffic over the local add traffic to avoid losses of transit packets. However, RPR addresses the overture of this feature by assessing and ensuring that the local node does not starve for long by setting an upper limit for the measured time interval by the access timer. In effect, a single-queue node exercising CM considers that congestion has set in, if the stipulated time of the access timer has expired, or

$$R_{fwd}(n) + R_{add}(n) \geq R_{th}^{CM}(low), \tag{5.16}$$

where $R_{th}^{CM}(low)$ represents the threshold value for the local outgoing rate, typically set at 80% of the link capacity C (note that in AM the local threshold is a device dependent parameter, i.e., 12.5% of the queue length of STQ, while in CM the local threshold is a rate-dependent parameter). The traffic monitor in the node also keeps measuring the number of active nodes N_{actv} that have transmitted at least one packet during the past aging interval T_{age}. If a single-queue node is found congested in the current aging interval, but was not so in the last one, the local fair rate $R_{loc}^{fair}(n)$ is estimated as

$$R_{loc}^{fair}(n) = R_{unres}/N_{actv}. \tag{5.17}$$

However, if the node remains congested for several aging intervals in succession, the local fair rate $R_{loc}^{fair}(n)$ is decided from a localized judgement involving the combined serviced rate of the node, i.e., $R_{fwd}(n)$ and $R_{add}(n)$. Thus, if $R_{fwd}(n) + R_{add}(n) < R_{th}^{CM}(low)$, then the link-capacity appears to be under-utilized and hence $R_{loc}^{fair}(n)$ is increased. However, if the combined rate $R_{fwd}(n) + R_{add}(n)$ is higher than a higher threshold value $R_{th}^{CM}(high)$ (typically set at 95% of the link capacity C), then $R_{loc}^{fair}(n)$ is decreased. Thus, in the CM scheme the maximum possible link-capacity utilization is kept bounded within the upper threshold $R_{th}^{CM}(high)$, thereby justifying that the scheme is truly conservative.

Finally, it may be noted that, during both AM and CM operations, RPR may also suffer performance degradation in the presence of unbalanced traffic distributions, leading to some instability in the network throughput. Several studies have reported on this problem and interested readers are referred to (Davik *et al.* 2005; Yuan *et al.* 2004) for further details.

5.11 Summary

In this chapter we presented the three optical networking standards: SONET/SDH, OTN, and RPR. Before presenting the SONET/SDH-based networks, the problem of network synchronization was discussed. The pre-SONET/SDH era network synchronization scheme, i.e., PDH, was discussed, bringing out its limitations for application in high-speed optical communication networks. Next, we presented the basic features of the master/slave HMS scheme for network synchronization, and described how this scheme is adopted in the SONET/SDH networks. Thereafter, the basic features of SONET framing schemes and the network operations using SONET-based network elements were described. The problem of poor link-capacity utilization in SONET while carrying the lower-rate packet-switched data streams (e.g, 10 Mbps, 100 Mbps, and 1 Gbps Ethernet LANs) was discussed, and the candidate DoS schemes – GFP, VCAT, and LCAS – for enhancing capacity utilization in SONET were presented.

Next, we presented the basic features of OTN as the convergent optical networking platform for carrying inclusively the SONET, as well as various data traffic streams, mainly the high-speed *x*GbE-based LAN traffic, across the optical metro/long-haul network. Finally, we presented the packet-switched metro ring standard, i.e., RPR, offering class-based services to the packet and circuit-switched traffic streams. The dual-queue RPR node architecture with primary and secondary transit-traffic queues was presented, followed by a description of the fairness algorithms used therein for the class-based MAC protocol.

EXERCISES

(5.1) What are the different levels of synchronization in a telecommunication network and in what sequence they are carried out in a receiving node? Discuss critically the differences between the bit synchronization in a point-to-point communication link and the network synchronization.

(5.2) What are the typical network synchronization schemes used in telecommunication networks? Which scheme is used in SONET/SDH networks and why?

(5.3) What are the different clock references in SONET? With a suitable example, illustrate how the clock references are used in a SONET ring and how does the network try to get around typical network failures.

(5.4) Consider two PCM bit streams with the frame durations of 125 μs arriving from two independent sources (nodes) at a third node, where the incoming bit streams have to be multiplexed. The system requirement dictates that the two incoming data streams must not slip from each other with an upper bound of one slip a day. Determine the clock-inaccuracy limit in terms of slips/frame for the incoming PCM tributaries.

(5.5) Assume that, in a two-channel PDH system, each multiplexed frame accommodates at the most five bits from each incoming data channel, along with two additional bits for realizing the pulse-stuffing mechanism. Sketch the multiplexed frame obtained using pulse-stuffing. Calculate the percentage variation in the incoming data rates (around the average per-channel data rate), that can be accommodated in the given PDH system.

(5.6) Discuss a suitable scheme, which can be used in a SONET/SDH-based ADM to sense the transmission rates of the incoming tributaries with respect to the local clock rate for deciding the actions of the pointer H3.

(5.7) Sketch a typical SONET connection over a path involving several links and sections along with terminal nodes, add-drop multiplexers, and digital crossconnect. Using this diagram, show how the various overheads for path/line/section are utilized in maintaining the connections.

(5.8) Give the maximum number of VTs that can be accommodated in an SPE of the basic SONET (STS-1) frame for VT1.5, VT2, VT3, and VT6, and determine the effective transmission rates (i.e., excluding the overhead bytes) for each VT_x.

(5.9) Calculate the capacity-utilization efficiency for DoS connections with and without VCAT for the following three Ethernet clients and justify the benefits of VCAT:

a) 10 Mbps Ethernet,
b) 100 Mbps Ethernet (fast Ethernet),
c) 1 Gbps Ethernet (gigabit Ethernet).

(5.10) Consider the following client networks for OTN:

a) GbE (1 Gbps),
b) OC-48 (2.488 Gbps),
c) OC-192 (9.953 Gbps),
d) 10GbE (10 Gbps),
e) 40GbE (40 Gbps),
f) 100GbE (100 Gbps).

Describe the basic mapping process, and sketch the overall mapping schemes to obtain the respective OTU_i's from each client network.

(5.11) What are the limitations of FDDI, DQDB, and SONET, and how are these limitations addressed in RPR-based metro ring networks. Justify that, with uniform traffic pattern in the network, RPR with its spatial-reuse feature of fiber rings can achieve a capacity utilization that is eight-times higher than that in an FDDI ring.

Part III

WDM Optical Networks

WDM Local-Area Networks

6

WDM-based broadcast-and-select transmission over an optical passive-star coupler (PSC) can significantly enhance the speed of optical LANs/MANs. This type of optical LAN/MAN, referred to as WDM LAN/MAN (or simply WDM LAN), can function using a variety of different network architectures. In particular, WDM LANs can transmit packets between two nodes using direct (i.e., single-hop) transmission or through intermediate nodes using multihop transmission, leading to two types of network architectures: single-hop and multihop. The nodes in WDM LANs can employ two types of transmitters as well as receivers: tunable transmitter (TT) or fixed transmitter (FT) and tunable receiver (TR) or fixed receiver (FR). This leads to four types of WDM transceiver configurations: TT-TR, TT-FR, FT-TR, and FT-FR. Among these four configurations, the first three can realize single-hop communication, while the fourth configuration generally leads to multihop networks. In this chapter, we describe various types of PSC-based WDM LANs and examine their salient performance features.

6.1 WDM in optical LANs

As discussed in Chapter 1, transmission capacity in optical networks can be enhanced significantly by using concurrent multi-wavelength transmissions supported by WDM technology. During the mid-1980s the benefit of WDM transmission was initially explored for LANs/MANs (or simply LANs) through the development of experimental testbeds and theoretical studies by several research groups. In a LAN-setting, WDM transmission can be employed over various physical topologies, such as bus, ring, star, and some possible combinations of these topologies. As discussed in Chapter 3, optical fiber buses suffer from the progressive losses at optical taps, while optical fiber rings with active nodes don't have this problem of bus topology. However, optical ring topology hasn't been an effective choice for WDM operation in LANs due to the complexity of node hardware, although it was extensively used later in the public-domain metro segment of optical networks. However, passive-star topology has been the favored choice for WDM LAN realization for some practical reasons. In particular, high-speed optical LANs can be readily realized by using WDM transmission through a passive-star coupler (PSC) placed at a central location of the network, offering its *natural* broadcasting feature, without the problem of optical buses delivering unequal power levels at the receiving nodes, nor with any need for complex optical add-drop functionality in WDM rings. This feature of optical broadcast topology with a PSC as a highly reliable central hub, enables easy implementation of multiple-access schemes, much needed in LAN architectures for handling bursty packet-switched traffic.

Further, a WDM LAN can utilize the transmission windows in optical fibers over the PSC-based passive-star topology through concurrent transmissions between the various node pairs, by using appropriate MAC protocols based on wavelength-division multiple access (WDMA) schemes (see Fig. 6.1). The WDMA-based concurrent communications between the node pairs is realized using *broadcast-and-select* transmission through the PSC,

Optical Networks. Debasish Datta, Oxford University Press (2021). © Debasish Datta.
DOI: 10.1093/oso/9780198834229.003.0006

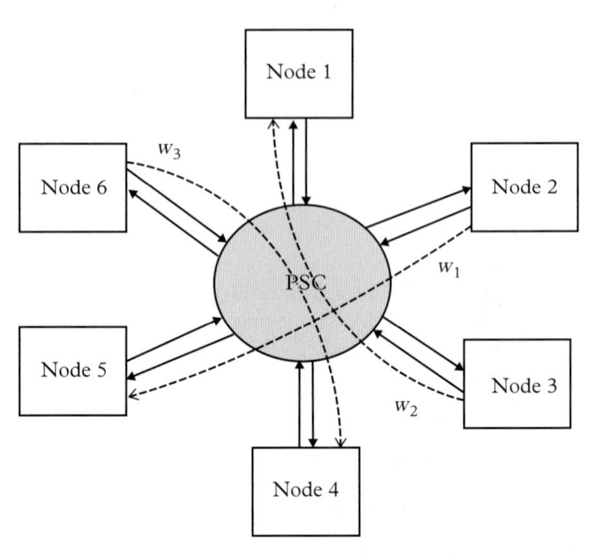

Figure 6.1 *Illustration of a WDMA-based six-node WDM LAN using broadcast-and-select transmission on three wavelengths* (w_1, w_2, *and* w_3) *over a passive-star topology.*

where several nodes broadcast data packets concurrently over non-overlapping wavelengths and the destined nodes set up their receivers at the desired wavelengths to select the intended data packets.

The PSC-based WDM LANs can be broadly divided into two categories: *single-hop* and *multihop* networks. In single-hop networks, either the transmitting or the receiving node, or both, must be tunable at the same wavelength during a packet transmission, so that the packet broadcast by a transmitting node over a given wavelength can be readily selected by the receiver of the destination node. Single-hop WDM LANs can have tunable transmitters (TTs) as well as tunable receivers (TRs) to employ this scheme. Otherwise, the transmitting node can have a TT, and before transmitting its packet can tune itself to the wavelength of the destined node using a fixed receiver (FR). Further, the transmitter can also have a fixed transmitter (FT) and when this node is transmitting, the destination node can tune to the desired transmitted wavelength using a TR. However, the network nodes using multiple FTs and FRs need to employ multihop connections. With a multihop connection, a packet transmitted by a source node may need to go through one or more intermediate nodes, if the transmit wavelengths at the source node don't match with any of the receiving wavelengths of the destination node; such networks are called multihop networks.

In view of the above, passive-star-based WDM LANs are also categorized in respect of the tunability of transceivers used in the nodes. In particular, each node in a WDM LAN can employ tunable or fixed transmitter(s) and receiver(s), leading to the four possible types of transceiver configurations:

- TT-TR: tunable transmitter(s) (TT), tunable receiver(s) (TR),
- TT-FR: tunable transmitter(s) (TT), fixed receiver(s) (FR),
- FT-TR: fixed transmitter(s) (FT), tunable receiver(s) (TR),
- FT-FR: fixed transmitter(s) (FT), fixed receiver(s) (FR).

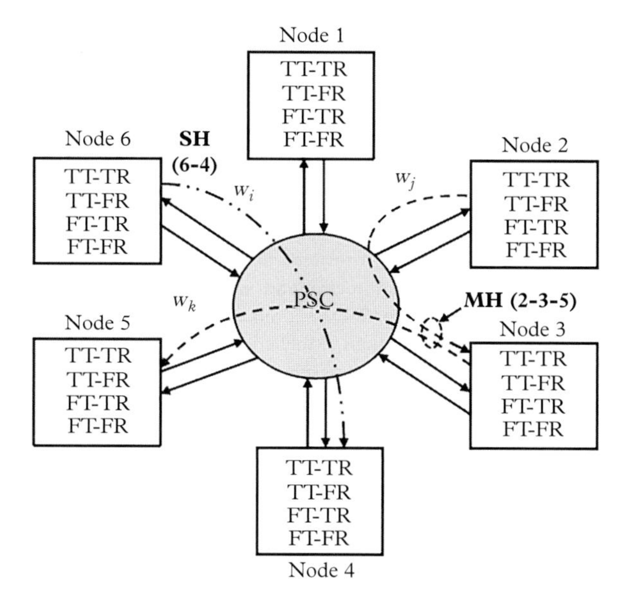

Figure 6.2 *Possible transceiver configurations in a WDM LAN using passive-star topology. Note that, single-hop connections (e.g., SH(6,4)) can be realized for each node pair by any one of the three transceiver configurations: TT-TR, TT-FR and FT-TR. With FT-FR transceiver configuration, some node pairs will have to set up multihop connections (e.g., MH(2-3-5)). However, with FT-FR^m or FT^m-FR (m representing the number of FTs/FRs per node) configuration, it is also possible to set up all single-hop connections in the network.*

Figure 6.2 illustrates the above four possible transceiver configurations of WDM LANs, realized over a PSC. WDM LANs using these transceiver configurations have been extensively studied using a variety of access schemes and protocols. The TT-TR configuration, belonging to the single-hop category, offers the most flexible operation. However, this scheme requires a significant amount of coordination, and might incur some delay while tuning the transmitter/receiver from one frequency to another from time to time. Such networks can be of two types. One of them uses pretransmission coordination (PC) using a specific shared channel (called a control channel) to arbitrate the transmission time and wavelength. The other type uses cyclically repeating pre-scheduled time frames, divided into some time slots, over which concurrent transmissions take place at non-overlapping wavelengths between multiple node pairs, thus obviating the need for PC. Of the two remaining configurations in the single-hop category, for the TT-FR case, the transmitter (i.e., a TT) of a source node can follow a scheduling scheme to tune itself to the specific wavelengths of the fixed receivers (FRs) at the destination nodes at appropriate time slots to establish single-hop communications. Further, in the FT-TR configuration, the destination node can adopt a scheduling scheme to tune its TR to the fixed wavelength of the FT at the transmitting node, thereby ensuring a single-hop communication.

The FT-FR configuration belonging to the multihop category avoids tuning delays, but with increased hardware, as each node usually needs to use multiple transmitters and receivers to set up communication with the remaining nodes. As mentioned earlier, in the FT-FR networks, since each pair of nodes cannot have the matching transmit and receive wavelengths, some node pairs may need multiple hops via some intermediate node(s) to set up a communication. However, if an FT-FR-based network has either multiple FRs or multiple FTs (i.e., FT-FR^m or FT^m-FR transceivers, with m equaling to the number of nodes N (or at least $N-1$), one can also realize single-hop connections in a WDM LAN.

Note that, although the WDM LANs were extensively studied using experimental testbeds and theoretical investigations, these networks could not see commercial deployment for various practical reasons, more so as the organizational LANs didn't have as much bandwidth requirements as the WDM LANs could offer, while needing more complex trasceiver designs. However, the experience gained through these studies has helped towards the subsequent developments in optical networks employing WDM transmission in the various segments of networking. For instance, the technologies for tunable lasers and optical filters, PSCs, optical switches, feasibility of various WDMA schemes, etc. used in WDM LANs, have eventually shown the path towards developing the technologies for today's optical WDM networks, in long-haul, metro, and access segments. With this spirit, we present in this chapter various types of WDM LANs that have been studied so far through experimental and theoretical investigations (Mukherjee 1992a; Mukherjee 1992b; Brackett, 1990). First, we present the noteworthy experimental developments that took place during the late 1980s and 1990s, and thereafter consider in detail the various network architectures studied theoretically using a wide range of design methodologies for single and multihop operations in WDM LANs.

6.2 Experiments on WDM LANs

Considering chronologically, some pioneering experiments were carried out during the mid-1980s on WDM-based networking in the British Telecom Laboratory (BTRL), AT&T Bell Laboratories, and Heinrich Hertz Institute (HHI), which were subsequently followed up by several other important investigations reported by Bell Laboratories, Bellcore, NTT, and various other organizations and universities.

The experiment carried out by BTRL (Payne and Stern 1986) used WDM transmission over an 8×8 passive-star network using mechanically tunable filters at the node receivers offering the feature of TRs. The filters used in the TRs were manually tunable over a range of 400 nm, each with a 10 nm bandwith. Thus the WDM transmission range was wide, though rather slow in changing the receiver wavelength due to mechanical tuning. The lasers were fixed-tuned (FTs) with the closest spacing of 15 nm, leading to a CWDM transmission over passive-star topology using an FT-TR transceiver configuration. This was the first proof-of-concept experimental demonstration of WDM-based packet-switched networking over a passive-star topology. Around the same time, Bell Laboratories demonstrated a point-to-point WDM link (Olsson *et al.* 1985), with a significant improvement in channel spacing (in the order of 1 nm), establishing the viability for DWDM transmission over optical fibers. Another noteworthy experiment in this direction was carried out at HHI, reporting passive-star-based video broadcasting using coherent optical communication technology (Bachus *et al.* 1986). Over a 64×64 (extendible upto 128×128) passive-star network, the experiment demonstrated the transmission of 10 wavelengths each at a rate of 70 Mbps with as small as 6 GHz channel spacing, which was indeed feasible due to the improved selectivity of coherent optical receivers. Moving forward with the confidence gained from these experiments, several other WDM-based networking projects were undertaken in various organizations and universities. We discuss some of these projects in the following.

6.2.1 LAMBDANET

The WDM-based LAN developed by the erstwhile Bellcore, named as LAMBDANET (Kobrinski *et al.* 1987; Goodman *et al.* 1990), used a PSC-based broadcast-and-select network employing one fixed-tuned laser (as an FT) at a unique wavelength in each node along with fixed-tuned multiple receivers, i.e., FRs (Fig. 6.3). Thus, LAMBDANET

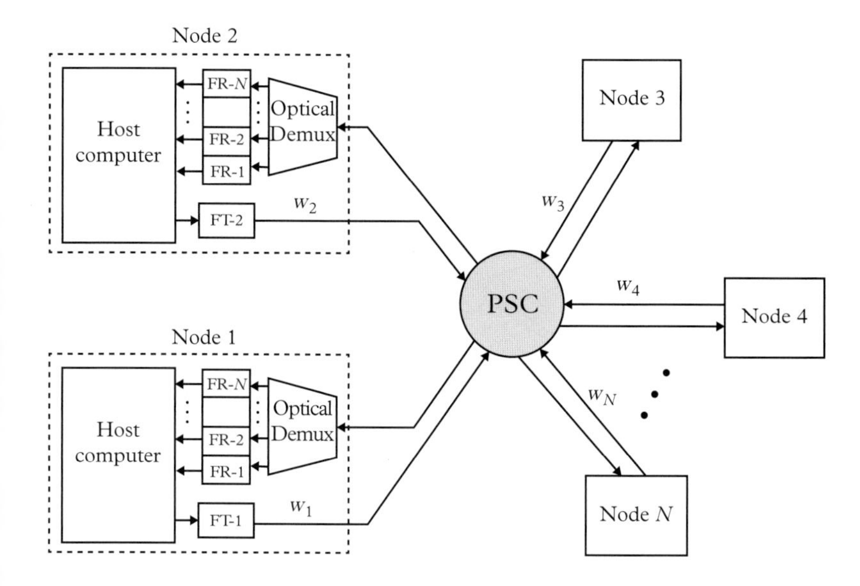

Figure 6.3 *LAMBDANET schematic (after Goodman et al. 1990).*

functioned with an FT-FR^m-based transceiver configuration at each node, with $m = N$ as the number of nodes. With 16 wavelengths (and hence 16 nodes), each LAMBDANET node could receive the signals concurrently from the remaining nodes (as well as its own transmitted signal), thereby supporting single-hop operation. Each node time-division multiplexed (TDM) its packets for the other nodes and transmitted the same using its unique fixed wavelength. At a receiving node, an optical demultiplexer using grating-based technology was used to select the received wavelengths from all the nodes concurrently, and pass them over to the bank of FRs, fixed-tuned at the specific transmit wavelengths of all the nodes (including its own transmit wavelength for error detection).

LAMBDANET used a 16 × 16 PSC with single-mode fibers and DFB lasers with 2-nm wavelength-spacing. The lasers could be modulated both with analog and digital signals (2 Gbps). The experiment demonstrated several interesting results. In particular, LAMBDANET offered a feasible non-blocking fully connected network architecture using WDM. This was also the first ever demonstration of a DWDM network with prototyped-and-packaged components and lasers. Furthermore, as an extension of the project, Bellcore set up a transmission record of 2 Gbps signals over 57 km with 16 wavelengths. Thus, LAMBDANET from Bellcore, as one of the leading projects, paved the way for future developments on WDM networks with novel network architectures, employing appropriate WDM technologies and MAC protocols.

6.2.2 FOX

Another experimental WDM network from Bellcore, the fast optical crossconnect (FOX), was developed for exploring the potential of WDM connectivity in parallel-processing-based computers (Arthurs *et al.* 1986; Cooper *et al.* 1988). In particular, FOX architecture realized high-speed communications between the processors and the shared memories through a crossconnect configuration using two PSCs (Fig. 6.4). As shown in the figure, the processors and the memory units were divided into two separate groups, with group P

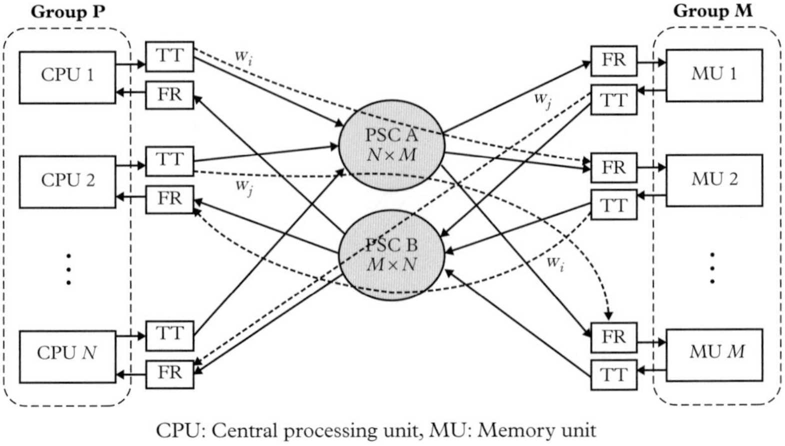

CPU: Central processing unit, MU: Memory unit

Figure 6.4 *FOX schematic. Note that, by using two PSCs, the network can support two-way concurrent transmissions on the same wavelength (after Arthurs et al. 1986).*

for processors and group M for memories. These two groups communicated with each other through the two PSCs (A and B) by using TT-FR-based transceivers at each node (processor or memory unit), thereby realizing a bidirectional crossconnect functionality between the two groups (N processors and K memory units) with single-hop connectivity.

The ($N \times M$) PSC (PSC A) realized the processor-to-memory (group P to group M) concurrent WDM connections, while the $M \times N$ PSC (PSC B) supported the memory-to-processor (group M to group P) concurrent WDM connections, both operating in broadcast-and-select mode. The TT in each processor could transmit to a memory unit by tuning to the receiving wavelength of the memory unit through PSC A. Then the TT in each memory unit could reach a processor in the opposite direction by tuning to the receiving wavelength of the processor through PSC B. Note that all these connections (group P to group M and vice versa) were realized with single-hop communication, indeed being feasible due to the tunability of the lasers. However, with limited number of wavelengths, the possibility of contentions came up, which was resolved by using a suitable contention-resolving protocol, following a binary exponential back-off algorithm. This experiment demonstrated, in particular, the need of fast-tunable lasers for realizing high-speed WDM LANs, which in effect triggered the need for further research on the fast-tunable WDM devices for high-speed WDM networks.

The concept of FOX was later extended by the Bellcore team to propose a WDM-based high-capacity packet-switching architecture, called the hybrid packet-switching system (HYPASS), where the receivers were also made tunable, implying the use of TT-TR configuration for its transceivers (Arthurs *et al.* 1986).

6.2.3 Rainbow

IBM demonstrated its WDM-based network, called Rainbow, using PSC with FT-TR-based transceivers (Dono *et al.* 1990). Rainbow was realized with 32 nodes communicating with each other at 200 Mbps on a target network-diameter of 55 km. Tunable optical filters, realized by Fabry–Perot etalons with piezo-electric control, were used at the TRs, and the FTs employed fixed-wavelength (unique for each node) DFB lasers. A representative block schematic of a Rainbow network is shown in Fig. 6.5. The connection setup process was in the slower category with the tuning time of optical filters being in the order of

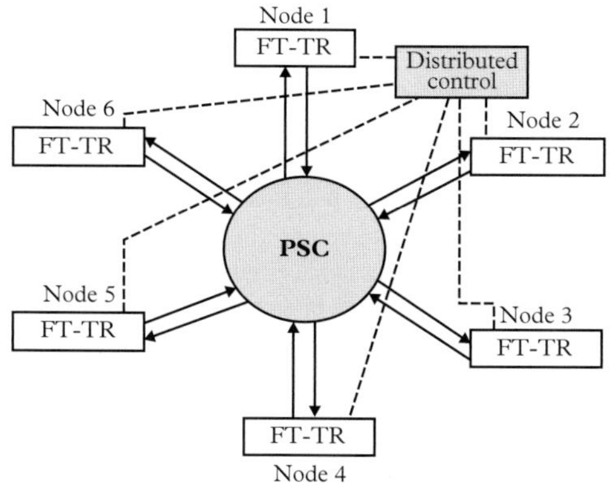

Distributed control mechanism:

- Source node FT transmits a packet repeatedly with the addresses of its own and the destination node, and keeps its TR tuned to the transmit wavelength of the destination node.
- TR in an idle node scans over entire wavelength range to check if it is the intended destination, knowing which, it tunes its TR at the desired wavelength and sends an acknowledgment (ack) to the source node.
- Having received the ack, the source node transmits its data packet.

Figure 6.5 *Rainbow schematic using FT-TR transceivers with PC-based distributed control mechanism (after Dono et al. 1990).*

$\simeq$10 μs/nm, and thus the network was considered to be more suitable for circuit-switched communications.

The Rainbow network employed a MAC protocol to tune the receiver filter of a destination node to the fixed transmit wavelength of the source node, before a packet was transmitted from the latter. In this PC-based MAC using distributed control mechanism – functionally an in-band receiver polling scheme – an idle receiver had to scan all the transmit wavelengths of the other nodes to know about the possible connection-setup requests addressed to itself. The connection-setup requests were sent by the source node on its transmit wavelength repeatedly in the form of a packet carrying the destination address. During this period, the receiver of the source node kept itself tuned to the transmit wavelength of the destination node, while waiting to receive an acknowledgment from the destination node. Upon receiving the acknowledgment, the source node would start its transmission, thus following a typical connection-setup procedure for circuit-switched connection. Since neither side (source or destination) used any additional wavelength to set up the connection, and the receiver had to scan the entire wavelength range in use for data transmission, this PC scheme was in the category of in-band receiver polling scheme. Note that the process of scanning over the optical transmission spectrum in a receiver to know about the connection-setup requests from the other nodes and the reception of the acknowledgment could get into a deadlock situation resulting from transmission errors; this possibility necessitated the use of a *time out* for the polling process of the network.

After experimenting with Rainbow, IBM came up with an improvised version, called as Rainbow-2, where the per-channel transmission speed was increased from 200 Mbps to 2 Gbps, but with a reduced network diameter in the range of 10–20 km. Rainbow-2 used the same MAC protocol as in its first version, and ensured an overall application-layer speed of 1 Gbps between the host computers.

6.2.4 TeraNet

TeraNet was an experimental WDM LAN developed in Columbia University, which offered a unique multiple-access scheme over PSC-based passive-star topology by employing WDMA along with multiple subcarrier frequencies assigned to each wavelength (Gidron and Acampora 1991). Thus, Teranet used a novel two-dimensional access scheme, realized with FT-TR-based optical transceivers and electronic subcarrier modems. Each node pair was assigned a unique combination of a wavelength plus a subcarrier frequency in the microwave range, leading to a combination of WDMA and subcarrier-frequency-division multiple access (SFDMA), together termed WD-SFDMA. The WD-SFDMA scheme used a more relaxed WDM channel spacing (as large as a few nanometers), and each wavelength within the large waveband around itself was intensity-modulated by multiple microwave subcarriers. The microwave subcarriers were modulated by the ingress baseband data, thereby utilizing the optical transmission window with fewer wavelengths. Thus, even with fewer optical channels (i.e., CWDM), and each of them carrying several subcarriers, TeraNet offered a feasible bandwidth-efficient transmission system with less-stringent WDM components. By using WD-SFDMA, Teranet could operate with a large number of channels (each channel being a combination of a wavelength and a subcarrier), offering the circuit as well as packet-switched services over the passive-star topology.

The WD-SFDMA transmission scheme took effect through the interfaces between the nodes and the passive-star optical topology, with the interfaces referred to as media interface units (MIUs). Thus, each MIU used a two-level modulation scheme, one at electrical level to modulate the microwave subcarriers with the ingress data, and the other at the next level to modulate the intensity of the FT lasers. The subcarrier modulation in MIUs used QPSK to transmit at 1 Gbps (i.e., with a symbol rate of 500 Megabauds/sec).

A representative block schematic of TeraNet is shown in Fig. 6.6, with its nodes offering both circuit and packet-switched services. Any node that has to offer packet-switched service is provisioned with p WD-SFDMA channels (e.g., $p = 2$ in Fig. 6.6a), while a node with a circuit-switched service is allocated one WD-SFDMA channel. Thus, each packet-switching node has p MIUs, while each circuit-switching node uses one MIU; a representative MIU block schematic is shown in Fig. 6.6b. A packet-switching node transmits/receives to/from p nodes directly and connects with the remaining nodes through some intermediate nodes, if necessary, thereby requiring multihop connections. One of the candidate logical topologies suitable for such multihop connectivity is ShuffleNet, which is a regular topology formed using a wrapped-up cylindrical structure with k columns and p^k rows (implying kp^k nodes), thereby needing kp^{k+1} WD-SFDMA channels (ShuffleNet is discussed later in this chapter in further detail).

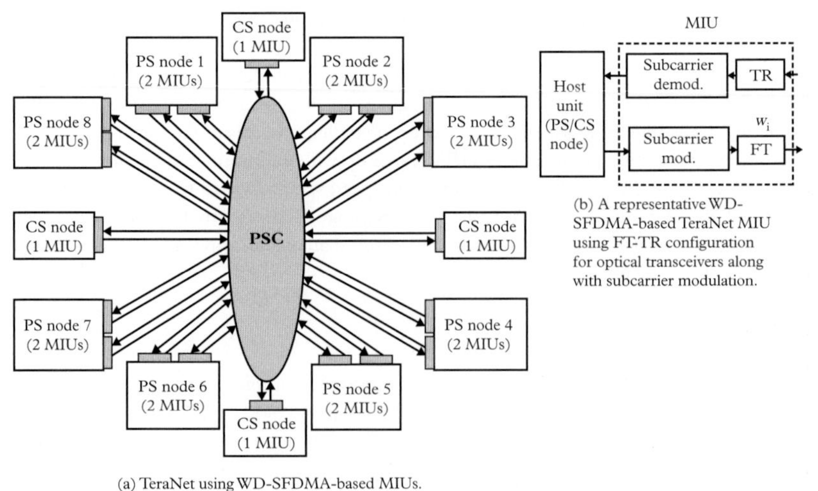

(a) TeraNet using WD-SFDMA-based MIUs.

(b) A representative WD-SFDMA-based TeraNet MIU using FT-TR configuration for optical transceivers along with subcarrier modulation.

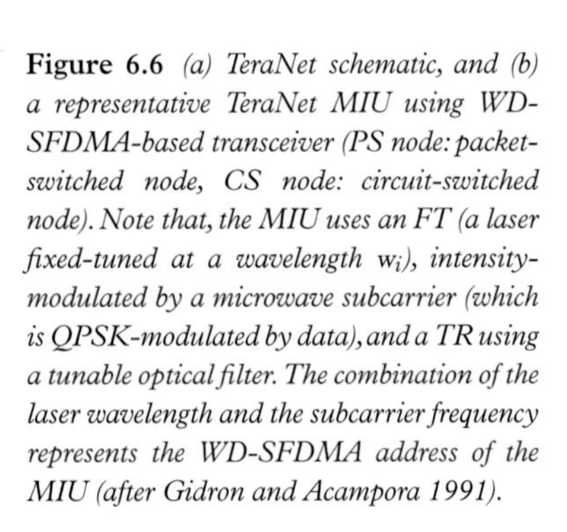

Figure 6.6 *(a) TeraNet schematic, and (b) a representative TeraNet MIU using WD-SFDMA-based transceiver (PS node: packet-switched node, CS node: circuit-switched node). Note that, the MIU uses an FT (a laser fixed-tuned at a wavelength* w_i*), intensity-modulated by a microwave subcarrier (which is QPSK-modulated by data), and a TR using a tunable optical filter. The combination of the laser wavelength and the subcarrier frequency represents the WD-SFDMA address of the MIU (after Gidron and Acampora 1991).*

The packet-switched network in TeraNet offered different classes of service – classes I, II, III, and IV, as in ATM-based networking. Class IV service was offered to carry out the control and management operations in the network, while the classes I through III services supported different types of data traffic with varying needs of quality of service (QoS). With appropriate transmission scheduling of WD-SFDMA channels and buffer allocation algorithm at the network nodes, the QoS requirements of various services were met.

6.2.5 STARNET

STARNET was also developed over a PSC-based passive-star topology at Stanford University with two coexisting logical subnetworks, one operating with a low speed (125 Mbps) handling the network control information, and the other providing a high-speed connectivity for the data traffic (Chiang *et al.* 1996). The low-speed subnetwork supporting the flow of control data between all the nodes managed the configuration of the high-speed subnetwork, so the latter could be dynamically reconfigured to optimize the bandwidth utilization. The low-speed subnetwork formed a logical ring over the passive-star through all the nodes, while the high-speed subnetwork could dynamically form a number of independently operating networks, which used multihop connections over the passive-star, whenever necessary. The network operation was realized by using some novel transceiver configuration, as discussed in the following.

Figure 6.7 illustrates the STARNET architecture, which employed FT-TR transceiver configuration for its nodes. Each STARNET node used a novel transmitter configuration, where one single laser operated as *two independent* transmitters for the two subnetworks. However, it used two independent receivers: main and auxiliary receivers for high and low-speed subnetworks, respectively. As shown in Fig. 6.7, the laser in the transmitter block, using a unique transmit wavelength, was subjected to a combined modulation scheme in a serial manner. In particular, the laser was first subjected to differential

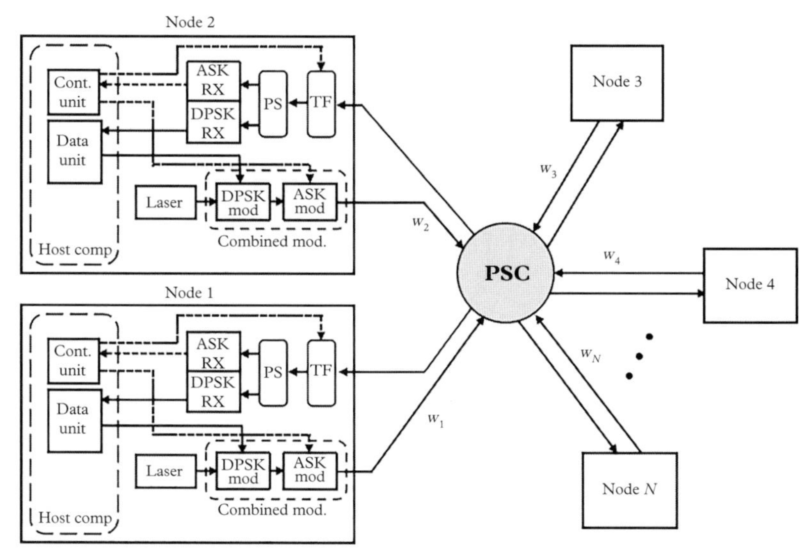

Figure 6.7 *STARNET schematic (after Chiang et al. 1996).*

phase-shift keying (DPSK) at the higher rate (2.5 Gbps) for the high-speed subnetwork, and the DPSK-modulated optical signal was subsequently passed through a modulator employing amplitude-shift keying (ASK) at the lower speed (125 Mbps) for the slower subnetwork exercising network control, both qualifying for coherent detection at the respective receivers. DPSK and ASK modulations were carried out using electro-optic modulators in a serial manner along with the ASK and DPSK receivers working in parallel. As evident from Fig. 6.7, the phase of the lightwave from the laser following DPSK modulation carries the high-speed (2.5 Gbps) data traffic, while its slowly varying envelope after ASK modulation carries the low-speed (125 Mbps) control information. Setting a lower-bound on the modulation depth for ASK (i.e., by using a non-zero light intensity for binary zero in the ASK signal) and the low-speed packets being more robust against noise, the receiver can strike a balance with the combined modulation scheme to obtain acceptable BERs for both ASK and DPSK modulation schemes.

The low-speed control subnetwork forms a ring over the PSC and operates following the FDDI protocol. This control subnetwork can configure more than one independent high-speed data networks (all these separate networks being collectively designated as a high-speed subnetwork for data) over the passive-star, by instructing the respective nodes to tune their receivers at appropriate wavelengths (see Fig. 6.8). As shown in the figure, two independent high-speed networks (solid lines) have been configured between some select nodes using the coordination carried out by the FDDI-based control subnetwork ring (dashed line).

The optical transmission system in STARNET employed heterodyne receivers, needing additional lasers as local oscillators, which in turn offered enhanced receiver sensitivity as compared to the non-coherent receivers employing direct detection. Subsequently, a simpler version, STARNET II, was examined using a direct detection scheme for ASK (control) signal and delay-demodulation scheme (weakly coherent) for DPSK (data) reception, thereby avoiding the need for more-complex heterodyne detection.

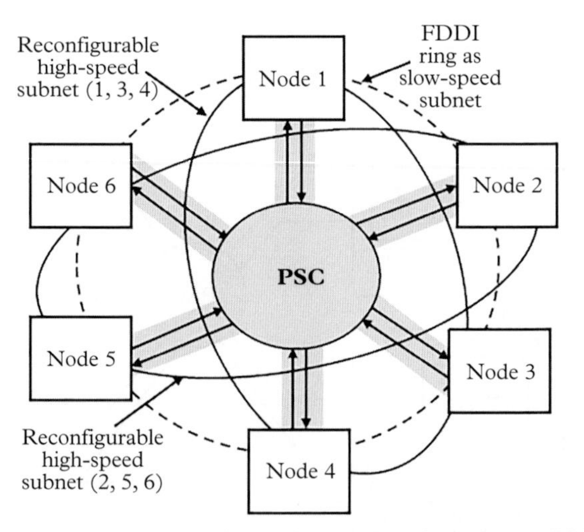

Figure 6.8 *Formation of slow-speed (logical FDDI ring, shown by dashed line) and reconfigurable high-speed logical subnetworks (shown by solid lines) in STARNET. Note that the nodes 1, 3, and 4 form one high-speed logical subnetwork, and the nodes 2, 5, and 6 form another coexisting high-speed subnetwork (after Chiang et al. 1996).*

6.2.6 SONATA

WDM networking between several groups of nodes was realized under the ACTS (advanced communications technologies and services) program of the European Union, by using a wavelength-routing device instead of a PSC. In this experiment, called the switchless optical network for advanced transport architecture (SONATA) (Bianco *et al.* 2001), large number of nodes were grouped into a number of clusters forming passive optical networks (PONs) (now a popular topology for optical access networks – see Chapters 4 and 7), and these PONs were interconnected using a centrally located passive wavelength-routing node (PWRN). Each node in this network used TT-TR transceiver configuration to transmit across the PWRN to provide full connectivity between all the users.

SONATA employed a WDMA/TDMA-based two-dimensional MAC protocol by choosing an appropriate wavelength and time slot for each connection. A block schematic of SONATA is shown in Fig. 6.9, where the transmit and receive blocks of the associated PONs are shown separately on the two sides of the PWRN for clarity (in reality, they are collocated). The PWRN is aided by wavelength converters (WCRs) in a feedback path and a control unit to set up the connections between the nodes by using appropriate wavelengths and time slots. Thus, the PWRN uses, on both sides, N ports to connect N PONs, K ports for K WCR arrays, and one port for the control unit. Each PON uses one unique wavelength and the nodes belonging to the same PON use different TDMA slots on the assigned wavelength for the PON. The WCR arrays are used dynamically to change (operated by the control unit) an incoming wavelength at an input port of the PWRN to another wavelength, which may become necessary for setting up a connection in the presence of non-uniform traffic patterns. While the optical amplifiers (OAs) are used to compensate for signal losses, mostly in splitters and the PWRN, the OA-based gates (AGs)

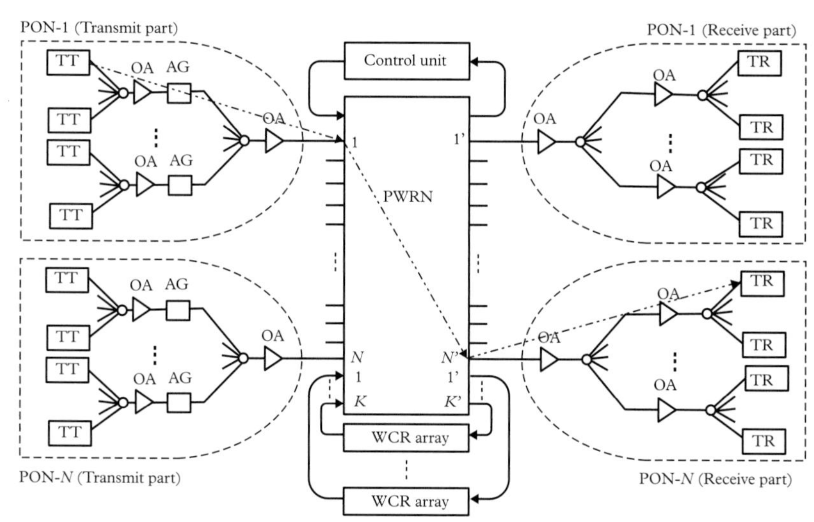

Figure 6.9 *SONATA schematic. Note that, the transmit and receive parts of a PON are shown separately on the two sides of the PWRN for clarity; in reality the transmitters and receivers are collocated, and the transmit and receive fibers also run together through same cable (after Bianco et al. 2001).*

are used to black out the idle periods during traffic flows (in between the packet bursts) to avoid noisy inputs at the receiver ends.

6.3 Single-hop WDM LANs

As described in the preceding section, the passive-star-based WDM LANs employing various transceiver configurations have been investigated extensively by using experimental testbeds. Several testbeds used single-hop architecture to implement WDMA-based concurrent transmissions between the node pairs. For example, Rainbow with FT-TR configuration employed a single-hop MAC protocol with a PC-based in-band polling scheme, while LAMBDANET with FT-FR^m configuration (i.e., with multiple FRs) used a simpler scheme for single-hop WDMA transmission. In TeraNet, multihop connections were employed with a two-dimensional multiaccess scheme (WD-SFDMA) to realize packet-switched networking. In this section, in order to get some insight into the performance of single-hop WDM LANs, we examine some useful MAC protocols that are applicable in such networks using the TT-TR transceiver configuration.

In single-hop WDM LANs, with or without PC for MAC implementation, one of the key issues turns out to be the capability of the transmitters and/or receiver filters to rapidly tune from one wavelength to another over the optical transmission spectrum, so that the incoming packets at every node can be transmitted on their arrival without much delay. The tuning delay of transmitters and receivers (filters) might be non-negligible compared to the packet durations, especially with short packet durations resulting from high transmission speeds ≥ 1 Gbps. In the following, we analyze two types of single-hop networks – the single-hop WDM LANs with and without PC – with the assumption that the transmitters and the filters can be tuned/retuned within a time duration that does not affect the performance significantly.

6.3.1 MAC using pretransmission coordination (PC)

MAC protocols in PC-based single-hop WDM LANs have been studied extensively by various research groups (Habbab *et al.* 1987; Sudhakar *et al.* 1991; Mehravari 1990; Mukherjee 1992a). In this section, we use the analytical models from (Habbab *et al.* 1987; Mehravari 1990) for examining some of the single-hop WDM LANs employing PC-based MAC protocols. In order to analyze these protocols, we assume that the network uses $M + 1$ wavelengths, $w_0, w_1, w_2, ..., w_M$ (say), with one specific wavelength (w_0) playing the role of control wavelength or channel (hereafter, we use wavelength and channel interchangeably in all forms of WDM networks). On w_0 each node transmits its control packet prior to the transmission of its intended data packet on one of the remaining wavelengths, called data wavelengths; this mechanism to coordinate between the transmitting and receiving nodes prior to the actual data-packet transmission leads to PC-based MAC. Each node, when not receiving any data packets from other nodes, keeps its receiver tuned to w_0 by default. Any node intending to transmit a data packet on the network, first tunes its transmitter to w_0 and broadcasts the control packet carrying the addresses of its own and the destination along with the index of the wavelength (i.e., j for wavelength w_j) on which its data packet will be transmitted. Having transmitted its control packet, the transmitting node immediately transmits its data packet (i.e., with no further delay), which is popularly known as the *tell-and-go* (TaG) scheme. Following (Habbab *et al.* 1987), we make the following assumptions to carry out the analysis.

- Network traffic follows Poisson distribution (i.e., infinite population model, see Appendix C), in both control and data channels.
- On the control channel, arrival requests come from all nodes. The traffic in the control channel is denoted as G, representing the average number of control packets arriving from all nodes during one control packet duration. The control packet duration is assumed to be unity, without any loss of generality.
- Each data packet is assumed to be of length L (i.e., L times the length of a control packet). Control channel traffic is assumed to be distributed uniformly into M data channels. Thus, the traffic in each data channel is given by $G_d = GL/M$.

In general, both control and data channels employ random media-access schemes to transmit the respective packets, and can use the same or different protocols. For example, both can use Aloha (i.e., pure Aloha) as the MAC protocol. We first consider this combination, i.e., Aloha/Aloha protocol for control/data packet transmissions. Then we consider a few other schemes and compare their performance. However, our discussions are confined to the class of MAC protocols that don't use carrier sensing (i.e., CSMA-based protocols are not used) owing to the small ratio of packet sizes (resulting from high-speed optical transmission) to the network propagation delay.

Aloha/Aloha protocol

In order to analyze the performance of the Aloha/Aloha protocol, we first note that a node intending to transmit a data packet must succeed in two subsequent contentions, one in the control channel and the next one in the data channel it has chosen for transmission. We need to determine the overall probability of success P_s in transmitting a data packet as the product of two independent probabilities: the probability $P_{s(con)}$ of successful transmission in the control channel and the probability $P_{s(data)}$ of successful transmission on the chosen data channel, i.e.,

$$P_s = P_{s(con)} P_{s(data)}. \tag{6.1}$$

Evaluation of $P_{s(con)}$ and $P_{s(data)}$ requires the estimates of the *vulnerable* (i.e., collision-prone) periods in control and data channels. For achieving successful transmission of a packet on a channel, no additional packet should arrive and collide during the transmission of the desired packet in the same channel, and the duration over which no additional packet should arrive to avoid collision in a given channel represents the vulnerable period. The vulnerable period τ_{con} in the control channel w_0 for the Aloha protocol is simply twice the control packet size, i.e., $\tau_{con} = 2$. As shown in Fig. 6.10, for a successful control packet CP_0 transmitted in the time interval $t \in [0, 1]$, τ_{con} lies in the time interval $t \in [-1, 1]$. However, the vulnerable period τ_{data} in a data channel (say, jth data channel) needs a critical observation, as illustrated in the same figure.

Note that, following the transmission of the control packet DP_0 on w_0 during $t = [0, 1]$, the chosen data channel (w_j, say) is used for data packet (DP_0) transmission during $t \in [1, L+1]$ using the TaG scheme. Hence, in the same data channel, no additional data packet should arrive and collide with DP_0. In the control channel, CP_a is the last permitted control packet to arrive before CP_0, which generates the data packet DP_a ending its transmission just at the moment when DP_0 begins transmission, thereby avoiding any collision. Similarly, CP_b represents the earliest permissible control packet after the transmission of CP_0, whose corresponding data packet DP_b begins transmission just after DP_0 completes its transmission. Hence, given that DP_0 needs to be successfully transmitted following the transmission of CP_0, no additional control packet should arrive demanding w_j

as the data channel during the time intervals $t \in [-L, -1]$ and $t \in [1, L]$, leading to a total vulnerable period $\tau_{data} = 2(L-1)$ for the data channel. Thus, one half of the vulnerable period, i.e., $L-1$ slots, would remain on the left side of the control packet CP_0, while the other half (i.e., the remaining $L-1$ slots) would appear on the right side of CP_0.

Using the values of vulnerable periods in the control and data channels, the overall probability of success P_s for a data packet transmission can be expressed as

$$P_s = P_{s(con)} P_{s(data)} = \exp[-2G] \exp\left[-\frac{2G(L-1)}{M}\right], \tag{6.2}$$

From the expression of P_s, one can write the per-channel throughput of the single-hop PC-based WDM LAN using Aloha/Aloha protocol as

$$S_d = G_d P_s = G_d \exp\left[-2G - \frac{2(L-1)G}{M}\right]. \tag{6.3}$$

With $G_d = GL/M$, S_d can be expressed as a function of per-channel traffic G_d as

$$S_d = G_d P_s = G_d \exp\left[-2\frac{G_d M}{L} - \frac{2(L-1)G_d}{L}\right]. \tag{6.4}$$

S_d can also be expressed as a function of the total network traffic G as

$$S_d = \frac{GL}{M} \exp\left[-2\frac{GL}{M}\left(1 + \frac{M-1}{L}\right)\right]. \tag{6.5}$$

Using Eq. 6.5, we obtain the maximum value for S_d at $G_d = L/[2(M-1+L)]$, given by

$$S_d(max) = \frac{L}{2e(M-1+L)}. \tag{6.6}$$

It appears from the above that, with $L \gg M$, $S_d(max)$ approaches $\frac{1}{2e}$, which is the maximum throughput of a single-channel Aloha network. Using the expression for S_d, we also obtain the packet transmission delay D experienced in the single-hop PC-based WDM LAN using Aloha/Aloha protocol as

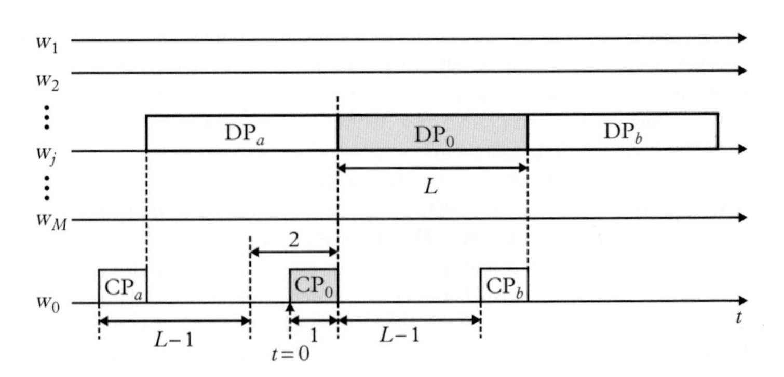

Figure 6.10 *Vulnerable periods in the control and data channels for TaG-based Aloha/Aloha protocol in PC-based single-hop WDM LAN.*

$$D = \frac{G_d}{S_d} = \exp\left[2\frac{GL}{M}\left(1 + \frac{M-1}{L}\right)\right]. \tag{6.7}$$

Overall network throughput can be estimated as $S = MS_d$, while the delay experienced in packet transmission at any node would be the same as the per-channel delay obtained in Eq. 6.7.

Slotted-Aloha/Aloha protocol

With slotted-Aloha/Aloha as the MAC protocol, the basic approach to evaluate the throughput and delay remains the same, though we need to have a re-look at the vulnerable periods for the control and data channels, as illustrated in Fig. 6.11. With slotted-Aloha as the control protocol, the vulnerable period for control packets is simply 1, i.e., $\tau_{con} = 1$. For data channels, using the similar observations as in the Aloha/Aloha protocol, with CP_0 as the control packet (Fig. 6.11), the vulnerable period on the left side of CP_0 lies in the time intervals $t \in [-(L-1), 0]$ and the vulnerable period on the right side of CP_0 lies in the time interval $t \in [1, L]$. This leads to the total vulnerable period for the data channel as $\tau_{data} = 2(L-1)$. Using these values for τ_{con} and τ_{data}, and following the same method as used in the Aloha/Aloha case, we obtain the per-channel throughput for slotted-Aloha/Aloha protocol as

$$\begin{aligned} S_d &= G_d P_s \\ &= G_d P_{s(con)} P_{s(data)} \\ &= G_d \exp\left[-G\right] \exp\left[\frac{-2(L-1)G}{M}\right] \\ &= G_d \exp\left[-2G_d\left(1 + \frac{M-2}{2L}\right)\right] \\ &= \frac{GL}{M} \exp\left[-2\frac{GL}{M}\left(1 + \frac{M-2}{2L}\right)\right]. \end{aligned} \tag{6.8}$$

Using Eq. 6.8, we obtain the maximum value for S_d at $G_d = L/[M + 2(L-1)]$, given by

$$S_d(max) = \frac{L}{e[M + 2(L-1)]}. \tag{6.9}$$

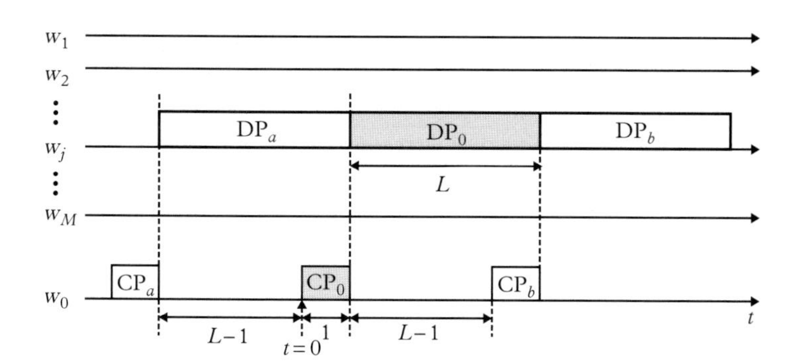

Figure 6.11 *Vulnerable periods in control and data channels for TaG-based slotted-Aloha/Aloha protocol in PC-based single-hop WDM LAN.*

With $L \gg M$, $S_d(max)$ reaches the throughput of single-channel slotted-Aloha, i.e., $\frac{1}{e}$. Using Eq. 6.8, we obtain the expression for D in the network, given by

$$D = \frac{G_d}{S_d} = \exp\left[2\frac{GL}{M}\left(1 + \frac{M-2}{2L}\right)\right]. \tag{6.10}$$

Slotted-Aloha/M-server-switch protocol

In this protocol, each node, with the arrival of an incoming data packet (at a time $t = t_0$, say) for transmission, starts monitoring the control channel to note the contents of the forthcoming control packets until $t = t_0 + L$. Note that the monitoring interval (from $t = t_0$ to $t_0 + L$) equals the length of a data packet. The information gathered from the control packets observed during the monitoring interval enables the node to determine which data channels are going to be free/busy during the next L slots after completing the monitoring process at $t = t_0 + L$. Thereafter, to indicate the destination node, the source node chooses at random one of the free (if any) channels and transmits its control packet using slotted-Aloha protocol, indicating therein the chosen channel. Following the transmission of the control packet, the node begins transmitting the data packet on the chosen channel. However, if all of the M channels are found to be in use, the node is blocked for data transmission, resembling a *queuing system* with M *servers* (each channel or wavelength representing a server) following the Erlang-B blocking model (see Appendix C). If P_B is defined as the blocking probability, then the throughput of a data channel can be expressed as (Mehravari 1990)

$$S_d = G_d P_{s(con)} P_{s(data)} = \frac{GL}{M} \exp(-G)(1 - P_B). \tag{6.11}$$

Note that multiple nodes may intend to use a particular free channel for packet transmission during the same interval leading to a contention, which is taken care of in the above expression by the probability of successful transmission $P_{s(con)} = \exp(-G)$ in the control channel following the slotted-Aloha model.

Next, we determine the expression for P_B, following the Erlang-B model for M servers. Note that the traffic arriving in each server (i.e., a channel) is governed by the control channel traffic G received in all of the L slots monitored by the node (i.e., LG), the probability of successful transmission for each control packet $\big(= \exp(-G)\big)$ and the number of servers M. The total control channel throughput over the L observed slots during the monitoring interval is $LG\exp(-G)$, which offers itself as the input traffic for the M servers. With these observations, and assuming that the input traffic, i.e., $LG\exp(-G)$, approximately follows a Poisson distribution, we express the Erlang-B blocking probability P_B as

$$P_B = \frac{\left[LG\exp(-G)\right]^M/M!}{\sum_{i=0}^{M}\left[LG\exp(-G)\right]^i/i!}. \tag{6.12}$$

Using the above expression for P_B in Eq. 6.11, we finally obtain the per-channel throughput S_d as

$$S_d = G_d \exp(-G)\left(1 - \frac{\left[LG\exp(-G)\right]^M/M!}{\sum_{i=0}^{M}\left[LG\exp(-G)\right]^i/i!}\right) \tag{6.13}$$
$$= \frac{GL}{M}\exp(-G)\left(1 - \frac{\left[LG\exp(-G)\right]^M/M!}{\sum_{i=0}^{M}\left[LG\exp(-G)\right]^i/i!}\right).$$

From the above expression of S_d, we express the average packet transmission delay D for the slotted-Aloha/M-server-switch protocol as

$$D = \frac{G_d}{S_d} = \frac{\exp(G)}{1 - \frac{\left[LG\exp(-G)\right]^M/M!}{\sum_{i=0}^{M}\left[LG\exp(-G)\right]^i/i!}}. \tag{6.14}$$

Performance of PC-based MAC protocols

Performance features of the PC-based MAC protocols for single-hop WDM LANs are compared using per-channel throughput and average delay in packet transmission for the respective networks. Figure 6.12 presents the plots of per-channel throughput vs. traffic for Aloha/Aloha and slotted-Aloha/Aloha protocols for the two combinations of L and M: $L = 10, M = 10$ and $L = 20, M = 10$. As evident from the plots, the slotted-Aloha/Aloha protocol offers better performance than the Aloha/Aloha protocol, and for both protocols, higher L offers higher peak throughput. The delay versus traffic plots for the two protocols

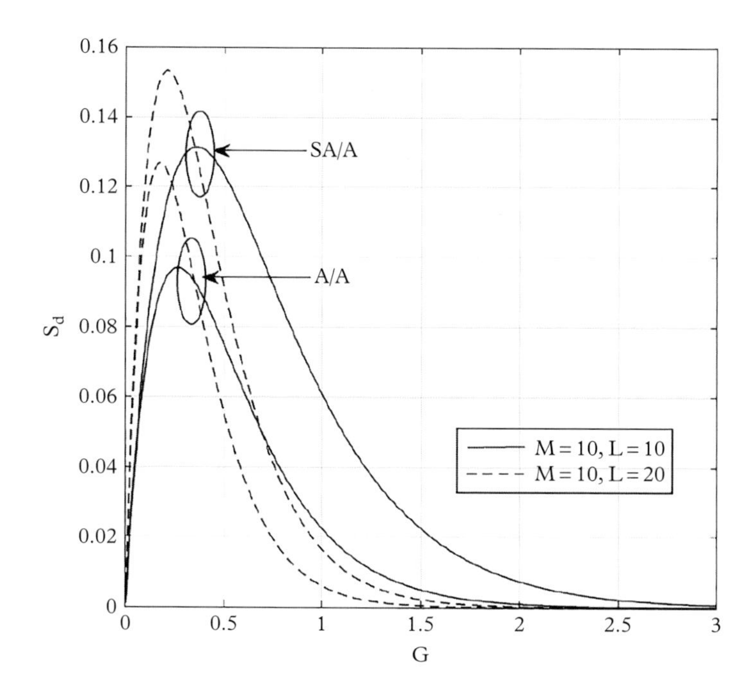

Figure 6.12 *Per-channel throughput S_d versus traffic G in the PC-based single-hop WDM LANs using Aloha/Aloha (A/A) and slotted-Aloha/Aloha (SA/A) protocols, ($L = 10, 20, M = 10$).*

are shown in Figure 6.13, where again the slotted-Aloha/Aloha offers better performance as compared to the Aloha/Aloha protocol.

Figure 6.14 presents the plots of throughput for slotted-Aloha/M-server-switch protocol for $L = 10, 20, 30, 40$, and $M = 10$, which shows that the throughput for this protocol

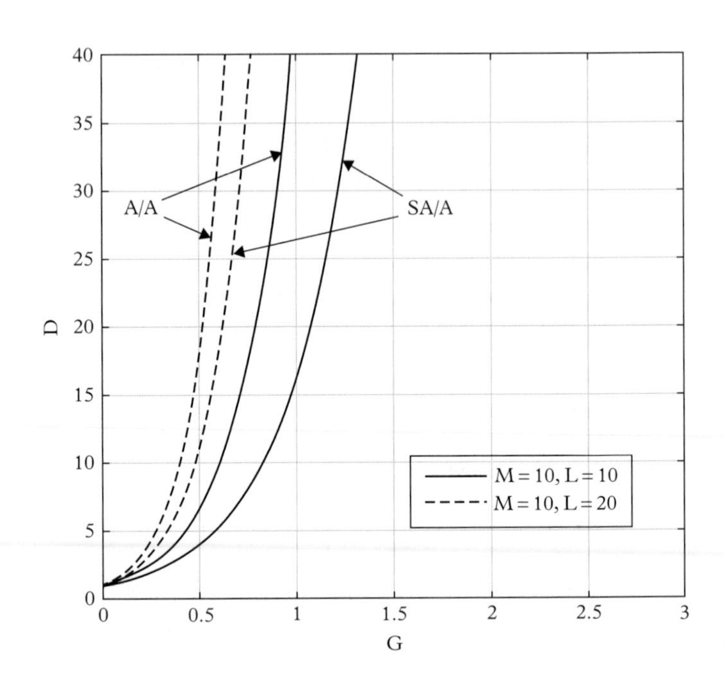

Figure 6.13 *Delay D versus traffic G in the PC-based single-hop WDM LANs using Aloha-Aloha (A/A), slotted-Aloha/Aloha (SA/A) protocols ($L = 10, 20, M = 10$).*

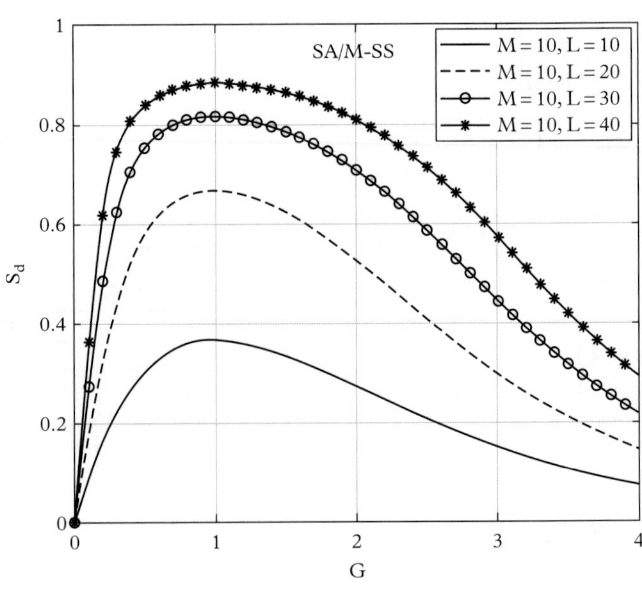

Figure 6.14 *Per-channel throughput S_d versus traffic G in the PC-based single-hop WDM LANs using slotted-Aloha/M-server-switch (SA/M-SS) protocols ($L = 10, 20, 30, 40, M = 10$).*

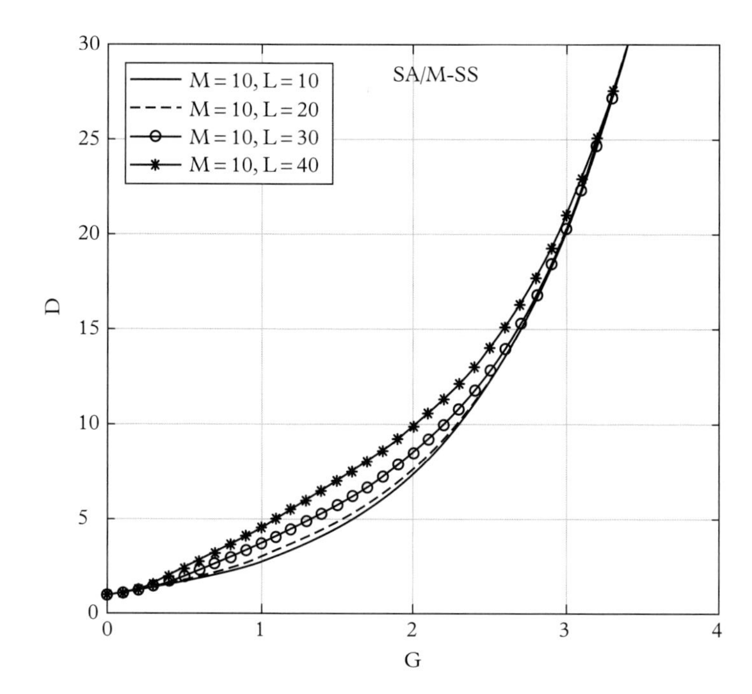

Figure 6.15 *Delay D versus traffic G in the PC-based single-hop WDM LANs using slotted-Aloha/M-server-switch (SA/M-SS) protocols* ($L = 10, 20, 30, 40, M = 10$).

outperforms the earlier two protocols. Delay performance for the slotted-Aloha/M-server-switch protocol is shown in Fig. 6.15, which again gives much better performance as compared to the other two single-hop PC-based MAC protocols.

Note that the analytical models used in the foregoing are based on the assumption that the tuning delay in the tunable transmitters and receivers is negligible as compared to the packet durations. This may not remain valid in high-speed networks, calling for the inclusion of tuning delay in the analytical models. Interested readers are referred to (Borella and Mukherjee 1996) and the other related works for more robust analytical results. Nevertheless, the analyses described here give an insight into the various MAC protocols along with the respective limits for the achievable throughput and delay in the PC-based single-hop WDM LANs.

6.3.2 MAC without PC

Without any PC between the nodes, a single-hop WDM LAN typically employs a two-dimensional time-wavelength assignment scheme (based on TDMA-WDMA) for realizing a viable MAC protocol. In particular, a synchronous time frame T_F with a given number of time slots (each time slot having a duration of T_s) is transmitted periodically. In each time frame, all the nodes get fair chances to transmit packets to the other nodes over a set of pre-assigned wavelengths using the available time slots, thus enabling concurrent transmission of packets over all the wavelengths in each time slot during the time frame. The time-wavelength assignment scheme can be realized in a number of ways, as discussed in the following.

Fixed time-wavelength assignment scheme

In this scheme, each node pair is assigned a specific wavelength over a given time slot in the time frame T_F. Consider a network with four nodes ($N = 4$) and three wavelengths ($M = 3$). Since four nodes will have $2 \times \binom{N}{2} = N(N-1) = 12$ possible connections between the node pairs, three wavelengths (w_1, w_2, w_3, say) will need $K = N(N-1)/M = 12/3 = 4$ time slots in a time frame (i.e., $T_F = 4T_s$). This scheme is illustrated in Table 6.1 with three rows representing the three wavelengths and four columns representing the four constituent time slots in the time frame T_F. Note that during a given time frame each node pair gets two time-wavelength *blocks* for both-way communications (i.e, node i to node j and node j to node i) in the time-wavelength assignment table.

The four time slots (four columns) in Table 6.1 are: time slot 1 $\in [t_0, t_0+T_s]$ (t_0 being an arbitrary starting point for the time frame $T_F = 4T_s$), time slot 2 $\in [t_0+T_s, t_0+2T_s]$, time slot 3 $\in [t_0+2T_s, t_0+3T_s]$, and time slot 4 $\in [t_0+3T_s, t_0+4T_s]$. This arrangement leads to $M \times K = 12$ time-wavelength blocks in the table to be filled up by the 12 possible connections between the node pairs of the four-node network. The network will have a $N \times N$ traffic matrix Γ with N diagonal elements being zero, thus leading to $N^2 - N$ (= 16 - 4 = 12 in the present example) elements representing the all possible traffic demands between node pairs. The 4×4 traffic matrix Γ is expressed as

$$\Gamma = \begin{bmatrix} \rho_{11} & \rho_{12} & \rho_{13} & \rho_{14} \\ \rho_{21} & \rho_{22} & \rho_{23} & \rho_{24} \\ \rho_{31} & \rho_{32} & \rho_{33} & \rho_{34} \\ \rho_{41} & \rho_{42} & \rho_{43} & \rho_{44} \end{bmatrix}, \tag{6.15}$$

with $\rho_{11} = \rho_{22} = \rho_{33} = \rho_{44} = 0$.

The time-wavelength assignment process with Table 6.1 is carried out as follows.

Step I: Start with the first row in Table 6.1, and the first row in Γ. Assign the first time-wavelength block of the table (i.e., wavelength w_1 and time slot 1) to the second element ρ_{12} (ρ_{11} being zero) in the first row of Γ. Carry on populating the subsequent time-wavelength blocks in the first row of the table for the remaining elements of the

Table 6.1 *Fixed assignment using TDMA-WDMA for four nodes and three wavelengths (channels) with frame time $T_F = 4T_s$.*

Wave-length	Time Slot 1	Time Slot 2	Time Slot 3	Time Slot 4
	$[t_0, (t_0+T_s)]$	$[(t_0+T_s), (t_0+2T_s)]$	$[(t_0+2T_s), (t_0+3T_s)]$	$[(t_0+3T_s), (t_0+4T_s)]$
w_1	node 1 → node 2	node 1 → node 3	node 1 → node 4	node 2 → node 1
w_2	node 2 → node 3	node 2 → node 4	node 3 → node 1	node 3 → node 2
w_3	node 3 → node 4	node 4 → node 1	node 4 → node 2	node 4 → node 3

Table 6.2 *Fixed assignment using TDMA-WDMA for four nodes and four wavelengths (channels) with frame time $T_F = 3T_s$.*

Wave-length	Time Slot 1	Time Slot 2	Time Slot 3
	$[t_0, (t_0 + T_s)]$	$[(t_0 + T_s), (t_0 + 2T_s)]$	$[(t_0 + 2T_s), (t_0 + 3T_s)]$
w_1	node 1 → node 2	node 1 → node 3	node 1 → node 4
w_2	node 2 → node 1	node 2 → node 3	node 2 → node 4
w_3	node 3 → node 1	node 3 → node 2	node 3 → node 4
w_4	node 4 → node 1	node 4 → node 2	node 4 → node 3

first row in Γ. In this example, the third block of the first row of the table completes the assignments for node 1 as source node (i.e., up to ρ_{14}). Hence the last block in the first row of the table is assigned to the first element of the next row of Γ, i.e., ρ_{21}.

Step II: Step I is continued until the last time-wavelength block of the table. Thus, the last-but-one matrix element ρ_{43} (ρ_{44} being zero) gets allocated to the last (right-bottom) block of the table.

One can generalize the above process after choosing the appropriate number of slots for the given set of wavelengths and number of nodes. Note that, with one additional wavelength, i.e., with $M = 4$, one can redo this assignment process, which will need 12/4 = 3 time slots (i.e., $T_F = 3T_s$), thereby reducing the delay incurred in packet transmission by one slot duration (Table 6.2).

Performance of fixed time-wavelength assignment schemes

The packet transmission delay in the fixed time-wavelength assignment schemes can be expressed as

$$D = T_F, \tag{6.16}$$

leading to the number of slots needed for a packet transmission, given by

$$E[s] = T_F/T_s, \tag{6.17}$$

which can be reduced by increasing the number of wavelengths, thereby indicating a tradeoff between delay and the number of wavelengths in the network. However, note that in this case the delay is deterministic in nature and hence without any significance of the $E[x]$ operator. Yet we retain this form in this case for uniformity of notation, as in the subsequent schemes the delay incurred by a packet would be governed by the contentions in the shared channels, thereby needing averaged estimates of the delay.

Partially-fixed time-wavelength assignment scheme

The rigid time-wavelength assignment scheme, as considered above, may lead to inefficient bandwidth utilization with a non-uniform traffic pattern between node pairs. In order to address this issue, one can make use of a partially fixed assignment scheme, where the possibility of collisions comes in. In such schemes, two or more node pairs may transmit on the same wavelength during the same time slot, leading to channel collision. In some other cases, a node while receiving packets from another node on a given channel, may receive packets from some other node on some other channel, and hence it cannot detect the latter signal, leading to what is called a receiver collision. However, if the collision-prone assignments are permitted judiciously to some specific node pairs having lower traffic demands, it can eventually lead to better bandwidth utilization in the network. We discuss these schemes in the following.

The assignment schemes that lead to receiver collisions are not considered here owing to their poor performance. We therefore consider below the schemes that encounter only channel collisions. As shown in Table 6.3, we consider a network with four nodes (i.e., $N = 4$), with $M = 3$ and $K = 3$, which offers only $M \times K = 9$ time-wavelength blocks for 12 elements of Γ. The shortfall of the block-spaces being $9 - 6 = 3$, three blocks are assigned with two concurrent transmissions. Transmissions in these blocks are likely to encounter collisions and the network needs to employ some random access protocol to address the problem of collision. This allocation scheme is known as a *destination allocation* scheme, as in a collision-prone time-wavelength block a given destination is allocated to receive signal from two sources on the same channel.

In another example, as shown in Table 6.4, we consider $M = 2$, while K remains the same (= 3), thereby offering six time-wavelength blocks for the assignment of 12 traffic matrix elements. Thus, in this case, all the six blocks in the time-wavelength assignment table are assigned 12/6 = 2 concurrent transmissions, thereby needing contention resolution protocol for all of them, although allowing only channel collisions. Note that this scheme

Table 6.3 *Partially-fixed assignment using TDMA-WDMA for four nodes and three wavelengths (channels) with frame time $T_F = 3T_s$. Note that three out of nine time-wavelength blocks are subjected to collision.*

Wave-length	Time Slot 1	Time Slot 2	Time Slot 3
	$[t_0, (t_0 + T_s)]$	$[(t_0 + T_s), (t_0 + 2T_s)]$	$[(t_0 + 2T_s), (t_0 + 3T_s)]$
w_1	nodes 1,3 → node 2	node 1 → node 3	node 1 → 4
w_2	node 2 → node 1	nodes 2,4 → node 3	node 2 → node 4
w_3	node 3 → node 4	node 4 → node 2	nodes 3,4 → node 1

Table 6.4 *Partially-fixed assignment using TDMA-WDMA for four nodes and two wavelengths (channels) with frame time $T_F = 3T_s$. Note that all time-wavelength blocks are subjected to collision.*

Wave-length	Time Slot 1	Time Slot 2	Time Slot 3
	$[t_0, (t_0 + T_s)]$	$[(t_0 + T_s), (t_0 + 2T_s)]$	$[(t_0 + 2T_s), (t_0 + 3T_s)]$
w_1	nodes 1,3 → node 2	nodes 1,2 → node 3	nodes 1,2 → node 4
w_2	nodes 2,3 → node 1	nodes 3,4 → node 2	node 3 → node 4, node 4 → node 1

cannot strictly be called a destination allocation scheme, as the block in the time-wavelength assignment table at the right-bottom corner (wavelength w_2, time slot 3) does not allocate the same destination for the two concurrent transmissions. The average packet transmission delay due to the collision-prone time-wavelength blocks exceeds the length of the time frame T_F, given by $D' > T_F$, accounting for possible multiple transmission attempts due to collisions.

Performance of partially fixed time-wavelength assignment schemes

Performance of the partially fixed time-wavelength assignment schemes will be governed by the estimate of $E[s]$ in the presence of collisions, by using a suitable model for D'. If there are Q collision-prone blocks out of $M \times K$ blocks in the time-wavelength assignment table, then the average packet transmission delay D' can be expressed as

$$D' = T_F(1 + \alpha_Q), \tag{6.18}$$

where α_Q is a collision-dependent factor, with $\alpha_Q = 0$ for $Q = 0$. The above expression implies that, without any collision-prone block, the average packet transmission delay is determined by the frame time only, i.e., $D = T_F$, while with all blocks subjected to collisions ($Q = MK$), the delay increases to $D' = T_F[1 + \alpha_{MK}]$. With $Q \leq MK$, when the collision-prone blocks are not identically affected by the collisions (due to non-uniform traffic and the number of nodes sharing a given channel), each collision-prone block may have a different value for α_Q (α_i, say), leading to a generalized expression for D', given by

$$D' = T_F\left(1 + \frac{\sum_{i=1}^{Q} \alpha_i}{Q}\right). \tag{6.19}$$

Using the above expression for D', we therefore obtain the expected number of slots $E[s]$ for a packet transmission as

$$E[s] = \frac{D'}{T_s} = \frac{T_F}{T_s}\left(1 + \frac{\sum_{i=1}^{Q} \alpha_i}{Q}\right). \tag{6.20}$$

Next, we evaluate α_i's for obtaining the value of $E[s]$ as follows. Assuming that the contending nodes in a given time-wavelength block (say, the ith block) transmit with a probability p_i of collision, the expected number of additional attempts (i.e., α_i) needed to transmit a packet successfully can be expressed as

$$\begin{aligned} \alpha_i &= \sum_{k=2}^{\infty} k p_i^{k-1}(1 - p_i) \qquad (6.21) \\ &= 2p_i(1 - p_i) + 3p_i^2(1 - p_i) + \cdots \\ &= (1 - p_i)[1 \times p_i + 3 \times p_i^2 + \cdots] \\ &= \frac{1 - (1 - p_i)^2}{1 - p_i}, \end{aligned}$$

where k is an integer representing the number of transmission attempts and p_i is expressed as

$$\begin{aligned} p_i &= 1 - \text{probability that only one node transmits} \qquad (6.22) \\ &= 1 - \binom{n}{1} p_{tx}(1 - p_{tx})^{n-1}, \end{aligned}$$

Table 6.5 *Comparison of delay performance for different time-wavelength assignment (WDMA-TDMA) schemes for non-PC-based single-hop WDM LANs.*

Number of wavelengths	$E[s]$ for four nodes		
	Fixed assignment of Table 6.2	Partially fixed assignment of Table 6.3	Partially fixed assignment of Table 6.4
4	3	—	—
3	—	4.5	—
2	—	—	7.5

with p_{tx} as the probability of transmission from any one of the n nodes. Assuming that all the n nodes are equally likely to transmit, we get $p_{tx} = 1/n$, and express p_i as

$$p_i = 1 - \left(1 - \frac{1}{n}\right)^{n-1}. \tag{6.23}$$

Note that, $\alpha_i = 0$ for contention-free time-wavelength blocks. Using the above expressions, we compare the performance of some non-PC based single-hop WDM LANs in Table 6.5 with uniform traffic pattern. As expected, with fewer wavelengths and more channel-sharing (i.e., with partially fixed assignment schemes), the expected number of time slots $E[s]$ increases for a successful packet transmission.

6.4 Multihop WDM LANs

PSC-based multihop WDM LANs have a different set of design aspects with respect to the single-hop versions (Acampora 1987; Hluchyj and Karol 1991; Mukherjee 1992b). Each node having a finite number of transmitters (FTs) and receivers (FRs), fixed at specific wavelengths, leads to varying number of hops to connect the other nodes. Thus some node pairs may be connected over a single hop, while other pairs may need two or more hops to establish a connection. Figure 6.16 shows the connections between the nodes in a six-node multihop WDM LAN, where we show that the *logical or virtual*[1] topology formed by the edges (connections using distinct wavelengths) between the nodes can significantly change with a larger number of FTs and FRs per node. For example, as shown in Fig. 6.16, with one FT and one FR (i.e., FT^1-FR^1 transceiver configuration) in each node, one can form a logical ring with six nodes using six wavelengths in the network. However, with FT^2-FR^2 transceiver configuration at each node, a logical mesh can be formed with enhanced connectivity between the nodes with 12 wavelengths, as more node pairs can now be provided with single-hop connection. In other words, the expected number of hops (averaged over all node pairs) is smaller in the 12-wavelength case, thereby ensuring better overall delay performance in the network.

[1] Note that *virtual* and *logical* topology would have the same significance throughout the book.

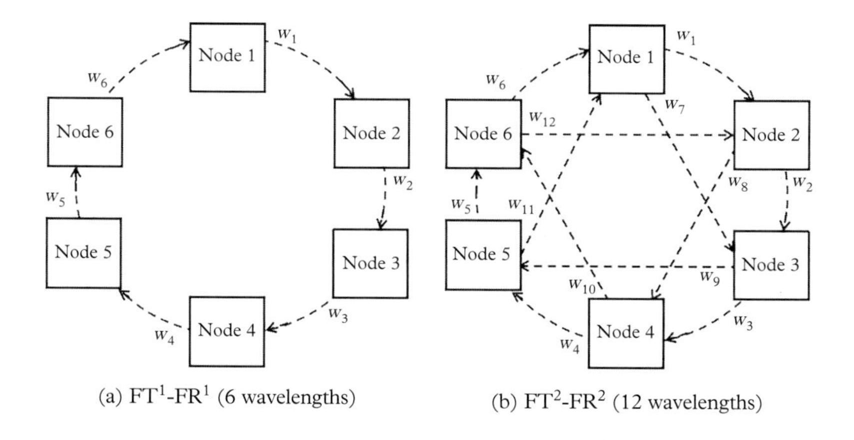

Figure 6.16 *Illustration of logical topologies in a six-node multihop WDM LAN using FT-FR transceiver configuration. The underlying physical topology for these logical topologies is based on a PSC, which is not shown in the diagram explicitly. Note that, the logical topology in part (a) is a ring with six wavelengths and FT^1-FR^1 transceiver configuration at each node, while in part (b) it is a mesh topology with 12 wavelengths and FT^2-FR^2 transceiver configuration at each node.*

The formation of logical topologies in multihop WDM LANs falls in two categories: regular topologies and the topologies without any specific regular formation. Regular topologies can be formed with some specific number of nodes along with a given logical degree for the nodes (i.e., the number of transceivers in each node), where routing and wavelength assignment schemes will follow certain specific rules and hence are easily implementable. However, in regular topologies, it may not be always possible to distribute traffic evenly over various wavelengths, leading to suboptimal network performance in respect of the delay and congestion in the network. However, for a given traffic distribution, one can design a suitable (generally non-regular) logical topology to optimize the network performance, but with more complex routing protocols and wavelength assignment schemes. In the following, we describe the salient features of the regular logical topologies.

With equal numbers of FTs and FRs (i.e.,FT^i-FR^j configuration with $i = j$), a network node may be considered to have a logical degree of $i = j = p$ (say). With this network setting and assuming that each node has the same logical degree p in an N-node network, a given node in the network can exhibit a *regular* connectivity graph to reach the other $N - 1$ nodes, resembling a p-ary directed *spanning tree*. Thus, every node in the network, as a *root node*, will have its own directed spanning tree governed by the logical degree p of the nodes, which in turn will provide the routing information for the root node. This feature of multihop WDM LAN is illustrated in Fig. 6.17, where after each hop, the reachability of the root node to other nodes increases p times. If the spanning tree is not *full*, all the leaf nodes will not be equidistant from the root node in respect of the number of hops h from the root node. In such cases, the number of hops between the farthest leaf and the root is called the *height* of the tree, given by $H = h_{max}$.

In an N-node multihop WDM LAN with a logical degree p and a tree height H, the number of nodes reached by the spanning tree (including the root node) would be exactly equal to the number of nodes N for a full spanning tree (i.e., all leaf nodes are H hops away from the root node), or else N will be upper-bounded by a polynomial in p, given by

$$N \leq 1 + p + p^2 + p^3 + \ldots + p^h + \ldots + p^H, \quad (6.24)$$

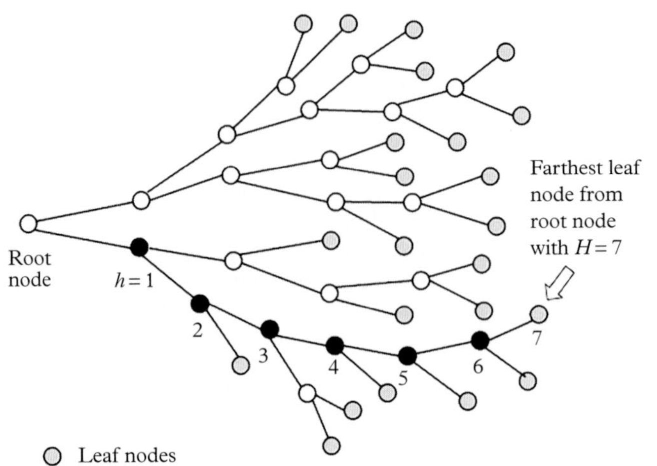

Figure 6.17 *Formation of spanning tree in a multihop WDM LAN with logical degree p =2.*

which can be simplified to obtain N in terms of H as

$$N \leq \begin{cases} \frac{p^{H+1}-1}{p-1}, & \text{for } p > 1 \\ H+1, & \text{for } p = 1. \end{cases} \tag{6.25}$$

In Eq. 6.25, equality will hold when the p-ary tree is full. This leads to a lower bound for H, given by

$$H = \begin{cases} \lceil \log_p[1 + N(p-1)] \rceil - 1, & \text{for } \ p > 1 \\ N - 1, & \text{for } p = 1. \end{cases} \tag{6.26}$$

Using a p-ary spanning-tree as the connectivity graph for each node as a source, one can express the expected number of hops $E[h]$ for a source node to reach any random destination node as

$$E[h] = \begin{cases} \frac{1}{N-1} \sum_{h=1}^{H} h n_p(h), & \text{for } p > 1 \\ \frac{N}{2}, & \text{for } p = 1, \end{cases} \tag{6.27}$$

where $n_p(h)$ represents the number of nodes reached at hop h from the root in a spanning tree with a logical degree p. Note that, $n_p(h)$ increases from p for $h = 1$ with a geometric progression (i.e., $n_p(h) = p^h$), but is upper-bounded by $[N - (p^H - 1)/(p-1)]$ at $h = H$ for an incomplete tree (as governed by Eq. 6.25). Using this observation on $n_p(h)$ in Eq. 6.27, one can simplify the expression for $E[h]$ as

$$E[h] = \frac{1}{N-1}\Big[p + 2p^2 + \cdots + hp^h + \cdots + (H-1)p^{H-1} + H\Big(N - \frac{p^H - 1}{p-1}\Big)\Big], \tag{6.28}$$

which can be further simplified for $p \geq 1$ as

$$E[h] = \begin{cases} \frac{p-p^{H+1}+NH(p-1)^2+H(p-1)}{(N-1)(p-1)^2}, & \text{for } p > 1 \\ \frac{N}{2}, & \text{for } p = 1. \end{cases} \quad (6.29)$$

The expression for $E[h]$ in Eq. 6.29 gives, in essence, the minimum-possible value for the average hop distance between any two nodes in a multihop WDM LAN, since in reality this hop distance may not always be realizable. In other words, it may not be always possible to assign the paths along the spanning tree, leading to some longer routes (i.e., taking a higher number of hops). In order to obtain the expected number of hops in a given network, we need to examine the logical topology based on the possible routes and wavelengths that can be allotted for all the connections in the network.

Next, we describe some well-known multihop networks with regular logical topologies, such as, ShuffleNet, Manhattan Street, Hypercube, and deBruijn networks, and examine their salient features.

6.4.1 ShuffleNet

A (p, k) ShuffleNet represents a *regular* network topology, where kp^k nodes (with k and p as integers) are arranged in k columns with each column having p^k nodes and each node having a logical degree p. The kth column is connected back to the first column, forming a connectivity graph which appears as a wrapped-around cylinder. An example (2,2) ShuffleNet is shown in Fig. 6.18, which has $k = 2, p = 2$, and $kp^k = 8$ nodes. When a multihop WDM LAN forms a (2,2) ShuffleNet as its logical topology, each node has to employ two FTs and two FRs on specific wavelengths, such that the desired connectivity graph is realized as shown in Fig. 6.18.

In the (2, 2) ShuffleNet, when a given node in a column is connected by single-hop transmission to another node in the next column (represented by a directed *arc* between these two nodes), it implies that one of the two FTs of the source node transmits at a wavelength, which is also the receiving wavelength of one of the two FRs at the destination node. Fixing of the wavelengths for FTs and FRs in all the nodes is governed by the connectivity graph of the given ShuffleNet topology, where the connections of the nodes between two successive columns follow a specific rule that mimics the *shuffling* of cards. For example, in the (2,2) ShuffleNet of Fig. 6.18, the directed connections (arcs) from the third and fourth nodes from the top in the first column (i.e., the nodes 2 and 3) are shuffled in between the directed arcs from the first and second nodes from the top in the first column (i.e., the nodes 0 and 1). This shuffling mechanism can be defined mathematically as follows. First, the nodes in each column are numbered from top to bottom as 0 through $p^k - 1$. Next, p directed arcs are drawn from node i on the given column to the nodes $j, j+1, ... j+p-1$ in the next column with the rule given by

$$j = (i \ mod \ p^{k-1}) \times p. \quad (6.30)$$

Note that, in the above rule, i or j doesn't represent a running node number, as used in Fig. 6.18. The above shuffling rule is applied for all the nodes in all the k columns to construct the complete topology of ShuffleNet. Using this interconnection rule of the topology, next we examine the salient features of ShuffleNet.

Topological characteristics and network performance

The performance of a (p, k) ShuffleNet can be assessed by evaluating the expected number of hops $E[h]$ between various node pairs. In order to evaluate $E[h]$ in ShuffleNet, one needs to examine its connectivity graph from a given node (as a source or root node) to all other nodes using the directed spanning tree, as discussed in the earlier section. For example, we consider the (2,2) ShuffleNet of Fig. 6.18, and draw the directed spanning tree for node 0 (as the root node) in the first column of the network, as shown in Fig. 6.19 (see the shaded part). It is evident from the figure that node 0 can reach nodes 4 and 5 in one hop over the wavelengths w_1 and w_2, and thereafter can reach nodes 1, 2, and 3 in two hops with the second hops transmitted over the wavelengths w_{10}, w_{11}, and w_{12}. Finally, the connections are wrapped around (i.e., repeated on the right side of column 2), and node 0 reaches the remaining nodes 6 and 7 either via node 1 or via node 3 using altogether three hops with the third hop transmitted either on the wavelengths w_3 and w_4 or on the wavelengths w_7 and w_8, respectively. This indicates that, in the present case there is a multiplicity of routes with the same hop distance (3), which is in a way good in the presence of traffic congestion from node 0. Note that the maximum number of hops (i.e., the tree height H) from the root

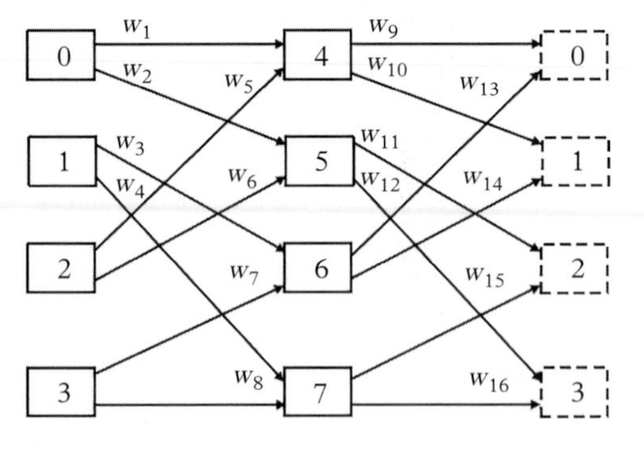

Figure 6.18 *(2,2) ShuffleNet. Note that between the first two columns, the two sets of arcs from the nodes 2 and 3 over the wavelengths w_5, w_6, w_7, and w_8 are shuffled (interleaved) in between the other two sets of arcs from the nodes 0 and 1 over the wavelengths w_1, w_2, w_3, and w_4. Similar shuffling of arcs takes place between column 2 and the last column (image of column 1) over the wavelengths w_9 through w_{16}.*

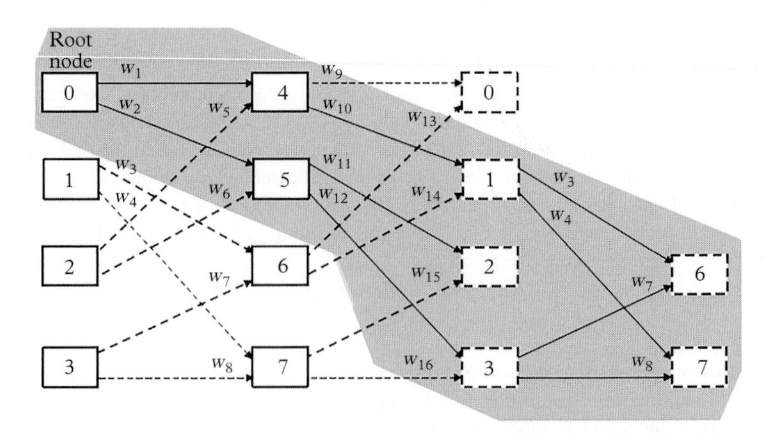

Figure 6.19 *Illustration of the directed spanning tree in a (2,2) ShuffleNet from node 0. Note that the nodes 6 and 7 can be reached through either node 1 or node 3 in the final hop. This type of multiple choices of routing helps in the presence of heavy traffic.*

node (node 0 in this case) in the spanning tree to reach all other nodes is three, which can be generalized for a (p, k) ShuffleNet as

$$H = 2k - 1. \tag{6.31}$$

In Fig. 6.18, we have considered that all the arcs have unique wavelengths, implying that the network will need kp^{k+1} (= 16 in this case, with $p = 2$, $k = 2$) wavelengths. With large number of nodes, having unique wavelengths for all the arcs may not be feasible, and the network may have to employ shared-wavelength transmission using appropriate multiple-access protocols.

Next, we generalize the above observations on the spanning tree in a ShuffleNet to find out the number of nodes n_h at a given hop distance h from a root node along its spanning tree. Typically, n_h keeps increasing with h having a geometric progression until the spanning tree covers $(k - 1)$ hops. This is better understood with higher values of k (i.e., with $k \geq 3$). For example, in Fig. 6.20 showing a (2,3) ShuffleNet, n_h increases geometrically with h up to $h = k - 1 = 2$ (i.e., two nodes at $h = 1$ and four nodes at $h = 2$); thereafter n_h at $h = k = 3$ and $h = k + 1 = 4$ increases to seven ($= p^k - 1$) and six ($= p^k - p$), respectively. Thus, for $h = k + i$ (with i = 0,1,2...), n_h can be expressed as $p^k - p^i$. It is instructive to verify this generalization for ShuffleNets with yet higher values of k. The variation of n_h with h varying from 1 to $2k - 1$ in a (p, k) ShuffleNet is presented in Table 6.6, leading to the governing equations, given by

$$n_h = \begin{cases} p^h, \text{ for } h = 1, 2, ..., k - 1 \\ p^k - p^{h-k}, \text{ for } h = k, k + 1, ..., 2k - 1. \end{cases} \tag{6.32}$$

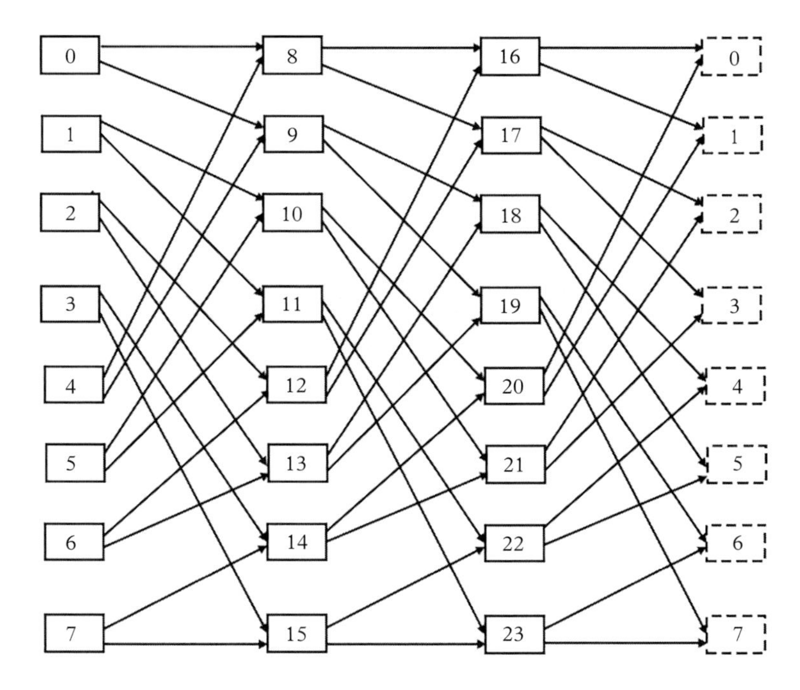

Figure 6.20 *(2,3) ShuffleNet.*

Table 6.6 *Number of hops h versus number of nodes h-hops away from the root node in the spanning tree of a (p, k) SuffleNet.*

Number of hops (h)	Number of nodes h hops away from root node
1	p
2	p^2
...	...
$k-1$	p^{k-1}
k	$p^k - 1$
...	...
$k+1$	$p^k - p$
$k+2$	$p^k - p^2$
...	...
$k+i$	$p^k - p^i$
...	...
$2k-1$	$p^k - p^{k-1}$

Using the above relations between n_h and h, one can express $E[h]$ for uniform traffic distribution in the network as

$$E[h] = \frac{\sum_{h=1}^{2k-1} hn_h}{\sum_{h=1}^{2k-1} n_h}, \tag{6.33}$$

where the numerator and denominator on the right-hand side are given by

$$\sum_{h=1}^{2k-1} hn_h = \sum_{h=1}^{k-1} hp^h + \sum_{h=k}^{2k-1} h(p^k - p^{h-k}) \tag{6.34}$$

$$\sum_{h=1}^{2k-1} n_h = \sum_{h=1}^{k-1} p^h + \sum_{h=k}^{2k-1} (p^k - p^{h-k}). \tag{6.35}$$

Using Equations 6.34 and 6.35 in Eqn.6.33, one can simplify $E[h]$ as

$$E[h] = \frac{kp^k(p-1)(3k-1) - 2k(p^k-1)}{2(p-1)(kp^k-1)}. \tag{6.36}$$

Next, we estimate the network throughput S of ShuffleNet under the assumption of uniform traffic from each node and uniform distribution of ingress traffic from a node to all other nodes. Furthermore, it is assumed that the queuing delay (at node buffers) is small compared to the transmission and propagation times of the packets. With these assumptions, the network performance is primarily determined by the expected number of hops $E[h]$ and per-channel transmission rate r_b. Since the network has kp^{k+1} wavelengths, the total number of bits transmitted per second concurrently over the entire network has an aggregate rate $R = r_b kp^{k+1}$, though on the average only one of the $E[h]$ packets would have the privilege of reaching the destination by a single hop. Therefore, the average value for the aggregate transmission rate is effectively reduced from R to $R/E[h]$. Hence, the throughput S of the entire network is expressed as

$$S = \frac{R}{r_b E[h]} = \frac{kp^{k+1}}{E[h]}. \tag{6.37}$$

Substituting for $E[h]$ from Eq. 6.36 in Eq. 6.37, we finally express S as

$$S = \frac{2kp^{k+1}(p-1)(kp^k-1)}{kp^k(p-1)(3k-1) - 2k(p^k-1)}. \tag{6.38}$$

This leads to the per-node throughput, given by

$$S_{node} = \frac{S}{kp^k} = \frac{2p(p-1)(kp^k-1)}{kp^k(p-1)(3k-1) - 2k(p^k-1)}. \tag{6.39}$$

Next, we present the performance of ShuffleNet in Fig. 6.21 using the plots of the expected number of hops $E[h]$ versus the number of nodes N, with the number of hops k as the parameter. As evident from the figure, for both cases (k = 2 and 3) $E[h]$ initially grows with the number of nodes (which also indicates increase in nodal degree p), and thereafter tends to saturate to a maximum value, which indeed remains below the maximum number of hops H ($= 2k + 1$ = 3 and 5 for k = 2 and 3, respectively). Note that, with ShuffleNet

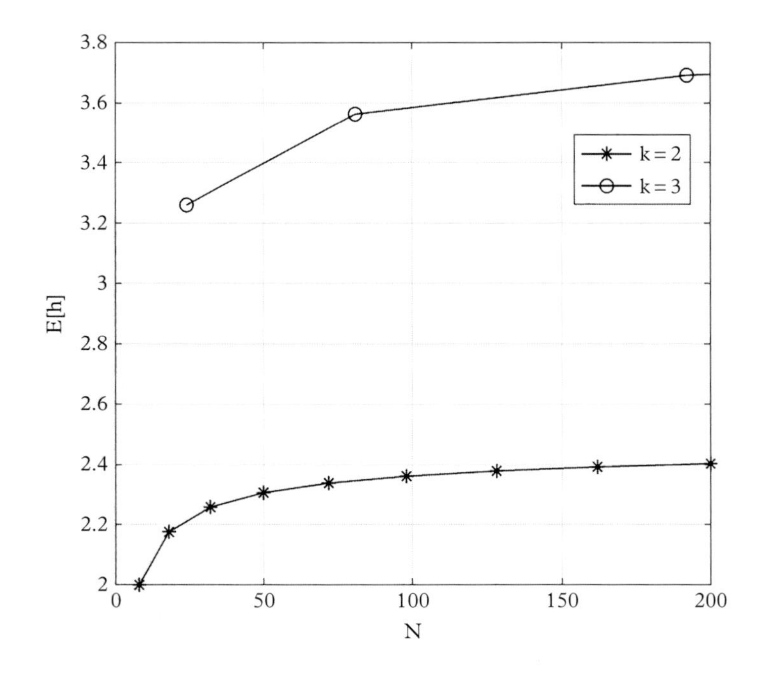

Figure 6.21 *Expected number of hops $E[h]$ versus number of nodes N in ShuffleNet. Note that, for each value of k, p is varied to obtain the permissible values for N.*

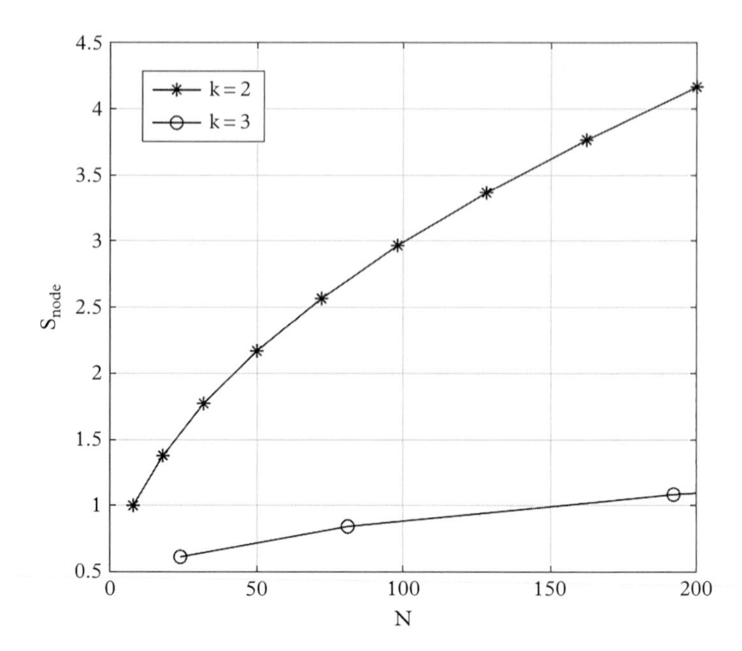

Figure 6.22 *Per-node throughput S_{node} versus number of nodes N in ShuffleNet. Note that, for each value of k, p is varied to obtain the permissible values for N.*

being a regular topology with certain discrete permissible values for the number of nodes (i.e., kp^k), the interpolated plots are of importance only at the specific points of abscissa, as shown by the *asterisks* and *circles* on the plots. Figure 6.22 presents the per-node throughput versus the number of nodes in ShuffleNet, where the throughput plots also exhibit saturation with increasing number of nodes.

Routing schemes

ShuffleNet needs to employ suitable routing schemes depending on the underlying constraints, such as number of nodes, number of transceivers in each node (which determines the number of available wavelengths in a network), traffic patterns (uniform, non-uniform, dynamic), etc. If the number of available wavelengths falls short of the number of arcs (kp^{k+1}) in the logical topology, the network needs to employ shared-wavelength transmission, implying that more than one node pair will use the same wavelength for packet transmission, thereby needing TDMA-based static or random wavelength-access schemes, such as Aloha, CSMA, etc.

Moreover, when a packet traverses intermediate nodes it may find contention over the desired transmit wavelength (i.e., the desired output port in the logical topology) and may have to wait in the node buffer. However, in high-speed networks, having large buffers for transit packets may not always be a preferred option, and hence the transit packet at a given node may have to be transmitted over some other wavelength, leading to a longer route. This type of routing is referred to as deflection routing which, with zero-buffer nodes, is popularly called *hot-potato* routing. The situation becomes more challenging when the traffic distribution becomes non-uniform and varies dynamically, calling for adaptive routing schemes. In the following, we first consider a simple routing scheme when the given ShuffleNet has distinct wavelengths for each arc, and uses the regular topological property to realize a fixed *self-routing* mechanism, so that each packet can be routed at any node based on its destination address.

Self-routing scheme The self-routing scheme falls under the class of fixed-routing schemes. In order to understand the self-routing scheme, we first consider an example (2,3) ShuffleNet, as shown earlier in Fig. 6.20. Each node in this 24-node ShuffleNet is assigned an address (c, r), with c and r representing the column and row coordinates, respectively. The column coordinate c represents an integer, i.e., $c = 0, 1, 2$ for first, second, and third columns, respectively. The row coordinate r represents a p-ary (in this example $p = 2$, implying binary coordinates for the rows) word $(r_2 r_1 r_0)$, ranging from (000) through (111), leading to $p^k = 8$ rows. For example, a node in the second column and third row of the (2,3) ShuffleNet will have the address $(c, r) = (1{,}010)$. Thus, in general a node with an address (c, r) in a (p, k) ShuffleNet will have $c \in (0, 1, \cdots, k-1, k)$ as the column coordinate labeled from the left to right end, and $r \in (0, 1, \cdots, p^k - 1)$ as the p-ary row coordinate labeled as 0 at the top to $p^k - 1$ at the bottom. The p-ary row coordinate can be expressed as

$$r = \sum_{j=0}^{k-1} r_j p^j = (r_{k-1}\ r_{k-2}\ \cdots\ r_j \cdots\ r_1\ r_0), \tag{6.40}$$

where r_j's represent the p-ary digits (i.e., $0 \leq r_i \leq p-1$). With this addressing scheme, for a source node in a (p, k) ShuffleNet located at an address (c_s, r_s), we express its row coordinate r_s as

$$r_s = \sum_{j=0}^{k-1} r_{s,j} p^j = (r_{s,k-1}\ r_{s,k-2}\ \cdots\ r_{s,0}), \tag{6.41}$$

where the p-ary variable r_j of Eq. 6.40 is renamed as $r_{s,j}$ to represent the address of the source node. For the source node, all p outgoing arcs will reach as many number of nodes at one-hop distance (i.e., on the next column around the cylindrically wrapped-up topology), with the following addresses:

$$(c+1) mod\ k, (r_{s,k-2}\ \ r_{s,k-3}\ \cdots\ r_{s,0}\ \ 0),$$

$$(c+1) mod\ k, (r_{s,k-2}\ \ r_{s,k-3}\ \cdots\ r_{s,0}\ \ 1),$$

$$\cdots$$

$$(c+1) mod\ k, (r_{s,k-2}\ \ r_{s,k-3}\ \cdots\ r_{s,0}\ \ (p-1)).$$

Next, consider a packet in a (p, k) ShuffleNet during its traversal from its source node (node s) to the destination node (node d), which has arrived at an intermediate node (node i) with an address (c_i, r_i). From this node, the packet should be routed over the appropriate arc (corresponding to a specific transmit wavelength) of node i, so that the packet reaches the next *desired* node with an address (c_{next}, r_{next}). Note that, governed by the ShuffleNet topology, from node i, the packet can go only to one of the p nodes on the next column C_{next} along the wrapped-up cylinder, given by,

$$c_{next} = (c_i + 1)\ mod\ k \tag{6.42}$$

where the *mod k* operation is essential to support the movement around the cylindrical topology. However, on column c_{next}, only one specific node out of the p nodes connected to node i must be chosen as the *next* node, so that the packet travels along the proper route to reach the desired next-hop node. In other words, r_{next} needs to be determined, following the desired route, as dictated by the ShuffleNet topology. By tracing the path traversed by such a route, one can observe that the packet from node i will be navigated hop by hop through successive columns, and the row of the node in each column will be chosen in accordance with the *column distance* D between node i and the destination node d, given by

$$\begin{aligned} D &= (k + c_d - c_i) \ mod \ k, \ \text{if} \ c_d \neq c_i \\ &= k, \ \text{if} \ c_d = c_i, \end{aligned} \tag{6.43}$$

where *mod k* operation again helps assess the true column distance in the cylindrical topology of ShuffleNet. Thus, from node i, the packet will travel hop by hop toward, the destination node d as follows:

◇ *the source node, i.e., node i, located at:*

$$(c_i, r_{i,k-1}, r_{i,k-2}, \cdots, r_{i,0}),$$

◇ *the node on the next column* c_{i+1}*, located at:*

$$(c_i + 1) \ mod \ k, \ \ (r_{i,k-2} \ \ r_{i,k-3} \ \cdots \ r_{i,0} \ \ r_{d,D-1}),$$

◇ *the node on the next-but-one column* c_{i+2}*, located at:*

$$(c_i + 2) \ mod \ k, \ \ (r_{i,k-3} \ \ r_{i,k-4} \ \cdots \ r_{i,0} \ \ r_{d,D-1} \ \ r_{d,D-2}),$$

$$\cdots \qquad \cdots \qquad \cdots$$

◇ *the destination node on column* c_d*:*

$$c_d, \ (r_{i,k-1-D} \ \ r_{i,k-D} \ \cdots \ r_{i,0} \ \ r_{d,D-1} \ \cdots \ r_{d,0}).$$

The traversal of the packet through the columns, as defined above, implies that, at each subsequent column the column coordinate of the respective node advances by one. However, in the respective row coordinate the second-most significant digit moves left-ward to the most significant position, followed by similar left-ward movements of the subsequent digits, vacating thereby the lowest significant digit. In the vacant least-significant digit, the digit $r_{d,D-1}$ is inserted after the first left-ward movement. Thereafter, in the subsequent step of movement, $r_{d,D-1}$ is replaced by $r_{d,D-2}$, and so on, until the packet reaches the destination with the lowest significant digit becoming $r_{d,0}$.

The foregoing observations indicate that, in a (p, k) ShuffleNet, for a packet at a given node (node i) located at (c_i, r_i) traveling toward its destination node (node d) located at (c_d, r_d), will be routed to the immediate next node located at (c_{next}, r_{next}), with c_{next} and r_{next} given by

$$c_{next} = (c_i + 1) \ mod \ k, \tag{6.44}$$

$$r_{next} = (r_{i,k-2} \ r_{i,k-3} \cdots r_{i,0} \ r_{d,D-1}), \tag{6.45}$$

and following this rule at each intermediate node, the packet will finally arrive at the destination node, with the column distance D between the source and destination nodes playing the pivotal role in navigating the packet all along.

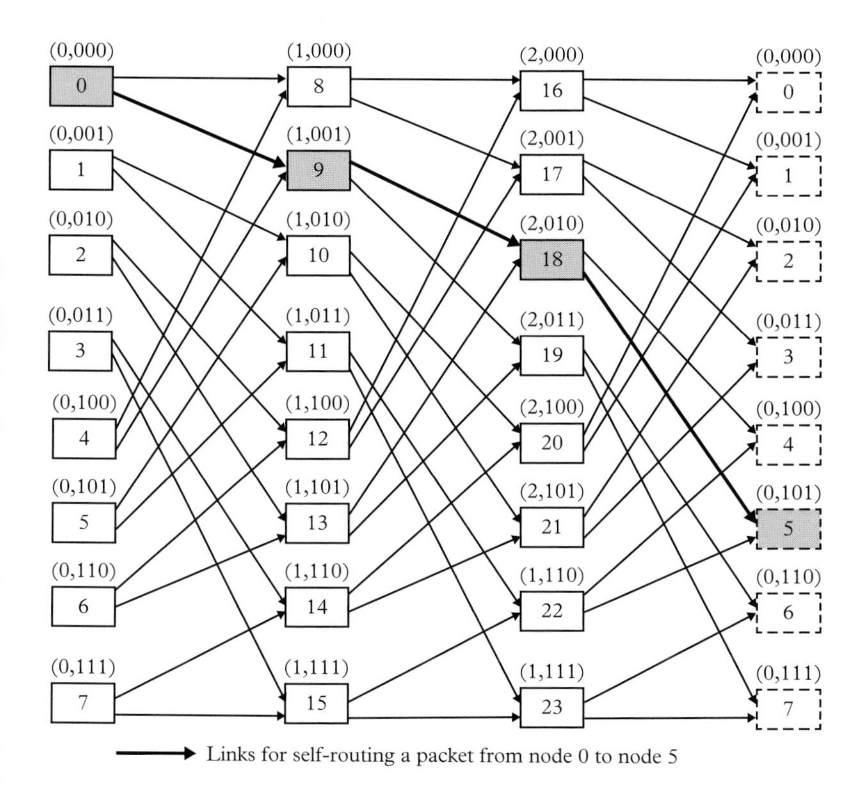

Figure 6.23 *An illustration of self-routing in a (2,3) ShuffleNet from node 0 to node 5 using the node sequence 0-9-18-5.*

In order to demonstrate the above routing scheme, let us consider again the (2,3) ShuffleNet in Fig. 6.23, where a packet is assumed to start from node 0 to reach its destination at node 5, following the route described by the node sequence 0-9-18-5 (according to the ShuffleNet connectivity shown in Fig. 6.23). Let us consider an intermediate node in this route, say node 9, and check whether the self-routing scheme routes the packet under consideration from node 9 to node 18. For the example under consideration with $p = 2$ and $k = 3$, the intermediate node is node 9 with the address $(c_i, r_i) = (1, 001)$, the destination node is node 5 with the address $(c_d, r_d) = (0, 101)$, and the node next to node 9 on the route needs to be determined from the address (c_{next}, r_{next}) following the above routing scheme (Eqs. 6.43 through 6.45). Thus, for the packet under consideration, in the first step, the value of D is estimated using Eq. 6.43 as $D = (k + c_d - c_i) mod\ 2 = (3 + 0 - 1) mod\ 2 = 2$. Next, from $r_i = 001$, we get $r_{i,2} = 0, r_{i,1} = 0, r_{i,0} = 1$. Using the values of D and the digits (in this example, binary digits or bits with $p = 2$) in r_i in Equations 6.44 and 6.45, we obtain the values of c_{next} and r_{next} as

$$c_{next} = (1 + 1)\ mod\ 3 = 2 \tag{6.46}$$

$$r_{next} = (r_{i,1}\ r_{i,0}\ r_{d,D-1}) = (r_{i,1}\ r_{i,0}\ r_{d,1}) = (010),$$

which implies that the next node has the address given by $(c_{next}, r_{next}) = (2, 010)$, as evident from Fig. 6.23. This address corresponds to node 18, which is indeed the next node in accordance with the desired ShuffleNet route, i.e., the node sequence 0-9-18-5.

Similar calculation shows that, finally the packet arrives at the destination node (node 5) in the next hop.

Adaptive-routing scheme As indicated earlier, one important aspect of any routing scheme in ShuffleNet is about how it responds to non-uniform traffic distribution and time-varying traffic. In practice, the traffic demands may change appreciably for some channels from time to time, while operating with the fixed self-routing scheme. With zero-buffer nodes, this may lead to losses and misrouting (deflection) of the transit packets, thereby degrading the network performance. There have been some useful studies on alternative routing schemes for ShuffleNet, that can adapt with the traffic variation and alleviate its impact on the network performance. We describe one such adaptive-routing scheme from (Karol and Shaikh 1991) without calling for any significant communication overhead, except some electronic buffers of nominal size at network nodes.

When the network operates without congestion, the adaptive routing scheme functions like the fixed-routing scheme. However, when the network encounters queuing in the buffer for some outgoing channel (output port) of a node, it *transfers* some part of the queued traffic (packets) to the buffer for some other outgoing channel carrying lesser traffic, often via a longer route. This leads to the deflection of packets over longer routes, but overall end-to-end delay might not increase as much due to the passage of the transferred packets possibly over a channel with less congestion. Moreover, ShuffleNet having multiple shortest paths (minimum number of hops) for longer routes, can also avoid additional hops, whenever the congestion takes place at a node where the transit packet appears as a *don't care* packet, implying that it can be routed through any one of the multiple shortest paths available at that node to reach the destination node. In view of the above, the present adaptive routing scheme categorizes the transit packets into two types and operates as follows:

1. **Single minimum-hop (SMH) packets:** SMH packets have one single minimum-hop path from the current node to the destination node. In ShuffleNet SMH packets take at the most k hops to reach the destination. An SMH packet is routed using the basic self-routing scheme. However, when the packet finds itself queued up in an output buffer at any node on the way with a queue size above a given threshold, the node moves this packet to the other available channel buffer with least queue length. However, such transfer of an SMH packet from one buffer to another is allowed only once during its complete traversal through the network. In order to implement this criterion, a binary status bit is *written* in the packet header, indicating whether it has been moved once from one buffer to another during its traversal through the network. Thereafter, in the event of encountering a full buffer again, the packet is dropped, and from the higher layer it is ensured that the overall network traffic does not exceed an upper limit, and the probability of packet loss is controlled within an acceptable limit.
2. **Multiple minimum-hop (MMH) packets:** MMH packets have multiple minimum-hop paths from the current node to the destination node, and hence are considered as don't care packets at the current node, as mentioned earlier. This empowers an MMH packet to choose any one of the multiple output ports of the current node without making any difference in respect of the hop count. Thus, the MMH packet is sent out through one of the multiple output ports (i.e., output channels) of the current node, whose buffer has fewer queued packets, thereby leading to an effective load balancing in the presence of congestion.

Note that, during the traversal of a given packet through the network, its status may change from one type to another. In particular, in the present routing scheme a packet during its journey can change its type in two possible ways. In one case, when an SMH packet is transferred to a longer path, the number of hops in its remaining journey increases thereby, with a possibility of getting converted into an MMH packet. In the other case, once an MMH packet reaches the destined column (but not the destined row), there remains a unique path with k more hops to reach the destination node, thus being converted to an SMH packet.

6.4.2 Manhattan street networks

Manhattan street network (MSNet) represents a regular mesh-configured, directed network topology that is similar to the geographical topology of the streets and avenues in Manhattan, USA (Maxemchuk 1985, 1987; Ayanoglu 1989; Lee and Kung 1995; Khasnabish 1989). The traffic in a $(U \times V)$ MSNet moves around two sets of orthogonally placed rings or toroids (U horizontal rings and V vertical rings) crisscrossing each other through the nodes, and the traffic in two successive rings, both in vertical and horizontal directions, move in opposite directions. Thus, as shown in Fig. 6.24, a (4×4) MSNet is a regular mesh topology using $(U + V) = 4 + 4 = 8$ unidirectional rings, where each of the $N = U \times V = 4 \times 4 = 16$ nodes is two-connected and hence operates with a fixed logical degree of 2. In other words, unlike ShuffleNet, each node in an MSNet of any size has a logical degree $p = 2$, and the number of wavelengths needed in MSNet is $2N$. Notwithstanding this structural limit, MSNets offer modular topology as the number of nodes can be added easily by increasing the number of rings in the horizontal or vertical direction. In an MSNet, both U and V are even numbers so that the counter-propagating rows/columns appear in pairs. Furthermore, in a perfect MSNet, $U = V = W$ (say) leading to a square topology with $N = U \times V = W^2$.

Topological characteristics and network performance

Consider a $(U \times V)$ MSNet, where each horizontal ring is represented as a row and each vertical ring as a column. Each node is assigned an address (r, c), with r and c representing row and column numbers ($r \in 0, 1, \cdots, U - 1$, and $c \in 0, 1, \cdots, V - 1$), respectively. For example, consider the (2,3) node in Fig. 6.24, which is placed at the intersection of row 2 and column 3. Owing to the *toroidal* nature of the topology (having no boundary in either the horizontal or vertical direction), the hop distance between rth and $(r + \Delta)$th rows will be the same as the hop distance between $(r+k)$th and $(r+k+\Delta)$th rows. Similarly, the hop distance between cth and $(c + \Delta)$th columns will be the same as the hop distance between $(c+k)$th and $(c+k+\Delta)$th columns. Thus, MSNet has the following characteristic features:

- the toroidal topology of MSNet is *isotropic*, which implies that every node has a similar set of connections with the rest of the nodes and thereby locally looks into the same topology around itself,
- MSNet is realized with wrapped-around connections, eliminating the *edge effect*, thereby reducing the route lengths. Such a boundary-less topology is referred to as a *manifold*.

As in ShuffleNet, one of the important performance features of MSNet is the expected number of hops $E[h]$ that a packet needs from a source node to traverse across the network to reach all the remaining nodes. Using the analytical model from (Khasnabish 1989), we examine this aspect in the following. Assume that there are n_h nodes that are reachable from

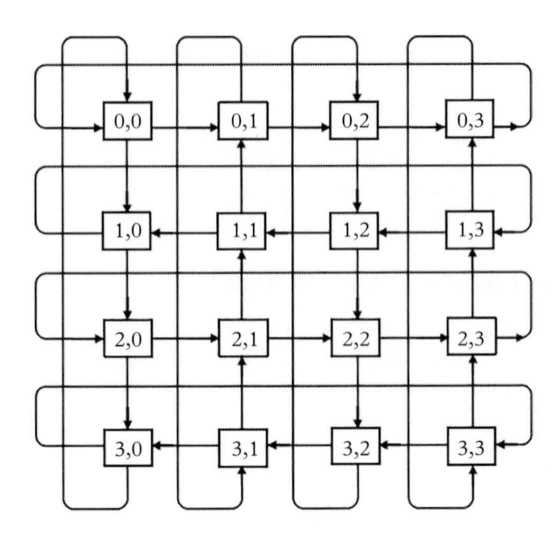

Figure 6.24 (4 × 4) *MSNet.*

a given source node using h hops. Note that, owing to the isotropic nature of the MSNet topology, n_h will be same for any source node. Hence, the probability P_h to reach a node at a hop distance h in an N-node ($N = U \times V$) MSNet with uniform traffic distribution can be expressed as

$$P_h = \frac{n_h}{N-1}. \tag{6.47}$$

Using the above expression for P_h, one can obtain the expected number of hops $E[h]$ as

$$E[h] = \sum_{h=1}^{H} hP_h = \frac{\sum_{h=1}^{H} hn_h}{N-1}, \tag{6.48}$$

where H represents the maximum number of hops (height) taken to reach the farthest node from a given source node. Assuming that $U = V = W$ (say) with W as an even number for a perfect MSNet, and thus with $N = W^2$, we observe that

$$H = \begin{cases} W, \text{ for } W = 2(2i+1) \\ W+1, \text{ for } W = 2(2i), \end{cases} \tag{6.49}$$

where i represents a non-zero positive integer. For $W = 2(2i+1)$ (implying that $W/2$ is an odd number), one can express n_h as

$$n_h = \begin{cases} 2, \text{ for } h = 1 \\ 4(h-1), \text{ for } 2 < h \le W/2+1 \\ 4(W-h+1), \text{ for } W/2+1 < h \le W-1 \\ 2, \text{ for } h = W. \end{cases} \tag{6.50}$$

For $W = 2(2i)$ (implying that $W/2$ is an even number), one can express n_h as

$$n_h = \begin{cases} 2, \text{ for } h = 1 \\ 4(h-1), \text{ for } 2 \leq h \leq W/2 \\ 4(W/2-1), \text{ for } h = W/2+1 \\ 4(W/2-1), \text{ for } h = W/2+2 \\ 4(W-h+1), \text{ for } W/2+2 < h \leq W \\ 2, \text{ for } h = W+1. \end{cases} \tag{6.51}$$

Using Eq. 6.48 through 6.51, we finally obtain $E[h]$ as

$$E[h] = \begin{cases} \frac{W^3/2+W^2-2W-2}{N-1}, \text{ for } W = 2(2i+1) \\ \frac{W^3/2+W^2-4}{N-1}, \text{ for } W = 2(2i). \end{cases} \tag{6.52}$$

From the above expressions for $E[h]$, we obtain the per-node throughput S_{node} (= number of wavelengths / (number of nodes $\times$ $E[h]$)) as

$$S_{node} = \frac{M}{W^2 E[h]} = \frac{2}{E[h]}, \tag{6.53}$$

where the factor of two in the numerator implies that $M = 2N = 2W^2$, as each node has two input/output ports (logical degree), each with one unique wavelength. Using the expression for $E[h]$ from Eq. 6.52 in Eq. 6.53, we obtain S_{node} as

$$S_{node} = \begin{cases} \frac{2(N-1)}{W^3/2+W^2-2W-2}, \text{ for } W = 2(2i+1) \\ \frac{2(N-1)}{W^3/2+W^2-4}, \text{ for } W = 2(2i). \end{cases} \tag{6.54}$$

Using S_{node}, the overall network throughput S is expressed as

$$S = NS_{node}. \tag{6.55}$$

With increasing W and having $N = W^2$, $E[h]$ asymptotically reaches a limiting value $\hat{E}[h]$, given by

$$\hat{E}[h] = W/2 + 1 = \sqrt{N}/2 + 1, \tag{6.56}$$

leading to the limiting values of per-node and network throughput as

$$\hat{S}_{node} = \frac{2}{\sqrt{N}/2+1} \tag{6.57}$$

$$\hat{S} = \frac{2N}{\sqrt{N}/2+1}.$$

Thus, with large W, one gets eventually a simple expression for network throughput estimate, given by $\hat{S} = 4\sqrt{N} = 4W$. A quick sanity check also leads to the same result, as

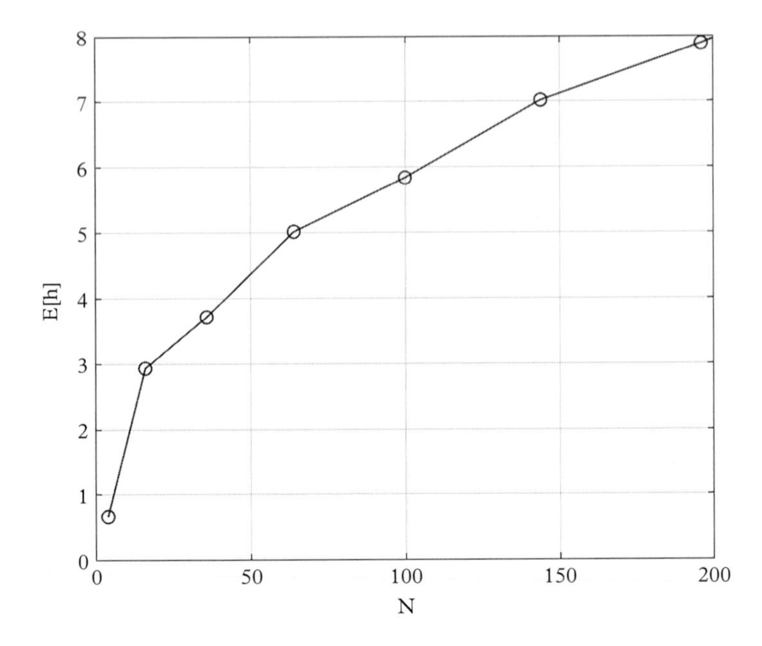

Figure 6.25 *Expected number of hops $E[h]$ versus number of nodes N in MSNet.*

follows. Note that the network as a whole has $2N$ concurrent transmissions from N nodes and each transmission can have an average hop count approximately equaling half of the network span, i.e., $W/2$. This leads to a network throughput estimate as $2N/(W/2) = 4W$, as obtained from Eq. 6.57 with large W.

Note that in MSNet, due to its isotropic connectivity through the toroidal manifold, the nodes do not usually employ any buffer, more so at high-speed optical transmission rates. In other words, delay caused by the detours taken by the packets due to deflections is more acceptable, when compared to the additional hardware complexity that needs to be incorporated in each node for the store-and-forward mechanism using buffers.

Figure 6.25 shows the plot of the expected number of hops $E[h]$ versus number of nodes N in MSNet (with $U = V$), which increases steadily with increasing number of nodes. This is expected because, unlike other multihop topologies, the logical degree of MSNet nodes is always fixed at two, thereby reducing the possibility of connections using fewer hops with the increasing number of nodes. For the same reason, as shown in Fig. 6.26, the per-node throughput of MSNet falls significantly with the increase in number of nodes.

Routing schemes

Routing in MSNet can be realized by using the available shortest paths through the two sets of rings. In the event of contention for a specific output port at a node, the packets can be deflected to the other port of the node, leading to deflection routing. Several routing schemes have been reported in the literature (Maxemchuk 1985, 1987; Ayanoglu 1989; Lee and Kung 1995), and we describe briefly one of these schemes in the following (Maxemchuk 1987).

Owing to the isotropic topology of MSNet, the routing operation can be facilitated by mapping the absolute address of a given destination node into a central address, i.e., (0,0),[2] and considering thereafter the incremental part of the absolute address of the source node

[2] Note that one can also choose a source node as the central node and allocate it a central address, i.e., (0,0) (Maxemchuk 1985).

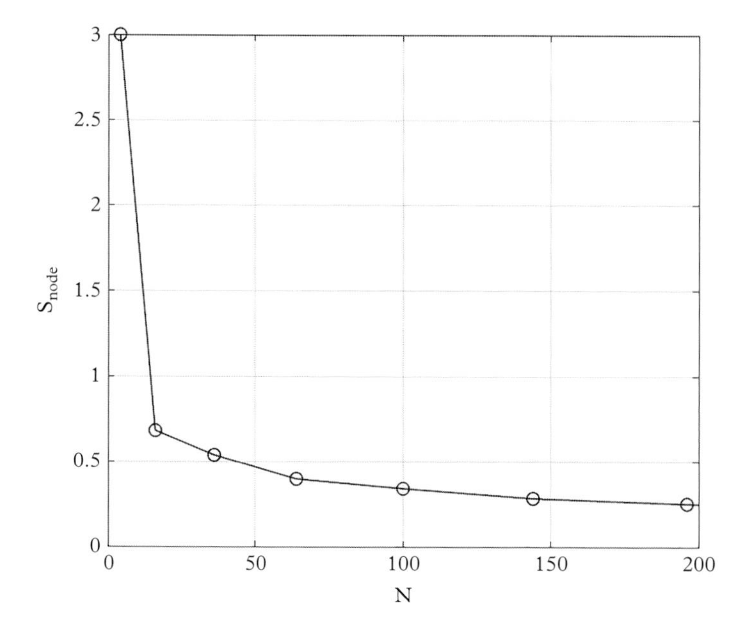

Figure 6.26 *Per-node throughput S_{node} versus number of nodes N in MSNet.*

(with respect to the destination node) as its relative address. In other words, the destination node can be considered as a central node (0,0), logically implying that the destination node is located at the intersection of the 0th row and the 0th column. Consequently, the other rows and columns are assigned numbers in the ranges of $[-(U/2-1), U/2]$ and $[-(V/2-1), V/2]$, respectively (U and V assumed to be even numbers). Thus, each of the remaining nodes, playing the role of candidate source nodes and represented by a generic source node address (r_s, c_s), can be assigned a relative address $(\hat{r}_s, \hat{c}_s)$, with $\hat{r}_s \in [-(U/2-1), U/2]$ and $\hat{c}_s \in [-(V/2-1), V/2]$. Based on this observation, for a given pair of source and destination nodes with the addresses (r_s, c_s) and (r_d, c_d) respectively, one can set the destination address (r_d, c_d) at the logical central address $(0,0)$, and determine the relative address $((\hat{r}_s, \hat{c}_s)$ of the source node (r_s, c_s) as

$$\hat{r}_s = \frac{U}{2} - \left[\left\{\frac{U}{2} - D_c(r_s - r_d)\right\} mod\, U\right] \tag{6.58}$$

$$\hat{c}_s = \frac{V}{2} - \left[\left\{\frac{V}{2} - D_r(c_s - c_d)\right\} mod\, V\right], \tag{6.59}$$

where the direction variables D_r and D_c are ± 1, depending on the directions of the links.

Note that, the destination node cannot be exactly at the center of the network (unless both U and V are odd numbers), and hence it is placed at the innermost corner of one of the four quadrants, such that the 0th row and the 0th column pass through this node itself. As shown in Fig. 6.27, the network would consist of four quadrants Q_1, Q_2, Q_3, and Q_4, and a given source node with a relative address $(\hat{r}_s, \hat{c}_s)$ will be located in:

1. quadrant Q_1, if $\hat{r}_s > 0$ and $\hat{c}_s > 0$,
2. quadrant Q_2, if $\hat{r}_s \leq 0$ and $\hat{c}_s > 0$,
3. quadrant Q_3, if $\hat{r}_s \leq 0$ and $\hat{c}_s \leq 0$,
4. quadrant Q_4, if $\hat{r}_s > 0$ and $\hat{c}_s \leq 0$.

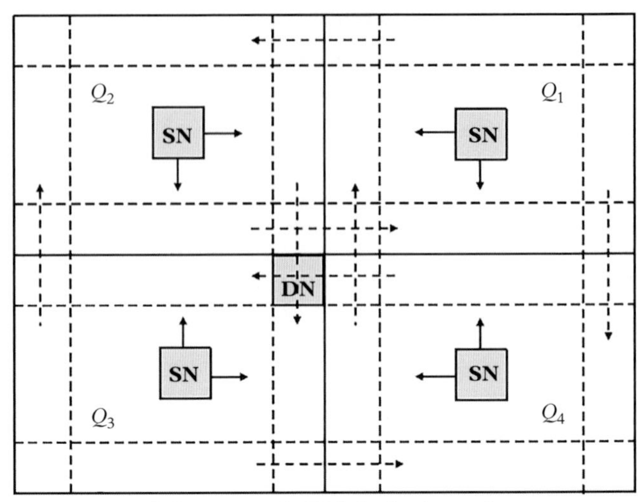

Figure 6.27 *Routing in MSNet using four-quadrant representation based on relative destination-centric addressing scheme, implying that the destination node (DN) is assigned (0,0) as its address. Note that the dashed arrows represent a typical set of directions of packet flow in the respective rows and columns. These directions should comply with the MSNet rule that any two neighboring rows or columns (including those around the outer edges, i.e., top/bottom or left/right) must have opposite directions of data flow.*

A packet from a source node in one of the above four quadrants should be transmitted toward the destination node (0,0), by using the links in appropriate horizontal/vertical directions, as shown by the solid arrows in Fig. 6.27. From the relative location of the source node, if located in quadrant Q_1, the preferred link should be directed downward or toward the left side. Similarly for the source nodes located in other three quadrants, the packets should move toward the centrally located destination node at (0,0) in appropriate directions. Thus, all packets start their inward traversals from the source node and move through intermediate node(s), if necessary, to eventually reach the destination node with the relative address of (0,0).

6.4.3 Hypercube networks

In a geometric sense, a hypercube is a closed space that can evolve through n-dimensional generalization, starting simply from a point (a hypercube with zero dimension, i.e., $n = 0$) through a line with $n = 1$ by moving the point, a square with $n = 2$ by moving the line, a cube (or a 3-cube) with $n = 3$ by moving the square, a tesseract (or a 4-cube) with $n = 4$ by moving the cube and so on. In networking, while seeking a suitable interconnection topology, we conceive the hypercube as a class of regular topologies (Li and Ganz 1992; Mukherjee 1992b; Bhuyan and Agarwal 1984) with each of its vertices representing a network node. In this section, we first consider one of its simple forms, binary hypercube (BHC), leading to a set of multihop network topologies, hereafter referred to as binary hypercube network (BHCNet). Next, we present a generalized version of BHCNet, called a generalized hypercube network (GHCNet).

BHCNet

A p-dimensional BHCNet is composed of $N = 2^p$ nodes, where each node is assigned a p-bit binary address, ranging from 0 to $(2^p - 1)$. The nodes in a BHCNet are connected following a specific interconnection pattern so that there exists a bidirectional logical link between two nodes if their binary addresses differ by only one bit. With the logical degree of the nodes as $p \geq 3$, BHCNets offer in general small network diameter, but the choices for number of nodes lack flexibility as the difference between any two successive values of N must differ by the power of 2.

Topological characteristics and network performance

The minimum hop distance between a given node pair in a BHCNet, say nodes i and j, is the number of bits that are different between the binary addresses of these two nodes, which is effectively the Hamming distance $H(i,j)$ between the node addresses. Figure 6.28 illustrates an 8-node BHCNet (i.e., $N = 8$) with a logical degree $p = \log_2 N = 3$. This BHCNet has a maximum hop distance $H = 3$ and from any given node the number of nodes with the hop distance $h = 1$ is given by $n_1 = 3$; similarly $n_2 = 3$, and $n_3 = n_H = 1$. With the probability P_h of reaching a destination node at a hop distance h as $n_h/(N-1)$ for uniform traffic pattern, we express the expected number of hops $E[h]$ for an 8-node BHCNet as

$$E[h] = \sum_{h=1}^{H} hP_h = \sum_{h=1}^{H} h\frac{n_h}{N-1} = 1 \times \frac{3}{7} + 2 \times \frac{3}{7} + 3 \times \frac{1}{7} = \frac{12}{7} \tag{6.60}$$

The above expression can be generalized for an N-node BHCNet with $N = 2^p$, p being an integer ≥ 3. For example, with $p = 4$ a 16-node BHCNet will have a logical degree $p = 4$, along with the same value (i.e., 4) for the maximum number of hops H between the node pairs. This topology can be drawn out by combining two 8-node BHCNets as shown in Fig. 6.29, where two identical 8-node BHCNets are placed side by side and every pair of nodes with the same binary address in the original 8-node BHCNets are connected by a single-hop arc. To differentiate the two nodes having the same address on both sides of the newly set up single-hop connection (shown by dotted arcs), (*i*) the binary address of the node from left BHCNet is appended with a 0 on the left side (i.e., on the left side of the most significant bit (MSB) of the original address) and (*ii*) the binary address of the node from right BHCNet is appended with a 1 on the left side of the MSB. For example,

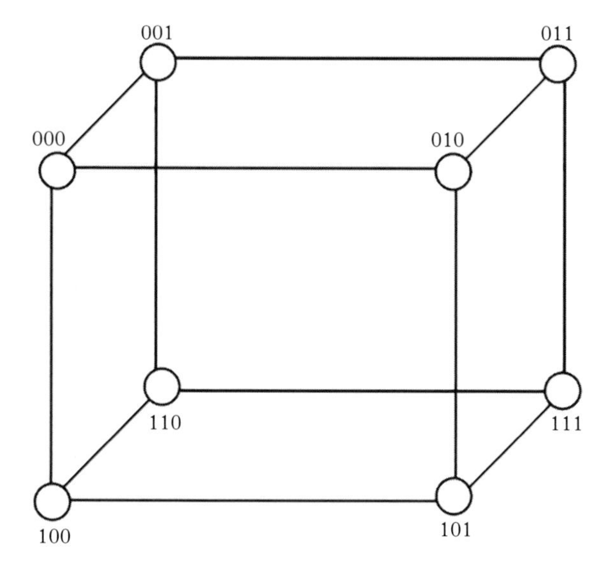

Figure 6.28 *8-node BHCNet with* $p = 3$.

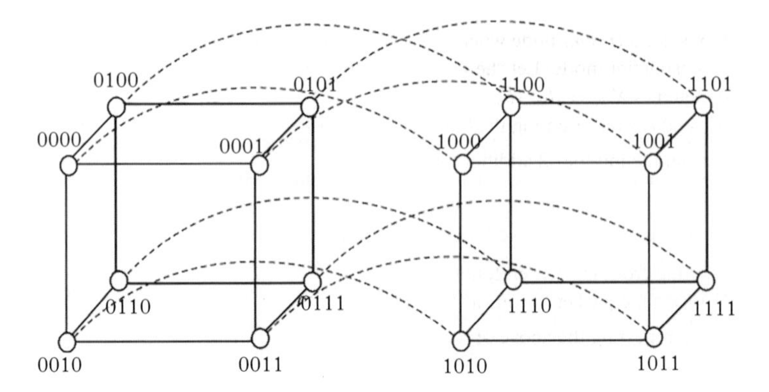

Figure 6.29 *16-node BHCNet; dashed lines indicate the links used to interconnect two 8-node BHCNets.*

the nodes with the address of (000) in both BHCNets now assume their new addresses as (0000) and (1000), and so on. This arrangement also complies with the original BHCNet node numbering rule, i.e., any two adjacent nodes having single-hop distance should have addresses differing only in one bit.

From the topology shown in Fig. 6.29, by observation one can easily find the values of n_h and P_h as

$$\begin{aligned} n_h &= 4, P_h = 4/15, \text{ for } h = 1 \\ n_h &= 6, P_h = 6/15, \text{ for } h = 2 \\ n_h &= 4, P_h = 4/15, \text{ for } h = 3 \\ n_h &= 1, P_h = 1/15, \text{ for } h = 4. \end{aligned} \tag{6.61}$$

Using these values of P_h and assuming uniform traffic pattern, we obtain $E[h]$ for a 16-node BHCNet as

$$E[h] = \sum_{h=1}^{H} hP_h = \sum_{h=1}^{H} h\frac{n_h}{N-1} = \frac{32}{15} \tag{6.62}$$

From the above two example cases (8-node and 16-node BHCNets), one can generalize the expression of $E[h]$ for N-node BHCNet by induction as

$$E[h] = \frac{N\log_2 N}{2(N-1)} = \frac{p \times 2^{p-1}}{2^p - 1}. \tag{6.63}$$

While BHCNet is a desirable topology with reasonable values for $E[h]$ and ease of routing with plenty of multiple minimum-path routes between the node pairs, its number of nodes increases exponentially with the logical degree p of the nodes.

Routing scheme

BHCNet follows a simple self-routing scheme, which is based on the node-numbering process in hypercubes. First, we number the output ports of each node (as a source node) in a manner such that the link i goes to the node whose binary address differs from that of the source node in the position i. With this framework, the self-routing scheme in a BHCNet is carried out as follows.

- ◇ Consider a *current* node where a packet under consideration needs to move toward its destination node. Let the addresses of the current and the destination nodes be denoted as $X_{cur} = \{b_{p-1}b_{p-2}\cdots b_1 b_0\}$ and $X_d = \{c_{p-1}c_{p-2}\cdots c_1 c_0\}$, respectively, with b_j's and c_j's representing the binary digits in the respective binary addresses.
- ◇ Obtain the modulo-2 addition of X_{cur} and X_d as Y, given by

$$Y = X_{cur} \oplus X_d. \tag{6.64}$$

- ◇ Starting from the left end, note the first non-zero (i.e., = 1) bit (say, ith bit) in Y. Forward the packet for onward transmission on the ith link.
- ◇ Keep repeating the above procedure until the packet reaches its destination.

GHCNet

GHCNet is a generalization of BHCNet with bidirectional logical links, offering richer connectivity between the nodes and more flexible choice for choosing the number of nodes in the network. The node addresses in GHCNet do not follow the fixed-radix numbering system, as used in BHCNet (binary addresses); instead, its constituent nodes are assigned addresses using a *mixed-radix* numbering system, which in turn brings in some useful features (Bhuyan and Agarwal 1984). In order to use this numbering system, the number of nodes N in a GHCNet is represented in a product form, given by

$$N = \prod_{k=1}^{r} m_k, \tag{6.65}$$

with m_k representing positive integers > 1 with $k \in [1, r]$. Any given node (say node i) can be assigned an r-tuple address X_i (instead of binary address used in BHCNet), given by

$$X_i = \{\alpha_r, \alpha_{r-1}, \cdots, \alpha_k, \cdots, \alpha_2, \alpha_1\}, \tag{6.66}$$

where α_k is defined as

$$\alpha_k = (0, 1, \cdots, m_k - 1), \text{ for } k = 1, 2, \cdots r, \tag{6.67}$$

and α_k for each k is assigned a specific radix. This r-tuple address X_i can be transformed into a mixed-radix-based decimal address $X_i^{(10)}$, given by

$$X_i^{(10)} = \sum_{l=1}^{k} w_k \alpha_k, \tag{6.68}$$

with w_k as the weight associated with each α_k, given by

$$w_k = \begin{cases} 1 \text{ for } k = 1 \\ \prod_{j=1}^{k-1} m_j \text{ for } k > 1. \end{cases} \tag{6.69}$$

Thus a node (say node i) in a GHCNet will be referred to by either an r-tuple X_i or by its equivalent mixed-radix-based decimal number $X_i^{(10)}$.

It would be worthwhile to visualize the above addressing scheme using a three-dimensional GHCNet. For example, consider that $r = 3$, indicating that the nodes will have

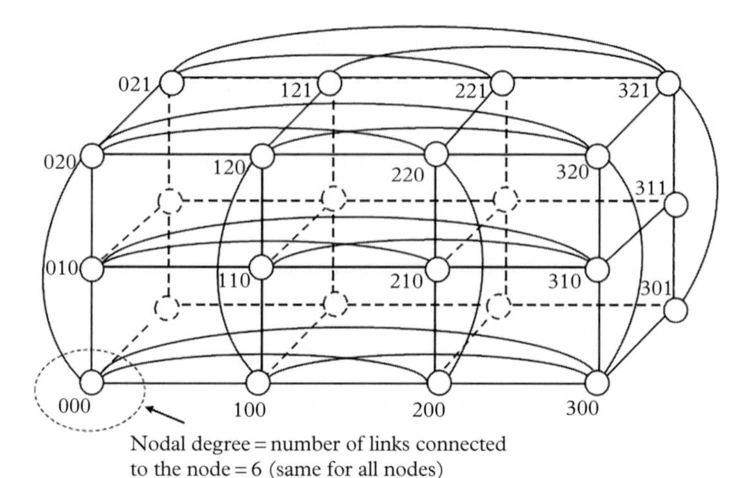

Figure 6.30 *24-node GHCNet with $r = 3$ and $m_1, m_2, m_3 = 2, 3, 4$, respectively; dashed lines and circles are used for some links and nodes on the rear plane but all the links for these nodes are not shown to avoid complication.*

three-tuple addresses, leading to a three-dimensional topology with $N = m_1 \times m_2 \times m_3$. Next, we assume that $m_1 = 2, m_2 = 3, m_3 = 4$, thereby implying that the three-tuple node address X_i for node i will be expressed as

$$X_i = \{\alpha_3, \alpha_2, \alpha_1\}, \tag{6.70}$$

with $\alpha_1 \in [0, m_1 - 1]$, $\alpha_2 \in [0, m_2 - 1]$ and $\alpha_3 \in [0, m_3 - 1]$, implying that $\alpha_1 = 0, 1$, $\alpha_2 = 0, 1, 2$, and $\alpha_3 = 0, 1, 2, 3$. This leads to a three-dimensional cubic structure having 2, 3, and 4 nodes in the three dimensions, as shown by the cube formation with $2 \times 3 \times 4 = 24$ nodes in Fig. 6.30. Using the values of m_j in Eq. 6.69, one can easily determine the values for w_j's as

$$w_1 = 1, w_2 = m_1 = 2, w_3 = m_1 m_2 = 6. \tag{6.71}$$

The decimal-based addressing of node i (with the mixed-radix-based three-tuple address as $X_i = \{\alpha_3, \alpha_2, \alpha_1\}$ will then be given by

$$X_i^{(10)} = w_3\alpha_3 + w_2\alpha_2 + w_1\alpha_1 = 6\alpha_3 + 2\alpha_2 + \alpha_1. \tag{6.72}$$

Thus the three-tuple addresses, (000) and (321), will correspond to the 0th and 23rd nodes, respectively, in the network topology shown in Fig. 6.30.

Topological characteristics and network performance

In order to understand the topological characteristics, we next consider the interconnection pattern used between the nodes in a GHCNet. The basic interconnection rule is a generalization of the rule used in BHCNet, i.e., each node pair connected by one hop (bidirectional logical link) should differ only by one element in their r-tuple addresses, so that there is only one "un-matched element" between the two r-tuple addresses. Also these two elements should only differ by one. As we shall see later, this feature will be helpful in facilitating the routing mechanism. The above interconnection rule can be expressed succinctly using the r-tuple addresses of the two connected nodes as follows:

◇ A given node (node i, say) located at the address $X_i = \{\alpha_r, \alpha_{r-1}, \cdots, \alpha_{k-1}, \alpha_k, \alpha_{k+1}, \cdots, \alpha_2, \alpha_1\}$ can be directly connected by one hop to another node (node j, say) at an address $X_j = \{\alpha_r, \alpha_{r-1}, \cdots, \alpha_{k-1}, \alpha'_k, \alpha_{k+1}, \cdots, \alpha_2, \alpha_1\}$, for all $k \in [1, r]$ provided that α'_k takes any possible integer value $\in [0, m_k - 1]$, *excepting* α_k. This leads to, as desired, a mismatch in only one position (i.e., at kth position) between the two r-tuples. Further, α_k and α'_k should only differ by one, i.e., $|\alpha_k - \alpha'_k| = 1$.

Using this rule, the GHCNet of Fig. 6.30 realizes one-hop connections between the candidate node pairs. In the figure, to avoid congestion, on the rear plane of the topology, dashed lines are used for some links and dashed circles are used for some nodes (without numbering), and the connections for the nodes (with dashed circles) are not shown. The salient features of the interconnection pattern are as follows.

- A GHCNet spans over r dimensions: $1, 2, 3, \cdots, k, \cdots, r$, with m_k nodes in kth dimension (axis). Each of the m_k nodes along kth axis is directly connected to all the remaining $(m_k - 1)$ nodes on the same axis.
- For any given node in a GHCNet, there are $(m_k - 1)$ links in the kth dimension (axis) going to as many neighbors on the same axis. For example, from the node at (000), there are 3 + 2 + 1 = 6 links going to the six neighbors on the three axes for three dimensions, and the same is true for all the nodes in the network. Thus, considering all the r dimensions and as many logical links emerging from the node under consideration, the total number of logical links connected to a node, i.e., the logical degree p of a node, is given by

$$p = \sum_{k=1}^{r} (m_k - 1). \tag{6.73}$$

- The total number of wavelengths required is governed by the product of p and the total number of nodes, given by

$$M = Np = \prod_{k=1}^{r} m_k \sum_{k=1}^{r} (m_k - 1). \tag{6.74}$$

- Since each link is connected to two nodes, the total number of links in a GHCNet is $N_L = pN/2$.
- The number of nodes at one-hop distance from a given source node in one of the r dimensions (say, kth dimension) = $(m_k - 1)$. This feature stems from the fact that there are exactly $(m_k - 1)$ nodes whose kth digits in their r-tuple addresses can differ by one from the r-tuple address of the given source node. Hence, the total number of such nodes at one-hop distance from the source node in all the r dimensions will be given by

$$n_1 = (m_1 - 1) + (m_2 - 1) + \cdots + (m_k - 1) + \cdots + (m_r - 1) = \sum_{k=1}^{r} (m_k - 1), \tag{6.75}$$

which obviously equals to p (Eq. 6.73). Similarly, the total number of nodes n_2 at two-hop distance from a given source node is given by

$$n_2 = \sum_{i=1}^{r}\sum_{j=1}^{r}(m_i - 1)(m_j - 1), i \neq j. \tag{6.76}$$

Generalizing the above expressions, the total number of nodes at a hop distance of h from a given source node can be expressed as

$$n_h = \sum_{i=1}^{r}\sum_{j=1}^{r}\sum_{k=1}^{r}\underbrace{(m_i - 1)(m_j - 1)(m_k - 1)\ldots}_{h \text{ factors, each representing one hop}}, i \neq j \neq k \neq \ldots, \tag{6.77}$$

where the right-hand side will have $\binom{r}{h}$ terms. Using this expression, and noting that the maximum value of h (i.e., H) equals r in GHCNet, one can express the expected number of hops $E[h]$ as

$$E[h] = \frac{\sum_{h=1}^{H=r} hn_h}{(N-1)}. \tag{6.78}$$

For a quick sanity check, let's consider an eight-node BHCNet (a special case of GHCNet) with $r = 3$. So, $m_1 - 1 = m_2 - 1 = m_3 - 1 = 1$, giving $n_h = 1\times\, ^3C_h = 3$, 3 and 1 for $h = 1, 2$, and 3 respectively. This leads to the expected number of hops for an eight-node BHCNet as

$$E[h] = \sum_{h=1}^{H} h\frac{n_h}{N-1} = 1 \times \frac{3}{7} + 2 \times \frac{3}{7} + 3 \times \frac{1}{7} = \frac{12}{7}, \tag{6.79}$$

which is the same as our earlier result for BHCNet, obtained by observation in Eq. 6.60. The expression for $E[h]$ in Eq. 6.78 is used to obtain the network and per-node throughputs (S, S_{node}, respectively), given by

$$S = \frac{M}{E[h]}, \quad S_{node} = \frac{S}{N}. \tag{6.80}$$

By using these results, we present in Figures 6.31 and 6.32 the plots of $E[h]$ and per-node throughput S_{node} for GHCNet versus number of nodes N. As evident from the plots, GHCNet with its flexibility in the logical degree of the nodes can achieve high per-node throughput and low delay as compared to other multihop WDM LANs discussed so far, but at the cost of increased number of transceivers per node.

Routing schemes

In order to implement routing in GHCNet, each element of all the r-tuple addresses of the network is converted into a binary word. Thus, the kth element of a node address is converted into a binary number having $\lceil \log_2 m_k \rceil$ binary digits or bits. The r-tuple address of each node can therefore be represented by B_r bits, given by

$$B_r = \sum_{k=1}^{r} \lceil \log_2 m_k \rceil. \tag{6.81}$$

Note that, for a given source destination pair, if the shortest route has h hops, then their addresses will also differ in h positions (i.e., h digits will be different out of r digits). Based on this feature, the routing scheme is carried out as follows.

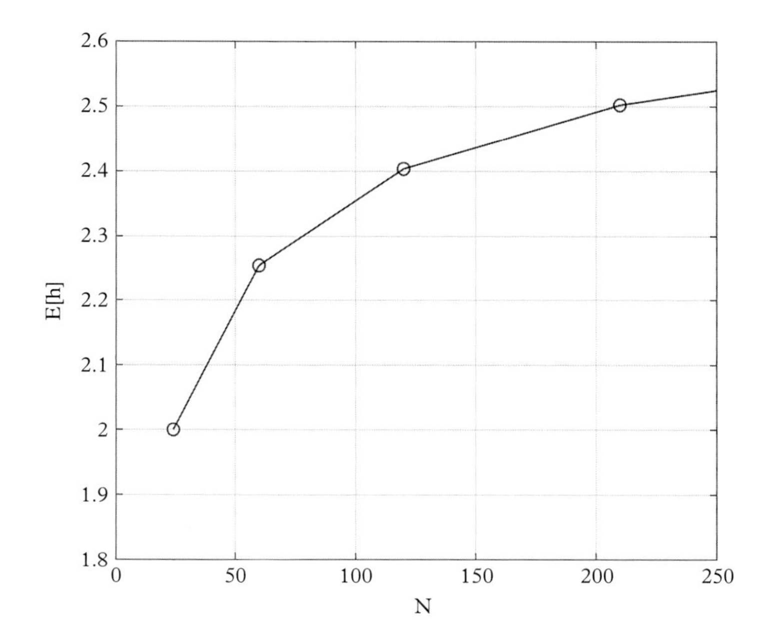

Figure 6.31 *Expected number of hops $E[h]$ versus number of nodes N in GHCNet. $r = 3$, (m_1, m_2, m_3) = (2,3,4), (3,4,5), (4,5,6), and (5,6,7) for the four results (circles on the plot from the left side) in the increasing order of number of nodes.*

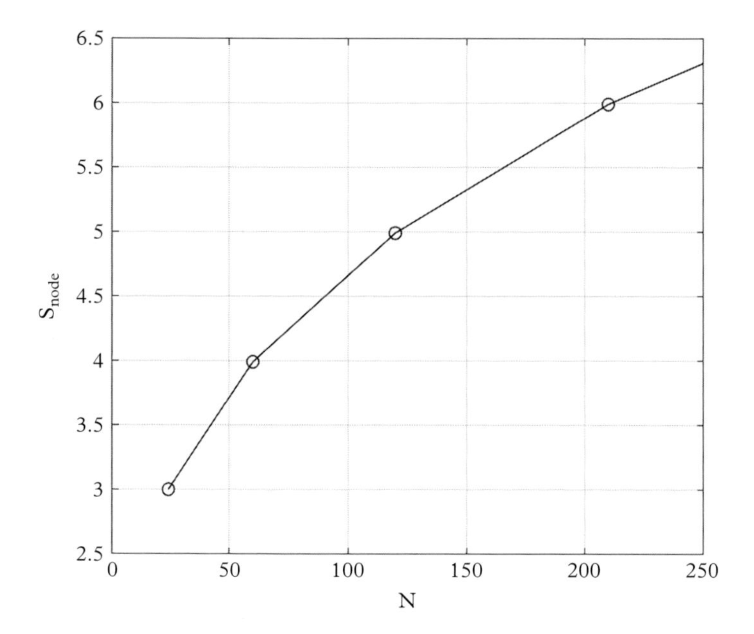

Figure 6.32 *Per-node throughout S_{node} versus number of nodes N in GHCNet. $r = 3$, (m_1, m_2, m_3) = (2,3,4), (3,4,5), (4,5,6), and (5,6,7) for the four results (circles on the plot from the left side) in the increasing order of number of nodes.*

- ◇ Consider a current node and compare its address X_{cur} digit by digit with the address of the destination node X_d. In other words, the pair of binary words representing the digits in same position of the addresses are subjected to modulo-2 addition, and the position(s) with non-zero result are found out. With h as the hop distance between the two nodes, the addresses will differ in h digits (corresponding to the Hamming distance between the two nodes).
- ◇ Pick up the first digit with a mismatch, say the jth digit, and route the packet in the direction (link) represented by the jth digit.
- ◇ Keep repeating the above procedure until the packet reaches its destination, where all digits of the packet's current address match all digits of the destination address.

Note that there are h disjoint paths of equal length of h hops if the Hamming distance is h for a given node pair. So there will be h alternate paths, which can be used when one of the next nodes corresponding to the available paths is busy. Thus one can choose readily an alternate path in GHCNet without deflection. In other words, the logical degree of nodes being large as compared to ShuffleNet and MSNet, GHCNet offers more alternate paths, but at the cost of an increased number of transceivers in each node.

6.4.4 de Bruijn networks

Another useful multihop WDM LAN with a *directed* regular topology is de Bruijn network (dBNet). In a (p, D) dBNet, each node is assigned an address using D digits, and each of these digits is chosen from a *set* of p integers, i.e., $0, 1, 2, \ldots, p-1$. For any given node, say node i, with an address $A_i = \{\alpha_1, \alpha_2, \cdots, \alpha_D\}$, the address of its neighboring nodes (i.e., the nodes reachable through one logical link) is obtained by a left-shift of the digits in A_i, which moves α_2 to the left-most position of the new address and the right most digit is replaced by one of the p digits. Thus, the address A_j of a neighboring node, say node j, is given by

$$A_j = \{\beta_1, \beta_2, \cdots, \beta_D\} = \{\alpha_2, \alpha_3, \cdots, \alpha_D, x\}, \tag{6.82}$$

where $\beta_D = x$ can assume any value from $(0, 1, 2, \cdots, p-1)$. However, for a node with all digits being the same (i.e., $\alpha_1 = \alpha_2 = \cdots = \alpha_D = \rho$, say), this left-shift operation will lead to a neighboring address, given by

$$A_j = \{\rho, \rho, \cdots, \rho, x\}, \tag{6.83}$$

which will result in self-looping around the node with $x = \rho$. These characteristics lead to (p, D) dBNet topologies with p^D nodes, where a few nodes (two nodes with $p = 2$) exhibit self-looping around themselves. An example (p, D) dBNet with $p = 2$ and $D = 3$ is shown in Fig. 6.33, wherein the nodes 0 and 7 at the addresses (000) and (111), respectively, exhibit self-looping.

Topological characteristics and performance

In a practical network-setting, self-looping around the nodes in dBNet graphs are removed, and the corresponding nodes become the exceptions having a logical degree of $p-1$, while the rest of the nodes have a logical degree of p. For example in the $(2, 3)$ dBNet shown in Fig. 6.33, with the removal of self-looping, nodes 0 and 7 will be left with a logical degree of 1, while the rest of the nodes will have a logical degree of 2. The interconnection scheme in

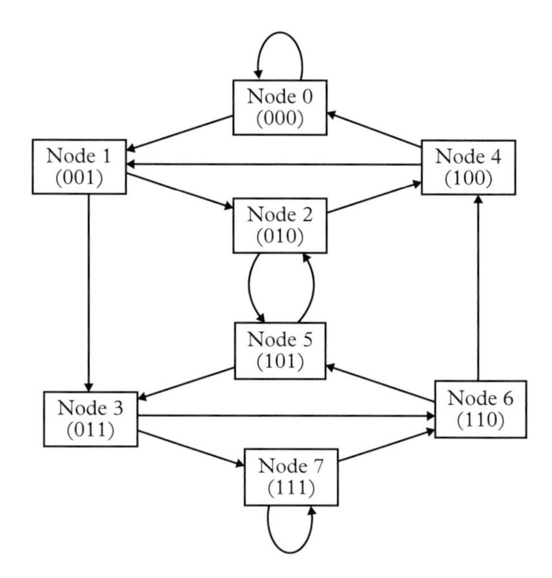

Figure 6.33 *(2,3) dBNet.*

dBNet leads to a maximum hop count H between a node pair, given by $H = D$. Thus, for the $(2, 3)$ dBNet shown in Fig. 6.33, the maximum hop count turns out to be 3.

Next we estimate the expected number of hops $E[h]$ in a dBNet. Noting that the network has $N = p^D$ nodes with D as the maximum hop count between a node pair, and following the spanning-tree model used earlier while obtaining Eq. 6.29, we express $E[h]$ in a (p, D) dBNet as

$$E[h] = \left[\frac{p - p^{H+1} + NH(p-1)^2 + H(p-1)}{(N-1)(p-1)^2}\right]_{H=D,\, N=p^D} \tag{6.84}$$
$$= \frac{p - p^{D+1} + p^D D(p-1)^2 + D(p-1)}{(p^D - 1)(p-1)^2}, \text{ for } p > 1.$$

Note that Eq. 6.84 gives a lower bound for $E[h]$ because in a dBNet some nodes (p nodes, in particular) will effectively have lesser logical degree ($< p$) due to self-looping around themselves. Using Eq. 6.84, the network and per-node throughputs are obtained as

$$S = \frac{M}{E[h]} = \frac{Np - n_{SL}}{E[h]}, \quad S_{node} = \frac{S}{N} = \frac{Np - n_{SL}}{NE[h]}, \tag{6.85}$$

with n_{SL} representing the number of nodes with self-looping, implying that $n_{SL} = p$.

Figure 6.34 presents the plot of the expected number of hops in dBNet versus number of nodes. The plot indicates that, the expected number of hops increases fast at lower values of N and thereafter attains an upper-bound with increasing number of nodes, but with an increase in the logical degree of nodes. Figure 6.35 shows the plot of per-node throughput versus number of nodes, which increases gradually with the increase in number of nodes.

Routing scheme

For dBnets having regular topology, one can again employ a self-routing scheme by utilizing the topological characteristics. We explain in the following the self-routing scheme for

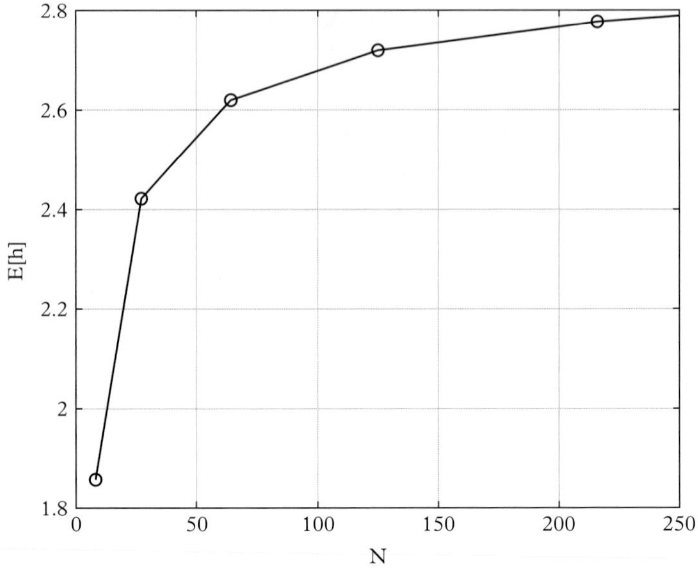

Figure 6.34 *Expected number of hops $E[h]$ versus number of nodes N in dBNet. $D = 3$, and p is increased for increasing the number of nodes N.*

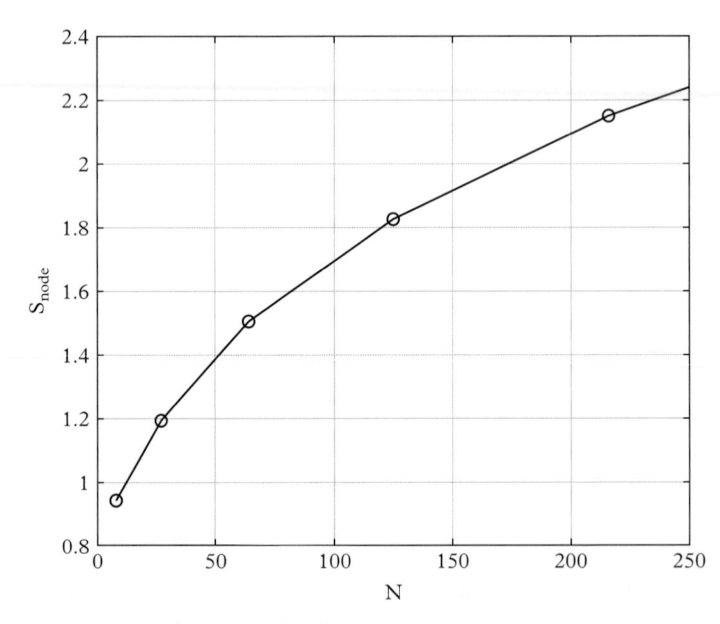

Figure 6.35 *Per-node throughput S_{node} versus number of nodes N in dBNet. $D = 3$, and p is increased for increasing the number of nodes N.*

dBNet (Sivarajan and Ramaswami 1991). Consider two nodes, node i and node j, with the addresses $A_i = \{\alpha_1, \alpha_2, \cdots, \alpha_D\}$ and $A_j = \{\beta_1, \beta_2, \cdots, \beta_D\}$, respectively, and assume that a packet has to be transmitted from node i to node j. The self-routing scheme is effected by using two operations *shift-and-match* followed by *merge* operation on the source and destination addresses, as follows.

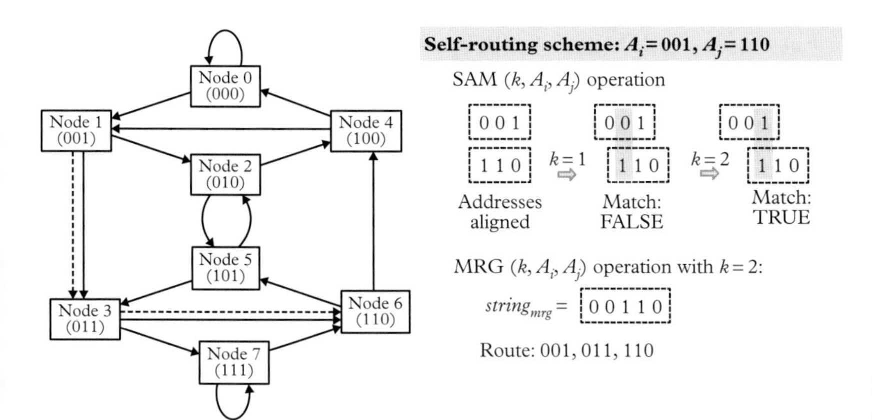

Figure 6.36 *Routing scheme in a (2,3) dBNet with eight nodes.*

- *Shift-and-match (SAM) operation:* Note that the network addresses are D-digit strings, with each digit having one of the p possible integer values. In this operation, the destination address is *shifted* with respect to the source address toward the right by $k \in [1, D-1]$ positions and thereafter the overlapping digits are compared (see Fig. 6.36). We denote this operation as $SAM(k, A_i, A_j)$, which takes effect as follows.
  - ◇ Start with $k = 1$ and execute $SAM(k, A_i, A_j)$. If the operation gives a match for all the overlapping digits, the result is TRUE, else the result is FALSE.
  - ◇ If the result is FALSE, increment k to $k + 1$, and execute $SAM(k, A_i, A_j)$. Repeat the process until a TRUE result is obtained and move to the next step. If no match is found even after $D - 1$ shifts, then also move to next step. Note that the match can take place at the most for $D - 1$ digits, or there can be no match at all.
- *Merge (MRG) operation:* Having executed the $SAM(k, A_i, A_j)$ operation, the MRG operation, denoted as $MRG(k, A_i, A_j)$, is carried out as follows.
  - ◇ In the case of a TRUE match, append the last $D - k$ digits of A_j on the right side of the D digits of A_i, thereby creating the merged string $string_{mrg} = \{\alpha_1, \alpha_2, \ldots, \alpha_D, \beta_{D-k+1}, \ldots, \beta_D\}$. If no match is found even after all the D shifts, append all the D digits of A_j on the right side of the D digits of A_i, creating thereby a $2D$-digit merged string, given by $string_{mrg} = \{\alpha_1, \alpha_2, \ldots, \alpha_D, \beta_1, \beta_2, \ldots, \beta_D\}$.
  - ◇ The merged string in both cases describes the complete route with first D digits as the source address, the next D digits counted from the second position as the next-hop node address and so on, finally with last D digits as the destination address.

We demonstrate the above routing scheme using an example taken from (2,3) dBNet of Fig. 6.36. Consider nodes 1 and 6 at the addresses (001) and (110) as the source and destination addresses A_i and A_j, respectively. The operation $SAM(k, A_i, A_j)$ is TRUE with $k = 2$. Hence $MRG(k, A_i, A_j)$ operation generates $string_{mrg}$ = (00110), indicating that the route follows the node sequence: 001 → 011 → 110.

6.4.5 Comparison of regular multihop topologies

In the preceding sections, we have considered some regular topologies for multihop WDM LANs, each one having its merits and limitations, but with the common simplicity of self-routing capability. We summarize in this section the salient features of these networks in respect of relevant network performance metrics.

First, we begin with ShuffleNet, which employs directed logical links and offers fixed logical degree of nodes in a given network, but can have different logical degrees of nodes for different network sizes. ShuffleNet offers alternate choices for routes between some of its node pairs, which are separated by large hop counts, thereby allowing alternate routes in presence of network congestion. Next we consider MSNet, which has two sets of rings (vertical and horizontal) with directed logical links. MSNet has a unique feature of the fixed logical degree of two for each node for any network size, along with the edge-less isotropic topology. Owing to its simplicity in node configuration, it has limited options for available routes between its node pairs, leading to large values of the expected number of hops, as compared to ShuffleNet. However, in ShuffleNet a given node pair can have a significant difference between the hop counts in two directions, giving an asymmetric delay between those node pairs. The choice for the number of nodes is restricted in both cases (and so also for all regular topologies), in ShuffleNet by the nodal degree p and the number of hops k, and in MSNet by the fixed logical degree (two) of nodes and the numbers of rows and columns.

The network topologies under hypercube classes (BHCNet and GHCNet) use bidirectional logical links. In BHCNet, the logical degree of nodes varies logarithmically with the network size (i.e., number of nodes). BHCNet has a small value for the maximum hop distance = $\log_2 N$. GHCNet offers a generalized version of BHCNet with a mixed-radix numbering system, allowing more flexible choices for the number of nodes. However, GHCNets while offering low values for the expected number of hops and high throughput, would require a large number of wavelengths with high logical degrees of nodes.

Next, we consider the main features of dBNet employing unidirectional logical topology. With a large logical degree of nodes, $E[h]$ in dBNet reaches an upper bound which is close to the theoretical limit in an arbitrary directed topology. However, in dBNet some of the nodes – the ones located at the addresses $(i, j, k, \dots)$ with $i = j = k = \dots$, have lower logical degree (precisely, by one) due to self-looping, as compared to the rest of the nodes in the network, thereby creating imbalance in traffic sharing between the logical links. Nevertheless, with larger network sizes, dBNet performs better than ShuffleNet, as the latter suffers from large hop counts between several node pairs. Further, the performance of dBNet comes closer to that of GHCNet, but needing fewer wavelengths.

To have an insightful understanding of the above multihop topologies, we next make an overall comparison between them in respect of some performance metrics for a specific network size $N = 64$, which happens to be a valid choice for all the four regular topologies, as shown in the following:

- ShuffleNet: $p = 2, k = 4, N = kp^k = 64$,
- MSNet: $U = V = W = 8, N = W^2 = 64$,
- GHCNet: $(m_1, m_2, m_3) = (8, 4, 2), N = m_1 \times m_2 \times m_3 = 64$,
- dBNet: $p = 4, D = 3, N = p^D = 64$.

With the number of nodes as the common platform, we next compare the above four topologies with 64 nodes in respect of the following metrics:

- ◇ expected number of hops $E[h]$,
- ◇ per-node throughput S_{node},
- ◇ logical degree of nodes p.

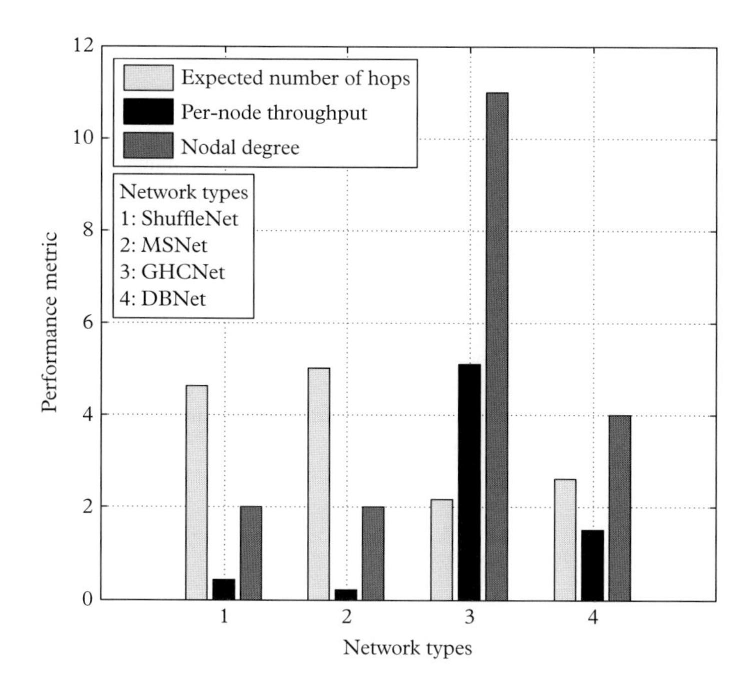

Figure 6.37 *Comparison of different multihop WDM LAN topologies with 64 nodes.*

The comparison is made using a bar chart in Fig. 6.37, where each network is characterized by three bars for the above three performance metrics. As evident from the figure, GHCNet offers the highest per-node throughput (=5.1) and the lowest expected number of hops (=2.16), but with the highest nodal degree (=11). In particular, the 64-node GHCNet requires 704 wavelengths (11 wavelengths in each of the 64 nodes) which is indeed exorbitantly high and needs a channel spacing as small as 35.5 GHz (=25, 000 GHz/704) in the 200 nm ($\simeq$25, 000 GHz) fiber transmission window around 1550 nm. However, this problem can be addressed through the shared use of wavelengths between the network nodes, unless the prevailing traffic increases the number of collisions, thereby reducing the throughput below an acceptable limit. However, one 64-node dBNet offers a per-node throughput of 1.5 (which is much higher as compared to the equivalent ShuffleNet or MSNet, but lower than the corresponding GHCNet), along with an average logical nodal degree of $\simeq$4, thus becoming a reasonably good choice (for this case study) among the other options, that are discussed here.

6.4.6 Non-regular multihop topologies

Multihop networks can also employ *non-regular* logical topologies instead of resorting to specific regular topologies, so that one can optimize the network performance for an arbitrary number of network nodes and non-uniform traffic flows between them. The design methodology should explore an appropriate logical topology, which can minimize the congestion, i.e., the maximum traffic that would flow in one (or more) logical link(s) in the network, or the network-wide average delay between the node pairs (Labourdette and Acampora 1991; Bannister *et al.* 1990; Iness *et al.* 1995). Such problems can be solved by using linear programming (LP), more precisely by using a mixed-integer LP (MILP), with appropriate network constraints formulated using linear expressions including equalities and/or inequalities along with a suitable objective function. Readers are referred, if necessary, to Appendix A for a primer on the LP-based optimization technique.

The problem of LP-based design optimization for multihop networks considers the network traffic as a set of multi-commodity flows over a graph with its vertices and edges representing the nodes and the logical links, respectively. Thus, a multihop WDM LAN is represented by a graph $G(N, L)$, with N as the set of vertices and L as the set of edges of the graph. Each edge l_j represents a logical link (with $l_j \in L$) between two adjacent network nodes in the logical topology. Note that a logical link can exist from one node to another if one of the transmit wavelengths of the source node matches with one of the receive wavelengths of the destination node. As mentioned earlier, the design problem is expressed as a set of linear equalities and/or inequalities along with a given objective function. The set of constraints provides the solution space, which is searched exhaustively to minimize the given objective function, representing the maximum traffic flow over a logical link (i.e., congestion). In order to formulate the problem, we first represent the traffic in the network by an $(N \times N)$ traffic matrix, given by

$$\Lambda = \{\lambda_{sd}\} \tag{6.86}$$

where λ_{sd} is the traffic representing the number of packets arriving at node s to reach node d during a given average packet duration, with $s, d \in N$. The traffic λ_{sd} may get split into multiple routes composed of specific logical links, with each route passing through a different set of nodes. Denoting the part of the traffic λ_{sd} flowing through a logical link between the nodes i and j $(i, j \in N)$ by ψ_{ij}^{sd}, one can express the total traffic flow from node i to node j as $\psi_{ij} = \sum_{s,d} \psi_{ij}^{sd}$ for all i, j, implying that the traffic ψ_{ij} consists of the contributions from all possible node pairs $\{(s, d)\}$. The LP formulation will select some of the ψ_{ij}'s for assigning a dedicated wavelength, corresponding to a logical link b_{ij} assuming a value of 1. For the node pairs that won't be selected by the LP formulation, b_{ij} would be made zero. We also assume that each node uses p transmitters (FTs) and p receivers (FRs), leading to a logical nodal degree of p. Therefore, the numbers of inbound as well as outbound channels at a node should be equal to p, i.e., $\sum_j b_{ij} = p$ for all i and $\sum_i b_{ij} = p$ for all j.

Using the above framework, the multihop topology design problem can be represented by an MILP formulation with b_{ij}'s as the binary variables and ψ_{ij}'s as the real variables. However, for large numbers of nodes, solving the MILP problem in *one go* turns out to be computationally complex. In order to ease out this hurdle, the overall design is divided into three smaller subproblems: logical connectivity problem (LCP), traffic-routing problem (TRP), and reconnection problem using branch-exchange operation (RP-BE). The first subproblem, LCP, is performed once at the beginning by using an integer LP (ILP) formulation with fewer constraints than the MILP, and thereafter the design revolves around recursively through TRP (an LP-based formulation) and RP-BE (a heuristic scheme) until the congestion (the maximum traffic observed in some logical link(s)) is brought down within a specific upper-bound (= 1, in the present context). We describe these subproblems in the following.

First subproblem: LCP

The first subproblem, LCP obtains an initial connectivity between the nodes based on the logical-degree constraint (p) of the nodes, where we determine the logical links b_{ij}'s between the node pairs with the objective of maximizing the number of single-hop connections in the network. This leads to an ILP problem governed by the following objective function and constraints.

LCP objective function LCP attempts to maximize the number of single-hop connections between the node pairs, which is described as

$$\text{maximize} \sum_{i,j} b_{ij}. \tag{6.87}$$

LCP constraints

- The logical degree of the nodes should be equal to a fixed number p, i.e.,

$$\sum_{j} b_{i,j} = p \quad \forall i, \tag{6.88}$$

$$\sum_{i} b_{i,j} = p \quad \forall j, \tag{6.89}$$

- A node pair can have, at most, one logical link, which implies

$$b_{i,j} = 0 \text{ or } 1, \forall i, j, i \neq j. \tag{6.90}$$

The above LP problem is solved by using the Simplex algorithm (see Appendix A), which in turn offers a set of b_{ij}'s with non-zero values (i.e., =1), leading to a primary version of the logical topology.

Second subproblem: TRP

In this subproblem, the result of LCP is used as the starting point. First, we make use of an LP formulation, where we set the values of b_{ij}'s obtained from LCP as *constants*. Consequently, the original ILP gets transformed into an LP. Next, this LP for TRP is carried out with an objective function and a few constraints as follows.

TRP objective function The LP for TRP runs with the objective function to minimize the maximum traffic flow (i.e., congestion) through a logical link, expressed as

$$\text{minimize } max_{i,j}\left[\psi_{ij}\right]. \tag{6.91}$$

Note that ψ_{ij} would exist only for the logical links having $b_{ij} = 1$; the remaining ψ_{ij}'s will be zero.

TRP constraints

- *Flow conservation at a node:* At a given node that operates as an intermediate node for the traffic flows between a given node pair (from node s to node d, say), the difference between the incoming and outgoing traffic flows should be zero, as happens in a solenoid for the magnetic flux lines. At the source node (i.e., node s) transmitting the traffic flows (corresponding to λ_{sd}), the difference would be equal to the ingress traffic from the same node, i.e., λ_{sd}, while for a destination node (i.e., node d) only receiving the traffic flows (corresponding to λ_{sd}), the difference would be the negative of the total traffic received, i.e., $-\lambda_{sd}$. This constraint, also referred to as *solenoidality* constraint for the given node pair (s, d), is expressed as

$$\sum_{j} \psi_{ij}^{sd} - \sum_{j} \psi_{ji}^{sd} = \begin{cases} \lambda_{sd} & \text{for } i = s \\ -\lambda_{sd} & \text{for } i = d \\ 0 & \text{otherwise.} \end{cases} \tag{6.92}$$

- *Total flow over a link:* The sum of the traffic flows from all source-destination pairs through the logical link i-j can be expressed as

$$\psi_{ij} = \sum_s \sum_d b_{ij} \psi_{ij}^{sd} \quad \forall i, j. \tag{6.93}$$

- *Nonnegativity of flow variables:* The traffic flow variables are physically unacceptable if they go negative, i.e.,

$$\psi_{ij}^{sd}, \psi_{ij} \geq 0 \quad \forall i, j, s, d. \tag{6.94}$$

Note that the above formulation does not put any constraint on the congestion, and hence the minimized value of $max_{i,j}[\psi_{ij}]$ may exceed the capacity (=1) of the corresponding logical link. In the next subproblem, we address this issue with the help of some heuristic reconnections for the under-utilized logical links by using the branch-exchange scheme (RP-BE).

Third subproblem: RP-BE

RP-BE is called for when TRP generates a congestion $max_{i,j}[\psi_{ij}]$ that exceeds the link capacity, i.e., 1. The values of the traffic flow variables obtained in TRP for each logical link are arranged in a list, termed a link list (*LL*) in the decreasing order of traffic flows from the top. Thereafter the BE operation is carried out between the links with the least utilization. RP-BE execution is a recursive process, which starts after the first round of TRP execution and formation of the initial *LL*. During RP-PE execution, the algorithm calls back TRP operation as and when needed, and each time after TRP is executed a fresh *LL* is formed. The iterations in RP-BE are carried out as follows.

(1) From *LL* choose two logical links from the bottom. The lower logical link of the two is denoted by l_L and the upper one is denoted by l_U. Assume that l_L is set up from node i_L to node j_L, and l_U is set up from node i_U to node j_U. Thus, we represent l_L and l_U as

$$l_U = (i_U \rightarrow j_U)$$
$$l_L = (i_L \rightarrow j_L)$$

(2) If there is any common node between the two selected logical links (i.e., all the four nodes are not distinct from each other),
   - ◇ then choose the next upper link as the new l_U (i.e., the one above the previous l_U) and l_L, and check whether the new pair of links has four distinct nodes. Repeat this process with the new l_U, if the above attempt fails. Continue to repeat this process with other links enlisted upward in *LL*, until two logical links are found with four distinct nodes. Name the latest pair of links again as l_U and l_L with l_U representing the new upper link.
   - ◇ carry out the BE operation between the same four distinct nodes by defining the two different (crisscrossed) logical links l'_L and l'_U as $l'_L = (i_L \rightarrow j_U)$, $l'_U = (i_U \rightarrow j_L)$.

(3) With the new connectivity, execute TRP operation again and find the congestion $max_{i,j}[\psi_{ij}]$. If the current congestion ≤ 1,
   - ◇ then jump to Step 6;
   - ◇ else go to the next step.

(4) If congestion (which is now more than 1) is larger than its last value,

- then dismantle l'_L and l'_U, restore l_L and l_U, keep these two links in LL as *marked* links, and go to Step 5;
- else accept l'_L and l'_U, *reorder* LL, and go back to Step 1.

(5) Leave the marked links and choose the next two logical links above the last two marked links from LL, denoting them as l_L and l_U (the last two links are retained for use in the subsequent TRP execution, but not referred to by the earlier names). Next, go back to Step 2. If all the links in LL are left out marked at this stage, the problem is considered to be *unsolvable*, and the algorithm stops. In this case, one can seek a possible solution with the next larger value of p.

(6) LL formed at this stage gives the final logical topology.

Case studies

In the following, we examine the above design approach through some illustrative case studies. First, we consider that a multihop WDM LAN needs to operate for a given nonuniform traffic with a number of nodes and the logical degree of the nodes that match with those of some regular topology (e.g., ShuffleNet, MSNet, etc.), and carry out the heuristic design as described above, producing a non-regular topology. Then we consider the same multihop WDM LAN with a regular topology (as indicated in the first case study) and the same nonuniform traffic, and run the TRP subproblem to distribute the non-unifiorm traffic over the regular topology, but skipping the first and third subproblems (i.e., LCP and RP-BE) as the values of b_{ij}'s are fixed by the regular topology under consideration. Thereafter, we examine whether the heuristic design with the non-regular topology can offer a better design. We also present a heuristic design of a network with a number of nodes, which cannot be matched with the number of nodes in the given regular topology.

For the regular topology, we consider a ShuffleNet with eight nodes ($N = 8$) and a nodal degree of two ($p = 2$), needing thereby 16 wavelengths. The traffic matrix for the network is nonuniform as given in Table 6.7, and the capacity of each logical link is 1. With these values of N and p, and the given non-uniform traffic matrix, we next determine the non-regular topology using the heuristic approach. In the first step, the LCP subproblem is executed using a Simplex-based LP-solver (CPLEX) leading to an initial logical topology with a set of logical links b_{ij}'s. Thereafter, the TRP subproblem is run with the input on the logical links obtained from the LCP results. The TRP results show that the traffic exceeds

Table 6.7 *Traffic matrix for an eight-node LAN.*

Nodes	0	1	2	3	4	5	6	7
0	0	.200	.100	.100	.200	.100	.129	.129
1	0.100	0	.200	.200	.100	.129	.100	.200
2	.100	.200	0	.100	.100	.100	.200	.100
3	.129	.129	.100	0	.200	.200	.200	.100
4	.200	.100	.100	.100	0	.100	.100	.100
5	.200	.100	.100	.100	.100	0	.100	.100
6	.200	.100	.200	.200	.100	.100	0	.100
7	.200	.100	.100	.100	.100	.200	.200	0

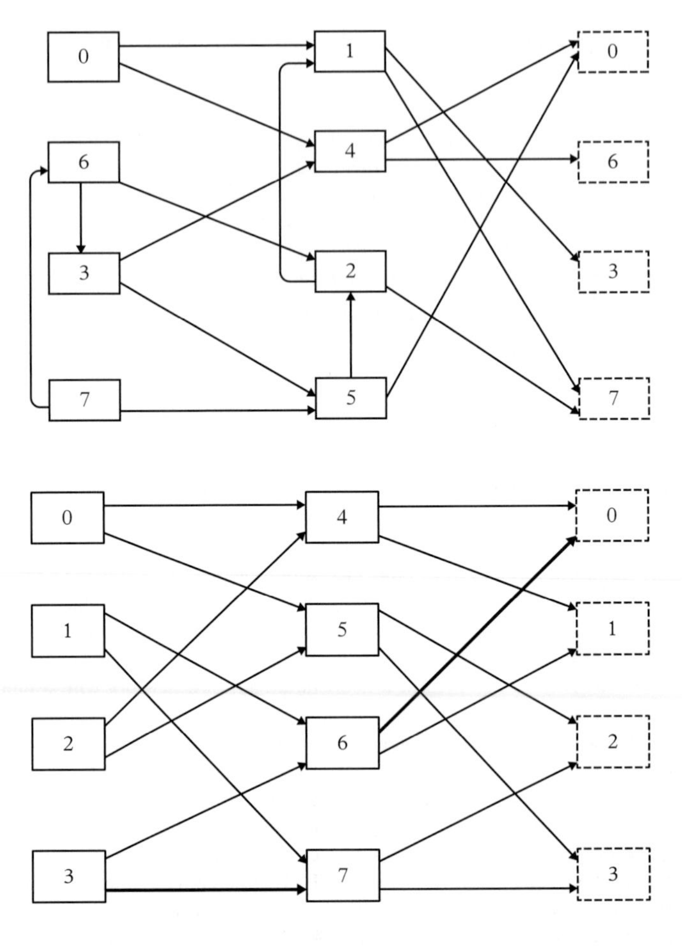

Figure 6.38 *Heuristic design of an eight-node multihop topology with nonuniform traffic pattern using LCP, TRP, and RP-BE, offering acceptable congestion (<1).*

Figure 6.39 *Traffic distribution using TRP in an eight-node ShuffleNet with nonuniform traffic pattern, resulting in unacceptable congestion in two logical links (thick lines).*

1 in some of the links, and hence the RP-BE subproblem is executed, where the TRP subproblem is called iteratively until an appropriate logical topology is reached with an acceptable congestion. Figure 6.38 shows the result of the above heuristic design, where the traffic in all the logical links is found to be less than 1, i.e., the network congestion remains within the acceptable limit.

In order to illustrate the merit of the above design over the equivalent regular topology, the traffic distribution over the eight-node ShuffleNet is estimated, where the LCP subproblem is skipped and the values of b_{ij}'s for eight-node ShuffleNet are fed as an input to the TRP before its execution. As indicated earlier, the RP-BE subproblem is also not used, as any BE operation would disturb the ShuffleNet topology. Having executed the TRP subproblem in this case, it is found that the eight-node ShuffleNet with the given nonuniform traffic pattern needs to support traffic flows exceeding 1 in two of its logical links (shown by the thick lines), thereby implying that the ShuffleNet topology fails to meet the capacity limit. This indicates the merit of the present design methodology offering a non-regular topology, although routing in such network will not be as simple as in ShuffleNet. We also present a case study with seven nodes with $p = 2$ (not matched with any (p, k) ShuffleNet), for which a design is obtained by the heuristic approach (using the same traffic matrix as given in Table 6.7, wherefrom the seventh row and seventh column are removed to obtain a 7×7

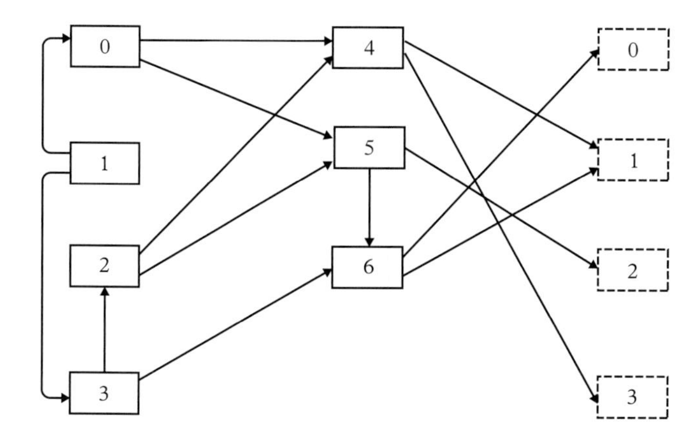

Figure 6.40 *Heuristic design of a seven-node topology with nonuniform traffic pattern.*

traffic matrix), and the resulting topology is shown in Fig. 6.40, where the traffic flows are distributed over the logical links with acceptable congestion.

6.5 Summary

In this chapter, we presented various types of WDM LANs realized over PSC-based passive-star topology using broadcast-and-select transmission. Several experimental testbeds – LAMBDANET, FOX, Teranet, STARNET, and SONATA – were presented with the description of their architectures and salient operational features. Thereafter, various types of WDM LANs were considered, falling in two broad categories: single-hop and multihop networks. These two categories of WDM LANs can be realized by employing different transceiver configurations for the network nodes, as explored in the experimental testbeds. In particular, transmitter(s) and receiver(s) in a node can be fixed-tuned or tunable, leading to the four possible transceiver combinations: TT-TR, TT-FR, FT-TR, and FT-FR. Out of the four transceiver configurations, the first three combinations (having at least the transmitter or the receiver tunable over the WDM transmission window) enable single-hop operation, while the fourth one (FT-FR) with multiple fixed-tuned transmitters and receivers generally leading to multihop operation. Single-hop networks are further classified into two categories with and without PC schemes. Several types of PC-based single-hop WDM LANs were described and analyzed in terms of throughput and delay performances. Single-hop WDM LANs without PC were also described and examined in terms of delay performance.

Next, some generic properties of multihop WDM LANs were examined in terms of the expected number of hops and throughput. Thereafter, we described some useful multihop WDM LANs with regular topologies – viz., ShuffleNet, MSNet, BHCNet, GHCNet, and dBNet – and estimated the expected number of hops and throughput for them. Subsequently, a comparison was made between these networks in terms of the basic network features: expected number of hops, numbers of transmitters and receivers, and per-node throughput. Finally, an ILP-cum-heuristic design methodology for non-regular topologies with non-uniform traffic pattern was presented, which can offer acceptable congestion for a given set of network parameters, e.g., logical degree of nodes, number of nodes, traffic matrix (nonuniform), but may not be feasible in an equivalent regular topology (in respect of the given number of nodes and their logical degree), albeit with a more complex routing mechanism.

EXERCISES

(6.1) Consider a single-hop WDM LAN using PC over 16 wavelengths. Assume the following: control packets consist of five bytes for source MAC address, five bytes for destination MAC address, $\log_2 M$ bits for wavelength indexing, with M as the number of wavelengths and two bytes for miscellaneous purposes.

a) Find the sizes of control and data packets, if the ratio of the data packet to the control packet sizes $L = 100$.

b) Consider two designs of the above network with $L = 100$ and 30. Calculate for both cases the maximum achievable per-channel throughput and the corresponding data-channel traffic and delay for the following MAC protocols:

i) Aloha/Aloha,

ii) slotted-Aloha/Aloha.

(6.2) From the analytical models for the single-hop WDM LANs developed for the three PC-based protocols, obtain an estimate for the upper-bound on the tuning delay τ_{TD} in tunable devices (transceivers) in the network, such that τ_{TD} doesn't affect the network latency significantly.

(6.3) Consider a spanning tree representing the minimum-hop paths from a source to all the destinations in a multihop 31-node WDM-LAN with a logical nodal degree of two. Evaluate the full-tree height and the expected number of hops.

(6.4) Consider a six-node passive-star multihop WDM LAN. For this network, form a unidirectional logical ring topology and determine the required number of wavelengths. Estimate the expected number of hops in this network to communicate between any two nodes. If each wavelength operates at 10 Gbps, evaluate the network throughput in Gbps.

(6.5) Consider a (p, k) ShuffleNet with $p = 3, k = 2$.

a) Draw the (3,2) ShuffleNet topology using the node numbering convention used for self routing.

b) How many wavelengths should this network have to set up concurrent transmissions for all possible connections?

c) Find out the number of hops from node 0 to all the other nodes in the network.

d) Calculate the expected number of hops in the network.

e) Trace a shortest-path route from node 0 to the highest-numbered node in the network. How many such shortest-path routes exist? Identify all of these paths.

(6.6) Consider a (3,4,5) GHCNet and find out the following:

a) number of nodes in the network,

b) number of wavelengths in the network for collision-free transmission,

c) logical nodal degree,

d) nodes reachable in one hop from node (3,2,1) (in mixed-radix representation),

e) path from node (0,0,0) to (4,3,2) using the GHCNet routing algorithm,

f) expected number of hops for this network.

(6.7) Consider a (2,4) dBNet and draw its topology. Find out the following:

a) number of nodes in the network,

b) number of wavelengths in the network for collision-free transmission, *c*) logical nodal degree,

d) route between two extreme nodes according to the self-routing algorithm,

e) expected number of hops for this network,

f) number of hops from node (0000) to all other nodes.

(6.8) Draw the topology of a (2,3) ShuffleNet with the number of nodes starting from 0. For this network,

a) trace the route from node 0 to node 23 following the self-routing scheme of ShuffleNet,

b) if both the outgoing links from the nodes 9 and 11 fail, then check whether node 23 would be reachable from node 0 – justify your answer by tracing the routes,

c) mention one possible situation, when node 18 would be unreachable from node 0 and justify your answer from the topology.

WDM Access Networks

7

In TDM PONs or simply PONs (presented in Chapter 4), the transmission capacity of optical fibers remains under-utilized with single-wavelength transmissions for upstream as well as downstream traffic. Use of WDM transmission in PONs can significantly enhance the overall network capacity and coverage area. WDM transmission in PONs can be realized in several possible ways, such as by using WDM exclusively or combining WDM with TDM, leading to WDM PONs and TDM-WDM PONs (TWDM PONs), respectively. Moreover, depending on the devices used, TWDM PONs can have different levels of flexibility for wavelength and bandwidth allocation to the ONUs. In this chapter, we present several physical configurations to realize WDM and TWDM PONs, and describe some useful wavelength and bandwidth allocation schemes for the latter. We also discuss briefly the needs of open access to the PON-based access networks for the various stakeholders, such as, service providers, network providers, and physical infrastructure providers, to ensure a fair, competitive, and collaborative ecosystem. Finally, we discuss briefly the roles of optical networking in the access segment of mobile networks.

7.1 WDM in access networks

PON-based optical access networks using TDM/TDMA transmission, while offering much higher bandwidth and network span as compared to the older versions (copper-based) of access networks, cannot utilize the bandwidth available in optical fibers. With unprecedented growth in network usage, optical access networks using TDM PONs are also going to get overloaded in foreseeable future. This limitation in TDM PONs (or simply PONs) can be addressed effectively by employing WDM transmission. There are a number of ways to realize WDM-enabled PONs, with appropriate devices and topological configurations (Kramer *et al.* 2012; Banerjee *et al.* 2005). One can develop PONs solely using WDM technology (i.e., without resorting to TDM transmission), or can employ WDM along with TDM transmission. The first category is named as WDM PON, while the second category is referred to as TDM-WDM PON (TWDM PON) or hybrid PON (HPON). However, depending on the devices used, a TWDM PON may or may not have the flexibility to allocate wavelengths as well as bandwidth (bytes) to ONUs dynamically. However, when such two-dimensional flexibility in wavelength and bandwidth allocation is employed, PONs can make much more effective use of the fiber bandwidth. Several studies have also been made on realizing long-reach TWDM PONs to extend the network coverage towards 100 km (Song *et al.* 2010). Furthermore, today optical networking is also playing a significant role in the access segment of mobile communication networks. In view of this evolving scenario, we present in this chapter the various network architectures of the WDM/TWDM PONs making use of appropriate optical device technologies, topological configurations, and wavelength and bandwidth allocation schemes.

Optical Networks. Debasish Datta, Oxford University Press (2021). © Debasish Datta.
DOI: 10.1093/oso/9780198834229.003.0007

7.2 WDM/TWDM PON architectures

One simple way to realize a WDM PON is to employ an AWG as an optical multiplexer/demultiplexer at the RN for the up/downstream transmission, respectively, over tree topology, where the OLT and ONUs use distinct wavelengths to realize WDM transmission. Another simple method – more so while upgrading an existing TDM PON – is to continue with OSP in the RN and use the OLT with WDM transmission on the downstream, while the ONUs use one wavelength on the upstream through TDMA-based transmission. However, the benefit of WDM can be more effectively utilized by using two stages of RN, where the first stage uses AWG or WSS and the second-stage RNs (multiple RNs) continue with OSPs thereby combining WDM along with TDM/TDMA over each wavelength (serving a group of users as in TDM PONs) to realize a TWDM PON architecture. The AWG-OSP configuration, while using AWG at the first-stage RN, imposes static allocation of wavelengths to each TDM group of users. However, the WSS-OSP configuration can realize flexible, and hence dynamic, wavelength and bandwidth allocations for the ONUs, but the first-stage RN loses its passivity owing to the use of WSS.

One can also use OSPs at both stages of a TWDM PON (i.e. OSP-only configuration) for the flexibility in wavelength and bandwidth allocation process (which is not realizable in an AWG-OSP configuration), while maintaining the passivity across the network (which is not realizable with WSS-OSP configuration), albeit with inferior power budget owing to increased power-splitting. There are also some other useful architectures for realizing WDM PONs, which can adapt different topological configurations, two of them being ring-and-stars and ring-and-trees topologies.

Besides the limitation of inflexible wavelength allocation, one important design issue for the TWDM PONs using the AWG-OSP configuration is that its ONUs use different (and fixed) wavelengths for different TDM groups of users. The ONUs used in this type of TWDM PONs are called *colored* ONUs, which become wavelength specific and cannot be procured in bulk. In this regard, the other option is to use *colorless* ONUs, generally for the WSS-OSP or OSP-only configurations, where the ONUs to be used for various users are identical with tunable lasers and optical filters, and the OLT in such PONs can tune the lasers and optical filters for setting up appropriate transmit/receive wavelengths in the ONUs. In the following, we describe these architectures for WDM and TWDM PONs with varying degree of optical hardware, design complexity, and flexibility in wavelength/bandwidth allocation schemes.

7.3 WDM PON using AWG

As mentioned earlier, in AWG-based WDM PONs, the AWG operates as a demultiplexer for the downstream transmission and as a multiplexer for the upstream transmission. Use of the AWG at RN relaxes the power budget because the power in one wavelength is fully directed to the desired ONU, without being divided in the OSP (at the RN) as in TDM PON. Figure 7.1 shows the physical configuration of a WDM PON using AWG at the RN.

The OLT in Fig. 7.1 uses N optical transmitters with unique wavelengths, which are multiplexed and transmitted on the downstream through a circulator towards the RN. The AWG at the RN demultiplexes and forwards the WDM signals received from the OLT into the respective distribution fibers to reach the ONUs. The ith ONU (ONUi) receives through a circulator the OLT transmission on wavelength w_i, which is predetermined and

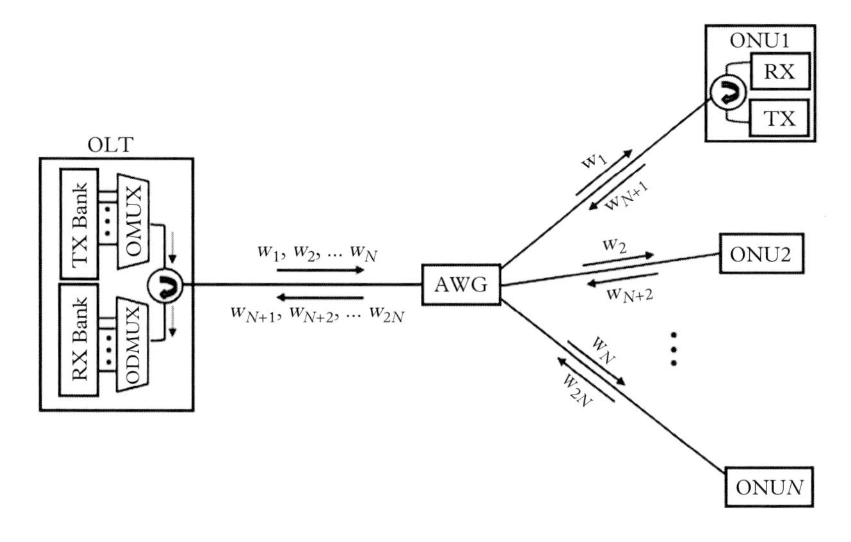

Figure 7.1 *WDM PON with the same number of wavelengths for upstream and downstream transmissions, with N being the number of ONUs, implying that each ONU has a unique pair of wavelengths for both directions of transmission.*

routed by the AWG to ONUi. The upstream transmitter in ONUi uses a wavelength w_i^{us} corresponding to an optical frequency f_i^{us} given by

$$f_i^{us} = f_i^{ds} \pm FSR, \tag{7.1}$$

i.e.,

$$\frac{1}{w_i^{us}} = \frac{1}{w_i^{ds}} \pm \frac{FSR}{c}, \tag{7.2}$$

where FSR is the free spectral range of the AWG, as defined in Chapter 2. The FSR-based fixation of the transmit wavelengths/frequencies allows the upstream signal from an ONU, say ONUi, to pass through the same AWG port connected to ONUi and reach the OLT through the same pair of circulators at both ends (ONU and OLT). All the upstream and downstream frequencies are placed within the respective FSRs of the AWG. The other alternative to this arrangement is to use a pair of AWGs at the RN with circulators on both sides, where the choice of the wavelengths will no longer be constrained by FSR, but must be non-overlapping for the upstream and downstream transmissions. An example architecture of WDM PON using this arrangement (dual-AWG RN) is shown in Fig. 7.2.

In AWG-based WDM PONs, the AWG remains constrained by an upper limit on the number of ports facing towards the distribution fibers, because with a large number of ports, the outer ports (toward the two edges of AWG) have to operate with weak optical power. In order to address this scalability problem in WDM PONs, one can use multiple stages of AWG, as shown in Fig. 7.3. The example case illustrated in the figure shows two stages of AWGs, with the first-stage AWG demultiplexing the downstream wavelengths into K wavebands, WB_i's, feeding into K second-stage AWGs. Each second-stage AWG finally demultiplexes the received waveband from the first-stage AWG into the constituent N/K wavelengths and forwards them to the respective ONU group (WB Gr.i's). On the upstream, the second stage AWGs receive the respective downstream wavelengths from the ONUs and forward them as wavebands to the first-stage AWG. The first-stage AWG multiplexes

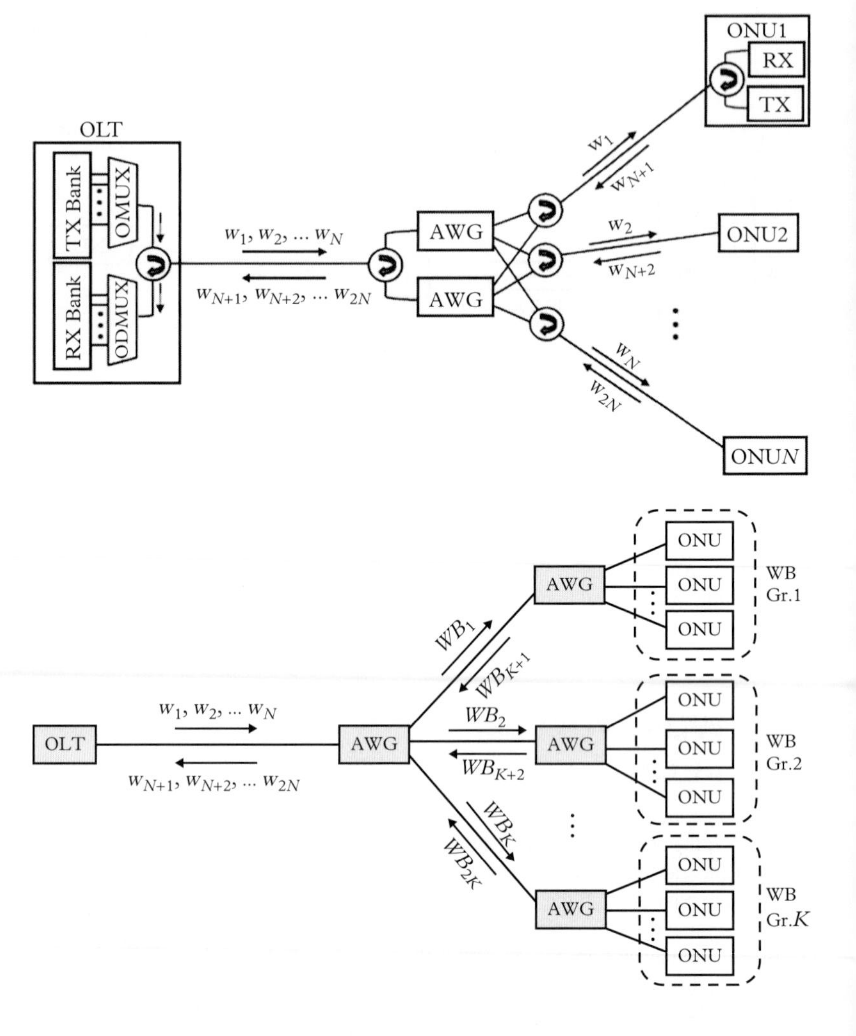

Figure 7.2 *WDM PON with the same number of wavelengths for upstream and downstream transmissions, and two AWGs at the RN.*

Figure 7.3 *WDM PON employing two stages of AWGs, with the first-stage operating over K wavebands WB_i's with $i \in [1, K]$. Each second-stage AWG receives one specific waveband from the first-stage AWG on the downstream and demultiplexes them into N/K wavelengths, along with the reverse functionalities (multiplexing) in the upstream direction at both stages of the AWGs. Note that, the OLT and ONUs use circulators (not shown), as in other architectures, to separate upstream and downstream signals.*

the wavebands received from the second-stage AWGs and forwards them to the OLT. The upstream wavelengths/wavebands are selected at the values separated from the downstream wavelengths/wavebands by the FSRs of the AWGs, so that the same AWG ports (looking toward ONUs) can work for both downstream and upstream signals.

7.4 WDM-upgrade of TDM PON

Some earlier versions of WDM PONs used one single upstream wavelength (typically 1330 nm), shared by all ONUs using TDMA through an OSP placed at the RN, while the downstream transmissions employed WDM from the OLT through the same RN, as shown in Fig 7.4. This type of architecture suited well during the early stage of WDM PON development, with marked asymmetry between the upstream (lower rate) and downstream

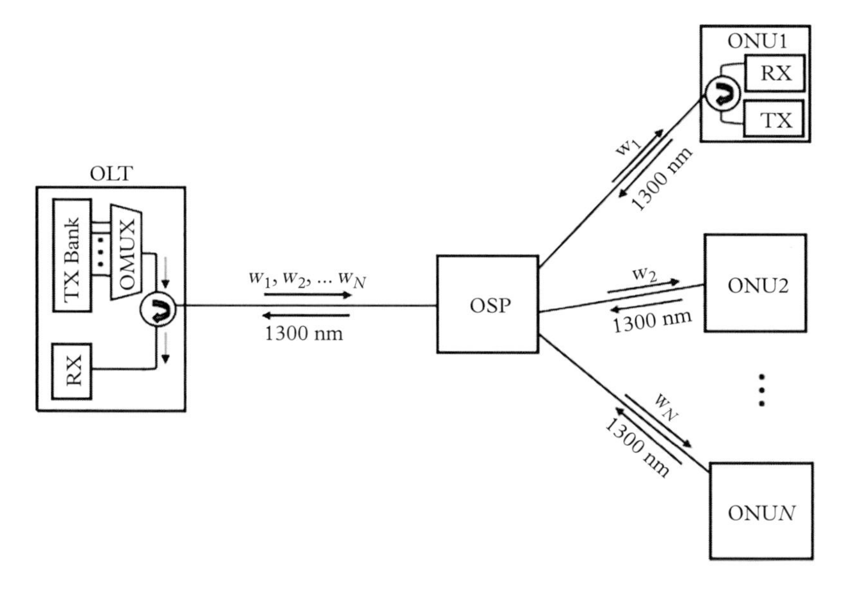

Figure 7.4 *WDM PON realized by upgrading the OSP-based TDM PON.*

(higher rate) traffic demands. However, due to the use of OSP at the RN, power budget in both directions remained constrained in this configuration as in TDM PONs.

7.5 WDM PON using loopback modulation at ONUs: RITE-Net

Another type of WDM PON architecture, called remote aggregation terminal equipment network (RITE-Net), was developed at Bell Laboratories with its ONUs using external modulation of an unmodulated optical signal sent from the OLT for upstream transmission (Frigo *et al.* 1994). In particular, the OLT used a tunable laser to send a downstream signal at a unique wavelength for each ONU through an AWG-based RN, followed by an interval of unmodulated light (viewed as a blank *optical chalkboard*) on the same wavelength, as shown in Fig. 7.5. At each ONU, the received signal was passed through an optical tap, with the data from the first part of the lightwave processed at the receiver, while the unmodulated part (i.e., the second part) of the signal is modulated at the external modulator by the ONU data, and transmitted through the second distribution fiber of the ONU to the AWG. Thereafter, the AWG was used to forward the upstream signals from all ONUs through the lower feeder fiber to the OLT.

7.6 WDM PON using spectral slicing of LEDs: LARNet

Another cost-effective version of WDM PON, called the local access router network (LARNet), was developed at Bell Laboratories (Zirngibl *et al.* 1995) by using LEDs as optical sources at the ONUs along with AWG-based RN (Fig. 7.6). The LED at an ONU offered a cheaper optical source (than a laser) with broad optical spectrum (BOS), which

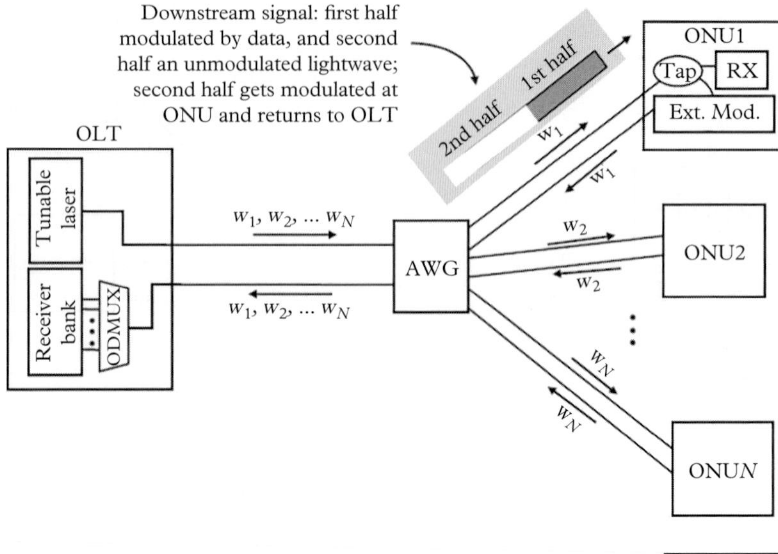

Figure 7.5 *WDM PON architecture using loopback modulation. The architecture is a generic form of the setup used in RITE-Net (after Frigo et al. 1994).*

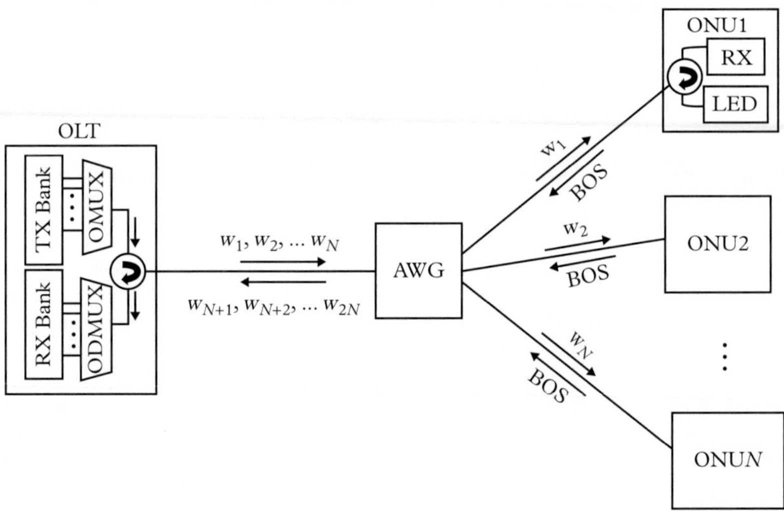

Figure 7.6 *WDM PON architecture using LEDs at ONUs transmitting broad optical spectrum (BOS) for the upstream, with spectral slicing realized in AWG. The architecture is a generic form of the setup used in LARNet (after Zirngibl et al. 1995.)*

- Downstream wavelengths: $w_1, w_2, \ldots w_N$: in 1550 nm window
- Upstream wavelengths: $w_{N+1}, w_{N+2}, \ldots w_{2N}$: in 1330 window, spectrally sliced by AWG from the BOSs transmitted by LEDs

was *sliced* by each AWG port facing toward the ONUs into a narrowband (sliced) optical spectrum. The spectral slice was unique for each AWG port receiving the LED signal from an ONU. In other words, each ONU effectively passed on a unique narrowband optical spectrum (with modulation) through the AWG toward the OLT, as shown schematically in Fig. 7.6. The downstream WDM transmission took place from the transmitter (laser) bank at the OLT. Note that although the use of LED reduced the ONU cost, owing to the slicing effect of AWG, upstream power was reduced significantly in the order of $1 : N$, with N as the number of ONUs. LARNet used the 1550 nm window for WDM transmission over the downstream, while a 1330 nm LED was used for spectral slicing at the AWG, which ensured non-overlapping slices of the BOS from each LED traveling in the feeder fiber without any interference.

7.7 Ring-and-stars topology: SUCCESS

A novel PON architecture with a migration path for the existing TDM PONs to gracefully adopt WDM transmission was explored at Stanford University, called the Stanford University access (SUCCESS) network (An *et al.* 2004). As shown in Fig. 7.7, SUCCESS was proposed with a ring-based feeder/collector connected to the star-based distribution segments using RNs with some special features to accommodate TDM as well as WDM transmissions. In particular, given a tree-based TDM PON with multiple trees *rooted* at the OLT (Fig. 7.7(a)), all the linear feeder fibers were reoriented to form a ring, where the distribution segments, in the form of stars, were connected to the ring-based feeder. The feeder ring using CWDM interconnected the OLT, all OSP-based RNs with nominal CWDM functionality (RN-TDMc) and the associated star-based distribution segments, to form a *ring-and-stars* topology, with in-built survivability against feeder-cut due to the ring-based connectivity (Fig. 7.7(b)).

The final step of migration in SUCCESS took place as illustrated in Fig. 7.7(c), where the ring-and-stars TDM/CWDM architecture was further improvised to realize a comprehensive TDM-cum-WDM PON architecture. In particular, the pre-existing TDM PONs continued to function as before, while the WDM PON functionality was incorporated gracefully using DWDM ring and two types of RNs: TDM-based RNs with nominal DWDM

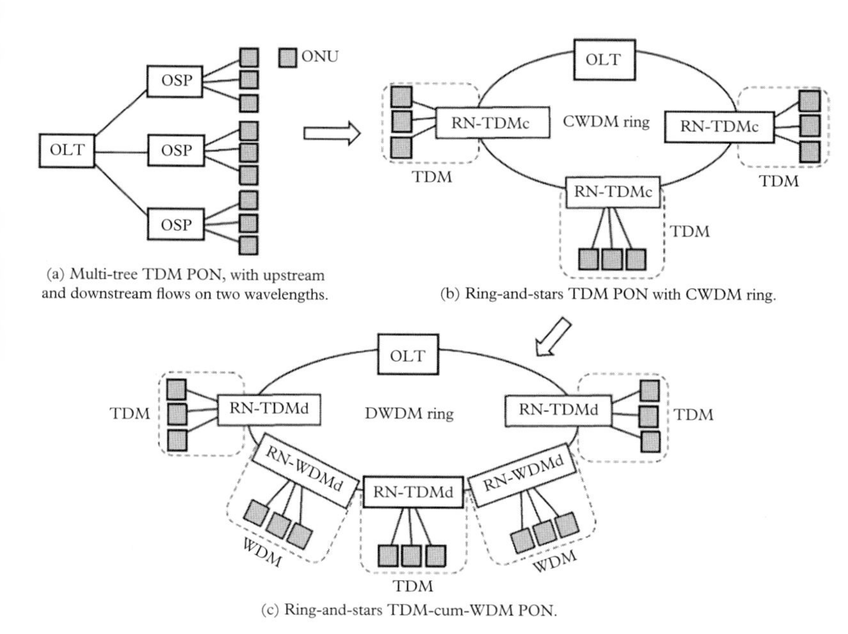

(a) Multi-tree TDM PON, with upstream and downstream flows on two wavelengths.

(b) Ring-and-stars TDM PON with CWDM ring.

(c) Ring-and-stars TDM-cum-WDM PON.

Figure 7.7 *Illustrating the process of migration from the multiple-tree (with common root) TDM PON architecture (a) to the ring-and-stars TDM PON architecture (b) and TDM-cum-WDM-PON architecture (c) as used in SUCCESS. RN-TDMc: RN for TDM PON with CWDM feeder ring, RN-TDMd: RN for TDM PON with DWDM feeder ring, RN-WDMd: RN for WDM PON with DWDM feeder ring.*

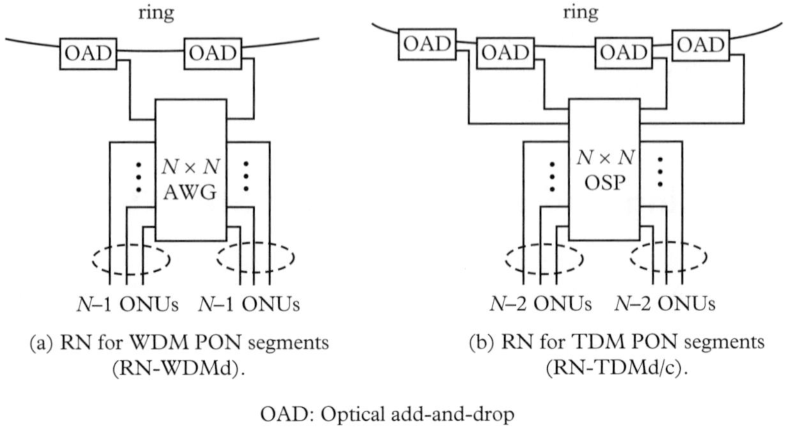

(a) RN for WDM PON segments (RN-WDMd).

(b) RN for TDM PON segments (RN-TDMd/c).

OAD: Optical add-and-drop

Figure 7.8 *RN configurations (RN-WDMd and RN-TDMd of Fig. 7.7) used in SUCCESS (after An et al. 2004).*

functionality (RN-TDMd's) and WDM-based RNs (RN-WDMd's). In the process of migration, SUCCESS also ensured cost-effective use of network sources. In particular, the ONUs were exempted from the use of individual optical sources. Instead, some centralized lasers located in the OLT, transmitted continuous-wave (i.e., unmodulated), bursts of light to the ONUs at the desired wavelengths which were intensity-modulated at the ONUs by the upstream data, using reflective semiconductor optical amplifiers (RSOAs), an idea much in line with RITE-Net as described earlier. Further, with the combined formation with WDM-based ring and stars, SUCCESS could also qualify for long-reach PONs.

Figure 7.8 presents the representative configurations of the two SUCCESS RNs for DWDM ring operation: RN-TDMd and RN-WDMd, as used in Fig. 7.7. In Fig. 7.8(a), the RN-WDMd uses an $N \times N$ AWG along with two optical add-drop (OAD) devices to carry out the bypass and add-drop functionalities on the transit and local outgoing/incoming wavelengths, respectively. Two upper ports of the AWG are used to connect to the feeder ring through the two OADs. The remaining $2(N-1)$ ports are used to connect the ONUs using distribution fibers. Then in Fig. 7.8(b), the RN-TDMd configuration used for the TDM groups of ONUs is illustrated, where an $N \times N$ OSP is used along with four OADs. The remaining $2(N-2)$ ports are used to connect the ONUs. The ONUs with the remote optical sources at the OLT are configured with an arrangement similar to that used in RITENet, thereby leading to a cost-effective ONU design. The OLT design also follows a similar approach to other versions of WDM PONs, with the banks of transmitters and receivers, along with the lasers to supply CW lightwaves for the ONUs as upstream carriers.

7.8 Ring-and-trees topology: SARDANA

A European project with seven partners explored a novel WDM-TDM PON architecture, called scalable advanced ring-based passive dense access network architecture (SARDANA) (Prat *et al.* 2010). SARDANA developed a PON based on a topology using ring-and-trees configuration, where the combination of one WDM ring and multiple TDM trees was explored as a solution toward metro-access integration. Figure 7.9 illustrates the architecture of SARDANA, where a WDM ring interconnects the OLT (as a CO) with the RNs, and each RN (functionally as an access node in the metro ring) connects to a TDM

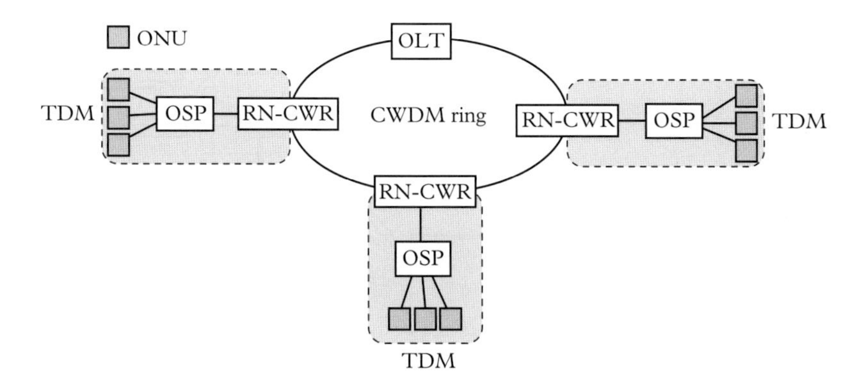

Figure 7.9 *SARDANA architecture using ring-and-trees topology (after Prat et al. 2010). RN-CWR: RN on CWDM ring.*

PON using splitter-based tree topology. SARDANA, with its capability to integrate metro and access, can effectively serve as a candidate architecture for long-reach PONs.

7.9 TWDM PONs

When a PON needs to support a large number of users, and at the same time each ONU doesn't need full-wavelength bandwidth, one can combine TDM and WDM over a tree topology leading to TWDM PON architecture. As mentioned earlier, a TWDM PON can typically employ two stages of RNs with an AWG at the first-stage RN from the OLT end, followed by a second stage of distribution using a number of OSPs, each of them connecting to a TDM-based (or more precisely, TDM/TDMA-based) group of ONUs. The first stage of RN can also use a WSS or an OSP with more flexibility in utilizing the available WDM channels (wavelengths) and bandwidth therein. We describe some of these TWDM PON architectures in the following.

7.9.1 Static TWDM PON using AWG and OSPs

In a two-stage TWDM PON using an AWG in the first-stage RN and a number of OSPs in the second stage, the ONUs (N ONUs, say) are divided into $G < N$ groups, and each group can use one unique wavelength, thereby needing $2G$ wavelengths for the upstream and downstream transmissions. With this arrangement, the network will need to employ G upstream wavelengths, and each upstream wavelength, supporting $K = N/G$ ONUs, will use TDMA as in TDM PONs. However, each of the G downstream wavelengths will use one-to-many TDM for the K ONUs. This arrangement is illustrated in Fig. 7.10, where with G wavelengths on up/downstream, the network uses G OSPs as the second-stage RNs. Looking from the AWG toward the ONUs, one can connect one OSP to each of the G AWG ports facing toward the ONUs and can thus extend the PON access to $K \times G = N$ ONUs. However, the AWG offers a *static wavelength allocation* to the ONU groups, while within a group of ONUs, TDMA slot allocations are carried out dynamically. Thus, there are effectively G static TDMA groups of ONUs (TDMA Gr. i's), with each group operating over a unique pair of wavelengths with TDMA-based DBA, and hence we call this type of PON a *static* TWDM PON. Note that, this architecture has been adopted by ITU-T in the 40-Gbps-capable next-generation PON standard (NG-PON2), which

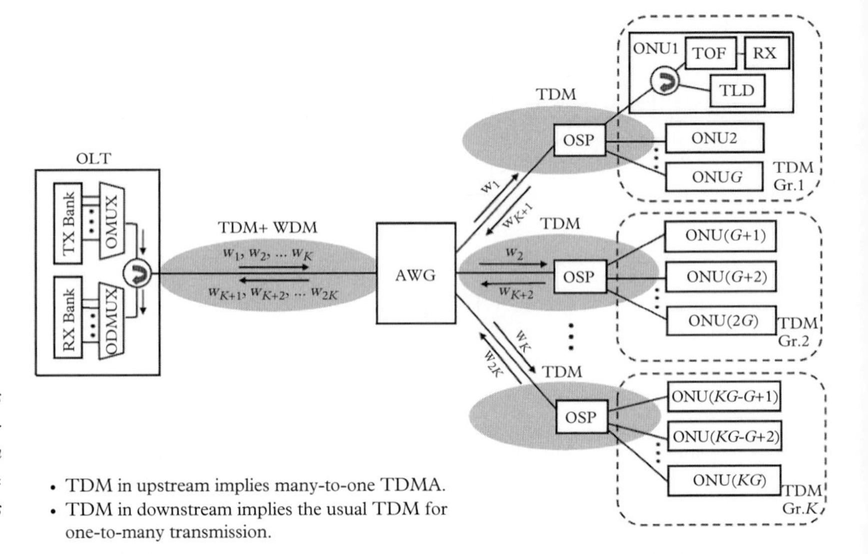

Figure 7.10 *TWDM PON with N ONUs and K wavelengths in each direction employing TDM/TDMA over WDM. Thus, each wavelength forms a TDM group of* $G = N/K$ *ONUs, with the ith group named as TDM Gr.i.*

provides convergence of all pre-existing PON standards (e.g., GPON, XG-PON, GEPON, 10G-EPON) (Nesset 2015).

As mentioned earlier, notwithstanding the advantage of improved power budget for using AWG in the first-stage RN (as compared to the architecture using OSPs at both stages), the AWG-OSP configuration with its static wavelength-routing functionality, denies any flexibility when some ONU group(s) starve for more bandwidth with respect to some other group(s) demanding much less bandwidth. The architectures described in the following address this issue.

7.9.2 Dynamic TWDM PON using WSS and OSPs

As mentioned earlier, the rigidity of AWG in TWDM PONs can be addressed by replacing the AWG in the first-stage RN by a WSS. Replacement of the AWG by a WSS in the first-stage RN enables dynamic wavelength allocation to ONU groups, though the electronic control on the WSS (from the OLT) makes the PON partly active. Nevertheless, with the AWG replaced by a WSS, the OLT can change the wavelength allocations for the ONU groups governed by the traffic variation across the ONUs, thereby achieving load-balancing across the wavelengths, leading to enhanced bandwidth utilization in the PON. In particular, in this WSS-enabled configuration, any overloaded ONU group attached to an OSP can be assigned multiple wavelengths, while multiple under-loaded ONU groups attached to respective OSPs can also share one single wavelength. However, in order to get this benefit, the ONUs must be colorless (by using tunable lasers and optical filters) to transmit/receive dynamically changing wavelengths for load-balancing (Das *et al.* 2012).

7.9.3 Dynamic TWDM PON using multiple stages of OSPs

In dynamic TWDM PONs one can also use OSPs at both the stages of the RNs, thereby ensuring fully passive realization of the network. Note that, the use of two stages of OSPs (or maybe more, if needed) is preferred to single-OSP topology for saving the fiber deployment cost, though both will lead to similar power budget for a given number of ONUs. Further, in order to utilize the full benefit of the OSP-only TWDM configuration, one must use colorless ONUs, so that any ONU across the entire PON can be arbitrarily allocated any wavelength for the upstream/downstream transmission. This configuration of dynamic TWDM has also been included by ITU-T in NG-PON2 standard (Nesset 2015). Thus, the NG-PON2 standard has been designed with an inclusive feature of supporting OSP-only and AWG-OSP configurations along with a mix-and-match of both, wherever needed (Wey *et al.* 2016).

However, owing to the use of OSPs, this configuration would have a more constrained power budget, with respect to the TWDM PONs using AWG/WSS-OSP configurations, more so for large number of ONUs and longer reach. The problem of constrained power budget in this configuration can, however, be taken care of, if necessary, by introducing OAs at appropriate locations, which need to be pumped remotely from the OLT end to ensure the passivity of the PON. We address this issue in the following.

7.9.4 TWDM PON using two stages of OSPs with remotely pumped EDFAs

As mentioned above, in order to realize long reach in TWDM PONs, the use of multiple stages of RNs, even with AWG/WSS-OSP combinations, might reduce the signal power received at both ends; the situation is likely to get more challenging when only OSPs are used at both stages of RNs. In such cases, one needs to include OAs (typically EDFAs in both directions) in the feeder segment, which can be remotely pumped from the OLT end, thereby retaining the passivity of the PON. One such realization of TWDM PON is shown in Fig. 7.11, where the EDFAs, marked as A and B in the figure, offer power gains for the downstream and upstream signals, respectively, and the ONUs make use of tunable laser diodes (TLDs) and tunable optical filters (TOFs) to function in colorless manner.

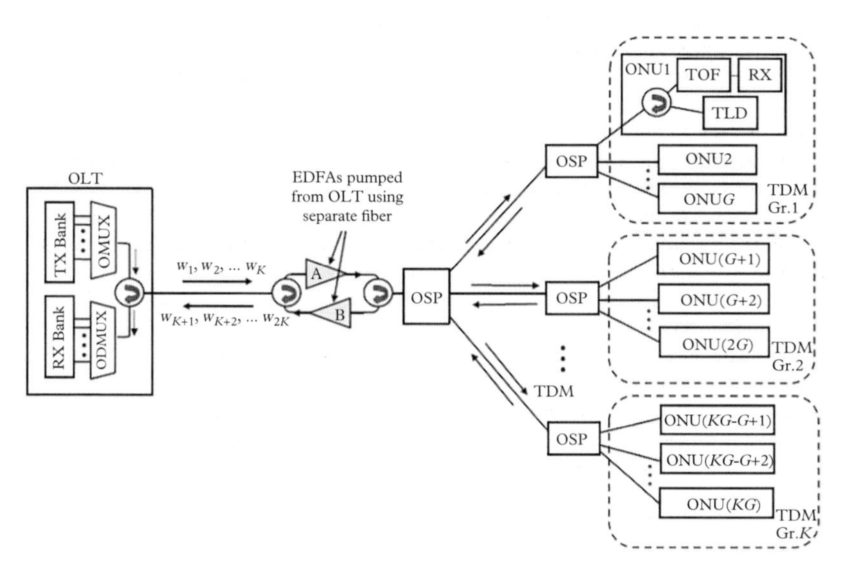

Figure 7.11 *TWDM PON using EDFAs, remotely pumped from the OLT. Note that, with the OSPs at two stages, wavelengths for the ONUs can be allocated with full flexibility, regardless of their physical connections with the OSPs. Note that, the transceivers in the ONUs are made tunable (and hence colorless) by using TLDs and TOFs to ensure flexible wavelength and bandwidth allocations.*

7.9.5 Wavelength and bandwidth allocation schemes for TWDM PONs

The TWDM PONs, as governed by the choice of the optical devices at the RNs (i.e., AWG-OSP, WSS-OSP, OSP-only), can have either static or dynamic wavelength allocations for the ONUs along with dynamic bandwidth (i.e., bytes) allocation on each wavelength. This leads to broadly two different categories of MAC protocols in TWDM PONs. In the static TWDM configuration (AWG-OSP), the PON practically gets divided into multiple TDM PONs, as many as the number of wavelengths used therein, with each wavelength serving one ONU group (attached to one OSP in the second stage of RNs) using TDM/TDMA transmission on down/upstream. For implementing dynamic wavelength and bandwidth allocation in WSS-OSP or OSP-only configuration, obviously the MAC protocols become more intricate, while offering better utilization of the spectrum in optical fibers. In the following, we describe these two types of schemes: the static wavelength and dynamic bandwidth allocation (SWDBA) scheme for the static AWG-OSP TWDM PONs and the dynamic wavelength and bandwidth allocation (DWBA) schemes for the dynamic WSS-OSP or OSP-only TWDM PONs.

Static wavelength and dynamic bandwidth allocation

The SWDBA schemes are realized over AWG-OSP configuration, wherein the AWG divides the entire network into several groups of ONUs with its wavelength-routing functionality. The OLT, as governed by the connectivity between the AWG ports and the OSPs, makes static allocations of the wavelengths to the ONUs, which in effect divides the ONUs into different groups depending on their connections to the OSPs, with each group using one wavelength and one OSP for upstream transmissions. Each wavelength group of the ONUs operates as a TDM PON, with a DBA scheme similar to the ones used in TDM PONs. However, as mentioned earlier, with the SWDBA scheme when the traffic in different ONU groups (operating on distinct wavelengths) vary, some wavelengths might remain under-utilized, while some other wavelengths might suffer from congestion.

Dynamic wavelength and bandwidth allocation

The dynamic TWDM PONs operating with a DWBA scheme need to employ an intelligent scheduling scheme concurrently in two dimensions: time and wavelength. Further, the TWDM PONs are in general designed to serve a large user population, often trying to reach sparsely located ONUs in remote areas. Thus, the DWBA schemes have to face additional challenges with large and widely varying RTTs for different ONUs. The problem of DWBA in dynamic TWDM PONs has been studied extensively with offline as well as online schemes (Kanonikas and Tomkos 2010; Dhaini et al. 2007; Kiaei *et al.* 2013; Buttabani *et al.* 2013). Some of the candidate DWBA schemes for dynamic TWDM PONs are as follows.

- Offline DWBA scheme: In an offline DWBA scheme, the OLT waits for the REPORTs from all ONUs during a cycle time, and thereafter assigns the free wavelengths to the received REPORTs.
- Online DWBA schemes:
  - Earliest finishing time (EFT): In EFT, the OLT responds immediately after each REPORT arrival, based on the earliest finishing time (EFT) of the ongoing upstream transmissions.
  - Latest finishing time (LFT): In LFT, the OLT responds immediately after each REPORT arrival on the basis of the latest finishing time (LFT) of the ongoing upstream transmissions.

- EFT with void-filling (EFT-VF): This scheme augments the EFT mechanism using VF in between the forthcoming (already committed by the OLT) upstream transmissions.
- LFT with void-filling (LFT-VF): This scheme augments the LFT mechanism by VF.
- Just-in-time (JIT): This scheme stands somewhere between the offline and the fully online schemes by using JIT-based allocations of wavelengths to the incoming requests.

In the following, we describe the above DWBA schemes belonging to the offline and online categories.

Offline DWBA scheme In an offline scheme, the OLT waits for an entire cycle time to collect the REPORTs from all ONUs and exercise its scheduling scheme thereafter for all the received requests. After receiving the REPORTs from the ONUs during a cycle time, the OLT can choose an appropriate grant-sizing scheme. While assigning wavelengths for a given REPORT request, the OLT typically chooses the wavelengths in the order they were freed in the previous cycle, thereby using the earliest-freed wavelength as the first assigned-wavelength and so on, along with a bandwidth (bytes) allocation policy so that the load among the wavelengths is distributed as uniformly as possible. However, this kind of allocation schemes appears to be a sort of micro-management of bandwidth utilization, as much bigger sacrifice is made through the waiting of the OLT during the entire cycle time. Thus, the offline schemes might offer high fairness in bandwidth allocation to the ONUs, but at the cost of low bandwidth utilization as the OLT keeps waiting through each cycle without any action. In the following, we address this issue while describing the online DWBA schemes, which becomes critical with large user population and network size.

Online DWBA schemes In online DWBA schemes, the execution of the EFT, LFT, EFT/LFT-VF, or JIT algorithm dealing with a large number of ONUs becomes time-critical, needing thereby an efficient procedural framework to find out the needed information-set *on the fly*. For this purpose, first we define some variables and parameters and formulate some useful relations between them. As discussed in Chapter 4 on TDM PONs, the network is assumed to operate with REPORT and GATE messages between the OLT and ONUs.

Overall, the ONUs in an OSP-only TWDM PON are divided *logically* into G groups. A group of ONUs, say gth group ($g \in [1, G]$) is assumed to share $M_g (\geq 1)$ upstream wavelengths out of M_{US} available upstream wavelengths in the network, such that

$$\sum_{g=1}^{G} M_g = M_{US}, \tag{7.3}$$

and each ONU group is assumed to have k_g ONUs, such that

$$\sum_{g=1}^{G} k_g = N, \tag{7.4}$$

with N representing the total number of ONUs in the TWDM PON.

The above segmentation of the upstream spectrum among the G ONU groups leads to a cost-effective design. In particular, the *tuning cost* of a laser in a transmitter can be two-fold. This implies that to make a laser tunable over the entire range of the optical spectrum may turn out to be *costly* in terms of both the tuning time and the associated optical hardware. However, it would usually suffice to make each ONU in an ONU group tunable over a segmented part of the entire optical spectrum (instead of giving access to the full spectrum), without much sacrifice in load-balancing across the network. Typically, $k_g > M_g$, and thus the given number (k_g) of ONUs in the ONU group g will be sharing a smaller number (M_g) of wavelengths through the TDMA-based scheduling algorithm, thereby ensuring only intra-group yet fair-enough load-balancing among the k_g ONUs. In order to formulate the procedural framework with these basic features, we next define some more variables and parameters for the dynamic TWDM PON as follows:

ONUi: ith ONU in the gth group of ONUs in the dynamic TWDM PON under consideration. Hence, ONUi can be allocated any one of the M_g wavelengths allocated for the gth ONU group,

$2\tau_i$: RTT for ONUi,

δ_{con}: transmission time for control (GATE/REPORT) messages,

AF_k^j: time instant when the first bit of the jth grant arrives at the OLT in a given cycle from an ONU in the ONU group g using wavelength $w_k \in M_g$ wavelengths,

AL_k^j: time instant when the last bit of the jth grant arrives at the OLT in a given cycle from an ONU in the ONU group g using wavelength $w_k \in M_g$ wavelengths,

FL_k: finishing time instant of the last-assigned grant transmission on the wavelength w_k arriving at the OLT in a given cycle from an ONU in the ONU group g using wavelength $w_k \in M_g$ wavelengths. Thus, FL_k can be expressed as

$$FL_k = \max_j \{AL_k^j\}, \forall w_k \in M_g. \tag{7.5}$$

In the OLT, when a wavelength and some bandwidth (number of bytes) are to be granted in response to a newly arrived REPORT from an ONU, say ONUi, the OLT checks the wavelength and bandwidth grants already made for the ongoing as well as yet-to-start transmissions from the set of wavelengths M_g in the ONU group g, and notes the respective values of AF_k^j's and AL_k^j's. From this information-set, the OLT determines the value of FL_k from Eq. 7.5, while ensuring that no overlap takes place between the new and old allocations in the time domain.

With the above framework, we assume that a new REPORT message has arrived at time t from ONUi to the OLT asking for a grant of β bytes. This REPORT needs to be attended to by the OLT with an appropriate wavelength assignment, say $w_x \in m_g$, along with the associated values of AF_x^n and AL_x^n, with n representing the new upstream grant to be assigned next in the given cycle under consideration. Thus, the newly arrived REPORT is represented by a set of four variables, given by a four-tuple report vector $RV_x = \{w_x, n, AL_x^n, AF_x^n\}$. The variables in RV_x must satisfy the following constraints.

1. *Wavelength allocation to ONUi*: the wavelength to be allocated to the new request, i.e., w_x, from ONUi in group g must remain within the wavelength set m_g assigned to ONUi, i.e.,

$$w_x \in M_g. \tag{7.6}$$

2. *Finishing-time allocation for the new grant at the OLT*: the finishing time AL_x^n observed at the OLT for the new grant should be ahead of the starting time AF_x^n by the grant size β and the duration of the REPORT message δ_{con}, i.e.,

$$AL_x^n = AF_x^n + \beta + \delta_{con}. \tag{7.7}$$

3. *Starting-time allocation for the new grant at the OLT*: the starting time AF_x^n has to be allocated immediately after the arrival of a new grant, or after the finishing time of the last, i.e., $(n-1)$th allocation, whichever takes place later, i.e.,

$$AF_x^n = max\{AL_x^{n-1}, t + 2\tau_i + \delta_{con} + \delta_{gt}\}, \tag{7.8}$$

where t represents the instant of time when the OLT sends the GATE message to the ONU, and $\delta_{gt} > 0$ represents the guard time, so that the two consecutive assignments do not collide due to unavoidable variations in system parameters, such as RTT, clock rate, etc.

The above components for RV_x are illustrated in Fig. 7.12 for a typical case, where an ONU, say ONUi, is assumed to have asked for a grant β, and has been accordingly allocated the starting and finishing times for the transmission of grant, as observed at the OLT. However, the choice of x and n will vary for the two schemes: EFT and LFT. We describe these schemes in the following without and with awareness of the possible voids left out between the consecutive allocations. As we shall see soon, the DWBA scheme shown in Fig. 7.12 illustrates the EFT scheme without the awareness of voids.

EFT scheme In the EFT scheme, the OLT scans through all the wavelengths $\{w_k\} \in m_g$ and assigns the wavelength w_x (say), which shows up with the earliest finishing time, i.e., $\min\{FL_k\}$. This implies that, among all the wavelengths in the wavelength group m_g, w_x represents the specific wavelength w_k with a k that corresponds to the lowest value for FL_k's. We denote the time slot for the new upstream grant β as the nth burst that does not overlap with the last $(n-1)$th (and earliest) burst transmission, to be transmitted over the wavelength w_x. Thus, w_x is chosen as

$$w_x = w_k, \arg\min_k\{FL_k\}, \tag{7.9}$$

and having picked up w_x as above, n is chosen as

$$n = j + 1, \arg\max_j\{AL_x^j\}. \tag{7.10}$$

As mentioned earlier, one can check now that the illustration shown in Fig. 7.12 follows in effect the online EFT scheme, if the wavelength w_x offers the EFT-based choice as above (note that, all wavelengths in the wavelength group M_g are not shown in the figure for simplicity). As evident from the diagram, it would suffice for the OLT to keep track of the finishing times of the latest transmission bursts on each wavelength from the set of M_g wavelengths assigned to the ONU group under consideration. However, as we shall see later, since the scheme doesn't look into the spectral utilization, some blank or void slots might remain left out and are never used thereafter. We discuss later a remedy for this problem in the void-aware versions of the EFT/LFT schemes.

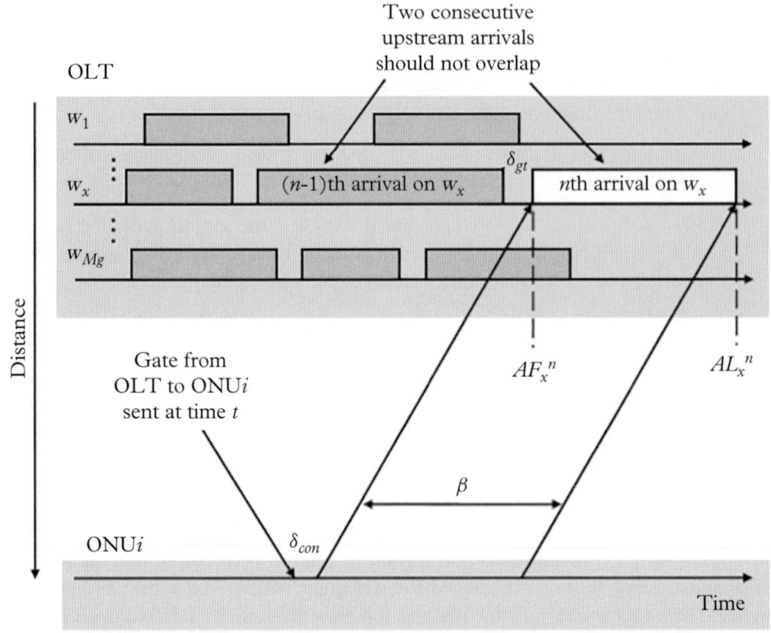

Figure 7.12 *An illustrative timing diagram for a typical DWBA scheme in a given cycle, where the OLT is shown to be dealing with m_g upstream wavelengths ($w_1, \cdots, w_x, \cdots, w_{M_g}$) for ONUi in group g. The other upstream wavelengths allocated by the OLT for the rest of the ONUs are not shown for simplicity. Note that, in the present example, the nth burst arrival at the OLT from ONUi takes place on wavelength w_x, satisfying all the constraints described by Eq. 7.6–7.8.*

LFT scheme The LFT scheme has a similar framework as used in the EFT scheme, but instead of considering the wavelength with the earliest finishing time of the ongoing transmissions, it goes for the wavelength that has the latest finishing time. Thus, in the illustration of Fig. 7.12, the LFT scheme will choose wavelength w_1, instead of w_x chosen in the EFT scheme. However, if the latest finishing time is higher than $t + 2\tau_c + \delta_{gt}$, then the assignment follows the EFT scheme. In an underloaded network, the LFT scheme would perform better than the EFT scheme as it advances the transmission of the grant in time. However, with increasing load both schemes would be bringing out similar results in respect of bandwidth utilization (Kanonikas and Tomkos 2010).

EFT-VF scheme In the EFT and LFT schemes, the voids once left behind in time are not filled up, leading to the under-utilization of bandwidth. This issue is addressed in the EFT-VF scheme as follows.

Note that the OLT cannot go back in time and fill in any void that has already *passed by* before the arrival of the new REPORT. However, it can always examine the forthcoming transmission bursts that have already been committed to the ONU under consideration and are yet to start. In doing so, it might discover some voids which are yet to be *visible* physically, and fill them in by the new request in advance, if the size of the new request fits into the void, leading to the VF-extensions of EFT and LFT. For the EFT scheme, we describe the mechanism in the following.

As discussed earlier, the EFT scheme makes a judicious choice of w_x and n based on its EFT-guideline. In the EFT-VF scheme, one needs to be more selective about the choice of $\{w_x, n\}$, by using the prior knowledge of the available set of voids V_i (say) for ONU-i. Note that, the voids need to be chosen with certain constraints, such that a given void must appear in time only after the arrival of the REPORT, and the size of the void should be able to accommodate the request. These constraints on V_i can be expressed as

$$V_i = \left\{AF_x^{j+1} - \max\left(AL_x^j, t + 2\tau_i + \delta_{con}\right) \geq \beta + \delta_{con}\right\}, \quad x \in M_g. \tag{7.11}$$

Having determined all the candidate voids by using the above expression, the EFT-VF scheme proceeds to allocate the grant with a wavelength w_x and slot index n as

$$w_x = \arg\min_x (AL_x^j \in V_i), \quad x \in M_g, \tag{7.12}$$

$$n = \arg\min_j (AL_x^j \in V_i) + 1, \quad x \in M_g. \tag{7.13}$$

This completes the formulation for the online EFT-VF scheme. Note that, the use of VF in the online LFT algorithm follows the same principle, and we don't elaborate further on this scheme. Notwithstanding the efficacy of the VF-based improvisation, one needs to check the complexity involved in its search process, which is indeed higher than the simple online EFT/LFT schemes as the number of voids is likely to exceed the number of wavelengths. Overall, with M wavelengths and N ONUs, the search space might increase from M for the EFT to $M + N$ for EFT-VF, assuming that, at a given time one grant remains pending at each ONU, which would in turn imply that the complexity will increase from $O(M)$ to $O(M + N)$. With some worthwhile improvisation, the complexity can however be reduced to $O(\log(M + N))$, as described in (Kanonikas and Tomkos 2010).

JIT scheme The JIT scheme makes a hybrid mix of offline and online schemes. In particular, after the beginning of a fresh cycle time, when an upstream wavelength becomes free, say at time t_1, the OLT responds just-in-time (JIT), i.e., immediately, to the REPORTs received from multiple ONUs upto that moment. The ONU REPORTs being available together at a time, the OLT carries out an appropriate bandwidth allocation scheme for multiple ONUs using TDMA, thereby ensuring fairness and load-balancing among them as in an offline scheme. The OLT keeps repeating the above bandwidth allocation process, as and when another wavelength becomes free.

Having allocated the grants in response to the REPORT messages from as many ONUs, the OLT suspends the allocation process until another wavelength becomes free. As soon as the next wavelength becomes free, say at time t_2, the OLT dynamically resumes the bandwidth allocation process for the REPORTs that arrived within the time interval $t_2 - t_1$. Thus, the JIT scheme keeps switching dynamically between idle and active states, while bringing in the fairness and load-balancing features of an offline scheme at every instant whenever a new wavelength becomes free.

7.10 Open access architecture

WDM-enabled PON-based optical access networks are now being transformed into an affordable high-bandwidth access solution across society. Hence, the business opportunities associated with the broadband WDM optical access networks need to be made *open* to various stakeholders in the domain of telecommunication networks for a fair, competitive, and collaborative ecosystem, while handling a large user population and the enormous bandwidth of optical fibers. One of these stakeholders is the service providers (SPs) who operate at the highest layer (application layer) of the network and hence remain directly in touch with the end-users. However, in the complex setting and hierarchy of the telecommunication networks, various other stakeholders engage themselves for making the network functional, such as network providers (NPs), physical infrastructure providers (PIPs), etc. This effectively leads to a class of access networks, termed *open access networks*, where different types of stakeholders need to be encouraged to step in and create competitive, collaborative, and cost-effective solutions for the end-users (Bhar *et al.* 2015). For example, the physical infrastructure in optical access networks, can be of two types – passive and active – where the tasks of fiber-laying and placement of RNs belong to the passive infrastructure category, while the various network equipment at the ONUs and OLT come under the active infrastructure category. The entire network infrastructure can be set up under one single NP, typically the local municipalities, or might even be shared by the private network NPs and PIPs. Even the SP, NP, and PIP may belong to one single administrative body in some cases. For a given SP, there can also be a number of NPs in a locality. With this kind of varying participation, open access networks need to set up business mechanisms for the fair scope of participation of all concerned.

In view of the above, we present a layered model for an open access network in Fig. 7.13, based on the various possible combinations of the SPs, NPs, and PIPs for different user groups (UGs). Note that, the roles of NPs and PIPs are fundamentally different from the roles of SPs, as SPs are directly accountable to the end-users. In general, NPs and PIPs can collaborate with each other, leading to an affordable infrastructure cost of the network. However, SPs need to take into consideration the service-level agreements (SLAs) with the users and comply with them with fairness. This needs strict policies so that the SPs are allocated appropriate chunks of bandwidth in the network, and the performance of the PON is acceptable to its UGs. Moreover, the security of the services across the SPs also need attention for a safe and secure communication for the end-users.

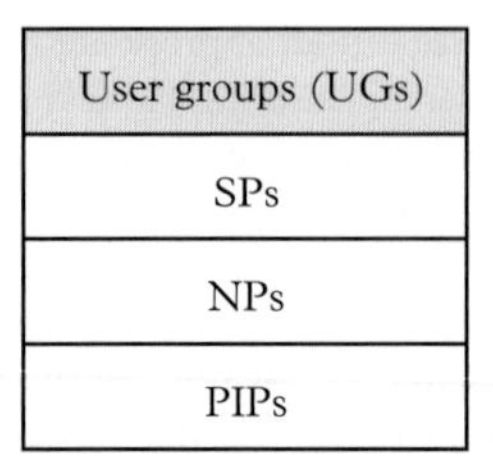

(a) Generic open access network model.

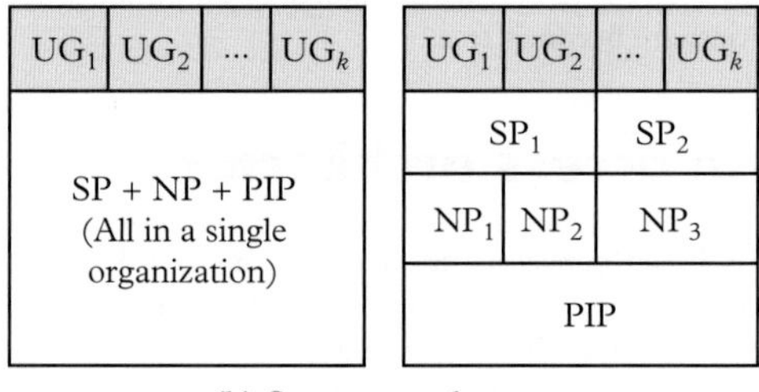

(b) Some example cases.

Figure 7.13 *Layered model for open access networks with various possible options.*

7.11 Optical networking in access segment of mobile networks

There has been a phenomenal growth in mobile data in the access segment due to the proliferation of various kinds of hand-held devices, such as smart phones. The backhaul connectivity using optical links in the access segment between the base stations (radio-access stations) in a mobile network and the associated metro-edge ring offer a high-capacity support to accommodate this increase in mobile data. In particular, at the user end of a mobile network, the base stations form the radio-access network, wherefrom each base station employs a back-haul optical link (could be copper or wireless too, but with much lower transmission capacity) to pass on the mobile data received from the end-users to an access node on the metro-edge ring. In one such realization used in the 4G mobile network, the base stations sample the microwave signals received from their end-users, and the sampled analog signals are used to modulate optical carriers therein. The modulated optical carriers (with analog modulation) are then transmitted using the *radio-over-fiber* optical links running from the base stations toward the metro-edge ring, thereby offering an analog and hence cost-effective backhauling solution.

Furthermore, owing to the overwhelming growth in the number of mobile users, cells in the cellular mobile networks need to be reduced in size to support larger per-user bandwidth, and more so for 5G networks. This in turn calls for large numbers of micro/nano/femto-cells, thereby needing as many base stations. Consequently, the backhauling cost needed with the increased number of base stations is escalated significantly. In order to address this issue, a pool of baseband units (BBUs) is set up, where the number of BBUs is kept lower than the number of base stations. The BBU pool with fewer BBUs (than the base stations) keeps communicating with the base stations by dynamically load-balancing the time-varying traffic, thereby offering optimized resource provisioning. While, the BBUs communicate with the respective access node on the metro-edge ring using the optical backhaul network, they make use of radio-over-fiber transmission through another set of optical fibers to connect with the radio units of the base stations. In order to facilitate this architectural design the radio heads (units) in the base stations are moved upfront, close to the antenna located on the top of the towers, while the necessary BBUs are placed at ground level. This leads to an efficient fronthaul design, making it easier for each base station to realize improved signal coverage and energy efficiency owing to the lower cable losses of microwave signals with the radio heads placed in proximity to the antenna elements.

7.12 Summary

In this chapter, we presented different versions of WDM-enabled PONs using tree topology – WDM and TWDM PONs – the latter with varying degrees of flexibility for the wavelength and bandwidth allocation process. First, the simplest WDM PONs were presented using AWG at the RN, with every ONU using unique wavelengths for upstream as well as downstream transmissions, and thus leading to inefficient bandwidth utilization of the optical fiber. Next, we described a scheme through which an existing TDM PON can be upgraded to operate as a WDM PON, where the downstream uses multiple WDM channels and the upstream transmissions are carried out using TDMA-based polling scheme as in TDM PONs. Following the introduction of these basic architectures, four WDM PON testbeds – RITENet, LARNet, SUCCESS and SARDANA – were presented with their salient operational features.

Next, we described the candidate architectures of TWDM PONs, where TDM/TDMA is introduced along with WDM transmission with enhanced bandwidth utilization. First, one of the possible architectures, named static TWDM PON, was discussed, using two stages of RNs with AWG as the first-stage RN followed by a number of OSPs as the second-stage RNs. Though the introduction of TDMA through OSPs could enhance the bandwidth utilization in the upstream, the wavelength allocation could only take place in the AWG in a static manner, thereby preventing the ONUs with more traffic in a given wavelength from sharing the unused bandwidth in some other wavelength(s). Next, a similar architecture with AWG replaced by WSS was described where flexibility in the wavelength assignment process could be realized, leading to a dynamic TWDM PON, although with the first-stage RN losing its passive nature. Then another version of dynamic TWDM PON with the most flexible and completely passive architecture was presented, which employs multiple stages of OSPs (i.e., without AWG/WSS), but with constrained power budget due to the power-splitting in the OSPs, a problem that can be compensated for by using remotely pumped OAs at appropriate locations. Thereafter, we presented a few wavelength and bandwidth allocation schemes for the two types (static and dynamic) of TWDM PONs and described their salient features. We also discussed briefly the issues concerning open access networks, which must encourage all stakeholders toward developing a fair, competitive, and collaborative ecosystem. Finally, we concluded the chapter with a brief discussion on the roles of optical access networks in the emerging mobile network architectures.

EXERCISES

(7.1) Consider three types of PONs using WDM transmission: (a) WDM PONs with single-stage RN using AWG, (b) TWDM PON using two stages of RN with AWG as the first-stage RN and multiple optical splitters at the second stage, and (c) TWDM PON using splitters only at RNs. Make an exhaustive comparison between these three options in respect of the power budget, cycle time, and hardware complexity of the OLT/ONUs.

(7.2) A long-reach WDM PON has to be set up to serve 64 ONUs over a span of 100 km by using AWG-based RN. If the transmit power of the optical source in the OLT is 3 dBm for each wavelength and the per-channel receiver sensitivity of the ONUs is −28 dBm for 1 Gbps downstream transmission, examine the feasibility of the power budget for downstream connection. Given: fiber loss = 0.2 dB/km, AWG insertion loss = 4 dB, connector loss = 1 dB.

(7.3) A TWDM PON has to be set up to serve 64 ONUs over a span of 40 km at 1 Gbps using eight wavelengths, and the network designer is given two options: (a) a static TWDM PON using AWG-OSP configuration, and (b) set up dynamic TWDM PON using OSP-only configuration. The OSPs are realized with multiple stages of 2×2 optical splitters as building blocks, with each of them having an insertion loss of 0.3 dB. If the transmit power of the optical source at the OLT is 3 dBm over each wavelength and the per-channel receiver sensitivity of the ONUs is −28 dBm, examine the feasibility of the downstream power budget in each case. Given: fiber loss = 0.2 dB/km, AWG insertion loss = 4 dB, connector loss = 1 dB.

(7.4) Consider a static TWDM PON with AWG-OSP configuration. The PON transmits at 1 Gbps (both ways) and supports 128 ONUs over 40 km with a fixed grant size of 10,000 bytes with a guard time of 1 μs. There are two options for the PON design in respect of the AWG and OSPs:

a) 1 × 16 AWG, 1 × 8 OSP,

b) 1 × 8 AWG, 1 × 16 OSP.

Determine the cycle time of a TDMA group for each design. Compare the two designs. Given: propagation time in optical fiber = 5 μs/km.

(7.5) Consider the design of a 128 ONU TWDM PON using two stages of RNs using OSPs only. The TWDM PON needs to cover a distance of 40 km and transmit at 1 Gbps both ways, with a transmit power of 3 dBm. The receivers have a sensitivity of −28 dBm while operating at 1 Gbps, and the OLT allocates a fixed grant size of 10,000 bytes for all ONUs and uses a guard time of 1 μs. If the design needs to ensure that the cycle time for the TDMA groups of ONUs remains confined within 1.5 ms, then estimate the number of wavelengths to be used and check whether any EDFA is needed to realize a feasible power budget. Given: fiber loss = 0.2 dB/km, connector loss = 1 km.

(7.6) A WDM PON uses one 1 : N AWG, which has a non-uniform power distribution $P(\theta)$ across its output ports, following a Gaussian distribution, given by

$$P(\theta) = \frac{1}{\sqrt{2\pi\sigma^2}} \exp\left(\frac{-\theta^2}{2\sigma^2}\right), \tag{7.14}$$

where θ is the angle made by the output ports with respect to the longitudinal axis passing through the input port. The angle θ takes up discrete values across the output ports and σ is determined by the overall AWG construction. Determine the expression for the power ratio PR (in dB) of the signal power levels at the central output port (with N assumed to be odd) and the outermost output port on one of two sides. Using this result, the network provider decides to connect the farthest ONU to the central AWG port, while the closer ONUs are connected to the outer ports of the AWG. If d_{max} and d_{min} are the distances of the farthest and closest ONUs from the RN respectively, determine the expression for the revised PR (as seen at the ONUs) with the above connection strategy for the ONUs.

(7.7) Develop a pseudo code to simulate the EFT and EFT-VF schemes for DWBA in a dynamic TWDM PON with N ONUs and compare the performance of the two schemes using computer simulation for a set of practical network specifications.

WDM Metro Networks

8

With overwhelming growth in traffic, optical metro networks have gone through major changes by employing WDM transmission. The legacy SONET/SDH-metro rings have mostly been upgraded using point-to-point WDM (PPWDM) transmission. Further, for cost-effective realization, WDM metro networks had to employ wavelength-routed optical networking (WRON), where a wavelength is bypassed optically at intermediate nodes. Moreover, the metro networks are in general split hierarchically into core and edge rings, with the metro-core ring interfacing with the long-haul backbone and the metro-edge ring interconnecting the metro-core ring with access segment. The PPWDM/WRON transmission is employed with each wavelength using circuit-switched SONET/SDH transmission, though the metro-edge rings can also use packet-switching for enhancing bandwidth utilization with bursty traffic from the access segment. In this chapter, we consider first the WDM metro networks using PPWDM/WRON-based rings and present their design methodologies using LP-based as well as heuristic schemes. Thereafter, we describe some packet-switched WDM ring testbeds, and examine the possible improvement in bandwidth utilization therein as compared to the circuit-switched WRON rings.

8.1 Metro networks: evolving scenario

Optical metro networks have traditionally been realized over ring topology for its simplicity and resilience. These networks have been using the SONET/SDH (hereafter, referred to as SONET only) technology with voice-optimized circuit-switched connectivity. For the public-domain metro networks, the SONET rings with ADM-based nodes have been readily available from the early days of optical networks, to the extent of being omnipresent, sprawling over large metro areas as well as covering long-haul networking spans using interconnected SONET rings. Thus, the SONET-based metro networks served well for the circuit-switched public telephone network (i.e., PSTN) along with an in-built capacity to carry the packet-switched traffic, albeit in circuit-switched conduits. The ability of the SONET standard to carry packet-switched traffic was further developed using various PoS/EoS technologies (discussed in Chapter 3). Subsequently, with the overwhelming rise in data traffic, corporate MANs, such as FDDI and DQDB, found some relevance, though without any availability for the common public, as needed for the metro networks in general. As described in Chapter 5, in response to these needs, the packet-based resilient ring network, i.e., RPR, was standardized for different classes of traffic, ranging from delay-critical circuit-switched services to the various delay-tolerant packet-switched data services. Thus, SONET and RPR offered viable solutions for the metro-ring networks, with RPR accommodating both delay-sensitive as well as delay-tolerant services with more efficient bandwidth utilization over fiber rings. However, so far the SONET rings with extensive deployment all over the world have remained by and large the main player in the metro domain.

However, gradually with the steady growth of data traffic, the SONET-based metro networks, positioned between the access and long-haul backbone networks, started

Optical Networks. Debasish Datta, Oxford University Press (2021). © Debasish Datta.
DOI: 10.1093/oso/9780198834229.003.0008

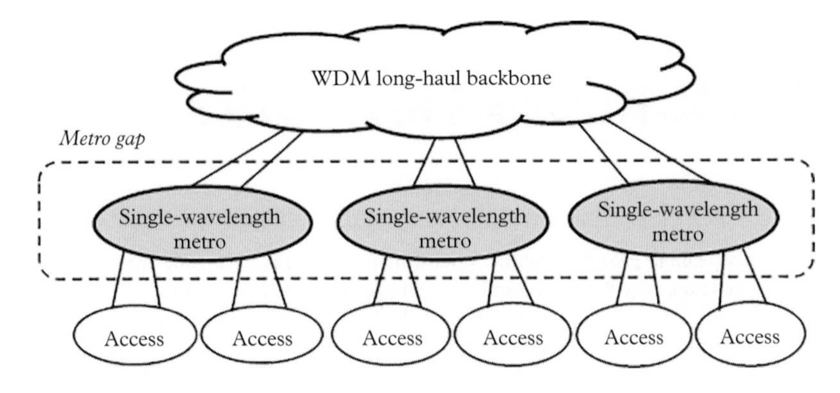

Figure 8.1 *Illustration of the metro gap between backbone and access segments.*

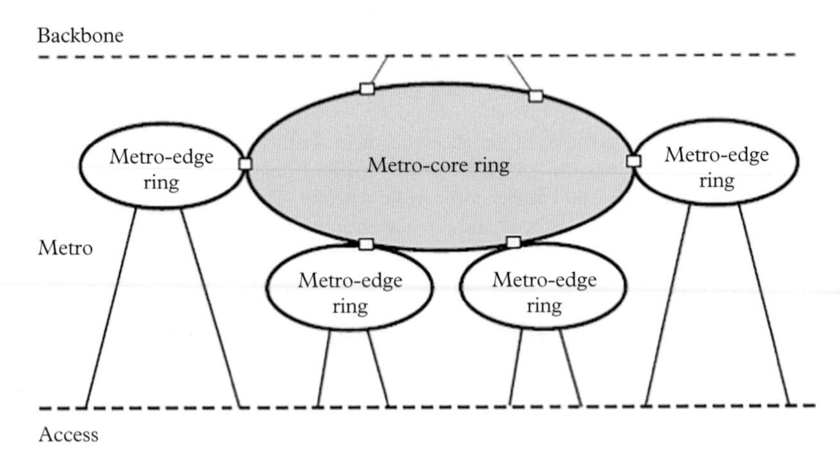

Figure 8.2 *Two-level hierarchy in metro-ring networks.*

facing serious bandwidth bottlenecks with single-wavelength transmission. In particular, the access networks started generating huge volumes of traffic with the emergence of various broadband access technologies, high-speed LANs, wireless access networks, PONs, etc. At this juncture, the capacity of long-haul networks was already available with much-enhanced capacity by the deployment of WDM technology. Thus, being sandwiched between the access segment with huge volume of traffic and the WDM-enabled long-haul networks, the bandwidth constraint of single-wavelength metro networks, popularly known as the metro gap (Fig. 8.1), turned out to be a major bottleneck. This, in effect, hindered the expanding access segment from tapping the bandwidth already available in the long-haul WDM backbone. In order to resolve the problem of the metro gap, introduction of WDM transmission over pre-existing metro networks became imperative.

Before getting into the details of WDM transmission in the metro rings, we discuss briefly how today's high-capacity WDM metro networks have been restructured using ring topology as the basic building block. In particular, the metro-ring networks are being split into two hierarchical levels: metro core and metro edge. In this configuration, multiple metro-edge rings, usually smaller in size than the metro-core rings, are interconnected by a larger metro-core ring (Fig. 8.2). Down the line of hierarchy, the metro-edge rings are connected with different types of access networks, such as ADSL, cable modem, cellular access, WiFi, GbE, 10GbE, PONs, and so on. However, each metro-core ring carries inter-edge-ring traffic and provides connection with the long-haul backbone network. Thus, the access networks get connected to the long-haul network through the metro-edge and metro-core rings.

8.2 Architectural options: PPWDM, wavelength routing, traffic grooming, circuit/packet switching

WDM transmission over a single optical fiber link between two terminal nodes is a simplistic scenario, where one can employ point-to-point transmission over each wavelength, leading to point-to-point WDM (PPWDM) transmission. However, in a network setting with several nodes interconnected by fiber links using a given topology (e.g., ring or mesh), one can provide other possible options, driven by the available WDM technology. Chronologically, in order to augment the capacity of metro-ring networks, the legacy SONET rings initially had to go through a basic upgrading process with PPWDM transmissions over each fiber link. However, the PPWDM links for a large number of wavelengths posed the problems of increased hardware and electronic processing at the network nodes. In particular, in a PPWDM ring, any intermediate node for a given connection has to regenerate the network traffic through OEO conversion for each wavelength and perform hop-by-hop electronic signal processing by using SONET-based ADMs. An example of one such PPWDM ring is shown in Fig. 8.3, operating with three wavelengths, w_1, w_2, and w_3. Each node in this example uses three ADMs with three sets of optical transmitters and receivers (i.e, three transceivers/transponders). For a large number of wavelengths, PPWDM networks must employ as many transceivers, resulting in increased hardware complexity. For example, 32 wavelengths with each of them transmitting at 40 Gbps would require 32 transmitters and 32 receivers (i.e., 32 transceivers) at each node in each direction, needing high-speed network equipment (NEs), e.g., IP routers, which will indeed be a complex hardware requirement along with large power consumption in the electronic devices used therein.

The hardware complexity of the nodes in a PPWDM ring could be reduced significantly by using a useful option, where the intermediate nodes can bypass the transit (i.e., passthrough) traffic carried by a wavelength in the optical domain itself. Such WDM ring networks, known as wavelength-routed rings, employ OADMs (i.e., optical add-drop multiplexers, see Chapter 2) at the network nodes. Note that, the wavelength-routing (WR) technique is also applied in long-haul networks over mesh topologies by using OXCs (i.e., optical crossconnects, see Chapter 2), a variant of OADMs made for mesh topology, and all forms of WR-based optical WDM networks over any possible physical topology are hereafter referred to as wavelength-routed optical networks (WRONs).

In a WRON-based ring, the transit traffic at a given node using OADM, can be bypassed optically without going through the electrical domain, while the traffic destined to this node can be dropped in, and the outgoing traffic from the same node can be added on the same wavelength. Thus, a traffic stream between a given node pair can be carried through several en route fiber links and intermediate nodes over a wavelength along the chosen path, commonly referred to as *lightpath*. The lightpaths, bypassed optically at intermediate nodes, are used to set up end-to-end all-optical connections between the

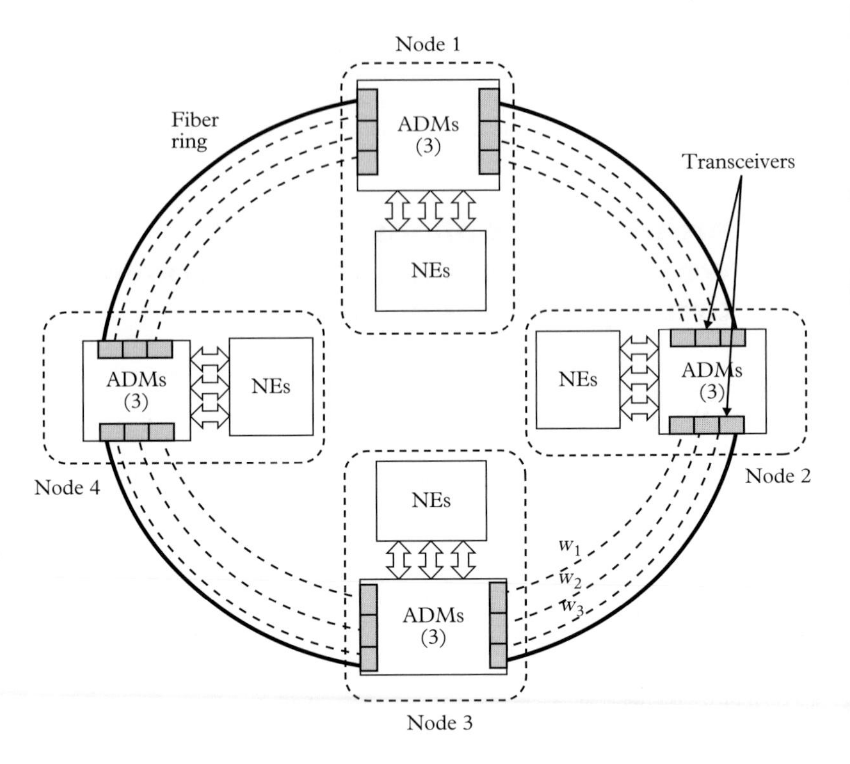

Figure 8.3 *Illustration of a SONET-over-PPWDM ring with three wavelengths, w_1, w_2, and w_3 (operation shown for one of the two counter-rotating rings). The transmitter and receiver (transceiver) of an ADM for each wavelength in one direction require two fiber ports (shown by shaded boxes on either side of ADMs). Every node performs OEO regeneration for each wavelength in each direction and carries out electronic processing in respective ADMs. Hence, for both rings, each node uses six transceivers in the three ADMs which in turn are connected to the respective NEs.*

source and destination nodes. Figure 8.4 illustrates an example four-node WRON ring. Note that, a node in a PPWDM ring (Fig. 8.3) uses one ADM for each wavelength and each ADM needs to have at least one electronic port, that should be compatible with the client traffic streams of PDH, IP, ATM, Ethernet, etc. to connect with the corresponding NE. However, in a WR ring (Fig. 8.4), the transit traffic is bypassed for some specific wavelength(s) at a given intermediate node, while the same node can handle the add/drop traffic using the rest of the wavelength(s) from/to the corresponding NEs. The configuration of the nodes used in WR rings is described in detail in the next section.

Next, we discuss some important features of the WR nodes from the viewpoint of circuit/packet switching and multiplexing options available in any network. In order to examine this issue, we bring in the notion of a network layer, called *optical* or *WDM layer*. The optical layer interconnects the NEs of the possible client (electronic) layers – PDH and SONET as circuit-switched clients, and IP, ATM, Ethernet etc. as packet-switched clients (Fig. 8.5) – with the node optics/optoelectronics consisting of WDM components and optical transceivers. For example, on the left side of the diagram in Fig. 8.5 (where we have not included ATM and other possible intermediate layers for simplicity), while moving downward through the SONET layer, one can visualize a traffic flow through the PDH-over-SONET-over-WDM stack as a three-layer circuit-switched connection using DCS (not shown) for PDH tributaries, PDH-to-SONET interface, ADM and WDM equipment, where the PDH layer functions as a client to the SONET layer and the SONET layer is a client to the optical layer. Further, through the same SONET layer, one can

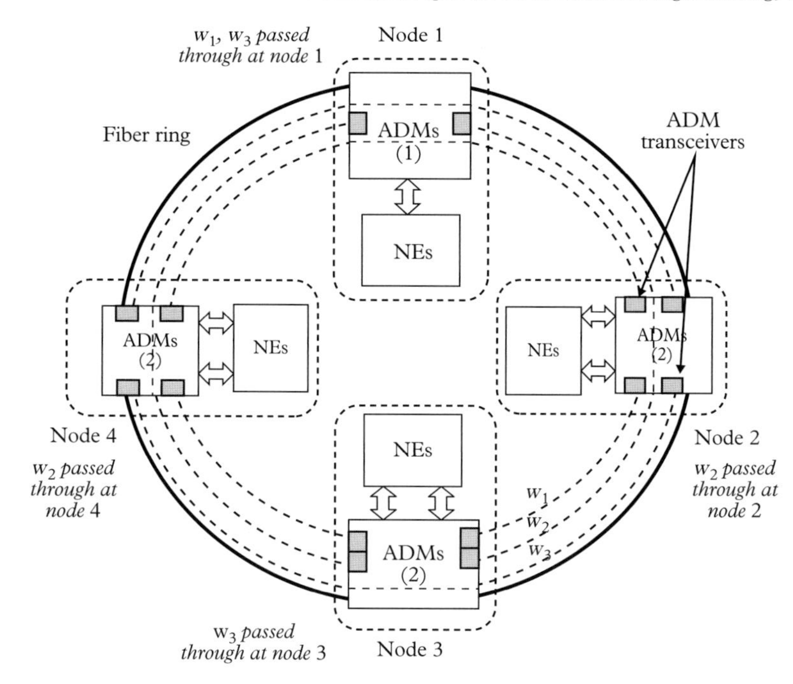

Figure 8.4 *Illustration of a WR ring (operation shown for one of the two counter-rotating rings, which can bypass as well as add-drop wavelengths, as necessary). Node 1 bypasses the wavelengths w_1 and w_3, and thus its only one ADM employs two transceivers for two directions and the ADM is connected to the respective NE to perform add-drop operation at w_2. However, each of the nodes 2, 3, and 4 bypasses one wavelength only (w_2 for nodes 2 and 4, and w_3 for node 3), and thus each of them uses four transceivers for two ADMs in each direction which are in turn connected to the respective NEs.*

visualize two more traffic flows between the packet-switched IP and Ethernet as client layers (functional from IP router and Ethernet switch) and the optical layer. This leads to two three-layered connections: IP-over-SONET-over-WDM and Ethernet-over-SONET-over-WDM, with IP and Ethernet operating through appropriate adaptation stages, i.e., EoS and PoS, respectively.

One noteworthy traffic-blending process takes place in the above example through the SONET equipment (i.e., in the corresponding ADM), where one circuit-switched traffic flow (PDH-over-SONET-over-WDM) is *circuit-multiplexed* with two packet-switched traffic flows (IP/Ethernet-over-SONET-over-WDM). In other words, concurrently over one single wavelength, three *circuits* (i.e., three sets of TDM slots or channels, through three physical ports of ADM) are time-multiplexed, to carry three different types of traffic streams (one being a circuit-switched synchronous traffic from/to PDH layer, and the other two as packet-switched asynchronous data traffic from/to IP and Ethernet layers). Note that, this process of circuit-multiplexing of different traffic flows plays a significant role in utilizing the capacity of the lightpaths in WRONs, popularly known as *traffic grooming* in the literature. The PDH-over-SONET-over-WDM circuit (i.e., some TDM slots in a SONET frame) carries circuit-switched legacy traffic, whose performance is traditionally assessed by the blocking probability. However, IP/Ethernet-over-SONET-over-WDM circuits (some other time slots of the same SONET frame) carry packet-switched traffic, the QoS for which is governed by the delay performance instead of the blocking probability. Thus, the three TDM circuits run concurrently through the same SONET ADM, albeit with different QoS needs. However, for the latter two circuits carrying packet-switched IP/Ethernet traffic, the network does not allow any packet-multiplexing at intermediate nodes along the SONET *path* (unlike in IP/Ethernet/ATM), leading to under-utilization of bandwidth at times. However, as shown in Fig. 8.5 on the right side of the diagram, the

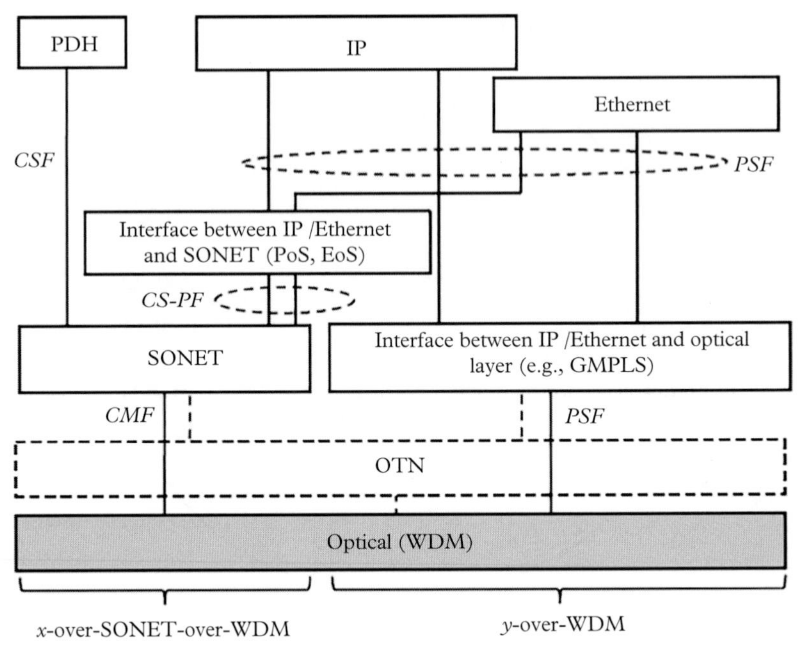

CSF: Circuit-switched flow, *PSF*: Packet-switched flow
CS-PF: Circuit-switched packet flow
CMF: Circuit-multiplexed flow (one *CSF* and two *CS-PFs*)

Figure 8.5 *Client layers working with the optical layer in a WDM ring.*

packet-based client-layer traffic, such as IP and Ethernet packets/frames (through GMPLS/carrier-Ethernet adaptation stage), can bypass SONET layer and directly reach the optical layer with fully packet-switched transmission, leading to the IP/Ethernet-over-WDM (two-layered) connections.

Thus the client layers in optical WDM networks can take effect from the respective NEs: IP router, Ethernet switch, ATM switch, DCS etc., in two possible ways: one being setup through SONET ADMs (x-over-SONET-over-WDM) using circuit-switched transmission of constant bit-rate PDH tributary or bursty packets, and the other using fully packet-switched transmission over WDM (y-over-WDM) without needing the SONET ADMs. Note that the second option does not allow the direct entry of PDH tributaries into WDM devices, i.e., y excludes the PDH tributaries, while x in the first option can accommodate both circuit-switched PDH tributaries as well as bursty packet streams, albeit with under-utilization of bandwidth for the circuit-switched packets. In Fig. 8.5, along with the above architectural layers, another layer representing the OTN-based convergence platform (see Chapter 5) is shown (with a dotted boundary) just above the optical layer, through which the high-speed network traffic flows (e.g., SONET, Ethernet) can be encapsulated appropriately for WDM transmission.

As discussed above, while introducing WDM in the existing optical networks, most of the early studies (Zhang and Qiao 1997; Gerstel *et al.* 1998a; Gerstel *et al.* 1998b; Wang *et al.* 2001) were directed toward utilizing WDM with the legacy circuit-switched SONET architecture (x-over-SONET-over-WDM) to carry the circuit-switched as well as the packet-switched traffic streams. However, the bandwidth utilization of the circuit-switched *bandwidth pipes* (OC-n) of SONET gradually turned out to be inefficient, as the major chunk of network traffic grew up in the form of bursty packet streams. Thus, with the huge growth of packet-switched traffic, the circuit-switched SONET rings faced challenges, more so at the metro-edge level. The metro-core rings need to interface large-bandwidth flows or fat pipes (e.g., OC-192 and higher ones) of long-haul backbones with their own moderate-bandwidth flows (e.g., OC-12, OC-48), where the burstiness of traffic flows is less visible. On the other hand, in the metro-edge rings, smooth packet flows are less likely, owing to the dominance of packet-based data traffic in the access segment. Thus, at the metro-edge level, WDM rings called for a need to migrate from the existing circuit-switched to an appropriate packet-switched architecture, wherever necessary, so as to carry the new generation of traffic with higher bandwidth utilization. This development motivated research on various possible versions of packet-switched WDM metro rings, such as, MAWSON, RingO, HORNET etc. (Herzog *et al.* 2004). In the following sections in this chapter, we therefore consider the WDM metro ring networks of two types: (*i*) SONET-over-WDM rings carrying circuit-switched traffic and (*ii*) packet-switched WDM rings using appropriate MAC protocols.

8.3 Circuit-switched SONET-over-WDM metro ring

As shown earlier, in Figures 8.3 and 8.4, the PPWDM and WR rings are examples of SONET-over-WDM rings. However, these figures indicate that the node architectures for PPWDM and WRON rings can differ significantly in respect of the ADM requirements. The PPWDM nodes, by using an ADM for each wavelength, can employ more flexible traffic grooming with access to all the time slots of all the wavelengths for time-multiplexing. However, a WRON node with OADM/ROADM can reduce the number of ADMs by offering the optical bypass mechanism for the wavelengths passing through the node, but without having any access to the empty time slots (if any) of the bypassed wavelengths.

For the PPWDM as well as WRON rings, the traffic streams from the client layers can access the available wavelengths in a node, with various possible combinations, e.g., PDH-over-SONET, ATM-over-SONET, IP-over-SONET, and Ethernet-over-SONET. Accordingly, appropriate NEs need to be interfaced with each ADM employed in a PPWDM ring. As shown in Fig. 8.6, in order to have an ATM-over-SONET-WDM and an

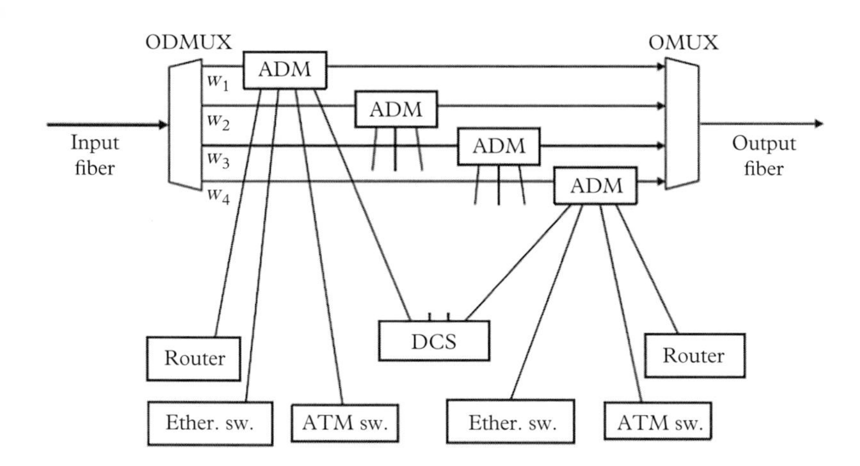

Figure 8.6 *Typical node using OADM in a PPWDM-based SONET ring with the client-layer NEs, such as routers, Ethernet and ATM switches, and DCS.*

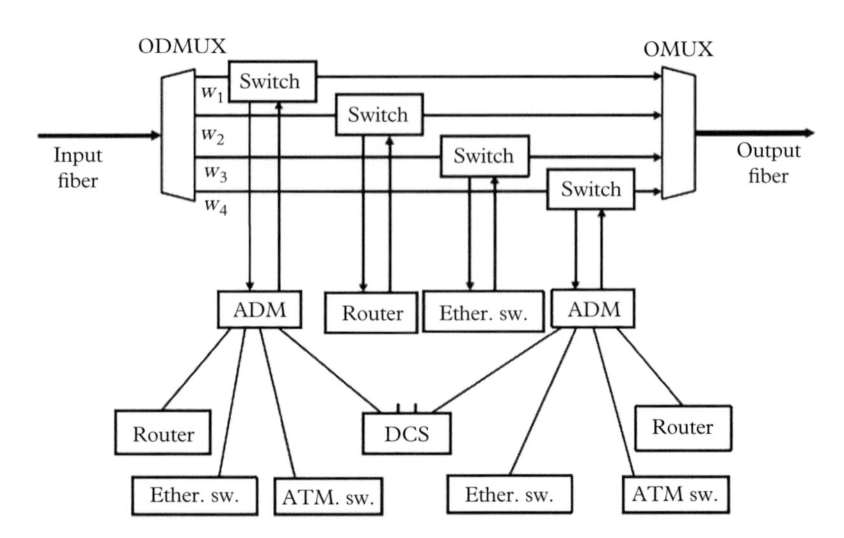

Figure 8.7 *Typical ROADM-based node in a WRON-based SONET ring with the client-layer NEs, where the ROADM employs optical demultiplexer (ODMUX), optical multiplexer (OMUX), and optical switches for dynamic reconfiguration of lightpaths. Note that some optical switches may be directly connected to the IP routers with appropriate interfaces using packet-switched transmission over some select wavelengths; however, this aspect is not considered in this chapter.*

IP-over-SONET-over-WDM connection in the PPWDM node, an ATM switch and an IP router have to be connected as client NEs with an ADM, and the ADM will have to choose appropriate wavelengths for onward transmission of the two traffic streams over PPWDM ring.[1] Moreover, for redistributing the traffic across the wavelengths, each node might also need a DCS. A similar set of NEs would also be needed with the ADMs operating with a ROADM at a WR node (Fig. 8.7), where the optical switches bring in the reconfigurability of the node in respect of add/drop/passthrough operations on the wavelengths.

Note that, such SONET-over-WDM rings would operate with semi-permanent connections, set up over a reasonably long period of time (days to months, typically), thereby practically handling nearly static traffic patterns. In the following, we examine the WDM metro networks using SONET-over-PPWDM/WRON architectures with static uniform traffic. Thereafter, we consider the design methodologies for the SONET-over-WRON rings with static nonuniform traffic.

8.3.1 PPWDM ring: uniform traffic

SONET-over-PPWDM ring networks have been studied extensively in the literature, for both static and dynamic traffic patterns (Ramaswami *et al.* 2010; Mukherjee 2006; Wang *et al.* 2001; Gerstel *et al.* 1998a; Gerstel *et al.* 1998b; Zhang and Qiao 1997). In this section, we consider static and uniform ingress traffic from the nodes to examine the basic design issues in such networks. As mentioned earlier, SONET-over-PPWDM ring networks employ OEO conversion for every wavelength at each node for each of the two counter-rotating unidirectional rings in clockwise (CW) and counter-clockwise (CCW) directions. Thus, each lightpath on each ring between two neighboring nodes (corresponding to a wavelength) is generated from the transmitter of one node and terminated at the receiver of the other node. So each node in a ring needs one transmitter and one receiver (i.e., one *transceiver*

[1] Note that, earlier in Chapter 2, the configurations of OADMs/ROADMs were discussed, but without involving the ADMs and client-layer NEs.

set) for each wavelength. If the network operates with M wavelengths on each ring, then for both CW and CCW rings each node will need N_{xr}^{node} transceivers and N_{adm}^{node} ADMs, given by

$$N_{xr}^{node} = 2M \tag{8.1}$$

$$N_{adm}^{node} = M. \tag{8.2}$$

Note that each ADM shown in the PPWDM node of Fig. 8.6 exhibits one half of its full functionality (for the left-to-right, i.e., CW ring), and in reality it will have its other half (not shown) operating for the same wavelength in the CCW ring. Assuming that there are N nodes operating over M wavelengths on each ring, the network as a whole will therefore need to employ N_{xr} transceivers and N_{adm} ADMs, given by

$$N_{xr} = N \times N_{xr}^{node} = 2NM \tag{8.3}$$

$$N_{adm} = N \times N_{adm}^{node} = NM. \tag{8.4}$$

For the present discussion, we assume that the ring has an even number of nodes, i.e., N is even. An example eight-node PPWDM ring is illustrated in Fig. 8.8, where we assume that each node sends out the same ingress traffic ρ, and this traffic is uniformly distributed to the rest of the nodes (i.e., to other seven nodes). In the present context (circuit-switched SONET traffic), the ingress traffic ρ is defined as the ratio of the number (n_L) of lower-rate ingress SONET traffic streams (typically, OC-3s) required to the number (n_C) of lower-rate tributaries that can be *groomed* into one WDM channel, i.e., $\rho = n_L/n_C$. For example, with OC-48 over each WDM channel (capable of grooming 16 OC-3s) in a metro ring, an ingress request of $n_L = 112$ (i.e., 112 SONET streams of OC-3) would result in a traffic $\rho = 112/16 = 7$, which would be flowing out to the rest of the nodes in the ring. Similarly, each node will also receive egress traffic of the same magnitude with the present assumption.

As a whole, the ring network is bidirectional with CW and CCW rings, though in Fig. 8.8 we show only the CW ring for simplicity. Each connection is set up on the CW or CCW ring (sometimes on both rings, with traffic splitting), governed by the criterion that the connection takes place over the shortest path. For example, a connection from node 1 to node 3 will be placed over the CW ring through node 2, as that would correspond to a shorter path, than the alternative path (i.e., the path through the nodes 1-8-7-6-5-4-3) over the CCW ring. With this framework, we next proceed to describe some heuristic design methodologies for the PPWDM ring.

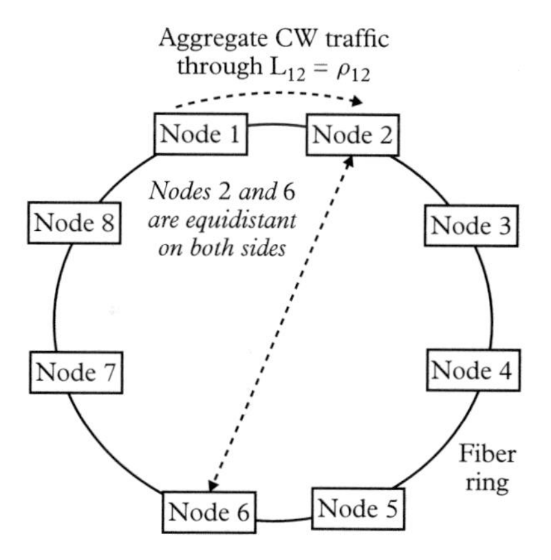

Figure 8.8 *Illustration of an eight-node PPWDM-based ring (for CW transmission) using shortest-path routing. Note that, based on shortest-path routing, as with nodes 2 and 6, there are several node pairs that will have the same path-cost over both sides of the ring. These node pairs will divide equally their traffic streams over the two sides using the CW and CCW rings. The CW ring, as shown in this figure, will thus carry 50% of the respective ingress traffic flows of these node pairs.*

Since all nodes are assumed to inject the same amount of ingress traffic ρ (with uniform distribution of the traffic to the remaining $N-1$ nodes), each link in each ring will carry the same traffic. Hence, due to the circularity of ring topology, it is possible to determine the traffic load on any of the two fiber rings by estimating the total load on any given fiber link between two adjacent nodes. We therefore consider the link from node 1 to node 2 (L_{12}) as the representative link and estimate its aggregate traffic ρ_{12} composed of the ingress traffic from all the node pairs on the CW ring.

Before enumerating the aggregate traffic ρ_{12} as defined above, we make some important observations. First we note that, on the left side of node 1, only the nodes that have a path-cost (number of links along the path) $\leq$ its maximum value (= half of the total number of links in the ring = 4 in this example), i.e., nodes 6, 7, 8, and 1, will send traffic to/through node 1 over the CW ring only. Further, these traffic components will have two categories, as discussed below.

- Node pairs (6-2, 7-3, 8-4, 1-5) have equal path-cost over both CW and CCW rings (each path covering half of the ring, i.e., four links). So nodes 6, 7, 8, and 1 will split their traffic components equally between CW and CCW rings. Hence, each of the CW traffic components through L_{12} for the node pairs 6-2, 7-3, 8-4, and 1-5, represented byρ_{12}^{62}, ρ_{12}^{73}, ρ_{12}^{84} and ρ_{12}^{15}, respectively, will be given by

$$\rho_{12}^{62} = \rho_{12}^{73} = \rho_{12}^{84} = \rho_{12}^{15} = \frac{\rho}{2 \times (N-1)} = \frac{\rho}{2 \times 7}. \tag{8.5}$$

- However, with the path-cost being < 4, the CW traffic components between the node pairs 7-2, 8-2, and 1-2 will be simply $\rho/(N-1) = \rho/7$. Hence, all the CW traffic components through L_{12} (*i*) to node 2 from the nodes 7, 8, and 1 (denoted as ρ_{12}^{72}, ρ_{12}^{82}, ρ_{12}^{12}), (*ii*) to node 3 from the nodes 8 and 1 (denoted as ρ_{12}^{83}, ρ_{12}^{13}), and (*iii*) to node 4 from node 1 (denoted as ρ_{12}^{14}) will be equal to $\rho/(N-1)$, i.e.,

$$\rho_{12}^{72} = \rho_{12}^{82} = \rho_{12}^{12} = \frac{\rho}{N-1} = \frac{\rho}{7} \tag{8.6}$$

$$\rho_{12}^{83} = \rho_{12}^{13} = \frac{\rho}{N-1} = \frac{\rho}{7} \tag{8.7}$$

$$\rho_{12}^{14} = \frac{\rho}{N-1} = \frac{\rho}{7}. \tag{8.8}$$

In view of the above, we can express the total CW traffic ρ_{12} through link L_{12} from node 1 to node 2 as a sum of traffic components of the above two categories as

$$\rho_{12} = \left[\rho_{12}^{62} + \rho_{12}^{73} + \rho_{12}^{84} + \rho_{12}^{15}\right] + \left[\left(\rho_{12}^{72} + \rho_{12}^{82} + \rho_{12}^{12}\right) + \left(\rho_{12}^{83} + \rho_{12}^{13}\right) + \left(\rho_{12}^{14}\right)\right] \tag{8.9}$$

Substituting the values for the respective traffic components, one gets

$$\rho_{12} = \left[4 \times \frac{\rho}{14}\right] + \left[\left(3 \times \frac{\rho}{7}\right) + \left(2 \times \frac{\rho}{7}\right) + \left(\frac{\rho}{7}\right)\right]. \tag{8.10}$$

As mentioned earlier, with the identical ingress traffic from each node along with its uniform distribution among all other nodes, the aggregate traffic in any link of the PPWDM ring (for both CW and CCW rings) would be same as ρ_{12}. By using this observation and generalizing Eq. 8.10 for N nodes (N being even in this example), we express the aggregate traffic over any of the links in an N-node PPWDM ring as

$$\begin{aligned}\rho_{link} &= \left(\frac{N}{2}\right)\frac{\rho}{2(N-1)} + \left(\frac{N}{2} - 1\right)\frac{\rho}{(N-1)} \\ &+ \left(\frac{N}{2} - 2\right)\frac{\rho}{(N-1)} + \ldots + \frac{2\rho}{(N-1)} + \frac{\rho}{(N-1)} \\ &= \left[\frac{\rho}{(N-1)}\right]\left[\frac{N}{4} + \frac{N(N-2)}{8}\right] = \frac{N^2\rho}{8(N-1)}.\end{aligned} \tag{8.11}$$

Therefore, the number of wavelengths M required to support the above traffic will be given by ρ_{link}, rounded-up to the next integer value, i.e.,

$$M = \lceil \rho_{link} \rceil. \tag{8.12}$$

Using the above result in Equations 8.1 and 8.2, we express the numbers of transceivers and ADMs, N_{xr}^{node} and N_{adm}^{node}, respectively, in a PPWDM node for both rings as

$$N_{xr}^{node} = 2M = 2\lceil \rho_{link} \rceil \tag{8.13}$$

$$N_{adm}^{node} = M = \lceil \rho_{link} \rceil. \tag{8.14}$$

Therefore, the total numbers of transceivers and ADMs, denoted as N_{xr} and N_{adm}, respectively, for the entire network can be expressed as (with N as an even number)

$$N_{xr} = 2N_{adm} = 2NM = 2N\lceil \rho_{link} \rceil. \tag{8.15}$$

Following a similar approach for an odd number of nodes one can also determine the expressions for M, N_{xr}, and N_{adm}.

It may be worthwhile to note that, for a 16-node PPWDM ring with 40 wavelengths, the network will have to deploy as many as 640 ADMs. Due to this considerable hardware requirement in PPWDM rings, WR rings are worth considering as a cost-efficient option, more so when DWDM transmission needs to be realized over metro networks. In the following section, we consider the problem of designing WRON rings with uniform traffic using ROADMs with the objective of keeping the number of wavelengths to a minimum.

8.3.2 WRON ring: uniform traffic

In this section, we present a heuristic design methodology for SONET-over-WRON ring for a uniform traffic pattern, as considered in the case of the PPWDM ring. With the uniform-traffic assumption, we consider again that the ingress traffic at each node is ρ, which is equally distributed to $(N-1)$ destination nodes, with $\lceil \rho/(N-1) \rceil$ as the traffic demand between each node pair (since in this case end-to-end connections are set up using one or more *full* wavelengths, i.e., lightpaths). The node pair for each connection needs to be provisioned with all-optical (i.e., single-hop) connection, optically bypassing all intermediate nodes (if any) by means of wavelength routing. Note that, with uniform traffic pattern and distribution, one can assume the connection between each node pair to be bidirectional, and hence the lightpaths are needed in pairs in two directions for each node pair by using the CW and CCW fiber rings.

To start with, we assume the traffic between a node pair as unity, ie., $\lceil \rho/(N-1) \rceil = 1$. This implies that each node pair is assumed to have a traffic demand, not exceeding one wavelength (i.e., one lightpath), which we represent by a bidirectional *arc* on a *circle* (Fig. 8.9). With this assumption, we determine the minimum number ($\zeta(N)$, say) of circles needed for N nodes, while ensuring that in each circle the constituent bidirectional arcs (or simply arcs) don't have any overlap. Having constructed $\zeta(N)$ circles, we multiply $\zeta(N)$ by the actual traffic demand $\lceil \rho/(N-1) \rceil$ to get the total number of wavelengths, given by

$$M = \left\lceil \frac{\rho}{N-1} \right\rceil \zeta(N). \tag{8.16}$$

Next, we estimate $\zeta(N)$ for setting up the single-lightpath connections for all node pairs by using an iterative algorithm. As mentioned above, if the node-pair traffic $\rho/(N-1) \leq 1$,

C1
i
a_{ij}
Bidirectional
WDM fiber ring
j
(a)
Circle
Arc

Iteration 1: Two nodes
One circle C_1 constructed with one arc (a_{ij})

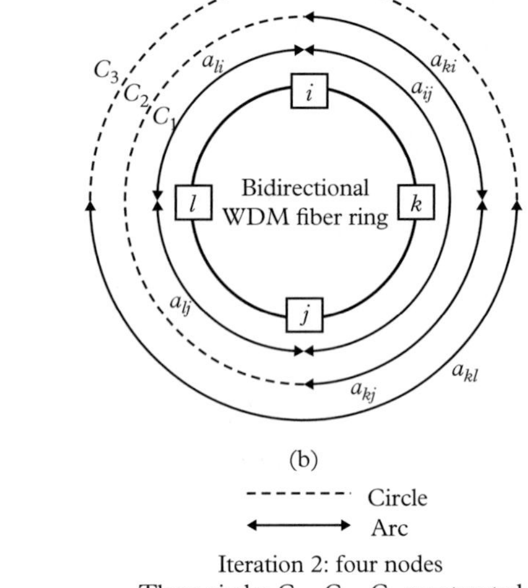

Iteration 2: four nodes
Three circles C_1 , C_2 , C_3 constructed with six arcs (a_{ij} , a_{li} , a_{lj} , a_{ki} , a_{kj} , a_{kl})

Figure 8.9 *Illustration of circle construction in SONET-over-WRON rings: (a) two-node ring, (b) four-node ring.*

then each circle will represent one wavelength, or else it might need a bundle of wavelengths. At this stage, we continue with this assumption (i.e., $\rho/(N-1) \leq 1$) and at the end, as already mentioned, convert the number of circles into the actual number of wavelengths.

The wavelength assignment scheme needs to ensure that *each circle once constructed gets maximally utilized, and a next circle is constructed only when the previous circle gets completely used up without any gap, or with a gap that cannot accommodate the remaining connections.* We carry out an iterative algorithm, starting with two nodes, and in every next iteration we add another pair to eventually develop a generalized expression for M. We carry three iterations with two, four, and six nodes and generalize the formulation thereafter by induction.

Iteration 1

In the first iteration, we consider two nodes (see Fig. 8.9(a)), i.e., nodes i and j (i.e., $N = 2 = N_1$, say). With this node pair, we set up an all-optical bidirectional connection represented by an *arc* a_{ij} on a circle C_1, with the arc representing a pair of lightpaths (one each on CW and CCW rings) partially filling up the circle, which can be laid either on the left or on the right side of the ring with respect to the diameter joining the two nodes. Without any loss of generality, let's assume that the arc is set up on the right side of the ring, as shown in Fig. 8.9(a). Thus, for the two given nodes on the ring, the number of circles $\zeta(2)$ needed to set up arc a_{ij} is just one, although the circle remains half-utilized.

Iteration 2

In the second iteration, we add two more nodes (nodes k and l, say) in the ring, interleaved between the earlier two nodes (Fig.8.9(b)). With the four nodes, it will require altogether ${}^4C_2 = 6$ arcs, which might require more circles. Out of the six arcs needed, we have already set up one arc using circle C_1 in the first iteration. We need to first utilize circle C_1 as much as possible and then construct more circles, when necessary. The circle construction process for these arcs is carried out as follows.

- We begin with arc a_{ij} already set-up in circle C_1, and identify the arcs to be introduced for connecting the two newly introduced nodes, i.e., the nodes k and l, with the earlier two nodes and also to set up connection between these two new nodes. In particular, we observe the following.
  - connections (arcs) needed for node k: a_{ki}, a_{kj}, a_{kl},
  - connections (arcs) needed for node l: a_{li}, a_{lj} (a_{lk} is same as a_{kl}).
- For node k, we have to introduce a new circle because the existing circle C_1 runs across this node (i.e., is optically bypassed). However, node l can utilize circle C_1. Therefore, we plug the arcs a_{li}, a_{lj} into the remaining gap of circle C_1. This realizes full utilization of C_1 (see Fig. 8.9(b)).
- Next we introduce a new circle C_2 and try to provision the remaining connections (arcs) without any overlap. We start with node k, which was deprived of any connection in circle C_1, and set up the arcs a_{ki}, a_{kj} on C_2.
- The remaining arc is to be set up between the nodes k and l, i.e., a_{kl}, which does not find any suitable gap in C_2. Hence, we construct the third circle C_3, and set up a_{kl} on C_3. No more connection is left in this iteration, leaving the circle C_3 partially filled. Note that, at this stage, altogether we have used three circles for four nodes.

Thus, after the second iteration with four nodes ($N = 4 = N_2$, say), one needs three circles C_1, C_2, C_3, as shown in Fig.8.9(b). Hence, the number of circles $\zeta(4)$ required for four nodes is three, which can be expressed recursively using $\zeta(2)$ and N_1 as

$$\zeta(4) = 3 = \left[\zeta(2) + \left(\frac{N_1}{2} + 1\right)\right] = 1 + \left(\frac{2}{2} + 1\right) \tag{8.17}$$

We validate this recursive relation in the next iteration.

Iteration 3

In the third iteration, we bring in two more nodes (nodes m and n, not shown in Fig. 8.9) interleaved between nodes i and l and nodes j and k, respectively, and assign arcs on the existing and new circles, if necessary. In particular, we need to set up altogether ${}^6C_2 = 15$ arcs using an adequate number of circles. Following the approach used in the second iteration, the circle construction process of the third iteration takes effect as follows.

- All the arcs already setup on the three circles (C_1, C_2, C_3) in the second iteration are retained. With this setting, we next identify the arcs to be introduced to connect the nodes m and n with the other nodes, including the connection between these two nodes as well. In particular, we observe the following.
  - connections (arcs) needed for node m: a_{mi}, a_{mj}, a_{mk}, a_{ml}, a_{mn},
  - connections (arcs) needed for node n: a_{ni}, a_{nj}, a_{nk}, a_{nl}.
- We observe that there are some gaps left out in the circles C_2 and C_3 which were constructed in the second iteration. The gap in C_2 is accessible to node n, but not to node m as C_2 runs through node m in the optical domain. So we setup arcs a_{mi} and a_{mj} on C_2. This exhausts the available gap in C_2.
- The gap available in C_3 is accessible to node m only, and using this gap we set up arcs a_{mk} and a_{ml} on C_3. This exhausts the available gap in C_3, and we need to introduce a next circle for setting up the remaining arcs.

- We introduce the fourth circle C_4 and try to provision the remaining connections. Going by the same approach, we set up arcs a_{ni} and a_{nj} on C_4. None of the remaining arcs can be plugged in C_4, and it calls for the next circle.
- Circle C_5 is constructed, and we set up the arcs a_{nk} and a_{nl} on C_5. C_5 cannot accommodate the left-out arc a_{mn}, and calls for another circle.
- Finally the last circle C_6 is constructed to set up arc a_{mn}. No more connection is left in this iteration leaving the circles C_4, C_5, and C_6 partially filled, and altogether we have constructed six circles for six nodes.

Note that the number of circles for this iteration ($\zeta(6)$) can again be expressed as

$$\begin{aligned}\zeta(6) &= 6 \\ &= \left[\zeta(4) + \left(\frac{N_2}{2} + 1\right)\right] \\ &= \left[\zeta(2) + \left(\frac{N_1}{2} + 1\right) + \left(\frac{N_2}{2} + 1\right)\right] \\ &= \left[1 + \left(\frac{2}{2} + 1\right) + \left(\frac{4}{2} + 1\right)\right].\end{aligned} \tag{8.18}$$

Generalization

Using the formation of Eq. 8.18, one can now make use of its underlying recursive behavior to obtain a generalized expression for the number of circles $\zeta(N)$ for N nodes as follows

$$\begin{aligned}\zeta(N) &= \left[1 + \left(\frac{2}{2} + 1\right) + \left(\frac{4}{2} + 1\right) + \cdots + \left(\frac{N-2}{2} + 1\right)\right] \\ &= 1 + 2 + 3 + \cdots + N/2 \\ &= \frac{N^2}{8} + \frac{N}{4}.\end{aligned} \tag{8.19}$$

Using the above expression for $\zeta(N)$, one can obtain the total number of wavelengths M for the N-node (N being even) SONET-over-WRON ring as

$$M = \left(\frac{N^2}{8} + \frac{N}{4}\right)\left\lceil \frac{\rho}{N-1} \right\rceil. \tag{8.20}$$

Considering that each node operates with $(N-1)$ arcs to reach as many nodes, the number of transceivers in a node (N_{xr}^{node}) is the product of the number of arcs and the number of wavelengths in each arc, given by

$$N_{xr}^{node} = (N-1)\left\lceil \frac{\rho}{N-1} \right\rceil, \tag{8.21}$$

thereby leading to the total number of transceivers required in the entire network (N_{xr}) as

$$N_{xr} = N \times N_{xr}^{node} = N(N-1)\left\lceil \frac{\rho}{N-1} \right\rceil. \tag{8.22}$$

In order to determine the number of ADMs needed in the WRON design, we consider the various circles used in the respective designs. For example in the two node example, one needs $N_{adm} = 2\lceil\rho\rceil$. For the four-node ring, one needs to count the number of ADMs for each circle. Thus, for the innermost circle there are three arcs and all are connected without any gap, leading to the need of 3 ADMs for each unit of traffic and hence $3\lceil\rho/3\rceil$ ADMs with the actual traffic. Further, the second circle has 2 arcs connected at node k, thus needing 3 ADMs for each unit of traffic. The outermost circle has only one arc, needing thereby 2 ADMs for unit traffic. Therefore, the total number of ADMs used in the four-node WRON ring is $N_{adm} = (3+3+2)\lceil\rho/3\rceil = 8\lceil\rho/3\rceil$. Similarly, for the six-node WRON ring, one can construct the circles and find out the number of ADMs.

Finally, note that the above provisioning scheme considers the minimization of the number of wavelengths (circles), which is relevant when the traffic is high enough, making the availability of adequate wavelengths a major issue. However, with moderate traffic demands when the upper limit of available wavelengths remains higher than the wavelength requirement, it becomes worthwhile to attempt to minimize the number of ADMs for cost minimization. We consider this design issue in the following for more realistic situations with *nonuniform* traffic patterns, by using LP-based design optimization as well as heuristic schemes.

8.3.3 WRON ring: nonuniform traffic

In the foregoing sections, we considered the problem of resource provisioning in WRON rings with identical ingress traffic in all nodes and uniform distribution of traffic from each node to others. While this approach gives a basic understanding of the resource-provisioning issues (e.g., the required number of wavelengths, transceivers, and ADMs), in a practical scenario the traffic ceases to be uniform between the nodes. The resource-provisioning task in such cases is more complex and is usually done off-line to set-up a network initially, which can be augmented from time to time with changed traffic demands. In general, such network design methodologies lead to optimization problems.

In view of the above, we consider two design methodologies in the following. First, we consider a design optimization technique based on LP formulation, which is, however, computationally intensive and may not scale up well for large networks. For large networks, one needs to employ heuristic schemes, which we discuss after describing the LP-based design methodologies.

LP-based design

The LP-based design methodology considers the network traffic as a set of multi-commodity flows over a graph with its vertices and edges representing the nodes and the fiber links of the physical topology, respectively. We therefore represent the WRON-based ring by a graph $G(V,E)$, with V as the set of vertices and E as the set of edges of the graph. Thus, in $G(V,E)$, each vertex represents a network node n (with $n \in V$) and each edge e_j represents a fiber link (with $e_j \in E$) between two adjacent network nodes, say node $(n-1)$ and node n.

The above N-node WRON ring would support M concentric circles, with each circle representing a distinct wavelength or WDM channel (as in Fig. 8.9). Furthermore, each circle (with a data rate, say OC-48) will consist of several *sub-circles*, representing the lower-rate SONET streams (e.g., OC-3), meaning thereby a grooming capacity of $48/3 = 16$. In other words, each circle (OC-48 frame in this example) will be able to groom or circuit-multiplex at the most G (= 16 in this example) smaller frames (or TDM slots) of OC-3 streams, corresponding to as many sub-circles.

In view of the above, the overall design should deal with the formation of these circles and their constituent sub-circles, while keeping the overall cost (number of ADMs) at a minimum. Thus, the design problem turns into an LP-based optimization (minimization) problem by using a set of linear constraints along with a linear objective function. The set of constraints provides the convex solution space, which is searched systematically by the LP-solver to minimize the objective function. Following the line of treatment in (Wang *et al.* 2001), we formulate the LP-based design problem for the SONET-over-WRON ring in the following.

In order to write the set of linear constraints and the objective function, we define the following parameters/variables:

- N: number of nodes in the ring,
- M: number of wavelengths or WDM channels $(w_1, w_2, ..., w_m, ..., w_M)$, represented by as many circles in the ring,
- G: maximum number of the lower-rate TDM slots of SONET (i.e., sub-circles) that can be groomed in the higher-rate TDM frame of SONET (i.e., circle) transmitted over a wavelength, thereby representing the *grooming capacity* of a wavelength,
- $\Gamma(\rho_{sd})$: traffic matrix, with ρ_{sd} as the traffic demand from node s to node d. Note that, ρ_{sd} represents the number of TDM slots to be provisioned as sub-circles over one or more circles (i.e., wavelengths) in CW and/or CCW rings $(\rho_{nn} = 0, \forall n)$,
- $TC_{sd}^{km,\delta}$: kth TDM slot (sub-circle) from node s to node d over a wavelength (circle) w_m along a direction δ, which is a binary variable $\in (0, 1)$, representing CW or CCW direction in the ring, respectively,
- N_{adm}^{nm}: number of ADMs operating on a wavelength w_m in node n. N_{adm}^{nm} is a binary variable $\in (0, 1)$,
- N_{adm}: total number of ADMs in entire network $= \sum_n \sum_m N_{adm}^{nm}$,
- ${}^{sd}O_{e_j}^{km,\delta}$: occupancy of kth TDM slot on wavelength w_m on edge e_j from node s to node d along a direction δ. ${}^{sd}O_{e_j}^{km,\delta}$ is a binary variable $\in (0, 1)$.

Figure 8.10 illustrates the function of a WRON node (node n, say), where a SONET TDM frame of a wavelength (circle) drops data of one sub-circle (i.e., a TDM slot) and also adds some data in the same sub-circle or TDM slot for onward transmission over the same wavelength. Thus the wavelength w_m goes through OEO conversion at node n, while the other wavelengths in the same node are bypassed in the optical domain.

The overall design of a WRON ring needs to use the given basic resources – M wavelengths or circles that can groom G lower-rate SONET streams or sub-circles, N nodes and the traffic matrix $\Gamma(\rho_{sd})$ – to find out an appropriate set of TDM connections $\{TC_{sd}^{km,\delta}\}$, so that the total number of ADMs in the network is minimized. As mentioned before, this design in turn amounts to solving an LP problem, defined by a linear objective function and a set of linear constraints (equalities/inequalities). The objective function and the constraints for the design problem are expressed as follows.

Objective function To minimize the total number of ADMs used for all operating wavelengths at all the network nodes, i.e.,

$$Minimize \sum_{n=1}^{N} \sum_{m=1}^{M} N_{adm}^{nm}. \tag{8.23}$$

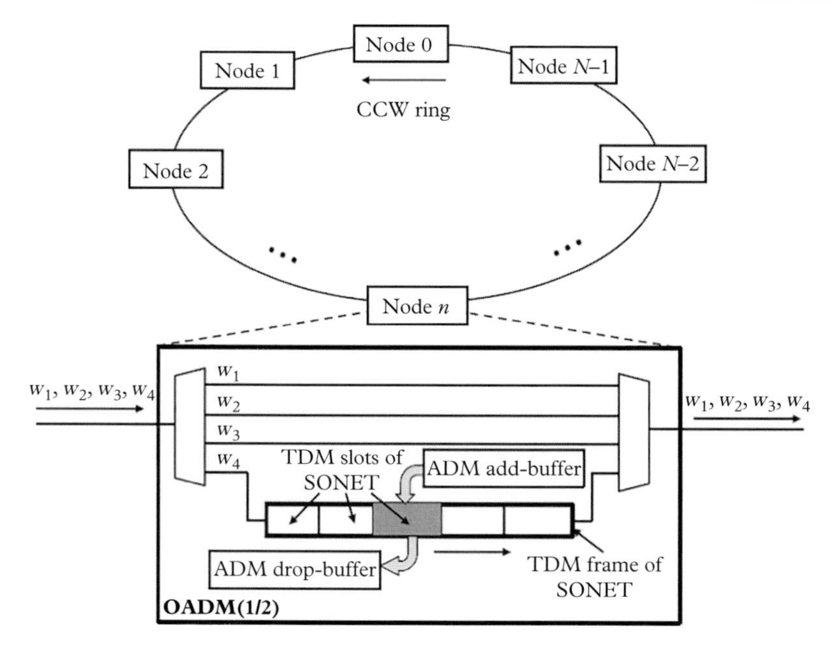

Figure 8.10 *Node configuration (node n, say) in an N-node WR ring using OADM. Each wavelength represents a circle, while the lower-rate TDM slots (groomed into a wavelength) represent the sub-circles. The shaded slot goes through the drop followed by the add operations on wavelength w_4, while the other three wavelengths, w_1, w_2, and w_3, are bypassed at node n using WR technology.*

Constraints

- *Traffic-load:* Between a given node pair sd (with s and d as the source and destination nodes, respectively), the traffic carried by all TDM slots ($TC_{sd}^{km,\delta}$, with $k = 1, 2, \ldots, G$) over all the wavelengths ($m = 1, 2, \ldots, M$) in two directions ($\delta = 0$ and 1) should be equal to the traffic load between the corresponding node pair (ρ_{sd}), i.e.,

$$\sum_m \sum_k \sum_\delta TC_{sd}^{km,\delta} = \rho_{sd} \qquad \forall(s, d). \tag{8.24}$$

- *Occupancy for a time slot in a wavelength over an edge along a particular direction (CW or CCW):* A given time slot k over a given wavelength w_m on an edge e_j in a specific direction δ can be occupied only by one single TDM circuit $TC_{sd}^{km,\delta}$ ($\in TC_{sd}$) between a node pair sd. One can express this occupancy constraint as

$$\sum_s \sum_d {}^{sd}O_{e_j}^{km,\delta} \leq 1 \qquad \forall(\delta, k, m). \tag{8.25}$$

- *Number of transmitted time slots:* Total number of time slots in a given SONET frame (connections) emerging from a source node (to all other nodes) on a wavelength toward a given direction (CW or CCW) should be upper-bounded by the product of the maximum number of available TDM slots per wavelength, i.e., G, and the number of ADMs (0 or 1) for the given wavelength in the given direction, i.e.,

$$\sum_e \sum_d TC_{sd}^{km,\delta} \leq GN_{adm}^{sm} \qquad \forall(\delta, s, m). \tag{8.26}$$

- *Number of received time slots:* Total number of time slots in a given SONET frame entering a receiving node on a given wavelength from a given direction (CW or CCW) should be upper-bounded by the product of the maximum number of available time slots per wavelength, i.e., G and the number of ADMs (0 or 1) for the given wavelength in the given direction, i.e.,

$$\sum_{e}\sum_{s} TC_{sd}^{km,\delta} \leq GN_{adm}^{dm} \qquad \forall(\delta, d, m). \tag{8.27}$$

- *Single-hop connections:* Every connection is set up using an all-optical connection between the given node pair with the following constraint

$$\sum_{s\neq n}\sum_{d\neq n} TC_{sd}^{km,\delta} \leq (1 - N_{adm}^{nm}) \qquad \forall(n, \delta, m, k). \tag{8.28}$$

  Note that this constraint is not applicable when the designer intends to employ traffic grooming (TG) at the intermediate nodes for a given connection, leading to more flexible design toward reducing the resource requirements. However, when one needs point-to-point connections between all node pairs, the above condition is invoked, but at the expense of larger number of wavelengths and ADMs.

With the above constraints and objective function, one can solve the LP-problem using Simplex-based LP solver. Some results of the above LP-based design for an example WRON ring are presented later and compared with the results obtained using the heuristic algorithm described in the following.

Heuristic designs

Several heuristic designs have been explored in the literature for SONET WDM metro rings (Zhang and Qiao 1997; Zhang and Qiao 2000; Chiu and Modiano 1998; Chiu and Modiano 2000; Gerstel *et al.* 1998a) employing efficient TG algorithms. The method, which we consider here from (Zhang and Qiao 1997; Zhang and Qiao 2000), addresses the design problem typically in two steps as follows.

1. *Sub-circle construction*: A number of sub-circles (SC, say) are constructed around the ring, with each sub-circle consisting of several *non-overlapping arcs* corresponding to the connection requests between various node pairs. As in the previous heuristic design of WRON rings carrying uniform traffic, each sub-circle, denoted now by ζ, supports a lower-rate SONET stream, and the connection request between a node pair comes always as an integral multiple of the data rate of a sub-circle. Thus each connection may need a bundle of arcs on the same number of sub-circles between a given node pair. Attempt is made through a suitable algorithm to construct the minimum number of sub-circles while supporting all the connection requests.
2. *Grooming sub-circles into circles*: Following the sub-circle construction, another algorithm is used to combine (groom) the sub-circles with their aggregate transmission rate approaching (but not exceeding) the capacity of a lightpath. Thus, the SC sub-circles constructed in the first step are groomed to form circles, each circle corresponding to a wavelength having a grooming capacity G. While grooming the sub-circles, an attempt is made so that the end nodes of the arcs in different sub-circles belonging to the same wavelength coincide as much as possible, thereby *minimizing the number of ADMs* in the network.

Before describing the algorithms, we define the relevant parameters and variables:

r_C: transmission rate of each circle (wavelength) assigned to a lightpath, with M as the number of available wavelengths on the fiber,

r_{SC}: transmission rate of a sub-circle; $r_C = Gr_{SC}$,

ρ_{sd}: number of arcs needed from source node s to destination node d,

h_{sd}: number of hops from source node s to destination node d over the shortest route, which is bounded by $N/2$ for a bidirectional ring. Thus, $\{h_{sd}\}$ will represent the *strides* or lengths of the arcs representing $\{\rho_{sd}\}$.

1. Sub-circle construction algorithm As before, the network traffic is represented by an $(N \times N)$ traffic matrix $\Gamma\{\rho_{sd}\}$, with $\rho_{ii} = 0$ and each off-diagonal non-zero element ρ_{sd} $(s \neq d)$ having a specific value of its stride h_{sd}. The algorithm operates iteratively with these connection requests to map them into a number of sub-circles as follows.

Step I: Arrange all the elements in $\Gamma\{\rho_{sd}\}$ (i.e., ρ_{sd}'s) in descending order of their strides (i.e., h_{sd}'s). Thus, the matrix element having the longest stride value is placed at the top of the ordered list, and the one with the lowest value of stride (i.e., the traffic element between the neighbouring pairs) is placed at the bottom. Hereafter, the algorithm operates iteratively.

Step II: Pick up the connection from the top of the list. If it is the first iterative cycle of the algorithm, then set up the arcs over the shortest route. If $k_{sd} = k_{ds}$, then all the arcs are bidirectional, else some are bidirectional and the rest are unidirectional. Each one of these arcs, set up in this process, is extrapolated by dotted lines to form a distinct sub-circle (as shown in Fig. 8.11(a)).

If this is the second or subsequent iterative cycle of the algorithm, check the following aspects for the arc (could be also multiple arcs on as many sub-circles, depending on the value of ρ_{sd}), and move to Step III.

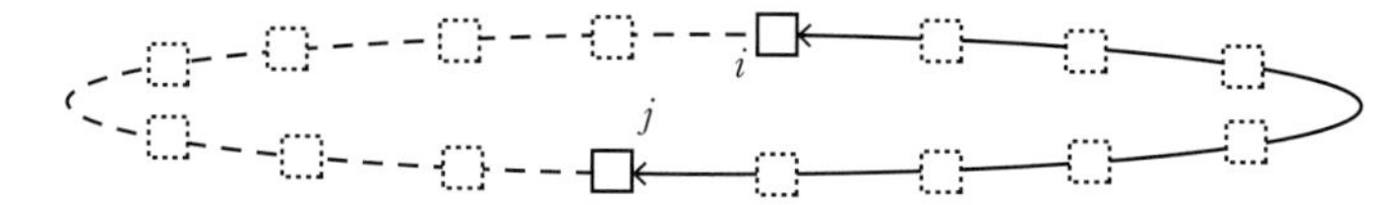

(a) A connection (arc) between the nodes i and j is set up on a sub-circle.

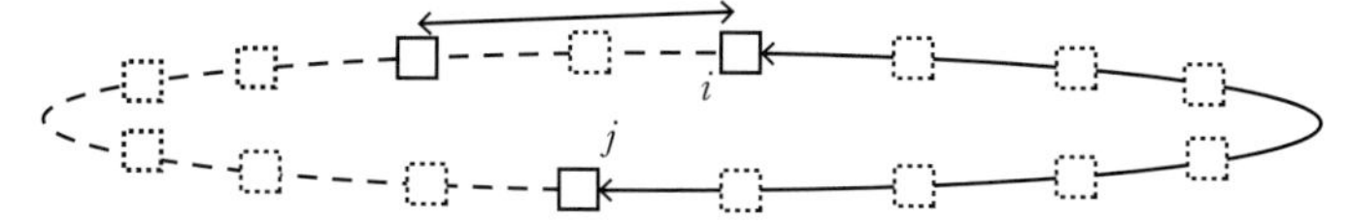

(b) An additional connection, on the same sub-circle of (a), which is contiguous with the previous arc, implying that the new connection does not create any additional gap.

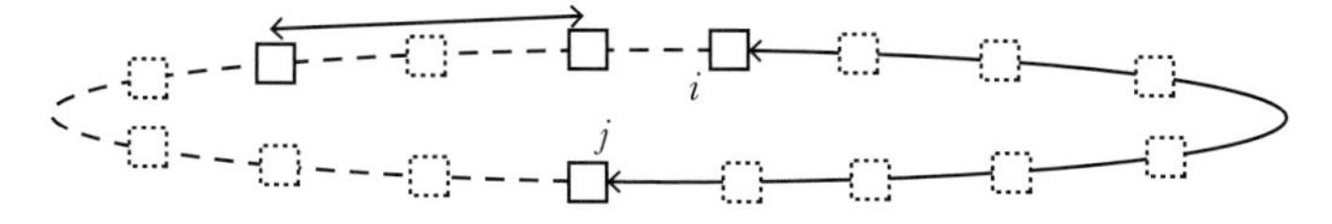

(c) An additional connection on the same sub-circle of (a), creating an additional gap.

Figure 8.11 *Formation of a sub-circle using arcs without and with gaps. The sub-circle in part (a) is obtained by setting up a bidirectional connection using an arc between the nodes s and d, and thereafter by extrapolating the arc over the remaining circular segment using a dashed line. Parts (b) and (c) illustrate the cases, where an additional connection is set up without and with a gap, respectively. The nodes that do not participate in any connection are shown using boxes with dotted outline, and the connections using arcs are shown with solid lines.*

(a) The new arc is a contiguous extension of some arc already set up over an existing sub-circle, as shown in Fig. 8.11(b).

(b) The new arc overlaps partially with some arc already set up over an existing sub-circle.

(c) The new arc can be placed on an already existing sub-circle without any overlap with the earlier arc(s) in the same sub-circle, but creates an additional gap in the sub-circle (i.e., two gaps on two sides).

Step III: If the observation (a) in Step II is satisfied, then the arc is *placed* on the same sub-circle, and the sub-circle is updated in terms of the *participating* nodes (represented by the small boxes with solid outlines in Fig. 8.11). Next, go to Step VI. Else, go to next step.

Step IV: If the observation (b) in Step II is true, then the other existing sub-circles are to be checked for possible contiguity between the present and the earlier arcs on those sub-circles. The arc will be set up as in Step III, if any sub-circle offers contiguous connection between the two arcs on the same sub-circle. If no existing sub-circle is found to accommodate the arc without overlap, then an *additional sub-circle* is introduced to set up the new connection (arc). Next, go to Step VI.

Step V: If the observation (c) is true for one or more of the existing sub-circles (implying that there exists no contiguity between the arc with any existing arc, i.e., condition (a) is not satisfied), then move this connection out of the present ordered list to a new list, named as *gap-maker List* (GML). Go to next step.

Step VI: If all the connection requests are tried out for setting up arcs and sub-circles, then go to the next step with the latest GML (with pending connections) along with the information on the constructed sub-circles, arcs, and the participating nodes therein. Else, remove the connection that has been allotted a sub-circle from the ordered list of connections, and go back to Step II.

Step VII: The pending connections for the gap maker(s) are set up in this step within the already created sub-circles with gap. However, additional sub-circle(s) may have to be set up when the number of sub-circles to be groomed into one wavelength (circle) exceeds the grooming factor G. The sub-circle construction algorithm is completed at this point. The total number of sub-circles obtained from this algorithm is denoted by SC (as defined earlier); this result is used as an input for the subsequent algorithm.

2. Sub-circle grooming algorithm In this algorithm, the SC sub-circles obtained from the last algorithm, are groomed (circuit-multiplexed) into bundles of sub-circles, i.e., circles, with each circle representing one unique wavelength. One can have two options for carrying out the grooming operation. In one of them, attempt is made to minimize the use of network resources by conserving the numbers of wavelengths/ADMs, where the connections between some node pairs may have to go for OEO conversion at intermediate nodes, thereby allowing multihop connections. In the other option, end-to-end all-optical connection is provided for each node pair using one or more lightpaths, which are bypassed at intermediate nodes (as considered in Section 8.3.2, for uniform traffic pattern). This option however does not carry out TG at any intermediate nodes.

Grooming option 1: Minimal use of resources with multihop connections, if necessary In this algorithm, the sub-circles obtained using the sub-circle construction algorithm are groomed to form circles with distinct wavelengths. The number of groomed sub-circles is denoted as SC_g. The initial value of SC_g is set to zero. We also define a variable M' given by

$$M' = \left\lceil \frac{SC - SC_g}{G} \right\rceil. \tag{8.29}$$

Thus, with $SC_g = 0$, the algorithm starts operating iteratively as follows, where SC_g and M' keep evolving with every iteration.

Step I: Mostly, each sub-circle will have multiple arcs ending at the source and destination nodes, called *participating nodes*. Thus, each sub-circle with multiple arcs will have several participating nodes. All these sub-circles with their constituent arcs are arranged in descending order of the number of participating nodes. Hence, a sub-circle (or multiple sub-circles) with the highest number of participating nodes will appear at the top of the list, called ordered sub-circle list (OSCL).

Step II: Compute M' as:

$$M' = \left\lceil \frac{SC - SC_g}{G} \right\rceil. \tag{8.30}$$

Next, we use M' to estimate the number M_{sc} of sub-circles in one of the possible circles to be formed, given by

$$M_{sc} = \left\lceil \frac{SC - SC_g}{M'} \right\rceil. \tag{8.31}$$

Since the number of groomed sub-circles would now go up by M_{sc}, the number of groomed sub-circles SC_g gets updated as

$$SC_g = SC_g + M_{sc}. \tag{8.32}$$

This updating mechanism indicates how many more sub-circles ($SC - SC_g$) will remain to be groomed in the next iterations to form the subsequent circles.

Step III: To form a circle from M_{sc} sub-circles, pick up *one sub-circle from the top* of the OSCL and the ($M_{sc} - 1$) *sub-circles from the bottom*. It is presumed (a heuristic presumption) that the topmost sub-circle with several arcs will have some gaps that are likely to get filled in by the shorter arcs in the sub-circles lying at the bottom. Having formed the circle, remove the sub-circles used in the circle from the OSCL. The pruned OSCL goes for the next iteration. Note that, some of the arcs from the bottom sub-circles may not find suitable gaps in the topmost sub-circles; in such cases these are kept aside as *ungroomed arcs* in each step for further consideration.

Step IV: Repeat Steps II and III with the latest list of sub-circles, and the latest value of SC_g, until the list gets exhausted with $SC_g = SC$ and $M' = 0$.

Step V: The arcs from the various sub-circles that could not be groomed are considered at this stage to create additional circles.

Step VI: Compute the number of circles created and the total number of participating ADMs in all the circles. The sub-circle grooming algorithm is completed at this point.

Grooming option 2: Single-hop connections Consider the same ordered list of sub-circles, i.e., the OSCL and carry out the following steps.

Step I: Pick up the first sub-circle and place it on a circle with a specific wavelength.

Step II: Check the other identical sub-circles below the previous sub-circle for an overlap in respect of the participating nodes. If available, groom this sub-circle into the already-created circle. If the number of such identical sub-circles is more than G, then introduce more circles. Repeat this procedure for similar other identical sub-circles, if present.

Step III: Pick up the other sub-circles with partial overlap with the previous circle(s), such that the participating nodes in the present sub-circles are same as the nodes already in the previous sub-circle(s). If so, and if the total number of sub-circles is less than G in some of the circles, then groom this sub-circle again in the previous circle(s).

Step IV: Continue the above process until all the sub-circles that are groomable into the previous circles get exhausted.

Step V: Assign additional wavelengths for the ungroomable sub-circles (if any).

Step VI: Count the total number of circles, giving the needed number of wavelengths. Count the number of participating nodes in each circle, the sum total of the participating nodes from all the circles giving the required number of ADMs. The algorithm is completed at this stage.

Case study

As a case study of the LP-based and heuristic designs presented in the foregoing, we consider a four-node WRON-based SONET-over-WDM ring. The traffic matrix $\Gamma = \{\rho_{sd}\}$ of the example four-node WRON ring is presented in Table 8.1.

For the heuristic algorithms, first we need to apply the sub-circle construction algorithm using the traffic matrix. We therefore form the OSCL for the connections and proceed as follows.

a) Sub-circle construction In Step I of sub-circle construction, first all the connections are arranged in the descending order of their strides h_{sd} as follows:

1. $h_{13} = h_{31} = 2, \rho_{13} = 2, \rho_{31} = 1,$
2. $h_{24} = h_{42} = 2, \rho_{24} = 2, \rho_{42} = 2,$
3. $h_{12} = h_{21} = 1, \rho_{12} = 1, \rho_{21} = 0,$
4. $h_{32} = h_{23} = 1, \rho_{23} = 3, \rho_{32} = 2,$

Table 8.1 *Traffic matrix for a four-node WDM ring. The first column and first row indicate the node numbers, and the matrix elements represent the traffic demands in terms of the required number of lower-rate SONET streams between the corresponding node pairs.*

Nodes	1	2	3	4
1	0	1	2	0
2	0	0	3	2
3	1	2	0	3
4	6	2	2	0

5. $h_{34} = h_{43} = 1, \rho_{34} = 3, \rho_{43} = 2,$
6. $h_{41} = h_{14} = 1, \rho_{41} = 6, \rho_{14} = 0.$

In Step II, arcs for the various connections are formed using appropriate sub-circles SCi's as follows:

1. $h_{13} = h_{31} = 2, \rho_{13} = 2, \rho_{31} = 1$. One bidirectional (BD) arc is created for 1-3/3-1 connection on $SC1$. Next, one unidirectional (UD) arc is created for 1-3 connection (overlapping with the arcs on $SC1$) on a new sub-circle, i.e., $SC2$.
2. $h_{24} = h_{42} = 2, \rho_{24} = 2, \rho_{42} = 2$. Two BD arcs for 2-4/4-2 connections (overlapping partially with $SC1$, $SC2$) are created on $SC3$ and $SC4$.
3. $h_{12} = h_{21} = 1, \rho_{12} = 1, \rho_{21} = 0$. One UD arc for 1-2 connection is created on $SC3$, having contiguity with the BD arc between 2-4.
4. $h_{32} = h_{23} = 1, \rho_{23} = 3, \rho_{32} = 2$. Needs two BD arcs for 2-3/3-2 connections, and one UD arc for 2-3 connection, which overlap with already introduced sub-circles. Hence, the two BD arcs are placed on two new sub-circles $SC5$ and $SC6$. Next, the remaining UD arc 2-3 is created on another new sub-circle $SC7$.
5. $h_{34} = h_{43} = 1, \rho_{34} = 3, \rho_{43} = 2$. Needs two BD arcs for connections 3-4/4-3, one UD arc for connection 3-4. Using the criterion of *no-overlap with contiguity*, the two BD arcs are placed on $SC1, SC2$. The remaining UD arc 3-4 is placed on $SC6$.
6. $h_{41} = h_{14} = 1, \rho_{41} = 6, \rho_{14} = 0$. Five UD arcs for connection 4-1 are placed on $SC1, SC2, SC3, SC4$, and $SC5$ using again the criterion of no-overlap and contiguity, while the last one is placed on $SC6$ with gap-making and hence moved to the GML.

The formation of the above sub-circles is illustrated in Fig. 8.12. As shown in the figure, the constituent arcs are placed on the respective sub-circles, forming a set of seven sub-circles, with six of them having contiguous arcs and one arc as the gap-maker on $SC6$, i.e.,

1. $SC1: 1 \leftrightarrow 3 \leftrightarrow 4 \rightarrow 1,$
2. $SC2: 1 \rightarrow 3 \leftrightarrow 4 \rightarrow 1,$
3. $SC3: 1 \rightarrow 2 \leftrightarrow 4 \rightarrow 1,$
4. $SC4: 2 \leftrightarrow 4 \rightarrow 1,$
5. $SC5: 2 \leftrightarrow 3 \rightarrow 4 \rightarrow 1,$
6. $SC6: 2 \leftrightarrow 3, 4 \rightarrow 1$ (gap maker),
7. $SC7: 2 \rightarrow 3.$

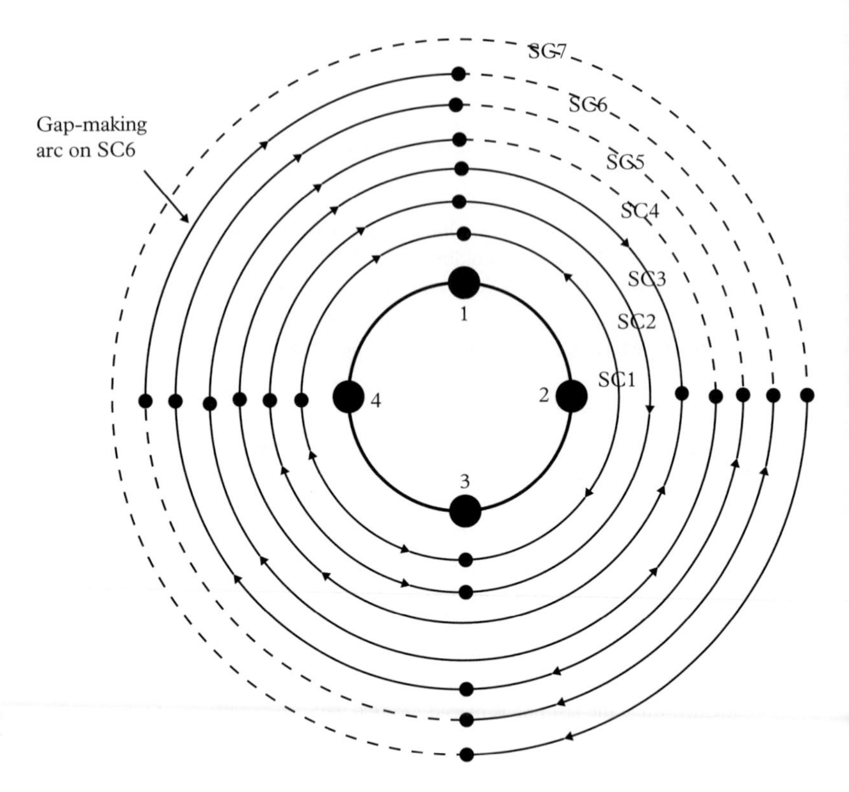

Figure 8.12 *Result of the sub-circle construction algorithm for a four-node WDM ring.*

Next, the task of sub-circle grooming is carried out on the seven sub-circles, $SC1$ through $SC7$, as follows.

b) Sub-circle grooming As described earlier, the sub-circle grooming can be carried out using two possible approaches. We carry out the grooming process with $G = 4$ for both approaches in the following.

Grooming option 1: Minimal use of resources with multihop connections, if necessary First, we arrange the sub-circles in the descending order of the number of participating nodes in a sub-circle to form the OSCL as follows.

1. $SC5$: $2 \leftrightarrow 3 \rightarrow 4 \rightarrow 1$,
2. $SC6$: $2 \leftrightarrow 3, 4 \rightarrow 1$ (gap maker),
3. $SC1$: $1 \leftrightarrow 3 \leftrightarrow 4 \rightarrow 1$,
4. $SC2$: $1 \rightarrow 3 \leftrightarrow 4 \rightarrow 1$,
5. $SC3$: $1 \rightarrow 2 \leftrightarrow 4 \rightarrow 1$,
6. $SC4$: $2 \leftrightarrow 4 \rightarrow 1$,
7. $SC7$: $2 \rightarrow 3$.

Note that the sequencing of SCi's with the same participating nodes can be arbitrary. With a grooming factor of 4 (i.e., $G = 4$), and seven sub-circles (i.e., $SC = 7$), we carry out the sub-circle grooming algorithm iteratively as follows.

- Initialize SC_g with $SC_g = 0$.
- Compute the following parameters and groom sub-circles iteratively:

  Iteration 1: Compute:

$$M' = \left\lceil \frac{SC - SC_g}{G} \right\rceil = \left\lceil \frac{7-0}{4} \right\rceil = 2, M_{sc} = \left\lceil \frac{C - C_g}{M'} \right\rceil = \left\lceil \frac{7-0}{2} \right\rceil = 4$$
$$SC_g = SC_g + M_{sc} = 0 + 4 = 4.$$

  Groom four sub-circles to form the first circle on wavelength w_1, say, with one from the top and the rest from the bottom: $SC5, SC3, SC4, SC7$ (which is feasible).

  Iteration 2: Consider the updated value of SC_g and compute:

$$M' = \left\lceil \frac{SC - SC_g}{G} \right\rceil = \left\lceil \frac{7-4}{4} \right\rceil = 1,$$
$$M_{sc} = \left\lceil \frac{SC - SC_g}{M'} \right\rceil = \left\lceil \frac{7-4}{1} \right\rceil = 3$$
$$SC_g = SC_g + M_{sc} = 4 + 3 = 7 = SC.$$

  Groom three sub-circles to form the second circle on wavelength w_2, say: $SC6, SC1, SC2$ (which is feasible).

  No further iteration is needed, as SC_g now equals SC, making $M' = 0$. Hence, finally the number of circles or wavelengths becomes two and the number of ADMs turns out to be eight. The formation of circles with this algorithm is illustrated in Fig. 8.13.

Grooming option 2: Single-hop connections In order to carry out the sub-circle grooming for all-optical connections between all node pairs, we first recall the output of the sub-circle construction algorithm for the same four-node ring in the following:

- $SC1$: $1 \leftrightarrow 3 \leftrightarrow 4 \rightarrow 1$,
- $SC2$: $1 \rightarrow 3 \leftrightarrow 4 \rightarrow 1$,
- $SC3$: $1 \rightarrow 2 \leftrightarrow 4 \rightarrow 1$,
- $SC4$: $2 \leftrightarrow 4 \rightarrow 1$,
- $SC5$: $2 \leftrightarrow 3 \rightarrow 4 \rightarrow 1$,
- $SC6$: $2 \leftrightarrow 3, 4 \rightarrow 1$ (gap maker),
- $SC7$: $2 \rightarrow 3$.

With the above list of sub-circles, the task of grooming them into circles is carried out as follows.

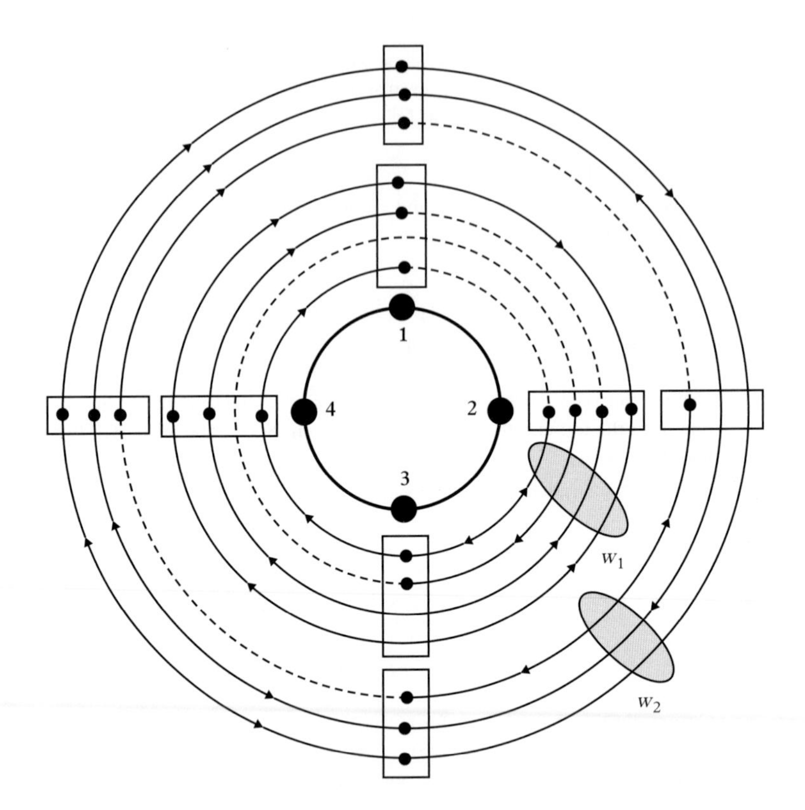

Figure 8.13 *Result of the heuristic algorithm for sub-circle grooming to minimize the numbers of ADMs/wavelengths (grooming option 1) for a four-node WDM ring considered for the sub-circle construction algorithm. Solution:* $M = 2, N_{adm} = 8$.

- The two sub-circles $SC1$ and $SC2$ have identical participating nodes, and hence are groomed into one wavelength, say w_1. Note that, with $G = 4$, we can groom the two sub-circles into one circle (wavelength), without exceeding the grooming limit.
- $SC3$ and $SC4$ differ from $SC1$ and $SC2$ in terms of participating nodes, while $SC4$ forms a subset in $SC3$. Hence, $SC4$ and $SC3$ are assigned the next wavelength, say w_2, and again with $G = 4$, this step of grooming also becomes permissible.
- Similarly $SC5$, $SC6$, and $SC7$ are assigned the next wavelength, say w_3, with a permissible grooming of three ($< G$) sub-circles into one circle.

Note that, with three wavelengths, the ADM requirements will be as follows:

- wavelength w_1: three ADMs at the nodes 1, 3, and 4,
- wavelength w_2: three ADMs at the nodes 1, 2, and 4,
- wavelength w_4: four ADMs at the nodes 2, 3, 4, and 1,

Thus, the single-hop all-optical heuristic design leads to three wavelengths and ten ADMs, and thus needs more resources than the earlier multihop design, using two wavelengths and eight ADMs.

Finally, we consider an LP-based multihop design for the same four-node WRON ring, but with a sparse traffic matrix shown in Table 8.2 for ease of computation. We also use

Table 8.2 *Traffic matrix for a four-node WDM ring for comparison of LP-based and heuristic solutions.*

Nodes	1	2	3	4
1	0	1	2	0
2	0	0	0	2
3	0	0	0	3
4	2	0	0	0

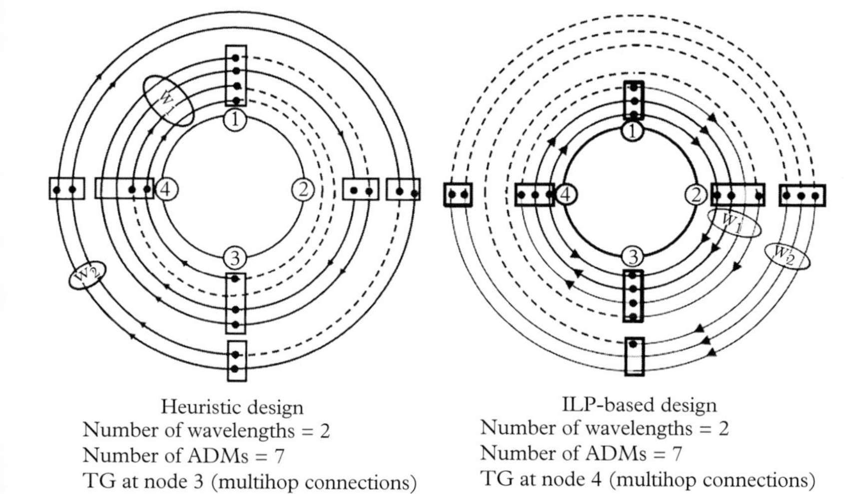

Figure 8.14 *Comparison of the designs obtained from the LP-based and heuristic methods for a four-node WDM ring with the traffic matrix given in Table 8.2.*

the multihop heuristic algorithm for the same network for a comparison between the results obtained from the two approaches. In particular, the LP-based design without any constraint on the single-hop connectivity between the nodes (i.e., without the constraint of Eq. 8.28) leads to the multihop design, which is compared with the heuristic design using the sub-circle construction and grooming steps (for the grooming option 1, allowing multihop connections). As shown in Fig. 8.14, the LP-based solution is close to the results obtained from the heuristic scheme, albeit with the traffic-grooming taking place at different nodes (i.e., at node 3 for the heuristic design and at node 4 for the LP-based design), while giving the same number of ADMs (seven) and wavelengths (two) in both cases. Note that, for the sparse traffic matrix used in this example case, the numbers of ADMs and wavelengths are found to be same for both LP-based and heuristic algorithms, though it cannot be generalized. Readers are encouraged to explore various other cases (with more populated traffic matrices) to see whether the LP-based solutions lead to better solutions.

8.3.4 Hub-centric WRON rings

Besides the two design methodologies discussed in the foregoing, some other design options have also been explored for the WRON rings. In one such design, one of the nodes in the ring is considered as a *hub* carrying out OEO conversion for all wavelengths, while the rest of the nodes would only transmit and receive at specific wavelengths to communicate with the hub. Any two non-hub nodes will therefore be communicating with each other through the hub by using two hops, while the hub as a node will communicate with all other nodes using a single hop. One can also use two or more hubs, and the other nodes can be connected to these hubs selectively, and the hubs can interconnect with each other using specific wavelengths. We discuss briefly in the following some of these hub-centric designs for the WRON rings.

Single-hub WRON ring

In a single-hub WDM ring using SONET transmission, the hub employs several ADMs on different wavelengths, while all other nodes practically operate as TMs on specific wavelengths. Formation of a typical single-hub WDM ring is shown in Fig. 8.15 using two bidirectional (CW and CCW) fiber rings.

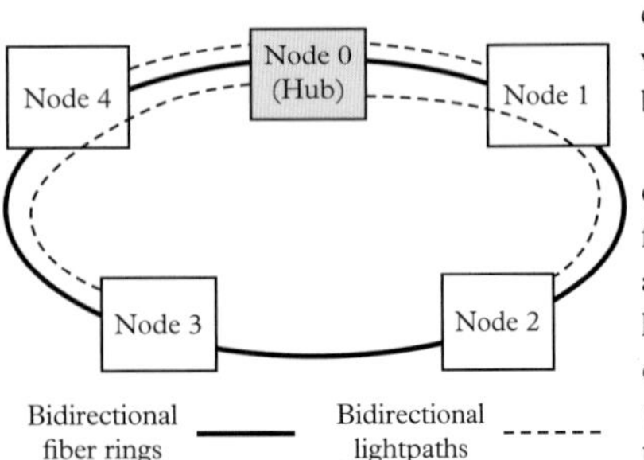

Figure 8.15 *Illustration of a single-hub 5-node WR ring.*

As shown in Fig. 8.15, each non-hub node is connected to the hub (node 0), by using one or more pairs of lightpaths on the two rings (i.e., bidirectional lightpaths on the two fibers), and these lightpaths are optically bypassed by using WR at intermediate nodes, if any. Thus, the ingress traffic at a given node addressed to all other nodes (including the hub) is exchanged through the hub, where the same wavelengths are used on disjoint paths over the ring on the two sides of the hub. Using this scheme with an odd number of nodes (i.e., N as an odd integer), and each node requiring l_i wavelengths for setting up all of its lightpaths, the total number of wavelengths can be expressed as

$$M = \left\lceil \frac{1}{2} \sum_{i=1}^{N-1} l(i) \right\rceil, \tag{8.33}$$

with the values for l_i estimated from the traffic matrix of the network, which takes into consideration the traffic grooming needed at each node to multiplex and send the incoming traffic flows toward the hub node.

Double-hub WRON ring

Double-hub WR rings employ two hubs, and each node is connected to both hubs by distinct lightpaths along with two or more lightpaths to connect the two hubs. A six-node double-hub WR ring is shown in Fig. 8.16, where the nodes 0 and 3 play the roles of two hubs.

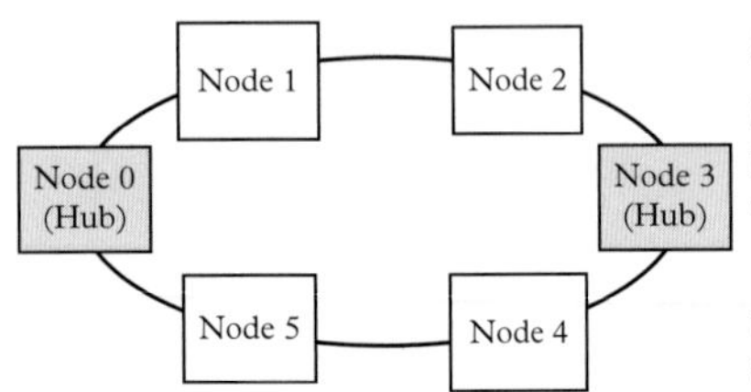

Figure 8.16 *Illustration of a double-hub six-node WR ring. Each node is connected to both hubs (nodes 0 and 3). Both hubs, being also network nodes, are connected to each other using two lightpaths.*

Assume that the two hubs are located at node 0 and node h in an N-node ring. With the two hubs, the ring is partitioned into two segments, one segment (segment A, say) with the nodes $0, 1, \cdots, (h-1)$, and the other segment (segment B, say) with the nodes $h, h+1, \cdots, (N-1)$. Positioning of the two hubs will depend primarily on the traffic pattern; for uniform traffic pattern the two hubs can be located at diametrically opposite locations for even number of nodes. With this network setting, one can adopt a similar approach as used for single-hub ring and find out the required number of wavelengths in the ring (Gerstel *et al.* 1998a).

8.3.5 Interconnected WRON rings

As discussed earlier, in order to address the problem of metro gap, metro rings are typically divided into two hierarchical levels: metro-core and metro-edge. Thus, a metro network would in general consist of a bigger metro-core ring surrounded by a number of smaller metro-edge rings, where the optical transmission is based on SONET-over-WDM architecture with a mix-and-match of WR and OEO conversion at intermediate nodes.

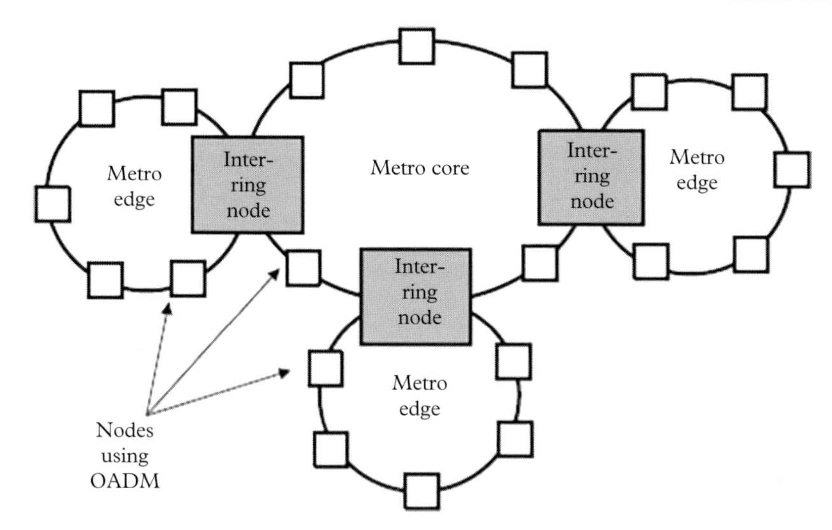

Figure 8.17 *Illustration of interconnected WDM rings, using all-optical or OEO-based interconnection.*

Interconnection of these WDM rings can be realized in different ways, such as by using all-optical connections between the adjacent rings or by using OEO-based interface. The interconnection between such WDM rings is illustrated in Fig. 8.17.

In OEO-based interconnection, wavelength reuse is possible in different rings, while all-optical interconnection can set up end-to-end lightpaths between the nodes located in different rings, but preventing wavelength reuse along with a more-constrained power budget owing to optical power losses at the intermediate OADMs and inter-ring node. The resource provisioning problem in such cases can be carried out using a hierarchical or an integrated approach. In hierarchical design, one can use the methods discussed in the foregoing sections for the constituent rings separately, and thereafter interconnect them with electrical or optical crossconnect. In this case, for electrical crossconnet, one can have flexibility for traffic grooming at the inter-ring node, but at the cost of electronic hardware. For all-optical connections between the rings, one needs more wavelengths, while the cost and power consumption remain lower in the inter-node ring. While provisioning end-to-end all-optical lightpaths over interconnected WDM rings along with intra-ring design, one can also frame an overall LP-based formulation leading to an optimum solution (Mukherjee 2006).

As mentioned above, the inter-ring nodes can be realized using different possible configurations, depending on the way the rings are interconnected with each other: all-optical interconnection, opaque interconnection through OEO conversion. The schematic diagram for an all-optical inter-ring node is shown in Fig. 8.18. The inter-ring node is placed between the rings A and B, and the block schematic shown in the figure represents functionally one half of the node, by connecting the CCW fiber of ring A to the CW fiber of ring B. The other half of the node (not shown for simplicity) operates for the CW fiber of ring A and the CCW fiber of ring B. The intra-ring CCW lightpaths arriving at the inter-ring node from ring A are either optically passed through the demultiplexer-switch-multiplexer combination to the output port of the inter-ring node connected to ring A itself or passed downward for local drop operation (and add operation in the reverse direction). In addition to the passthrough and add-drop lightpaths, the inter-ring CCW/CW lightpaths arriving at the inter-ring node from ring A/B for moving over to ring B/A are switched to the

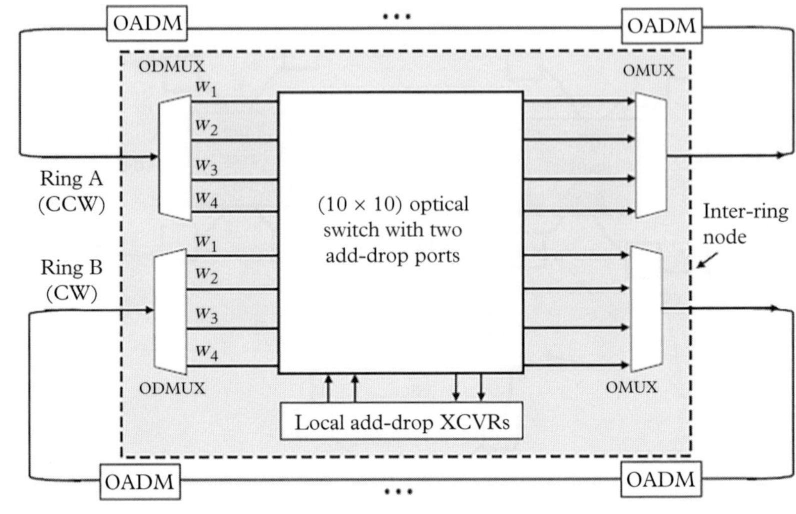

Figure 8.18 *Illustration of interconnected WDM rings, with all-optical inter-ring node.*

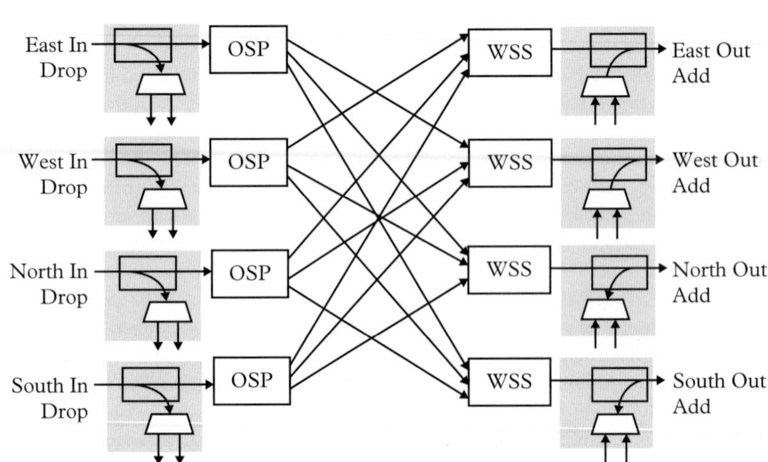

Figure 8.19 *Illustration of an MD-ROADM using OSPs and WSSs. Add-drop operation takes place at input/output (east/west/north/south) ports using optical coupler and multiplexer/demultiplexer.*

appropriate output ports of the optical switch that can take the lightpaths to the multiplexer connected to ring B/A. Similar operations take place also between the CW-A and CCW-B rings in the other half of the inter-ring node.

The inter-ring nodes can also be realized by extending the regular ROADMs (with a nodal degree of two) to form MD-ROADMs, as described in Chapter 2. A typical MD-ROADM using OSPs and WSSs as an inter-node ring to interconnect two WDM rings is shown in Fig. 8.19, with the east and west ports to be used for one ring and the north and south ports for the other. We shall discuss MD-ROADMs in further detail in Chapter 9, while addressing the node configurations needed for the migration from the interconnected-rings to the mesh-connected topology for long-haul WDM backbone networks.

Though interconnecting two rings using a single inter-ring node offers a simple mechanism for expanding the networking coverage, the connectivity between the nodes lying in the two rings may get throttled in the inter-ring node. Further, the single-node-based

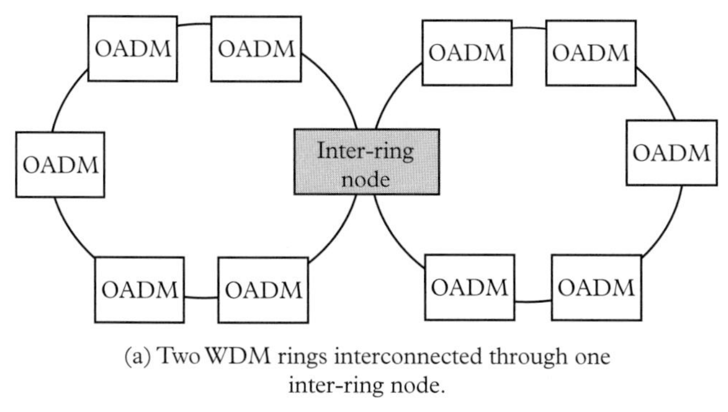

(a) Two WDM rings interconnected through one inter-ring node.

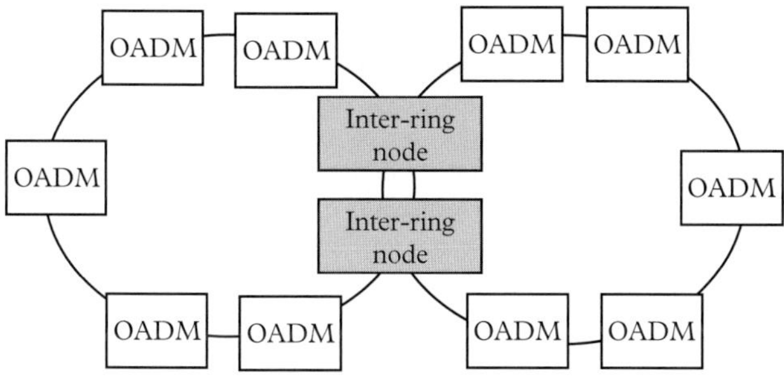

(b) Two WDM rings interconnected through two inter-ring nodes.

Figure 8.20 *WDM rings interconnected through (a) one and (b) two inter-ring nodes.*

interconnection loses communication between the nodes in the two rings, when the inter-ring node fails. Thus, one can have inter-ring connectivity through one or two inter-ring nodes as shown in Fig. 8.20, where two rings are shown to have been interconnected through one (in part (a)) and two (in part (b)) inter-node rings, and one can use OEO, all-optical, or mix-and-match of both to interconnect the two rings. One can also use four inter-ring nodes for enhancing further the survivability of the inter-ring connections between two SONET-over-WDM rings, and we discuss such issues later in Chapter 11, dealing with the survivability of optical networks.

8.4 Packet-switched WDM metro ring

As mentioned earlier, WDM metro rings have been designed and deployed using the pre-existing circuit-switched SONET rings. However, these circuit-switched SONET-over-WDM networks have not been flexible enough to carry the growing traffic in the metro segment due to the unprecedented increase in the packet-switched traffic. For single-wavelength metro networks, we described earlier (in Chapter 3) the RPR-based ring architecture employing packet-switched transmission. Similarly, for WDM metro ring networks, a variety of packet-switched architectures have been studied in various research groups, such as, MAWSON (Summerfield 1997), RingO (Carena *et al.* 2004), HORNET (Shrikhande *et al.* 2000), WONDER (Antonio *et al.* 2008), SMARTnet (Rubin and Hua 1995), and

RINGOSTAR (Herzog *et al.* 2005). A comprehensive review of most of these packet-switched WDM metro networks can be found in (Herzog *et al.* 2004).

Design of the packet-switched WDM metro rings makes use of some of the features that were used earlier in single-wavelength LANs/MANs, such as in FDDI, DQDB, or RPR, along with the devices needed for WDM transmission. For the SONET-over-WDM rings, we considered the problem of setting up circuit-switched connections through *permanent/semi-permanent* TDM slots as *circuits* over multiple wavelengths. However, for the packet-switched WDM rings, each node needs to use an appropriate MAC protocol to access the available wavelengths on a *packet-by-packet* basis. This requirement led to the developments of several testbeds for the packet-switched metro WDM ring networks with varying node configurations and MAC protocols. Before going into the details of the specific networks, we briefly describe below some of the typical schemes for the MAC protocols in these networks: slotted WDM ring, multitoken WDM ring, and meshed WDM ring.

- *Time-slotted WDM ring*: Time-slotted (or simply slotted) WDM ring networks employ TDMA along with WDM, where the transmission time in each wavelength is divided into fixed-size time slots for packet transmission. Each node employs a TT (or multiple FTs) and an FR, as in the WDM LANs described in Chapter 6. To transmit a packet in a time slot, each node uses a TDMA-based MAC protocol over a wavelength to which the receiver of the destined node is tuned. This scheme can achieve high bandwidth-utilization from WDM wavelengths with uniform traffic between the network nodes. With non-uniform and time-varying traffic demands, the network needs to add extra features to the MAC protocol to make a dynamic allocation of multiple slots to the nodes having more data to transmit. In general, such networks use collision-free (or collision-avoidance) schemes for MAC protocols.
- *Multitoken WDM ring*: Unlike slotted rings, multi-token rings do not employ TDMA slots. Each wavelength in a multi-token ring operates in a way similar to FDDI. Thus, on each wavelength a token moves around the entire ring. Whenever the token on a given wavelength reaches a node, the node can hold the token for a duration governed by the MAC protocol and transmit its own packet; thereafter the node releases the token for onward transmission. On each wavelength, the protocol operates in a distributed manner to ensure fairness for all the nodes.
- *Meshed WDM ring*: Meshed rings utilize additional (dark) fibers existing in fiber cables realizing direct connections for some node pairs along with the ring topology. This arrangement changes the ring topology to a *mesh-like* topology, thereby giving enhanced performance by avoiding electronic processing at the intermediate nodes for distant connections.

With the above introduction to the possible networking strategies, we next consider some of the packet-switched WDM metro-ring testbeds that have been developed by various research groups.

8.4.1 MAWSON

One of the early studies on packet-switched WDM metro rings has been on a network called *metropolitan area wavelength switched optical network* (MAWSON), where the time-slotted WDM ring was used for the MAC protocol (Summerfield 1997). The number of operating wavelengths (M) in MAWSON was the same as the number of nodes (N) in the network, i.e., $N = M$. Each node employed $(M-1)$ FTs and one FR (i.e., FT $^{M-1}$-FR configuration),

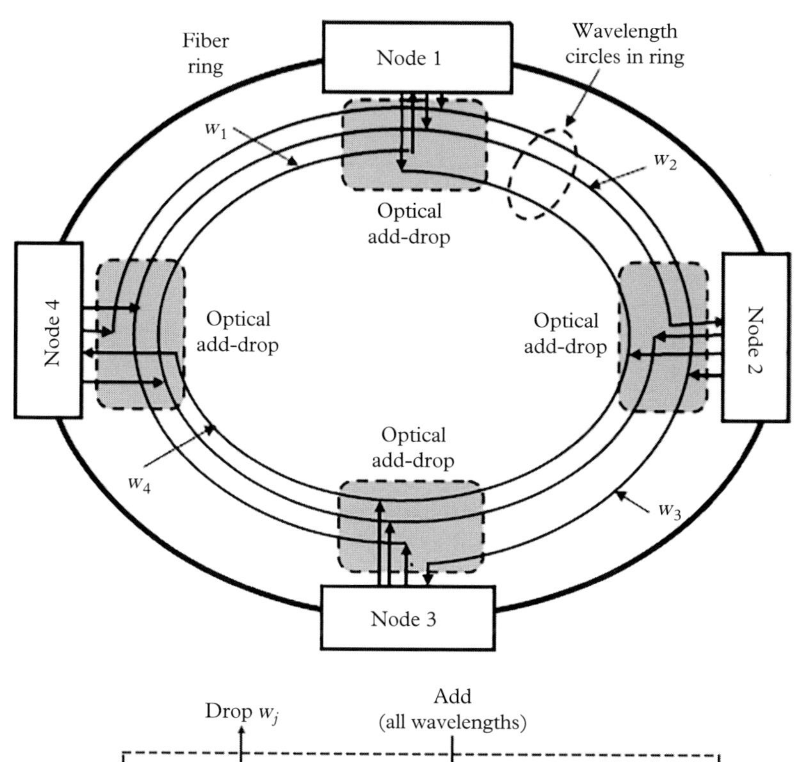

Figure 8.21 *MAWSON schematic for four nodes.*

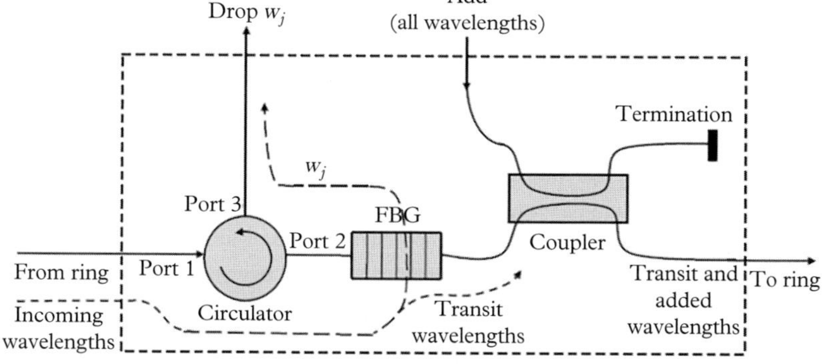

Figure 8.22 *MAWSON node configuration (after Summerfield 1997).*

with the FR set up at a fixed wavelength (unique for each node) among the M wavelengths operating over the ring. Further, each node used an OADM for connecting to the fiber ring. Thus, a node intending to transmit to a specific destination node with a receiving wavelength j, had to choose the transmitter tuned at the wavelength j. A four-node MAWSON ring is illustrated in Fig. 8.21.

The node configuration in MAWSON is shown in Fig. 8.22. The OADM operation is realized in each node by using an FBG to reflect a specific wavelength into the local node, which goes around a circulator for the local *drop* operation. The FBG passes on the other $(M - 1)$ wavelengths to a coupler to combine them with the ingress traffic of the node for onward transmission. In particular, the ingress data packets modulate appropriate FTs (not shown), and the modulated outputs of the FTs are *added* or combined with the

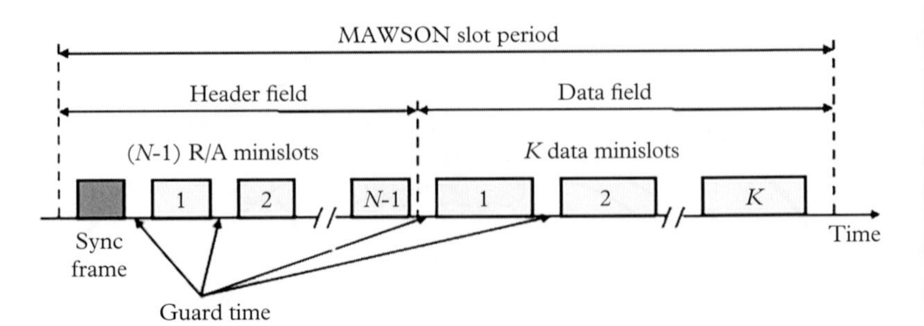

Figure 8.23 *Slot structure in MAWSON (after Summerfield 1997).*

FBG output in the coupler and forwarded as the composite WDM signal to the output fiber port of the node. Note that, normally one FT becomes active during a given time slot, implying that only one wavelength is availed by a node during a time slot to reach the corresponding destination. However, for multicast operation, multiple wavelengths can be availed for onward transmission. Since each node can receive data packets on only one wavelength, multiple source nodes cannot send data to a single destination during the same time slot to avoid collisions. This requires an appropriate MAC protocol based on TDMA. MAWSON uses a request/allocation(R/A)-based scheme to realize a collision-free MAC protocol, as described in the following.

Any node in MAWSON, say node i, when intending to transmit data to another node, say node j, has to *request* for bandwidth (i.e., time slots) on the receiver wavelength of node j (i.e., on wavelength j) and wait for the *allocation* from node j, which must be transmitted by node j on the receiver wavelength of node i, i.e., wavelength i. Thus, the MAC scheme in MAWSON is based on an R/A protocol (RAP), and to execute the RAP the entire transmission time along the fiber ring is divided into fixed-size time slots for all wavelengths. Each fixed-size time slot on a wavelength has a header and payload (data) fields, and these time slots are kept synchronized across all the M wavelengths. The slot structure of MAWSON is illustrated in Fig. 8.23. In each slot, header and payload fields are composed of a number of *minislots*. The header field of each slot starts with its first minislot as the preamble for synchronization (sync frame). The next part of the header field consists of $(N-1)$ R/A minislots for executing the RAP operation, as each node needs to interact with the remaining $(N-1)$ nodes for exchanging R/A information. The header field in each slot is followed by the data field with a maximum number (K) of minislots, called data minislots (DMSs).

Each of the $(N-1)$ minislots in a slot header has two parts for performing two different tasks. When a node, say node i, needs to send data (i.e., DMSs) to any one of the other $(N-1)$ nodes, say node j, it requests on wavelength j for a number (k) of DMSs, with $k \in (0, K)$, thereby needing a binary number with $\lceil log_2(K+1) \rceil$ bits. Then, in response to the request, when allocation has to be made from node j to node i on wavelength i, it would require a field of K bits with its nth bit made 1 if the nth DMS has been allocated to node i. Thus, the $(N-1)$ minislots on each receiving wavelength, carrying the R/A information contents (i.e., $\lceil log_2(K+1) \rceil$ bits for the request and K bits for the DMS allocation), will form the main part of the header field of a slot. Furthermore, each of the $(N-1)$ minislots will have a few more fields as shown in Fig. 8.24. In particular, after a preamble (sync) field, each minislot will have a few bits indicating the beginning of field (BOF), followed by all the R/A bits (as described above), finally ending up with the bits indicating the end of field (EOF).

Overall, MAWSON architecture employs a simple node configuration without any need for a carrier-sensing mechanism (or channel inspection, discussed later), while its RAP

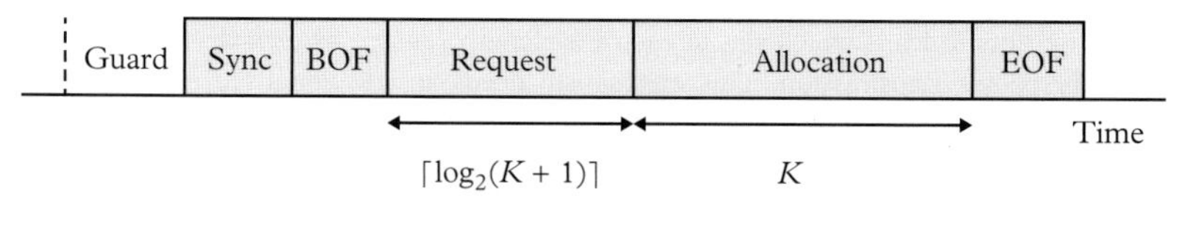

Figure 8.24 *Minislot structure in a MAWSON slot header (after Summerfield 1997).*

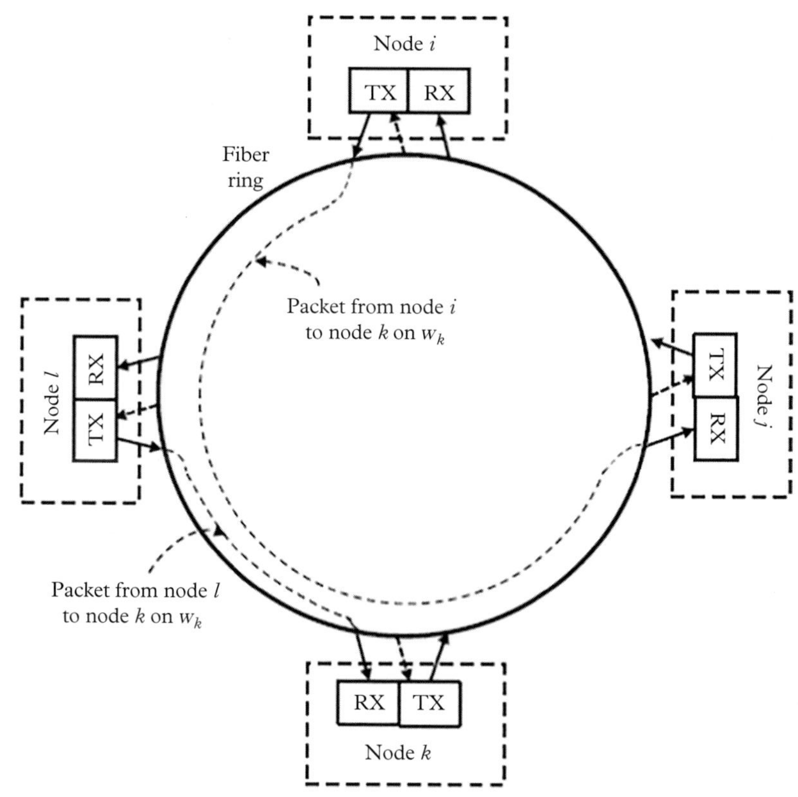

Figure 8.25 *Basic network configuration of RingO.*

protocol needs about 10% overhead and in practice can provide a throughput approaching 70% of the full capacity. Moreover, the RAP scheme introduces some MAC latency, as the sequential transmission of the request (source to destination) and the allocation (destination to source) of information bits call for at least one round-trip propagation delay around the ring.

8.4.2 RingO

Another packet-switched WDM metro ring, called *ring optical network* (RingO) (Carena *et al.* 2004), also used time-slotted WDM ring architecture. The basic network configuration of RingO is shown in Fig. 8.25. Note that, the RingO configuration apparently looks similar to that of MAWSON in respect of destination stripping (i.e., packets being dropped at the destination node) using a unique drop wavelength for each node. However, RingO employs *a posteriori* channel inspection, which requires to read the headers of the passing-by slots to learn which slot in each wavelength (except its own drop wavelength) is free for packet transmission. This mechanism needs an additional receiver module in each node for all wavelengths to help the transmitter, as shown in Fig. 8.25 by an additional *inward arrow* from the ring to the transmitter block of each node.

In contrast with RingO, MAWSON used the R/A-based interactive MAC (i.e., RAP), thereby having no need to carry out channel inspection, but with increased latency. Gradually, RingO evolved into three versions, with the first two versions using one unidirectional ring and the same MAC protocol with the number of wavelengths being exactly the same as the number of nodes (i.e., $N = M$). However, the node architecture in the second version was more compact and simpler. The nodes in both these versions of RingO use multiple transmitters (laser array) and one receiver, all being fixed-tuned at specific wavelengths (implying FT $^{M-1}$-FR transceiver configuration). We describe next the first and second versions of RingO, and subsequently discuss briefly how better scalability could be achieved by employing bidirectional rings in the third version.

As mentioned above, RingO employed channel inspection on the desired wavelength before transmission of packets to the destined node for realizing a collision-free MAC protocol over the synchronized time slots. The node configuration used in the first version of RingO is shown in Fig. 8.26. Every node uses an EDFA at its input for signal amplification followed by an optical demultiplexer. The demultiplexer drops the wavelength (drop wavelength) to which the node receiver is fixed-tuned and forwards the other wavelengths to the 90/10 splitters for channel inspection and onward transmission of the passthrough traffic. Each splitter drops 10% of its input power toward a specific photodiode of the photodiode array for channel inspection. However, the passthrough outputs of the splitters (with 90% of the input power) are multiplexed using an optical multiplexer and the multiplexed output is forwarded to a power combiner after delaying the passthrough signal in an optical delay line to absorb the processing time in the node controller. The local data signal modulates the multiplexed laser output(s) (typically one laser output at a time, or more when multicasting is needed) from the laser array using an external modulator and pass on the modulated light to the power combiner, where the modulator output is added with the multiplexed passthrough lightwaves for onward transmission on the ring. The node controller, having derived the information on the wavelength availability in a given time slot (through channel inspection), selects the appropriate laser(s) in the laser array for external modulation.

The MAC protocol in RingO (both versions), along with channel inspection, employs a number of buffers, called virtual output queues (VOQs). Generally, each VOQ (not shown) is kept for one of the $(N-1)$ possible destination nodes (or for a group of destination nodes) in an N-node RingO network. The nodes in RingO using channel inspection are prone to unfairness, as the upstream nodes will tend to *grab* the empty slots in a given wavelength. There are various schemes to address this problem, such as synchronous or asynchronous round robin schemes. However, in order to keep the complexity low, the packet at the head of the line (HOL) in the longest VOQ is transmitted on the respective wavelength, provided an empty slot is available. This gives a suboptimal delay performance, but with a simpler implementation.

The second version of RingO employs the same MAC protocol as the first version, while the node configuration uses a more compact hardware, as shown in Fig. 8.27. In the first step after EDFA-based pre-amplification, 10% of the incoming power is dropped using one 90/10 splitter to an AWG (operating as a demultiplexer) for channel inspection, thereby avoiding multiple splitters (one for each wavelength), as used in the first version. Thereafter, an FBG along with a circulator drops the desired wavelength, which makes the node hardware more

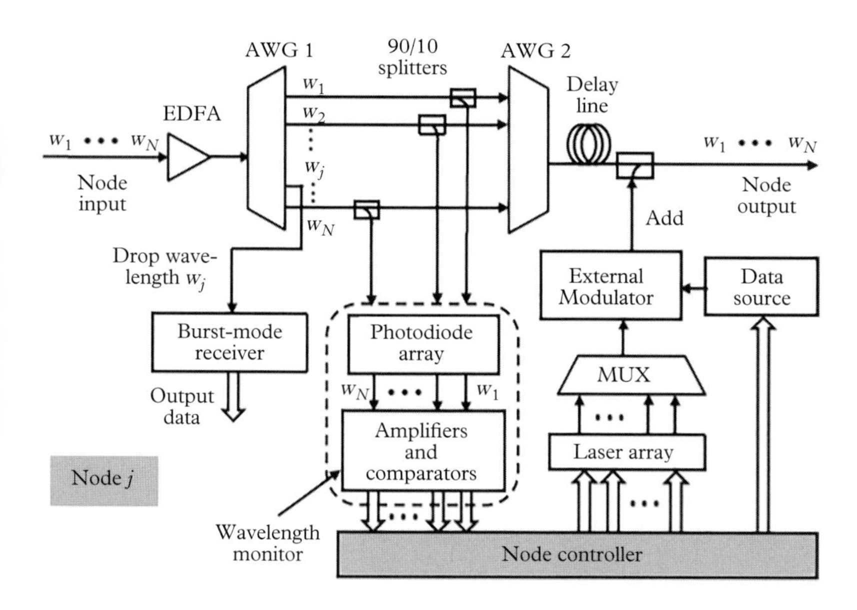

Figure 8.26 *RingO node (node j with w_j as the drop wavelength) configuration for unidirectional ring (first version) (Carena et al. 2004, ©IEEE).*

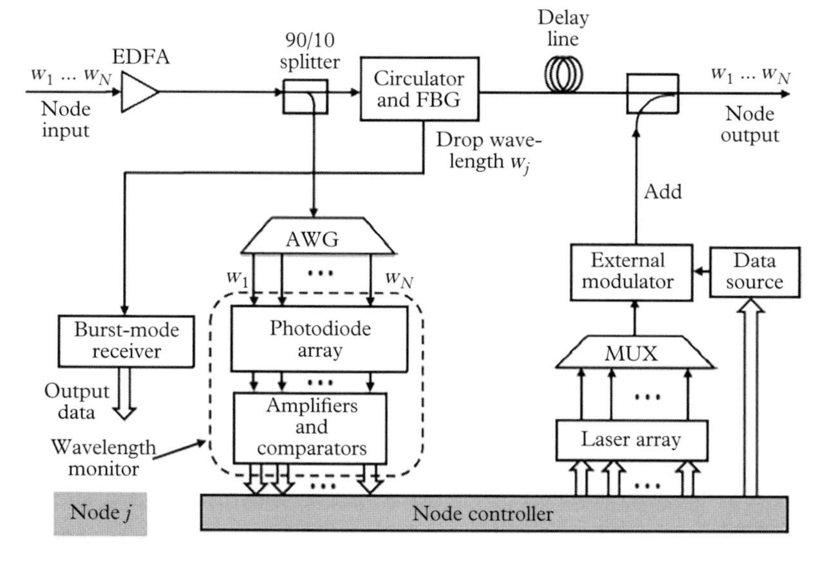

Figure 8.27 *RingO node (node j with w_j as the drop wavelength) configuration for unidirectional ring (second version) (Carena et al. 2004, ©IEEE).*

compact and robust. The remaining part of the node in this version is kept the same as that used in the nodes of the first version.

The first two versions of RingO employed unidirectional ring with the number of nodes being equal to the number of wavelengths (i.e., $N = M$), thus lacking network scalability. The third version of RingO was developed with bidirectional rings with a scalable architecture in terms of the number of nodes, i.e., one could choose $N > M$, while using almost similar node hardware with channel inspection. The modified nodes in the third version have access to two counter-propagating rings, and these two rings are in practice configured through an optical loopback mechanism into a *folded bus* (Carena *et al.* 2004). The two-ring approach can realize a distinct feature of spatial reuse with $N > M$, where the same wavelength can be concurrently used over the network with non-overlapping routes. With the folded-bus feature, this node configuration leads to a significant gain in throughput with large number of nodes.

8.4.3 HORNET

Another noteworthy testbed for packet-switched WDM ring, called *hybrid optoelectronic ring network* (HORNET), was developed by employing again the time-slotted WDM channels, with each node using the TT-FR transceiver configuration (Shrikhande *et al.* 2000). As in MAWSON and RingO, every node in HORNET has a unique drop wavelength to receive data, and hence any node, while intending to communicate to another node, must transmit at the drop wavelength of the destination node. The transmitter in every node, along with the local ingress data packet, also transmits an electrical subcarrier modulated (using frequency-shift keying, i.e., FSK) by the bits of the packet header on the drop wavelength of the destination node over a given time slot. The subcarrier frequencies – unique for each node, and hence N in number (for N nodes) – are placed beyond the highest frequency of the baseband electrical data spectrum. Each wavelength (w_j, say) in HORNET, which is a drop wavelength for node j, is associated with a unique subcarrier frequency (f_{cj}, say), i.e., the wavelengths $w_1, w_2, ..., w_N$ correspond to the electrical subcarriers $f_{c1}, f_{c2}, ..., f_{cN}$, respectively.

When a HORNET node transmits a data packet on a wavelength w_j to reach the destination node j, it also transmits concurrently the corresponding modulated subcarrier f_{cj} on the wavelength w_j during the same time slot. However, when a node has no data to transmit to node j it does not transmit its subcarrier on w_j. The local ingress data packet is combined electrically with the subcarrier, and the combined electrical signal is used to modulate the transmitting laser in the TT. In order to carry out the various steps of operations, the node configuration is divided into three functional blocks: *slot manager*, *smart drop*, and *smart add* blocks. We discuss below the node function in further details using the HORNET node configuration shown in Fig. 8.28.

As shown in Fig. 8.28, each HORNET node receives the incoming optical WDM signal in its slot manager block from the upstream nodes in the ring and splits the same with a 90/10 ratio, with the 10% power directed to the subcarrier receiver which employs photodetection and demodulation of all the incoming subcarriers. At the subcarrier receiver in the slot manager block, after photodetection, all the data packets from different wavelengths get translated in the electrical domain, thereby colliding in time domain. Nevertheless, the subcarriers being spectrally distinct and placed at frequencies higher than the highest frequency of data spectrum (governed by the bit rate of data streams), remain distinguishable and hence separable using a bank of bandpass filters. Thus, the subcarrier receiver separates by bandpass filtering and performs demodulation for N subcarriers concurrently and produces N parallel baseband data signals from the N subcarriers. Consider a given node, say node j, for the following discussion.

The demodulator for the subcarrier frequency f_{cj} corresponding to the drop wavelength w_j of node j employs FSK demodulation and extracts the overhead bits for the incoming data packet on the same wavelength. If no node transmits any packet to node j during the time slot under consideration, the FSK receiver output of node j becomes zero (no output) indicating that the node has no signal to receive in the corresponding time slot. The waveforms of other subcarrier frequencies (after respective bandpass filtering) are *treated* like binary amplitude-shift keying (ASK) waveforms for the detection of presence/absence of the subcarrier (though the subcarriers, if present, are FSK-modulated (with constant envelope) by the respective transmitters). Thus, these subcarrier waveforms are forwarded

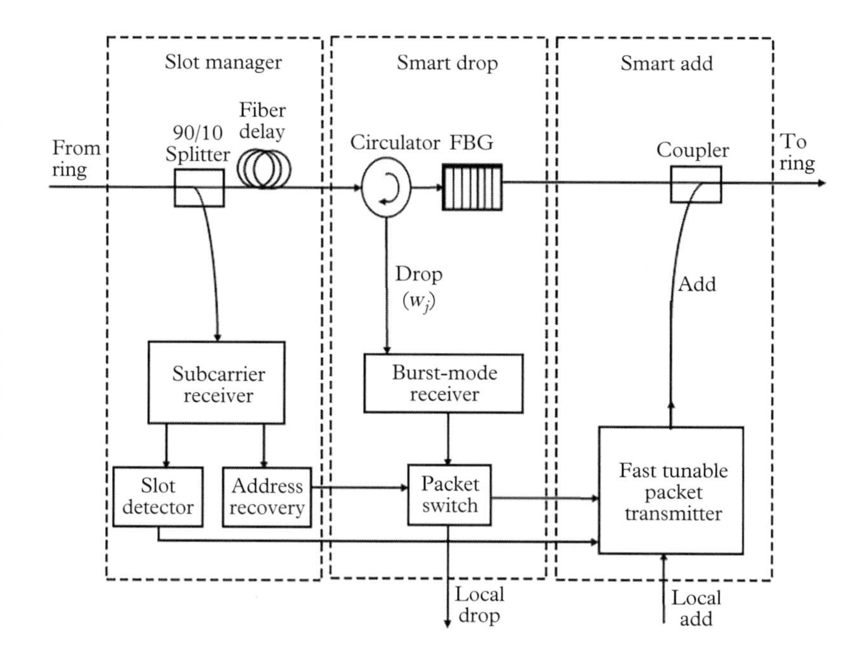

Figure 8.28 *HORNET node configuration (Shrikhande et al. 2000, ©IEEE).*

to ASK demodulators using envelope detection to check the occupancy/availability of the corresponding wavelengths for transmission from node j. If the desired wavelength ($\neq w_j$) is free, the transmitter of node j can choose to transmit its packet over the same wavelength on the current time slot. Thus, the slot manager block decides the actions to be taken by the node during the time slot under consideration.

Having dropped 10% of the incoming optical power at the port connected to the subcarrier receiver of the slot manager, the 90/10 splitter passes the remaining 90% of the received power onto the right-side port. Thereafter, the WDM signal from the right-side port of the splitter is delayed by a small duration τ by using a fiber delay line. The delayed WDM signal is forwarded to a circulator-FBG combination in the smart drop block for dropping down the drop wavelength w_j for reception at the burst-mode receiver (BMR) and passing on the rest of the wavelengths towards the next stage, i.e., the smart add block. Note that, the smart drop block ensures (as in MAWSON and RingO) destination-stripping of the packets addressed to this node. During the time duration τ spent in the fiber delay line (adjusted to the processing time in slot manager plus some guard time), the subcarrier receiver processes the FSK and ASK demodulator outputs. The FSK demodulator output enables a packet switch to direct the received data packet from the BMR output to the local host. However, the $(N-1)$ ASK demodulator outputs are processed at the slot detector block and its output is passed onto the TT in the smart add block in identifying whether the current time slot is free on the drop wavelength w_k (say) of its desired destination node, i.e., node k.

If a packet is ready for transmission on the same wavelength, it is transmitted from the local host in the current time slot. Before modulating the laser of TT, header of the packet in the form of FSK-modulated subcarrier (corresponding to w_k, i.e., with the frequency f_{ck}) is added to the data bits in electrical domain in the transmitter located in the smart add block. This composite electrical signal, with the concurrent control bits carried by the subcarrier

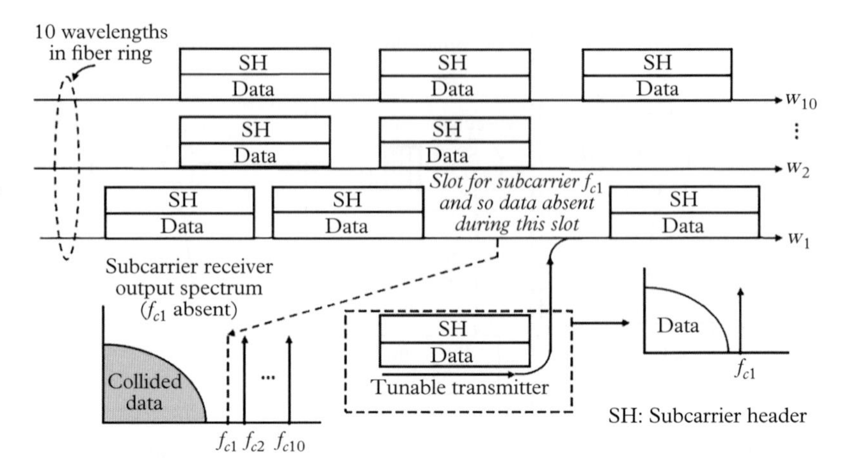

Figure 8.29 *Illustration of HORNET MAC operation at some node, say node j, with w_j as the drop wavelength and f_{cj} as the subcarrier frequency for sending the control information. The received subcarrier spectrum indicates that the subcarrier f_{c1} is absent in the slot manager section, implying that the node under consideration (i.e., node j) can send its packet to node 1 on wavelength w_1 (Shrikhande et al. 2000, ©IEEE).*

and the data bits in the baseband form during the same time slot, is used to modulate the intensity of the tunable laser in TT. The laser output is added optically at the combiner with the transit optical signals from upstream nodes (coming from the FBG output of the smart drop section) for onward transmission. When this slot reaches the destination node (i.e., node k), the particular subcarrier at f_{ck} uniquely foretells (before the node can take action on the data packet owing to the latency introduced between slot manager and the remaining part of the node by using the fiber delay line) about the incoming data bits, thereby executing channel inspection needed for executing its MAC protocol.

Different MAC protocols have been examined for HORNET, such as slotted ring, variable slots with collision-avoidance-based CSMA (CSMA/ CA) scheme and a few others (Wonglumsom *et al.* 2000; White *et al.* 2000). For all these variations, a channel inspection mechanism based on the subcarrier multiplexing remains the central functionality for MAC. The basic subcarrier-based slot management scheme described in the foregoing can be used for both slotted ring with fixed packet sizes or with variable packets, by employing channel inspection with *a posteriori* channel selection at a transmitting node.

The HORNET MAC operation using the subcarriers is illustrated in Fig. 8.29, where we consider node j as an example case with its drop wavelength and subcarrier frequency as w_j and f_{cj}, respectively. The illustration shows that the slot manager in node j finds from the ASK demodulator outputs that the subcarrier f_{c1} is missing and hence also the data in that slot. So the slot manager instructs the smart add block to avail the passing-by time slot on w_1, if it intends to send any packet to node 1. However, the smart drop section drops the incoming packet, if any on the wavelength w_j (drop wavelength for node j), leading to destination stripping. If the FSK demodulator at f_{cj} in slot manager block identifies the presence of f_{cj}, the slot manager block demodulates the carrier and processes the respective header to decide whether it should forward the received packet to the local host or the smart add block for onward (multihop) transmission. With the information from the slot manager block, the smart add block uses an appropriate scheme so that the packets waiting in the queues for transmission from the local node are sent out with the objective of maximizing the network throughput.

8.4.4 RINGOSTAR

RINGOSTAR (Herzog *et al.* 2005) is an augmented version of the packet-switched RPR network, a single-wavelength bidirectional metro ring, as described earlier in Chapter 3. In RINGOSTAR, in addition to the bidirectional ring realizing the RPR-centric operation at each node, some selected nodes are provisioned with WDM transmission. These WDM-enabled nodes use additional fibers and some centralized WDM components to get interconnected with each other, thereby forming a star topology. This leads to a combined ring-and-star topology, where logically RINGOSTAR uses two subnetworks, called ring subnetwork and star subnetwork, as illustrated in Fig. 8.30. The star subnetwork appears like the *spokes in a wheel* offering shorter routes on multiple wavelengths for distant node pairs. The nodes in the star subnetwork, are called ring-and-star homed (RSH) nodes, while the rest of the nodes are called ring-homed (RH) nodes. The RH nodes employ destination stripping in the electronic domain using OEO conversion, as in RPR. The RSH nodes, in addition to destination stripping, also strip off or remove some of the transit packets and forward them over the star subnetwork, to offer a shorter route. Thus each RSH node mimics the role of a destination node in respect of packet stripping. This functionality is termed *proxy stripping*, and the nodes where this function is carried out, i.e., the RSH nodes, are also called *proxy-stripping* nodes.

The proxy-stripping feature in RINGOSTAR is illustrated in Fig. 8.31, where we consider node 3 as a source node transmitting to node 5 and node 2 as another source node transmitting to node 8. For a packet going from node 3 to node 5, it takes the CCW ring for availing the shortest route. So the RSH node on its route (i.e., node 4) doesn't strip the packet and forwards it to node 5. However, for a packet going from node 2 to node 8, the transmission takes place on the CW ring upto the nearest RSH node (node 1, in this case). At node 1 the received packet from node 3 is stripped off from the ring (a case of proxy-stripping operation), and thereafter transmitted over the WDM-based star subnetwork to reach node 7, bypassing the peripheral ring subnetwork segment consisting of nodes 12,

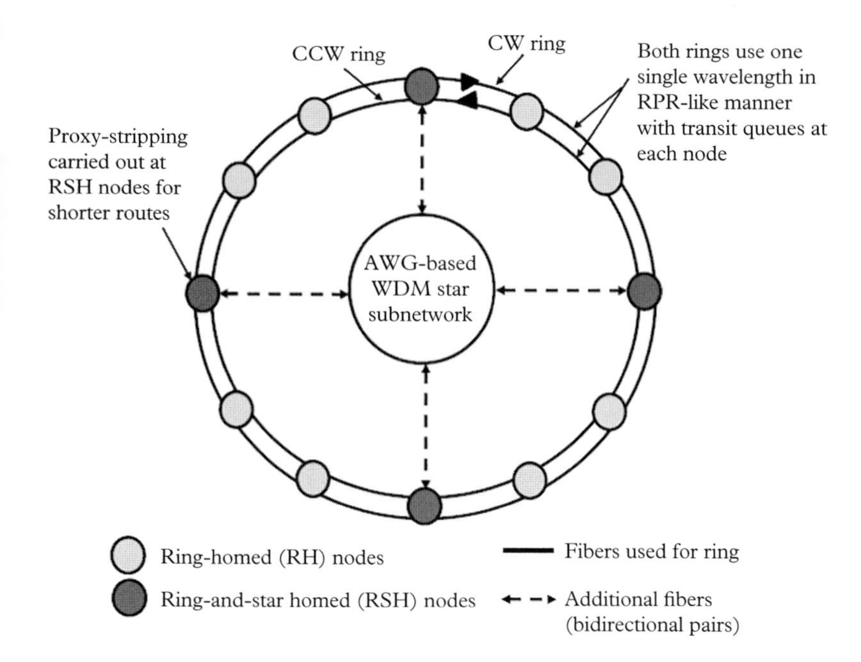

Figure 8.30 *RINGOSTAR architecture using ring and star subnetworks (after Herzog et al. 2005).*

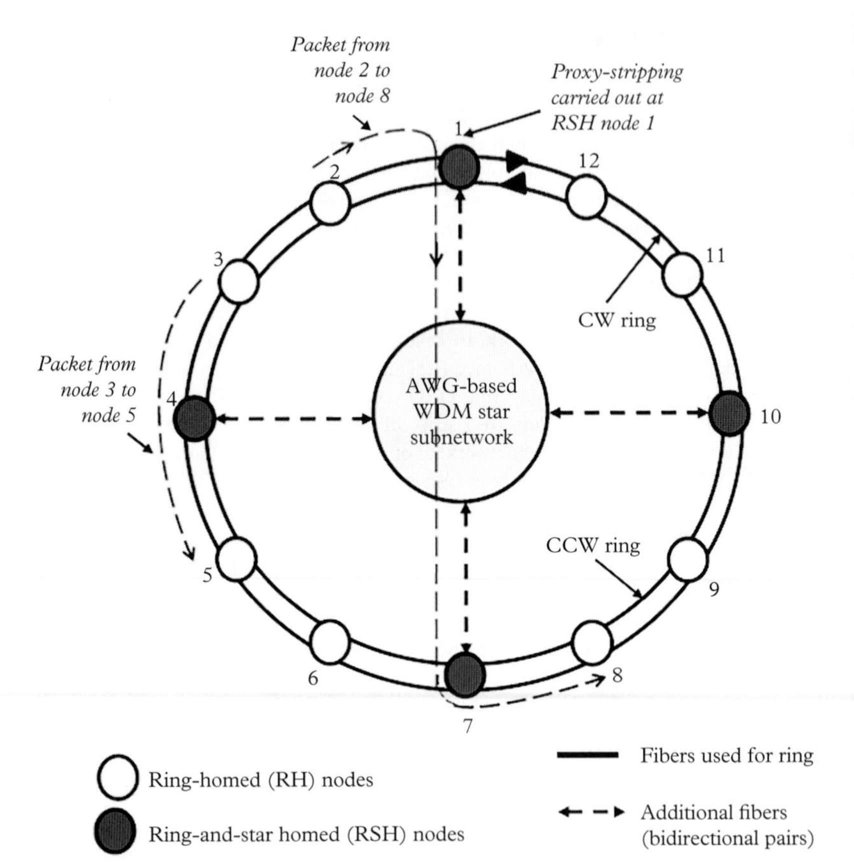

Figure 8.31 *Proxy-stripping example in RINGOSTAR (after Herzog et al. 2005).*

11, 10, and 9. Subsequently, the packet is reinserted from node 7 into the ring subnetwork on the CCW ring on the shortest route to reach its destination, i.e., node 8, where finally the destination stripping is carried out for the packet. As a result, this packet avoids delay in the transit queues and eases out the release of ingress packets at the nodes 12, 11, 10, and 9 (note that, as discussed in Chapter 3, in RPR the transit traffic can choke the flow of ingress traffic at intermediate nodes). Thus the proxy-stripping functionality using WDM transmission at selected nodes (along with the basic destination-stripping feature at each node) leads to more efficient reuse of ring bandwidth and enhances the overall network performance significantly.

The WDM-based star subnetwork interconnecting the RSH nodes is realized by using a centrally located AWG, as shown in Fig. 8.32. The basic properties of AWG, as described in Chapter 2, are utilized to realize the star subnetwork. In particular, the AWG has $D \times D$ input-output ports, which are connected using optical fibers to the RSH nodes through a number of power combiners and splitters, along with the OAs inserted in between the splitters/combiners and AWG to compensate for the power losses in the passive devices. In particular, the RSH nodes are divided into D groups, and hence the number of RSH nodes is given by

$$N_{RSH} = DS, \tag{8.34}$$

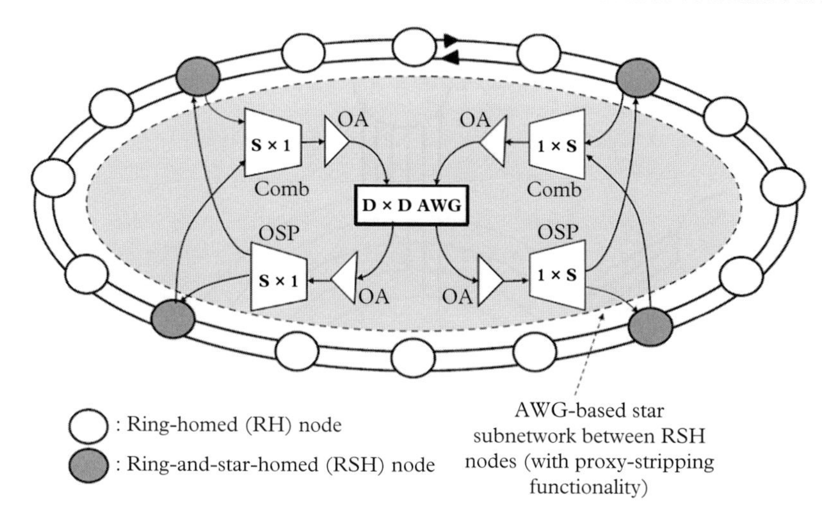

Figure 8.32 *RINGOSTAR physical topology: single-wavelength-based ring subnetwork and WDM-based star subnetwork with $D = 2$ and $S = 2$ (Herzog et al. 2005, ©IEEE). OA: optical amplifier, Comb: optical combiner, OSP: optical splitter.*

where S (> 1) represents the number of RSH nodes in each of the D groups. The total number of nodes N is therefore expressed as

$$N = N_{RSH} + N_{RH} = DS + N_{RH}, \qquad (8.35)$$

with N_{RH} as the number of RH nodes, which can only perform destination stripping and are not connected to the star subnetwork. Outputs of all the RSH nodes belonging to a given group are combined using a ($S \times 1$) power combiner and the combiner output is connected at one of the D inputs of the $D \times D$ AWG. In Fig. 8.32, with $D = 2$ and $S = 2$, $N_{RSH} = 4$, and thus each group has two RSH nodes, which are connected to a 2×1 combiner, and the combiner output is connected to one of the D ($= 2$) input ports of the AWG at the center through the respective OA. The D ($= 2$) AWG output ports are connected to the same number of (1×2) splitters (since $S = 2$) through OAs, and the output of the splitters are connected back to the receivers of the two RSH nodes belonging to the same group.

8.5 Bandwidth utilization in packet and circuit-switched WDM rings

As described in the foregoing sections, the testbeds for the packet-switched WDM metro rings, by and large, employed time-slotted transmission scheme over the WDM ring topology, where each node uses TT-FR or FT^M-FR configuration for the transceivers. A typical logical topology formed in many of the packet-switched WDM rings (e.g., MAWSON, HORNET) is illustrated in Fig. 8.33, where each node as a destination receives packets on a unique wavelength from all other nodes.

As shown in Fig. 8.33, the packets from the source nodes for a given destination are time-multiplexed on the same wavelength in the well-defined time slots. The six-node ring in the figure thus employs six wavelengths, with each wavelength carrying the packets from five nodes in a time-multiplexed manner. Each node uses one queue with a buffer, which is assumed to be large enough to prevent any loss of the incoming packets from the host. In this section, we examine the salient performance features of the packet-switched WDM ring networks and compare them with the circuit-switched WDM ring networks in respect of bandwidth utilization.

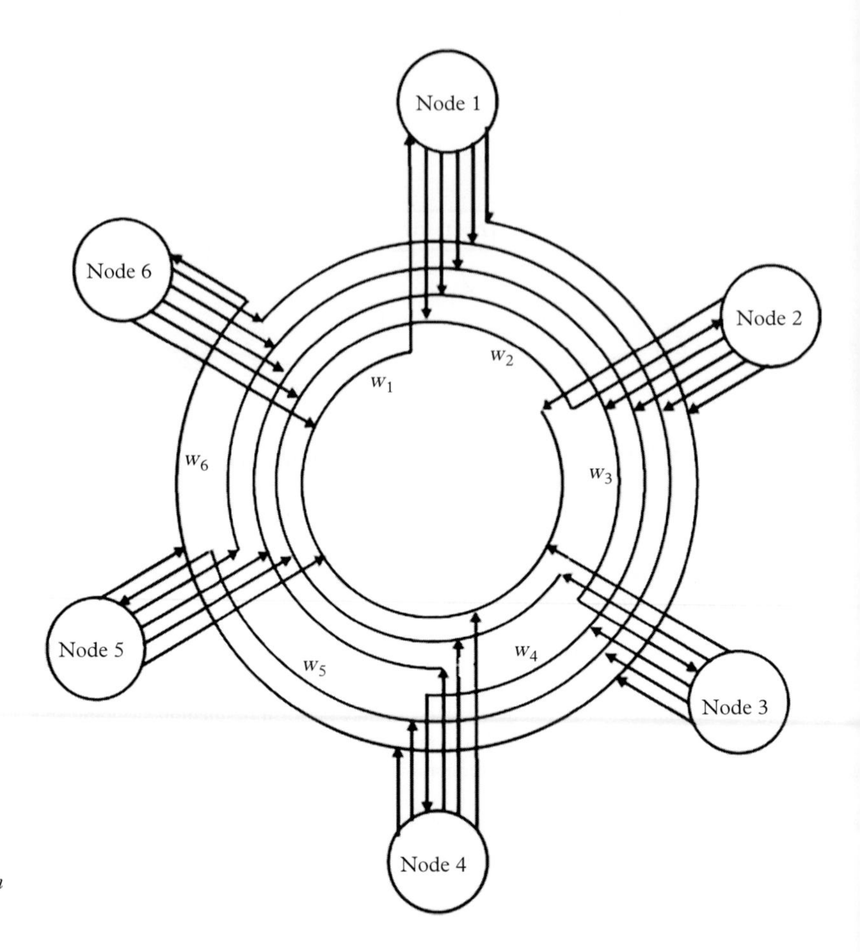

Figure 8.33 *Logical topology formation over a time-slotted WDM ring.*

First, we consider the MAC issues in packet-switched WDM rings with unidirectional transmission, where for a given node, while receiving the packets from the other nodes, the farthest transmitting node enjoys the highest privilege. Figure 8.34 illustrates this issue. As evident from the figure, the destination node generates the stream of free slots for all the remaining nodes (as the candidate source nodes) and hence the *nearest downstream node* can grab as many slots as it wants before any other downstream nodes can avail the same. In this manner, the nodes farthest (actually a neighbor in the reverse direction) to the destination node gets the least chance to avail a free time slot, thereby creating an unfairness in the MAC protocol. The designer needs to examine how the ring would scale to a bigger size with larger number of nodes, without increasing the unfairness in respect of access delay beyond an acceptable limit.

As shown in Fig. 8.34, consider a receiving wavelength w_k, assigned to node k, which is shared by all other nodes for packet transmission to reach node k. The queuing model to analyze such a transmission scheme can be based on $M/D/1$ queuing systems with *vacation* (see Appendix C), where the service time turns out to be deterministic as all the slots are of fixed length. However, for a given transmitting node, say node i, the vacation times appear randomly from the upstream nodes with higher privilege, and gets manifest as increased medium-access delay for node i, while waiting to avail a free slot, that has not been availed by

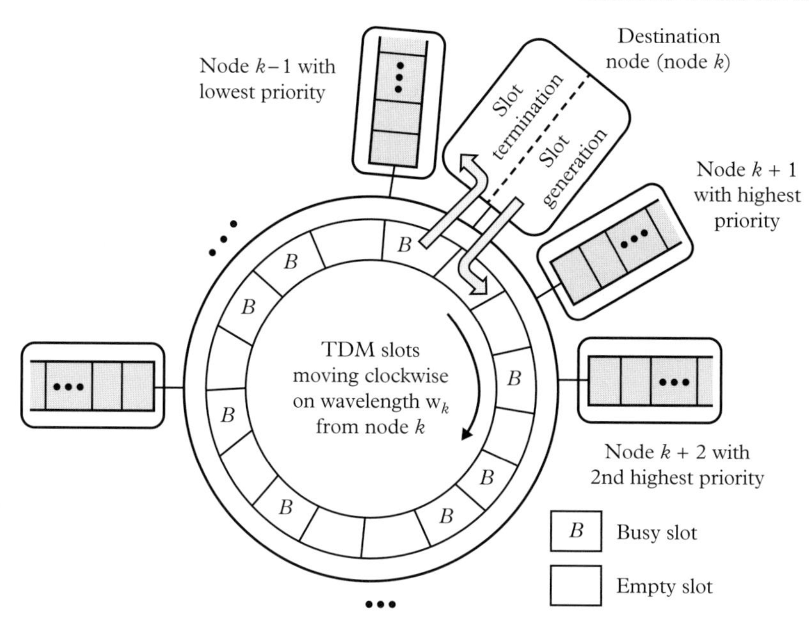

Figure 8.34 *Illustration of slot-generation/termination functionalities of a destination node (node k) on wavelength w_k in a time-slotted packet-switched WDM ring.*

the upstream nodes. The unfairness in the MAC protocol, thus introduced, can be addressed by forcing the higher-privilege nodes to skip slot(s) with a specified probability, which should decrease for the downstream nodes closer to the destination node. However, this would raise the average delay of the network to some extent, while reducing the unfairness, and one needs to see how the overall delay is affected as the network scales up to larger size with more nodes and larger diameter.

Next, we proceed to assess the packet- and circuit-switched (SONET-based) WDM rings in respect of *bandwidth utilization*. The primary cause for bandwidth under-utilization in SONET for bursty traffic is the multiplexing scheme used. In SONET, the smaller-rate streams like OC-3 are time-division multiplexed into a higher-rate OC frame keeping the packet flows within the OC-3 pipes undisturbed without allowing any migration of packets from one OC-3 to another. Therefore, the packets inside an OC-3 pipe cannot be reprovisioned to any other OC-3 pipe even if the latter has empty TDM slots (while being independently compressed in time domain to realize TDM-based circuit-multiplexing), which results in wastage of bandwidth and makes the SONET-based WDM ring inefficient in handling bursty traffic. Figure 8.35 illustrates this issue of bandwidth under-utilization in SONET-based WDM ring. As shown in the figure, even if a lower-rate stream goes empty, the data packets from the other lower-rate streams cannot avail the empty space (i.e., the available bandwidth therein).

In order to estimate the bandwidth utilization in packet and circuit-switched WDM rings, we consider an N-node N-channel WDM ring, where the ingress packets into the network is assumed to be uniformly distributed over all the N channels and every such channel gets the traffic uniformly from $(N-1)$ nodes. With this framework, bandwidth-utilization of any one channel will be the same as that of the whole network, and hence we will consider any arbitrary channel (channel k, say) from the N channels in the ring for estimating the bandwidth utilization in both types of networks, i.e., time-slotted packet-switched and circuit-switched WDM rings. Denoting the packet arrival rate (uniform for all

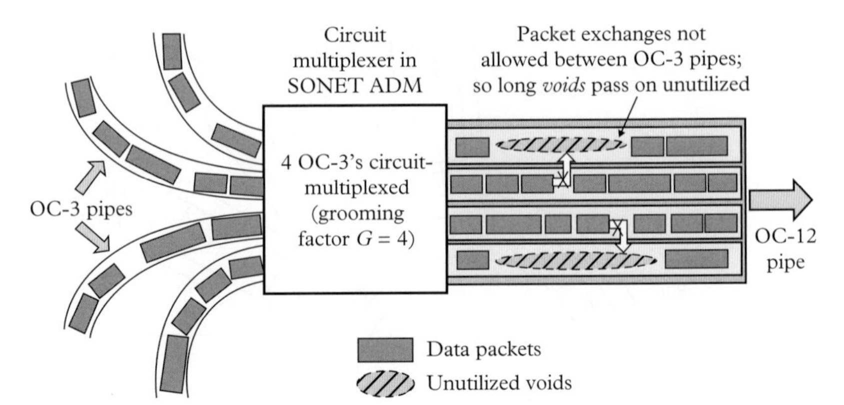

Figure 8.35 *Bandwidth under-utilization in SONET-based WDM ring networks with bursty traffic.*

channels) in a particular channel (w_k) by Λ, the arrival rate from a given source node i on that channel is given by

$$\lambda_{ik} = \Lambda/(N-1). \tag{8.36}$$

For evaluating the bandwidth utilization of the packet-switched ring in the kth channel (and so also for all channels, and hence for the network as well), we need to find the probability that the slot (generated from the destination node) is empty (i.e., not utilized) in the same channel even after passing through all the downstream nodes, i.e., all the queues of the corresponding nodes have been empty. Using the basic property of queuing systems in steady state, the probability that the buffer of node i would be empty, is assumed to be

$$P_i(0) = 1 - x_{ik}, \tag{8.37}$$

where x_{ik} is the traffic at the ith node addressed to node k, arrived within the comprehensive service time defined as the sum of fixed service time i.e., $E[X] = 1$ for the slot duration and the average *vacation time* $E[V(i)]$ for a particular destination. With t_s as the slot duration, we therefore express x_{ik} as

$$x_{ik} = \lambda_{ik}(1 + E[V(i)])t_s. \tag{8.38}$$

The average vacation time $E[V(i)]$ at node i can be evaluated as follows. Consider that node i has been able to grab a free slot on the mth attempt, implying thereby that it has missed $(m-1)$ slots as they were used already by the upstream nodes. The probability of grabbing the mth slot for transmission by node i will imply a vacation period $V(i)$ of $m-1$ slots. Assuming that all the nodes transmit independently, the probability of $V(i)$ will follow geometrical distribution, given by

$$P[V(i) = m-1] = q_i^{m-1}(1-q_i), \tag{8.39}$$

where q_i is the probability that node i fails an attempt, i.e., encounters a busy slot. Thus, the probability q_i represents the probability that one of the preceding (upstream) nodes has used the time slot. Hence, $(1-q_i)$ is in effect the probability that node i can transmit, while no node preceding node i has intended to transmit. One can therefore express $(1-q_i)$ as

$$1 - q_i = \rho_{ik} \prod_{j=1}^{i-1} (1 - \rho_{jk}), \tag{8.40}$$

where ρ_{jk} ($= \lambda_{jk} t_s$) represents the traffic from node jth to node k. Hence, the expression for q_i is obtained as

$$q_i = 1 - \rho_{ik} \prod_{j=1}^{i-1} (1 - \rho_{jk}). \tag{8.41}$$

Using Eq. 8.39 in Eq. 8.41, the average vacation time $E[V_i]$ at node i on the kth channel can be expressed as

$$E[V(i)] = \sum_{m-1=0}^{\infty} (m-1) q_i^{m-1} (1 - q_i) = q_i/(1 - q_i), \tag{8.42}$$

which can be used in Eq. 8.38 to obtain the estimate of x_{ik}. Finally, using x_{ik}'s, the under-utilization factor α_{pkt} for the packet-switched WDM ring can be expressed as

$$\alpha_{pkt} = \prod_{i=1}^{N-1} (1 - x_{ik}). \tag{8.43}$$

Therefore, the bandwidth utilization u_{pkt} for the packet-switched WDM ring is expressed as

$$u_{pkt} = 1 - \alpha_{pkt}. \tag{8.44}$$

Next, we consider the case of the circuit-switched SONET-over-WDM ring. Taking cue from the earlier discussion, we consider the packet arrival rate at each node for a particular destination, which is same as in Eq. 8.36. In reality, the packets in G OC-3 pipes are compressed into OC-n ($n > 3$) pipe with the effective arrival rate given by

$$\lambda_{ckt} = \frac{\Lambda}{(N-1)}. \tag{8.45}$$

Each packet is served through an OC-n pipe, thereby making the service time, given by

$$t_p^{ckt} = r_b L, \tag{8.46}$$

with L as the packet length and r_b as the bit rate of OC-n bit stream. Therefore, the effective traffic intensity (x_{ckt}) on each OC-3 is the bandwidth utilization factor for circuit-switched WDM ring (u_{ckt}), given by

$$u_{ckt} = x_{ckt} = \lambda_{ckt} t_p^{ckt}. \tag{8.47}$$

Equations 8.44 and 8.47 provide the necessary formulas to compare the bandwidth utilization in both types of switching options in WDM ring networks. The plots of bandwidth utilization in both cases for a six-node (and hence using six wavelengths) WDM ring network, as obtained from these two equations, are shown in Fig. 8.36. As the OC-3 pipe capacity is 155.52 Mbps, and the capacity of each wavelength is considered to be 2.5 Gbps (precisely, 2.48832 Gbps), we get the grooming capacity as $G = 16$. The plots of bandwidth utilization for the two networks in Fig. 8.36 indicate indeed the marked superiority of packet-switched WDM ring over the circuit-switched WDM ring in terms of bandwidth utilization.

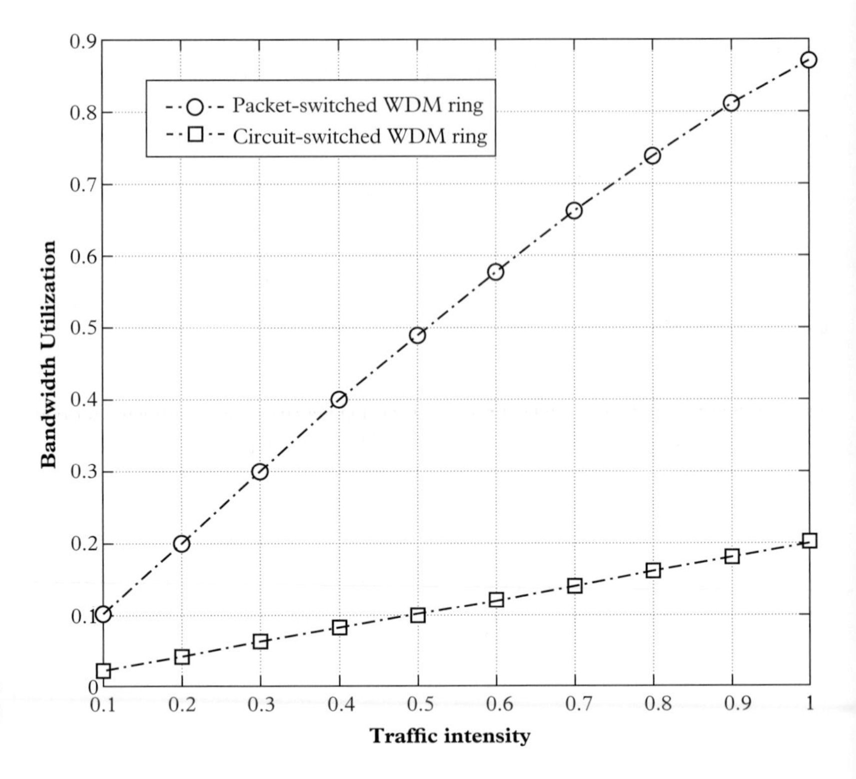

Figure 8.36 *Comparison of bandwidth utilization between circuit- and packet-switched architectures in WDM ring networks.*

Besides the issues discussed in respect of the MAC protocol and bandwidth utilization in packet-switched WDM rings, one also needs to check the physical-layer issues. In the packet-switched ring architecture, the transmitted packets on each wavelength traverse through the intermediate nodes which are based on passive optical devices, thereby causing insertion losses for the passthrough signal. Furthermore, for header-reading, a small part of the received optical power is tapped at every downstream node, thereby causing a progressive signal loss.

To get an insight into this issue, we consider HORNET as an example. Note that, in HORNET, the control and data segments in a given slot are likely to get contaminated by different magnitudes of noise components. The tapped signal ($\simeq$10% of received power) with subcarrier-modulated control bits in the slot-manager block will have smaller power level, but with the shot noise components adding up from all the wavelengths. However, the data signal filtered and dropped down in the smart-drop block will be stronger ($\simeq$90% of the received power) and free from the shot noise contributions from the other wavelengths, thereby being contaminated by the shot noise generated just from the power of the received wavelength only. This will make the SNR superior (higher) for the data bits retrieved in the smart drop block as compared to the control bits retrieved in the slot manager block.

The difference in the SNR between the control and data bits in HORNET will make the control bits more vulnerable to bit errors (higher BER). However, the packet-error rate (PER) being the deciding factor, and with $PER = L \times BER$, the control segment in a packet having much fewer bits than the data segment can manage with higher BER for the same target PER. This issue can be further addressed by transmitting the control segment at a slower rate (while remaining within the TDM slot), as the lower-rate transmission reduces the receiver bandwidth, thereby decreasing the noise proportionately. Thus, the designer needs to make use of the interplay between the various relevant parameters, e.g., tapping ratio in the slot-manager module, control bit rate, laser power, receiver sensitivity, etc., so that the ring can host an adequate number of nodes over the required network span with acceptable PER values for data as well as control packets. Similar assessment of the other packet-switched WDM rings would also be useful, to ensure the feasibility of the respective networks.

8.6 Summary

In this chapter, we presented various forms of optical WDM metro networks, using ring topology. Several candidate architectures of WDM ring networks were considered for both circuit- and packet-switched transmission schemes. Hierarchical splitting of metro networks into core and edge rings and the suitability of circuit/packet switching for the two levels (core/edge) were discussed.

For the circuit-switched WDM ring networks, we considered two basic categories, PPWDM and WRON, operating with the TDM-based circuit-switched SONET/SDH transmission. First, we presented heuristic design methodologies for PPWDM and WRON rings with uniform traffic pattern between the node pairs and estimated the number of wavelengths and ADMs required in the network. For non-uniform traffic, we presented some design methodologies using LP-based as well as heuristic schemes. Subsequently, the WRON-based rings using single-hub and double-hub logical topologies were examined, followed by a brief discussion of interconnected WDM ring networks.

For the packet-switched WDM ring networks, some of the testbeds developed by various research groups were described, such as MAWSON, RingO, HORNET, and RINGOSTAR. Thereafter, we presented a generic mathematical model to assess the performance of packet- and circuit-switched WDM ring networks in respect of bandwidth utilization. Using the analytical model a comparison of the circuit and packet-switched WDM rings was made in terms of bandwidth utilization in fiber rings, with the packet-switched WDM rings showing significantly better bandwidth utilization for bursty traffic.

EXERCISES

(8.1) Discuss, using an appropriate layered architecture, how the packet- and circuit-switched traffic flows are circuit-multiplexed in SONET-over-WDM ring networks. What is the limitation of the circuit-switched transmission of packet-switched traffic and why does this practice continue to exist?

(8.2) Consider an N-node mesh-configured WRON having bidirectional fiber links with a given set of lightpaths, where a node pair is allocated only one lightpath in each direction. Assume that a cut-set C_i bisects the network into two parts with $N - n_i$ and n_i nodes. If the cut-set consists of f_i fiber pairs and l_i lightpaths, then find out the expression for the lower-bound for the number of wavelengths M_{min} needed in the entire network. Using this model, show that for an N-node ring (i.e., two-connected mesh) with N as an odd integer, $M_{min} = \frac{N^2-1}{8}$.

(8.3) Repeat the above exercise for WRON rings with an even number of nodes and determine the lower-bound for the number of wavelengths in the network.

(8.4) Consider a six-node bidirectional ring with the following traffic matrix (integer multiples of OC-12 connection).

$$\Gamma = \begin{bmatrix} 0 & 7 & 7 & 2 & 5 & 6 \\ 9 & 0 & 5 & 9 & 7 & 6 \\ 7 & 4 & 0 & 5 & 4 & 6 \\ 8 & 8 & 6 & 0 & 4 & 9 \\ 3 & 7 & 5 & 4 & 0 & 3 \\ 9 & 6 & 7 & 5 & 4 & 0 \end{bmatrix}.$$

Find out the numbers of ADMs and wavelengths by using a multihop heuristic design to support the traffic demand with a grooming capacity of four in each wavelength

(8.5) Repeat the above exercise using a multihop LP-based method. Compare the results obtained from the LP formulation and the heuristic scheme.

(8.6) Consider that a WRON-based ring with five nodes employs single-hub configuration, where the numbers of OC-48 lightpaths needed from the various nodes to the hub node have been estimated from the traffic matrix for the WRON as follows (node 0 is the hub):

a) from node 4 to hub: 2,

b) from node 3 to hub: 4,

c) from node 2 to hub: 1,

d) from node 1 to hub: 3.

Determine the number of wavelengths needed in the network. Also calculate the number of OC-48 transceivers to be provisioned in the hub.

(8.7) Consider a WRON-based ring with six nodes, whose virtual topology looks like a *circle* with several *arcs* as lightpaths representing the end-to-end connections between all non-adjacent node pairs. With this network-setting, sketch the schematic diagram of an OADM for the add-drop and passthrough operations.

(8.8) In the six-node WRON ring of the above exercise, assume that the optical transmitters transmit 15 dBm of optical power at 2.5 Gbps for each lightpath and the receivers have a sensitivity of −23 dBm, and the ring has a circumference of 30 km with equal distances between the adjacent nodes. Trace the longest lightpath in the ring and check the feasibility of the power budget for the same. Given: connector loss = 1 dB, demultiplexer loss = 3 dB, multiplexer loss = 3 dB, fiber loss = 0.2 dB/km.

(8.9) Discuss critically the realizability of the various packet-switched WDM metro networks, particularly in respect of their scalability of physical size, node hardware, and latency.

(8.10) Consider a five-node packet-switched WDM metro ring based on HORNET architecture. With the HORNET-node configuration presented in Fig. 8.28, formulate the power budget for the data packets with a ring circumference of 15 km. If the transmit power from the lasers is 15 dBm and the receiver sensitivity for the data packets is −27 dBm at 1 Gbps, estimate the worst-case power budget and comment on its feasibility. Given: insertion and connecting loss in the input splitter (in addition to the 10% tapping loss) = 1 dB, insertion loss in FBG (including circulator) = 2 dB, insertion loss in the output combiner = 2 dB, coupling and insertion loss in the output combiner = 1.5 dB, connector loss = 1 dB. Assume equal distance between adjacent nodes.

WDM Long-Haul Networks

9

Ever-increasing traffic in the long-haul optical networks called for WDM transmission, all the more, as compared to the metro and access segments. Further, although the interconnected WDM rings served the needs of long-haul networks initially, mesh-connected WDM networks evolved as a natural and more effective solution with enhanced connectivity and resilience against network failures. In this chapter, we first discuss the basic design challenges and the candidate node configurations for the WRON-based mesh networks. Thereafter, we present various offline design methodologies for this class of networks using LP-based and heuristic schemes. The impact of wavelength conversion in mesh-configured WRONs is also examined analytically and the gain in bandwidth utilization is assessed in WRONs with full wavelength conversion as compared to those not using any wavelength conversion. Finally, we present some online routing and wavelength assignment schemes that can be used for the operational WRONs.

9.1 Wavelength-routing in long-haul networks

The WDM long-haul networks have certain characteristic features, in contrast with the underlying metro or access segments. In general, the metro segment employs fiber rings or interconnected fiber rings, while the access segment can use tree, star, or a mix-and-match of tree/star and ring topologies using optical fibers. As compared to these regular physical topologies, the WDM long-haul networks employ preferably mesh topologies (or more precisely, incomplete mesh topologies) with enhanced connectivity and failure resilience, which have evolved from the interconnected rings used in old networking setups. The design of a WDM long-haul network is governed by its mesh topology with large number of nodes, preferably employing a wavelength-routing technique for the en route traffic flows to reduce the load of IP routers in the electrical domain. This class of wavelength-routed optical networks (i.e., WRONs[1]) would be more challenging than the WRONs over ring topology, with numerous options for setting up connections across the mesh topology using appropriate routes and wavelengths, along with increased transmission impairment in the long routes.

Furthermore, while addressing the above issues, the overall design should be cost-effective in respect of capital and operational expenses and be compliant with the QoS requirements of different classes of traffic flows. In other words, one needs to come up with designs that make conservative use of network resources (transceivers, wavelengths, OEO regeneration, etc.), while not compromising with the network performance in respect of the

[1] Note that, we have already dealt with WRONs in Chapter 8 on WDM metro networks, but using ring topology, where the physical degree of a node or simply the nodal degree (i.e., the number of bidirectional fiber links connected to the node) has been limited mainly to just two, or a few more for the nodes interconnecting multiple ring networks. However, for the WRONs with a mesh topology, the physical degree of a node would be variable and larger, thereby calling for more complex node configurations.

Optical Networks. Debasish Datta, Oxford University Press (2021). © Debasish Datta.
DOI: 10.1093/oso/9780198834229.003.0009

network capacity and QoS parameters. The plethora of such underlying issues makes the design of long-haul WRONs quite challenging and leads to a class of optimization problems.

9.2 Node configurations in mesh-connected WRONs

Traditionally, SONET/SDH has been the legacy optical-networking platform for metro and long-haul networks, even after the introduction of WDM transmission. In the metro segment, the network formation has been mainly realized using a ring or interconnected rings. These networks have been set up using the nodes with a physical degree (or nodal degree) of two within a ring, and multi-degree inter-ring nodes with a nodal degree of four to interconnect two rings. When there has been a need to interconnect three or more rings at an inter-ring node for extending a metro network or for realizing a mesh-like long-haul network, the nodal degree of the inter-ring nodes needed to be enhanced to yet-higher values. For WDM networks, this task has been carried out by extending the ROADMs with a degree of two to form multi-degree ROADMs (MD-ROADMs) having degrees $D > 2$ with additional hardware.

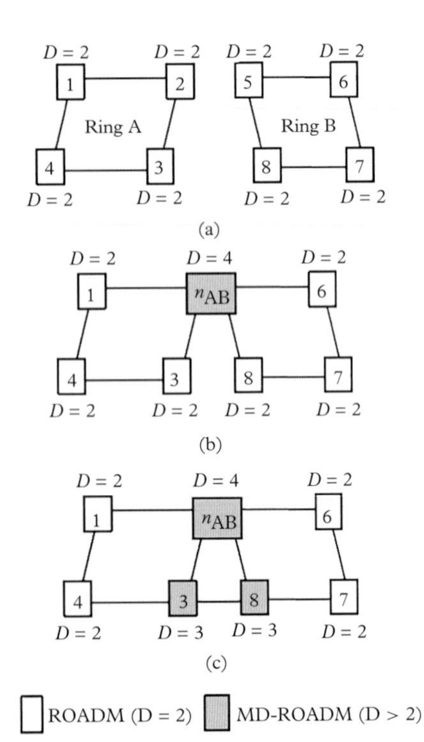

Figure 9.1 *Illustration of transforming two interconnected rings to a mesh network by using MD-ROADMs.*

Figure 9.1 illustrates the above transformation process with some typical examples demonstrating the need of MD-ROADMs with varying degrees $D \geq 3$. Fig. 9.1(a) presents two four-node WDM rings (Ring A and Ring B) each using four regular ROADMs with a nodal degree $D = 2$. In Fig. 9.1(b) the two four-node WDM rings are interconnected to form a bigger network by bringing together the nodes 2 and 5 and replacing them by an inter-ring node of degree four, renamed as node n_{AB}. Fig. 9.1(c) shows another example, wherein one needs more connectivity in the network than what is shown in (b) with the additional link between nodes 3 and 8. This requires two MD-ROADMs of degree three for nodes 3 and 8. One can realize MD-ROADMs using optical splitters along with WSSs offering colorless and directionless functionality, as discussed in Chapter 2. However, with large nodal degree, the MD-ROADM configuration involving optical splitters doesn't scale up well, thereby needing full-fledged optical crossconnects, i.e., OXCs, also introduced briefly in Chapter 2. We discuss below the various options for OXC realization in further detail. Note that, one can also interconnect rings with multiple linear fiber links to approach toward a mesh configuration (see Exercise 9.1)

When a WRON is being set up afresh with its nodes needing high nodal degrees, one can directly opt for OXCs using optical multiplexers/demultiplexers along with optical switches realized with electro-optic, MEMS-based, or other forms of optical switching elements. This type of OXCs carries out the switching operation of the incoming lightpaths as a whole in a transparent manner. However, in practice, one can also use appropriate mix-and-match of electrical and optical hardware within an OXC with respective pros and cons. We discuss these variations in OXC configurations in the following.

In general, the OXCs used in WDM mesh networks can realize switching in three possible ways: all-optical switching, electrical switching with OE-EO conversion, and 1300 nm optical switching with OEO-OEO conversion. The OXC configuration with all-optical switching has already been discussed in Chapter 2 (Fig. 2.80). Further, the OXC configuration with electronic switching through OE-EO conversion is shown in Fig. 9.2, where the received lightpaths are demultiplexed and transformed into electrical signals using OE conversion. Thereafter switching is carried out in the electrical domain. The switch outputs are passed through EO conversion and multiplexed into the respective output ports. The add-drop functionality is carried out by the electrical switch fabric.

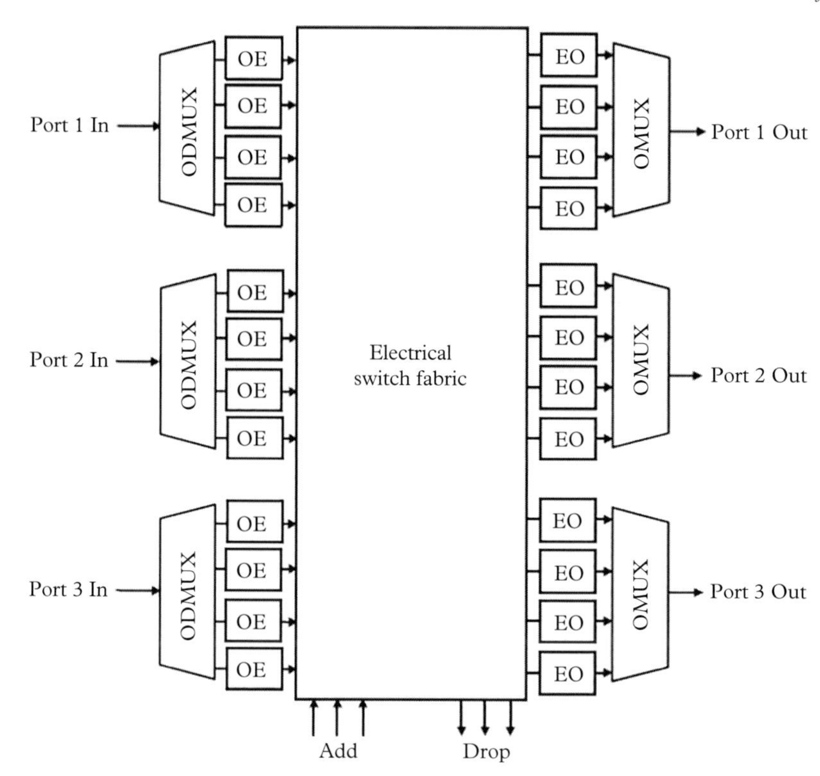

Figure 9.2 *An OXC with OE-EO conversion and electrical switching for four-wavelength WDM inputs. OMUX: optical multiplexer, ODMUX: optical demultiplexer.*

Figure 9.3 shows the OXC, which carries out optical switching on a single wavelength, i.e., 1300 nm using OEO-OEO conversion. Hence, the incoming lightpaths, after being demultiplexed, are all passed through the transceivers (transponders), and each transponder converts its input wavelength (say w_i) into an electrical signal, which is subsequently transmitted on 1300 nm toward the switch input port. Thus, regardless of the associated wavelength, each incoming lightpath gets transformed into a 1300 nm optical signal, and the optical switch fabric needs to switch all the optical signals arriving at its input ports on the same wavelength, i.e., 1300 nm. Having been switched to the appropriate switch output ports, the 1300 nm signals are once again passed through the OEO-based transponders to get converted into appropriate wavelengths and multiplexed out through the output ports of the OXC. In this case, the add-drop signals are also passed through the optical switch operating at 1300 nm, and they also go through the similar transponders for both add as well as drop operations.

As discussed in Chapter 2, MD-ROADMs typically use a combination of optical splitters and WSSs (see Fig. 2.79). This configuration can be modified by replacing the optical splitters on the input end by WSSs, which makes the node scalable by preventing the power-splitting losses at input ports. Figure 9.4 shows this WSS-WSS configuration for MD-ROADMs/OXCs, which also offers wavelength-flexibility in add-drop operation, thereby making the node colorless in respect of add-drop functionality. However, notwithstanding these features, the natural broadcasting functionality offered by the power-splitting operation in the splitter-WSS configuration wouldn't be feasible in the WSS-WSS configuration.

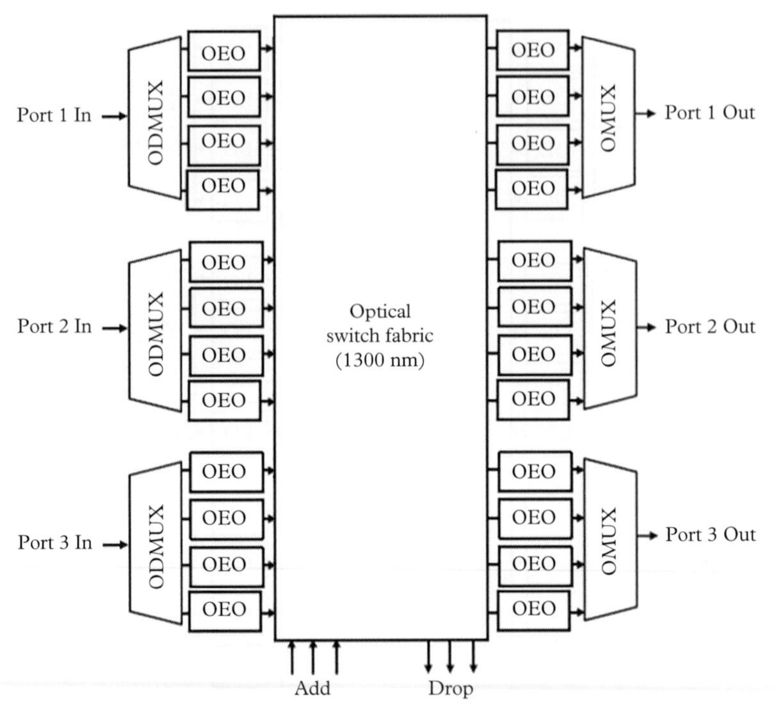

Figure 9.3 *An OXC with OEO-OEO conversion and optical switching at 1300 nm for four-wavelength WDM inputs. OMUX: optical multiplexer, ODMUX: optical demultiplexer.*

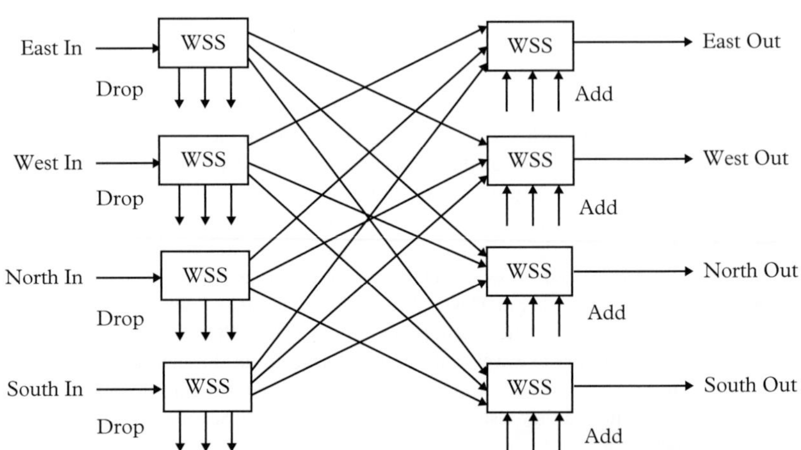

Figure 9.4 *MD-ROADM with WSSs on both sides, offering colorless add-drop functionality.*

9.3 Design issues in long-haul WRONs

As in metro WRONs, the design methodology of long-haul WRONs needs to use appropriate resource-provisioning (dimensioning) schemes for a set of traffic demands between the node pairs in a given network, but with a non-regular (incomplete) mesh topology. The design is initially carried out offline for setting up a network afresh (i.e., *greenfield* network) or for augmenting an existing network setting. Besides the offline design, which is important to start the network operation, one also needs to design some computationally efficient online routing and wavelength assignment (RWA) scheme with a *dynamic* resource allocation mechanism to deal with *time-varying* traffic patterns and unpredictable network failures.

The option for end-to-end all-optical transmission (without or with wavelength conversion) avoiding OEO conversion and electronic processing at the intermediate nodes becomes a more sought-after feature for long-haul mesh-configured WRONs, having much larger size and higher nodal degrees, as compared to the ring-based metro networks. However, some limited options for OEO conversion, if provided along with OXCs, can allow more traffic to be groomed (multiplexed) at intermediate nodes for efficient utilization of the fiber bandwidth. However, in practice it is hard to support all-optical connections between all node pairs due to the limited number of transceivers and wavelengths. These issues make the offline design of the long-haul WRONs a critical optimization problem, governed by numerous constraints, such as available network resources, traffic matrix, congestion in the network, etc.

The offline design methodology for WRONs using one single LP-based formulation becomes computationally complex, more so with large networks (Krishnaswami and Sivarajan 2001; Mukherjee 2006; Ramaswami *et al.* 2010). Hence, the overall design problem is preferably broken into smaller subproblems, where each subproblem is solved using a smaller LP-based formulation or a heuristic scheme. Owing to the constraints on the numbers of available wavelengths and per-node transceivers, some selected node pairs (typically with large traffic and/or having high-priority for all-optical communications) are connected using all-optical transmission, thereby creating a set of end-to-end virtual connections (VCs). These VCs are routed through specific paths, leading to virtual paths (VPs), and the VPs, once assigned specific wavelengths, lead to the creation of lightpaths. The VPs (i.e., VCs with specific paths) constitute a virtual topology (VT), which would usually look different from the underlying physical topology. The VPs, when assigned appropriate wavelengths, lead to a lightpath topology (LT), which is similar to VT, but with specific wavelength assignments (WAs) made for all the VPs.

Figure 9.5 illustrates the significance of a VC and the respective VP in Fig. 9.5(b) for a six-node WRON (shown in Fig. 9.5(a)) and the corresponding lightpath in Fig. 9.5(c)

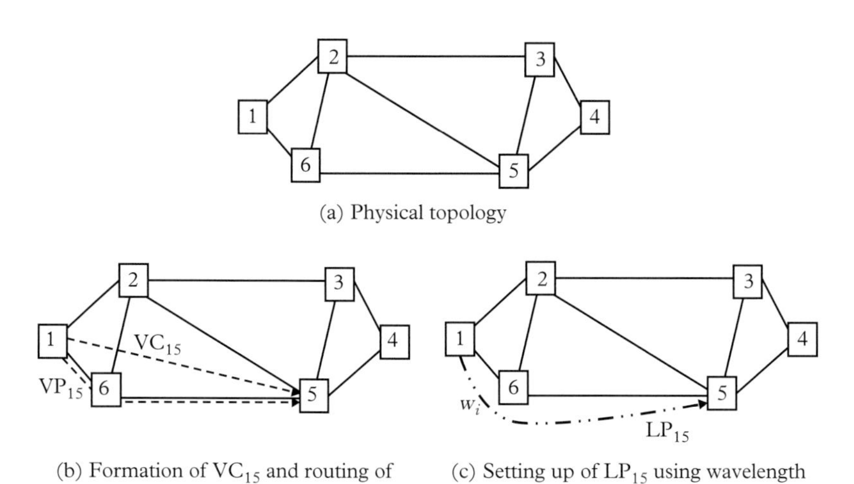

Figure 9.5 *Formation of an example VC (VC_{15}) from node 1 to node 5, the corresponding VP (VP_{15}) and lightpath (LP_{15}), in a six-node mesh-configured WRON (shown in part (a)), where all nodes operate with OXCs realizing the wavelength-routing functionality. In part (b), VC_{15} is shown as a direct connection needed from node 1 to node 5, which is routed through node 6 to set up VP_{15}. In part (c), VP_{15} is assigned the wavelength w_i to form the corresponding lightpath LP_{15}. Note that LP_{15} is optically bypassed (i.e., wavelength-routed) by the OXC at node 6.*

after the assignment of a wavelength (w_i, say). As mentioned above, all the VPs lead to the formation of a VT, while the respective lightpaths form an LT. The design problem addressing the formation of a VT is referred to as VT design (VTD), while the overall design of LT is referred to as lightpath topology design (LTD). The traffic demands between the node pairs that cannot be assigned lightpaths due to limited resources are usually groomed with the traffic of the node pairs already rewarded with direct (i.e., single-hop) lightpaths. Thus, finally the designer will have to use the LTD for setting up the necessary multihop connections using concatenated lightpaths with OEO regeneration at the appropriate intermediate nodes.

Once a WRON with a mesh topology starts operating with the initial offline design, it needs to carry out an appropriate online RWA scheme, which would require fast algorithms to set up and tear down lightpaths dynamically in accordance with the time-varying traffic and other unforeseen situations, such as network failures. The task of online RWA is carried out by a *control plane* of the network using appropriate protocols that run concurrently with the data-transporting part of the network, generally referred to as the *data plane*, the latter operating through a set of lightpaths provisioned across the physical topology. The control plane is typically realized over the same physical topology of the network by using a dedicated wavelength (or a fraction of a specific wavelength). The chosen wavelength for the control plane *must visit* every node *electronically* (i.e., must undergo OEO conversion at each node), so that the control information is shared with all nodes for the online RWA and various other control and management operations.

Figure 9.6 illustrates an embedded control plane using one single (or part) wavelength over the given physical topology in a WRON, where the data traffic flows concurrently through the data plane (not shown explicitly) involving the same nodes and the physical topology through a set of lightpaths. As indicated earlier, initially, the data plane starts operating using the set of lightpaths obtained from the offline LTD, which keeps evolving with time while responding to the time-varying network traffic and unpredictable situations, such as network failures and various management issues. Note that, the control plane may or may not use the complete physical topology, but it should reach each node at least once

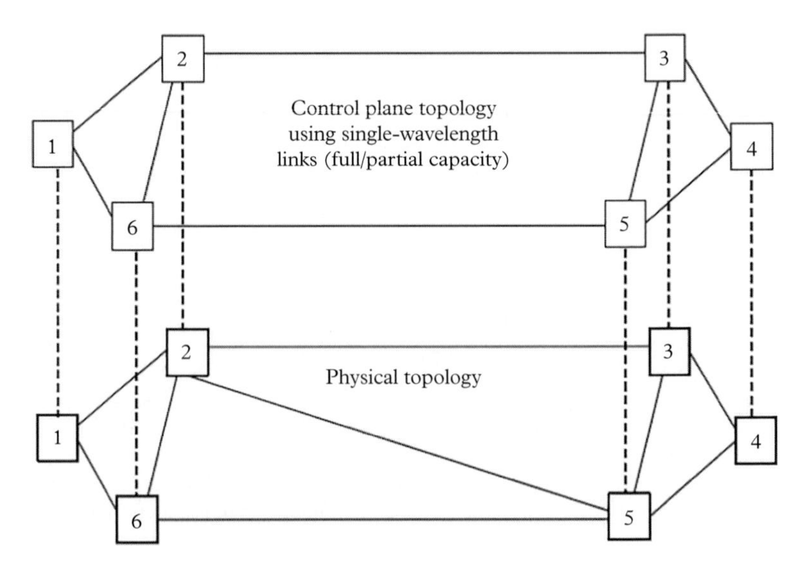

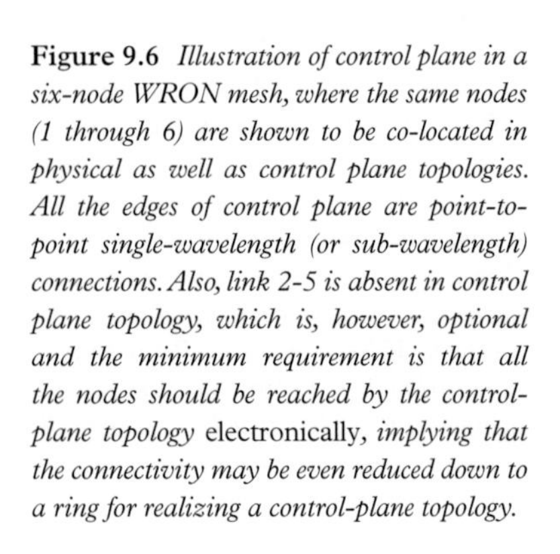

Figure 9.6 *Illustration of control plane in a six-node WRON mesh, where the same nodes (1 through 6) are shown to be co-located in physical as well as control plane topologies. All the edges of control plane are point-to-point single-wavelength (or sub-wavelength) connections. Also, link 2-5 is absent in control plane topology, which is, however, optional and the minimum requirement is that all the nodes should be reached by the control-plane topology* electronically, *implying that the connectivity may be even reduced down to a ring for realizing a control-plane topology.*

electronically, implying that its minimum topology should preferably be an OEO-based ring passing through all the nodes.

In order to appreciate further the role of a control plane in a WRON, let's also have a look into the dynamics of the WRON traffic. In particular, though a major part of the traffic in an operational WRON would vary slowly over time with long connection-holding times, typically ranging over weeks, months, or beyond, there would also be a fraction of the traffic that will vary with much shorter holding times, leading to what is known as *network churn* (Simmons 2014). The control plane should therefore sense and reprovision the network resources to address such traffic dynamics from time to time.

In view of the above challenges, the control plane in a WRON must be agile enough to address comprehensively the various issues in real time. Several standards for control planes in WRONs have been developed by the international bodies, such as generalized multiprotocol label switching (GMPLS) from IETF, and automatic switched optical network (ASON) from ITU-T, which operate with a distributed control framework. Further, a centralized framework of control plane in optical WDM networks is currently emerging, which uses the software-defined network (SDN) architecture. We consider these issues in further detail in Chapter 12, dealing specifically with the control and management of optical networks.

9.4 Offline design methodologies for long-haul WRONs

As indicated earlier, the offline design methodologies of mesh-configured WRONs, being complex in nature, are in practice divided into a few subproblems. We discuss one such design methodology using two subproblems as follows.

MILP-VTD and ILP/heuristic-WA

- *First subproblem* (MILP-VTD): The first subproblem in this approach employs a mixed-integer LP (MILP) formulation, which considers both integer and real variables. The MILP would typically set up VPs between some selected node pairs along specific paths or routes, leading to a VTD, but without any WA for the VPs. Further, while carrying out the MILP-VTD with the given resource constraints, the residual traffic demands between the left-out node pairs without dedicated VPs also get groomed into the designed VPs.
- *Second subproblem* (ILP/heuristic-WA): In the second subproblem, the task of WA is carried out for the VPs obtained from the first subproblem, transforming each VP into a lightpath (or more) with specific wavelength(s). Thus, the VTD obtained in the first subproblem is now transformed into an LTD by carrying out WA. The task of WA can be accomplished by using graph-coloring technique based on an integer LP (ILP) or by using some suitable heuristic scheme.

In the following, we describe the two subproblems in further detail.

9.4.1 MILP-VTD

We formulate the MILP problem following the line of treatment presented in (Ramaswami and Sivarajan 1996). In particular, using the MILP we seek to find a VTD consisting of a set of VPs, for a given physical topology with the given numbers of transmitters and receivers in each node. However, we don't place any upper limit on the number of wavelengths, and let the number of wavelengths be eventually determined through the MILP execution and next subproblem (i.e., WA). Before describing the objective function and the constraints for the MILP, we define the relevant network parameters and variables as follows.

MILP parameters:

N: number of nodes in the network.

e_{lm}: a binary number, which equals one if there exists a bidirectional fiber link between the nodes l and m, and zero otherwise. Thus, $\{e_{lm}\}$ represents the adjacency matrix for the given physical topology.

Λ: traffic matrix $\{\lambda^{sd}\}$, with λ^{sd} as the traffic (i.e., the number of packets arriving during an average packet duration) from a source node s to a destination node d.

d_{ij}: propagation delay through the shortest path between a node pair (i,j) through the physical topology. If a VP exists between this node pair, then d_{ij} for this VP will be the sum of the propagation delays of all of its constituent fiber links (e_{lm}'s).

d_{max}: maximum allowable propagation delay between any node pair across the network.

Δ_i^{TX}: number of transmitters in node i, i.e., logical out-degree of node i.

Δ_i^{RX}: number of receivers in a node i, i.e., logical in-degree of node i.

For each nod pair (i,j), the shortest path through appropriate links (i.e., e_{lm}'s) of the physical topology has to be *pre-computed* (using a shortest-path algorithm, e.g., Dijkstra's algorithm) and the total propagation delay for each shortest path ($= d_{ij}$) needs to be provided as an input to the LP-solver. The physical degree of a node depends on the number of bidirectional fiber links connected to it, while the logical out-degree or in-degree of a node is determined by the number of VPs that can emerge from or sink into the node, respectively. In the present problem, we assume that, $\Delta_i^{TX} = \Delta_i^{RX} = \Delta \; \forall i$ for simplicity, i.e., the logical nodal degree is same for outgoing and incoming traffic. Next, we define the variables used in the MILP.

MILP variables:

k: index of a VP between a node pair, upper-bounded by the number of transceivers in the two nodes, i.e., $k = 0, 1, \ldots, \Delta - 1$.

$b_{ij}(k)$: a binary variable, which equals one, if there exists the kth VP from node i to node j, and zero otherwise. Note that the set of $b_{ij}(k)$'s with given directions defines the *directed* VT of the WRON, which is to be determined through the execution of the MILP.

α: delay *relaxation* factor (to be multiplied with d_{max}) to ease out the route selection process, as a higher value of α would allow more routes to become feasible.

$\lambda_{ij}^{sd}(k)$: traffic from node s to node d over the kth VP from node i to node j.

$\lambda_{ij}(k)$: total traffic through the kth VP from node i to node j.

λ_{max}: $max_{ij}\lambda_{ij}(k)$ = traffic through the VP from node i to node j, that turns out as the VP carrying the maximum traffic (referred to as *congestion*) among all VPs in the WRON.

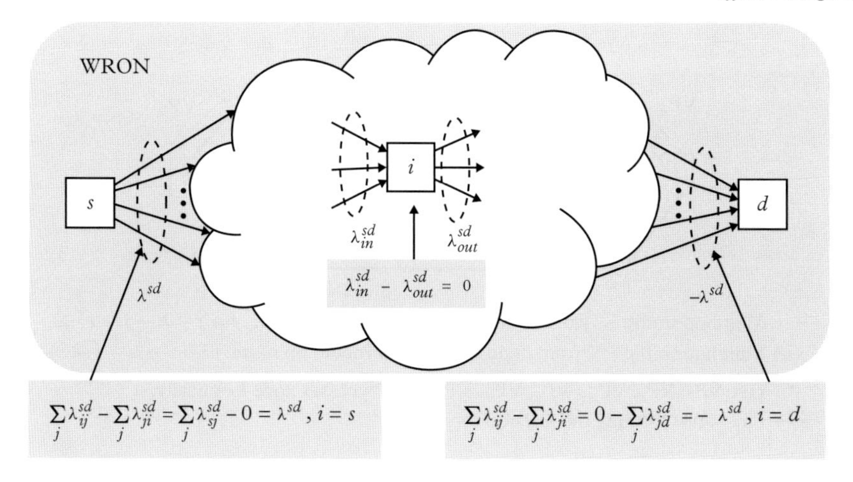

Figure 9.7 *Illustration of flow conservation at WRON nodes. Nodes s and d are the source and destination nodes, respectively, for a traffic flow λ^{sd}. Node i is one of the intermediate nodes for the same connection located inside the WRON, where a part of λ^{sd} (λ_{in}^{sd}, say) enters node i from various possible directions. Similarly, λ_{out}^{sd} represents the part of λ^{sd}, that exits from node i in different directions and must be equal to λ_{in}^{sd}. The illustration shows how the flow conservation at source, destination, and intermediate nodes takes effect in accordance with Eq. 9.2.*

MILP objective function:

The MILP-based VTD is carried out with an objective to minimize the traffic flow through the VP carrying the maximum traffic (congestion), leading to the following objective function:

$$\text{minimize}[\lambda_{max}]. \tag{9.1}$$

MILP constraints:

- *Flow conservation:* At an intermediate node in a VP, the difference between the incoming and outgoing traffic flows should be zero. However, at the source node of the VP, the difference would be the same as the outgoing traffic from the same node. Similarly, at the destination node of the VP, the difference would be the negative of the traffic flowing in it (see Fig. 9.7). This constraint (also referred to as solenoidality, see Chapter 7) is therefore expressed for all s, d, i as

$$\sum_j \lambda_{ij}^{sd}(k) - \sum_j \lambda_{ji}^{sd}(k) = \begin{cases} \lambda_{sd}, \text{ for } i = s \\ -\lambda_{sd}, \text{ for } i = d \\ 0, \text{ otherwise.} \end{cases} \tag{9.2}$$

- *Total traffic flow in a VP:* Through a VP, multiple traffic flows can be passed through by packet-multiplexing, leading to TG. Thus, the total traffic through the kth VP from node i to node j can be expressed as

$$\lambda_{ij}(k) = \sum_{s,d} \lambda_{ij}^{sd}(k), \ \forall\ i, j, k. \tag{9.3}$$

  Moreover, the traffic component $\lambda_{ij}^{sd}(k)$ of λ^{sd} can exist if and only if $b_{ij}(k) = 1$; hence

$$\lambda_{ij}^{sd}(k) \le b_{ij}(k)\lambda^{sd}, \ \forall\ i, j, s, d, k. \tag{9.4}$$

- *Maximum traffic flow through a VP:* The traffic flow through the kth VP, i.e., $\lambda_{ij}(k)$, would be always limited within λ_{max}, i.e.,

$$\lambda_{ij}(k) \le \lambda_{max}, \ \forall\ i, j, k. \tag{9.5}$$

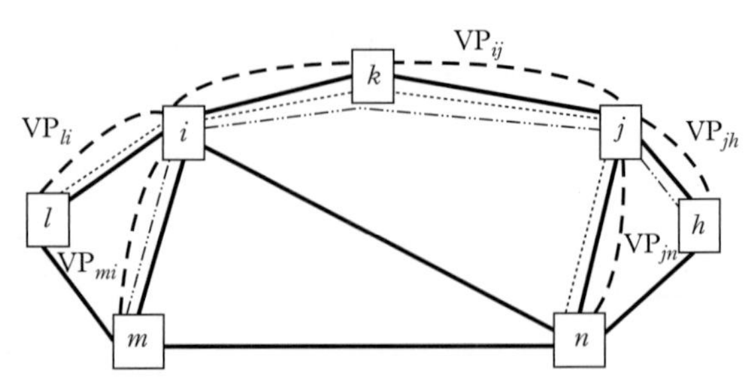

Figure 9.8 *Example of two multihop traffic flows passing through (i.e., groomed into) one common VP. In particular, λ_{ln} is carried by three VPs: VP_{li}, VP_{ij}, VP_{jn}, and λ_{mh} is carried by three lightpaths: VP_{mi}, VP_{ij}, VP_{jh}. Thus, VP_{ij} carries out TG for both the traffic flows, λ_{ln} and λ_{mh} together, along with its own traffic from node i to node j (i.e., λ_{ij}), for which it has originally been set up.*

- *Average propagation delay:* The propagation delay along the route of a VP is constrained by an upper bound on the average propagation delay with due weightage of the traffic carried. This constraint is expressed as

$$\sum_{ijk} \lambda_{ij}^{sd}(k) d_{ij} \leq \lambda^{sd} \alpha d_{max}, \ \forall \ s, d. \tag{9.6}$$

- *Values of $b_{ij}(k)$'s:* A given VP, say kth VP, from node i to node j, may or may not exist, leading to a binary constraint on $b_{ij}(k)$ (as defined earlier), given by

$$b_{ij}(k) = 0, 1 \quad , \ \forall \ i, j, k. \tag{9.7}$$

- *Numbers of transmitters and receivers:* The total number of connections emerging from a node is limited by the number of transmitters in the node, while the total number of connections terminating at a node is limited by the number of receivers in the node. One can express these two constraints as

$$\sum_{j,k} b_{ij}(k) \leq \Delta \quad , \ \forall \ j \text{ (for transmitters)}, \tag{9.8}$$

$$\sum_{i,k} b_{ij}(k) \leq \Delta \quad , \ \forall \ j \text{ (for receivers)}. \tag{9.9}$$

- *Non-negativity constraint:* All traffic flows must be non-negative, leading to the constraints, given by

$$\lambda_{ij}^{sd}(k), \lambda_{ij}(k), \lambda_{max} \geq 0, \ \forall \ i, j, s, d, k. \tag{9.10}$$

Note that the use of binary variables $b_{ij}(k)$'s along with the real-valued traffic flows implies that the present formulation represents an MILP problem. However, it may not be always possible to guarantee binary values for all $b_{ij}(k)$'s. One possible way to address this issue is to relax the MILP formulation (called as LP relaxation) by letting $b_{ij}(k)$ to assume any real value in the range [0, 1]. Having realized a set of real values for $b_{ij}(k)$, one can approximate the real $b_{ij}(k)$'s to the nearest integers (i.e., 0 or 1) through *rounding-up/down* operation, while limiting the total number of rounded-up $b_{ij}(k)$'s for each node to Δ with preference to those carrying higher traffic (Ramaswami *et al.* 2010). Transforming the variables $b_{ij}(k)$'s accordingly, one can execute the same optimization program (MILP) again with the binary values for $b_{ij}(k)$'s supplied as constants, thereby converting the MILP problem into a simpler LP problem.

9.4.2 ILP/heuristic wavelength assignment

The problem of WA for the VPs can be carried out using the ILP-based graph-coloring approach (Oki 2013; Zang *et al.* 2000) or by using heuristic schemes. Several heuristic WA schemes have been reported in the literature (Simmons 2014; Zang *et al.* 2000; Mokhtar and Azizoglu 1998) with varying degrees of complexity and spectrum utilization. In the following, first we describe an ILP-based graph-coloring approach and then discuss some of the heuristic schemes.

ILP-WA using graph-colouring In the graph-coloring problem (GCP), first an *auxiliary graph* $G_a(U, L)$ is formed with U nodes and L edges. Each VP obtained from the first subproblem with a specific route is considered as a node in $G_a(U, L)$. Thus, U nodes in $G_a(U, L)$ represent as many VPs obtained from the first subproblem. If any two VPs in the original physical topology or graph of the WRON share even a single optical fiber link, then the two corresponding nodes in the auxiliary graph are *connected* using an *edge*. In other words, for each node pair in $G_a(U, L)$, say the nodes u and u', an edge $l_{u,u'}$ is set up, if the two corresponding VPs share any common fiber link along their routes in the physical topology; else they are left *disconnected*. In order to carry out the task of WA, GCP assigns *wavelengths* or *colors* to all the nodes in $G_a(U, L)$, such that no two nodes, if connected by an edge, have the same color. Having generated $G_a(U, L)$ coloring of the nodes (i.e., WA) in $G_a(U, L)$ is carried out using an ILP as follows.

Assume that the VPs of the first subproblem (i.e., the nodes of $G_a(U, L)$) are to be assigned wavelengths from a pool of M wavelengths $w_1, w_2, \ldots, w_m, \ldots, w_M$, following the constraints of the ILP-based GCP, with an objective of using the minimum number of wavelengths. Before stating the objective function and the constraints for the ILP, we define the following variables.

ILP variables:

$p_u^{w_m}$: a binary variable, which equals one, if wavelength w_m is assigned to a node u in $G_a(U, L)$, and zero otherwise.

q^{w_m}: a binary variable, which equals one, if wavelength w_m has been used *at least once* in $G_a(U, L)$, and zero otherwise.

ILP objective function: Minimize the total number of wavelengths used (Z, say), i.e.,

$$\text{Minimize } Z = \sum_{m=1}^{M} q^{w_m}. \tag{9.11}$$

LP constraints:

- *One wavelength for each node*: Each node in $G_a(U, L)$ must be allocated only one wavelength, i.e.,

$$\sum_{m=1}^{M} p_u^{w_m} = 1, \ \forall m \in M. \tag{9.12}$$

- *Wavelengths for adjacent nodes*: Two adjacent (i.e., connected by an edge) nodes, u and u' in $G_a(U, L)$, must not have the same wavelengths, i.e.,

$$p_u^{w_m} + p_{u'}^{w_m} \leq q^{w_m}, \ \forall l_{u,u'} \in L, \forall m \in M. \tag{9.13}$$

Recall that $l_{u,u'}$ exists only when u and u' are connected by an edge in $G_a(U, L)$.

- *Ordered assignment of wavelengths*: Wavelengths are assigned with the ascending order of m, i.e.,

$$q^{w_m} \geq q^{w_{m+1}}, \ \text{for } m \in (1, 2, \ldots, M-1). \tag{9.14}$$

- *Binary constraints*: As defined earlier, the variables $p_u^{w_m}$ and q^{w_m} are binary in nature, making the GCP an ILP, i.e.,

$$p_u^{w_m} \in (0, 1), \ \forall m \in M, \ \forall u \in U \tag{9.15}$$

$$q^{w_m} \in (0, 1), \ \forall m \in M. \tag{9.16}$$

After running the ILP, one gets the solutions to the WA problem, i.e., the values of $p_u^{w_m}$'s $\forall u, m$ and the total wavelength count Z from the minimized objective function.

Furthermore, note that the traffic in some VPs might exceed unity, implying that they should be provisioned multiple lightpaths. For example, when the traffic λ_{ij} in a VP from node i to node j exceeds unity, then it should be allotted $n_{ij}^{LP} = \lceil \lambda_{ij} \rceil$ lightpaths, which may or may not be link-disjoint. First, these LPs must be allocated separate nodes in $G_a(U, L)$, and secondly if they have partial or full overlap, they need to be connected with additional edge(s). We illustrate the possible scenarios using an example four-node physical topology in Fig. 9.9.

As shown in Fig. 9.9(a), the given four-node physical topology (solid lines) has a VT shown by dashed lines. Fig. 9.9(b) shows the auxiliary graph $G_a^1(U, L)$ for the case when all the VPs carry traffic which are ≤ 1, wherein u_5 (representing VP_{24}) is connected to u_4 (representing VP_{23}) as well as u_6 (representing VP_{43}) due to the overlapping edges. In Fig. 9.9(c), the auxiliary graph $G_a^2(U, L)$ becomes different from $G_a^1(U, L)$, as two of its VPs have traffic exceeding unity. As evident from the illustration in Fig. 9.9(c), VP_{12} with $\lambda_{12} = 2.5$ will need $\lceil 2.5 \rceil = 3$ lightpaths leading to the three nodes u_{1a}, u_{1b}, u_{1c} in the auxiliary graph $G_a^2(U, L)$ (instead of u_1 in $G_a^1(U, L)$), with three interconnecting edges between themselves (as they are assumed to run on the same route). Similarly, VP_{24} with $\lambda_{24} = 1.7$ will need $\lceil 1.7 \rceil = 2$ nodes, u_{5a} and u_{5b} with one interconnecting edge between themselves, along with other additional edges to connect with the relevant nodes in $G_a^2(U, L)$. In particular, u_{1a}, u_{1b}, u_{1c} form a triangle owing to their complete overlap on the same path, but the triangle remains isolated from the rest, as these VPs don't overlap with any other VP. However, u_{5a}, u_{5b}, while being interconnected to each other (as they overlap on the same path), also must have connecting edges with u_4 and u_6 due to the respective overlapping edges.

Subsequently, in both cases, having formed the appropriate auxiliary graphs, one can use the same in the above ILP-GCP formulation (Equations 9.11 through 9.16) to carry

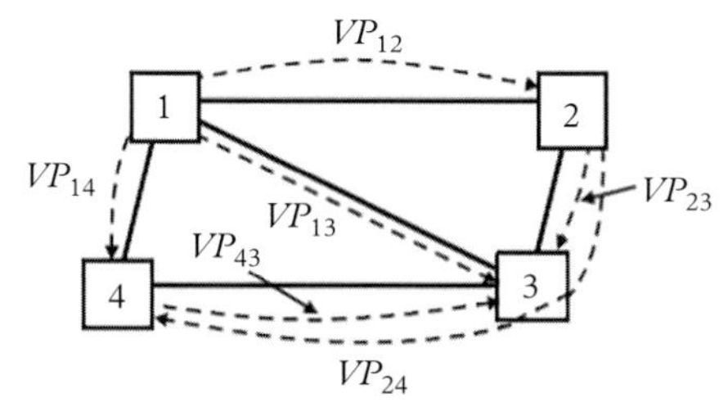

(a) Four-node WRON graph with VPs in dashed lines

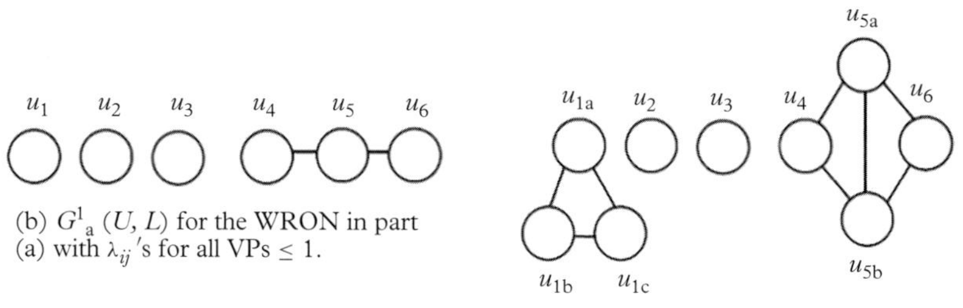

(b) $G^1_a\ (U, L)$ for the WRON in part (a) with λ_{ij}'s for all VPs ≤ 1.

(c) $G^2_a\ (U, L)$ for the WRON in part (a), with λ_{ij}'s for two VPs > 1., i.e., λ_{12} in $\text{VP}_{12} = 2.5$, λ_{24} in $\text{VP}_{24} = 1.7$.

Figure 9.9 *Illustration of graph coloring in a four-node WRON. The circles marked as u_k's in parts (b) and (c) are the nodes in the auxiliary graphs $G^1_a(U,L)$ and $G^2_a(U,L)$, respectively, corresponding to the traffic λ_{ij}'s in the VPs in two different cases. In part (b) $\lambda_{ij} < 1$ in all VPs, and hence each VP is mapped to a unique node in the auxiliary graph. Thus, in part (b) $VP_{12}, VP_{13}, VP_{14}, VP_{23}, VP_{24}, VP_{43}$ are mapped to $u_1, u_2, u_3, u_4, u_5, u_6$, respectively. However, in part (c) $\lambda_{12} = 2.5$ and $\lambda_{24} = 1.7$, while all other λ_{ij}'s are < 1. Thus, in part (c), VP_{12} is mapped to three nodes in $G^2_a(U,L)$, i.e., u_{1a}, u_{1b}, u_{1c}, and similarly VP_{24} is mapped to u_{5a}, u_{5b}, while other nodes in $G^2_a(U,L)$ remain the same as in $G^1_a(U,L)$, albeit with different set of edges between the respective nodes.*

out the task of WA. Thereafter, the traffic in the VPs with traffic >1 can be divided into the respective multiple lightpaths assigned to them, though this would necessitate provisioning additional transceivers in the respective nodes. However, this approach might lead to a suboptimal grooming of traffic, and for further refinement of the design one can get back to the original MILP (converted to LP) with modified $b_{ij}(k)$'s as constants and unequal numbers of transceivers in the nodes to get an improved traffic distribution, leading to lessened congestion in the network.

Heuristic-WA Several heuristic schemes have been studied for WA in WRONs. In the following, we describe some of these schemes, which can be used to assign wavelengths to the VPs obtained from the first subproblem.

- *Most-utilized wavelength first* (MUWF): In the MUWF algorithm, an attempt is made to route the connection on the most-utilized wavelength first, which amounts to the maximum possible *packing* of traffic into the available wavelengths and hence also referred to as PACK algorithm. In particular, all the wavelengths are placed in an ordered list with the most-used wavelength at the top and the least at the bottom. Using this list, the assignment of wavelength for a given route is made from the top.
- *Least-used wavelength first* (LUWF): The LUWF algorithm attempts to distribute the network traffic on the least-used wavelength first, ensuring thereby nearly uniform load for all wavelengths. Hence, all the wavelengths are placed in an ordered list with the least-used wavelength at the top and the most-used wavelength at the bottom (i.e., in the reverse order as compared to MUWF). Using this list, the assignment of wavelength for a given route is made from the top. Thus, the algorithm ensures that the traffic gets spread over all the wavelengths, and hence it is also referred to as the SPREAD algorithm.

- *Exhaustive search for wavelengths* (ESW): In the ESW algorithm, all the available wavelengths are considered and the shortest path for each wavelength is found out. From this set of wavelengths, the one that gives the overall shortest route from among all the wavelength-specific shortest routes found in the first step is assigned for the requested connection.
- *Random fit (assignment) of wavelengths* (RFW): RFW algorithm simply assigns to a given connection a wavelength randomly chosen from the available wavelengths.
- *Fixed fit (assignment) of wavelengths* (FFW): In this algorithm, all the wavelengths are arranged in a list with a fixed order and the first available wavelength from the top is assigned for each requested connection. This algorithm is thus also referred to as the *first-fit of wavelengths* algorithm, bearing incidentally the same acronym, i.e., FFW.

As evident from the above discussion, FFW and RFW appear to be the fastest algorithms, and the FFW certainly gives better bandwidth utilization. Other algorithms are time-consuming, with MUFW and ESW being the slowest, but ensuring the best resource utilization. In the following, we describe the steps of the FFW algorithm combined with the results on the values of b_{ij} from MILP, to carry out the WA process.

1. Enlist the VPs with descending order of their propagation delays d_{ij}'s. Note that each VP once set up by the MILP, gets assigned a route from the delay constraint.
2. Enlist sufficient number of wavelengths in some order (we don't consider any upper limit for the number of wavelengths, as the FFT algorithm itself eventually uses a minimum number of wavelengths).
3. Assign the first wavelength from the wavelength list to the longest VP. The FFW algorithm along with the consideration to the length of the route helps in reusing wavelengths and thus conserves the wavelength resource efficiently.
4. Assign wavelengths to all the VPs down the list using FFW algorithm to complete the WA process.

As indicated earlier in the ILP-GCP method, for the above heuristic scheme one also needs to assign the VPs having $\lambda_{ij}(k)$'s > 1 more wavelengths, the numbers being equal to $\lceil \lambda_{ij} \rceil$, and thereafter split the traffic flows accordingly over the respective lightpaths and modify the transceiver requirements in the respective nodes.

Case study on MILP-VTD

In the following, we present a case study on an offline VTD for a given WRON using MILP. Using the MILP-VTD subproblem with CPLEX software, VTs are computed for a six-node physical topology representing a WRON (see Fig. 9.10), to be provisioned for

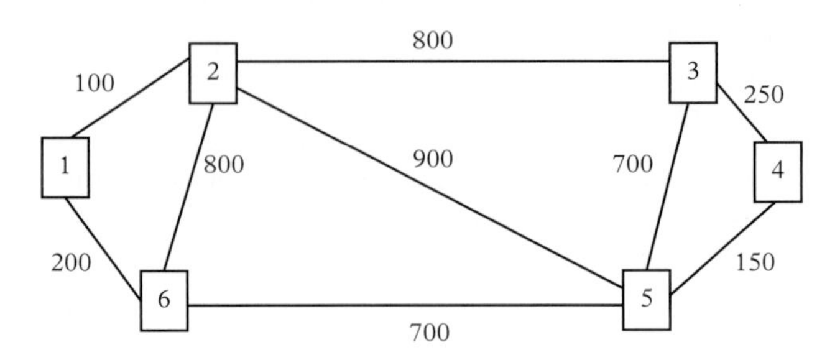

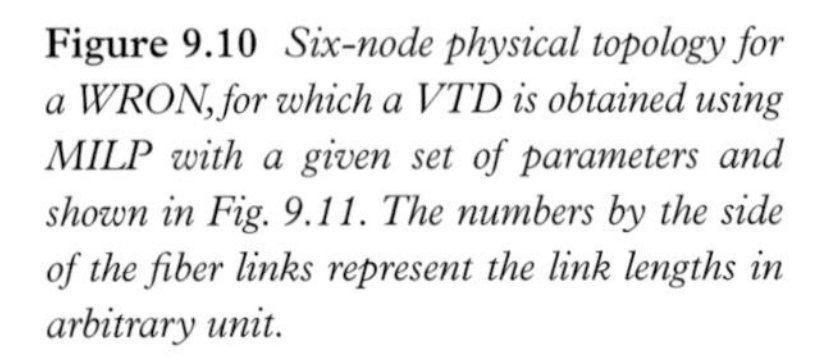
Figure 9.10 *Six-node physical topology for a WRON, for which a VTD is obtained using MILP with a given set of parameters and shown in Fig. 9.11. The numbers by the side of the fiber links represent the link lengths in arbitrary unit.*

Table 9.1 *Traffic matrix for the six-node WRON of Fig.9.6.*

Nodes	1	2	3	4	5	6
1	0.000	0.467	0.334	0.500	0.169	0.724
2	0.478	0.000	0.962	0.464	0.705	0.145
3	0.281	0.827	0.000	0.491	0.995	0.942
4	0.827	0.436	0.391	0.000	0.902	0.153
5	0.292	0.382	0.421	0.716	0.000	0.895
6	0.447	0.726	0.771	0.538	0.869	0.000

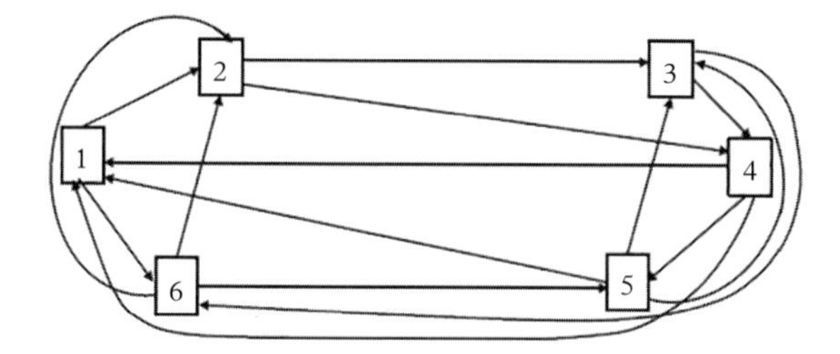

Figure 9.11 *VTD for the six-node physical topology of Fig. 9.10, with $\Delta = 2$, $\alpha = 2$, $k = 0, 1$, resulting in a congestion $\lambda^{max} = 1.189$. Note that, the design creates 12 VPs, which can be used in the ILP-WA for assigning wavelengths to them.*

Table 9.2 *MILP results for congestion λ^{max} in the six-node WRON of Fig. 9.10, with $k \in (0, \Delta - 1)$.*

Δ	1	2	3	4	5
MILP ($\alpha = 1.5$)	X	1.231	0.434	0.230	0.146
MILP ($\alpha = 2$)	8.844	1.189	0.434	0.228	0.146
MILP ($\alpha \to \infty$)	8.567	1.189	0.434	0.228	0.146

Table 9.3 *MILP results for congestion λ^{max} in the six-node WRON of Fig. 9.10, with $k = 0$.*

Δ	1	2	3	4	5
MILP ($\alpha = 2$)	8.844	2.379	1.302	0.910	0.728
MILP ($\alpha \to \infty$)	8.567	2.379	1.302	0.910	0.728

the traffic matrix given in Table 9.1. Figure 9.11 shows one VT designed with $\Delta = 2$ and $\alpha = 2$, with a congestion $\lambda_{max} = 1.189$. Tables 9.2 and 9.3 present the values of congestion for a few combinations of Δ and α, with two different cases of k-values. As expected, with higher values for Δ and α, the congestion decreases significantly. Further, as k is reduced, the congestion increases significantly, as shown in Table 9.3.

Using the VTD shown in Fig. 9.11 for the WRON shown in Fig. 9.10, the task of WA can be carried out using the ILP-GCP formulation (Equations 9.11 through 9.16) presented earlier, and the readers are encouraged to carry out WA for this problem and examine the results in terms of the number of wavelengths used and the number of times each wavelength has been reused in the design. The same exercise can be repeated by using the heuristic WA, and the results can be compared with those obtained from the ILP-GCP method.

9.5 Wavelength conversion in long-haul WRONs

The techniques for wavelength conversion in WRONs were discussed in Chapter 2. In this section, we discuss how the wavelength converters (WCRs) can be used in the nodes of a WRON. Typically, WCRs are used selectively in OXCs, particularly when wavelength-continuous connection requests become infeasible due to increased traffic, less connectivity and limited resources (mainly the number of available wavelengths and transceivers).

In Fig. 9.12, we present a WRON node configuration, where each output port of the optical switch is passed through a WCR, offering the node an absolute flexibility with full wavelength conversion. In Fig. 9.13, we present the share-per-link configuration with WCRs being available to each outgoing link in a node, i.e., with one WCR for each optical multiplexer (OMUX). The third configuration, the cheapest one of all, is shown in Fig. 9.14, where the node employs only one WCR, which is shared by the entire node, leading to the share-per-node configuration.

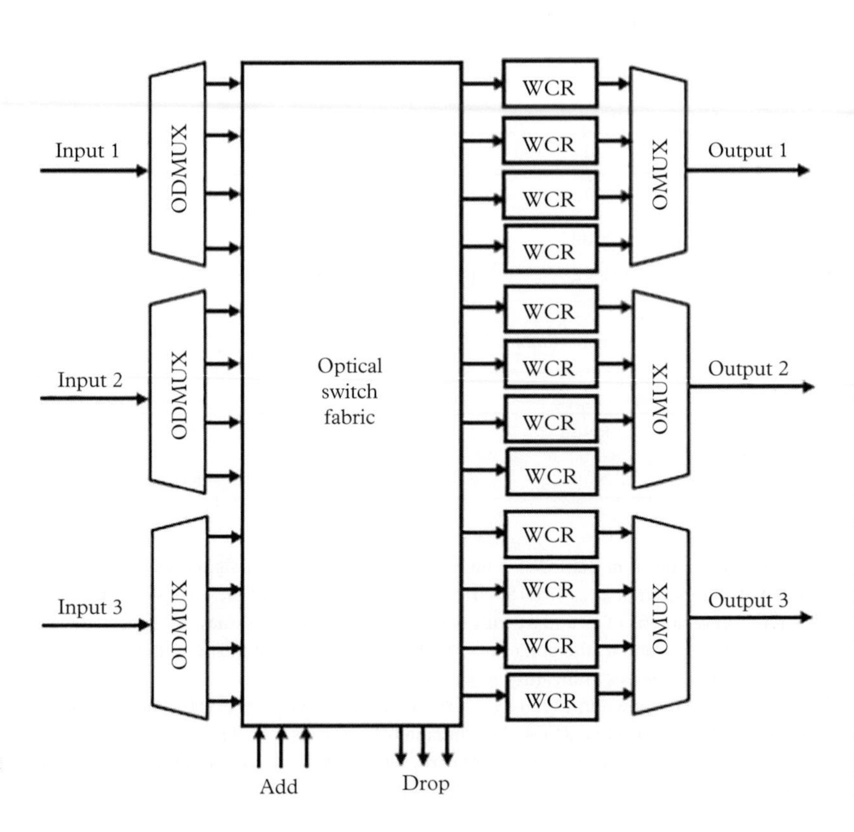

Figure 9.12 *WRON node configuration with full wavelength conversion.*

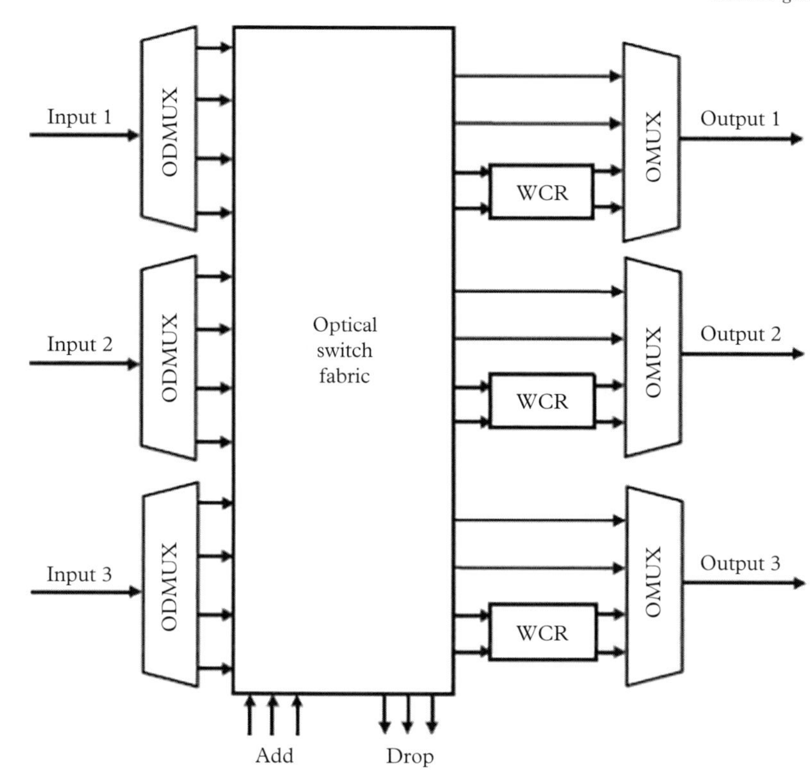

Figure 9.13 *WRON node configuration with per-link sharing of WCRs.*

9.6 Wavelength utilization with wavelength conversion

In long-haul WRONs, wavelength is viewed as one of the significant resources, which needs to be increased in number to accommodate more traffic. However, a close look into the network would reveal that the utilization of each wavelength can be reduced in a large network with fewer wavelengths. In particular, while seeking to set up a long-haul lightpath traversing through a number of optical fiber links, finding a wavelength-continuous lightpath would become much harder with fewer wavelengths. In such cases, judicious use of WCRs in WRON nodes is expected to reduce the probability of blocking for all-optical connections $P_{b(ao)}$ and thereby improve the wavelength utilization significantly. However, note that the notion of $P_{b(ao)}$ in WRONs doesn't relate to the traditional blocking probability model (Erlang B) in a telephone network, as the latter is based on the stochastic processes of arrival and departure of telephone calls, in contrast with the problem of resource utilization in the present context without any notion of stochastic process.

In the next three subsections, following the models used in (Barry and Humblet 1996), we examine the impact of wavelength conversion on wavelength utilization for a specified upper limit for $P_{b(ao)}$ in WRONs, so that a network operator can decide on the number of wavelengths and the associated requirements (e.g., the number of transceivers,

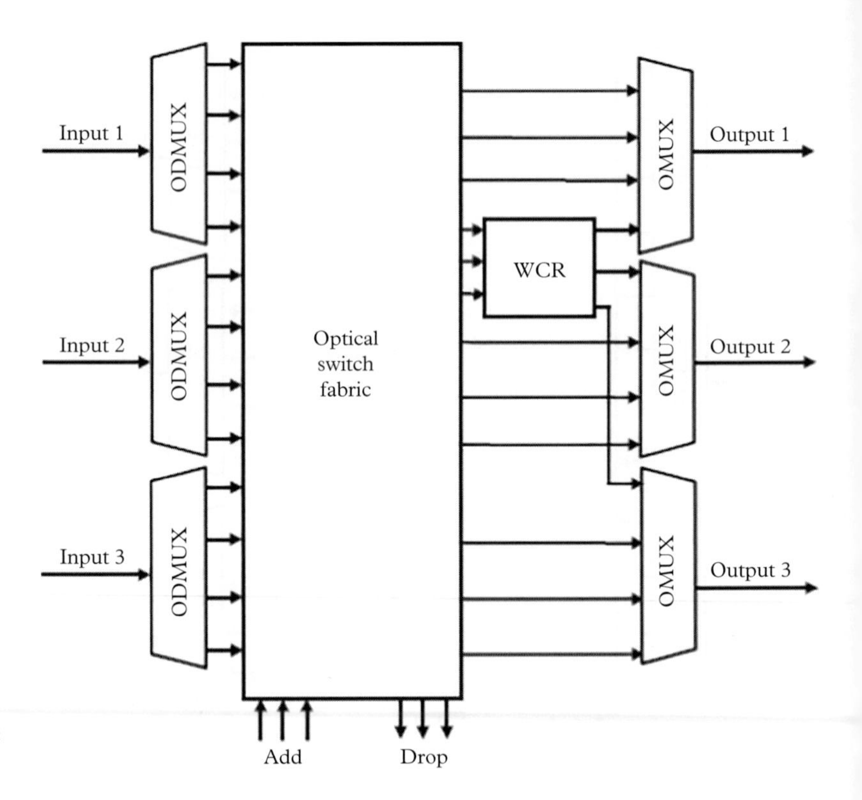

Figure 9.14 *WRON node configuration with per-node sharing of WCRs.*

OA bandwidth, use of WCR wherever necessary, etc.) while planning for the network installation (or subsequent augmentation).

9.6.1 WRONs with full wavelength conversion

For the evaluation of $P_{b(ao)}$ in this case, consider an example linear segment in a six-node WRON mesh in Fig. 9.15(a) showing an all-optical connection (AOC) from node 4 to node 1, which can be a single-wavelength lightpath when at least one wavelength is free in all the fiber links along the three-hop path, and will need wavelength conversion at some intermediate node(s) if different wavelengths are free in different fiber links. Fig. 9.15(b) presents a generalized scenario of part (a), where a candidate AOC is to be set up between a pair of nodes that are H hops away from each other, and the concerned nodes can use full wavelength conversion to improve wavelength utilization in the WRON.

With reference to the example of AOC in a WRON shown in Fig. 9.15, we denote the wavelength utilization factor by γ_{fc} for full wavelength conversion (i.e., when all nodes employ full wavelength conversion) with M as the number of operating wavelengths on each fiber link across the network. For a given value of γ_{fc}, the probability that all the wavelengths are used up in a hop can be obtained as

$$P_1 = \gamma_{fc}^M, \tag{9.17}$$

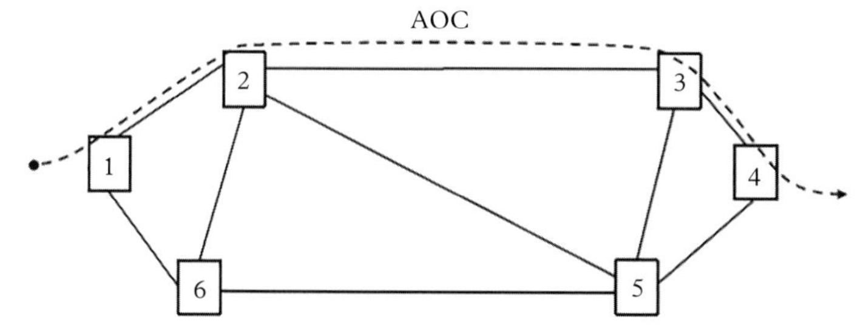

(a) An all-optical connection (AOC) with three hops from node 1 to node 4

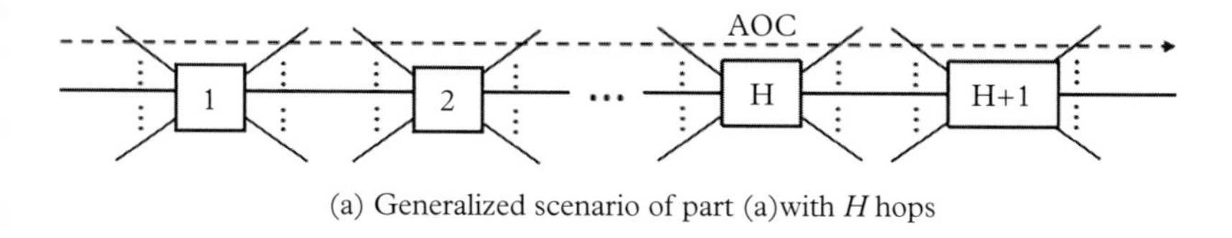

(a) Generalized scenario of part (a)with H hops

Figure 9.15 *Illustration of all-optical connection (AOC) in a WRON. Wavelength utilization can be improved across the network along the connection path by using wavelength conversion at some/all nodes.*

where we assume that the wavelength occupancies in the fiber links are independent of each other. This model leads to the probability that at least one wavelength is free in each one of the H hops along the chosen all-optical path, given by

$$P_2 = (1 - P_1)^H = (1 - \gamma_{fc}^M)^H. \tag{9.18}$$

Note that, the probability that not a single wavelength is free on all the hops along the path, i.e., $(1 - P_2)$, will represent the blocking probability $P_{b(ao)}$, as defined earlier. We therefore express $P_{b(ao)}$ as

$$P_{b(ao)} = 1 - P_2 = 1 - (1 - \gamma_{fc}^M)^H. \tag{9.19}$$

Using the above relation, we finally express γ_{fc} with full wavelength conversion in terms of H, M, and a specified $P_{b(ao)}$, as

$$\gamma_{fc} = \left[1 - (1 - P_{b(ao)})^{1/H}\right]^{1/M} \approx \left(\frac{P_{b(ao)}}{H}\right)^{1/M}. \tag{9.20}$$

The above expression indicates, for a given value of $P_{b(ao)}$, how one should choose the network parameters/resources (i.e., H and M), while setting up the WRON with a given value of γ_{fc}. Figure 9.16 shows the variation of γ_{fc} with M for different values of H and with $P_{b(ao)} = 0.001$, indicating how with larger number of wavelengths, the wavelength utilization factor increases, while the effect of H is not felt much due to full wavelength conversion.

9.6.2 WRONs without wavelength conversion

Without any wavelength conversion throughout the network, we denote the bandwidth-utilization ratio in the network as γ_{nc}. With a similar framework as in the previous case, one can express the probability that a given wavelength will be free on all the H hops as

$$P_3 = (1 - \gamma_{nc})^H. \tag{9.21}$$

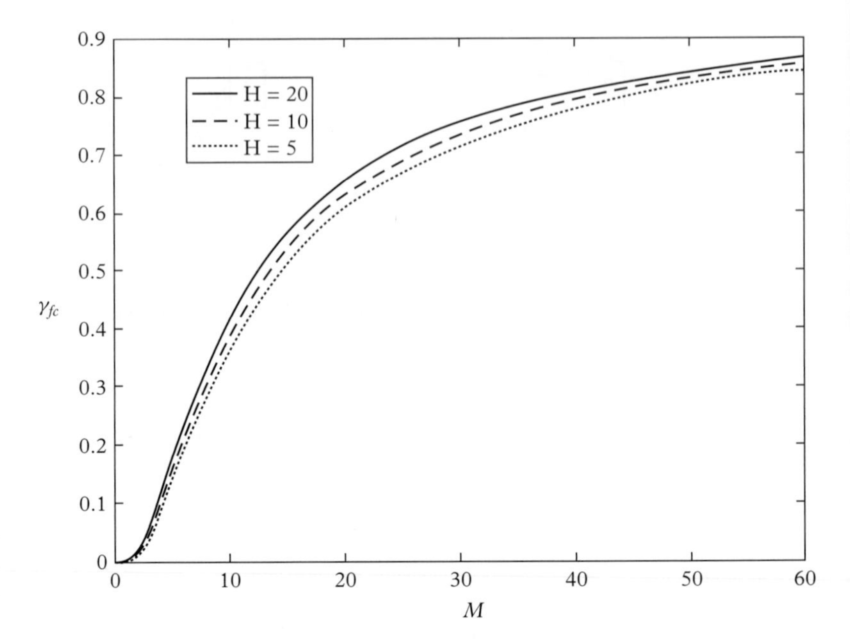

Figure 9.16 *Plots of wavelength utilization factor versus number of wavelengths in WRONs with full wavelength conversion. Given:* $P_{b(ao)} = 0.001$.

Hence, the probability that the given wavelength will be busy in at least one hop (out of H hops) can be expressed as

$$P_4 = 1 - P_3 = 1 - (1 - \gamma_{nc})^H. \tag{9.22}$$

Note that, the probability that each one of the M wavelengths will be busy on at least one of the H hops along the path, i.e., P_4^M will represent the probability $P_{b(ao)}$. We therefore express $P_{b(ao)}$ without wavelength conversion as

$$P_{b(ao)} = P_4^M = \left[1 - (1 - \gamma_{nc})^H\right]^M. \tag{9.23}$$

Using the above relation, we finally express the bandwidth utilization factor without wavelength conversion in terms of H, M, and a specified $P_{b(ao)}$, as

$$\gamma_{nc} = 1 - \left(1 - P_{b(ao)}^{1/M}\right)^{1/H}. \tag{9.24}$$

With $(1-x)^\alpha \approx 1 + \alpha \ln(1-x)$ for small x and α (i.e., $\ll 1$), one can simplify the above expression as

$$\gamma_{nc} \approx -\frac{1}{H} \ln\left(1 - P_{b(ao)}^{1/M}\right). \tag{9.25}$$

Note that, γ_{nc} bears the same significance as γ_{fc}, but with a smaller value in the absence of wavelength conversion. The variation of γ_{nc} versus M for different values of H is presented in Fig. 9.17 with $P_{b(ao)} = 0.001$. A comparison of the plots with and without wavelength

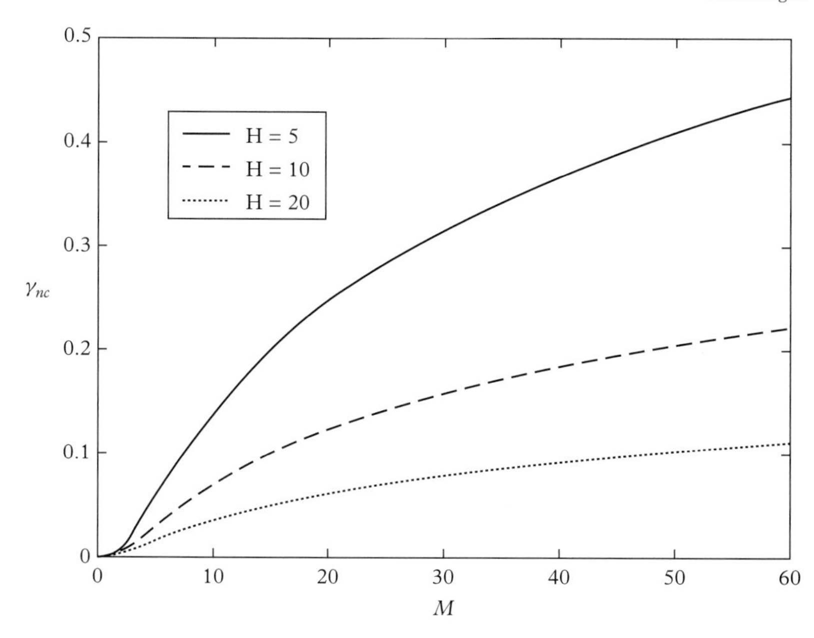

Figure 9.17 *Plots of wavelength utilization factor versus number of wavelengths in WRONs without any wavelength conversion. Given:* $P_{b(ao)} = 0.001$.

conversion shows that the wavelength utilization becomes a strong function of H without wavelength conversion.

9.6.3 Comparison of WRONs with and without wavelength conversion

Using full wavelength conversion, one can enhance the resource utilization in a WRON. We therefore express the ratio of the utilization factors with and without wavelength conversion as the achievable gain G_{WC} through wavelength conversion, given by

$$\begin{aligned} G_{WC} &= \frac{\gamma_{fc}}{\gamma_{nc}} \qquad (9.26) \\ &= \frac{\left[1-(1-P_{b(ao)})^{1/H}\right]^{1/M}}{1-\left(1-P_{b(ao)}^{1/M}\right)^{1/H}} \\ &\approx \frac{H^{(1-1/M)}P_{b(ao)}^{1/M}}{-\ln\left(1-P_{b(ao)}^{1/M}\right)}. \end{aligned}$$

In Fig. 9.18, we present the variation of the gain G_{WC} with the number of wavelengths M for different values of H with a given value of $P_{b(ao)}$ (= 0.001). It is evident that, for larger values of H, the gain increases sharply with M at lower values of M, thereby indicating a strong effectiveness of wavelength conversion for large networks with fewer wavelengths. However, for a given H, even though G increases initially with M, it subsequently reaches a maximum achievable value, and thereafter with higher M it decreases and settles down to some value > 1, determined by H. The latter effect is attributed to the fact that, with large

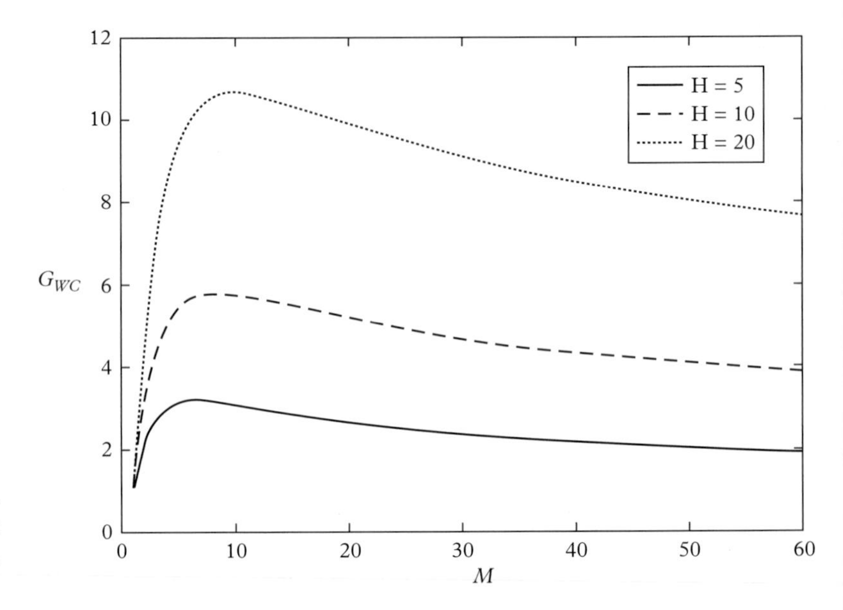

Figure 9.18 *Plots of wavelength utilization gain versus number of wavelengths in WRONs by using full wavelength conversion. Given:* $P_{b(ao)} = 0.001$.

number of wavelengths, it gets easier to find a free path, thereby playing down the impact of wavelength conversion on $P_{b(ao)}$.

9.7 Online RWA for operational WRONs

In an operational WRON, the task of RWA has to be carried out online and hence needs a different set of considerations as compared to the offline designs. Having carried out the offline dimensioning in a long-haul WRON, one gets an estimate of the resources needed on a long-term basis. However, to run the network efficiently, the provisioned resources need to be allocated, released, and reallocated for the incoming connection requests using a fast online RWA to make the best use of the resources. Note that the online RWA problem is relevant for the metro segment as well, though with fewer constraints, as the network size is much smaller in metro networks and connections are to be set up over regular topologies, such as rings or a few interconnected rings.

A variety of online RWA algorithms for WRONs have been reported in the literature, and the interested readers are encouraged to go through (Simmons 2014; Zang *et al.* 2000) for an overview of the various options. We briefly describe here some of the candidate schemes that are relevant for long-haul WRONs. For the online RWA in WRONs, as the requests for connections keep arriving at each node, the RWA protocol needs to decide the path and the wavelength for each connection in real time. The task of online RWA is in general carried out in two steps: routing in the first step using a single-hop all-optical path (lightpath), followed by a WA scheme for the same path in the next step. If a single-hop connection cannot be supported due to inadequate SNR (i.e., higher BER) at the receiving node, the connection can be broken into concatenated lightpaths with the OEO-based regeneration at appropriate intermediate node(s). Further, if the available route is not wavelength-continuous, then one can go for wavelength conversion, provided any WCR is available at the right intermediate node, or else can go for OEO-based regeneration at the same node. In the following, we discuss some online all-optical RWA schemes, where the options for multihop connections or wavelength conversion (if needed) can be accommodated as well with appropriate modifications.

Online routing

There are a number of candidate online routing schemes to carry out the RWA in WRONs. We describe below the following three online RWA schemes:

(1) fixed-path routing,
(2) alternative-path routing,
(3) dynamic-path routing,

with the first two schemes belonging to the *constrained* routing category, while the third one operates with an unconstrained scheme.

(1) *Fixed-path routing*

In the fixed-path routing scheme (which is the simplest of the above three schemes), for each node pair a fixed path or route is kept pre-computed (as a constraint) in the source node, based on the shortest-path criterion. Whenever a request arrives at a node, the predefined route is allocated immediately. Thus the allocation of route takes place instantaneously, though the chosen routes may not be the best ones because the traffic keeps changing with time, with the possibility of passing the route through an already congested link or links. The lack of adaptability of this scheme with the time-varying traffic scenario may cause connection blocking, even when some other routes remain available for the connection in demand.

(2) *Alternative-path routing*

In the alternative-path routing scheme, a number of alternative paths, say P paths, are kept as pre-computed multiple options (and hence with less stringent constraint as compared to fixed-path routing) for each node for all possible destination nodes. When a connection request arrives at a node, one of these P paths is assigned for the requested connection, where the choice of the path is made using some criterion based on the current network status. Typically, one can choose the path which has the least congestion at that instant, or at least they do not share a congested route of the network. However, this search will take a while as compared to the fixed-path routing scheme. Note that this search is not global, but from within the pre-computed set of P paths only, which indeed saves some time but may not provide the best route that could be obtained from a global search following the arrival of the request (which is considered next in the dynamic-path routing). However, one can incorporate some criteria while pre-computing the candidate paths (P) for a node pair. One of them ensures that all the P paths for a source-destination pair are link-disjoint. This also helps in setting up a backup path to make the primary path survivable (this issue is addressed in Chapter 11). Further, to improve the route-selection process, the network can periodically update its information base on the network status and pre-compute accordingly the paths, thereby enabling itself to choose the paths more efficiently for each node pair.

(3) *Dynamic-path routing*

In the dynamic-path routing scheme, each node carries out online path-computation, based on the current network-state information, and hence it is also known as *adaptive unconstrained routing* scheme. The source node needs to set up a lightpath over the *shortest* available path with a predefined cost parameter within a time duration not exceeding a specified limit. In order to find out the appropriate path on real-time basis, i.e., within the given time constraint, the node needs to *prune* the network, so that only the available resources are taken into consideration during search operation by removing the unavailable resources temporarily from the record of the available resources. Note that this search process being carried out across the entire network, takes a longer time than that used in the alternate-path routing scheme, and hence the network pruning becomes a necessity to make the search feasible within the given time limit. Having pruned the network temporarily, a number of shortest paths (K shortest paths) are computed and the least-loaded path is chosen from the set of K shortest paths.

To make a choice from the above routing options, one should note that the fixed-routing option offers the simplest and fastest scheme, though being unaware of the current network status it may lead to increased congestion at times. The dynamic-routing scheme is the best in respect of adaptability and hence ensures the best network-bandwidth utilization, though it will take the longest search time, with the possibility of exceeding the time limit for the search operation at times. The alternate-path routing scheme comes between the two extremes (fixed and dynamic), in respect of both searching speed and network resource utilization, and thus usually makes a preferred choice.

Online wavelength assignment

Having found the route for a connection, the network employs one of the candidate WA schemes, such as MUWF (or PACK), LUWF (or SPREAD), RFW, ESW, or FFW, as discussed earlier for the offline design. For the online RWA scheme, indeed a simple search mechanism is preferred. Thus, in general, the FFW scheme provides a favourite choice of online WA scheme in a practical networking scenario, unless the wavelength utilization factor becomes the foremost issue.

9.8 Summary

In this chapter we dealt with the long-haul WRONs over mesh topology. First, we presented the major design issues and candidate node configurations for wavelength-routing. Thereafter, some offline design methodologies were examined for resource provisioning in WRONs using candidate algorithms. To carry out the task in a computationally efficient manner, the designs are generally split into component subproblems, which in turn use a mix-and-match of LP-based optimization techniques and heuristic algorithms. We presented one such design methodology, where the first subproblem employs MILP-based VTD leading to a set of VPs. In the second subproblem of this design, the VPs can be assigned wavelengths either by using an ILP-based graph-coloring approach or some heuristic algorithm, such as one of the MUWF, LUWF, RFW, ESW, or FFW schemes.

Next, the role of wavelength conversion in WRONs was discussed and the candidate node architectures were presented with three possible options of wavelength conversion: the nodes with full wavelength conversion, share-per-link wavelength conversion, and share-per-node wavelength conversion. An analytical model was presented to determine the wavelength utilization in WRONs with full wavelength conversion, as well as without any wavelength conversion. Using this framework, an expression for the achievable gain in wavelength utilization in a WRON using full wavelength conversion (with respect to similar WRON, but not employing any wavelength conversion) was determined with a specified upper limit for the blocking probability of all-optical connections. Finally, some online RWA schemes were described with various options for the online routing schemes: fixed-path routing, alternative-path routing, and dynamic-path routing, along with the heuristic WA schemes, already described for the offline designs.

EXERCISES

(9.1) As illustrated in the example shown in Fig.9.1, sketch some other candidate topological interconnections between multiple WDM rings to form mesh-like topology and specify the requirements of the interconnecting nodes in the proposed interconnected-ring networks.

(9.2) Consider an MD-ROADM used in Fig.9.1 at node n_{AB} and sketch its configuration using optical demultiplexers, multiplexers, and switches. Sketch alternate configurations for the same MD-ROADM by using (i) optical splitters and WSSs and (ii) WSSs only. Comment on the basic differences between these configurations.

(9.3) Draw the configuration for a MEMS-based OXC that can support three input and three output fibers with 16 wavelengths on each fiber, such that any four wavelengths from/to each fiber can be dropped/added through tunable transceivers (i.e., with logical nodal degree of four). Specify the sizes of the optical demultiplexers and multiplexers, optical switch (MEMS), and the number of tunable transceivers needed in the OXC.

(9.4) Consider an N-node mesh-configured WRON offering a set C of connections $\{c_{s,d}\}$ with s and d representing the source and destination nodes for the connection $c_{s,d}$. Assume that the connection $c_{s,d}$ employs $l(s,d)$ fiber links. If the WRON has E bidirectional fiber links, determine the expression for the lower-bound on number of wavelengths needed in the network in terms of the number of nodes, total internodal fiber links used for the connections in C and the average physical degree of a node in the network.

(9.5) Consider a 3×3 OXC for four wavelengths using MEMS in Spanke's switch architecture and sketch the OXC configuration. Estimate the gain of an EDFA as the loss-compensating amplifier placed after the OXC. Given: connector loss = 1 dB, loss in optical demultiplexer/multiplexer = 3 dB, loss in a MEMS element = 1 dB.

(9.6) Consider an N-node WRON mesh with a given traffic matrix. The network vendor can afford Δ_T transceivers for the entire network and decides to provision them for the nodes in accordance with the traffic handled by each node. Using this provisioning scheme, develop an expression for the number of transceivers Δ_i for each node, while ensuring that each node has at least one transceiver to remain connected with the network.

(9.7) Consider the seven-node WDM mesh network shown in Fig. 9.19 with two tunable transmitters and two tunable receivers per node and two wavelengths per fiber.

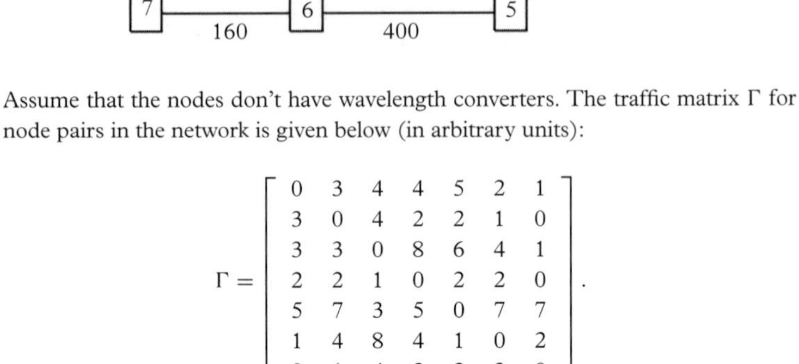

Figure 9.19 *Seven-node WRON topology for Exercise 9.7.*

Assume that the nodes don't have wavelength converters. The traffic matrix Γ for node pairs in the network is given below (in arbitrary units):

$$\Gamma = \begin{bmatrix} 0 & 3 & 4 & 4 & 5 & 2 & 1 \\ 3 & 0 & 4 & 2 & 2 & 1 & 0 \\ 3 & 3 & 0 & 8 & 6 & 4 & 1 \\ 2 & 2 & 1 & 0 & 2 & 2 & 0 \\ 5 & 7 & 3 & 5 & 0 & 7 & 7 \\ 1 & 4 & 8 & 4 & 1 & 0 & 2 \\ 0 & 1 & 4 & 2 & 2 & 3 & 0 \end{bmatrix}.$$

Determine a set of lightpaths to be constructed using an algorithm *maximizing the single-hop traffic* with the largest traffic considered first, where if two traffic elements have the same value, the first traffic element obtained in the row-wise sequence (i.e., beginning with the topmost row from left to right and then moving downward to the next row) is selected. Show the RWA for the lightpaths obtained by using the shortest-path and first-fit schemes for routing and WA, respectively.

(9.8) Evaluate the wavelength utilization ratio in a WRON mesh using 10 wavelengths with full and without wavelength conversion for five-hop connections for a blocking probability of 0.001. Using these results, estimate also the gain in wavelength utilization that can be achieved by using full wavelength conversion in a network which currently doesn't employ any wavelength conversion.

(9.9) In a given WRON, two client nodes (IP routers) are connected to two OXCs at two different WRON nodes, with four intermediate nodes between them. In the WRON, each fiber between any two nodes supports four wavelengths. Consider that 0.7 is the probability that a wavelength is being used on a fiber for an ongoing connection.

a) Find the probability that a new connection request is blocked when the nodes are not equipped with wavelength converters.

b) Find the probability that a new connection request is blocked when the nodes are equipped with wavelength converters, having full wavelength-conversion capability.

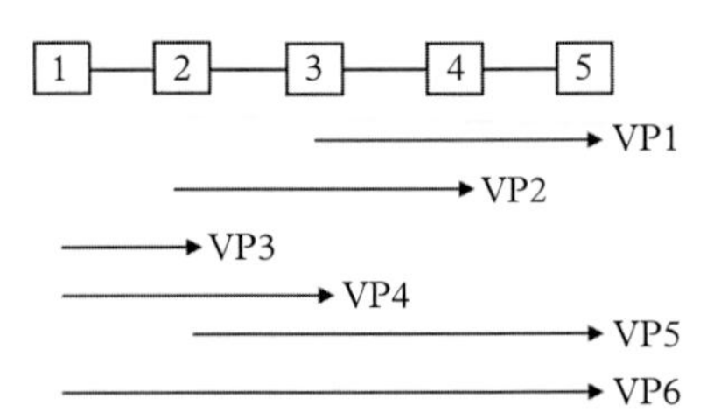

Figure 9.20 *Five-node linear WRON for Exercise 9.10.*

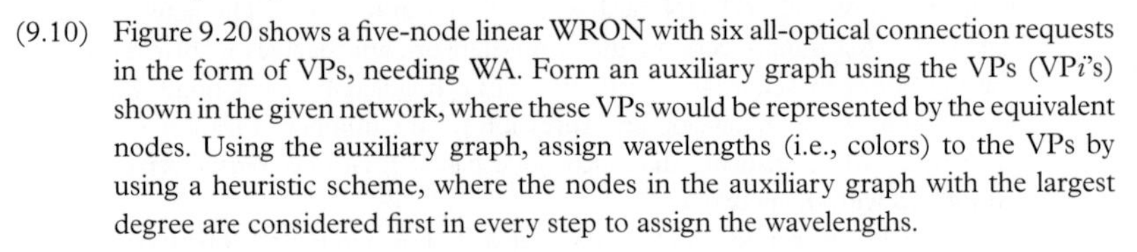

(9.10) Figure 9.20 shows a five-node linear WRON with six all-optical connection requests in the form of VPs, needing WA. Form an auxiliary graph using the VPs (VP*i*'s) shown in the given network, where these VPs would be represented by the equivalent nodes. Using the auxiliary graph, assign wavelengths (i.e., colors) to the VPs by using a heuristic scheme, where the nodes in the auxiliary graph with the largest degree are considered first in every step to assign the wavelengths.

Part IV

Selected Topics in Optical Networks

Transmission Impairments and Power Consumption in Optical Networks

10

The physical layer in optical networks suffers from various types of transmission impairments due to the non-ideal passive and active devices used therein. For example, the losses in passive optical devices and fiber links, noise generated in optical receivers and amplifiers, spectral spread of optical sources, dispersion and nonlinear phenomena in optical fibers, crosstalk in optical switches, etc. can degrade the quality of the received signal at the destination nodes, thereby increasing the receiver BER beyond acceptable limits. Further, as in any other form of telecommunication networks, the power consumption in various active devices across optical networks keeps increasing with the growth of network traffic, speed, and size, thereby demanding power-aware designs of the network elements and protocols as well. In this chapter, first we examine the impact of various transmission impairments in optical networks, followed by some candidate impairment-aware designs for different networking segments. Finally, we present some power-aware design approaches for optical networks.

10.1 Physical-layer issues in optical networks

Even with optical fibers offering enormous bandwidth and extremely low loss, the physical layer in optical networks needs a judicious design approach to control the transmission impairments from the non-ideal devices for successful end-to-end transmission. Losses in various passive WDM devices, noises in optical amplifiers and receivers, crosstalk from switches, spectral spread of optical sources, and linear and nonlinear impairments in fibers can degrade the signal quality, increasing the BER beyond acceptable limits. The variety and magnitude of transmission impairments increase significantly in a WDM network, as compared to its single-wavelength version. The impact of transmission impairments (or simply impairments) in the physical layer may vary for different network segments, and calls for a careful approach to examining them and designing the transmission system accordingly. For example, a lightpath over a long route may get contaminated by large noise and crosstalk, thereby producing a high BER, and hence this may have to be blocked or broken at some intermediate node for OEO-based regeneration. Thus, the impairment-aware design of optical networks may have to look into the cross-layer issues, whenever necessary.

Furthermore, while designing the physical layer of an optical network, one needs to be aware of the power consumption across the network. Power consumption has all along been an important issue in communication systems and the researchers in this area have made great strides over the past three centuries. Customarily, the energy or power consumption

Optical Networks. Debasish Datta, Oxford University Press (2021). © Debasish Datta.
DOI: 10.1093/oso/9780198834229.003.0010

(henceforth, we use power and energy consumptions interchangeably depending on the context) efficiency of communication systems is described using a parameter, defined as η_{EpBpD} = energy per bit per distance (in joules per bit per 1000 km of link length, abbreviated as Jpbp1000), with the reference distance set at 1000 km (Tucker 2011a). To begin with, the first trans-Atlantic wireless communication system set up during the 1860's needed an $\eta_{EpBpD} \simeq 0.1$ Jpbp1000. This parameter made great progress with time (over the next $\simeq$ 100 years), and during the years around 1950, the early undersea coaxial communication systems came up with an η_{EpBpD} as low as 10^{-5} Jpbp1000 = 10 μJpbp1000. It took only another half century (in the years around 2010), to bring down this number to 1.1 nJpbp1000 by using optically amplified dense-WDM optical networks operating with 40 wavelengths, each operating at 40 Gbps – a stupendous achievement of mankind indeed! However, it is worthwhile to note that, from the physical-layer perspective, the efficiency parameter η_{EpBpD} has a close link with the quality of transmission (QoT), as the spectral efficiency η_S ($\equiv$ network speed) in a digital communication system (expressed in joules per bit per sec) can be enhanced by using multilevel modulation schemes, such as QAM, which might in turn degrade η_{EpBpD} by pushing up the energy consumed per bit.

In view of the above, we discuss in this chapter the issues concerning transmission impairments and power/energy consumption in optical networks, over different network segments or scales (e.g., access to long-haul).

10.2 Impact of transmission impairments in optical networks

In order to explore the impairment-aware design approaches for optical networks, we first examine the basic transmission systems in terms of the achievable BER, both for single-wavelength and WDM systems. In the following, we therefore present analytical models to estimate BER for different types of optical transmission systems, followed by discussions on the power and rise-time budgets in optical WDM networks.

10.2.1 BER in optical receivers

In this section, we examine the BER performance of optical receivers in different operating conditions, while using direct detection of intensity modulated lightwaves (i.e., for IM-DD systems, as discussed in Chapter 2). One of these operating conditions concerns with the fundamental performance limit (known as *quantum limit*) for optical receivers, when the receiver offers the best possible IM-DD performance, operating as an ideal *photon counter*. The next scenario that we discuss deals with the optical receivers that work with the thermal and shot noise components, and represent a practical class of optical receivers used in single-wavelength transmission links. Thereafter, we consider all-optical WDM links (lightpaths) taking into account all the relevant noise components, including thermal and shot noise, ASE noise from optical amplifiers, and crosstalk components from the relevant optical devices. The underlying physical phenomena and the theoretical models to analyze these noise components are described in Appendix B.

BER in quantum limit

In a practical optical receiver, BER is governed by all kinds of noise components: thermal noise, shot noise, and various other noise components arising from the use of WDM devices and OAs (for WDM systems), along with pulse spreading due to various dispersion mechanisms in optical fibers. However, it is instructive to understand how an optical receiver would behave in an ideal condition, when the receiver operates as a ideal photon counter under *quantum limit* in an IM-DD system.

Quantum limit operation of an optical receiver implies that the receiver is capable of *counting* the number of photons arriving at the photodiode for a perfectly rectangular optical pulse. In other words, while operating in the quantum limit, the receiver is assumed to have the perception to *see the quantized version of light*. Such an assumption is however hypothetical, as a practical receiver will be contaminated by various noise components, and moreover the current impulses created by the arriving photons will be smeared due to the finite receiver bandwidth, thereby preventing the receiver from deciphering the number of photons from the noise-contaminated current profile with the overlapping smeared-impulses generated by the received photons (see Fig. B.1 in Appendix B). However, even with this ideal ambience in the receiver, there will yet remain a finite probability of error, i.e., nonzero BER, due to uncertainty in the photon emission process in the transmitter itself. We examine this issue in the following.

In general, error in the decision made in a binary receiver can take place if a transmitted one (or a zero) is received as a zero (or a one). Thus, the BER in a binary receiver is expressed as

$$\text{BER} = P(0)P(1/0) + P(1)P(0/1), \tag{10.1}$$

with $P(0)$ and $P(1)$ as the probabilities of transmission for binary zero and one, respectively. $P(1/0)$ and $P(0/1)$ are the conditional probabilities for detecting a one when a zero has been transmitted and for detecting a zero when a one has been transmitted, respectively. Customarily, we assume that the probabilities of transmission for zeros and ones are the same and hence equal to half, such that we can write BER as

$$\text{BER} = \frac{1}{2}\big[P(1/0) + P(0/1)\big]. \tag{10.2}$$

With this expression of BER we next proceed to find out the conditional probabilities $P(1/0)$ and $P(0/1)$ in an optical receiver operating in quantum limit.

First, we consider an ideal receiver as described above, and assume that the receiver receives an optical power P_1 during the reception of a binary one and zero power (i.e., $P_0 = 0$) during the reception of a binary zero, implying that the source gets completely extinguished (i.e., switched off) during a zero transmission. The power received during the binary one reception in the bit interval T would be related to the average number of photons Λ_1 received in the same interval, given by

$$P_1 = \frac{\Lambda_1 hf}{T}, \tag{10.3}$$

with f as the frequency of the received photons. However, the actual number of photons (k, say) received during the interval T will be random and Poisson distributed (due to the photoemission uncertainties in the optical source itself, see Appendix B) with the Poissonian probability $P(k/\Lambda_1)$, given by

$$P(k/\Lambda_1) = \frac{\Lambda_1^k \exp(-\Lambda_1)}{k!}. \tag{10.4}$$

Using Eq. 10.4, we therefore obtain $P(0/1)$ as

$$P(0/1) = P(k = 0/\Lambda_1) = \exp(-\Lambda_1). \tag{10.5}$$

Next, in order to estimate $P(1/0)$ we note that, while receiving a binary zero, the optical source remains completely switched off. Hence, the receiver being an ideal photon counter will never receive any photon *for certain*, implying thereby that, $P(1/0) = 0$. Using these observations on $P(0/1)$ and $P(1/0)$, we express the BER under quantum limit as

$$\text{BER} = \frac{1}{2}\big[\exp(-\Lambda_1) + 0\big] = \frac{\exp(-\Lambda_1)}{2}. \tag{10.6}$$

Hence, the number of photons needed during the reception of a binary one, for a specified BER, can be expressed as

$$\Lambda_1 = \ln\left(\frac{1}{2\,\text{BER}}\right). \tag{10.7}$$

For a BER of 10^{-9}, we obtain $\Lambda_1 = 21$ from the above expression, implying that at least 21 photons/bit are needed during the reception of a binary one to ensure a BER of 10^{-9} in an optical receiver operating in quantum limit. The average received power P_{av}^{QL} needed in a receiver operating in quantum limit can be expressed as

$$P_{av}^{QL} = \frac{1}{2}(P_1 + P_0) = \frac{P_1}{2} = \frac{\Lambda_1 hf}{2T} = \frac{\Lambda_1 hcr}{2w}, \tag{10.8}$$

where, as mentioned earlier, we have assumed $P_0 = 0$ implying perfect laser extinction during zero transmission, and used Eq. 10.3 for P_1. Further, r represents the transmission rate (bit rate), given by $r = 1/T$, with T as the bit interval. Note that the average number of received photons/bit is $\Lambda_{av} = \Lambda_1/2$, which is independent of the bit rate r, while the corresponding value of the average received power P_{av}^{QL} is proportional to r. This is expected naturally, as the same number of photons (and hence energy) must be received during a shorter value of T at higher bit rate r, thereby making the required power proportionately higher with higher transmission speed. For example, with the BER requirement of 10^{-9}, while $\Lambda_{av} = 10.5$ photons/bit regardless of the bit rate, the value of P_{av}^{QL} changes from -59 dBm to -49 dBm as the bit rate increases from 1 Gbps to 10 Gbps at $w = 1550$ nm.

BER with shot and thermal noise components

An optical receiver, in practice, will have thermal and shot-noise components, when operating in a single-wavelength transmission system, or operating in a WDM system without any significant impairment from the ASE noise or crosstalk components. In this section, we evaluate the BER for such optical receivers.

As discussed in Appendix B, with a photodiode as current source, our analysis will be based on the current-source models for the signal and noise components. The receiver current $i(t)$ is sampled at appropriate instants of time within each bit interval, and the sampled current in each bit interval, denoted as i_k, is compared with a threshold current I_{TH} to take the decision. We therefore express two possible outcomes of i_k, i.e., i_{k1} and i_{k0}, respectively for binary one and zero receptions, as

$$i_{k1} = I_1 + i_{n1}, \text{ for binary one reception} \tag{10.9}$$

$$i_{k0} = I_0 + i_{n0}, \text{ for binary zero reception,} \tag{10.10}$$

where I_1 and I_0 are the DC components (or mean values) of i_{k1} and i_{k0}, respectively. I_1 and I_0 can be expressed as

$$I_1 = R_w P_1 + I_{BD}, \text{ for binary one reception} \tag{10.11}$$

$$I_0 = R_w P_0 + I_{BD}, \text{ for binary zero reception.} \tag{10.12}$$

In the above expressions, I_{BD} represents a combination of the omnipresent currents in the photodiode due to background illumination and dark current, and R_w is the photodiode responsivity (as defined in Chapter 2). The terms i_{n1} and i_{n0} in Equations 10.9 and 10.10 represent the sampled values of the receiver noise currents for binary one and zero receptions, respectively, following zero-mean Gaussian distributions, given by

$$p_1(i_{n1}) = \frac{1}{\sqrt{2\pi\sigma_1^2}} \exp\left(\frac{-i_{n1}^2}{2\sigma_1^2}\right) \tag{10.13}$$

$$p_0(i_{n0}) = \frac{1}{\sqrt{2\pi\sigma_0^2}} \exp\left(\frac{-i_{n0}^2}{2\sigma_0^2}\right). \tag{10.14}$$

The respective noise variances σ_1^2 and σ_0^2 are expressed as

$$\sigma_1^2 = \sigma_{th}^2 + \sigma_{sh1}^2 \tag{10.15}$$

$$\sigma_0^2 = \sigma_{th}^2 + \sigma_{sh0}^2, \tag{10.16}$$

with σ_{th}^2 and σ_{shi}^2 ($i = 1, 0$ for binary one and zero receptions, respectively) as the thermal and shot-noise variances (following the models described in Appendix B), given by

$$\sigma_{th}^2 = 4k\theta B_e/R \tag{10.17}$$

$$\sigma_{sh1}^2 = 2q(R_w P_1 + I_{BD})B_e \tag{10.18}$$

$$\sigma_{sh0}^2 = 2q(R_w P_0 + I_{BD})B_e, \tag{10.19}$$

with B_e as the electrical bandwidth of the receiver, R as the input resistance of the receiver preamplifier, and θ representing the receiver temperature in °K.

Using the above statistical models for the signal and noise components, we consider the two cases of signal reception: binary one and zero. For the reception of binary one, the receiver makes an error if $i_{k1} < I_{TH}$, i.e., when

$$R_w P_1 + I_{BD} + i_{n1} < I_{TH} \tag{10.20}$$

$$\text{or, } i_{n1} < -R_w P_1 + I'_{TH},$$

with $I'_{TH} = I_{TH} - I_{BD}$. Using the above condition, we therefore express the conditional probability $P(0/1)$ as

$$P(0/1) = \frac{1}{\sqrt{2\pi\sigma_1^2}} \int_{-\infty}^{-R_w P_1 + I'_{TH}} \exp\left(\frac{-i_{n1}^2}{2\sigma_1^2}\right) di_{n1} \tag{10.21}$$

$$= \frac{1}{2}\text{erfc}\left(\frac{R_w P_1 - I'_{TH}}{\sqrt{2}\sigma_1}\right).$$

Similarly, for the reception of binary zero, the receiver makes an error if $i_{k0} > I_{TH}$, i.e., when

$$R_w P_0 + I_{BD} + i_{n0} > I_{TH} \quad (10.22)$$
$$\text{or, } i_{n0} > I'_{TH} - R_w P_0,$$

which leads to the expression of $P(1/0)$, given by

$$P(1/0) = \frac{1}{\sqrt{2\pi\sigma_0^2}} \int_{I'_{TH}-R_w P_0}^{\infty} \exp\left(\frac{-i_{n0}^2}{2\sigma_0^2}\right) di_{n0} \quad (10.23)$$
$$= \frac{1}{2}\text{erfc}\left(\frac{I'_{TH} - R_w P_0}{\sqrt{2}\sigma_0}\right).$$

Using $P(0/1)$ and $P(1/0)$ from Equations 10.21 and 10.23 in Eq. 10.2, we obtain the BER as

$$\text{BER} = \frac{1}{4}\left[\text{erfc}\left(\frac{R_w P_1 - I'_{TH}}{\sqrt{2}\sigma_1}\right) + \text{erfc}\left(\frac{I'_{TH} - R_w P_0}{\sqrt{2}\sigma_0}\right)\right]. \quad (10.24)$$

In order to minimize the BER, the threshold current I'_{TH} needs to be optimized using the log-likelihood criterion; however, a more straightforward approach can be adopted with a close-to-optimum result by equating the arguments of the two terms in the parentheses on the right-hand side of the above expression as

$$\frac{R_w P_1 - I'_{TH}}{\sqrt{2}\sigma_1} = \frac{I'_{TH} - R_w P_0}{\sqrt{2}\sigma_0}, \quad (10.25)$$

leading to the expression of I'_{TH}, given by

$$I'_{TH} = \frac{\sigma_0 R_w P_1 + \sigma_1 R_w P_0}{\sigma_1 + \sigma_0} \quad (10.26)$$
$$\text{i.e., } I_{TH} = \frac{\sigma_0 R_w P_1 + \sigma_1 R_w P_0}{\sigma_1 + \sigma_0} + I_{BD}.$$

Substituting the value of I_{TH} from Eq .10.26 in Eq. 10.24, we finally obtain the expression for the BER as

$$\text{BER} = \frac{1}{2}\text{erfc}\left[\frac{R_w(P_1 - P_0)}{\sqrt{2}(\sigma_1 + \sigma_0)}\right] = \frac{1}{2}\text{erfc}\left(\frac{Q}{\sqrt{2}}\right), \quad (10.27)$$

where Q represents the quality factor of the receiver and should be higher for lower BER. For example, Q should be $\simeq$ 6 and 7 to ensure a BER = 10^{-9} and 10^{-12}, respectively. Note that the numerator of Q represents the signal-current swing between the binary one and zero levels (i.e., the eye-opening at the decision-making point of the receiver), while the factor $(\sigma_1 + \sigma_0)$ in the denominator represents the average of the root-mean-squared (rms) values of receiver noise for one and zero receptions, respectively. The threshold, as governed by Eq. 10.26, is biased toward the binary zero level as the noise variance of zero is less than that of one, but it falls exactly halfway between the two signal levels when the receiver becomes dominated by the thermal noise, making $\sigma_1^2 \simeq \sigma_0^2 \simeq \sigma_{th}^2$. At very high bit rates, the receiver input resistance R needs to be brought down to achieve the required bandwidth

($B_e \simeq 0.7r$), as input capacitance C in the receiver circuit cannot be brought down below a certain limit and $B_e \propto \frac{1}{RC}$. This makes the thermal noise variance $\sigma_{th}^2 = 4k\theta B_e/R$ quite large at high bit rate, even exceeding the shot noise components; under this situation the receiver is said to be operating in the *thermal-noise limit*. However, when the received optical power is fairly large, the receiver might get dominated by shot noise and thus operate in the *shot-noise limit*.

Note that, the above BER analysis didn't consider the effect of fiber dispersion and other pulse-spreading phenomena in optical transmitter and receiver circuits. However, one simple yet effective means to address this issue would be to replace $(P_1 - P_0)$ in the numerator of Q by $\alpha_D(P_1 - P_0)$ with $\alpha_D(\leq 1)$ representing the reduction factor of the eye-opening (observed at the input of the receiver decision-making circuit, see Fig. 2.48) caused by pulse-spreading mechanisms. Impact of fiber dispersion along with other pulse-spreading phenomena in optical fiber links is considered later, in the context of system design, based on the power and rise-time budgets of optical fiber links. Furthermore, if necessary, the receiver BER can be significantly improved (reduced) by using FEC codes (e.g., Reed–Solomon codes), albeit with a nominal increase in the transmission rate.

BER of a lightpath in WDM mesh

Optical receivers in WDM mesh networks, particularly in long-haul WRONs, need to operate in the presence of additional noise components, over and above those (thermal and shot noise processes) we considered above. Figure 10.1 illustrates a typical long-haul WRON segment, where a given lightpath LP_{24} traverses from node 2 to node 4 through an intermediate node (node 3). During the course of its traversal across the network segment, the lightpath encounters losses in the passive devices of the intermediate nodes and fiber links, and gets contaminated by the ASE noise from the OAs and crosstalk from the OXCs. With the analytical models presented in Appendix B, one can estimate the various noise components along with the received signal, which are in turn used to evaluate the receiver BER, following the same BER model as described above. In the following, we first demonstrate this approach for a simple example and generalize the procedure thereafter.

Considering the example illustrated in Fig. 10.1, we observe that the lightpath LP_{24} goes through node 3 (as an intermediate node), the links l_1 and l_2, and the OAs used therein. In

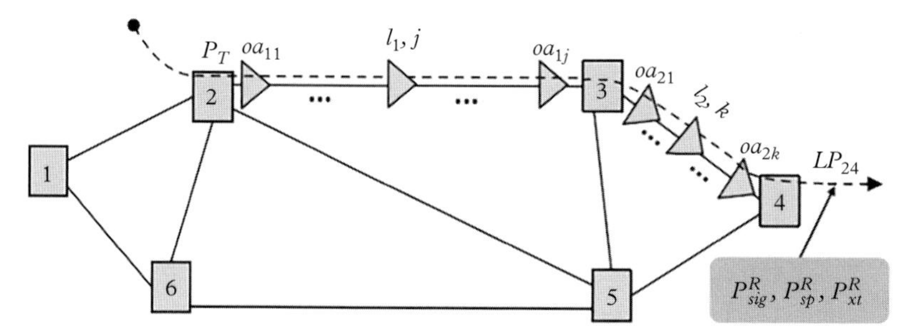

Lightpath LP_{24} is set up from node 2 to node 4 through node 3, using the links l_1 and l_2 with j and k as the numbers of OAs employed in these links, respectively.

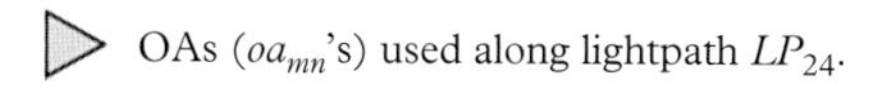

Figure 10.1 *Illustration of a lightpath (LP_{24}) set up between two nodes (node 2 to node 4) in a six-node WRON. In all the nodes and OAs traversed through, LP_{24} gets contaminated by crosstalk/noise and incurs loss/gain in addition to the losses in fiber links, and their cumulative effect finally shows up at the receiving node (node 4) as P^R_{sp} (ASE noise power) and P^R_{xt} (crosstalk power) along with the signal power P^R_{sig}. Note that, all OAs are actually provisioned in pairs for the two bidirectional fibers used in each link, but shown for one direction only for the fiber links used in the lightpath under consideration. For the remaining fiber links, OAs are not shown.*

order to estimate the received signal power P^R_{sig}, the accumulated ASE power P^R_{sp}, and the accumulated crosstalk power P^R_{xt} at node 4, we evaluate step-by-step the impairments that lightpath LP_{24} undergoes along its route from node 2 to node 4.

In Fig. 10.2, we illustrate how the signal transmitted from node 2 on lightpath LP_{24} of Fig. 10.1 encounters various types of impairments on one of its two links, i.e., link l_1. Denoting the fiber length spanned by link l_1 by F_1, the number of OAs in the link can be obtained as $j = \lfloor \frac{F_1}{\Delta F} + 0.5 \rfloor + 1$, where ΔF is the separation between two consecutive OAs. Note that, the OA on link l_1 immediately following node 2 (i.e., oa_{11}) compensates for the losses incurred in the same node by a passthrough lightpath, while oa_{12} through oa_{1j} compensate for the losses in the respective preceding fiber segments as in-line OAs. For simplicity of the analysis, we assume that the link lengths are integral multiples of ΔF. Since all in-line OAs compensate for the fiber segments of same length ΔF, we can write $G_{oa_{12}} = G_{oa_{13}} = \cdots = G_{oa_{1j}} = 1/L_{\Delta F} = G_{oa_{IL}}$ (say). With this network setting, we express the gains $G_{oa_{1i}}$'s of the OAs in link l_1 as

$$G_{oa_{11}} = \frac{1}{L_{dm}L_{sw}L_{mx}} \tag{10.28}$$

$$G_{oa_{1i}} = G_{oa_{IL}} = \frac{1}{L_{\Delta F}} \quad \text{for } i = 2, 3, \ldots, j, \tag{10.29}$$

with L_{dm}, L_{sw}, and L_{mx} representing the insertion losses in the ODMUX, switch, and OMUX (see Fig. 10.3), respectively, and $L_{\Delta f}$ as the fiber loss in each fiber segment ΔF. For simplicity, we also assume that the losses in a given type of optical device (ODMUX, switch, and OMUX) used in all nodes are identical (i.e., L_{dm}, L_{sw}, and L_{mx} are the same at all nodes), though in reality they would be different, and one can accommodate this variation

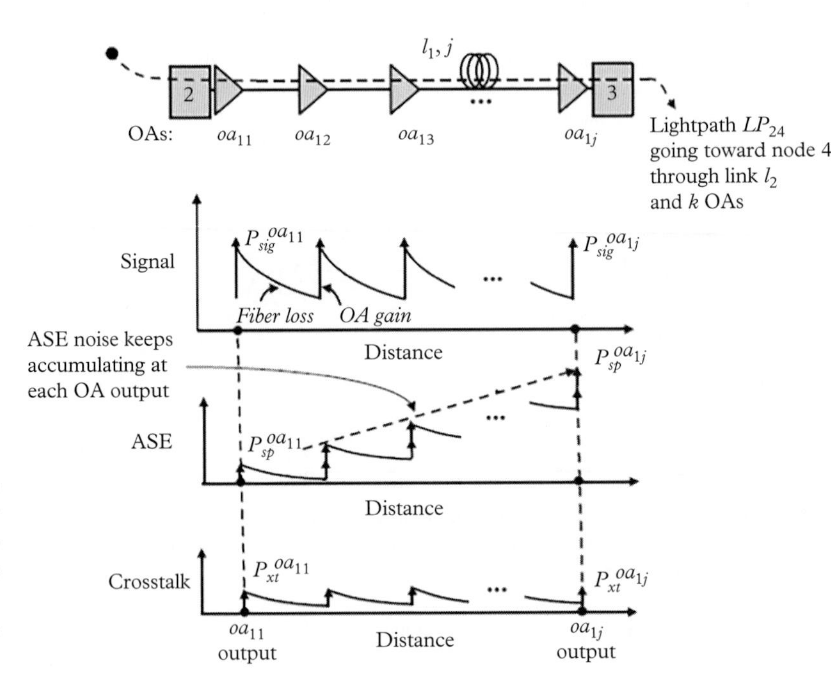

Figure 10.2 *Illustration of how the signal, ASE noise, and crosstalk components grow/decay along the fiber link l_{23} of lightpath LP_{24}, shown in Fig.10.1, where each node uses an OXC. Note that $P^{oa_{11}}_{sig} = P^{oa_{1j}}_{sig}$ and $P^{oa_{11}}_{xt} = P^{oa_{1j}}_{xt}$, while the ASE noise keeps accumulating at each OA output.*

in the present formulation itself. Further, the (insertion) losses in connectors used with the above devices are considered to be included in the respective device losses.

With the above network setting (Fig. 10.1), we enumerate in the following the episodes of signal losses/gains, and generation/growth/decay of ASE noise and crosstalk components along the lightpath LP_{24} at relevant locations (i.e., at the outputs of the nodes and OAs).

(1) *At the output of source node, i.e., node* 2: The transmitted power P_T from one of the transmitters of node 2 passes through the optical switch and OMUX with the losses L_{sw} and L_{mx}, respectively (like the locally added lightpath (AL) in the OXC shown using a dashed line in Fig. 10.3). In the process, the signal also gets contaminated by a switch crosstalk (we consider that the hetero-wavelength crosstalk components in OMUX/DEMUX are much less than the switch crosstalk components) with a crosstalk-to-signal ratio ρ_{xt}. Thus, the lightpath exits node 2 with the signal power $P_{sig}(2)$, given by

$$P_{sig}(2) = P_T L_{sw} L_{mx}. \tag{10.30}$$

Next, we evaluate the accompanying crosstalk component $P_{xt}(2)$ contributed by the switch of OXC at the output of node 2 by an incoming signal from a preceding node (say, node 6 in Fig. 10.1). We assume that in the link between the nodes 6 and 2, there are n OAs (not shown in the figure), with OA_{61} as the node-loss compensating OA following node 6 and $(n-1)$ in-line OAs. With this consideration, we express $P_{xt}(2)$ at the output of node 2 as

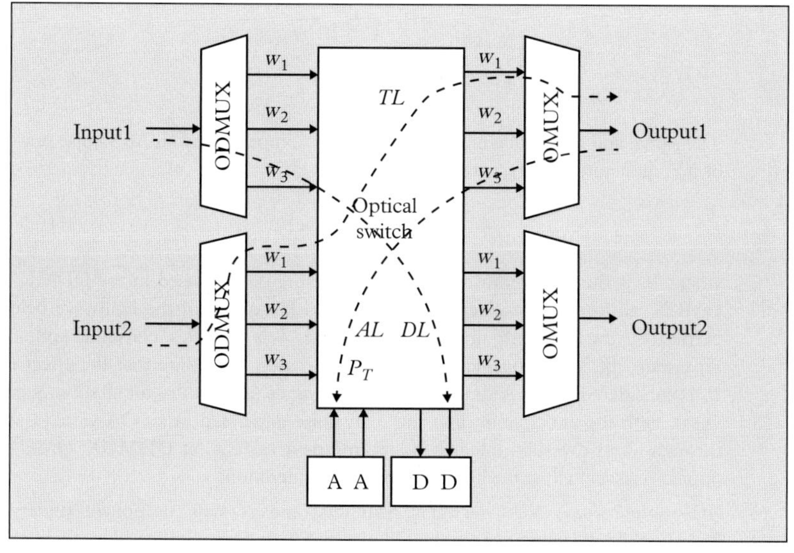

AL: Added lightpath (loss in OXC $= L_{sw} L_{mx}$)
DL: Dropped lightpath (loss in OXC $= L_{dm} L_{sw}$)
TL: Passthrough or transit lightpath (loss in OXC $= L_{dm} L_{sw} L_{mx}$)

Figure 10.3 *Illustration of losses in an OXC. Note that a locally added lightpath (denoted by the dashed line AL in the OXC) undergoes a loss $L_{sw}L_{mx}$ through the optical switch and OMUX. Similarly, a dropped lightpath (denoted as DL) undergoes a loss $L_{dm}L_{sw}$ through the ODMUX and optical switch, and a passthrough or transit lightpath (denoted as TL) incurs a loss of $L_{dm}L_{sw}L_{mx}$ through the ODMUX, switch, and OMUX.*

$$\begin{aligned} P_{xt}(2) &= \rho_{xt} \underbrace{P_T L_{sw} L_{mx}}_{\text{From node 6}} G_{oa_{61}} \underbrace{(L_{\Delta F} G_{oa_{IL}})^{n-1}}_{\text{GLP(1)=1}} \\ &\quad \times \underbrace{L_{dm} L_{sw} L_{mx}}_{\text{Loss in node 2}} \\ &= \rho_{xt} P_T L_{sw} L_{mx} \times \frac{1}{L_{dm} L_{sw} L_{mx}} \times L_{dm} L_{sw} L_{mx} \\ &= \rho_{xt} P_T L_{sw} L_{mx}, \end{aligned} \tag{10.31}$$

where GLP(1) represents the gain-loss product (=1) for a fiber segment of length ΔF and the following loss-compensating OA (see Eq. 10.29). Further, $G_{oa_{61}}$ is substituted by $G_{oa_{11}}$ using Eq. 10.28 (note that, both OAs have the same gain, as OXCs in all the nodes are assumed to be having same loss).

(2) *At the output of* oa_{11}: The OXC output from node 2 is amplified in oa_{11} by $G_{oa_{11}}$ to produce a signal output $P_{sig}^{oa_{11}}$, given by

$$\begin{aligned} P_{sig}^{oa_{11}} &= G_{oa_{11}} P_{sig}(2) \\ &= \frac{1}{L_{dm} L_{sw} L_{mx}} \times L_{sw} L_{mx} P_T = \frac{P_T}{L_{dm}}. \end{aligned} \tag{10.32}$$

The above signal is accompanied by an amplified crosstalk component with a power $P_{xt}^{oa_{11}}$, given by

$$\begin{aligned} P_{xt}^{oa_{11}} &= \rho_{xt} P_T L_{sw} L_{mx} \times G_{oa_{11}} \\ &= \rho_{xt} P_T L_{sw} L_{mx} \times \frac{1}{L_{dm} L_{sw} L_{mx}} \\ &= \frac{\rho_{xt} P_T}{L_{dm}}. \end{aligned} \tag{10.33}$$

The signal and crosstalk components will be accompanied by an ASE noise power of $p_{sp}^{oa_{11}}$ generated at oa_{11}, given by

$$p_{sp}^{oa_{11}} = n_{sp} G_{oa_{11}} hf B_o, \tag{10.34}$$

where B_o is the bandwidth of the optical filtering devices used in the ODMUX, OMUX, and optical receivers, f is the central frequency of the lightwave being carried by the lightpath under consideration, h is Planck's constant, and n_{sp} represents the ionization factor in the EDFA-based OA. Note that the effect of B_o is considered on the ASE noise power in advance (and so also for all subsequent OAs), with the anticipation that the ASE noise generated in an OA would pass through all of the subsequent bandlimiting devices (e.g., at ODMUX, OMUX, optical receiver) along the lightpath under consideration.

(3) *At the output of* oa_{12}: Next, the signal, ASE noise, and crosstalk components undergo losses in the following fiber segment of length ΔF, and get amplified by oa_{12} by the amount equaling the inverse of the loss in the preceding fiber segment (Eq. 10.29). However, oa_{12} also adds an ASE noise power, denoted as $p_{sp}^{oa_{12}}$. Hence, the power levels for signal, ASE, and crosstalk components at oa_{12} output can be expressed as

$$P_{sig}^{oa_{12}} = P_{sig}^{oa_{11}} \underbrace{L_{\Delta F} G_{oa_{12}}}_{\text{GLP(1)=1}} = P_{sig}^{oa_{11}} = \frac{P_T}{L_{dm}} \tag{10.35}$$

$$P_{sp}^{oa_{12}} = p_{sp}^{oa_{11}} L_{\Delta F} G_{oa_{12}} + p_{sp}^{oa_{12}} = p_{sp}^{oa_{11}} + p_{sp}^{oa_{12}} \tag{10.36}$$

$$P_{xt}^{oa_{12}} = P_{xt}^{oa_{11}} L_{\Delta F} G_{oa_{12}} = P_{xt}^{oa_{11}} = \frac{\rho_{xt} P_T}{L_{dm}}, \tag{10.37}$$

where $p_{sp}^{oa_{12}} = n_{sp}(G_{oa_{12}} - 1)hfB_o$.

(4) *At the output of* oa_{1j}: The computational process used in Step 3 is repeated for all the remaining OAs. Note that, with the identical in-line OAs used to compensate the losses $L_{\Delta F}$ in equal-length fiber segments, one can write $G_{oa_{12}} = G_{oa_{13}} = \cdots = G_{oa_{1j}} = G_{oa_{22}} = G_{oa_{23}} = \cdots = G_{oa_{2k}} = G_{oa_{IL}}$, with $G_{oa_{IL}}$ representing the gain of an in-line OA, and hence $p_{sp}^{oa_{12}} = p_{sp}^{oa_{13}} = \cdots = p_{sp}^{oa_{1j}} = p_{sp}^{oa_{22}} = p_{sp}^{oa_{23}} = \cdots = p_{sp}^{oa_{2k}} = p_{sp}^{oa_{IL}}$, with $p_{sp}^{oa_{IL}}$ representing the ASE noise generated in an in-line OA. We therefore express the powers of signal, ASE noise, and crosstalk components at the output of oa_{1j} as

$$\begin{aligned} P_{sig}^{oa_{1j}} &= P_{sig}^{oa_{12}} \underbrace{(L_{\Delta F} G_{oa_{IL}})^{j-1}}_{\text{GLP(1)=1}} \\ &= P_{sig}^{oa_{12}} = P_{sig}^{oa_{11}} = \frac{P_T}{L_{dm}} \end{aligned} \tag{10.38}$$

$$\begin{aligned} P_{sp}^{oa_{1j}} &= P_{sp}^{oa_{12}} (L_{\Delta F} G_{oa_{IL}})^{j-1} + (j-2) p_{sp}^{oa_{IL}} \\ &= P_{sp}^{oa_{12}} + (j-2) p_{sp}^{oa_{IL}} \\ &= p_{sp}^{oa_{11}} + \underbrace{p_{sp}^{oa_{12}}}_{=p_{sp}^{oa_{IL}}} + (j-2) p_{sp}^{oa_{IL}} \\ &= p_{sp}^{oa_{11}} + (j-1) p_{sp}^{oa_{IL}} \end{aligned} \tag{10.39}$$

$$\begin{aligned} P_{xt}^{oa_{1j}} &= P_{xt}^{oa_{12}} (L_{\Delta F} G_{oa_{IL}})^{j-1} \\ &= P_{xt}^{oa_{12}} = P_{xt}^{oa_{11}} = \frac{\rho_{xt} P_T}{L_{dm}}, \end{aligned} \tag{10.40}$$

implying that the ASE noise builds up steadily with the number of OAs while the signal and crosstalk powers remain the same due to the loss-gain compensation in the fiber-OA pairs (as illustrated earlier in Fig. 10.2).

(5) *At the output of node* 3: The output from oa_{1j} will pass through node 3 (an intermediate node, for which oa_{1j} is a pre-amplifier) with the same loss parameters as in node 2, but with an additional crosstalk component from node 3 (worst case), thereby giving the powers for the signal, ASE noise, and crosstalk components at the output of node 3 as

$$P_{sig}(3) = L_{dm} L_{sw} L_{mx} P_{sig}^{oa_{1j}} \tag{10.41}$$

$$P_{sp}(3) = L_{dm} L_{sw} L_{mx} P_{sp}^{oa_{1j}} \tag{10.42}$$

$$P_{xt}(3) = L_{dm} L_{sw} L_{mx} P_{xt}^{oa_{1j}} + \rho_{xt} P_T L_{sw} L_{mx}. \tag{10.43}$$

(6) *At the output of* oa_{21}: The signal, ASE noise, and crosstalk components from node 3 will next be amplified by $G_{oa_{21}}$ along with the additional ASE noise component from oa_{21}, leading to

$$P_{sig}^{oa_{21}} = G_{oa_{21}} P_{sig}(3) \tag{10.44}$$

$$\begin{aligned}
&= \underbrace{G_{oa_{21}} L_{dm} L_{sw} L_{mx}}_{\text{GLP(2)=1}} P_{sig}^{oa_{1j}} \\
&= P_{sig}^{oa_{1j}} = P_{sig}^{oa_{11}} = \frac{P_T}{L_{dm}}
\end{aligned}$$

$$\begin{aligned}
P_{sp}^{oa_{21}} &= G_{oa_{21}} P_{sp}(3) + p_{sp}^{oa_{21}} \\
&= G_{oa_{21}} L_{dm} L_{sw} L_{mx} P_{sp}^{oa_{1j}} + p_{sp}^{oa_{21}} \\
&= P_{sp}^{oa_{1j}} + \underbrace{p_{sp}^{oa_{21}}}_{=p_{sp}^{oa_{11}}} = \{p_{sp}^{oa_{11}} + (j-1)p_{sp}^{oa_{IL}}\} + p_{sp}^{oa_{11}} \\
&= 2p_{sp}^{oa_{11}} + (j-1)p_{sp}^{oa_{IL}}
\end{aligned} \tag{10.45}$$

$$\begin{aligned}
P_{xt}^{oa_{21}} &= G_{oa_{21}} P_{xt}(3) \\
&= G_{oa_{21}} (L_{dm} L_{sw} L_{mx} P_{xt}^{oa_{1j}} + \rho_{xt} P_T L_{sw} L_{mx}) \\
&= P_{xt}^{oa_{1j}} + \rho_{xt} P_T / L_{dm} = 2P_{xt}^{oa_{11}} = \frac{2\rho_{xt} P_T}{L_{dm}},
\end{aligned} \tag{10.46}$$

where GLP(2) represents the loss-gain product (=1) of a node followed by its loss-compensating OA (see Eq. 10.28).

(7) *At the receiver of the destination node, i.e., node* 4: The output of oa_{21} will thereafter go through similar episodes, as undergone in the fiber link between the nodes 2 and 3. Having reached node 4 (after traversing through $(k-1)$ in-line OAs), the incoming lightpath LP_{24} will be dropped, following the dashed lightpath DL in the OXC of node 4 into a receiver, as illustrated in Fig. 10.3. Therefore, one can express the power of the received (dropped) signal at node 4 as

$$\begin{aligned}
P_{sig}^R &= P_{sig}^{oa_{21}} \underbrace{(\text{GLP}(1))^{k-1}}_{=1} L_{dm} L_{sw} \\
&= P_{sig}^{oa_{21}} L_{dm} L_{sw} = P_{sig}^{oa_{11}} L_{dm} L_{sw} \\
&= \frac{P_T}{L_{dm}} \times L_{dm} L_{sw} = P_T L_{sw}.
\end{aligned} \tag{10.47}$$

Similarly, we express the ASE noise and crosstalk components received at node 4 as

$$\begin{aligned}
P_{sp}^R &= P_{sp}^{oa_{21}} (\text{GLP}(1))^{k-1} L_{dm} L_{sw} + \\
&\quad (k-1) p_{sp}^{oa_{IL}} L_{dm} L_{sw} \\
&= \{2p_{sp}^{oa_{11}} + (j-1)p_{sp}^{oa_{IL}}\} L_{dm} L_{sw} + (k-1) p_{sp}^{oa_{IL}} L_{dm} L_{sw} \\
&= \{2p_{sp}^{oa_{11}} + (j+k-2)p_{sp}^{oa_{IL}}\} L_{dm} L_{sw}
\end{aligned} \tag{10.48}$$

$$\begin{aligned}
P_{xt}^R &= P_{xt}^{oa_{21}} L_{dm} L_{sw} (\text{GLP}(1))^{k-1} + \underbrace{\rho_{xt} P_T L_{sw}}_{\text{Crosstalk at node 4}} \\
&= P_{xt}^{oa_{21}} L_{dm} L_{sw} + \rho_{xt} P_T L_{sw} \\
&= \frac{2\rho_{xt} P_T}{L_{dm}} \times L_{dm} L_{sw} + \rho_{xt} P_T L_{sw} = 3\rho_{xt} P_T L_{sw},
\end{aligned} \tag{10.49}$$

where the last term on the right side in the first row of P_{xt}^R represents the crosstalk generated in the switch of node 4 (i.e., $\rho_{xt} P_T L_{sw}$), which is equal to $P_{xt}(2)/L_{mx}$, as in this case the switch crosstalk component is dropped at the receiver without going through the OMUX stage of the OXC.

(8) *Generalization:* Following the above steps, one can find out the expressions for the signal, ASE noise, and crosstalk components at the receiving end of any lightpath in a WRON. Note that, the example lightpath we have considered has one intermediate node (at node 2), and presence of more intermediate nodes in a lightpath will need the replication of the method used for node 2. Further, the WRON considered in the present example uses a mesh topology, while the given procedure can be used for other topologies, such as in ring where the OXCs will have to be replaced by ROADMs.

Having evaluated the received signal, ASE noise, and crosstalk components at the destination node, one needs to determine the *beat noise* variances (using the method described in Appendix B), and use them along with other noise components to evaluate the BER for a given lightpath from Eq. 10.27. We consolidate these steps as follows.

From Eq. B.42 (Appendix B), we express the total receiver noise variances $\sigma^2_{wdm_i}$ for binary one and zero receptions as

$$\sigma^2_{wdm_i} = \sigma^2_{sh_i-wdm} + \sigma^2_{s_i-sp} + \sigma^2_{sp-sp} + \sigma^2_{s_i-xt} + \sigma^2_{th}, \tag{10.50}$$

where $\sigma^2_{sh_i-wdm}, \sigma^2_{s_i-sp}, \sigma^2_{sp-sp}, \sigma^2_{s_i-xt}$, and σ^2_{th} represent the variances of the shot noise, signal-spontaneous (spontaneous noise implying ASE) beat noise, spontaneous-spontaneous beat noise, signal-crosstalk beat noise, and thermal noise, respectively. In $\sigma^2_{sh_i-wdm}$, $\sigma^2_{s_i-sp}$, and $\sigma^2_{s_i-xt}$, the suffix $i = 1$ or 0 for binary one or zero reception, respectively, as these variances are signal-dependent. We express these signal-dependent components of $\sigma^2_{wdm_i}$ as (see Appendix B)

$$\sigma^2_{sh_i-wdm} = 2qR_w(P^R_{sig_i} + P^R_{sp} + P^R_{xt})B_e \tag{10.51}$$

$$\sigma^2_{s_i-sp} = 4R^2_w P^R_{sig_i} P^R_{sp}\left(\frac{B_e}{B_o}\right) \tag{10.52}$$

$$\sigma^2_{s_i-xt} = 2\eta_p\eta_s R^2_w P^R_{sig_i} P^R_{xt}, \tag{10.53}$$

while the signal-independent spontaneous-spontaneous and thermal noise variances are given by

$$\sigma^2_{sp-sp} = \left(R_w P^R_{sp}\right)^2 \frac{B_e(2B_o - B_e)}{B^2_o} \tag{10.54}$$

$$\sigma^2_{th} = \frac{4k\theta B_e}{R}, \tag{10.55}$$

with R as the input resistance of the receiver preamplifier and θ representing the receiver temperature in °K (as in Eq. 10.17). Note that the crosstalk-crosstalk beat noise components are ignored as their contributions are much smaller than the contributions from the remaining noise components. Using the above noise variances and Eq. 10.27, we finally express the BER for a lightpath (for IM-DD transmission) at the destination receiver as

$$\mathrm{BER}^{wdm} = \frac{1}{2}\mathrm{erfc}\left[\frac{R_w\{P^R_{sig}(1) - P^R_{sig}(0)\}}{\sqrt{2}(\sigma_{wdm_1} + \sigma_{wdm_0})}\right] = \frac{1}{2}\mathrm{erfc}\left(\frac{Q^{wdm}}{\sqrt{2}}\right), \tag{10.56}$$

where $P^R_{sig}(1)$ and $P^R_{sig}(0)$ represent the received optical power for binary one and zero transmissions, respectively, and $P^R_{sig}(0) = \epsilon P^R_{sig}(1)$ with ϵ as the laser extinction factor in the transmitter, and $P^R_{sig} = \{P^R_{sig}(1) + P^R_{sig}(0)\}/2 = P^R_{sig}(1)(1+\epsilon)/2$. As mentioned earlier, for the single-wavelength transmission links, the impact of fiber dispersion can also be taken into account in this case by using the reduction factor α_D for the receiver eye-opening in the numerator of Q^{wdm} in Eq. 10.56.

10.2.2 Power and rise-time budgets

Any communication link set up through an optical network, be it a single-wavelength or WDM network, needs to satisfy two basic physical-layer constraints. One of them dictates that the received optical power at the destination node must be higher than what is required to achieve an acceptable BER (for digital transmission), preferably along with a *system margin* of a few dB to take care of unforeseen variations in source power and the losses/gains experienced in the en route devices. The other constraint is imposed on the pulse-spreading, which needs to be satisfied so that the pulses do not spread out beyond a given limit. In other words, the received pulses due to various dispersion mechanisms in fiber and bandlimiting electronic circuits on both ends must not have a large *rise time* (or fall time), such that the decision circuit can *sample* the full amplitude of the received signal pulses during the respective bit intervals. These two constraints are popularly known as *power* and *rise-time budgets* in optical communication links, which are also collectively called the link budget, ensuring the needed QoT for the point-to-point communication links in an optical network. As we shall see soon, the rise-time budget, if not adhered to, may also have a deleterious effect on the power budget of the system.

Figure 10.4 illustrates the power and rise-time budgets for a binary optical communication link (a lightpath in WDM networks) using optical fibers, connectors, splices, and lossy passive devices used in the nodes and OAs, if any. As illustrated in Fig. 10.4(a), the power budget in the link can be formulated by a linear expression (using the power levels and gains/losses in dBm and dB scales, respectively) equating the transmit power P_T^{dBm} with the fiber/device losses L_i^{dB}'s and the amplifier gains G_j^{dB}'s through the link along with the needed received power P_R^{dB} and a system power margin P_M^{dB}. Thus, one can express the power budget as

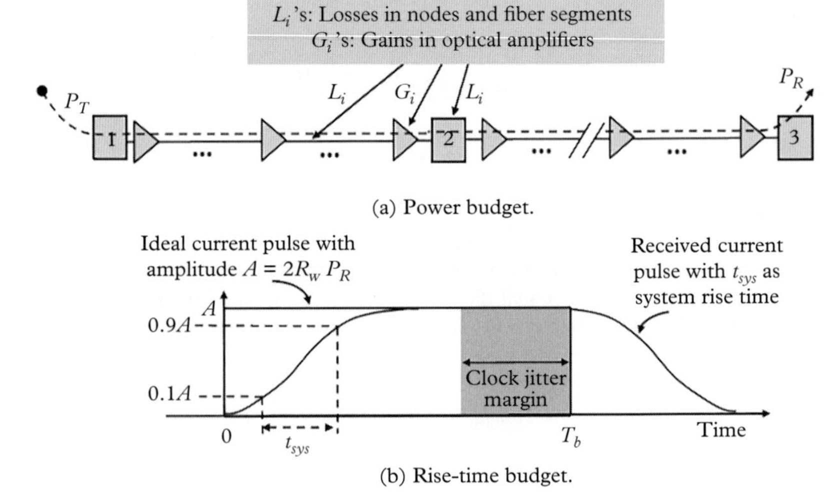

Figure 10.4 *Illustration of link budgets using a lightpath in a WRON traversing from node 1 to node 3 via node 2 with a transmit power P_T, which is received at node 3 with a power P_R. If P_R is the average received power, the amplitude of the received current pulse is given by $A = 2R_wP_R$. The received power P_R must be adequate to satisfy the required BER at the receiver, leading to the power budget. Also, the system rise time t_{sys} is the time taken by the received pulse amplitude to reach from 10% to 90% of the pulse amplitude A, and should be less than the bit interval T_b with a margin left out for clock jitter.*

$$P_T^{dBm} = \sum_i L_i^{dB} + \sum_j G_j^{dB} + P_M^{dB} + P_R^{dBm}, \tag{10.57}$$

where one needs to take into account the various noise components in the receiver (as described earlier) while determining the value for P_R^{dBm} for a specified receiver BER.

The rise-time budget in a link, as shown in Fig. 10.4(b), needs to consider all the underlying phenomena contributing to the system rise time t_{sys} (which also approximately equals the fall time) of the received electrical pulse at the input of the decision-making circuit in the receiver (see Fig. 2.48). As shown in Fig. 10.4(b), the system rise time is measured as the time interval during which the received pulse (at the input of the decision-making circuit, and hence in electrical domain) rises from 10% to 90% of its maximum value. Thus, at the decision-making point in the receiver, all kinds of dispersion-induced pulse-spreading components $t_D(i)$'s from the fiber dispersions along with the transmitter and receiver rise times, t_{tx} and t_{rx} respectively, manifest themselves together, leading to t_{sys}, where the constituent components are customarily added up on the root-mean-squared basis. Thus, one can obtain an estimate of t_{sys}, given by

$$t_{sys} = \sqrt{t_{tx}^2 + \sum_i t_D^2(i) + t_{rx}^2}, \tag{10.58}$$

where the transmitter rise time t_{tx} depends on the combined switching speed of the source and the associated driver circuit, dispersion-induced components $t_D(i)$'s depend on the type of fiber, link length, laser linewidth and bit/symbol rate, and the receiver rise time t_{rx} depends on the combined effect of the speed of photodiode and the electrical bandwidth of the front-end receiver circuit. The system rise time t_{sys} should not, in general, exceed ~70 % of the bit interval T_b, with the remaining part kept desirably *flat* to absorb the problem of timing jitter in the recovered clock. Note that if the system rise time exceeds the prescribed upper limit the receiver eye-opening factor α_D (as discussed earlier for Equations 10.27 and 10.56) might get reduced, thereby increasing the receiver BER. This will, in turn, necessitate a stronger control on the fiber dispersion mechanisms and the rise times of transceiver circuits and devices, along with increased transmit power, if necessary.

10.3 Impairment-aware design approaches

Design of an optical WDM network needs to consider a wide range of issues, right from the QoT in physical-layer to teletraffic performance (e,g., delay, throughput) at higher layers. Physical-layer centric issues would include the availability of quality network resources and network topology, along with the specified QoT (typically BER), that may be constrained by the limitations of the constituent non-ideal networking devices. Each network segment, such as access, metro, or long-haul network, uses different topological formations (e.g., tree, ring, mesh), along with varying transmission techniques and devices. For example, PONs using tree-based topology will pose different physical-layer challenges as compared to the ones faced in WRON-based metro rings and long-haul mesh networks. This will need separate physical-layer design considerations for different network segments to make the overall network function with the expected QoT. In the following, we consider some of the candidate optical network architectures and discuss their underlying impairment issues that might impact the QoT of the network significantly.

10.3.1 Impairment awareness in PONs

Currently, PONs are the most popular form of optical access networks, where there are some unique challenges in the physical layer. In particular, PONs, both for the single-wavelength (one wavelength in each direction) or WDM transmission, have some common design issues due to their basic physical topology (i.e., tree), while the WDM/TWDM PONs have some additional issues from their different possible configurations using AWG/WSS and OSPs. In the tree topology used in a PON, different ONUs are located mostly at different distances from the OLT. Thus the OLT receives packet bursts with varying power levels (different for different ONUs), as the transmitters of ONUs cannot be tailor-made with different transmit-power levels.

For example, if the receiver at the OLT is set with a high threshold for receiving high-power packets, the lower-power packets will be prone to high BER as many of its binary ones (contaminated by noise) might fall below the threshold. Therefore, an OLT receiver needs to dynamically learn and adjust the needed threshold for the packets with varying power levels, failing which the receiver BER may get significantly affected (increased). Note that, although at the user-end different ONUs might receive different power levels from the OLT, at a given ONU the received power level doesn't vary from burst (of packets) to burst. As discussed earlier in the chapters on PON, in order to address the problem of varying power level received from different ONUs, the receiver in the OLT makes use of a special type of receiver, called burst-mode receiver (BMR), which can dynamically change the threshold with varying power levels of the received packets, albeit with the help of some alternating preamble (PA) bits (i.e., 101010... pattern) attached as headers in the transmitted packets (Fig. 10.5).

Note that, even though the threshold values are estimated dynamically for each packet, the threshold estimates are bound to get contaminated by the receiver noise. Using the alternating PA bits (which also facilitates clock recovery), the receiver gets a reasonable estimate of the threshold, which can be made more accurate at the expense of longer PA field as the longer duration of the PA field helps in averaging out the receiver noise contaminating the threshold estimate. The statistical nature of the decision threshold with a finite length of PA field degrades the BER in a BMR, and the BER model in such receivers also gets more complex analytically. In the following, we describe an analytical method (Eldering 1993) for estimating the BER in a BMR in the presence of threshold uncertainty.

As mentioned above, in a BMR the threshold current I_{TH}, estimated from the PA, is a random variable. The threshold is assumed to follow a Gaussian statistic, with a mean threshold current $\bar{I}_{TH}$ and a variance σ^2_{TH}, given by (see Eq.10.26 for the mean threshold current)

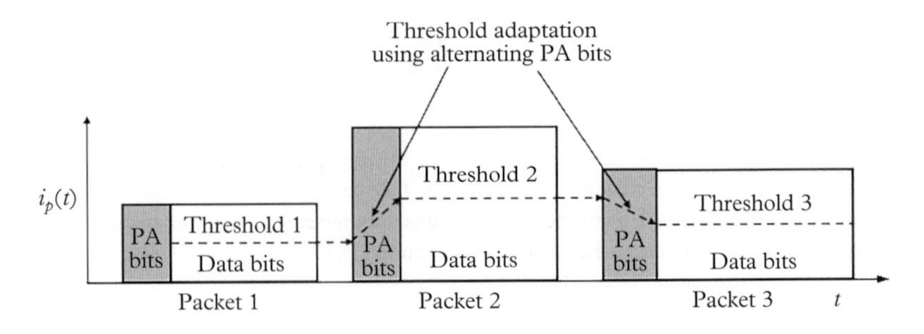

Figure 10.5 *Illustration of BMR operation in a PON. Three received packets of different optical power levels produce photocurrent $i_p(t)$ of different amplitudes at the receiver, thereby needing three different values of the decision threshold for the three packets. The BMR dynamically* learns *the correct threshold from the past threshold value by charging or discharging through the alternating bits of the preamble (PA) attached to each packet by using an adaptive electronic circuit.*

$$\bar{I}_{TH} = \frac{I_1\sigma_0 + I_0\sigma_1}{\sigma_1 + \sigma_0} + I_{BD} \tag{10.59}$$
$$\sigma_{TH}^2 = \frac{\sigma_1^2 + \sigma_0^2}{2n_{pa}},$$

where n_{pa} represents the length of the PA field in terms of the number of bits, implying that the variation of the threshold current reduces with the increase in the PA length. Therefore, the Gaussian distribution of the estimated threshold current I_{TH} is expressed as

$$p(I_{TH}) = \frac{1}{\sqrt{2\pi\sigma_{TH}^2}} \exp\left[\frac{-(I_{TH} - \bar{I}_{TH})^2}{2\sigma_{TH}^2}\right]. \tag{10.60}$$

In order to obtain the receiver BER in the presence of threshold uncertainty, one needs to average the conditional BER for a given arbitrary threshold I_{TH} over the statistics of the threshold current of Eq. 10.60. We therefore express the average BER in a BMR as

$$\text{BER}_{bmr} = \int_{-\infty}^{\infty} \text{BER}(I_{TH})p(I_{TH})dI_{TH}, \tag{10.61}$$

where the conditional BER(I_{TH}) for a given I_{TH} can be obtained from Eq. 10.24 (or its modified form when using OA in TWDM PON) to determine the final value of BER_{bmr}. Computation of BER_{bmr} from the above expression indicates that, for a given number of PA bits, the transmitter has to pay a power penalty, typically a few dB, (i.e., needs to increase the transmit power by the same amount), to ensure a BER close to that achieved in a normal receiver operating with fixed received power and decision threshold.

Furthermore, in WDM PONs using AWGs, ONUs encounter a nonuniform power distribution across the AWG output ports, typically following a Gaussian distribution. As a result, the central output port of AWG receives the strongest optical signal while the output ports toward the edges (on both sides of the central port) operate with weaker optical signal. Hence, the ONUs located at places nearer to the RN should be connected to the edge-ports, while the far-away ONUs should preferably be connected to the AWG output ports in the central region. Moreover, any output port of an AWG might encounter crosstalk from adjacent channels (incident on the adjacent output ports) owing to the angular spread of lightwaves over the adjacent output ports caused by the spectral spread of the lasers. In long-reach PONs, these problems may get compounded by the ASE noise coming from the OA used in the feeder segment. All these problems – the tapered power distribution, crosstalk across the AWG output ports, and the ASE noise from the OA – can be addressed together by combining the approach reported in (Ratnam *et al.* 2010) with the analytical model described earlier to evaluate BER for WDM networks in the presence of ASE noise. Further, a dynamic BER-aware heuristic scheme can be adapted, based on such an analytical model, so that whenever a new ONU is introduced, it is given access to an appropriate output port of the AWG (Shi *et al.* 2013) to ensure the best possible BER at the OLT receiver for the new ONU.

10.3.2 Impairment awareness in long-haul WRONs

As described earlier, in long-haul WRONs, lightpaths undergo a wide range of impairments, e.g., signal losses, various noise and crosstalk components along with pulse spreading due to fiber dispersion and transceiver rise times. The impact of all these impairments that determine the BER of a lightpath (reflected in Q^{wdm} of Eq. 10.56 along with α_D), can be included in the lightpath-provisioning algorithms in WRONs by making a prior *BER-check* (Ramamurthy *et al.* 1999a). In particular, one can choose a lightpath with a prior online estimate of BER and set up the lightpath if it satisfies the BER constraint, as shown schematically in the flow diagram of Fig. 10.6. Although this approach can increase connection blocking, the lightpaths that are allowed would be meeting the BER constraint offering smooth functioning of the physical layer with acceptable BER performance. Further, in order to get around the problem of increased BER, one can adopt a *translucent* transmission scheme, where the long lightpaths with higher BER are broken at appropriate intermediate nodes and 3R processing is employed therein through OEO conversion (Ramamurthy *et al.* 1999b), thereby refreshing the optical signals for acceptable BER at the intermediate and destination nodes.

In addition to the above impairments, WDM systems are also prone to the fiber nonlinearities due to the large aggregate power carried by the WDM channels in optical fibers, especially due to the FWM-induced crosstalk in long-haul networks (see Chapter 2). Hence, the design of such networks should also be aware of the impact of the FWM-induced impairments and the resource provisioning in the network needs to be made judiciously to keep the deleterious effects of fiber nonlinearities under control. We explain in the following some of the candidate methods to address the problem of FWM-induced crosstalk in WDM networks.

One possible scheme to completely get rid of FWM interferences is to make the channel spacings between consecutive channels in a WDM network different from each other, as indicated in Chapter 2 (Fig. 2.19(a)) (Forghieri *et al.* 1995). We describe this scheme in the following.

Consider that the frequency grid in a DWDM network allots the central frequencies of the WDM frequency slots (WDM channels) as $f_1, f_2, \ldots, f_i, \ldots, f_M$ in a frequency grid with a reference frequency f_0, such that

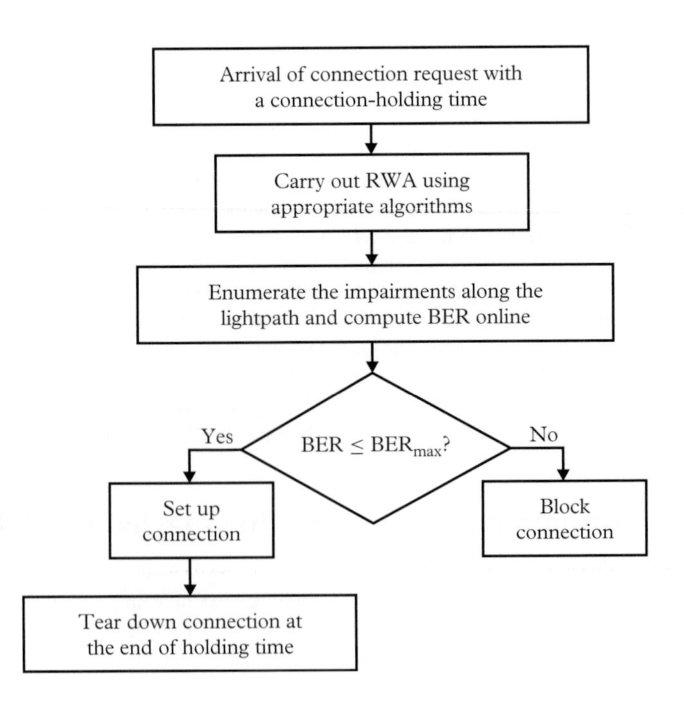

Figure 10.6 *Impairment-aware RWA scheme in WRONs.*

$$f_i = f_0 + n_i \Delta f, \tag{10.62}$$

where n_i is an integer $\in [1, M]$, representing the WDM slot or channel number in the WDM frequency grid with M frequency slots. As explained in Chapter 2, an FWM component generated from the channel frequencies f_i, f_j, and f_k would have the frequency f_{ijk}, given by

$$\begin{aligned} f_{ijk} &= f_i + f_j - f_k \\ &= (f_0 + n_i \Delta f) + (f_0 + n_j \Delta f) - (f_0 + n_k \Delta f) \\ &= f_o + (n_i + n_j - n_k)\Delta f \\ &= f_o + n_{ijk} \Delta f, \end{aligned} \tag{10.63}$$

with $n_{ijk} = n_i + n_j - n_k$. The above expression implies that, for any choice of i, j, and k, if n_{ijk} doesn't coincide with the index of any of the operational WDM channels in a link, then the resulting FWM component won't interfere with the parent (operational) WDM channels. This in turn will lead to the basic requirement that

$$n_{ijk} \notin (n_1, n_2, \cdots, n_M) \; \forall i, j, k \in [1, M], k \neq i, j, \tag{10.64}$$

which would imply that the channel separation between two channel frequencies must be unique. In other words, the separation for any pair of channels will differ from the separation of any other channel pair. The channel selection process from an M-channel WDM grid following above condition would eventually lead to an ILP problem, where the total bandwidth used in the fiber by all the assigned channels must be minimized so that the required bandwidth doesn't exceed the available transmission window of optical fibers. However, the ILP turns out to be computationally NP-complete, and one therefore needs to use suitable heuristic methods for the solution, e.g., Golomb ruler (Forghieri *et al.* 1995).

Note that, notwithstanding the capability of the above approach to completely avoid FWM interference in the system, the utilization of the fiber transmission window may get reduced significantly. Note that the separation Δf of two adjacent WDM channels needs to be $\geq \Delta f_{wdm}(min)$ (governed by the modulation bandwidth and laser linewidth, see Eq. B.58 in Appendix B). With this lower bound on channel separation along with the condition imposed by Eq. 10.64, the number of available channels within the fiber transmission window may fall short of the expected number of channels in a DWDM system.

Another practical means to get around the problem of FWM interference has been to control its magnitude rather than avoiding it completely, while keeping the channel spacings uniform (Adhya and Datta 2012). This approach would be much desirable, as any lightpath would get contaminated by other noise components, and it should be enough if the FWM-induced noise components don't shoot up beyond the aggregated noise components from the other physical phenomena. We describe this method in the following.

A close look at the FWM generation process in a WDM link reveals that the FWM frequencies are produced more in the central region of the spectrum, while at the edges on both sides there appear much fewer FWM components. This phenomenon is illustrated in Fig. 10.7 for a WDM system with equispaced WDM channels (optical frequencies), where the strength of the FWM components (i.e. total power of the FWM components generated at a given frequency – shown as an *envelope*) is much larger at the central channels. Hence, the lightpaths that suffer from lesser noise contributions from the shot and beat noise components, i.e., the shorter lightpaths, can withstand higher FWM interference (crosstalk) and can therefore be assigned the wavelengths in the central channels. On the other hand,

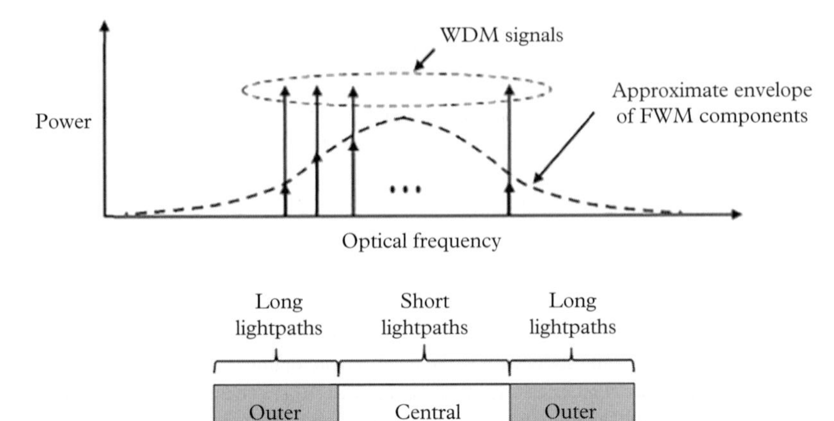

Figure 10.7 *Distribution of FWM components in a WDM system with equally spaced channel frequencies.*

Figure 10.8 *FWM-aware wavelength allocation in WRONs.*

the longer lightpaths, being contaminated by more noise and crosstalk components, can be assigned the wavelengths in the channels located at the two edges of the WDM window, so that the total noise power for the lightpaths in both parts of the spectrum doesn't differ much, thereby ensuring higher as well as minimal variation in the overall optical SNR of lightpaths across the entire network. The basic principle of this FWM-aware wavelength assignment (WA) scheme is illustrated in Fig. 10.8.

In order to execute the above WA scheme, the algorithm needs to know in advance the average length k_{av} (i.e., the number of fiber links) of the lightpaths, *which are yet to be set up*, so that the FWM-aware algorithm can identify a given lightpath as a long (length $\geq k_{av}$) or a short (length $< k_{av}$) one. This requirement leads to a *chicken-and-egg* problem, and one can get around this difficulty by first carrying out a *dry-run* of lightpath provisioning *without FWM awareness*, such as the MILP-VTD followed by the first-fit WA scheme, as described in Chapter 9. From the prior dry-run, an *approximate* estimate of the average length of lightpaths is obtained using the lengths of the preliminary lightpaths. Thereafter, one can run the same design process again, but by replacing the first-fit algorithm (or any other FWM-unaware WA scheme used in the dry run) with the FWM-aware wavelength assignment scheme, as discussed above. The approach is summarized as follows.

(1) FWM-unaware lightpath design is carried out (as a dry-run) using one of the methods described in Chapter 9 and the preliminary lightpath lengths are computed.

(2) Average lightpath length k_{av} is estimated from the above step.

(3) Lightpath design is carried out once again with the above FWM-aware wavelength assignment scheme using the estimated value of k_{av} from Step 2.

10.4 Power consumption in optical networks

The last few decades have witnessed outstanding growth in telecommunication networking in respect of its reach and capacity, with a plethora of networking technologies deciding the overall network architecture and cost. However, the unprecedented growth of today's Internet in terms of its geographical spread and bandwidth-intensive applications, has put an additional burden on the overall networking cost from the power/energy consumed in every segment of the network, along with the harmful carbon emissions. As per the ongoing global developments, the total power consumption in networks is expected to exceed the 10% mark of the 2010-global electricity supply, thereby calling for power-aware optical network designs. Though much progress has been made over the last three centuries (as mentioned in the beginning of this chapter), yet much more remains to be done in this regard, as our needs to communicate over the network have grown significantly and continue to grow with time. It is understood that today the power consumption in networks, or broadly in the area of information and communication technology (ICT), is relatively smaller than the total world-wide power consumptions in other areas ($\leq 10\%$), but in foreseeable future this fraction is going to increase, thereby needing the network designs to become far more power-aware. In this direction, extensive research is being carried out world over, exploring suitable power-efficient networking practices (Tucker 2011a; Tucker 2011b; Heddeghem *et al.* 2016; Kani 2013; Dixit *et al.* 2014; Valcarenghi *et al.* 2012) that would be able to carry out the task of *greening* the network, commensurate with the available power/energy.

Different network settings, including topological formations, network architectures, and device technologies, may lead to variations in the overall power consumptions in the various segments of optical networks – access, metro, long-haul – calling for segment-specific measures to enhance the overall power efficiency. For example, the long-haul backbones with mesh topology can enhance power efficiency by optimally reducing the routing/switching and transmission hardware used at the network nodes and employing *sleep-mode* options in these network elements (NEs) while remaining idle. Further, in PONs used in the access segment, substantial power-saving may be possible by employing sleep-mode operation in ONUs, as the major power consumption in a PON takes place in the large number of ONUs at the user-end, which also remain idle for some time interval in each DBA cycle. In fact, with the optical access networks increasingly penetrating into the society, one needs to make the design of PONs much power-efficient, more so for the ONUs used at the user-end. The available information (Dixit *et al.* 2014) indicates that the access networks consume about 80–90% of the total power consumed by today's Internet, and the ONUs consume 60% of the power consumed in PONs. In the following, we discuss the power-consumption issues and some of the schemes that can be adopted for power-efficient operations of optical networks.

10.4.1 Power-awareness in PONs

To make a PON power-efficient, one needs to explore the possible methods to save power in its OLT and all the ONUs. However, ONUs being large in number, the total power saving attainable in all the ONUs together far exceeds what one can achieve in the OLT, and thus most of the research on this topic has been carried out on making the ONUs more power-efficient.

Power-efficient operational modes in ONUs can be broadly categorized into two types: *adaptive link rate* (ALR) mode and *low-power modes* (LPMs). In ALR mode, standardized for GPON, the link rates from the ONUs are decided from a few given choices, according to the traffic flow from the user-end. The LPM-based solutions can adopt various possible schemes (suitable for EPONs and GPONs), where the hardware in ONUs are turned off to varying degrees. We describe some of the LPM-based schemes in the following.

LPM-based power-saving schemes in PONs

PONs can employ a variety of LPM-based schemes as follows.

- *Doze scheme*: In this scheme, when in low-power state, the transmitter in an ONU is fully turned off, but the receiver remains on. However, the transmitter remains in sync with the clock of the receiver, so that when the transmitter needs to be turned on, it can quickly become synchronized with the OLT from the recovered clock of the receiver, and resume upstream transmission.
- *Deep-sleep scheme*: In the low-power state of this scheme, the transmitter and the receiver in an ONU are completely turned off with substantial power saving. However, the ONU needs a long time to wake up and resume transmission (3–5 ms). Further, the ONU wakes up only after any activity-detection from the user-end, for which a nominal hardware remains on. This scheme becomes useful for low traffic with long inter-arrival times (typically, much exceeding the DBA cycle times). However, the sleeping ONU might lose downstream data, as it wakes up only from the activity-detection stimuli from the customer end or after the expiry of a locally set time limit. The problem of long wakeup time from deep sleep can be addressed by using an appropriate prediction algorithm, that can help in deciding the appropriate time to initiate the wakeup process in advance (Bhar *et al.* 2016).
- *Cyclic-sleep scheme*: In this scheme, the transmitter and receiver in an ONU go through cycles of sleep and wakeup periods. After waking up, if an ONU sees that it has to communicate, it resumes transmission/reception, or else it goes back to the sleep state for the transmitter and receiver.

The OLT and ONUs in a PON need to choose some of the above schemes to make the PON operate with LPM in a power-efficient manner. In general, the OLT comes to know from the ONUs about their capabilities and can negotiate with the ONUs for selecting an appropriate scheme for LPM operation through the control and management (C/M) plane messages of PON. In this process, the OLT and ONUs can jointly choose and switch back and forth between the worthwhile LPM schemes over time.

Next, we consider an example case from (Valcarenghi *et al.* 2012) to get an insight into the power-efficiency measures for PONs. First, we consider a TDM PON working with LPM using the deep-sleep scheme with a cycle time of T_C. Also consider that, within a given cycle time, the ONU under consideration was sleeping, but wakes up owing to some activity detection with a wakeup time overhead T_{oh} and thereafter resumes transmission and transmits for a duration T_{tx}, as granted by the OLT. With P_{ac} and P_{sl} as the power consumptions in an ONU during the active (including T_{tx} and T_{oh}) and sleeping ($T_C - T_{oh} - T_{tx}$) periods respectively, we express the energy consumptions E^{ac} and E^{sl}_{lpm} in the ONU operating without any power-saving measure and with the sleep-based LPM, respectively, as

$$E^{ac} = P_{ac}T_C \tag{10.65}$$

$$E^{sl}_{lpm} = P_{ac}(T_{oh} + T_{tx}) + P_{sl}(T_C - T_{oh} - T_{tx}). \tag{10.66}$$

If the PON has N nodes, then $T_{tx} \approx T_c/N$ with the assumption that the guard time between the data packets from the ONUs in a cycle time is negligibly small (i.e., $\ll T_{tx}$). Using this assumption and the above expressions, we express the overall power/energy saving ratio η_{pon} in the PON as

$$\begin{aligned}\eta_{pon} &= \frac{E^{ac} - E^{sl}_{lpm}}{E^{ac}} \\ &= \frac{P_{ac}T_C - \left[P_{ac}(T_{oh} + T_C/N) + P_{sl}(T_C - T_{oh} - T_C/N)\right]}{P_{ac}T_C} \\ &= 1 - \left(\frac{T_{oh}}{T_C} + \frac{1}{N}\right) - \frac{P_{sl}}{P_{ac}}\left(1 - \frac{T_{oh}}{T_C} - \frac{1}{N}\right) \\ &= \left(\frac{N-1}{N} - \rho_T\right)(1 - \rho_P),\end{aligned} \tag{10.67}$$

where $\rho_P = P_{sl}/P_{ac}$ and $\rho_T = T_{oh}/T_C$ represent the power-saving and time-saving ratios, respectively, implying that both factors play important roles in attaining power efficiency in a PON using a deep-sleep-based LPM scheme. As expected, the smaller the values of ρ_T and ρ_P, the larger the energy-saving efficiency, and the effects of both determine together the efficiency in a product form.

Figure 10.9 shows some representative plots of η_{pon} versus ρ_T for different values of ρ_P in a PON with $N = 20$. The plots indicate that a target value for η_{pon} can be achieved for different combinations of ρ_P and ρ_T, and one needs to see to what extent each of these two parameters can be adjusted for a given value of η_{pon}. For example, to have an η_{pon} of 40%, one can have $\rho_T = 20\%$ and $\rho_P = 47\%$ or $\rho_T = 40\%$ and $\rho_T = 22\%$, and the designer needs to make a choice for the option that makes it easier to realize in the ONU hardware. The other candidate schemes can be similarly modeled to estimate the values for the respective power saving ratios. Furthermore, one should note that, while operating in sleep-based schemes, the wakeup time would increase the delay experienced in the network (leading to a cross-layer issue), and hence the OLT needs to judiciously choose the pros and cons of the possible schemes while opting for a candidate power-saving option.

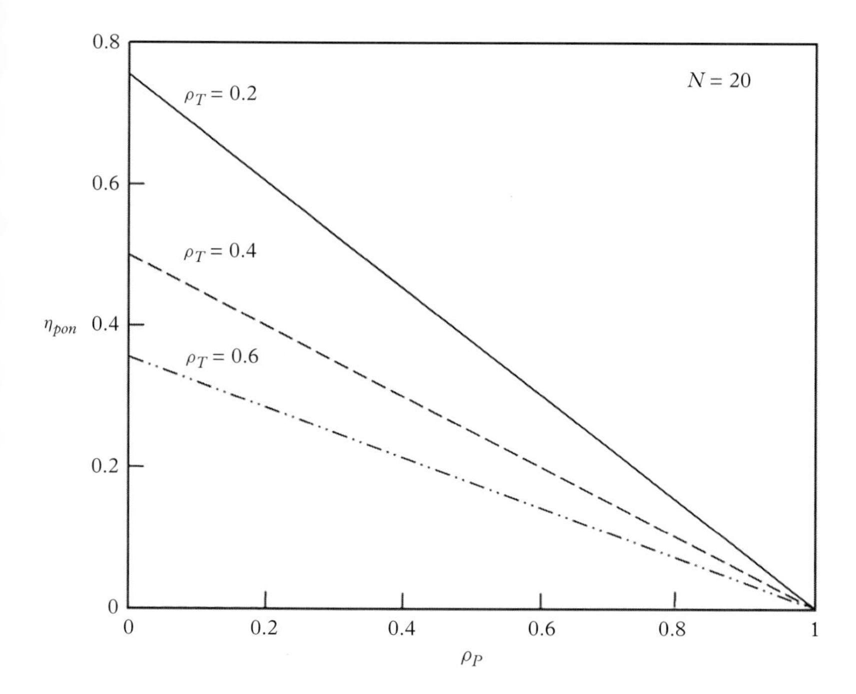

Figure 10.9 *Variation of power/energy efficiency η_{pon} in PON with power-saving ratio ρ_P for different values of time-saving ratio ρ_T.*

10.4.2 Power-awareness in long-haul WRONs

In order to get an insight into the power-consumption issues in the long-haul WRONs, we follow the model presented in one of the survey papers on this topic (Heddeghem *et al.* 2016). For this purpose, we illustrate in Fig. 10.10 a WRON-based long-haul network, indicating the usage of various network elements in a given lightpath.

The WRON in Fig. 10.10 shows an example lightpath-based connection L_{24} between the nodes 2 and 4 with the participating NEs along its route. These NEs include IP routers (IPRs), OXCs, and OAs and all of them contribute to the power consumption in the network for the shown lightpath-based connection. Note that the IPR at node 3 (an intermediate node for lightpath LP_{24}) does not participate in the passthrough operation of the lightpath. The average total power consumed by all such lightpaths is modeled by using the following network parameterts/variables:

- P_x: power consumed in an NE in a WRON: OXC, IP router, optical transceiver, and OA, represented by $P_{ox}, P_{ip}, P_{xr}, P_{oa}$, respectively,
- Γ: total traffic in the network in Gbps,
- $E[h_x]$: average hop count for a given connection between the NEs of type x operating in a specific network layer. For example, IPRs and OXCs would belong to different layers, thereby having different hop counts. Thus the hop count h_{ip} for IPRs would be just 1 for an end-to-end lightpath, while the hop count h_{ox} for OXCs for the same lightpath can be > 1 if the lightpath is optically bypassed by one or more intermediate nodes. However, a multihop connection with two concatenated lightpaths (say) will have $h_{ip} = 2$,
- Q_x: Power rating of a given NE of type x, given by its power consumption per unit bandwidth (precisely, transmission rate), i.e., p_x/r_x in watts/Gbps, with p_x as the transmitted power in watts and r_x as the transmission (bit) rate in Gbps,
- Q_x^T: Total power consumed by all the NEs of the same type x across the network. Note that Q_x represents the basic unit of power consumption in an NE of type x. Assuming a proportional dependence on Γ and $E[h_x]$ for the total power consumed Q_x^T, one can get its estimate by multiplying Q_x with Γ and $E[h_x]$ as

$$Q_x^T = Q_x \Gamma E[h_x]. \tag{10.68}$$

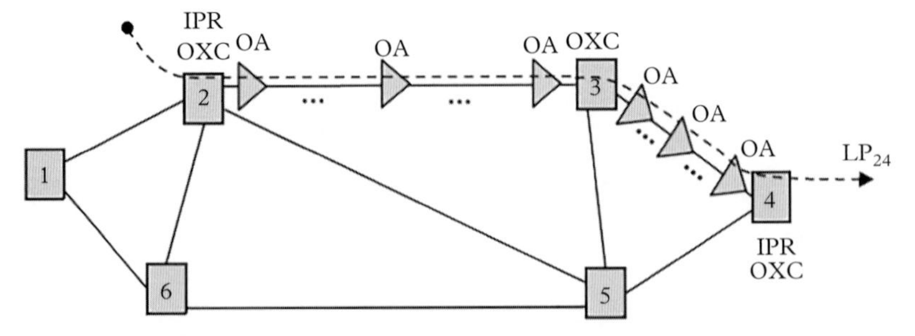

Figure 10.10 *Power-aware WRON model. Note that, while most NEs (OXCs, IPRs, and OAs) along the considered lightpath LP_{24} participate in its end-to-end transmission/routing, the IPR at node 3 doesn't participate in the passthrough operation of the same lightpath at node 3 and hence the same IPR (i.e., the IPR at node 3) is not shown explicitly for the functioning of lightpath LP_{24}.*

Furthermore, the variable Q_x^T needs to be multiplied by some appropriate multiplication factors, governed by the NEs' overhead, protection, and over-provisioning requirements. We describe these factors in the following.

- η_y^x: a multiplication factor for Q_x^T in an NE of type x accounting for the three different aspects (y's): its external overhead (e.g., cooling, power provisioning, heating of IC chips, etc.), failure protection, and over-provisioning, given by η_{eo}^x, η_{pr}^x, and η_{op}^x, respectively.

  In practice, η_{eo}^{ip}, η_{eo}^{ox}, η_{eo}^{xr} are $\simeq 2$ for external overhead, while $\eta_{eo}^{oa} = 1$, as most of the optical amplifiers are deployed outside the network node without any active cooling system. η_{pr}^x's are typically set at 2 in optical layer for full protection, i.e., $\eta_{pr}^{ox} = \eta_{pr}^{xr} = \eta_{pr}^{oa} = 2$. However, for IPRs, generally no such protection is used, implying that $\eta_{pr}^{ip} = 1$.
- α_x: a multiplication factor for Q_x^T accounting for the transceiver ports, in the type x of NE under consideration. In general $\alpha_x = 2$ for IPR, OXC, and optical transceiver, because for these NEs the transceiver capacity is necessary on both sides of the hop. However, α_{oa} for OAs in each hop would be equal to $2(\lfloor \frac{F}{\Delta F} + 0.5 \rfloor + 2)$, with ΔF being the separation between two successive OAs ($\simeq 80$ km, typically).
- W_x: a weighting factor for each type of NE to take care of the additional power consumption in the NE, besides the power required for optical transmission, switching, and amplification. The typical values of W_x are as follows (Heddeghem *et al.* 2016).
  - ◇ IPR: $W_{ip} = 10$,
  - ◇ OXC: $W_{ox} = 1$,
  - ◇ Transceiver: $W_{xr} = 10$,
  - ◇ OA: $W_{oa} = 2$.

Using the above observations, we express the total power consumption in all the NEs of a given type x across the entire network as

$$P_x = \eta_{eo}^x \eta_{pr}^x \eta_{op}^x Q_x^T \alpha_x = \eta_{eo}^x \eta_{pr}^x \eta_{op}^x \Gamma(E[h_x] + b_x) Q_x \alpha_x W_x, \tag{10.69}$$

where b_x is a binary variable; $b_{ip} = 1$, as an IPR, needs to connect client NE incurring additional power consumption, while $b_{ox} = b_{xr} = b_{oa} = 0$. Using these definitions, we express the power consumption in the entire network as

$$\begin{aligned} P_{net} &= \sum_x P_x = P_{ip} + P_{ox} + P_{xr} + P_{oa} \qquad (10.70) \\ &= \eta_{eo}^{ip} \eta_{pr}^{ip} \eta_{op}^{ip} \Gamma(E[h_{ip}] + 1) Q_{ip} \alpha_{ip} W_{ip} + \\ &\quad \eta_{eo}^{ox} \eta_{pr}^{ox} \eta_{op}^{ox} \Gamma E[h_{ox}] Q_{ox} \alpha_{ox} W_{ox} + \\ &\quad \eta_{eo}^{xr} \eta_{pr}^{xr} \eta_{op}^{xr} \Gamma E[h_{xr}] Q_{xr} \alpha_{xr} W_{xr} + \\ &\quad \eta_{eo}^{oa} \eta_{pr}^{oa} \eta_{op}^{oa} \Gamma E[h_{oa}] Q_{oa} \alpha_{oa} W_{oa} \\ &= \sum_x \underbrace{\eta_{eo}^x \eta_{pr}^x \eta_{op}^x \Gamma(E[h_x] + b_x) Q_x \alpha_x W_x}_{\text{4 terms: } x \in ip,ox,xr,oa}. \end{aligned}$$

It appears from Eq. 10.70 that, any candidate mechanism to reduce P_{net} must bring down some of the terms on its right-hand side, such as $\eta_{eo}^x, \eta_{pr}^x, \eta_{xr}^x, E[h_x], Q_x$, and α_x, by some improvement factor β_y^x (say). Thus, we express the improved value of P_x for a given type x of NE as P_x^{imp}, given by

$$P_x^{imp} = \frac{\eta_{eo}^x}{\beta_{eo}^x} \frac{\eta_{pr}^x}{\beta_{pr}^x} \frac{\eta_{op}^x}{\beta_{op}^x} \frac{\Gamma}{\beta_\Gamma^x} \left(\frac{E[h_x]}{\beta_h^x} + b_x \right) \frac{Q_x \alpha_x W_x}{\beta_Q^x}, \tag{10.71}$$

with all β_y^x's being ≥ 1 necessarily. However, the achievable values for β_y^x's for different cases (combinations of x and y) would differ from each other, depending on the means adopted for the respective improvement schemes. Next, by taking the ratio of P_x to P_x^{imp}, we obtain the comprehensive improvement factor for an NE of type x as

$$\beta_{comp}^x = \frac{P_x}{P_x^{imp}} = \beta_{eo}^x \beta_{pr}^x \beta_{op}^x \beta_\Gamma^x \frac{E[h_x] + b_x}{E[h_x]/\beta_h^x + b_x} \beta_Q^x. \tag{10.72}$$

Using the above results, we finally express the total power reduction factor β_{net} in the network as

$$\begin{aligned} \beta_{net} &= \frac{\sum_x P_x}{\sum_x P_x^{imp}} \qquad (10.73) \\ &= \frac{\sum_x \eta_{eo}^x \eta_{pr}^x \eta_{xr}^x \Gamma (E[h_x] + b_x) Q_x \alpha_x W_x}{\sum_x \frac{\eta_{eo}^x}{\beta_{eo}^x} \frac{\eta_{pr}^x}{\beta_{pr}^x} \frac{\eta_{op}^x}{\beta_{op}^x} \frac{\Gamma}{\beta_\Gamma^x} \left(\frac{E[h_x]}{\beta_h^x} + b_x \right) \frac{Q_x \alpha_x W_x}{\beta_Q^x}}. \end{aligned}$$

In light of the above model, we next discuss the various possible schemes that can reduce the power consumption in long-haul WRONs.

Power-saving schemes in long-haul WRONs

In particular, the candidate power-saving schemes should be able to enhance the values of some or all of the improvement factors β_y^x's (i.e., $\beta_{eo}^x, \beta_{pr}^x, \beta_{op}^x, \beta_\Gamma^x, \beta_h^x$, and β_Q^x) above unity, for the various network elements (NEs) in WRONs. We discuss these schemes briefly in the following.

Improvement factor for external overhead β_{eo}^x*:* The power consumption in an NE has typically an external overhead factor arising from the cooling system, and other associated overheads. The corresponding overhead factor β_{eo}^x can be reduced by using the cooling systems with improved power usage effectiveness (= the ratio of the total power consumed to the useful power consumed, and hence should be less for better performance), IC chips with high temperature-tolerance (so that less cooling is required), and efficient power provisioning for the NE. High-temperature operations of IC chips and hard discs without any failure are being explored for this purpose. Further, the uninterrupted power supply (UPS) units can be chosen judiciously to avoid light-loading, as these units operate with much lower efficiency with light loads, while offering considerably high efficiency ($\simeq 90\%$) near full load.

Improvement factor for protection mechanism β_{pr}^x*:* Protection of traffic against network failures is an essential part in any network to meet the SLAs, which is realized by using alternate arrangements causing additional energy consumption. Long-haul WRONs, as discussed in Chapter 11, employ dedicated or shared lightpaths for this purpose, implying about 50% utilization of resources (for dedicated protection), which in turn leads to the worst-case (maximum) value of the protection factor as $\beta_{pr}^x \simeq 2$ for all

NEs (i.e., for all x's). However, the protection factor can be reduced by using some discreet usage of the network resources as follows.

- *Sleep mode:* One of the useful means is to use the additional resources that give protection in sleep modes, where the NEs in the backup paths can go through the spells of *sleep-and-wakeup* (SAW) phases. This can be realized effectively by dissociating the backup paths from the primary paths and letting all NEs in the backup paths go through the SAW phases. Interventions of sleeping phases by the *wakeup calls* ensures that each backup path does not remain unused for a long time in the event of a failure that needs its service.
- *Power-aware routing:* One can also weigh between the shortest-path routes and the paths which may be somewhat longer but with lower energy consumption, leading to power-aware routing.
- *SLA-aware routing:* Another useful higher-layer perspective is to differentiate the connection setup algorithms for varying SLAs; in other words lightpaths with cheaper SLAs can be set up without protection, while the connections with costlier SLAs are set up with protected lightpaths.

Improvement factor for over-provisioning β^x_{op}: Network operators need to over-provision networks to absorb the occasional increase in traffic flows (traffic bursts) across the network above their average values and also to support traffic scale-up with time. Note that the additional resources used for this purpose are different from those used to set up backup paths to recover from the failures. Typically, the over-provisioning factor β^x_{op} is upper-bounded by 2, and can be reduced by sleep-mode operation. The sleep-mode operation in this case won't be needing wakeup calls (as in the SAW scheme), that are otherwise needed to address the unpredictable failure events on a *nearly-real-time* basis.

Improvement factor for network traffic β^x_Γ: Reduction in traffic volume can be realized by using data compression (i.e., source coding), which effectively reduces the redundancy in the incoming data stream, as in a multimedia signal. Since today's network traffic is dominated by already-compressed multimedia traffic, it becomes hard to further compress such data streams for long-haul WRONs. However, traffic flows in the long-haul WRONs can be reduced effectively by using caching at the source, destination, and intermediate proxy-servers. The content-distribution network (CDN) also provides effective reduction of traffic flows, by providing the requested data from a nearer server, thereby reducing the physical spans of the data flows.

Improvement factor for power-rating (i.e., power/rate ratio) in NEs β^x_Q : Improvement in the power/rate ratio Q_x is a fundamental aspect of the hardware design, involving better chassis design and utilization in the equipment, more power-efficient electronic design, and sleep-mode operations for all of its subsystems driven by static as well as dynamic sleep/wakeup functionalities.

Improvement factor for the number of hops β^x_h: Setting up single-hop connections by using end-to-end lightpaths definitely saves the electrical power consumption in IP routers at all intermediate nodes in a WRON. However, for long lightpaths the impairments suffered along the path might increase the BER above an acceptable level, thereby opting for OEO conversion at intermediate nodes, but with more power consumption. However, OEO conversions at intermediate nodes help in traffic-grooming, leading to a possible decrease in the number of wavelengths, which in turn can reduce the number of transceivers. Thus, a judicious design is needed to balance these conflicting issues.

Using the above framework and appropriate techniques for improving the multiplication factors η_y^x's, an estimate for the network-wide improvement factor β_{net} can be made. Without going into further details of the various specifications of the NEs and the individual multiplication factors, we summarize here the crux of the results reported in (Heddeghem *et al.* 2016).

The entire exercise to find the overall improvement factor β_{net} was carried out using two approaches: moderate-effort (ME) reduction, and best-effort (BE) reduction in power consumption. In the ME-based approach, the candidate solutions were sought with the approaches that are operationally feasible with the technologies available in practice (reference year being 2010), while the BE-based approach made use of challenging solutions which are expected to be feasible in course of time. The ME-based approach predicted an overall reduction factor of 2.3, which was expected to improve to ~6 in 10 years' time, driven by the dynamics of Moore's law in respect of the future energy-efficient CMOS technologies. However, the traffic growth in the next 10 years was predicted to be 10 times, thereby *overshadowing* all the predicted benefits through the ME-based method. However, the BE-based method predicted an improvement factor of 32 (64, with the futuristic CMOS-technology improvement), which was expected to beat the 10-times growth of power consumption due to the traffic scale-up in the 10-year time frame.

10.5 Summary

In this chapter, we dealt with two physical-layer issues in optical networks: transmission impairments and power consumption. Having described the underlying challenges from the transmission impairments and power consumption in optical networks, we presented analytical models to evaluate BER in single-wavelength and WDM transmission systems. In particular, we examined the BER for optical receivers operating under quantum limit as well as in the presence of various noise components encountered in single-wavelength systems. We extended this analysis to the lightpaths in WRONs with the accumulated noise and crosstalk components from the en route optical amplifiers and switches. Following the BER analysis, we presented the link design issues, including the system power budget constrained by the receiver BER, transmitted power, and losses/gains encountered in the link, and the rise-time budget constrained by the transmission rate, fiber dispersion components, and rise times of the transceivers used in the link.

Next, we described some impairment-aware design approaches for the PONs as well as long-haul WDM networks. For PONs, we examined the BER performance of the BMR-based optical receivers in the OLTs, leading to a framework to assess the interplay between the length of preamble field in the transmitted packets, receiver BER, and additional transmit power needed to combat the threshold uncertainty in BMRs. For the long-haul WRONs, making use of the BER model for the lightpaths, an impairment-aware lightpath setting-up methodology was described. For combating with the fiber nonlinearity from FWM, two approaches were presented, one avoiding FWM completely though with reduced spectrum utilization in optical fibers and the other controlling FWM interference within a tolerable limit, but without sacrificing the spectrum utilization.

Next, we described some power-aware design approaches for the PONs and long-haul WRONs. For PONs, we described various power-aware modes for ONU operation and presented an estimate of power-saving ratio in the ONUs using the LPM scheme based on the deep-sleep mechanism. Finally, for the long-haul WRONs, we presented a comprehensive approach to modeling the power consumption in the various network elements and discussed the possible techniques to reduce power consumption across the network.

EXERCISES

(10.1) How in practice is the power/energy-consumption efficiency defined in telecommunication networks? Discuss how the efficiency has evolved over the last three centuries. Also, discuss the conflicting roles between the ways to enhance power-consumption efficiency and spectrum utilization in the high-speed WDM networks.

(10.2) Indicate the major differences between the shot and thermal noise components in optical receivers. Write down the expression for the shot-noise variance in an APD-based optical receiver and discuss how one could optimize the receiver performance (in respect of SNR) in the presence of APD excess noise.

(10.3) Consider that a PIN-based binary optical receiver, operating at 10 Gbps with an incoming optical signal at 1.55 μm, offers a BER of 10^{-9} with a received optical power of −18 dBm. Determine how inferior it is (in dB) with respect to an equivalent receiver operating in quantum limit. Discuss how one can improve the BER of this receiver. (Given: Planck's constant = 6.63×10^{-34} joule-second).

(10.4) Consider an optical communication link operating between two cities, using WDM transmission over the C band of a single-mode fiber. If the link uses 32 wavelengths, each operating at 10 Gbps, calculate the effective transmission capacity (in Gbps) of the link. Estimate the maximum allowable value of chromatic dispersion for each wavelength, if the link length is 100 km and the laser linewidth is 100 MHz. Ignore polarization mode dispersion.

(10.5) A 1550 nm optical communication link transmits over 70 km of single-mode optical fiber at 10 Gbps without any optical regeneration. The laser at the transmitting end emits a power of 3 dBm and the fiber has a loss of 0.2 dB/km. The optical fiber link needs to splice multiple fiber spools, each having a length of 30 km. Each splicing incurs a splicing loss of 0.1 dB and each of the two connectors at the transmitting/receiving ends has a loss of 1 dB. If the receiver has a sensitivity of −18 dBm, estimate the system-power margin, if any.

(10.6) Consider the optical fiber communication link of the last exercise, and estimate its system rise-time. Comment on the result. Given: fiber dispersion = 8 ps/nm/km, transmitter rise time = 10 ps, laser linewidth = 100 MHz.

(10.7) Discuss the needs of preamble bits in the data bursts transmitted in a PON. Discuss the pros and cons of using long preamble fields in the data bursts in respect of the power penalty in BMRs, clock jitter, and bandwidth utilization.

(10.8) Consider a TDM PON with the maximum distance between the OLT and ONUs as 20 km, where the RN uses an optical splitter with a splitting ratio of 1:32, which is realized using 2×2 optical splitters as the building block. Assuming that the upstream transmission takes place at 1 Gbps, verify whether the BMR at the OLT would ensure a BER $\leq 10^{-9}$. Given: PIN responsivity = 0.8 A/w, receiver input resistance = 50 ohms, receiver temperature = 27°C, BMR power penalty = 2 dB,

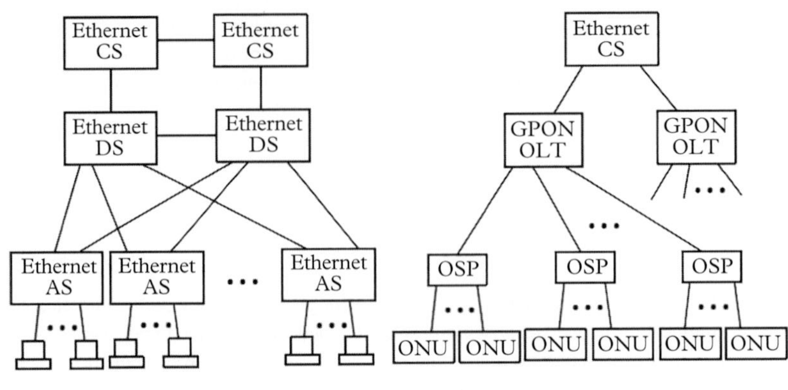

Figure 10.11 *Ethernet and GPON designs.*

transmit power from ONU = 5 dBm, laser extinction = 5%, insertion loss in 2 × 2 optical splitter = 0.3 dB.

(10.9) Consider a building complex, which needs to be provisioned either by Ethernet or GPON for about 2000 end-users. As shown in Fig. 10.11, the GPON design supports two OLTs connected to one core Ethernet switch from the top, and each OLT can support 40 GPON ports and each port can connect to 32 ONUs using OSPs. However, the Ethernet design employs three levels of switching: core, distribution, and access, with two core switches, two distribution switches and 16 access switches, with each access switch being able to support 150 switched ports. Thus, both designs employ a degree of fault tolerance at the switching levels, where the Ethernet design can accommodate at most $16 \times 150 = 2400$ end-users, and the GPON design can support at most $2 \times 40 \times 32 = 2560$ ONUs, thereby ensuring that both designs can serve close to but not less than 2000 end-users. The power consumptions in the respective network elements are as follows:

Ethernet core switch: 836 W,

Ethernet distribution switch: 725 W,

Ethernet access switch: 970 W,

GPON OLT: 2500 W,

GPON ONU: 10 W.

Calculate the power consumptions in both networks and explain, using the results, why GPON would be *preferred* as a green architecture by the service providers.

(10.10) A TDM PON operates with 32 ONUs, with each ONU operating in deep-sleep LPM to save power consumption. If the power-saving and time-saving ratios in the ONUs are 20% and 40%, respectively, determine the overall power-saving ratio in the PON. Repeat the calculation when the values of power-saving and time-saving ratios are reversed.

Survivability of Optical Networks

11

Today's telecommunication network, with the unprecedented dependence of the entire society on it, needs to deliver services that can survive against unpredictable failures across the network. Recovery from failures can be made using various possible means, broadly categorized in two types: protection and restoration schemes. Realization of these schemes varies for the different network segments: access, metro, and long-haul networks. The protection schemes are proactive in nature and need more resources, while assuring fast recovery from failure. However, the restoration schemes are reactive in nature, as in such schemes a network starts exploring the possible alternate connections after any failure takes place, and hence offer slower recovery while needing fewer resources. In this chapter, we present various protection and restoration schemes, as applicable to the respective network segments – SONET/SDH, PON and WDM/TWDM PON, WDM-based ring, and long-haul networks using wavelength-routing – and discuss the underlying issues for the implementation of these schemes.

11.1 Network survivability

Telecommunication networks offering a wide range of services to society must be reliable, and every attempt needs to be made to make it survivable against failures that may be encountered from various unforeseen situations. The survivability measures can be of different categories depending on the network SLAs, and hence different networks may have to employ different schemes to ensure that the network services are recovered within the committed recovery time following the occurrences of failures. In optical networks, failures can be of various types, ranging from simple link failures due to fiber-cuts, failures of active devices in a node, or even the failure of an entire node. When such failures take place, the network should be ready to recover promptly, so that the network services remain available with a specified percentage availability P_A, typically set at 99.999%, which in turn leads to a downtime of $(1 - P_A/100) \times 365 \times 24 \times 60 \simeq 5$ minutes in a year's time, not a trivial commitment indeed! There can also be large-scale disasters affecting one or more nodes and all the connected links, but these are beyond the scope of the present deliberation.

The most effective way to address network failures is to pre-configure protection or backup paths for the working connections, and switch over to the protection path soon after a working path fails. To address the problem of multiple failures occurring in the same region without much time gap, the service recovery for one failure needs to take place before the occurrence of the next failure. This leads to two more performance metrics concerning network survivability: *mean time to repair* a failure (τ_R) and *mean time between failures* (τ_F). Obviously, the former must be much smaller compared to the latter (i.e., $\tau_R/\tau_F \ll 1$) to avoid multiple failures at the same time. To make the task easier, it is customary to break down the network into multiple zones (i.e., satisfying the constraint $\tau_R/\tau_F \ll 1$ with increased τ_F in

Optical Networks. Debasish Datta, Oxford University Press (2021). © Debasish Datta.
DOI: 10.1093/oso/9780198834229.003.0011

smaller zones) and employ concurrent survival strategies in different zones, thereby helping to limit the number of failures needing attention at a given time.

Survivability of networks is generally addressed by different network layers in different ways, though they need to converge eventually in time. In respect of the physical layer of optical networks, survivability needs to be ensured in all hierarchical segments of the network, needing specific measures for the access, metro, and long-haul segments. Note that, these three segments are formed using different physical topologies; typically, long-haul optical networks use (incomplete) mesh topology, metro optical networks go for ring topology, and optical access networks in general employ tree topology. The mesh topology used in a long-haul backbone, being richly connected, can employ a variety of recovery schemes using backup paths for the protection of working paths against network failures with additional resources, or then can try to discover alternate paths following the occurrence of a failure. The ring topology in metro networks represents two-connected mesh topology and hence the recovery schemes in metro networks are more constrained as compared to those used in mesh-connected backbones. Finally, the tree topology or its variants used for the optical access segment, i.e., in PONs, is yet another variety, where the feeder fiber is the *weakest* or the most vulnerable part of the network, and needs some arrangement at least to protect against feeder cuts. Furthermore, in PONs one can also address the problems of failures in the OLT as well as in the distribution segment. Thus, the task of making an optical network survivable against failures needs to be accomplished for the different segments of the network using various recovery mechanisms.

11.2 Protection vs. restoration

For a prompt recovery after a failure, a network can employ some proactive mechanism, known as a *protection* scheme, which uses pre-configured protection or backup paths for the working connections. Although this scheme needs additional network resources, its pre-configured protection paths ensure that the networks recover fast and in a predictable manner. However, with limited additional resources, the network may not be able to set up pre-configured protection paths for recovery from failure and hence would try to *discover* alternate paths for the failed connections in a reactive manner. This type of recovery mechanism is called *restoration* scheme. A restoration scheme may require longer time for a failed connection to get re-established, and the network might even fail to set up the alternate path within the time limit specified in the SLA. Thus, the task of failure recovery is carried out in different ways for the protection and the restoration schemes, the former being pre-planned (i.e., proactive) offering faster recovery and the latter resorting to an exploratory (i.e., reactive) and hence slower approach. A network can also use a mix-and-match of both schemes for different types of connections having different SLAs.

Recovery time in an optical network is expected to scale with the number of nodes and area covered by the network. The access and metro networks, generally employing regular topologies (e.g., tree and ring), can use more readily the protection schemes, ensuring fast recovery time (typically $\leq$ 50 ms), while the long-haul backbone networks tend to use protection and/or restoration schemes, with recovery times for restoration schemes increasing at times to a few seconds or even minutes. Overall, as indicated earlier, the network services with the support of protection/restoration schemes for failure-recovery, should remain available with the prescribed availability constraint (i.e., typically with $P_A \geq$ 99.999%, amounting to a down time of $\leq$ 5 minutes a year) along with the criterion that $\tau_r/\tau_F \ll 1$.

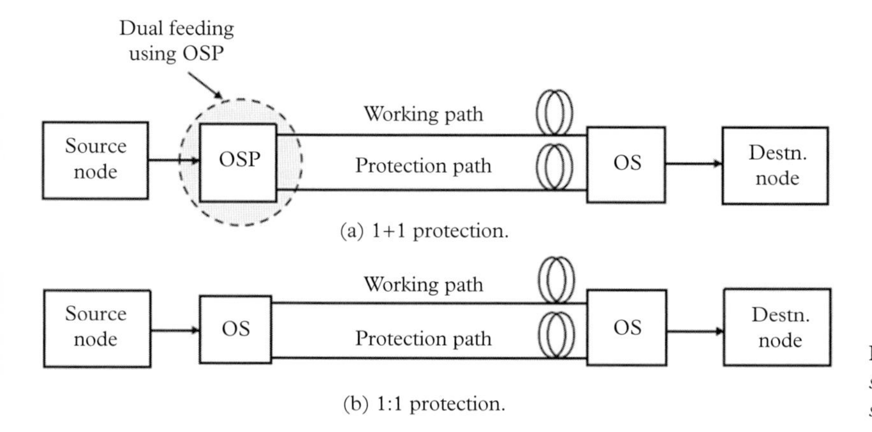

Figure 11.1 *1+1 and 1:1 protection schemes. OS: optical switch, OSP: optical splitter.*

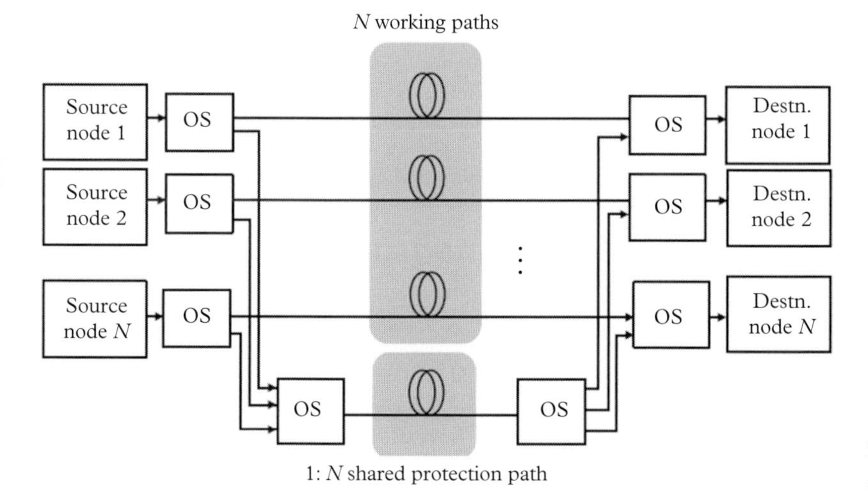

Figure 11.2 *1:N protection scheme with shared protection path. OS: optical switch.*

Protection-based recovery schemes can have different types of backup arrangements: 1+1, 1:1, and 1:N schemes, as shown in Figures 11.1 and 11.2. In Fig. 11.1, for a 1+1 dedicated protection scheme, as shown in part (a), a working path is set up along with a pre-configured protection path, and the source node keeps transmitting on both the paths by *dual-feeding* using an optical splitter (OSP). The receiving node through an optical switch (OS) listens to the working path, and when the working path fails, the OS in the receiving node switches to the protection path. As a result, the recovery process is very fast as it does not need any signaling protocol between the two nodes. Consequently, the data loss incurred remains minimal, except during the time interval that the receiver takes to detect the link failure and also the time needed to reconfigure the OS of the receiving node.

For 1:1 shared protection scheme, as shown in Fig. 11.1(b), the source node uses a switch to transmit only on primary path, and the backup path does not carry any signal when the primary path works, and hence the optical transmitter consumes less power as the power need not be split up into two paths (a 3 dB transmit-power advantage over 1+1 scheme). As and when the primary path fails in a 1:1 scheme, and the failure message reaches the source and destination nodes, the OS at the source end switches to the protection path and the source node starts transmitting through the same. The receiving node also reconfigures its OS to receive the signal sent over the protection path, and the overall recovery process needs more time as compared to the 1+1 scheme. Another advantage of 1:1 scheme (besides power saving) is that the protection path can be used for carrying low-priority data when not in use for protection. When the protection path is in use for protecting its original working path affected by a failure, the low-priority traffic flows get blocked. Another merit of this protection scheme is that a protection path can serve more than one working path in practice, as discussed in the following.

In particular, one can also opt for another scheme, called 1:N shared protection scheme (Fig. 11.2), where for N working paths, one protection path is provisioned, assuming that not more than one of the N working paths will fail at the same time. In this case also, the shared-protection path can carry low-priority data traffic. Furthermore, in the various protection schemes, one can switch the failed path over to an entirely different path with disjoint links and nodes; this type of protection is termed *path switching*. However, the protection path can also retain some of the unaffected links and get around only the failed link through some other link or links; this type of protection is called *line switching*.

As mentioned earlier, in restoration schemes the recovery process is exploratory in nature using an appropriate signaling scheme, and hence takes much longer time as compared to the protection schemes, which might lead to the possibility of data losses. Overall, the choice for a recovery scheme in a given network is governed by the cost of additional network resources and an upper limit for the recovery time in the event of a network failure.

11.3 Survivability measures in SONET/SDH networks

SONET/SDH is the oldest form of single-wavelength optical networking standard for metro and long-haul segments all over the world. In this section, we describe the survivability measures that have been used in SONET/SDH networks and have stood the test of time. In particular, we describe in the following different types of protection schemes used in single-wavelength SONET-based networks (hereafter, we discuss the protection schemes with reference to SONET only, as in Chapter 5).

With the two-connected mesh topology, protection of SONET rings can be arranged using two basic schemes, which were also adopted later for the SONET-over-WDM rings. In particular, the SONET-based networks operate as *self-healing* rings by using the following schemes:

- unidirectional path-switched ring (UPSR),
- bidirectional line-switched ring (BLSR).

These SONET protection schemes are required to recover from the failure in $\simeq$ 50 ms, including the failure detection and recovery phases. The basic difference between the UPSR and BLSR schemes lies in the mechanisms that these schemes employ for setting up duplex connections between the node pairs in a ring. The UPSR scheme uses the same direction along the ring for setting up the duplex connections (i.e., the *to-and-fro* connections remain unidirectional, thereby justifying its name), making one connection longer than the other, excepting the node pairs (each node in a SONET ring being an ADM), which are equidistant on the two sides of the ring. However, the BLSR schemes set up duplex connections bidirectionally between a node pair on the shortest path and hence on the same segment of the ring, thereby having no differential propagation delay between the two connections for a node pair. Note that, when a link/node failure takes place in a ring, more than one connection might be torn down and the network needs to concurrently employ recovery processes for all of them. We describe below the UPSR and BLSR schemes in further detail, along with a useful mechanism to realize survivable interconnection between two adjacent SONET rings.

11.3.1 Unidirectional path-switched ring

In the UPSR scheme, SONET employs two fibers carrying optical signals in opposite directions: clockwise (CW) and counter-clockwise (CCW). One of the two rings (say, the CW ring) employs working or primary paths for all connections (TDM slots), while the other ring (CCW ring) provides protection or backup paths for all the working connections. Thus, working paths always flow only in the CW direction, and protection paths only in the CCW direction. Figure 11.3 illustrates the UPSR scheme, which operates on the path layer of SONET, implying that only the terminal nodes of a given path participate in setting up the protection path when a link/node failure takes place.

Consider the connections between nodes 5 and 1 in Fig. 11.3. The working path from node 5 to node 1 takes one TDM slot on the shorter segment on the CW ring, while the working path from node 1 to node 5 takes a TDM slot on the longer segment on the CW ring itself, thus making the connections unidirectional. The corresponding pair of protection paths are concurrently set up on the CCW ring, and thus again being unidirectional as

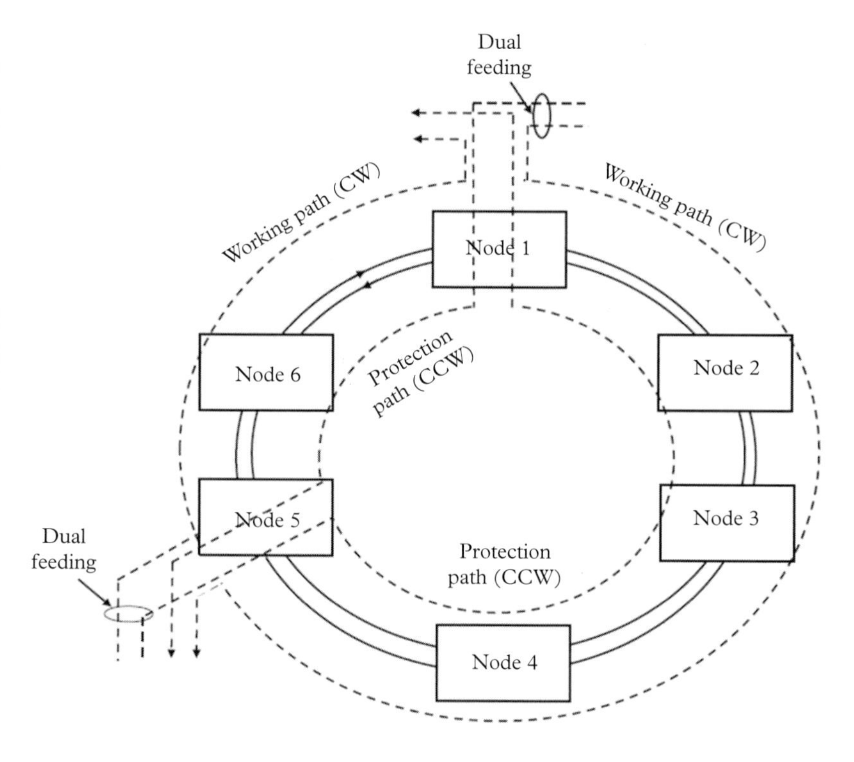

Figure 11.3 *UPSR scheme in SONET. Note that the nodes in SONET rings use ADMs, and the working and protection paths are* dual-fed *from the source node in the CW and CCW rings by using an optical splitter (not shown) for 1+1 protection. However, depending on whether there has been a failure along the working path, the receiving node chooses one of the two signals (from the working or the protection path) by using an optical switch (not shown).*

the working paths, albeit in the opposite direction. In order to examine the protection mechanism, consider the connection from node 5 to node 1, with node 5 *dual-feeding* the transmit signal using an optical splitter (not shown) into the working and protection fibers in two different directions to reach node 1. Node 1 keeps receiving signal from node 5 through the working path (CW) as well as from the protection (CCW) path. In a normal situation (i.e., in absence of any failure), the signal coming through the working path is selected by node 1 using an optical switch. However, if any link/node failure takes place in the working path, the signal from the working path doesn't arrive at node 1. Having detected the loss of signal (or a weak signal), node 1 reconfigures its optical switch (not shown) to receive the signal from the protection path, leading to a fast 1+1 protection against link/node failure.

Though the UPSR scheme is fast, it makes inefficient use of bandwidth available in the fiber rings while using path switching for the failed connection. In particular, in the UPSR scheme a duplex connection consumes the required bandwidth (TDM slot) all over the ring. Moreover, as mentioned earlier, the two-way connections will have differential propagation delay between the two directions of transmission. However, the nodes are simple, where only the receiver at the destination node needs to respond in a 1+1 protection scheme for link/node failures. In a voice-dominated metro ring connected down to the access segment, most connections are between one-and-many in nature, as the CO-node needs to be reached by most of the other nodes over the ring. In such networks, UPSR offers a simple and useful solution.

11.3.2 Bidirectional line-switched ring

There are two BLSR schemes: BLSR2 and BLSR4. Both BLSR schemes carry out 1:1 protection by switching the affected point-to-point link only (and hence in line layer), i.e., they don't switch the entire path of a failed connection as in UPSR, thereby needing a complex signaling scheme as compared to UPSR. However, the BLSR schemes make better use of the fiber bandwidth and hence are useful in SONET rings supporting high transmission speeds, network traffic, and size.

The BLSR2 scheme employs two fiber rings in CW and CCW directions, and in both fibers 50% capacity is kept reserved for protection purpose. As shown in Fig. 11.4, although using two fibers in two directions, BLSR2 *logically* allows four separate traffic flows. The working paths get 50% capacity of CW and CCW rings and the remaining 50% capacity of each ring is kept reserved for protection purpose, while the working and protection paths are not dual-fed from the source nodes, thereby realizing 1:1 protection. For a duplex connection between a node pair, unlike in UPSR, both (to-and-fro) working paths are set up through the shortest path on the CW and CCW rings, thus setting up the bidirectional traffic flows for the two connections on two rings in the same segment of the ring with zero differential propagation delay (unlike UPSR). When a link fails, BLSR2 restores the affected connections by using the protection capacity in the two fibers on the opposite segment (i.e., opposite to the failure-affected working segment) of both rings, and hence it is referred to as *ring switching*. Further, as the bidirectional connections are set up on the same shortest path and ring segment, BLSR2 can reuse the bandwidth on the two rings in the remaining segment as none of the duplex connections engage the needed TDM slot over the entire ring.

The BLSR4 scheme uses four fibers, as shown in Fig. 11.5. The two outermost rings are used for the working paths, with one employing CW and the other employing CCW transmission. Using the two rings (CW and CCW), to-and-fro working paths for a duplex connection are set up along the shortest paths on the same segment of the ring, thus doing away with the differential delay experienced in UPSR. This feature of BLSR4 is similar to BLSR2, albeit with full capacity of the fibers for the working and protection paths. As shown

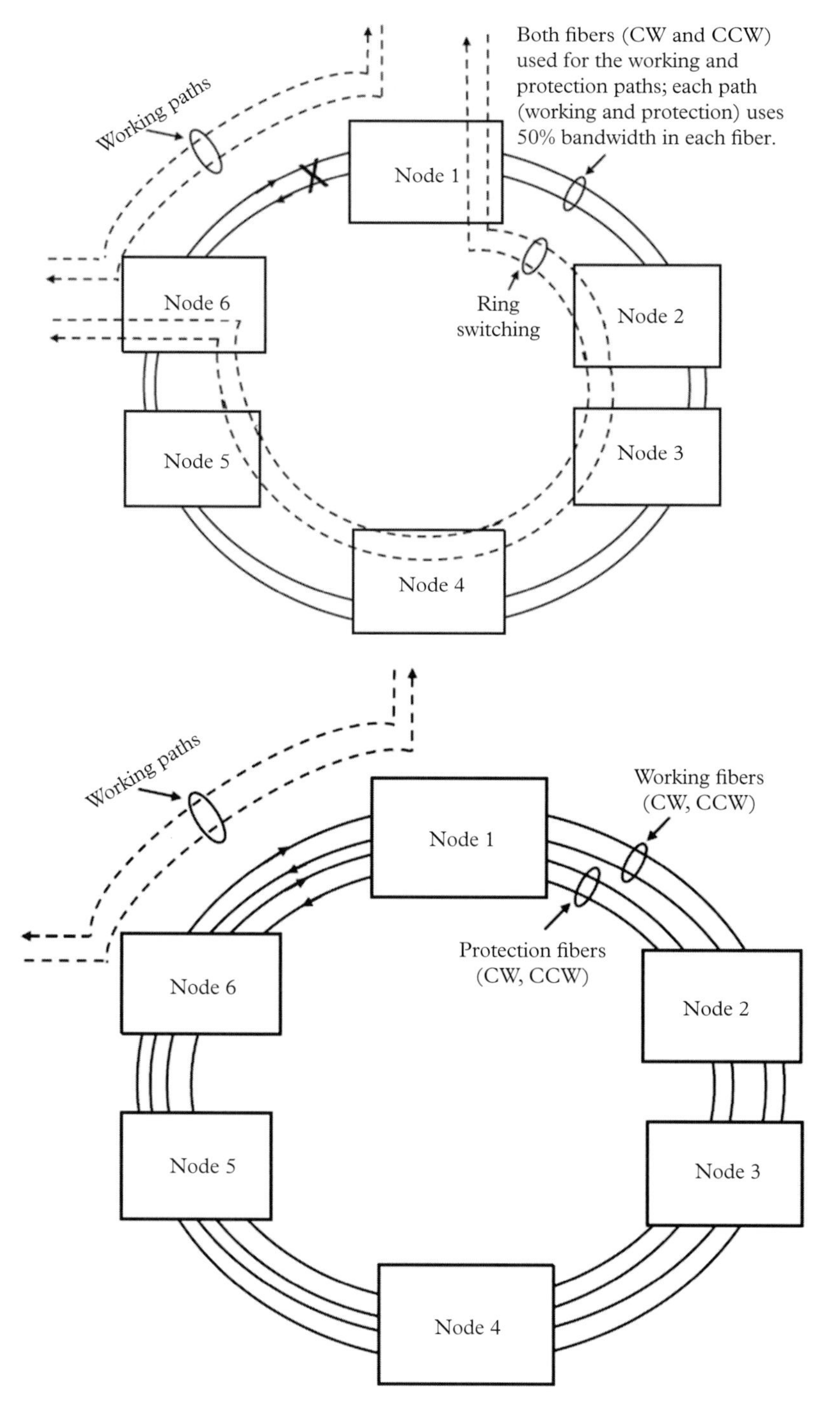

Figure 11.4 *BLSR2 protection scheme in SONET, where two physical rings operate as four* logical *rings, with each physical ring accommodating the working as well as protection traffic flows on a 50:50 capacity-sharing basis.*

Figure 11.5 *BLSR4 scheme in SONET.*

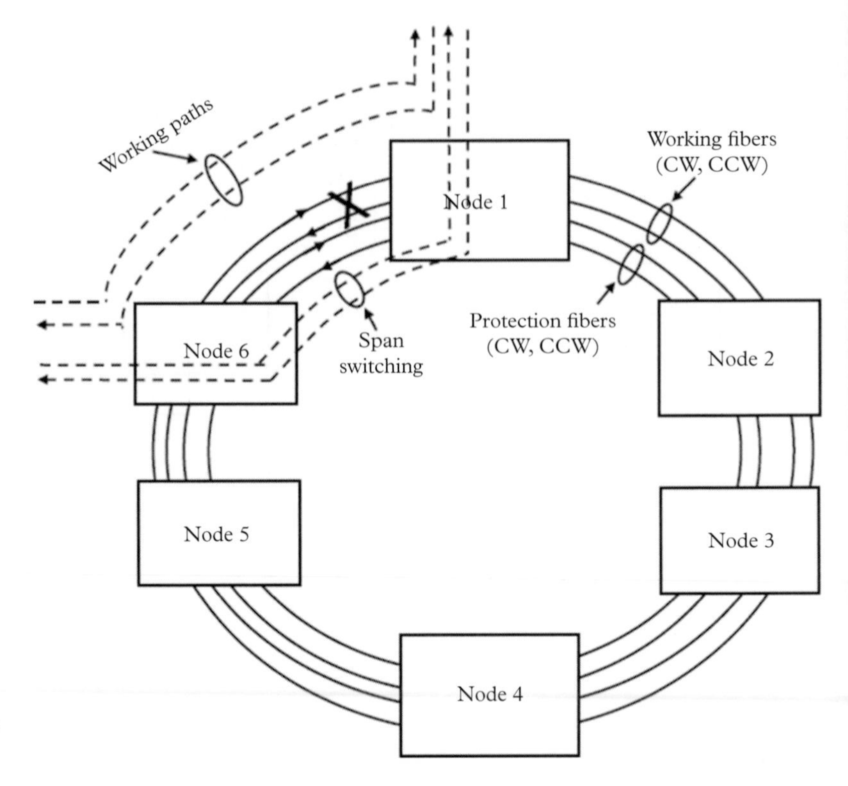

Figure 11.6 *BLSR4 protection scheme in SONET using span switching.*

in Fig. 11.5, from node 6 to node 1, the working path is set up on the CW ring using the shortest path, while from node 1 to node 6, the working path is set up on the next inner ring with CCW transmission also using the shortest path and hence on the same segment of the ring. As a result, besides zero differential delay between the to-and-fro connections, the remaining segments of both rings are available for setting up other connections (as in BLSR2), leading to high bandwidth utilization in the two working fiber rings. The other two rings in BLSR4 are provisioned for protection, and again one employs CW and the other employs CCW transmission, thereby offering similar benefits as in the working fibers.

In the event of link failure in the working fibers (i.e., in the outermost rings), the protection paths are set up on the protection fibers with the CW-based working connection moved to CW-based protection ring, and CCW-based working connection moved to CCW-based protection ring on the same segment of the ring. This type of BLSR4 protection is called as *span switching*, as shown in Fig. 11.6. However, if all the four links fail (e.g., fiber-bundle cut) at a point, then the protection fibers in the opposite segment of the ring are used to set up the protection paths (Fig. 11.7), and this type of BLSR4 scheme is referred to as ring switching, as in BLSR2, but with the full capacity of fibers being available for the working/protection paths.

So far, we have discussed BLSR schemes, dealing with link failures, typically due to fiber cuts. However, the situation becomes more complex in BLSR, when a node fails (though less likely, in general) in a ring, as each node on two sides of the failed node *presumes* that the link between itself and the failed node has undergone a link failure. In this situation,

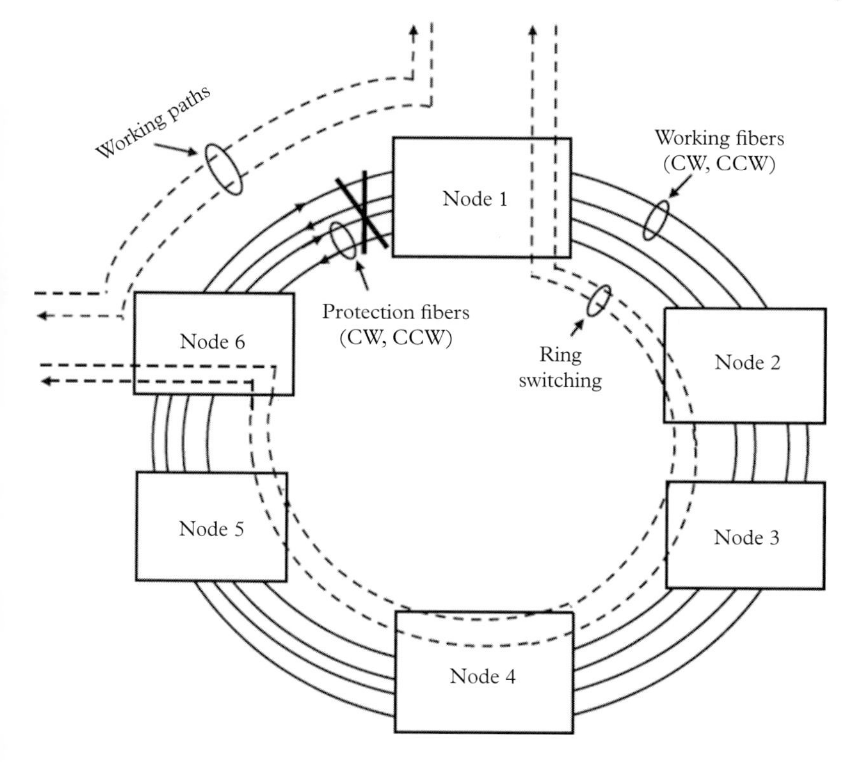

Figure 11.7 *BLSR4 protection scheme in SONET using ring switching.*

the failure detection process takes longer due to the need for a time-consuming signaling process to resolve the issue, and eventually the failure recovery is realized by keeping the failed node isolated from rest of the network using *loopback switching* carried out at the two neighboring nodes.

11.3.3 Survivable interconnection of SONET rings

When two SONET rings are interconnected, the interconnection between them can be made survivable by engaging two nodes from each ring to set up the interconnection. With an interconnection using one node from each ring, the network becomes vulnerable when one of the two nodes or the interconnecting link fails. Figure 11.8 illustrates a survivable interconnection scheme of two rings involving four participating nodes for the needed interconnection, so that from each of the two interconnecting nodes in a given ring, there are always two disjoint paths (i.e., dual access) to reach the other ring. As shown in the figure, a working path WP from node 1 of ring A to node 10 of ring B gets split into two working paths, WP1 and WP2, at node 4 of ring A using a *drop-and-continue* operation, which makes the working path survivable even if one of the two interconnecting links between the rings A and B fails. In similar manner, the protection path for WP from node 1 to node 10 needs to be set up also, which should be link-disjoint with WP in ring A and node-disjoint with the node that performs the drop-and-continue operation in ring A (i.e., not involving node 4 for drop-and-continue operation). Thus, the required protection path for WP can be set up from node 1 through nodes 7, 6, and 5 in ring A, with node 5 performing the drop-and-continue operation (not shown).

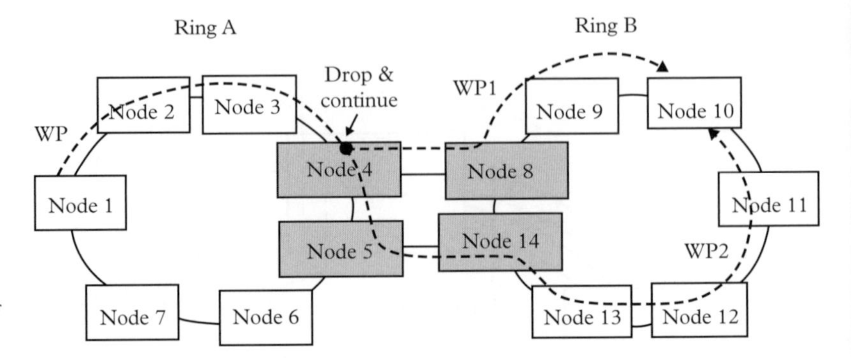

Figure 11.8 *Survivable interconnection of SONET rings.*

Similar interconnection can also be realized for SONET-over-WDM rings, where the drop-and-continue mechanism at the inter-ring nodes can be implemented either in an all-optical manner by using OSPs or by using OEO conversion. However, the OEO-based drop-and-continue mechanism will permit wavelength reuse in the two rings, while the optical drop-and-continue operation, though offering all-optical connections across the two rings, might call for more wavelengths and restrict the power budget for the inter-ring lightpaths due to power-splitting.

11.4 Survivability measures in PONs

Survivable PON designs are in general based on 1:1 protection schemes, which can be categorized into three types: feeder-protection scheme using protection feeder fiber (type A), feeder-cum-OLT protection scheme using protection feeder fiber along with backup OLT (type B), and the third one extending the second scheme for a comprehensive protection scheme (type C), taking care also of the distribution segment (ITU-T GPON convergence 2005; Kantarci and Moufta 2012; Kim *et al.* 2003). Note that, we discussed two broad classes of PONs in Chapters 5 and 7. One of them is TDM PON, or simply PON, employing TDM with two wavelengths in two directions. The other class uses WDM or TDM-WDM (TWDM) transmissions in both directions to realize larger capacity and reach. In the following, we describe some of the useful protection schemes for PON, WDM PON, and TWDM PON.

11.4.1 PON protection schemes for feeder and OLT

Figure 11.9 presents the schematic diagrams for the first two protection schemes for PONs. As shown in the figure, type A uses a protection feeder fiber connecting the OLT and the RN (passive splitter/combiner) in parallel with the working feeder fiber. In a normal situation, an optical switch (OS) connects the OLT with the working feeder fiber. In the event of a cut in the working feeder fiber, the OS connects the OLT with the protection feeder fiber, bypassing the cut working feeder fiber. Type B has a backup OLT which can also be switched into operation, when necessary, by using another OS connected to both the OLTs. The other side of this OS is connected to the next OS on its right side supporting feeder fiber protection. All of these protection schemes are 1:1 in nature, as the OLT and the backup OLT don't employ dual-feeding into the working and protection fibers.

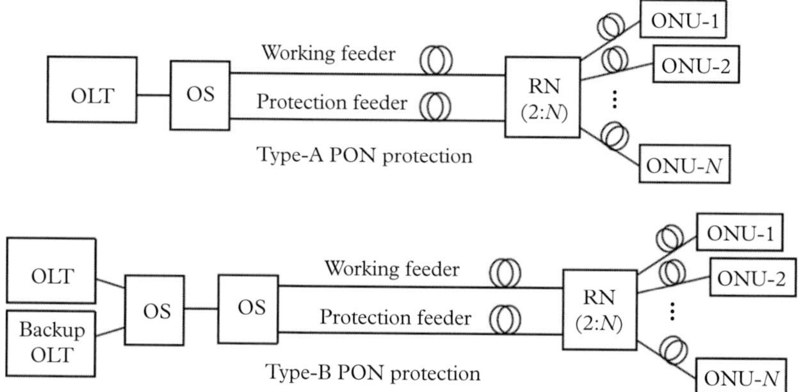

Figure 11.9 *TDM PON protection schemes: types A and B. Note that, for simplicity, the circulators needed to separate the incoming and outgoing signals at various points in a PON (see Chapters 4 and 7) are not shown explicitly in this diagram and the following ones on the PON protection schemes.*

Survivability of the above protection schemes for PONs can be assessed by using an analytical model, based on a parameter called *asymptotic unavailability* (or simply unavailability) associated with each network element used in a PON (Chen and Wosinska 2007). In particular, for all PON elements, such as OLT, feeder fiber, OSP, OS, distribution fibers, and ONUs (and WDM devices in WDM/TWDM PONs), the unavailability parameter U_i of an element i is defined as the product of its mean failure rate r_F^i and the mean time to repair τ_R^i, given by

$$U_i = r_F^i \tau_R^i = \frac{\tau_R^i}{\tau_F^i}, \tag{11.1}$$

where $\tau_F^i = 1/r_F^i$ represents the mean time between two consecutive failures for the PON element i. To get an insight into the unavailability parameters, consider the case of an optical fiber. Typically, one kilometer of optical fiber has a failure rate $r_F^{OF} = 570 \times 10^{-9}$ per hour (Chen and Wosinska 2007), implying that in a period of one year an optical fiber of length 1000 km is likely to fail ($r_F^{OF} \times 1000$ per hour) $\times$ (365×24 hours) $= (570 \times 10^{-9} \times 1000) \times (365 \times 24) = 4.993 \approx 5$ times. Further, for the optical fibers the mean time to repair is typically found to be $\tau_R^{OF} = 6$ hours, leading to the asymptotic unavailability of an optical fiber, given by

$$U_{OF} = r_F^{OF} \tau_R^{OF} = (570 \times 10^{-9}) \times 6 = 3.42 \times 10^{-6}/\text{km}. \tag{11.2}$$

Similarly, the values of U_i for the OLT, ONU, OSP, and OS (and AWG or WSS for WDM PON) are also available from the network providers. Using these unavailability parameters, we present the connection unavailability model for a PON in the following.

Considering the PON with feeder protection (i.e., Type A protection, see Fig. 11.9), we present a connection unavailability model in Fig. 11.10, where the unavailability parameter U_i of each element is represented by a box. These boxes are interconnected from the OLT on the left end to a given ONU on the right end by using a mix of series/parallel connections, depending on the protection mechanism for the respective devices. For example, with type A protection, U_{OLT} (for the OLT) and U_{OS} (for the OS following the OLT) are joined using a serial connection, while following the OS there appears a pair of parallel segments, with each arm of this pair having $U_i = U_{FF}$ (for the respective feeder fiber). Thereafter, the two U_{FF}'s are connected to the serially connected segment of U_{OSP}

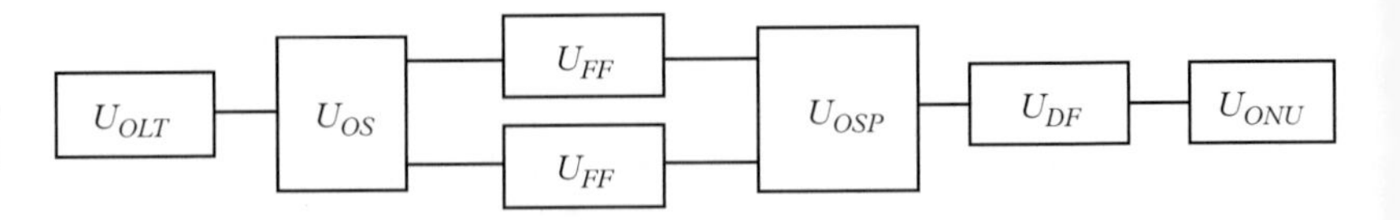

Figure 11.10 *Connection unavailability model for a TDM PON using type A protection.*

(for the RN), U_{DF} (for the distribution fiber) and U_{ONU} (for the ONU). Note that, for any serial connection, the unavailability parameters will add up, while for the segments in parallel the unavailability parameters will be multiplied. Using this framework, we express the connection unavailability U^A_{pon} of an ONU in a PON using Type A protection as

$$U^A_{pon} = U_{OLT} + U_{OS} + (U_{FF})^2 + U_{OSP} + U_{DF} + U_{ONU}, \tag{11.3}$$

which would increase for an unprotected PON with single feeder fiber as

$$U^{UP}_{pon} = U_{OLT} + U_{OS} + U_{FF} + U_{OSP} + U_{DF} + U_{ONU}. \tag{11.4}$$

In the above expressions, $U_{FF} = l_{FF}U_{OF}$ and $U_{DF} = l_{DF}U_{OF}$, with l_{FF} and l_{DF} as the lengths of the feeder fiber and the longest distribution fiber in kilometer. Using the above framework, one can readily obtain the connection unavailability U^y_{xPON} for any given PON configuration (i.e., PON, WDM PON, TWDM PON) with varying protection schemes, with x and y representing the type of PON and the protection scheme adopted therein, respectively (see Exercises 11.4 and 11.5).

11.4.2 Comprehensive PON protection scheme

For comprehensive protection (type C), one can have different schemes with varying complexity and performance (Kantarci and Moufta 2012; Kim *et al.* 2003; Nadarajah *et al.* 2006; Chen and Wosinska 2007). Figure 11.11 illustrates one candidate scheme for comprehensive protection of PONs from (Kantarci and Moufta 2012). This scheme is an extended version of a type B scheme, with the distribution segment of the PON also being protected using a *pair-wise* fail-safe operation between the ONUs. With N ONUs, the PON divides them into $N/2$ pairs (N being assumed to be an even number), the two members in each pair (of adjacent/neighboring ONUs) helping each other in the event of a fiber-cut in the distribution segment. For example, in the ONU pair 1 consisting of ONU-1 and ONU-2, if ONU-2 gets disconnected from the RN due to a cut in its distribution fiber, then ONU-1 helps out as its OSP sends a part of the received downstream signal to ONU-2 also. After the failure detection (from signal loss), the OS in ONU-2 gets configured to switch the downstream signal for ONU-2 (received by the OSP of ONU-1) to the transceiver (XCVR) of ONU-2, bypassing the failed distribution fiber. The upstream signal also follows the same route, but on a different wavelength with the OSPs operating as combiners in the reverse direction.

Some other schemes for distribution-segment protection have been reported in (Nadarajah *et al.* 2006; Kim *et al.* 2003). As proposed in (Nadarajah *et al.* 2006), between the ONUs a LAN is set up for inter-ONU communications by using two alternative schemes. One of them *emulates* the LAN between ONUs through the OLT, by involving both upstream and downstream communications using subcarriers. Any signal loss detected through this LAN at any ONU's LAN receiver helps in locating the distribution fiber that has been cut, which

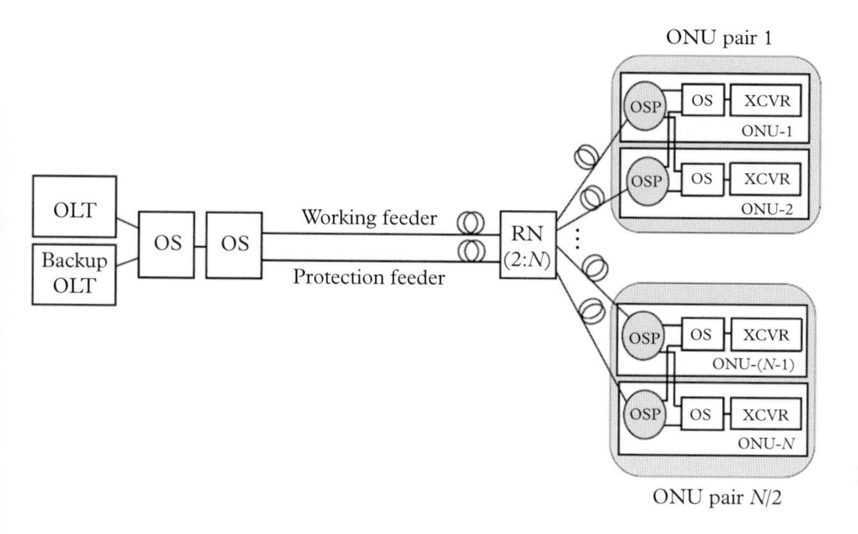

Figure 11.11 *An example of comprehensive TDM PON protection scheme.*

is followed up by establishing an alternate path. In the other scheme, the RN (a 1:N OSP as splitter/combiner) is replaced by an $(N+1) \times (N+1)$ passive star, with loopback fibers to each ONU. The loopback arrangement sets up the LAN, and helps protect the distribution segment.

11.4.3 Protection scheme for WDM PONs

Protection schemes for WDM PONs with tree-based topologies also use similar approaches as used in PONs (Chen and Wosinska 2007; Ruffini *et al.* 2012; Chan *et al.* 2003). However, WDM PONs being able to support a longer network reach with higher capacity, implementations of the respective protection schemes would have additional features and be desirably comprehensive, covering the OLT, feeder, and distribution segments. We describe a useful protection scheme for WDM PONs in the following.

An example protection scheme for WDM PON is shown in Fig. 11.12, where N ONUs are paired into $M = N/2$ groups as in the comprehensive scheme discussed earlier for PONs, but with some additional features (Chen and Wosinska 2007). As before, the feeder fiber and the OLT are kept protected against failure, with the two OSs along with a backup OLT as well as a protection feeder fiber. For WDM operation, the RN employs an AWG and M OSPs and the optical spectrum is split into two wavebands: waveband 1 and waveband 2. Each waveband is subdivided into two sub-bands: A and B in waveband 1, and C and D in waveband 2, wherefrom sub-bands A and C are selected for upstream transmission from ONU ends, and sub-bands B and D are used for downstream reception at ONUs. Each sub-band has M wavelengths, i.e., as many as the number of ONU pairs.

Each ONU pair in this scheme has an interconnection between them through two OSs and a fiber link. For the upstream traffic from an ONU pair, one of the wavelengths from sub-band A and one of the wavelengths from sub-band C are chosen for the two ONUs, with sub-band A serving the odd-numbered ONUs, and sub-band C serving the even-numbered ONUs. Similarly, for the downstream traffic, the wavelengths from the sub-bands B and D are chosen for the ONUs in each pair. An optical waveband filter (WBF) in each ONU in a pair selects from the downstream signal the appropriate wavelength (from sub-band B or

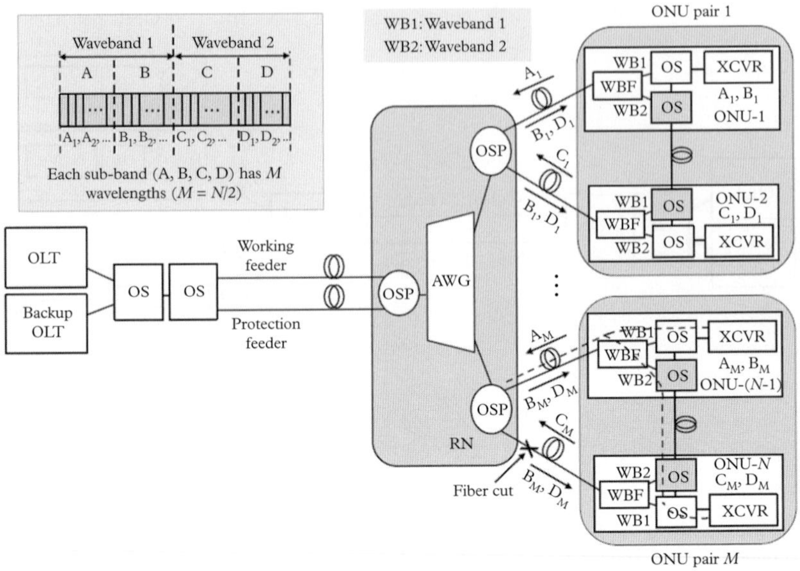

Figure 11.12 *Example of a comprehensive WDM PON protection scheme.*

D, with B for the odd-numbered ONU and D for the even-numbered ONU) to pass on the filtered signal through an OS to the XCVR. Similarly, for the upstream transmission, each WBF chooses the sub-band A or C using appropriate WBF ports facing the XCVR side of an ONU. Note that, there are two OSs in each ONU, and the OS directly connected to the XCVR is in white color, and the other OS facing toward the next ONU in the pair is shaded in gray color. In normal situation (i.e., without any failure in the distribution fiber), the OSs in white are used. The shaded OSs come into play when one of the distribution fibers for the same ONU pair gets cut.

The resilience to the failures in the feeder fiber (fiber cut) and the OLT is well understood from the previous discussions on the protection schemes for PONs. Therefore, next we examine the possible failures in the distribution fibers. Consider that the distribution fiber between RN and ONU-N has been cut. The WBF in ONU-($N-1$) would as usual receive both the incoming wavelengths B_M and D_M, and pass them on to the upper and the lower output ports, respectively, and the reception of D_M at ONU-($N-1$) will be useful for ONU-N when its distribution fiber gets cut. The wavelength B_M would go to the WBF port connected to the transceiver (XCVR) of ONU-($N-1$) through the OS in white, and the wavelength D_M would go to the WBF port connecting to the OS in gray which is connected to the gray-shaded OS of ONU-N via a local fiber link between the ONU pair. When the distribution fiber of ONU-N gets cut, the gray OS of ONU-N would configure itself to receive the signal coming from the gray OS of ONU-($N-1$) through the WBF of ONU-(N-1). Thereafter, the output of white OS in ONU-N would switch the incoming signal from its gray OS toward the XCVR of ONU-N for reception. The upstream signals from the two ONUs are transmitted at wavelength A_M (from ONU-($N-1$)) and at wavelength C_M (from ONU-N) and passed on to the OSP at the RN following the same path as followed by the received signals, albeit in the opposite direction. The dashed lines between the lowest OSP in the RN and the Mth ONU pair (Nth and ($N-1$)th ONUs) illustrates the signal paths when the distribution fiber connected to ONU-N gets cut.

11.4.4 Protection scheme for TWDM PONs

The protection scheme that we discuss here for TWDM PON using AWG-OSP configuration is partly similar to the one discussed for WDM PON. Figure 11.13 shows an example protection scheme for TWDM PONs (Chen and Wosinska 2007), where the features of disjoint working and protection feeders and the backup OLT are common with the last protection scheme discussed for WDM PON. There is also the interconnection between every pair of adjacent ONUs, but with some differences due to the TDM-based operation at the user-end. Unlike the previous configuration, each ONU in an ONU-pair uses one OSP (instead of WBF) at the input followed by one OS. The OSP in each ONU in an ONU pair is crossconnected with the OS in the other ONU of the same ONU pair for 1:1 protection of distribution-fiber failure. When a distribution fiber fails, the other distribution fiber, with the help of the OSP, offers connection to the isolated ONU owing to the failure (cut) of its distribution fiber.

For example, consider the Mth ONU pair (i.e., ONU-($N-1$) and ONU-N) in Fig. 11.13, and assume that the distribution fiber connected to ONU-N has been cut. In this situation, after failure detection, the 2:1 OS in ONU-N switches to its upper port looking up toward ONU-($N-1$), instead being configured toward the Nth distribution fiber (through the neighboring OSP). Thus, the signal meant for ONU-N arrives at the OSP in ONU-($N-1$) and moves down (along the dashed line) to the OS of ONU-N. Through this OS, ONU-N picks up the right TDM slot from the received signal for its XCVR for further processing. Similarly, the OS in ONU-($N-1$) also picks up its own TDM slot, and the XCVR of ONU-(N – 1) receives its own signal, as shown by another dashed line between the OS and the XCVR in ONU-($N-1$).

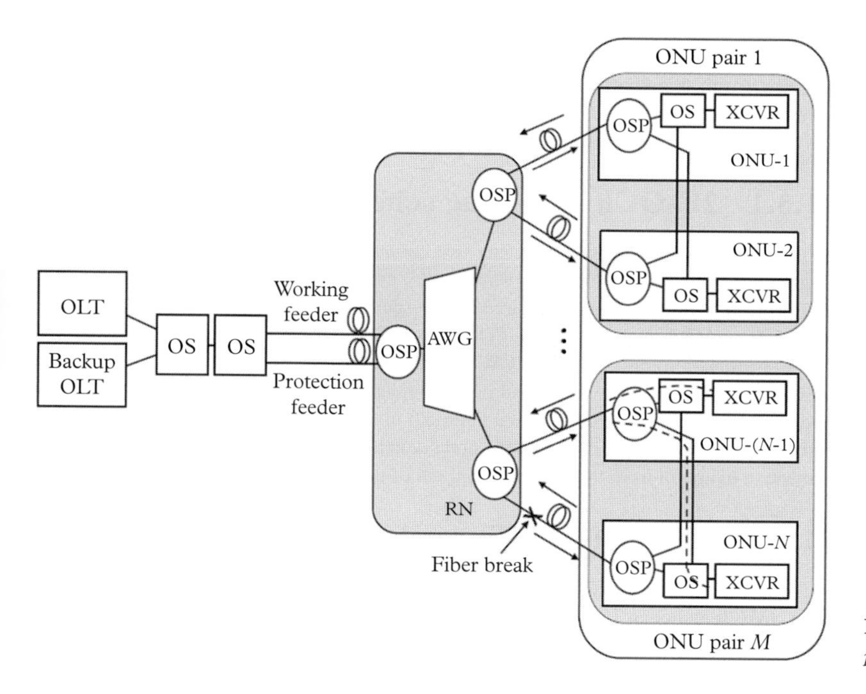

Figure 11.13 *Example of a TWDM PON protection scheme.*

11.5 Survivability measures in WDM metro networks

Survivability mechanisms in WDM ring networks make use of the basic concepts employed in the protection schemes for SONET using two or four fiber rings. However, instead of TDM slots in SONET, the granularity of traffic goes up to at least a wavelength, i.e., an optical channel (OCh), or an optical multiplex section (OMS), i.e., a bundle of wavelengths in a fiber. As in SONET, the protection schemes in WDM rings are also categorized in two basic types – dedicated and shared protection schemes – and each type can employ one of the multiple options as follows.

- *Dedicated protection* using two fiber rings, with two options:
  - two fiber rings with OCh switching (2F-OCh-DP-Ring);
  - two fiber rings with OMS switching (2F-OMS-DP-Ring).
- *Shared protection* using two or four fiber rings, each with two options:
  - two fiber rings with OCh switching (2F-OCh-SP-Ring);
  - two fiber rings with OMS switching (2F-OMS-SP-Ring);
  - four fiber rings with OCh switching (4F-OCh-SP-Ring);
  - four fiber rings with OMS switching (4F-OMS-SP-Ring).

OCh-based schemes of both types offer higher flexibility because protection of each wavelength (OCh) can be carried out independently, while the OMS-based schemes are required at times when an entire set of multiplexed wavelengths (OMS) needs to be switched for the protection of several lightpaths. In the following, we describe two of the OCh-based protection schemes, one using two fiber rings with dedicated protection and the other using four fibers with shared protection. With the knowledge of basic concepts used in these two schemes (along with the UPSR and BLSR schemes), the interested readers can learn about the other possible protection schemes from (Li *et al.* 2005).

11.5.1 2F-OCh-DP-Ring scheme

In the 2F-OCh-DP-Ring scheme using two counter-propagating (CW and CCW) fiber rings, the working and protection wavelengths or lightpaths are set up in the working and protection fibers in opposite directions, respectively. Thus, much like in UPSR (see Fig. 11.3), each lightpath (instead of TDM slots in SONET) from a source node is dual-fed (for 1+1 protection) into CW and CCW rings through its ROADM (instead of the ADM in a SONET node) for the working and protection lightpaths, respectively, and the receiving node uses an OS to select the working lightpath in normal condition. When a failure takes place in the working path, the receiving node switches to the protection fiber to receive the protection lightpath from the opposite segment of the ring.

11.5.2 4F-OCh-SP-Ring scheme

Figure 11.14 illustrates the 4F-OCh-SP-Ring scheme in a six-node WDM ring using ROADMs in its nodes. In this protection scheme, as in BLSR4, two counter-propagating fibers (CW and CCW) are allocated for the working lightpaths and the other two counter-propagating fibers are reserved for the protection lightpaths (without dual feeding), thereby offering 1:1 protection. With four fibers, the 4F-OCh-SP-Ring scheme can carry out span

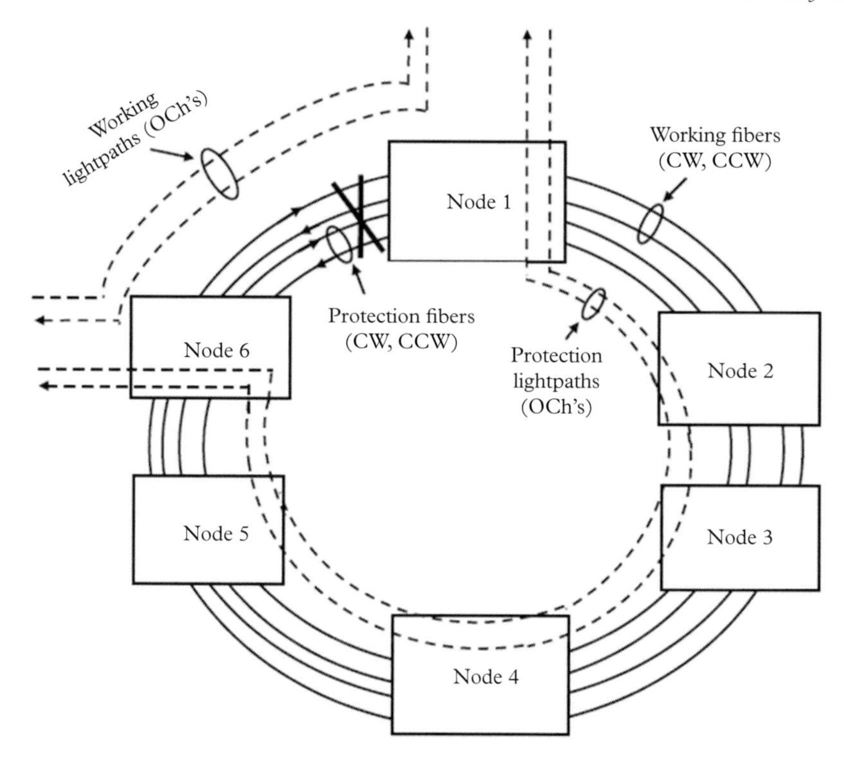

Figure 11.14 *Operation of a 4F-OCh-SP-Ring protection scheme in a six-node WDM ring. The illustration in the figure demonstrates the case of ring switching due to a fiber-bundle cut between the nodes 6 and 1.*

as well as ring switching for lightpaths (instead of TDM slots in BLSR4). As in BLSR4, with the failure in working fibers only, span switching takes place, while with the failure of all four fibers (i.e., fiber-bundle cut), the 4F-OCh-SP-Ring scheme needs to employ ring switching, as shown in Fig. 11.14.

Note that, in the case of WDM rings also, the recovery time needs to be within 50 ms as in SONET rings. For the earlier scheme, i.e., in 2F-OCh-DP-Ring, with 1+1 protection, the recovery is fast as in UPSR, while for the present scheme, i.e., in 4F-OCh-SP-Ring scheme with shared protection, the recovery time would be longer and needs to be kept under the acceptable limit, thereby limiting the size of the ring in terms of the coverage area and number of nodes. Nevertheless, the present protection scheme (as in BLSR schemes of SONET) will be free from the problem of differential delay in full-duplex connections, thus offering better QoS than 2F-OCh-DP-ring. Further, for interconnected SONET-over-WDM rings, one can extend the above schemes using inter-ring nodes as discussed earlier for the SONET rings (Fig. 11.8).

11.6 Survivability measures in long-haul WDM networks

The long-haul WDM networks are in general set up in the form of an incomplete mesh, so they do not conform to any regular topological configuration. Thus, the design of survivable long-haul mesh-configured WDM networks using wavelength routing (i.e., WRONs) is complex and has to use different approaches from those used in the access and metro segments. For employing protection-based failure recovery of all connections in such networks, every lightpath across the network should be provisioned with appropriate protection lightpaths using one of the candidate schemes discussed earlier: 1+1, 1:1, and 1:N schemes. One could also employ restoration-based recovery schemes with some nominal over-provisioning of network resources, but without keeping any resource reserved for the protection of the working lightpaths.

One can solve the associated design problems for the survivable mesh-configured WRONs using LP-based formulations or heuristic approaches, though the LP-based approaches may not be favored in large networks due to the computational complexity. The LP-based design methodologies for WDM LAN, metro, and long-haul WRONs have been dealt with in detail in the respective chapters in this book, and by extending those approaches one can also obtain appropriate designs in these cases. In this direction, there has been extensive research on protection and restoration schemes for mesh-configured WRONs employing LP (Caenegem *et al.* 1998; Ramamurthy and Mukherjee 1999a; Ramamurthy and Mukherjee 1999b; Sahasrabuddhe *et al.* 2002; Ramamurthy *et al.* 2003), which have also been used subsequently to validate various useful heuristic schemes. However, LP-based treatments for failure recovery are beyond the scope of this book, and the interested readers are referred to (Sahasrabuddhe *et al.* 2002; Ramamurthy *et al.* 2003). In the present deliberation, we describe some heuristic design methodologies for mesh-configured survivable WRONs. Before describing these designs, we present below some performance metrics, that will serve in assessing the designs of survivable WRONs and offer a broad perspective of the underlying issues.

11.6.1 Upper-bounds of traffic scale-up factor in survivable WRONs

A useful performance metric of a WRON design for a given traffic matrix is the allowable traffic scale-up factor α (> 1), by which the entire traffic matrix might increase over time. The upper-bound for α in a survivable WRON, is its maximum possible value α^{max}, called the upper-bound of the traffic scale-up factor (or simply upper-bound, in the present context) that could be supported without allowing the network traffic to exceed the overall network capacity. In the following, we use a model from (Sahasrabuddhe *et al.* 2002) to determine analytically the expressions for α^{max} in WRONs using different criteria. As and when a network encounters a failure, the permissible value of α^{max} gets more constrained due to the loss of network capacity (or resources) caused by the failure. Thus, while making the network survivable against failures, it becomes worthwhile to assess the value of α^{max} in a given WRON in the event of a link failure. Note that, as assumed implicitly so far, we consider in the following that a WRON in general experiences one link failure at a time, implying that the characteristics of failures in the network follow the criterion, given by $\tau_R/\tau_F \ll 1$ (Section 11.1).

Upper-bound from reduced cut-set due to link failure

One viewpoint to assess the value of the upper-bound α^{max} is to use the *cut-set* of the network graph when a link fails (i.e., we consider the case of single link failure, implying that $\tau_R/\tau_F \ll 1$). For this purpose, we represent the physical topology of a WRON by a graph $G = (V, E)$, as used in Chapter 9, with V representing the vertices or the nodes and E representing the edges or the fiber links. In G, a fiber cut will break several lightpaths, and all of them should be recovered by using the lightpaths kept reserved for protection (dedicated or shared) or IP-based restoration scheme. In order to find the upper-bound α^{max} using cut-set-based model, we first bisect the graph into two parts: V_1 and V_2, using a cut-set consisting of C fiber links. Assuming that p and q represent two nodes in V_1 and V_2, respectively, we express the total traffic between V_1 and V_2 across C links before the failure as

$$\Lambda_{V_1V_2} = \sum_{p\in V_1} \sum_{q\in V_2} \Lambda_{pq}, \tag{11.5}$$

which would flow through C links of the cut-set. In the above expression, Λ_{pq} represents the total traffic (in bits/second) from node p in V_1 to node q in V_2. Next, we presume that, among C links in the cut-set, one fiber link has failed due to fiber cut, leading to $(C-1)$ links to connect the two sets of nodes in V_1 and V_2. Hence, $\Lambda_{V_1V_2}$ would flow through $(C-1)$ fiber links after the failure with average congestion $\hat{\rho}_c$ across each link in the reduced cut-set, given by

$$\hat{\rho}_c = \frac{\Lambda_{V_1V_2}}{\eta_u M(C-1)}, \tag{11.6}$$

where M represents the number of wavelengths in the WRON and $\eta_u(\leq 1)$ is the average usage of a wavelength in the fiber links across the network. Note that the value of $\hat{\rho}_c$ would vary between different cut-sets. We denote the maximum value of $\hat{\rho}_c$ as $\hat{\rho}_c(max)$ (estimated from the $\hat{\rho}_c$'s for all possible cut-sets), which should not exceed the transmission rate r of a lightpath. In other words, the traffic scale-up factor α must not increase beyond the ratio $r/\hat{\rho}_c(max)$. Thus, one can express the upper-bound of the cut-set-based model as

$$\alpha_1^{max} = \frac{r}{\hat{\rho}_c(max)}. \tag{11.7}$$

Note that, the above model is generic in nature and thus applies for both protection and restoration schemes.

Upper-bound from the per-node transmitters and receivers in protection schemes

Another way to estimate the upper-bound for the protection schemes is to use the transmit and receive capacities at all the nodes engaged for the working paths. Assume that node i has Δ_{tx}^i transmitters and Δ_{rx}^i receivers for the working paths, and each one of them sources and sinks traffic flows, respectively, at a data rate r. Then, the maximum traffic the node can source is $r\Delta_{tx}^i$, and the maximum traffic the node can sink is $r\Delta_{tx}^i$. This leads to the fact that the outgoing traffic from node i, i.e., $\sum_{j\in V, j\neq i} \Lambda_{ij}$, into the network should not exceed $r\Delta_{tx}^i$. Similarly, the incoming traffic to node i, i.e., $\sum_{j\in V, j\neq i} \Lambda_{ji}$, from the network should not exceed $r\Delta_{rx}^i$. Therefore, one can express the upper-bound α_2^{max} of the transceiver-constrained traffic scale-up factor for a protection scheme as

$$\alpha_2^{max} = min\left[\frac{r\Delta_{tx}^i}{\sum_{j\in V, j\neq i} \Lambda_{ij}}, \frac{r\Delta_{rx}^i}{\sum_{j\in V, j\neq i} \Lambda_{ji}}\right]. \tag{11.8}$$

Upper-bound with disconnected transceivers due to a link failure in restoration schemes

Upper-bound for the traffic scale-up factor can also be calculated from the reduced resources (transceivers) due to link failure in the restoration-based WRONs. Note that the overall resources in a restoration-based WRON (i.e., without reserving any resource for protection) might be somewhat more than the resources consumed by the working paths in protection-based WRONs, but would be less than the total number of resources used for the working as well as protection lightpaths in the protection-based WRONs. Thus, for the restoration-based WRONs, we represent the numbers of transmitters and receivers in a node (node i, say) by $\Delta^i_{tx}(res)$ and $\Delta^i_{rx}(res)$, respectively, such that $\Delta^i_{tx}(res) > \Delta^i_{tx}$ and $\Delta^i_{rx}(res) > \Delta^i_{rx}$. Further, the transceivers in all nodes are considered to be available for the recovery process, in accordance with the restoration-based survivable WRONs.

If d_i is the physical degree of node i, then one can configure the node to transmit from the $\Delta^i_{tx}(res)$ transmitters over d_i fiber links. If one of these links fails, then the transmitters of the source node attached to the failed link will become disconnected from the network, reducing the aggregate transmitted traffic in bits/second from node i into the network from $r\Delta^i_{tx}(res)$ to $\approx r\Delta^i_{tx}(res)(d_i - 1)/d_i$ (assuming that each link attached to the node gets a nearly equal share of the available transmitters). Similarly, when a link fails, one of the receivers of the destination node attached to the failed link will become defunct, reducing the aggregate received traffic at node i from $r\Delta^i_{rx}(res)$ to $\approx r\Delta^i_{rx}(res)(d_i - 1)/d_i$. As mentioned earlier, in a restoration-based recovery scheme, all the lightpaths would be working paths between the node pairs. Hence, the above transmitted/received traffic at node i will represent the maximum transmission capacity of node i in the event of the link failure, and this capacity must not be exceeded by the actual traffic. In other words, the actual transmitted traffic $\sum_{j\in V, j\neq i} \Lambda_{ij}$ from node i and the actual received traffic $\sum_{i\in V, i\neq j} \Lambda_{ji}$ at node i should not exceed $r\Delta^i_{tx}(res)(d_i - 1)/d_i$ and $r\Delta^i_{rx}(res)(d_i - 1)/d_i$, respectively. Hence, the upper-bound of the traffic scale-up factor for a restoration scheme using the entire network resources can be expressed as

$$\alpha_3^{max} = min\left[\frac{r\left[\Delta^i_{tx}(res)\frac{d_i-1}{d_i}\right]}{\sum_{j\in V, j\neq i}\Lambda_{ij}}, \frac{r\left[\Delta^i_{rx}(res)\frac{d_i-1}{d_i}\right]}{\sum_{j\in V, j\neq i}\Lambda_{ji}}\right]. \tag{11.9}$$

Upper-bounds for protection and restoration schemes

In the foregoing, we have obtained three upper-bounds: $\alpha_1^{max}, \alpha_2^{max}$, and α_3^{max}, from three different considerations. As mentioned earlier, α_1^{max} is governed by the cut-sets and applicable to both protection and restoration-based designs. However, α_2^{max} is only applicable to protection-based designs with the numbers of transceivers corresponding to the working paths only. However, α_3^{max} is based on the reduced resources due to link failures in restoration schemes, where the numbers of transceivers will correspond to the total available resources (i.e., no resources will be reserved for protection paths). Using these observations, we obtain more specific expressions for the upper-bounds α_P^{max} and α_R^{max} for the protection and restoration schemes, respectively, as

$$\alpha_P^{max} = min[\alpha_1^{max}, \alpha_2^{max}], \tag{11.10}$$

$$\alpha_R^{max} = min[\alpha_1^{max}, \alpha_3^{max}]. \tag{11.11}$$

One can use these expressions as the performance indicators while examining the possible designs of survivable mesh-configured WRONs.

11.6.2 Protection-based design of survivable WRONs

Next, we focus on the protection-based WRON design with an objective to minimize the total number of wavelengths used for working and protection lightpaths. In LP-based problem formulation, one can express this objective as

$$minimize\left[\sum_{k}(n_k^{wlp} + n_k^{plp})\right], \tag{11.12}$$

with n_k^{wlp} and n_k^{plp} representing the numbers of wavelengths used for the working and protection lightpaths, respectively, for the kth connection. However, we describe here the steps of a heuristic approach based on the MILP-VTD and FF-based WA scheme (see Chapter 9).

1. In the first step of the heuristic scheme, the MILP-VTD is carried out with the numbers of transmitters and receivers in each node reduced to half of the available numbers. In other words, if the actual numbers of transmitters and receivers in node i is 2 Δ_{tx}^i and 2 Δ_{rx}^i, then we execute the current MILP-VTD for the working paths only by using Δ_{tx}^i and Δ_{rx}^i ($\forall i$), transmitters and receivers, respectively.
2. The execution of MILP-VTD returns a set of primary or working lightpaths $b_{ij}(k)$'s, which are subsequently duplicated over the link-disjoint next-shortest paths for obtaining the protection lightpaths. As discussed in Chapter 9, the MILP-VTD scheme runs to minimize the maximum congestion in the network links, which implicitly addresses our present objective of maximizing the traffic scale-up factor.
3. Having carried out the above design for the working as well as the protection lightpaths using the respective transceivers, the task of WA is carried out using the FFW scheme (or any other scheme, as discussed in Chapter 9) while keeping the protection lightpaths disjoint from the respective working lightpaths.

Finally, one needs to estimate α_1^{max} and α_2^{max} from Equations 11.7 and 11.8 using the groomed traffic obtained from the MILP-VTD and evaluate α_P^{max} using Eq. 11.10 to assess the design.

11.6.3 Restoration-based design of survivable WRONs

The heuristic scheme for the restoration-based design follows a similar approach as above, albeit without any resource reservation for protection, and with the assumption that the WRON has been over-provisioned with some reasonable additional resources which are less than the resources that would have been reserved for the protection paths in the WRON if using the protection scheme. The necessary steps of the design are as follows.

1. Arbitrarily, one of the fiber links in the given WRON is considered to have failed, and the MILP-VTD scheme is carried out, with the failed link removed from the physical topology. Note the lightpath congestion.
2. The above step is repeated for all other fiber links as failed fiber links, considered one at a time.
3. Find out α_1^{max} and α_3^{max} from Equations 11.7 and 11.9 using the groomed traffic obtained from the MILP-VTD and evaluate α_R^{max} using Eq. 11.11 to assess the design.

11.6.4 Recovery time for survivable WRONs

The recovery time for survivable WRONs would indicate whether the recovery scheme used in the network would be acceptable for the given SLA. The recovery process in a WRON depends on the specific recovery scheme used (protection or restoration) and various processing delays in network elements and fiber links. In the following we use the models from (Sahasrabuddhe *et al.* 2002) to estimate the recovery times for the protection and recovery schemes in a WRON.

Recovery time in protection scheme:

The recovery time needed to set up the protection path in a protection scheme starts from the instant of link failure, and ends with the final setting-up of the protection lightpath. For 1+1 protection (dedicated protection), setting-up of the protection path doesn't need any reconfiguration of the switch at the source node as the transmit power is already split and dual-fed over the working and protection paths. However, the destination node must receive the failure message from the node closest to the failed link and thereafter should reconfigure its OS to listen to the already-set-up and active protection path to resume the signal reception. This indeed makes the data loss minimal. However, for 1:1 or 1:N protection scheme, readiness to transmit through an alternate path must be realized through optical switching at the source and destination ends, as well as by reconfiguring all the en route OXCs. This takes additional steps, making the recovery time longer than that for 1+1 protection scheme. We consider here the case of the 1:1 protection scheme and define the relevant parameters of the protection scheme in the following.

τ_{det}^{P}: time needed to detect a link failure at the adjacent node connected to the failed fiber link.

τ_{msg}: message processing time in a node.

τ_{prop}^{i}: propagation time on ith fiber link.

τ_{oxc}: time needed to reconfigure the switch in an OXC, which can be as large as 7–10 ms for MEMS-based switching. It could be much smaller ($\simeq$ 10 ps) for electro-optic switches, albeit with higher optical crosstalk. WRONs in general don't need the agility of a packet-switched network, and hence the nodes in a WRON use circuit switching, for which MEMS-based switching would be the preferred choice.

h_{sf}: number of hops between the source node s and the node adjacent to the failed link (node f, say).

h_{sd}^{pp}: number of hops between the source and destination nodes (nodes s and d, respectively) along the protection lightpath.

Next, we enumerate the recovery process following the 1:1 protection scheme. First, the link failure is detected in the node adjacent (nearest) to the failed link, which takes the failure detection time τ_{det}^{P}. Then this node sends alarm messages to the source and destination nodes through the control-plane network. Upon receiving the message the source node processes the message, needing the message processing time τ_{msg}, and sends a message to the destination node through the protection path with the necessary reconfigurations carried out at the en route OXCs. The destination node (which by this time has presumably received the failure message), upon receiving the message from the source node, confirms receipt of the message through the protection path, thereby completing the recovery process. Further, the transmission time of the alarm message is considered to be negligible as compared to the other delay elements. With this framework, we express the recovery time τ_{REC}^{prot} for 1:1 protection scheme as

$$\tau_{REC}^{prot} = \tau_{det}^{P} + \sum_{i=1}^{h_{sf}} \tau_{prop}^{i} + (h_{sf}+1)\tau_{msg} + (h_{sd}^{pp}+1)\tau_{oxc} + 2\sum_{j=1}^{h_{sd}^{pp}} \tau_{prop}^{j} + 2h_{sd}^{pp}\tau_{msg}, \tag{11.13}$$

where the six terms on the right side of the equation bear the following implications. The first term is the time needed to detect a link failure at the nearest node (node k, say) connected to the failed fiber link and the second term represents the propagation time taken by the failure message to travel from node k to the source node s. The third term is the cumulative (hop-by-hop) failure-message processing time at all the en route nodes on the path between the nodes k and s. The fourth term represents the cumulative OXC-reconfiguration time at all the en route nodes on the protection path between the source and destination nodes s and d. The fifth term accounts for the back-and-forth total propagation time between the nodes s and d along the protection path, once for sending the reconfiguration signal and next for the acknowledgment. The sixth term gives the back-and-forth cumulative message-processing time between the nodes s and d along the protection path, once for sending the reconfiguration message and the next for acknowledgment message (all messages are assumed to use the same duration).

Note that, with a link failure, several working lightpaths are likely to get torn down, and the network needs to carry out the protection-based recovery process for all concurrently, and one should note the longest recovery time for the worst-affected lightpath for a given failed link, $\tau_{REC}^{prot}(k)$, say. Thereafter, this process should be repeated for all possible link failures and finally check whether the longest of these recovery times, i.e., $max\{\tau_{REC}^{prot}(k)\}$, remains within the upper limit set by the SLA.

Recovery time in restoration scheme involving IP layer

Recovery process in a restoration scheme, operating at IP layer, starts after the failure detection at the destination nodes of all the lightpaths carried by the failed link, instead of the nodes attached to the failed link. This detection process takes much longer, as all the destination nodes have to service the interrupt at IP layer due to the failure. Thus, the failure detection time for the restoration case τ_{det}^{R} is much longer than its counterpart τ_{det}^{P} in the protection scheme. Having detected the failure, the destination nodes broadcast the link-update message to all other nodes, where the broadcast algorithm needs a long time in the order of milliseconds. Having received the broadcast message through the network, the IP routers in the nodes recompute their routing table, needing typically a time of ~200 ms. With this framework of the restoration scheme, we next define the relevant parameters for the restoration scheme:

τ_{det}^R: time needed to detect a link failure at a destination node, whose reception has been affected by the failure.

τ_{bcast}: time needed by a router to run the broadcast algorithm.

τ_{route}: time needed by a router to recompute its routing table.

τ_{prop}^i: propagation time on ith fiber link.

h_{df}: number of hops between a destination node d of a lightpath and the node adjacent to the failed link (node f, say).

h_{dn}: number of hops between a destination node of a lightpath and the farthest node (say, node n) in the network (as each node needs to recompute its routing table).

Using the above definitions, we express the recovery time τ_{REC}^{rest} for a lightpath in a WRON using the restoration scheme as

$$\tau_{REC}^{rest} = \tau_{det} + \sum_{i=1}^{h_{df}} \tau_{prop}^i + \sum_{j=1}^{h_{fn}} \tau_{prop}^j + (h_{fn} + 1)\tau_{bcast} + \tau_{route}, \tag{11.14}$$

where the first term is self-explanatory, the second and third terms are the propagation times from a lightpath's destination node (node d) to the node adjacent to the failed link (node f) and between node f and the farthest node therefrom (node n, say), respectively. The fourth term is the cumulative running time of the broadcast-algorithm at all the nodes to reach the farthest node, and the fifth term represents the re-computation time for routes at the routers. Finally, in order to assess the compliance of the restoration scheme with the SLA, one needs to compute the recovery time of the network as above for each link failure at a time, and ensure that the longest recovery time remains within the acceptable limit.

11.7 Convergence of survivability schemes in multiple layers

In this section, we address the convergence issues between the candidate layers in an optical network, as these layers might start trying concurrently to recover from a failure. In optical networks, the optical layer serves as the lowest layer and so far we have mostly discussed about the survivability schemes employed in this layer. However, when a failure takes place, the event of failure is sensed at the network-centric layers: layer-1, layer-2, and layer-3, and the recovery effort can start in these layers thereafter. Since the actions may not progress with similar agility in each layer, the network should ensure some convergence of the multi-layer attempts to come out of the failure. In other words, if the recovery takes place earlier in one layer, the other layers should know that and act accordingly (Shiragaki *et al.* 2001; Pickavet *et al.* 2006). We discuss this issue in the following.

To start with, for example, consider the layer-1 in a SONET-over-WDM network, which involves in reality a few layers/sublayers for recovery mechanism. While looking up from the bottom of the network protocol stack, one can see some nested protocols (sublayers) within the physical layer of optical network, which use lightpaths (needing setting-up/tearing-down/recovery processes) in WDM transmission and higher-rate SONET channels (OC-48, OC-192, etc.) carried by each lightpath, with each of them being groomed with several

SONET-based lower-rate OC channels (OC-3, OC-12, etc.). Usually, the recovery mechanism of WDM or optical layer (OL) gets mingled with that of self-healing SONET (another sublayer of layer-1) using SONET-over-WDM transmission, as the recovery of a lightpath (OCh) or waveband-path (OMS) is tantamount to the recovery of the constituent lower-rate optical channels (OCs) as well. However, layer-2 (typically, Ethernet, ATM, MPLS, RPR, etc.) and layer-3 (IP) would employ concurrently their own recovery mechanisms, remaining mostly unaware of others' ongoing recovery mechanisms. Notwithstanding this apparent independence, a network must be sure that one recovery mechanism does not have any conflict with the others, and eventually the recovery is achieved with appropriate inter-operability between the layers in shortest possible time.

To understand the inter-operability of multiple survivability schemes, let's consider, for example, the optical layer (OL) and one more candidate layer on the top of OL in an optical WDM network. Note that, as discussed in Chapter 9, there can exist several architectural formations in optical WDM networks: IP-over-SONET-over-WDM, IP-over-WDM, Ethernet-over-SONET-over-WDM, Ethernet-over-WDM, and so on. In order to get an insight, we consider here the case of IP-over-WDM, implying that over the WDM layer (i.e., OL), IP layer would run, and hence the survivability mechanisms of both layers would attempt to recover from the failed state, say due to a link failure. Given this situation, the two survivability schemes may inter-operate in two possible ways: parallel and serial modes of operation, as illustrated in Fig. 11.15. We describe these schemes in the following.

In the parallel mode, both layers start working toward recovery after the failure detection in the respective layers, and the layer that realizes the recovery first, informs the other layer to

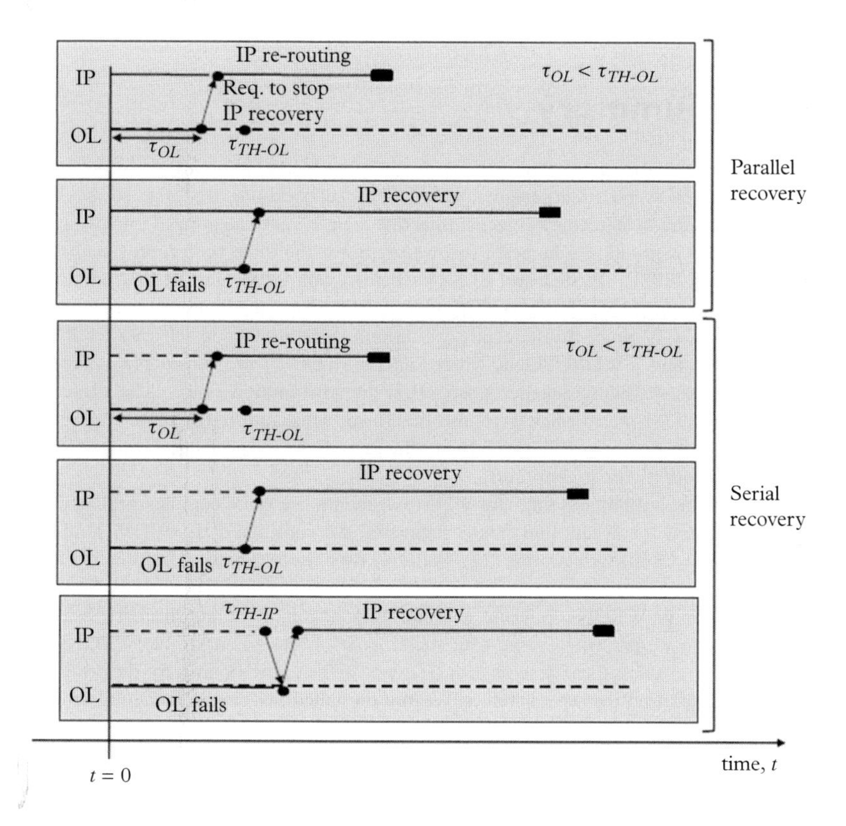

Figure 11.15 *Convergence scenarios for multilayer recovery schemes.*

stop the recovery process of the latter and carry out further steps, if necessary. For example, with a dedicated protection scheme, the OL realizes the recovery much earlier than the IP layer. Generally, the OL sets up a threshold time for recovery (τ_{TH-OL}), and informs the IP layer to stop the IP recovery process if the recovery time $\tau_{OL} < \tau_{TH-OL}$. The IP layer, having received this message from the optical layer, stops its recovery operation, and carries out appropriate re-routing as the working path for the affected connection(s) get replaced by the protection path. Thereafter, the original working path is repaired and takes over the earlier connection status, followed up by further re-routing once again. The last step of this example recovery-case makes the scheme revertible in nature. However, the network can also employ a non-revertible recovery scheme, where even after the repair of the failed link or node, the connection continues to use the protection path, and hence the second re-routing process in IP layer is obviated; this makes the non-revertible recovery scheme a preferable option for restoration-based recovery schemes.

In the serial mode of multi-layer recovery attempts, the OL being the fastest in general, starts the recovery process with a given threshold (upper limit) of the recovery time. If the OL can recover before/within the threshold time, it notifies the IP layer immediately after the recovery. The IP layer then completes the task of re-routing, knowing that the recovery has already been achieved in the OL within the time limit. When the OL fails to accomplish the recovery within the given time limit, it passes on a message after the expiry of the time limit to the IP layer indicating that it has failed to achieve recovery, and thereafter the IP layer proceeds to achieve the recovery on its own. The IP layer can also employ a threshold of waiting time and send a message to the OL, which in turn reconfirms to the IP layer its incapability of completing the protection process, following which the IP layer completes the recovery process.

11.8 Summary

In this chapter, we presented various survival mechanisms used in optical networks for the three hierarchical segments: access, metro, and long-haul backbone networks. To start with, we introduced the basic recovery mechanisms that are generally categorized into two types: protection and restoration schemes. First, we presented the three protection schemes for SONET rings: UPSR, BLSR2, and BLSR4 schemes, with UPSR and BLSR2 employing two counter-propagating (CW and CCW) rings and BLSR4 using four fiber rings, with two CW and two CCW rings. Subsequently, we described the protection schemes for PONs, WDM PONs, and TWDM PONs, where various schemes were considered addressing varying degrees of failures: feeder fiber, OLT, and distribution fibers. The protection schemes for WDM metro rings were also described, which utilize the basic concepts of SONET protection schemes, but with the traffic granularity for protection being at least a lightpath or more (lightpath bundles).

Ensuring survivability of long-haul WDM backbones is a more complex problem than those considered for access and metro segments. We considered the WDM long-haul networks using wavelength-routing (i.e., WRON) over mesh topology and described the underlying protection/restoration issues. First, we defined some upper-bounds for the traffic scale-up factor in WRONs, offering a broad perspective of the survivability issues in WRONs. Then we presented some heuristic schemes to design survivable WRONs and discussed how to assess these designs with reference to the upper-bounds for the protection and restoration-based recovery schemes. We also presented two simple models to estimate the recovery times in the mesh-configured long-haul WRONs for the protection and IP-based restoration schemes. Finally, we discussed briefly the convergence issues between the survivability schemes, that may be employed at the same time by the network in different layers.

EXERCISES

(11.1) The problem of survivability against failures in access and metro segments is generally addressed using protection-based designs. On the other hand, the long-haul mesh networks can employ protection and/or restoration schemes to ensure network survivability. Justify this practice with reference to the basic features and constraints of the respective network segments.

(11.2) Discuss the pros and cons of the UPSR and BLSR2/4 schemes in SONET rings. Explain how the upper-bound on recovery time (50 ms) influences the design of SONET rings. By using a suitable illustration, describe a scheme to recover from a node failure in the BLSR2 scheme.

(11.3) Consider that two SONET rings using UPSR-based protection scheme are interconnected using dual access, thereby involving four nodes (with two nodes from each ring). Sketch the two interconnected rings and illustrate the protection scheme that can be used for inter-ring connections. Discuss how the devices used for the protection scheme might influence the scalability of the ring network.

(11.4) As indicated in the section on PON protection, the survivability of a PON can be assessed in terms of the unavailability of its connections to the ONUs, by using the concept of asymptotic unavailability of the network components, given by $U_x = \tau_R \times$ failure rate, with x representing the various PON elements/devices. Using this model and the following values of U_x's, estimate the worst-case availability of the ONU connections in the unprotected TDM PON shown in Fig. 4.2(c). Assume that the PON has a feeder length of 15 km and the distribution fibers are all within 5 km.

Asymptotic unavailabilities:

OLT: $U_{OLT} = 5.12 \times 10^{-7}$,

ONU: $U_{ONU} = 5.12 \times 10^{-7}$,

optical fiber: $U_F = 3.42 \times 10^{-6}$/km,

remote node (RN) using optical splitter (OSP): $U_{RN} = U_{OSP} = 10^{-7}$,

optical switch: $U_{OS} = 4 \times 10^{-7}$ (used in the next problem).

(11.5) Repeat the above problem with a PON using the Type-A protection scheme as shown in Fig. 11.9.

(11.6) Compare the two-fiber OCh-DP-Ring and the four-fiber OCh-SP-Ring protection schemes used for WDM ring networks, in respect of the recovery time and differential delay in full-duplex connections. Discuss the implications of these differences while setting up a WDM ring network.

(11.7) Consider a long-haul WRON with its lightpaths carrying data at 10 Gbps across the network, where each fiber can carry 40 wavelengths. Imagine that the network has been bisected using a cut-set C generating two parts having 10 fiber links in the cut-set. With this bisected view, one finds that the WRON has suffered a link failure, which happens to be one of the 10 fiber links in the cut-set. The total traffic following across the cut-set before the link failure was found to be 1000 Gbps, which needs to be accommodated with the 9 out of 10 fiber links due to the link failure. In this scenario, if the average usage of the wavelengths is 60%, estimate the upper-bound for the traffic scale-up factor in the network after the link failure.

Optical Network Control and Management

12

The task of network control and management is generally realized in two logical planes – control and management – which collaboratively operate to ensure smooth, secure, and survivable traffic flow in the data plane of the network. Some of the functionalities are realized mainly in the control plane, which needs real-time execution, such as recovery from network failures, network reconfiguration due to traffic variation, etc. Some other functionalities deal with the performance monitoring, configuration management, network security, accounting and billing, etc., which are less time-sensitive and addressed by the management plane. In this chapter, we first discuss the philosophy of the multiple-layer abstraction of telecommunication networks, including control, management, and data planes, and thereafter describe various network control and management techniques used in optical networks: operation, administration and management (OAM) in SONET, generalized multiprotocol label switching (GMPLS), automatically switched optical network (ASON), and software-defined optical networking (SDON) in WDM networks.

12.1 Multiple-plane abstraction of telecommunication networks

In telecommunication networks, control and management functionalities are the basic needs for ensuring efficient and reliable operation. These functionalities become more complex when the network span and bandwidth increase, as in optical WDM networks. The management functionalities are required to serve the needs of clients, such as assuring network performance, security, fault tolerance, accounting, billing, etc., through appropriate management software embedded in the network nodes. However, routing and re-routing of connection requests, recovery of connections in the events of failures, etc. are the main issues addressed by the control functionality. In other words, control and management operations are conceived as two logical entities or *planes*, known as control and management planes, operating in collaboration across the entire network, ensuring efficient data communication through data (or transport) plane.

The boundary between the control and management planes is somewhat *smeared* in some situations, at times making it difficult to portray them as independent planes. Broadly speaking, the management plane manages the overall network operation with the control plane carrying out some of its time-sensitive tasks. An example to illustrate this would be worthwhile. Consider that there has been a fiber-cut in a long-haul WDM network. This is a failure in the transmission system that should be detected and all the lightpaths flowing through the cut-fiber must be re-routed, complying with the SLAs of the clients (e.g., downtime of the network), which are in turn available from the management-plane information base. While the new routes assigned to the affected lightpaths may not be the best routes (i.e., with least cost), this arrangement will prevent disruption of connections for

Optical Networks. Debasish Datta, Oxford University Press (2021). © Debasish Datta.
DOI: 10.1093/oso/9780198834229.003.0012

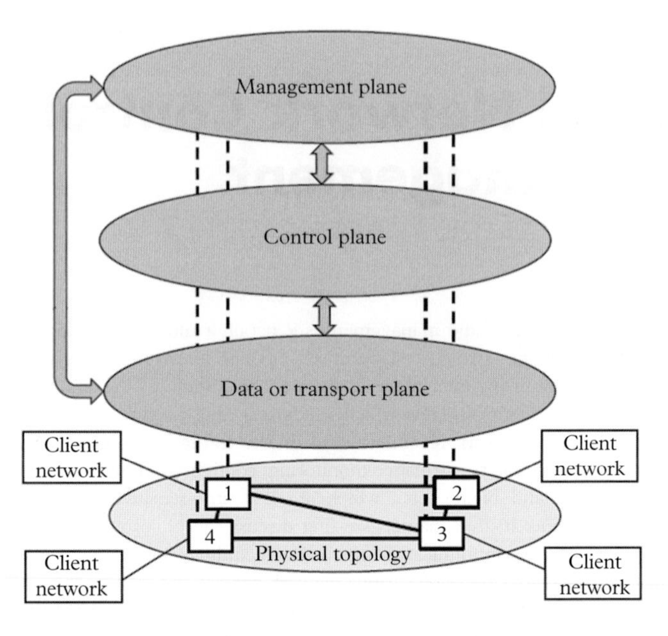

Figure 12.1 *Three-plane abstraction of a long-haul network. The three planes (data, control, and management) operate through the physical topology, and each client network can communicate with the three planes through the node (NE), to which the client network is attached in the physical topology.*

the clients, which used the cut-fiber. Further, repair of the fiber would start, following the detection of the fiber-cut and be carried out typically in a few hours. From this example, it is evident that the re-routing (reconfiguration) of connections is to be carried out on a real-time basis, and thus this task would be performed by the control plane, but in compliance with the SLAs from the management plane. However, the *field work* needed to repair the fiber-cut cannot be done immediately and will be addressed by the management plane in due course.

In Fig. 12.1, we present an abstract view of the control and management operations for a network, consisting of three planes: data (transport) plane, control plane, and management plane. The data plane takes care of the data transmission across the entire network once the connections are set up. The management plane interacts with the clients, maintains an information base for the network elements (NEs), and carries out its tasks of performance monitoring, configuration management, security, accounting and billing, etc. over time and also passes on the necessary input to the control plane for real-time execution of the time-sensitive tasks. Collaborating with the management plane, the control plane carries out configuration/reconfiguration of connections driven by the dynamic traffic variation and occurrences of failures across the network. The records of all NEs are maintained in the management information base (MIB) belonging to the management plane. Typically, a client interacts with the management plane through the attached node by using in-band or out-of-band data communication.

As indicated in Chapter 9, for a WDM network, generally a separate (or a shared) wavelength is allocated for such communications across the network, thereby forming a logical topology for the control plane. This logical topology may be different from the physical topology as well as from the logical topology of the data plane, but it must visit each node electronically (minimum configuration being generally a ring). For direct interactions (usually slower ones) between the data and management planes, management information flows may take place as shown in Fig. 12.1 by the direct double-sided arrow between the management and data planes, which would bypass the control plane logically. However, there are several interactions between the management and data planes (faster ones) that must go through the control plane, such as re-routing of connections in the event of a link failure or while coping with traffic variations.

12.2 Framework of control and management planes

Operation of control and management planes in a network can be based on a centralized or a distributed framework. The major aspects that need attention for controlling and managing an operational network include:

- network performance,
- fault tolerance,
- configuration and reconfiguration of NEs (nodes) and connections,
- network security,
- accounting and billing.

In the following, first we discuss the salient features of network management in further detail.

Performance management: Performance management in a network deals with monitoring the performance of network and checking whether the QoS parameters committed to by the service providers are being complied with. Further, this entity of network management should be in *dialogue* with the clients and network administrators to ensure smooth running of the network.

Fault management: Fault management in a network ensures that faults occurring anywhere in the network are detected promptly, and appropriate actions are taken thereafter. As mentioned earlier, part of this task (arrangement of alternate paths for the disrupted connections until the necessary repair is made) needs to be carried out on a real-time basis by the control plane. However, the repair work needed for the network failure (failed node/link) is undertaken through a maintenance group and may take some time. After the repair work is completed, the network may have to go for connection reconfiguration, which is also carried out through the control plane.

Configuration management: Configurations of NEs need to be recorded in the MIB and managed throughout the network to ensure that the network operation is carried out efficiently. Governed by the QoS parameters guaranteed to the clients (available in MIB), the connections running through the network are configured/reconfigured from time to time. Configuration management deals with these tasks across the entire network. However, the execution of connection configuration/reconfiguration, being real-time in nature, is passed on to the control plane.

Security management: Network security management deals with the authentication of users, encryption of data, partitioning the network in multiple domains, etc.

Accounting management: Accounting management deals with the billing records of all the NEs in the network.

Guided by the management plane and the evolving network scenario, the control plane acts at each node following the policy-based directives, and accordingly the NEs in the data plane forward the data packets through the data plane of the network. In particular, the routing of data packets is carried out at each node by the control plane by using an appropriate mechanism, whenever

- a new connection request comes in,
- an existing connection is removed,
- failures occur.

When a new connection request arrives or an existing connection needs re-routing, the control plane carries out the route discovery through exchanges of messages between the relevant nodes. Following this step, the data plane starts data transmission from source to destination node. The reverse process is carried out for tearing down a connection, and the resources used by the departing connection are released for other connection requests.

In the following, we describe the operation of control and management planes using the illustrative diagram in Fig. 12.2. In the figure, we consider a linear segment from a network with three nodes: nodes 1, 2, and 3. Node 2 being in the middle between nodes 1 and 3 has to forward traffic from both sides, and also to source and sink data from and to itself, respectively. Roles of the control and management planes along with the data plane are illustrated for node 2. The management-plane entity of node 2 is shown at the top to perform its functions (performance, configuration, fault, security, and accounting/billing management) using the inputs from the clients through the control plane along with its MIB and appropriate protocol. The necessary actions are transformed into messages to be shared with control plane for real-time actions, and other messages are passed on to the network management staff through a graphic user interface (GUI), whenever necessary. The network maintains the element management system (EMS) for the records of all NEs, typically distributed into the servers kept at some selected nodes, which are shared by the management-plane entity of each node through control-plane-aided communications. The control and management-plane entities interact so that the control plane can carry out configuration/reconfiguration tasks using its control protocol and dynamically-maintained

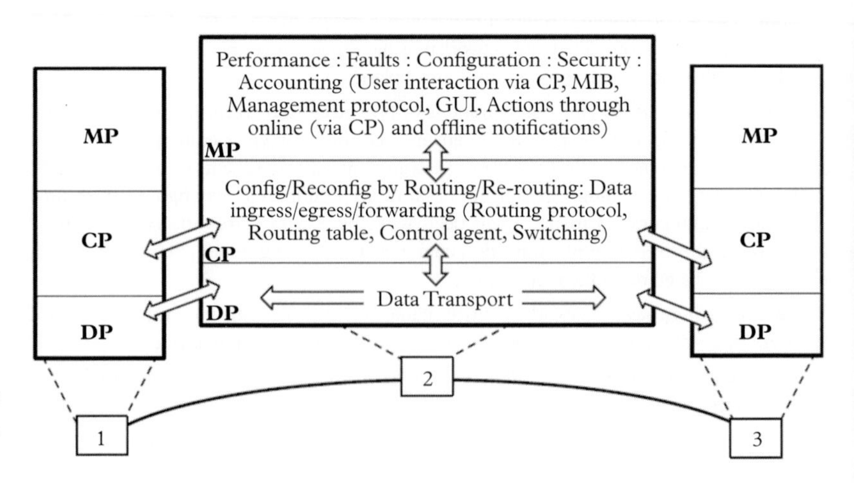

Figure 12.2 *Operations of the management and control planes; MP: management plane, CP: control plane, DP: data plane.*

routing tables, leading to the forwarding and add/drop operations on the transit and ingress/egress traffic flows, respectively.

12.3 Control and management of optical networks

The control and management schemes in metro and long-haul optical networks depend on the respective physical-layer configurations. In particular, single-wavelength SONET/SDH standards (henceforth, to consider SONET only, as in Chapter 5) have used time-tested control and management schemes, popularly known as operation and management (OAM) systems, which have been generally operated over ring topologies. However, for WDM mesh networks, the protocols for control and management mechanisms have evolved over the years, leading to the generalized multiprotocol label switching (GMPLS), automatically switched optical network (ASON) and optical version of software-defined network (SDN), i.e., SDON.

The development of GMPLS has evolved from the pre-existing multiprotocol label switching (MPLS) protocol for packet-switched networks, undertaken by IETF, while ASON has been developed by ITU-T with a set of standard interfaces to interconnect multiple network clusters or domains. However, both of them employ distributed control and management across the network. SDN has been a more recent development and is still evolving, employing centralized control and management in contrast with ASON and GMPLS.

Some of the salient features of these standards for WDM networking are worth mentioning at this stage. For example, GMPLS is applied between the nodes within a network cluster or a domain, where there exists a common address space with path-computation capability, generally using the NEs from the same vendor (or compatible vendors) within a single administrative jurisdiction. Examples of such network domains include any autonomous system (network) or a routing zone of a network using interior gateway protocol. Consequently, GMPLS remains confined within a given networking domain, enjoying a trusted environment. Thus, within a given networking domain, GMPLS operates in all the NEs, while specific interfacing techniques are needed to interconnect multiple networking domains. In such cases, one needs the ASON-based inter-domain networking, which remains oblivious to what goes on within the individual network domains, while interconnecting multiple domains by using appropriate network interfaces, called as *reference points*. Thus, ASON uses these reference points to interconnect multiple networking domains, which might be operating internally using GMPLS or any other proprietary control and management protocols. Hence, for the inter-domain networks, one can visualize a protocol suite, such as, IP-over-GMPLS/ASON-over-WDM, that would be able to operate across the boundaries between different GMPLS domains.

Overall, in metro and long-haul optical networks, within a SONET-based network jurisdiction, the tasks of OAM are carried out independently, while for similar operations in the SONET-over-WDM networks, the optical or WDM layer has to run in parallel with OAM of SONET through the WDM devices. However, for the IP-over-WDM networks (i.e., without SONET), GMPLS carries out the tasks of network control and management, albeit within a network domain, while ASON as an umbrella provides an inclusive inter-domain architecture to operate with GMPLS and other candidate networking-domains as the client layers. Furthermore, SDN/SDON offers an evolving class of networking solutions, providing a centralized (as compared to the others offering distributed control and management) and programmable network architecture with much flexibility and enhanced resource utilization. In the following sections, we describe the salient operational features of the control and management planes in optical networks, based on OAM, GMPLS, ASON, and SDN/SDON.

12.4 Operation, administration and management in SONET

OAM functionalities in SONET take effect through the various overhead bytes (POH, LOH, and SOH bytes), as defined in Chapter 5. Using these bytes, the SONET OAM module in every node carries out a wide range of tasks: the connection (path) setup/tear-down, alarming, performance monitoring, data communication for network maintenance, etc., along with automatic protection switching in the event of network failures. In the following, we revisit the functions of the SONET overhead bytes and describe how the SONET nodes work hierarchically through the three layers – path, line, and section – and execute the various tasks of OAM to keep the network operational.

For a given end-to-end connection between a node pair (ADMs in general, and TMs for terminal nodes in linear segments) in SONET, the path is the longest connection span and hence represents the highest SONET layer between the two nodes, and then follow the line and the section layers with decreasing order of hierarchy. Figure 12.3 illustrates an example scenario (a more detailed version of Fig. 5.11), where a path is set up between two nodes using the SONET overhead bytes. As shown in the figure, the SONET nodes at the two ends of the path need to exchange PDH signals. There are three intermediate nodes which also participate in carrying this path from source to destination. The nodes at the two ends play the role of PTE, and operate (insert/extract) with all the three sets of overhead bytes (POH, LOH, SOH), while the intermediate nodes work (insert/extract) with LOH and SOH (or only SOH) bytes, depending on the roles the intermediate nodes play (e.g., as LTE or STE). Note that, for the path under consideration, an ADM at an intermediate node operating in the LTE mode can add/drop lower-rate SONET streams, while an ADM at an intermediate node operating in the STE mode performs as a regenerator only. As in any layered-network architecture, each layer of a node communicates to the peer layer of the next relevant node (not necessarily the adjacent node, excepting the section layer) by processing the specific header information, and passing it up/down to the next layer.

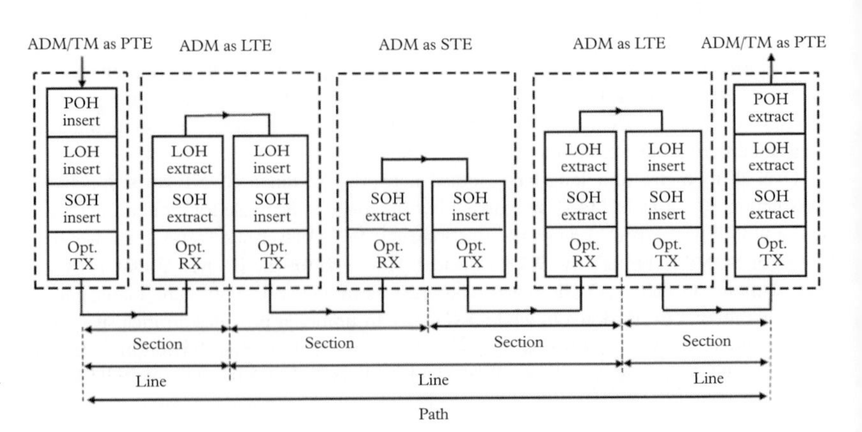

Figure 12.3 *Illustration of SONET OAM functionalities.*

For an STS-1 frame transmission, while setting up a path from a node on one end (say, the left end in Fig. 12.3), the path layer in SONET typically transforms 28 T1 frames into seven VT-1.5s (see Chapter 5) along with the POH bytes to form an STS-1 SPE and passes this on to the underlying line layer. The POH field carries out its share of OAM functionalities by using some of its constituent bytes as follows:

- *path trace* byte used to check the integrity or temporal continuity of the path by inserting (at source) and reading (at destination) a periodically repeated bit stream from frame to frame,
- *path BIP* byte used for error monitoring (BER check) at path level,
- *path signal label* used to identify the type of the payload,
- *path status* byte used to convey the status of a path by sending an error count from the destination to the source node.

By inserting/extracting the above POH bytes, the path layer carries out all the services needed between the path-terminating equipment (i.e., PTEs) at the respective end nodes, thereby continuously assessing the performance of the connection and ensuring protection-switching for the path set-up between the two nodes. Note that the POH bytes are only processed (inserted or extracted) at the terminal nodes (PTEs) of a connection.

After the insertion of payload and POH bytes in the SONET frame at the source node, the line layer inserts the LOH bytes, and through this process carries out its share of OAM functionalities as follows:

- *payload pointer* bytes used for synchronization,
- *line BIP* bytes used for error-monitoring at the line level,
- *APS* bytes used for automatic protection switching,
- *line data com* bytes used for the line maintenance with alarms, monitoring and control,
- *orderwire* byte used for voice communication at the line level.

At the next level, the section layer inserts an SOH field which operates between any two adjacent SONET nodes for signal regeneration and associated tasks. The SOH bytes operate as follows:

- *framing* bytes used for delimiting the frames,
- *STS-1 ID* byte used for STS-1 frame number as an identification in an STS-N frame,
- *section BIP* byte used for error monitoring at the section level,
- *orderwire* byte used for voice communication at the section level,
- *section data com* bytes used for section maintenance with alarms, monitoring, and control.

After the insertion of SOH bytes, the entire TDM frame is scrambled (see Fig.5.8), and the scrambled electrical signal is converted to its optical equivalent and transmitted over the fiber to the next node.

As the transmitted signal from the source node traverses its designated path, the intermediate nodes, playing the role of LTE, extract and reinsert the LOH and SOH bytes, while the intermediate nodes playing the role of STE extract and reinsert the SOH bytes only. Finally the destination node extracts all overhead bytes, processes them to accomplish end-to-end communication over the concerned path. For a path in the reverse direction, similar operations are carried out from the right PTE to the left PTE through the intermediate nodes (as LTE/STE), thereby setting up a duplex SONET connection between the two nodes. Thus, by using the overhead bytes of the three layers, SONET carries out all the tasks of OAM at the relevant nodes along the path.

12.5 Generalized multiprotocol label switching

GMPLS provides a set of network control and management protocols, standardized by IETF, for WDM-based packet-switched optical networks (Banerjee *et al.* 2001a; Banerjee *et al.* 2001b; Mannie 2004). In particular, GMPLS provides a generalization of the multiprotocol label switching (MPLS) protocol to offer packet-switching capabilities over time, wavelength, waveband, and space (fiber). In order to describe the functionalities of GMPLS, we first make a quick review of MPLS protocols.

12.5.1 MPLS: basic features

Owing to the unprecedented growth of Internet and emergence of wide range of applications, including multimedia communications, cloud computing, social networking, e-commerce, etc., the hop-by-hop packet forwarding mechanism of IP networks gradually turned out to be inadequate to support the necessary QoS guarantees. With this backdrop, the concept of MPLS was born in late 1990s for deployment in the packet-switched networks, offering traffic-engineering (TE) capabilities and virtual packet networking (VPN) services, while ensuring minimal increase in resource costs in the existing networking infrastructure. Although MPLS can work with different packet-switched networks, we discuss in the following the operation of MPLS protocols when adopted in IP networks.

The MPLS layer, functioning between the IP and link layers, is often referred to as layer 2.5, where the hop-by-hop packet forwarding scheme of IP is replaced by *tunnel-like* virtual circuit-switched connections across the network. In particular, at an edge router of a core network, the IP packets having a matching set of prefixes in their destination addresses are assigned a *label*. The label is added as an encapsulation over the IP header (later to be prepended with the layer-2 header) and all are sent over a common route, which is set up across the network, prior to the transmission of the packets. For example, if three IP packets have the destination addresses *a.b.c.*1, *a.b.c.*2, and *a.b.c.*5, then they are represented by a group address as *a.b.c.x* (belonging to a specific subnet), which traverse along a common route across the network. Such a group of IP packets with the matching prefixes, while demanding the same class of service (COS), represents a *class*, collectively represented as a *forward equivalent class* (FEC).[1]

All the packets belonging to the same FEC are assigned a common *label*, which is placed before the header of each IP packet belonging to the same FEC. Note that, after the transmission of these labelled IP packets, their MPLS label keeps changing at each intermediate node (each input label being replaced by an outgoing label), and these labels are pre-defined before the transmission of the packets using an MPLS protocol, called label distribution protocol (LDP) (discussed later). When the labelled packets belonging to an FEC are received by the edge router of the destined subnet, the MPLS labels are stripped

[1] Note that this abbreviation (FEC) is also used for forward error-correcting codes.

off from all packets and thereafter these packets are forwarded to the respective destination nodes corresponding to their IP addresses.

The edge routers located at the boundary of an MPLS network are referred to as label edge routers (LERs), while the routers located inside the core network are called label-switching routers (LSRs). As mentioned above, for a given FEC, the MPLS labels at the edge nodes (i.e., at LERs) and at the intermediate nodes (i.e., LSRs) are set up using LDP, before any MPLS packet belonging to that FEC is forwarded by an LER. Having left the LER for onward journey, the packet goes through LSR(s) to reach the destination node in a pre-configured path, called a label-switched path (LSP). In particular, by inspecting the label of an incoming packet, each LSR assigns a new label to the packet for the next hop, as already pre-assigned in all the LSRs by LDP. Once the LSP for an FEC is set up using LDP, the subsequent routing of all the MPLS packets belonging to the same FEC becomes much faster as compared to the IP networks, as the intermediate nodes need not run any routing algorithm (excepting when the labels are set up by LDP). Further, an MPLS packet can be assigned multiple labels along the path, implying that it might carry multiple LSPs within itself to complete its journey. The corresponding LSP labels are *stacked* one after another in the MPLS packet, necessitating the use of a stack bit in the MPLS header, as explained in the following.

MPLS header placement in a packet is shown in Fig. 12.4. As shown in the figure, the MPLS header has in total 32 bits, placed between the layer-2 and IP headers, followed by the payload. The left-most 20 bits are reserved for a label, the following three bits for COS to carry out traffic engineering (discussed later), the next bit (stack bit) represents the bottom of the label stack, and the last eight bits are used to indicate the time to leave (TTL) to avoid infinite circulation of any packet in the network. The stack bit is one for the oldest entry in the label stack, and zero for the other entries. The MPLS packet in part (b) is shown to have multiple labels stacked together between layer-2 and layer-3 headers.

Note that, between the same pair of source and destination LERs, the MPLS packets belonging to different FECs (due to different COSs) can be assigned different routes while setting up the labels using LDP, and thus the network can avoid congestion on a single route,

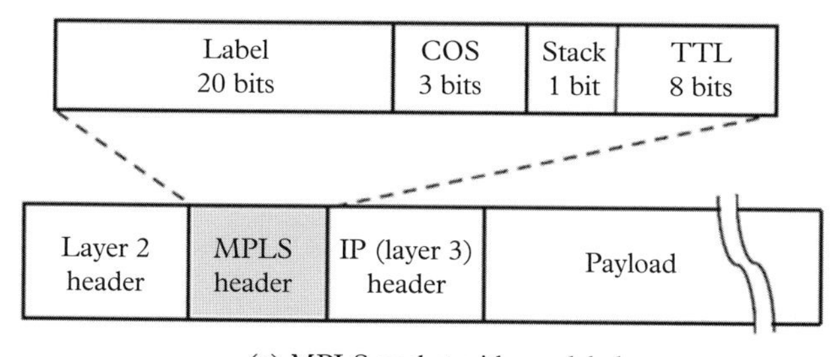

(a) MPLS packet with one label.

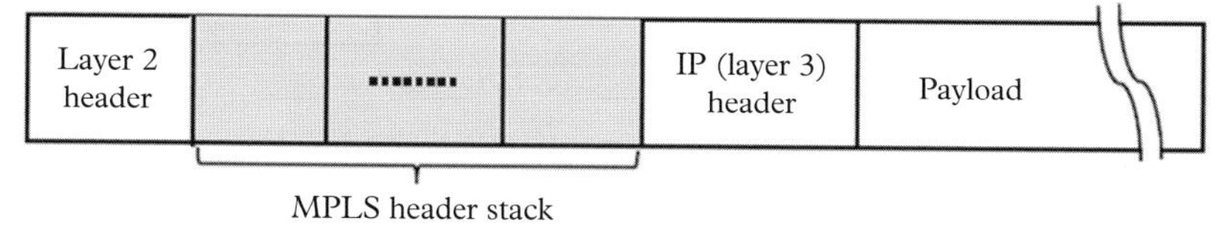

(b) MPLS packet with multiple labels.

Figure 12.4 *MPLS packets with one and multiple labels.*

leading to TE. Further, the use of *tunnel-like* LSPs enables the MPLS networks to offer the VPN service, whenever needed.

As evident from the foregoing, an MPLS network needs to employ an appropriate LDP to set up the LSPs. There are several candidate protocols, which can use either some modified versions of the existing IP protocols or some new ones, e.g., resource reservation protocol with traffic-engineering-based extension (RSVP-TE) and constrained-routing-based label distribution protocol (CR-LDP). The basic functions of LDP are illustrated in Fig. 12.5, where the use of the LDP scheme enables the LERs and LSRs to exchange the necessary information for *binding* or *connecting* all the labels assigned to the LERs/LSRs along each LSP, thereby enabling fast hop-by-hop packet forwarding using label-switching at each LSR across a given MPLS domain. Note that, the process of creating LSPs using LDP remains decoupled from the packet-forwarding mechanism in LSRs across the network.

An example scenario of MPLS networking is illustrated in Fig. 12.6, where the data as well as the legacy voice (TDM) services are shown connected through an MPLS domain using stacked labels over the select links. Note that the packet traffic (IP/Ethernet)

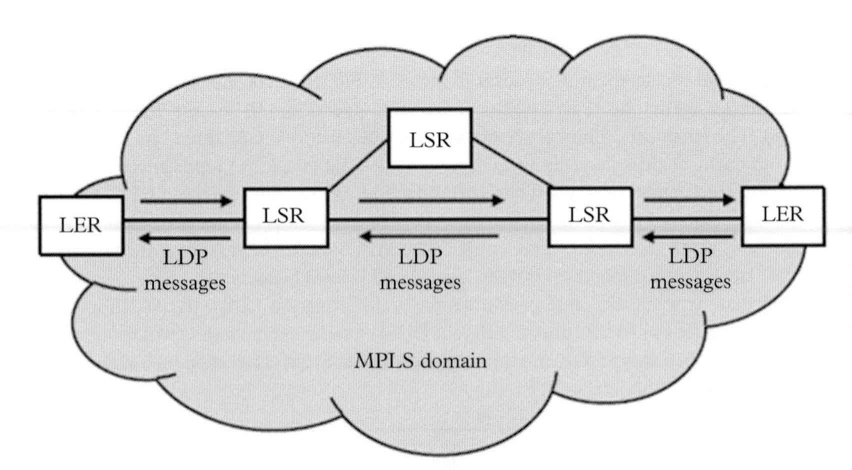

Figure 12.5 *Illustration of MPLS LDP.*

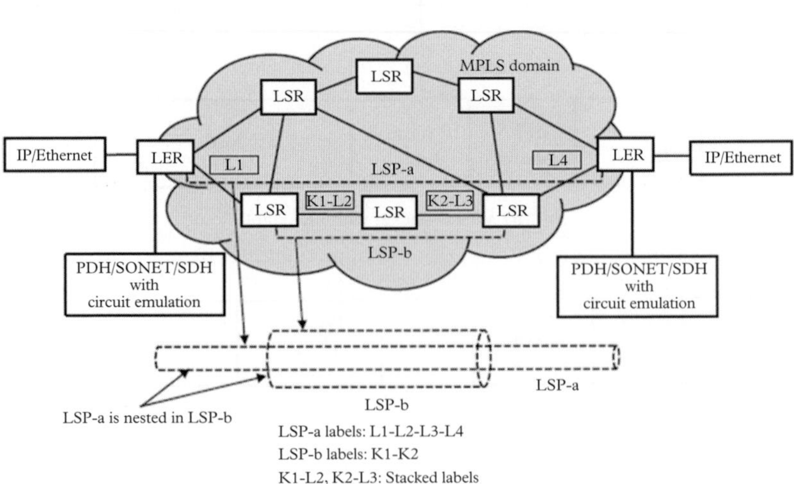

Figure 12.6 *Illustration of MPLS LSPs, wherein two LSPs are nested by using label stacks.*

enters directly into the LERs, while the circuit-switched flows are packetized using circuit emulation and fed into the LER. As shown in the figure, there are two LSPs over a common set of links, leading to the nested LSPs using stacked labels.

12.5.2 GMPLS architecture

The motivation for developing the GMPLS standard for IP-over-WDM networks was to generalize the concept of MPLS label, so that the ingress packets in an optical network could be assigned labels corresponding to the various possible granularities of optical transmission capacity. The granularity was offered over a wide range, such as a full wavelength or a waveband from the optical transmission window, time slots over a given wavelength, and one or more optical fibers from a fiber bundle. In other words, given the multi-dimensional capacity of optical fibers, the aim was to use an appropriate mechanism in GMPLS protocols to offer the resources as much as possible in a flexible manner, for realizing the best-possible network performance. Thus, GMPLS was designed to extend the LSP-based virtual circuit-switching technique of MPLS to support various switching capabilities in WDM networks with appropriate interfaces in its routers as follows:

- packet-switch capable (PSC) interface,
- time-slot-switch capable (TSC) interface,
- lambda(wavelength)-switch capable (LSC) interface,
- fiber-switch capable (FSC) interface.

This multi-dimensional switching capability in WDM networks led to the notion of nested or hierarchical LSPs, where the LSPs of different interfaces were nested one after another. Thus, the GMPLS label request during the label-binding process for an LSP needs to carry a type field, including multiple values: packet, PDH(ANSI/ETSI), SONET, wavelength, waveband, and fiber, for interfacing GMPLS router with IP routers, ADM/DCS, OADM, and OXC playing their respective roles. A block schematic of a GMPLS router is presented in Fig. 12.7, illustrating all the four switching capabilities of GMPLS.

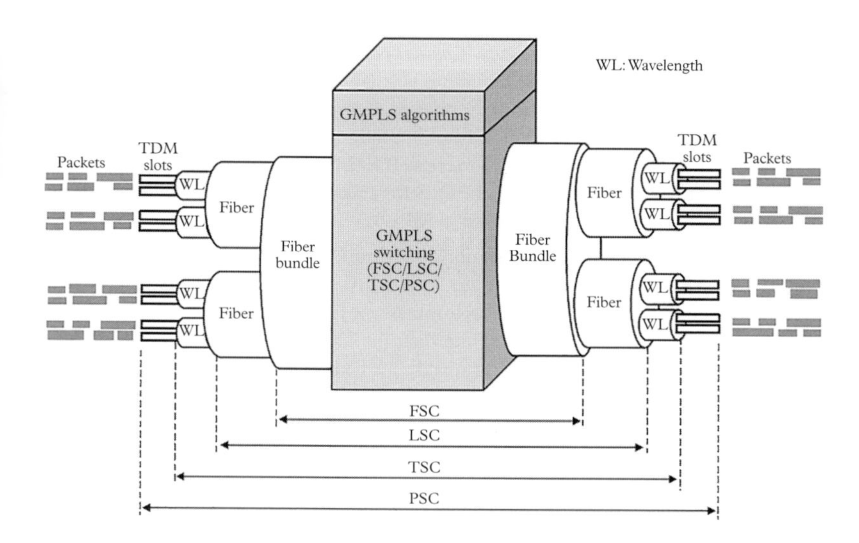

Figure 12.7 *Block schematic of a GMPLS router with four switching capabilities: FSC, LSC, TSC, and PSC.*

12.5.3 GMPLS protocols

The GMPLS protocol generalizes the basic concept of the MPLS protocol using the concept of a generalized label (G-label) with variable length, where the notion of label is far more broadened, and the operation of the network is carried out solely through the GMPLS-based control plane. Further, the LSPs set-up in MPLS networks are unidirectional in nature, while the GMPLS LSPs are all *bidirectional* conforming to the fact that all optical links are bidirectional. The control plane in GMPLS is kept independent of the data plane, which simplifies the control and management operations throughout the network.

Unlike MPLS, there are five different ways that a G-label may be assigned: fiber label representing a fiber in a fiber bundle, wavelength label representing a specific wavelength in a given fiber, waveband level representing a band of consecutive wavelengths from the WDM wavelength range, TDM label representing a time slot in a wavelength, and finally MPLS label generically representing packet-switched client networks.

The three major tasks carried out by GMPLS protocols are *routing, LSP establishment*, and *link management*. The routing scheme uses OSPF-TE and IS-IS-TE protocols (discussed later) for auto-discovery of the network topology and advertisement of the resource availability, with necessary enhancements for link protection and route discovery for backup in a shared-risk link group. The signaling scheme for LSP establishment uses RSVP-TE and CR-LDP protocols, with the necessary enhancements for the establishment of bidirectional LSPs, label-exchanging for non-packet (TDM) networks, use of suggested labels for faster label establishment and waveband support. The link management scheme uses link management protocol (LMP) for control channel management, link-connectivity verification, identification of link properties of adjacent nodes, and fault isolation. In the following we discuss all the components of GMPLS protocols in further details.

- *Routing*

  The routing scheme in GMPLS networks is designed using the existing routing protocols of IP: open shortest path first (OSPF) and internet system to internet system (IS-IS), with the necessary extensions. OSPF and IS-IS are basically link-state protocols, where the router in an area distributes the information on its connectivity with the rest of the network by using the state of its links (active interfaces) with the neighbors along with the underlying *cost* of sending data over those links. As the acronym OSPF indicates, *open shortest path first* implies the open or active link from the node that offers the minimum cost (shortest path). However, through this auto-discovery scheme, IP routers operate with a localized vision of the network, while remaining oblivious to the cost of the entire path from the ingress to the egress nodes, thus lacking the ability to exercise TE. In particular, such routing schemes cannot choose a route to ensure the QoS for a requested connection, thereby calling for the TE-based extension as used in MPLS.

  TE: Employing the TE constraints on the routing scheme leads to the placement of LSPs on pre-computed routes as in MPLS, so that the chosen routes satisfy the QOS needs. Only after such routes are found out, is the signaling process started to set up the LSPs. Note that, the link-state information (i.e., whether a link is up or down) doesn't qualify for setting up the LSPs; the active links must have sufficient bandwidth available to check whether the active link would qualify as a TE-qualified link (TE-link) or not. One of the ways to implement this mechanism is to use TE constraint on the OSPF algorithm. The extended version of the OSPF algorithm for this task is termed as constraint-based shortest path first (CSPF) algorithm, which can find out the paths with adequate bandwidth along with specified QoS limits (e.g., delay and delay-jitter parameters) through the *flooding* operation across the network. The database of all the TE links (TED) is computed by the path computation algorithm, which is eventually used to generate a TE-based network graph composed of all the TE links. GMPLS routing makes use of this graph for building up information for the LSP establishment process.

During the formation of TED, control operations remain confined to the channels of the control plane, while all the TE links that are advertised, correspond to the channels of data plane for setting up the LSPs.

Inter-domain routing: GMPLS routing, discussed so far, is used for a GMPLS network under a routing area or domain. In order to distribute the routing information across different domains, one needs to do it judiciously so that the routing scheme remains scalable. One possible way to address this problem has been to decide upon a few inter-domain (edge-to-edge) TE links through which the traffic can flow across any adjoining pair of GMPLS domains. ASON, as discussed later, comes forward with intelligent interfaces between different network domains to address the problems of inter-domain connectivity.

- *LSP establishment*
  Setting-up a unidirectional LSP in a GMPLS network takes effect by using the G-label, where all the necessary requirements of an LSP are specified. In particular, the process starts with a message sent from an ingress node, using a G-label request with the LSP type (representing specific switching need), payload type, signal type, protection requirement, and *suggested label*. The concept of suggested label is a unique feature in GMPLS, which is needed to overcome the problem of the long set-up time of optical switches. We explain the role of suggested label in the following.

  Suggested label: While setting up a unidirectional LSP, an upstream GMPLS node suggests a label to its immediate downstream node for onward transmission. The downstream node can refuse the suggestion and propose another label as may be deemed appropriate based on the availability, and the upstream node must re-configure itself accordingly (Fig. 12.8). However, the downstream node may also agree to the suggested label from the upstream node to receive the signal if it has the appropriate resource available to receive signal from the upstream neighbor. This mechanism helps (when the suggestion is agreed upon by the downstream node), as the suggested label can meanwhile be used by the upstream node to pre-configure itself with the proposed label.

  For bidirectional LSPs, the same procedure can be repeated in both directions. However, this process can be made more efficient by using an integrated approach to set up LSPs in both directions (Banerjee *et al.* 2001b).

- *Link management*
  The notion of a link (or a channel) in GMPLS networks is a connectivity between two adjacent nodes, which could imply a fiber link as a whole, a wavelength on a fiber, or a TDM slot in a wavelength. Whatever it may be, a given channel may be assigned to transport data or control traffic, referred to as a data or control channel. Control channels would carry messages for purposes such as, signaling, routing, and other tasks to be performed within the jurisdiction of control plane, while the data channels would only carry data packets. Thus, the identities of all the channels between adjacent node pairs need to be identified as either control or data channels

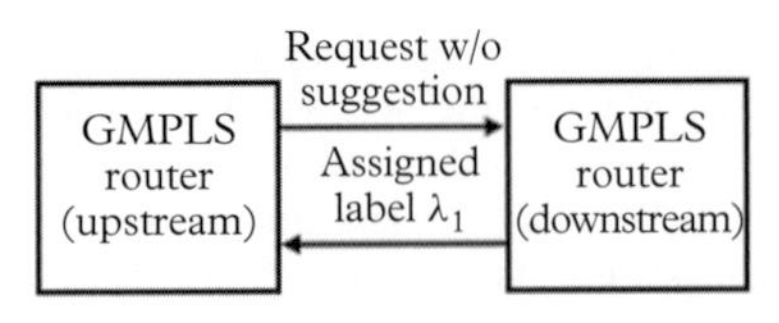

(a) Label assignment without any suggested label.

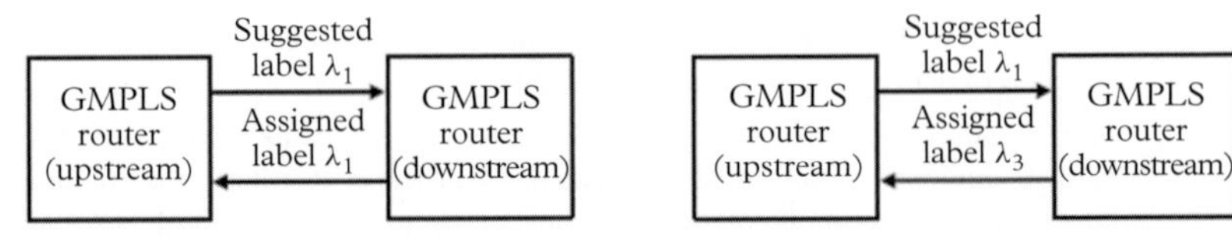

(b) Label assignment using suggested label, which is agreed upon by downstream router; upstream router saves time.

(c) Label assignment using suggested label, which is refused by downstream router and a different label is assigned.

Figure 12.8 *Illustration of suggested label in GMPLS network.*

in the nodes (LSRs/LERs). This task is carried out by a protocol called the link management protocol (LMP). In particular, LMP has a number of crucial functionalities to be carried out between the adjacent nodes: control channel management, data link discovery and capability exchange, data link connectivity verification, and fault management. In the following, we describe the above functionalities of LMP.

Control channel management: In the first step of control channel management, the neighboring nodes go through an initialization process, by activating the control channels and exchanging their identities through the activated control channels. Once the control channels get activated and the identities and capabilities are exchanged between the nodes, the connectivity is set up and managed thereafter by the control channel management process using the *Hello* protocol.

Data link discovery and capability exchange: This functionality of LMP helps LERs and LSRs to discover the data links between the adjacent neighbors through a sequence of message exchanges. Thereafter, messages are exchanged between the nodes for sharing the specific features of data links.

Data link connectivity verification: Verification of status of the data links are needed to ensure that they continue to remain functional, and operator intervention would be needed as and when link failure takes place.

Fault management: When a fault occurs in a data link, the concerned downstream node for the faulty link detects the problem (non-receipt of data or a fall in received signal quality). Thereafter, the link is isolated and the location of the fault is notified to the source node. Fault localization becomes important for supporting protection/restoration of the link (see Chapter 11).

One more important functionality of LMP is known as *link bundling*, which becomes worthwhile to reduce the burden on the routing protocol. In GMPLS-enabled optical networks, there might be a number of links between a given node pair, and often it may not be needed to refer to these links separately. Such links can be bundled together as a single entity and collectively referred to as a link bundle. The agreement on link bundling can be reached by LMP through the exchange of link identifiers, thereby registering such bundles as single TE links.

12.6 Automatically switched optical network

ASON represents the optical network architecture developed by ITU-T community for long-haul WDM optical networks, using the combination of management, control, and data planes, along with certain special features (Jajszczyk 2005; ITU-T ASON G.8080 2006). Essentially, the long-haul networks covering large areas need to accommodate different administrative areas and carrier networks, and the carrier networks in different areas might use NEs from different network vendors. This diversity of long-haul networks leads to the formation of various *domains*, based on geographical areas, administrative jurisdictions, carriers, and network vendors. In large networks with such diversities, ASON offers the much-needed multi-domain operation facilitating the deployment of multiple carriers, which can in turn deploy NEs from multiple vendors in different areas. In the following, we describe the ASON architecture employing different types of network interfaces between the various domains.

12.6.1 ASON architecture

By and large, the ASON architecture follows the generic representation as shown earlier in Fig. 12.1, but with the necessary multi-domain functionalities and multiple-layer formation within the data and control planes (Jajszczyk 2005). The data plane operating with multiple standards (e.g., circuit-switching for SONET, packet-switching for IP, virtual circuit-switching for ATM/MPLS, and GMPLS) call for different switching options, thereby leading to standard-specific *layers* in control plane. An example multi-domain network operating with multiple carriers and equipment from different vendors is illustrated in Fig. 12.9 with ASON interfaces.

As shown in Fig. 12.9, the example network formation has two carrier networks from Carrier A and Carrier B, and each carrier network is composed of two subnetworks, resulting into four domains: Domains A1 and A2 of Carrier A, and Domains B1 and B2 of Carrier B. Within a given domain, the links between any two adjacent nodes use some intra-domain communication scheme, while the links between different domains employ inter-domain communication scheme. Thus, in ASON the NEs in multiple domains interact through three specific interfaces: internal network-network interface (I-NNI), external network-network interface (E-NNI), and user-network interface (UNI). UNI represents a bidirectional signal interface between the user (requesting a service from a client network) and the control plane entity of ASON in a given node in the neighborhood of the user.

When multiple domains operate, the relationship between the domains may remain administratively disjoint, or not in the cases when both have trust-based relation and/or use equipment from the same vendor. In the latter case, ASON will have separate control plane entities for two domains, but connected through an internal arrangement using I-NNI as the interface. Otherwise, they get connected through external interface, i.e., E-NNI. In Fig. 12.9, a UNI is located as a reference point between a user from a client network (bottom left) and an NE of Domain A2 of Carrier A. Note that Carrier A operates with two administrative domains: Domain A1 and Domain A2. These two domains, though under the same carrier, use NEs from different network vendors and hence interfaced through an external interface, i.e., E-NNI. Domain A2 and Domain B1, operating under two different carriers, are also interfaced through E-NNI. However, in the network area of Carrier B, Domains B1 and B2 use NEs from the same vendor and have trust-based relation, thereby using I-NNI as the interface between them.

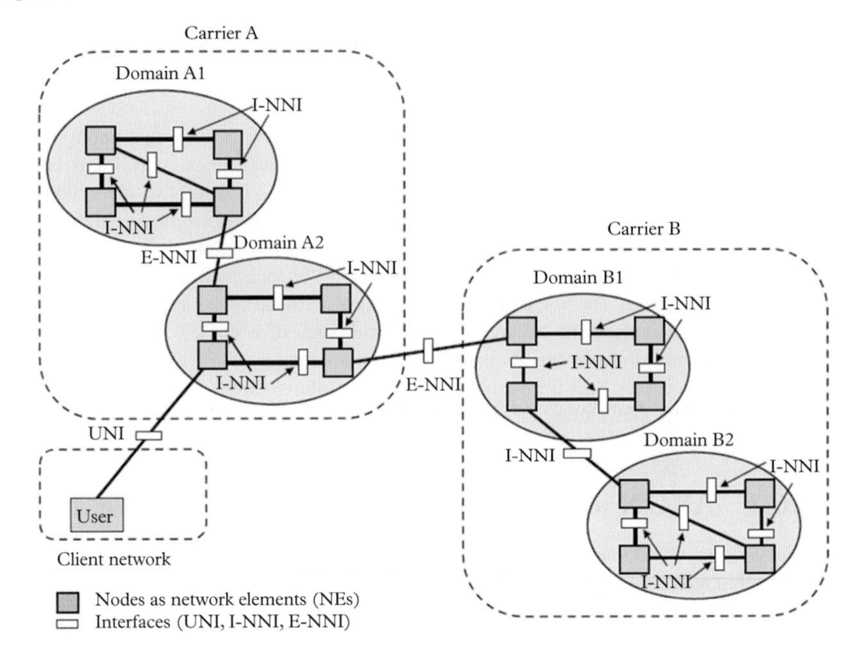

Figure 12.9 *Example multi-domain network operating with ASON.*

ASON offers three main connection options: unidirectional point-to-point, bidirectional point-to-point, and unidirectional point-to-multi-point connections. Another connection option – an asymmetric connection – can also be realized as two unidirectional connections or as a special setting of a bidirectional connection. In respect of holding time (service duration) of connections, one can choose one of the three types: permanent connection, switched connection, and soft-permanent connection. The permanent connections are typically meant for long holding times ranging from months to years, and hence do not need any support of automatic reconfiguration functionality from control plane. In other words, the permanent connections are set up directly by the management plane manually. However, the switched connections are set up on demand for the source-destination pairs, by using the automatic routing and switching capability of the control plane. The soft-permanent connection is a hybrid mix of the last two cases, where the source and destination users from the respective clients set up permanent connections with the respective edge nodes of the network at two ends, and the internal routing and switching between the two edge nodes of the network are handled by the control plane.

12.6.2 ASON protocols

With the above architectural framework, ASON messages are exchanged across the network for executing the automated functionalities which, mostly being real-time in nature, come under the purview of the control plane. Major operational features of ASON control plane include: *discovery scheme, routing scheme, call and connection control scheme, signaling scheme,* and *survivability measures*. In the following, we describe all these functionalities of the ASON control plane (King *et al.* 2008).

- *Discovery*
  For the networking functionalities to take effect, each node needs to carry out three different types of discovery processes: *neighbour discovery, resource discovery,* and *service discovery*. The discovery processes are carried out through exchanges of messages between the nodes with the relevant identity attributes: neighbour, resource, and service. The discovery messages are initiated from each node carrying its own identity, while the acknowledgment messages return the identity from the other end. The overall discovery process works as follows.

  Neighbor discovery: The neighbor discovery process first builds up the adjacency matrix, which is essential to set up any connectivity from a given node. Toward this goal, a given node needs to first assess the *physical adjacency*. Next comes the *layer adjacency*, which indicates the association between the two ends of a logical link for a given layer, which in turn helps in carrying out the routing scheme. The third adjacency, known as the *control adjacency*, indicates the association of the control-layer entities in two neighboring NEs.

  Resource discovery: The resource discovery process enables a node to explore the network topology and the resources of the NEs along with their capabilities. However, due to the hierarchical nature of the ASON control mechanism, details of the topology formation in a domain may remain hidden to the nodes in the other domain.

  Service discovery: The service discovery process verifies the service capabilities of various administrative domains of a network. The service capability implies that the COS, routing schemes, etc. are supported by different domains.

- *Routing*
  The control plane in ASON employs hierarchical routing scheme and operates with a source-based step-by-step approach. In the multi-domain architecture of ASON, not every node in a domain gets to know about the topology of any other domain, and a specific node (or nodes) would play the role of *routing controller* (RC) and helps the other nodes in the domain for inter-domain connections. The RC of a domain has the knowledge of its own domain topology with all other nodes in the same domain. Further, the RC knows how its own domain is connected to the other domains at the RC level, but it doesn't have the knowledge of the internal topology of the other domains. The knowledge of the RCs of the other domains is acquired by an RC in a given domain through the discovery processes discussed earlier.

  The routing process within a given domain operates with the internal routing scheme, which may vary from domain to domain. For communication across multiple domains, the ingress node gets the request from the client and approaches its RC. Next, from this RC, the RCs of other domains are reached, so that a call can be routed through several domains. Within a given domain, the routing is carried out using a proprietary routing scheme through the I-NNI interfaces, and for the connections going beyond the domain under consideration gets routed through the domain RC using the E-NNI interface. The routing scheme across multiple domains is illustrated in Fig. 12.10.

- *Call and connection control*
  Calls and *connections* are two functionalities in ASON that are carried out as two separate processes, where end-to-end call messages precede the corresponding connection messages. Call is viewed as a mechanism to coordinate the relevant NEs to *associate* the two end points of a requested connection, so that a service can be

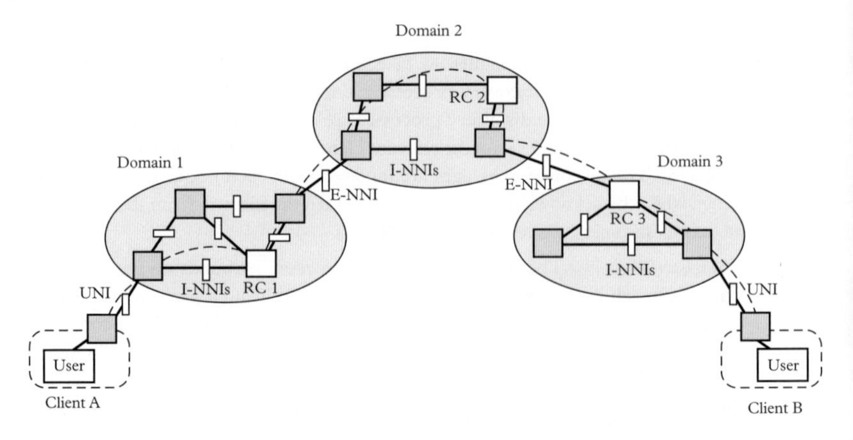

Figure 12.10 *Multidomain routing scheme in ASON using route controllers (RCs).*

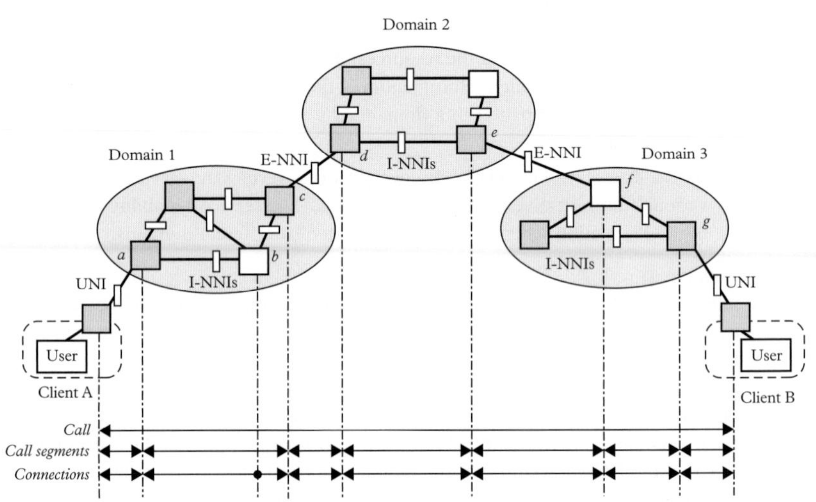

- Route for the call is assumed to pass through the node-sequence {*a-b-c-d-e-f-g*}
- Node *b*, being an internal node in Domain 1, does not *interact* with call segment.

Figure 12.11 *An example of call and connection control schemes in ASON.*

offered to the caller and the called ends, and be maintained for the necessary duration. However, connection is the subsequent process, where having associated the end points using call mechanism, appropriate concatenated (if necessary) network links are engaged to actually set up the connection so that the transport of data takes place between the two end points. The two processes are illustrated in Fig. 12.11, where call and connection schemes work between two client networks separated by the three domains, Domains 1, 2, and 3. The call passes through a chosen route: through the node sequence *a-b-c-d-e-f-g* in the three domains. Note that the call segments are shown between the edge nodes only, while the connections are set up using all en route nodes. Thus, the call segments don't involve any internal node (node *b* being the only internal node in this example), while the connections involve all the en route nodes (i.e., including node *b* as well).

- *Signaling*
  Signaling is an elementary component of all the control plane tasks, transporting all the control messages over the control plane for setting up, tearing down, maintaining, and restoring all the optical connections or lightpaths. Generally, *common-channel signaling* is a well-accepted and scalable signaling scheme used across various network domains, so that the signaling network remains disjoint with the transport network (data plane). Automatic discovery and routing are supported by signaling schemes, which enable the network to remain less dependent on manual intervention.
- *Survivability measures*
  ASON realizes survivability through collaborative involvement of its three planes. Guided by the SLAs, the protection of some of the selected connections are set up by the management plane. In the event of a failure, the control plane operates to use the protected paths for the failed connections. For unprotected connections, the control plane attempts to recover the connection by re-routing through the available spare capacity in the network (i.e., employs some restoration scheme, as discussed in Chapter 11). The re-routing service falls in two categories: hard and soft re-routing. The hard re-routing category takes effect in the event of failure, while soft re-routing is employed for optimizing traffic distribution across the network, network maintenance, and some planned network engineering from time to time.

12.7 Software-defined networks

Software-defined networking (SDN) is a relatively recent concept with the unique feature of centralized and programmable operations that can control and manage different types of networking media and technologies: wireline (e.g., Ethernet, optical fiber) and wireless networks, across various possible networking segments (Thyagaturu *et al.* 2016; Nunes *et al.* 2014; Kreutz *et al.* 2015; Cao *et al.* 2017). In particular, the networking community around the world has been engaged in research over the years to bring out a networking paradigm that can accommodate the three much-desired important aspects: programmability of networking functions, centralization of network control, and complete separation of data and control planes. The centralized operation of the control plane empowers SDN with a unique potential to unify the diverse range of NEs/nodes, vendors, and administrative domains. However, the control plane in an SDN is spared from looking into the specific details of any NE. Instead, it uses an *abstraction* of all the features of diverse NEs to map them into a set of higher-level parameters understood by the applications run by the network operators. In the following, we describe the rationale behind this philosophy.

In any given network, the operators need to configure the NEs by transforming the higher-level networking policies into equipment-specific settings in the NEs which may vary between different network vendors. Such configuration/reconfiguration tasks, with time-varying traffic and occasional network failures, need to be carried out on a real-time basis, and hence in an automated manner. As discussed earlier, GMPLS can address this problem, albeit with extended and hence complex versions of the pre-existing IP/MPLS protocols. However, SDNs, by using the centralized control and abstraction of low-level NE features into higher-level network parameters, can carry out such operations readily with enhanced network performance and resource utilization.

12.7.1 SDN architecture

The operation of an SDN-based network is illustrated in Fig. 12.12. As shown in the figure, the control plane runs a network operating system (NOS), which carries out the abstraction process to map the various settings of the NEs into some abstract parameters that are *understood* by the application layer. The details of the settings in NEs are received through southbound interface (SBI), which are converted by the NOS into a set of abstract parameters and passed on to the application plane through northbound interface (NBI). The control plane may be operated by one or more interconnected controllers located at some selected nodes to ensure efficient and reliable control operation. There can be a hierarchical connectivity between multiple controllers, where one of them might play a *coordinating role*, or all might collaborate with each other using a *horizontal framework*. However, the underlying NEs in the data plane remain oblivious to the number of controllers, and stay attached to, and recognize, the control plane as a single entity.

On the top of the control plane, the SDN operates using a number of applications from the application plane through the NBI by using a northbound protocol: *re*presentational *s*tate *t*ransfer (REST). NBI works with a client-server model, where the controller and the application plane play the roles of server and client, respectively. Using NBI and the inputs from the application layer, network operators interact dynamically through the control plane with the NEs. Since the settings of NEs are mapped into an abstract information set understood by the control plane, the SDN can run the network with a diverse range of network infrastructures seamlessly without needing any details of the specific NEs.

The management plane works on the administrative issues along with slower or non-real-time tasks, and interacts accordingly with all the other planes. For example, the management plane interacts with the application plane on the SLAs to be followed by the network. For the control plane, interactions with the management plane take place for the slower actions to be undertaken in the network, such as NE setting (not the dynamic node reconfigurations), repair works after link/node failures (with backup paths set up on a real-time basis by the control plane), etc. In the data plane, initial network settings and maintenance of NEs are carried out through management plane instructions, some of them being handled offline and hence manually.

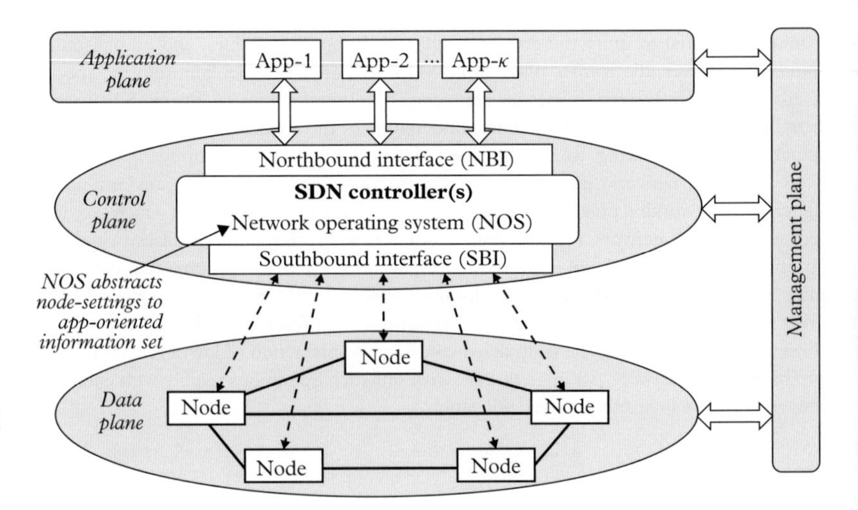

Figure 12.12 *SDN architecture. Note that, the control plane may operate with one SDN controller (or more) with the necessary interconnections between themselves.*

As shown in Fig. 12.12, the SDN needs to employ an appropriate protocol between the control and data planes through SBI. There have been a number of choices available for such a protocol: OpenFlow protocol, path computation element protocol (PCEP), network configuration protocol (NETCONF), etc. These protocols are, however, element-specific; for example a south-bound message from the OpenFlow interface will not comply with the NETCONF interface at the NE. OpenFlow being a popular protocol for the purpose, we discuss its basic features in the following.

12.7.2 OpenFlow protocol

The OpenFlow protocol was being explored even before research on SDN started. The primary objective of the original OpenFlow protocol was for encouraging the network researchers to run experimental protocols on a part or *slice* of an operational network in a campus (mainly with Ethernet switches), so that their new ideas on the networking protocols could be tried out without interfering with the parent operational network (McKeon *et al.* 2008). Before the development of OpenFlow protocols, researchers in the networking community felt rather *ossified* within the legacy framework of commercial switches and routers with hidden details, as their own ideas remained untested on a practical networking platform. Thus, the OpenFlow activity started in a few academic campuses over packet-switched LANs by exploiting some common features (use of flow tables to forward an incoming packet) of the commercial switches and routers (nodes/NEs), and later this idea was extended to include circuit-switched traffic as well, for use in a larger platform such as the SDN.

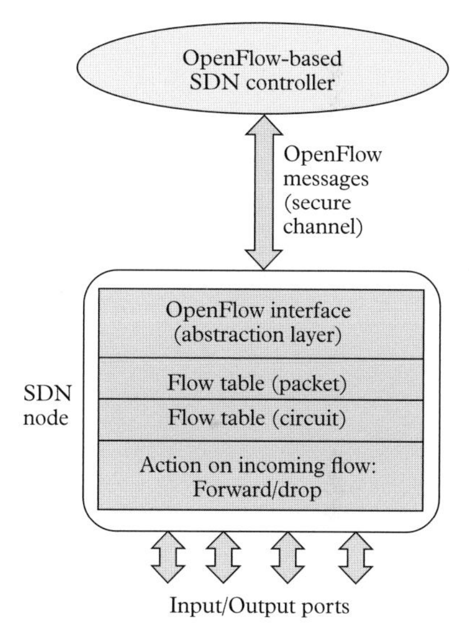

Figure 12.13 *Functional diagram of OpenFlow protocol in SDN.*

The OpenFlow protocol (or more precisely the extended version of the OpenFlow protocol) used in the SDN control plane, views the packet- and circuit-switched traffic streams commonly as *flows*. The operation of the OpenFlow protocol is illustrated in Fig. 12.13, where the control plane abstracts the task of each node in the form of *flow tables* for the packet- and circuit-switched traffic. When asked by a node for an incoming packet (not finding any match in the existing flow table), the NOS in the control plane communicates through SBI with the OpenFlow interface (abstraction layer) of the node, and provides the necessary flow-table information. Then the node updates the flow table and forwards the packet (packet/circuit-switched) using the updated flow table. Thus, centrally, the control plane decides on the access to the network, routing of flows, TE, etc. across the entire network. In other words, the execution of the control-plane tasks is orchestrated across the network in a centralized manner by using the OpenFlow-based flow tables set up in the SDN-enabled nodes.

The SDNs enabled by the dual-flow-based (packet and circuit flows) OpenFlow protocol can operate with a versatile control framework on all kinds of optical networks (circuit/packet-switched, fixed/flexible-grid[2]), and can offer support for the networking needs in all network segments. For example, with the datacenters located in the optical long-haul, metro, or access networks using the framework of OpenFlow-based SDN, the cloud-based services can reach the end-users efficiently through the concatenation of circuit- and packet-switching network domains, irrespective of their locations. Thus, the data can flow seamlessly from the datacenters to the users or other datacenters through the packet- and circuit-switched network domains, utilizing the dual-flow support from the OpenFlow protocol.

Figure 12.14 illustrates the flow tables used by the OpenFlow protocol in an SDN node, where part (a) shows a flow table for the packets of the packet-switched flows with

[2] Flexible-grid optical WDM networks are popularly known as elastic optical networks (see Chapter 14).

Input port	VLAN ID	Layer-2 (Ethernet) SA, DA, Type	Layer-3 (IP) SA, DA, Type	Layer-4 (TCP) Source, Destination

(a) Flow table for packets.

Input/ output port	Input/ output wavelength	VCG	Start TDM slot	Signal type

(b) Flow table for circuits.

Figure 12.14 *Flow tables in an SDN node for packets and circuits.*

the layer-2, layer-3, and layer-4 headers, typically representing Ethernet, IP, and TCP, respectively, along with the input port number of the node and VLAN ID. Thus, an incoming packet for a packet-switched flow with matching values for these fields with one such flow in the flow table for the packets, would follow the corresponding flow along the respective route, and accordingly the actions would be taken to forward the packet to an appropriate port for onward transmission. Similarly, Fig. 12.14(b) shows the flow table for the packets of circuit-switched flows, the flow table consisting of layer-0 and layer-1 information, corresponding to a wavelength and a TDM slot, respectively, along with a virtual concatenation group (VCG) and input/output port numbers of the node. Hence, when a packet belonging to a circuit-switched flow arrives at the given node and finds a match with the wavelength, TDM, VCG, and the assigned ports of the flow table, it gets forwarded to the appropriate output port of the node for onward transmission. When an incoming packet doesn't find a match in the existing flow tables, the node under consideration forwards its query to the controller (control node) to seek the necessary routing information and wait for the response from the controller for onward transmission through the appropriate output port of the node. We discuss below some of the messages used by the OpenFlow protocol between the NOS in the controller and the forwarding node under consideration through SBI.

Initially to set up a connection between the controller and a node, *hello* messages are used in both directions between the NOS of the controller and the abstraction layer of the node through SBI. Once the connection is set up, both keep communicating by using *echo request/reply* messages to check the connection status (i.e., alive/dead), which also helps the controller to assess the latency of the corresponding connectivity. The controller also asks for the identity and the various capabilities of each node by using the message, called *features*, following which the configuration parameters of the nodes are exchanged and the abstraction is set up using *configuration* messages.

The SDN controller needs to collect the needed information from the nodes by using *read state* messages, and it can also modify the OpenFlow tables in the nodes by using *modify state* messages. Further, *packet in* and *packet out* messages are used in the controller for executing the forwarding operations in the nodes by using the flow tables. However, in the reverse direction each node keeps informing its status (i.e., up or down) to the controller by using *port status* messages.

12.7.3 SDN virtualization

The idea of network *virtualization* was explored while designing the OpenFlow protocol, so that multiple virtual networks can coexist and operate independently in a given physical network. Such virtual segmentation of a given network infrastructure enables the network users with the privilege of carrying out experiments with their new ideas on an independent part or *slice* of a running network. In the process of network virtualization in an SDN, some specific node is chosen to play an advisory role over the others, and is called as network hypervisor. By using network hypervisors, SDNs can realize a range of slice-specific and inter-slice network functions, including routing, switching etc. In effect, a network hypervisor creates multiple virtual network slices that remain practically isolated from each other.

12.7.4 SDN Orchestration

In a practical scenario, a functional SDN may employ multiple vendors, technologies, administrative regions, etc., which must work together (orchestrate) to ensure that the network runs smoothly from end to end, in spite of the heterogeneity of the network domains. The procedure that achieves this collaborative SDN operation in the background is called *orchestration*. Orchestration can be of two types: *multi-domain* and *multi-layer*. In the following, we discuss these two types of SDN orchestration.

In SDNs using multiple domains, operating with heterogeneous technologies, vendors, and administrative regions, one needs to ensure best utilization of resources, while setting up connections across different network domains. Multi-domain orchestration helps in achieving this object by using appropriate routing and bandwidth/spectrum allocation algorithms between different domains using a *horizontal framework*. Typically, each domain operates with a domain-specific controller, and these controllers interact with each other horizontally (i.e., without using any higher-level orchestrator) by using a *flat* control protocol to set up a connection across various domains (Fig. 12.15). As shown in the figure, the

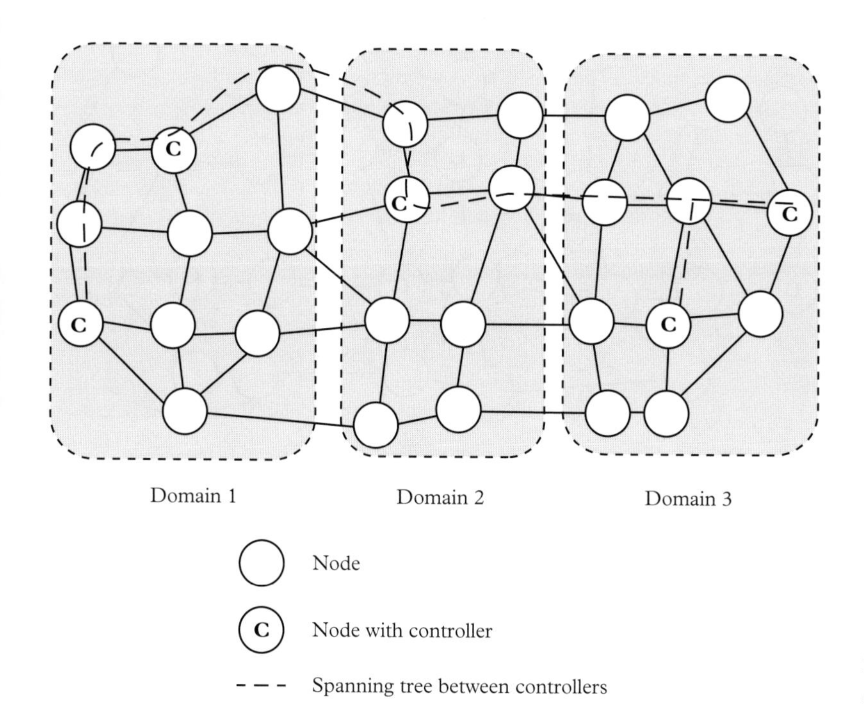

Figure 12.15 *Multi-domain control in SDN using flat orchestration.*

network is divided into multiple domains, and each domain employs one or more controllers that interact between themselves, forming a horizontal or flat architecture. Note that these controllers are physically collocated with some selected nodes, and besides working for all the nodes in their respective domains, they also need to interact between themselves to share and update (synchronize) control information on a periodic basis. One possible means of carrying out this task is as shown in Fig. 12.15, where the controllers are connected through a spanning tree which might use full or partial capacity of the constituent point-to-point optical links. Note that the connections are to be point-to-point with OEO conversion at each controller, so that information can be shared between all controllers in the electrical domain. In order to make the architecture survivable against failures (link/node), one needs to set up backup connections for the spanning-tree links using an appropriate design methodology.

However, multi-layer orchestration is realized by using a hierarchical (vertical) coordination, where on the top (logically) of the domain-specific SDN controllers, another node plays the role of an orchestrator and operates notionally as an *umbrella* over the underlying SDN controllers (Fig. 12.16). As evident from the example, the orchestrator, like the controllers, is also collocated at some select node in the network and coordinates with all the controllers across all the domains, and forms a star-based logical connection-set to carry out the task of orchestration. In this case as well, the design methodology should address the survivability of the control plane along with a backup orchestrator, because with a failed orchestrator, the coordination between controllers would cease to exist.

In setting up an SDN control-plane framework for multi-domain horizontal orchestration, one needs to find appropriate physical locations for the controllers (collocated within

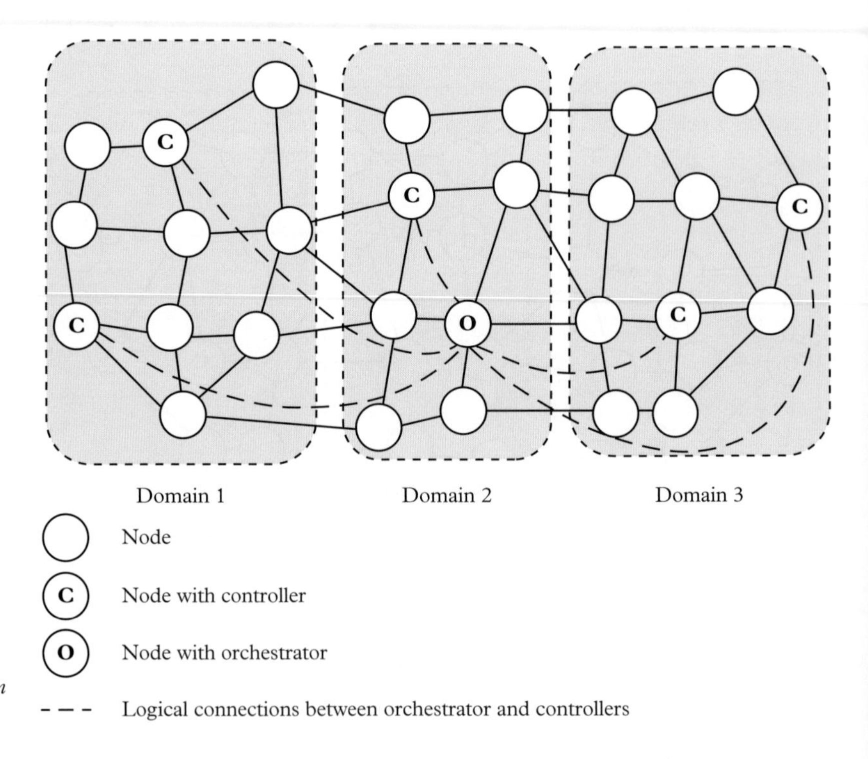

Figure 12.16 *Multi-domain control in SDN using hierarchical orchestration.*

select nodes), and a survivable control-plane topology to interconnect them. The choices should be made judiciously, so that the traffic between the controllers and the network nodes, as well as the inter-controller traffic used to synchronize the controllers, experience a latency not exceeding a specified upper-bound. Similarly, for multi-layer orchestration, the locations of the controllers at the lower layer and the orchestrator at the higher layer are to be determined again with a guarantee for the control-plane latency to remain confined below the specified upper-bound. Over-provisioning of network resources for control plane would slow down the data plane, while under-provisioning would prevent the control plane from being adequately agile. Thus, designing an appropriate orchestration framework for SDN control plane turns out to be an optimization problem, which can be addressed using LP-based formulations and/or heuristic algorithms. The interested readers are referred to (Lourenço *et al.* 2018) for these design methodologies.

12.7.5 A queuing-theory perspective of SDN

The performance of an SDN cluster using a centralized control plane can be assessed by using the model of *network of queues with feedback*, based on Jackson's theorem (see Appendix C). In this section, we present one such model (Mahmood *et al.* 2015) that deals with one control node supporting a cluster of packet-forwarding nodes (hereafter simply referred to as nodes), as shown in Fig. 12.17. Each node (node i in the figure) in the cluster receives packets from outside (ingress traffic) as well as from the other nodes in the cluster, along with the packets from the control node. The packets coming from the control node to a given node include the answers to the queries from the node itself and also the updates from the other nodes coming via the control node. The model assumes that the arrivals and service times are independent in each node across the network. Further, the arrival process and service durations are assumed to follow Poisson and exponential distributions, respectively. These assumptions along with Jackson's theorem allow us to model each node as an $M/M/1$ queue, thereby simplifying the analysis significantly.

The network cluster consists of N nodes with λ_i and μ_i representing the external arrival rate and the service rate for the ith node, respectively. From each node, say node j, a fraction

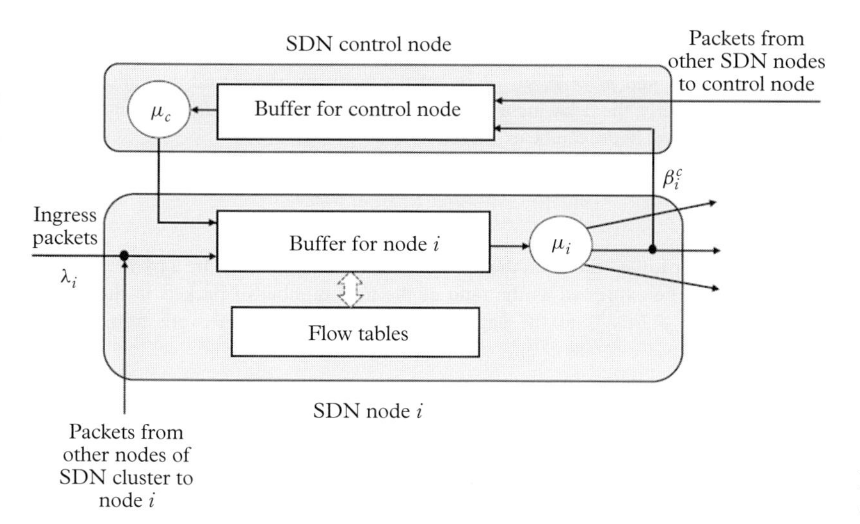

Figure 12.17 *Queuing model for an SDN cluster with one control node.*

α_{ji} of its departing traffic goes to node i. Further, from each node, say node i, a fraction β_i^c of the arriving traffic λ_i is forwarded to the control node (node c, say) for necessary information. The total arrival rate at the control node is assumed to be Γ_c with a service rate of μ_c. The traffic arriving at node i from outside and the other nodes directly (excluding the traffic from the control node) is represented by γ_i, while the overall traffic arriving at node i including the traffic from the control node and others is denoted by Γ_i. With this framework, we express the input arrival rate γ_i at node i arriving from outside as well as the other nodes of the cluster as

$$\gamma_i = \lambda_i + \sum_{j=1,\ j\neq i}^{N} \alpha_{ji}\gamma_j. \tag{12.1}$$

As mentioned above, from the control node each node would receive answers to the queries from itself as well as the updates from the other nodes, as necessary, through the control node. Thus the overall arrival rate Γ_i at node i can be expressed as

$$\Gamma_i = \gamma_i + \beta_i^c\lambda_i + \sum_{j=1,\ j\neq i}^{N} \alpha_{ji}\beta_j^c\lambda_j, \tag{12.2}$$

where $\alpha_{ji} = 1$ or 0, indicating whether the update from node j involves node i or not. Furthermore, at the control node the total arrival rate is represented by Γ_c, expressed as

$$\Gamma_c = \sum_{i=1}^{N} \beta_i^c\lambda_i. \tag{12.3}$$

Using the above framework, and assuming that the network runs in steady state (implying that the traffic at each node is upper-bounded by unity), one can express the expected number of packets at node i and the control node, represented by $E[N_i]$ and $E[N_c]$ respectively, as

$$E[N_i] = \frac{\rho_i}{1-\rho_i}, \quad E[N_c] = \frac{\rho_c}{1-\rho_c}, \tag{12.4}$$

where $\rho_i = \Gamma_i/\mu_i$ and $\rho_c = \Gamma_c/\mu_c$ represent the traffic in node i and the control node, respectively. Similarly, the delay incurred in the respective nodes, i.e., $E[T_i]$ for node i and $E[T_c]$ for the control node, can be expressed as

$$E[T_i] = \frac{1}{\mu_i - \Gamma_i}, \quad E[T_c] = \frac{1}{\mu_c - \Gamma_c}. \tag{12.5}$$

Next, by using Little's theorem, the average delay $E[T_{net}]$ incurred by a packet entering the network can be expressed as the ratio of the total number of packets in the system $(= \sum_{i=1}^{N} E[N_i] + E[N_c])$ to the aggregate arrival rate in the network from outside $(= \sum_{i=1}^{N} \lambda_i)$. Thus, we obtain $E[T_{net}]$ as

$$E[T_{net}] = \frac{\sum_{i=1}^{N} E[N_i] + E[N_c]}{\sum_{i=1}^{N} \lambda_i}. \tag{12.6}$$

Further, the average delay incurred by a packet in each node would have two components, one of them being for the packets having direct flow through the node itself and the other for the packets which need to be forwarded to the control node, and thereafter traverse through the node again for the second time. Thus one can express the two types of delay incurred in a node as

$$\begin{aligned} T_i^{dir} &= T_i, \text{ with probability } 1 - \beta_i^c \\ &= \underbrace{T_i}_{\text{1st visit}} + T_c + \underbrace{T_i}_{\text{2nd visit}}, \text{with probability } \beta_i^c. \end{aligned} \tag{12.7}$$

Using the above expressions, we therefore express the average comprehensive delay in node i as

$$\begin{aligned} E[T_i^{comp}] &= (1 - \beta_i^c)E[T_i] + \beta_i^c(2E[T_i] + E[T_c]) \\ &= (1 + \beta_i^c)E[T_i] + \beta_i^c E[T_c] \\ &= \frac{1 + \beta_i^c}{\mu_i - \Gamma_i} + \frac{\beta_i^c}{\mu_c - \Gamma_c}. \end{aligned} \tag{12.8}$$

It is instructive to note that the size of the SDN cluster, i.e., N, should not be increased to the extent that Γ_c is forced to approach μ_c, failing which the control node will not be able to remain in a steady state and will thereby cease to have any centralized control over the SDN nodes in the cluster.

12.7.6 Software-defined optical networks

In order to run optical networks with SDN-based control and management, hereafter referred to as software-defined optical networks (SDONs) (Channegowda *et al.* 2013; Bhaumik *et al.* 2014; Thyagaturu *et al.* 2016), all the optical devices and modules used in the network nodes – optical transceivers, various WDM devices, optical switching elements, ROADMs, OXCs, etc. – need to be programmable by the network controller with appropriate abstraction of all the device parameters. These devices are typically circuit-switched (and can also be packet/burst-switched in future) and need to be flexibly configured with time-varying traffic by the control plane NOS (through SBI), so that different types of optical signals (with varying bit rates, modulation schemes, wavelengths, etc.) can be transmitted across the network. As discussed earlier, all these devices will have to interact with the controller using specific set of messages through SBI. We discuss below some of the relevant optical devices that need to be controlled by the SDON control plane automatically in a programmable manner.

Optical transceivers and switching elements

Optical transmitters used in the existing optical networks, with bit rates not exceeding 10 Gbps, employ IM of lasers or LEDs while the corresponding receivers employ DD (non-coherent) of the received optical signal using photodetectors. These IM-DD transceivers (see Chapter 2) are simple to design and can be controlled relatively easily as compared to their higher-speed versions. However, the optical transceivers operating at higher bit rates employ multi-level modulation schemes, such as QPSK, QAM, and OFDM, which need to employ coherent or nearly coherent modulation/demodulation techniques. In the optical networks operating under the SDON framework, the control plane needs to employ an appropriate abstraction mechanism, so that its NOS can relate the application-layer messages with the transceiver parameters, such as modulation scheme, bit rate, wavelength, etc.

Further, in the forthcoming optical networking scenario, when elastic optical networking (EON) is adopted for flexible-grid (different optical connections assigned different bit rates and hence different bandwidths) operation, this abstraction process has to go down deeper into the technology used in the physical layer of the network. In particular, the NOS should be able to abstract the parameters of the bandwidth-variable transceivers and switching devices (e.g., ROADMs, OXCs, see Chapter 14) to be used in EONs into the higher-level application layer messages and vice versa.

Non-SDN optical NEs

One major concern with using the OpenFlow-based control will arise from the legacy WDM-based NEs (ROADMs and OXCs), which do not have any SDON-interfacing abstraction layer installed therein and hence cannot be controlled by OpenFlow-based SDON controllers. This problem is addressed by making the SDON controller backward-compatible through the retro-fitting devices. This arrangement is illustrated in Fig. 12.18, where a sublayer is introduced between the SBI and the native interface of the legacy NEs for the conversion of the messages in both directions. As shown in Fig. 12.18, the legacy devices are by default designed to operate with a simple network management protocol (SNMP) interface, and hence the retro-fitting abstraction layer is designed to bridge between SBI and the SNMP interface, thereby giving an end-to-end compatibility between OpenFlow-based controller and the legacy optical devices. As shown in the figure, having been retro-fitted with the control layer of SDON, such legacy optical nodes can be readily inserted into the network to interwork with the neighboring SDON nodes.

12.7.7 SDN/SDON experiments

In optical WDM networks with their underlying GMPLS/ASON operation, SDN can be used to map all the features of the WDM-based NEs (e.g., ROADM, OXC, WSS) in terms of various switching capabilities (e.g., packet, TDM slot, wavelength, fiber) into a set of abstracted information by the NOS at each WDM node. Thereafter, the network operators can centrally use various applications through the application layer interface and operate

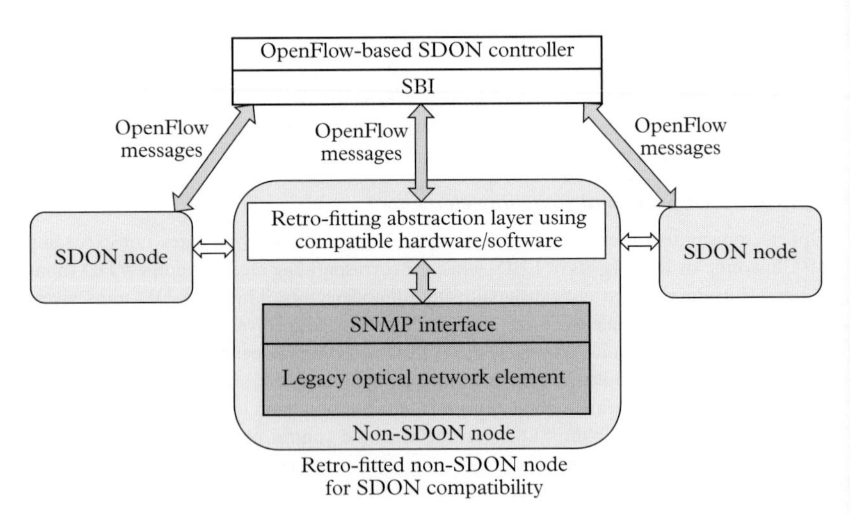

Figure 12.18 *Representative block schematic of a retro-fitted non-SDON node operating alongside the neighboring SDON nodes within an SDON.*

the network efficiently in a dynamic manner. Several experimental studies in this direction have been reported in the literature (Gudla *et al.* 2010; Liu *et al.* 2011; Liu *et al.* 2013; Channegowda *et al.* 2013; Patel *et al.* 2013). We discuss briefly some of these developments in the following.

Laboratory testbeds

In the experimental testbed for SDN/SDON reported in (Gudla *et al.* 2010) from Stanford University, the circuit- and packet-switched functionalities in optical networks were controlled using the OpenFlow protocol, realized by using a programmable hardware platform. In particular, the testbed employed NetFPGA, a hardware platform that used field-programmable gate array (FPGA) on a peripheral component interconnect (PCI) card installed in the host computers, the latter playing the role of OpenFlow-enabled packet switch. A representative block schematic of the network testbed is presented in Fig. 12.19. As shown in the figure, the two host computers using NetFPGA are programmed to function as packet switches, i.e., as Ethernet switch or IP router, to switch/route the incoming packets to the output ports. These two packet switches (PS1 and PS2) are interconnected through two optical networking devices: AWG and WSS, representing a simplified version of a WDM-based optical networking domain. In reality, an optical networking domain would be an interconnected set of several WDM NEs (e.g., ROADMs, OXCs). All the packet switches (only two are shown as an example case) and optical NEs are connected to the OpenFlow-based SDN/SDON controller, which in turn controls all of them. We describe below an example connection scenario (Fig. 12.19), as reported in (Gudla *et al.* 2010).

Consider that two connections are to be set up in the SDON shown in Fig. 12.19; one between user A and the video server, and the other between user B and the same video server. The users A and B are seeking to watch two different video-streaming programs from the same video server across the optical networking domain. For the connection demanded by client A from the video server, a video request is sent from the computer of user A to PS1. Similarly, user B also sends a request to PS1 but for another video program from the same video server. Let's consider one of the two demands, say the one from user A to the video server through PS1. The request from user A goes to PS1, which is subsequently forwarded

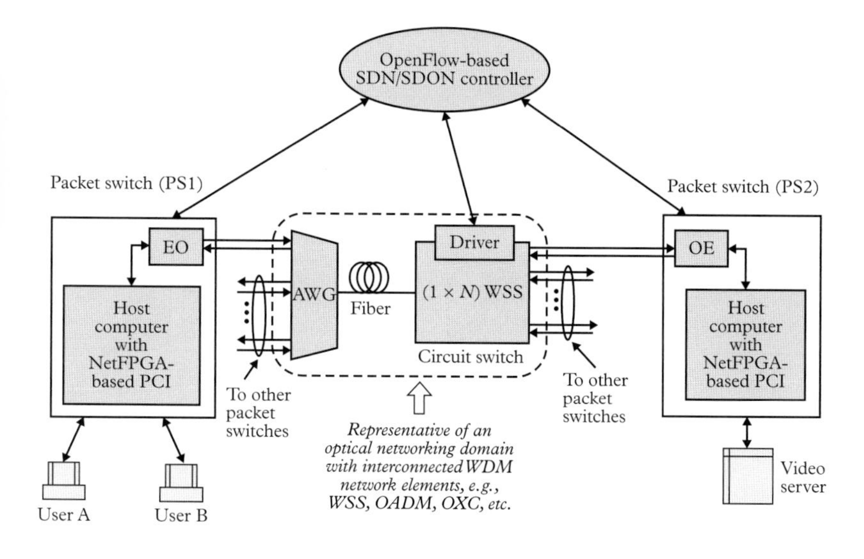

Figure 12.19 *Representative block schematic of the SDON testbed reported in (Gudla et al. 2010).*

to the OpenFlow-based SDN controller for setting up a bidirectional connection through the WSS and AWG (note that, if the needed flow table does already exist, the referral to the SDN controller is not needed). The SDN controller then sets up the flow tables in PS1 and PS2 for the connection, and thus far the operation goes on with the OpenFlow functioning using the packet flow model. Finally, the video server transmits the video traffic (packets) across the connection set up by the OpenFlow-based controller using a circuit-switched flow model. Thus two packet switches (PS1 and PS2) and one circuit switch (WSS along with AWG) work together, as governed by the SDN controller, to set up a video-streaming connection using both packet and circuit-switched NEs, but still conforming to the QoS requirements of the video connection. Similarly, the connection between user B and the video server through PS1 is also set up through the SDN controller.

Following the above experiment, a similar investigation on SDN/SDON was reported in (Liu *et al.* 2011), where an OpenFlow-based WDM network was realized as a proof-of-concept with four OXCs. The optical domain was again flanked by two packet-switched IP domains, and the test traffic flowed between the two packet-switched domains through the optical network in the middle, thereby setting up connections across the packet as well as circuit-switched domains by using the OpenFlow-based SDN controller. Thereafter, the same research group, along with two others, went on together to do a large-scale field trial of an OpenFlow-based multi-granularity optical WDM network (Liu *et al.* 2013), which we discuss later, in the section on multinational experiments.

In another experimental testbed reported in (Channegowda *et al.* 2013), an OpenFlow-based SDN/SDON was emulated using packet-switched layer-2 domains (e.g., Gigabit Ethernet) as well as circuit-switched optical network domains, for setting up connections between the servers and users (Fig. 12.20). The optical segment of the network testbed employed two different domains. One of them used bandwidth-variable OXCs, transmitters and receivers for realizing optical networking with flexible grid over optical fiber spectrum (i.e, EON), while the other domain employed the traditional fixed spectral grid using ROADMs with fixed-rate transceivers. The users were offered connections to the video servers across the two circuit-switched optical networking domains flanked by two packet-switched layer-2 domains, which were offered unified control by using OpenFlow-based SDN/SDON controller. GMPLS control plane was also included with appropriate abstraction into the SDN/SDON control plane, so that GMPLS-enabled networking elements could be unified under the umbrella of one single OpenFlow-based SDN/SDON controller.

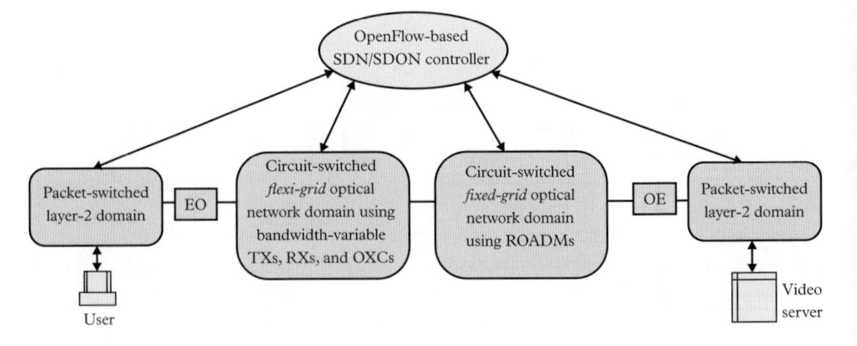

Figure 12.20 *Representative block schematic of the flexigrid SDON testbed reported in (Channegowda et al. 2013).*

Multinational experiments

We consider in the following two noteworthy multinational projects, viz., OFELIA (Europe and Brazil) and OpenFlow-based UCP (Japan, China, and Spain), undertaken in the recent past to carry out experiments on SDN/SDON-based large-scale networks.

OFELIA – European nations and Brazil

A large-scale SDN testbed development activity was undertaken as a multinational project, named as *OpenFlow in Europe-linking infrastructure and applications* (OFELIA) (Gerola *et al.* 2013), offering a unique multi-domain SDN platform, where the researchers could explore their networking ideas and test them through the virtualization of the network into slices. The network was designed to operate over ten sub-testbeds or *islands*, located in various European nations and Brazil. The virtualization capability of the entire network was realized using a software tool called *virtual topologies generation in OpenFlow* (VeRTIGO). The VeRTIGO tool could form and isolate the network slices precisely, thereby enabling the participating researchers to conduct innovative experiments independently within the respective network slices, residing across the underlying physical layer. This large-scale network needed orchestration to move the traffic through multiple and heterogeneous network domains using a federal philosophy, which was realized using an orchestration framework called *federated testbed for large-scale infrastructure experiments* (FELIX). FELIX was able to serve as a federal framework, orchestrating a huge range of computing and networking resources distributed over wide spread SDN islands, which were in turn connected via multi-domain *transit network services* (TNS). Figure 12.21 presents a representative block schematic of the multinational networking testbed used in the OFELIA project, consisting of ten islands interconnected through the TNS-based multi-domain transit networking infrastructure. As shown in the figure, the researchers from various SDN islands could form their slices and operate them independently by using FELIX and VeRTIGO as the orchestrator and virtualizer, respectively. FELIX and VeRTIGO were colocated in some nodes physically, from where they could operate across the entire network in a dynamic manner.

The overall project was conceived for two broad categories of services: data-domain services (user-centric) and infrastructure-domain services (network-centric). Data-domain services mostly dealt with efficient on-demand data and video transmissions, real-time pre-processed data (satellite) delivery over geographically distant locations, by configuring dynamically the NEs across the network with the required bandwidth, delay, and other QoS guarantees. The infrastructure domain services focused on migration of virtual infrastructures in an efficient manner, which was helpful for realizing cloud-based services and disaster management. For example, SDN-enabled data mobility to a remote datacenter was indeed a help to a user to access the needed services from a remote location with a quality experienced at the user's home location. Furthermore, OFELIA also demonstrated the feasibility of integrating the existing GMPLS networks with OpenFlow-based SDN domains, and thus created a practical and unique platform to experiment with.

OpenFlow-based UCP – Japan, China, and Spain

As mentioned earlier, a field trial of OpenFlow-based unified control plane (UCP) for multi-layer multi-granularity SDON was carried out at a large scale by a team of researchers from Japan, China, and Spain, for verifying the feasibility and overall efficiency of the SDN concept (Liu *et al.* 2013). The network accommodated various traffic granularities: packet, burst, and wavelength with the respective switching domains, i.e., IP, OBS, and OCS

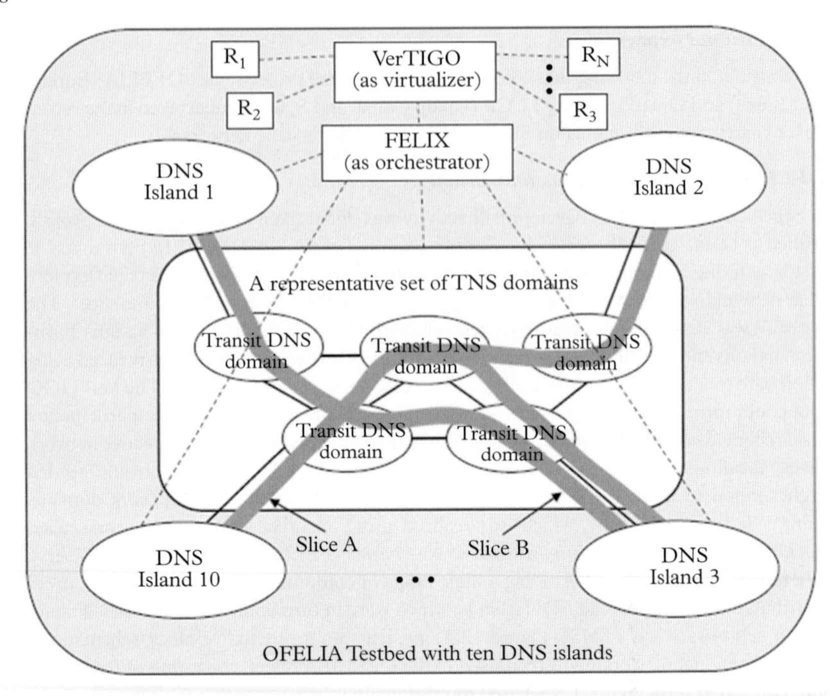

• Virtualization using Slice A between Islands 2, 10, and 3
• Virtualization using Slice B between Islands 1 and 3
• R_1, R_2 ... R_N: Researchers physically operating from various islands (not shown explicitly), while their intended slices are being created/deleted by VeRTIGO

Figure 12.21 *Representative block schematic of the OFELIA testbed reported in (Gerola et al. 2013).*

(OCS domain being termed as wavelength-switching network (WSON)). In particular, the project developed OpenFlow-based IP routers (OF-R), OpenFlow-based OBS routers (OF-OBS-ER/CR), with ER and CR representing edge and core routers, and OpenFlow-based ROADMs with virtual Ethernet interface, which could communicate with centralized NOS-based controller using OpenFlow protocol and SBI. The overall SDN performance was evaluated by measuring the network latencies for dynamic end-to-end cross-layer path creation and restoration through measurements on the network spanning over Tokyo, Beijing, and Barcelona, where the overall latency of the network was found to be primarily governed by the NOS-processing and the propagation times of the control messages across the network.

Industry initiatives and open network foundation

Considering the potential of SDN/SDON, several equipment manufacturers started working toward the development of the needed technology (Bhaumik *et al.* 2014; ONF 2011). For example, Cisco worked on open networking environment (ONE) using REST and Java for the NBI, and ONE platform kit (onePK) for the SBI operations, while Juniper developed Contrail as an SDN-based solution for cloud computing with the necessary features of network virtualization and orchestration (Enterprise Networking Planet 2013; Bhaumik *et al.* 2014). Further, Google developed Espresso (Yap *et al.* 2017) as the SDN-based Internet peering edge routing infrastructure, that is programmable and integrated with global traffic system. Similarly, other companies engaged in research on networking such as, HP, Alcatel-Lucent, and Huawei, also started working toward a variety of solutions for SDN.

Furthermore, in 2011 an operator-led consortium, known as the Open Networking Foundation (ONF), was launched, which has been engaged in taking the SDN movement further through providing the candidate network operators the freedom and flexibility to develop dynamic and cost-effective SDN/SDON solutions. Through the ONF initiatives, an open-source next-generation SDN (NG-SDN) platform has been formed to offer hardware-independent programmable networking solutions. In order to follow the ongoing developments in this direction, the interested readers are referred to (ONF 2011).

12.8 Summary

In this chapter, we presented the fundamental concepts of network control and management functionalities, followed by the descriptions of the various standards used in optical networks. To begin with, we explained the generic multiple-plane abstraction of telecommunication networks using the three logical planes – control, management, and data planes – and explained how the control and management planes work together in a collaborative manner to ensure smooth, secure, and survivable traffic flow in the data plane. In particular, we described the functionalities of the control plane needing real-time execution, such as recovery from network failures, network reconfiguration with time-varying traffic, etc., and functionalities of the management plane that are less time-sensitive in nature, such as performance monitoring, configuration management, network security, accounting and billing, etc.

Next, we described the functioning of the OAM scheme in SONET-based optical networks using the overhead bytes. For WDM networks, we presented the control and management schemes realized using the GMPLS, ASON, and SDN/SDON protocols, with the SDN/SDON exercising the control of the network in a centralized manner, in contrast with its predecessors (GMPLS, ASON) using distributed network control. GMPLS architecture was explained with a brief introduction to the MPLS protocol, followed by the GMPLS protocols using various levels of switching capabilities: PSC, TSC, LSC, and FSC. Next, ASON architecture was described with the features of multidomain optical networking using appropriate network interfaces: UNI, I-NNI, and E-NNI. Finally, the principle of SDN-based networking with centralized network control was presented with its basic architecture using OpenFlow protocol, along with the SDN-specific features, such as network virtualization and orchestration. A queuing-theoretic perspective of the SDN-based centralized control was presented using a model based on the *open network of queues with feedback*. Thereafter, we described how the concept of SDN is being adapted to realize SDON functionalities for optical networks and presented some noteworthy investigations that have been carried out in the recent past and are also being conducted currently on the SDN/SDON-based networking.

EXERCISES

(12.1) Using a suitable illustration, describe how the SONET overhead bytes carry out the tasks of OAM.

(12.2) Discuss the pros and cons of the distributed and centralized control planes in telecommunication networks.

(12.3) Considering the coexistence of the control (distributed) and data planes in WRONs, as shown in Fig. 9.6, propose and draw a suitable 3×3 node configuration for an OXC-based mesh-connected WRON with a logical nodal degree (in/out) of 2. Each node needs to perform the operations of both the planes, with the control information being handled by using one unique wavelength out of eight wavelengths available in the network. Assume that the control-plane topology follows the physical topology of the network. Also indicate how the node configuration will change if the control plane uses a ring topology.

(12.4) Mention the driving forces behind the development of the MPLS protocol and explain the role of FEC in MPLS protocol. Consider an MPLS network with five ingress packets arriving with the destination (IP) addresses a.b.c.3, a.b.c.5, a.b.c.7, a.b.d.4, and a.b.d.9 at one of its edge nodes (i.e., at an LER). Explain with a suitable illustration how the LER will treat them by using their CoSs and destination addresses to form the respective FECs and set up the corresponding LSPs.

(12.5) Describe the role of suggested labels in setting up unidirectional LSPs in a GMPLS network. Discuss a suitable method for setting up bidirectional LSPs in GMPLS networks.

(12.6) Describe how ASON is used to interconnect multiple domains of GMPLS networks and roles of the necessary ASON interfaces therein. Illustrate the interdomain routing between various GMPLS domains using ASON interfaces.

(12.7) Describe with a suitable illustration how the centralized control operation is realized in SDN. Explain the roles of the flow tables in OpenFlow protocol in controlling the flows of circuit as well as packet-switched traffic.

(12.8) Consider a cluster of N SDN nodes operating with one control node (see Fig. 12.17). Each SDN node has an ingress arrival rate of λ, of which on average, 30% of the packets are forwarded to the control node. The control node serves these requests from the SDN nodes with a service rate of μ. Using the concept of open network of queues with feedback (see Section 12.7.5), determine the maximum possible value of N that can be accommodated in the SDN cluster, so that the control node can remain in the steady state.

(12.9) Explain how SDN orchestration helps in realizing multi-domain networks, and mention the basic differences between the horizontal and vertical frameworks of SDN orchestration. Describe, with suitable illustration, how SDN operates through a multi-domain network using a horizontal framework.

(12.10) Explain, with suitable diagram, how the legacy WDM nodes can be retro-fitted in an SDON. Construct an example SDON schematic for setting up circuit- and packet-switched connections.

Datacenter Networks

13

In today's society, datacenters play a significant role by providing data storage and computing power needed to carry out a wide variety of tasks, from web mail, web searching, online transactions, and social networking for individuals to big-data processing tasks at organizational level, including machine learning, high-performance computing, etc. Datacenters typically consist of a large number of interconnected servers, ranging from hundreds to hundreds of thousands or even more, hence needing bandwidth-efficient networking between themselves, with energy-efficient operation. Use of optical switching in intra-datacenter networks can significantly improve the overall performance of datacenters with enhanced speed and reduced energy consumption. In this chapter, we present various types of datacenters, using electrical, optical, and hybrid switching technologies. Finally, we present some heuristic methods to design the WRON-based long-haul networks for hosting datacenters at suitable locations, and the popularity-based replication strategy of various objects therein.

13.1 Datacenters: evolving scenario

In today's networking scenario, data-storage capacity and computing power need to be available from the network as basic resources for running a wide range of computer applications at various levels, such as webmail service, web searching, online transactions, social networking for individual users, and big-data processing tasks carried out at the organizational level, such as, machine learning, high-performance computing, climate forecasting, etc. Our day-to-day life, both on personal and professional fronts, has never been so dependent on access to the wide range of data from the network. In order to offer such services, it is now found convenient for a service-providing organization to keep the necessary data resources and the computing power at some central location within its own premise or elsewhere, with some other third-party organization providing such facilities on a rental basis. These centralized facilities with large data-storage capacity and computing power are popularly known as datacenters.

Typically, a datacenter consists of a large number of servers, as large as thousands to hundreds of thousands, or even more, which must be interconnected with a bandwidth-efficient internal network, leading to the formation of a complex *intra-datacenter network*, or simply *datacenter network* (DCN).[1] The servers in datacenters would be accessed by the users and other datacenters through the supporting long-haul/metro/access networks, while remaining oblivious to the locations of the datacenters. This leads to the notion of having the data sources and the computing power available from *somewhere* in the network, viewed as a *cloud*. For small organizations, such cloud-based services are generally offered

[1] Henceforth, *DCN* will be used as an abbreviation for the internal networking aspects within a datacenter, while *datacenter* will represent the entire setup, viewed from outside as a resource by the users and network operators.

Optical Networks. Debasish Datta, Oxford University Press (2021). © Debasish Datta.
DOI: 10.1093/oso/9780198834229.003.0013

by datacenters owned by a third party extending such services to several organizations, while big organizations (e.g., Google, Microsoft, Amazon, etc.) would generally set up their own datacenters in multiple locations, to provide services to users all over the world.

Centralized placement of large number of servers in datacenters leads to better management and maintenance, along with improved flexibility in provisioning the resources in the presence of dynamic business demands. The servers in a datacenter need to be interconnected using a judiciously designed DCN, which should operate with adequate bandwidth, fault tolerance, and power/energy efficiency. However, often the datacenters may have to be connected to some other datacenters through a supporting network, when necessary. For example, in various organizations, in order to reach users over widespread locations and also for resilience to network failures, multiple datacenters are generally used in different locations for storing the same objects/contents, and data replication needs to be carried out between these datacenters from time to time through the supporting network. With this arrangement, users from the client nodes should be able to access one of the datacenters for a given object/content, governed by its proximity to a datacenter hosting the desired object.

In DCNs, the transmission media would typically be optical or a mixture of electrical and optical links. Further, DCNs can also employ wireless connectivity between the switches by using millimeter-wave transmission. The task of switching in DCNs can be realized using electrical, electrical-cum-optical, or all-optical switching devices, depending on the network size and bandwidth requirement. In the following, we first describe various candidate architectures and the design issues for the DCNs using these three switching techniques, and thereafter describe the approaches for designing long-haul WDM networks hosting the datacenters.

13.2 Datacenter networks: architectural options

DCNs in general can broadly employ two possible architectures: switch-centric and server-centric. A typical DCN would consist of a large number of servers, the number ranging from a few thousands to as large as hundreds of thousands (to even a million), organized in rows of vertically stacked server racks, and the entire set of servers in all the racks is interconnected using various suitable topologies (Xia *et al.* 2017; Shuja *et al.* 2016; Kachris *et al.* 2013; Chen *et al.* 2013; Al-Fares *et al.* 2008), which could employ server- or switch-centric architectures. In switch-centric architectures, each server-rack in a DCN uses a switching unit which is placed on the top of that rack and is popularly called top-of-the-rack (ToR) switch. When a DCN employs a tree-based switching topology, the ToR switches are also called edge switches, as they are placed at the network *edges*, with the servers connected as *leaf* nodes to those switches. In these topologies, the edge/ToR switches of all the racks are connected hierarchically to a number of switches at different higher levels following an appropriate topology. However, some DCNs don't use tree-based topologies, and in such networks the servers also participate in switching along with fewer and smaller switching units (as compared to the switch-centric architectures) by using multiple networking ports or cards, leading to server-centric architectures.

Some of the known topologies that have been explored for DCNs include switch-centric architectures, such as tree, fat tree, virtual layer 2 (VL2), Aspen tree, etc., and server-centric topologies, such as DCell, BCube, etc., all using electrical switching equipment transmitting mostly over single-wavelength fiber-optic links (Xia *et al.* 2017). Due to the overwhelming growth in the size of datacenters, more effective DCN topologies are being explored with the partial support of optical switching. Some of these studies explored hybrid technologies with electrical and optical-switching units operating in parallel to meet the bandwidth requirements, such as c-Through and Helios. Testbeds for DCNs with all-optical switching have also been developed to utilize the full capacity of optical transmission and switching technologies, such as OSA, Mordia, LIONS, etc.

Some alternative support can also be realized alongside the wired DCN infrastructure by adding wireless links between the ToRs using millimeter-wave communication, as the unpredictable propagation losses experienced by millimeter-wave frequencies in the open air would be practically nonexistent for in-house wireless links in a DCN. Furthermore, in the millimeter-wave range one can realize (with small-size antenna elements) highly directive antennas with phased-array configuration offering beam-steering capability, which would offer one additional dimension to the switching capability between the ToR switches.

It therefore appears that one can have purely electrically switched wired DCNs with copper and optical fibers as transmission media, or can realize DCNs using an appropriate mix-and-match of electrical, optical, and beam-steering-based millimeter-wave switching, as governed by the needs of DCN size, traffic characteristics, and the acceptable network latency. With a high-level view of the possible alternatives, one can thus visualize the optically switched paths as bandwidth-intensive *highways*, while referring to the steerable millimeter-wave wireless links (though with much less bandwidth as compared to the optical paths) as the wiring-free and agile *flyways* within a DCN. However, in the context of optical networking, the roles of wireless transmission links are beyond the scope of the subsequent discussions, and the interested readers could refer to (Hamza *et al.* 2016; Zhou *et al.* 2012).

Next, we discuss some of the specific functional features that need to be taken into consideration while designing DCNs. In a DCN, several online applications are run with real-time or nearly real-time service requirements, while the big data-processing jobs are executed in batch mode with less sensitivity to network latency, such as machine learning and high-performance computing. In the presence of such wide-ranging tasks running concurrently, DCNs in general exhibit highly time-varying traffic pattern between the servers; as reported by Microsoft in (Greenberg *et al.* 2009), the traffic in DCN might cyclically move through 50–60 different patterns in a day, with each pattern maintained for about 100 seconds at its 60th percentile. The design of DCNs must address this variability in traffic pattern, while choosing the physical topology and the protocols running in the DCN.

Furthermore, one of the major issues in DCNs is the need of end-to-end high-speed communication between the constituent servers with adequate bandwidth. For example, with a given web-search query, it will be required for the search engine in a DCN to resort to concurrent communications with several other servers in the same DCN, some of them being located in different racks. Further, for high-volume scientific computing tasks, the DCN must engage multiple servers to execute parallel computing to speed up the computation process. Overall, the DCN must ensure that an adequate bandwidth is available between its own servers, regardless of their physical proximity, to engage in high-speed *conversations*, while executing a given task without exceeding a tolerable latency limit. Note that, when two servers are placed in the same rack under the same edge switch, they can use the full port (switch port) bandwidth between themselves, which may not be true for the servers located in different racks, owing to possible congestion in the switches at higher levels.

In order to address the above issues, the design of DCNs needs to consider two network parameters: bisection bandwidth and oversubscription ratio. Any given DCN topology (graph) can be bisected in two parts with minimal number of link-cuts, each part having the same or a similar number of nodes, and one can find out the set of links (cut-set) bridging the two parts and their aggregate bandwidth, called bisection bandwidth; this would give an estimate of the speed that the servers from the two sides can avail for the inter-server communication. The need for oversubscription comes in with an objective to reduce the overall cost of DCNs, which implies that the DCN remains oversubscribed by its servers in respect of the available bisection bandwidth. More specifically, a server on a given rack, while connected to its own edge switch with a dedicated (unshared) bandwidth/speed (typically 10 Gbps using Gigabit Ethernet), may not manage the same speed (10 Gbps) while communicating to another server belonging to a different rack due to the bandwidth bottleneck at the higher-level switching element. Thus, the oversubscription ratio is defined as the ratio of the server port-bandwidth to the available bandwidth at the same server while communicating to another server on a different rack in the DCN. For example, in a tree-based topology, a given server may practically communicate at 400 Mbps through its 1 Gbps (1000 Mbps) port, thereby being constrained by an oversubscription ratio of (1000 Mbps to 400 Mbps) = 2.5:1. The oversubscription ratio improves as the bisection bandwidth approaches the server transmission bandwidth (i.e., port speed).

Overall, DCNs have a number of architectural options using electrical-switching equipment, which can be augmented partially or replaced fully by optical switching devices to improve the DCN performance. Hence, in the following, we present first a few electrically switched DCNs, followed by the sections on the DCNs employing hybrid (i.e., electrical-cum-optical) as well as all-optical switching techniques.

13.2.1 DCNs using electrical switching

In this section, we examine some of the DCN architectures using only electrical switching, though the transmission between the switching equipment is mostly realized using optical fiber links. In particular, electrical transmission over coaxial cables becomes infeasible at high transmission rates for link lengths exceeding a few meters (e.g., 10 Gbps transmission can be supported up to 7 meters). As mentioned earlier, several physical topologies have been explored for DCN realization using electrical switching, which can be categorized as

- switch-centric connectivity, e.g., tree, fat tree (Xia *et al.* 2017; Al-Fares *et al.* 2008), virtual layer-2 (VL2) (Greenberg *et al.* 2009), Aspen tree (Walraed-Sullivan *et al.* 2013), Jellyfish (Sing-la *et al.* 2012),
- server-centric connectivity, e.g., DCell (Guo *et al.* 2008), BCube (Guo *et al.* 2009), FiConn (Li *et al.* 2009).

In the following we describe some of these DCNs (tree, fat tree, VL2, Aspen tree, DCell, and BCube) and discuss the salient features of the respective networks.

Switch-centric connectivity: tree

In a simple tree topology, as shown in Fig. 13.1, all the edge/ToR switches are connected to a core switch. From the core switch the access to the Internet can be provided by using a router. However, for dedicated connections with other datacenters, the core switch may have to be connected through layer-2 tunnels over the Internet backbone.

Physical topology of the tree, as shown in Fig. 13.1, uses two levels of switching with its root lying in the core switch. As a result, any communication between two servers belonging to two different racks needs to be routed through the core switch at the root, which may

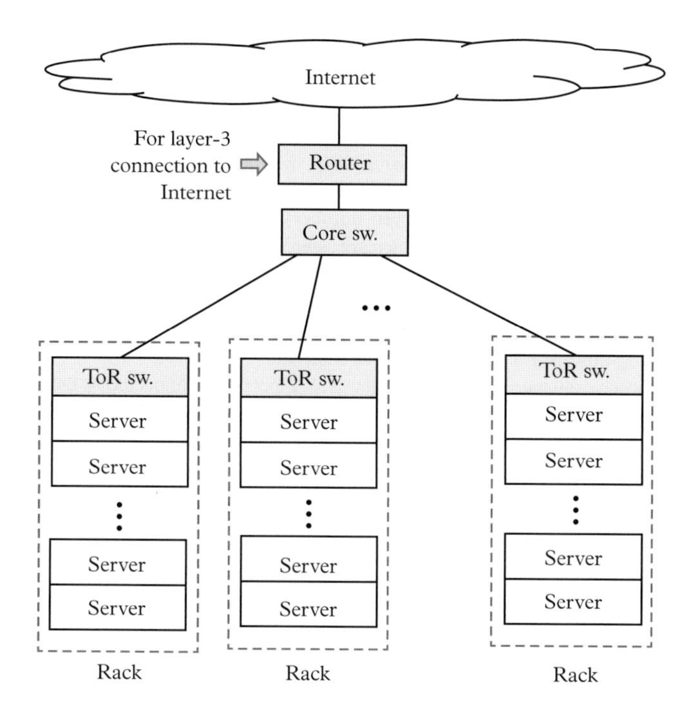

Figure 13.1 *Tree topology for DCN with layer-3 Internet connectivity using a router.*

cause a significant bandwidth bottleneck. Thus, implementation of such two-level DCNs in large scale (in the order of thousands and more) becomes mostly infeasible with a reasonable oversubscription ratio. Although this problem can be partially alleviated by increasing the number of levels in the tree, the problem of having one single root (i.e., one core switch) can be better addressed by using suitable topological variations of the basic tree topology, as discussed in the following.

Switch-centric connectivity: fat tree

As compared to the simple tree topology with single root, fat-tree topology uses multiple roots with as many core switches and additional switching levels for enhanced connectivity, offering a full port-bandwidth between all possible pairs of servers (i.e., full bisection bandwidth). In particular, its multiple core switches carry out the routing of some of the inter-rack traffic between themselves at the highest level. Further, all inter-rack communications are not routed through the core switches owing to the introduction of an additional switching level – aggregate switching level – between the edge and core switching levels. The fat-tree topology, resembling a folded-Clos non-blocking topology, was proposed in (Leiserson 1985) as a class of non-blocking switching networks for interconnecting the processors of a general-purpose parallel-computing based super-computer.

As mentioned above, a traditional fat tree uses three levels of switching – edge, aggregate and core – with the edge switches lying at the bottom of the hierarchy to interconnect with the servers. The aggregate switches are placed just above the edge switches, and the core switches are placed at the next higher level, i.e., the topmost level of the hierarchy. The core switches are connected to the Internet using appropriate routing devices. All switches are same in respect of the port count (radix) and port speed, and hence can be procured in bulk as a commodity switch, except the core switches, which will have to employ additional ports to get access to the Internet. Note that some novel variations of the traditional fat-tree topology have also been studied by various research groups and are discussed in the following sections.

A traditional non-blocking fat tree using three switching levels, as shown schematically in Fig. 13.2, is called a k-ary fat tree, where the edge and aggregate switches are arranged into $k/2$ groups or pods (k being an even number), with each pod comprising $k/2$ edge switches and $k/2$ aggregate switches, which are *completely connected* between themselves and can be bisected downward with null cut-set. Note that, each switch in all levels employs exactly k ports, i.e., with the same switch radix k. However, for each core switch, all of its k ports connect downward with the aggregate-level switches, while it will have a few additional port(s) looking upward for connecting to the Internet.

For the switches at aggregate and edge levels, $k/2$ ports are connected to the switches of the upper level and $k/2$ ports are connected to the switches of the lower level (or the servers in the case of level 1, i.e., edge level). The $k/2$ downward ports of each edge switch are connected to $k/2$ servers, implying thereby that the network employs $(k/2)^2 = k^2/4$ servers per each pod and hence $k^2/2 \times k/2 = k^3/4$ servers in total. In a given pod comprising the edge and aggregation switches, each edge switch uses all of its $k/2$ upward ports to get connected to all $k/2$ aggregate switches, and each aggregate switch similarly gets connected to all $k/2$ edge switches, thus giving a complete connectivity (and hence fault-tolerant connections) between the switches of the two levels within each pod. Further, at aggregate level all $k^2/2$ switches have in total $k^2/2 \times k/2 = k^3/4$ ports looking upward to connect to the core switches. Since each core switch has k switching ports (looking downward) to get connected with the aggregate-level switches, the core level must employ $(k^3/4) \div k = k^2/4$

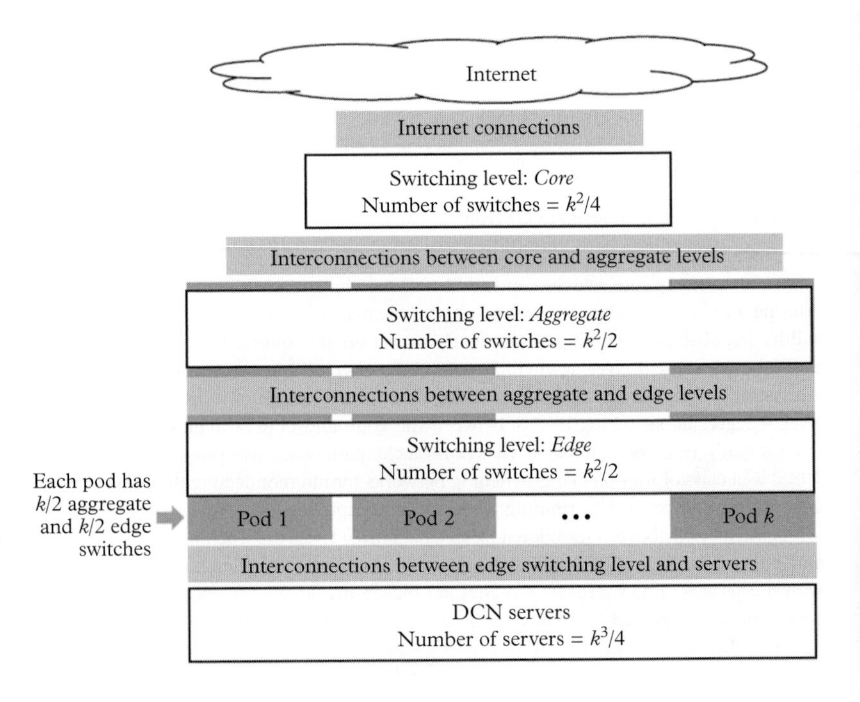

Figure 13.2 *Block schematic for a k-ary fat tree with $k^2/4$ core switches (roots) and $k^3/4$ leaf nodes (servers), where each switch has k ports. However, the core switches have additional ports for external connections with Internet.*

switches, so that each of the $k/2$ upper ports of an aggregate switch can get connected to one core switch. We summarize these features of a k-ary fat tree in the following.

- Number of switching levels $n = 3$.
- Number of aggregate and edge switches $= k^2/2$. For larger n, this expression is generalized as $k^{n-1}/2^{n-2}$.
- Number of core switches $= k^2/4$ $(= k^{n-1}/2^{n-1}$, for n levels of switching).
- Total number of servers $= k^3/4$ $(= k^n/2^{n-1}$, for n levels of switching).
- Total number of switches = number of edge switches + number of aggregate switches + number of core switches $= (k^2/2)\times 2+k^2/4 = 5k^2/4$ $(= (k^{n-1}/2^{n-2})\times (n-1) + k^{n-1}/2^{n-1} = (2n-1)k^{n-1}/2^{n-1}$, for n levels of switching).
- Switches at each level are identical (except the switches at core level, as these switches should have some more ports for accessing the Internet), which permits the use of commodity switches at edge and aggregate levels.

Next, we illustrate the above features of fat tree using a 4-ary topology, as shown in Fig. 13.3. The 4-ary fat-tree topology employs $k^2/2 = 8$ edge switches, eight aggregate switches and $k^2/4 = 4$ core switches, the total number of switches being $5k^2/4 = 20$, where the network supports $k^3/4 = 16$ servers. Note that, in a fat tree with extensive cabling requirement between these switches, the wiring complexity would increase significantly with large number of servers. Further, in spite of full bisection bandwidth, fat tree takes a considerable amount of time to reconfigure the network and restore the connection(s) affected by the internal failures. We discuss this issue in the following.

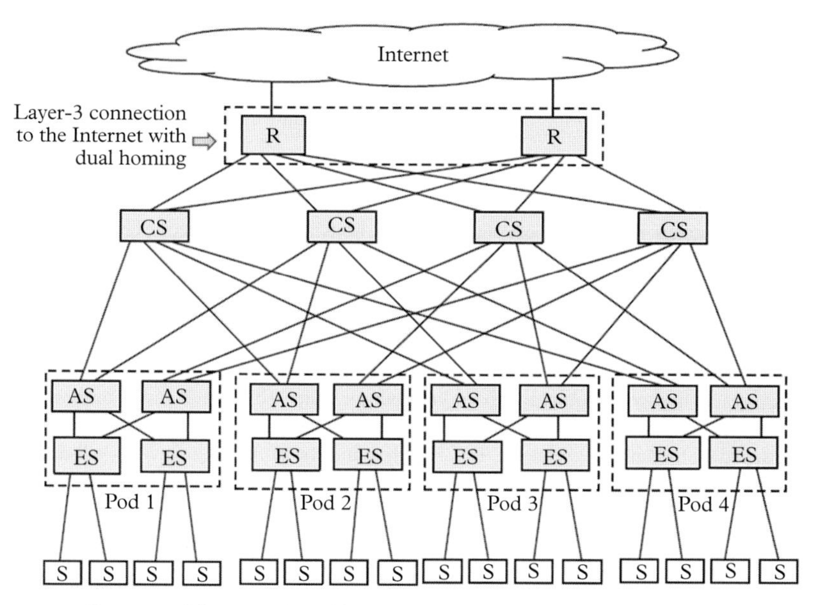

Figure 13.3 *Traditional k-ary (k = 4) fat-tree topology for DCN.*

Consider the case of a single-link failure in a fat-tree-based DCN for one of the links connecting an edge switch attached to some destination server. Following the failure, the network controller has to send back the failure message to the respective source server(s) (there could be multiple sources sending packets to the same destination), so that the source(s) can re-route the packets, avoiding the failed link, which is possible in a fat tree. However, in the meantime, a large number of packets from the sources might arrive at the input end of the failed link and get dropped. The time that is taken by the source(s) to react is non-negligible, as during the reverse journey of the failure message all the way, each switch up/down the hierarchical path has to compute the desired route to deliver the message to the sources. Such disruptions in the network may slow down the DCNs significantly. As we shall see later, there are some topologies, such as Aspen tree, that can address this issue with adequate fault tolerance, albeit at the cost of increased hardware.

Switch-centric connectivity: VL2

Virtual layer 2 or VL2, a DCN architecture explored by Microsoft, also employs a multi-rooted hierarchical topology, as in a fat tree, with the same number of switching levels, i.e., ToR (or edge), aggregate, and core. The topmost switching level in VL2 was termed as the intermediate switching level in the original paper from Microsoft (Greenberg *et al.* 2009); however, we continue here with the generic name (i.e., core level), for uniformity of nomenclature. Despite using multi-rooted tree topology, VL2 has some important differences with the fat-tree topology. In particular, VL2 realizes its connectivity between the core and aggregate levels with each core switch being connected to *all* the aggregate switches, and vice versa. Thus, each aggregate switch also gets connected to all core switches, whereas in fat tree each *pod* remains connected from aggregate level to all core switches. However, each ToR switch is connected to only two vertically adjacent aggregate switches, although the connections between the switches at these two vertically adjacent levels are realized with higher transmission speed, as compared to the connections between the servers and the ToR switches. Furthermore, VL2 uses some novel addressing schemes for its servers and employs a unique set of protocols, enabling the DCN to support dynamic internal traffic with high agility, while avoiding hot-spot formations. We explain below the salient features of VL2 in further detail.

Assume that k_A is the total number of ports in each aggregate switch, with $k_A/2$ ports for the links connecting with the core switches and $k_A/2$ ports for the links connecting with the ToR switches. Since each port of an aggregate switch connects with one core switch, the number of core switches $S_C = k_A/2$. Similarly, if each core switch has k_C ports (downward), then the number of aggregate switches is $S_A = k_C$, for full connectivity between a core switch to all aggregate switches. Furthermore, each ToR switch has only two links connecting two aggregate switches, implying that there are $S_T = S_A/2$ ToR switches. However, as indicated earlier, the number of links between ToR and aggregate switches being small, each of these links uses higher speed, compared to the links between each ToR switch and its own servers. The speed-up ratio (i.e., the speed of the links between the ToR and aggregate switches to the speed of the links between the ToR switches and the underlying servers) will depend on the number of servers under each ToR switch, to avoid bandwidth bottleneck. Thus, if there are N_S^T servers connected to each ToR switch, and each link between server and ToR switch operates at a speed of r Gbps, then each uplink from a ToR switch to an aggregate switch should have a speed equaling $N_S^T \times r/2$ Gbps. Figure 13.4 shows an example VL2 realization, where the DCN gets connected to the Internet using multiple routers (not shown explicitly) for reliable Internet access.

From Fig. 13.4, it appears that, with 20 servers below each ToR switch and 1-Gbps speed on each of the links between a ToR switch and its servers, VL2 would use two links from

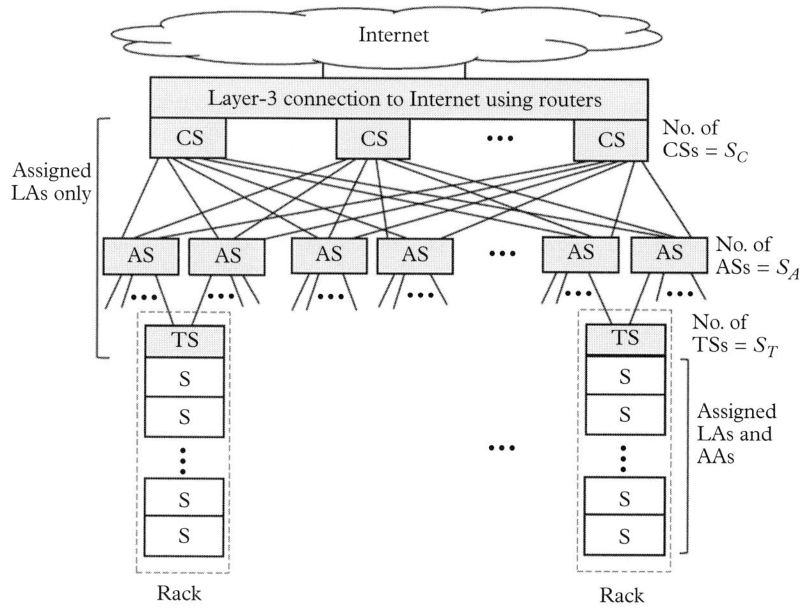

Figure 13.4 *VL2 topology for DCN with layer-3 connectivity with the Internet using routers (after Greenberg et al. 2009).*

each ToR switch to connect to two aggregate switches, each with a speed of $20 \times 1/2$ Gbps = 10 Gbps, and these two links would therefore be ensuring connections to the next higher level (aggregate level) without any bandwidth bottleneck. Similarly, all the links between the aggregate and core switches will also transmit at 10 Gbps. Furthermore, due to the bipartite connections between the core and aggregate switches, one gets $S_C k_C = S_A k_A/2$ and the total number of servers in the network becomes $N_S = S_T N_S^T = S_A N_S^T/2$, with all the optical links between the three switching levels transmitting at $N_S^T \times r/2$ Gbps (= 10 Gbps in the present example).

With the above architectural settings, VL2 doesn't incur oversubscription but uses less wiring than the fat tree, by using higher transmission speed between the ToR and aggregate switches. However, in order to get around the problem of traffic agility and the associated issues – e.g., hot-spot formations, impact of one service on another, fragmentation of address space, etc. – the operation of VL2 is orchestrated by creating a sense of *virtualization* for each service, thereby making all the servers associated with a given service communicate with an *illusion* of being connected to a single non-interfering Ethernet (layer-2) switch; hence the name VL2.

Addressing and routing mechanisms in VL2 play important roles for achieving the above objectives. In particular, VL2 uses a unique addressing scheme with two types of IP addresses: local addresses (LAs) and application addresses (AAs). Location-specific IP addresses, i.e., LAs, are assigned, as usual, to all switches and the network interfaces of the servers, with which in turn the DCN operates with the IP-based layer-3 routing protocols. However, all the applications running different collaborative tasks in various servers across the DCN are assigned a set of application-specific addresses, i.e., AAs, which don't change even if the locations of the servers offering the services (applications) are changed. When some servers are assigned a given application to serve for, they are allocated an AA. VL2 employs a *directory system*,[2] which receives the AA from the server requesting the service of a specific application. In turn, the directory system, when requested by the source server for an application (service) with an AA, maps the AA to the appropriate LA and the request for the application goes to the appropriate application server, based on its LA. This enables the application servers to operate with a notion (to the servers requesting the service) that they all belong to a single AA space (IP subnet) corresponding to a specific application. With this addressing scheme, the requesting servers remain unaware (agnostic) of how and when the necessary application servers change their locations owing to virtual machine migration or reprovisioning (due to overload). Therefore these servers can get around the scaling bottleneck of the address resolution protocol (ARP) and dynamic host control protocol (DHCP) in large Ethernets.

The LAs and AAs are used as follows. While routing packets from one server to another, the source server encapsulates the packets with AA addresses. Having reached the parent ToR, the AAs are encapsulated by the LA of the destination ToR (using directory system) since the switching network (the switches at different levels of DCN hierarchy) don't understand AAs. Having delivered the packets to the destination ToR, the LA addresses are stripped off and the packets are routed according to the AA address to the respective servers in the destined rack. Thus servers under a given application service get to *feel* that they belong to a subnet slated for the application service they are engaged in, regardless of their LAs. However, the switched network remains oblivious to the AAs and operates the routing mechanism exclusively based on LAs only. This dichotomy in the addressing scheme of VL2 makes the network more efficient in respect of routing through the switching levels of the network, as well as in bringing together the servers engaged in a given application, without the need for looking into the network architectural details.

Furthermore, in order to address the traffic agility more efficiently, VL2 uses the *valiant load balancing* (VLB) scheme, where the traffic flow between any pair of servers is distributed over all possible paths randomly without adhering to any centralized routing algorithm or traffic engineering mechanism. Propagation time across a datacenter being small, the diversity of paths (between long and short paths) created in the VLB scheme doesn't add much to the latency, while alleviating the problems of localized congestion due to the agile traffic pattern. The VLB scheme also leads to a graceful resilience to link failures in VL2, since all paths are not likely to fail simultaneously. VL2 also utilizes layer-3 features, such as equal-cost multipath (ECMP), in spreading the traffic over multiple equal-cost subpaths. Thus, VLB and ECMP together help the entire network to operate without forming traffic hot spots.

Switch-centric connectivity: Aspen tree

Aspen tree is another hierarchical tree topology, which offers some inherent resilience to failures in DCNs (Walraed-Sullivan *et al.* 2013). As discussed earlier, in a traditional fat tree, it is possible to reconfigure/re-route the connections affected by link failure, but its restoration time is large and increases significantly with increase in the DCN size. Aspen tree, with some additional hardware, can bring down the restoration time in the event of a link failure.

The name of this topology is coined from nature: Aspen trees, usually seen to grow in the northern hemisphere in groups, cannot easily be destroyed due to their *resilient root system* below the soil. Aspen tree topology, proposed for DCN, also has an in-built localized

[2] For the operational details of the directory system and the necessary setup, readers are referred to (Greenberg *et al.* 2009).

survival mechanism in the event of link failures, and hence the name. In particular, the Aspen-tree topology for DCNs represents a generic multi-rooted fat-tree topology, where some redundancy is introduced with additional hardware at some switching level(s). This enables the network to *react* to the failures *locally*, obviating the need for a *global* (and hence long and complex) recovery process across the network involving all switching levels.

An (n, k) Aspen tree generically represents a multi-rooted tree with a set of n switching levels (with $n \geq 3$), using k-port switches at all the levels. However, an Aspen tree differs from the traditional fat tree in respect of the connectivity between the vertically adjacent layers. In particular, the switches in any two adjacent layers, say ith and $(i-1)$th layers, can employ more interconnections as compared to a traditional fat tree, offering thereby localized recovery paths against a link failure. We describe below the basic design concepts of Aspen tree.

In order to explain the salient features of Aspen tree, we define below some parameters, followed by the relations between them.

k: number of switch ports, presuming that all switches are commodity switches with the same number of switching ports, excepting the core switches.

L_i: ith level of switching, with $i = 1, 2, \ldots, n$. The nth level is the topmost (core) switching level, with all of its k ports communicating with the switches at the $(n-1)$th level (with some additional ports for connecting to the Internet).

S: total number of switches in a level, except the core switching level. The core switching level L_n has $S/2$ switches, as in a traditional fat-tree topology.

p_i: number of pods in level L_i. A pod is defined as a group of switches in one or more level(s). For all levels (except L_n) while looking downward, there are vertical bisections with null cut-set, thereby dividing the cut portions of the level(s) into pods. The core level, i.e., L_n, has only one pod, i.e., $p_n = 1$, as the switches at this level have crisscrossed downward connections spanning over all the pods in the next lower level, barring any bisection with null cut-set. Furthermore, at level L_1, each switch represents itself as a pod with null cut-sets on both sides. Pods can also be formed by grouping switches from the vertically adjacent layers.

m_i: number of switches in pod p_i. As mentioned above, $m_1 = 1$, as each switch in level L_1 has its independent downward connections only to the underlying servers in its own rack, and hence perfectly bisectable from the other L_1 switches.

r_i: *responsibility* count of a switch in layer L_i, implying that a switch in layer L_i would connect to r_i pods in the next underlying layer L_{i-1}.

c_i: number of connections from a switch at layer L_i to each pod in layer L_{i-1}. Note that, $c_i = 1$ for a traditional fat-tree.

Figure 13.5 shows an example Aspen tree with four levels and four ports per switch, and hence termed as (4,4) Aspen tree. This Aspen tree offers some fault-tolerance feature, which we will discuss later, after developing below a generic analytical model for the Aspen-tree topology by using the above parameters.

The number of switches S in a level is the product of the number of pods in the given level and the number of switches in each pod in that level (except the core level (L_n), where the number of switches is $S/2$, as per the basic features of fat trees), i.e.,

$$p_i m_i = S, 1 \leq i \leq (n-1); \quad p_n m_n = S/2. \tag{13.1}$$

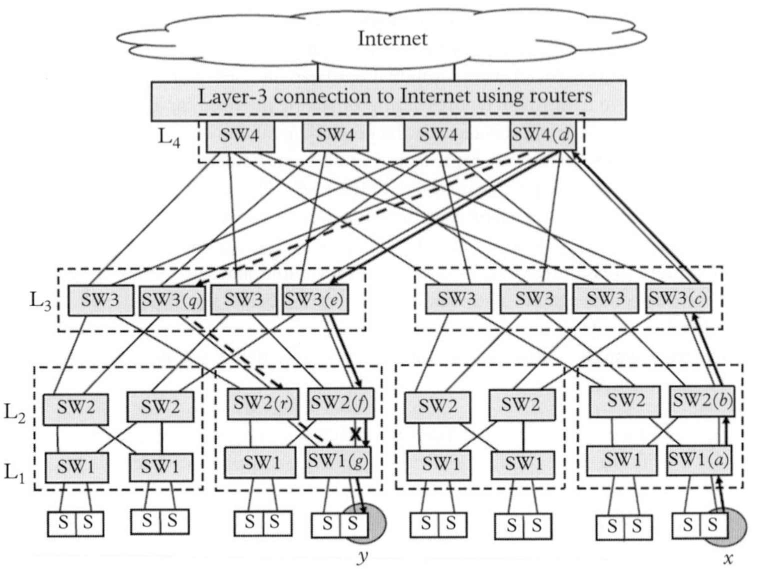

Figure 13.5 *A (4,4) Aspen tree with 16 servers employing an FTV = (1,0,0) (after Walraed-Sullivan et al. 2013). As implied by the FTV elements, the redundancy (one from each switch in L_4) takes effect between levels 4 and 3. The solid arrows (→) represent the original path from x to y (a-b-c-d-e-f-g), and the dashed arrows represent the alternate path from x to y (a-b-c-d-q-r-g), when the link f-g fails (marked as ×).*

Fault tolerance at a given layer, say L_i, depends on how many switches in the next underlying layer L_{i-1} are connected with each switch in layer L_i. This number is simply the number of down-side ports of each switch, i.e., $k/2$, which would be the product of r_i and c_i, i.e.,

$$r_i c_i = k/2, 2 \leq i \leq (n-1); \quad r_n c_n = k. \tag{13.2}$$

As we shall see later, r_i and c_i have an interplay, as their product must be equal to $k/2$. In other words, for increasing the fault tolerance one might increase c_i, which will in turn reduce the number of pods p_i, eventually leading to a decrease in the number of servers that can be supported for a given value of k. Further, note that the above constraint doesn't apply to the core and edge levels. There exists also a relation between two vertically adjacent levels, say level L_i and level L_{i-1}. In particular, the set of pods at a higher level, say L_i, should *cover* all pods in the next lower level, i.e., level L_{i-1}, i.e.,

$$p_i r_i = p_{i-1}, 2 \leq i \leq n, \tag{13.3}$$

which implies that, from the higher level L_i, if we take one switch from each pod, i.e., p_i switches from p_i pods, then each of the downward $p_i r_i$ connections from them (each switch having r_i downward links) will reach one pod at level L_{i-1}. Note that, the above constraint does not apply for $i = 1$, as there is no level below L_1.

An important feature of Aspen tree is the controllable fault tolerance, which can be assessed by using the downward connectivity variable c_i, as defined earlier (Eq. 13.2). Note that in a traditional fat tree, a switch at a given level (except layer L_1) is connected downward to only one switch of each pod in the next layer, making $c_i = 1$. In Aspen tree, this count can be more than one, implying that the switch at the level L_i can have more than one connection to a pod at the level below, by virtue of which the fault tolerance takes effect, i.e., one can have $c_i > 1$, for $n \geq i \geq 2$ in an Aspen tree.

In view of the above, the fault tolerance of an Aspen tree can be designed on a per-level basis by enhancing the values of c_i for different levels. The redundancy in the design can be quantified comprehensively using a metric, termed as fault tolerance vector (FTV). The FTV is an $(n-1)$-tuple vector $(\phi_n, \phi_{n-1}, \cdots, \phi_j, \cdots, \phi_2)$, with its ith element expressed as

$$\phi_i = c_i - 1, n \geq i \geq 2, \tag{13.4}$$

where ϕ_i represents the excess connection count from a switch in level L_i to a pod in level L_{i-1}. Note that ϕ_i is exactly zero for a traditional fat tree. This leads to the fact that any level in an Aspen tree, say ith level with $n \geq i \geq 2$, having the same connection pattern downward as in a fat tree will have $\phi_i = c_i - 1 = 1 - 1 = 0$. Any additional downward connection from this level would make ϕ_i a positive integer representing the in-built redundancy of the same level. With this observation, one can express FTV in an Aspen tree as

$$\text{FTV} = (\phi_n, \phi_{n-1}, \cdots, \phi_2) = (c_n - 1, c_{n-1} - 1, \cdots, c_2 - 1). \tag{13.5}$$

From the above relations, we next estimate the number of switches needed in all the switching levels. Using Eq. 13.3, one can develop a set of recursive relations for the number of pods in level L_i, given by

$$\begin{aligned} p_1 &= p_2 r_2 \qquad (13.6)\\ p_2 = p_3 r_3 \Rightarrow p_1 &= (p_3 r_3) r_2 \\ &\cdots \\ p_{n-1} = p_n r_n \Rightarrow p_1 &= (p_n r_n) r_{n-1} \cdots r_3 r_2 \\ p_n = 1 \Rightarrow p_1 &= r_n r_{n-1} \cdots r_3 r_2, \end{aligned}$$

leading to the generalized expression for p_i, given by

$$p_i = \prod_{j=i+1}^{n} r_j. \tag{13.7}$$

Next we proceed to find out the expression for the number of switches in each switching level. Noting that all the levels (except the core level L_n) have the same number of switches S, and in level L_1, each switch is independent as a pod with $m_1 = 1$, we express S using Eq. 13.1 as the number of switches in edge level L_1 as

$$S = m_1 p_1 = p_1 = \prod_{j=2}^{n} r_j = r_n \prod_{j=2}^{n-1} r_j. \tag{13.8}$$

Using Eq. 13.2, we express r_j and r_n in terms of k, c_j and c_n as

$$r_j = \frac{k}{2c_j}, 2 \leq i \leq (n-1); \quad r_n = \frac{k}{c_n}. \tag{13.9}$$

Using the above expressions for r_j and r_n, we further simplify the expression for S in Eq. 13.8 as

$$S = p_1 = \frac{k}{c_n} \prod_{j=2}^{n-1} \frac{k}{2c_j} = \frac{k^{n-1}}{2^{n-2}} \prod_{j=2}^{n} \frac{1}{c_j} = \frac{k^{n-1}}{2^{n-2}} \times \frac{1}{CEF}, \tag{13.10}$$

and express the number of servers in the network N_S as

$$N_S = \frac{k}{2} \times S = \frac{k}{2}\left[\frac{k^{n-1}}{2^{n-2}} \times \frac{1}{CEF}\right] = \frac{k^n}{2^{n-1}} \times \frac{1}{CEF}, \tag{13.11}$$

where CEF represents the connection-enhancement factor, which depends on the redundancy introduced in the Aspen-tree design by choosing the proper element values of FTV (ϕ_i's). It is evident from Eq. 13.11 that, in order to increase CEF, one needs to compromise in terms of the number of servers. To get around this problem, one can opt for a higher value for k (i.e., bigger switches), thereby restoring the DCN capacity (reduced due to the introduction of fault tolerance) in respect of the number of servers.

Next we return to the (4,4) Aspen tree shown earlier, in Figure 13.5, to explain the role of FTV in further detail. As mentioned earlier, the (4,4) Aspen tree has 16 servers with four-port switches, and it employs an FTV = (1,0,0), implying that one additional per-switch connection is set up between the switches at the levels L_4 and L_3 for fault tolerance. This implies that $CEF = c_2c_3c_4 = 1 \times 1 \times 2 = 2$, with $n = k = 4$, leading to the number of servers N_S, given by

$$N_S = \frac{k^n}{2^{n-1}} \times \frac{1}{CEF} = \frac{4^4}{2^3} \times \frac{1}{2} = 16. \tag{13.12}$$

Note that, N_s would increase from 16 to 32 with FTV = (0,0,0) (i.e., with $CEF = 1 \times 1 \times 1 = 1$). We next explain, using this network setting, how additional connections between levels L_4 and L_3 lead to *localized* fault tolerance, though reducing the size of the network in terms of the number of servers that can be supported in the network (i.e., from 32 to 16).

As indicated by an FTV of (1,0,0) in the (4,4) Aspen tree under consideration (Fig. 13.5), its left-most element is $\phi_4 = c_4 - 1 = 2 - 1 = 1$, implying that the highest level (i.e., L_4) should have one excess downward connectivity from each of its four switches to each pod in the next level (i.e., L_3), as compared to a (4,4) Aspen tree with FTV = (0,0,0). In a (4,4) Aspen tree with FTV = (0,0,0), a switch at L_4 (i.e., the core switch) connects to the next lower level, i.e., at L_3, with only one link reaching each pod, and hence the four ports of each core switch at L_4 can reach four pods at level L_3 (i.e., can reach two more pods, in contrast with the (4,4) Aspen tree with FTV = (1,0,0)), thereby doubling the number of servers in the network. Thus, in the (4,4) Aspen tree with FTV = (1,0,0) each switch in the topmost level reaches one pod in the next level through two links, which in turn gives the fault tolerance in the network, but with the reduced (halved in this case) number of servers. In the following, we explain the implication of fault tolerance in the (4,4) Aspen tree with an FTV = (1,0,0) by using an example case of inter-server communication between two servers belonging to two different racks (i.e., under two different ToR switches).

As shown in Fig. 13.5, the connection from a source server (x) to a destination server (y) is routed through the path *a-b-c-d-e-f-g*, traversing through the topmost level (L_1). However, as the link *f-g* gets out of order, the route must be redirected through an alternate path, if available. As evident from the figure, there exist redundant connections between L_4 and L_3, as each switch in L_4 connects to two switches of each pods of L_3, with $C_4 = 2$ and $\phi_4 = 1$.

Thus, with the failed link, the failure message needs to visit the core switch in the backward direction to set up the connection through the path *a-b-c-d-q-r-g*, thereby bypassing the failed link. Note that, the redundant connection being available at the topmost level, the failure message has to pass through half the network only. The latency of the path recovery can be further enhanced by providing redundancy in the lower levels, though the topmost level should be the first choice as it covers all the server racks. In other words, one should start increasing the element values in FTV from the left end only. However, with the increase in redundancy and having a fixed number of ports per switch (i.e., k), the number of servers will be reduced significantly (Eq. 13.11), and one can get around this issue by using a larger value for k, thereby leading to increased hardware.

Server-centric connectivity: DCell

DCell (Guo *et al.* 2008) is also based on a hierarchical topology, however of a different type, where the servers use multiple networking ports and thus get engaged in switching operation as well. The DCell topology is realized recursively using small switches (mini-switches), with each of them connecting to one group or cell of servers. Being enabled by multiple ports, the servers from different cells are connected using direct links between themselves. Thus, the higher-level switches are not needed for interconnecting multiple cells, and the network can make use of smaller switches with lower cost, albeit with additional network ports in the servers. We explain below the recursive methodology used in forming the DCell-based DCNs.

A DCell network can be built recursively in multiple steps for different levels, e.g., DCell_i with i levels, starting from a basic cell, called Dcell_0, by using one mini-switch and a group of servers. Thus, Dcell_0 is built around one single mini-switch interconnecting a number of servers, say n servers. The next level of DCell, i.e., DCell_1 is formed by using n Dcell_0's, which get interconnected through their constituent servers with multiple network ports (two in the case of DCell_1). We describe this process in the following with an example DCell_1 formation procedure.

Figure 13.6 illustrates the formation of a DCell_1 network using five, i.e., $(n+1)$ DCell_0's, assuming that $n = 4$ for each DCell_0, i.e., each DCell_0 consists of four servers with each server having two networking ports. Given one such DCell_0, say $\text{DCell}_0[0]$, each one of its four servers can be connected to another server belonging to a different DCell_0, by using the second port (one port is already used to connect to the parent mini-switch) over a point-to-point link. Thus, the four servers of $\text{DCell}_0[0]$ can be connected to four other servers, belonging to four additional DCell_0's, designated as $\text{DCell}_0[1]$, $\text{DCell}_0[2]$, $\text{DCell}_0[3]$, and $\text{DCell}_0[4]$. In this manner, each of the additional four DCell_0's will be connected to the remaining four other DCell_0's through their respective servers. Hence, in total, five (i.e., $n+1$) DCell_0's will be interconnected, thereby forming the next level of DCell, i.e., DCell_1. In this example, with $n = 4$, we can therefore have a DCell_1 of $n \times (n+1) = 4 \times 5 = 20$ servers, using five mini-switches. If each mini-switch has eight ports (i.e., $n = 8$), then the DCN at DCell_1 level will have $8 \times 9 = 72$ servers with $n+1 = 9$ mini-switches. Overall, each DCell_0 is conceptualized as a *node* with n servers, and $n+1$ such nodes are interconnected using the servers having two networking ports, thereby forming the next higher level DCell, i.e., DCell_1 for $n(n+1)$ servers.

Following the above steps of recursion, one can build a DCell_2 network by considering each DCell_1 again as one single node having $n = 72$, implying that each of its nine DCell_0 networks employs an eight-port switch interconnecting eight servers. Hence, DCell_2 can have $n \times (n+1) = 72 \times 73 = 5{,}256$ servers. The next level of DCell, i.e., DCell_3 will therefore be able to support $5{,}256 \times 5{,}257 = 2{,}763{,}792 \approx 2.8$ million servers! Theoretically, DCell scales doubly exponentially with the degree of the servers, i.e., n. However, to realize this

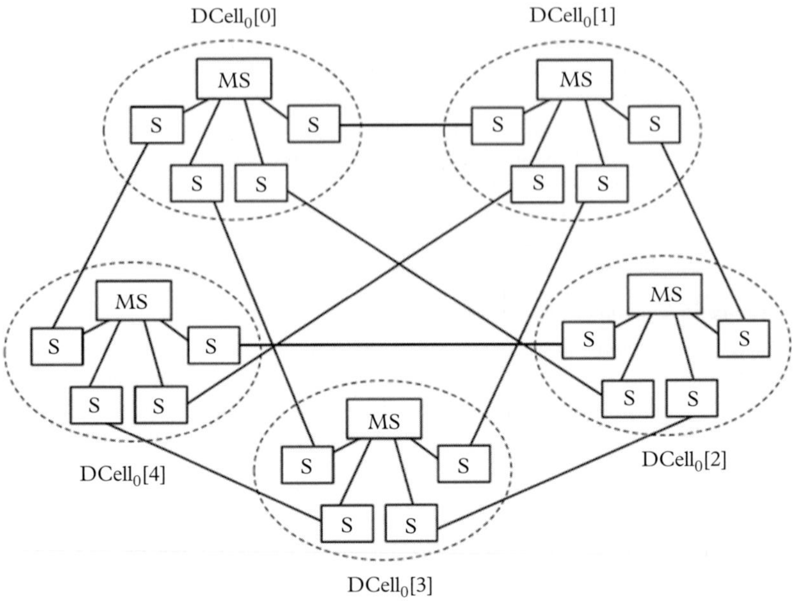

Figure 13.6 *$DCell_1$-based DCN formation using five $DCell_0$'s as the constituent nodes at the basic level (after Guo et al. 2008). Note that each $DCell_0$ employs one mini-switch (MS) connected to four servers (S), and thus the entire DCN supports 20 servers.*

connectivity, the servers must have an adequate number of ports. For example, in a $DCell_2$ network, each server must connect to a server from one of the other 72 constituent $DCell_1$'s and also to its $DCell_0$ mini-switch as well as one neighboring $Dcell_0$ server within the parent $Dcell_1$, needing thereby three ports. Thus, in effect the servers will have to play an important role in carrying out the task of switching in the DCN.

In view of the above, one can utilize the flexibility at the starting point of the design in respect of the size of the mini-switch. For example, using mini-switches with four ports, one can have a $DCell_1$ with 20 servers with two server ports, a $DCell_2$ with $20 \times 21 = 420$ servers with three server ports, and a $DCell_3$ with $420 \times 421 = 176{,}820$ servers with four server ports. However, with mini-switches having eight ports, one can have $DCell_3$ with the number of servers exceeding the million mark (2.8 million), again with four server ports. It therefore needs discretion to decide the mini-switch size and the level of a DCell to reach the desired size of the network (in terms of the number of servers) with a feasible port count in each server. Furthermore, DCell employs a distributed routing scheme, called as DCell fault-tolerant routing (DFR), that achieves a near-optimal routing solution utilizing the server-centric architecture of DCell and can handle a wide range of failures.

Server-centric connectivity: BCube

BCube is a modular DCN architecture with the feature of graceful performance degradation with failures (Guo *et al.* 2009). This feature makes the architecture highly suitable for relocatable shipping-container type DCNs, which needs to migrate from one location to another in a sealed container and hence in less serviceable form, thereby demanding more resilience to failure encountered due to mobility. Like DCell, BCube is also a server-centric DCN, but with more participation from the switches (mini-switches) in the network

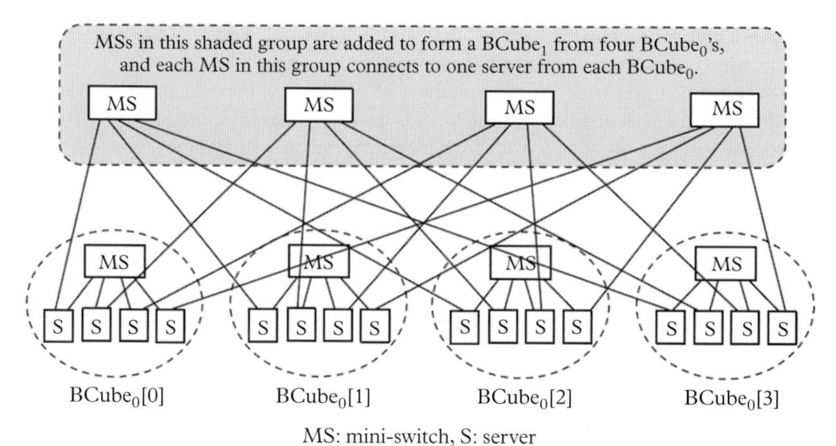

Figure 13.7 *$BCube_1$-based DCN formation using four $BCube_0$'s as the constituent blocks at the basic level (after Guo et al. 2009). Note that each $BCube_0$ employs one mini-switch connected to four servers, and thus the entire DCN supports 16 servers.*

operation. Each server typically has four networking ports, which are supplemented by several mini-switches to form the network topology.

BCube, like DCell, follows a recursive design approach, with its basic or initial topology being the same as in DCell. Thus, the basic BCube network, called as $BCube_0$, employs a number of servers connected to a mini-switch. However, the recursive process, which scales up the network further, is different from that of DCell. Consider that a $BCube_0$ network uses one n-port mini-switch with n servers. The next higher-level BCube, i.e., $BCube_1$ can be constructed using n $BCube_0$'s, where n additional n-port mini-switches are used at a higher level with each of the n ports of each additional mini-switch being connected to one server from each of the n $BCube_0$'s. In doing so, each server must have an additional networking port. We describe below this design procedure with suitable examples.

Figure 13.7 illustrates the formation of an example $BCube_1$ using four-port mini-switches, with each of its four servers having two networking ports. Unlike in DCell, no two servers in BCube are directly connected, thereby needing fewer server ports as compared to DCell. As shown in the figure, the four identical $BCube_0$ networks: $BCube_0[0]$, $BCube_0[1]$, $BCube_0[2]$, and $BCube_0[3]$, are interconnected using four additional four-port mini-switches. Each of the four ports of a newly added mini-switch is connected to one server from each of the four different $BCube_0$'s, so that one server from each $BCube_0$ can get connected directly through the higher-level mini-switch. Thereafter, three more servers will be left from each group, which will, likewise, need three more mini-switches to have direct connectivity, thereby using all the four additional mini-switches at $BCube_1$ level. As a result, all the $4 \times 4 = 16$ servers of the four $BCube_0$'s need to have two networking ports, and from a given $BCube_0$ each server is connected to a unique new mini-switch. Thus, in $BCube_1$ we now have eight mini-switches with four ports, and 16 servers with each server having two ports.

The above procedure is recursively continued to construct the next-higher levels of BCube network. To illustrate this process, we assume that we have the above $BCube_1$ network, which has already been developed from a $BCube_0$ with four servers. We consider this $BCube_1$ network as one single unit and replicate the same in three more units, giving four $BCube_1$ networks in total: $BCube_1[0]$, $BCube_1[1]$, $BCube_1[2]$, and $BCube_1[3]$. Next, we connect all the servers of the four $BCube_1$'s to an appropriate number of additional four-port mini-switches placed at the next higher level. Note that each $BCube_1$ has 16 servers in contrast with four servers in each BCube_0, and BCube_2 will therefore have $4 \times 16 = 64$ servers, which must be connected to the additional four-port mini-switches. The additional mini-switches will also have the same number of ports (four ports in this example), so that the cost benefit of commodity switch holds good for large-scale procurement. Hence, in order to connect 64 servers with four-port mini-switches in BCube_2, one will need $64 \div 4 = 16$ (i.e., 4^2) mini-switches, and each server will be using three (= the level number of the BCube under construction, i.e., two for BCube_2 + 1 = 2 + 1) networking ports. Therefore, with n-port switches at the kth level of BCube, by induction, we can generalize the required number of switches as n^k. In other words, with the mini-switches of n ports with n servers in the basic BCube_0 level, if we need to construct a BCube at kth level, i.e., BCube_k, we will need n^k mini-switches, each with n ports, and the entire network will support n^{k+1} servers, with each server having $k + 1$ ports.

Note that, in a BCube_k, switches do not interconnect directly, and the same is the case between the servers. The servers with multiple networking ports operate as relay-nodes in the network. The BCube-based DCNs scale up reasonably well. For example, with four-port mini-switches, a BCube_3 can accommodate $4^{3+1} = 256$ servers, each server using four ports. If the switch-port count is increased to eight, the number of servers would scale up to $8^{3+1} = 4096$ servers, with each server using again four ports, and with 16 ports this number will be $16^{3+1} = 65{,}536$. With the same 16-port switches, for a BCube_4 the server count will reach the million mark (i.e., 1,048,576), with each server using five ports. With eight-port switch it will require a BCube_5 for the server count to cross the million mark (2,097,152), with each server using six ports. Note that, BCube is superior to DCell in respect of architectural modularity, and notwithstanding the double-exponential scalability (of DCell) the traffic in DCell gets imbalanced. Further, a trade-off is realized in BCube between the server complexity and the number of switches, which is cost-effective as the switches used are small in size. Moreover, BCube adopts a fault-tolerant source-routing protocol, called BCube source routing (BSR) protocol, which can operate without coordination between the switches and without involving the servers in the routing process, and scales well with load-balancing capability.

13.2.2 DCNs using electrical-cum-optical switching

Judicious use of optical switching along with the electrical-switching elements not only alleviates the bandwidth bottleneck in a DCN, but also helps in reducing the power consumption. The basic philosophy of accommodating optical switching technology in a DCN is illustrated in Fig. 13.8. As shown in the figure, a separate optical-networking plane is added using a passive optical switch in parallel with the original electrically switched tree-based network between the ToR switches, which reduces the task of switching in the electrical domain significantly. In the following, we describe some of these DCN architectures using hybrid (electrical and optical) switching technologies, such as, c-Through (Wang *et al.* 2010) and Helios (Farrington *et al.* 2010). However, as we shall see later, the optical switching devices, while providing high-speed alternate paths to large-volume data flows, might add to the network latency owing to the sluggish reconfiguration process in the MEMS-based optical switches, and hence need to be used with discretion.

c-Through

c-Through is a DCN architecture using a hybrid packet and circuit-switching (HyPac) scheme, where the packet switching takes place in the electrical domain and relatively *stationary* packet flows are circuit-switched in the optical domain (Wang *et al.* 2010).

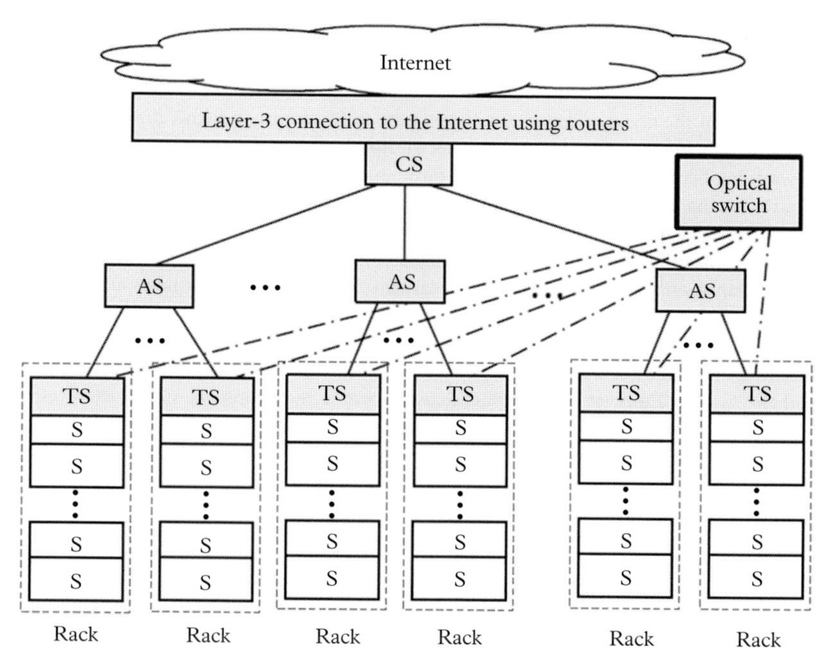

Figure 13.8 *Generic scheme for HyPac realization (after Wang et al. 2010).*

Notionally, the HyPac-based DCN proposed in (Wang *et al.* 2010) follows the generic architecture, shown in Fig. 13.8, where the electrical packet-switching is carried out over a tree topology with the ToR switches placed at the edges of the tree, while the optical switching is typically employed in parallel between the ToR switches using a centralized MEMS-based optical switch. As mentioned earlier, switch reconfiguration in MEMS is a slow process ($\simeq$ 10 ms), and hence only the relatively stationary or stable inter-rack packet flows are preferably transmitted through the MEMS switch. If the packet flows remain stationary over durations that are large enough (referred to as *elephant* flows) as compared to the MEMS switching time, the packet transmissions in optical domain takes place almost in a circuit-switched manner. By contrast, the bursty packets (referred to as *mouse* flows) are switched by the electrically packet-switching devices, typically Ethernet switches. Note that, although each of the electrical and optical parts of the DCN appears to have a single root, their combination effectively leads to a *multi-rooted tree*. Thus, owing to the diversion of elephant flows through the optical plane employing high-speed transmission through the optical switch, overall bandwidth-constraint of the network gets relaxed significantly.

In order to make the circuit-switching optical plane effective in the HyPac scheme, the servers engaged in inter-rack communications need to ensure enough data packets in the buffers before transmission, so that the durations of the optical-packet flows become appreciably large (i.e., ensuring elephant flows) as compared to the reconfiguration time of the MEMS. Or else, with bursty packet transmissions (i.e., mouse flows), the MEMS-based switch will keep spending a considerable amount of time for frequent reconfigurations, thereby wasting the bandwidth of the optical plane of the network. Such hybrid networks need additional traffic-control schemes, so that the packets from a source server for a given task are appropriately split for onward transmission over the electrically packet-switched and optically circuit-switched media.

In view of the above, aided by an appropriate traffic control scheme, the bursty inter-rack mouse flows are moved through the electrical-switching plane, while the stable elephant flows going between the racks are routed through the optical switch. However, the traffic-control mechanism carrying out the traffic-estimation task adds some latency over and above the delay involved in the switch reconfiguration process. Nevertheless, this latency problem is alleviated by the all-optical transmission (thus avoiding the packet-processing and transmission time overheads in electrical switches) of the elephant flows along with a considerable amount of savings in respect of wiring complexity and power consumption.

The HyPac architecture was realized using a testbed, called c-Through, by using an *emulated* optical switch on an electrical switch, as illustrated in Fig. 13.9, with 16 servers and two Gigabit Ethernet (GbE) switches. As shown in the figure, one GbE switch (GbE switch 1 in the figure) is logically configured into four parts, operating as four separate VLANs, with each VLAN connected to a group of four servers from each virtual rack (VR-A, VR-B, VR-C, and VR-D). Each of the four VLANs – VLAN-A, VLAN-B, VLAN-C, and VLAN-D – are used for intra-rack traffic for the respective virtual racks. The ports of the second GbE switch (GbE switch 2 in the figure), operating at the higher level (core) of the electrically switched tree topology, are connected to one free port from each VLAN-part of the first GbE switch (thereby requiring four ports from the second switch), so that each logical rack is connected to the next higher-level switch to form a tree with electrical switching.

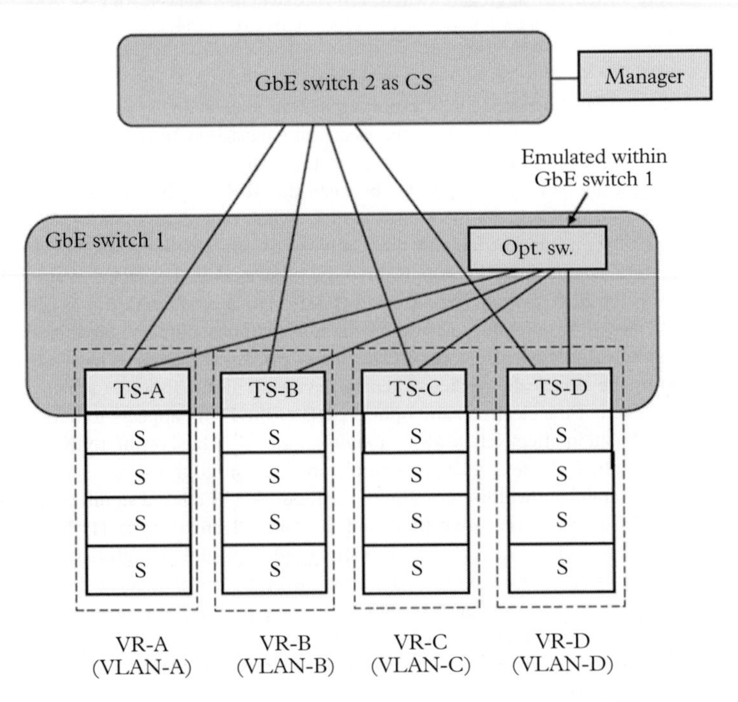

Figure 13.9 *Block schematic of c-Through testbed with an* emulated *optical switch (after Wang et al. 2010). Four VLANs (VLAN-A, VLAN-B, VLAN-C, and VLAN-D) are configured within Gigabit Ethernet (GbE) switch 1, which logically creates four virtual ToR switches (TS-A, TS-B, TS-C, and TS-D) supporting four virtual racks of servers: VR-A, VR-B, VR-C, and VR-D. The optical switch is emulated within GbE switch 1 itself. The core switch (CS) operates from GbE switch 2, along with the network manager.*

Further, within GbE switch 1, its internal switching fabric is utilized to emulate the optical switching (equivalent to the core level) between the four racks belonging to the four VLANs, by allowing high-speed transmission at 4 Gbps (which is four times the transmission rate in a GbE switch). The reconfiguration time of the optical switching system is emulated by delaying the packet transmissions by an amount equaling to the reconfiguration time of the optical switch along with the re-synchronization times needed at the optical transceivers. As and when the manager of the network, operating from GbE switch 2, finds that there are enough inter-rack packets between a pair of racks qualifying for circuit switching, the respective packet flows are transmitted through the emulated optical switch with prior switch reconfiguration.

The testbed performance was examined for several applications, such as virtual machine (VM) integration, Hadoop, and fast Fourier-transform (FFT) computation based on message-passing interface (MPI), with encouraging results on the overall network throughput. For example, while executing a VM integration task with an oversubscription ratio of 40:1 and an input buffer of 128 KB for c-Through, the completion time was found to decrease from around 280 seconds for electrical network to 120 seconds for the c-Through network. However, as expected, with an oversubscription ratio of 10:1, this difference shrinks down with 75 seconds for electrical network and 50 seconds for c-Through. Similar results were also observed with the other tasks.

Helios

Helios also employed a HyPac-based combination of electrical packet-switching and optical circuit-switching devices in realizing hybrid DCN as in c-Through. However, Helios realized the optical-switching functionality by using MEMS-based optical switch (instead of emulated optical-switching in c-Through), along with the option for WDM transmission (Farrington *et al.* 2010).

Figure 13.10 presents the schematic diagram of a Helios-based DCN. As shown in the figure, the network employs two levels, one with several pods at lower level (edge), and the other with core switches. The core switches are of two types. Some of the core switches

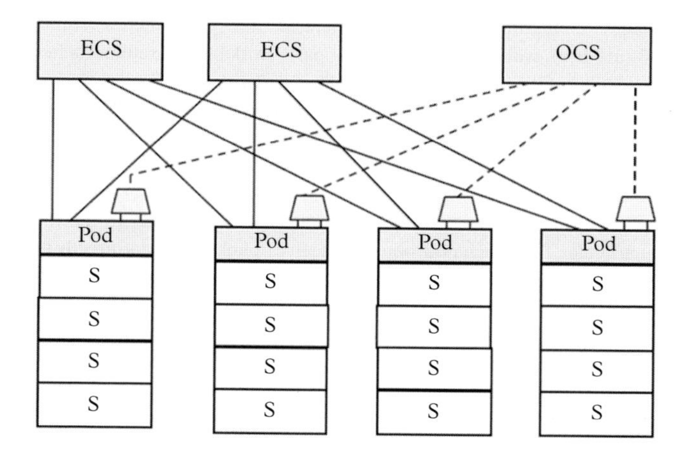

Figure 13.10 *Representative block schematic for a Helios network (after Farrington et al. 2010). Each pod with eight ports (four transmitters and four receivers) supports four uplinks for the core-level switches (ECSs) and four downlinks for the servers. Further, from each pod, two optical uplinks are wavelength-multiplexed using two distinct wavelengths to provide one WDM uplink to the optical core switch (OCS). Note that, in the actual testbed (shown later), Helios makes use of multiple OCSs to enhance the capacity of optical plane furthermore.*

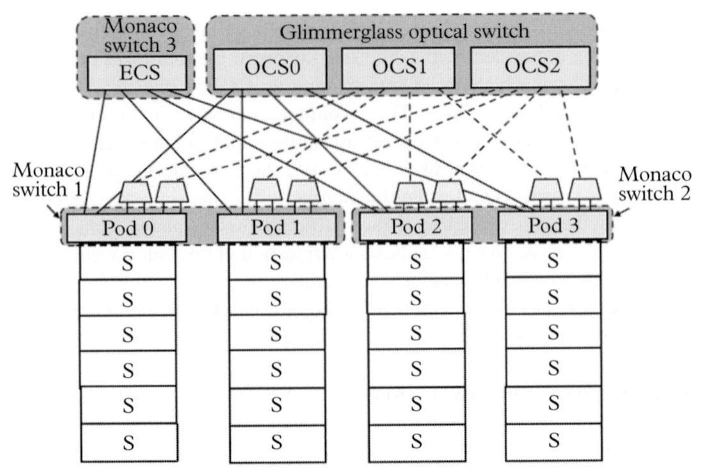

Figure 13.11 *One of the possible configurations of Helios prototype (after Farrington et al. 2010). Four pods are configured on two 24-port Monaco 10 Gbps Ethernet switches, while one electrical core switch (ECS) is configured with four ports of another Monaco Gigabit Ethernet switch. Three optical core switches (OCSs) are configured in 64-port Glimmerglass optical-switching device, with some spare ports left out for future use.*

are electrically switched, called electrical core (packet) switches (ECSs) responding to the bursty-packet traffic (i.e., mouse flows), while the rest are optical core switches (OCSs) (Fig. 13.10 shows only one OCS, while in practice there can be more) carrying the more stable component of the traffic (i.e., elephant flows) between the servers in a circuit-switched manner. The pods can employ electrical multiplexing (TDM) and/or optical multiplexing (WDM) to enhance the effective capacity of the links reaching the OCSs.

One of the configurations of the Helios prototype is shown in Fig. 13.11, with 24 servers interconnected using a 64-port Glimmerglass optical switch and three 24-port 10 Gigabit Monaco Ethernet (10GbE) switches, along with another 48-port Dell Ethernet switch (not shown) for control purpose. The 64-port optical switch is partitioned into three (or more for other possible configurations) virtually separated four-port switches (some ports left unused for future use), with all the four virtual parts working as circuit switches. Further, two 24-port Monaco GbE switches are virtually configured into four pods (pod 0, 1, 2, and 3) at the bottom layer, whose six downlinks connect to $6 \times 4 = 24$ servers. The third Monaco GbE switch is used as an ECS (whose four ports are used for the configuration under consideration) at the higher level, to which each pod connects using one of its six uplinks.

From each pod, the rest of the uplinks, i.e., five uplinks, connect to the three optical switches. In particular, from each pod, one of these five uplinks connects directly to one OCS (i.e., OCS0). Thus, the four ports of OCS0 get connected to four pods. The remaining four uplinks from each pod are grouped into two pairs, and the two ports of each pair employ two transceivers on wavelengths 1270nm and 1290 nm. These two wavelengths are optically multiplexed/demultiplexed into/from a single optical super-link (dashed lines) to connect one of the remaining two OCSs (OCS1 and OCS2). Thus, at each pod two WDM signals (each with two wavelengths) are transmitted/received on two super-links to connect with OCS1 and OCS2. The control switch, not shown in the diagram, executes the necessary network control operations in the background.

As indicated for c-Through, traffic measurements are necessary in Helios as well. The control module in Helios carries out this job by using *flow counters*. By using flow counters, it classifies the packet flows in the two categories, mouse flows ($\leq$ 15 Mbps) and elephant flows ($>$ 15 Mbps). With the elephant flows at the core level, it forms a traffic matrix and accordingly configures the OCSs. Mouse flows at core level are packet-switched through the ECS. As and when the OCSs need reconfiguration, prior notifications are sent to the servers from the controller to avoid packet losses.

Performance analysis of Helios prototypes was carried out in terms of throughput as a function of time and of traffic stability. With some engineering of the physical-layer issues – debouncing of packets during circuit reconfigurations and removal of over-restrictive noise-cleaning mechanisms in optical receivers – throughput could be improved significantly. Further, the throughput exhibited considerable increase with more stable traffic flows through optical switches, as this type of traffic flow required fewer optical switch reconfigurations over the observation period.

13.2.3 All-optical DCNs

DCNs can use all-optical physical topologies with WDM transmission for large-scale datacenters, leading to considerable improvement in speed and power consumption. There have been several investigations on all-optical DCNs, such as OSA (Chen *et al.* 2014), Mordia (Porter *et al.* 2013), DOS (Ye *et al.* 2010), LIONS (Yin *et al.* 2013a), OSMOSIS (Minkenberg *et al.* 2006), and FISSION (Gumaste *et al.* 2013). In this section we present some of these all-optical DCNs, which utilized some of the basic concepts of WDM-based LANs, MANs, and PONs, that we discussed in earlier chapters.

OSA

Optical switching architecture (OSA) was developed as an all-optical DCN, which employed wavelengths as well as wavebands (contiguous band of wavelengths) over a WDM-based optical-switching network between its servers (Chen *et al.* 2014). Like its predecessors (c-Through and Helios), OSA also differentiates between mouse and elephant flows, albeit in a different manner. In particular, it employed multihop WDM connections by concatenating multiple lightpaths through OEO conversion at intermediate ToRs for bursty mouse flows, thereby offering scope for packet multiplexing (traffic grooming) in the electrical domain at intermediate ToRs. However, single-hop all-optical lightpaths were set up in OSA for stable and large-volume elephant flows. Computer simulation of OSA architecture and small prototype implementation and measurements from the simulation and experiments indicate promising features of the OSA architecture, such as high bisection bandwidth and flexibility in the underlying WDM lightpath topology adapting with time-varying traffic demands.

A basic schematic of the OSA architecture is shown in Fig. 13.12, where the ToR switches over the L server racks are interconnected using a wide range of passive optical devices: optical multiplexers/demultiplexers and power combiners, WSSs, circulators, and MEMS-based optical switching matrix (OSM) at the highest level (i.e., as a core switch) of the two-level tree-configured optical interconnect hierarchy. Each ToR switch has M optical transceiver ports, transmitting and receiving M distinct wavelengths, viz., $w_1, w_2, \cdot\cdot, w_M$. The transmitted wavelengths from a ToR are multiplexed in an optical multiplexer, and thereafter a bandwidth-variable WSS (BV-WSS) (see Chapters 2 and 14) converts the wavelength-multiplexed signal from the multiplexer into k appropriate wavebands, where the wavebands may have varying number of wavelengths. Note that, each ToR has M downlinks to connect to M servers, and M servers can choose to transmit M wavelengths as optical uplinks, which are subsequently grouped into $k < M$ wavebands through the respective WSS.

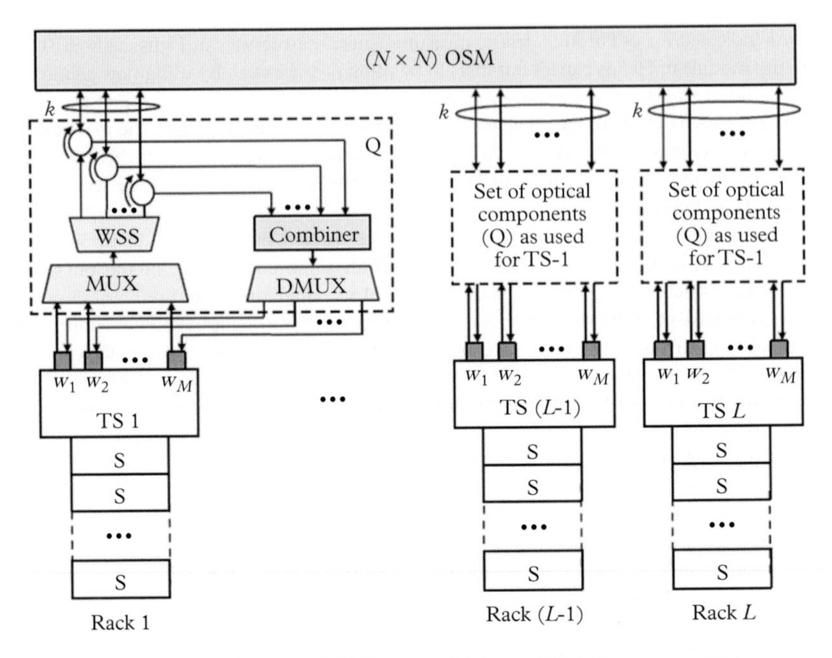

Figure 13.12 *OSA architecture (after Chen et al. 2014). Number of racks $L = N/k$.*

These k wavebands created at the WSS output are thereafter forwarded through an optical circulator to one of the N input ports of the $N \times N$ OSM, with $N = k \times L$. The circulator is needed at this point, as the received signal from the same port of the OSM needs to be isolated from the transmitted signal and directed toward the receivers of the same ToR switch. All of the k wavebands from a ToR switch reach k ports of the $N \times N$ OSM (with $N > k$), with the MEMS mirrors in the OSM being oblivious to the received wavelengths in the wavebands. These incoming wavebands are therefore switched to the desired (pre-configured) output ports of the optical switch and forwarded to the destined ToR switches through the circulators, optical power combiners, and demultiplexers. Overall features of the network is described by the following parameters:

- number of racks (= number of ToRs): L,
- number of wavelengths in OSA: M,
- number of servers in each rack (= number of transceivers in each ToR): M,
- OSM input/output ports: $N \times N$,
- WSS input/output ports: $1 \times k$,
- per-ToR input/output ports at OSM: k,
- structural constraints: $k < M$ and $N = kL$

As mentioned above, each ToR offers k ports from its WSS to OSM and vice versa, represented by as many distinct wavebands, which get connected to k ports of the OSM. Thus the nodal degree of a ToR switch (for a rack) as seen by the OSM is k, while its number of transceiver ports is $M (> k)$. Hence, $L = N/k$ ToR switches can concurrently transmit through the OSM, and each ToR switch can communicate directly (single-hop all-optical connection) with k ToR switches in the network. In order to increase the connectivity between the servers from different racks, one can adopt the concept of multihop connections used in WDM LANs (Chapter 6), where two nodes can have communications via other intermediate nodes with OEO conversions therein, albeit using multiple passages through the same OSM. Every attempt would be made for provisioning the elephant flows with single-hop all-optical connections, while the mouse flows could be transmitted using multihop connections, if necessary, with OEO conversions at intermediate ToR switches.

An example architecture of OSA, named OSA-2560 supporting 2560 servers, was simulated for performance measurement (Chen *et al.* 2014). OSA-2560 employed a MEMS-based OSM with 320 ports (i.e., with a 320×320 non-blocking OSM), $N =$ 80 ToRs, and $M = 32$ wavelengths. The nodal degree of each ToR was considered to be $k = 4$, and hence with 80 ToRs, it required $4 \times 80 = 320$ ports for the optical switch, justifying the size of the 320×320 OSM. Further, with $k = 4$ and $M = 32$, each ToR required 32 transceivers, followed by one 32×1 multiplexer, one 1×4 WSS and four circulators for onward transmission. Further, each ToR used 32 downlinks to connect with 32 servers, and hence 32 servers could choose 32 wavelengths as optical uplinks, which were subsequently grouped into four wavebands. For reception of the switched outputs, each ToR required 4×1 combiner, followed by a 1×4 demultiplexer. With this architecture, a ToR could communicate on four wavebands to four other ToRs using single-hop connectivity. Furthermore, with the reconfiguration capability of WSS, OSA could control the speed of each waveband in the range of [0, $M \times 10$ Gbps]. Hence, with $M = 32$, each ToR could choose a speed for each of its four optical ports (wavebands), varying with the quantum of 10 Gbps, i.e., $0, 10, 20, \cdots, 320$ Gbps. The switch configuration and the corresponding network topology were dynamically controlled by the OSA manager (not shown explicitly in the figure). However, the overall latency of OSA-2560 would depend on the overhead time needed for traffic estimation and switch reconfiguration processes. The above OSA configuration was simulated and performance measurements were reported in (Chen *et al.* 2014). We discuss the important simulation results in the following.

The reported measurements were based on some typical traffic patterns, such as MapReduce, and some real datacenter traffic matrices. One of the important performance metrics examined from these measurements is the bisection bandwidth of the network. It was observed that OSA-2560 can offer high bisection bandwidth, reaching about 60% or even 100% of non-blocking network for different traffic patterns. However, the time overhead for the traffic estimation and the switch reconfiguration process from one to another traffic pattern was indeed non-negligible (290 ms), out of which the traffic estimation itself took 161 ms. Thus, there remains a large scope for improving the associated algorithms and the switching technology to bring down this overhead.

On a smaller scale, an OSA testbed was also developed, which employed eight Optiplex Dell servers, emulating virtually a large number of end-hosts (servers) for the DCN as follows. In particular, each Optiplex Dell server could run four virtual machines (VMs), emulating four virtual servers (VSs) as the end-hosts of the network. Thus, each Optiplex Dell server realized a virtual rack (VR) of four VSs, thereby configuring 32 VSs for eight VRs from eight Optiplex Dell servers. Furthermore, to have eight ToR switches for the eight VRs, four Dell PowerEdge servers were used. Each one of these four servers was used to emulate two VMs, giving altogether eight VMs operating as eight virtual ToR (VToR) switches for the eight VRs. Each VToR used six network interface cards (NICs), one of which went to the network OSA manager, one was connected to one VR accessing the VSs underneath (logically) the same VToR. The remaining four NICs were connected through optical transceivers, optical multiplexers, WSS, and circulator to reach the OSM. On the receiving side of the same ToR, circulator outputs were fed into optical demultiplexers, with the demultiplexed outputs reaching the receiver ports of the respective transceivers. This arrangement was replicated for each VToR, thus completing the entire testbed setting. A Linux server was used over and above all the above Dell servers, functioning as the OSA manager, which controlled the OSM and all the VToRs using Ethernet connections and the WSSs through RS-232 serial ports.

The optical part of the OSA used the following devices. In particular, each VToR used 1×4 CoAdna WSS, along with a 32-port Polaris series-1000 OSM overlooking all the eight VToRs of the entire testbed from the highest level (core). The wavelengths used by the VToR transceivers were 1545.32 nm, 1544.53 nm, 1543.73 nm, and 1542.94 nm, corresponding to the channel numbers 40 through 43 of the 100 GHz ITU-T grid.

Notwithstanding the large bisection bandwidth and scalability of the architecture, managing the bursty time-varying mouse flows with sluggish OSM was indeed a challenge in OSA. As the experimental results indicate, the mouse flows got affected in OSA as the *stability time* of the traffic reduced below 5 seconds. Using a typical production center dataset, it was found that, as the stability time of the flows reduced from 9 seconds to 2 seconds, the mouse flows got affected by 1% to 4.5% (4.5 times deterioration). In other words, when OSA had to go through more frequent reconfigurations, higher fractions of mouse flows were affected with increased latency. A few possible remedies could alleviate this problem as follows.

One possible remedy proposed by the authors was to reserve a few wavelengths along with some OSM and WSS ports as static circuits (in the form of some suitable topologies) for mouse flows, so that the bursty flows can be assigned the needed routes instantaneously, albeit at the cost of those ports being blocked for the stable connections. The other alternatives would be to set up a ring or a star network between the ToRs with some reserved ports therein or to use an additional electrical switch connecting each ToR switch.

Mordia

Mordia stands for *microsecond optical research datacenter interconnect architecture*, realizing WDM-based optical switching in a DCN using WSSs, to bring down the switch reconfiguration time by three orders from $\simeq$ 10 ms (in MEMS-based DCNs) to $\simeq$ 10 μs (Porter *et al.* 2013). A block schematic of the Mordia testbed using WSS-based WDM ring for 24 ToR switches (TSs) is shown in Fig. 13.13. The latency incurred in the OSM used in OSA and its predecessors is addressed in Mordia using the WSS-based distributed optical

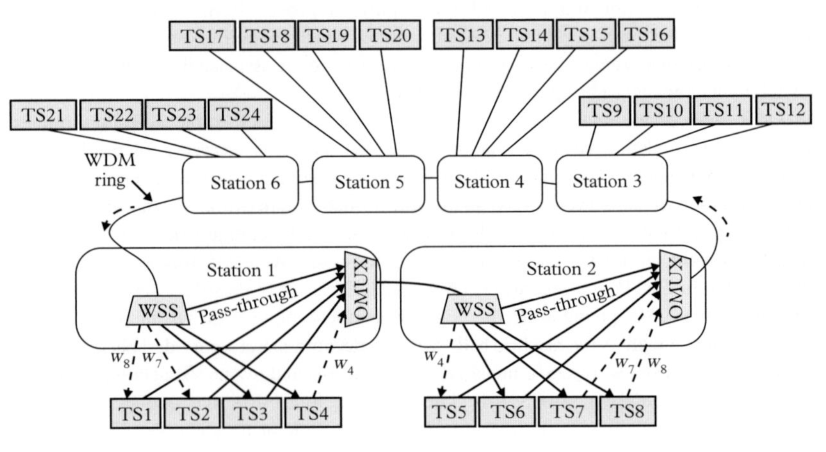

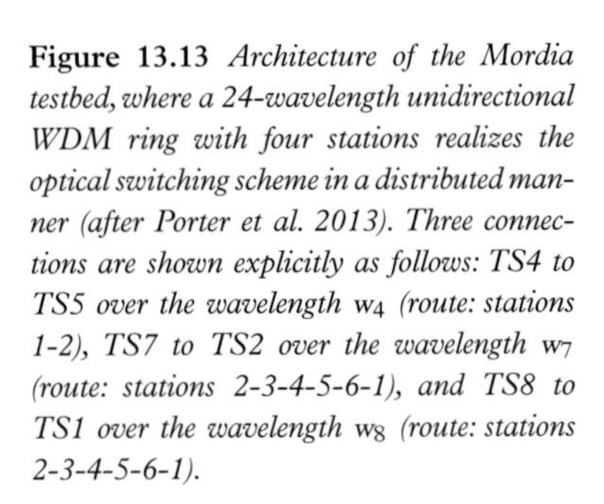
Figure 13.13 *Architecture of the Mordia testbed, where a 24-wavelength unidirectional WDM ring with four stations realizes the optical switching scheme in a distributed manner (after Porter et al. 2013). Three connections are shown explicitly as follows: TS4 to TS5 over the wavelength* w_4 *(route: stations 1-2), TS7 to TS2 over the wavelength* w_7 *(route: stations 2-3-4-5-6-1), and TS8 to TS1 over the wavelength* w_8 *(route: stations 2-3-4-5-6-1).*

switching technique along with a novel traffic scheduling scheme. The WSSs used in Mordia can be built with liquid-crystal technology (LCoS), offering fast reconfiguration process ($\simeq$ 10 μs). As we shall see later, unlike the *centralized* switching scheme used in OSA employing OSM, Mordia uses a *distributed* switching scheme across the DCN. The TSs in Mordia are clustered into several groups, called *stations*, which are interconnected through a unidirectional WDM ring, and the connection between the TSs is controlled by using WSS-enabled nodes at each station.

Before going into the operational details of the Mordia architecture, we have a re-look at the hybrid and all-optical DCN architectures discussed so far. In particular, we note that, in hybrid (c-Through and Helios) architectures, the networking protocols employed some traffic analysis followed by the optical switch configuration process. In effect, both of them would search for some hot spots in terms of the connections through an optical switch to let the elephant flows pass through the switched all-optical paths. Such schemes, referred to as *hot-spot scheduling* (HSS), were needed to overcome the impact of sluggishness of the MEMS-based optical switches on bursty (mouse) traffic flows. However, in the Mordia ring, the switching time of WSSs being in the order of 10 μs, the traffic-scheduling scheme didn't have to look for highly stable flows, such as in HSS-based switching schemes using flow durations exceeding 100 ms. Effectively, with WSS as the switching device at each station, the overall latency in Mordia scaled down to the order of $\sim$ 100 μs, thereby allowing the bursty mouse flows to be routed through the optical switching system with satisfactory delay performance.

Along with the WSS-based optical switching, Mordia employed a different type of traffic management scheme, called traffic matrix switching (TMS), in place of HSS, so that the network could maximize the benefit of the faster switching hardware. In particular, the controller in Mordia operated in the background through the TSs to estimate the traffic from the servers and used much shorter *schedules* of packet flows instead of the long ones used in HSS, and configured the agile WSSs accordingly (Mordia testbed realized 11.5 μs of switching time). Scheduling of the WSS-based switching system across the ring was carried out in a cyclic manner, with the cycle time being much higher than the WSS reconfiguration time. Assuming that there are N TSs in the network, each TS would use $(N - 1)$ output buffers where the packets due for transmission arrive from the servers and wait until the beginning of the next cycle. However, the cycle duration might change with time, governed by the controller which sets up the cycles in accordance with the traffic dynamics.

The data packets in Mordia, queued up in all the TS buffers across the ring during an ongoing cycle, formed a short-term *traffic matrix*, which was scheduled to be switched through the WSS in the next cycle; hence the name TMS. The respective TS transceivers used appropriate transmit wavelengths for the transmission through the WSS during the next cycle. The cycle was segmented into a number of time slots, and during each time slot the data packets corresponding to a part of the traffic matrix (a sub-matrix) were transmitted through a specific WSS configuration, which had to be set up prior to the respective transmission. Thus, in between two consecutive time slots in a cycle, there was an interval for WSS reconfiguration, called the *night* interval, when the WSSs got reconfigured and hence the transmissions from the TSs were inhibited during these intervals. Duration of the slots was typically kept at least one order higher than the night interval, so that the time spent during WSS reconfiguration couldn't affect the network latency. Thus, the entire traffic matrix for a cycle time was partitioned into several sub-matrices, and each sub-matrix was scheduled for one specific time slot in the cycle time along with one specific WSS-configuration, with pauses during the short night intervals in between the consecutive slots.

After the completion of a given cycle time (say, previous cycle), all the sub-matrices were transmitted in the next cycle (say, present cycle) over their respective time slots through the optical switch. The buffers storing the transmit packets in the previous cycle were not allowed to become saturated (to prevent overflow), implying an underlying assumption that the network was not driven beyond its capacity.

The Mordia ring testbed, as shown in Fig. 13.13 employed six stations interconnected through the WDM ring, and each station hosted four TSs, leading to a 24-TS Mordia DCN. Each station employed one WSS and one optical multiplexer (OMUX), and each of them (WSS and OMUX) was connected to all the four TSs belonging to the same station. In a given station, each TS used a unique wavelength, and thus the four wavelengths from the four TSs were multiplexed along with the passthrough wavelengths of WSS for onward transmission on the counter-clockwise WDM ring.

The Mordia stations shown in Fig. 13.13 were implemented in effect by using four rack-mounted *trays*, with three trays having six stations (two each) and the fourth tray carrying a FPGA-based control board operating as the traffic scheduler. Figure 13.13 illustrates three candidate connections configured using the WSSs and OMUXs over the WDM ring by using three distinct wavelengths. For example, the wavelength w_4 connects TS4 to TS5, located at the stations 1 and 2, respectively, while the wavelengths w_7 and w_8 connect TS7 and TS8 located at station 2 to TS1 and TS2, located at station 1. Thus, over the WDM ring, with the help of the traffic-matrix scheduler, all-optical connections could be established during the respective time slots in a given cycle, with a switching latency of 11.5 μs.

The Mordia testbed can be scaled up physically over C-band using $\simeq$ 40 wavelengths, although the corresponding control plane must speed up accordingly for the scheduling purpose. Further expansion is also possible by stacking multiple WDM rings, but with the possibilities of connection blocking due to the limited port counts in the available WSS devices, which can however be addressed with tunable transmitters, at the cost of increased cost. With all these arrangements, the authors of Mordia gave an estimate of the maximum number of TSs to be 704, which would be effectively supporting *optically circuit-switched* high-speed connections between $\simeq$ 28,000 servers with the assumption that each TS would be connected to 40 servers.

LIONS

LIONS stands for *low-latency interconnect optical network switch*, and it can be used for DCNs as well as high-performance computing systems (Yin *et al.* 2013a). Like its former version, named as datacenter optical switch (DOS) from the same research group (Ye *et al.* 2010), the central enabling optical device used in an N-node LIONS is an $(N + R) \times (N + R)$ AWG, which offers a static routing matrix between its N input/output ports, where the routing flexibility is brought in by using electrical buffers placed on the loopback path between R additional output/input ports in the AWG.

Figure 13.14 presents a block schematic of an N-node LIONS. Note that, the nodes in LIONS represent the ToR switches of the DCN. At the core level of the network, AWG is used as the central routing/switching element, along with the other supporting optical and electrical hardware. As shown in the figure, each ToR switch or node is spilt into two parts, one part with the transmitter (TX) and the other part with the receiver (RX), which are shown on the input and output sides of the AWG, respectively (though the TX and RX of each ToR switch are physically collocated inside the same network element). The TX in each ToR switch employs an optical channel adapter (OCA) using an optical label generator (OLG), packet encapsulation (PE), and electro-optic conversion (lasers, typically) to send the incoming packets to the core switching system including the AWG, label extractor (LE), and tunable wavelength converter (TWC). The AWG output is forwarded to the optical receivers belonging to the various ToR switches, employing optical demultiplexer

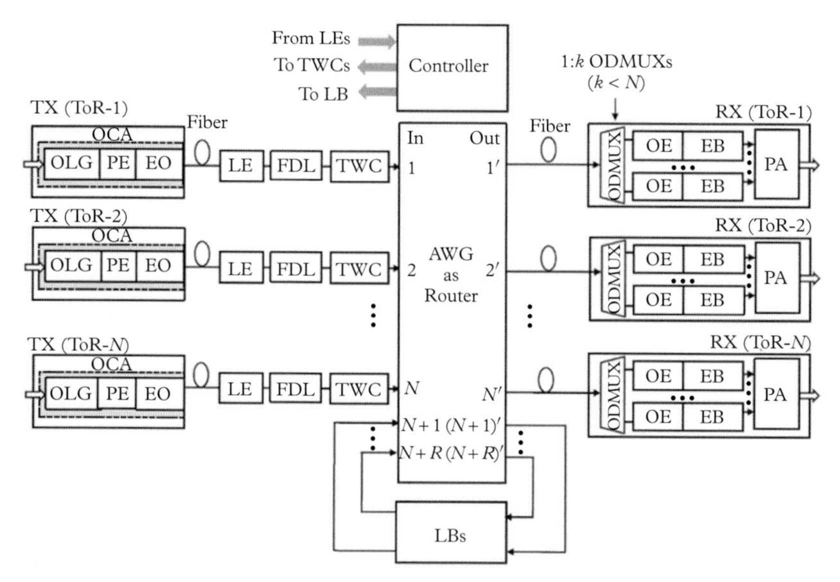

Figure 13.14 *LIONS architecture with N nodes (ToR switches) (after Yin et al. 2013). OLG: optical label generator, PE: packet encapsulation, EO: electro-optic converter (laser), LE: label extractor, TWC: tunable wavelength converter, ODMUX: 1 × k optical demultiplexer with k < N, OE: opto-electric converter (optical receiver), EB: electrical buffer, PA: packet adapter, LB: loopback buffer (electrical).*

(ODMUX) followed by optical receiver (OE converter), electrical buffer (EB), and packet adapter (PA) to retrieve the received data packets. The loopback buffers (LBs), realized in the electrical domain with the necessary OE/EO conversions are used to resolve output port contentions, which we discuss later.

Ideally, the number of receivers in each ToR switch should be the same as the number of nodes N (precisely $N-1$). However, in that case, the entire DCN needs $\sim N^2$ optical receivers for OE conversion, which would prevent the architecture from being scalable. In view of this, each node employs k ($< N$) receivers, preceded by a $1:k$ ODMUX. The received packets in electrical form are kept in buffers (EBs), and subsequently passed on to the PAs to obtain the packets destined to the receiving ToR switch.

The unique feature of this architecture is the AWG-based routing scheme and electrical buffering in the loopback path of the AWG using LBs. In particular, when an optical packet arrives at the central switching system, including AWG preceded by LE and TWC, the LE extracts the header and passes the label on to the controller block. In the controller block, the optical label is converted into electrical domain and processed to get the destination address and the packet length. Thereafter, the controller maps the destination address to the appropriate output port of the AWG, and determines the appropriate wavelength for onward transmission through AWG and whether the packet needs to wait in the loopback buffer (LB). Accordingly, messages are sent to the TWC and LB, so that the respective TWC and LB function accordingly. During the time spent by the controller in carrying out its task, the optical packet is delayed in the FDL, the delay being fixed for all incoming packets. Note that, in LIONS, the latency in switching system comes from the controller processing time and the tuning time overhead of the TWCs (typically in the range of nanoseconds), the latter being much faster than the switching systems used in OSA and Mordia. However, the LBs in LIONS add to the overall delay, when the packets are sent to the LBs through the AWG output ports $N+1$ through $N+R$ due to output port contentions.

The architecture of the electrical LBs in the loopback path play a significant role in LIONS. In particular, the static routing-configuration in AWG is relaxed and made dynamic by introducing the loopback path. The loopback operation can be executed using three possible buffer configurations: shared loopback buffer (SLB), distributed loopback buffer (DLB), and mixed loopback buffer (MLB). Before describing the buffering schemes, we note that, Fig. 13.14 shows R ports on both sides of the AWG, kept for loopback functionality. In SLB, $R = 1$, i.e., AWG has $(N + 1)$ ports on input and output sides, implying that the packets encountering contentions from anyone of the N input ports are directed to one single output port, i.e., the $(N+1)$th port, for loopback entry into the AWG through $(N+1)$th input port. The optical signal exiting from the $(N+1)$th output port of the AWG are optically demultiplexed into specific wavelengths, which are subsequently passed through OE converters, for being stored in the shared electrical buffers. Upon hearing from the buffer controller, the stored packets are passed through the EO converters at the desired wavelength and forwarded to the $(N + 1)$th input port of the AWG. From the $(N + 1)$ input port, having waited in the buffer for the stipulated time estimated by the controller, these packets are routed to the desired output port of the AWG. Figure 13.15 illustrates the mechanism of SLB, as described above.

The DLB scheme is more complex than SLB in respect of hardware, and obviously better in performance in terms of latency. In DLB (not described here, and the interested readers are referred to (Yin *et al.* 2013a) for further details), the AWG uses $2N$ input/output ports, where N ports are used for the actual data inputs, while the remaining N ports (instead of only one additional port in SLB) are used to resolve the contention problem, with distributed

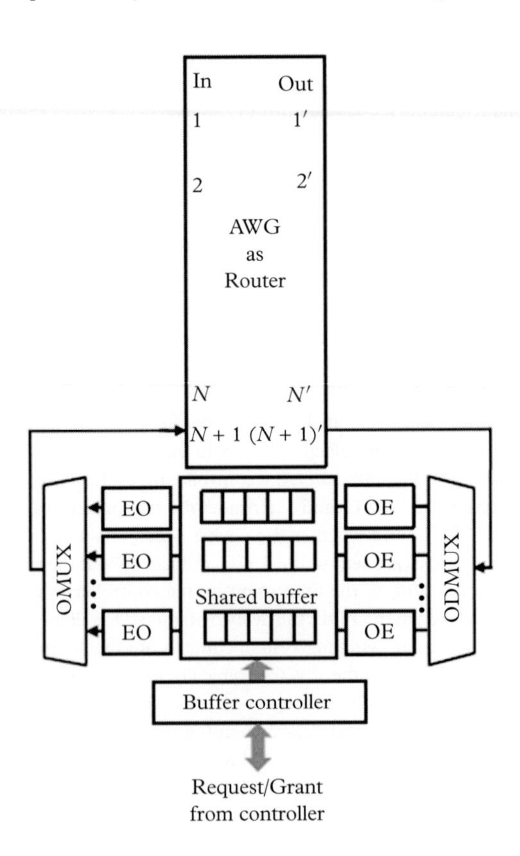

Figure 13.15 *SLB schematic for LIONS (after Yin et al. 2013).*

and independent buffers for each of the AWG input ports. The third option, i.e., MLB, uses a mix-and-match of both SLB and DLB, and the details of its functionality and hardware configuration can be found in (Yin *et al.* 2013a).

Although out of the three buffering schemes, DLB performs the best, in respect of hardware any loopback buffering scheme turns out to be a challenging option for all three loopback schemes, when the network scales up to a large size. In order to get around this problem, an alternative version of DOS/LIONS architecture was examined, by using an all-optical *negative acknowledgment* (AO-NACK) scheme for the packets lost in contention for the desired output ports in the AWG. We describe this architecture – LIONS with AO-NACK – in the following, with its illustration presented in Fig. 13.16.

As shown in Fig. 13.16, the AWG output from the output port $(N+1)'$, the deflected packets are passed on to the AO-NACK generator through circulator C_4. The output of the AO-NACK generator passes through the AWG through circulator C_4 again through the same AWG port in the opposite direction and reach the sender's input port of AWG. Thereafter, with the help of the circulators C_3, C_2, and C_1, the AO-NACK reaches the right source (ToR switch) and gets converted into the electrical domain by the OE converter to extract the NACK message. The message in turn informs the ToR switch that the packet got blocked and needs retransmission.

An exhaustive simulation package of large-scale LIONS was developed, and the performance was measured in terms of the network latency and throughput. Further, a four-node testbed was developed for LIONS, and measurements were made to validate the simulation model developed for the network architecture. Overall, it was found that LIONS with AO-NACK performs well, and thus offers a practical solution to resolve the extensive hardware issue in DLB-based LIONS without much significant difference in the results. A proof-of-concept experiment on LIONS based on AO-NACK was reported in (Yin *et al.* 2013a) using packet durations of 204.8 ns with a distance of 20 meters between hosts and the switch, where error-free retransmissions using AO-NACK mechanism were realized successfully at 10 and 40 Gbps.

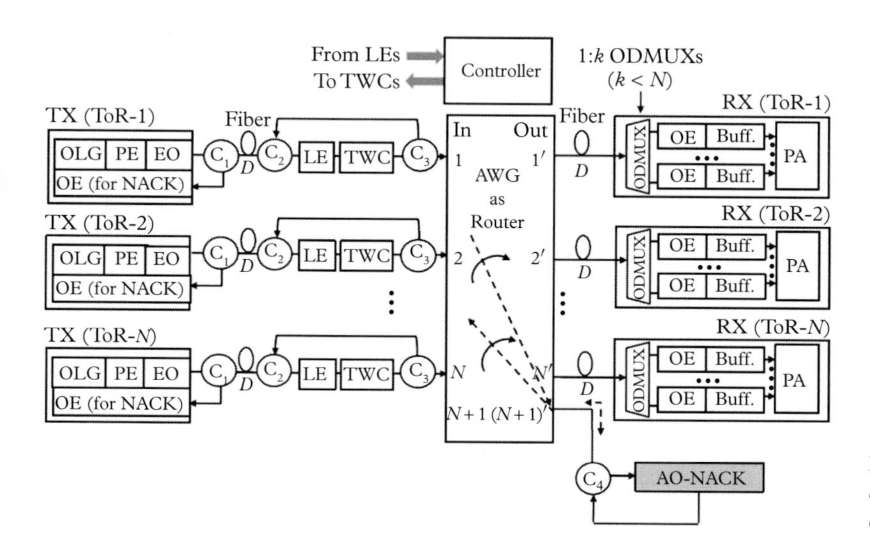

Figure 13.16 *LIONS with AO-NACK. All circulators (C_1, C_2, C_3, and C_4) operate in the clockwise direction (after Yin et al. 2013).*

13.2.4 DCNs using optical switch and passive stars over fat trees

So far we have discussed the electrical DCNs followed by the recent developments in hybrid and all-optical DCN architectures, such as c-Through, Helios, OSA, Mordia, and LIONS. In this section, we describe a candidate hybrid architecture that promises high scalability in respect of the number of servers, along with high-speed interconnections between the servers. In the following, we present the basic schematic of this architecture, named *optical switch and passive-stars over fat trees* (OSPS-FT), using one MEMS switch and a number of PSCs along with small traditional electrical DCNs using fat tree (FT) topology or its variants. Each PSC-based network is configured as a multihop WDM LAN using an appropriate virtual topology, such as ShuffleNet, GHCNet, dBNet, etc.

As shown in Fig. 13.17, the OSPS-FT architecture uses one $N \times N$ MEMS switch on the top. Below the MEMS switch there are three levels of PSCs, and the MEMS switch is connected only to the PSCs at the lowest level (level 1). One PSC is placed at the highest level (level 3) denoted as L3-PSC, which is connected only to N_2 level-2 PSCs (L2-PSC), and at the bottom of the PSC-hierarchy there are N_1 level-1 PSCs (L1-PSCs) connected to specific L2-PSCs as well as to the MEMS switch, thereby needing $N_1 \times N_2$ L1-PSCs in total. Each L1-PSC is connected downward to L k_{FT}-ary FTs, with each FT consisting of S_{FT} servers. Thus, with each of the N_2 L2-PSCs (i.e., L2-PSC-1 through L2-PSC-$N2$) being connected downward to N_1 L1-PSCs, and the L3-PSC being connected downward to N_2 L2-PSCs, all of the $N_1 \times N_2 \times L$ FTs get interconnected through the three levels of PSCs, leading to the total number of servers in the OSPS-FT architecture as $S_{total} = N_1 N_2 L S_{FT}$. Since each L1-PSC connects to one pair of input/output ports of the $N \times N$ MEMS switch, we obtain $N = N_1 N_2$ and hence the total number of servers becomes $S_{total} = N_1 N_2 L S_{FT} = N L S_{FT}$.

With the above network-setting, OSPS-FT routes its traffic through different candidate paths, governed by the nature of traffic flows (*elephant* and *mouse* flows), and the connectivity between the PSCs. In particular, this architecture routes the elephant-category traffic from

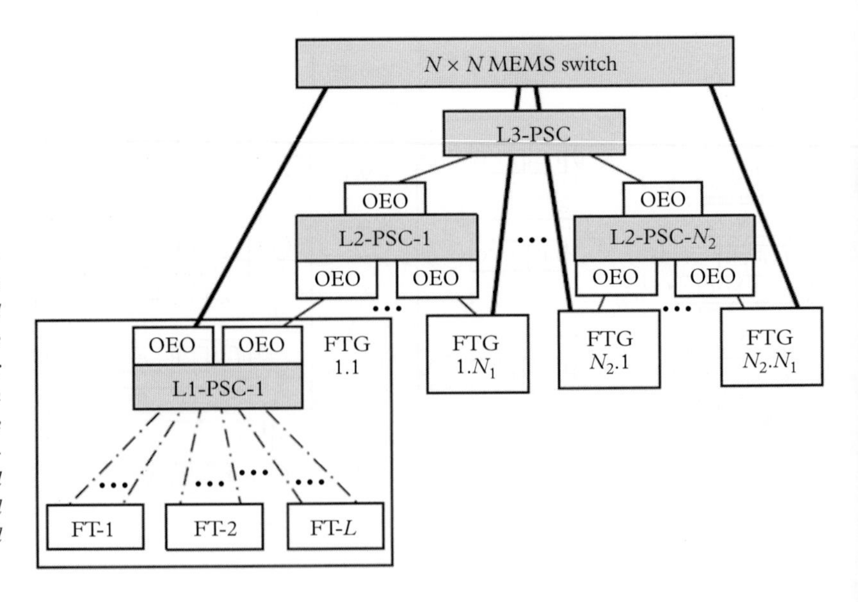

Figure 13.17 *Hybrid DCN using OSPS-FT architecture. Total number of servers $S_{total} = N_1 N_2 L S_{FT} = N L S_{FT}$. Thick solid lines represent the elephant traffic flows, while the thin solid lines are the connections for the mouse traffic flows. In the example design discussed in the text, each PSC at the three levels employs multihop WDM LAN following (p, k) ShuffleNet topology, with p and k representing the logical nodal degree and the number of hops in the respective virtual topology.*

the ToR switches of each FT through the respective L1-PSCs and MEMS switch, while bypassing the level-2 and level-3 PSCs. The traffic flows through this route, being of elephant category, are not affected much by the long reconfiguration time ($\simeq$ 10 ms) of the MEMS switch. On the other hand, all the PSCs at the three levels support together the mouse-category traffic flows between the FTs, and for them the routing takes place in the electrical domain (enabling fast-switching) through the multihop WDM LANs embedded in the PSCs.

As mentioned above, each of the PSCs at three levels operates as a multihop WDM LAN, and as shown in Fig. 13.17 each L1-PSC gets connected downward to a group of L FTs, denoted as FTG(i,j); thus the ith L2-PSC (i.e., L2-PSC-i) gets connected downward through N_1 L1-PSCs to the FT groups FTG$(i,1)$ through FTG(i,N_1). Note that the FTs under the same L1-PSC (parent L1-PSC) can connect with each other using their parent L1-PSC. However, the packets from an FT under a given L1-PSC can reach another FT belonging to a different L1-PSC through either one L2-PSC or through a combination of two L2-PSCs and the L3-PSC. Further, each PSC at levels 1 and 2, while connecting to a PSC at the next higher level or to the MEMS switch (from the PSCs at level 1 only), needs to use one of its ports using OEO conversion (with multiple FTs and FRs) for ShuffleNet routing, which also facilitates wavelength reuse in different multihop WDM LANs and helps in realizing feasible optical power budget across the DCN. Moreover, with the optically opaque (OEO) interconnections between the PSCs, the OEO ports of different PSCs with multiple FTs and FRs can run independently the self-routing algorithm of the respective ShuffleNets, thereby ensuring high-speed interconnections between the servers for mouse flows. The task of traffic classification needs to be carried out in the source servers or in the respective ToR switches in each FT. In order to visualize the design and functioning of OSPS-FT architecture, we consider an example case in the following.

- Looking upward from the server level, first we consider the basic electronic module of the DCN as 8-ary FT (i.e., $k_{FT} = 8$) with the numbers of servers and core switches, S_{FT} and N_{FT}^{CS} respectively, given by

  ◇ $S_{FT} = \frac{k_{FT}^3}{4} = 128$,

  ◇ $N_{FT}^{CS} = \frac{k_{FT}^2}{4} = 16$.

- Next, each L1-PSC is chosen as a (3,3) ShuffleNet (i.e., $p = 3$ and $k = 3$, see Chapter 6) supporting $N_{L1}^{SN} = kp^k = 3 \times 3^3 = 81$ ports (nodes). Hence, with $N_{FT}^{CS} = 16$ for each FT, each (3,3) ShuffleNet (embedded in each L1-PSC) can connect downward at the most with four 8-ary FTs engaging its $16 \times 4 = 64$ nodes (or corresponding PSC ports), as it must use two additional ports to connect with the parent L2-PSC and the MEMS switch. Thus, from the remaining $81 - 64 = 17$ ports, two ports remain engaged to connect to the parent L2-PSC and MEMS switch, thereby leaving $81 - (64 + 2) = 15$ ports unused. Note that the next-lower available size of ShuffleNet cannot be used in this case, as it doesn't offer adequate number of ports (i.e., < 66). With this choice of parameters, one can thus express the number of FTs that can be supported by each L1-PSC as

  ◇ $L = \left\lfloor \frac{N_{L1}^{SN}-2}{N_{FT}^{CS}} \right\rfloor = \left\lfloor \frac{81-2}{16} \right\rfloor = 4$, which confirms what we assumed in the beginning. Note that, the ShuffleNet nodes used at level 1 to connect with the ShuffleNets at the level 2 will be using transceivers with appropriate wavelengths to comply with the ShuffleNets to which they will be interfacing. Furthermore, the ShuffleNet nodes used at level 2 to connect with the ShuffleNets at the levels 1 and 3 will have two sets of such transceivers with appropriate wavelengths.

In practice, 128-port PSCs would be the nearest available size (in the form of 2^x, built with x stages of 2×2 optical splitters as the building block) which can be used in this case, but with some spare ports remaining unused. In other words, the design of the ShuffleNet from the PSC would follow a port-utilization ratio of 81:128, implying that 81 ports would be the number of ShuffleNet nodes out of 128 ports, from which $64 + 2 = 66$ ports would be utilized by the DCN. However, one can indeed improve the utilization of the PSC ports by varying the size of the FTs and ShuffleNets. One can also explore other multihop regular/non-regular topologies for the WDM LANs.

- Each L2-PSC is chosen as a (4,2) ShuffleNet (i.e., $p = 4$ and $k = 2$) with 32 ports to connect downward to 31 L1-PSCs and 1 port for connecting with L3-PSC at the higher level. Thus, the number of L1-PSCs that can be accommodated by each L2-PSC is given by

  ◇ $N_1 = 32 - 1 = 31.$

  In this case, one can use PSCs exactly with $2^5 = 32$ ports, without leaving any unused ports.

- L3-PSC is chosen as a (2,3) ShuffleNet (i.e., $p = 2$ and $k = 3$) having 24 ports, thereby connecting to 24 L2-PSCs. Hence, we obtain the number of L2-PSCs as

  ◇ $N_2 = 24.$

  However, the nearest available PSC port count is 32, which is used at this level, leaving thereby eight unused ports.

- The above design leads to the total number of servers in the OSPS-FT architecture as $S_{total} = N_1 N_2 L S_{FT} = 31 \times 24 \times 4 \times 128 = 380,928$. The choices of N_1 and N_2 lead to the total number of L1-PSCs, $N_{L1(total)}$, given by

  ◇ $N_{L1(total)} = N_1 N_2 = 31 \times 24 = 744 = N,$

  implying that the design needs an $N \times N$ MEMS switch with $N = 744$, which is within the realizable values of the port counts ($\simeq 1000$) for the MEMS switches (Kim *et al.* 2003). Further, the largest PSCs used in the design have 128 ports, implying that the signal splitting loss will be $10 \log_{10} 128 + \log_2 128 \times 1 = 28$ dB (assuming an insertion loss of 1 dB in each 2×2 optical splitter used as a building block in the PSC). Hence, by using lasers with 15 dBm transmit power and a receiver sensitivity of -18 dBm at 10 Gbps, it would be possible to realize a power budget satisfying the BER constraint in each fiber link (not exceeding a few tens of meters, along with a pair of connectors on two sides) of all the ShuffleNets. This aspect can be readily checked by using the impairment models described in Chapter 11 (see Exercise 13.8).

- Finally, we examine the latency of the OSPS-FT architecture for various possible traffic flows. The latency of the elephant flows will naturally be dominated by the reconfiguration time of the MEMS switch ($\simeq$ 10 ms), as the delay incurred in the underlying L1-PSCs and the FTs at the two ends would be much smaller. However, for the mouse flows, the end-to-end delay would be upper-bounded by the delay incurred by the packets which will have to flow through the PSCs at all the three levels (i.e., worst-case delay). The delay of these packets will have two components: the multihop packet transmission delay incurred in the three ShuffleNet-based WDM LANs and the delay incurred in the two FTs at two ends. We estimate the worst-case delay D_{WLAN}^{max} of the mouse flows in WDM LANs as

$$D_{WLAN}^{max} = 2D_{L1}^{max} + 2D_{L2}^{max} + D_{L3}^{max}, \tag{13.13}$$

where D_{L1}^{max}, D_{L2}^{max}, and D_{L3}^{max} represent the worst-case delay components incurred at the ShuffleNets at the levels $L1$, $L2$, and $L3$, respectively. Note that, with small propagation delay along the fiber links in the in-house DCNs, each delay component will be the product of the maximum number of hops (= $2k - 1$) in the respective (p, k) ShuffleNets and the packet transmission time τ_P (ignoring queuing delay in the ShuffleNet nodes). Using this observation and the above expression, we express D_{WLAN}^{max} as

$$\begin{aligned} D_{WLAN}^{max} &= [2(2 \times 3 - 1) + 2(2 \times 2 - 1) + \\ &\quad (2 \times 3 - 1)]\tau_P \\ &= 21\tau_P. \end{aligned} \tag{13.14}$$

With a typical packet size of 1200 bytes (=9600 bits) transmitted at 10 Gbps, we get $\tau_P = 9600 \times 10^{-10} = 0.96$ μs, leading to $D_{WLAN}^{max} = 21 \times 0.96$ μs $= 20.16$ μs, which is indeed much less than the MEMS delay and would comply well with the typical requirements in a DCN for mouse flows. However, note that the overall delay in both cases would be increased by the traffic analyzer module for classifying the ingress traffic in the elephant and mouse categories.

Furthermore, the PSC-based multihop WDM LANs and the FTs can have various other choices. For example, PSCs can employ GHCnet, dBnet, or even non-regular virtual topology (see Chapter 6), while the FTs can use VL2, Aspen tree, or some other topology. Optimum choices for S, L, N_1, and N_2 can be made through an exhaustive search using an appropriate heuristic scheme or an LP-based design, based on appropriate objective function and relevant constraints. Further, note that, the use of multiple levels of PSCs has been a necessity to avoid large-sized (i.e., with large port count) PSCs, as larger PSCs would significantly impact the power budget in the optical links used therein. One can also increase the number of levels of PSCs beyond three to increase the size of the DCN while keeping the latency and power budget within the acceptable limits.

As we see from the above case study, the OSPS-FT architecture can be employed for a reasonably large DCN, where the multi-level (three levels of PSCs and one MEMS in the present design) optical plane offers large bandwidth, while the electrically switched FTs being smaller in size, can avoid excessive wiring and hardware in electrical domain.

13.3 Networking with datacenters in long-haul WDM networks

While setting up a long-haul WDM network along with datacenters placed at different locations across the network, the overall network design deserves some special consideration. For example, in order to ensure *low-latency* access of data to the end-users in client networks, the objects in a datacenter often need to be replicated in several other datacenters, located at different places across the network. Multiple datacenters with the same objects, located at different places, will also provide redundancy against network failures. Furthermore, large datacenters need to be located at the places where land is available at affordable prices along with adequate power supply, the latter constraint needing proximity of the large datacenters to the power stations.

Special attention is also needed to protect the networks hosting datacenters from large-scale disasters, which may be natural or man-made, e.g., earthquake, tsunami, hurricane, tornado, weapons of mass destruction, electromagnetic pulse attacks, etc. Network failures caused by such incidents are much larger in scale as compared to the simple link or node failures from the fiber cuts or malfunctioning of network components, as an entire area may get devastated resulting in the failure of one or more nodes hosting the datacenters and all the adjoining fiber links around those nodes. One way to address such large-scale problems is to first identify appropriate groups of nodes and links at various geographical locations, which are likely to be affected in the event of such disasters, popularly known as shared-risk groups (SRGs). Figure 13.18 illustrates an eight-node WRON with six SRGs and three datacenters (DC1, DC2, and DC3), where the nodes 1, 2, 6, 7, and the links connected to them form the respective SRGs (i.e., four SRGs), while some nodes being close to each other together form one SRG collectively, such as the node pairs (3,8) and (4,5) along with the attached links from the two SRGs. As shown in Fig. 13.18, all network nodes are connected to the respective local client networks and hence we simply call them client nodes (CN*i*'s) in the following discussion.

Having identified the SRGs, one needs to have an estimate of the number of datacenters that the network would be able to host, and find out the objects that are highly popular and should be replicated in several other datacenters and the objects that may need fewer

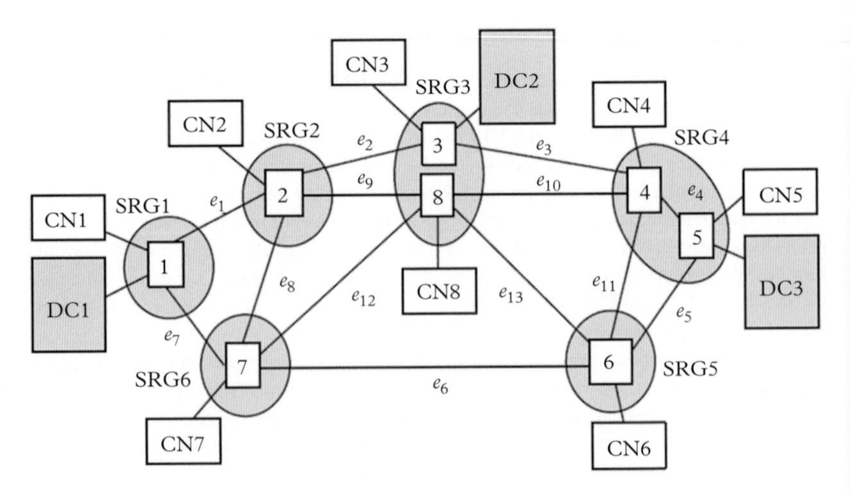

Figure 13.18 *Formation of SRGs in an eight-node WRON. CN: client network, DCi: ith datacenter. All nodes are connected to the respective CNs, while some of them are hosting datacenters, i.e., DC1, DC2, and DC3.*

replications. Further, the supporting long-haul WDM network (assumed to be a WRON) needs to be provisioned with protection (backup) lightpaths preferably for most of the working (primary) all-optical connections, so that the protection lightpaths don't pass through the same SRGs on their routes as the ones on the working lightpaths, i.e., the working and protection lightpaths, should be on the *SRG-disjoint* routes. Moreover, attempts should be made that the working and protection lightpaths traverse through minimum number of SRGs.

Due to extremely low loss and enormous bandwidth, today the optical WDM networks (with ring, inter-connected rings, or mesh topology at different scales, e.g., metro and long-haul networks) appear to be the most suitable networking platform for interconnecting datacenters and client nodes. However, such networks might differ at times from the conventional long-haul carrier networks in respect of topological characteristics. For example, some datacenters might be remotely located (in search of affordable land prices and adequate power supply), which will, in turn, have to be connected to the rest of the network with the other datacenters and client nodes over fairly long distances, exceeding hundreds of kilometers, at times without any intermediate node to regenerate the WDM signals using OEO conversion. This feature indeed makes the long-haul connections between the remote datacenters and the other nodes much costlier than what is experienced in typical carrier networks, calling for parallel and proprietary network infrastructure. However, both practices can coexist, and trade-offs are necessary in this respect on whether to use only the existing carrier networks or to go for proprietary long-haul WRONs in a given situation, which will however remain connected to the carrier WRONs for giving access to the end-users. In the following, we present a simplistic design methodology for setting up a networking infrastructure, hosting datacenters using long-haul WRONs.

Several studies have been made on the design of WRON-based carrier networks, where datacenters are placed at appropriate locations (nodes) ensuring latency-aware access to the client nodes and other datacenters as well, both using LP-based formulations and heuristic schemes. However, to get a quick insight, we consider a heuristic scheme, which breaks the overall problem into four steps. A rigorous approach considering the entire design as an LP-based optimization problem can also be explored, though feasible in general for small networks (as seen earlier in Chapters 8 and 9), and the interested readers are referred to (Dong *et al.* 2011) for an exhaustive treatment on this topic.

In the first step of the four-step design procedure, we determine the possible locations of datacenters with a given number (K, say) of datcenters using an *SRG-aware* heuristic scheme. Next, we assess the popularity of different contents or objects O_i's, based on which the objects are to be replicated in various datacenters. Thereafter, the appropriate number and locations of datacenters are chosen (from the K datacenters already found out in the first step) for the replication of each object, based on the object popularity. Finally, the design wraps up by setting up the SRG-aware working and protection lightpaths between all the client nodes and datacenters with a given network traffic matrix and the physical topology of the WRON under consideration.

13.3.1 Estimation of datacenter locations

In order to determine the possible locations of the datacenters for a given value of K, we first consider that each node in an N-node network is a candidate node for datacenter placement, and determine the sum of the SRG-aware distances from this node to all other nodes, where the SRG-aware distance covered by a chosen route (shortest path in terms of the link lengths) is expressed as the number of SRGs (n_{sd}^{srg}) crossed by the route under

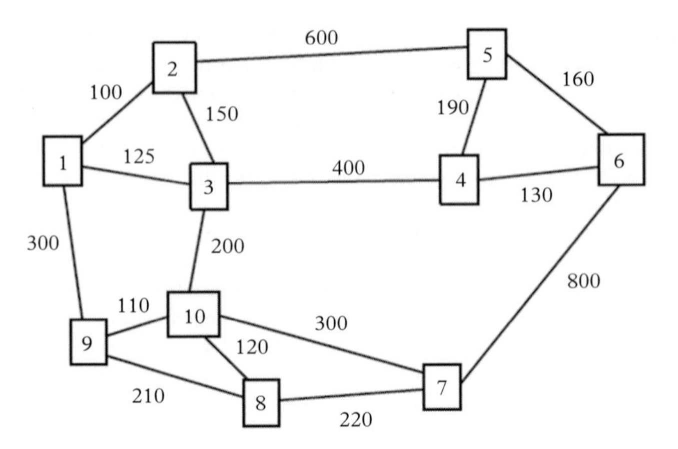

Figure 13.19 *Ten-node WRON as an example case to study problems related to the DC placement, object replication therein and disaster-resilient lightpath provisioning. Note that, in the ensuing case study, we will assume that each node represents an SRG.*

consideration. Thus, for a given node d as a candidate datacenter, we express the sum SUM_d of the SRG-aware routing distances for node d as

$$SUM_d = \sum_{s \neq d} n_{sd}^{srg}, \tag{13.15}$$

for all nodes in the network. Having obtained SUM_d for all nodes in the network, we arrange them in the ascending order of SUM_d. Thereafter, if $K = 1$, we choose the node on the top as the node to host a datacenter. If $K = 2$, we consider the first two nodes from the top as the nodes to host two datacenters, and so on. Having gone through this exercise, we maintain this ordered list of datacenters, named as OL_{DC}, which we use later while assigning multiple datacenters to some objects on the basis of their popularity levels.

To illustrate the above scheme, we consider a 10-node WRON as shown in Fig. 13.19 with $K = 3$. For this network, by applying the above method we obtain the datacenter locations for different values of K as

1. $K = 1$, datacenter hosted by: node 3,
2. $K = 2$, datacenters hosted by: nodes 3, 10,
3. $K = 3$, datacenters hosted by: nodes 3, 10, 2.

Note that, the datacenter locations also need to satisfy other diverse constraints, such as affordable land prices and proximity to the power stations particularly for large datacenters. However, for simplicity, we considered only the issue of SRG-awareness in the present computation, while the other factors, if necessary, can be incorporated in the cost model for more realistic estimates.

13.3.2 Popularity estimates for objects

As mentioned earlier, object or content replication in multiple datacenters plays an important role, and depends primarily on the popularity of the contents. An object request from a client node should go to the nearest datacenter hosting the requested object, which not only reduces the access latency, but also leaves more space/bandwidth for other nodes in the network and reduces power consumption with shorter transmission distances.

Following the line of treatment in (Dong *et al.* 2011), the objects are ranked in accordance with their *access rates* or *popularity*, i.e., how frequently a given object is asked for by the client nodes. Viewing this process statistically, one can form an *object popularity distribution*, which typically follows a pattern, known as Zipf's distribution. In other words, one can rank a given object with an index $i \in [1, Q]$, called as popularity index, which is smallest (= 1) for the most popular object. Thus, the probability ϕ_i to access an object O_i with a popularity index i will be *inversely related* to i. In other words, $i = 1 = i_{min}$ represents the most popular object with the highest access probability and $i = Q = i_{max}$ represents the least popular object with the lowest access probability. Following the Zipf's distribution, this statistical feature is captured by the access probability ϕ_i for an object O_i, and expressed as

$$\phi_i = \frac{\beta}{i} \tag{13.16}$$

$$\beta = \left[\sum_{i=1}^{Q} \frac{1}{i}\right]^{-1},$$

such that the sum of the probabilities = $\sum_i^Q \phi_i = 1$. To illustrate the method, we assume $Q = 4$ (i.e., $i = 1$ through 4), and obtain the probability estimates as a percentage, called the percentage popularity $P_i = \phi_i \times 100\%$, as

$i = 1$: $P_1 = 48\% = P_{max}$ ($\phi_1 = \phi_{max} = 0.48$)
$i = 2$: $P_2 = 24\%$ ($\phi_2 = 0.24$),
$i = 3$: $P_3 = 16\%$ ($\phi_3 = 0.16$),
$i = 4$: $P_4 = 12\% = P_{min}$ ($\phi_4 = \phi_{min} = 0.12$).

13.3.3 Replication of objects in datacenters

With the given values of Q and ϕ_i's ($\phi_{max}, \cdots, \phi_i, \cdots, \phi_{min}$, for $i \in [1, Q]$), the number of datacenters where the replication needs to be carried out for the ith object can be estimated by using a simple linear model, as shown in Fig. 13.20 (Dong *et al.* 2011). The number of replicas needed for the ith object, i.e., R_i, can be expressed as a linear function of ϕ_i in the interval $\phi_i \in [\phi_{min}, \phi_{max}]$, given by

$$R_i = m\phi_i + c, \phi_i \in [\phi_{min}, \phi_{max}], \tag{13.17}$$

with m as the slope of the linear plot of R_i versus ϕ_i and c as the intercept on the R_i axis by extrapolating the linear plot to $\phi_i = 0$.

Note that the linear plot in Fig. 13.20 is applicable only in the valid range of $\phi_i \in [\phi_{min}, \phi_{max}]$). We assume that the least popular object, i.e. the object Q with $\phi_i = \phi_Q = \phi_{min}$, will have $R_Q = 1$ (in general, it can be > 1 though, but with the minimum allocation among all other objects) and the most popular object, i.e., object 1 with $\phi_i = \phi_1 = \phi_{max}$ will be replicated at all the K datacenters with $R_1 = K$. Making use of these criteria in Eq. 13.17, we obtain

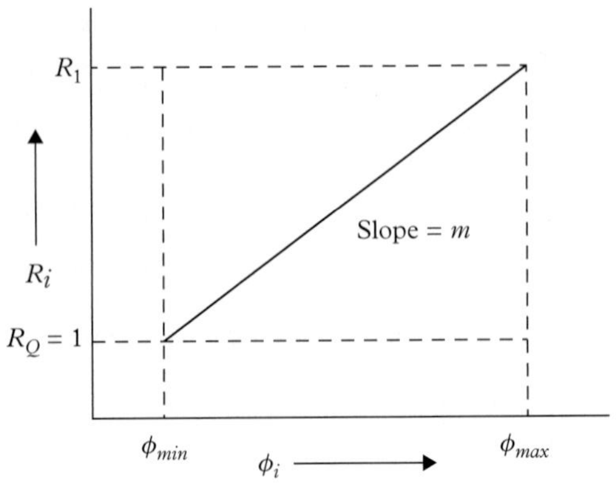

Figure 13.20 *Illustration of the linear model for the number R_i of datacenters required for a given popularity ϕ_i of an object.*

$$m = \frac{R_1 - R_Q}{\phi_{max} - \phi_{min}} = \frac{K-1}{\phi_{max} - \phi_{min}} \tag{13.18}$$

$$c = R_i - m\phi_i = 1 - m\phi_{min} = \frac{\phi_{max} - K\phi_{min}}{\phi_{max} - \phi_{min}}. \tag{13.19}$$

Substituting the expressions of m and c in the earlier expression of R_i, we finally obtain R_i as

$$R_i = \left\lceil \frac{K-1}{\phi_{max} - \phi_{min}}\phi_i + \frac{\phi_{max} - K\phi_{min}}{\phi_{max} - \phi_{min}} \right\rceil. \tag{13.20}$$

Using the above expression and $Q = 4$, we get the values of R_i's for the present example (i.e., for the WRON in Fig. 13.19 with $K = 3$) as $R_1 = 3$, $R_2 = 2$, $R_3 = 2$, and $R_4 = 1$, where R_1 and R_4 have been already preset as 3 and 1, respectively. Having known the values of R_i's as above, we next find out the specific datacenter location(s) out of K datacenters for each object. In particular, by recalling the ordered list OL_{DC} of K datacenters, we carry out the task of object placements in the following steps.

1. Assign all objects to the datacenter on the top of the list OL_{DC}, as each object should be available at least in one datacenter, and that datacenter should have the minimum cost.
2. Next consider the popularity list for the objects with the values of R_i for all popularity indices $i \in [1, Q]$, which is also arranged in ascending order of i (which means descending order of popularity ϕ_i). Replicate all objects in the next datacenter from OL_{DC}, except the least popular object (or objects, if more than one object has a popularity index $i = Q$, i.e., if $R_Q > 1$) at the bottom of the popularity list.
3. Repeat Step 2 with the next datacenter in OL_{DC}, while excluding the second least-popular object(s) from the bottom of the popularity list.
4. Keep repeating Step 3, until all objects get replicated to the required number of datacenters as determined earlier by the values of R_i's.

With $K = 3$ and $Q = 4$, one can obtain the object replications as

1. object 1 (most popular) with $R_1 = 3$ is placed at all of the 3 datacenters, located at the locations $L1$ (node 3), L_2 (node 10) and L_3 (node 2),
2. object 2 with $R_2 = 2$ is placed at the datacenters located at $L1, L_2$,
3. object 3 with $R_3 = 2$ is placed at the datacenters located at $L1, L_2$,
4. object 4 with $R_4 = 1$ is placed at the datacenters located at $L1$.

13.3.4 Designing WRONs hosting datacenters

Disaster-resilience in a network hosting datacenters needs to be realized by using protection lightpaths that must be SRG-disjoint with the working lightpaths, and this needs to be ensured all the more for the lightpaths between the datacenters. However, we simplify the approach by making all protection paths SRG-disjoint with the respective working paths. The present approach addresses this aspect to arrive at a simple heuristic solution to the problem.

In order to set up the lightpaths as above, in the first step, one needs to identify the various possible SRGs and form an SRG information matrix (SIM). From a given physical topology, represented by a graph $G(N, E)$ with N representing the nodes and e_i's $\in E$ representing the edges (i.e., fiber links), one can form a SIM as follows. First, the SRG domains identified in the given WRON (as shown earlier in Fig. 13.18) are numbered along with all the links in the graph (which are already numbered in E). Next, we define the association of a route r_i with an SRG, say SRG_j, by a binary SIM element ψ_{ij}, defined as

$$\begin{aligned} \psi_{ij} &= 1, \quad \text{if } r_i \text{ goes through } SRG_j, \\ &= 0, \quad \text{otherwise.} \end{aligned} \tag{13.21}$$

For the example shown in Fig. 13.18, the links $e_2, e_3, e_9, e_{10}, e_{12}, e_{13}$, and nodes 3 and 8 form SRG_3. Thus, for a given route from node 1 to node 4 via node 3 (say, route r_i) passing through SRG_2 and SRG_3, one obtains $\psi_{i2} = 1$ and $\psi_{i3} = 1$, leading to the cumulative value of ψ_{ij}'s as $\psi_i^C = \psi_{i2} + \psi_{i3} = 2$. However, to estimate the overall cost ($COST_i$, say) of the ith route, we take into account the propagation delay as well, and express $COST_i$ as

$$COST_i = (\psi_i^C + 1)d_i, \tag{13.22}$$

where d_i represents the total distance covered by route i along its links, and SRG-aware cost ψ_i^C is incremented to $\psi_i^C + 1$, to ensure that even when the route does not pass through any SRG (i.e., $\psi_i^C = 0$) it has a cost governed by the propagation delay of the route.

Next, we consider the overall traffic matrix (in Gbps) $\Gamma = [\lambda_{sd}]$ in the network, with λ_{sd} as the traffic from node s to node d, wherein all kinds of traffic flows (i.e., client to datacenter, datacenter to client, datacenter to datacenter, client to client) are included. With this network setting, we describe the SRG-aware routing scheme in the following.

1. Arrange the traffic elements in $\Gamma = [\lambda_{sd}]$ in descending order. The highest data rate is to be assigned a lightpath first.
2. For the first traffic element in the ordered list, find the k shortest routes.

3. For each of the chosen k routes (say, route i), compute the cost, i.e., $COST_i$.
4. Sort all the k routes with ascending order of the costs (i.e., the topmost element with the least cost will be the most-preferred path).
5. Choose the topmost path in the list as the primary path, and assign the wavelength using the FFW algorithm. If any wavelength is not available, move to the next choice and follow the FFW algorithm again for wavelength assignment. For simplicity, we assume that available wavelengths are adequate in number.
6. Choose the next shortest path for the backup lightpath and follow the same procedure as in Step 5.
7. Repeat the above steps for all the traffic elements.

Note that, in practice, the transceiver constraints might limit the number of lightpaths being set up between the node pairs and some node pairs might need more lightpaths (or even sub-lightpaths). In such cases, one can modify the algorithm in line with the heuristic approaches discussed in Chapter 9 on long-haul WRONs. Further, the proposed design approach can also be extended to realize power-aware routing by including the power consumption in networking devices in the cost metric. Overall, we kept the design procedure simple here to get an insight of the problem and encourage the readers to extend the given scheme to accommodate various practical aspects and assess the overall design.

A representative result of the above design procedure carried out for the network shown earlier in Fig. 13.18 is illustrated in Fig. 13.21, with the traffic matrix given in Table 13.1 and the link lengths in km beside the respective optical fiber links in the figure. The traffic matrix has been synthesized from the four distinct components: client network (CN) at a node to a datacenter at another node (CN-DC) and the reverse (DC-CN), datacenter to datacenter located at different nodes (DC-DC), and CN to CN located at different nodes (CN-CN), with the following average bandwidth distribution:

- DC-DC: 4 Gbps
- DC-CN: 3 Gbps
- CN-DC: 2 Gbps
- CN-CN: 1 Gbps,

leading to the traffic matrix (with the DC locations as determined earlier), as given in Table 13.1.

Thus, each element in the traffic matrix represents the composite bandwidth/speed requirement between a node pair considering the contribution from all the four possible components: DC-DC, DC-CN, CN-DC, and CN-CN, where a node pair without any datacenter associated with it shows up with the minimum bandwidth demand. However, a node pair with each of them supporting a datacenter, comes up with the largest bandwidth demand. The figure shows five illustrative sets of working-cum-protection paths for some select connections (to avoid clumsiness) as follows (with the assumption that each node represents an SRG):

◇ Node 2 → node 3 – primary path: 2-3, backup path: 2-1-3, transmission rate: 10 Gbps (OC-192),

◇ Node 10 → node 2 – primary path: 10-3-2, backup path: 10-9-1-2, transmission rate: 10 Gbps (OC-192),

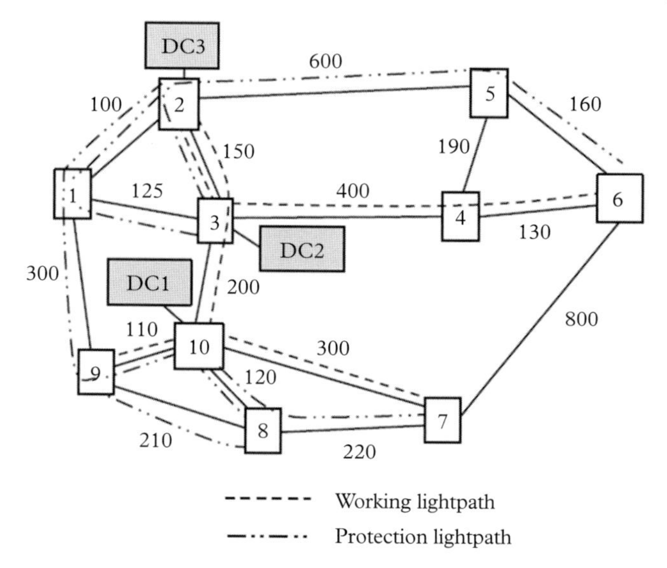

Figure 13.21 *Formation of lightpaths in SRG-aware WRON hosting datacenters using the network shown earlier in Fig.13.18. The working and protection lightpaths for a few connections are shown using solid and dashed lines, respectively.*

Table 13.1 *Composite traffic matrix in Gbps for the 10-node WRON of Fig.10.19.*

Nodes	1	2	3	4	5	6	7	8	9	10
1	0	3	3	1	1	1	1	1	1	3
2	4	0	10	4	4	4	4	4	4	10
3	4	10	0	4	4	4	4	4	4	10
4	1	3	3	0	1	1	1	1	1	4
5	1	3	3	1	0	1	1	1	1	3
6	1	3	3	1	1	0	3	1	1	3
7	1	3	3	1	1	1	0	1	1	3
8	1	1	1	1	1	1	1	0	1	3
9	1	3	3	1	1	1	1	1	0	3
10	4	10	10	4	4	4	4	4	4	0

- Node 10 → node 9 – primary path: 10-9, backup path: 10-3-1-9, transmission rate: 4 Gbps (rounded up to OC-192 at 10 Gbps),
- Node 10 → node 7 – primary path: 10-7, backup path: 10-8-7, transmission rate: 4 Gbps (rounded up to OC-192 at 10 Gbps),
- Node 3 → node 6 – primary path: 3-4-6, backup path: 3-2-5-6, transmission rate: 4 Gbps (rounded up to OC-192 at 10 Gbps).

On the other hand, the connections, like node 4 to node 6, or node 7 to node 9, will have only a traffic of 1 Gbps (as none of these nodes hosts any datacenter), rounding up to 2.5 Gbps (OC-48). Thus, with the above bandwidth requirements, by rounding up the requirements to the nearest OC-n speeds, the design will call for two different transmission rates at OC-48 and OC-192, implying multi-rate transmissions over the WRON. The present approach, though heuristic in nature, gives an insight into the SRG-aware lightpath setting-up process in WRONs hosting datacenters, which can be readily improved using the various schemes discussed earlier in Chapter 9 dealing with long-haul WRONs. For more exhaustive SRG-aware design approach, the interested readers are referred to (Habib *et al.* 2012).

13.4 Summary

In this chapter, we examined various types of datacenters, employing electrical, electrical-cum-optical (hybrid), and optical switching architectures for setting up DCN between the servers. Furthermore, we also considered the design methodologies for the long-haul WRON backbones, where the datacenters need to be placed at suitable locations with variety of objects.

For electrically switched architectures, we considered two broad categories: switch-centric and server-centric architectures. In particular, we described some noteworthy switch-centric architectures: traditional fat tree, VL2, and Aspen tree, all being based on the hierarchical tree topology. The architecture of the fat-tree-based DCN was discussed in detail as a benchmark, with its strictly non-blocking feature, albeit with its wiring complexity preventing scalability. VL2 architecture, as an improvisation of the fat-tree architecture, was described, and its special features, e.g., VLB and LA/AA-based addressing scheme, were explained. Further, a fault-tolerant switch-centric DCN architecture, known as Aspen tree, was described in detail, where the network can recover from the failures in a localized manner by using additional connectivity between the vertically adjacent switching layers, thereby ensuring faster recovery as compared to the DCNs using traditional fat-tree topology. Next, two server-centric electrically switched architectures, DCell and BCube, were presented, where the servers participate in interconnecting each other with the help of smaller electrical switches.

Next, two noteworthy hybrid switching architectures for intra-DCN were presented, c-Through and Helios, where parallel optical-switching planes are used to supplement the electrically switched part of the DCN for improving the overall performance, including bisection bandwidth, network latency, and power consumption. Subsequently, we described some of the datacenter testbeds using all-optical switching for intra-DCN, including OSA, Mordia, and LIONS, with OSA and LIONS employing centrally switched topology and Mordia using WSS-based distributed optical switching through WDM ring network. A novel hybrid intra-DCN architecture, referred to as OSPS-FT, was presented for high-capacity scalable datacenters employing switched passive-star couplers over electrically-switched fat trees. Finally, we presented some heuristic approaches to designing long-haul WRONs with SRG-aware RWA for hosting datacenters at suitable locations, along with a popularity-based replication strategy of various objects that need to be placed in the selected datacenters.

EXERCISES

(13.1) Consider a DCN using a 6-ary fat tree architecture and find out the following:

a) number of servers,

b) number of ToR switches,

c) number of aggregate switches,

d) number of core switches.

Taking the two extreme ToR switches at the two ends, trace the path between them and justify that the connection is non-blocking.

(13.2) Consider that a DCN employs Aspen tree as its physical topology with four levels of switching with each switch having six ports. Calculate the following:

a) number of ToR switches for the three FTVs: (0,0,0), (0,2,0), and (2,2,2),

b) number of servers for the three FTVs: (0,0,0), (0,2,0), and (2,2,2).

Sketch the Aspen tree topology with four levels and six-port switches with the FTV of (2,2,2) and illustrate how the number of servers is significantly reduced to achieve a high level of fault tolerance.

(13.3) A server-centric DCN has to be designed following DCell$_n$ topology with the basic DCell network (i.e., DCell$_0$) having one mini-switch connected to five servers. Design and draw the DCell$_0$ and DCell$_1$ topologies. If the overall DCell$_n$ topology needs to accommodate at least 800,000 servers, determine the value for n.

(13.4) A DCN using fat-tree topology employs 48-port Ethernet switches as the building block. Assume that the links in the DCN carry an average traffic of 0.75 Erlang, and it is desirable that the probability of packet loss in the switches, each of which uses a shared buffer for all ports, remains below 10^{-6}. Assuming that the packet arrivals and service durations follow Poisson and exponential distributions, respectively, estimate the needed buffer size in the switches. Also indicate how the estimated buffer size would perform, when the arrival processes at the switch ports become non-Poissonian with heavy-tail distributions. Given: average packet size = 1500 byte.

(13.5) Explain why the two sets of addresses, location and application addresses (LA and AA), are used in a VL2 network. Justify how the valiant load-balancing (VLB) in VL2 addresses the challenges of the traffic agility and network failures.

(13.6) Why does Helios employing HyPac switching classify the traffic flows into *mouse* and *elephant* categories and distribute these traffic streams between the optical and electrical switching units? Discuss how one should decide the threshold between the mouse and elephant traffic flows.

(13.7) Consider the OSA architecture for an all-optical DCN, and determine the end-to-end path loss for an all-optical connection between any two ToR switches. Using this estimate, determine the transmit power needed from each transmitter if the receiver sensitivity for each wavelength is −18 dBm at 10 Gbps. Given: loss in optical multiplexer/demultiplexer = 3 dB, loss in WSS = 3 dB, loss in circulator = 2 dB, loss in combiner = 2 dB, loss in MEMS switch = 4 dB, connector loss = 1 dB.

(13.8) Consider an OSPS-FT architecture of DCN, where the lowest level of the network uses 4-ary fat trees in electrical domain and a MEMS-based optical switch at the top. Below the MEMS switch, for the three levels of PSC-based multihop WDM LANs, the DCN uses three types of ShuffleNet as follows:

- ◇ (2,3) ShuffleNets at level 1 (bottom-most level of PSCs)
- ◇ (4,2) ShuffleNet at level 2,
- ◇ one (2,3) ShuffleNet at level 3 (top-most level of PSCs).

Using the above design guideline, construct and sketch the entire OSPS-FT architecture and evaluate the number of servers the DCN can support, the number of PSCs at levels 1 and 2 and the size of the MEMS switch. Also specify the sizes of the PSCs needed at each level and check whether the power budget of the network would be feasible, if each optical link has a length $\leq$ 15 m and all transmitters transmit 10 dBm of optical power at 10 Gbps with a receiver sensitivity of -18 dBm. Given: connector loss = 1 dB, fiber loss = 0.2 dB/km, insertion loss in each 2×2 optical splitter in a PSC = 0.5 dB.

(13.9) In a long-haul WRON mesh, six contents/objects need to be replicated at multiple DCN locations, governed by their respective popularity indices. Following Zipf's distribution, estimate the probabilities of accessing the objects using their respective popularity indices, i.e., 1 through 6. Assume that there are six DCNs placed at as many locations across the WRON with an ordered list, where the DCN locations are arranged in ascending order of their mean distances from the other nodes. Using the popularity indices of the objects and the location-based ordered list of the DCNs, determine the locations of the DCNs, where each object has to be placed/replicated.

Elastic Optical Networks

14

In WDM networks using fixed frequency grid, transmission rates might vary for different connections, leading to inefficient bandwidth utilization in optical fibers with lower-rate connections set up over wide frequency slots. In elastic optical networks (EONs), the frequency grid is made flexible, thereby improving the effective network capacity. A flexible frequency grid consists of smaller frequency slots, and a transmitting node can use multiple frequency slots using suitable modulation techniques, e.g., optical OFDM, Nyquist-WDM, and optical arbitrary waveform generation. However, this requires bandwidth-variable transceivers and other devices to set up variable-rate connections across the network. In this chapter, first we discuss the design challenges in EONs and the evolving technologies for the network elements. Thereafter, we present some offline (LP-based and heuristic) design methodologies for EONs leading to routing and spectral allocation (RSA) for the required connections. Finally, we present some online fragmentation-aware RSA schemes for the operational EONs.

14.1 Challenges in fixed-grid WDM networks

The WDM networks discussed so far have been considered to employ a fixed grid of optical frequencies, which has evolved over the last few decades from the channel spacing in the order of a few nanometers (each nanometer being equivalent to $\simeq$ 125 GHz) down to the currently used ITU-T standard of 50 GHz ($\simeq$ 0.8 nm). With the emerging needs in today's telecommunication networks, the peak bandwidth demands for connections started growing beyond 100 Gbps, while across the network span from time to time the demands for transmission rates remained variable from subchannel rates (e.g., $\leq$ 10 Gbps) to as large as a few Tbps, referred to as superchannel rates. Having visualized these ensuing challenges, some interim solutions were explored earlier with mixed transmission rates of 10, 40, and 100 Gbps over the fixed 50 GHz grid. However, optical transmission at the rates of 10 and 40 Gbps led to under-utilization of channel bandwidth. On the other hand, 100 Gbps transmission over 50 GHz was realized with a spectral efficiency of 100 Gbps/50 GHz = 2 bps/Hz, by using coherent digital modulation techniques, such as quadrature phase-shift keying (QPSK). Thereafter, it turned out to be difficult to accommodate the connection requests with the transmission rates exceeding 100 Gbps within the 50 GHz frequency grid for long distance connections. In particular, transmission at the rates > 100 Gbps within 50 GHz bandwidth needed high-order digital modulation schemes, e.g, QAM with multiple levels of carrier amplitude as well as phase, thus ensuring higher spectral efficiency than QPSK. This class of multi-level modulation schemes with densely populated signal space (constellation), notwithstanding their high spectral efficiency, made the transmitted signal very prone to transmission impairments (e.g., noise, crosstalk, signal losses, and dispersion) due to the reduced *distance* between the *constellation points* in the *signal space*. Thus, use of these modulation schemes at high transmission rates were found to be unsuitable for optical transmissions across long-haul WDM networks.

Optical Networks. Debasish Datta, Oxford University Press (2021). © Debasish Datta.
DOI: 10.1093/oso/9780198834229.003.0014

Various possible ways to address the above challenges have been explored by using the notion of a *flexible* or *elastic* spectral grid over a new class of optical WDM networks, known as elastic optical networks (EONs) (Jinno 2017; Gerstel *et al.* 2012; Jinno *et al.* 2009). By using flexible grid in EONs, attempts are being made to minimize the under-utilization of bandwidth for low-rate transmissions, while ensuring that the transmission at high rates exceeding 100 Gbps also becomes feasible using a mechanism called multiple-carrier (multicarrier) communication (Goldsmith 2005), wherein each carrier (also called a subcarrier) carries a smaller bandwidth of information while the aggregate transmission bandwidth can be chosen to be small or large, governed by the number of carriers chosen for concurrent transmission. Since each carrier carries information with lower bandwidth, it encounters less noise (per carrier) and hence can offer the desired optical reach. In the following section, we discuss the basic concept of elastic or flexible optical spectrum by means of multicarrier communication.

14.2 Elastic optical spectrum

As mentioned above, in EONs all the transceivers operate with flexible transmission rates, thereby enabling the use of variable optical transmission bandwidth for the connections across the network. This in turn necessitates the use of bandwidth-variable optical network elements, such as modulators, optical filters, switches, and multiplexers/demultiplexers, across the entire network. Optical spectrum utilization with such an arrangement increases significantly as compared to fixed-grid WDM networks.

The notion of elastic or flexible grid in EON is realized by dividing the entire optical transmission spectrum into narrower *frequency slots* (FSs) as compared to the 50 GHz standard, and allocating a variable number of slots for a given connection, as per its transmission-rate requirement. Figure 14.1 illustrates how EON operates using flexible

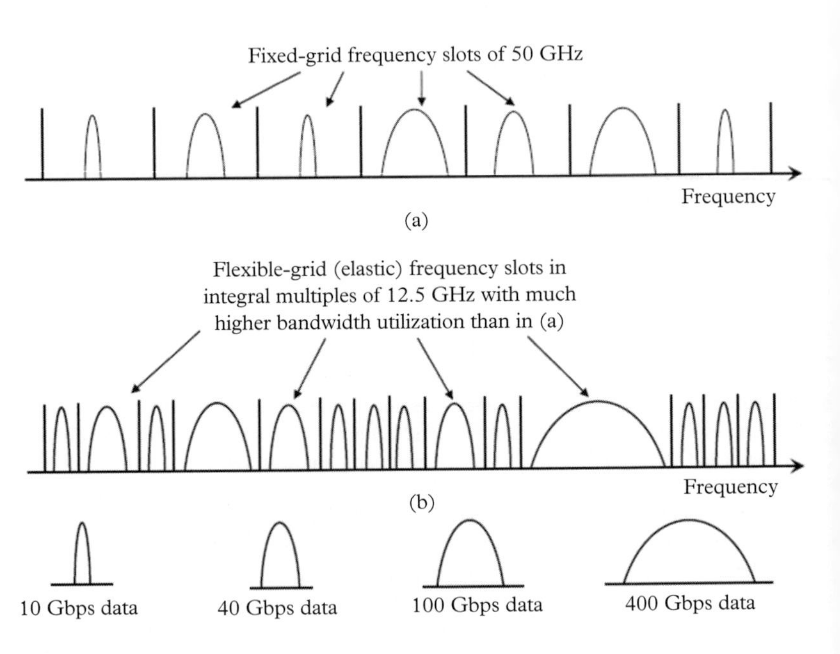

Figure 14.1 *Illustration of the fixed-grid and elastic (flexi-grid) optical spectra.*

transmission bandwidth (part (b)) as compared to the fixed-grid WDM networks (part (a)). As evident from Fig. 14.1(b), EON with flexible grid composed of narrower FSs can effectively *pack* more traffic demands into a given transmission window. As mentioned earlier, any attempt to transmit at higher rate using a single carrier within a given bandwidth becomes error-prone over long distance. In EON, concurrent transmission of multiple carriers with smaller bit rates over narrow FSs emerged as the possible solution to this problem. Such multicarrier-based transmission techniques have already been in use in wireless communication systems, which eventually found a strong candidature in optical networks. By using variable number of carriers over a bundle or band of as many narrow FSs, it becomes possible in EONs to transmit with flexible or elastic transmission rates, while with the per-carrier bandwidth (i.e., the width of an FS) remaining small, one can get around the problem of higher BER while transmitting at a high aggregate transmission rate (i.e., sum of the transmission rates of the bundled carriers).

Typically, for each FS, the input binary data stream in an EON transmitter is used to carry out quadrature-phase-shift keying (QPSK) on an optical or electrical carrier (the electrical carrier is later translated up to a suitable optical carrier), leading to a $12.5 \times 2 = 25$ Gbps transmission. Two parallel QPSK signals can be further multiplexed in the optical domain by transmitting two orthogonal polarizations of light, leading to the transmission of dual-polarized QPSK (DP-QPSK) signal at 50 Gbps for each FS, which are again combined (multiplexed) together to form optical signal with multicarrier modulation offering flexible transmission rates in multiples of 50 Gbps. However, to further exploit the bandwidth of each FS, one can use higher-level modulations for each FS, such as quadrature-amplitude modulation (QAM) schemes, but constrained by the optical reach of the connection due to poorer error performance of QAM with respect to QPSK.

14.3 Multicarrier transmission in EON

In order to realize multicarrier-based elastic spectrum, EONs are evolving toward employing optical carriers with a narrow transmission bandwidth (i.e., FS) of 12.5 GHz, which is one-fourth of the currently used 50 GHz standard for channel spacing (ITU-T G.694.1 2002). Using the FSs of 12.5 GHz, an EON transmitter can set up bandwidth-variable end-to-end optical connections, i.e., lightpaths with variable speed, hereafter referred to as *elastic optical paths* (EOPs), by selecting appropriate number of carriers, which are assigned contiguous FSs. This leads to the multicarrier transmission, where each carrier is modulated using a bandwidth-efficient digital modulation scheme. By using such a modulation scheme, it is also possible to extract a *two-fold flexibility* of the transmission speed, where one can aggregate more carriers as well as pack information in each carrier as much as possible, albeit constrained by the optical reach. For example, an optical carrier can employ QPSK modulation over 12.5 GHz FS to realize a 25 Gbps transmission, while by adopting 16 QAM the same carrier can transmit at a rate of 50 Gbps. Thus, eight contiguous FSs, each using QPSK, can have an aggregate transmission rate of 200 Gbps, while it can be raised to 400 Gbps by using 16 QAM for each subcarrier, but with a reduced optical reach in the second case.

In the following, we illustrate the above features of EON using Fig. 14.2. As shown in part (a), the entire spectrum is divided into narrow FSs of width Δf, and any transmitting node can select one or a band of some contiguous FSs for transmitting its ingress traffic. Further, guard bands (Δf_G) are inserted between any two consecutive bands of FSs to make sure that the EOPs corresponding to those bands don't interfere with each other in the spectral domain. Δf_G is preferably set as an integral multiple of Δf, i.e., $\Delta f_G = n_G \Delta f$, with $n_G = 1$, typically. A possible scenario of concurrent reception of multiple bands is also illustrated in

(a) Spectral bands consist of a variable number of narrow slots (i.e., FSs), separated by guard bands. $\Delta f_G = 2\Delta f$, Δf = slot width (channel spacing).

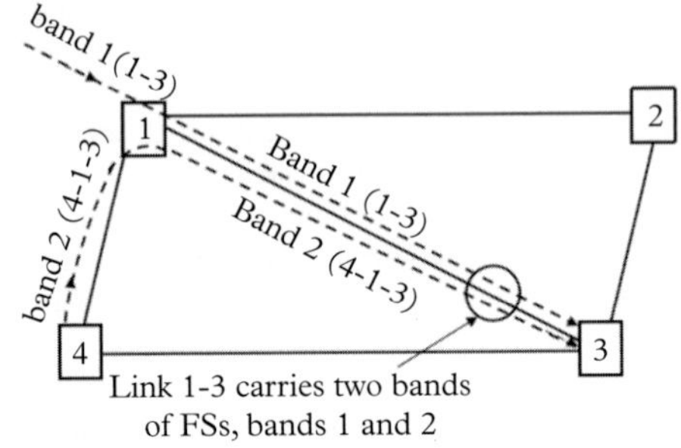

(b) Band multiplexing: two different spectral bands (1 and 2) get multiplexed over one fiber link, and received at the same node (node 3) distinctly, with the separation of the guard band.

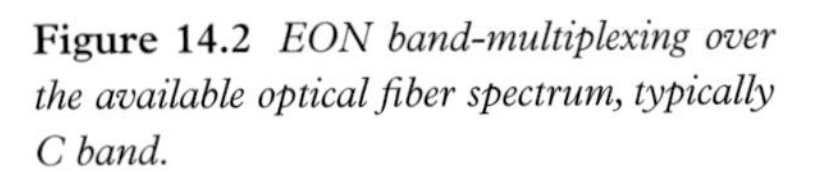
Figure 14.2 *EON band-multiplexing over the available optical fiber spectrum, typically C band.*

Fig. 14.2(b), where the data traffic from node 1 flows to node 3 over band 1 and the data traffic from node 4 to node 3 flows via node 1 over band 2. Thus, node 3 has to receive both the spectral bands (bands 1 and 2) through the common fiber link between the nodes 1 and 3. So far, a variety of multicarrier transmission schemes have been explored for EONs, many of them being centered around the well-known concept of *orthogonal frequency-division multiplexing* (OFDM).

The multiple carriers in OFDM are chosen to be *orthogonal* to each other over a symbol duration (a group of bit intervals), and packed densely with overlapping spectra, thereby providing a highly efficient and flexible utilization of transmission bandwidth (Goldsmith 2005). The spectral efficiency of OFDM over non-OFDM schemes is illustrated in Fig. 14.3, where one can see that with overlapping spectra, OFDM can pack its carriers with narrower FS width, which is in fact half of the FS width used in the non-OFDM-based multicarrier systems. Theoretically, each individual spectrum is a *sinc* function (when rectangular pulses are used as the modulating baseband signal) with decaying-but-oscillating tails spreading far into several adjacent FSs (not shown in the diagram for simplicity). However, the spectral nulls of each sinc function beyond its parent FS fall exactly on the central peaks of all the adjacent FSs. However, even with this kind of overlapping spectrum, the orthogonality (the *condition of orthogonality* is discussed in the next section) between the chosen carriers in OFDM helps in extracting the information on each carrier successfully at the receiving end.

The optical version of OFDM (OOFDM) can be realized using several candidate schemes with varying degrees of hardware complexity and performance (Armstrong 2009; Shieh *et al.* 2008; Gerstel *et al.* 2012; Chandrasekhar and Liu 2012). Some of the options for realizing OOFDM in EONs are as follows:

- intensity-modulation (IM) of light using electrical OFDM signal and direct-detection (DD) of the intensity-modulated optical signal (IMDD-OOFDM),

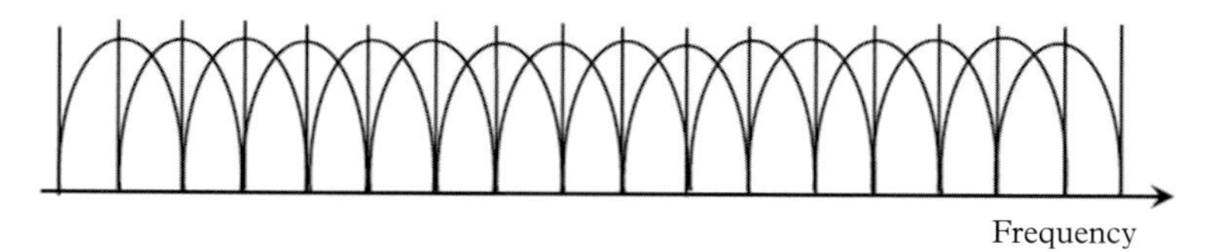

(a) Orthogonal multicarrier transmission spectrum with overlap.

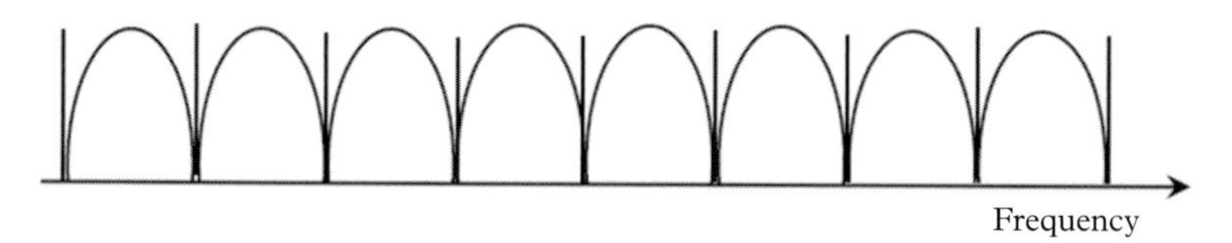

(b) Non-orthogonal multicarrier transmission spectrum without overlap.

Figure 14.3 *Comparison of multicarrier communication systems using OFDM and non-OFDM transmission schemes. Note that in the given example, the non-OFDM system is able to transmit eight carriers, while the OFDM system can accommodate 16 carriers within the same spectral window.*

- upconversion of electrical OFDM to OOFDM using amplitude modulation of laser (UC-OOFDM),
- all-optical realization of OOFDM (AO-OOFDM),
- electro-optic generation of optical OFDM (EO-OOFDM) with I/Q modulation carried out in the optical domain.

Some other options for optical multicarrier transmissions have also been explored: Nyquist-WDM transmission (Sinefield *et al.* 2013) and transmission systems using optical arbitrary waveform generation (OAWG) (Gerstel *et al.* 2012). In the following, we describe the implementation schemes for these multicarrier-based transmission systems for EONs.

14.3.1 OOFDM transmission

Before explaining the possible transmission schemes for OOFDM, we describe below the salient features of the electrical OFDM technique. Thereafter, the basic concept of OFDM is utilized to implement the OOFDM transceivers using appropriate schemes.

Electrical OFDM basics

A typical electrical OFDM (or simply OFDM) transmitter realized in the electrical domain is illustrated in Fig. 14.4, where an incoming binary data stream is passed through several stages of signal processing before the final OFDM waveform is generated. First, the transmitter employs a bits-to-QAM-symbol mapper, where the incoming data bits are grouped into k-bit words (using a serial-to-parallel converter (SP)), and each k-bit word, say the nth k-bit word, is mapped into the corresponding complex-valued QAM symbol pair $X[n]$ (representing in-phase/quadrature-phase (I/Q) components) from the M-QAM (with $M = 2^k$) signal constellation. Each complex-valued QAM symbol pair $X[n]$ spans over a QAM symbol period T_s^{QA}, given by

$$T_s^{QA} = kT, \tag{14.1}$$

with T representing the bit interval of the incoming binary data stream.

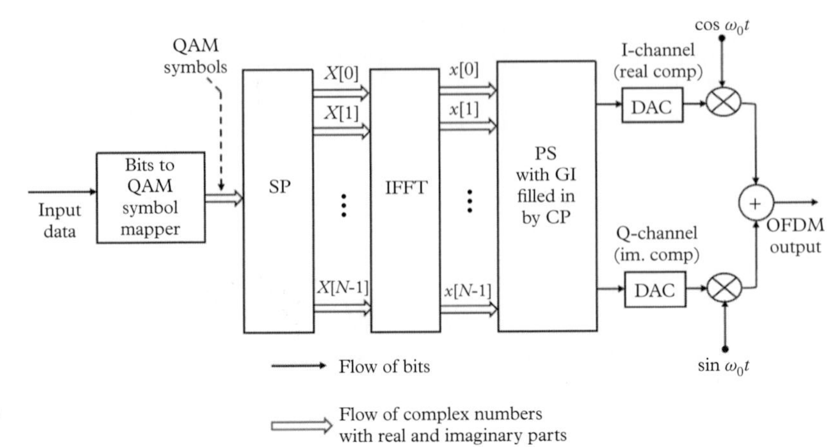

Figure 14.4 *Electrical OFDM transmitter. GI: guard interval, CP: cyclic prefix.*

Having received the complex QAM symbols ($X[n]$'s) from the bits-to-symbol mapper, the SP converter passes them in parallel to the digital signal-processing (DSP) block carrying out inverse fast Fourier transform (IFFT), which in turn generates the complex discrete-time signal samples $x[n]$'s, given by

$$x[n] = \frac{1}{\sqrt{N}} \sum_{0}^{N-1} X[i] \exp\left(\frac{j2\pi ni}{N}\right), 0 \leq n \leq N-1. \tag{14.2}$$

These samples are passed through parallel-to-serial converter (PS), and being complex in nature, are separated into I and Q channels carrying the real and imaginary components, along with guard times between two consecutive OFDM symbols. The guard times are introduced between adjacent OFDM symbols in wireless communications systems to combat multipath fading, where the signal at the end of each OFDM symbol is folded back in the preceding guard time with a *cyclic prefix* to alleviate the problem of signal spreading and ease out the problem of synchronization. In EONs, the use of guard time and cyclic prefix would have similar implications but due to different physical phenomena: various fiber-dispersion mechanisms creating pulse spreading. These I/Q outputs are passed through digital-to-analog converters (DACs) producing the analog waveforms (real and imaginary parts) at baseband level, which are subsequently passed through the respective multipliers to get multiplied with the cosine and sine carriers, following which the multiplier outputs are added up to produce the final OFDM signal.

For a given input bit stream, each $X[n]$ is generated from k consecutive bits (i.e., k-bit word) and N such $X[n]$'s are available from the bits-to-QAM-symbol mapper during kN bit intervals implying that each OFDM symbol will span over the duration T_s^{OF}, given by

$$T_s^{OF} = NT_s^{QA} = NkT_b. \tag{14.3}$$

Further, for the orthogonality to be satisfied, the spacing between any two OFDM carriers, say f_j and f_k, must be an integral multiple of the inverse of the OFDM symbol period, i.e.,

$$|f_j - f_k| = m\left(\frac{1}{T_s^{OF}}\right), \quad m = \text{an integer } (\neq 0). \tag{14.4}$$

This relation stems from the fact that, when two sinusoidal carriers $x_j(t)$ and $x_k(t)$ having the frequencies f_j and f_k, respectively (satisfying the condition of Eq. 14.4), are multiplied and their product is integrated over a given time interval T_s^{OF}, one gets the normalized (to the integration interval T_s^{OF}) value of the product ψ_{jk} as

$$\begin{aligned}\psi_{jk} &= \frac{1}{T_s^{OF}} \int_0^{T_s^{OF}} x_j(t)x_k(t)dt \qquad (14.5)\\ &= \frac{\sin[\pi(f_j - f_k)T_s^{OF}]}{\pi(f_j - f_k)T_s^{OF}} = \text{sinc}[\pi(f_j - f_k)T_s^{OF}].\end{aligned}$$

Note that ψ_{jk} goes to zero when $|f_j - f_k|$ becomes equal to m/T_s^{OF}, with m as an integer $\neq 0$, leading to the condition of *orthogonality*. However, with $j = k$, the value of ψ_{jk} becomes unity, leading to the condition of *orthonormality*. Thus, if any of the N incoming OFDM carriers in an OFDM receiver, say jth carrier, is multiplied with a locally generated carrier in the OFDM receiver with a frequency $f_k = f_j$ and integrated over T_s^{OF}, the receiver gets back the amplitude of the incoming carrier by virtue of the orthonormality condition. However, the other incoming OFDM carriers with $f_j \neq f_k$ are all reduced to zero after the above multiply-and-integrate operation owing to the condition of orthogonality between the incoming and locally generated carrier waveforms. The entire idea of OFDM is built upon this concept of orthonormality vis-a-vis orthogonality.

Following the above properties of OFDM signal, and using the reverse mechanism (with respect to the OFDM transmitter), the OFDM receiver is constructed as shown in Fig. 14.5, where the received OFDM signal is split into two arms and multiplied by the I/Q-phase RF carriers and then lowpass-filtered. The filtered outputs from the I/Q-channels are quantized using analog-to-digital converters (ADCs) and the cyclic prefix is removed after the serial-to-parallel (SP) conversion. The SP outputs are transformed back using the DSP-based FFT block (*implicitly* carrying out the orthonormality/orthogonality-based signal processing), giving back the baseband QAM symbols, which are subsequently passed through PS and mapped finally into the binary bit stream.

While providing a spectrally efficient multicarrier transmission scheme, OFDM signal suffers from a problem arising from the interference between the coherent carriers within each OFDM symbol, leading to a random amplitude fluctuation of the aggregate OFDM

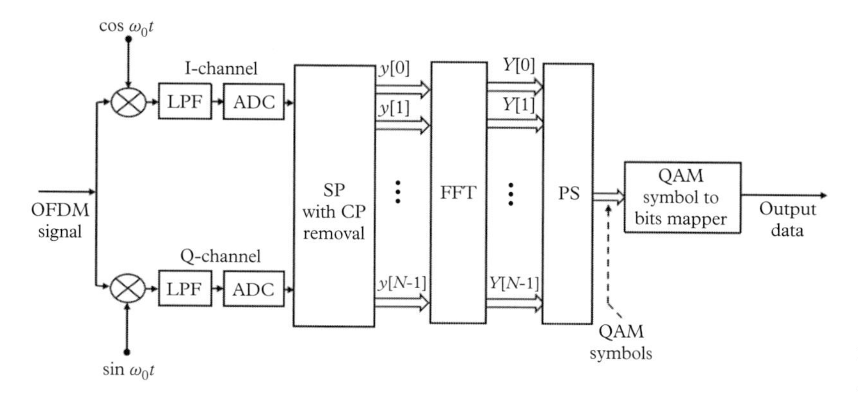

Figure 14.5 *Electrical OFDM receiver. LPF: lowpass filter.*

waveform. This phenomenon is measured by the peak-to-average power ratio (PAPR) within an OFDM symbol, and occurrence of high PAPR may become critical in highly packed high-power optical spectrum, pushing the optical fibers into the nonlinear region. However, a small amount of fiber dispersion can prevent the occurrence of large PAPR, thereby reducing the impact of PAPR-induced fiber nonlinearity.

OOFDM schemes

The basic schemes described above for OFDM modulation and demodulation are transformed into their equivalent schemes appropriate for generating OOFDM. As mentioned earlier, there are multiple options for the generation and reception of OOFDM signal: IMDD-OOFDM, UC-OOFDM, AO-OOFDM, and EO-OOFDM. In the following, we describe these three mechanisms for OOFDM.

IMDD-OOFDM

IMDD-OOFDM offers a straightforward realization of OOFDM, where the electrical version of OFDM generated at a radio frequency (RF) (using the transmitter configuration of Fig. 14.4) is used to modulate directly a laser (or an external modulator), leading to a light output whose intensity (not its phase, frequency, or amplitude) is modulated (IM) with the RF-based OFDM. At the receiving end, the incoming optical signal is received at a photodiode for direct detection (DD), and the photodetected electrical signal is passed through the typical OFDM receiver configuration, as shown earlier in Fig. 14.5.

Note that, due to the non-coherent detection used in IM-DD systems, the receiver sensitivity becomes inferior as compared to the OOFDM systems employing coherent detection, and hence may not be a preferred choice for long-haul WDM networks. However, for the EON-based PONs this scheme can be useful with simpler and hence cheaper implementations.

UC-OOFDM

In UC-OOFDM, the basic electrical OFDM modulation is carried out using an intermediate frequency at RF. The electrical OFDM is next *up-converted* using amplitude modulation (instead of IM, as used in IMDD-OOFDM) of a laser diode (LD) at MZI giving a suppressed-carrier double-sideband modulation of optical carrier by the electrical OFDM. Figure 14.6 illustrates the configurations of UC-OFDM transmitter and receiver in parts (a) and (b), respectively. The signal spectra at MZI input and output ports (i.e., at the points $\boldsymbol{x}, \boldsymbol{y}$) are shown in the shaded inset of the figure. Following the MZI, one of the two sidebands (the higher sideband, i.e., ω_2, is chosen in the present example) is filtered out to produce the upconverted version of the electrical OFDM at the optical filter output (i.e., the spectrum shown for the point $\boldsymbol{z}$).

At the receiving end, the demodulation process at intermediate frequency (IF) is preceded by the down-conversion from the optical to electrical domain through coherent heterodyne detection using the local laser playing the role of the local oscillator, as shown in Fig. 14.6(b). The IF carrier is extracted using a carrier recovery scheme (not shown) and used for coherent demodulation of the in-phase and quadrature-phase (in the I and Q arms) using multipliers. Subsequently, the OFDM demodulation is carried out in the electrical domain by using DSP-based FFT block as discussed earlier for electrical OFDM (i.e., the way similar to that shown in Fig. 14.5).

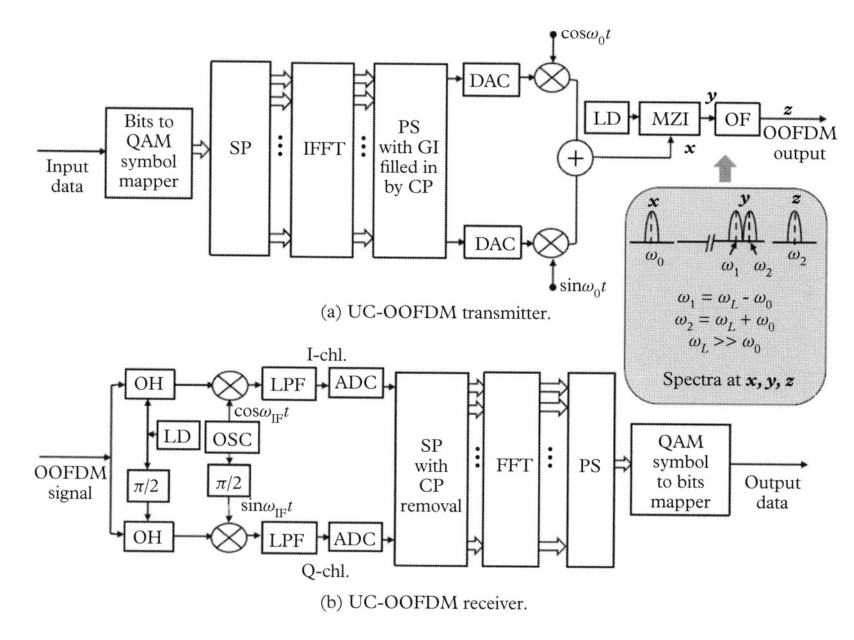

Figure 14.6 *Block diagrams for UC-OOFDM transmitter and receiver. LD: laser diode, GI: guard interval, CP: cyclic prefix, OF: optical filter, OSC: oscillator, OH: optical hybrid, ω_L = transmit laser (angular) frequency.*

Note that coherent heterodyne detection can also be carried out using intradyne detection where the local laser operates at a frequency in the received optical signal band, requiring thereby some additional signal-processing tasks. Further, though not shown in the figure, the receiver needs to control the polarization of local laser so that the SOP (i.e., state of polarization) of the local laser matches with the SOP of the incoming lightwave.

AO-OOFDM

The block diagrams for AO-OOFDM transmitter and receiver are illustrated in Fig. 14.7, where both modulation and demodulation are carried out in the optical domain. In the transmitter (part (a)), the output of the laser diode (LD) is passed through a multiple-carrier generator, producing optical carriers with the frequencies needed for generating OOFDM waveform. Note that, while realizing the multicarrier generator, one needs to ensure the needed orthogonality between the multiple optical carriers for successful OFDM generation. One possible method to realize this feature is to amplitude-modulate the laser with a periodic pulse train at a frequency of $1/T_s^{OF}$ by using an external modulator, such that the pulse-modulated laser output generates lightwave harmonics separated exactly by a frequency spacing of $1/T_s^{OF}$, thereby ensuring their orthogonality over the QAM-symbol interval T_s^{OF}. The output of the multicarrier generator is demultiplexed using an optical demultiplexer, and thereafter all the optical carriers at the demultiplexer output undergo single-carrier modulation by the complex (I/Q) inputs from the bits-to-QAM-symbol mapper, where each optical carrier is modulated in amplitude and phase (for realizing QAM), by using complex Mach–Zehnder modulators (CMZMs). The modulated carriers are next combined together through a power combiner, leading to the desired OOFDM signal.

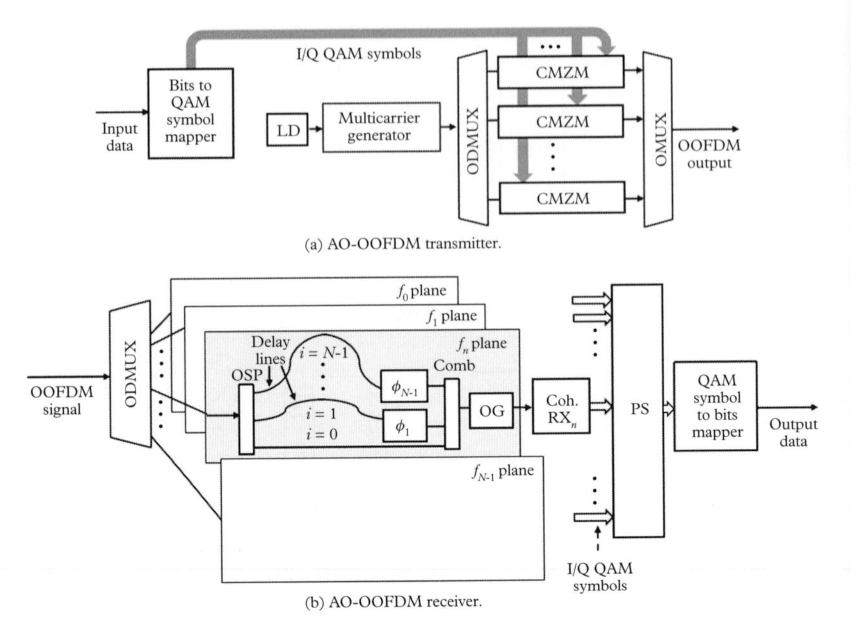

Figure 14.7 *AO-OOFDM transceiver block diagrams, with nth plane in the receiver (corresponding to the nth carrier) shown in details. OG: optical gate, Coh. RX_n: coherent optical receiver for nth plane (for simplicity, local laser for coherent detection is not shown), Φ_i's = phase shifts in the optical phase shifters, OSP: optical power splitter, Comb: optical power combiner.*

At the destination node, the received optical signal is demultiplexed into its N constituent frequency components, and each of these N components undergoes optical signal-processing on a frequency *plane* or *channel* (see Fig. 14.7(b)) (Sanjoh *et al.* 2002). The task of optical signal-processing is carried out by following the basic definition of the discrete Fourier transform (DFT) $X[n]$ for the discrete-time signal $x[i]$'s (see Eq. 14.2), given by

$$X[n] = \frac{1}{\sqrt{N}} \sum_{0}^{N-1} x[i] \exp\left(-\frac{j2\pi ni}{N}\right), 0 \leq n \leq N-1. \tag{14.6}$$

In essence, we need to get back the QAM symbol in the form of $X[n]$ from $x[i]$'s for each plane (i.e., each carrier frequency) through optical signal-processing. This amounts to splitting each of the N demultiplexer outputs (corresponding to one of the N carriers) into N *arms* on the respective plane by using one $1 \times N$ optical splitter (OSP), where the optical signal in each arm of the given plane passes through an optical delay line and a phase shifter (corresponding to each term of the summation in Eq. 14.6). Thus, in the nth plane, the magnitude of delay in the ith arm will be iT_s^{OF}/N and the following phase shifter in the same arm will introduce a phase shift $\Phi_i^n = 2\pi ni/N$, with $0 \leq i \leq N-1$.

The delayed and phase-shifted components from all the N arms on each plane are combined in an optical combiner to form a composite optical signal, whose *complex envelope* carries the desired $X[n]$. The combined optical signal from each plane is subsequently passed through an optical gate and a coherent optical receiver carrying out the opto-electronic conversion, thereby extracting the respective value of $X[n]$ from the I/Q channels (not shown explicitly). Optical gating (OG) is required to enable the DFT-processing of the combined signal in each plane at the *right time*. These QAM symbols for all values of n (from N planes) are thereafter passed through the PS converter followed by the QAM-symbol-to-bit mapper to finally obtain the output data stream. Thus, the overall optical hardware employs N planes for N carriers, and each plane employs N *delay-and-phase-shift* arms to perform the needed signal-processing task in the optical domain. Note that, the scheme presented here does not have any provision for a guard interval, which can however be introduced with additional hardware.

In AO-OOFDM, multiple carriers generated from the single laser must satisfy the orthogonality conditions of OFDM, mandating that the lasers in the transmitter and the receiver should be coherent, and the receiver should also employ coherent detection (shown as Coh. RX_n in the figure) to get back the baseband signal. This is why AO-OOFDM systems are also referred to as coherent WDM (Co-WDM) systems in the literature (Shieh *et al.* 2008).

EO-OOFDM

The block diagrams of transmitter and receiver for EO-OOFDM are shown in Fig. 14.8. In the transmitter (Fig. 14.8(a)), analog version of the OFDM I/Q waveforms at the baseband level are generated using electrical signal processing. The I/Q waveforms are next used to modulate the laser output in I and Q channels separately (unlike the double-sideband modulation of laser output in UC-OOFDM), and combined thereafter to produce the OOFDM signal. In other words, unlike UC-OOFDM, in this case two separate I/Q channels are employed in optical domain itself (note the $\pi/2$ phase shifter inserted in the CMZM's Q-channel). Thus, the generation of baseband OFDM signal (i.e., the complex envelope) is carried out in the electrical domain, whereas the final I/Q modulation of the optical carrier is realized in the optical domain.

The receiver (Fig. 14.8(b)) carries out the inverse set of operations after converting the optical signal into the electrical counterpart through coherent detection for I/Q channels, which are next passed through the lowpass filters, ADCs, and SP converter, followed by the necessary DSP-related blocks (FFT and PS) and QAM-symbol-to-bits mapper to get back the desired data stream.

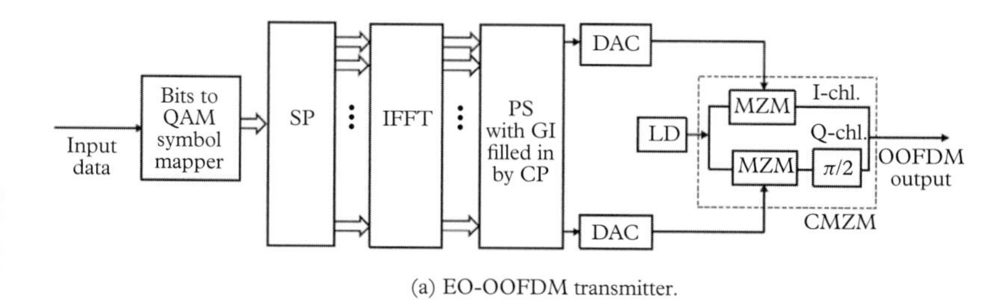

(a) EO-OOFDM transmitter.

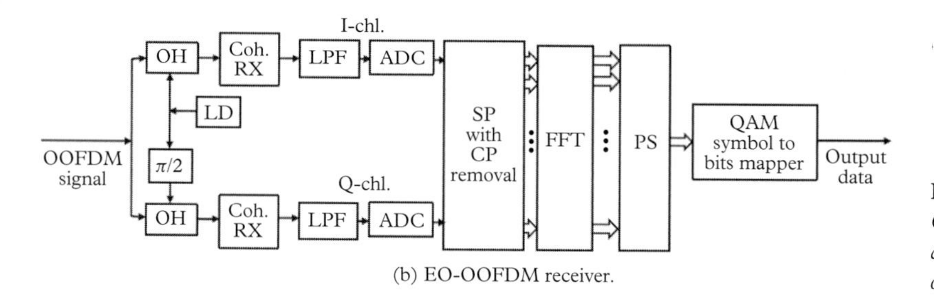

(b) EO-OOFDM receiver.

Figure 14.8 *Block diagrams for EO-OOFDM transmitter and receiver. LD: laser diode, OH: optical hybrid, Coh. RX: coherent optical receiver.*

14.3.2 Nyquist-WDM Transmission

Nyquist-rate transmission is a classic benchmark for digital communication systems, implying that the transmission rate over a channel with a given bandwidth is equal to the Nyquist rate, which is determined by the channel bandwidth itself. For the lowpass (i.e., baseband) channels, the rate is twice the channel bandwidth, while for the bandpass channels (as in WDM channels) the rate is just equal to the channel bandwidth. Thus, a Nyquist-rate WDM system refers to a WDM link, where the per-channel bit-rate approaches the Nyquist rate. For example, when a WDM channel with a channel spacing and hence a channel bandwidth $\Delta f = 50$ GHz can transmit at a rate $r = 50$ Gbps, the system is said to be Nyquist-rate WDM (or more precisely, Nyquist-rate DWDM system due to 50 GHz channel spacing). This implies that, over a WDM channel, if the transmission system achieves the ratio $\rho = \Delta f/r = 1$, then it qualifies as a Nyquist-rate WDM transmission system. Broadly, any WDM system that has a ρ that is close to but > 1 (typically, in the range between 1 and 1.2) is called a *quasi-Nyquist-rate* WDM system, while the systems with $\rho < 1$ are referred to as *super-Nyquist-rate* WDM systems.

In view of the above, a per-channel transmission rate of 10 Gbps in a WDM system with 50 GHz channel spacing (or bandwidth) gives $\rho = 50/10 = 5$, and hence this WDM transmission system doesn't even qualify as a quasi-Nyquist-rate WDM system. However, a 40 Gbps per-channel transmission rate in a WDM system with 50 GHz channel spacing will practically represent a quasi-Nyquist-rate WDM system with $\rho = 50/40 = 1.25$, and a 100 Gbps per-channel transmission rate in a WDM system with the same channel spacing would indeed be a super-Nyquist-rate WDM system with $\rho = 50/100 = 0.5$. Attempting a super-Nyquist-rate WDM transmission (using multilevel modulation schemes) with the transmission rates much above 100 Gbps over a 50 GHz WDM grid may not be feasible as the transmission impairments would increase BER with the reduced distances between the constellation points in signal space. Moreover, such systems would call for high-speed DSP chips, although similar super-Nyquist-rate (i.e., with $\rho \leq 0.5$) transmission may not be as challenging over smaller channel spacings. Thus, with 12.5 GHz as the reduced-size FS, quasi/super-Nyquist-rate WDM transmissions become feasible with much lower demands on high-speed electronics, and therein comes the candidature of such WDM transmission systems for the EONs using 12.5 GHz FSs.

Note that, OFDM allows spectral overlaps between the adjacent carriers without any interference (and hence offers high spectral efficiency), but at the cost of high-speed ADC/-DAC and DSP hardware. However, for efficient spectral packing, non-OFDM multicarrier quasi-Nyquist/super-Nyquist WDM systems must employ tight per-channel spectrum using appropriate optical filtering, while avoiding the ADC/DAC and DSP hardware of OFDM. We discuss below the implementation aspects of the Nyquist/quasi-Nyquist-rate WDM (hereafter referred to simply as Nyquist-WDM) systems for EONs.

As mentioned above, while avoiding the use of high-speed ADC/DAC and DSP chips in OOFDM, the main challenge of Nyquist-WDM systems is to prevent the spectral overlap between adjacent WDM channels. This, in turn, requires implementation of optical filters for each channel with *sharp roll-off*, which can use precisely designed photonic devices, such as photonic spectral processors with specified optical transfer functions (Sinefield *et al.* 2013). Figure 14.9 presents the schematic block diagrams for the Nyquist-WDM transmitter and receiver.

As shown in Fig. 14.9(a), independent lasers are used for generating multiple carriers, obviating the need of phase-locked operation. This configuration with independent lasers also remains free from the PAPR problem experienced in OOFDM-based systems. The laser outputs are passed through the bank of multi-level complex Mach–Zehnder modulators

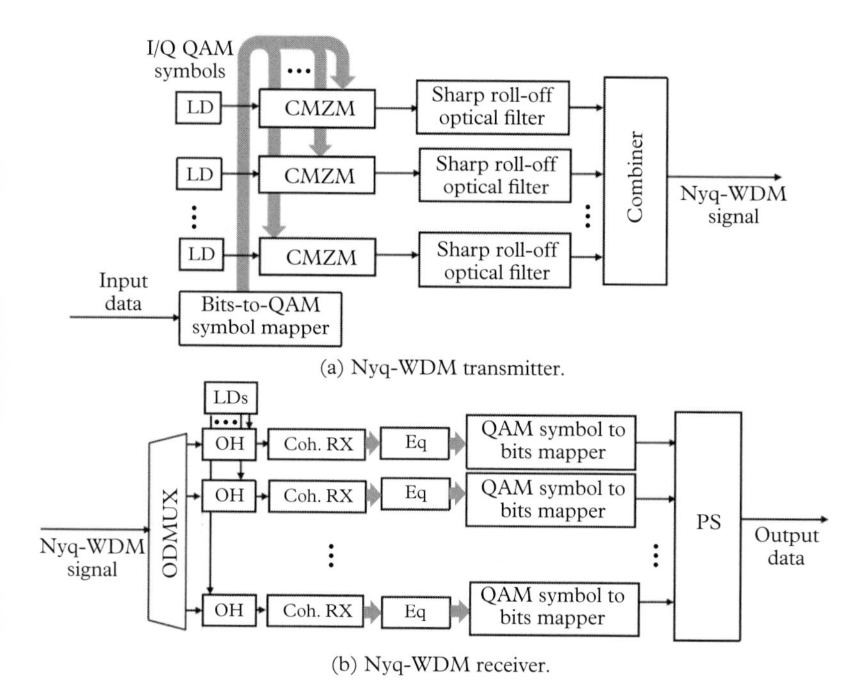

Figure 14.9 *Block diagrams for Nyq-WDM transmitter and receiver. Eq: equalizer, Coh. RX: coherent optical receiver.*

(CMZMs) with the respective I/Q modulating signals from the bits-to-QAM-symbol mapper for quasi-Nyquist or Nyquist transmission, but with the spectral tails on both sides of each FS. Hence the modulator outputs are passed through tight optical filters with sharp roll-off, which in turn create oscillating tails (like the *sinc* function) in the time domain, causing inter-symbol interference (ISI) in the receiver. However, the ISI created through this process (in order to save inter-carrier interference (ICI)) is managed at the receiver by using equalization (Eq blocks in the figure) with a tapping length in accordance with the filtering-induced tails in the time domain.

The receiver for the Nyquist-WDM signal (Fig. 14.9(b)), upon receiving the optical signal, demultiplexes it into the constituent carriers from the multiplexed spectrum and passes them through as many coherent optical receivers aided by the local LDs and OHs. The receiver outputs after equalization are passed through QAM-symbol-to-bits mappers and a PS converter to retrieve the desired data stream. Along with this arrangement, sparing an additional 10% bandwidth for each FS can control the ISI problem within practical limits (Chandrasekhar and Liu 2012). In other words, for a per-channel transmission rate of r, keeping a channel spacing $\Delta f \simeq 1.1 \times r$ (i.e., with $\rho = 1.1$) would serve the purpose. If necessary, further improvement in BER performance can be achieved through FEC codes.

14.3.3 OAWG-based multicarrier transmission

Arbitrary waveform generation in an OAWG scheme for EONs implies that the transmitter can employ multicarrier transmission over the available FSs, where each carrier can choose a different modulation format exhibiting different or arbitrary waveform, as determined by the constellation of the chosen modulation scheme (Gerstel *et al.* 2012). However, the transmission system does not mandate any orthogonality condition to be satisfied by the

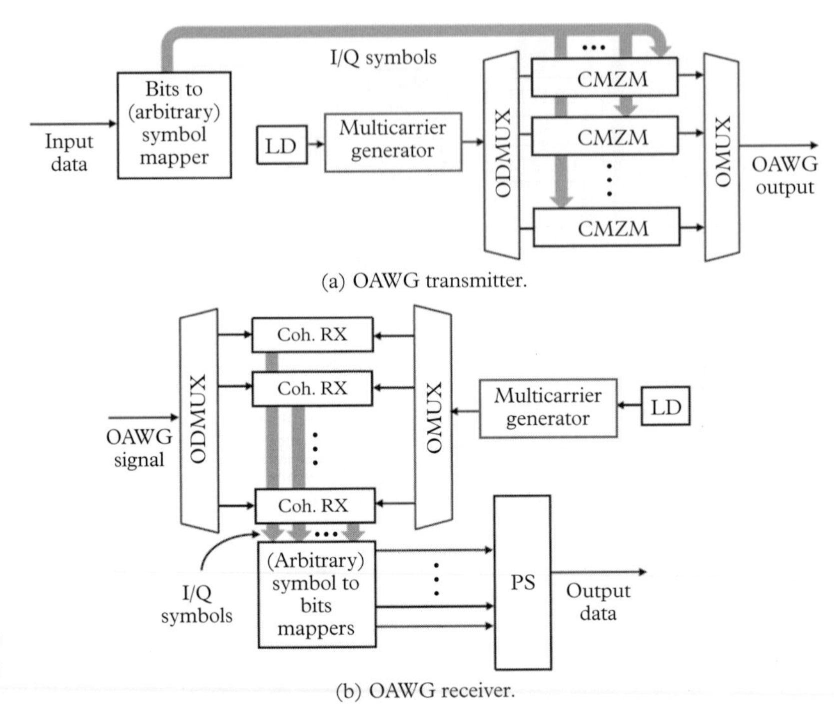

Figure 14.10 *Block diagrams for OAWG transmitter and receiver. Coh. RX: coherent optical receiver.*

optical carriers, although the lasers used (in transmitters as well as receivers) must be coherent to allow coherent modulation/demodulation of m-ary PSK/QAM signals with varying constellation size, e.g., QPSK, 8PSK, or 16QAM.

Figure 14.10 presents the block diagrams for the transmitter and receiver used in OAWG-based transmission systems. As shown in part (a), the transmitter makes use of one laser which is amplitude-modulated with a periodic pulse train to generate equally-spaced spectral lines fitting with the central frequencies of the FSs of EON (as in AO-OOFDM). These optical carriers are demultiplexed and are subjected to different modulation formats and transmission rates (as needed by the traffic demands) spanning over one FS (single-channel) to several FSs (superchannel).

On the receiving side, the received OAWG signal is demultiplexed and treated with coherent detection (local LDs and OHs not shown), thereby delivering the m-ary PSK/QAM symbols, which are subsequently mapped into the corresponding data bits by the symbol-to-bits mapper and serialized through the PS converter into the desired data stream. Overall, OAWG promises a highly flexible transmission scheme, where, by and large, other transmission schemes (OFDM or non-OFDM) can also fit in, leading to a versatile transmission framework.

14.4 Bandwidth-variable network elements for EON

The flexibility of an EON greatly depends on how its constituent network elements are implemented and can adapt to the time-varying bandwidth requirements of the incoming

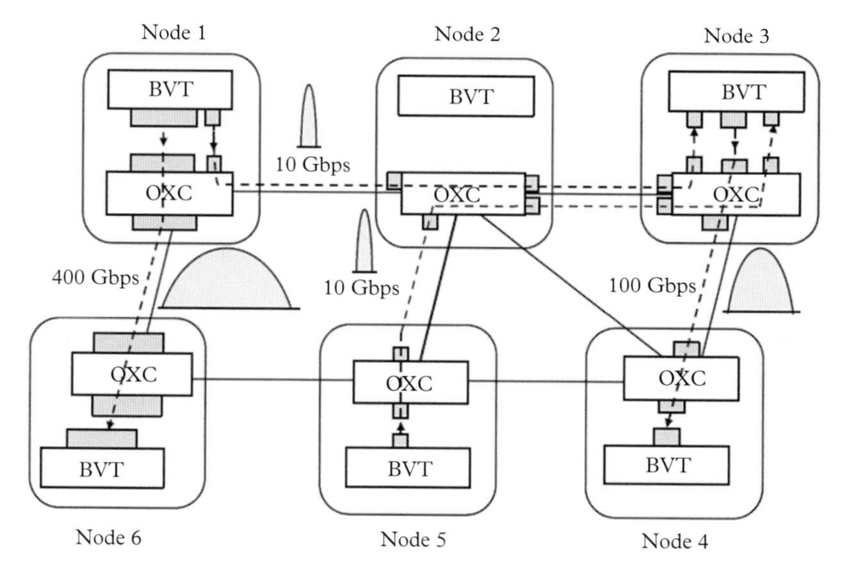

Figure 14.11 *EON operation using BV network elements with 10, 100, and 400 Gbps connections across a six-node network. Note that node 2 doesn't originate/terminate any connection, and hence operates only as a cross-connect for pass-through connections between other nodes; hence the BVT of node 2 does not participate with the pass-through connections. While the number of transponder ports in a BVT is fixed, each port in a BVT can configure itself with the bandwidth commensurate with the connections (dashed lines) being set up. Similarly, the ports of the OXCs also participate with variable bandwidth.*

connection requests. The typical network elements in an EON include bandwidth-variable (BV) transmitters and receivers, commonly called as BV-transceivers or BV-transponders (BVTs) (which are realized using the transmitters and receivers, as described in the foregoing), bandwidth-variable ROADMs and OXCs (BV-ROADMs and BV-OXCs), and their constituent devices, such as BV optical filters, BV-WSSs, etc. In the following, first we describe at a higher level how these network elements work together to set up BV connections across EONs, then we illustrate how the various network elements, e.g., BVTs, BV-WSSs, BV-ROADMs, and BV-OXCs, are realized in practice.

Figure 14.11 illustrates a six-node EON operating with BVTs and BV-OXCs to set up end-to-end connections with EOPs. A number of EOPs are established using respective BV ports of each BVT, which are routed through the BV-OXCs as follows:

(1) one EOP from node 1 to node 6 at 400 Gbps over the direct link 1-6 (i.e., without any intermediate node),
(2) another EOP from node 1 to node 3 at 10 Gbps over the link 1-2-3 via node 2 as the intermediate node,
(3) one EOP from node 5 to node 3 at 10 Gbps over the link 5-2-3 via node 2 as the intermediate node,
(4) one EOP from node 3 to node 4 at 100 Gbps over the direct link 3-4.

The bandwidth associated with each connection is shown symbolically by the width of the concerned ports (represented by the *width* of the shaded boxes attached to each BVT and OXC) and the corresponding optical spectra. Note that, although two EOPs pass through node 2, its BVT does not participate in these passthrough connections. In a dynamic situation, bandwidth of the connections going from/through/to the nodes would vary with time, thereby implying that the respective ports of BVTs and OXCs will have to tune themselves to the required band of FSs. In the following we describe the operations of these devices and discuss how such functionalities are realized in practice. Note that, in Chapter 2, we discussed a wide range of devices for WDM networks with fixed-grid applications, while we deferred the discussion on the devices for EON to this chapter, for better appreciation of the mechanisms in the context of bandwidth-variable operations of EONs.

14.4.1 Sliceable BVT

So far we have discussed about the BVTs for use in EONs, carrying out the necessary tasks of modulation and demodulation schemes using flexible bandwidth. However, more efficient realizations of BVTs in respect of hardware utilization have received much attention in the recent years, which are referred to as sliceable-BVTs (S-BVTs).

With a close look at the BVTs, one can visualize that through one single port of a BVT-based transmitter it is also possible to transmit multiple bands of FSs *logically* representing different EOPs (as independent data flows) onto the corresponding output fiber. For example, a 400 Gbps transponder, while transmitting a 100 Gbps flow, need not waste its remaining 400 − 100 = 300 Gbps hardware capacity, and can transmit an additional one or more logically separate flows to some other destination node(s) through the same fiber port, although not exceeding 300 Gbps. Thus, the same transmitter port becomes capable to *slice* its 400 Gbps transmission capacity into multiple slices of independent flows to different destination nodes with each slice carrying an integral multiple of one FS (12.5 GHz), thereby with much-enhanced utilization of the transmitter hardware (López *et al.* 2014; Zhang *et al.* 2015; Sambo *et al.* 2015).

We illustrate the basic features of S-BVTs from the network-centric viewpoint in Fig. 14.12 with some example connection scenarios in a six-node EON. As shown in the figure, at node 1 there arrives four connection (or flow) requests (A, B, C, and D) with different bandwidth demands. Node 1 has three physical ports, ports P, Q, and R (i.e., the physical nodal degree is three), each port being capable of transmitting a maximum bandwidth of 400 Gbps. The connection request A wants to reach node 6 through port R of node 1 with a transmission rate of 200 Gbps, while the connection requests B, C, and D want to exit through port P to reach three different destination nodes with transmission rates

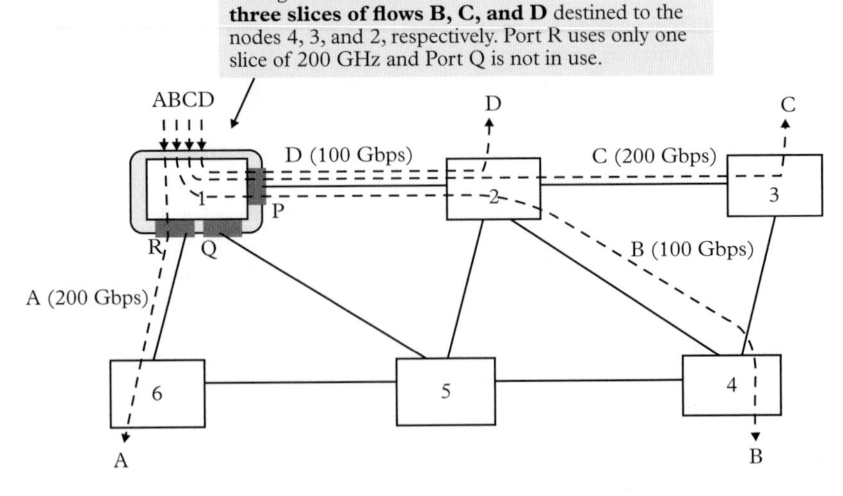

Figure 14.12 *Sliceability features of S-BVTs. Note that the node under consideration, i.e., node 1, has three ports and hence three S-BVTs.*

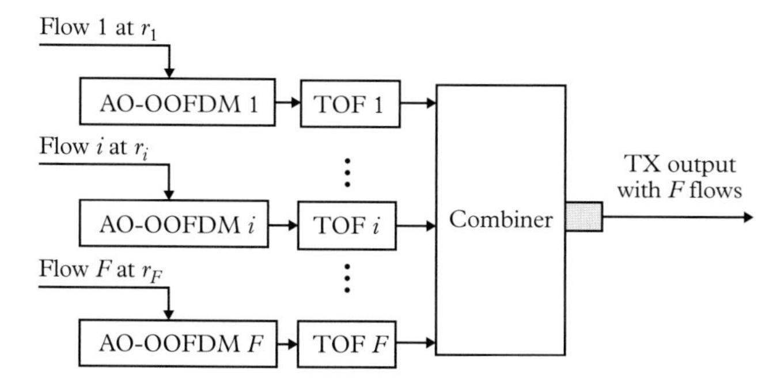

Figure 14.13 *Typical S-BVT transmitter using AO-OOFDM with the sliceability features. The transmitter in an S-BVT uses F AO-OOFDM as modulators, modulated by F independent flows, and each flow has a bit rate r_i, not exceeding the maximum capacity of the modulators, i.e., r_{max}. The receiver in an S-BVT carries out similar operations in the reverse direction.*

of 100 Gbps, 200 Gbps, and 100 Gbps, respectively. Thus, port P of node 1 connected to the fiber reaching node 2 uses three slices of flows: flow B at the rate of 100 Gbps reaching node 4 through node 2, flow C at the rate of 200 Gbps reaching node 3 through node 2, and flow D at the rate of 100 Gbps reaching directly to node 2. Port P is therefore using the *sliceability* feature of the transmitter using three slices of flows, summing up to the full capacity (i.e., 400 Gbps in this example) of the port. However, port R of node 1 connected to the fiber reaching node 6 is using the partial capacity (200 Gbps out of 400 Gbps), and the remaining capacity is available for use, whenever necessary. Port Q of node 1 reaching node 5 is not in use at the moment.

One candidate realization of S-BVT, as illustrated in Fig. 14.13, employs F AO-OOFDM-based modulators, each modulator generating one slice of spectrum for its own input data stream for onward transmission through the output port. Thus, each modulator is modulated by an independent flow with a bit rate of r_i, with r_i not exceeding the full capacity of the modulator, i.e., r_{max}. Each modulator output is passed through a tunable optical filter (TOF_i), and the TOF outputs are optically combined in the combiner to realize the frequency-multiplexed slices or FSs at the output port, with each slice corresponding to one flow with a variable rate r_i corresponding to the proportionate number of FSs. The slices from the output port carry different independent flows, which reach the respective destination nodes after exiting from the OXC of the source node (shown later in Fig. 14.14) on the specific output fiber and traverse through the chosen routes in the network. However, for this to happen successfully across the network, all the intermediate and destination nodes should be tuned appropriately to pass through or receive the corresponding slice, which is practically an EOP with variable data rate.

Therefore, the total transmission rate r_{tx}^{tot} from the S-BVT transmitter can be expressed as

$$r_{tx}^{tot} = \alpha_1 r_1 + \cdots + \alpha_i r_i + \cdots + \alpha_F r_F, \tag{14.7}$$

where α_i is 1 if the ith input port of the combiner receives a flow, else zero. The maximum possible value of r_i that a modulator can support is r_{max} and hence the maximum speed the entire transmitter can support is given by

$$r_{tx}^{max} = \max\{r_{tx}^{tot}\} = F \times r_{max}, \tag{14.8}$$

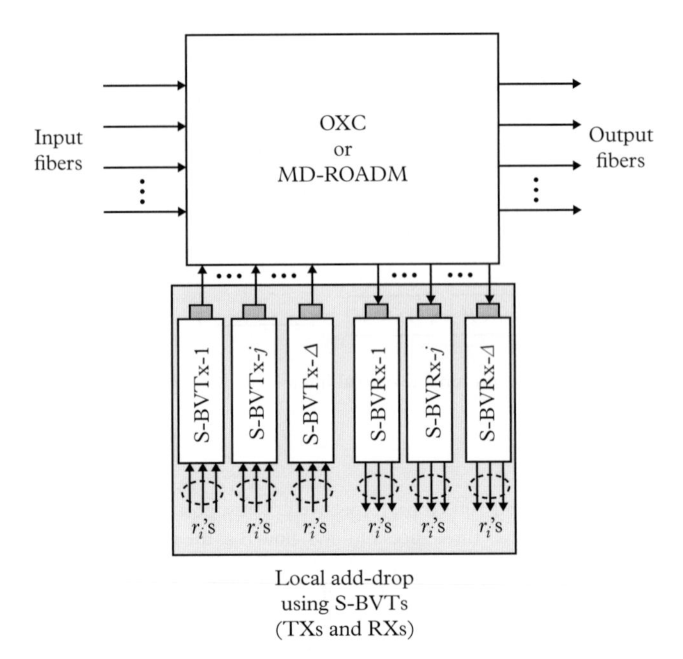

Figure 14.14 *Typical S-BVT node. SBVTx-i and SBVRx-i are the transmitter and receiver of the ith S-BVT in the node.*

which will be mapped into an equivalent number of FSs, governed by the constellation sizes used by each carrier in the modulators. The receiver in an S-BVT will carry out a similar operation in the reverse direction.

A candidate S-BVT-based node configuration, shown in Fig. 14.14, illustrates how the S-BVT transmitters (SBVTx-j) and receivers (SBVRx-j) are connected with the rest of the node using OXC or MD-ROADM. Note that the pair of SBVTx-j and SBVRx-j constitute the jth S-BVT of the node. Thus, each node would have a number of transceiver ports, and hence a number of such S-BVTs, depending on the physical nodal degree. Furthermore, as discussed later, all the optical elements in an OXC or MD-ROADM also need to be bandwidth-variable, while forwarding the pass-through EOPs with different FS allocations. Note that, the overall connection set-up task will therefore need a global coordination for the tuning of all the optical devices across the network involved in setting up the EOP, and this task can be better realized using a centralized control plane, such as by embedding an SDON-based (see Chapter 12) control framework in EONs.

14.4.2 BV-WSS

In EONs, while setting up an EOP around a central frequency f_n with a given bandwidth $m\Delta f$, all the network elements along the EOP must tune to the same frequency band. The BV-WSS is an important building block to realize the bandwidth variability along with switching for BV-ROADMs and BV-OXCs. Designs of the fixed-grid WSSs, ROADMs, and OXCs have been standardized using a number of techniques based on various combinations of power splitters, fixed-tuned filters, optical switches, AWGs, MEMS, etc. However, the variability in the bandwidth of the optical filters in WSS was not attempted in the earlier configurations. One of the key elements that can be used in basic BV-WSS configuration

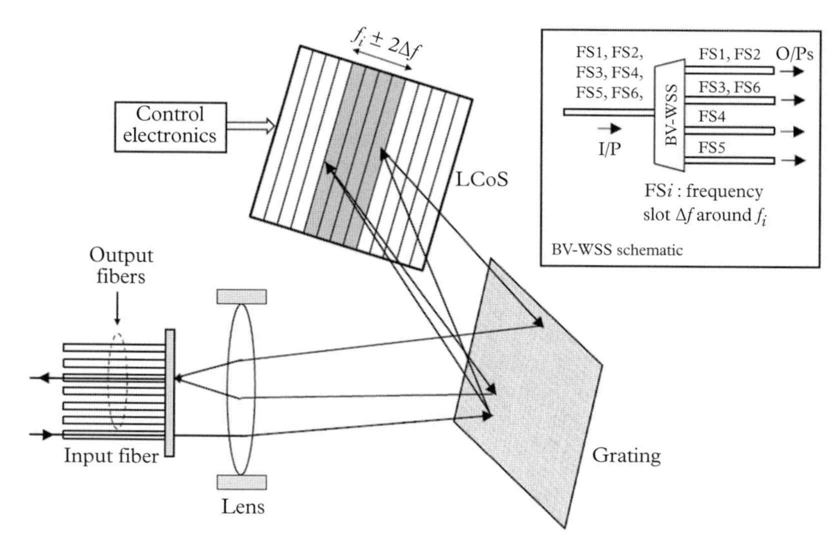

Figure 14.15 *Functional diagram of a BV-WSS using LCoS and reflection grating; block schematic of the BV-WSS is shown in the inset.*

for the variability of bandwidth is a silicon-based planar device, i.e., liquid crystal on silicon (LCoS) (Xie *et al.* 2017; Strasser and Wagener 2010), which was originally developed for display purposes. LCoS exhibits the role of a beam-steering element, as in MEMS, when used in a traditional WSS along with an additional control on the bandwidth of the light beam being steered.

Figure 14.15 presents the functional diagram of a BV-WSS using LCoS and reflection grating. As shown in the figure, light from the input fiber is first incident on the planar reflection grating. The incident light gets diffracted by the grating toward the LCoS, which is a planar liquid crystal array placed on a CMOS substrate. The CMOS-based processing-backplane empowers each LCoS pixel to get controlled by an electronic signal, so that the number of pixels reflecting the incoming light from the grating can be controlled from time to time. The LCoS plane is arranged as vertical columns of pixels, and each pixel column can be turned around its vertical axis to steer the incident light beam with frequencies within a specific FS. Thus turning a specific column toward a specific output fiber reflects a specific FS to the same output port. Hence, as more contiguous pixel columns are simultaneously rotated toward a desired output fiber port, a larger number of contiguous FSs are directed to the desired output port of the BV-WSS, leading to the bandwidth-variable connectivity. For carrying out the entire task, focusing the light beams onto the desired planes (grating and LCoS) is an important function, and the lens system carries out the task both ways. Thus, as shown in the figure, if the BV-WSS needs to switch four FSs around f_i (say), then the four contiguous columns of LCoS pixels should turn together around the vertical line corresponding to f_i toward the desired output fiber port of the BV-WSS.

Note that, in the configuration shown in Fig. 14.15, the grating operates in the reflection mode. However, one can also realize the same functionality by using a transmission grating with two focusing systems (lenses) on two sides. Further, as in usual WDM devices, similar operation can also be realized over planar optics, thereby giving more stable operation against physical stress and vibrations.

14.4.3 BV-ROADMs and BV-OXCs

BV-ROADMs are realized by using BV-WSSs and other supporting components. Figure 14.16 presents a typical configuration of a simple BV-ROADM with two ports (note that two such units are needed for a bidirectional ring) using one BV-WSS and an optical power combiner. Moreover, as shown in Fig. 14.17, one can also employ multiple BV-WSSs along with optical power splitters for multi-degree (nodal degree) BV-ROADMs with north-south-east-west ports to interconnect two ring networks.

As shown in Fig. 14.18, BV-OXCs are also realized in the same manner as in BV-ROADMs, albeit with a larger number of ports. Note that, due to the bandwidth variability, OXC and ROADM are driven to use the same basic component, i.e., BV-WSS, unlike those used in fixed-grid WRONs, where MEMS-based/electro-optic/thermo-optic switches are used in place of LCoS-based BV-WSSs. There are also some other mechanisms to realize BV optical filters, such as those using optical transversal filters (Swekla and MacDonald 1991) and thermo-optic switches (Xie *et al.* 2009), which might also be useful in realizing BV devices for EON.

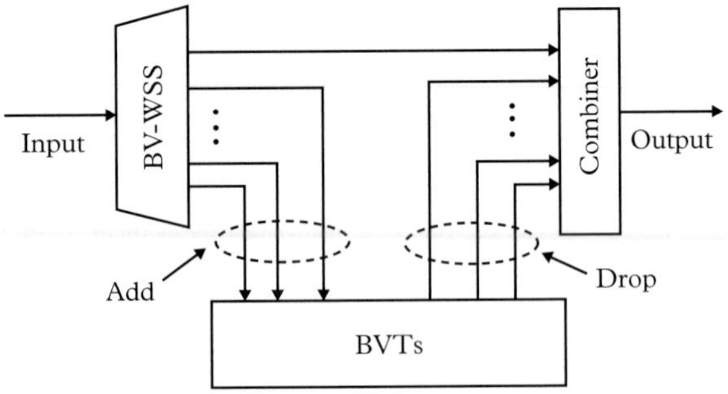

Figure 14.16 *Block diagram of a typical BV-ROADM for ring networks. Note that all the output ports of BV-WSS are bandwidth-variable and compatible with the ports of all the transponders in the BVT.*

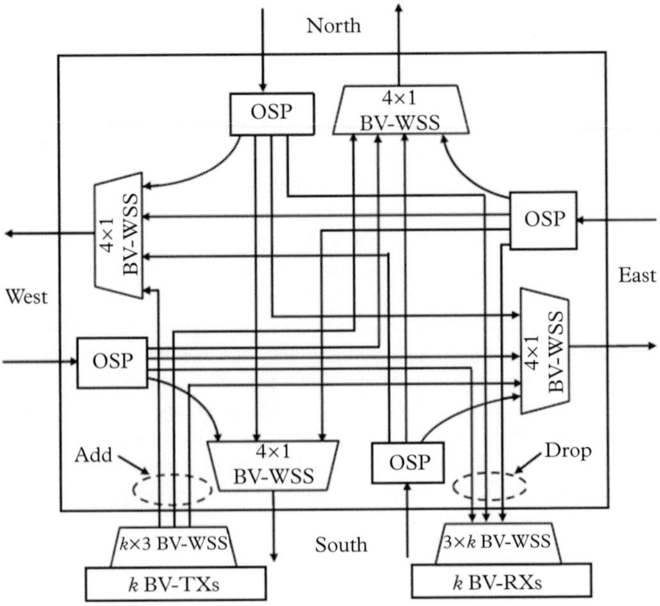

Figure 14.17 *Block diagram of a typical four-port (north-south-east-west) BV-ROADM for interconnecting ring networks. BV-TX: BV transmitter, BV-RX: BV receiver; a pair of BV-TX and BV-RX makes a BVT.*

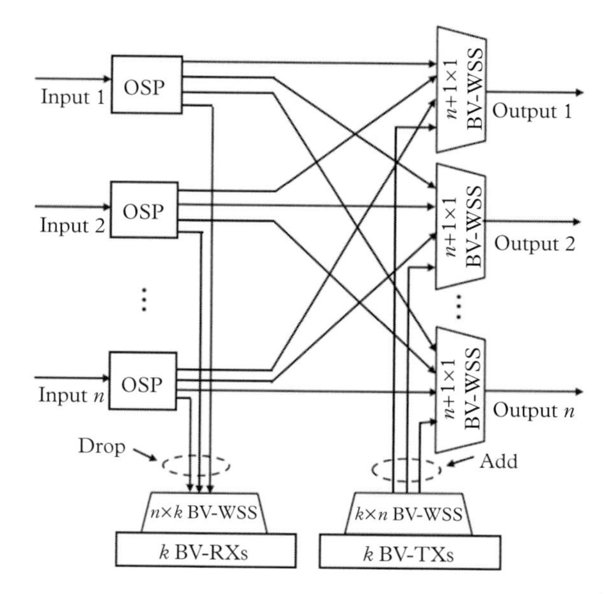

Figure 14.18 *Block diagram of a typical BV-OXC.*

14.5 Design issues in EONs

The task of designing and operating EONs has some basic commonalities with that of fixed-grid WRONs described in Chapters 8 and 9. However, owing to the flexibility of spectrum assignment to each connection, EON designs get more complex and hence more challenging. We discuss the EON design issues in the following.

In order to utilize the available transmission spectrum in optical fibers, any given connection needs to be assigned a band of FSs, depending on the bandwidth request. Thus, the connection setup problem in EON needs a suitable route as well as a band of FSs (each one, typically, with a bandwidth of 12.5 GHz), instead of the full-channel capacity (of 50 GHz) in DWDM WRONs. Further, this band could be as small as one FS or a number of FSs, demanding an aggregate speed exceeding 100 Gbps at times. Thus, in EONs one needs to carry out an appropriate routing and spectral assignment (RSA) scheme (Christodoulopoulos *et al.* 2011; Klinkowski and Walkowiak 2011; Capucho and Resendo 2013; Wang *et al.* 2012; Yin *et al.* 2013; Thießen and Çavdar 2014), instead of the RWA algorithms that are used in WRONs. The RSA schemes in EONs need to address some unique issues for connection requests across the network, as discussed in the following.

In order to utilize the available optical spectrum efficiently, an RSA scheme has to ensure spectral contiguity of the FSs chosen for a connection. In other words, all the FSs should be placed contiguously in the optical spectrum (i.e., without any gap in between), so that the band of requested FSs is not fragmented. Note that each fragmented part of a band will need guard bands on either sides, leading to inefficient spectral utilization. The entire band of the FSs chosen for a connection should also be available continuously in all the fiber links in a given route. Thus, the *spectral contiguity* and the *spatial continuity* over the links in the chosen route of the available FSs play a pivotal role in any RSA scheme.

RSA schemes can be executed with a given estimate of the overall traffic matrix, when the network needs to be set up or augmented, and in such cases an RSA turns out to be an *offline problem*. However, as time passes, connections keep departing/arriving in a

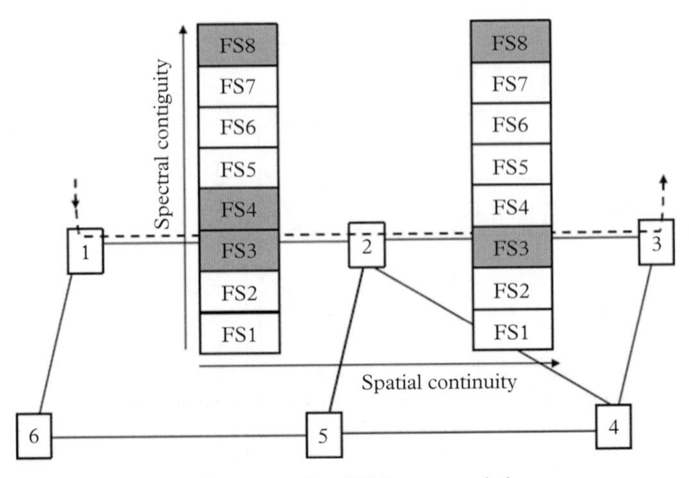

Figure 14.19 *Illustration of the RSA constraints in EON arising from the requirements of spectral contiguity and spatial continuity. Note that the connection request for three FSs from node 1 to node 3 through the route 1-2-3 is going to be successful with both spectral contiguity and spatial continuity for the FSs: FS5, FS6, FS7. However, if the connection request required four slots, the connection would have been blocked.*

dynamic manner, and thus keeping the continuity and contiguity constraints satisfied for subsequent connection requests becomes an *online problem* (Wang *et al.* 2012; Wang and Mukherjee 2012; Wang and Mukherjee 2013; Christodoulopoulos *et al.* 2013; Ba *et al.* 2016; Singh and Jukan 2017). In particular, ongoing arrival and departure of connections with different bandwidth requirements can pose the problem of fragmented spectrum, leaving out scattered *holes* of one or more spectral slots (FSs) over the optical spectrum on various links in the network. The RSA schemes for offline provisioning might also face the problem of spectral fragmentation, which can however be addressed without any time-sensitivity. However, while handling dynamic traffic, the impact of fragmentation becomes more critical, with increase in connection blocking and consequent reduction in spectrum utilization.

Overall, the offline RSA problem can be visualized as a two-dimensional resource-provisioning problem. As illustrated in Fig. 14.19, consider a linear segment of the given six-node network, along which one needs to set up a connection between the nodes 1 and 3 via node 2, with a requirement of three out of eight FSs in the spectrum. As evident from the example, FS5, FS6, and FS7 are the three adjacent (contiguous) FSs that are available in link 1-2 and the same set of slots is also available in link 2-3, thereby satisfying the constraint of spatial continuity as well. Thus, considering the three participating nodes and the two links connecting them, one can conceive a two-dimensional frame, where the vertical dimension with eight slots represents the spectrum availability in each link and the horizontal dimension with two links stands for the spatial allocation of the FSs in the associated links. In the given example, the connection request under consideration happens to find the three available slots (FS5, FS6, FS7) which are free in horizontal (spatial) as well as vertical (spectral) dimensions, thereby allowing the network to set up the connection. When the network fails to find the same set of contiguous FSs in each link along the candidate paths, i.e., in both dimensions, it encounters a connection blocking. We summarize below the design challenges as follows:

- spectral contiguity of FSs over the available transmission spectrum for a given connection,
- spatial continuity of each FS in a connection across all the connecting fiber links,

- assignment of FSs and routes for the connections, so that the network bandwidth (FSs) is utilized maximally.

In the following section, we describe some of the candidate methodologies to carry out the offline RSA in EONs.

14.6 Offline RSA in EONs

As discussed in the foregoing, the offline RSA problem in EONs needs to address several issues while provisioning the network resources. Typically, the resources include a physical topology, the numbers of FSs available in the fiber links and transceivers in each node across the network, and one has to *measure* the cost of the design appropriately for a given traffic matrix. One important cost-measure is the estimate of the total number of FSs used across the network for a given traffic matrix, which the design must minimize. In the following, we consider an ILP-based offline approach that attempts to minimize the usage of FSs, leading to high spectral utilization.

Consider a given EON topology, represented by an N-node graph $G = (V, E)$, with V representing the network nodes (vertices) and E representing the fiber links (edges) of the network. The carrier in every FS (Δf) is assumed to carry a base transmission rate r_B, though one can also use different values for the transmission rate by choosing different levels of modulation. As mentioned earlier, with QPSK modulation for $\Delta f = 12.5$ GHz, r_B will become 25 Gbps for each FS while, by using 16 QAM for each FS, one can increase r_B to 50 Gbps. However, this variability will also affect the optical reach of the intended connections. For simplicity, we don't consider this issue, and assume that the base rate r_B transmitted over an FS (Δf) can span across the given network without any constraint coming from the optical reach. With this assumption (i.e., with fixed r_B for each FS across the entire network), an incoming connection request with a demand for a rate r_i would require $\mu_i = \lceil r_i/r_B \rceil$ carriers over as many FSs that must be contiguous in the spectrum and continuous across the fiber links along the desired route.

Note that, in the MILP problem considered for the long-haul WRONs in Chapter 9, the objective was to minimize the congestion, which implicitly assumed a packet-centric queuing model. However, in Chapter 8 for SONET/SDH-based WDM metro networks, an ILP problem was formulated with the objective of minimizing the number of ADMs, while establishing connections with given capacities following a circuit-switched traffic model. In the present problem, we adopt an ILP-based approach following (Capucho and Resendo 2013), where the objective function is based on the minimization of the cost measured as the bandwidth (number of FSs) usage in the network. For simplicity, we assume that each node is equipped with an adequate number of transceivers, thereby imposing no constraint on the logical nodal degree (number of outgoing/incoming EOPs) in the ILP formulation. If the traffic exceeds the total available capacity, it may not be possible to set up all connections, and the ILP would fail to offer a feasible solution, indicating that the network needs more resources (typically, more FSs in the present problem setting).

In order to formulate the ILP problem for the offline RSA, we assume that the FSs are indexed from the lower end of the optical fiber spectrum (i.e., the FS with the minimum frequency is assigned the minimum index value = 1), and the network has a given physical topology and traffic matrix in terms of the number of FSs between the node pairs. With this framework, we define below the relevant variables and constants that are needed to execute the ILP problem.

ILP variables and constants:

$e(ij)$: a binary variable indicating whether there exists a physical bidirectional link or edge between a node pair (i,j). $e(ij) = 1$, when a physical link exists between the node pairs (i,j), else $e(ij) = 0$.

n_G: number of FSs in a guard band, which is kept free, following the band of assigned slots for each connection (EOP), except the last one in the spectrum. Typically, $n_G = 1$.

n_{sd}: an integer representing the traffic demand in terms of the number of FSs from a node s (source node) to a node d (destination node). Hence, the traffic matrix is given by $\Gamma = [n_{sd}]$, with $s, d = 1, 2, \cdots, N$, and $n_{ii} = 0, \forall i$.

μ_{sd}: an integer representing the modified traffic demand including the guard band n_G, i.e., $\mu_{sd} = n_{sd} + n_G$.

SW: available spectral window on each fiber link, which is an integral multiple of the size of an FS.

M_{FS}: an integer representing the total number of FSs in SW, i.e., $SW = M_{FS}\Delta f$.

M: an integer representing the modified value of M_{FS} to include the effect of n_G, i.e., $M = M_{FS} + n_G$. This is needed to generalize the FS allocation model. However, the Mth FS is a hypothetical FS (never used), but helps in using the same formula for all FS allocations, including the ones those go upto M_{FS}th FS.

m: an integer representing the index of the FSs in SW, varying from 1 to M.

u: an integer representing the index of the first FS of a band of FSs allocated for a connection (EOP). Thus, for a connection from node s to node d, $m = u$ for the first FS and $m = m_{max} = u + \mu_{sd} - 1 + n_G$ for the last FS. With $n_G = 1$, $m_{max} = u + \mu_{sd}$.

f_1: the lowest or the starting carrier frequency of SW in the first FS (for $m = 1$), which spans over the interval $(f_1 \pm \Delta f/2)$.

f_M: the highest or the last carrier frequency of SW in Mth FS, which spans over the interval $(f_M \pm \Delta f/2)$.

ρ, R^k: ρ is one of the k-shortest routes between a node pair in the network; $\rho \in R^k$ with R^k representing the set of all k-shortest routes for all the node pairs in the network.

R^k_{sd}: the set of k-shortest routes between the node pair (s, d), with $R^k_{sd} \in R^k$.

δ^{ij}_ρ: a binary variable, which is 1 if link $e(ij)$ falls on route ρ between node s and node d; else $\delta^{ij}_\rho = 0$.

β^ρ_{sd}: distance traversed by an EOP through the fiber links in route ρ from node s to node d.

$Q^{sd}_{\rho,u}$: a binary occupancy variable, which is 1 when the uth FS is the first FS for a connection from node s to node d on route ρ; else $Q^{sd}_{\rho,u} = 0$.

U^{ij}_m: a binary occupancy variable, which is 1 when the mth slot on link e_{ij} is used, else $U^{ij}_m = 0$.

Using the above variables and constants, next we define the objective function and various constraints for the ILP.

ILP Objective function

In the present problem, the optical spectrum usage across the network (in all links) is considered as the cost Z of the network. Thus, Z is defined as the number of FSs assigned to the connections (EOPs) between all node pairs, with a weighting for the distance covered through the fiber links that each EOP traverses along a given route. Thus, we express Z as

$$Z = \sum_{sd}\sum_{\rho}\sum_{u} \mu_{sd}\beta^{\rho}_{sd}Q^{sd}_{\rho,u}, \tag{14.9}$$

where the summation over (s, d) pairs includes the traffic demands μ_{sd} from all node pairs, the factor β^{ρ}_{sd} takes into account the effect of the distances along the routes chosen (available shortest path) for the various (s, d) pairs, and the summation over u for all $Q^{sd}_{\rho,u}$'s (0 or 1) helps to pick up (in the summation) the appropriate block of FSs along all the constituent links of the chosen route for each EOP. Therefore, the objective function of the ILP can be expressed as

$$\text{minimize}\{Z\} \tag{14.10}$$

ILP constraints:

- *Uniqueness of a route for a node pair:* A given node pair (s, d) (s and d representing the source and destination nodes, respectively) should have only one unique route (no splitting allowed), and an FS, once allocated as the first FS for the connection on the same route, cannot be allocated to any other connection along the same route. In other words, for a given (s, d) pair, only one of the possible $Q^{sd}_{\rho,u}$'s for all ρ's and all u's can be 1, and the rest will be 0, i.e.,

$$\sum_{\rho,u} Q^{sd}_{\rho,u} = 1, \forall s, d. \tag{14.11}$$

- *Selection of the first FS in a route:* The first FS in a connection (i.e., uth FS on a given route ρ from node s to node d) must have a large enough number ($= \mu_{sd}$) of free and contiguous FSs ahead of itself, for all links on the same route (i.e., all permissible values of $m \geq u$ and $m(max) = (u + \mu_{sd}) \leq M$). In other words, the FS farthest from uth FS should not fall beyond the available spectral window SW. With this criterion, and given the definitions of the binary occupancy variables $Q^{sd}_{\rho,u}$ and U^{ij}_{m} with $m \geq u$ and $m(max) = (u + \mu_{sd}) \leq M$), a few cases may arise as follows.
  1. Link l_{ij} falls on route ρ. So with $Q^{sd}_{\rho,u} = 1$, all the assigned FSs following the uth FS on $l_{ij} (\in \rho)$ are occupied and hence U^{ij}_{m}'s are 1 with $u \leq m \leq m_{max} \leq M$. Thus, $Q^{sd}_{\rho,u} = U^{ij}_{m} = 1$. Note that, U^{ij}_{m} is route-agnostic.
  2. Link l_{ij} doesn't fall on route ρ. Yet one can have a situation where $Q^{sd}_{\rho,u} = 1$ as well as $U^{ij}_{m} = 1$, since U^{ij}_{m} is route-agnostic (as observed above). Thus, again in this case $Q^{sd}_{\rho,u} = U^{ij}_{m} = 1$.
  3. Link l_{ij} doesn't fall on route ρ. However, for a given value of u, the uth FS is not allocated for the connection on ρ (i.e., $Q^{sd}_{\rho,u} = 0$), although the mth FS (with m being in the range $[u, u + \mu_{sd}]$) is allocated elsewhere with $U^{ij}_{m} = 1$. Thus, $Q^{sd}_{\rho,u} < U^{ij}_{m}$.
  4. If neither the uth FS is allocated to any route for any node pair, nor is the mth slot allocated in any link across the network, then $Q^{sd}_{\rho,u} = U^{ij}_{m} = 0$.
  5. If $Q^{sd}_{\rho,u} = 1$, then U^{ij}_{m} cannot be zero by definition (and hence must be 1). Thus $Q^{sd}_{\rho,u}$ can never be greater than U^{ij}_{m}.

The above cases can be collectively represented as

$$Q^{sd}_{\rho,u} \leq U^{ij}_{v}, \forall s, d;\ u = 1, \cdots, M - \mu_{sd}, u \leq v \leq (u + \mu_{sd}). \tag{14.12}$$

- *Availability of the requested FSs for a connection:* In order to assign the requested number of FSs (μ_{sd}) for a connection between the node pair s, d on a route ρ passing through a link e_{ij}, one needs to satisfy

$$\sum_{sd} \sum_{\rho} \sum_{u} \delta^{ij}_{\rho} Q^{sd}_{\rho,u} - U^{ij}_{v} \leq 0, \text{ for } u \leq v \leq (u + \mu_{sd}). \tag{14.13}$$

- *Upper limit on the total number of allocated slots in a link:* Given a spectral window $SW = M \times \Delta f$, the total number of slots allocated over a link from all possible connections must be upper-bounded by the maximum number of available slots M, i.e.,

$$\sum_{v} U^{ij}_{v} \leq M, \forall e_{ij} \in E. \tag{14.14}$$

Solving the above ILP formulation using a Simplex algorithm, one can find out the FS allocations for all the node pairs and also the minimized value for the objective function Z. The overall spectrum utilization can be expressed as

$$\eta_{SU} = \frac{Z}{Z_{max}}, \tag{14.15}$$

with Z_{max} as the total number of FSs available on all the fiber links across the network. Z_{max} is estimated as the product of the number of FSs in each fiber link and the total number N_E of fiber links (or edges $\in E$) in the network, i.e.,

$$Z_{max} = MN_E. \tag{14.16}$$

The above ILP-based design can be made more flexible by allowing spectral-splitting on multiple routes for the connection requests that cannot be accommodated in one single route owing to the unavailability of contiguous slots. Such flexibility is expected to offer higher spectral utilization in the network, though needing more transceivers at the network nodes. In such cases, to make the design compliant with the limited resources, one may have to add a constraint from the available number of transceivers in each node, given by

$$\sum_{d} \sum_{\rho} \sum_{m} Q^{sd}_{\rho,m} \leq \Delta, \forall s \tag{14.17}$$

$$\sum_{s} \sum_{\rho} \sum_{m} Q^{sd}_{\rho,m} \leq \Delta, \forall d,$$

along with appropriate modifications in the ILP to include the feature of spectral-splitting.

In order to make the above design impairment-aware, one can also pre-compute the BER (extending the models presented in Chapter 10) for each candidate route and select only those satisfying the upper limit of BER, which might however reduce the spectral utilization across the network. Furthermore, the impairment-aware ILP-formulation can be made more flexible and efficient in respect of bandwidth utilization, by taking into account the interplay between the optical reach and the per-FS transmission rate for each connection. For example, for a given FS, a higher-level modulation scheme (e.g., by using $N_C > 4$ for

N_C-QAM with N_C as the QAM constellation size; note that $N_C = 4$ for QPSK) can be chosen for realizing a higher per-FS transmission rate for the shorter connection distances, which will require fewer FSs as compared to QPSK. On the other hand, the long-haul connections will have to operate with simpler modulation schemes, e.g., QPSK or some N_C-QAM with lower value of N_C, to satisfy the required BER criterion, however requiring larger number of FSs as compared to the higher-level modulation schemes for the same bandwidth demand. This design flexibility is expected to improve the overall bandwidth utilization in the network along with impairment awareness, and can be accommodated in the above ILP-based formulation, by pre-computing and identifying the candidate routes based on their respective BER values for different multi-level modulation schemes.

The offline design can also be carried out using heuristic schemes, when the network size becomes too large, and hence, increases the computational complexity of the ILP problem. In the following, we discuss briefly one such heuristic scheme. In the first step of this scheme, the traffic-matrix elements (if given in terms of Gbps) are converted into the number of required FSs. Next, the traffic demands (numbers of FSs) for different connections are listed in a descending order, and the required FSs are assigned to each traffic element in the same order. Thereafter, determine a set of k shortest routes for each node pair using a shortest-path algorithm (Dijkstra's algorithm, say). With this ordered list, and the sets of the shortest routes, the heuristic algorithm is carried out as follows.

(1) Consider the largest traffic element (number of FSs) from the top of the ordered list and choose the shortest available route.

(2) Assign the FSs using the first-fit algorithm with FS positions indexed from the left (lower frequency end), i.e., fit them into the left-most available contiguous FSs over all the links in the route. If the spectral contiguity and spatial continuity are not feasible, try the next available shortest routes one by one.

   a) If the allocation is successful go to Step 3 (not applicable for the first connection), else go to the next step;

   b) Explore the already-provisioned connection(s) that would block the present connection, and move the same connection(s) to the next shortest path(s), and try to set up the present connection under consideration. If this is not feasible, the present connection request is considered to be blocked.

(3) Remove the top element from the ordered list and go back to Step 1. If the list is empty, the algorithm is deemed to have been completed.

In this scheme one can also include BER-awareness and go for traffic-splitting when the entire band of FSs in a connection request is not available along the fiber links across the network. Having obtained the RSA for a given traffic matrix, one can determine the performance of the scheme by estimating the spectral utilization in the network, blocking due to the limited resources, and spectral fragmentation ratio ρ_{frag} across the network, defined as

$$\rho_{frag} = \sum_{l} \rho_{frag}(l) = \sum_{l} \frac{N_{free}^{T}(l) - N_{free}^{LB}(l)}{N_{free}^{T}(l)}, \tag{14.18}$$

where $\rho_{frag}(l)$ is the fragmentation ratio in a link l, with $N_{free}^{T}(l)$ and $N_{free}^{LB}(l)$ represents the total number of free FSs and the number of FSs in the largest band of free FSs in link l. ρ_{frag} gets minimized in the ILP-based offline design through the objective function, while one needs to check the same for the designs obtained using heuristic schemes (see Exercise 14.9). Moreover, ρ_{frag} would keep changing for an operational network, and one needs to employ an appropriate online RSA scheme to keep the fragmentation ratio within a reasonable limit.

Case study

For a case study, we consider next a six-node EON, as shown in Fig. 14.20, where the numbers beside the fiber links indicate the representative distances in arbitrary units. For this network, the ILP-based design is carried out to examine the salient features of the design with varying traffic scale-up factor α_{SF}, with a starting traffic matrix given in Table 14.1.

Figure 14.21 presents the plots of spectrum utilization η_{SU} versus the traffic scale-up factor α_{SF} for the six-node EON shown in Fig. 14.20, operating with the traffic matrix given in Table 14.1 with the number of transceivers = 5 and four values for $M_{FS} = 40, 60, 80, 100$, and $n_G = 1$. Note that, with $\Delta < 5$, the ILP won't be able to produce a feasible solution, while with larger values for Δ one can explore traffic-splitting, whenever needed. The plots for $M_{FS} = 40$ and 60 indicate that, with the increase in α_{SF} for each M_{FS}, η_{SU} exhibits proportionate increase, and thereafter *stops* at some maximum value corresponding to the maximum allowable values of α_{SF}, 3 and 4.5, respectively. In other words, beyond these values of α_{SF}, ILP doesn't return any feasible solution due to the paucity of FSs, and hence $M_{FS} = 60$ allows higher α_{SU} as compared to $M_{FS} = 40$. With the next higher values of M_{FS} (i.e., with $M_{FS} = 80$ and 100), η_{SU} saturates beyond $\alpha_{SF} = 6$ and hence are not captured in the given plots. Thus, in order to accommodate higher α_{SF}, one can run the ILP with larger values of M_{FS} and assess the required number of FSs to design the network for a projected increase (scale-up factor) in traffic over a given time frame.

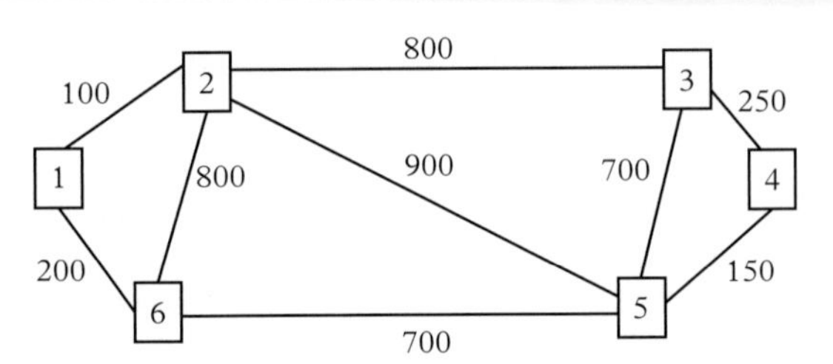

Figure 14.20 *Six-node EON for the case study with the distances of the links expressed in kilometers.*

Table 14.1 *Traffic matrix for the six-node EON of Fig.14.19.*

Nodes	Node 1	Node 2	Node 3	Node 4	Node 5	Node 6
Node 1	0	2	6	4	3	2
Node 2	4	0	7	2	6	4
Node 3	5	5	0	3	7	2
Node 4	2	4	3	0	5	7
Node 5	3	5	2	4	0	2
Node 6	2	6	5	7	3	0

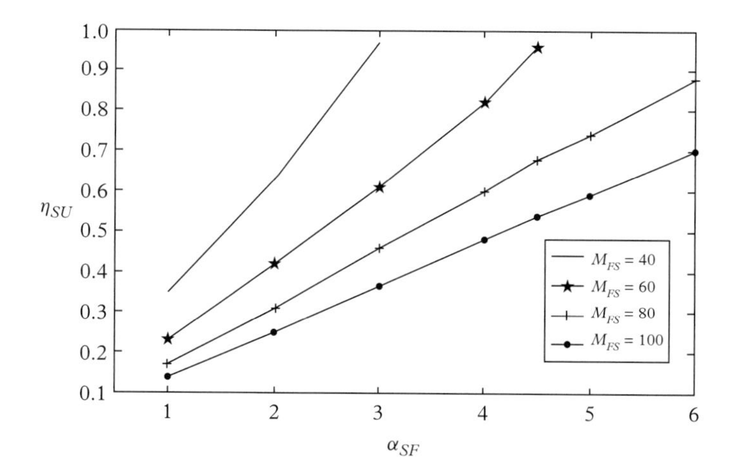

Figure 14.21 *Plots of spectrum utilization η_{SU} versus traffic scale-up factor α_{SF} for the network in Fig.14.20, using the traffic matrix of the network given in Table 14.1. $M_{FS} = 40, 60, 80, 100$, and $n_G = 1$.*

14.7 Online RSA in EONs

Operational EONs need to employ an online RSA algorithm for time-varying traffic in a dynamic manner. As described above, for offline RSA, the problem of spectral fragmentation is addressed once for all by allocating appropriate sets of FSs for each connection using an optimum or a nearly optimum (heuristic) scheme. However, having started with an offline provisioning, an operational network keeps going through dynamic changes in the traffic pattern, which in turn creates *spectral holes* in the fiber links, leading to the spectral fragmentation, and makes it difficult to satisfy the contiguity and continuity criteria for all connection (i.e., EOP) requests across the network. Thus, while setting up an EOP, the required number of FSs may be available in fragmented parts of the optical spectrum along a given route, which needs to be avoided as much as possible by using appropriate online RSA schemes (Wang *et al.* 2012; Wang and Mukherjee 2012; Wang and Mukherjee 2013; Christodoulopoulos *et al.* 2013; Ba *et al.* 2016; Singh and Jukan 2017). These schemes can be categorized in two broad types:

- proactive defragmentation,
- reactive defragmentation.

In proactive schemes, the network employs some general network-wide policy that minimizes the probability of fragmentation of the available optical spectrum. In reactive schemes, appropriate actions are taken as and when a requested connection is unable to find a contiguous set of FSs over the desired route. We describe some of these schemes in the following.

14.7.1 Proactive defragmentation

In proactive defragmentation schemes, the network usually goes by some centralized spectral allocation policy in advance or keeps on making some periodic spectral adjustments, so that when a new connection request arrives, the possibility of accessing the required contiguous/continuous band of FSs is enhanced. We discuss below two such schemes: spectral partitioning and periodic spectral cleaning.

Spectral partitioning

One of the candidate proactive defragmentation schemes is realized by *partitioning* the optical spectrum for large and small bandwidth requirements. For example, an incoming connection request can be allocated the requested number of FSs from the lower end of the optical spectrum (i.e., first-fit), if the number of FSs is below a specified threshold. Or else, the requested number of FSs can be allocated from the higher end of the optical spectrum (i.e., last-fit). The partitioning scheme, called as soft partitioning scheme, is illustrated in Fig. 14.22 with a threshold value for the number of FSs, N_{th}, corresponding to the threshold bandwidth demand of $N_{th}\Delta$. The value of N_{th}, for deciding a connection request in favor of the first-fit or the last-fit FS, can be decided through prior simulation (as a dry run) for an expected traffic pattern for different values of N_{th}. Using the results of simulation, one can choose the optimum value for N_{th}, for which the probability of connection blocking is minimized. Notwithstanding this value of N_{th} determined from the dry run, FS allocation will continue from both ends of the spectrum towards making the maximum possible utilization of the optical spectrum, without considering $N_{th}\Delta$ as a hard partition. However, the connection requests with higher bandwidth can also be allocated the FSs in the lower end of the spectrum and the reverse for the connection requests with lower bandwidth, depending the dispersion characteristics of the fiber links and the EOP lengths.

One can also rigidly divide the entire spectrum into multiple segments (two or more) for different ranges of bandwidth requirements, and the incoming connection requests can be allocated FSs in the respective spectral segments as governed by the bandwidth requests on a first-fit or last-fit basis. Figure 14.23 illustrates one such scheme called as rigid partitioning scheme, where the optical spectrum is divided into two parts (binary partitioning) at an optical frequency $f = K\Delta$, and the connection requests with bandwidth demands $\leq K\Delta$ are allocated a set of first-fit (or last-fit) FSs from the lower segment, while the rest of the connections are similarly placed on the higher spectral segment. In this case also, the lower-

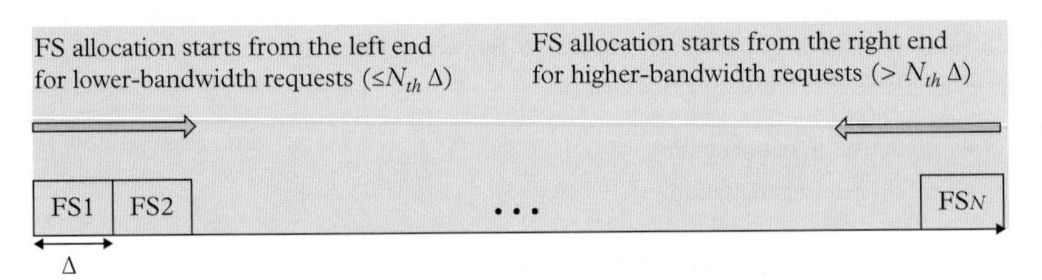

Figure 14.22 *Illustration of the centralized defragmentation scheme with soft spectral partitioning.*

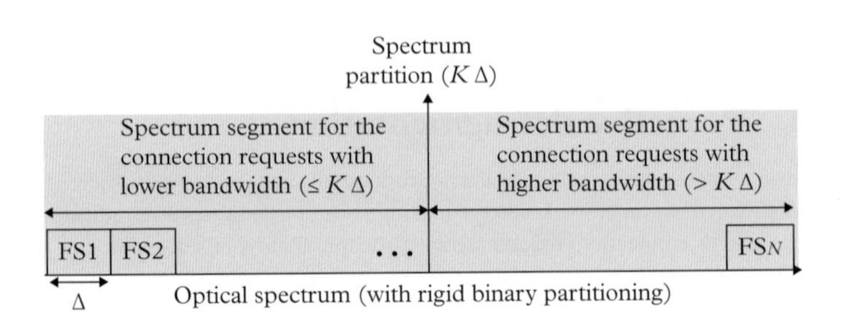

Figure 14.23 *Illustration of the centralized defragmentation scheme with rigid (binary) spectral partitioning. One can use more partitions depending on the traffic characteristics.*

and higher-bandwidth requests can be swapped to higher and lower segments, depending on the fiber dispersion characteristics and the EOP lengths.

Periodic spectral cleaning

Another proactive scheme employs periodic adjustments of the spectral allocations for the ongoing EOPs in a *hitless* manner (i.e., without any disruption), so that the spectrum occupancy in the network links *moves* toward one or both end(s) of the spectrum without crossing over any occupied FS in any fiber link across the network. Note that the scattered occupancies of the FSs with holes on both sides in the optical spectrum prevent the availability of a long contiguous stretch of FSs. Hence, such FSs can be shifted *on the fly* towards the two ends of the spectrum, thereby *cleaning* and consolidating the optical spectrum with more availability for the new EOP requests. Following such a cleaning operation, whenever a new EOP request arrives at a node, it has a fair chance of finding the needed contiguous stretch of FSs along the desired route. One such hitless periodic spectral-cleaning scheme is illustrated in Fig. 14.24.

However, the periodic cleaning operation should not be carried out too frequently, as it can impact the connections going through spectral shifting owing to the retuning of transmitters, receivers, and all the associated network elements at the intermediate nodes. In particular, such shifting or retuning operations might take a *non-negligible* time as compared to the durations of the ongoing connections, and during this time interval the signal received at the destination node will incur losses due to mistuning. The shifting time required for a given FS will increase with the amount of spectral shift (i.e., $k\Delta$, with k as a non-zero integer) needed for the spectral consolidation. Further, the task of consolidating the spectrum on one link by shifting the FSs may not always be doable for the other links, through which the same EOPs would pass, as the FSs in the other links might not be moveable without disruptions.

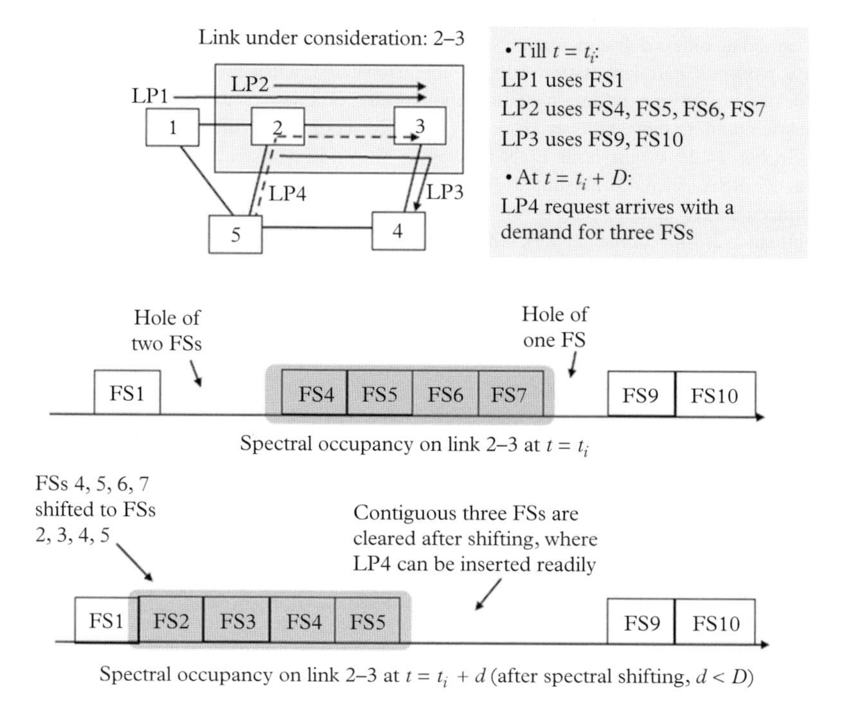

Figure 14.24 *Illustration of periodic spectral cleaning using hitless shifting by retuning of the devices involved in the existing EOPs. For simplicity, we assume that the guard band $n_G = 0$ in this and in all subsequent illustrations of the online defragmentation schemes.*

In order to reduce the time for retuning during spectral shifting, shifting all the FSs to one specific end of spectrum might take a longer time in some cases, resulting in an increase in the disruption time. However, the disruption time can be minimized by shifting the FSs selectively to appropriate sides. For example, in Fig. 14.24 by shifting FS4 through FS7 to the right side (being nearer to FS9) will reduce the disruption time in this case. In general, the FSs to be shifted can be partly shifted to the left end and the remaining FSs can be shifted to the right end simultaneously. In some cases, this parallelism would lead to the minimization of the maximum retuning times of all the FSs, thereby minimizing the disruption time due to the retuning process. However, in practice, the shifting operation may impact the existing EOPs during the retuning process, more so when the FSs are to be moved over a large frequency range. In order to minimize this effect, before starting the shifting operation for the FSs of the existing EOP, a stand-by connection can be made by using additional FSs (if available) for the existing EOP. This leads to a make-before-break (MbB) arrangement for the old EOPs, where the additional FSs are used to *make* the connection using new FSs, followed by the *breaking* of the transmission over the original FSs.

14.7.2 Reactive defragmentation

The reactive defragmentation schemes also need to be hitless in nature, so that reactive actions do not disrupt the existing connections, i.e., the FSs while being moved to accommodate new connections do not cross over any other FS in use. Hence, these schemes are also called non-disruptive schemes, while the non-hitless schemes would disrupt the ongoing connections and are not preferred in EONs. However, note that, even in hitless cases, there might be some impairments arising from the finite tuning times of all the optical devices participating in an EOP. We discuss some of the hitless reactive schemes in the following.

Hitless spectral adjustment

A candidate hitless defragmentation scheme is illustrated in Fig. 14.25, which is also known as a *push-pull* defragmentation scheme. In this scheme, when faced with a connection blocking due to spectral fragmentation, the blocking of a new connection can be avoided by attempting to move hitlessly an already allocated set of FSs on the fly.

In Fig. 14.25, a nine-node EON topology is shown in part(a), using which the hitless reactive RSA scheme is illustrated. As shown in Fig. 14.25(b), at time t_1, four all-optical connections are already set up between the node pairs 1-2, 1-3, 3-4, and 4-5 along the linear segment 1-2-3-4-5 of the network following shortest-hop-count criterion. The connections have the following FS allocations:

- node pair 1-2, one FS (FS1),
- node pair 1-3, two FSs (FS2 and FS3),
- node pair 3-5, one FS (FS1),
- node pair 4-5, one FS (FS2).

With these connection settings, at time $t = t_2$ ($> t_1$) the connection 1-2 departs (Fig. 14.25(c)). Thereafter, at $t = t_3$ ($> t_2$) there arrives a new request for a connection between the node pair 1-5 for one FS to be set up on the same linear segment, providing the shortest hop-count. Given the earlier settings, the connection can only be set up on FS4 without interfering with other connections. However, for this purpose, the connections 3-5 and 4-5 are moved up to the higher FSs, and thus FS1 becomes free for the new connection 1-5. This movement in spectral domain (retuning of the transceivers and the intermediate

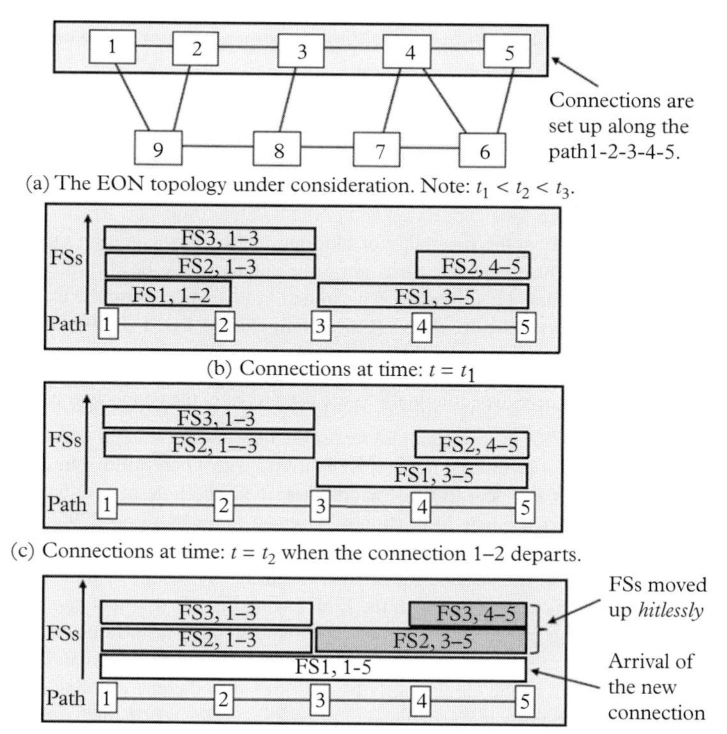

Figure 14.25 *Illustration of hitless online RSA, where all the connections in the example EON are set up over the linear segment 1-2-3-4-5 on the basis of shortest-path criterion, and the connection 1-3 uses two FSs while the rest of the connections use only one FS. Note that, due to hitless movement of the FSs for the connections 3-5 and 4-5, the new connection 1-5 doesn't have to use any additional FS. Without the upward movement of the FSs for the connections 3-5 and 4-5 carried out in this scheme, the connection 1-5 would require FS4, leading to inefficient bandwidth utilization.*

network elements) becomes hitless for other connections because when the FSs are moved up, they *do not clash or interfere* with any other connection, as shown in Fig. 14.25(d). As discussed earlier in the cases of other proactive schemes, in this scheme also, the RSA mechanism might have to make use of MbB operation to avoid the impact of retuning on the ongoing EOPs.

A useful algorithmic approach for the reactive hitless spectral adjustment of the occupied FSs has been proposed by (Singh and Jukan 2017), where an optimum solution is reached by constructing an *auxiliary graph* with each EOP representing a node in the same graph. Much like the auxiliary graph of the graph-coloring algorithm used for the wavelength assignment in WRONs (Chapter 9), the nodes in the present auxiliary graph are connected by an edge if they (the existing and new EOPs) have any spectral overlap on the common route during the residual holding times of the existing EOPs and the ensuing holding time of the new EOP request. Using the auxiliary graph and the concerned nodes therein, a possible retuning of the existing EOPs is attempted by using a *maximum independent set* (MIS) algorithm to accommodate the new EOP request, and the new requests that fail to find a solution from this attempt are blocked. However, in order to implement this algorithm, one may again have to take resort to the MbB scheme during the retuning process for minimizing connection disruptions.

Hitless spectral adjustment with holding-time-aware reordering of FSs

Adding the awareness of holding times of the connections in progress to the reactive schemes can alleviate the difficulties faced during the retuning process. While adjusting the spectrum occupancy in a link, if an EOP that has to leave the system after a long holding time is given a band of FSs in the middle of the spectrum (thereby leaving spectral holes on both sides), then setting up a new EOP with large bandwidth-demand becomes difficult. However, if the ongoing long-holding-time connection had been allocated toward the leftmost or rightmost available bands of FSs, the possibility of blocking for the forthcoming EOPs could have been minimized. This effectively turns out to be similar to the head-of-the-line blocking experienced in queues. This situation is illustrated in Fig. 14.26 using the plots of residual holding time (τ_i) versus the FS indices (FS_k) for the existing EOPs (EOPk). In the example case, we assume that there are 10 FSs available in the optical spectrum.

Fig. 14.26(a) shows the case where FSs are allocated to the requested EOPs without any awareness of the holding times of the EOPs. Hence, as the earliest-to-depart EOP (i.e., EOP2 using FS2, in this example) leaves the network, it creates a one-FS hole in the middle which cannot be utilized if a new EOP arrives with a bandwidth demand exceeding one FS. Further, if the new EOP demands three FSs, the hole at FS2 will continue to remain unutilized, as the new EOP cannot be set up either by using FS9 and FS10 (two contiguous slots only), or by using FS2 (one slot only). However, in Fig. 14.25(b) the FSs are shown to have been allocated to the EOPs with the knowledge of their holding times, and their allocations are ordered with the EOP having the shortest holding-time (EOP2 in this example) being placed on the rightmost FS and so on. This arrangement will ensure that the departure of the very next EOP (EOP2 in this case) will lead to an increase in the available bandwidth to the contiguously placed three FSs, thereby enabling the setting up of the new connection (with the demand of three FSs), which was denied in the earlier case

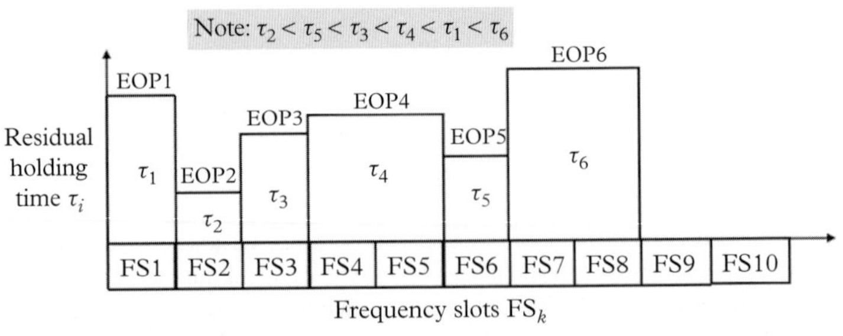

(a) τ_i vs. FSs, with the FSs allocated without any awareness of τ_i's.

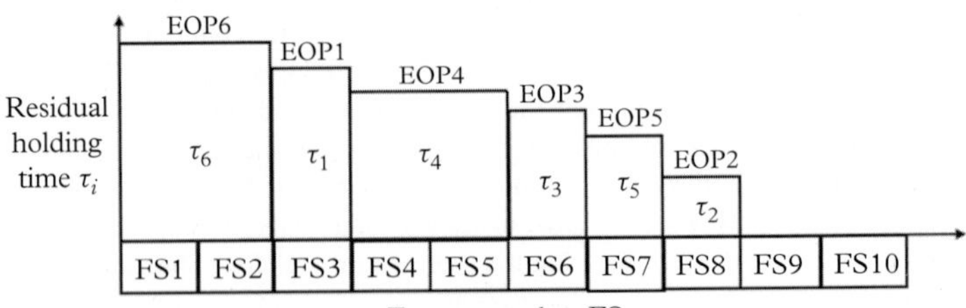

(b) τ_i vs. FSs, wherein the FS indices increase with decreasing τ_i values, following the holding-time-aware FS allocation policy.

Figure 14.26 *Illustration of holding-time-aware FS allocation for avoiding spectral fragmentation in online RSA.*

(Fig. 14.25(a)). Thus, we observe that an orderly placement of the FSs of the existing and new EOPs with the awareness of their holding times can enhance the accessible bandwidth leading to reduced blocking for the incoming EOP requests. The major issue in carrying out the orderly placement of the FSs in this scheme lies in the fact that such spectral adjustments, being in the need to retune the FS allocations on the fly when a new EOP request comes, might cease to remain hitless in some situations and should therefore be avoided in such instances.

14.8 Summary

In this chapter, we presented the ongoing developments in the area of bandwidth-efficient EONs. First, we presented the basic concept of flexible frequency grid used in EONs, explaining how by using a flexible grid with smaller FSs one can realize higher bandwidth utilization in optical WDM networks. Next, we considered various multicarrier modulation techniques: OOFDM, Nyquist-WDM, and OAWG. Before describing the various schemes to implement OOFDM, we presented the basics of electrical OFDM scheme, which was subsequently adopted in the various OOFDM schemes: IMDD-OOFDM, UC-OOFDM, AO-OOFDM, and EO-OOFDM. Further, we described the multicarrier modulation and demodulation schemes based on the principles of Nyquist-WDM and OAWG. Special types of bandwidth-variable devices needed for setting up connections across an EON, such as BVT, S-BVT, BV-WSS, BV-ROADM, and BV-OXC, were described along with the discussions on their special features.

Next we discussed the design issues in EONs, where each end-to-end connection across the network would require spectral contiguity and spatial continuity of the FSs in the fiber links along the route chosen for the connection. Driven by these constraints, the offline design methodologies for EONs were presented using ILP and heuristic schemes leading to the offline RSA for a given traffic matrix, and possible improvisation of the designs were discussed using cross-layer considerations. The online RSA schemes were presented for the two broad categories: proactive and reactive schemes. Two proactive online RSA schemes were presented, using the notions of soft and rigid partitioning of the available optical spectrum in the network. Finally, some reactive hitless online RSA schemes were presented, with some of them being based on the spectral adjustment of the existing EOPs to accommodate a new EOP, and the other scheme using an additional spectral adjustment mechanism by re-ordering the FSs of the existing EOPs, based on the holding times of the existing EOPs and the new EOP request.

EXERCISES

(14.1) Consider the versions of OOFDM transceivers that need to employ coherent optical transmission. Justify the need of coherent transmission in the respective transceivers and identify the components that would need special care to ensure coherent operation of the transceivers.

(14.2) Consider an all-optical OOFDM receiver (Fig. 14.7) to be used in EON. Indicate the roles of the components that are used to carry out the task of signal processing in the optical domain for OFDM demodulation.

(14.3) You are given four transmission systems for optical WDM links using a 50 GHz transmission grid, which have to operate at 10 Gbps, 40 Gbps, 100 Gbps, and 400 Gbps. Grade them as Nyquist, quasi-Nyquist, super-Nyquist, or non-Nyquist systems and justify your statement using the basic criterion of Nyquist transmission.

(14.4) In a given EON, five connection requests have been made with the speeds $r_1 = 10$ Gbps, $r_2 = 20$ Gbps, $r_3 = 40$ Gbps, $r_4 = 100$ Gbps, and $r_5 = 400$ Gbps. Each FS has the width of 12.5 GHz and the width of the guard bands is the same as one FS. Further, BPSK modulation is used for each of the FSs, which are combined using OFDM when multiple FSs are to be transmitted. Estimate the number of FSs needed for the five connection requests.

(14.5) Consider the above exercise with QPSK-OFDM transmission and estimate the number of FSs needed for the five connection requests.

(14.6) Consider the four-node EON, as shown in Fig. 14.27, with the link lengths indicated beside the respective links in arbitrary units. Assume that the width of each FS is 12.5 GHz, and that each connection uses a guard band of one FS, appended at the end of its allocated FS(s). The network needs to set up all connections as end-to-end all-optical lightpaths. Carry out an offline RSA for the EON using the heuristic scheme described in 14.6, where each link can carry at the most eight FSs in each direction. The traffic matrix Γ in terms of transmission rates (in Gbps) is given by:

$$\Gamma = \begin{bmatrix} 0 & 50 & 25 & 60 \\ 25 & 0 & 50 & 60 \\ 25 & 30 & 0 & 30 \\ 25 & 50 & 30 & 0 \end{bmatrix}.$$

From the spectral assignment determine the spectral utilization in the network and the probability of blocking due to resource exhaustion.

(14.7) Scale up the traffic in the above exercise by a factor of 1.5 and carry out the offline RSA again. Compare the results obtained in the two cases.

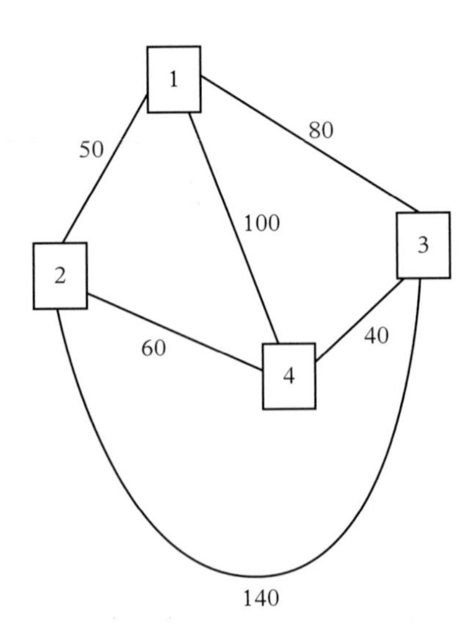

Figure 14.27 *Four-node EON physical topology for Exercise 14.7.*

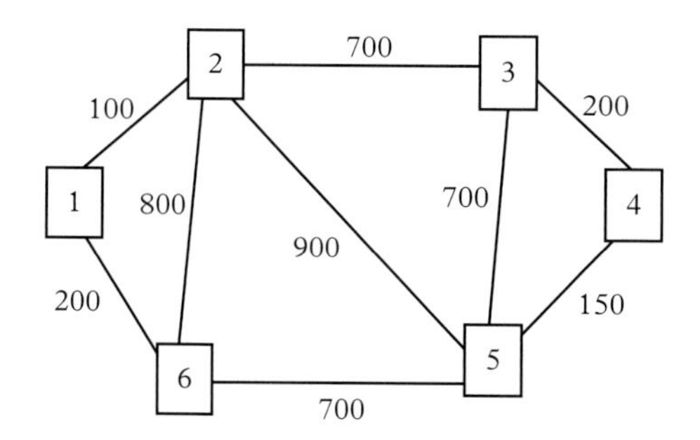

Figure 14.28 *Six-node EON physical topology for Exercise 14.9.*

(14.8) Consider the six-node EON shown in Fig. 14.28 with the following connection requests.

Node pair	FSs requested	Selected path
1, 2	2	$1 \rightarrow 2$
1, 3	2	$1 \rightarrow 2 \rightarrow 3$
4, 6	2	$4 \rightarrow 5 \rightarrow 6$
2, 4	1	$2 \rightarrow 3 \rightarrow 4$
1, 4	1	$1 \rightarrow 2 \rightarrow 3 \rightarrow 4$

Each connection in the EON uses one FS as the guard band, appended after the requested FS(s). The total number of FSs available in each link (bidirectional) is 10.

a) Carry out the spectral allocation and illustrate the same on the network topology.

b) Calculate the spectral fragmentation for each fiber link and the average spectral fragmentation across the entire network.

Optical Packet and Burst-Switched Networks

15

In order to improve bandwidth utilization in circuit-switched WRONs, various techniques for optical packet-switching (OPS) have been studied, although needing complex technologies, such as header extraction, processing and insertion, packet alignment for optical switching, etc. An intermediate solution between the WRONs and OPS networks – optical burst-switched (OBS) network – has been explored, where several packets are clubbed together to form optical bursts, which are transmitted with the headers sent as control packets ahead of the respective bursts. In OBS networks, one can get around the challenges of OPS networks, while improving the bandwidth utilization as compared to the WRONs. In this chapter, we first present the node architectures, followed by the header-processing schemes and switch designs for the OPS networks. Next, we present the basic concepts of OBS networking and describe the necessary network protocols, including the burst assembly scheme, just-enough time signaling, resource-reservation, and routing schemes.

15.1 Bandwidth granularity in WDM networks

WDM networks using wavelength-routing, i.e., WRONs, as discussed in Chapters 8 and 9, use optical circuit-switching (OCS) with long holding times, typically ranging from days to months. However, these high-speed lightpath-based connections often go underutilized, resulting in poor bandwidth-utilization efficiency. In other words, the bandwidth granularity of the entire network turns out to be rather coarse, while using high-speed lightpaths operating at bit rates ranging from a few Gbps to 100 Gbps. This limitation of WRONs necessitated thorough studies on whether the switching technology in WDM networks could perform the electrical *store-and-forward* switching mechanism of IP networks in the optical domain itself.

As described in Chapter 8, for metro networks the packet-switched operation was explored in the optical domain over WDM rings; however, such networks are not scalable enough for long-haul optical mesh topologies. Later on, as discussed in Chapter 14, one practical remedy to this problem evolved through the development of EON technologies using flexible frequency grid. However, long before the studies on EONs started, the research community in the area of optical networks explored the feasibility of optical packet-switching (OPS) on WDM networks, attempting IP-like packet-switching in the optical domain itself. Subsequently, the research on OPS networks was followed up by some useful studies on a novel and less-complex variant of the OPS networks, known as optical burst-switched (OBS) networks.

Though the technology for OPS functionalities in a node seemed highly challenging, various research groups moved on to develop novel OPS-based node architectures for

Optical Networks. Debasish Datta, Oxford University Press (2021). © Debasish Datta.
DOI: 10.1093/oso/9780198834229.003.0015

WDM networks (Guillemot *et al.* 1998; Carena *et al.* 1998; Papadimitriou *et al.* 2003; Pattavina 2005; Xu *et al.* 2001). The main challenges in developing OPS networks were faced while attempting to *mimic* the store-and-forward mechanism of IP networks during the passage of optical packets through the OPS nodes. In particular, each OPS node needed to receive, hold the incoming optical packets in the buffers for synchronization, perform wavelength conversion, and extract and reinsert their headers, all to be carried out on the fly in the optical domain. To get around these challenges of the OPS nodes, several studies were carried out toward realizing the OBS networks, wherein a number of optical packets were combined together at the source node to form optical bursts with longer duration. The optical bursts formed at the source nodes were transmitted with their headers sent as control packets ahead of the actual bursts carrying the payload. In effect, the OBS networks promised an intermediate solution between the packet-switched OPS networks with the finest bandwidth granularity and the OCS-based WRONs with the coarse granularity of full-capacity wavelengths (Qiao and Yoo 1999; Battestilli and Perros 2003; Verma *et al.* 2000). In this chapter, we describe the basic architectures and operational features of the OPS and OBS networks.

15.2 OPS networks

In order to carry out the packet-forwarding operation in an OPS node, an optical switch would play a central role, where the switching has to be carried out in a synchronous manner, such that all the packets at the input ports can be moved simultaneously to the desired output ports. For the synchronous switching, all the packets transmitted across the network should have the same duration, equaling a specific slot interval, along with some guard times before/after the packets within each slot. Further, before carrying out switching, the input packets at each node have to be aligned in time using variable fiber delay lines (FDLs) to get ready for the synchronous switching operation. To make the operation of the nodes compatible with the IP networks with variable packet durations, asynchronous node architectures have also been explored. In this class of OPS nodes, the variable-size packets are divided into a number of fixed-duration slots (with padding if necessary in the last slot). In such nodes, the terminal behavior would appear to be asynchronous, while the internal switching mechanism of the node is to be carried out synchronously with aligned time slots.

Besides the switching operation, feasibility of the OPS nodes will depend greatly on the successful accomplishment of various other challenging tasks, such as synchronization, extraction and re-insertion/insertion of the headers from/into the packets optically, and dynamic wavelength conversion on the packets. We describe below the various possible architectures of OPS nodes and the challenges faced therein.

15.2.1 OPS node architecture: generic form

A generic form of OPS nodes is shown in Fig. 15.1. The $N \times N$ node, operating with M wavelengths, uses three cascaded stages between an input column of demultiplexers and an output column of multiplexers, for carrying out synchronous contention-free optical switching of the incoming packets, along with the local add-drop operation.

The first stage of the OPS node (following the demultiplexer column) taps out signals for the header processing and control unit (HPCU). The HPCU establishes alignment of the received packets (for synchronized switching) using FDLs, and control the tunable wavelength converters (TWCs) in the first stage and the optical switch in the second stage, based on the extracted header information. The HPCU-controlled TWCs ensure that the

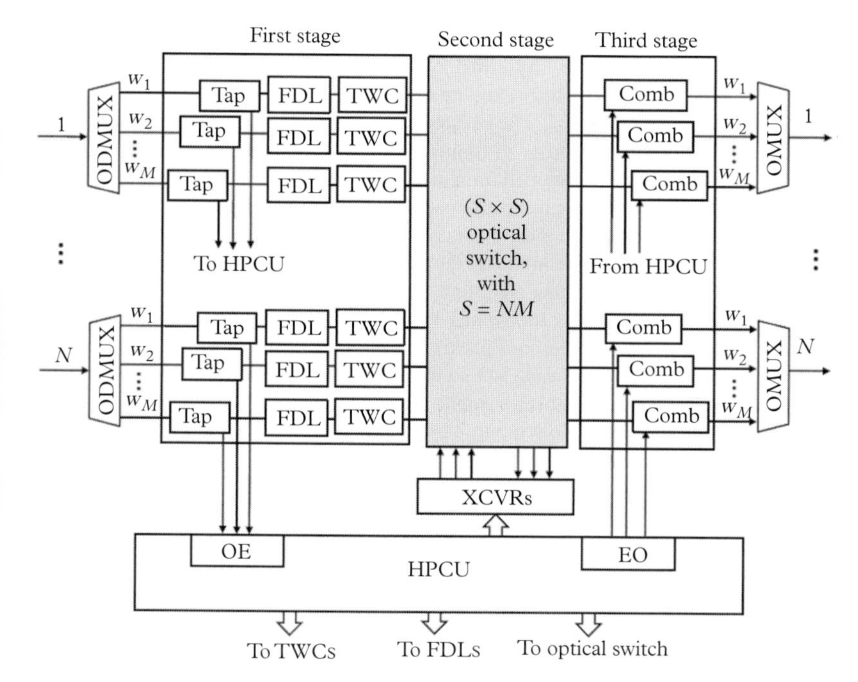

Figure 15.1 *$N \times N$ OPS node architecture: a generic form. OMUX: optical multiplexer, ODMUX: optical demultiplexer, Comb: optical combiner, XCVR: optical transceiver, FDL: fiber delay line, TWC: tunable wavelength converter, HPCU: header processing and control unit.*

incoming packets assume suitable wavelengths to avoid contentions in the switch. For the second stage, the HPCU configures the $S \times S$ optical switch ($S = NM$) during the right time slot, such that the time-aligned packets with appropriate wavelengths are switched in *one go* by the optical switch fabric to the desired output ports. In the third stage, the outputs of the optical switch are combined with the modified headers from the HPCU and are finally passed through the multiplexers to reach the output ports of the node with appropriate time slots and wavelengths. Headers extracted from the first stage are processed at the HPCU in the electrical domain through EO conversion of the tapped signals, and in turn the HPCU manages the node control functionalities for various subsystems operating in all the three stages as well as the local transceivers (XCVRs) for add-drop operation. The old headers of the pass-through packets are stripped off in the TWCs using gating functionality of the SOAs used therein.

As shown in Fig. 15.1, each input port of the $N \times N$ node receives M wavelengths, which are demultiplexed and passed on to M optical taps, to extract the header of each packet and forward the payload to an FDL. Each optical tap typically employs a 10:90 power-splitting ratio with 10% power going to the HPCU, while the rest of the power (i.e., 90%) is forwarded toward the optical switch through the FDL and TWC. The tapped (10%) optical power for the HPCU is first converted into the electrical domain through OE conversion, and thereafter passed on for further processing. The remaining 90% optical power of the received packet coming out from the tap first goes through an FDL, where the packet is delayed for packet alignment (needed for synchronized switching) as well as for the time spent during the header processing in the HPCU. The delayed packets are then passed through the TWCs for wavelength conversion (with input from the HPCU), if necessary, which ensures that the received packets are carried by suitable wavelengths and reach the desired output ports of the optical switch, without any contention. As mentioned earlier, the TWCs are also used to erase the old headers of all the pass-through packets by switching off the SOAs in the TWCs during the appropriate time intervals, such that the updated headers can be reinserted at the third stage in the passthrough packets. The packets addressed to the node under consideration (i.e., the local node) are dropped by the optical switch toward the receivers in the XCVRs, while for the packets to be transmitted from the XCVRs, fresh headers are generated by the HPCU and added later with the respective packets in the third stage. We discuss below the header generation and insertion/reinsertion process in further details.

Following the EO conversion in the HPCU, the header of each incoming packet is used, along with an appropriate routing algorithm, to determine the output port of the optical switch through which the packet should be forwarded as a pass-through packet or dropped locally. Further, for the pass-through packets, at the HPCU the header is changed and converted into optical domain using EO conversion and combined with the payload (i.e., without the old header, that has been stripped off at the respective TWC) coming out from the optical switch at the combiner input. The packets with the received payloads and the changed headers are next forwarded from the combiners to the respective output ports of the node. The locally added packets through the optical switch from the transmitters (in XCVRs) are also combined with the respective headers through the same process as used for the passthrough packets, albeit with fresh headers. The optical switch keeps getting configured/reconfigured by HPCU, as governed by the received header bytes of the incoming packets as well as by the locally added packets.

Overall, the OPS networks function using appropriate protocols based on IP-over-WDM architecture, with hop-by-hop forwarding of packets. However, the existing IP routing protocols will need some modifications for the segmentation of the IP packets into fixed-duration slots at the edge nodes. Further, in order to avoid contentions in FDLs, the OPS nodes can also adopt deflection routing (i.e., routed to a switch output port, different from the desired one), reducing thereby the packet losses, but at the cost of increased delay. In the following, we describe in further detail the important functional blocks used in OPS nodes and the challenges therein.

15.2.2 Header processing, TWC, and FDL units

One of the critical design issues in the OPS nodes (Fig. 15.1) is the successful extraction of headers from the incoming packets and reinsertion of modified headers on the passthrough packets with precise timing. Further, the FDLs should be able to change dynamically for synchronous switching and the TWCs need to employ appropriate dynamic wavelength-conversion technology for each input port of the node, so that the incoming packets at various input ports, contending for the same output of the switch, do not spectrally interfere with each other. Furthermore, FDLs can also be used as the first-in first-out (FIFO)-based buffers placed typically inside the switching stages and/or at the switch output for avoiding contentions. We discuss below the various design issues of these subsystems used in the OPS nodes in further detail.

Header processing

Header processing plays a significant role in the OPS nodes for routing the incoming packets to appropriate destination nodes, following which the other devices in the node start carrying out the respective tasks. In the first step of header processing, the node needs to extract the header from the received optical packet at each port of the node. For successful header extraction, the HPCU needs a minimum amount of tapped optical power to detect the packet header, with a BER not exceeding a specified upper limit. This in turn reduces the power for the original packet moving toward FDL. Note that this tapping loss for header extraction takes place at each node along the chosen route, which in turn progressively degrades the signal quality of the packet received at the destination node. The optical amplifiers used at each node to compensate for this loss (and other losses in a node) add ASE noise, reducing the optical SNR. Thus the tapping power for the header extraction at each node cannot be increased much (typically kept within 10%) without making the payload of a packet vulnerable to the bit errors at the destination node.

The conflicting needs of optical power levels between the header and payload can be alleviated by using a lower transmission rate for the headers, as a header uses much fewer bits as compared to the payload. This strategy will in turn reduce the receiver bandwidth of the OE unit in the HPCU block, leading to lower receiver noise. The reduction of receiver noise would need lower received optical power for achieving the same SNR that would be needed to ensure the specified BER. This arrangement would consequently demand less power-tapping at the optical taps for header extraction, thereby leaving a larger share of power for the payload.

For the header transmission, beside combining the header and payload one after another in time domain, one can also use electrical subcarrier modulation for the header bits, while keeping the payload unmodulated in electrical domain (i.e., as baseband signal). In other words, the modulated (by header bits) subcarrier waveform is multiplexed with the baseband payload signal electrically, leading to subcarrier multiplexing (SCM) in the electrical domain, where the subcarrier frequency f_{SC} is chosen to be higher than the highest frequency ($\approx r$, the per-channel bit rate) of the baseband spectrum of the payload. The electrically multiplexed subcarrier and the baseband payload, occupying non-overlapping electrical spectrum (with $f_{SC} > r$), are used to modulate the laser intensity, thereby leading to a comprehensive intensity-modulated (IM) optical signal with the subcarrier component of the modulating signal carrying the header and the baseband counterpart carrying the payload. Since the header and payload remain spectrally non-overlapping and hence transmitted concurrently through SCM, the overall packet transmission time is reduced. However, the spectral contents from the modulated subcarrier (lying beyond the spectral region occupied by the payload, but overlapped in time domain) need to be removed before the header reinsertion, which can be accomplished at the TWC stage, by optically filtering out the subcarrier spectrum (sidebands) prior to carrying out the wavelength conversion.

Regarding the header-insertion (for the locally added packets) or the header-reinsertion (for the pass-through packets) process, precise timing of insertion is extremely important, as the header inserted into the outgoing packet should not temporally overlap with the payload. This problem is addressed by keeping a safe time-gap between the header and the payload. For the SCM-enabled header extraction/reinsertion process, this problem is nonexistent, as the header and the payload would occupy non-overlapping electronic spectra and thus can fully or partially overlap in the time domain. Note that, a similar SCM-based header transmission scheme was also used in HORNET, a packet-switched WDM metro ring discussed earlier in Chapter 8.

TWC schemes

TWCs in an OPS node can be realized in several ways, as described in Chapter 2, such as by using OE regeneration, SOA-based gating, MZI-based interferometry with SOAs, and fiber nonlinearity. In the present context, one needs to make a choice for an appropriate method that suits the OPS functionalities. In particular, SOA-based methods are fast and can operate as a gating device as well for header removal, although contributing ASE noise. Optical fibers with Kerr nonlinearity can generate FWM components, thereby generating a new optical wavelength while needing a minimum length of fiber for nonlinear interaction between the input lightwaves. As we shall see below, FDLs already use fiber segments, and

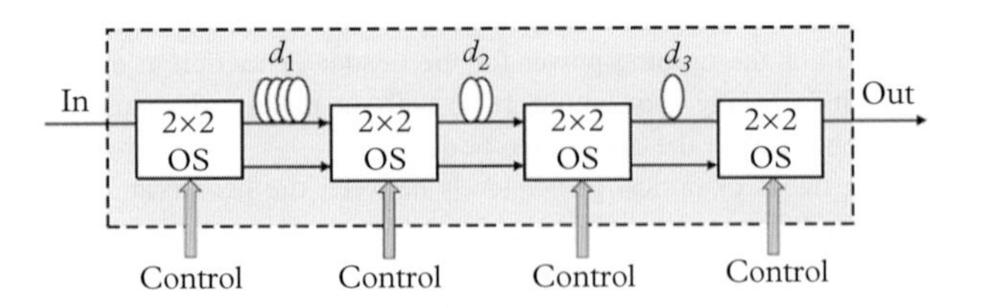

Figure 15.2 *Example three-stage FDL configuration offering programmable delay. 2 × 2 OS: 2 × 2 crossbar optical switch. With the binary-exponential setting of the delay elements at consecutive stages, the values of the delay elements become: $d_1 = d, d_2 = d/2, d_3 = d/4$. With k stages of delay elements, the kth delay element would therefore offer a delay $d_k = d/2^{k-1}$.*

hence the use of fibers for TWCs also can make the node hardware very bulky. The OE regeneration is straightforward and can be operated with high speed and flexibility. With these pros and cons of the candidate methods, the SOA-based realizations stand as the most useful choice for the TWCs used in the OPS nodes.

FDL configuration

FDLs are realized using fiber coils, which should be long enough to delay the optical packets by the required duration. For example, consider a 1000-byte optical packet with a duration of 800 ns (implying a packet size of 8000 bits transmitted at 10 Gbps), which finds an overlap with another (earlier) packet over a duration of 50 ns in the switch input port, and therefore needs to be delayed by the same amount for slot synchronization. With the velocity of lightwave through silica fiber as 2×10^8 m/sec, a fiber delay of 50 ns will need a fiber length $= (50 \times 10^{-9}) \times (2 \times 10^8) = 10$ m.

The delay requirements for the incoming packets in an OPS switch might vary with time, governed by their misalignments in time and the header-processing time, with the second component usually having a fixed value. Thus, with the need of variable delay from FDLs, one needs to realize programmable FDLs by using multiple fiber coils connected in cascade through 2 × 2 optical switches (OSs) where the OSs operate as *crossbar* switches. The programmable FDL, as shown in Fig. 15.2, uses three fiber coils, offering the delays of d_1, d_2, and d_3, and four 2 × 2 crossbar optical switches. By configuring these switches in various combinations of cross/bar states, one can have a total delay $D = \alpha d_1 + \beta d_2 + \gamma d_3$, with α, β, and γ representing three binary variables. Thus, the example configuration can offer a minimum delay $D_{min} = 0$ when α, β, and γ are all set to zero by configuring the first and the fourth OSs (from the left side) in cross state, and the two switches in the middle in bar state. However, the maximum delay $D_{max} = d_1 + d_2 + d_3$ is realized by configuring all the OSs in bar state. Typically, the values of delay in the FDL elements are set up with binary-exponential values, i.e., $d_1 = d$ (say), $d_2 = d/2$, $d_3 = d/4$, and so on.

15.2.3 Optical switching schemes

The optical switch in an OPS node is the core functional block, which needs to reconfigure much faster than the switches typically used in circuit-switched WRONs, such as MEMS. The switching device should be chosen so that the switch reconfiguration time is much less than the packet durations (fixed time slots, in this case). For example, an incoming packet having 1000 bytes, i.e., 8000 bits, would have a duration of $\tau = 800$ ns for a per-channel transmission rate of 10 Gbps, and hence an optical switching element in an OPS node should have a switch reconfiguration time, preferably not exceeding $\tau_{sw} = \tau/10 = 80$ ns. Moreover, the switching for the packets present at all input ports is to be carried out synchronously during the same slot interval.

Besides the need of synchronous switching, the optical switch fabric in an OPS node must ensure minimal internal blocking. Blocking in optical switch will result in packet losses, which will in turn lead to retransmissions, causing increased latency in the network. In an OPS node, contentions can take place in its switch when more than one input packet intends to reach the same output port at the same time with the same wavelength. In such cases, the switch can get around the problem of contention by utilizing all the three available dimensions: time, wavelength, and space. In this regard, the TWCs used preceding the switch can help for wavelength conversion, while the dimension of time can be availed by using additional FDLs at the output side of the switch fabric. Further, at a higher level, the node can use the third dimension (i.e., space) by adopting deflection routing, where the packets might reach out of sequence with an increased latency.

There are several possible switching devices and associated schemes that may be considered for the OPS nodes, with varying degree of suitability. As mentioned earlier, MEMS-based switches do not qualify for packet switching owing to their large reconfiguration time. Some viable options include electro-optic, thermo-optic, liquid-crystal switches, and gating devices based on SOAs. The electro-optic and thermo-optic switches have small values for the switching time, typically in the range of 3–5 ns. However, the electro-optic switches have high insertion losses (8–10 dB) and moderate crosstalk performance (25–30 dB), along with non-negligible PDL. PDL in electro-optic switches can be brought down using higher drive voltage, but at the cost of switch reconfiguration speed. However, the thermo-optic switches consume high power needing external cooling and need large die-area, leading to poor integration for large switches. The liquid-crystal switches also offer fast switching ($\simeq$ 5 ns), while being capable of operating with polarization diversity. Further, the crosstalk performance of liquid-crystal switches is excellent (50 dB). SOA-based gating modules are indeed fast and also provide necessary power amplification to compensate the losses incurred in the passive devices of a node. In addition to the OPS node architectures using the above switching devices, one can also use an AWG with large port count as a static switch along with TWCs on the output side, where the AWG is provisioned with some spare input/output ports used for feeding back the contending packets using FDLs. We describe in the following some of the useful switching schemes that can be used in the OPS nodes.

Wavelength-routed switching scheme

One of the synchronous OPS node architectures, developed in the *keys to optical packet switching* (KEOPS) project from Europe (Guillemot *et al.* 1998), was realized using the wavelength-routed switching (WRS) scheme at each input port of the switch, as shown in Fig. 15.3. Each input/output port in the switch (which functions as the second stage of the node architecture shown in Fig. 15.1) deals with a single wavelength at a time, and the switch operates using $S \times S$ input/output ports, with $S = N \times M$ for an $N \times N$ node and M wavelengths.

As shown in Fig. 15.3, at each input port of the switch, the received optical signal coming from a given TWC in the first stage of the node (Fig. 15.1) is passed through the respective optical demultiplexer in column A of Fig. 15.3. The demultiplexers in this column work as wavelength routers forwarding their input packets to the appropriate FDLs. Thus, the incoming packet at each demultiplexer of column A is routed to one of the S multiplexers (column B) through an FDL, where a given route from a demultiplexer output port of column A delays the packet by a specific amount, varying between zero and $(S-1)d$, with d as the delay incurred in one unit of FDL.

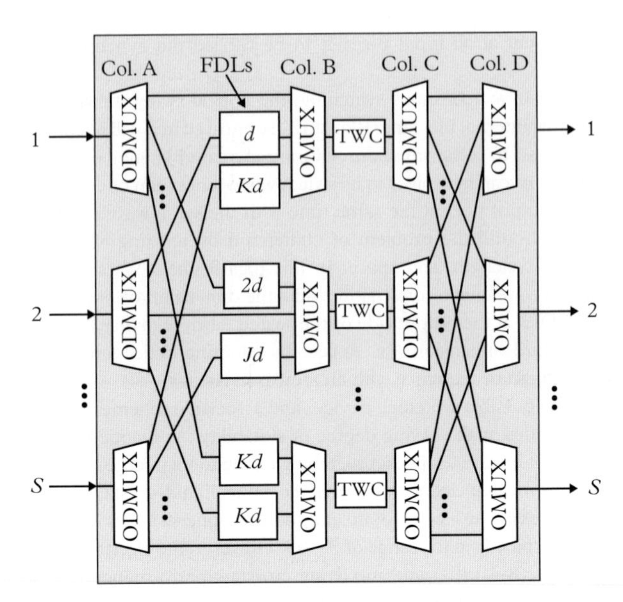

Figure 15.3 *An $S \times S$ switch architecture using WRS configuration for an $N \times N$ OPS node, operating with M wavelengths (as in KEOPS). $S = N \times M$, d = delay in one unit of FDL, and $K = S - 1, J = S - 2$.*

The choice of delay (in FDLs) for each incoming packet is determined by the column of TWCs in the first stage of the node (Fig. 15.1) preceding the demultiplexers of column A in Fig. 15.3, because the changed wavelengths (if necessary) at these TWCs will decide the specific output ports of the demultiplexers of column A, through which the input packets are routed. In the following column of multiplexers in column B, each input port of a multiplexer will receive one packet at a time with a specific delay and a wavelength, which subsequently goes for another column of TWC. The TWC outputs next get demultiplexed (routed) by the demultiplexers of column C to the final column of multiplexers (column D) with a fixed interconnection between columns C and D to eventually reach the desired output port of the switch with the specific wavelength during the right time slot.

Another OPS switch architecture was explored in the KEOPS project employing broadcast-and-select (BS) scheme, wherein the similar optical multiplexing/demultiplexing (OMUX/ODMUX) elements, FDLs, and TWCs (as in WRS) were used along with a number of optical power splitters (OSPs). One of the OSPs was used as the broadcasting device for the combined input WDM signals into various delay elements (FDLs), which were thereafter passed through an interconnection of one column of OSPs, two columns of SOAs as gating devices, and three columns of OMUX/DEMUX elements. The interested readers are referred to (Guillemot *et al.* 1998) for further details of the BS-based optical packet-switching scheme.

AWG-based scheme with recirculation buffers

Another switching scheme for the OPS nodes has been explored by using AWG as the switching device (Xu *et al.* 2001; Hunter *et al.* 1999). We discuss here this switching scheme, developed in the *wavelength switch optical packet network* (WASPNET) project, conducted by a collaboration of British universities (Hunter *et al.* 1999). The WASPNET switch made use of an AWG along with recirculating buffers, as shown in Fig. 15.4. In particular, the switching scheme employs an over-sized AWG with a large input/output port count ($L+S$, say), that exceeds the number of actual switch ports (S) by an integer L. These additional

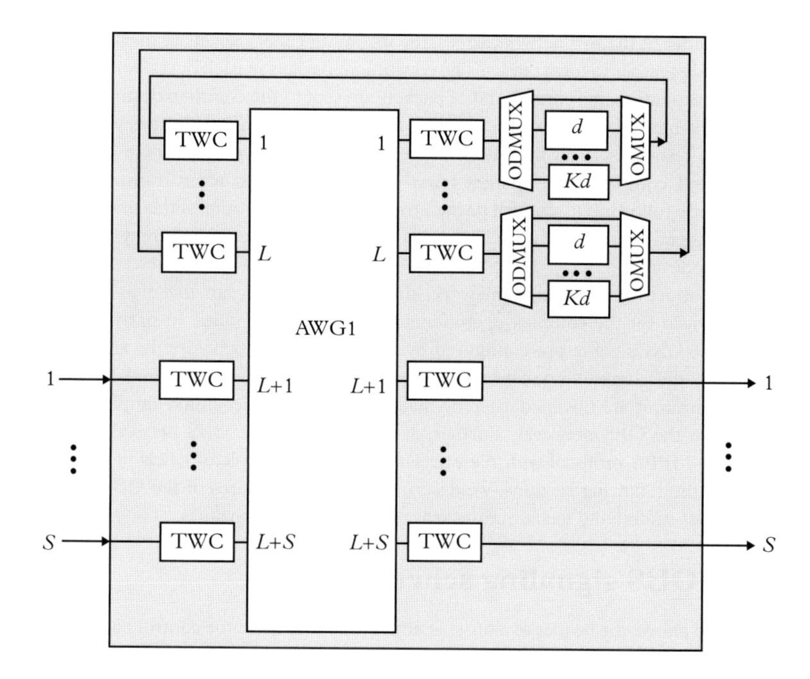

Figure 15.4 *$S \times S$ switch architecture for OPS node, using AWG with recirculation buffers. d = delay in the smallest element of FDL, and K is an integer.*

L ports in the AWG are used for recirculation of the contending input packets around the AWG itself to eventually route the packets to the desired output ports (i.e., the lower S output ports) by means of a wavelength-selective buffering scheme using TWCs, FDLs, and demultiplexer-multiplexer pairs. As shown in Fig. 15.4, another column of S TWCs is also used at the final stage following the AWG output ports for a reason that we explain below.

Overall, for an $S \times S$ switch, the present scheme uses one $(L+S) \times (L+S)$ AWG, $2(L+S)$ TWCs, L demultiplexer-multiplexer pairs, and KL fixed FDLs with varying delay options from d to Kd (with d as the delay in the smallest FDL element and K as an integer). As indicated above, the packets from the S lower output ports of the AWG are passed once more through a column of S additional TWCs to obtain the final output signals. This column of TWCs is necessary as the AWG operates internally with a larger number of wavelengths ($> M$) generated in the input TWCs, to use the $(L + S)$ ports for packet-forwarding/recirculation, and hence some of the outgoing wavelengths from the AWG may have to be converted to the network-compliant wavelengths before reaching the final switch output ports.

15.3 OBS networks

OBS networks provide a balance between the coarse bandwidth granularity of OCS-based WRONs and the packet-based fine granularity of OPS networks, while using simpler node hardware as compared to the OPS nodes (Qiao and Yoo 1999). In OBS networks, a source node at the network edge aggregates the packets received from its clients, having a common destination node at some other location at the network edge. Thus, with a pair of source and destination nodes at the two network edges, the incoming client packets at the source node are aggregated together to form bursts. The constituent packets in each burst, being destined to the same destination, the entire burst of packets are sent to the common destination node with one single header. The header, called a control packet, is sent in advance, which keeps reserving the resources at all the intermediate (core) nodes (with electronic conversion, processing, and control) for its parent burst. Hence, this burst, when transmitted from the source node following its control packet, traverses the same route as the control packet through the already configured switches at the intermediate nodes and thereby reaches the destination node smoothly.

With the above operational framework, the OBS networks can make use of OXCs without any need for the complex operational features of OPS nodes. In particular, in an OBS node the OXCs being pre-configured by the control packets before the arrival of the parent bursts, the complex optical header extraction/reinsertion process and FDL-based slot-synchronization are obviated, thereby making the OBS nodes much simpler than the nodes used in the OPS networks. Further, at a higher level the OBS networks conform mostly to the MPLS protocol with the wavelengths taking the role of labels of the MPLS networks. In the following sections, we describe the salient features of the OBS networks and the various underlying techniques to realize the OBS functionalities.

15.3.1 OBS signaling schemes

As mentioned above, the header of a burst in an OBS network, i.e., the control packet, is sent to the destination node, ahead of the payload (i.e., the burst) *with a timing offset*, say Δ. In the signaling scheme that we discuss now, the control packet, sent over a specific wavelength w_c with a given value of Δ, is received at each intermediate node and processed electronically therein. Using the control-packet information at each intermediate node, an appropriate route is selected and the OXC is configured and reserved for a time interval on a wavelength with a specific internal path for forwarding the forthcoming data burst (corresponding to the control packet) to the appropriate output port. Thus, the data burst moves along the same path as set up and traversed by the control packet. However, there might be contentions in an OXC in getting access to the required output port, while reserving the time slot for the input data packet. This problem can be addressed by using FDLs and TWCs based on the control packet information (received in advance), and hence without any need for their real-time switch reconfiguration as required in the OPS nodes. However, the optical switches used in the OBS nodes should be agile enough, such that the OXCs therein can reconfigure with adequate time margin between the arrivals of the control packets and the respective bursts.

Figure 15.5 illustrates the operation of a typical OBS network, where the control packet from a source node s to a destination node d is transmitted ahead of the corresponding data burst with an offset time Δ. The offset time is estimated such that the sum of the header processing and switch reconfiguration times at each node on the route is $\leq \Delta$ (i.e., at the nodes 1, 2, and d, in the example shown in Fig. 15.5), thereby ensuring that the data burst finds the en route OXCs ready for forwarding/receiving, whenever it reaches any intermediate/destination node. If τ_{cpp} and τ_{swc} represent the control-packet processing and switch reconfiguration times, respectively, then one can express Δ as

$$\Delta \geq k\delta, \tag{15.1}$$

where δ is given by

$$\delta = \tau_{cpp} + \tau_{swc}, \tag{15.2}$$

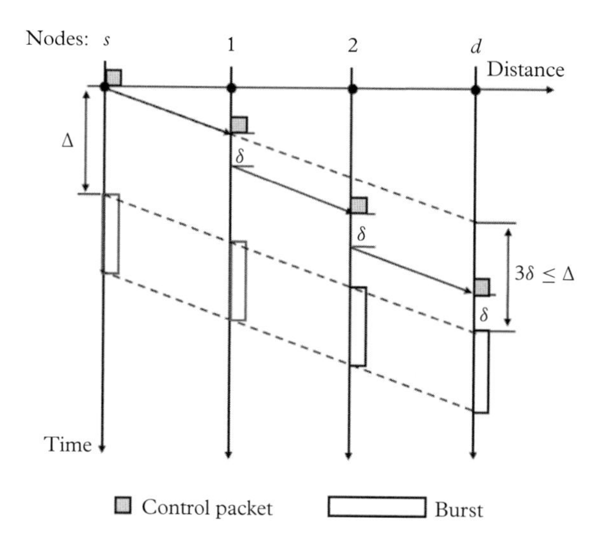

Figure 15.5 *An illustration of OBS network operation over a linear segment with s and d as the source and destination nodes at the network edges and the nodes 1 and 2 as the intermediate nodes on the chosen route.* $\Delta \geq 3\delta = 3(\tau_{cpp} + \tau_{swc})$*. Note that,* $\Delta = 3\delta$ *for the JET signaling.*

and k is an integer representing the number of nodes traversed by the burst, including the intermediate and destination nodes. In the present example, with $k = 3$ (i.e. the nodes 1, 2, and d) one therefore needs to set up the offset time as $\Delta \geq 3\delta$. When the offset time is specifically set up as $\Delta = 3\delta$ ($\Delta = k\delta$, in general), the OBS network is said to be operating with a *just-enough time* (JET) signaling scheme.

Besides the JET scheme, a few other signaling schemes have also been explored in OBS networks (Battestilli and Perros 2003). In one of these schemes, the offset time Δ is estimated statistically (Verma *et al.* 2000). In particular, as soon as a burst is formed at the source node, the control packet is sent out. However, the transmission of the corresponding burst is delayed until a token is generated within the node itself following a Poisson arrival process with a pre-assessed mean arrival rate (and hence mean interarrival time). Thus the offset time Δ becomes a random variable around the mean interarrival time of the Poisson arrival process. The statistical nature of the offset time staggers the arrival instants of the bursts at the core nodes, reducing thereby the possibilities of contention. However, the JET-based scheme stands as the simplest yet efficient signaling scheme for the OBS networks, and we therefore describe below the various functionalities of OBS networks, primarily based on the JET signaling scheme.

15.3.2 OBS nodes with JET signaling

A typical OBS node architecture using JET signaling is shown in Fig. 15.6. The control packets are sent over a fixed wavelength w_c from all nodes across the network, while the selection of wavelengths for bursts are made from the source (edge) nodes. At the intermediate and destination nodes, the control packet is demultiplexed by the OXC and sent to the control-packet processing and control unit (CPPCU) for the processing of control packet, OXC reconfiguration, and retransmission of the modified control packet through OE and EO conversions. Using the results of control-packet processing, the OXC is informed about the needed time slot and wavelength for the forthcoming burst corresponding to the control packet being processed, and accordingly the optical switch in the OXC is scheduled with the necessary configuration for the reserved time slot and wavelength. The

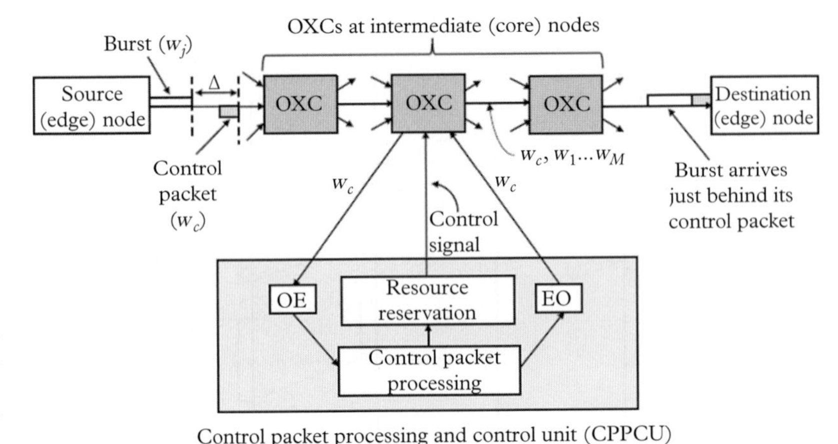

Figure 15.6 *OBS networking with OXC-based nodes using a JET signaling scheme, where the CPPCU is shown for only one node. OXCs are in general provisioned with FDLs and TWCs (not shown here) to reduce packet losses in OBS nodes.*

corresponding burst, on arrival at the node, is switched immediately to the appropriate output port for onward transmission.

In the OBS nodes shown in Fig. 15.6, the control packet received at a node might find that the needed time slot on the specified wavelength is not free, implying that when the corresponding burst arrives at the node, it might get lost. In order to prevent the possible burst loss, an FDL can be inserted with a limited capacity before/after the switch in an OXC; or else one can go for deflection routing to avoid the packet loss, but at the cost of increased delay in both cases. Use of TWCs and FDLs within the OXC is the best option to avoid possible contentions in OXCs. However, with the use of FDLs and TWCs, the OBS reservation schemes need to be designed appropriately at each node. In the following, we describe the operation of OBS networks based on the various schemes and technologies used in the network nodes.

15.3.3 Burst assembly schemes

The length of a burst is a key parameter in OBS networks, which needs to be decided in the source node at the network edge. The burst assembly (or aggregation) scheme, employed at a source node, needs to collect the ingress packets (with the same destination node) over a certain time interval using a preset timer, and ensure that the burst size is confined within some lower and upper limits. Note that, the longer bursts, once lost in contention on their routes toward destinations, result in to the losses of large number of packets, while the shorter bursts (each with its own control packet) lead to a large number of control packets with reduced bandwidth utilization. The time interval, during which the ingress packets are collected for aggregation, also should not exceed an upper limit. With a long time interval for aggregation, the packets that arrive earlier and wait in the buffer, suffer more delay as compared to the packets that arrive toward the end of the aggregation interval. Thus, the bursts can be classified based on the QoS demands, with each class having a specific combination of the three burst aggregation parameters: aggregation time interval, and the lower and upper limits on the burst size.

15.3.4 Resource reservation schemes

Each OBS node along the route of a transmitted burst needs to reserve the resources in its OXC for the needed input-output connectivity along with the FDLs and TWCs (if used) during the necessary time interval. Further, the allocated resources for the burst must be released, immediately after the burst leaves the node. The process of resource reservation and release, collectively referred to as the resource reservation scheme, can be realized in various ways. We discuss two useful resource reservation schemes in the following.

In one of these schemes, called *advanced reservation* (AR), the resource reservation in all the nodes on a route is made, soon after the control packet arrives (say, at time t_0) and the processing of the control packet is carried out at $t_0+\delta$. By using the values of the offset time Δ and the duration of the forthcoming burst τ_B from the control packet, the node under consideration estimates the time t_D when the corresponding burst would leave the same node. Thus, t_D can be expressed as

$$t_D = t_0 + \Delta + \tau_B. \tag{15.3}$$

Using the estimated value of t_D, the node next reserves the OXC resources for the specific switch configuration during the time interval $[t_0+\delta, t_D]$. Note that, although the actual burst arrives only at $t_A = t_0 + \Delta$, the OXC resources remain reserved over a longer interval, i.e., right from $t_0+\delta$ therefore blocking the switch capacity in advance. Fig. 15.7(a) illustrates the operation of the AR scheme with reference to a given intermediate node in an OBS network.

In another scheme, called the *delayed reservation* (DR) scheme (illustrated in Fig. 15.7(b)), the resource reservation in OXC is delayed by $\Delta - \delta$ seconds. In other words, the reservation is made from the expected arrival time t_A until the instant t_D when the packet leaves the node. Thus, the reservation is made only during the interval $[t_A, t_D]$, thereby keeping the interval $\Delta - \delta$ free, which can be utilized for the other subsequent control packet arrivals. In both schemes, the resources reserved in the OXC are released immediately after the burst leaves the node, i.e., at $t = t_D$.

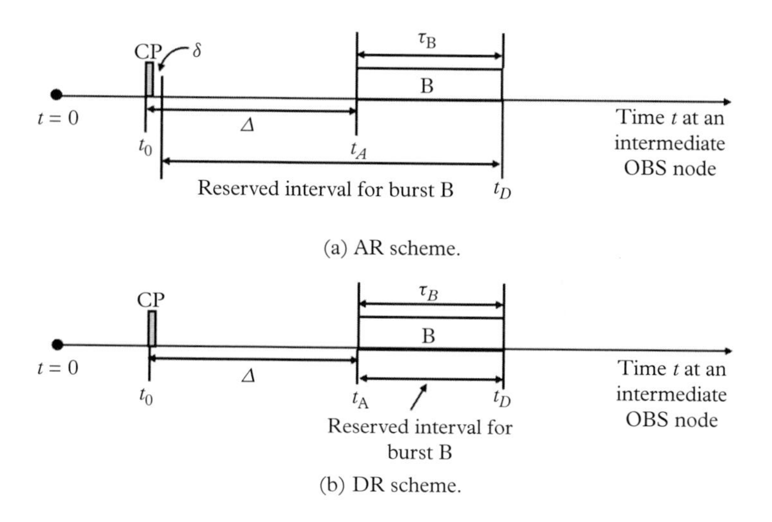

Figure 15.7 *Illustration of the AR and DR schemes at intermediate nodes in an OBS network.*

Next, in Fig. 15.8, some example cases are shown, illustrating when a burst may or may not get lost in the DR scheme at a given intermediate node in an OBS network. The figure presents the three cases of arrivals for two consecutive bursts, B1 and B2, wherein we make use of the following variables.

t_{01}, t_{02}: arrival times of the control packets at the node under consideration for the bursts B1 and B2, respectively.

Δ_1, Δ_2: offset times for the bursts B1 and B2, respectively.

t_{A1}, t_{A2}: arrival times of the bursts B1 and B2, respectively, at the node under consideration.

τ_{B1}, τ_{B2}: durations of the bursts B1 and B2, respectively.

t_{D1}, t_{D2}: departure times of the bursts B1 and B2, respectively, at the node under consideration. Thus, $t_{D1} = t_{A1} + \tau_{B1}$ and $t_{D2} = t_{A2} + \tau_{B2}$.

In Fig. 15.8, for all the three cases (parts (a), (b), and (c)), the control packets CP1 of the burst B1 arrives earlier than the control packet CP2 for the burst B2. Hence, the node makes the reservation for B1 before it receives the control packet CP2 for B2. However, in parts (a) and (b), the node can find an available time slot for burst B2 as well without any contention. In particular, in part (a), B2 arrives and leaves earlier than B1, i.e., $t_{A1} > t_{A2}+\tau_{B2}$, and in part (b), B2 arrives after B1 has left, i.e., $t_{A2} > t_{A1}+\tau_{B1}$. However, in the example case of part (c) the expected duration to be spent by B2 overlaps with that of B1, i.e., $t_{A1} < t_{A2} < t_{A1}+\tau_{B1}$, and thus B2 gets dropped as B1 has already been provisioned by its control packet CP1 before the arrival of the control packet CP2 for B2.

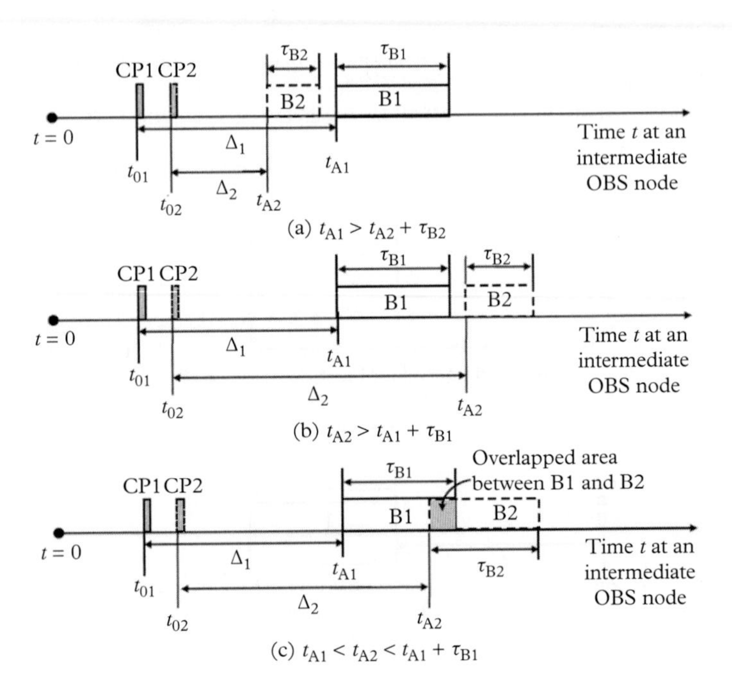

Figure 15.8 *Different types of burst-arrival cases in the DR scheme.*

15.3.5 Use of FDLs and TWCs in OBS nodes

As mentioned earlier, one can use FDLs and TWCs in OBS nodes to address the contentions between the incoming bursts to be switched to the same output port of the optical switch used in the OXC. The use of FDLs and TWCs in an OBS node for this purpose is illustrated in Fig. 15.9.

As shown in the figure, the FDLs and TWCs together can be used to resolve the contention, which is accomplished with the help of the CPPCU module of the node. However, although the OBS node follows the similar usage of FDLs and TWCs as in an OPS node, it doesn't need to reconfigure them on a real-time basis. Regarding the use of TWC, the control packet also has to be changed accordingly, so that the next node in the route is informed about the changed wavelength of the forthcoming burst wavelength. In the following, we illustrate the role of FDLs in achieving improved resource reservation process with reduced burst losses.

Figure 15.10 illustrates the role of an FDL in alleviating the problem of contention in OBS node, using DR-based JET signaling. With reference to the case shown in Fig. 15.8(c), the FDL to be used for burst B2 must provide a minimum delay d_{min} to prevent the loss of B2, because in the absence of the FDL, B2 would have been dropped to avoid collision with B1. Since the FDL delays are quantized into discrete values (see Fig. 15.2), in reality the FDL delay should be chosen such that $d_{FDL} \geq d_{min}$.

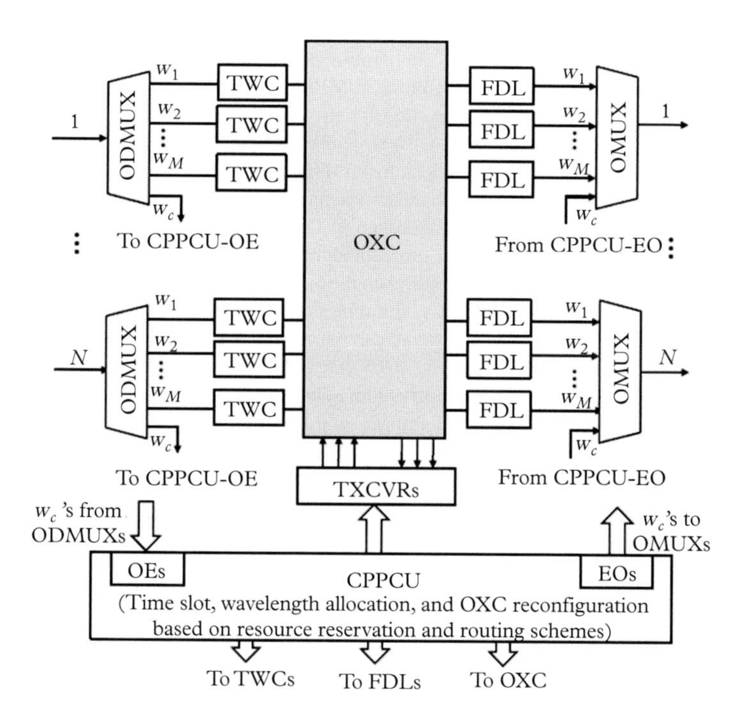

Figure 15.9 *OBS node architecture with FDLs and TWCs.*

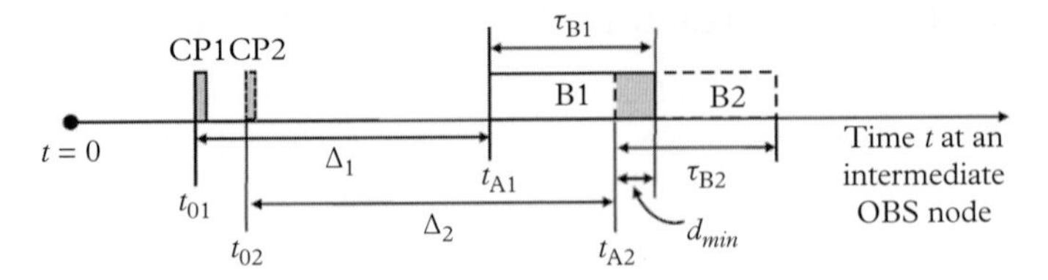

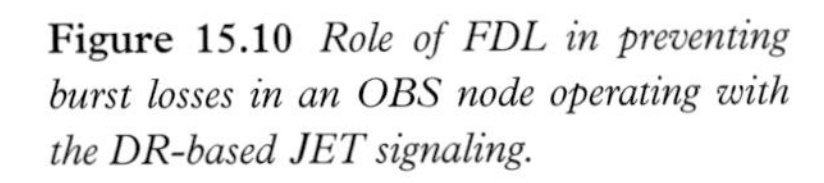

Figure 15.10 *Role of FDL in preventing burst losses in an OBS node operating with the DR-based JET signaling.*

15.3.6 Routing schemes in OBS networks

Functionally, the OBS networks come close to the MPLS networks, with the optical bursts *resembling* the forward equivalent classes (FECs) of ingress MPLS packets (see Chapter 12) at the source nodes. In the MPLS networks, first a label-distributing message (like a control packet in OBS networks) is sent through the network using a suitable label-distributing protocol (LDP) to set up the label-switched path (LSP) from a source node to a destination node. This leads to a synergy between MPLS and OBS networks employing a similar protocol suite, such as IP-over-MPλS, where the collected IP packets at the edge are clubbed into optical bursts (as a group/class of packets) and sent through the OBS network over a wavelength (λ representing a wavelength) playing the role of the label in MPLS networks. However, appropriate modifications are needed to adopt the basic philosophy of the IP-over-MPLS routing in OBS networks in the form of a suitable protocol suite, popularly termed IP-over-MPλS protocol.

With the IP-over-MPλS protocol, the control packets move across OBS network in the same manner as the LDP packets move in MPLS networks, for setting up the labels (wavelengths in the case of OBS networks) for the intermediate nodes along the chosen LSP (lightpath in the case of OBS networks) in a hop-by-hop manner. Note that the lightpaths may assume different wavelengths on different links, governed by the resource reservation scheme used in the nodes and wavelength availability therein. Thus, for each burst, an end-to-end connection is established in OBS networks, following the principle of the hop-by-hop packet forwarding protocol along with an appropriate OBS signaling scheme. This task is accomplished by setting up the output wavelengths using TWCs, input-output switch connectivity in the OXCs, and time slots using FDLs at each node from the burst information. Note that the burst would carry several nuggets of information, including, offset time, burst duration, class of service (CoS), etc., which will enable the nodes to exercise the underlying schemes for resource reservation. Following the control packet with an offset time, the burst would move along the lightpath set up by the control packet, just as an FEC would flow through the MPLS network after its labels have been successfully set up for the LSP. The control packets, after completing the resource reservation task and routing in each node, are modified for the next hop and retransmitted on the control wavelength w_c.

In reality, the routing of control packets in OBS networks would be more complex than what is done in an MPLS network in respect of the resource reservation scheme, due to the intricate optical device technologies and the contentions in limited-size FDLs, while the flow of FECs along LSPs would be similar to the flows of the optical bursts along the assigned lightpaths. Overall, the FDLs, TWCs, and OXCs along with the signaling scheme in use, will need to be *integrated* with the IP/MPLS protocol, leading to an efficient IP-over-MPλS protocol, for carrying out successful end-to-end burst communications in OBS networks.

For JET-based OBS networks, one can also offer class-based services by varying the offset time for different classes of traffic. Note that, setting a higher value for Δ, a source node can ensure less contention, thereby offering lower burst loss probability, as the control packet can *buy* more time to find a contention-free time interval at intermediate nodes. In other words, one can ensure lower burst losses with longer offset time, albeit with an increase in overall delay. However, when a burst gets lost due to contention, the same burst needs to be retransmitted, which also increases the overall delay, and usually the increase in delay due to larger offset time tends to be lower than the increase in delay due to burst retransmissions.

15.3.7 Latency and burst losses

One can get an insight into the OBS networks through some estimates of the network latency and probability of burst losses. In the following, we assess these two performance metrics, presuming that the OBS nodes don't use FDLs.

In order to get an estimate of the network latency, we note that, it is the end-to-end packet delay, rather than the burst delay, that deserves attention in the OBS networks, as different packets in a burst spend different times in a queue to form a burst at the source node, and once formed and transmitted, the burst and hence its constituent packets have a smooth journey through the intermediate nodes, albeit with some propagation delay. However, in reality, when a burst gets blocked/lost due to contention, a bunch of packets are lost in one go, which subsequently need retransmission, incurring additional delay. At the moment, we won't consider this component of latency, as it involves higher-layer protocols. With this framework (Verma *et al.* 2000), we next find out the various components of the average end-to-end delay D_P across the network for a packet in a burst. One of the components of D_P takes place at the source node, as a given packet in a burst needs to wait until the entire burst is formed, and thus the average waiting period $\bar{\tau}_{BQ}$ for the first packet is the largest of all the packets in the burst. The next component of D_P at the source node is the average offset time $\bar{\Delta}$, while the third component with the present assumptions (i.e., without FDLs) would be simply the average propagation time $\bar{\tau}_{prop}$ through the network, as the core nodes, being pre-configured, pass on the incoming bursts through the OXCs without incurring any delay at the nodes. Finally, one needs to consider the average burst duration $\bar{\tau}_B$ (transmission time) to get an estimate of D_P. Hence, the average end-to-end packet delay D_P for the first packet in a burst can be expressed as the sum of these four components, given by

$$D_P = \bar{\tau}_{BQ} + \bar{\Delta} + \bar{\tau}_{prop} + \bar{\tau}_B. \tag{15.4}$$

However, as and when the bursts encounter contentions at the intermediate nodes, the delay increases due to the use of FDLs for contention resolution, implying that the above expression would give us a lower-bound for D_P, albeit with higher packet losses.

In order to get an estimate of the probability of burst losses at the core nodes, one can employ a simple queuing model as follows (Verma *et al.* 2000). As before, presuming that the OBS nodes don't use any FDL while TWCs are available with full wavelength-conversion capability, we represent the probability of burst losses P_{BL} (synonymous with blocking probability) at the core nodes by using the $M/M/c/c$ queuing model (Erlang B loss model), as

$$P_{BL} = \frac{\left(\frac{\lambda}{\mu}\right)^M / M!}{\sum_{i=0}^{M} \left(\frac{\lambda}{\mu}\right)^i / i!} = \frac{\rho^M / M!}{\sum_{i=0}^{M} \rho^i / i!}, \tag{15.5}$$

where λ represents the burst arrival rate at a core node, μ is the inverse of the average duration of the optical bursts, ρ is the traffic intensity, and M is the number of wavelengths available in the network. As evident from Eq. 15.5, the burst losses can be brought down significantly by increasing the number of wavelengths. Further, the burst losses can be brought down by using the FDLs, and thus the above expression represents the worst-case estimate of P_{BL} in the OBS networks.

15.4 Summary

In this chapter, we presented the packet/burst-centric versions of optical WDM networks, OPS and OBS networks, both promising higher bandwidth utilization in optical fibers as compared to the circuit-switched WRONs. First, we presented the OPS network with its packet-based bandwidth granularity offering maximum bandwidth utilization from optical fibers, while calling for packet alignment, tunable wavelength conversion, header extraction/insertion and processing, all to be carried out in real time using complex hardware. We presented a generic form for the OPS node architecture and the devices to be used therein to perform the above tasks. Then, we described a few useful switch architectures used in the testbeds developed for the OPS networks in KEOPS and WASPNET projects.

Next, we discussed the basic concepts of OBS networks, and described a typical node architecture to realize the OBS functionalities. Thereafter, we described the necessary protocols for OBS networks, including burst assembly scheme, just-enough time (JET) signaling, resource-reservation, and routing schemes. Finally, with some simplifying assumptions, we analytically evaluated the performance of OBS networks in terms of the network latency and probability of burst losses at core nodes.

EXERCISES

(15.1) Consider a 1200 byte optical packet to be switched at an OPS node with a bit rate of 40 Gbps. What should be the maximum allowable switching time for the OPS node? Comment on the suitability of the various possible optical-switching technologies for the OPS node.

(15.2) Consider an OPS node operating with the configuration shown in Fig. 15.1 and express its total insertion loss in terms of the losses incurred in their constituent optical devices. Assuming the given losses in various constituent building blocks as given below, estimate the insertion loss in the node and indicate whether one should use a post-amplifier following the node, and, if so, what should be the amplifier gain. Given: insertion loss in demultiplexer/multiplexer = 3 dB, power tapping for header processing = 10%, insertion loss in power tap = 0.5 dB, insertion loss in FDL (for largest delay) = 5 dB, insertion loss in the optical switch = 12 dB, insertion loss in optical combiner = 2 dB, connector loss = 1 dB. Assume that the TWCs (using SOAs) don't contribute to the node insertion loss.

(15.3) Consider a WDM network operating with both WRON-based optical circuit-switching (OCS) as well as OBS, where for the control operation one wavelength is kept reserved, and the OCS and OBS parts of the network are assigned three and two wavelengths, respectively. At each node, the control wavelength from the adjacent nodes undergoes OE-EO conversions for sharing OCS/OBS control information with the rest of the node. Draw a suitable configuration for a 3×3 node with the given network-setting, that will be able to cater to the needs of both OCS and OBS operations. Discuss on the feasibility of independent control operations of the OCS and OBS segments of the network with one control wavelength and the switching technologies that would be appropriate for the OCS and OBS operations.

(15.4) Sketch a block schematic for a programmable FDL, where an optical packet needs to be delayed with a variable delay from zero to 0.7 μs in steps of 0.1 μs. Design the FDL.

(15.5) Consider that an optical burst has to reach from its ingress node to the egress node through five intermediate nodes by using JET protocol. The total length of the chosen route through the intermediate nodes across the network is 200 km along the optical fiber links, and each node in the network needs 100 ns to process the incoming control packet for the subsequent burst transmission, while the network transmits over each wavelength at a bit rate of 10 Gbps.

a) Calculate the necessary offset time for the control packet with respect to the burst for which it is being sent.

b) Assume that the average burst size in the network is 10,000 bytes, and the first packet in a burst in the ingress node has to wait with an average waiting time of 10 μs before the burst carrying this packet is transmitted. Estimate the average delay incurred by the first packet in a burst for the given pair of ingress and egress nodes.

(15.6) An OBS network is realized with its nodes using TWCs, but without any FDL. The network operates with five wavelengths, each transmitting at 10 Gbps. Consider a core node in this network, where optical bursts arrive at a rate of 50,000 bursts/second with an average burst size of 125,000 bytes. Estimate the burst-loss probability at the node using the $M/M/c/c$ queuing model.

(15.7) An OBS network operates with the DR-based JET signaling by using FDLs. Assume that the two control packets CP1 and CP2 have arrived one after another with a difference of 0.8 μs (CP1 has arrived earlier). The offset times for CP1 and CP2 are 0.2 μs and 0.25 μs, respectively. The durations of the corresponding data bursts, i.e., B1 (for CP1) and B2 (for CP2), are 1 μs and 1.2 μs. Determine the minimum length of the FDL needed to prevent contention between the two data bursts.

Appendix A
Basics of Linear Programming and the Simplex Algorithm

A.1 Background

Linear programming (LP) offers an optimization technique, where the constraints and objective function are linearly related with the variables of a given problem. From a wider perspective, LP is seen as a subclass of a larger class of optimization problems, called convex optimization. Thus, for an LP-based optimization problem, one needs to ensure that the solution space is *convex* in nature in the given n-dimensional problem space. In this appendix, we first discuss the background and salient features of LP problems, and then present the basic analytical steps for solving LP problems using the well-known Simplex algorithm.

Historically, the problem of finding a solution to a given set of linear inequalities was first addressed by J. Fourier way back in 1826, and subsequently many mathematicians worked on various aspects of the problem. Much later, in 1939, L. Kantorovich proposed an approach to solve LP problems for minimizing the cost of an army, which remained unnoticed for several years. Thereafter, interest in LP problems went up significantly following the work of G. B. Dantzig in 1947, who developed a general LP formulation for the US Air Force, and later invented the well-known Simplex algorithm which could address a wide range of LP problems (Dantzig 1963). Around the same time, T. C. Koopmans also worked independently in this area, showing how LP could be used for analysing the classical problems of economics. Eventually, in 1975 Kantorovich and Koopmans went on to win the shared Nobel prize in economics. However, Dantzig's work was not considered for Nobel, as his work seemed more mathematical in nature as compared to the problems in economics, and mathematics was not listed as a subject for the Nobel prize (Vanderbei 2008).

Subsequently, the LP problem was shown to be solvable in polynomial time by L. Khachiyan (Khachiyan 1979), and thereafter N. Karmarkar (Karmarkar 1984) introduced a novel interior-point method to solve LP problems. Dantzig's Simplex algorithm gained huge popularity and found applications in a wide range of problems involving LP formulation. Simplex has since been applied in many professional software packages and is used extensively for solving problems in a wide range of areas: operation research, economics, telecommunication networks, etc.

A.2 Geometric interpretation

First, we explain the concept of convexity using geometric examples. The property of convexity of a closed region or space implies that the given region should not have any *indentation* on its boundary, nor should it have any hole within, as shown in Fig. A.1. In part (a) of the figure representing a convex region, one can find a point within the same region by linearly interpolating any other two points. Thus, if the points $\mathbf{u}_1$ and $\mathbf{u}_2$ belong to the

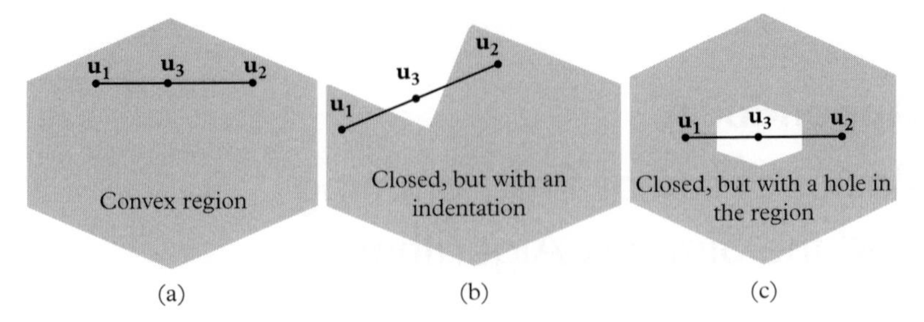

Figure A.1 *Illustration of the property of convexity.*

given convex region, then $\mathbf{u}_3 = \lambda\mathbf{u}_1 + (1-\lambda)\mathbf{u}_2$ (with $0 \leq \lambda \leq 1$) will also lie within the same region. However, with the examples shown in Fig. A.1(b) and (c), this will not hold true as $\mathbf{u}_3$ falls within the dented portion or in the hole of the closed region, respectively. This example scenario can be readily stretched to appreciate that the property of convexity will also hold true in any n-dimensional space. An LP problem seeks to find a solution in a given convex region defined by a set of linear constraints in the form of straight lines, planes, or hyperplanes (depending on the value of n) as the boundaries of the region, where a linear objective function, also a line, plane, or a hyperplane for a given cost function, needs to be maximized or minimized to achieve the optimum solution.

In the following, first we illustrate the basic features of LP using a simple two-dimensional example, and thereafter generalize the concept for larger dimensions.

Consider an optimization problem with the following linear constraints and objective function involving two variables x_1 and x_2.

Constraints:

$$x_1 + 2x_2 \leq 7 \tag{A.1}$$

$$4x_1 + x_2 \leq 9, \tag{A.2}$$

as well as,

$$x_1 \geq 0 \tag{A.3}$$

$$x_2 \geq 0, \tag{A.4}$$

with the last pair of inequalities (Equations A.3 and A.4) representing the non-negativity constraints on the variables (which are also linear).

Objective function:

$$\textit{Maximize} \quad z = 4x_1 + 2x_2 \tag{A.5}$$

Before we proceed, we replace the first two inequalities of Equations A.1 and A.2 by equalities, resulting in two straight lines in the (x_1, x_2) plane, as shown in Fig. A.2. Further, the two non-negative constraints of Equations A.3 and A.4, when converted to equalities, would represent the x_1 and x_2 axes, respectively. The shaded area bounded by the two straight lines by the equalities obtained from Equations A.1 and A.2 and the x_1 and x_2 axes

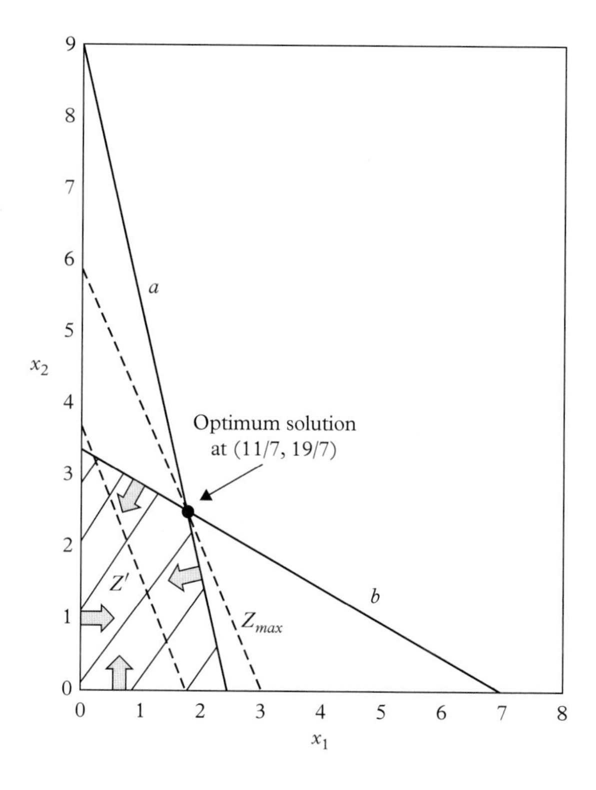

Figure A.2 *Geometric illustration of an example LP problem to maximize a given objective function, as specified in Equations A.1 through A.5. The lines a and b with the arrows directed toward the origin indicate the regions of inequalities given by Equations A.1 and A2, respectively. Similarly, the lines aligned with the x and y axes with inward arrows (into the first quadrant) represent the constraints imposed by Equations A.3 and A.4. The shaded region represents the convex space bounded by the given constraints. z' and z_{max} are the plots of the objective function for two different values, and z_{max} corresponds to the plot of the objective function that passes through the appropriate corner point, offering the optimum solution at (11/7, 19/7).*

representing the two equalities obtained from Equations A.3 and A.4 represents the *convex* region in the first quadrant, while the objective function z obtains its maximum possible value of $z_{max} = 82/7$ in the convex region, while touching a specific *corner point* of the convex region at $(x_1, x_2) = (11/7, 19/7)$, leading to the desired optimum solution. Note that another plot of the objective function with a lower value (i.e., z') fails to offer the needed solution. We describe below the significance of this result in further detail.

As shown in Fig. A.2, the shaded convex region bounded by the given set of linear constraints has some corner points or extrema, and the objective function for a given cost represents a straight line, which is not parallel to any of the lines representing the constraints. Hence, if the cost is increased from a low value, the line representing the cost function moves up, and as it is not parallel to any of the constraint lines (which is a degenerate case), it can only touch one single corner point at the top, for the maximization of cost. While the cost line, for some other values of the cost, can run through any other corner point, it does not maximize the cost, but pierces the convex region through an infinite number of points in the continuum (see z' in Fig. A.2). However, the latter case with the cost line running through the convex space has no significance in the present problem, as we seek a unique solution that would maximize the cost. Hence, as and when the cost line is placed in such a manner that it touches only one corner point at the top (for maximization problem),[1] one can be sure that the optimum solution has been reached. This leads to the fact that the

[1] Note that, one could also seek to *minimize* the objective function in a different problem setting.

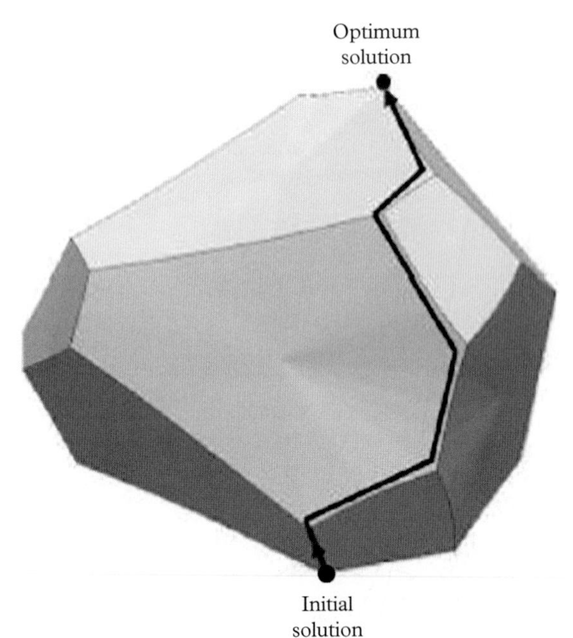

Figure A.3 *Linear optimization: an example using a 3D polyhedron as a convex solution space.*

solution to an LP problem is located at one of the various corner points of the convex space, and the task of an efficient LP solver is to find an algorithmic method that will ensure that the objective function starts with a suitable initial value and reaches the appropriate corner point with the least computational effort.

The graphical solution in Fig. A.2 and the discussion of the same give us some insight into the LP-based optimization problem, but such a method cannot be used in a practical problem with a large number of variables and constraints. In order to solve even this small two-variable problem computationally, one needs to first convert all the inequalities mathematically into equalities with some additional variables. For example, one has to convert the first constraint, given by $x_1 + 2x_2 \leq 7$, to $x_1 + 2x_2 + x_3 = 7$ with $x_3 \geq 0$ as an artificial variable (typically referred to as a slack or surplus variable, depending on the type of inequality ($\leq$ or $\geq$, respectively). For the two constraints, this conversion process will eventually lead to four variables, i.e., the two original variables (x_1 and x_2) and two more slack/surplus variables (x_3 and x_4) to convert the two inequalities into two equalities, while the constraints in Equations A.3 and A.4 are taken care of by considering only the non-negative solutions for (x_1, x_2). The four-dimensional solution space to this problem will now represent a *polyhedron* (instead of a planar region as observed in Fig. A.2) as the convex region of feasible solutions, from which the optimum solution needs to be obtained by using the objective function.

In order to get some additional insight, we present a three-dimensional (3D) polyhedron as an *analog of such problems* in Fig. A.3. Note that, although with more than three variables it is not possible to visualize the problem in 3D, we still consider this analogy for gaining an insight into the Simplex algorithm. In particular, the two-dimensional planes or surfaces of the 3D polyhedron will be replaced by *hyperplanes* in a higher-dimension polyhedron (i.e., with dimensions > 3). Note that all the vertices of such a polyhedron would be similar to the corner points of the shaded convex space in Fig. A.2. As illustrated in Fig. A.3, an automated computational algorithm, like Simplex, will start from one of the vertices as the starting solution of the problem and move along the edges of the polyhedron through the other vertices or corner points to reach the appropriate vertex or corner point as the optimum solution (given by the coordinates of the corresponding vertex in the diagram). For example, while solving the problem considered in Fig. A.2 through Simplex, initially all the original variables (x_1, x_2) are set to zero making the solution assume the form of $(x_1, x_2, x_3, x_4) = (0, 0, x_3, x_4)$. This sets x_3 and x_4 equal to the right-hand-side numbers of the respective constraint inequalities (i.e., 7 and 9), and the initial solution $(0, 0, 7, 9)$ corresponds to one of the vertices of the respective four-dimensional polyhedron. Thereafter, using this initial solution and the given objective function, Simplex *discovers* a judicious path to move through some *select edges and vertices* (all vertices being possible solutions) and eventually reach the specific vertex corresponding to the optimum solution ($x_1 = 11/7, x_2 = 19/7$, as in Fig. A.2), by following a systematic computational algorithm.

A.3 Algebraic framework

While the above examples give an insight into convex optimization problems, in practice the constraints and variables are large in number, and thus one needs to follow a systematic computational method based on the framework of linear algebra. Using the line of treatment in (Hadley 1961), we present below the basic algebraic framework for the LP problem.

Generalizing the earlier example represented by Equations A.1 through A.4, the linear constraints in an LP problem, m in number, can be expressed in a general form as

$$a_{i1}x_1 + a_{i2}x_2 + \cdots + a_{ij}x_j + \cdots + a_{iu}x_u \;\; \{\leq = \geq\} \;\; d_i, \quad i = 1, ...m, \tag{A.6}$$

where a_{ij}'s and d_i's are constants. As indicated in the above expression, each constraint can be of one of the three types: two possible inequalities ($\leq$ and $\geq$) and equality. We need to find the values of the variables, x_1, ..., x_u that have to satisfy the non-negativity constraint, given by

$$x_i \geq 0, \quad i = 1, ..., u, \tag{A.7}$$

and maximize (or minimize) a cost function z, expressed as

$$z = c_1x_1 + c_2x_2 + ... + c_ux_u. \tag{A.8}$$

As discussed in the previous (geometric) example, the inequalities in the constraints can be converted into equalities by using slack and surplus variables absorbing the $\geq$ and $\leq$ inequalities, respectively. Therefore, we will need to add one additional slack/surplus variable to each constraint governed by each inequality (say, with v such constraints), giving rise to $u + v = n$ variables in total. However, the constraints governed by the equalities will not need any introduction of slack or surplus variables. With these observations, one can rewrite the constraints in Eq. A.6 as

$$\mathbf{Ax} = \mathbf{d} \tag{A.9}$$

where $\mathbf{A}$ is a rectangular matrix with m rows and $n = u + v$ columns ($v \leq m$), given by

$$\mathbf{A} = \begin{bmatrix} a_{11} & a_{12} & \cdots & a_{1u} & 1 & 0 & \cdots & 0 & 0 \\ a_{21} & a_{22} & \cdots & a_{2u} & 0 & 1 & \cdots & 0 & 0 \\ \cdots & \cdots & \cdots & \cdots & \cdots & \cdots & \cdots & \cdots & \cdots \\ \cdots & \cdots & \cdots & \cdots & \cdots & \cdots & \cdots & \cdots & \cdots \\ \cdots & \cdots & \cdots & \cdots & \cdots & \cdots & \cdots & \cdots & \cdots \\ a_{w-1,1} & a_{w-1,2} & \cdots & a_{w-1,u} & 0 & 0 & \cdots & -1 & 0 \\ a_{w1} & a_{w2} & \cdots & a_{wu} & 0 & 0 & \cdots & 0 & -1 \\ \cdots & \cdots & \cdots & \cdots & \cdots & \cdots & \cdots & \cdots & \cdots \\ \cdots & \cdots & \cdots & \cdots & \cdots & \cdots & \cdots & \cdots & \cdots \\ \cdots & \cdots & \cdots & \cdots & \cdots & \cdots & \cdots & \cdots & \cdots \\ a_{m1} & a_{m2} & \cdots & a_{mu} & 0 & 0 & \cdots & 0 & 0 \end{bmatrix} \quad \text{(A.10)}$$

and $\mathbf{x}$ and $\mathbf{d}$ are the column vectors, expressed as[2]

$$\mathbf{x} = (x_1, x_2, \cdots, x_u, x_{u+1}, \cdots, x_{u+v}) \quad \text{(A.11)}$$

$$\mathbf{d} = (d_1, d_2, \cdots, d_m). \quad \text{(A.12)}$$

In $\mathbf{A}$, the first two rows represent two examples of constraints with the $\leq$ inequality having positive coefficients $(+1)$ for slack variables, while the $(w-1)$th and wth rows indicate two examples of constraints with the $\geq$ inequality having negative coefficients (-1) for surplus variables. Note that, this arrangement ($+1$ coefficients for slack variables and -1 coefficients for surplus variables) ensures that the slack and surplus variables remain non-negative. Further, the mth row represents an example constraint with equality, with zero coefficients for all slack and surplus variables that are introduced by the other constraints with inequalities. However, it's convenient mathematically to convert all the constraints in the form of $\leq$ using simple manipulations. For example, by multiplying both sides by -1, one can transform a constraint from a $\geq$ form to a $\leq$ form. An equality-based constraint can also be transformed into a pair of constraints based on $\leq$ and $\geq$. For example, the following equality

$$a_{i1}x_1 + a_{i2}x_2 + \cdots + a_{in} = d_i \quad \text{(A.13)}$$

can be transformed as a pair of $\leq$ and $\geq$ constraints, given by

$$\begin{aligned} a_{i1}x_1 + a_{i2}x_2 + \cdots + a_{in} &\leq d_i \\ a_{i1}x_1 + a_{i2}x_2 + \cdots + a_{in} &\geq d_i. \end{aligned} \quad \text{(A.14)}$$

In view of the above, we continue below with the $\leq$ inequalities for all the constraints, expressed as

$$a_{i1}x_1 + a_{i2}x_2 + \cdots + a_{ij}x_j + \cdots + a_{im}x_m \;\; \{\leq\} \;\; d_i, \quad i = 1, ...m, \quad \text{(A.15)}$$

[2] Note that we don't use the transpose notation $(.)^T$ for expressing the column vectors in terms of their elements (i.e., $(x_1, x_2, \cdots, x_u, x_{u+1}, \cdots, x_{u+v})^T$ for $\mathbf{x}$) for notational simplicity. However, while carrying out row-column multiplications the transport notation is used explicitly for the sake of clarity, as in Eq. A.19.

leading to the matrix $\mathbf{A} = \{a_{ij}\}$ for the m original variables and m slack variables given by

$$\mathbf{A} = \begin{bmatrix} a_{11} & a_{12} & \dots & a_{1m} & 1 & 0 & \dots & 0 & 0 \\ a_{21} & a_{22} & \dots & a_{2m} & 0 & 1 & \dots & 0 & 0 \\ \dots & \dots & \dots & \dots & \dots & \dots & \dots & \dots & \dots \\ \dots & \dots & \dots & \dots & \dots & \dots & \dots & \dots & \dots \\ \dots & \dots & \dots & \dots & \dots & \dots & \dots & \dots & \dots \\ a_{m-1,1} & a_{m-1,2} & \dots & a_{m-1,m} & 0 & 0 & \dots & 1 & 0 \\ a_{m1} & a_{m2} & \dots & a_{mm} & 0 & 0 & \dots & 0 & 1 \end{bmatrix}. \tag{A.16}$$

In the above expression, the m columns with $i \in (m+1, 2m)$ (i.e., with $n = 2m$) form an $m \times m$ identity matrix. Overall, the non-negativity (feasibility) criterion continues to exist with all the variables of $\mathbf{x}$, i.e.,

$$\mathbf{x} \geq \mathbf{0}. \tag{A.17}$$

With $u = m$ and $n = 2m$, the cost function z of Eq. A.8 at this stage can be generalized using the vectors, $\mathbf{x} = (x_1, x_2, ..., x_m, ..., x_{2m})$ and $\mathbf{c} = (c_1, c_2, ..., c_m, ..., c_{2m})$, as

$$\begin{aligned} z &= \mathbf{cx} \\ &= \sum_{j=1}^{2m} c_j x_j \\ &= c_1x_1 + c_2x_2 + \cdots + c_m x_m + 0 \times x_{m+1} + \cdots + 0 \times x_{2m} \\ &= c_1x_1 + c_2x_2 + \cdots + c_m x_m \\ &= \sum_{j=1}^{m} c_j x_j, \end{aligned} \tag{A.18}$$

where the values of the c_j's in $\mathbf{c}$ with $j > m$ correspond to the slack variables and hence are equal to zero, i.e., $c_{m+1} = c_{m+2} = \cdots = c_{2m} = 0$. In other words, the slack variables being artificial in nature, don't make any contribution to the cost function.

Note that, $\mathbf{A}$ is a rectangular $(m \times 2m)$ matrix with m independent constraints (rows), and hence the rank of $\mathbf{A}$ is m. Thus, there are m independent column vectors in $\mathbf{A}$, known as *basis vectors*, each with m elements and all other column vectors of $\mathbf{A}$, called non-basis vectors, can be expressed as linear combinations of the basis vectors. Each set of m basis vectors forms an $m \times m$ basis matrix $\mathbf{B}$, and leads to a *basic solution*, designated as $\mathbf{x}_B$, having only m elements.

Hence, if we want to construct a $2m$-element solution for $\mathbf{Ax} = \mathbf{d}$ from an m-element basic solution $\mathbf{x}_B$, then m elements or variables of the $2m$-element solution would be the same as the elements of $\mathbf{x}_B$, and the remaining m elements (which are non-basic variables) of the $2m$-element solution can be made zero, so that the product of the ith row of $\mathbf{B}$, i.e., $\mathbf{r}_{Bi}$ and $\mathbf{x}_B$ equals to d_i, which in turn will be equal to $\mathbf{r}_{Ai}\mathbf{x}$ with $\mathbf{r}_{Ai}$ as the ith row of $\mathbf{A}$, i.e.,

$$\begin{aligned} \mathbf{r}_{Bi}\mathbf{x}_B &= (r_{B1}, r_{B2}, \cdots, r_{Bm})(x_{B1}, x_{B2}, \cdots, x_{Bm})^T \\ &= (a_{i1}, a_{i2}, \cdots, a_{im}, \cdots, a_n)(x_1, x_2, \cdots, x_m, \cdots, x_{2m})^T \\ &= \mathbf{r}_{Ai}\mathbf{x} = d_i, \end{aligned} \tag{A.19}$$

and

$$\mathbf{B}\mathbf{x}_B = \mathbf{d}. \tag{A.20}$$

Further, one can also express $\mathbf{Ax}$ as

$$\mathbf{Ax} = \mathbf{B}\mathbf{x}_B + \mathbf{B}_{NB}\mathbf{x}_{NB} = \mathbf{d}, \tag{A.21}$$

where $\mathbf{B}_{NB}$ is the non-basis matrix in $\mathbf{A}$ and $\mathbf{x}_{NB}$ represents the non-basis vector in $\mathbf{x}$. Further, with $\mathbf{x}_{NB} = \mathbf{0}$, Eq. A.21 gets simplified to Eq. A.20.

With the above framework, one can choose several possible combinations of m linearly independent columns out of $2m$ columns in $\mathbf{A}$ to form an acceptable $\mathbf{B}$. Thus, one can have ${}^{2m}C_m = \frac{2m!}{m!m!}$ formations of basis matrices $\mathbf{B}$ from $\mathbf{A}$. However, all of the m-element basic solutions ($\mathbf{x}_B$'s) obtained from these $\mathbf{B}$'s may not satisfy the non-negativity (i.e., feasibility) criterion. In order to solve the problem, one needs to choose only those basic solutions from this set, where all the m elements are non-negative, which are called as basic feasible solutions (BFSs). Note that, this solution will remain confined in the all-positive segment in the $2m$-dimensional convex space, defined by $\mathbf{Ax} = \mathbf{d}$ (as in the first quadrant of the two-variable LP illustration, as shown in Fig. A.2). With reference to the example illustrated in Fig. A.3, this will imply that, for a given polyhedron with $\frac{2m!}{m!m!}$ corner points in the $2m$-dimensional space, one needs to find the specific corner point or BFS that will maximize (or minimize) the objective function, thereby leading to the optimum solution. The Simplex algorithm is used to accomplish this task in a step-by-step manner.

A.4 Simplex algorithm

The first important step of the Simplex algorithm is to obtain an initial BFS. In subsequent steps the algorithm moves from the initial BFS to the next BFS and so on, eventually reaching the optimum solution in minimum number of steps. We discuss next the first step, where one finds an initial BFS through a simple observation.

A.4.1 Determining the initial BFS

A BFS would have no more than m non-zero elements corresponding to the rank m of $\mathbf{A}$ (Eq. A.16). So we first seek a BFS $\mathbf{x}_B$ (as defined in Eq. A.20), hereafter called an initial BFS, which will have m elements as the basic variables from a candidate basis matrix $\mathbf{B}$. Note that the sub-matrix of $\mathbf{A}$ with its last m columns (i.e., on the right side) and m rows forms a candidate $m \times m$ basis matrix $\mathbf{B}$. Hence, the easiest way to get the initial BFS $\mathbf{x}_B$, is to set the original variables (i.e., $x_1, x_2, ..., x_m$) at zero complying with the feasibility criterion (i.e., x_i's ≥ 0), which would in turn leave the m slack variables as the m nonzero elements forming the intended initial BFS $\mathbf{x}_B$. Thus, setting $x_1 = x_2 = \cdots = x_m = 0$, we obtain the initial BFS $\mathbf{x}_B$ for Eq. A.9 (with $\mathbf{A}$ given by Eq. A.16) as

$$\begin{aligned} \mathbf{x}_B &= (x_{B1}, x_{B2}, \cdots, x_{Bm}) \\ &= (x_{m+1}, x_{m+2}, \cdots, x_{2m}) \\ &= (d_1, d_2, \cdots, d_m), \end{aligned} \tag{A.22}$$

leading to a feasible solution $\mathbf{x}$ to $\mathbf{Ax} = \mathbf{d}$ corresponding to the initial BFS $\mathbf{x}_B$, given by

$$\begin{aligned}\mathbf{x} &= (0, 0, \cdots, 0, x_{B1}, x_{B2}, \cdots, x_{Bm}) \\ &= (\mathbf{0}, \mathbf{x_B}).\end{aligned} \tag{A.23}$$

Having obtained the initial BFS ($\mathbf{x}_B$) and the corresponding feasible solution $\mathbf{x}$ for $\mathbf{Ax} = \mathbf{d}$, we next estimate its cost z from Eq. A.18. Using Eq. A.18 and noting that the cost of the ith basic variable $c_{Bi} = c_{m+i}$, the cost z of the initial BFS can be expressed in a form involving c_{Bi}'s, given by

$$\begin{aligned}z &= \sum_{j=1}^{2m} c_j x_j \\ &= c_1 \times 0 + c_2 \times 0 + \cdots + c_m \times 0 + \sum_{i=1}^{m} c_{Bi} x_{Bi} \\ &= \sum_{i=1}^{m} c_{Bi} x_{Bi} \\ &= 0,\end{aligned} \tag{A.24}$$

with all c_{Bi}'s being zero as the slack variables have zero cost. However, the cost z will keep changing as the algorithm moves through the subsequent steps. Using the initial BFS with zero cost, we next move on to obtain the next possible BFS with higher value of z (for the maximization problem). This becomes the next step of the Simplex algorithm.

Note that, the above initial BFS is a utopian solution from a realistic viewpoint as it forces all the original variables of $\mathbf{x}$ to zero, thereby also producing a zero cost. Notwithstanding this aspect, the initial BFS serves as a useful starting point of the Simplex algorithm, as the algorithm is capable (as we shall soon see) of moving from any given BFS to another candidate BFS with higher cost value (for maximization problem), and thereafter systematically can reach the optimum BFS (as shown in the polyhedron example in Fig. A.3) in the minimum number of steps. Note that, in Fig. A.3, the initial solution at the bottom corner of the polyhedron *symbolically* represents the initial BFS of the Simplex algorithm. Having determined the initial BFS, one can thereafter judiciously transform one specific basis column vector in the basis matrix $\mathbf{B}$ into a non-basis column vector, while converting a non-basis column vector from the rest of $\mathbf{A}$ into a basis column vector, and thus obtain a different basis matrix with another set of basis column vectors (or simply basis vectors) leading to a changed BFS with improved cost. The algorithm keeps repeating this process iteratively until z is maximized (or minimized for the minimization problem).

A.4.2 Moving from one BFS to another with improved cost

As mentioned above, to change the BFS from the initial BFS, one needs to select one column of $\mathbf{A}$ which is not in $\mathbf{B}$ (i.e., select a non-basis column vector from $\mathbf{A}$) and replace it by a basis column vector while changing a basis column vector in $\mathbf{B}$ to a non-basis column vector. The choice of the column vectors has to be made in a way such that the cost function z increases or decreases as governed by the LP objective, i.e., cost maximization or minimization, respectively, and no element in the new basic solution becomes negative.[3] The transformation realized by using the above method will form a different basis matrix, leading to a new BFS. In effect, this operation should move the initial corner point (corresponding to the initial BFS) in the *convex solution space* to the next corner point or BFS, ensuring the change of the cost function in the appropriate direction, while maintaining the non-negativity criterion for all the elements (with at least one element being non-zero) in the new BFS. This process should continue by moving the initial BFS through the appropriate BFSs in a step-by-step manner, until the desired BFS is reached, ensuring the maximum value for the objective function. The essential steps to move from the initial or an intermediate BFS to the next candidate BFS will follow a generic transformation process using a judicious *column-replacement scheme.* In the following, we describe this scheme in further details.

Column-replacement scheme

First, we express the initial BFS $\mathbf{x}_B$, corresponding to the initial basis matrix $\mathbf{B}$, as

$$\mathbf{B}\mathbf{x}_\mathbf{B} = \mathbf{d}, \tag{A.25}$$

implying that $\mathbf{x}_B = \mathbf{B}^{-1}\mathbf{d}$. Next, we consider arbitrarily one of the non-basis columns of $\mathbf{A}$, say $\mathbf{a}_k$, which is located outside $\mathbf{B}$, and assume that $\mathbf{a}_k$ will have to be replaced by a basis column vector while making the rth basis column vector of $\mathbf{B}$ (i.e., $\mathbf{b}_r$) a non-basis column vector (see Fig. A.4).

By definition, $\mathbf{a}_k$ being at the moment a non-basis vector, can be expressed as a linear combination of the basis vectors in $\mathbf{B}$, given by

$$\mathbf{a}_k = \sum_{i=1}^{m} p_{ik}\mathbf{b}_i = p_{rk}\mathbf{b}_r + \sum_{i=1, i\neq r}^{m} p_{ik}\mathbf{b}_i. \tag{A.26}$$

Note that the index i in the above expression represents the ith column in $\mathbf{B}$ and hence the $(m+i)$th column in $\mathbf{A}$. Thus, $\mathbf{b}_r$ in $\mathbf{B}$ is the same as $\mathbf{a}_{m+r}$ in $\mathbf{A}$. Also note that the vector $\mathbf{b}_r$ is deliberately taken out of the summation on the right-hand side as its status is going to

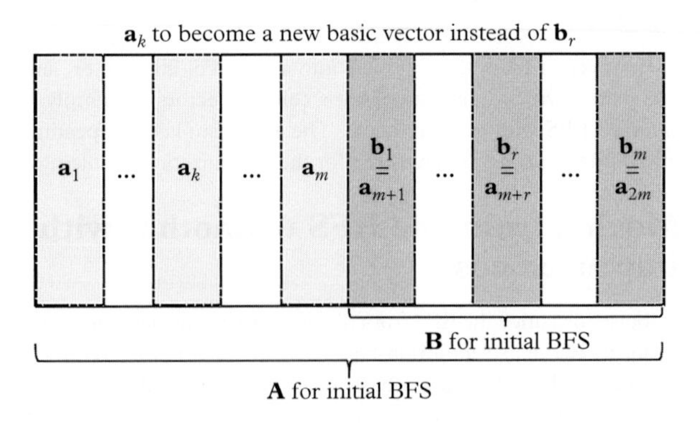

Figure A.4 *Column-replacement scheme in the Simplex algorithm.*

[3] Hereafter, we shall be discussing only on the maximization problem. The minimization problem is a *dual* of the maximization problem and can be solved using appropriate modification of the maximization problem.

change from a basis to a non-basis vector after the column-replacement process. Using the above expression, $\mathbf{b}_r$ can be expressed as

$$\mathbf{b}_r = \frac{\mathbf{a}_k}{p_{rk}} - \sum_{i=1, i\neq r}^{m} \frac{p_{ik}}{p_{rk}} \mathbf{b}_i. \tag{A.27}$$

Using $\mathbf{b}_r$ from the above expression, we express $\mathbf{Bx}_B$ as

$$\begin{aligned}
\mathbf{Bx}_B &= x_{B1}\mathbf{b}_1 + x_{B2}\mathbf{b}_2 + \cdots + x_{Br}\mathbf{b}_r + \ldots + x_{Bm}\mathbf{b}_m \qquad \text{(A.28)}\\
&= x_{B1}\mathbf{b}_1 + x_{B2}\mathbf{b}_2 + \cdots + x_{Br}\left(\frac{\mathbf{a}_k}{p_{rk}} - \sum_{i=1, i\neq r}^{m} \frac{p_{ik}}{p_{rk}}\mathbf{b}_i\right)\\
&\quad + \cdots + x_{Bm}\mathbf{b}_m\\
&= \sum_{i=1, i\neq r}^{m} x_{Bi}\mathbf{b}_i + x_{Br}\left(\frac{\mathbf{a}_k}{p_{rk}} - \sum_{i=1, i\neq r}^{m} \frac{p_{ik}}{p_{rk}}\mathbf{b}_i\right)\\
&= \sum_{i=1, i\neq r}^{m} \left(x_{Bi} - x_{Br}\frac{p_{ik}}{p_{rk}}\right)\mathbf{b}_i + \frac{x_{Br}}{p_{rk}}\mathbf{a}_k.
\end{aligned}$$

Given the fact that the left-hand side in the above expression is $\mathbf{Bx}_B$, it turns out that on the right-hand side, in the last line, all $\mathbf{b}_i$'s with $i \neq r$ continue to remain as basis vectors, while $\mathbf{a}_k$ (previously a non-basis vector) now assumes the role of a basis vector instead of $\mathbf{b}_r$. Hence, the coefficients of $\mathbf{b}_i$'s in the summation term, i.e., $\left(x_{Bi} - x_{Br}\frac{p_{ik}}{p_{rk}}\right)$ represent the new basic variables $\hat{x}_{Bi}$'s for $i = 1, 2, \cdots, m$, with $i \neq r$, and the coefficient of $\mathbf{a}_k$, i.e., $\frac{x_{Br}}{p_{rk}}$ in the second term represents the new basic variable $\hat{x}_{Br}$ for $i = r$. Thus, the coefficients of $\mathbf{b}_i$'s and $\mathbf{a}_k$ together constitute the new basic solution after the column replacement. However, for the new basic solution to become feasible, i.e., to become a new BFS, all of its elements must become non-negative, leading to the requirements, given by

$$\begin{aligned}
x_{Bi} - x_{Br}\frac{p_{ik}}{p_{rk}} &\geq 0, \quad \text{for} \quad i \neq r, \qquad \text{(A.29)}\\
\frac{x_{Br}}{p_{rk}} &\geq 0.
\end{aligned}$$

Note that $x_{Bi} \geq 0$, as x_{Bi}'s are the elements of a BFS and hence must be non-negative. However, $p_{rk} \neq 0$, as $\mathbf{b}_r$ can contribute to $\mathbf{a}_k$ only with a non-zero coefficient for itself in Eq. A.26. Presuming that $p_{rk} > 0$, the requirement for feasibility of the new solution can be stated as

$$\frac{x_{Bi}}{p_{ik}} - \frac{x_{Br}}{p_{rk}} \geq 0, \text{ with } p_{ik} > 0. \tag{A.30}$$

The above results imply that if the feasibility condition for basic solutions is fulfilled while transforming $\mathbf{a}_k$ into a basis vector instead of $\mathbf{b}_r$, one can get a set of new basis vectors $(\hat{\mathbf{b}}_1, \hat{\mathbf{b}}_2, \cdots, \hat{\mathbf{b}}_m)$ forming the modified basis matrix $\hat{\mathbf{B}}$, given by

$$\begin{aligned}
\hat{\mathbf{b}}_i &= \mathbf{b}_i \quad \text{for} \quad i \neq r, \qquad \text{(A.31)}\\
\hat{\mathbf{b}}_r &= \mathbf{a}_k,
\end{aligned}$$

which will in turn offer a new BFS $\hat{\mathbf{x}}_B$, given by

$$\hat{x}_{Bi} = x_{Bi} - x_{Br}\frac{p_{ik}}{p_{rk}} \quad \text{for} \quad i \neq r, \qquad \text{(A.32)}$$
$$\hat{x}_{Br} = \frac{x_{Br}}{p_{rk}}.$$

Consequently, the initial cost function z is changed to $\hat{z}$, where c_{Br} will be replaced by c_k due to the replacement of $\mathbf{b}_r$ by $\mathbf{a}_k$ in the basis vector set. We express $\hat{z}$ in terms of z, the elements of initial BFS $\mathbf{x}_B$, and p_{ik}'s, as

$$\hat{z} = \sum_{i=1}^{m} c_{Bi}\hat{x}_{Bi} = \sum_{i=1,i\neq r}^{m} c_{Bi}\left(x_{Bi} - x_{Br}\frac{p_{ik}}{p_{rk}}\right) + c_k\frac{x_{Br}}{p_{rk}}. \qquad \text{(A.33)}$$

Since $c_{Bi}(x_{Bi} - x_{Br}p_{ik}/p_{rk}) = 0$ for $i = r$, we simplify the above expression by including its zero-valued rth term within the summation (i.e., within $\sum$ on the right-hand side) to express z as

$$\begin{aligned}\hat{z} &= \sum_{i=1}^{m} c_{Bi}\left(x_{Bi} - x_{Br}\frac{p_{ik}}{p_{rk}}\right) + c_k\frac{x_{Br}}{p_{rk}} \qquad \text{(A.34)}\\ &= \sum_{i=1}^{m} c_{Bi}x_{Bi} - \frac{x_{Br}}{p_{rk}}\sum_{i=1}^{m} c_{Bi}p_{ik} + c_k\frac{x_{Br}}{p_{rk}}\\ &= z - \frac{x_{Br}}{p_{rk}}(z_k - c_k),\end{aligned}$$

where we define a new column-specific cost function z_k, given by

$$z_k = \sum_{i=1}^{m} c_{Bi}p_{ik}. \qquad \text{(A.35)}$$

Note that, z_k is the weighted sum of c_{Bi}'s using the coefficients p_{ik}'s used earlier in expressing $\mathbf{a}_k$ in terms of $\mathbf{b}_i$'s in Eq. A.26. Further, from Eq. A.34, it is evident $\frac{x_{Br}}{p_{rk}}(z_k - c_k)$ would play a significant role to assess whether $\hat{z}$ has changed from z in the right direction towards the optimum solution after the column replacement. In particular, $\frac{x_{Br}}{p_{rk}}$ being finite and non-negative, the sign of $(z_k - c_k)$ will eventually determine the selection of column $\mathbf{a}_k$ from $\mathbf{A}$ (outside $\mathbf{B}$) for the column replacement. We make use of this observation later for moving the BFS towards the optimum solution.

Impact of column replacement

Having obtained the conditions for column replacement, one needs to assess the impact of column replacement on the overall LP problem. In order to carry out this exercise, we consider some arbitrary column vector, say $\mathbf{a}_j$ in $\mathbf{A}$, and examine how this column gets affected by the column-replacement process involving $\mathbf{a}_k$ and $\mathbf{b}_r$. For this purpose, we first express the column vector $\mathbf{a}_j$ in terms of the column vector $\mathbf{b}_r$ as

$$\mathbf{a}_j = \sum_{i=1}^{m} p_{ij}\mathbf{b}_i = p_{rj}\mathbf{b}_r + \sum_{i=1,i\neq r}^{m} p_{ij}\mathbf{b}_i. \qquad \text{(A.36)}$$

Next, in Eq. A.36 we substitute $\mathbf{b}_r$ from Eq. A.27, to incorporate the impact of column replacement, leading to the expression of $\hat{\mathbf{a}}_j$, given by

$$\begin{aligned}
\hat{\mathbf{a}}_j &= p_{rj}\left(\frac{\mathbf{a}_k}{p_{rk}} - \sum_{i=1,i\neq r}^{m} \frac{p_{ik}}{p_{rk}}\mathbf{b}_i\right) + \sum_{i=1,i\neq r}^{m} p_{ij}\mathbf{b}_i \qquad (A.37)\\
&= \sum_{i=1,i\neq r}^{m} \underbrace{\left(p_{ij} - p_{ik}\frac{p_{rj}}{p_{rk}}\right)}_{\hat{p}_{ij}, i\neq r} \underbrace{\mathbf{b}_i}_{\hat{\mathbf{b}}_i, i\neq r} + \underbrace{\frac{p_{rj}}{p_{rk}}}_{\hat{p}_{rj}} \underbrace{\mathbf{a}_k}_{\hat{\mathbf{b}}_r}\\
&= \sum_{i=1}^{m} \hat{p}_{ij}\hat{\mathbf{b}}_i.
\end{aligned}$$

Equation A.37 implies that all $\mathbf{a}_j$'s are now modified as above, where the column vectors $\mathbf{b}_i$'s in $\mathbf{B}$ are changed to $\hat{\mathbf{b}}_i$'s in $\hat{\mathbf{B}}$ (actually only one column in $\mathbf{B}$, i.e., $\mathbf{b}_r$, is changed to $\mathbf{a}_k$ to form $\hat{\mathbf{B}}$ from $\mathbf{B}$). Consequently, all the associated coefficients (i.e., p_{ij}'s) for $\mathbf{a}_j$ are modified from p_{ij}'s to $\hat{p}_{ij}$'s as follows:

$$\begin{aligned}
\hat{p}_{ij} &= p_{ij} - p_{ik}\frac{p_{rj}}{p_{rk}} \quad \text{for} \quad i \neq r \qquad (A.38)\\
\hat{p}_{rj} &= \frac{p_{rj}}{p_{rk}}.
\end{aligned}$$

Note that, for $j = k$, we obtain from Eq. A.37

$$\begin{aligned}
\hat{p}_{ik} &= \left[p_{ij} - p_{ik}\frac{p_{rj}}{p_{rk}}\right]_{j=k} = 0 \quad \text{for} \quad i \neq r \qquad (A.39)\\
\hat{p}_{rk} &= \left[\frac{p_{rj}}{p_{rk}}\right]_{i=k} = 1,
\end{aligned}$$

implying that now the kth column vector (i.e., $\mathbf{a}_k$) in the modified $\mathbf{A}$ ($\hat{\mathbf{A}}$, say), i.e., ($\hat{p}_{1k}$, $\hat{p}_{2k}, \cdots \hat{p}_{r,k-1}, \hat{p}_{rk}, \hat{p}_{r,k+1}, \cdots \hat{p}_{mk}$) becomes equal to $(0, 0, \cdots, 0, 1, 0 \cdots, 0)$ with its rth element being the only non-zero element equaling unity. As we shall see later, Eq. A.38 will serve as an importatnt mathematical tool to evaluate the values of $\hat{p}_{ij}$'s after each step of column replacement to transform all the basis vectors in the initial $\mathbf{B}$ corresponding to the initial BFS to the ones located in the other columns in $\mathbf{A}$, thereby leading to the final BFS.

Next, we recall the column-specific cost function z_j for column j (as defined in Eq. A.35 for column k) and examine the impact of column replacement on $(z_j - c_j)$. The updated value of z_j after the column replacement, denoted by $\hat{z}_j$, is expressed from its definition as

$$\hat{z}_j = \sum_{i=1}^{m} \hat{c}_{Bi}\hat{p}_{ij} = \sum_{i=1,i\neq r}^{m} \hat{c}_{Bi}\left(p_{ij} - p_{rj}\frac{p_{ik}}{p_{rk}}\right) + \hat{c}_{Br}\frac{p_{rj}}{p_{rk}}. \qquad (A.40)$$

Note that, $p_{ij} - p_{rj} \times (p_{ik}/p_{rk}) = 0$ with $i = r$, $\hat{c}_{Bi} = c_{Bi}$ for $i \neq r$, and $\hat{c}_{Br} = c_k$ due to the column replacement, which allows us to replace $\sum_{i=1,i\neq r}^{m}$ simply by $\sum_{i=1}^{m}$ in the above expression and simplify $\hat{z}_j$ as

$$\hat{z}_j = \sum_{i=1}^{m} c_{Bi}\left(p_{ij} - p_{rj}\frac{p_{ik}}{p_{rk}}\right) + c_k\frac{p_{rj}}{p_{rk}} \tag{A.41}$$

$$= \underbrace{\sum_{i=1}^{m} c_{Bi}p_{ij}}_{z_j} - \sum_{i=1}^{m} c_{Bi}p_{rj}\frac{p_{ik}}{p_{rk}} + c_k\frac{p_{rj}}{p_{rk}}$$

$$= z_j - \frac{p_{rj}}{p_{rk}}\left(\sum_{i=1}^{m} c_{Bi}p_{ik} - c_k\right).$$

Recognizing that $\sum_{i=1}^{m} c_{Bi}p_{ik} = z_k$, we obtain $\hat{z}_j$ as

$$\hat{z}_j = z_j - \frac{p_{rj}}{p_{rk}}(z_k - c_k). \tag{A.42}$$

Using this expression, we obtain the updated version of $z_j - c_j$, i.e., $\hat{z}_j - c_j$, as

$$\hat{z}_j - c_j = (z_j - c_j) - \frac{p_{rj}}{p_{rk}}(z_k - c_k). \tag{A.43}$$

As evident from the above, the column-replacement scheme needs to compute the value of $(z_j - c_j)$ for each column in $\mathbf{A}$ and examine their modified values. In particular, for any column vector in $\mathbf{A}$, say column j, after a column replacement for column k, the change in $\hat{z}$ with respect to z must be positive (for maximization problem), which is ensured by a positive increment in $z_j - c_j$. Thus, the changes that occur for $(\hat{z}_j - c_j)$'s after a column-replacement step will have to be examined, and the value of j (i.e., jth column) for which $(\hat{z}_j - c_j)$ would ensure maximum positive increment (for the maximization problem), would qualify for the next column replacement, i.e., column j will assume the role of column k for the non-basis to basis vector transformation. We consolidate below the first and subsequent steps of this algorithm to reach the optimum solution.

Step-by-step column replacement

To begin with, we consider the first step of column replacement by moving the initial BFS to the next BFS, and thereafter generalize the first step for carrying out the subsequent steps.

1. In the first column-replacement step from the initial BFS, all z_j's will be zero, since all c_{Bi}'s are zero at this stage. Hence, one should choose a j for which $(z_j - c_j) = -c_j$ is most negative, as this will cause the largest positive increment in z $(= \hat{z} - z)$ when the corresponding column vector *enters* the set of basis vectors by column replacement. Denote this j as k, and go to Step 3.
2. For an intermediate BFS, one needs to check whether $(\hat{z}_j - c_j)$ for all values of j have assumed non-negative values. *If yes, then the transformation process is complete and the corresponding solution is optimum*; else, go to the next step.
3. Having chosen the appropriate j as k, one needs to choose appropriate value of r, i.e., column $\mathbf{b}_r$ in the original basis vector set. In other words, for a chosen value of k, one must choose a specific r, so that no element in the next BFS goes negative. Using Eq. A.30, we express the non-negativity criterion for the BFS as

$$\hat{x}_{Bi} = x_{Bi} - x_{Br}\frac{p_{ik}}{p_{rk}} \geq 0 \quad \text{for} \quad i \neq r, \tag{A.44}$$

which implies that one should choose the value of r, governed by the condition, given by

$$\frac{x_{Br}}{p_{rk}} = \min_i \left\{ \frac{x_{Bi}}{p_{ik}}, p_{ik} > 0 \right\}, \tag{A.45}$$

4. Compute the modified values of the necessary variables with the values chosen for k and r by using Equations A.32, A.38, A.42 and A.43 as follows:

Elements of modified BFS $\hat{\mathbf{x}}_B$:

$$\begin{aligned} \hat{x}_{Bi} &= x_{Bi} - x_{Br}\frac{p_{ik}}{p_{rk}} \quad i \in (1, m; \neq r) \\ \hat{x}_{Br} &= \frac{x_{Br}}{p_{rk}} \end{aligned} \tag{A.46}$$

Modified cost function $\hat{z}$:

$$\hat{z} = z - \frac{x_{Br}}{p_{rk}}(z_k - c_k) \tag{A.47}$$

Modified values of the coefficients $\hat{p}_{ij}$'s:

$$\begin{aligned} \hat{p}_{ij} &= p_{ij} - p_{ik}\frac{p_{rj}}{p_{rk}} \quad i \in (1, m+1, \neq r), \quad j \in (0, 2m) \\ \hat{p}_{rj} &= \frac{p_{rj}}{p_{rk}}, \quad j \in (0, 2m). \end{aligned} \tag{A.48}$$

Values of $\hat{z}_j - c_j$:

$$\hat{z}_j - c_j = (z_j - c_j) - \frac{p_{rj}}{p_{rk}}(z_k - c_k). \tag{A.49}$$

5. Return to Step 2.

In the Simplex algorithm, the above computational task to reach the optimum solution is executed customarily by bringing all the variables and parameters into a tabular format, called *tableau*, which keeps evolving from one BFS to another until the optimum solution is reached. In the following, first we describe the tableau formation process, and thereafter explain the use of tableau to execute the algorithm in each step of the transformation process.

A.4.3 LP solution using tableau formation

As described in the foregoing, the entire computational process, from the initial BFS to the final BFS (i.e., optimal solution), has to be carried out in a step-by-step manner. The first step of computation starts with the initial BFS and a matrix $\mathbf{P}_{ext} = \{p_{ij}\}$ constructed by extending the matrix $\mathbf{A}$ with one additional column on the left side and one additional row at the bottom. The additional column on the left side consists of the elements of the initial BFS $\mathbf{x}_B = \mathbf{d} = (d_1, d_2, \cdots, d_m)$ and the cost z. The additional row at the bottom consists of z (z being the common element for the additional row and column) and the column-wise values of $z_j - c_j$'s. Having constructed $\mathbf{P}_{ext}$, it is presented in a tabular form, named as Tableau 1. In each of the subsequent steps of column-replacement, a modified version of $\mathbf{P}_{ext}$ ($\hat{\mathbf{P}}_{ext}$, say) is obtained, which is represented by a modified tableau.

In the following, we first describe how Tableau 1 is constructed. Next, we make use of Tableau 1 to carry out the first column-replacement process and construct therefrom the next tableau, i.e., Tableau 2. Finally, we generalize the subsequent column-replacement steps needed to reach the optimal solution.

Forming the first tableau

As mentioned above, the matrix $\mathbf{P}_{ext}$ is formed by extending $\mathbf{A}$, by adding the elements of the initial BFS $\mathbf{x}_B$, the cost z, and the column-wise cost values $(z_j - c_j)$'s in the new column and row. First, we express $\mathbf{P}_{ext}$ as

$$\mathbf{P}_{ext} = \begin{bmatrix} p_{10} & p_{11} & \cdots & p_{1m} & \cdots & p_{1,2m} \\ p_{20} & p_{21} & \cdots & p_{2m} & \cdots & p_{2,2m} \\ \cdots & \cdots & \cdots & \cdots & \cdots & \cdots \\ \cdots & \cdots & \cdots & \cdots & \cdots & \cdots \\ \cdots & \cdots & \cdots & \cdots & \cdots & \cdots \\ p_{m0} & p_{m1} & \cdots & p_{mm} & \cdots & p_{m,2m} \\ p_{m+1,0} & p_{m+1,1} & \cdots & p_{m+1,m} & \cdots & p_{m+1,2m} \end{bmatrix}. \tag{A.50}$$

The $(m+1)$ elements in the first column of $\mathbf{P}_{ext}$ (which is the additional column inserted on the left side of $\mathbf{A}$) are governed by the initial BFS and the corresponding cost function z, given by

$$\begin{aligned} p_{10} &= x_{B1} = d_1 \\ p_{20} &= x_{B2} = d_2 \\ &\cdots \\ p_{m0} &= x_{Bm} = d_m \\ p_{m+1,0} &= z = 0. \end{aligned} \tag{A.51}$$

The elements of the last row (which is inserted as the additional row below $\mathbf{A}$) as

$$\begin{aligned} p_{m+1,0} &= z = 0 \\ p_{m+1,1} &= z_1 - c_1 = -c_1 \\ p_{m+1,2} &= z_2 - c_2 = -c_2 \\ &\cdots \\ p_{m+1,j} &= z_j - c_j = -c_j \\ &\cdots \\ p_{m+1,m} &= z_m - c_m = -c_m \\ p_{m+1,m+1} &= z_{m+1} - c_{m+1} = 0 \\ &\cdots \\ p_{m+1,2m} &= z_{2m} - c_{2m} = 0, \end{aligned} \tag{A.52}$$

where z and z_j's for all j's are zero for the initial BFS. The rest of the elements of $\mathbf{P}_{ext}$, i.e., p_{ij} for $i = 1$ to m and $j = 1$ to $2m$ are same as a_{ij} of $\mathbf{A}$ for $i = 1$ to m and $j = 1$ to $2m$, i.e.,

$$p_{ij} = a_{ij} \quad i \in (1, m), \quad j \in (1, 2m). \tag{A.53}$$

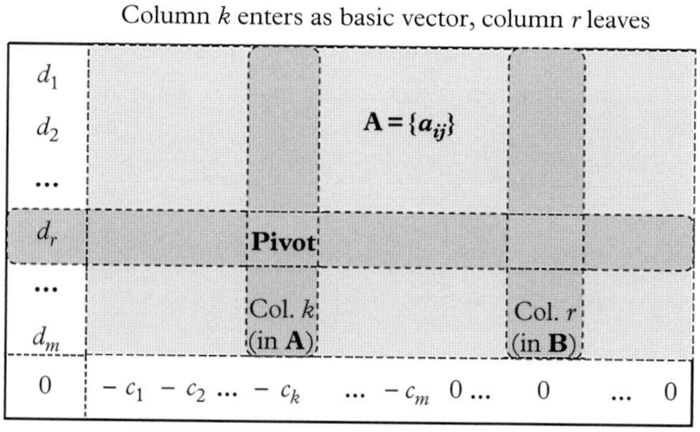

Column k enters as basic vector, column r leaves

d_1						
d_2			$\mathbf{A} = \{a_{ij}\}$			
...						
d_r		**Pivot**				
...		Col. k (in **A**)		Col. r (in **B**)		
d_m						
0	$-c_1$ $-c_2$ ...	$-c_k$	... $-c_m$	0 ...	0	... 0

Figure A.5 *Tableau 1 for the initial BFS, presenting* $\mathbf{P}_{ext}$ *in tabular form.*

Equations A.51 through A.53 are next used to form Tableau 1 corresponding to the initial BFS, as shown in Fig. A.5. We make use of Tableau 1 to determine the next BFS and the corresponding $\hat{\mathbf{P}}_{ext}$, leading to Tableau 2. Thereafter, we generalize the overall procedure.

Moving to the next BFS using Tableau 1 and generalization using subsequent tableaus

As discussed earlier, the task of moving to the next BFS is carried out by column replacement in $\mathbf{P}_{ext}$, leading to $\hat{\mathbf{P}}_{ext}$. In particular, with the initial BFS, i.e., $\mathbf{x}_B = \mathbf{d}$, one has to choose the appropriate column, called the entering column (column k, say), such that $p_{m+1,k} = z_k - c_k = -c_k$ has the most negative (i.e., with largest magnitude) value in the lowest row of $\mathbf{P}_{ext}$ (see Step 1 in the column-replacement process). Next, the column k in $\mathbf{A}$, i.e., $\mathbf{a}_k$, should enter the basis matrix, and an appropriate column in $\mathbf{B}$, say $\mathbf{b}_r$ should become a non-basis vector, while ensuring that the next solution $\hat{\mathbf{x}}_B$ doesn't have any negative element. The value of r is determined using Eq. A.45 and the values of $\mathbf{x}_{Bi}$'s and p_{ik}'s. The positions of the kth column in $\mathbf{A}$ and the rth column in $\mathbf{B}$ (i.e., the $(m+r)$th column in $\mathbf{A}$) are illustrated in Tableau 1, as shown in Fig. A.5. The crossover element between the $(k+1)$th column and the rth row of $\mathbf{P}_{ext}$ is called a *pivot*. Note that, the $(k+1)$th column of $\mathbf{P}_{ext}$ is the kth column of $\mathbf{A}$. The implication of the name pivot will become clear in course of the following discussion.

Next, with the appropriately chosen columns (i.e., column k and column r), we proceed to evaluate $\hat{\mathbf{x}}_B$, $\hat{z}$, $\hat{p}_{ij}$'s and $\hat{z}_j - c_j$'s. With the new set of values, we get the next version of $\mathbf{P}_{ext}$, i.e., $\hat{\mathbf{P}}_{ext}$, given by

$$\hat{\mathbf{P}}_{ext} = \begin{bmatrix} \hat{p}_{10} & \hat{p}_{11} & \cdots & \hat{p}_{1m} & \cdots & \hat{p}_{1,2m} \\ \hat{p}_{20} & \hat{p}_{21} & \cdots & \hat{p}_{2m} & \cdots & \hat{p}_{2,2m} \\ \cdots & \cdots & \cdots & \cdots & \cdots & \cdots \\ \cdots & \cdots & \cdots & \cdots & \cdots & \cdots \\ \cdots & \cdots & \cdots & \cdots & \cdots & \cdots \\ \hat{p}_{m0} & \hat{p}_{m1} & \cdots & \hat{p}_{mm} & \cdots & \hat{p}_{m,2m} \\ \hat{p}_{m+1,0} & \hat{p}_{m+1,1} & \cdots & \hat{p}_{m+1,m} & \cdots & \hat{p}_{m+1,2m} \end{bmatrix}, \tag{A.54}$$

where $\hat{p}_{ij}$'s are obtained from Equations A.46 through A.49 as follows:

$$\hat{p}_{i0} = \hat{x}_{Bi}, \quad i \in (1, m) \tag{A.55}$$
$$\hat{p}_{m+1,0} = \hat{z},$$

where

$$\hat{x}_{Bi} = x_{Bi} - x_{Br}\frac{p_{ik}}{p_{rk}}, \quad i \in (1, m; \neq r) \tag{A.56}$$

$$\hat{x}_{Br} = \frac{x_{Br}}{p_{rk}}$$

$$\hat{z} = z - \frac{x_{Br}}{p_{rk}}(z_k - c_k);$$

$$\hat{p}_{ij} = p_{ij} - p_{ik}\frac{p_{rj}}{p_{rk}}, \quad i \in (1, m+1; \neq r), j \in (0, 2m) \tag{A.57}$$

$$\hat{p}_{rj} = \frac{p_{rj}}{p_{rk}}, \quad j \in (0, 2m)$$

$$\hat{p}_{m+1,j} = \hat{z}_j - c_j, \quad j \in (1, 2m),$$

where

$$\hat{z}_j - c_j = (z_j - c_j) - \frac{p_{rj}}{p_{rk}}(z_k - c_k). \tag{A.58}$$

Note that, all the elements in $\mathbf{P}_{ext}$ go through the same transformation formula, and thus eventually the pivot-centric transformation of $\mathbf{P}_{ext}$ for all of its rows turns out to be a *Gaussian elimination method* for each step of column replacement. Through this process, the pivot element goes to unity and all other elements in the same column become zero, and accordingly the remaining elements in the matrix are transformed.

Using $\hat{\mathbf{P}}_{ext}$, we then form the next tableau, i.e., Tableau 2, as shown in Fig. A.6. The remaining column-replacement steps are carried out recursively as follows to reach the optimum solution.

1. Find the column (say column k) of the last $\hat{\mathbf{P}}_{ext}$ with the most negative value for $\hat{p}_{m+1,k} = \hat{z}_k - c_k$, implying that column k will enter in the basis matrix.
2. Determine the appropriate value of r, using the same method of computation as used in Eq. A.45.
3. Compute all the elements of $\hat{\mathbf{P}}_{ext}$ and form the new tableau.
4. Check whether or not all $(z_j - c_j)$'s are non-negative for the columns 1 to m. If negative values are there, go back to Step 1. Else, the algorithm is complete with the desired BFS.

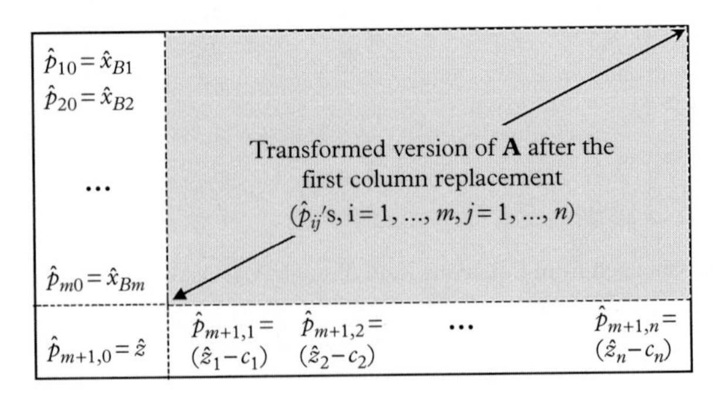

Figure A.6 *Tableau 2 for transformation by second column replacement.*

Following the above tableau-based computational steps, we present in the following a simple case study to illustrate the Simplex algorithm.

A.4.4 Case study

For the present case study, we recall the same two-variable LP problem, which was illustrated earlier geometrically using two-dimensional plots of the linear constraints. The constraint equations used in the same example are recalled below as

$$x_1 + 2x_2 \leq 7 \tag{A.59}$$

$$4x_1 + x_2 \leq 9, \tag{A.60}$$

along with the objective function

$$\max\{z = 4x_1 + 2x_2\}, \tag{A.61}$$

and the non-negativity criterion on the variables, given by

$$x_1 \geq 0 \tag{A.62}$$

$$x_2 \geq 0. \tag{A.63}$$

In order to determine the optimum solution using the Simplex algorithm, we first convert the inequalities into equalities using slack variables (x_3 and x_4) as

$$x_1 + 2x_2 + 1 \times x_3 + 0 = 7 \tag{A.64}$$

$$4x_1 + x_2 + 0 + 1 \times x_4 = 9, \tag{A.65}$$

leading to the matrix form of Eq. A.9 (i.e., $\mathbf{Ax} = \mathbf{d}$), with

$$\mathbf{A} = \begin{bmatrix} 1 & 2 & 1 & 0 \\ 4 & 1 & 0 & 1 \end{bmatrix}, \tag{A.66}$$

$$\mathbf{x} = (x_1 \ x_2 \ x_3 \ x_4), \tag{A.67}$$

and

$$\mathbf{d} = (7, 9). \tag{A.68}$$

The above equations, with the assumption that $x_1 = x_2 = 0$ at the initial stage, lead to the initial BFS $\mathbf{x}_B$, given by

$$\mathbf{x}_B = (x_{B1} \ x_{B2}) = (x_3 \ x_4) = (d_1 \ d_2) = (7, \ 9). \tag{A.69}$$

Using the above results, we next form $\mathbf{P}_{ext}$ for the present case study, as

$$\mathbf{P}_{ext} = \begin{bmatrix} 7 & 1 & 2 & 1 & 0 \\ 9 & 4 & 1 & 0 & 1 \\ 0 & -4 & -2 & 0 & 0 \end{bmatrix}, \tag{A.70}$$

leading to Tableau 1 as shown in Fig. A.7.

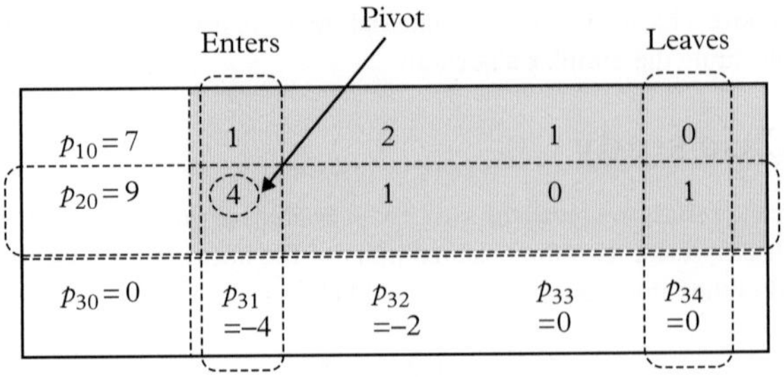

Pivot

Enters | Leaves

	Enters			Leaves
$p_{10}=7$	1	2	1	0
$p_{20}=9$	(4)	1	0	1
$p_{30}=0$	p_{31} =−4	p_{32} =−2	p_{33} =0	p_{34} =0

Figure A.7 *Tableau 1 for the present case study.*

As evident from Eq. A.70, second column of Tableau 1 from left is the most sensitive column with largest negative element (= −4) in the last row, and hence this column should enter as a basis vector by replacing a suitable column (column r of **B**, say) among the existing basis vectors represented by the fourth and fifth columns of Tableau 1. Going by the procedure set up in the foregoing, we use Eq. A.45 to carry out the search for an appropriate value for r as follows.

- Compute $\frac{x_{Bi}}{p_{ik}}$ with $k = 1$:
  For row 1: $i = 1$, $\frac{x_{B1}}{p_{11}} = 7/1 = 7$
  For row 2: $i = 2$, $\frac{x_{B2}}{p_{21}} = 9/4$ (x_{B2} corresponds to $r = 2$ in **B**)
- $\min\{\frac{x_{Bi}}{p_{i1}}\}$ is realized with $r = 2$, i.e., column 2 of **B** or column 5 (i.e., $j = 4$ in $\mathbf{P}_{ext}$ of Tableau 1. Hence column 5 of Tableau 1 should be replaced. Thus, column 2 of Tableau 1 *enters* as a basis vector and column 5 *leaves* as a basis vector and becomes a non-basis vector.

Having determined the entering and leaving columns for the basis matrix, we need to compute the changes effected by the column replacement on all the elements of the modified version of $\mathbf{P}_{ext}$, i.e., in $\hat{\mathbf{P}}_{ext}$. Using Equations A.55 through A.58, all the elements in $\hat{\mathbf{P}}_{ext}$ are calculated from $\mathbf{P}_{ext}$ as:

$$\begin{aligned} \hat{p}_{i0} &= 19/4, 9/4, 9, \quad i \in (1,3) \\ \hat{p}_{i1} &= 0, 1, 0, \quad i \in (1,3) \\ \hat{p}_{i2} &= 7/4, 1/4, -1, \quad i \in (1,3) \\ \hat{p}_{i3} &= 1, 0, 0, \quad i \in (1,3) \\ \hat{p}_{i4} &= -1/4, 1/4, 1, \quad i \in (1,3), \end{aligned} \tag{A.71}$$

leading to the formation of Tableau 2, as shown in Fig. A.8. Note that, in Tableau 2 the second element from the top of column 1, being in the row of the pivot element of Tableau 1 (which is now converted to unity after the first column replacement), assumes an intermediate value of the first basic variable for the new BFS.

The above results imply that, after the transformation, the third column of Tableau 2 is left with only one negative element with a value of −1. Further, in Tableau 2 column 2 is now (010) with the pivot of the first step becoming unity. The other elements of $\mathbf{P}_{ext}$ also get changed accordingly. As mentioned earlier, this implies that the above transformation has in effect executed a Gaussian elimination method on the matrix $\mathbf{P}_{ext}$ around the pivot. Furthermore, the last element in column 2 of Tableau 2, i.e., $\hat{p}_{31} = 0$, implying that the cost of this column has increased from −4 to 0, which is indeed desirable and has been achieved through the column-replacement operation.

$\hat{p}_{10}=\frac{19}{4}$	0	7/4	1	−1/4
$\hat{p}_{20}=\frac{9}{4}$	1	1/4	0	1/4
$\hat{p}_{30}=9$	$\hat{p}_{31}=0$	$\hat{p}_{32}=-1$	$\hat{p}_{3,3}=0$	$\hat{p}_{3,4}=1$

Pivot: 7/4; Enters: column 3; Leaves: column 4

Figure A.8 *Tableau 2 for the present case study.*

$\hat{p}_{10}=\frac{19}{7}$	0	1	4/7	−1/7
$\hat{p}_{20}=\frac{11}{7}$	1	0	−1/7	2/7
$\hat{p}_{30}=\frac{82}{7}$	$\hat{p}_{31}=0$	$\hat{p}_{32}=0$	$\hat{p}_{33}=4/7$	$\hat{p}_{34}=6/7$

Figure A.9 *Tableau 3 for the present case study.*

In the next iteration, as indicated above, we find that the most negative element in the last row of Tableau 2 is -1 in its column 3. By using again Eq. A.45, we find that column 4 of Tableau 2 should be replaced, considering the element located at the crossing of its row 1 and column 3 as the pivot. Using the same method as used for the formation of Tableau 2, we obtain Tableau 3 as shown in Fig. A.9, where the elements of the last row are non-negative, indicating that we have reached the final BFS as the optimum solution given by

$$(x_1, x_2) = (11/7, 19/7) \tag{A.72}$$

It would be instructive to get back to the earlier geometric illustration (Fig. A.2) of the same optimization problem considered in the present case study to demonstrate the Simplex algorithm. Recall that, in Fig. A.2, the linear plots of the constraint equations were shown as equalities, where the shaded region represented the convex space of the problem. In the Simplex algorithm the initial BFS was located at $(x_1, x_2, x_3, x_4) = (0, 0, 7, 9)$ in the 4D space, which for the 2D diagram in Fig. A.2 was located at the origin, i.e., at $(x_1, x_2) = (0, 0)$ (slack variables being of no significance in 2D diagram). Subsequently, in the Simplex algorithm the initial BFS moved through two steps of column replacement to the optimum solution given by $(x_1, x_2) = (11/7, 9/7)$, which is also the location of optimum solution obtained geometrically in Figure A.2.

Practical variations of LP problems

So far we have discussed the basic features of the Simplex algorithm, confined to the maximization problem, which was improved later to increase the computational efficiency. The LP problems may also have to optimize the solution based on the minimization of an objective function. In reality, the two problems are dual and the minimization problem can be carried out by making use of the framework developed for the maximization problem. Further, when the LP constraints have $\geq$ inequalities, the surplus variables (with negative coefficients) come in, preventing the basis matrix from becoming an identity matrix. This necessitates the introduction of artificial variables, and the LP matrix increases in size, needing some additional computations.

Some practical LP problem may at times need to minimize the maximum (*min-max*) of some linear combination of its variables as the objective function, which in effect moves away from the linear to the non-linear problem domain. These problems can also be solved within the framework of the Simplex algorithm, but at the cost of increased computational steps. In such cases, one needs to express the min-max objective function as

$$minimize \quad max_{i=1}^{m}\{z_i\}. \tag{A.73}$$

The above objective function is next modified with a new variable β as

$$\beta - z_i \geq 0, \quad \forall i \in [1, \cdots, m], \tag{A.74}$$

and

$$minimize \quad \beta. \tag{A.75}$$

Thus, in effect one needs to generate some more constraint equations, including β and the associated surplus variables for each new constraint, which are next included in the original LP formulation. This eventually transforms the nonlinear constraint into a set of linear constraints, but increasing the computational complexity due to the increased size of the matrix $\mathbf{P}_{ext}$ and the corresponding versions of tableaus.

Furthermore, some practical LP problems deal with only integer variables, leading to the integer LP (ILP) problems, while the LPs involving both the integer and floating-point variables are known as mixed-integer LP (MILP) problems. Practical LP solvers using the Simplex algorithm include a wide range of options, serving as efficient software packages: CPLEX, GLPK, etc. The analytical approach described in this appendix should help develop an insight into the underlying computational task in an LP solver, and the readers are referred to (Vanderbei 2008; Dantzig 1963; Hadley 1961) for further details.

Appendix B
Noise Processes in Optical Receivers

B.1 Noise sources in optical communication systems

The sources of noise processes observed in optical receivers originate from a wide range of devices, including photodetectors and receiver circuits, optical sources, optical amplifiers, etc. The two major noise components include additive thermal noise and a signal-dependent noise, known as *shot noise*. The thermal noise is generated in the receiver itself without any dependence on the received signal. However, the origin of the shot noise is in the optical source itself, where the emission of photons takes place in a non-deterministic manner. Though the average number of photons emitted from an optical source (LED or laser) during the transmission of a binary one in a given bit interval is determined by the driving current, the actual number of photons emitted during the bit interval and their emission times remain unpredictable, thereby making the emission process random in nature. Similarly, an optical source with incomplete extinction during binary zero transmission will also lead to non-deterministic photon emissions, though much less in number as compared to the emission during binary one transmission.

The emitted light from an optical source, after traversing through the optical fiber link(s) and lossy optical devices gets attenuated and the optical signal received at the photodetector has a lower power and hence fewer photons. These photons, due to their random emission process at the source end, arrive at the receiver also at random instants of time along with the non-deterministic photon count, which are thereafter photodetected to produce an equivalent electrical signal. The root cause of this two-dimensional randomness (in terms of the received photon count and the photon arrival times) in the received optical signal (and hence also in the photodetected electrical signal) lies therefore in the unpredictable photon-emission process in the optical source itself.

The other possible noise processes observed in optical receivers, particularly in WDM networks, include the ASE noise components from the optical amplifiers that the signal passes through. Furthermore, the optical switches, multiplexers, demultiplexers, and filters add interference from the other co-propagating lightwaves, which manifest themselves as crosstalk components and add to the overall noise variances, thereby degrading the SNR at the optical receiver. Furthermore, when the fibers are driven into nonlinear range of operation, FWM-induced crosstalk also causes degradation of the SNR.

The power spectra of optical sources are governed by the phase noise associated with the emitted lightwaves. Hence, the spectrum of optical sources needs to be analyzed for an estimate of their spectral width (or linewidth). As discussed in Chapter 2, the spectrum of lasers determines the extent of dispersion suffered by the optical signal, especially in the single-mode optical fibers. Further, in coherent optical communication systems, laser coherence (related to the linewidth) plays a significant role in determining the receiver BER.

In the following we discuss the statistical features of the various noise and crosstalk components in optical receivers, which are used in Chapter 10 to evaluate the impact of transmission impairments on the physical-layer performance of optical communication systems.

B.2 Shot noise

In order to have an insight into the shot noise statistics in an optical receiver, we first consider the incidence of a rectangular optical pulse of power P spread over a time interval T at the receiving surface of a *pin* photodiode (as a photodetector). Hence, the average number of photons received by the photodiode will be given by

$$\Lambda = \frac{PT}{hf}, \tag{B.1}$$

with h as Planck's constant and f as the frequency of the photons or the associated lightwave. Given the average number Λ of photons arriving in the bit interval T, the exact number of photon arrivals k (say) will be random, governed by the quantum-mechanical nature of photon emission. Moreover, the photon emission times at the source, and hence also the photon arrival times at the receiver, would be memoryless. Thus, k would follow a Poisson distribution, just as in a queue receiving a memoryless arrival process (see Appendix C). Hence, the probability of k for a given Λ can be expressed as the Poisson distribution, given by

$$P(k/\Lambda) = \frac{\Lambda^k \exp(-\Lambda)}{k!}. \tag{B.2}$$

Each of these k photons would generate a photoelectron (hole–electron pair) upon arrival i.e., at $t_j \in [0, T]$. In a photodiode, to obtain the number of photoelectrons, k should be replaced by ηk due to the photon absorption process in the photodiode, with η as its quantum efficiency (see Chapter 2). Each photoelectron, while crossing through the photodiode and the load resistance, would generate a photocurrent impulse $h_p(t)$ with an area of electronic charge q. The shape of $h_p(t)$ will be governed by the transit time of the charge carriers in the photodiode, the parasitic circuit elements therein, and the input impedance of the following electronic circuit in the receiver, mainly the equivalent input resistance and capacitance of the receiver preamplifier.

Figure B.1 illustrates a sample photocurrent variation in response to the arrivals of photons in a photodiode in a given interval T, with each photon generating one impulse $h_p(t-t_j)$ at a random arrival instant t_j.[1] With this visualization of the photodetection process, we express the photocurrent $i_p(t)$ as a sum of k such randomly occurring impulse functions at time instants t_j's, i.e., $\{h_p(t - t_j)\}$, given by

$$i_p(t) = \sum_{i=0}^{k} h_p(t - t_j), \tag{B.3}$$

where $h_p(t)$, as indicated above, represents a current impulse generated by the flow of an electronic charge q across the photodiode, and hence is integrable to an area equaling to the electronic charge q, i.e.,

$$\int_0^\infty h_p(t)dt = q. \tag{B.4}$$

[1] In reality, k photons will be able to generate ηk photoelectrons, with η (< 1) as the quantum efficiency of the photodiode. In the immediate discussion, we carry on with the assumption that $\eta = 1$, and include later the effect of η on the photodetection process.

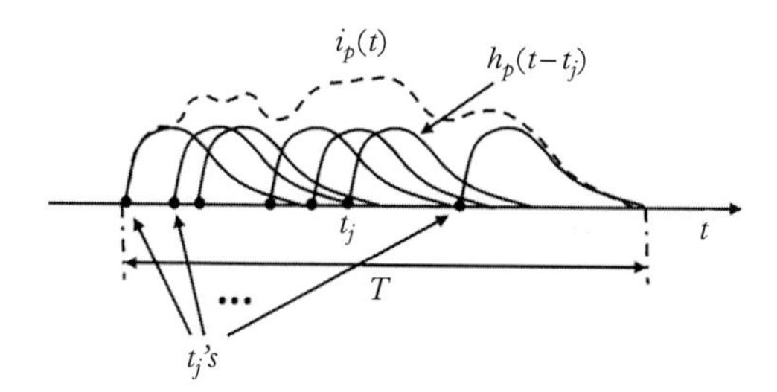

Figure B.1 *Illustration of a typical variation of photocurrent $i_p(t)$ with time, developed from the photons arriving in a photodiode.*

For the present case of photon arrivals during a bit interval T, the upper limit for the integration can be set at T for practical purposes, though the tails of last few arrivals might spill over the bit boundary.

Next, by using the basic model of $i_p(t)$, we proceed to find out the spectral and other relevant properties of the photocurrent. First, we note that the variable k being Poisson-distributed, its mean and variances would be equal, given by

$$E[k] = \sigma_k^2 = \Lambda. \tag{B.5}$$

Further, the photocurrent being the superposition of independently arrived identical impulses, would practically follow a Gaussian probability density function (pdf). However, the Gaussian assumption might not work well when the impulses are sharp and few in number within the interval T, thereby preventing adequate overlap of the adjacent impulses; nevertheless, we assume that the received power $(= \frac{\Lambda T}{hf})$ is not so small, and hence carry on with the Gaussian assumption. With this observation, we next determine the average and autocorrelation function for $i_p(t)$, which will finally enable us to examine its spectral behavior.

The ensemble average of $i_p(t)$ (which would also be the time average, given that the process can be considered ergodic), we note that, $i_p(t)$ has two dimensions of randomness in k and t_j, where k is a Poisson-distributed count variable, while t_j's can be shown to be uniformly distributed in a time interval of duration T (i.e., the duration of a bit) with a probability density function $p_i(t) = 1/T$ (Davenport and Root 1958; Gagliardi and Karp 1995). Therefore, the ensemble average of $i_p(t)$ can be obtained by averaging it over the statistics of k and t_j as

$$\begin{aligned} E[i_p(t)] &= E_{k,j}[i_p(t)] \\ &= E_{k,j}\Big[\sum_{j=1}^{k} h_p(t-t_j)\Big] \\ &= E_k\Big[E_j\Big[\sum_{j=1}^{k} h_p(t-t_j)\Big]\Big] \\ &= E_k\Big[\int_0^T dt_1 \int_0^T dt_2 \ldots \int_0^T dt_j \ldots \int_0^T dt_k \\ &\quad \times p(t_1)p(t_2)\ldots p(t_j)\ldots p(t_k)\sum_{j=1}^{k} h_p(t-t_j)\Big]. \end{aligned} \tag{B.6}$$

Noting that, $p(t_j) = 1/T \in [0, T]$, $E[k] = \Lambda$ and $\int_0^T h_p(t - t_j)dt = q$ for any t_j, we simplify $E[i_p(t)]$ as

$$E[i_p(t)] = E_k\left[\int_0^T \frac{dt_1}{T}\int_0^T \frac{dt_2}{T}\cdots\int_0^T \frac{dt_j}{T}\cdots\int_0^T \frac{dt_k}{T} \times \sum_{j=1}^{k} h_p(t - t_j)\right] = E_k\left[k \times \frac{q}{T}\right] = \frac{q}{T}E[k] = \frac{q\Lambda}{T} = q\lambda = I, \tag{B.7}$$

where $\lambda = \Lambda/T$ is the arrival rate of photons at the photodiode in the optical receiver and I represents the ideal value of the average photocurrent under the assumption of $\eta = 1$.

Following the similar statistical framework, the autocorrelation function of $i_p(t)$ can be expressed as

$$R(\tau) = E_{k,j}[i_p(t)i_p(t-\tau)] = E_{k,j}\left[\sum_{j=1}^{k} h_p(t - t_j)\sum_{l=1}^{k} h_p(t - \tau - t_l)\right], \tag{B.8}$$

where, for the second summation, the suffix has been changed from j to l, to distinguish between the terms with $j = l$ and $j \neq l$. We denote the sum of terms with $j = l$ as $S_{j=l}$ and the sum of the terms with $j \neq l$ as $S_{j\neq l}$, so that

$$R(\tau) = E_k\big[S_{j=l} + S_{j\neq l}\big], \tag{B.9}$$

with k^2 terms in total, where $S_{j=l}$ will have k terms and $S_{j\neq l}$ will have $k^2 - k$ terms. By using the statistical features of t_j and k as before and going through some algebraic simplifications (Davenport and Root 1958; Gagliardi and Karp 1995), we obtain $S_{j=l}$ and $S_{j\neq l}$ as

$$S_{j=l} = \frac{k}{T}\int_0^T h_p(t)h_p(t-\tau) = \frac{k}{T}R_h(\tau) \qquad S_{j\neq l} = (k^2 - k)\left(\frac{q}{T}\right)^2, \tag{B.10}$$

where $R_h(\tau)$ represents the autocorrelation function for $h_p(t)$. Noting that $E[k^2] = (E[k])^2 + \sigma_k^2 = \Lambda^2 + \Lambda$, we obtain $R_p(\tau)$ as

$$R(\tau) = E_k\left[\frac{k}{T}R_h(\tau) + (k^2 - k)(\frac{q}{T})^2\right] = \lambda R_h(\tau) + (\lambda e)^2. \tag{B.11}$$

At this stage, as mentioned earlier, we introduce a necessary correction factor in Equations B.7 and B.11, to replace λ by $\eta\lambda$, because the actual number of photoelectrons contributing to the photocurrent will be fewer by the factor η due to the absorption loss in the photodiode. Hence, the average value and its autocorrelation function of $i_p(t)$ are now modified as

$$I_p = \eta q\lambda = \eta I = \frac{\eta q}{hf}P = R_w P \tag{B.12}$$
$$R_p(\tau) = \eta\lambda R_h(\tau) + I_p^2,$$

where R_w is the responsivity of photodiode (as defined in Chapter 2). The expression of I_p is η times the ensemble average I, which is expected due to the ergodic assumption for the photocurrent. Similarly, $R_p(\tau)$ has a constant term equaling I_p^2, and the other term is proportional to the autocorrelation function of the impulse response $h_p(t)$ of the photodiode. This leads us to visualize the photocurrent $i_p(t)$ as a noise process with a DC component I_p along with a zero-mean noise component, which is recognized as the shot-noise current $i_{sh}(t)$ (also referred to as filtered-Poisson process) with a Gaussian probability density function. We therefore express $i_p(t)$ as a sum of its mean and random components, given by

$$i_p(t) = I_p + i_{sh}(t), \tag{B.13}$$

where the shot noise component $i_{sh}(t)$ will have an autocorrelation function given by

$$R_{sh}(\tau) = R_p(\tau) - I_p^2 = \eta\lambda R_h(\tau). \tag{B.14}$$

Using the above expression, we next obtain the spectral density of photocurrent by taking the Fourier transform of the autocorrelation function (using the Weiner–Kintchine theorem) as

$$S_{sh}(f) = FT[R_{sh}(\tau)] = \eta\lambda|H_h(f)|^2, \tag{B.15}$$

with $H_h(f)$ representing the Fourier transform (and hence the frequency response) of $h_p(t)$, which in turn leads to the overall spectral density of the photocurrent as

$$S_p(f) = \eta\lambda|H_h(f)|^2 + I_p^2\delta(f), \tag{B.16}$$

with $H_h(0) = \int_0^\infty h_p(t)dt = q$. Figure B.2 illustrates a representative plot of $S_p(f)$ with its DC component $I_p^2\delta(f)$ and the continuous spectrum following the profile of $|H_h(f)|^2$. For a practical receiver, the frequency response of a photodiode will be much wider than the receiver bandwidth B_e and hence the continuous part of the spectrum will practically remain flat at a value of qI_p ($= \eta q\lambda$, see Eq. B.12) over the two-sided electronic bandwidth $[-B_e, B_e]$. As a result, the shot noise power (also the variance as it has a zero mean) would

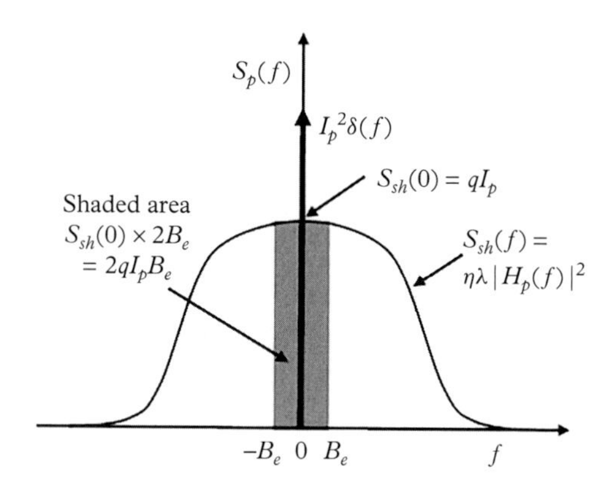

Figure B.2 *Spectral density $S_p(f)$ of photocurrent. Note that, the strength of the impulse function at $f = 0$ represents the DC or the* signal component *of $i_p(t)$, while the continuous spectrum represents the* zero-mean shot-noise *process, with its shaded area giving the* shot-noise variance *observed at the receiver input, band-limited by electrical filtering.*

be given by the power captured by a rectangle of height $\eta\lambda|H_h(0)|^2 = \eta\lambda q^2 = qI_p$ and width $2B_e$, given by

$$\sigma_{sh}^2 = qI_p \times 2B_e = 2qI_pB_e = 2qR_wPB_e. \tag{B.17}$$

Equations B.12 and B.17 expressing the average and variance of the photocurrent play a significant role in formulating the statistical model of optical receivers, along with the other noise components as described in the following.

B.3 Thermal noise

Thermal noise is omnipresent in all kinds of communication receivers, and it is generated from the random movements of charge carriers due to various physical phenomena occurring within the receiver. In general, the power P_{th} associated with thermal noise is expressed as

$$P_{th} = 4k\theta B_e, \tag{B.18}$$

where k is the Boltzman constant and θ is the receiver temperature in °K.

In general, the thermal noise follows the Gaussian distribution and can be modeled as either an electrical voltage or a current source. In other words, the above noise power can be thought of being generated from a noisy electrical source, that can be conceived as a Thevenin voltage source or a Norton current source, as shown in Fig. B.3, where the equivalent voltage or current source delivers the noise power into a load resistance R. As shown in Fig. B.3(a), the voltage source model fits well into the receivers, where the input amplifier is a voltage amplifier with high input resistance R (functioning as a load resistance to the source) attempting to extract maximum voltage in a noisy ambience from the input device (say, from an antenna) as a voltage source with a small internal resistance r_v ($\ll R$) in series. On the other hand, as shown in Fig. B.3(b), there are cases where the input amplifier operates as a current-to-voltage amplifier (i.e., a transimpedance amplifier) with low input resistance R, receiving the input signal in the form of current from a noisy current source (e.g., a photodiode, which we justify in the following) having a high internal resistance r_i ($\gg R$) in shunt. Thus, the noise power P extracted by a load resistance R from a noisy source can be represented either as $P = \overline{v_n^2}/R$ (with $r_v \ll R$ for a noisy voltage source), or as $P = \overline{i_n^2}R$ (with $r_i \gg R$ for a noisy current source), determined by the nature of the source under consideration.

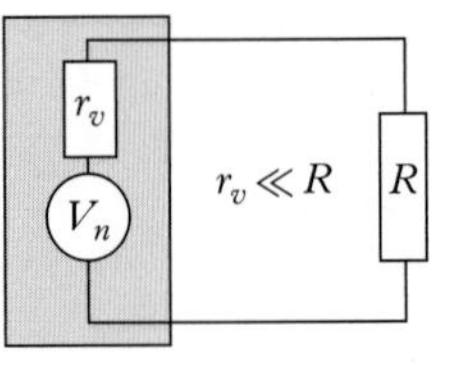

(a) Noise process as a voltage source.

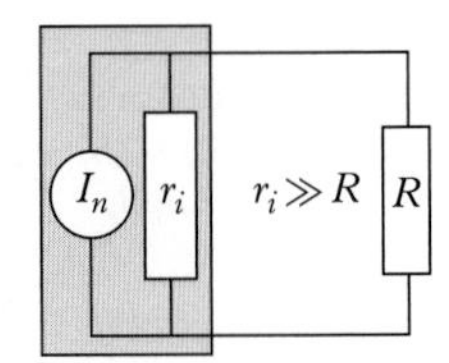

(b) Noise process as a current source.

Figure B.3 *Sources of noise processes viewed as (a) voltage and (b) current source, following Thevenin's and Norton's equivalent circuits, respectively. Note that, in optical receivers, the current-source model is used which adds up with shot noise current $i_{sh}(t)$ to represent the total receiver noise current in single-wavelength transmission system. In WDM receivers, other noise components come in, as discussed later.*

Next, we note that, the photocurrent generated in a photodiode is determined by the photon arrival rate (i.e., the received optical power), and hence doesn't change if the load resistance R is changed, while the voltage across R changes with R. In other words, the photodiode functions as a current source (Norton's model, as shown in Fig. B.3(b)) at the input of an optical receiver, with an input amplifier with low input resistance to extract current from the photodiode. The low input resistance of the transimpedance amplifier, in effect, competes with the high internal resistance of the photodiode (in shunt) to draw the maximum possible current from the inherent current source in the photodiode, just as the high input resistance of a voltage amplifier competes with the low internal resistance of the input voltage source (in series) to extract maximum voltage from the voltage source. Hence, with the input transimpedance amplifier (preamplifier) having low input resistance, the thermal noise source in an optical receiver will manifest itself practically as a current source with the *current variance* expressed as

$$\sigma_{th}^2 = \overline{i_n^2} - \underbrace{(\overline{i_n})^2}_{=0} = \frac{P}{R} = \frac{4k\theta B}{R}, \tag{B.19}$$

where $\overline{i_n}$ is considered to be zero, as the thermal noise has a zero mean. The thermal-noise variance σ_{th}^2 being dimensionally the same as the shot-noise variance (both having the dimension of current-squared, see Eq. B.17) adds up with σ_{sh}^2 to give the total receiver noise variance due to the thermal and shot noise components. Note that, for the receivers receiving signal with voltage amplifiers having high input resistances compared to the input devices, the thermal noise will follow the Thevenin model (Fig. B.3(a)) and manifest itself as a voltage source with the voltage variance of $4k\theta BR$.

B.4 ASE noise

Optical amplifiers used in the optical communication systems, while amplifying the input signals, add noise generated from the ASE. In a long-haul network, a connection set up on a wavelength may have to pass through multiple optical amplifiers, each of them contributing some amount of ASE noise, which co-propagate with the signal along the rest of the path. At the receiving node, along with the signal these noise components enter the receiver through a band-limiting optical filter followed by a photodiode. The signal, as well as the noise components, while being photodetected, interact nonlinearly with each other following the *square-and-average* functionality of the photodiode (see Chapter 2), thereby producing beat frequencies, or beat noises. These beat noise components enter the receiver following the photodiode, if they fall within the electronic bandwidth B_e of the receiver. As a result, when optical amplifiers are used in an optical link, the overall electrical noise process in the optical receiver can become more complex than what we discussed in the earlier sections.

Next, we describe the beat-noise generation process following the model in (Olsson 1989), and examine the electrical noise variances in the receiver contributed by the ASE noise in EDFAs as the optical amplifiers. We carry out the analysis for one single amplifier placed before the receiver, while the analysis can be readily applied to the accumulated ASE noise from multiple optical amplifiers, by enumerating the ASE contributions along the signal path and treating the total accumulated ASE noise as one single entity (see Chapter 10).

Consider an optical amplifier, through which a signal (lightwave) at a given frequency f_s (or wavelength w_s) passes, with a power gain G, and is then received by an optical receiver having its optical filter tuned to f_s with an optical bandwidth B_o. The amplifier adds ASE noise to the signal, and the received ASE noise power P_{sp} at the filter output in the receiver can be expressed as

$$P_{sp} = n_{sp}(G-1)hfB_o = N_{sp}B_o, \tag{B.20}$$

where n_{sp} represents the ionization factor in the amplifier and $N_{sp} = n_{sp}(G-1)hf$ is the one-sided power spectral density of the ASE noise.

The electromagnetic wave $E_s(t)$ associated with the received optical signal can be simply expressed as (i.e., without any phase noise and initial phase as epoch, which are of lesser significance at this stage)

$$E_s(t) = \sqrt{2GP_{in}}\cos(2\pi f_s t), \tag{B.21}$$

with P_{in} as the optical power at the optical amplifier input, which is contaminated by ASE noise of power P_{sp}. The received ASE noise remains bandlimited within the bandwidth B_o around f_s (i.e., within $[f_s - B_o/2, f_s + B_o/2]$) by the optical filter at the receiver input, which is assumed to have an ideal rectangular transfer function around f_s. Hence, as shown in Fig. B.4, the ASE noise spectrum assumes a rectangular shape with the one-sided power spectral density N_{sp}, such that the ASE noise power at the filter output becomes $P_{sp} = N_{sp}B_o = n_{sp}(G-1)hfB_o$, as defined in Eq. B.20.

For the convenience of analysis, the filtered rectangular ASE noise spectrum is split into $2J+1$ *spectral slices*, each with a bandwidth of $\delta f = \frac{B_o}{2J+1}$ (Fig. B.4), and the noise in each spectral slice is represented by a sinusoid, with a power of $\sqrt{2N_{sp}\delta f}$ at a frequency $(f_s + j\delta f)$ for $j = 0, \pm 1, \ldots \pm J$. Thus, the field associated with the ASE noise received at the photodiode is represented as a sum of sinusoids, given by

$$E_{sp}(t) = \sum_{j=-J}^{J} \sqrt{2N_{sp}\delta f}\cos\{2\pi(f_s + j\delta f)t + \phi_j\}, \tag{B.22}$$

where ϕ_j is a random phase (epoch) associated with each term, uniformly distributed over the interval $[0, 2\pi]$, making each term a stationary process. Further, all ϕ_j's are independent of each other, which is an essential feature of the noise model preventing each spectral slice and its mirror image around the central frequency f_s, i.e., $f_s \pm j\delta f$ from behaving as the sidebands of a modulated carrier.

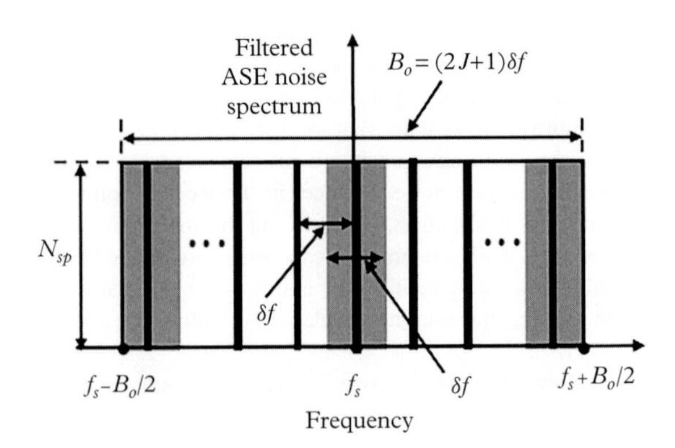

Figure B.4 *Power spectral density of the filtered ASE noise with B_o as the bandwidth of the optical filter centered around f_s. The entire spectrum is split into $2J+1$ spectral slices (shaded stripes) separated by δf. Each spectral slice of bandwidth δf has a power of $N_{sp}\delta f$, and is modeled as a sinusoid having an amplitude of $\sqrt{2N_{sp}\delta f}$ at a frequency located at the center of the slice (i.e., at $f_s + j\delta f, j = 0, \pm 1, \ldots, \pm J$) along with an arbitrary phase.*

Using $E_s(t)$ and $E_{sp}(t)$, we next express the photocurrent using the square-and-average functionality of the photodiode, given by

$$i_p(t) = R_w\overline{[E_s(t) + E_{sp}(t)]^2}. \tag{B.23}$$

Utilizing the models for $E_s(t)$ and $E_{sp}(t)$ from Equations B.21 and B.22 in Eq. B.23, $i_p(t)$ is expressed as

$$\begin{aligned} i_p(t) &= R_w\overline{[E_s^2(t) + E_{sp}^2(t) + 2E_s(t)E_{sp}(t)]} \\ &= I_p + \underbrace{A\overline{\left[\sum_{j=-J}^{J} \cos(2\pi f_s t)\cos\{2\pi(f_s + j\delta f)t + \phi_j\}\right]}}_{i_{s-sp}(t)} \\ &\quad + \underbrace{B\overline{\left[\sum_{j=-J}^{J} \cos\{2\pi(f_s + j\delta f)t + \phi_j\}\right]^2}}_{i_{sp\text{-}sp}(t)} \\ &= I_p + i_{s\text{-}sp}(t) + i_{sp\text{-}sp}(t), \end{aligned} \tag{B.24}$$

where I_p, A, and B are given by

$$\begin{aligned} I_p &= R_w G P_{in} \\ A &= 2R_w\sqrt{2GP_{in} \times 2N_{sp}\delta f} = 4R_w\sqrt{GP_{in}N_{sp}\delta f} \\ B &= 2R_w N_{sp}\delta f. \end{aligned} \tag{B.25}$$

In Eq. B.24, the first term I_p on the right-hand side represents the signal component of the photocurrent. The second term $i_{s-sp}(t)$ with the coefficient A represents the sum of all possible beat-noise components resulting from the product of the signal and each spontaneous noise (ASE) term, and is called the signal-spontaneous (*s-sp*) beat-noise current. The third term $i_{sp-sp}(t)$ with the coefficient B represents the sum of all beat-noise terms produced through the product of each pair of spontaneous terms, called the spontaneous-spontaneous (*sp-sp*) beat-noise current. Generally, the *s-sp* terms are stronger, as the signal is stronger than the ASE noise power in a spectral slice, while *sp-sp* terms are indeed weaker, although large in number. We examine the statistical features of these random beat noise currents, $i_{s\text{-}sp}(t)$ and $i_{sp\text{-}sp}(t)$, in the following.

The expression for $i_{s\text{-}sp}$ current can be simplified as

$$\begin{aligned} & i_{s\text{-}sp}(t) \\ &= \frac{A}{2}\overline{\left[\sum_{j=-J}^{J} \cos\{2\pi(2f_s + j\delta f)t + \phi_j\} + \sum_{j=-J}^{J} \cos(2\pi j\delta ft + \phi_j)\right]} \\ &= \frac{A}{2}\sum_{j=-J}^{J} \cos(2\pi j\delta ft + \phi_j)], \end{aligned} \tag{B.26}$$

where the high-frequency terms in the form of $\cos\{2\pi(2f_s + j\delta f)t + \phi_j\}$ are averaged out to zero over the effective time constant of the photodiode (equivalent to the inverse of the overall speed of the photodiode, which is way above the time period of the lightwave frequency). Note that the individual components of the sum in the final expression of $i_{s-sp}(t)$ have equal amplitude and zero average and ϕ_j's are random over $[0, 2\pi]$. Hence, $i_{s-sp}(t)$ will have a zero mean and the total power will be the sum of the powers in each component, given by

$$\overline{i^2_{s\text{-}sp}} = \frac{1}{2} \times \frac{A^2}{4} \times (2J+1) = 2R_w^2 P_{in} N_{sp} G B_o, \tag{B.27}$$

where $2J+1$ has been substituted by $\frac{B_o}{\delta f}$. Note that the above *s-sp* beat-noise will be white Gaussian in nature due to the large number of the spectral slices (theoretically infinite with $\delta f \to 0$) with identical power and identically distributed random phases, albeit band-limited in the two-sided bandwidth B_o due to the prior optical filtering. With these observations, we express the two-sided spectral density N_{s-sp} noise as

$$N_{s\text{-}sp} = 2R_w^2 P_{in} N_{sp} G, \ \text{ for } f \in [-B_o/2, B_o/2]. \tag{B.28}$$

In general, the optical bandwidth B_o will be larger than the electronic bandwidth B_e of the receiver (to avoid signal truncation due to laser frequency drifts and misalignments), and thus the receiver will eventually have a *s-sp* variance $\sigma^2_{s\text{-}sp}$ as the product of the two-sided spectral density $N_{s\text{-}sp}$ and the two-sided electronic bandwidth $2B_e$, given by

$$\sigma^2_{s\text{-}sp} = N_{s\text{-}sp} \times 2B_e = 4R_w^2 P_{in} N_{sp} G B_e. \tag{B.29}$$

Considering that the output of the amplifier goes directly to the receiver, the received signal power P^R_{sig} and the received ASE noise power P^R_{sp} can be expressed as

$$P^R_{sig} = P_s = P_{in} G \tag{B.30}$$

$$P^R_{sp} = P_{sp} = N_{sp} B_o. \tag{B.31}$$

Using the above expressions, we finally express σ^2_{s-sp} from Eq. B.29 in terms of the received signal and ASE noise powers as

$$\sigma^2_{s\text{-}sp} = 4R_w^2 P^R_{sig} P^R_{sp} B_e / B_o. \tag{B.32}$$

Next, we consider the *sp-sp* beat noise current i_{sp-sp}. From Eq. B.24, we express i_{sp-sp} as

$$\begin{aligned} i_{sp\text{-}sp}(t) & \qquad\qquad (B.33) \\ &= B\overline{\left[\sum_{j=-J}^{J} \cos\{2\pi(f_s + j\delta f)t + \phi_j\}\right]^2} \\ &= B\overline{\sum_{j=-J}^{J} \cos\{2\pi(f_s + j\delta f)t + \phi_j\} \sum_{k=-J}^{J} \cos\{2\pi(f_s + k\delta f)t + \phi_k\}} \\ &= B\overline{\sum_{j=-J}^{J} \cos\beta_j \sum_{k=-J}^{J} \cos\beta_k} \ \text{(say)} \\ &= \frac{B}{2}\overline{\sum_{j=-J}^{J}\sum_{k=-J}^{J} \{\cos(\beta_j + \beta_k) + \cos(\beta_j - \beta_k)\}} \\ &= \frac{B}{2}\sum_{j=-J}^{J}\sum_{k=-J}^{J} \cos(\beta_j - \beta_k), \end{aligned}$$

wherein again the high-frequency terms in the form of $\cos(\beta_j + \beta_k)$ are averaged out to zero. Substituting back the values of β_j and β_k, we simplify i_{sp-sp} as

$$i_{sp\text{-}sp}(t) = \frac{B}{2} \sum_{j=-J}^{J} \sum_{k=-J}^{J} \cos\{(k-j)2\pi\delta f t + (\phi_j - \phi_k)\}. \tag{B.34}$$

In Eq. B.32, there will be $2J + 1$ terms with $j = k$, sum of which will represent a DC component of $i_{sp\text{-}sp}(t)$, given by

$$I_{sp\text{-}sp} = \frac{B}{2} \times (2J+1) = R_w N_{sp}(G-1)B_o. \tag{B.35}$$

The remaining terms will be spread over $f \in [-B_o, B_o]$ (i.e., over twice the optical bandwidth B_o) at different spectral slices of width δf. In the spectral slices located at higher frequencies, the number of terms will decrease, finally vanishing at $f = \pm B_o$. As $\delta f \to 0$, the discrete sum of spectral lines will eventually become a continuous spectrum with a triangular shape (as expected from the spectral convolution resulting from the product of the two noise-sums), with its maximum value occurring at $f = 0$, and the spectral triangle going down to zero on both sides at $f = \pm B_0$. Thus, $i_{sp\text{-}sp}$ will follow a zero-mean Gaussian distribution with the triangular spectral density, from which the electronic receiver following the photodiode will receive the noise power lying within the electronic bandwidth $[-B_e, B_e]$.

Further, from Eq. B.34, it is evident that, all the terms with $j - k = 1$ will have the form $\frac{B}{2}\cos\{2\pi\delta f t + (\phi_j - \phi_{j-1})\}$, and fall in the spectral slice $\in [\delta f/2, 3\delta f/2]$, i.e., in the spectral slice number $j = 1$ (hence in the spectral slice on the adjacent-right side of the central slice). The number of such terms with $j - k = 1$ would be $2J$. The same number of terms will also fall in the slice number $j = -1$ (i.e., for $f \in [-3\delta f/2, -\delta f/2]$) for $j - k = -1$ (hence in the spectral slice on the adjacent-left side of the central slice). Further, a given term in $j = 1$ slice would have a twin in $j = -1$ slice, adding up coherently; each one of such $2J$ pairs will have an amplitude equaling $2 \times \frac{B}{2} = B$. Thus the total power in each of these two spectral slices would be the sum of the powers of the $2J$ pairs, each pair having a power of $B^2/2$. Thus, in the two spectral slices together from both sides of the central slice ($f \in [-\delta f/2, \delta f/2]$), the total power will be given by

$$\begin{aligned} Q &= 2J \times \frac{B^2}{2} \\ &\approx \frac{B_o}{\delta f} \times \frac{(2R_w^2 N_{sp}\delta f)^2}{2} \\ &= 2R_w^2 N_{sp}^2 B_o \delta f. \end{aligned} \tag{B.36}$$

Equation B.36 implies that, the *sp-sp* power in Q (for $|j-k| = 1$) gets distributed over a frequency interval of $2\delta f$, and with $\delta f \to 0$ (i.e., $J \to \infty$), $\frac{Q}{2\delta f}$ becomes the power spectral density $N_{sp\text{-}sp}(f)$ for $i_{sp\text{-}sp}$ at $f = 0$, i.e.,

$$N_{sp\text{-}sp}(0) = \frac{Q}{2\delta f} = R_w^2 N_{sp}^2 B_o, \tag{B.37}$$

along with the terms for $j = k$ appearing in the central slice, producing in effect the DC component $I_{sp\text{-}sp}$ as given in Eq. B.35 (not to be considered in the continuous spectral density $N_{sp-sp}(f)$). As mentioned above, the spectrum $N_{sp\text{-}sp}$ will linearly fall on both sides of $f = 0$ to zero at $f \pm B_o$, with a negative slope $\zeta = -N_{sp-sp}(0)/B_o$ for $0 \le f \le B_o$, and with a positive slope of the same magnitude for $-B_o \le f \le 0$ which can be expressed as

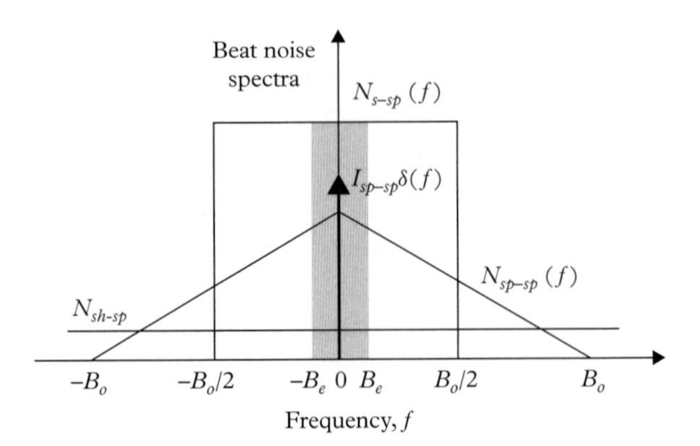

Figure B.5 *Spectra of the beat and shot-noise components generated in electrical domain from the ASE noise.*

$$N_{sp\text{-}sp}(f) = \zeta f + N_{sp\text{-}sp}(0), \text{ for } 0 \leq f \leq B_o \tag{B.38}$$
$$= -\zeta f + N_{sp\text{-}sp}(0), \text{ for } -B_o \leq f \leq 0.$$

Integrating the above spectrum in the interval $[-B_e, B_e]$, we obtain the variance of the *sp-sp* noise as

$$\sigma^2_{sp\text{-}sp} = R_w^2 N_{sp}^2 (2B_o - B_e) B_e, \tag{B.39}$$

which can be expressed in terms of the received ASE power $P_{sp}^R = N_{sp} B_o$ as

$$\sigma^2_{sp\text{-}sp} = R_w^2 \left(\frac{P_{sp}^R}{B_o}\right)^2 (2B_o - B_e) B_e = \left(R_w P_{sp}^R\right)^2 \frac{B_e(2B_o - B_e)}{B_o^2}. \tag{B.40}$$

Figure B.5 illustrates the representative spectral behavior of the various noise components resulting from the ASE noise. The rectangular spectrum over the frequency range $[-B_o/2, B_o/2]$ represents the *s-sp* continuous spectrum, which would be filtered by the electronic bandwidth, leading to the respective noise variance. The vertical arrow at $f = 0$ represents the DC component of the *sp-sp* noise, while the triangular continuous spectrum represents the continuous spectrum of the *sp-sp* noise, which is integrated over the two-sided electronic bandwidth $2B_e$ leading to the variance of the *sp-sp* noise component.

Furthermore, the total incident ASE noise at the photodiode will also lead to a shot noise power with a variance of $\sigma^2_{sh\text{-}sp}$ (in addition to the shot noise generated by the signal itself, as discussed earlier), given by

$$\sigma^2_{sh\text{-}sp} = N_{sh\text{-}sp} \times 2B_e = 2qR_w P_{sp}^R B_e. \tag{B.41}$$

B.5 Crosstalk

In WDM networks, optical crosstalk is originated in optical switches, filters and multiplexing/demultiplexing devices, which once generated at a node keeps co-propagating with the signal. Thus, like the ASE noise accumulation, all the crosstalk components generated at different intermediate nodes eventually lead to an accumulated crosstalk at the receiving end, which in turn creates beat-noise components through the nonlinear photodetection process. The basic difference in this case lies in the fact that, the crosstalk components, being a fraction of some other signal, do not have wideband spectrum as observed for the ASE noise. Crosstalk components can be generated for a given signal (lightwave) from the other lightwaves in the same channel, while passing through a non-ideal switching device, and this type of crosstalk is called homo-wavelength crosstalk. Also, there can be hetero-wavelength or inter-channel crosstalk components which are generated from the spectral tails encroaching from adjacent channels into the channel under consideration in non-ideal filtering/multiplexing devices. However, be it homo-wavelength or hetero-wavelength crosstalk, their spectra overlapping with the signal spectrum (fully for the home-wavelength case and partially for the other) of the lightpath under consideration are photodetected along with the desired signal received at the destination node, leading to the signal-crosstalk (s-xt) beat-noise components.

Consider a WDM receiver receiving an optical signal on a lightwave with a power P_s, along with a number of crosstalk components with powers $P_{x1}, \ldots, P_{xk}, \ldots, P_{xK}$. With these signal and crosstalk components, the total received electromagnetic field $E(t)$ at the photodiode can be expressed as

$$E(t) = \underbrace{\sqrt{2P_s}\cos\{2\pi f_s t + \psi_s(t) + \phi_s\}}_{signal} + \underbrace{\sum_{k=1}^{K} \sqrt{2P_{xk}}\cos\{2\pi f_s t + \psi_{xk}(t) + \phi_{xk}\}}_{crosstalk}, \tag{B.42}$$

where f_s, $\psi_s(t)$ are the frequency, time-varying phase, and random phase offset (epoch) for the signal component of $E(t)$. The crosstalk lightwaves in the second term are expanded around f_s with $\psi_{xk}(t)$ and ϕ_{xk} as the respective time-varying and epoch phase components, with $\psi_{xk}(t)$ taking into account the respective modulations in the crosstalk components. Using this representation, and following the same procedure as used in determining the ASE-caused beat noises, we assume again a zero-mean Gaussian model for the s-xt beat noise, with the variance $\sigma^2_{s\text{-}xt}$, given by

$$\begin{aligned} \sigma^2_{s\text{-}xt} & \\ &= \eta_p\eta_s \sum_{k=1}^{K} \overline{[R_w\sqrt{2P_s}\sqrt{2P_{xk}}\cos\{\underbrace{(\psi_s(t) - \psi_{xk}(t))}_{\Delta\psi_{s\text{-}k}(t)} + \underbrace{(\phi_s - \phi_{xk})}_{\Delta\phi_{s\text{-}k}}\}^2]} \\ &= \eta_p\eta_s \sum_{k=1}^{K} 4R_w^2 P_s P_{xk}\overline{\cos^2\{\Delta\psi_{s\text{-}k}(t) + \Delta\phi_{s\text{-}k}\}} \\ &= 2\eta_p\eta_s \sum_{k=1}^{K} R_w^2 P_s P_{xk} = 2\eta_p\eta_s R_w^2 P_s P_{xt}, \end{aligned} \tag{B.43}$$

where $P_{xt} = \sum_{k=1}^{K} P_{xk}$ is the total crosstalk power, η_p is the polarization mismatch factor (typically equaling half on the average), and the average of $\cos^2\{\Delta\psi_{s-k}(t) + \Delta\phi_{s-k}\}$ is also assumed to be half. Another factor η_s is introduced as the signal and crosstalk bit streams being uncorrelated, temporal overlap of a signal bit (one) with a crosstalk bit (one) will be uncertain with a probability of a half. However, for the worst-case estimate, one can assume the two mismatch factors (η_p and η_s) to be unity. The worst-case accumulated crosstalk components will also lead to a shot-noise variance, given by

$$\sigma^2_{sh\text{-}xt} = 2qR_w P_{xt} B_e. \tag{B.44}$$

B.6 Total receiver noise

Total noise in an optical receiver would comprise the various components, as discussed in the foregoing. However, with single-wavelength transmission without any use of optical amplifiers, the receiver noise would be simply given by the thermal and shot noise components, with its variance given by

$$\sigma^2_{sw} = \sigma^2_{sh} + \sigma^2_{th}, \tag{B.45}$$

where the shot-noise variance would differ for the binary zero and one receptions. For WDM systems, the receiver noise variance needs to include the beat noise from the ASE and crosstalk components (if any) along with the thermal and shot noise components, and hence expressed as

$$\sigma^2_{wdm} = \sigma^2_{sh\text{-}wdm} + \sigma^2_{s\text{-}sp} + \sigma^2_{sp\text{-}sp} + \sigma^2_{s\text{-}xt} + \sigma^2_{th}, \tag{B.46}$$

where $\sigma^2_{sh\text{-}wdm}$ will consist of three components from the signal, ASE and crosstalk components (see Equations B.17, B.41, and B.44), from which the signal-induced components will be signal-dependent and hence will vary for binary zero and one receptions. Similarly, $\sigma^2_{s\text{-}sp}$ and $\sigma^2_{s\text{-}xt}$ will also be signal-dependent and vary accordingly from bit to bit. In all of these computations, we have assumed the worst-case scenario, where the polarization of signal lightwave is assumed to be the same as those of ASE or crosstalk lightwaves. Note that there is another type of beat noise, crosstalk-spontaneous beat noise, which we have ignored due to its insignificant contribution to the entire noise process. Furthermore, when the optical fibers are driven into the nonlinear region while carrying large number of wavelengths, the WDM links might suffer from crosstalk, e.g., from the FWM interferences, which also need to be addressed, if necessary, while evaluating the beat noise components.

B.7 Laser phase noise

Phase noise in lasers plays a central role in determining its spectral width. While the structural features and the material decide overall spectral envelope of a laser (Fig. 2.38), in a given mode of a multimode laser or in a single-mode laser, the spectral shape is controlled fundamentally by the quantum-mechanical nature of the photon emission process in a laser.

Consider a lightwave that has been generated after the population inversion of electrons from the valence band to the conduction band, and assume that this lightwave moves along the active layer and grows coherently by stimulated emission. During this passage of the coherently growing lightwave, some spontaneous photons will also drop from the higher to the lower energy level without having any phase relation with the passing-by stimulated lightwave. These spontaneous emissions will disturb the linearly increasing phase (with time) of the coherent lightwave, by creating some abrupt phase drifts (jumps) of random magnitude and sign.

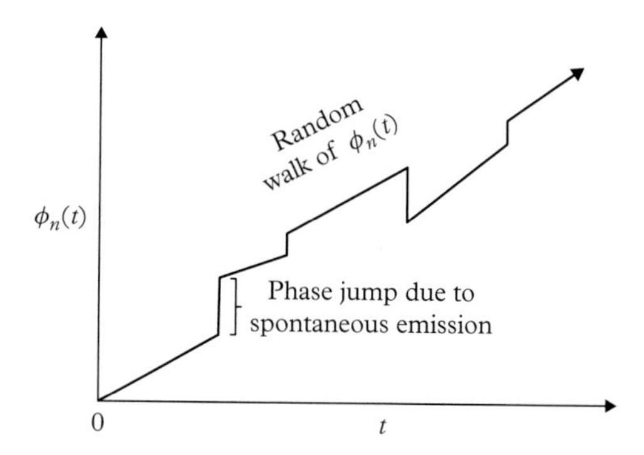

Figure B.6 *Illustration of random walk of phase noise $\phi_n(t)$ with time.*

Driven by the above phenomenon, the instantaneous phase of the resultant lightwave will move away from the linear path that it would have followed in the absence of the spontaneous emissions. Thus, as shown in Fig. B.6, the instantaneous phase $\phi_n(t)$ of the stimulated lightwave will evolve with time following a *random walk* process, whose derivative would represent the laser frequency noise, given by $\mu_n(t) = \frac{1}{2\pi}\frac{d\phi_n(t)}{dt}$. In reality, the phase noise process $\phi_n(t)$ would behave as a Weiner process, with the underlying frequency noise $\mu_n(t)$ becoming a zero-mean white noise process. This phase/frequency noise will eventually manifest as a finite spectral width (or linewidth) in the laser spectrum. In single-mode fibers, the linewidth will lead to the chromatic dispersion causing pulse broadening. Furthermore, the presence of laser phase noise would affect the receiver performance in coherent optical communication systems, the extent of which will depend on the modulation schemes used in the system. In the following, we follow an analytical model used in (Salz 1986) and examine the phase noise process to estimate its impact on the laser spectrum.

Experimental observations indicate that the underlying frequency noise in a laser follows a white spectrum, but follows a $1/f$-pattern at very low frequencies (in $\sim$ MHz region), as compared to the high-speed modulating data streams. Thus, in the optical communication systems, we assume that a single-mode semiconductor laser, dealing with much higher modulation speeds (as compared to the frequencies in the $1/f$ region of laser spectrum), would practically be affected by a zero-mean white Gaussian frequency noise $\mu_n(t)$ with a two-sided spectral density ξ (say). We therefore express the phase noise $\phi_n(\tau)$ as an integrated version of $\mu_n(t)$ over an observation interval τ (typically a bit/symbol interval in an optical receiver), given by

$$\phi_n(\tau) = 2\pi \int_0^\tau \mu_n(t)dt. \tag{B.47}$$

Next, we determine the variance of the phase noise $\sigma^2_{\phi_n}$. Note that, $\phi_n(\tau)$ being practically a filtered version of the zero-mean white Gaussian noise $\mu_n(t)$, will also be Gaussian with a zero mean. Hence, $\sigma^2_{\phi_n}$ would be same as the mean-squared value of $\phi_n(\tau)$, i.e., $\sigma^2_{\phi_n} = E[\phi_n^2(\tau)]$. From Eq. B.47, we take the frequency-domain route to find out the variance $\sigma^2_{\phi_n}$. The integral in Eq. B.47 implies that $\phi_n(\tau)$ is obtained from $\mu_n(t)$ by passing it through an *integrate-and-dump* filter with a duration of its impulse response $h_{ID}(t)$ as τ, i..e., $h_{ID}(t) = u(t) - u(t-\tau)$, with $u(t)$ as the unit step function of time. Hence, the transfer function of $h_{ID}(t)$, i.e., its Fourier transform, can be expressed by a sinc function, given by

$$H_{ID}(f) = \text{FT}[h_D(t)] = \tau\left[\frac{\sin(\pi f\tau)}{\pi f\tau}\right] = \tau\,\text{sinc}(\pi f\tau). \tag{B.48}$$

Using the above expression for $H_{ID}(f)$ and recognizing that $E[\phi_n^2(\tau)]$ can be expressed as the filtered version of the spectral density of $\mu_n^2(t)$ through $H_{ID}(f)$, we obtain $\sigma_{\phi_n}^2$ as

$$\sigma_{\phi_n}^2 = E[\phi_n^2(t)] = \xi \int_{-\infty}^{\infty} |H_{ID}(f)|^2 df = 4\pi^2 \xi \tau, \tag{B.49}$$

leading to the pdf of the zero-mean Gaussian phase noise $p(\phi_n)$, given by

$$p(\phi_n) = \frac{1}{\sqrt{2\pi\sigma_{\phi_n}^2}} \exp\left(\frac{-\phi_n^2}{2\sigma_{\phi_n}^2}\right). \tag{B.50}$$

With the above statistical model of $\phi_n(\tau)$, we next represent the lightwave emitted from a laser as an electromagnetic wave $E_s(t)$, given by

$$E_s(t) = A\cos\{2\pi f_s t + \phi_n(t) + \varphi)\}, \tag{B.51}$$

where A and f_s are the amplitude and frequency of $E_s(t)$, $\phi_n(t)$ is the phase noise in $E_s(t)$, and φ is a random phase term (epoch) uniformly distributed in $[0, 2\pi]$ to make $E_s(t)$ a stationary process. The autocorrelation function $R_{E_s}(\tau)$ of $E_s(t)$ can be expressed as

$$\begin{aligned} &R_{E_s}(\tau) \\ &= E[E_s(t)E_s(t-\tau)] \\ &= E[A\cos\{2\pi f_s t + \phi_n(t) + \phi)\}A\cos\{2\pi f_s(t-\tau) + \phi_n(t-\tau) + \phi)\}. \end{aligned} \tag{B.52}$$

Ignoring the $\cos(4\pi f_s T)$ as before (i.e., averaged out to zero), we obtain $R_{E_s}(\tau)$ as

$$\begin{aligned} R_{E_s}(\tau) &= \frac{A^2}{2}\int_{-\infty}^{\infty} \cos\Delta\phi_n p(\Delta\phi_n) d(\Delta\phi_n) \\ &= \frac{A^2}{2}\exp(-2\pi^2\xi\tau)\cos(2\pi f_s\tau), \end{aligned} \tag{B.53}$$

where $\Delta\phi_n = \phi_n(t) - \phi_n(t-\tau) = 2\pi\int_{t-\tau}^{t} \mu_n(t)dt$. The phase noise being a stationary process, $\phi_n(t) - \phi_n(t-\tau)$ would have the same variance as $\phi_n(\tau) - \phi_n(0) = \phi_n(\tau)$. Thus, the pdf $p(\phi_n)$ of ϕ_n from Eq. B.50 is used for the pdf of $\Delta\phi_n$ in Eq. B.53, to obtain the expression for $R_{E_s}(\tau)$. Next, by taking Fourier transform of $R_{E_s}(\tau)$, we express the laser spectrum as

$$S_{laser}(f) = \frac{A^2}{4\pi^2\xi}\left[\frac{1}{1+\left(\frac{f+f_s}{\pi\xi}\right)^2} + \frac{1}{1+\left(\frac{f-f_s}{\pi\xi}\right)^2}\right], \tag{B.54}$$

with $\int_{-\infty}^{\infty} S_{laser}(f)df = A^2/2$. The above spectrum is known as a Lorentzian spectrum, as shown in Fig. B.7, with its 3 dB linewidth B_L given by

$$B_L = 2\pi\xi, \tag{B.55}$$

implying that the spectral density of the frequency noise $\mu_n(t)$ is directly related to the laser linewidth, i.e.,

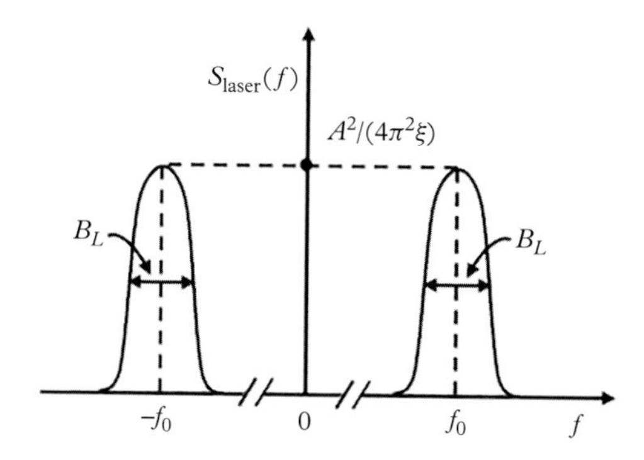

Figure B.7 *Representative Lorentzian spectrum of lasers.*

$$\xi = \frac{B_L}{2\pi}. \tag{B.56}$$

The linewidth estimated in the foregoing would be increased further for the lightwaves carrying modulation, and the optical filters receiving a modulated lightwave must have an adequate bandwidth to avoid any signal loss due to spectral truncation. For estimating this bandwidth, we first assume that the modulation bandwidth of an ideal laser (i.e., with zero linewidth) is B_M such that at least 95% of the modulated power falls within $f_s \pm B_M/2$. With this definition, we express the overall bandwidth B_L^M of the modulated lightwave as the root-mean-squared value of B_L and B_M (Kazovsky 1986), given by

$$B_L^M = \sqrt{B_L^2 + B_M^2}. \tag{B.57}$$

In practical systems, the laser frequency might drift with time, and the frequency of a procured laser for a given WDM slot may not fall exactly in the centre of the slot. With due considerations to these unavoidable situations, it is customary to set up the channel spacing Δf_{WDM} of a given WDM frequency grid, with the necessary allowances, given by

$$\Delta f_{\text{WDM}}(\text{min}) = B_L^M + \Delta f_M, \tag{B.58}$$

where Δf_M represents the extra margin that one needs to include in the channel spacing to avoid any significant signal loss during optical filtering and/or interferences (crosstalk) from the adjacent channels.

Appendix C
Basics of Queuing Systems

C.1 Queuing: network perspective

Telecommunication networks need to deal with incoming service requests with a finite set of resources (e.g., available time slots, frequencies, wavelengths, network connectivity), thereby having some chance of failing to serve the service requests with the desired QoS. For example, a telephone call, as it arrives at a telephone exchange, the call is either set up (i.e., serviced) or refused (i.e., blocked) depending on the availability of transmission resources from that node. However, a request for sending a data packet from a computer to another can wait for a while without the *go/no-go* (i.e., *connect/block*) notion used for the telephone calls. Service quality in the first case is generally measured by the *probability of blocking*, while in the second case the *delay* experienced by the packet while waiting in the system gives a measure of the network performance. While looking from an ingress point in the given network, both systems are viewed as *queuing systems*; one without any buffer to handle real-time service requests, while the other using a buffer of appropriate size to prevent data loss. Circuit-switched networks fall in the category dealing with the blocking probability, while packet-switched networks are assessed using the amount of delay incurred by a packet in the network.

C.2 Basic configurations and definitions

A basic queuing system is represented as a buffer followed by a server, as shown in Fig. C.1 (a). Customers send service requests (messages, packets, calls) to the queue with an arrival rate λ, and the average value of service durations τ is represented by $E[\tau] = 1/\mu$, with μ as the service rate of the server. A queuing system might also employ multiple servers to speed up the performance, as shown in Fig. C.1 (b). The arrival instants at the queue input can be a random or a deterministic process. Similarly, the service durations for the incoming service requests might also fall in the same two categories, i.e., random or deterministic.

A generic queuing system is represented by the $A/B/c/K/Z/P$ notation, where A defines the arrival process, B defines the process representing the service durations, c stands for the number of servers, K denotes the buffer size, Z is the queuing (or servicing) discipline, and P represents the population size. With the first-in first-out (FIFO) queuing discipline and infinite population, the notation for the queuing systems is usually simplified as $A/B/c/K$, and in the following deliberations, we will be focussing on the variations of this type of queuing system. The specific symbols used for A and B depend on the statistical nature of the corresponding processes (interarrival times and service durations), and are in general defined as follows.

> M: Memoryless process, applicable for the interarrival times as well as the service durations, following exponential distribution. As we shall see later, the number of

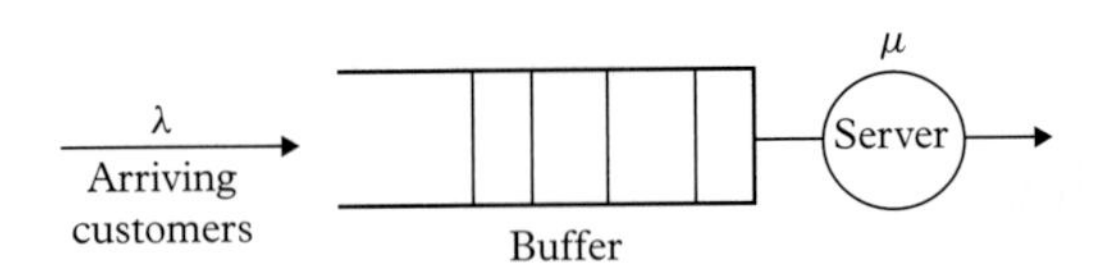

(a) Single-server queuing system.

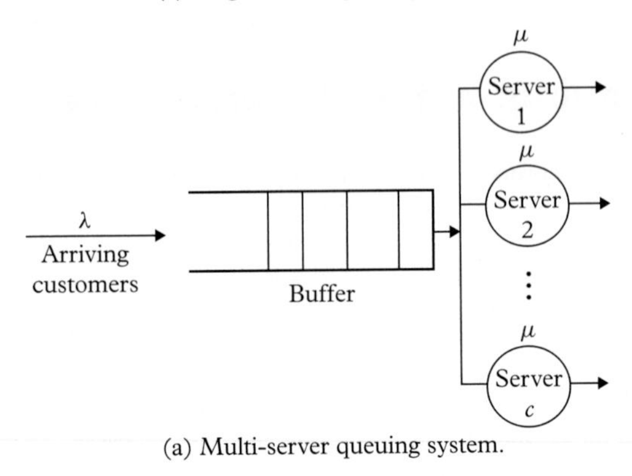

(a) Multi-server queuing system.

Figure C.1 *Illustration of basic queuing systems: (a) with single server, (b) with multiple servers.*

arrivals in a given time interval will follow Poisson distribution for an exponentially distributed interarrival process.

G: General distribution, applicable for some arrival processes and service durations that are not memoryless. For such processes one needs to obtain an appropriate distribution by taking into account of the underlying events representing the process under consideration.

D: Deterministic distribution for interarrival times or service durations. In this case, the service durations or interarrival times remain constant over time.

When the arrival process and the service durations are memoryless, the single-server queuing system is denoted as a $M/M/1/K$ system. These queuing systems are simply denoted as $M/M/1$ system, when the buffer size is infinitely large. An understanding of the $M/M/1$ system helps significantly in studying various other queuing systems, such as, $M/M/1/K, M/M/c/K, M/M/c/c, M/G/1, M/D/1$, and so on. Waiting time or delay in a queuing system for a packet (generically, a *customer*), say the ith packet, is expressed as the sum of the waiting time in the buffer and the service time, given by

$$T_i = W_i + \tau_i, \tag{C.1}$$

where W_i represents the waiting time spent by the ith packet in the buffer of the queue and τ_i is the service time or duration of the same packet. As mentioned earlier, for the queues without buffer, the performance is assessed by the blocking probability or the percentage of the customers that are denied service. For both types of network, the basic queuing performance is governed by a fundamental variable, called *traffic intensity* (also referred to as *utilization*), defined as $\rho = \lambda/\mu$.

C.3 Little's theorem

One fundamental aspect of a queuing system is its behavior under the steady state, generally described by a theorem, called Little's theorem, which takes effect only if $\lambda < \mu$. Consider that, in a queuing system, $N(t)$ customers are present at a given time instant t, and each of these customers undergoes some delay T_i in the buffer as well as in the server. Little's theorem states that, in the steady state, the average number $E[N]$ of customers in the system can be related to the average value of the delay $E[T]$ incurred by a customer as

$$E[N] = \lambda E[T], \tag{C.2}$$

which implies that, during the average *duration of stay* (= $E[T]$) of a customer in the system, the number of customers that arrive is equal to the product of the arrival rate λ and $E[T]$. The form of Little's theorem is quite generic in nature and hence also applies for the average waiting time W in the buffer and the average number of customers waiting therein $E[N_b]$, i.e.,

$$E[N_b] = \lambda E[W]. \tag{C.3}$$

One can arrive at these relations analytically, though we won't go into the details of the proof, and the readers are referred to (Bertsekas and Gallager 2003; Leon-Garcia 1994) for further details. As we shall see in the following, Little's theorem serves as an extremely useful tool for examining various features of queuing systems.

C.4 Poisson arrival process

As mentioned earlier, in a memoryless arrival process, the number of arrivals over a given time interval follows a Poisson distribution, while the interarrival times follow an exponential distribution. Though the Poisson distribution is derivable from the binomial distribution, its relation to a memoryless arrival process is of profound importance in queuing systems and can be derived by using a few fundamental assumptions on the arrival process. We carry out this exercise using a simple illustration in Fig. C.2, where we assume a time interval $[0, t + \Delta t]$ with Δt being a very small duration as compared to t.

As shown in Fig. C.2, we assume that k customers arrive with service requests during the time interval $[0, t + \Delta t]$ in a queuing system. Our objective is to find out the probability that k packets have arrived in the time interval $[0, t]$. Before proceeding further, we make three assumptions as follows.

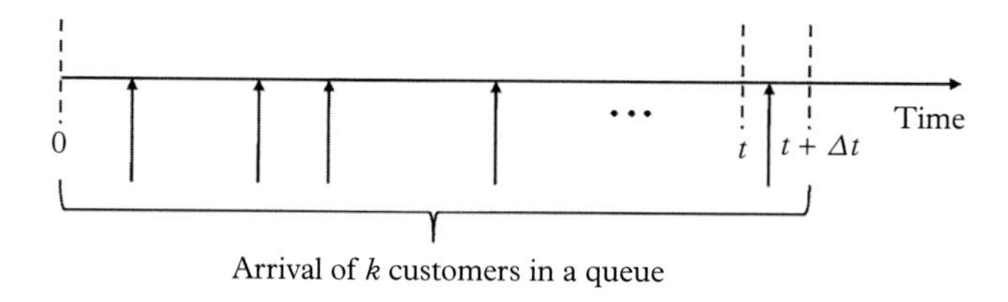

Figure C.2 *Arrival of k customers in a queue within a time interval* $[0, t + \Delta t]$.

(1) Probability that one arrival (i.e., $k = 1$) takes place during the short time interval $[t, t + \Delta t]$ is given by

$$P\{1 \in [t, t + \Delta t]\} = \alpha \Delta t + o(\Delta t), \tag{C.4}$$

where α is a constant determined by the arrival process, and $o(t)$ becomes negligible as $\Delta t \to 0$, implying thereby

$$\lim \ \Delta t \to 0 \ \frac{o(\Delta t)}{\Delta t} = 0. \tag{C.5}$$

Equations C.4 and C.5 imply that, as Δt increases, the assumption that one single arrival takes place in the interval $[t, t + \Delta t]$ with a probability of $\alpha \Delta t$ becomes *increasingly inaccurate* in the order of $o(\Delta t)$. However, when Δt is reduced, this error (i.e., $o(\Delta t)$) reduces at a faster rate than Δt itself. Thus, as we keep decreasing Δt, the probability $P\{1 \in [t, t + \Delta t]\}$ is *more and more accurately* represented by $\alpha \Delta t$.

(2) Probability of more than one arrival in $[t, t + \Delta t]$ is $o(\Delta t)$. This assumption implies that, as $\Delta t \to 0$, the probability of two or more arrivals become zero. Hence, the probability of zero arrival in the interval $[t, t + \Delta t]$ can be expressed as $P\{0 \in [t, t + \Delta t]\} = 1 - (\alpha \Delta t + o(\Delta t))$, which would approach $1 - \alpha \Delta t$ as $\Delta t \to 0$.

(3) Number of arrivals in *non-overlapping* time intervals are statistically independent, which sets up the memoryless property of the underlying random process.

With the above assumptions, we next consider that k customers have arrived during the time interval $[0, t + \Delta t]$ and proceed to evaluate $p_k(t) = P\{k \in [0, t]\}$, representing the probability that k customers have arrived in the time interval $[0, t]$. As shown in Fig. C.2, the time interval $[t, t+\Delta t]$ is broken into two adjoining intervals: $[0, t]$ and $[t, t+\Delta t]$. Since the arrivals in these two non-overlapping intervals are independent, one can express $p_k(t + \Delta t)$ in the sum-of-product form as

$$\begin{aligned} p_k(t + \Delta t) = & \ P\{k \in [0, t]\}P\{0 \in [t, t + \Delta t]\} \\ & + P\{k - 1 \in [0, t]\}P\{1 \in [t, t + \Delta t]\} \\ & + P\{k - 2 \in [0, t]\}P\{2 \in [t, t + \Delta t]\} + \cdots, \end{aligned} \tag{C.6}$$

where the third and all subsequent terms on the right-hand side will vanish with Δt approaching zero, as dictated by the assumption 2. Using this observation on Eq. C.6 and the foregoing assumptions, we simplify the above expression as

$$\begin{aligned} p_k(t + \Delta t) &= P\{k \in [0, t]\}(1 - \alpha \Delta t) \\ & \quad + P\{k - 1 \in [0, t]\}\alpha \Delta t \\ &= p_k(t)(1 - \alpha \Delta t) + p_{k-1}(t)\alpha \Delta t, \end{aligned} \tag{C.7}$$

leading to a difference equation, given by

$$p_k(t + \Delta t) - p_k(t) = -\alpha \Delta t\big[p_k(t) - p_{k-1}(t)\big]. \tag{C.8}$$

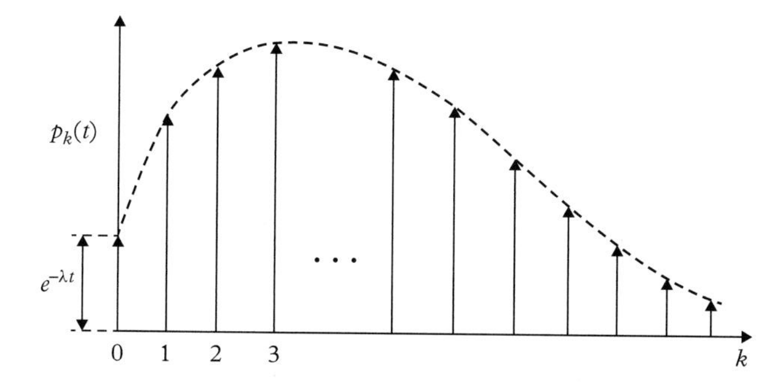

Figure C.3 *Poisson distribution $p_k(t)$.*

With $k = 0$, the second term on the right-hand side in the above expression would vanish, leading to a simple first-order differential equation with $\Delta t \to 0$, given by

$$\frac{dp_0(t)}{dt} = -\alpha p_0(t). \tag{C.9}$$

The above equation, along with the obvious initial condition that $p_0(0) = 1$, would lead to the exponential distribution for $p_k(t)$, expressed as

$$p_0(t) = \exp(-\alpha t). \tag{C.10}$$

Using the above expression for $p_0(t)$ in Eq. C8, one can easily obtain the solutions for (i) one arrival in the time interval $[0, t]$ as $p_1(t) = \alpha t \exp(-\alpha t)$ and (ii) two arrivals in the time interval $[0, t]$ as $p_2(t) = \frac{(\alpha t)^2}{2} \exp(-\alpha t)$. Through subsequent recursions, one can obtain eventually the expression for $p_k(t)$ as

$$p_k(t) = \frac{(\alpha t)^k}{k!} \exp(-\alpha t). \tag{C.11}$$

The above expression represents the Poisson distribution as shown in Fig. C.3, implying that the number of arrivals k (also called the count variable) in a given interval t will be Poisson-distributed for a memoryless arrival process. Note that the distribution extends from $k = 0$ to ∞, implying thereby that the above model represents infinite population of customers. In most practical cases, the population is not infinite, but the results using the infinite-population model gives much useful insight of queuing systems.

The mean and variance of the Poisson arrival process can be obtained from the probability distribution function $p_k(t)$. The mean arrival count Λ in time t can be expressed as

$$\Lambda = E[k] = \sum_{0}^{\infty} kp_k(t) = \sum_{0}^{\infty} kp_k(t)\frac{(\alpha t)^k}{k!} \exp(-\alpha t) = \alpha t, \tag{C.12}$$

implying that our original constant α in reality is equal to the Poisson arrival rate, i.e., $\alpha = \Lambda/t = \lambda$, thereby leading to the expression for the Poisson distribution in terms of its arrival rate λ, given by

$$p_k(t) = \frac{(\lambda t)^k}{k!} \exp(-\lambda t). \tag{C.13}$$

Using a similar approach, the mean-squared value of the count variable k, denoted as $E[k^2]$, can be expressed as

$$E[k^2] = \sum_0^\infty k^2 p_k(t) = \sum_0^\infty k^2 \frac{(\lambda t)^k}{k!} \exp(-\lambda t) = (\lambda t)^2 + \lambda t = \Lambda^2 + \Lambda, \quad \text{(C.14)}$$

implying that the variance of the count variable k, denoted as σ^2, would be given by

$$\sigma^2 = E[k^2] - (E[k])^2 = (\Lambda^2 + \Lambda) - \Lambda^2 = \Lambda. \quad \text{(C.15)}$$

Thus, for the Poisson arrival process, the variance is equal to the mean, which is feasible *dimensionally* as k represents a dimensionless count variable.

For a given Poisson distribution, one can find the statistics of the interarrival times as follows. Consider a time interval $[0, t]$, during which the probability of zero arrival for a Poisson arrival process is $p_0(t) = \exp(-\lambda t)$. Note that $p_0(t)$ represents the probability that no arrival takes place in $[0, t]$, or more precisely no arrival takes place *at least* during the interval $[0, t]$, and the interarrival time T_{ia} (say) could also be more than t. Using this observation, we express the probability that the interarrival time $T_{ia} \geq t$ as

$$P\{k = 0 \in T_{ia} \geq t)\} = p_0(t) = \exp(-\lambda t), \quad \text{(C.16)}$$

which, in effect, is complementary to the cumulative distribution function (CDF) of t, denoted by $P_{ia}(t)$. Therefore, we express $P_{ia}(t)$ as

$$P_{ia}(t) = 1 - P\{k = 0 \in T_{ia} \geq t)\} = 1 - \exp(-\lambda t). \quad \text{(C.17)}$$

Having obtained the CDF of t as above, the pdf $p_{ia}(t)$ of t is expressed as

$$p_{ia}(t) = \frac{dP_{ia}(t)}{dt} = \frac{d}{dt}\big[1 - \exp(-\lambda t)\big] = \lambda \exp(-\lambda t), \quad \text{(C.18)}$$

which represents an exponential distribution as shown in Fig. C.4. The results on $p_k(t)$ and $p_{ia}(t)$ together lead to an important feature of the memoryless Poisson arrival process. In particular, the Poisson distribution for the count variable (a discrete variable) and the exponential pdf of interarrival time (a continuous variable) stand together to represent a

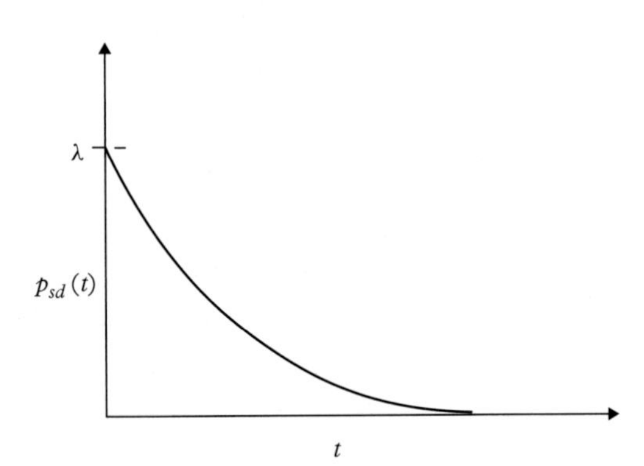

Figure C.4 *Exponential distribution.*

memoryless arrival process, be it the case of the arrival of telephone calls at a telephone exchange, or the arrival of packets in a switch or router, or the birth process in a nation.

Note that the memoryless service duration τ (call or packet duration) would also follow an exponential distribution, governed by the service completion rate $\mu = 1/E[\tau]$. Therefore, the pdf for the service durations $p_{sd}(\tau)$ can be expressed as

$$p_{sd}(\tau) = \mu \exp(-\mu\tau). \tag{C.19}$$

C.5 Markov process

The Markov process represents a random process which finds extensive use in modeling queuing systems. A random process $X(t)$ is called a Markov process if the future of $X(t)$, with the given present, is independent of the past. In other words, given that t_i represents the present time instant of observation for a Markov process $X(t)$ and t_{i+1} is the immediate next (future) time instant of observation, with $t_1 < t_2 < \cdots t_i < t_{i+1}$, one can write

$$\begin{aligned} &P[X(t_{i+1}) = x_{i+1}/X(t_i) = x_i, \cdots, X(t_1) = x_1] \\ &= P[X(t_{i+1}) = x_{i+1}/X(t_i) = x_i], \end{aligned} \tag{C.20}$$

where the second line expresses the memorylessness of the queuing system.

When $X(t)$ is discrete and integer-valued in nature (e.g., the number of packets in a queue), then the Markov process $X(t)$ is called a Markov chain, represented by $N(t)$, with the value of $N(t)$ called the state of the queue. Markov chains can be of two types, continuous-time and discrete-time. In a continuous-time Markov chain, the state transitions can take place at any time, while in the discrete-time Markov chains the state transitions take place at discrete instants of time. In our subsequent deliberations, we shall be focusing on the continuous-time Markov chain. In order to have an insight into the queuing systems that we consider here, we describe in the following some of the basic features of the continuous-time Markov chain.

For continuous-time Markov chains, the arrival process naturally follows a Poisson distribution, as it ensures that the interarrival times are continuous random variable and memoryless in nature. The joint probability that the state of a continuous-time Markov chain $N(t)$ assumes the $i + 1$ values, $n_{i+1}, n_i, \cdots, n_2, n_1$, at the $i + 1$ arbitrary time instants $t_{i+1} < t_i < \cdots t_2 < t_1$, can be expressed as

$$\begin{aligned} &P[N(t_{i+1}) = n_{i+1}, N(t_i) = n_i, \cdots, N(t_1) = n_1] \\ &= P[N(t_{i+1}) = n_{i+1}/N(t_i) = n_i] \cdots P[N(t_2) = n_2/N(t_1) = n_1], \end{aligned} \tag{C.21}$$

ensuring that, what happens next will only depend on what happened in the immediate past and not on the other previous states.

In practice, $N(t)$ would start from an arbitrary initial state at $t = 0$ and move through the subsequent states with the respective transition probabilities $p_{ij}(t)$'s. In each state $N(t)$ would spend some time, that is exponentially distributed (if memoryless), eventually reaching a steady with $p_{ij}(t) = p_{ij}$.

Further, we consider another important aspect of $X(t)$ – its stationarity. Often it is important to know the transition probability of $X(t)$ from one state to another over a given time interval t, given by $P[X(t_0 + t) = j/X(t_0) = i]$ for $t \geq 0$. The stationarity of the transition probability is ensured through the condition, given by

$$P[X(t_0 + \tau) = j/X(t_0) = i] = P[X(\tau) = j/X(0) = i] = p_{ij}(\tau), \text{ say,} \tag{C.22}$$

implying that the time difference and not the absolute times will determine the transition probability under stationary condition. Obviously, in a zero time interval a state cannot change implying $p_{00}(0) = 1$, and similarly $p_{ij}(0) = 0$ with $i \neq j$. These observations lead to the fact that the matrix $\{p_{ij}(\tau)\} = P(\tau)$ (say) is an identity matrix for $\tau = 0$, given by

$$P(0) = \{p_{ij}(0)\} = I. \tag{C.23}$$

In steady state, $N(t)$ would move around from one state to another in one or multiple steps. One can estimate the m-step probability $p_{ij}(m)$ for a continuous-time Markov process to reach from state i to state j, by using the one-step probabilities $p_{kl}(1)$'s, where the states k and l are the intermediate states, thereby leading to a state-transition matrix.

C.6 Single-server queues: *M/M/1* and *M/M/1/K*

As mentioned earlier, the $M/M/1$ system represents a single-server queue with infinite buffer that follows a FIFO discipline, where the interarrival times and service durations are exponentially distributed. Thus, the single-server queuing system, as shown earlier in Fig. C.1(a), becomes an $M/M/1$ queue using an infinite buffer with exponentially distributed interarrival times (Poisson arrivals) and service durations. Figure C.5 shows the state-transition diagram of an $M/M/1$ queue, where the state $N(t)$ of the queuing system represents the number of customers in the buffer and server, and the state of the queue increments or decrements by one for each arrival or departure of customers, respectively. With the arrival and service (i.e., departure) rates denoted by λ and μ, every increment of state is indicated by a *right arrow* marked by λ, and every decrement in state is indicated by a *left arrow* marked as μ. In steady state with $\lambda < \mu$, for a given state, say state i, the probability of increment of the state by one in a short time interval $[t, t + \Delta t]$ should be the same as the probability of decrement by one in the state during the same time interval.

As evident from the state-transition diagram, in the time interval $[t, t + \Delta t]$, the queuing system can enter state i with a probability $\lambda\Delta t$, if there is one arrival in the queue, while being in state $i - 1$ with a probability p_{i-1}. On the other hand, the system can also enter state i, if there is one departure from the queue with a probability $\mu\Delta t$, when the queue is in state $i + 1$ with a probability p_{i+1}. This implies that, given the memoryless arrival and service completion processes during $[t, t + \Delta]$, the total probability of entering state i, P_i^{reach}, would be the sum of the two probabilities $p_{i-1} \times \lambda\Delta t$ and $p_{i+1} \times \mu\Delta t$, i.e.,

$$P_i^{enter} = \lambda p_{i-1}\Delta t + \mu p_{i+1}\Delta t, \text{ for } i = 1, 2, \cdots \tag{C.24}$$

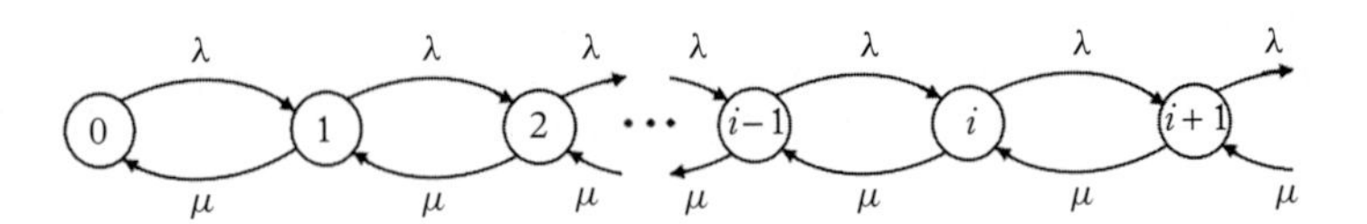

Figure C.5 *State-transition diagram for M/M/1 queuing system.*

Similarly, the queue might leave state i if there is an arrival causing a transition to the state $i+1$ or there is a departure causing a transition to the state $i-1$, with the total probability of leaving state i, P_i^{leave}, given by

$$P_i^{leave} = (\lambda + \mu)p_i \Delta t, \text{ for } i = 1, 2, \cdots \tag{C.25}$$

During the steady-state operation, these two probabilities, P_i^{enter} and P_i^{leave}, should be equal (i.e., $P_i^{enter} = P_i^{leave}$), leading to the *global balance equation*, given by

$$\lambda p_{i-1} + \mu p_{i+1} = (\lambda + \mu)p_i \text{ for } i = 1, 2, \cdots \tag{C.26}$$

For state 0 ($i = 0$), the above equation gets pruned to a simpler one as $N(t)$ being a count of packets cannot go below zero. Hence, only the process of entering state 0 from state 1 due to a departure (μp_1) and the process of leaving state 0 to state 1 due to an arrival (λp_0) keep balancing each other in the steady state, leading to

$$\lambda p_0 = \mu p_1. \tag{C.27}$$

This leads to the relation between p_1 and p_0, given by

$$p_1 = \frac{\lambda}{\mu} p_0 = \rho p_0, \tag{C.28}$$

where ρ, as defined earlier, represents the traffic intensity in Erlang for the queuing system. Using this relation in Eq. C.26 with $i = 1$, one obtains

$$\lambda p_0 + \mu p_2 = (\lambda + \mu)\rho p_0, \tag{C.29}$$

leading to

$$p_2 = \rho^2 p_0. \tag{C.30}$$

Using further recursive iterations for Eq. C.26 and generalization, eventually we get

$$p_i = \rho^i p_0. \tag{C.31}$$

For the normalization of the state probabilities, we sum up all p_i's to 1 and obtain

$$\sum_{i=0}^{\infty} p_i = 1 = (1 + \rho + \rho^2 + \cdots)p_0 = \frac{p_0}{1 - \rho}, \tag{C.32}$$

leading to the expression for p_0, given by

$$p_0 = 1 - \rho. \tag{C.33}$$

Using Equations C.31 and C.33, the probability p_i is obtained as

$$p_i = (1 - \rho)\rho^i. \tag{C.34}$$

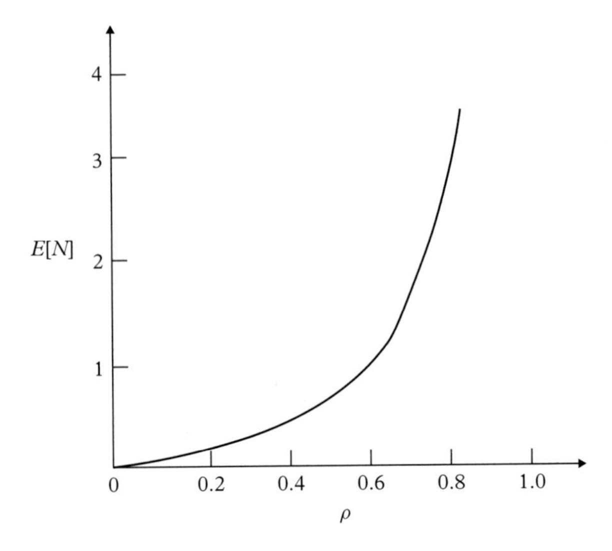

Figure C.6 *Plot of $E[N]$ versus ρ for $M/M/1$ queuing system.*

Equation C.34 offers an extremely useful relation, which helps in assessing the performance of a given queuing system (i.e., $M/M/1$ queue). First, using Eq. C.34, we obtain the average number of packets in the system $E[N]$ as

$$E[N] = \sum_{i=0}^{\infty} ip_i = \sum_{i=0}^{\infty} i(1-\rho)\rho^i = \frac{\rho}{1-\rho}. \tag{C.35}$$

As evident from the plot of $E[N]$ vs. ρ shown in Fig. C.6, the average number of packets in the queuing system remains reasonably low for lower values of ρ (below $\simeq 0.5$), but increases rapidly beyond this range. With ρ approaching unity (i.e., with the packet arrival rate λ reaching the service rate μ), $E[N]$ shoots up toward infinity making the average packets in the system and the delay arbitrarily large. This is in conformity with our earlier conjecture that the given queuing system can remain in steady state as long as ρ remains below unity (i.e., $\lambda < \mu$).

Next, using Little's theorem, we obtain the average delay $E[T]$ in the system (including the buffer and server) as

$$E[T] = \frac{E[N]}{\lambda} = \frac{1}{\lambda}\left(\frac{\rho}{1-\rho}\right) = \frac{1}{(\mu-\lambda)} = \frac{E[N]+1}{\mu}, \tag{C.36}$$

which also shows a similar trend as for $E[N]$, but with a pedestal of delay equaling to the average service duration $E[\tau] = 1/\mu$ (Fig. C.7).

The delay incurred in the buffer by the packets can be estimated from $E[T]$ by subtracting the delay in the server $E[\tau] = 1/\mu$ from the same, given by

$$E[W] = E[T] - E[\tau] = \frac{1}{\mu-\lambda} - \frac{1}{\mu} = \frac{\rho}{\mu(1-\rho)}. \tag{C.37}$$

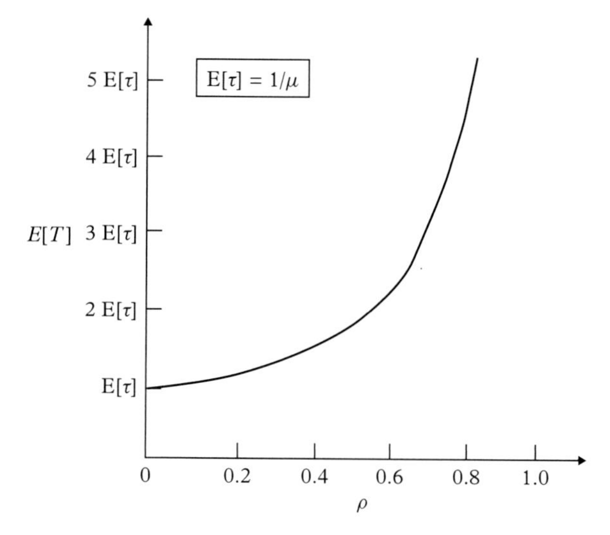

Figure C.7 *Plot of $E[T]$ versus ρ for $M/M/1$ queuing system.*

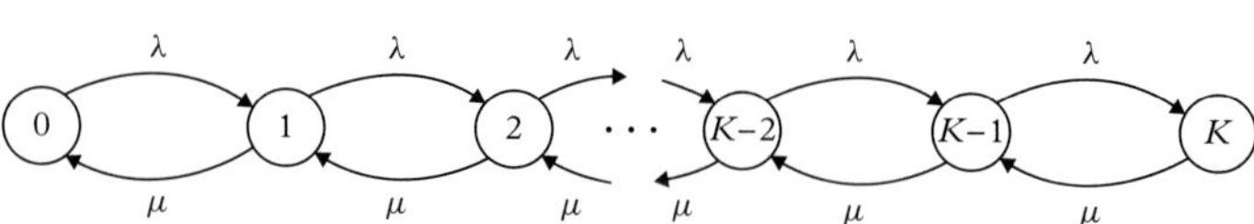

Figure C.8 *State-transition diagram of $M/M/1/K$ queuing system.*

Using Eq. C.37 and Little's theorem, we therefore express the average buffer occupancy $E[N_b]$ as

$$E[N_b] = \lambda E[W] = \frac{\rho^2}{1-\rho}. \tag{C.38}$$

In the foregoing, we considered the $M/M/1$ queuing system with infinite buffer, while in real life the buffer would have a finite size, leading to $M/M/1/K$ queuing system, where K is the number of packets the system can accommodate at the most (including the buffer and server). Due to the finite size of the buffer, the queue states are limited in the range $[0, K]$ and hence the Markov chain gets upper-bounded by the state K, as shown in Fig. C.8.

According to the state-transition diagram of Fig. C.8, the arrival rate would be zero if the system has $N(t) = K$ packets, i.e.,

$$\begin{aligned} \lambda_n &= \lambda, \text{ for } n \leq K-1 \\ &= 0, \text{ for } n \geq K. \end{aligned} \tag{C.39}$$

With this observation on the $M/M/1/K$ Markov chain, the global balancing equation of $M/M/1$ queue (Eq. C.26) gets modified as

$$\begin{aligned} \lambda p_{i-1} + \mu p_{i+1} &= (\lambda+\mu)p_i, \text{ for } i = 1, 2, \cdots, K-1 \\ \lambda p_0 &= \mu p_1 \\ \lambda p_{K-1} &= \mu p_K. \end{aligned} \tag{C.40}$$

Following similar steps as in the $M/M/1$ queue with the above expressions, one can obtain p_i as

$$p_i = p_0\rho^i \text{ for } i \leq K \quad (C.41)$$
$$= 0 \text{ for } i > K.$$

Using the normalizing condition, i.e., with $\sum_{i=0}^{K} p_i = 1$, we obtain the expression for p_0 as

$$p_0 = \frac{1-\rho}{1-\rho^{K+1}}, \quad (C.42)$$

leading to the final expression for p_i, given by

$$p_i = \frac{(1-\rho)\rho^i}{1-\rho^{K+1}} \text{ for } i \leq K \quad (C.43)$$
$$= 0 \text{ for } i > K.$$

Note that, due to the finite size of the buffer, the queue state can never grow beyond K even when λ approaches μ. However, with $\rho = 1$, the above expression of p_i apparently becomes indeterminate and is evaluated using L'Hospital's rule as $p_i(\rho \to 1) = \frac{1}{K+1}$.

For $\rho < 1$, i.e., with $\lambda < \mu$, the average number of packets in the queuing system is obtained as

$$E[N] = \sum_{i=0}^{K} ip_i = \sum_{i=0}^{K} i\frac{(1-\rho)\rho^i}{1-\rho^{K+1}} \quad (C.44)$$
$$= \frac{(1-\rho)}{1-\rho^{K+1}} \sum_{i=0}^{K} i\rho^i$$
$$= \frac{\rho}{1-\rho} - \frac{(K+1)\rho^{K+1}}{1-\rho^{K+1}}.$$

With $\rho = 1$, all states of the queue assume equal probability $(= \frac{1}{K+1})$, leading to $E[N] = \sum_{i=0}^{K} \frac{i}{K+1} = K/2$.

In an $M/M/1/K$ system, incoming packets are denied entry when the buffer gets fully occupied with a probability p_K. This in turn reduces the actual arrival rate into the queue from λ to λ_K, given by

$$\lambda_K = \lambda(1 - p_K), \quad (C.45)$$

which implies that, in queuing systems with limited buffer, the traffic carried by the queue becomes less than the traffic offered. Following the above expression, the carried traffic ρ_C in the present case can be expressed in terms of the offered traffic ρ as

$$\rho_C = \rho(1 - p_K). \quad (C.46)$$

Further, using Little's theorem with the above expression from λ_K, one can obtain the system delay $E[T]$ as

$$E[T] = \frac{E[N]}{\lambda_K} = \frac{E[N]}{\lambda(1-p_K)}, \quad (C.47)$$

albeit with packet losses in the finite buffer.

C.7 Multiple-server queues: *M/M/c* and *M/M/c/c*

In this section, we first consider a multiple-server queuing system with an infinite buffer, denoted as $M/M/c$. The state-transition diagram of $M/M/c$ queuing system is shown in Fig. C.9, where the arrivals take place with the rate λ, while the service rate varies from $c\mu$ to μ depending on how many of the servers remain busy at a given instant of time. The global balance equations of the system can be divided into two types: (*i*) when not all servers are busy, and (*ii*) when all servers are busy.

When not all servers are busy (i.e., $i < c$), the global balance equations are expressed as

$$(\lambda + i\mu)p_i = \lambda p_{i-1} + (i+1)\mu p_{i+1} \text{ for } c > i \geq 1 \qquad \text{(C.48)}$$
$$\lambda p_0 = \mu p_1.$$

However, when all servers are busy (i.e., $i \geq c$), the global balance equation is expressed as

$$c\mu p_i = \lambda p_{i-1} \text{ for } i \geq c. \qquad \text{(C.49)}$$

Following the same procedure as used before for the $M/M/1$ system, we obtain from Eq. C.48 the recursive expression of p_i as $p_i = \frac{\lambda}{i\mu}p_{i-1}$ for $c \geq i \geq 0$, leading to

$$p_i = \frac{1}{i!}\left(\frac{\lambda}{\mu}\right)^i p_0 = \frac{\alpha^i}{i!}p_0, \text{ for } c \geq i \geq 0, \qquad \text{(C.50)}$$

where $\alpha = \lambda/\mu$. Note that, in a $M/M/c$ queuing system with c servers, the traffic intensity in the queue becomes $\rho = \frac{\lambda}{c\mu} = \frac{\alpha}{c}$. Next, from Eq. C.49 we obtain the expression for p_i for $i \geq c$ as

$$p_i = \frac{\lambda}{c\mu}p_{i-1} = \rho p_{i-1} \text{ for } i \geq c. \qquad \text{(C.51)}$$

Using the above expression and $p_c = \frac{\alpha^c}{c!}p_0$ from Eq. C.50, we obtain p_{c+1} as

$$p_{c+1} = \rho p_c = \frac{\alpha^c}{c!}\rho p_0. \qquad \text{(C.52)}$$

Generalizing the expression of p_{c+1} further through recursive iterations of Eq. C.51, we express p_i for $i \geq c$ as

$$p_i = \frac{\alpha^c \rho^{i-c}}{c!}p_0 \text{ for } i \geq c. \qquad \text{(C.53)}$$

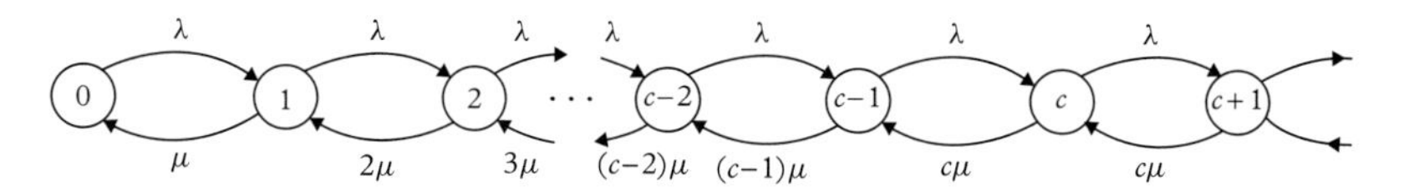

Figure C.9 *State-transition diagram of M/M/c queuing system.*

Using the normalizing condition on p_i's (i.e., $\sum_{i=0}^{\infty} p_i = 1$, using p_i's for $c \geq i \geq 0$ and $i \geq c$ from Equations C.50 and C.53, respectively), we obtain the expression for p_0 as

$$p_0 = \left[\sum_{i=0}^{c-1} \frac{\alpha^i}{i!} + \frac{\alpha^c}{c!(1-\rho)} \right]^{-1}. \tag{C.54}$$

An important measure of the $M/M/c$ queue is the probability P_W that an arrived customer has to wait in the queue before being served. We express P_W, as the sum of the probabilities p_i for $i \in [c, \infty]$, given by

$$\begin{aligned} P_W = P[N \geq c] &= \sum_{i=c}^{\infty} \frac{\rho^{i-c}\alpha^c}{c!} p_0 \\ &= \frac{\alpha^c p_0}{c!} \sum_{k=0}^{\infty} \rho^k \\ &= \frac{\alpha^c p_0}{c!(1-\rho)} \\ &= \frac{\alpha^c}{c!(1-\rho)} \left[\sum_{i=0}^{c-1} \frac{\alpha^i}{i!} + \frac{\alpha^c}{c!(1-\rho)} \right]^{-1}, \end{aligned} \tag{C.55}$$

which is known as the Erlang-C formula.

Next, we determine the average number of packets that would be waiting in the buffer. Note that the number of packets waiting in the buffer (excluding the servers) for a given queue state $i > c$ is $i - c$. We therefore express the average number of packets $E[N]$ waiting in the buffer as

$$E[N_b] = \sum_{i=c}^{\infty} (i-c)p_i = p_c \sum_{k=0}^{\infty} k\rho^k = \frac{\rho p_c}{(1-\rho)^2} = \frac{\rho P_W}{1-\rho}. \tag{C.56}$$

The average delay incurred by a packet in the buffer can be expressed as $E[W] = E[N_b]/\lambda$, using which we obtain the total delay in the system $E[T]$ as

$$\begin{aligned} E[T] &= E[W] + E[\tau] \\ &= \frac{E[N_b]}{\lambda} + \frac{1}{\mu} \\ &= \frac{1}{\lambda} \frac{\rho P_W}{(1-\rho)} + \frac{1}{\mu} \\ &= \frac{P_W}{\mu c(1-\rho)} + \frac{1}{\mu}. \end{aligned} \tag{C.57}$$

A multiple-server queue *without buffer*, receiving traffic with memoryless interarrival times and service durations, is denoted as $M/M/c/c$ system, implying that the system with c servers can hold (serve) at the most c customers at a time. This type of queuing system typically resembles telephone exchanges, where if all the servers (outgoing trunks) are busy, any new call request gets blocked. Figure C.10 shows the state-transition diagram of an $M/M/c/c$ queuing system whose number of states are upper-bounded by c.

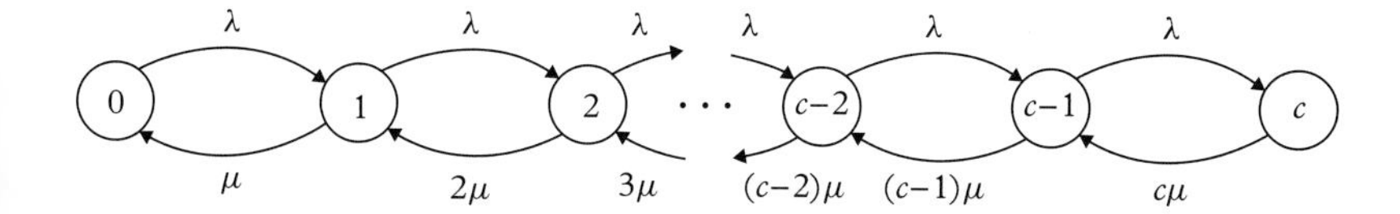

Figure C.10 *State-transition diagram of $M/M/c/c$ queuing system.*

Note that the global balance equations in this case will be the same as in $M/M/c$ system for $c \geq i \geq 0$, and hence are reproduced below from Eq. C.48 as

$$\begin{aligned}(\lambda + i\mu)p_i &= \lambda p_{i-1} + (i+1)\mu p_{i+1} \text{ for } c > i \geq 1 \\ \lambda p_0 &= \mu p_1,\end{aligned} \tag{C.58}$$

while the transitions expressed by Eq. C.49 will be non-existent. By using the above expressions recursively, as before, we obtain p_i as

$$p_i = \frac{\alpha^i}{i!} p_0, \text{ for } c \geq i \geq 0. \tag{C.59}$$

With this expression for p_i and using the normalizing condition $\sum_{i=0}^{c} p_i = 1$, we obtain p_0 as

$$p_0 = \left[\sum_{i=0}^{c} \frac{\alpha^i}{i!} \right]^{-1}. \tag{C.60}$$

Using the above expression for p_0, we finally obtain the probability P_B $(= p_c)$ that all the servers would be busy as

$$P_B = p_c = \frac{\alpha^c p_0}{c!} = \frac{\alpha^c / c!}{\sum_{i=0}^{c} \frac{\alpha^i}{i!}}. \tag{C.61}$$

This relation is known as the Erlang-B formula, representing the blocking probability in a telephone exchange with c outgoing trunks. The actual arrival rate in the queue can therefore be expressed as $\lambda_a = \lambda(1 - P_B)$. Using this expression along with Little's theorem, we obtain the average number $E[N]$ of customers in the system (also called carried traffic) as

$$E[N] = \lambda_a E[\tau] = \frac{\lambda}{\mu}(1 - P_B). \tag{C.62}$$

C.8 M/G/1 queue

So far we have considered the queuing systems with memoryless interarrival times as well as service durations, thus both following the exponential distribution. In this section, we consider $M/G/1$ systems representing single-server queues, where the interarrival times are exponentially distributed, but the service durations follow some general distribution. In order to estimate the performance of this type of systems, we first look into the waiting time of a packet in the system, including the delay due to the earlier-arrived packets in the buffer as well as the packet, if any, which is being serviced at present and is left with a *residual* service time. Note that when the service times don't remain memoryless (i.e., non-exponential) the

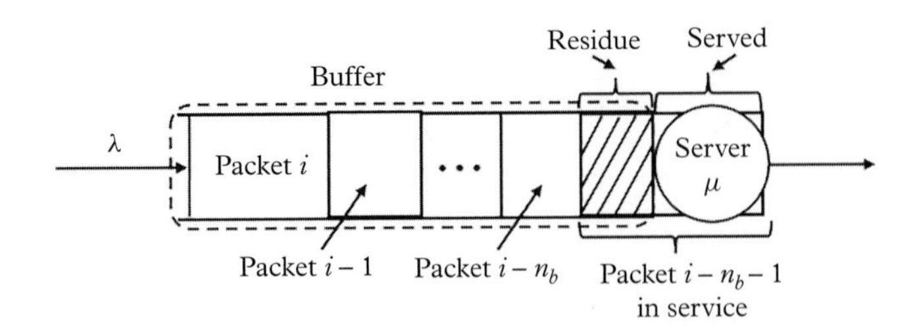

Figure C.11 *Snapshot of a single-server queue with infinite buffer, where packet i has arrived at a time when n_b packets are waiting in the buffer, and the oldest packet $i - n_b - 1$ is being served in the server with a residual time left behind in the buffer.*

average waiting time $E[W]$ in the buffer and the total delay $E[T]$ for each packet would be intrinsically different from what we obtained earlier for $M/M/1$ systems.

In view of the above, we first consider an arrival of a packet (packet i, say) in a single-server queue with infinite buffer. Packet i finds, on its arrival, that there are n_b packets waiting in the buffer for service, while the oldest packet (i.e., the packet $i - n_b - 1$) in the queue is being serviced in the server with a *residual time* R_i left behind in the buffer. Figure C.11 presents a snapshot of the queue, when packet i arrives.

The waiting time of packet i, as shown in Fig. C.11, can be expressed as

$$W_i = \sum_{j=i-n_b}^{i-1} \tau_j + R_i. \tag{C.63}$$

where τ_j's represent the durations of the packets seen by packet i ahead of itself in the buffer. From the above expression, one can express the average waiting time in the buffer $E[W]$ as

$$\begin{aligned} E[W] &= E\left[\sum_{i-n_b}^{i-1} E[\tau_j/n_b]\right] + E[R'] \\ &= E[\tau]E[n_b] + E[R'] \\ &= \frac{E[n_b]}{\mu} + E[R'], \end{aligned} \tag{C.64}$$

where $E[R']$ represents the average value of R_i's, and $E[R']$ will have two components, one when the queue is empty (i.e., zero delay) with the probability of p_0 and the other when the queue is not empty with a probability $1 - p_0$. Thus, we express $E[R']$ as

$$E[R'] = p_0 \times 0 + (1 - p_0) \times E[R], \tag{C.65}$$

where $E[R]$ represents the average residual time when the system has at least one packet. Using the fact that, when the system is not empty, then $1 - p_0 = \lambda E[\tau]$, we express $E[R']$ as

$$E[R'] = (1 - p_0)E[R] = \lambda E[\tau]E[R]. \tag{C.66}$$

Substituting $E[R']$ from Eq. C.66 and using Little's theorem for $E[n_b]$ (i.e., $E[n_b] = \lambda E[W]$) in Eq. C.64, we obtain $E[W]$ as

$$\begin{aligned} E[W] &= \frac{E[n_b]}{\mu} + E[R'] \\ &= \frac{\lambda E[W]}{\mu} + \lambda E[\tau]E[R] \\ &= \rho(E[W] + E[R]), \end{aligned} \tag{C.67}$$

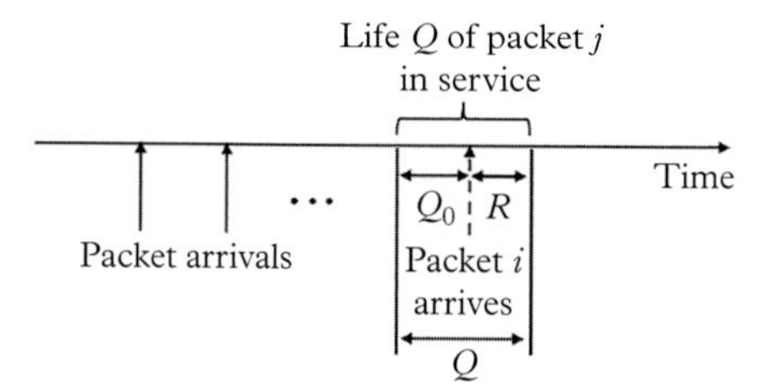

Q_0: Age of packet in service, i.e., already-served part of Q when packet i arrives
R: Residue of Q, yet to be served when packet i arrives

Figure C.12 *Illustration of the life, age, and residue of the oldest (in-service) packet j, i.e., the packet $i-n_b-1$ in Fig.C.11, when packet i arrives in the queue.*

leading to the expression for $E[W]$ in terms of E[R] and ρ, given by

$$E[W] = \frac{\rho E[R]}{1-\rho}. \tag{C.68}$$

Equation C.68 implies that the average waiting time of a packet in the buffer will be proportional to the average residual time of a packet, which would follow a *general* pdf in the case of $M/G/1$ queuing systems.

In order to evaluate $E[R]$, we follow the analytical model from (Kleinrock 1975), as illustrated in Fig. C.12 for the single-server queue considered earlier in Fig. C.11. The packets in the buffer are arranged following FIFO discipline, where packet j (i.e., packet $i-n_b-1$ in Fig. C.11) with a duration of Q, called its *life*, is found being served, when packet i arrives and is placed behind packet $i-1$. In particular, Fig. C.12 illustrates that the packet i has arrived in the queue at a specific instant of time, when packet j of life Q has already been served for a time interval Q_0 (called its *age*) and is left with the residual time R_j, and hence needs to wait for the time W_j with an expression in the form of Eq. C.63, derived earlier for the ith packet.

With the definitions of the life Q, age Q_0, and residue R_j of the in-service packet (or, simply R), we next consider the CDF and pdf of Q as $F_Q(q)$ and $f_Q(q)$, given by

$$\begin{aligned} F_Q(q) &= P[Q \le q] \\ f_Q(q) &= \frac{dF_Q(q)}{dq}. \end{aligned} \tag{C.69}$$

Similarly, we define the CDF and pdf of R as $F_R(r)$ and $f_R(r)$, given by

$$\begin{aligned} F_R(r) &= P[R \le r] \\ f_R(r) &= \frac{dF_R(r)}{dr}. \end{aligned} \tag{C.70}$$

Next, we formulate the probability of the event that the arrival instant of the packet i in the queue falls within the life $Q=q$, which is the service duration of packet j (i.e., the in-service packet $i-n_b-1$ in Fig. C.12). We denote the pdf of arrival instant of the packet falling within the service time or life Q as $f_A(q)$. Hence, the probability that the packet arrives during a small interval $[q, q+\delta q]$ within the life time Q of the packet that is being serviced will be $f_A(q)dq$. Intuitively, the probability $f_A(q)dq$ will be higher for higher values of q (that is for longer life) as well as for higher probability of occurrence of the segment of q, i.e., $f_Q(q)dq$.

In other words, one can model $f_A(q)dq$ as the probability that is proportional to the product $qf_Q(q)dq$, given by

$$f_A(q)dq = Kqf_Q(q)dq, \tag{C.71}$$

where K is the constant of proportionality. Integrating $f_A(q)dq$ over all possible values of q to unity, we obtain

$$\int_0^\infty f_A(q)dq = 1 = \int_0^\infty Kqf_Q(q)dq = KE[Q], \tag{C.72}$$

implying that $K = 1/E[Q]$, with $E[Q]$ representing the expected (mean) value for Q. Recalling that Q represents the service time of the packets, in the above expression we can substitute $E[Q]$ by $E[\tau]$, leading to

$$f_A(q) = Kqf_Q(q) = \frac{qf_Q(q)}{E[\tau]}. \tag{C.73}$$

Next, we find out the conditional probability that, in a given life time $Q = q$, the residual time $R \leq r$, i.e., $P[R \leq r/Q = q]$ (with $R \leq Q$, as residual time can not be longer than life). Assuming that r would be uniformly distributed over the given $Q = q$, we can express this probability as

$$P[R \leq r/Q = q] = \frac{r}{q}. \tag{C.74}$$

Using the above expression, we obtain the conditional probability that $r \leq R \leq r + dr$ given that $q \leq Q \leq q + dq$, as

$$\begin{aligned} P[r \leq R \leq r + dr, q \leq Q \leq q + dq] &= \frac{dr}{q} \times \frac{qf_Q(q)}{E[\tau]}dq \\ &= \frac{f_Q(q)}{E[\tau]}drdq. \end{aligned} \tag{C.75}$$

Next, we express the probability $f_R(r)dr$ by integrating the above probability over all possible values of $q \in [r, \infty]$, i.e.,

$$\begin{aligned} f_R(r)dr &= \int_{q=r}^\infty \frac{f_Q(q)dq}{E[\tau]}dr \\ &= \frac{dr}{E[\tau]}\int_{q=r}^\infty f_Q(q)dq \\ &= \frac{1 - F_Q(r)}{E[\tau]}dr, \end{aligned} \tag{C.76}$$

leading to the unconditional pdf of R, given by

$$f_R(r) = \frac{1 - F_Q(r)}{E[\tau]}. \tag{C.77}$$

Using the above result, we express the Laplace transform of $f_R(r)$ as $F_R(s) = \frac{1 - F_Q(s)}{sE[\tau]}$, with a factor of s in the denominator as $f_R(r)$ has an implicit multiplying-factor of a unit-step function $u(r)$ since the residual life time can assume only non-negative values. Using this transform-domain representation, one can determine any statistical moment, say the nth moment $E[R^n]$, of the residual life time R from $F_R^n(0) = [\frac{d^n F_R(s)}{ds^n}]_{s=0} = (-1)^n E[R^n]$ (Kleinrock 1975), from which, for the present problem, we need to estimate the first moment (i.e., the expected or average value) of the residual time R as $E[R] = -F_R^1(0)$. We therefore express the average value of R as

$$E[R] = -[F_R^1(s)]_{s=0} = -\left[\frac{d}{ds}\left\{\frac{1 - F_Q(s)}{sE[\tau]}\right\}\right]_{s=0} = -\left[\frac{-sF_Q^1(s) - 1 + F_Q(s)}{s^2 E[\tau]}\right]_{s=0}. \tag{C.78}$$

Noting that $F_Q(0) = 1$ and $F_Q^1(s)$ can be expressed in the form of $sP(s)$ with $P(s)$ as a polynomial of s, the function within the parentheses in the last line of the above equation becomes indeterminate at $s = 0$. We therefore apply L'Hospital's rule in Eq. C.78 to obtain $E[R]$ as

$$E[R] = \left[\frac{F_Q^2(s)}{2E[\tau]}\right]_{s=0} = \frac{F_Q^2(0)}{2E[\tau]} = \frac{E[\tau^2]}{2E[\tau]}, \tag{C.79}$$

where $E[\tau^2] = F_Q^2(0)$ represents the mean-squared value (i.e., the second moment) of the packet duration or life. Using Eq. C.79 in Eq. C.68, and substituting $E[\tau]$ by $1/\mu$, we therefore obtain $E[W]$ as

$$E[W] = \frac{\rho E[R]}{1-\rho} = \frac{\rho E[\tau^2]}{2E[\tau](1-\rho)} = \frac{\lambda E[\tau^2]}{2(1-\rho)}. \tag{C.80}$$

Expressing $E[\tau^2]$ as $(E[\tau])^2 + \sigma_\tau^2$ with σ_τ^2 as the variance of τ_i's, we obtain another form of $E[W]$ as

$$E[W] = \frac{\lambda[(E[\tau])^2 + \sigma_\tau^2]}{2(1-\rho)} = \frac{\rho(1 + C_\tau^2)}{2(1-\rho)} E[\tau], \tag{C.81}$$

where $C_\tau^2 = \sigma_\tau^2/(E[\tau])^2$ represents the coefficient of variation of the service times τ_i's. The above equation is known as the Pollaczek-Khinchin (PK) formula for queuing systems. The average delay $E[T]$ is obtained from the PK formula as

$$E[T] = E[W] + E[\tau] = \frac{\rho(1 + C_\tau^2)}{2(1-\rho)} E[\tau] + E[\tau]. \tag{C.82}$$

It would be worthwhile to see how this expression leads to our earlier results on the average waiting time for $M/M/1$ queues. Noting that the variance of exponential service times equals the square of the average, we get $C_\tau = 1$, leading to the expression of $E[T]$ as

$$E[T] = \frac{\rho(1+C_\tau^2)}{2(1-\rho)}E[\tau] + E[\tau] \tag{C.83}$$
$$= \frac{\rho}{\mu(1-\rho)} + \frac{1}{\mu}$$
$$= \frac{1}{\mu - \lambda},$$

which is same as the result obtained for $M/M/1$ queues in Eq. C.36. However, for $M/D/1$ queues, $C_\tau = 0$, and hence $E[T]$ is expressed as

$$E[T] = \frac{\rho}{2(1-\rho)}E[\tau] + E[\tau] = \frac{\rho}{2\mu(1-\rho)} + \frac{1}{\mu} = \frac{2-\rho}{2\mu(1-\rho)}, \tag{C.84}$$

where $E[\tau] = 1/\mu$ is in reality a constant representing a fixed packet size.

In some single-server queuing systems, the server might have to serve multiple arrival processes with different levels of priority, coming to the server from different buffers. In other words, while serving the packets of one arrival process (say $X(t)$), the server may receive a packet from another arrival process $Y(t)$ with higher priority. Owing to the higher priority for $Y(t)$, the server would complete the servicing of the current packet of $X(t)$, keep the other packets of $X(t)$ waiting in the buffer and offer service for the packet(s) arrived with higher priority, i.e., from $Y(t)$, and resume servicing $X(t)$ once the packet(s) in the buffer of $Y(t)$ are served.[1] In this situation, for $X(t)$ the server is said to have gone on *vacation* to offer service for $Y(t)$. An $M/G/1$ queue with vacation can be used to model the media-access or switching/routing protocols at the network nodes, where a lower-priority arrival process $X(t)$ has to compromise on its access to the server as and when demanded by another higher-priority arrival process $Y(t)$. The average time delay for the packets from the lower-priority process can be derived extending the $M/G/1$ queuing model, by accommodating the additional delay caused by the higher-priority arrivals. The expression of $E[W]$ for $X(t)$ in the queuing system following the model of $M/G/1$ with vacation (under the influence of the higher-priority arrivals) can be obtained as (Bertsekas and Gallager 2003)

$$E[W] = \frac{\lambda E[\tau_X^2]}{2(1-\rho)} + \frac{E[\tau_Y^2]}{2E[\tau_Y]}, \tag{C.85}$$

where $E[\tau_X^2]$ is the mean-squared value of the service time for the lower-priority process $X(t)$, and $E[\tau_Y]$ and $E[\tau_Y^2]$ are the mean and mean-squared values, respectively, for the service times of the higher-priority process $Y(t)$.

C.9 Network of queues

In a practical network, the queues used in the network nodes remain interconnected through the network topology, leading to a *network of queues*. There could be situations where the packets never come back to a queue already visited once, leading to a cascade of queues without any feedback. Such networks of queues are generally referred to as *tandem network of queues* (TNQ). However, the packets in a network might visit a queue multiple times due to a feedback mechanism, as governed by the network connectivity and protocols. Such

[1] There can also be *preemptive* schemes where the server would switch immediately to serve any new arrival of $Y(t)$, leaving behind the packet in $X(t)$ with incomplete service.

networks of queues are known as *networks of queues with feedback*, which are broadly divided into two categories: *open networks of queues with feedback* and *closed networks of queues*. In the first category, customers can arrive from outside and exit from the network as well, while for the second category, as the name indicates, customers move around the network without any ingress or egress traffic. We describe some of these networks of queues in the following.

C.9.1 Tandem networks of queues

To get an insight, we consider an example TNQ in Fig. C.13, where two single-server queues, ξ_1, ξ_2, are considered with infinite buffers, employing two servers with μ_1 and μ_2 as the service rates, respectively. Poisson arrival of packets takes place in the first queue ξ_1 at a rate of λ. In general, such networks of queues are described by a joint state vector $\mathbf{n}(t)$ at time t, given by $\mathbf{n}(t) = (N_1(t), N_2(t))$, where $N_1(t)$ and $N_2(t)$ are the states (i.e., number of customers in the first and the second queues at time t, respectively). For simplicity of notation, we represent $\mathbf{n}(t)$, $N_1(t)$, and $N_2(t)$ simply as $\mathbf{n}$, N_1, and N_2, respectively.

As shown in Fig. C.14, the TNQ of Fig. C.13 can pass through various possible state vectors, among which some will be terminal or boundary states, e.g., the states $\mathbf{n} = (0,0), (0,i_2), (i_1,0)$, with $i_1, i_2 > 0$. Thus, the terminal states would occur when one or both of the two queues become empty. Other (non-terminal) states would be represented by $\mathbf{n} = (N_1, N_2) = (i_2, i_2)$ with $i_1, i_2 > 0$. As before, equating the incoming and outgoing transition rates of the states, the global balance equations for the terminal states can be expressed as

$$\begin{aligned} \lambda p(0,0) &= \mu_2 p(0,1) \\ (\lambda + \mu_1)p(i_1,0) &= \lambda p(i_1 - 1, 0) + \mu_2 p(i_1, 1) \\ (\lambda + \mu_2)p(0,i_2) &= \mu_1 p(1, i_2 - 1) + \mu_2 p(0, i_2 + 1). \end{aligned} \tag{C.86}$$

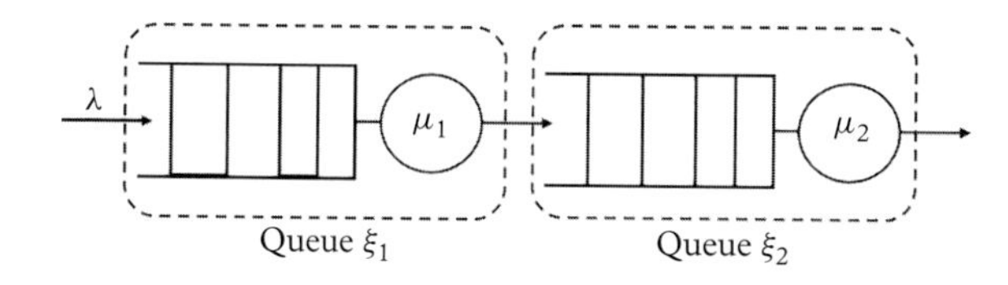

Figure C.13 *Tandem network of queues.*

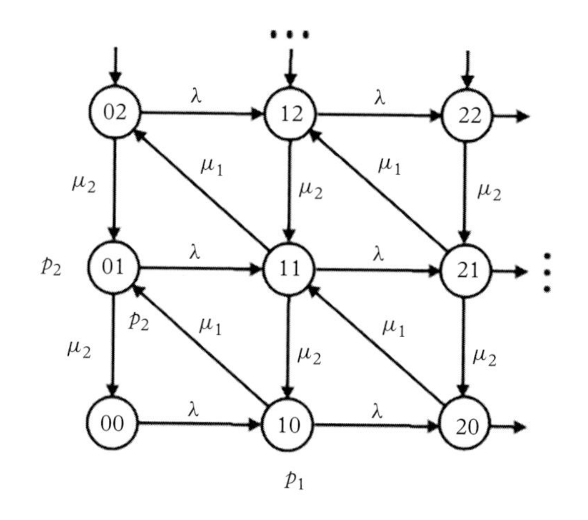

Figure C.14 *State-transition diagram of a TNQ with two queues, as shown in Fig. C.13. Note that, the two bits (from left to right) in the circles (network states) represent the number of customers in the two queues ξ_1 and ξ_2, i.e., i_1 and i_2, respectively.*

Similarly, one can write the global balance equation for a non-terminal state vector $\mathbf{n} = (i_1, i_2)$ as

$$(\lambda + \mu_1 + \mu_2)p(i_1, i_2) = \lambda p(i_1 - 1, i_2) + \mu_2 p(i_1, i_2 + 1) \\ + \mu_1 p(i_1 + 1, i_2 - 1), \tag{C.87}$$

where the left-hand side represents the total outward transition rate from the state (i_1, i_2) due to arrival/departure of packets, and the first, second, and third terms on the right-hand side are the transition rates into state (i_1, i_2) from the states $(i_1 - 1, i_2)$, $(i_1, i_2 + 1)$, and $(i_1 + 1, i_2 - 1)$, respectively. By using recursion on the above global balancing equations and applying the condition of normalization on the state probabilities, i.e.,

$$\sum_{i_1=1}^{\infty} \sum_{i_2=1}^{\infty} p(i_1, i_2) = 1, \tag{C.88}$$

one can obtain the solution for the state probability $p(i_1, i_2)$ as

$$p(i_1, i_2) = (1 - \rho_1)\rho_1^{i_1}(1 - \rho_2)\rho_2^{i_2}, \tag{C.89}$$

with $\rho_1 = \lambda/\mu_1$ and $\rho_2 = \lambda/\mu_2$, implying that the tandem network remains stable with $\rho_1, \rho_2 < 1$. Note that, the solution has come out in *product form* with its factors looking similar to that obtained in the $M/M/1$ queuing systems. In reality, the first queue is an $M/M/1$ queue with its state probability, given by $p_1(i_1) = (1 - \rho_1)\rho_1^{i_1}$, which is the same as the first factor on the right side of the above expression, and would sum up to unity for all values of i_1, with a given value of i_2. Using this observation and summing up $p(i_1, i_2)$ from Eq. C.89 over all possible values of i_1, we obtain the state probability $p_2(i_2)$ of the second queue ξ_2 as

$$\begin{aligned} p_2(i_2) &= \sum_{i_1=0}^{\infty} (1 - \rho_1)\rho_1^{i_1}(1 - \rho_2)\rho_2^{i_2} \\ &= (1 - \rho_2)\rho_2^{i_2} \underbrace{\sum_{i_1=0}^{\infty} (1 - \rho_1)\rho_1^{i_1}}_{=1} \\ &= (1 - \rho_2)\rho_2^{i_2}. \end{aligned} \tag{C.90}$$

By using Equations C.90 and C.89, one can therefore express $p(i_1, i_2)$ as

$$p(i_1, i_2) = (1 - \rho_1)\rho_1^{i_1}(1 - \rho_2)\rho_2^{i_2} = p_1(i_1)p_2(i_2), \tag{C.91}$$

which is an important result implying that the joint probability of the two queues in tandem is the product of the state probabilities of the two independent queues. In other words, the numbers of packets in the two queues in a two-stage TNQ at any instant of time are independent variables. This result can also be obtained by using Burke's theorem (Burke 1956), which states that (*i*) the departure process of any system using $M/M/1, M/M/c$ or $M/M/\infty$ queue is Poisson-distributed with the departure rate equaling to the Poissonian arrival rate λ, and (*ii*) at any time t (say), the state of the system $N(t)$ is independent of the sequence of packet-departure instants prior to t. Note that, the above result in product form for the joint state probabilities can be generalized for any number of queues in tandem, wherein the queue with the highest traffic intensity (i.e., with the lowest service rate) would constrict the traffic flow in the network, thereby becoming the *bottleneck point* of the network.

C.9.2 Networks of queues with feedback

In the networks of queues with feedback, open as well as closed networks, packets can visit a queue multiple times which in turn makes the resulting input arrival process non-Markovian as the past and present traffic flows get mixed up, thereby making the analysis very complicated as compared to the tandem networks of queues. However, the analysis of networks of queues with feedback can be simplified by using an extremely useful theorem, known as *Jackson's theorem* (Jackson 1957; Bertsekas and Gallager 2003; Leon-Garcia 1994). In the following, in order to prove Jackson's theorem, we consider the case of open networks of queues with feedback (ONQ-FB), and the same approach can also be extended for closed networks of queues.

Figure C.15 illustrates a simple form of ONQ-FB, where, from the output of the first queue ξ_1, a fraction (probability) p_{11} of the outgoing packets are fed back to the same queue, increasing the total arrival rate to queue ξ_1 from γ to $\gamma + p_{11}\gamma_1$. With this notion of feedback mechanism, we next consider a generic ONQ-FB with K nodes, each operating as a single-server queue with infinite buffer and following a FIFO discipline. In a representative queue (queue ξ_i at node i, say) in the ONQ-FB, packets arrive from outside the network following an independent Poisson arrival process. Assume that, in queue ξ_i, external packets arrive at a rate γ_i, with $\gamma_i > 0$ for at least one queue. Once a packet is serviced in ξ_i, it might go to another queue, say ξ_j, with a probability p_{ij} or exit the network with the probability of $p_{ix} = 1 - \sum_{j=1}^{K} p_{ij}$. With this framework, one can express the total arrival rate in queue ξ_i as

$$\lambda_i = \gamma_i + \sum_{j=1}^{K} \lambda_i p_{ji}, \text{ for } i = 1, \ldots, K. \tag{C.92}$$

Next, we define a K-tuple network state vector at a time t (t being not explicitly included in the notation, as before) as

$$\mathbf{n} = (N_1, N_2, \ldots, N_{i-1}, N_i, N_{i+1}, \ldots, N_K), \tag{C.93}$$

with N_i representing the number of packets in queue ξ_i. With this framework, one can state Jackson's theorem as

$$p(\mathbf{n}) = p_1(N_1)p_2(N_2)\ldots p_K(N_K), \tag{C.94}$$

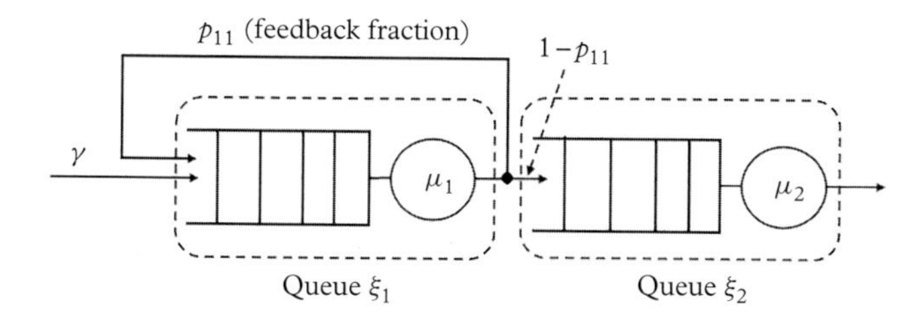

Figure C.15 *Illustration of an open network of two queues with feedback.*

implying that the probability of the state vector $p(\mathbf{n})$ can be expressed as a product of *apparently* independent probabilities $p_i(N_i)$'s of the states N_i's in K *equivalent* queues, which in turn helps in representing the given ONQ-FB as a TNQ of as many equivalent queues, albeit with modified arrival and service rates. This is a significant simplification of a complex scenario, as in this form the entire network is represented as a TNQ, from which it becomes straightforward to find out the overall network performance in terms of the number of customers waiting in the network and the delay incurred therein in an additive form of the respective components (number of customers or delay values) in each queue.

In the following, we present a proof of Jackson's theorem assuming that all the queues in the network follow $M/M/1$ queuing model. Note that even though the queues receive Poisson arrivals from the external world (at least one queue must receive external arrivals), the actual inputs to all queues may not be memoryless owing to the feedback path(s) from within the network. The proof presented here can also be extended for $M/M/c$ queues, although the theorem won't work with finite buffers causing possible packet losses therein (Jackson 1957).

In the first step to prove Jackson's theorem, we arbitrarily choose a queue, say queue ξ_i, in the given ONQ-FB. The network state vector is represented by $\mathbf{n} = (N_1, N_2, \ldots, N_{i-1}, N_i, N_{i+1}, \ldots, N_K)$, as defined earlier. When queue ξ_i happens to have $N_i + 1$ packets (i.e., one additional packet, as compared to state $\mathbf{n}$) and the other queues continue to have the same number of packets as in state $\mathbf{n}$, the state vector of the network is denoted as $\mathbf{n}^{i+}$, given by

$$\mathbf{n}^{i+} = (N_1, N_2, \ldots, N_{i-1}, N_i + 1, N_{i+1}, \ldots, N_K). \tag{C.95}$$

During the operation of the network, the state vector can move from state $\mathbf{n}^{i+}$ to state $\mathbf{n}$, with a packet leaving queue ξ_i and exiting the network altogether with a transition rate $\phi(\mathbf{n}^{i+}, \mathbf{n})$, given by

$$\phi(\mathbf{n}^{i+}, \mathbf{n}) = \mu_i\left(1 - \sum_{j=1}^{K} p_{ij}\right) = \mu_i p_{ix}. \tag{C.96}$$

Similarly, when ξ_i has $N_i - 1$ packets, the network state vector is denoted as $\mathbf{n}^{i-}$, given by

$$\mathbf{n}^{i-} = (N_1, N_2, \ldots, N_{i-1}, N_i - 1, N_{i+1}, \ldots, N_K), \tag{C.97}$$

and the state vector can move from state $\mathbf{n}^{i-}$ to state $\mathbf{n}$, with a transition rate $\phi(\mathbf{n}^{i-}, \mathbf{n})$, given by

$$\phi(\mathbf{n}^{i-}, \mathbf{n}) = \gamma_i. \tag{C.98}$$

Following similar notations and approach, when the network is in a state $\mathbf{n}^{j+i-}$ with $N_i - 1$ packets in ξ_i, and $N_j + 1$ packets in ξ_j, the network state vector is represented by

$$\begin{aligned}\mathbf{n}^{j+i-} \\ = (N_1, N_2, \ldots, N_{i-1}, N_i - 1, N_{i+1}, \ldots, N_{j-1}, N_j + 1, N_{j+1}, \ldots, N_K),\end{aligned} \tag{C.99}$$

and a transition from state $\mathbf{n}^{j+i-}$ to state $\mathbf{n}$ would take place with a transition rate $\phi_F(\mathbf{n}^{j+i-}, \mathbf{n})$, given by

$$\phi(\mathbf{n}^{j+i-}, \mathbf{n}) = \mu_j p_{ji}, \tag{C.100}$$

which is an internal (i.e., within the network) transfer of a packet from queue ξ_i to queue ξ_j.

With the above definitions and observations, we next proceed to prove Jackson's theorem as follows. In essence, as we observe from the foregoing discussions, we need to establish that the given product form of the state probability $p(\mathbf{n})$ defines a steady state of the given ONQ-FB. For this purpose, one therefore needs to establish that

$$\lim \Delta t \to 0 \frac{p(\mathbf{n}; t + \Delta t) - p(\mathbf{n}; t)}{\Delta t} = 0, \tag{C.101}$$

where $p(\mathbf{n}; t)$ and $p(\mathbf{n}; t + \Delta t)$ represent the probabilities of the state vector $\mathbf{n}$ at time t and $t + \Delta t$, respectively. In order to establish the above criterion, we express Eq. C.101 in terms of an equivalent difference equation, given by

$$\begin{aligned} p(\mathbf{n}; t + \Delta t)\Delta t & \\ = & \underbrace{\left[1 - \left(\sum_{i=1}^{K} \gamma_i + \sum_{i=1}^{K} \mu_i\right)\right]}_{\phi(\mathbf{n},\mathbf{n})} p(\mathbf{n}; t)\Delta t \\ & + \sum_{i=1}^{K} \underbrace{\mu_i p_{ix}}_{\phi(\mathbf{n}^{i+},\mathbf{n})} p(\mathbf{n^{i+}(t)})\Delta t + \sum_{i=1}^{K} \underbrace{\gamma_i}_{\phi(\mathbf{n}^{i-},\mathbf{n})} p(\mathbf{n^{i-}(t)})\Delta t \\ & + \sum_{i=1}^{K}\sum_{j=1}^{K} \underbrace{\mu_j p_{ji}}_{\phi(\mathbf{n}^{j+i-},\mathbf{n})} p(\mathbf{n^{j+i-}(t)})\Delta t + o(\Delta t), \end{aligned} \tag{C.102}$$

with $o(\Delta t)$ as the error term that vanishes as $\Delta t \to 0$. The underbraced parts of the four terms in the right-hand side of the above expression indicate the following:

(1) $\phi(\mathbf{n}, \mathbf{n})$: transition rate for the network to continue in state $\mathbf{n}$ in spite of the possible transitions,

(2) $\phi(\mathbf{n}^{i+}, \mathbf{n})$: transition rate for the network to move from state $\mathbf{n}^{i+}$ to state $\mathbf{n}$ (see Eq. C.96),

(3) $\phi(\mathbf{n}^{i-}, \mathbf{n})$: transition rate for the network to move from state $\mathbf{n}^{i-}$ to state $\mathbf{n}$ (see Eq. C.98),

(4) $\phi(\mathbf{n}^{j+i-}, \mathbf{n})$: transition rate for the network to move from state $\mathbf{n}^{j+i-}$ to state $\mathbf{n}$ (see Eq. C.100).

With $\Delta t \to 0$, and presuming that the steady state can be arrived at, we remove the time dependence of the probabilities and divide both sides by Δt to recast the above expression, implying that eventually we need to prove that

$$\begin{aligned} p(\mathbf{n}) & \\ = & \left[1 - \left(\sum_{i=1}^{K} \gamma_i + \sum_{i=1}^{K} \mu_i\right)\right] p(\mathbf{n}) + \sum_{i=1}^{K} \mu_i p_{ix} p(\mathbf{n^{i+}}) \\ & + \sum_{i=1}^{K} \gamma_i p(\mathbf{n^{i-}}) + \sum_{i=1}^{K}\sum_{j=1}^{K} \mu_j p_{ji} p(\mathbf{n^{j+i-}}). \end{aligned} \tag{C.103}$$

Dividing both sides by $p(\mathbf{n})$ in the above equation and moving $\sum_{i=1}^{K} \gamma_i + \sum_{i=1}^{K} \mu_i$ to the left-hand side, we obtain

$$\sum_{i=1}^{K} \gamma_i + \sum_{i=1}^{K} \mu_i = \sum_{i=1}^{K} \mu_i p_{ix} \underbrace{\frac{p(\mathbf{n}^{\mathbf{i}+})}{p(\mathbf{n})}}_{T1} + \sum_{i=1}^{K} \gamma_i \underbrace{\frac{p(\mathbf{n}^{\mathbf{i}-})}{p(\mathbf{n})}}_{T2} + \sum_{i=1}^{K} \sum_{j=1}^{K} \mu_j p_{ji} \underbrace{\frac{p(\mathbf{n}^{\mathbf{j}+\mathbf{i}-})}{p(\mathbf{n})}}_{T3}. \tag{C.104}$$

The three factors $T1$, $T2$, and $T3$ in the above equation are evaluated using the product form of state probabilities as

$$\begin{aligned} T1 &= \frac{p(\mathbf{n}^{i+})}{p(\mathbf{n})} \\ &= \frac{p_1(N_1)p_2(N_2)\dots p_i(N_i+1)\dots p_K(N_K)}{p_1(N_1)p_2(N_2)\dots p_i(N_i)\dots p_K(N_K)} \\ &= \frac{p_i(N_i+1)}{p_i(N_i)} = \frac{\rho_i^{N_i+1}(1-\rho_i)}{\rho_i^{N_i}(1-\rho_i)} = \rho_i = \frac{\lambda_i}{\mu_i}, \end{aligned} \tag{C.105}$$

$$\begin{aligned} T2 &= \frac{p(\mathbf{n}^{i-})}{p(\mathbf{n})} \\ &= \frac{p_1(N_1)p_2(N_2)\dots p_i(N_i-1)\dots p_K(N_K)}{p_1(N_1)p_2(N_2)\dots p_i(N_i)\dots p_K(N_K)} \\ &= \frac{p_i(N_i-1)}{p_i(N_i)} = \frac{\rho_i^{N_i-1}(1-\rho_i)}{\rho_i^{N_i}(1-\rho_i)} = \frac{1}{\rho_i} = \frac{\mu_i}{\lambda_i}, \end{aligned} \tag{C.106}$$

$$\begin{aligned} T3 &= \frac{p(\mathbf{n}^{j+i-})}{p(\mathbf{n})} \\ &= \frac{p_1(N_1)p_2(N_2)\dots p_i(N_i-1)\dots p_i(N_j+1)\dots p_K(N_K)}{p_1(N_1)p_2(N_2)\dots p_i(N_i)\dots p_K(N_K)} \\ &= \frac{p_i(N_i-1)p_j(N_j+1)}{p_i(N_i)p_j(N_j)} = \frac{\rho_i^{N_i-1}(1-\rho_i)\rho_j^{N_j+1}(1-\rho_i)}{\rho_i^{N_i}(1-\rho_i)\rho_j^{N_j}(1-\rho_j)} \\ &= \frac{\rho_j}{\rho_i} = \frac{\mu_i\lambda_j}{\mu_j\lambda_i}. \end{aligned} \tag{C.107}$$

Substituting for $T1$, $T2$, and $T3$ in Eq. C.104, we obtain

$$\begin{aligned} \sum_{i=1}^{K} \gamma_i + \sum_{i=1}^{K} \mu_i &= \sum_{i=1}^{K} \mu_i p_{ix} \frac{\lambda_i}{\mu_i} + \sum_{i=1}^{K} \gamma_i \frac{\mu_i}{\lambda_i} + \sum_{i=1}^{K} \sum_{j=1}^{K} \mu_j p_{ji} \frac{\mu_i \lambda_j}{\mu_j \lambda_i} \\ &= \sum_{i=1}^{K} p_{ix}\lambda_i + \sum_{i=1}^{K} \frac{\gamma_i \mu_i}{\lambda_i} + \sum_{i=1}^{K} \sum_{j=1}^{K} \frac{\mu_i \lambda_j p_{ji}}{\lambda_i}. \end{aligned} \tag{C.108}$$

Recognizing that in the steady state the total external arrival rate into the network in all queues must be equal to the total departure rate from all network queues, i.e., $\sum_{i=1}^{K} \gamma_i = \sum_{i=1}^{K} p_{ix}\lambda_i$, we obtain from the above expression

$$\sum_{i=1}^{K} \mu_i = \sum_{i=1}^{K} \frac{\gamma_i \mu_i}{\lambda_i} + \sum_{i=1}^{K}\sum_{j=1}^{K} \frac{\mu_i \lambda_j p_{ji}}{\lambda_i}. \tag{C.109}$$

Note that the above expression is the outcome of utilizing the product-form for the state probabilities while finding $T1$, $T2$, and $T3$.

However, the above equality in Eq. C.109 can also be obtained in a different manner as well. Note that the second term on the right-hand side of the above expression itself can be split *without* making use of the product-form state probabilities as

$$\sum_{i=1}^{K}\sum_{j=1}^{K} \frac{\mu_i \lambda_j p_{ji}}{\lambda_i} = \sum_{i=1}^{K} \frac{\mu_i}{\lambda_i} \sum_{j=1}^{K} \lambda_i p_{ji}. \tag{C.110}$$

Noting that $\sum_{j=1}^{K} \lambda_i p_{ji}$ represents the internally fed-back traffic into queue ξ_i, we express $\sum_{j=1}^{K} \lambda_i p_{ji}$ as

$$\sum_{j=1}^{K} \lambda_i p_{ji} = \lambda_i - \gamma_i, \tag{C.111}$$

because the total arrival rate into the queue ξ_i is λ_i, including the external arrival rate γ_i and the feedback traffic. Using the above expression in Eq. C.110, we obtain

$$\sum_{i=1}^{K}\sum_{j=1}^{K} \frac{\mu_i \lambda_j p_{ji}}{\lambda_i} = \sum_{i=1}^{K} \frac{\mu_i}{\lambda_i}(\lambda_i - \gamma_i) = \sum_{i=1}^{K} \mu_i - \sum_{i=1}^{K} \frac{\mu_i \gamma_i}{\lambda_i}, \tag{C.112}$$

leading to

$$\sum_{i=1}^{K} \mu_i = \sum_{i=1}^{K} \frac{\gamma_i \mu_i}{\lambda_i} + \sum_{i=1}^{K}\sum_{j=1}^{K} \frac{\mu_i \lambda_j p_{ji}}{\lambda_i}, \tag{C.113}$$

which is identical to Eq. C.109, an equality that was earlier obtained from Eq. C.104 by the use of state probabilities expressed in the product form, thereby establishing Jackson's theorem.

Next, we illustrate the implications of Jackson's theorem in ONQ-FBs with an example shown in Fig. C.16. As shown in the figure, we consider an ONQ-FB having three $M/M/1$ queues ξ_1, ξ_2, and ξ_3, with service rates μ_1, μ_2, and μ_3, respectively, where from the output of queue ξ_2 a fraction p_2 is fed back to queue ξ_1. Further, from queue ξ_1, a fraction p_1 of its outgoing customers exit the network, and thus queue ξ_2 receives a fraction $(1 - p_1)$ of the output packets from queue ξ_1. The external arrival takes place only at queue ξ_1 with an arrival rate of γ.

With the above network setting, one can relate the total input arrival rate λ_1 at queue ξ_1 with the external arrival rate γ as

$$\lambda_1 = \gamma + p_2(1 - p_1)\lambda_1, \tag{C.114}$$

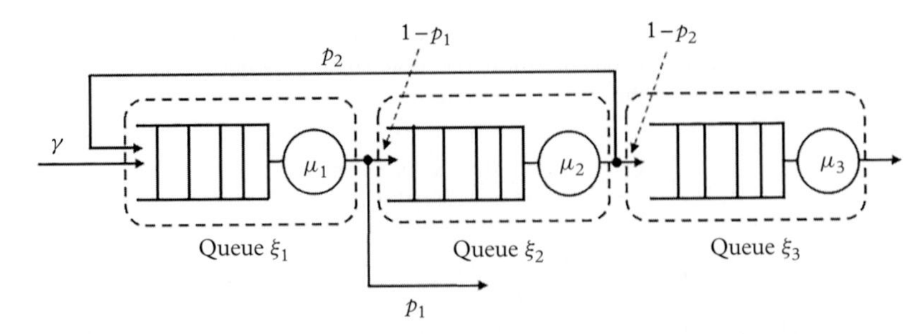

Figure C.16 *Example ONQ-FB with three queues.*

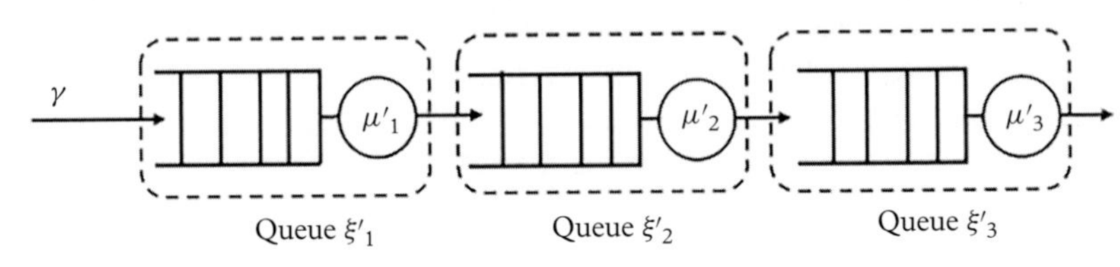

Figure C.17 *Equivalent tandem network for the ONQ-FB shown in Fig.C.16.*

leading to the expression of λ_1, given by

$$\lambda_1 = \frac{\gamma}{1 - p_2(1 - p_1)}. \tag{C.115}$$

Following a similar approach, one can express the arrival rates for the queues ξ_2 and ξ_3 as

$$\lambda_2 = (1 - p_1)\lambda_1 = \frac{(1 - p_1)\gamma}{1 - p_2(1 - p_1)} \tag{C.116}$$

$$\lambda_3 = (1 - p_2)\lambda_2 = \frac{(1 - p_1)(1 - p_2)\gamma}{1 - p_2(1 - p_1)}. \tag{C.117}$$

Using the values of λ_1, λ_2 and λ_3 for the three queues, we express the respective traffic intensities as

$$\rho_1 = \frac{\lambda_1}{\mu_1} = \frac{\gamma}{\mu_1[1 - p_2(1 - p_1)]} = \frac{\gamma}{\mu'_1} \tag{C.118}$$

$$\rho_2 = \frac{\lambda_2}{\mu_2} = \frac{(1 - p_1)\gamma}{\mu_2[1 - p_2(1 - p_1)]} = \frac{\gamma}{\mu'_2} \tag{C.119}$$

$$\rho_3 = \frac{\lambda_3}{\mu_3} = \frac{(1 - p_1)(1 - p_2)\gamma}{\mu_3[1 - p_2(1 - p_1)]} = \frac{\gamma}{\mu'_3}. \tag{C.120}$$

Following Jackson's theorem, we therefore express the probability of the network state vector $\mathbf{n}$ in the product form as

$$\begin{aligned} p(\mathbf{n}) &= p_1(N_1)p(N_2)p(N_3) \\ &= \rho_1^{N_1}(1 - \rho_1)\rho_2^{N_2}(1 - \rho_2)\rho_3^{N_3}(1 - \rho_3), \end{aligned} \tag{C.121}$$

leading to the equivalent TNQ as shown in Fig. C.17.

The product-form representation of $p(\mathbf{n})$ in Eq. C.121 and the equivalent TNQ shown in Fig. C.17 imply that the three equivalent queues (ξ'_1, ξ'_2, and ξ'_3) operate with the same external Poisson arrival rate γ, but with the modified service rates μ'_1, μ'_2, and μ'_3, so that the number of customers (N_1, N_2, and N_3) remain invariant. With this representation along with Equations C.119 through C.121, the modified service rates of ξ'_1, ξ'_2, and ξ'_3 are therefore expressed as

$$\frac{\gamma}{\rho_1} = \mu'_1 = \mu_1[1 - p_2(1 - p_1)] \tag{C.122}$$

$$\frac{\gamma}{\rho_2} = \mu'_2 = \frac{\mu_2[1 - p_2(1 - p_1)]}{1 - p_1} \tag{C.123}$$

$$\frac{\gamma}{\rho_1} = \mu'_3 = \frac{\mu_3[1 - p_2(1 - p_1)]}{(1 - p_1)(1 - p_2)}. \tag{C.124}$$

From the above results, we obtain the total average number of customers in the network as the sum of the average numbers of customers in all three queues, given by

$$\begin{aligned} E[N] &= E[N_1] + E[N_2] + E[N_3] \\ &= \frac{\rho_1}{1 - \rho_1} + \frac{\rho_2}{1 - \rho_2} + \frac{\rho_3}{1 - \rho_3}. \end{aligned} \tag{C.125}$$

Note that, although the third original queue ξ_3 is not involved in the feedback mechanism, its equivalent queue ξ_3 is driven to operate with a modified service rate (i.e., $\mu'_3 \neq \mu_3$) because its arrival rate γ is different from the actual arrival rate of the original third queue ξ_3. Using Little's theorem in Eq. C.110, we finally obtain the total delay incurred between the input of ξ_1 and the output of ξ_3 as

$$\begin{aligned} E[T] &= \frac{1}{\gamma} E[N] \\ &= \frac{1}{\gamma}\left[\frac{\rho_1}{1 - \rho_1} + \frac{\rho_2}{1 - \rho_2} + \frac{\rho_3}{1 - \rho_3}\right]. \end{aligned} \tag{C.126}$$

Finally, note that, although the equivalent queues with tandem connectivity will give appropriate estimates for the average number of customers and the corresponding values of the delay incurred in the network, their statistical distributions won't match with those of the original queues, as the arrival rates in these queues will no longer remain Poissonian.

Bibliography

A. S. Acampora, "A multichannel multihop local lightwave network," *IEEE GLOBECOM 1987*, Tokyo, Japan, November 1987.

A. Adhya and D. Datta, "Evaluation of accumulated FWM power at lightpath ends in wavelength-routed optical networks," *IEEE/OSA Journal of Optical Communications and Networking*, vol. 4, no. 4, pp. 314–325, April 2012.

G. P. Agrawal, *Nonlinear Fiber Optics*, 3rd edition, Elsevier, 2001.

M. Al-Fares, A. Loukissas, and A. Vahdat "A scalable, commodity data center network architecture," *ACM SIGCOMM 2008*, Seattle, USA, August 2008.

V. Alwayn, *Optical Network Design and Implementation*, Cisco Press, 2004.

F.-T. An, K. S. Kim, D. Gutierrez, S. Yam, E. Hu, K. Shrikhande, and L. G. Kazovsky, "SUCCESS: a next-generation hybrid WDM/TDM optical access network architecture," *IEEE/OSA Journal of Lightwave Technology*, vol. 22, no. 11, pp. 2557–2569, November 2004.

ANSI Standard X3T9.5, "Fiber distributed digital interface (FDDI)," 1989.

ANSI Standard X3T12, "Synchronous optical network (SONET)," 1995.

A. Antonio, A. Bianco, A. Bianciotto, V. De Feo, J. M. Finochietto, R. Gaudino, and F. Neri, "WONDER: A resilient WDM packet network for metro applications," *Elsevier Journal of Optical Switching and Networking*, vol. 5, no.1, pp. 19–28, March 2008 .

Y. Arakawa and A. Yariv "Quantum well lasers – gain, spectra, dynamics," *IEEE Journal of Quantum Electronics*, vol. 22, no. 9, pp. 1887–1899, September 1986.

J. Armstrong, "OFDM for optical communications," *IEEE/OSA Journal of Lightwave Technology*, vol. 27, no. 3, pp. 189–204, February 2009.

E. Arthurs, J. M. Cooper, M. S. Goodman, H. Kobrinski, M. Tur, and M. P. Vecchi, "Multiwavelength optical cross-connect for parallel-processing computers," *Electronic Letters*, vol. 24, pp. 119–120, 1986.

E. Ayanoglu, "Signal flow graphs for path enumeration and deflection routing analysis in multihop networks," *IEEE GLOBECOM 1989*, Dallas, USA, November 1989.

S. Ba, B. C. Chatterjee, S. Okamoto, N. Yamanaka, A. Fumagalli, and E. Oki, "Route partitioning scheme for elastic optical networks with hitless defragmentation," *IEEE/OSA Journal of Optical Communications and Networking*, vol. 8, no. 6, pp. 356–370, June 2016.

E.-J. Bachus, R.-P. Braun, C. Caspar, E. Grossman, H. Foisel, K. Hermes, H. Lamping, B. Strebel, and F. J. Westphal, "Ten-channel coherent optical fiber communication," *Electronics Letters*, vol. 22, pp. 1002–1003, 1986.

R. Ballart and Y-C Chiang, "SONET: now it's the standard optical network," *IEEE Communications Magazine*, vol. 27, no. 3, pp. 8–15, March 1989.

A. Banerjee, J. Drake, J. P. Lang, B. Turner, K. Kompella, and Y. Rekhter, "Generalized multiprotocol label switching: an overview of routing and management enhancements," *IEEE Communications Magazine*, vol. 39, no. 1, pp. 144–150, January 2001a.

A. Banerjee, J. Drake, J. P. Lang, B. Turner, D. Awduche, L. Berger, K. Kompella, and Y. Rekhter, "Generalized multiprotocol label switching: an overview of signaling enhancements and recovery techniques," *IEEE Communications Magazine*, vol. 39, no. 7, pp. 144–151, July 2001b.

A. Banerjee, Y. Park, F. Clarke, H. Song, S. Yang, G. Kramer, K. Kim, and B. Mukherjee, "Wavelength-division multiplexed passive optical network (WDM-PON) technologies for broadband access: a review," *OSA Journal of Optical Networking*, vol. 4, no. 11, pp. 737–758, November 2005.

J. A. Bannister, L. Fratta, and M. Gerla, "Topological design of the wavelength-division optical network," *IEEE INFOCOM 1990*, San Francisco, USA, June 1990.

R. A. Barry and P. A. Humblet, "Models of blocking probability in all-optical networks with and without wavelength changers," *IEEE Journal on Selected Areas in Communications*, vol. 14, no. 5, pp. 858–867, June 1996.

R. A. Barry, V. W. S. Chan, K. L. Hall, E. S. Kintzer, J. D. Moores, K. A. Rauschenbach, E. A. Swanson, L. E. Adams, C. R. Doerr, S.G. Finn, H. A. Haus, E. P. Ippen, W. S. Wong, and M. Haner, "All-optical network consortium – ultrafast TDM networks," *IEEE Journal on Selected Areas in Communications*, vol. 14, no. 5, pp. 999–1013, June 1996.

T. Battestilli and H. Perros, "An introduction to optical burst switching," *IEEE Communications Magazine*, vol. 41, no. 8, pp. S10–15, August 2003.

J. Bellamy, *Digital Telephony*, John Wiley, 3rd edition, 2003.

D. Bertsekas and R. Gallager, *Data Networks*, Prentice Hall, 2nd edition, 2003.

C. Bhar, N. Chatur, A. Mukhopadhyay, G. Das, and D. Datta, "Designing a green optical network unit using ARMA-based traffic prediction for quality of service-aware traffic," *Springer Photonic Network Communications*, vol. 32, no. 3, pp. 407–421, December 2016.

C. Bhar, G. Das, A. Dixit, B. Lannoo, M. V. D. Wee, D. Colle, D. Datta, M. Pickavet, and P. Demeester, "A green open access optical distribution network with incremental deployment support," *IEEE/OSA Journal of Lightwave Technology*, vol. 33, no. 19, pp. 4079–4092, October 2015.

P. Bhaumik, S. Zhang, P. Chowdhury, S.-S. Lee, J. H. Lee, and B. Mukherjee, "Software-defined optical networks (SDONs): a survey", *Springer Photonic Network Communications*, vol. 28, no. 1, pp. 4–18, 2014.

L. N. Bhuyan and D. P. Agarwal, "Generalized hypercube and hyperbus structures for a computer network," *IEEE Transactions on Computers*, vol. c-33, no. 4, pp. 323–333, April 1984.

A. Bianco, E. Leonardi, M. Mellia, and F. Neri, "Network controller design for SONATA: A large-scale all-optical WDM network," *IEEE ICC 2001*, Helsinki, Finland, June 2001.

M. S. Borella and B. Mukherjee, "Efficient scheduling of nonuniform packet traffic in a WDM/TDM local lightwave network with arbitrary transceiver tuning latencies," *IEEE Journal on Selected Areas in Communications*, vol. 14, pp. 923–934, June 1996.

C. A. Brackett, "Dense wavelength division multiplexing networks: principles and applications," *IEEE Journal on Selected Areas in Communications*, vol. 8, no. 6, pp. 948–964, August 1990.

S. Bregni, "A historical perspective on telecommunications network synchronization," *IEEE Communications Magazine*, pp. 158–166, June 1998.

P. J. Burke, "The output of a queuing system," *Operations Research*, vol. 4, pp. 699–704, 1956.

A. Buttabani, M. De Andrade, and M. Tornatore, "A multi-threaded dynamic bandwidth and wavelength allocation scheme with void filling for long reach WDM/TDM PONs," *IEEE/OSA Journal of Lightwave Technology*, vol. 31, no. 8, pp. 1149–1157, April 2013.

B. V. Caenegem, W. V. Parys, F. D. Truck, and P. Demeester, "Dimensioning of survivable WDM networks," *IEEE Journal on Selected Areas in Communications*, vol. 16, no. 7, pp. 1146–1157, September 1998.

S. A. Calta, S. A. de Veer, E. Loizides, and R. N. Strangewayes, "Enterprise systems (ESCON) architecture – system overview," *IBM Journal of Research and Development*, vol. 36, no. 4, pp. 535–541, July 1992.

X. Cao, N. Yoshikane, I. Popescu, T. Tsuritani, and I. Morita, "Software-defined optical networks and network abstraction with functional service design," *IEEE/OSA Journal of Optical Communications and Networking*, vol. 9, no. 4, pp. C65–75, April 2017.

J. H. L. Capucho and L. C. Resendo, "ILP model and effective genetic algorithm for routing and spectral allocation in elastic optical networks," *IEEE Microwave and Optoelectronics Conference (IMOC) 2013*, Rio de Janerio, Brazil, August 2013.

A. Carena, V. D. Feo, J. M. Finochietto, R. Gaudino, F. Neri, C. Pogilione, and P. Poggiolini, "RingO: an experimental WDM optical packet network for metro applications," *IEEE Journal on Selected Areas in Communications*, vol. 22, no. 8, pp. 1561–1571, October 2004.

A. Carena, M. D. Vaughn, R. Gaudino, M. Shell, and D. J. Blumenthal, "OPERA: an optical packet experimental routing architecture with label swapping capability," *IEEE/OSA Journal of Lightwave Technology*, vol. 16, no. 12, pp. 2135–2145, December 1998.

D. Cavendish, K. Murakami, S-H. Yun, O. Matsuda, and M. Nishihara, "New transport services for next-generation SONET/SDH systems," *IEEE Communications Magazine*, vol. 40, no. 5, pp. 80–87, May 2002.

K. Chaitanya and D. Datta, "Gain dynamics of EDFA chains in wavelength-routed optical networks," *IEEE ANTS 2009*, Delhi, India, December 2009.

T. J. Chan, Y. C. Ku, C. K. Chan, L. K. Chen, and F. Tong, "A novel bidirectional wavelength division multiplexed passive optical network with 1:1 protection," *OFC 2003*, Atlanta, USA, March 2003.

C. Chandrasekhar and X. Liu, "OFDM based superchannel transmission technology," *IEEE/OSA Journal of Lightwave Technology*, vol. 30, no. 24, pp. 3816–3823, December 2012.

M. Channegowda, R. Nejabati, and D. Simeonidou (Invited), "Software-defined optical networks technology and infrastructure: enabling software-defined optical network operations," *IEEE/OSA Journal of Optical Communications and Networking*, vol. 5, no. 10, pp. A274–282, October 2013.

K. Chen, A. Singla, A. Singh, K. Ramachandran, L. Xu, Y. Zhang, X Wen, and Y. Chen, "OSA: an optical switching architecture for data center networks with unprecedented flexibility," *IEEE Transactions on Networking*, vol. 22, no. 2, pp. 498–511, April 2014.

M. Chen, H. Jin, Y. Wen, and V. C. M. Leng, "Enabling technologies for future data center networking: a premier," *IEEE Network*, vol. 27, no. 4, pp. 8–15, July/August 2013.

J. Chen and L. Wosinska "Analysis of protection schemes in PON compatible with smooth migration from TDM-PON to hybrid WDM/TDM-PON," *OSA Journal of Optical Networking*, vol. 6, no. 5, pp. 514–526, 2007.

A. H. Cherin, *Introduction to Optical Fibers*, McGraw-Hill, 1983.

T. K. Chiang, S. K. Agrawal, D. T . Mayweather, D. Sadot, C. F. Barry, M. Hickey, and L. G. Kazovsky, "Implementation of STARNET: a WDM computer communications network," *IEEE Journal on Selected Areas in Communications*, vol. 14, no. 5, pp. 824–839, June 1996.

A. L. Chiu and E. H. Modiano, "Reducing electronic multiplexing costs in unidirectional SONET/WDM ring networks via efficient traffic grooming," *IEEE GLOBECOM 1998*, Sydney, Australia, November 1998.

A. L. Chiu and E. H. Modiano, "Traffic grooming algorithms for reducing electronic multiplexing costs in WDM ring networks," *IEEE/OSA Journal of Lightwave Technology*, vol. 18, no. 1, pp. 2–12, January 2000.

I. Chlamtac, A. Ganz, and G. Karmi, "Lightpath Communications: an approach to high-bandwidth optical WAN's," *IEEE Transactions on Communications*, vol. 40, no. 7, pp. 1171–1182, July 1992.

A. R. Chraplyvy, "Optical power limits in multi-channel wavelength-division multiplexed systems due to simulated Raman scattering," *Electronic Letters*, vol. 20, no. 2, pp. 58–59, 19th January 1984.

K. Christodoulopoulos, I. Tomkos, and E. A. Varvarigos, "Elastic bandwidth allocation in flexible OFDM-based optical networks," *IEEE/OSA Journal of Lightwave Technology*, vol. 29, no. 9, pp. 1354–1366, May 2011.

K. Christodoulopoulos, I. Tomkos, and E. A. Varvarigos, "Time-varying spectrum allocation policies and blocking analysis in flexible optical networks," *IEEE/OSA Journal on Selected Areas in Communications*, vol. 31, no. 1, pp. 13–25, January 2013.

T. Clark, *Designing Storage-Area Networks*, Addison-Wesley, Reading, 1999.

C. Clos, "A study of non-blocking switching networks," *Bell System Technical Journal*, vol. 32, no. 2, pp. 406–424, March 1953.

J. Cooper, J. Dixon, M. S. Goodman, H. Kobrinski, M. P. Veechi, E. Arthurs, S. G. Menocal, M. Tur, and S. Tsuji, "Nanosecond wavelength switching with a double-section distributed feedback laser," *CLEO 1988*, Anaheim, USA, April 1988.

J. A. Crossett and D. Krisher, "SONET/SDH network synchronization and synchronization sources," *IEEE GLOBECOM 1992*, Orlando, USA, December 1992.

D. G. Cunningham, "The status of the 10-Gigabit Ethernet standard," *ECOC 2001*, Amsterdam, Netherlands, September-October 2001.

J. D'Ambrosia, "40 Gigabit Ethernet and 100 Gigabit Ethernet: the development of a flexible architecture," *IEEE Communications Magazine*, pp. S8–14, March 2009.

G. B. Dantzig, *Linear Programming and Extensions*, Princeton University Press, 1963.

G. Das, B. Lanoo, A. Dixit, D. Colle, M. Pickavet, and P. Demeester, "Flexible hybrid WDM/TDM PON architectures using wavelength selective switches," *Elsevier Journal of Optical Switching and Networking*, vol. 9, no. 2, pp. 156–169, April 2012.

W. B. Davenport and W. L. Root, *An Introduction to the Theory of Random Signals and Noise*, McGraw-Hill, 1958.

F. Davik, A. Kvalbein, and S. Gjessing, "Improvement of resilient packet ring fairness," *IEEE GLOBECOM 2005*, St. Louis, USA, November-December 2005.

F. Davik, M. Yilmaz, S. Gjessing, and N. Uzun, "IEEE 802.17 resilient packet ring tutorial," *IEEE Communications Magazine*, vol. 42, no. 3, pp. 112–118, March 2004.

E. Desurvire, *Erbium-Doped Fiber Amplifiers: Principles and Applications*, John Wiley, New York, 1994.

A. R. Dhaini, C. M. Assi, M. Maier, and A Shami, "Dynamic wavelength and bandwidth allocation in hybrid TDM/WDM EPON networks," *IEEE/OSA Journal of Lightwave Technology*, vol. 25, no. 1, pp. 277–286, January 2007.

A. Dixit, B. Lanoo, D. Coolie, M. Pickavet, and P. Demeester, "Energy efficient dynamic bandwidth allocation for Ethernet passive optical networks: overview, challenges, and solutions," *Elsevier Journal of Optical Switching and Networking*, vol. 18, no. 2, pp. 169–179, November 2015.

X. Dong, T. El-Gorashi, and J. M. H. Elmirghani, "Green IP over WDM networks with data centers," *IEEE Journal of Lightwave Technology*, vol. 29, no. 12, pp. 1861–1880, June 2011.

N. R. Dono, P. E. Green, K. Liu, R. Ramaswami, and F. F. K. Tong, "A wavelength division multiple access network for computer communication," *IEEE Journal on Selected Areas in Communications*, vol. 8, pp. 983–993, August 1990.

T. Durhuus, B. Mikkelsen, C. Joergensen, S. L. Danielsen, and K. E. Stubkjaer, "All-optical wavelength conversion by semiconductor optical amplifiers," *IEEE/OSA Journal of Lightwave Technology*, vol. 14, no. 6, pp. 942–954, June 1996.

R. Dutta, A. E. Kamal, and G. E. Rouskas, Ed., *Traffic Grooming for Optical Networks: Foundations, Techniques, and Frontiers*, Springer, 2008.

N. Economides, "Telecommunications regulation: an introduction," New York University Working Paper No. 2451/26175, pp. 48–76, 26 June, 2013.

F. Effenberger, D. Cleary, O. Haran, G. Kramer, R. D. Li,, M. Oron, and T. Pfeiffer, "An introduction to PON technologies," *IEEE Communications Magazine*, pp. S17–25, March 2007.

C. A. Eldering, "Theoretical determination of sensitivity penalty for burst mode fiber optic receivers," *IEEE Journal of Lightwave Technology*, vol. 11, no. 12, pp. 2145–2149, December 1993.

Enterprise Networking Planet "Cisco and Juniper reveal networking strategies," http://www.enterprisenetworkingplanet.com/datacenter/cisco-and-juniper-reveal-sdn-strategies.html, 2013.

D. J. Farber, "A ring network," *Datamation*, vol. 26, no. 2, pp. 44–46, February 1975.

N. Farrington, G. Porter, S. Radhakrishnan, H. H. Bazzaz, B. Subramanya, Y. Fainman, G. Papen, and A. Vahdat, "Helios: A hybrid electrical/optical switch architecture for modular data centers," *ACM SIGCOMM 2010*, New Delhi, India, August-September, 2010.

C. H. Foh, L. Andrew, E. Wong, and M. Zukerman, "Full-RCMA: A high utilization EPON," *IEEE Journal on Selected Areas in Communications*, vol. 22, no. 8, pp. 1514–1524, October 2004.

F. Forghieri, R. W. Tkach, and A. R. Chraplyvy, "WDM systems with unequally spaced channels," *IEEE/OSA Journal of Lightwave Technology*, vol. 13, no. 5, pp. 889–897, May 1995.

H. Frazier and H. Johnson "Gigabit Ethernet: from 100 to 1000 Mbps," *IEEE Internet Computing*, pp. 24–31, January-February 1999.

N. J. Frigo, P. P. Iannone, P. D. Magill. T. E. Darcie, M. M. Downs, B. N. Desai, U. Koren, T. L. Koach, C. Dragone, H. M. Tresby, and G. E. Bodeep, "A wavelength-division multiplexed passive optical network with cost-shared components," *IEEE Photonics Technology Letters*, vol. 6, no. 11, pp. 1365–1367, November 1994.

R. M. Gagliardi and S. Karp, *Optical Communications*, John Wiley, 2nd edition, 1995.

M. W. Geis, T. M. Lyszczarz, R. M. Osgood, and B. R. Kimbal, "30 to 50 ns liquid-crystal optical switches," *Optics Express*, vol. 18, no. 18, August 2010.

M. Gerola, R. D. Corin, R. Riggio, F. De Pellegrini, E. Salvadori, H. Woesner, T. Rothe, M. Suñe, and L. Bergesio, "Demonstrating inter-testbed network virtualization in OFELIA SDN experimental facility," *IEEE INFOCOM 2013, Workshop on Computer Communication*, Turin, Italy, April 2013.

O. Gerstel, M. Jinno, A. Lord, and S. J. B. Yoo, "Elastic optical networking: a new dawn for the optical layer?" *IEEE Communications Magazine*, vol. 50, no. 2, pp. S12–20, February 2012.

O. Gerstel, R. Ramaswami, and G. Sasaki, "Cost-effective traffic grooming in WDM rings," *IEEE INFOCOM 1998*, San Francisco, USA, March-April 1998a.

O. Gerstel, P. Lin, and G. Sasaki, "Wavelength assignment in a WDM ring to minimize cost of embedded SONET rings," *IEEE INFOCOM 1998*, San Francisco, USA, March-April, 1998b.

R. Gidron and A. S. Acampora, "Design and implementation of a distributed switching node for a multihop ATM network," *IEEE GLOBECOM 1991*, Phoenix, USA, November 1991.

A. Goldsmith, *Wireless Communications*, Cambridge University Press, USA, 2005.

M. S. Goodman, J. L. Gimlett, H. Kobrinski, M. P. Vecchi, and R. M. Bulley, "The LAMBDANET multiwavelength network: architecture, applications, and demonstrations," *IEEE Journal on Selected Areas in Communications*, vol. 8, no. 6, pp. 995–1004, August 1990.

P. E. Green, *Fiber Optic Networks*, Prentice Hall, 1993.

A. Greenberg, J. R. Hamilton, N. Jain, S. Kandula, C. Kim, P. Lahiri, D. A. Maltz, P. Patel, and S. Sengupta, "VL2: a scalable and flexible data center network," *ACM SIGCOMM 2009*, Barcelona, Spain, August 2009.

V. Gudla, S. Das, A. Shastri, N. McKeown, L. G. Kazovsky, and S. Yamashita, "Experimental demonstration of OpenFlow control of packet and circuit switches," *OFC 2010*, San Diego, USA, March 2010.

C. Guillemot, M. Renaud, P. Gambini, C. Janz, I. Andonovic, R. Bauknecht, B. Bostica, M. Burzio, F. Callegati, M. Casoni, D. Chiaroni, F. Clerot, S. L. Danielsen, F. Dorgeuille, A. Dupas, A. Franzen, P. B. Hansen, D. K. Hunter, A. Kloch, R. Krahenbuhl, B. Lavigne, A. Le Corre, C. Raffaelli, M. Schilling, J.-C. Simon, and L. Zucchelli, "Transparent optical packet switching: the European ACTS KEOPS project approach," *IEEE/OSA Journal of Lightwave Technology*, vol. 16, no. 12, pp. 2117–2134, December 1998.

A. Gumaste, B. M. K. Bheri, and A. Kshirsagar, "FISSION: flexible interconnection of scalable systems integrated using optical networks for data centers," *IEEE ICC 2013, Symposium on Optical Networks and Systems*, Budapest, Hungary, June 2013.

C. Guo, G. Lu, D. Li, H. Wu, X. Zhang, Y. Shi, C. Tian, Y. Zhang, and S. Lu, "BCube: a high performance, server-centric network architecture for modular data centers," *ACM SIGCOMM 2009*, Barcelona, Spain, August 2009.

C. Guo, H. Wu, K. Tan, L. Shi, Y. Zhang, and S. Lu, "DCell: a scalable and fault-tolerant network structure for data centers," *ACM SIGCOMM 2008*, Seattle, USA, August 2008.

I. M. I. Habbab, M. Kavehrad, and C.-E. W. Sundberg, "Protocols for very high speed optical fiber local area networks using a passive star topology," *IEEE/OSA Journal of Lightwave Technology*, vol. 5, no. 12, pp. 1782–1794, December 1987.

M. F. Habib, M. Tornatore, M. De Leenheer, F. Dikbiyik, and B. Mukherjee, "Design of disaster-resilient optical datacenter networks," *IEEE/OSA Journal of Lightwave Technology*, vol. 30, no. 16, pp. 2563–2573, August 2012.

G. Hadley, *Linear Programming*, Addison-Wesley, 1961.

A. S. Hamza, J. S. Deogun, and D. R. Alexander, "Wireless communication in data centers: a survey," *IEEE Communications Surveys and Tutorials*, vol. 18, no. 3 pp. 1572–1595, 3rd quarter 2016.

J. Hecht, *Understanding Fiber Optics*, Prentice Hall, 5th Ed., 2005.

W. V. Heddeghem, B. Lanoo, D. Coolie, M. Pickavet, and P. Demeester, "A quantitative survey of the power saving potential in IP-over-WDM backbone networks," *IEEE Communications Surveys and Tutorials*, vol. 18, no. 1, pp. 706–731, 1st quarter 2016.

M. Herzog, M. Maier, and M. Reisslein, "Metropolitan area packet-switched WDM networks: a survey on ring systems," *IEEE Communication Surveys and Tutorials*, vol. 6, no. 2, pp. 2–20, 2nd quarter, 2004.

M. Herzog, M. Maier, and A. Wolisz "RINGOSTAR: an evolutionary AWG-based WDM upgrade of optical ring networks," *IEEE/OSA Journal of Lightwave Technology*, vol. 23, no. 4, pp. 1637–1651, April 2005.

M. J. Hluchyj and M. J. Karol, "Shufflenet: an application of generalized perfect shuffles to multihop lightwave networks," *IEEE/OSA Journal of Lightwave Technology*, vol. 9, no. 10, pp. 1386–1397, October 1991.

D. J. Hunkin and G. W. Litchfield, "An optical fibre section in the Cambridge digital ring," *Proceedings of IEE*, vol. 130, no. 5, September 1983.

D. K. Hunter, M. H. M. Nizam, M. C. Chia, I. Andonovic, K. M. Guild, A. Tzanakaki, M. J. O'Mahony, J. D. Bainbridge, M. F. C. Stephens, R. V. Penty, and I. H. White, "WASPNET: a wavelength switched packet network," *IEEE Communications Magazine*, vol. 37, no. 3, pp. 120–129, March 1999.

IEEE Standard 802.ah, "Ethernet in first mile (EFM)," September 2004.

IEEE Standard 802.6, "Distributed queue dual bus (DQDB) subnetwork of a metropolitan area network," December 1990.

IEEE Standard 802.17, "Resilient packet ring (RPR)," June 2004.

J. Iness, S. Banerjee, and B. Mukherjee, "GEMNET: A generalized, shuffle-exchange-based, regular, scalable, and modular multihop network based on WDM lightwave technology," *IEEE/ACM Transactions on Networking*, vol. 3, no. 4, pp. 470–476, August 1995.

H. Ishio, J. Minowa, and K. Nosu, "Review and status of wavelength-division-multiplexing technology and its application," *IEEE/OSA Journal of Lightwave Technology*, vol. 2, no. 4, pp. 448–463, August 1984.

ITU-T Recommendation G.7041, "Generic framing procedure," August 2016.

ITU-T Recommendation G.8080, "Architecture for the automatically switched optical network," June 2006.

ITU-T Recommendation G.984.1, SG 15, "Gigabit-capable passive optical networks (G-PON): general characteristics," March 2003.

ITU-T Recommendation G.984.2, SG 15, "Gigabit-capable passive optical networks (G-PON): physical media dependent (PMD) layer specification," March 2003.

ITU-T Recommendation G.984.3, SG 15, "Gigabit-capable passive optical networks (G-PON): transmission convergence (TC) layer specification," July 2005.

ITU-T Recommendation G.984.4, SG 15, "Gigabit-capable passive optical networks (G-PON): ONT management and control interface specification," June 2005.

ITU-T Recommendation G.987, "10 Gigabit-capable passive optical networks (10G-PON)," June 2012.

ITU-T Recommendation G.803, "Architecture of transport networks based on the synchronous digital hierarchy (SDH)," March 2000.

ITU-T Recommendation G.694.1, "Spectral grids for WDM applications: dense WDM frequency grid," June 2002.

ITU-T Recommendation G.694.2, "Spectral grids for WDM applications: coarse WDM frequency grid," December 2003.

ITU-T Recommendation G.709, "Interfaces for the optical transport network," June 2016.

J. R. Jackson, "Networks of waiting lines," *Operations Research*, vol. 5, no. 4, pp. 518-521, August 1957.

R. Jain, "Performance analysis of FDDI token ring networks," *ACM SIGCOMM 1990*, Philadelphia, USA, September 1990.

A. Jajszczyk, "Automatically switched optical networks: benefits and requirements," *IEEE Communications Magazine*, vol. 43, no. 2, February 2005.

V. Jayaraman, Z.-M. Chuang, and L. A. Coldren, "Theory, design, and performance of extended tuning range semiconductor lasers with sampled gratings," *IEEE Journal of Quantum Electronics*, vol. 29, no. 6, pp. 1824–1834, June 1993.

M. Jinno, "Elastic optical networking: roles and benefits in beyond 100-Gb/s era," *IEEE/OSA Journal of Lightwave Technology*, vol. 35, no. 5, pp. 1116–1124, May 2017.

M. Jinno, H. Takara, B. Kozicki, Y. Tsukishima, Y. Sone, and S. Matsuoka, "Spectrum-efficient and scalable elastic optical path network: architecture, benefits, and enabling technologies," *IEEE Communications Magazine*, vol. 47, no. 11, pp. 66–73, November 2009.

C. Kachris, K. Kanonakis, and I. Tomkos "Optical interconnection networks in data centers: recent trends and future challenges," *IEEE Communications Magazine*, vol. 51, no. 9, pp. 39–45, September 2013.

J. Kani, "Power saving techniques and mechanisms for optical access networks systems," *IEEE/OSA Journal of Lightwave Technology*, vol. 31, no. 4, February 2013.

B. Kantarci, and H. T. Moufta, "Availability and cost-constrained long-reach passive optical network planning," *IEEE Transactions on Reliability*, vol. 61, No. 1, pp. 113–124, March 2012.

K. Kanonakis and I. Tomkos, "Improving the efficiency of online upstream scheduling and wavelength assignment in hybrid WDM/TDMA EPON networks," *IEEE Journal on Selected Areas in Communications*, vol. 28, No. 6, pp. 838–848, August 2010.

K. C. Kao and G. A. Hockham, "Dielectric-fiber surface waveguides for optical frequencies,"*Proceedings of IEE*, vol. 113, no. 7, pp. 1151–1158, 1966.

N. Karmarkar, "A new polynomial-time algorithm for linear programming," *Combinatorica*, vol. 4, no. 4, pp. 373-395, 1984.

M. J. Karol and S. Z. Shaikh, "A simple adaptive routing scheme for congestion control in ShuffleNet multihop lightwave networks," *IEEE Journal on Selected Areas in Communications*, vol. 93, no. 7, pp. 1040–1051, September 1991.

S. V. Kartalopoulos, *Next Generation SONET/SDH: Voice and Data*, Wiley-IEEE Press, January 2004.

L. G. Kazovsky, "Impact of laser phase noise in optical heterodyne communication systems," *Journal of Optical Communications*, vol. 7, no. 2, pp. 66-78, June 1986.

G. Keiser, *Optical Fiber Communications*, McGraw-Hill, 4th edition, 2008.

S. Keshav, *An Engineering Approach to Computer Networking: ATM Networks, the Internet, and the Telephone Network*, Addision-Wesley, 1997.

L. Khachiyan, "A polynomial algorithm for linear programming," *Doklady Akademii Nauk SSSR*, 224 (5), pp. 1093–1096, 1979.

B. Khasnabish, "Topological properties of Manhattan Street networks," *Electronics Letters*, vol. 25, no. 20, pp. 1388–1389, 28 September, 1989.

M. S. Kiaei, L. Meng, C. M. Assi, and M. Maier, "Efficient scheduling and grant sizing methods for WDM PONs," *IEEE/OSA Journal of Lightwave Technology*, vol. 28, no. 13, pp. 1922–1931, July 2013.

Y.-m. Kim, J. Y. Choi, J.-h. Ryou, Hyun-mi Baek, Ok-sun Lee, Hong-shik Park, Minho Kang, Gwan-joong Kim, and J.-h. Yoo, "Cost-effective protection architecture to provide diverse protection demands in Ethernet passive optical network," *ICCT 2003*, Beijing, China, April 2003.

J. Kim, C. J. Nuzman, B. Kumar, D. F. Lieuwen, J. S. Kraus, A. Weiss, C. P. Lichtenwalner, A. R. Papazian, R. E. Frahm, N. R. Basavanhally, D. A. Ramsey, V. A. Aksyuk, F. Pardo, M. E. Simon, V. A. Lifton, H. B. Chan, M. Haueis, A. Gasparyan, H. R. Shea, S. C. Arney, C. A. Bolle, P. R. Kolodner, R. R. Ryf, D. T. Neilson, and J. V. Gates, "1100 × 1100 port MEMS-based optical crossconnect with 4-dB maximum loss," *IEEE Photonics Technology Letters*, vol. 15, no. 11, pp. 1537-1539, November 2003.

D. King. Y. Lee, H. Xu, and A. Farrel, "Path computation architectures overview in multi-domain optical networks based on ITU-T ASON and IETF PCE," *IEEE NOMS Workshop 2008*, Salvador, Brazil, April 2008.

L. Kleinrock, *Queueing Systems, Volume I: Theory*, John Wiley, 1975.

M. Klinkowski and K. Walkowiak, "Routing and spectrum assignment in spectrum sliced elastic optical path network," *IEEE Communication Letters*, vol. 15, no. 8, pp. 884–886, August 2011.

Z. Knittl, *Optics of Thin Film Filters*, John Wiley, 1976.

H. Kobrinski, R. M. Bulley, M. S. Goodman, M. P. Vecchi, C. A. Brackett, L. Curtis, and J. L. Gimlett, "Demonstration of high capacity in the LAMBDANET architecture, a multiwavelength optical network," *Electronic Letters*, vol. 23, no. 16, pp. 824–826, 30 July 1987.

G. Kramer, M. D. Andrade, R. Roy, and P. Chowdhury "Evolution of optical access networks: architectures and capacity upgrade," *Proceedings of IEEE*, vol. 100, no. 5, pp. 1188–1196, May 2012.

G. Kramer, B. Mukherjee, and G. Pesavento, "IPACT: a dynamic protocol for an Ethernet PON (EPON)," *IEEE Communications Magazine*, vol. 40, no. 2, pp. 74–80, February 2002.

D. Kreutz, P. E. Verissimo, and S. Azodolomolky, "Software-defined networking: a comprehensive survey," *Proceedings of IEEE*, vol. 103, no. 1, pp. 14–76, January 2015.

R. Krishnaswami and K. Sivarajan, "Design of logical topologies: a linear formulation for wavelength-routed optical networks with no wavelength changers,"*IEEE/ACM Journal of Networking*, vol. 9, no. 2, pp. 186–198, April 2001.

J.-F. P. Labourdette and A. S. Acampora, "Logically rearrangeable multihop lightwave networks," *IEEE Transactions on Communications*, vol. 39, no. 8, pp. 1223–1230, August 1991.

H. G. Lang, *A Phone of Our Own: the Deaf Insurrection Against Ma Bell*, Gallaudet University Press, 2000.

W.-T. Lee and L.-Y. Kung, "Binary Addressing and Routing Schemes in the Manhattan Street Network," *IEEE/ACM Transactions on Networking*, vol. 3, no. 1, pp. 26–30, February 1995.

J.-Y. Lee, I-S. Hwang, A. Nikoukar, and A. T. Liem, "Comprehensive performance assessment of bipartition upstream bandwidth assignment schemes in GPON," *IEEE/OSA Journal of Optical Communications and Networking*, vol. 5, no. 11, November 2013.

C. E. Leiserson, "Fat-trees: universal networks for hardware-efficient supercomputing," *IEEE Transactions on Computers*, vol. C-34, no. 10, pp. 892–901, October 1985.

A. Leon-Garcia, *Probability and Random Processes for Electrical Engineering*, Pearson, 2nd ed., 1994.

B. Li and A. Ganz, "Virtual topologies for WDM star LANs: the regular structure approach," *IEEE INFOCOM 1992*, Florence, Italy, May 1992.

D. Li, C. Guo, H. Wu, K. Tan, Y. Zhang, and S. Lu, "Ficonn: using backup port for server interconnection in data centers," *IEEE INFOCOM 2009*, April 2009.

M.-J. Li, M. J. Soulliere, D. J. Tebben, L. Nederlof, M. D. Vaughn, and R. E. Wagner, "Transparent optical protection ring architectures and applications," *IEEE/OSA Journal of Lightwave Technology*, vol. 23, no. 10, pp. 3388–3403, October 2005.

L. Liu, T. Tsuritani, I. Morita, H. Guo, and J. Wu, "OpenFlow-based wavelength path control in transparent optical networks: a proof-of-concept demonstration," *ECOC 2011*, Geneva, Switzerland, September 2011.

L. Liu, D. Zhang, T. Tsuritani, R. Vilalta, R. Casellas, L. Hong, I. Morita, H. Guo, J. Wu, R. Martinez, and R. Munõz, "Field trial of an OpenFlow-based unified control plane for multilayer multigranularity optical switching networks," *IEE/OSA Journal of Lightwave Technology*, vol. 31, no. 4, pp. 506–514, February 2013.

V. López, B. de la Cruz, Ó. G. de Dios, O. Gerstel, N. Amaya, G. Zervas, D. Simeonidou, and J. P. Fernandez-Palacios, "Finding the target cost for sliceable bandwidth variable transponders," *IEEE/OSA Journal of Optical Communications and Networking*, vol. 6, no. 5, pp. 476–485, May 2014.

R. B. R. Lourenço, S. S. Savas, M. Tornatore, and B. Mukherjee, "Robust hierarchical control plane for transport software-defined networks," *Elsevier Journal of Optical Switching and Networking*, vol. 30, pp. 10–22, November 2018.

Y. Luo and N. Ansari, "Limited sharing with traffic prediction for dynamic bandwidth allocation and QoS provisioning over Ethernet passive optical networks," *OSA Journal of Optical Networking*, vol. 4, no. 9, pp. 561–572, September 2005.

K. Mahmood, A. Chilwan, O. Østerbø, and M. Jarschel, "Modelling of OpenFlow-based software-defined networks: the multiple node case," *IET Networks*, Vol. 4, No. 5, pp. 278–284, 2015.

T. H. Maiman, "Stimulated optical radiation in Ruby," *Nature*, vol. 187, pp. 493–494, August 1960.

E. Mannie, Ed., "Generalized multi-protocol label switching (GMPLS) architecture," RFC 3945, Oct. 2004.

N. F. Maxemchuk, "Regular mesh topologies in local and metropolitan area networks," *AT&T Technical Journal*, vol. 64, no. 7, pp. 1659–1685, September 1985.

N. F. Maxemchuk, "Routing in the Manhattan street network," *IEEE Transactions on Communications*, vol. 35, no. 5, pp. 503–512, May 1987.

M. P. McGarry, M. Reisslein, and M. Maier, "Ethernet passive optical network architectures and dynamic bandwidth allocations", *IEEE Communications Surveys and Tutorials*, vol. 10, no. 3, 3rd quarter, 2008.

N. McKeown, T. Anderson, H. Balakrishnan, G. Parulkar, L. Peterson, J. Rexford, S. Shenker, and J. Terner, "OpenFlow: enabling innovation in campus networks," *Computer Communication Review*, vol. 32, no. 2, pp. 69–74, 2008.

N. Mehravari, "Performance and protocol improvements for very high speed optical fiber local area networks using a passive star topology," *IEEE/OSA Journal of Lightwave Technology*, vol. 8, no. 4, pp. 520–530, April 1990.

R. M. Metcalfe and D. R. Boggs, "Ethernet: distributed packet switching for local computer networks," Technical Report, Xerox Palo Alto Research Center, California, USA, 1975.

R. M. Metcalfe, D. R. Boggs, C. P. Thacker, and B. W. Lampson, "Multipoint data communication system with collision detection," United States Patent 4063220, December 1977.

C. Minkenberg, F. Abel, P. Muller, R. Krishnamurthy, M. Gusat, P. Dill, I. Iliadis, R. Luijten, B. R. Hemenway, R. Grzybowski, and E. Schiattarella, "Designing a crossbar scheduler for HPC applications," *IEEE Micro*, vol. 26, no. 3, pp. 58–71, May-June 2006.

A. Mokhtar and M. Azizoglu, "Adaptive wavelength routing in all-optical networks," *IEEE/ACM Transactions on Networking*, vol. 6, no. 2, pp. 197–206, April 1998.

B. Mukherjee, "WDM-based local lightwave networks, Part I: single-hop systems," *IEEE Network*, vol. 6, no. 3, pp. 12–27, May 1992a.

B. Mukherjee, "WDM-based local lightwave networks, Part II: multihop systems," *IEEE Network*, vol. 6, no. 4, pp. 20–32, July 1992b.

B. Mukherjee, *Optical WDM Networks*, Springer, USA, 2006.

N. Nadarajah, E. Wong, M. Attygalle, and A. Nirmalathas "Protection switching and local area network emulation in passive optical network," *IEEE/OSA Journal of Lightwave Technology*, vol. 24, no. 5, pp. 1955–1967, May 2006.

D. Nesset (invited), "NG-PON2 technology and standards," *IEEE/OSA Journal of Lightwave Technology*, vol 33, no. 5, pp. 1136–1143, March 2015.

B. A. A. Nunes, M. Mendonca, X.-N. Ngyen, K. Obracza, and T. Turletti, "A survey of software-defined networking: past, present and future of programmable networks," *IEEE Communication Surveys and Tutorials*, vol. 6, no. 3, 3rd quarter 2014.

E. Oki, *Linear Programming and Algorithms for Communication Networks*, 2013, CRC Press.

N. A. Olsson, "Lightwave systems with optical amplifiers," *IEEE/OSA Journal of Lightwave Technology*, vol. 7, no. 7, pp. 1071–1082, July 1989.

N. A. Olsson, J. Hegarty, R. A. Logan, L. F. Johnson, K. L. Walker, L. G. Cohen, B. L. Kasper, and J. C. Campbell, "68.3 km transmission with 1.37 Tbit km/s capacity using wavelength division multiplexing of ten single-frequency lasers at 1.5 μm," *Electronics Letters*, vol. 21, no. 3, pp. 105–106, 31 January 1985.

M. J. O'Mahony, C. Politi, D. Klonidis, R. Nejabati, and D. Simeonidou, "Future Optical Networks," *IEEE/OSA Journal Of Lightwave Technology*, vol. 24, no. 12, pp. 4684–4696, December 2006.

Open Networking Foundation (ONF), http://www.opennetworking.org/, 2011.

H. Ono, A. Mori, K. Shikano, and M. Shimazu, "A low-noise and broadband erbium-doped tellurite fiber amplifier with a seamless amplification band in the C- and L-bands," *IEEE Photonics Technology Letters*, vol. 14, no. 8, pp. 1073–1075, August 2002.

G. I. Papadimitriou, C. Papazoglou, and A. S. Pomportsis, "Optical switching: switch fabrics, techniques, and architectures," *IEEE/OSA Journal of Lightwave Technology*, vol. 21, no. 2, pp. 384-405, February 2003.

A. Parekh and R. Gallager, "A generalized processor sharing approach to flow control in integrated services networks," *IEEE/ACM Transactions on Networking*, vol. 1, no. 3, pp. 344–357, June 1993.

A. N. Patel, P. N. Ji, and T. Wang, "QoS-aware optical burst switching in OpenFlow based software defined optical networks," *ONDM 2013*, Brest, France, April 2013.

A. Pattavina, "Architectures and performance of optical switching nodes for IP networks," *IEEE/OSA Journal of Lightwave Technology*, vol. 23, no. 3, pp. 1023–1032, March 2005.

D. B. Payne and J. R. Stern, "Transparent single-mode fiber optical networks," *IEEE/OSA Journal of Lightwave Technology*, vol. 4, no. 7, pp. 864–869, July 1986.

M. Pickavet, P. Demeester, D. Colle, D. Staessens, B. Puype, L. Depre, and I. Lievens, "Recovery in multilayer optical networks," *IEEE/OSA Journal of Lightwave Technology*, vol. 24, no. 1, pp. 122–134, January 2006.

G. Porter, R. Strong, N. Farrington, A. Forencich, P. Chen-Sun, T. Rosing, Y. Fainman, G. Papen, A. Vahdat, "Integrating microsecond circuit switching into the data center," *ACM SIGCOMM 2013*, Hong Kong, August, 2013.

J. Prat, J. Lazaro, P. Chanclou, R. Soila, A. M. Gallardo, A. Teixeira, G. M. TosiBeleffi, and I. Tomkos, "Results from EU project SARDANA on 10G extended reach WDM PONs," *OFC 2010*, San Diego, USA, March 2010.

C. Qiao and M. Yoo, "Optical burst switching (OBS)- a new paradigm for an optical Internet," *Journal of High Speed Networks*, vol. 8, no. 1, pp. 69–84, January 1999.

B. Ramamurthy, D. Datta, H. Feng, J. P. Heritage, and B. Mukherjee, "Impact of transmission impairments on the teletraffic performance of wavelength-routed optical networks," *IEEE/OSA Journal of Lightwave Technology*, vol. 17, no. 10, pp. 1713–1723, October 1999a.

B. Ramamurthy, H. Feng, D. Datta, J. P. Heritage, and B. Mukherjee, "Transparent vs. opaque vs. translucent wavelength-routed optical networks," *OFC 1999*, San Diego, USA, February 1999b.

S. Ramamurthy and B. Mukherjee, "Survivable WDM mesh networks, Part I – Protection," *IEEE INFOCOM 1999*, New York, USA, March 1999a.

S. Ramamurthy and B. Mukherjee, "Survivable WDM mesh networks, Part II – Restoration," *IEEE INFOCOM 1999*, New York, USA, March 1999b.

S. Ramamurthy, L. Sahasrabuddhe, and B. Mukherjee, "Survivable WDM mesh networks," *IEEE/OSA Journal of Lightwave Technology*, vol. 21, no. 4, pp. 870–883, April 2003.

R. Ramaswami and K. Sivarajan, "Design of logical topologies for wavelength-routed optical networks," *IEEE Journal on Selected Areas in Communications*, vol. 14, no. 5, pp. 840–851, June 1996.

R. Ramaswami, K. Sivarajan, and G. Sasaki, *Optical Networks: A Practical Perspective*, 3rd edition, Morgan Kaufmann, 2010.

J. Ratnam, S. Chakraborti, and D. Datta, "Impact of transmission impairments on demultiplexed channels in WDMPONs employing AWG-based remote nodes," *IEEE/OSA Journal of Optical Communications and Networking*, vol. 2, no. 10, pp. 848–858, October 2010.

E. G. Rawson, "The Fibernet II Ethernet-compatible fiber-optic LAN," *IEEE/OSA Journal of Lightwave Technology*, vol. 3, no. 3, pp. 496–501, June 1985.

E. G. Rawson and R. M. Metcalfe, "Fibernet: Mulitimode optical fiber for local computer networks," *IEEE Transactions on Communications*, vol. 26, no. 7, pp. 983–990, July 1978.

J. Roese, R.-P. Braun, M. Tomizawa, and O. Ishida, "Optical transport network evolving with 100 Gigabit Ethernet," *IEEE Communications Magazine*, vol. 48, no. 3, pp. S28–34, March 2010.

F. E. Ross, "FDDI – a tutorial," *IEEE Communications Magazine*, vol. 24, no. 5, pp. 10–17, May 1986.

I. Rubin and H.-K. Hua, "SMARTNet: An all-optical wavelength-division meshed-ring packet-switching network," *IEEE GOLBECOM 1995*, Singapore, November 1995.

M. Ruffini, D. Mehta, B. O'Sullivan, L. Quesada, L. Doyle, and D. B. Payne, "Deployment strategies for protected long-reach PON,"*IEEE/OSA Journal of Optical Communications and Networking*, vol. 4, no. 2, pp. 118–128, February 2012.

L. Sahasrabuddhe, S. Ramamurthy, and B. Mukherjee, "Fault management in IP-over-WDM networks: WDM protection versus IP restoration," *IEEE Journal on Selected Areas in Communications*, vol. 20, no. 1, pp. 21–33, January 2002.

J. Salz, "Modulation and detection for coherent lightwave communications," *IEEE Communications Magazine*, vol. 24, no. 6, pp. 38–49, June 1986.

N. Sambo, P. Castoldi, A. D'Errico, E. Riccardi, A. Pagano, M. S. Moreolo, J. M. Fàbrega, D. Rafique, A. Napoli, S. Frigerio, E. H. Salas, G, Zervas, M. Nolle, J. K. Fischer, A. Lord, and J. P. F.-P Giménez, "Next generation sliceable bandwidth variable transponders," *IEEE Communication Magazine*, vol. 53, no. 2, pp. 163–171, February 2015.

H. Sanjoh, E. Yamada, and Y. Yoshikuni, "Optical orthogonal frequency division multiplexing using frequency/time domain filtering for high spectral efficiency up to 1bit/s/Hz," *OFC 2002*, Anaheim, USA, March 2002.

M. Scholten, Z. Zhu, E. Hernandez-Valencia, and J. Hawkins, "Data transport applications using GFP," *IEEE Communications Magazine*, vol. 40, no. 5, pp. 96–103, May 2002.

J. Senior, *Optical Fiber Communications*, Prentice Hall, 2nd edition, 1996.

F. Shehadeh, R. S. Vodhanel, C. Gibbons, and M. Ali, "Comparison of gain control techniques to stabilize EDFAs for WDM networks," *OFC 1996*, San Jose, USA, February-March 1996.

L. Shi, A. Nag, D. Datta, and B. Mukherjee, "New concept in long-reach PON planning: BER-aware wavelength allocation," *Elsevier Journal of Optical switching and Networking*, vol. 10, no. 4, pp. 475–480, November 2013.

W. Shieh, H. Bao, and Y. Tang, "Coherent optical OFDM: theory and design," *Optics Express*, vol. 16, no. 2, pp. 841–859, January 2008.

T. Shiragaki, S. Nakamura, M. Shinta, N. Henmy, and S. Hasegawa, "Protection architecture and applications of OCh shared protection ring," *Springer Optical Networks Magazine*, vol. 2, no. 4, pp. 48–58, July/August 2001.

K. V. Shrikhande, I. M. White, D. Wonglumsom, S. M. Gemelos, M. S. Rogge, Y. Fukashiro, M. Avenarius, and L. G. Kazovsky, "HORNET: A packet-over-WDM multiple access metropolitan area ring network," *IEEE/OSA Journal on Selected Areas in Communication*, vol. 18, no. 10, pp. 2004–2016, October 2000.

J. Shuja, K. Bilal, S. A. Madani, M. Othman, R. Ranjan, P. Balaji, and S. U. Khan, "Survey of techniques and architectures for designing energy-efficient data centers," *IEEE Systems Journal*, vol. 10, no. 2, pp. 507–519, June 2016.

J. M. Simmons, *Optical Network Design and Planning*, Springer, 2014.

D. Sinefield, S. Ben-Ezra, and D. M. Maron, "Nyquist-WDM filter shaping with a high resolution colorless photonic spectral processor," *Optics Letters*, vol. 38, no. 17, pp. 3268–3271, September 2013.

S. K. Singh and A. Jukan, "Efficient spectrum defragmentation with holding-time awareness in elastic optical networks," *IEEE/OSA Journal of Optical Communications and Networking*, vol. 9, no. 3, pp. B78-89, March 2017.

A. Sing-la, C.-Y. Heng, L. Popa, and P. B. Godfrey, "Jellyfish: networking data centers randomly," *USENIX/ACM Conference on Network System Design Implementation*, San Jose, USA, April 2012.

K. Sivarajan and R. Ramaswami, "Multihop lightwave networks based on de Bruijn networks," *IEEE INFOCOM 1991*, Bal Harbour, USA, April 1991.

M. K. Smit and C. Van Dam, "PHASAR-based WDM-devices: Principles, design and applications," *IEEE Journal of Selected Topics in Quantum Electronics*, vol. 2, no. 2, pp. 236–250, June 1996.

R. G. Smith, "Optical power handling capacity of low loss optical fibers as determined by stimulated Raman and Brillouin scattering," *Applied Optics*, vol. 11, no. 11, pp. 2489–2160, November 1972.

J. Sömmer, S. Gunreben, F. Feller, M. Kohn, A. Mifdaoui, D. Saß, and J. Scharf, "Ethernet – a survey on its fields of application," *IEEE Communications Surveys and Tutorials*, vol. 12, no. 2, pp. 263–284, 2010.

H. Song, B.-H. Kim, and B. Mukherjee, "Long-reach optical access networks: a survey of research challenges, demonstrations, and bandwidth assignment mechanisms," *IEEE Communications Surveys and Tutorials*, vol. 12, no. 1, pp. 112–123, 1st quarter 2010.

C. E. Spurgeon, *Ethernet: the Definitive Guide*, O'Reilly Media, vol. 1, February 2000.

T. A. Strasser and J. L. Wagener, "Wavelength-selective switches for ROADM applications," *IEEE Journal of Quantum Electronics*, vol. 16, no. 5, pp. 1150–1157, September-October 2010.

G. N. M. Sudhakar, N. D. Georganas, and M. Kavehrad, "Slotted Aloha and reservation Aloha protocols for very high-speed optical fiber local area networks using passive star topology," *IEEE/OSA Journal of Lightwave Technology*, vol. 9, No. 10, pp. 1411–1422, October 1991.

Y. Suematsu, "Dynamic single-mode lasers," *IEEE/OSA Journal of Lightwave Technology*, vol. 32, no. 6, pp. 1144–1158, March 2014.

Y. Suematsu and S. Arai, "Single-mode semiconductor lasers for long-wavelength optical fiber communications and dynamic semiconductor lasers," *IEEE Journal on Selected Areas in Quantum Electronics*, vol. 6, no. 6, pp. 1436–1448, November/December 2000.

M. A. Summerfield, "MAWSON: a metropolitan area wavelength switched optical network," *APCC 1997*, Sydney, Australia, November 1997.

B. E. Swekla and R. J. MacDonald, "Optoelectronic transversal filter," *Electronics Letters*, vol. 27, no. 19, 26 July 1991.

C. Thießen and Ç. Çavdar, "Fragmentation-aware survivable routing and spectrum assignment in elastic optical networks," *IEEE RNDM 2014*, Barcelona, Spain, January 2014.

A. Thyagaturu, A. Merican, M. P. McGarry, M. Reisslein, and W. Keller, "Software defined optical networks (SDONs): a comprehensive survey," *IEEE Communications Surveys and Tutorials*, vol. 18, no. 4, pp. 2738–2786, 4th quarter 2016.

F. A. Tobagi and M. Fine, "Performance of unidirectional broadcast local area networks: Expressnet and Fasnet," *IEEE Journal on Selected Areas in Communications*, vol. SAC-1, no. 5, pp. 913–926, November 1983.

D. Tolmie and J. Renwick, "HIPPI: simplicity yields success," *IEEE Network*, vol. 7, no. 1, pp 28–32, January 1993.

P. Tran-Gia and Th. Stock, "Approximate performance analysis of the DQDB access protocol," *ITC Specialist Seminar*, Adelaide, Australia, 1989.

R. S. Tucker, "Green optical communications – Part I: energy limitations in transport," *IEEE Journal on Selected Topics in Quantum Electronics*, vol. 17, no. 2, pp. 245–260, March/April 2011a.

R. S. Tucker, "Green optical communications – Part II: energy limitations in networks," *IEEE Journal on Selected Topics in Quantum Electronics*, vol. 17, no. 2, pp. 261–274, March/April 2011b.

L. Valcarenghi, D. P. Van, P. G. Raponi, P. Castoldi, D. R. Campelo, S.-W. Wong, S.-H. Yen, L. G. Kazovsky, and S. Yamashita "Energy efficiency in passive optical networks: where, when and how?" *IEEE Network*, vol. 26, no. 6, pp. 61–68, November-December 2012.

R. J. Vanderbei, *Linear Programming: Foundations and Extensions*, Springer, 3rd edition, 2008.

S. Verma, H. Chaskar, and R. Ravikanth, "Optical burst switching: a viable solution for terabit IP backbone," *IEEE Network*, vol. 14, no. 6, pp. 48–53, November-December 2000.

M. Walraed-Sullivan, A. Vahdat, and K. Marzullo, "Aspen trees: balancing data center fault tolerance, scalability and cost," *ACM SIGCOMM 2013*, Santa Barbara, USA, December 2013.

Y. Wang, X. Cao, Q. Hu, and Y. Pan, "Towards elastic and fine-granular bandwidth allocation in spectrum-sliced optical networks," *IEEE/OSA Journal of Optical Communications and Networking*, vol. 4, no. 11, pp. 906–917, November 2012.

G. Wang, M. Kaminsky, K. Papagiannaki, T. S. Eugene Ng, M. A. Kozuch, and M. Ryan, "c-Through: part-time optics in data centers," *ACM SIGCOMM 2010*, New Delhi, India, August-September 2010.

R. Wang and B. Mukherjee, "Spectrum management in heterogeneous bandwidth networks," *IEEE GLOBECOM 2012*, Anaheim, USA, December 2012.

R. Wang and B. Mukherjee, "Provisioning in elastic optical networks with non-disruptive defragmentation," *IEEE/OSA Journal of Lightwave Technology*, vol. 31, no. 15, pp. 2491–2500, August 2013.

J. Wang, W. Cho, V. Rao Vemuri, and B. Mukherjee, "Improved approaches for cost-effective traffic grooming in WDM ring networks: ILP formulations and single-hop and multihop connections," *IEEE/OSA Journal of Lightwave Technology*, vol. 19, no. 11, pp. 1645–1653, November 2001.

A. Werts, "Propagation de la lumiere coherente dans les fibres optiques," *L'Onde Electrique*, 46, pp. 967–980, 1966.

J. Wei, Q. Cheng, R. V. Penty, I. H. White, and D. G. Cunningham, "400 Gigabit Ethernet using advanced modulation format: performance, complexity, and power dissipation," *IEEE Communications Magazine*, pp. 182–187, February 2015.

J. S. Wey, D. Nesset, M. Valvo, K. Gorbe, H. Roberts, Y. Luo, and J. Smith (Invited), "Physical layer aspects of NG-PON2 standards - Part 1: optical link design," *IEEE/OSA Journal of Optical Communications and Networking*, vol. 8, no. 1, pp. 33–42, January 2016.

I. M. White, M. S. Rogge, K. Shrikhande, Y. Fukashiro, D. Wonglumsom, F.-T. An, and L. G. Kazovsky, "Experimental demonstration of a novel media access protocol for HORNET: a packet-over-WDM multiple-access MAN ring," *IEEE Photonics Technology Letters*, vol. 12, no. 9, pp. 1264–1266, September 2000.

P. J. Winzer, "Beyond 100G Ethernet," *IEEE Communications Magazine*, pp. 26–30, July 2010.

D. Wonglumsom, I. M. White, K. Shrikhande, M. S. Rogge, S. M. Gemelos, F.-T. An, Y. Fukashiro, M. Avenarius, and L. G. Kazovsky, "Experimental demonstration on an access point for HORNET: a packet-over-WDM multiple-access MAN," *IEEE/OSA Journal Lightwave Technology*, vol. 18, no. 12, pp. 1709–1717, December 2000.

W. Xia, P. Zhao, Y. Wen, and H. Xie, "A survey on data center networking (DCN): infrastructure and operations," *IEEE Communications Surveys and Tutorials*, vol. 19, no. 1, pp. 640–656, 1st quarter 2017.

D. Xie, D. Wang, M. Zhang, Z. Liu, Q. You, Q. Yang, and S. Yu, "LCoS-based wavelength-selective switch for future finer-grid elastic optical networks capable of all-optical wavelength conversion," *IEEE Photonics Journal*, vol. 9, no. 2, April 2017.

N. Xie, T. Hashimoto, and K. Utaka, "Ultimate-low-power-consumption, polarization-independent, and high-speed polymer Mach–Zehnder thermo-optic switch," *OFC 2009*, San Diego, USA, March 2009.

L. Xu, H. G. Perros, and G. Rouskas, "Techniques for optical packet switching and optical burst switching," *IEEE Communications Magazine*, vol. 39, no. 1, pp. 136–142, January 2001.

M. Yamada, A. Mori, K. Kobayashi, H. Ono, T. Kanamori, K. Oikawa, Y. Nishida, and Y. Ohishi, "Gain-flattened tellurite-based EDFA with a flat amplification bandwidth of 76 nm," *IEEE Photonics Technology Letters*, vol. 10, no. 9, pp. 1244–1246, September 1998.

K. K. Yap, M. Motiwala, J. Rahe, S. Padgett, M. Holliman, G. Baldus, M. Hines, T. Kim, A. Narayanan, A. Jain, V. Lin, C. Rice, B. Rogan, A. Singh, B. Tanaka, M. Verma, P. Sood, M. Tariq, M. Tierney, D. Trumic, V. Valancius, C. Ying, M. Kallahalla, B. Koley, and A. Vahdat, "Taking the edge off with Espresso: scale, reliability and programmability for global Internet Peering," *ACM SIGCOMM 2017*, Los Angeles, USA, August 2017.

A. Yariv and P. Yeh, *Photonics: Optical Electronics and Modern Communications*, 6th edition, 2007, Oxford University Press, 2007.

X. Ye, Y. Yin, P. Mejia, and R. Proietti, "DOS: a scalable optical switch for datacenters," *ANCS 2010*, La Jolla, USA, October, 2010.

Y. Yin, R. Proietti, X. Ye, C. J. Nitta, V. Akella, and S. J. B. Yoo, "LIONS: an AWGR-based low-latency optical switch for high-performance computing and data centers," *IEEE Journal of Selected Topics in Quantum Electronics*, vol. 19, no. 2, March-April 2013a.

Y. Yin, H. Zhang, M. Zhang, M. Xia, Z. Zhu, S. Dahlfort, and S. J. B. Yoo, "Spectral and spatial 2D fragmentation-aware routing and spectral assignment algorithms in elastic optical networks," *IEEE/OSA Journal of Optical Communications and Networking*, vol. 5, no. 10, pp. A100–106, October 2013b.

S. J. B. Yoo, "Wavelength conversion technologies for WDM network applications," *IEEE/OSA Journal of Lightwave Technology*, vol. 14, no. 6, pp. 955–966, June 1996.

S. J. B. Yoo, "Optical Packet and Burst Switching Technologies for the Future Photonic Internet," *IEEE/OSA Journal of Lightwave Technology*, vol. 24, no. 12, pp. 4468–4492, December 2006.

T. Yoshida, H. Asakura, H. Tsuda, T. Mizuno, and H. Takahashi, "Switching characteristics of a 100-GHz-spacing integrated 40-λ 1×4 wavelength selective switch," *IEEE Photonics Technology Letters*, vol. 26, no. 5, pp. 451–453, March 2014.

P. Yuan, V. Gambrioza, and E. Knightly, "The IEEE 802.17 media access protocol for high-speed metropolitan-area resilient packet rings," *IEEE Network*, vol. 18, no. 3, pp. 8–15, May-June 2004.

H. Zang, J. Jue, and B. Mukherjee, "A review of routing and wavelength assignment approaches for wavelength-routed optical WDM networks," *Springer Optical Networks Magazine*, vol. 1, no. 1, pp. 47–60, 2000.

X. Zhang and C. Qiao, "Scheduling in unidirectional WDM rings and its extensions," *SPIE Proceedings, vol. 3230, All Optical Communication Systems: Architecture, Control and Network Issues III*, pp. 208–219, October 1997.

X. Zhang and C. Qiao, "An effective and comprehensive approach to traffic grooming and wavelength assignment in SONET/WDM rings," *IEEE Transactions on Networking*, vol. 8, no. 5, pp. 608–617, October 2000.

J. Zhang, Y. Zhao, X. Yu, J. Zhang, M. Song, Y. Ji, and B. Mukherjee, "Energy-efficient traffic grooming in sliceable-transponder-equipped IP-over-elastic optical networks," *IEEE/OSA Journal of Optical Communications and Networking*, vol. 7, no. 1, pp. A142–152, January 2015.

X. Zhou, Z. Zhang, Y. Zhu, Y. Li, S. Kumar, A. Vahdat, B. Y. Zhao, and H. Zheng "Mirror on the ceiling: Flexible wireless links for data centers," *ACM SIGCOMM 2012*, Helsinki, Finland, August 2012.

M. Zirngibl, C. H. Joyner, L. W. Stulz, C. Dragone, H. M. Presby, and I. P. Kaminow, "LARnet, a local access router network," *IEEE Photonics Technology Letters*, vol. 7, no. 2, pp. 1041–1135, February 1995.

M. Zuckerman and P. Potter, "The DQDB protocol and its performance under overloaded traffic conditions," *ITC Specialist Seminar*, Adelaide, Australia, September 1989.

Abbreviations

Note: Some of the abbreviations in the book are used with multiple implications (i.e., with different expanded versions) in different contexts. For example, FEC is used for "forward error correction" in the context of reliable communication in the physical layer, while it is also used as "forward equivalent class" in the MPLS protocol at higher network layer.

Abbreviation	Expanded version
AA:	Active area (for laser diodes)
AA:	Allocation address (for VL2 datacenter)
AAL:	ATM adaptation layer
AC:	Access control
ACTS:	Advanced communications technologies and services
ADC:	Analog-to-digital converter
ADM:	Add-drop multiplexer
ADSL:	Asymmetric digital subscriber line
AG:	Amplifier-based optical gate
AL:	Active layer (for laser diodes)
AL:	Added lightpath (for WRONs)
AL:	Arbitrated loop (for DQDB protocol)
ALR:	Adaptive link rate
AM:	Aggressive mode
ANSI:	American National Standard Institute
AOC:	All-optical connection
AO-NACK:	All-optical negative acknowledgment
AO-OOFDM	All-optical OOFDM
APD:	Avalanche photodiode
APON:	ATM PON

APS	Automatic protection switching
AR:	Advanced reservation
ASE:	Amplified spontaneous emission
ASK:	Amplitude-shift keying
ASON:	Automatically-switched optical network
ATM:	Asynchronous transfer mode
AT&T:	American Telegraph and Telephone
AWG:	Arrayed-waveguide grating
B-CIR:	Class B with committed information rate
BD:	Building distributor
BE:	Best effort
B-EIR:	Class B with excess information rate
BER:	Bit-error rate
BFS:	Basic feasible solution
BHCNet:	Binary hypercube network
BIP:	Byte-interleaved priority
BLSR:	Bidirectional line-switched ring
BMR:	Burst-mode receiver
BoF:	Beginning of field
BOS:	Broad optical spectrum
BS:	Broadcast and select
BSR:	BCube source routing
BTRL:	British telecommunication research laboratory
BV:	Bandwidth variable
BVT:	Bandwidth-variable transponder
CCIR:	Internal Radio Consultative Committee
CCITT:	Internal Telegraph and Telephone Consultative Committee
CCW:	Counter-clockwise
CD:	Campus distributor
CDC:	Countdown counter
CDF:	Cumulative distribution function
CDN:	Content distribution network

CEF: Connection-enhancement factor

CGM: Cross-gain modulation

cHEC: Core header error control

C/M: Control and management

CM: Conservative mode

CN: Client network

CN_i: Client node (*i*th)

CO: Central office

CoS: Class of service

CPLEX: Simplex-based LP-solver software

CPPCU: Control packet processing and control unit

CRC: Cyclic redundancy check

CR-LDP: Constrained-routing-based LDP

CSMA/CD: Carrier-sense multiple access with collision detection

CSP: Constraint-based shortest path

CVD: Chemical vapor deposition

CW: Clockwise

CWDM Coarse WDM

DAC: Digital-to-analog converter

DAS: Direct attached storage

DBA: Dynamic bandwidth allocation

dBNet: de Bruijn network

DBR: Distribute Bragg reflector

DCF: Dispersion-compensating fiber

DCN: Intra-datacenter network

DCS: Digital crossconnect system

DD: Direct detection

DED: Demand-driven excess distribution

DFB: Distributed feedback

DFR: DCell fault-tolerant routing

DFT: Discrete Fourier transform

DH: Double-heterojunction

DLB:	Distributed loopback buffer
DMS:	Data minislots
DoS:	Data over SONET
DOS:	Datacenter optical switch
DP-QPSK:	Dual-polarized quadrature phase-shift keying
DP:	Data packet
DPSK:	Differential phase-shift keying
DQ:	Distributed queue
DQDB:	Distributed-queue dual bus
DQPSK:	Differential quadrature phase-shift keying
DR:	Delayed reservation
DS1:	Digital signal 1
DSF:	Dispersion-shifted fiber
DSL:	Digital subscriber line
DVI:	Digital visual interface
DWBA:	Dynamic wavelength and bandwidth allocation
DWDM	Dense WDM
EB:	Electrical buffer
ECL:	Emitter-coupled logic
ECS:	Electrical core switch
ECSA:	Exchange carrier standard association
ED:	Ending delimiter
EDFA:	Erbium-doped fiber amplifier
EDTFA:	Erbium-doped tellurite fiber amplifier
EED:	Equitable excess distribution
EFM:	Ethernet in the first mile
EFT:	Earliest finishing time
EFT-VF:	EFT with void filling
EMS:	Element management system
E-NNI:	External network-network interface
EOF:	End of field/frame

EON: Elastic optical network

EO-OOFDM: Electro-optic OOFDM

EOP: Elastic optical path

EoS: Ethernet over SONET

EPON: Ethernet PON

ERF: Earliest request first

ESCON: Enterprise serial connection

ESS: Electronic switching system

ESW: Exhaustive search for wavelengths

ETSI: European Telecommunications Standards Institute

FA: Frequency alignment

FBG: Fiber Bragg grating

FC: Frame control (in FDDI)

FC: Fiber channel (in SAN)

FCS: Frame check sequence

FD: Floor distributor

FDDI: Fiber-distributed digital interface

FDL: Fiber delay line

FDM: Frequency-division multiplexing

FEC: Forward error correction (for error-free communication)

FEC: Forward equivalent class (for MPLS protocol)

FELIX: Federated testbed for large-scale infrastructure experiment

FFT: Fast Fourier transform

FFW: First-fit wavelength

FICON: Fiber connection

FIFO: First-in first-out

FOX: Fast optical crossconnect

FPGA: Field programmable gate array

FPI: Fabry–Perot interferometer

FR: Frame relay (for network protocol)

FR: Fixed receiver (for WDM node)

FS: Frequency slot (for EON)

FS:	Frame status (for FDDI)
FSC:	Fiber switch capable
FSK:	Frequency-shift keying
FSR:	Free spectral range
FT:	Fixed transmitter (for WDM network)
FT:	Fourier transform
FTTB:	Fiber-to-the-building
FTTC:	Fiber-to-the-curb
FTTH:	Fiber-to-the-home
FTV:	Fault-tolerant vector
FWHM:	Full-width half-maximum
FWM:	Four-wave mixing
GbE:	Gigabit Ethernet
GCP:	Graph coloring problem
GEM:	Generalized encapsulation method
GFF:	Gain-flattening filter
GFP:	General framing procedure
GFP-F:	GFP-frame
GFP-T:	GFP-transparent
GHCNet:	Generalized hypercube network
GMII:	Generalized media-independent interface
GML:	Gap-maker list
GMPLS:	Generalized multiprotocol label switching
GPON:	Gigabit PON
GPS:	General process sharing
GRC:	Global request counter
GTC:	GPON transmission convergence
GUI:	Graphic-user interface
GVD:	Group velocity dispersion
HDLC:	High-level data link control
HE:	Head end
HFC:	Hybrid fiber coax

HHI: Heinrich Hertz Institute
HIPPI: High-performance parallel interface
HMS: Hierarchical master-slave
HOL: Head of the line
HORNET: Hybrid optoelectronic ring network
HPCU: Header processing and control unit
HSS: Hotspot scheduling
HSSG: Higher-Speed Studies Group
HyPac: Hybrid packet and circuit switching
HYPASS: Hyprid packet-switching system
ICI: Inter-carrier interference
ICT: Information and communication technology
IDN: Integrated digital network
IEEE: Institute of Electrical and Electronic Engineering
IET Institute of Engineering and Technology
IETF: Internet Engineering Task Force
IF: Intermediate frequency
IFFT: Inverse fast Fourier transform
IL: Insulation layer (for laser diodes and LEDs)
IL: Insertion loss (for lossy devices)
ILP: Integer linear programming
IM: Intensity modulation
IM-DD: Intensity modulation – direct detection
IMDD-OOFDM Optical OFDM using IM-DD
I-NNI: Internal network-network interface
IP: Internet protocol
IPACT: Interleaved-polling with adaptive cycle time
IPG: Interpacket gap
ISDN: Integrated service digital network
ISI: Inter-symbol interference
IS-IS: Intermediate system to intermediate system

ITU:	International telecommunication (telegraph) union
IXC:	Inter-exchange carrier
JET:	Just enough time
JIT:	Just in time
LA:	Local address
LAN:	Local-area network
LARNet:	Local access router network
LATA:	Local access and transport area
LB:	Loopback buffer
LC:	Late counter
LCAS:	Link-capacity allocation scheme
LCoS:	Liquid crystal on silicon
LCP:	Logical connectivity problem
LD:	Laser diode
LDP:	Label distribution protocol
LEAF:	Large effective-area fiber
LE:	Label extractor
LED:	Light-emitting diode
LER:	Label edge router
LFT:	Latest finishing time
LFT-VF:	LFT with void filling
LG:	Limited grant
LIONS:	Low-latency interconnect optical network switch
LLID:	Logical link ID
LMP:	Link management protocol
LO:	Local oscillator
LOH:	Line overhead
LP:	Linear programming
LPM:	Low-power mode
LRF:	Longest request first
LSC:	Lambda-switch capable

LSP: Label-switched path
LSR: Label-switching router
LT: Lightpath topology
LTD: Lightpath topology design
LTE: Line-terminating equipment
LUWF: Least-utilized wavelength first
MAC: Media-access control
MAN: Metropolitan-area network
MAWSON: Metropolitan-area wavelength-switched optical network
MbB: Make before break
MCVD: Modified chemical vapor deposition
MDI: Media-dependent interface
MDTF: Multilayer dielectric thin film filter
ME: Moderate effort
MEMS: Micro-electro-mechanical system/switch
MIB: Management information base
MILP: Mixed-integer linear programming
MIS: Maximum independent-set
MIU: Media interface unit
ML: Metallization layer
MLB: Mixed loopback buffer
MMH: Multiple minimum-hop
MPCP: Multipoint-control protocol
MPI: Message passing interface
MPLS: Multiprotocol label switching
MQW: Multiple quantum well
MRG: Merge
MSB: Most significant bit
MSNet: Manhattan street network
M-STFQ: Modified STFQ
MUWF: Most-utilized wavelength first
MZI: Mac–Zehnder interferometer

NA:	Numerical aperture
NBI:	Northbound interface
NE:	Network element/equipment
NETCONF:	Network configuration protocol
NIC:	Network interface card
NOS:	Networking operating system
NP:	Network provider
NZ-DSF:	Non-zero DSF
OA:	Optical amplifier
OAD:	Optical add drop
OADM:	Optical add-drop multiplexer
OAM:	Operation, administration, and management
OAWG:	Optical arbitrary waveform generation
OBS:	Optical burst switching
OCA:	Optical channel adaptor
OCh:	Optical channel
OCS:	Optical core switch (for carrying out optical switching)
OCS:	Optical circuit-switching (one of the switching techniques in optical domain)
ODMUX:	Optical demultiplexer
ODUk:	Overhead channel data unit
OEO:	Optical-electronic-optical
OFDM:	Orthogonal frequency-division multiplexing
OFELIA:	OpenFlow in Europe-linking infrastructure and applications
OL:	Optical layer
OLG:	Optical label generator
OLT:	Optical line terminal
OMCC:	ONU management and control channel
OMCI:	ONU management and control interface
OMS:	Optical multiplex section
OMUX:	Optical multiplexer

ONQ-FB: Open network of queues with feedback

ONU: Optical networking unit

OOFDM: Optical version of OFDM

OPS: Optical packet switching

OPUk: Overhead channel payload unit

OS: Optical switch

OSA: Optical switching architecture

OSM: Optical switching matrix

OSP: Optical splitter

OSPF: Open shortest path first

OTDM: Optical time-division multiplexing

OTN: Optical transport network

OXC: Optical crossconnect

PA: Preamble (in a packet burst)

PA: Pre-arbitrated (for DQDB)

PA: Packet adapter (in DCN)

PAPR: Peak-to-average power ratio

PBS Polarization beam splitter

PC: Polarization control (in coherent optical receivers)

PC: Pretransmission coordination (in WDM LAN)

PCBd: Physical control block downstream

PCEP: Path computation element protocol

PCI: Peripheral component interconnect

PCM: Pulse-code modulation

PCS: Physical-coding sublayer

pdf: probability density function

PDH: Plesiochronous digital hierarchy

PDL: Polarization-dependent loss

PDU: Protocol data unit

PER: Packet-error rate

PHY: Physical layer

PIP:	Physical infrastructure provider
PK:	Pollaczek-Khinchin
PLI:	Payload length indicator (in GPON)
PLI:	Payload identifier (in GFP)
PLOAM:	Physical-layer operation, administration and management
PLOU:	Physical-layer overhead for upstream
PMA:	Physical medium attachment
PMD:	Polarization-mode dispersion (for optical fibers)
PMD:	Physical-medium dependent (for FDDI, Gigabit Ethernet)
POH	Path overhead
PON:	Passive Optical Network
PoP:	Point of Presence
POTS:	Plain old telephone service
PP:	Point-to-point
PPWDM:	Point-to-point WDM
PRC:	Primary reference clock
PRS:	Primary reference source
PS:	Parallel-to-serial converter
PSC:	Packet-switch capable (for GMPLS protocol)
PSC:	Passive-star coupler (optical device used for optical broadcast)
PSK:	Phase-shift keying
PSTN:	Public switched-telephone network
PTE:	Path-terminating equipment
PTF:	Power transfer function
PTI:	Payload type indicator
PTQ:	Primary transit queue
PWRN:	Passive wavelength-routing node
QA:	Queue arbitrated
QAM:	Quadrature amplitude modulation
QoS:	Quality of service
QoT	Quality of transmission
QPSK:	Quadrature phase-shift keying

R/A: Request/allocation

RAP: Request allocation protocol

RC: Routing controller

REST: Representational state transfer

RF: Radio frequency

RFW: Random-fit wavelength

RH: Ring homed

RingO: Ring optical network

RIP: Refractive index profile

RITE-Net: Remote integration of terminal equipment

rms: Root-mean-squared

RN: Remote node

ROADM: Reconfigurable optical add-drop multiplexer

RP-BE: Reconnection problem using branch-exchange operation

RPR: Resilient packet ring

RSA: Routing and spectral assignment

RSH: Ring and star homed

RSOA: Reflective semiconductor optical amplifier

RSVP-TE: Remote reservation protocol with traffic engineering

RTT: Round-trip time

RWA: Routing and wavelength assignment

RX: Receiver

SAM: Shift and match

SAN: Storage-area network

SARDANA: Scalable advanced ring-based passive dense access network architecture

SAW: Sleep and wake-up

SBI: Southbound interface

SBS: Stimulated Brillouin scattering

S-BVT: Sliceable bandwidth-variable transmitter

SC: Sub-circle

SCB: Single copy broadcast

SCM: Subcarrier modulation/multiplexing
SD: Sequence delimiter
SDH: Synchronous digital hierarchy
SDN: Software-defined network
SDON: Software-defined optical network
SF: Switch fabric
SFD: Start frame delimiter
SFDMA: Subcarrier frequency-division multiple access
SLA: Service-level agreement
SLB: Shared loopback buffer
SMH: Single minimum-hop
SMT: Station management
SN: Switching node
SNMP: Simple network management protocol
SNR: Signal-to-noise ratio
SOA: Semiconductor optical amplifier
SOH: Section overhead
SONATA: Switchless optical network for advanced transport architecture
SONET: Synchronous optical network
SOP: State of polarization
SP: Strict priority (for protocols)
SP: Service provider (for network access)
SP: Serial-to-parallel converter (for OFDM in EON)
SPE: Synchronous payload envelope
SPM: Self-phase modulation
SPSFT: Switched passive-star over fat trees
SRG: Shared risk group
SRS: Stimulated Raman scattering
STE: Section-terminating equipment
STFQ: Start time fair queuing
STM: Synchronous transport module
STP: Shielded twisted pair

STQ: Secondary transit queue

STS: Synchronous transport signal

SUCCESS: Stanford University access network

SWDBA: Static wavelength and dynamic bandwidth allocation

TaG: Tell-and-go

TC: Transmission convergence

TCA: TC adaptation

T-CONT: Transmission container

TDI: Transmission for the Deaf, Inc.

TDM: Time-division multiplexing

TDMA: Time-division multiple access

TE: Terminating end (for optical fiber bus)

TE: Transverse electric (for electromagnetic fields)

TE: Traffic-engineering (for MPLS, GMPLS, SDN)

TG: Traffic grooming

tHEC: Type header error control

THT: Token holding time

TIR: Total internal reflection

TM: Transverse magnetic (for electromagnetic fields)

TM: Terminal multiplexer (for SONET)

TMS: Traffic matrix switching

TNS: Transit network service

TNQ: Tandem network of queues

TOF: Tunable optical filter

ToR: Top of the rack

TR: Tunable receiver

TRT: Token rotation time

TSC: Time-slot-switch capable

TT: Tunable transmitter

TTL: Time to leave

TRP: Traffic-routing problem

TTRT: Target token rotation time

TWC:	Tunable wavelength converter
TX:	Transmitter
UC-OOFDM	Upconverted OOFDM
UCP:	Unified control plane
UD:	Unidirectional
UNI:	User-network interface
UOB:	Unidirectional optical bus
UPS:	Uniterrupted power supply
UPSR:	Unidirectional path-switched ring
URQ:	Upstream request queue
UTP:	Unshielded twisted pair
VC:	Virtual connection
VCAT:	Virtual concatenation
VCG:	Virtual concatenation group
VCI:	Virtual channel identifier
VCSEL:	Vertical cavity surface-emitting laser
VeRTIGO:	Virtual topologies generation in OpenFlow
VLB:	Valiant load balancing
VM:	Virtual machine
VOQ:	Virtual output queue
VP:	Virtual path (for WRON)
VP:	Virtual port (for ONUs in PONs)
VPN:	Virtual private network
VR:	Virtual rack
VS:	Virtual server
VT:	Virtual topology (for WLAN, WRON, DCN)
VT:	Virtual tributary (for SONET)
VTD:	Virtual topology design
VToR:	Virtual ToR
WA:	Wavelength assignment
WAN:	Wide-area network

WASPNET: Wavelength-switched optical packet network

WBF: Waveband filter

WCR: Wavelength converter

WDM: Wavelength-division multiplexing

WDMA: Wavelength-division multiple access

WED: Weighted excess distribution

WFQ: Weighted fair queuing

WR: Wavelength routing

WRON: Wavelength-routed optical network

WRS: Wavelength-routed switching

WSS: Wavelength-selective switch

XCVR: Transceiver

XPM: Cross-phase modulation

Index